MISSION CO
LEARNING RESOURC

DATE DUE

APR 15 1994			
GAYLORD			PRINTED IN U.S.A.

Ultrastructure, Macromolecules, and Evolution

LAWRENCE S. DILLON

Texas A & M University
College Station, Texas

PLENUM PRESS · NEW YORK AND LONDON

Library of Congress Cataloging in Publication Data

Dillon, Lawrence S
Ultrastructure, macromolecules, and evolution.

Includes index.
1. Cytology. 2. Ultrastructure (Biology) 3. Evolution. I. Title.
QH581.2.D54 574.87 80-20550
ISBN 0-306-40528-8

© 1981 Plenum Press, New York
A Division of Plenum Publishing Corporation
233 Spring Street, New York, N.Y. 10013

All rights reserved

No part of this book may be reproduced, stored in a retrieval system, or transmitted, in any form or by any means, electronic, mechanical, photocoping, microfilming, recording, or otherwise, without written permission from the Publisher

Printed in the United States of America

QH 581.2 D54

Preface

Thus far in the history of biology, two, and only two, fundamental principles have come to light that pervade and unify the entire science—the cell theory and the concept of evolution. While it is true that recently opened fields of investigation have given rise to several generalizations of wide impact, such as the universality of DNA and the energetic dynamics of ecology, closer inspection reveals them to be part and parcel of either of the first two mentioned. Because in the final analysis energy can act upon an organism solely at the cellular level, its effects may be perceived basically to represent one facet of cell metabolism. Similarly, because the DNA theory centers upon the means by which cells build proteins and reproduce themselves, it too proves to be only one more, even though an exciting, aspect of the cell theory.

In fact, if the matter is given closer scrutiny, evolution itself can be viewed as being a fundamental portion of the cell concept, for its effects arise only as a consequence of changes in the cell's genetic apparatus accumulating over geological time. Or, if one wishes, the diametrically opposite standpoint may be taken. For, if current concepts of the origin of life hold any validity, the evolution of precellular organisms from the primordial biochemicals must have proceeded over many eons of time prior to the advent of even the most primitive cell. Hence, evolution is the single basic principle, and the cell, in this light, is secondary, merely serving as the instrument that carries out the processes.

No matter which point of view is found more acceptable, it is clear that these two fundamentals of biology are so closely interrelated that in any treatise that transcends narrow taxonomic boundaries, discussion of either to the exclusion of the other makes meaningless that which holds great significance and tedious much that is exciting.

Nor can the macromolecules always be sorted into neat, discrete units and kept separated from descriptions of the cell parts. For, as the building blocks of protoplasm, they can no more be segregated from consideration of cellular morphology than can bricks, lumber, and concrete be eliminated from a discussion of building construction. Hence, this topic joins those two that form the warp and woof of biological fabric, much like the pile in a Persian rug of intricate design.

The exposition of these three intertwined subjects from the broadest possible comparative approach, then, is the purpose of this book. Factual matters necessarily comprise the greater part of the text. But the facts indeed would be empty if from them no questions arose concerning their significance or if no problems for future investigation were revealed. For at this level and with this approach the newer and more classical aspects of biology can be seen to be but adjacent cells in a complex organism of a single science, and, as in those cells, the interfaces are found to interdigitate in a most intricate fashion.

Like its predecessor in the trilogy, this second member is not merely a review of the literature but more properly is to be considered an analytical synthesis from a diversity of viewpoints. As a consequence of its broad, in-depth approach, a number of novel aspects have come to light, so that new terms for several organelles have been introduced, the lysosomes and centriolar derivatives in particular having received such treatment. Moreover, an occasional regrouping of certain biochemicals has proven essential, and, where present information was insufficient, the need for such revision or redefinition has been indicated for other classes of macromolecules or cell structures. Similarly in those few instances that are supported by adequate researches, hypotheses concerning the functioning of such organelles as the flagellum have received analysis, and alternative concepts have been advocated. That phylogenetic theories and those of the origin of the cell parts have also received comparable treatment should, of course, be a foregone conclusion, for classification and phylogenesis are a direct consequence of evolutionary considerations.

Far too many persons have contributed to the completion of this volume than could possibly be mentioned here, but fortunately those who have made available electron and light micrographs or other illustrative material can be acknowledged appropriately with the figures. Among those who have been especially generous in this manner are Drs. Hilton H. Mollenhauer, of the U.S. Department of Agriculture, College Station, F. S. Sjöstrand, University of California at Los Angeles, E. M. Mandelkow and M. Mandelkow, Max Planck Institute, Heidelberg, and T. Eda, Teikyo University, so that especially warm thanks are expressed to them. Finally, my wife, as always, has collaborated in the preparation of the manuscript and artwork and in researching the literature, interpreting data, and polishing the phraseology.

LAWRENCE S. DILLON

Contents

1

Biological Membranes

Although none is completely satisfactory, the very best cell part with which to begin a discussion of the structure and evolutionary history of the cell probably is the membrane (Frey-Wyssling, 1955). In the first place, it is the only organelle aside from the genetic mechanism that is universally present in living things—even the viruses are provided with a membranous capsid that covers the virion. Second, the presence of a membranous envelope endows cellular organisms with one of their most characteristic features. Because of it, living creatures have the ability to absorb materials from their surroundings, even against a gradient, that is, when the concentration of the given ion or compound in the organism is manyfold greater than that in the medium. In contrast, in nonliving systems, even when a membrane is present, the movement of chemicals is always from the greater concentration to the lesser. Although certain colloidal particles, such as coacervates and proteinoid microspheres, also possess a limited capacity for concentrating material against a gradient in biological fashion, the ability varies with dilution, composition of the milieu, and other factors without influence in living systems. Moreover, many organelles other than the plasmalemma of the eukaryote and prokaryote cells alike are constructed of membranes.

In addition, a major generalization has prevailed in the literature that has attempted to ascribe a single pattern of molecular structure to all membranes regardless of their site within the cell. Because specialized aspects of these ubiquitous structures are more profitably discussed with the organelles of which they are a part, present considerations are confined first to general features and then to two all-important organelles of the cell, the plasma and nuclear membranes.

1.1. MEMBRANES IN GENERAL

As a group, membranes of cells average 75 Å in thickness but range from as little as 50 Å to as much as 200 Å, but in light of recent findings, the lower limits and average cited are probably subject to correction (Sjöstrand, 1978). In comparing single-cell sections of mouse kidney and pancreas by electron microscopy, Sjöstrand (1963a) reported two width classes of membranes. The thinner one, measuring from 50 to 60 Å, was confined to mitochondria and endoplasmic reticula, and the thicker one, ranging between 90 and 100 Å, included the plasma membrane and that surrounding zymogen granules. In cross section under the electron microscope, natural membranes typically have the appearance of what has become known as the "unit membrane."

1.1.1. The Unit-Membrane Concept

In essence, the original unit-membrane concept advanced by J. D. Robertson (1959, 1960) was a modification, based on electron microscopic and X-ray diffraction data, of an older theory of biological membrane structure (Danielli and Davson, 1935). It proposed that all such membranes are constructed over a single molecular pattern, regardless of the species or site within the cell from which they are derived. Among the most convincing data that support the theory is the consistency of their appearance in electron micrographs (Figure 1.1A,B). In the plasma membrane, as well as in those from the endoplasmic reticulum, nucleus, Golgi apparatus, and other organelles, three layers appear to be present when suitably stained. The two outside layers are uniformly dark (electron dense or opaque), while the intermediate one is consistently pale (electron transparent). To explain this configuration, the unit-membrane concept proposed that biological membranes consist of two parallel layers of phospholipid molecules, held together by van der Walls' forces. The constituents of the respective layers face in opposite directions, so that the electron-transparent nonpolar carbon chains lie adjacent to form the central clear region. Thus the two electron-dense faces consist of the polar moieties that become electron opaque with osmium staining (Figure 1.1B). In addition, each face was conceived as being covered with a monomolecular layer, the thinner external (distal, or E) face consisting of proteins and the thicker internal (proximal, plasmal, or P) one being constructed variously of proteins or carbohydrates. While the general pattern was believed to be universal, the actual chemical nature of the phospholipids and proteins was considered to vary both from species to species and from organelle to organelle in the same cell.

The Unitary Concept. In turn, the unit-membrane concept leads into another generalization, termed the unitary theory of cell structure (Robertson, 1967). According to this view, the eukaryotic cell consists of three "phases." Phase 1, the greatest portion of the cell, is the nucleocytoplasmic matrix, which

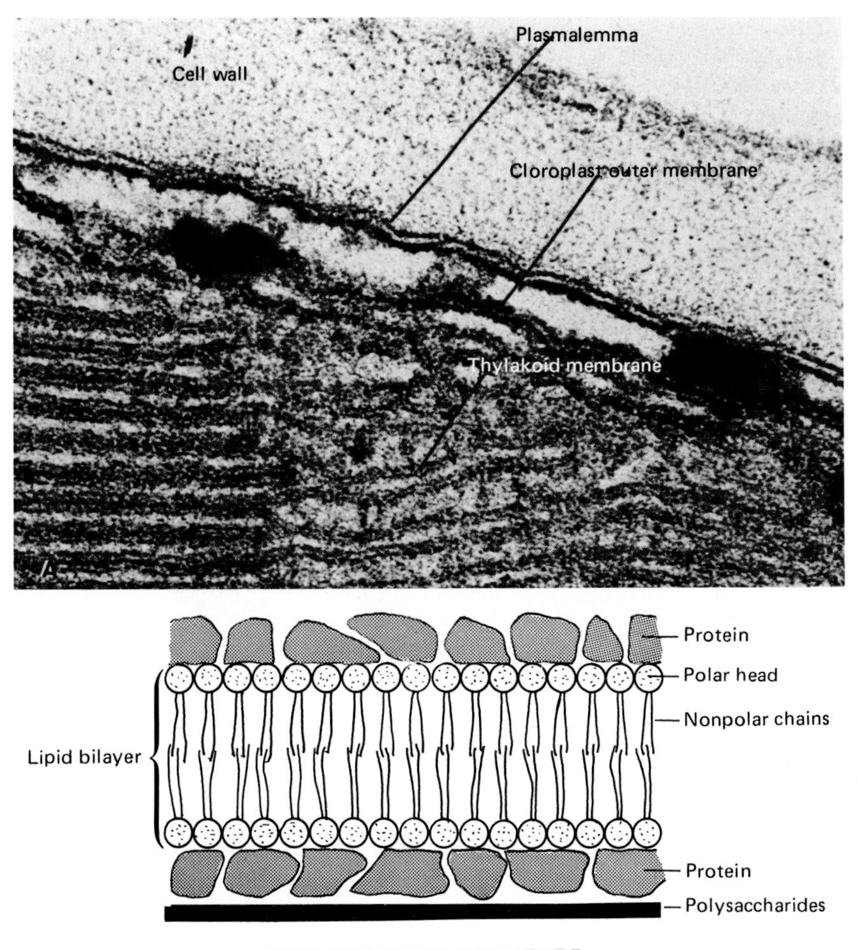

Figure 1.1. The unit-membrane concept. (A) Transverse section of a corn-leaf cell, 1 month old, 200,000×. Each of the three kinds of membrane shown appears to consist of two dense layers enclosing a transparent one. Careful comparisons, however, show that no two are exactly alike. (B) The transparent inner layer is accepted as being comprised of a lipid bilayer, whereas the dense ones are made up of protein. (A, courtesy of Crane and Hall, 1972.)

is made into an integrated whole by the pores of the nuclear membrane. Phase 2, the membranous portion, is made continuous from the plasma to the nuclear membrane by means of the endoplasmic reticulum, Golgi apparatus, and mitochondrial membranes (Daniels, 1964). Phase 3 is an external fraction brought into the interior of the cell by invagination, as when solid matter is taken in during phagocytosis or liquid by pinocytosis (Section 1.2.1). This concept thus makes clear that the original internal surface of the plasma membrane, even in its derivatives, is always adjacent to the nucleocytoplasmic matrix and that the E face consistently contacts substances from the milieu, even during phagocytotic and secretory activities and also during translocation to another region or organelle. This polarity of the membrane phase is viewed as constant in this concept, not the physical continuity of the several organelles.

1.1.2. Present Status of the Concept

As further investigations into the nature of biological membranes reached fruition, the validity of certain aspects of the original unit-membrane concept began to be questioned. Possibly the first paper to raise some doubts was that cited earlier (Sjöstrand, 1963a), which pointed out the absence of uniformity in thickness of the membranes from different types of organelles (Figure 1.1). Some other data that in one way or another can be reconciled with the concept only with difficulty are briefly summarized below

Protein-to-Lipid Ratios. In reevaluating the theory in light of biochemical composition, Korn (1966) disclosed serious problems that were encountered in correlating his data with the theoretical requirements of the unit-membrane concept. He pointed out that if all membranes are similar, their protein-to-lipid ratios should approximate one another closely. Instead he reported extensive variation in the ratios in the instances investigated. Among bacteria, for example, he showed that in *Streptococcus faecalis* the molar ratio of amino acid to phospholipid was 14.2:1.0 (Shockman *et al.*, 1963), in *Bacillus licheniformis,* 19.6:1.0 (Salton and Freer, 1965), and in *B. megaterium,* 22.6:1.0 (Weibull and Bergström, 1958). Furthermore, cholesterol was absent in all these bacteria, while it was consistently present in vertebrate membranes. Among the vertebrate tissues for which membrane analyses were available, he found similar wide divergence in the proportions of the constituents. In erythrocytes the molar ratio of amino acid to phospholipid to cholesterol approximated 16:1:1 (Maddy and Malcolm, 1965), whereas in myelin a ratio of only 2.4:1:0.67 had been reported (O'Brien and Sampson, 1965).

Another aspect analyzed by Korn presents similar difficulties. If, as the unit-membrane concept proposes, two monomolecular layers of protein cover the surfaces of a bimolecular layer of lipids, the ratio of surface area of the lipids to that of the proteins should be unity. In calculating the area ratios, Korn assumed amino acids on the average to occupy 17 $Å^2$, phospholipid molecules 70 $Å^2$, and cholesterol molecules 38 $Å^2$. Among bacteria, the results indicated that the protein-to-lipid surface area ratio varied from 3:1 to more than 5:1, and in erythrocytes 2.5:1. Hence, as Korn pointed out, there is an excess of protein simply to provide a covering layer, if the lipids are arranged as hypothesized in a bimolecular fashion.

Chemical Composition of Lipids. As summarized in that same study,

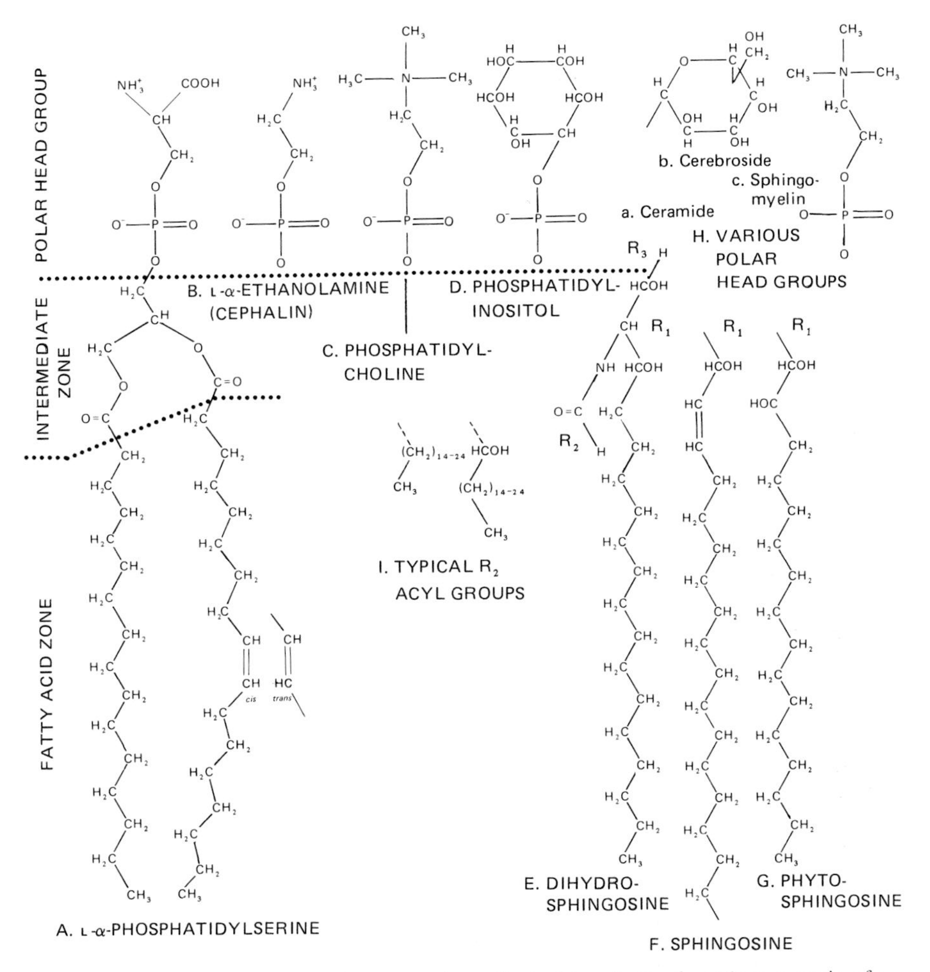

Figure 1.2. Molecular structure of some membrane lipids. Many lipids of membranes consist of a polar head group (A–D, H), to which two fatty acid chains typically are attached by structures located in the intermediate zone (A). The ceramide (a) is illustrated by E.

the chemical composition of the lipids also deviates widely from taxon to taxon. Among bacteria, no steroids occur in the membranes (Kaneshiro and Marr, 1965; Op den Kamp *et al.*, 1965), and in such species as *Azotobacter agilis* and *Escherichia coli*, the lipids are entirely represented by a single phospholipid, phosphatidylethanolamine (Figure 1.2B). On the other hand, that same phospholipid and two others are present in the membranes of *Bacillus megaterium*; the two additions include phosphatidylglycerol and lysylphosphatidylglycerol, the latter in small quantities. In contrast, most metazoan membranes contain cholesterol (Figure 1.3A), but the percentage present is subject to extensive fluctuation from tissue to tissue. Whereas in myelin (O'Brien and

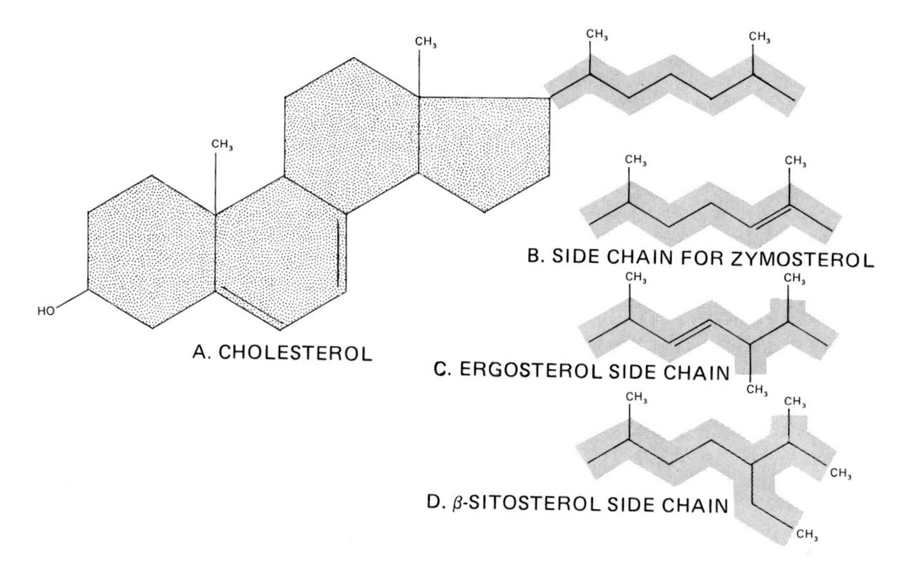

Figure 1.3. Sterols of membranes. Typically sterols have a complex molecular structure, similar to that shown in coarse stippling (A); various side chains, shown in dark stippling, are appended to the representative structure.

Sampson, 1965) and erythrocytes (Ways and Hanahan, 1964) this steroid is the most abundant lipid, providing 25% of the entire fraction, in mitochondria and microsomes (Dallner *et al.*, 1965) it is found only sparingly, with phosphatidylcholine (Figure 1.2C) providing approximately half. Nevertheless, in spite of the wide fluctuation in composition, reexamination of the above data from a different point of view exposes an evident trend in the lipids—from a simpler condition in bacteria to a more complex one among metazoans. Consequently, although too few analyses are as yet available to permit more than a suggestion of the possibility at this time, it is not unlikely that an evolutionary sequence in the lipid components will eventually be found to exist from the lower to higher taxa.

Water Relationships. Another source of doubt as to the validity of the unit-membrane concept derived from the relationship of water to the other components. Because the central layer was theorized to consist of lipids and therefore to be hydrophobic and the outer lamellae wettable and hydrophilic, the moisture content of membranes should be largely concentrated in the latter rather than in the former. Several studies intimated that water may actually be distributed in the opposite fashion. One such investigation on the amoeba, *Pelomyxa carolinensis* (Brandt and Freeman, 1967), used the electron microscope to examine some effects of various concentrations of $CaCl_2$ and NaCl upon the plasma membrane. If the concentration of $CaCl_2$ was maintained at a uniform low level while that of NaCl was increased, the membrane's diameter increased from a mean of 94 Å to one of 140 Å; raising the $CaCl_2$ level without lowering that of NaCl then restored the membrane to its original thickness. Because the outer layers remained unaltered during these experiments, the investigators concluded that their observations resulted in part from the redistribution of water and electrolytes *within* the central lamella. Hence, water was evidenced to be present in the latter, at least under certain conditions.

Other Conflicting Data. A different approach supporting similar conclusions was employed by Branton and Moor (1964). By use of freeze-etching techniques on cells of the onion root tip, they found that the nuclear membrane did not conform to all theoretical implications of the unit-membrane concept. With these procedures, after the cell has been fractured with the ultramicrotome knife, the exposed surface is "etched" by permitting water to evaporate for a time before a carbon replica is made for the actual examination. If water is distributed as outlined above to accord to the theory's demands, shrinkage should be greater in the outer lamellae than in the inner; however, the opposite results were obtained. Because of these and earlier comparable data, several workers (e.g., Sjöstrand, 1963b; Lucy and Glauert, 1964) suggested that membranes may consist of a phospholipid matrix in which protein globules are embedded, or may be a combination of micelles and lamellae (see Kavanau, 1965, for a review of this concept). Nevertheless, the freeze-etch studies, it must be emphasized, did suggest a three-layered condition for the membranes examined; only the sequence of the constituent molecules was questioned.

The Current View of the Unit-Membrane Concept. More recently, a study of the trilamellar arrangement of membrane structure was made, in which the arrangement of the proteins was particularly stressed (Green, 1972; Green and Brucker, 1972). In part, these papers showed that the trilamellar appearance of sectioned membranes viewed under the electron microscope was the result of preparatory techniques. It is essential that objects prepared for electron microscopic study be thoroughly dehydrated before exposure to the extremely high vacuum employed in such instruments. One of the results of such dehydration, it was shown, was to induce the lipid fraction of the membrane to clump into a "water hose" configuration (Figure 1.8B), so that the bimolecular layer no longer remains (Figure 1.4). Today, although the unit membrane still exists as a working concept in a few quarters, especially in theoretical accounts or in those based on synthetic membranes (Fox and Keith, 1972), it has been replaced in biological researches on the subject by other models, such as those described in the following section.

1.1.3. Other Concepts of Membranes

Although innumerable models and varieties of membrane structure have been proposed since the unit-membrane concept first began to lose favor, the problem of the molecular organization of membranes is still not completely

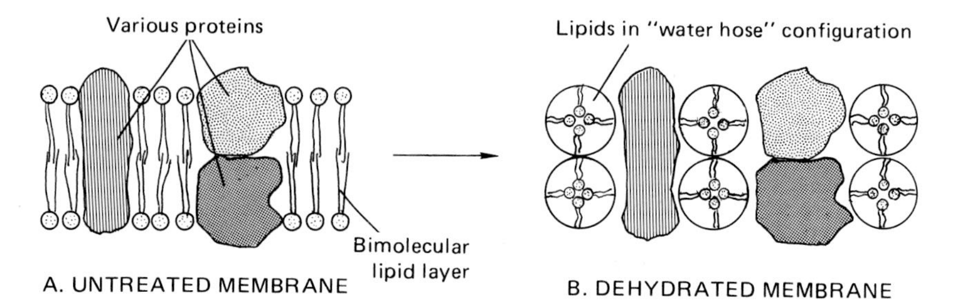

Figure 1.4. Effect of dehydration on membrane structure. According to Green (1972) dehydration during preparation for ultrastructural study induces lipids in membranes to assume a "water hose" configuration, with polar heads inward and the fatty acid chains outward.

resolved. To recount the details of all the models that have been advocated during the intervening years would require an unjustifiable amount of space, so only the more important and current proposals receive attention below. Out of the maze of data and ideas that have emerged from the numerous studies of the membrane, it is becoming apparent that all fall into a limited number of major categories.

Liquid Models of Membranes

The Fluid Mosaic Concept. Among the many recent models of membrane structure at the molecular level is the fluid mosaic concept (Singer, 1972, 1977; Singer and Nicolson, 1972). Actually this is a modification of the Wallach-Zahler (1966) concept, which proposed the arrangement of proteins and lipids but did not refer to fluidity. In this, as in most current models, the double layer (bilayer) of phospholipids remains, but here it is viewed as discontinuous and fluid—more specifically, it is considered to be a two-dimensional viscous solution whose constituents are highly oriented. The proteins, however, no longer are thought to cover the surfaces of the phospholipid bilayer but are treated as being embedded in them. That is to say, the proteins are considered to float within the liquid medium or sometimes to penetrate through it (Figure 1.5A). In addition, there may be extrinsic, or "soluble," proteins that float on the surface of the membranes; those that are contained in whole or part within the membranes are then referred to as intrinsic (Green, 1972). In some instances, certain of the phospholipid members of the matrix may interact with given proteins. This type of structure is dynamic and is believed to apply to most cellular membranes, including the plasma membrane and those of mitochondria, endoplasmic reticulum, Golgi material, and chloroplasts. Others such as myelin sheaths of neurones or the capsids of small viruses, nonetheless may be rigid, so all membranes are not perceived as being identical.

The thermodynamics of certain aspects of membrane structure are particularly clearly phrased in Singer and Nicolson's (1972) explanation of the two types of noncovalent interactions that occur, hydrophobic and hydrophilic. The first variety of this class of interactions is defined as those thermodynamic factors that sequester hydrophobic and nonpolar groups away from water. Such sequestering is demonstrated by the immiscibility of such nonpolar substances as oils and other hydrocarbons with water. For example, at 25°C an expenditure of 2.6 kcal of free energy is required to transfer 1 mole of methane from a nonpolar medium, such as petroleum, to water. Similar requirements for free energy in comparable large amounts doubtlessly are involved in many other interactions of nonpolar bodies or parts, such as among proteins contained in aqueous solutions in which the nonpolar amino acid residues are considered to be largely confined to the interiors of the macromolecules, out of contact with the water.

Hydrophilic interactions, the second variety, are similar to the first except that they proceed in the opposite direction; they are defined as those thermody-

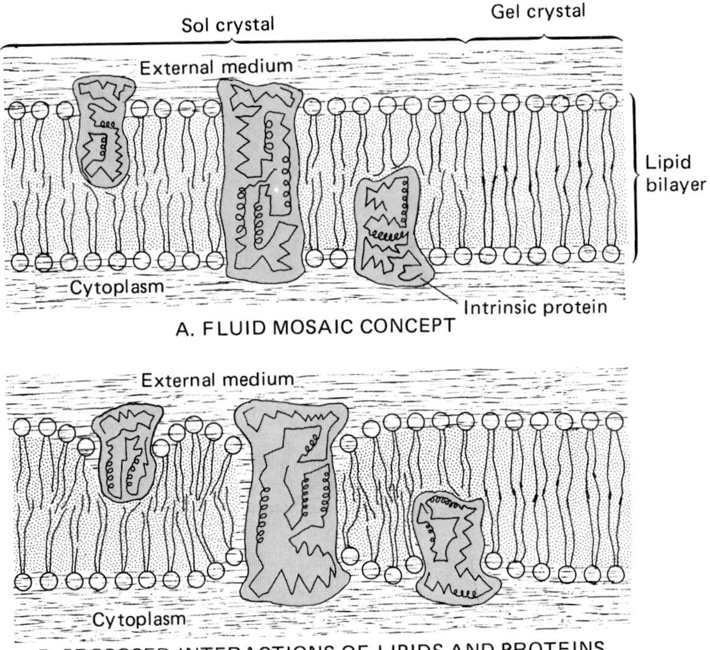

Figure 1.5. The fluid mosaic concept. (A) In this concept, the lipid bilayer is viewed as discontinuous, being interrupted by the presence of proteins that penetrate one or both layers. (B) A modification of the above, proposed here, suggests how the lipids may interact upon contact with proteins.

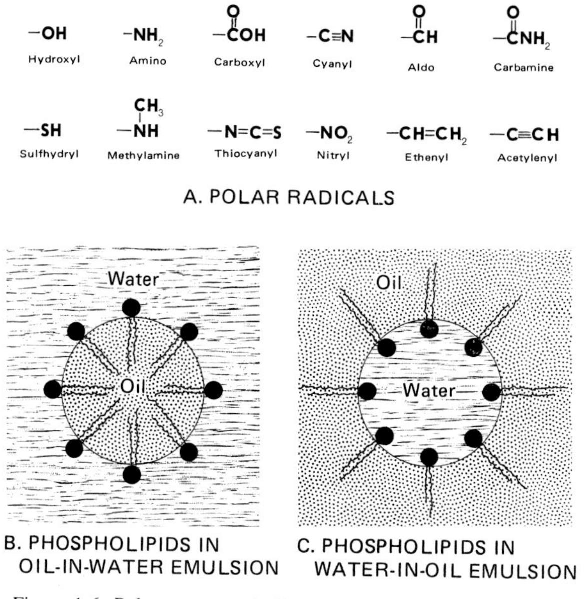

Figure 1.6. Polar groups and effects of polarity in various media.

namic factors responsible for the sequestering of ionic and polar groups from nonpolar environments. For instance, at 25°C the transfer of zwitterionic* glycine from water to the hydrocarbon acetone requires 6.0 kcal of free energy. Thus almost all of the ionic amino acid residues of proteins are considered to be in contact with water and usually are exposed on the exterior of the molecule, insofar as X-ray crystallographic studies can discern.

Polarity Aspects. Because polar radicals include OH^-, $COOH^-$, NH_2^-, and the others illustrated (Figure 1.6A), all amino acid molecules are polar at the end where the peptide bonds are formed. In addition, serine, cysteine, threonine, aspartic and glutamic acids, lysine, arginine, and tyrosine (Table 1.1) have polar radicals at their free ends. Thus a problem may be perceived to exist with the concept.

This problem involves the orientation of the polar and nonpolar amino acids in their respective environments. In their diagrams and discussions, Singer and Nicolson suggested that most protein molecules penetrate only half-way through the phospholipid bilayer, leaving one such layer undisturbed. However, the large number of amino acids with polar free ends usually present in protein molecules makes it extremely unlikely that only those protein strands that are almost exclusively nonpolar would be immersed in the phospholipid milieu, leaving the ionic and polar portions concentrated in the aqueous medium above the surface of the bilayer (Figure 1.5A). While perhaps a few pro-

*Bearing both a positive- and a negative-charged radical.

teins might have their amino acids distributed sufficiently clustered into polar and nonpolar strands to accord with such an arrangement, that condition must be rare. Because the structural organization of proteins is primarily that demanded by their activities on specific substrates, restrictions imposed by membrane components necessarily must be secondary. As protein structure is primarily functional, the phospholipids of membranes more than likely respond to the presence of the protein, rather than the constituents of the latter being adapted to those substances. Moreover, the bilayer is viewed as being in a fluid condition, whereas the proteins are solids. Thus it would seem likely that the phospholipids are polarized in relation to the inserted protein molecules, perhaps as suggested in Figure 1.5B. However, another concept that provides a better solution to this problem has been presented recently, which receives attention later.

Amino Acid Sequences and Polarity. A comparison of polarity in soluble and membrane proteins has recently been made (Vanderkooi and Capaldi, 1972; Capaldi, 1974), using primary structural data of 206 representatives of the soluble type and 9 from membranes. In making the analysis, the three alcohol amino acids (serine, threonine, and tyrosine), which are relatively

Table 1.1
The 20 Amino Acids of Proteins and Their Standard Abbreviations

Amino acid	3-Letter abbrevia-tion	1-Letter abbrevia-tion	Amino acid	3-Letter abbrevia-tion	1-Letter abbrevia-tion
Alanine	Ala	A	Methionine	Met	M
Arginine	Arg	R	Phenylalanine	Phe	F
Asparagine	Asn	N	Proline	Pro	P
Aspartic acid	Asp	D	Serine	Ser	S
Cysteine	Cys	C	Threonine	Thr	T
Glutamic acid	Glu	E	Tryptophan	Trp	W
Glutamine	Gln	Q	Tyrosine	Tyr	Y
Glycine	Gly	G	Valine	Val	V
Histidine	His	H	Either Asn or Asp	-	B
Isoleucine	Ile	I	Either Glu or Gln	-	Z
Leucine	Leu	L	Either Ile or Leu	-	J
Lysine	Lys	K			

weakly polar, were combined into an intermediate category with two weakly nonpolar amino acids (glycine and histidine). Half of this group were counted with polar amino acids and the other half with nonpolar. The soluble proteins were found to have a range from 36 to 55% polarity, with a mean of 44%; in contrast, the nine membrane proteins ranged from 29 to 41% polarity, with a mean of 36%.

Although the percent differences undoubtedly are of much significance, there is a more meaningful aspect of molecular polarity that needs attention. The question appears to be whether or not the lower numbers of polar amino acids present in membrane proteins result from their being concentrated in the proportionately smaller exposed portion and, accordingly, their relative absence from those long segments of the protein chain that are embedded within the lipid bilayer. Unfortunately too few amino acid sequences of membrane proteins have been established as yet to resolve the problem fully; actually only two such proteins have been analyzed in this manner, and of this pair, one is incomplete.

Cytochrome c Oxidase. The incomplete sequence is from cytochrome *c* oxidase, a protein of mitochondrial membranes that has been suggested to serve as a proton pump (Wikström, 1977; Krab and Wikström, 1978; Wikström and Krab, 1978). Although other roles have been proposed (Moyle and Mitchell, 1978), the primary structure of the segment of this substance that has been determined (Tanaka *et al.*, 1977) is of great interest in showing a degree of evolutionary kinship to the β chain of hemoglobin. As aligned by the investigators cited, 11 sites of the 109-residue sequence are found to be identical in the two chains, whereas random chance alone would lead to a mean of 6 identities. One of the peptides, however, is a portion of the membrane protein, cytochrome *c* oxidase, while the other is a segment of a soluble protein. According to the polarity concepts enumerated in the foregoing discussion, the former should contain a smaller fraction of polar amino acids than the latter, but in fact quite the opposite condition prevails. Polar residues, indicated in Table 1.2 by a "P" above or below the symbol for the amino acid, account for a total of 36% of the first peptide and only for 28% of the second. But a greater discrepancy exists between conceptual requirements and actual. Rather than having long sections of the molecule consisting largely of nonpolar amino acids and others of polar amino acids that could be immersed in the lipid bilayer or exposed in the aqueous medium, respectively, the polar and nonpolar components are intermixed rather uniformly. In fact, greater uniformity exists in the membrane-embedded protein, for the longest series of nonpolar amino acids is six units in length; two such series occur, one from sites 81 to 86 and the other from sites 96 to 101. In addition, four sections consist of five nonpolar residues. The soluble hemoglobin chain, however, has two sections of eight nonpolar components, the first from sites 9 to 16 and the second from sites 31 to 38, and, in addition, has a sequence of seven such constituents occupying sites 83 to 89 and one of six from 67 to 72, not to mention several that are five sites in length.

Table 1.2
The Primary Structure of the Heme a Segment of Cytochrome c Oxidase and a Comparison with a β Chain of Human Hemoglobin [a]

						5					10					15					20					25		
							P		P	P	P		P		P						P	P		P		P		
Cytochrome oxidase	NH_3	S	H	G	S	H	E	T	D	E	E	F	D	A	R	W	V	T	Y	F	N	K	P	D	I	D	A	W
Hb β-chain segment		V	H	L	T	P	E	E	K	S	A	V	T	A	L	W	G	K	V	N	V	D	E	V	G	G	E	A
							P	P	P									P		P		P	P				P	

			30					35					40					45					50					55					60				
	P		P	P			P						P				P		P			P				P			P	P		P	P				
Cyto	E	L	R	K	G	M	N	T	L	V	G	Y	N	L	V	P	E	P	K	I	I	D	A	A	L	R	A	C	R	R	L	N	D	F	A	S	A
Hb	L	G	R	L	L	V	V	Y	P	W	T	Q	R	F	F	E	S	F	G	D	L	S	T	P	D	A	V	M	G	N	P	K	V	K	A	H	G
			P									P	P			P				P					P					P		P		P			

	65					70					95					80					85					90					95					100	
		P			P			P	P	P					P	P							P	P		P				P	P						
Cyto	V	R	I	L	E	V	V	K	D	K	A	G	P	H	K	E	I	Y	P	Y	V	I	Q	E	L	R	P	T	L	N	E	L	G	I	S	T	P
Hb	K	K	V	L	G	A	F	S	D	G	L	A	H	L	D	N	L	K	G	T	F	A	T	L	S	E	L	H	C	D	K	L	H	V	D	P	E
	P	P							P						P	P		P								P				P	P				P		P

				105				109	
	P	P				P	P		
Cyto	E	E	L	G	L	D	K	V	COOH
Hb	N	F	R	L	L	G	N	V	
	P		P				P		

[a] Based on Tanaka *et al*. (1977).

A Bacterial Lipoprotein. The second membrane protein whose primary structure has been established is a lipoprotein from the outer membrane of *E. coli* (Braun and Bosch, 1972); moreover, a peptide extension on the precursor of the molecule has had its sequence determined (Inouye *et al*., 1977). This segment (Table 1.3) shows a high degree of specialization in that only 10% of its 20 sites contain a polar amino acid, namely sites 2 and 5, where lysine occurs. The actual lipoprotein, which has a fatty acid chain attached to the cysteine in site 1, consists of only 58 amino acid residues, of which 28 are polar; thus 48% of its constituents are polar, in spite of this being a membrane protein. In fact, this is one of the most characteristic members of this class, and according to Inouye *et al*. (1977), it is probably the most thoroughly investigated. Additionally, the polar amino acids are uniformly distributed throughout the molecule, so that definite polar and nonpolar segments are lacking. Accordingly, there appears little to substantiate the concept, based on purely theoretical considerations, that membrane proteins differ strongly from soluble ones in the distribution of polar amino acids.

The Question of Fluidity. The near-universality of a fluid (or better, a liquid crystalline) state in membranes has been sharply questioned (Oldfield,

Table 1.3
Amino Acid Sequence of the Lipoprotein of E. coli Outer Membrane and Precursor Extender[a,b]

```
            5         10        15        20         5        10        15        20
      P     P                                     P   P   P P       P   P     P   P
NH3 M K A T K L V L G A V I L G S T L L A G C S S N A K I D E L S S D V Q T L N A K

           25        30        35        40        45        50        55    58
      P       P P   P     P   P   P     P P P     P   P P P   P         P   P P
    V D E L S N D V N A M R S D V Q A A K D D A A R A N E R L D N M A T K Y R K COOH
```

[a] Based on Braun and Bosch (1972) and Inouye *et al.* (1977).
[b] Precursor extension is italicized. P, polar amino acid. Fatty acid chain is attached to the initial cysteine of the protein molecule.

1973). Nuclear magnetic resonance experiments using deuterium indicated that, at the usual temperatures of culturing the bacteria, most of the lipids were in a state of mobility similar to the crystalline gel state of lecithin (Metcalfe *et al.*, 1972; Oldfield *et al.*, 1972; Warren and Metcalfe, 1977). Part of the conceptual difficulty is that solid or gel crystalline regions in membranes might limit the movement of substances through them. Indeed, Krasne *et al.* (1971) demonstrated that valinomycin's enhancement of potassium ion transport ceased when the temperature was lowered sufficiently fo freeze the lipid bilayer. However, under the same conditions the antibiotic gramicidin A continued to function normally. Accordingly, it was proposed that the molecules of the latter substance form a channel through the bilayer, as discussed in a later section. Thus a completely fluid crystalline structure for membranes may not be requisite *in vivo* for transport processes to occur.

That a liquid state could exist in membranes under optimal growth conditions for many organisms has been amply confirmed by the measured degree of movement permitted the membrane constituents (Eisenberg and McLaughlin, 1976; Fahey and Webb, 1978). For instance, one lipid molecule was found to be able to exchange places with a neighbor at a rate of 10^7 times per second and could move laterally within its own layer 2×10^{-8} cm^2 per second; however, the rate of lateral diffusion has been determined to vary with the concentration of phospholipid (Fisher and James, 1978). In contrast, to exchange places with a neighbor in the opposing molecular array in the bilayer, that is, a process known by the inelegant but descriptive term "flip-flop," requires hours or perhaps days and may not actually occur in biological membranes (Kornberg and McConnell, 1971; Johnston *et al.*, 1975). Free movement also was reported to be provided for each individual molecule of the membrane, so that any one could rotate on its axis at rates that have not been ascertained. The long chains of the fatty acid moieties appear also to undergo changes in configuration (Lee, 1975; Albrecht *et al.*, 1978), the distal tips being more active than the residues nearer the polar portion (Eletr and Keith, 1972). States of the membrane proteins are also influenced by the energization of the membrane (Williams, 1978). That fluidity can be a property of a membrane is strongly evidenced by the plasmalemma of a flagellate, one end of the cell of which rotates during swimming, as described more fully in Chapter 4, Section 4.2 (Tamm, 1979).

Nevertheless, the physical stage of membranes is sensitive to temperature changes in the environment, and accordingly undergoes transitions from liquid crystalline to gel as temperatures are lowered. However, the exact temperature at which the alteration in state takes place varies with the lipid and the configuration of the double-bonded carbons in the fatty acid moieties. Carbon-to-carbon double bonds are fixed at ambient temperatures usually in the *cis* arrangement rather than in the *trans* (Figure 1.2A). These introduce a bend of 30° in the chain as indicated in the Figure 1.7, which has the effect of lowering the

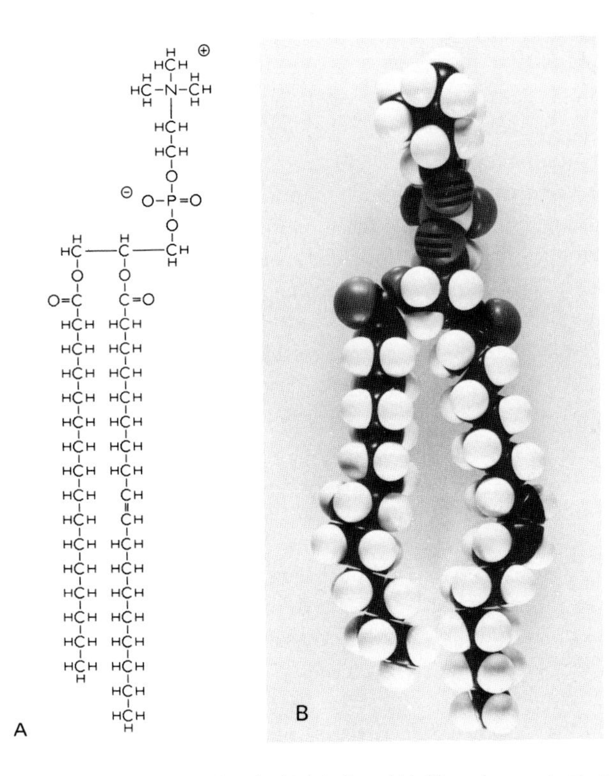

Figure 1.7. Molecular structure of a phosphatidylcholine. (A) The primary structure. (B) A CPK space-filling model of the same substance. The right-hand chain contains a bend resulting from the *cis* arrangement of the double bond between carbons 9 and 10; the other bends noted result from single bonds in the gauche configuration. (Courtesy of Eisenberg and McLaughlin, 1976, and *BioScience*.)

temperature at which transition to the gel state occurs. Membranes taken near the hoof of a reindeer, where they are exposed to severe cold, were found to contain many double bonds in the fatty acid chains, while those from the more protected thigh region contained few (Fox, 1972). Nonetheless, as numerous organisms spend their entire lives at temperatures close to 0°C, including the seaweeds and plankton of Arctic and Antarctic seas and even metazoans that dwell in the abysmal regions of the oceans, biological membranes obviously must function in the gel state that doubtlessly ensures from such frigid conditions. Fish from polar regions are known to contain eight or more glycoproteins in their blood (Osuga and Feeney, 1978; Osuga *et al.*, 1978), but whether similar molecules are present in their cell membranes has not been established. Temperature decreases have been shown to affect membrane lipid metabolism in *Acholeplasma* (Christiansson and Wieslander, 1978) and in mouse LM cells (Gilmore *et al.*, 1979a,b).

Asymmetry in the Lipid Bilayer. Although throughout the entire foregoing discussion of membrane structure the constituents of the basic bilayer have been treated merely as lipids, it should not be supposed that only a single variety of this class of biochemicals is present. Three, four, or even more major types commonly occur, accompanied in many cases by several varieties of lesser abundance. Currently much interest is being displayed in the manner in which these components are distributed within given membranes, for in some instances distribution may be asymmetric or nonrandom. The mammalian erythrocyte probably represents the most thoroughly documented case of asymmetric distribution of the lipid constituents of the membrane. In the distal monolayer is located most of the phosphatidylcholine and sphingomyelin, while in the proximal monolayer is found most of the phosphatidylserine and phosphatidylethanolamine (Figure 1.2B; Bretscher, 1972; Zwall *et al.*, 1973; Kahlenberg *et al.*, 1974; Dallner, 1977; van Deenan *et al.*, 1977). Further, within a given monolayer the constituents may be nonrandomly dispersed, so that one type of lipid may predominantly surround one species of protein, and another type may associate chiefly with a second. However, this second kind of asymmetry is difficult to ascertain and has not been satisfactorily documented.

The proteins of membranes also may be asymmetrically distributed (Gebhardt *et al.*, 1977), as well as the carbohydrates (Sturgess *et al.*, 1978). In erythrocytes the protein spectrin has been shown to be located on the surface of the proximal monolayer, while other proteins are confined to the outer. Also, in the plasma membrane of rabbit intestinal cells, leucine β-naphthylamidase was demonstrated to be confined to the outer face (Takesue and Nishi, 1978). Similarly, in the sarcoplesmic reticular membrane of rabbit fast muscle, ATPase was found to be located across the membrane with both internal and external polar regions exposed, calsequestrin was situated on the inner face, and much of the glycoprotein was on the outer surface (Hidalgo and Ikemoto, 1977).

How this nonrandom differentiation of inner and outer portions of a membrane is achieved has not been established, but a number of molecular mechanisms have been proposed. One suggestion is that specific enzymes are involved in the processes (Bretscher, 1973), while the electric potential that exists across a membrane has been advocated by others (McLaughlin and Harary, 1974). The configuration of the lipid molecules themselves is a third possibility that has received attention (Israelachvili and Mitchell, 1975), and a fourth conceives that lipid–protein interactions are active (Schafer *et al.*, 1974). While the condition may not be typical of all membranes, however, it has been firmly established as occurring in the erythrocyte, the discs of the outer segment of the retinal rods (Litman and Smith, 1974), and the capsid membrane of certain viruses (Schafer *et al.*, 1974, among others). In contrast, it has been demonstrated as being absent from the membranes of microsomes and Golgi vesicles of the liver (Sundler *et al.*, 1977).

Miscellaneous Membrane Models. In addition to the Wallach–Zahler model of membrane structure that has become widely accepted today, there are a number of others that have never gained substantial support. However, only those concepts of recent vintage that seem likely to receive future attention are reported here, as a fuller history of membrane structure is to be found in Wallach (1972) and Kotyk and Janáček (1977), where concise reviews are provided.

A Micellar Model. A rather early modification of the unit-membrane concept was in part based on micelles (Lucy and Glauert, 1964), minute globules of phospholipids having the polar moieties arranged around the surface (Figure 1.8A). The interior of the membrane is viewed as a mixture of such micelles with proteins and glycoproteins in the form of a quasi-two-dimensional array. This random array, then, is thought to be coated on each surface with longer protein molecules, as in the unit-membrane concept.

A model similar to the micellar in some particulars has been suggested to apply to cases in which two or more adjacent membranes of like type fuse to one another. Such "fusion membranes" are especially typical of the myelin sheath that forms around neurones, but they also occur in organelles, such as in mitochondria of mammalian heart cells and in the chloroplasts of diverse algae. Rather than being in globular micelles, however, the phospholipids are believed to form a concentric "water hose" configuration. This conformation is viewed as being an elongate cylinder comprised of inverted phospholipids, having the polar heads arranged around a central lumen while the fatty acid moieties project distally (Figure 1.8B). A monolayer of lipids and proteins then overlies the cylinder, many such cylinders and coats lying side by side to form the membranous sheet.

The Bimodal-Protein Model. The bimodal-protein model of protein structure (Vanderkooi and Green, 1970; Green and Brucher, 1972) is somewhat

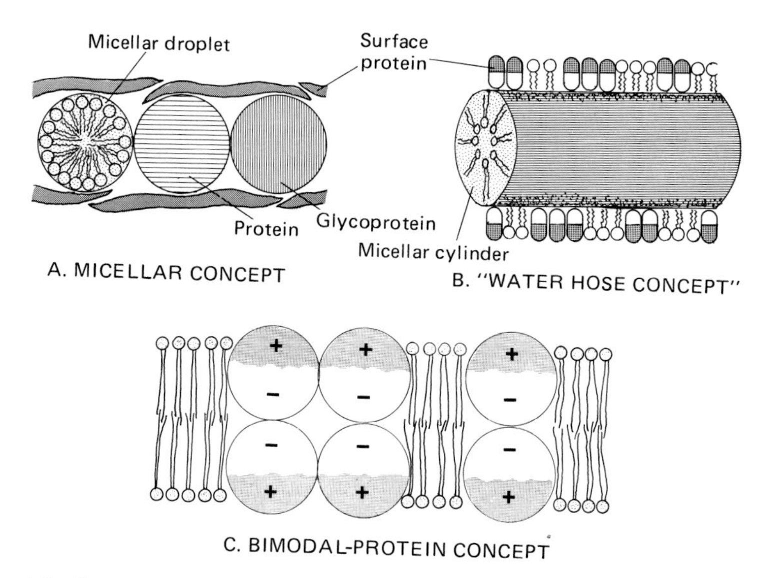

Figure 1.8. The micellar and bimodal-protein concepts. (A) In the micellar concept, the phospholipids are arranged in droplets with the polar moieties situated around the surface. These, together with protein and glycoprotein molecules, are arranged in sheets between two layers of surface proteins. (B) In the water hose variation of the micellar concept, the polarized phospholipids form continuous cylinders placed between complex layers of polar proteins and phospholipids. (C) The bimodal-protein model resembles the fluid mosaic pattern, but the proteins are conceived to occur in pairs, with polar portions externally and nonpolar parts in contact.

akin to both the fluid mosaic concept, which it antedates, and the repeating-unit model of Green and Perdue (1966; Green, 1972). Its resemblance to the former is in its consisting of a bimolecular layer of lipids in which proteins are scattered and to the latter in its proteins being conceived to be repetitious. From both it differs in that the proteins are believed to be bimodal, so that each species of protein occurs in pairs (Figure 1.8C), one member being located on the P surface of the membrane, the other on the E. The bimodality arises conceptually from each molecule's consisting of a polar portion externally and a nonpolar portion internally. In addition the latter hydrophobic portions are thought to be able to become charged with contrasting electric potentials, thus explaining asymmetry of charges sometimes noted in biological membranes.

Proteinaceous-Membrane Concepts

An Interdigitated Model. During the years when the unit-membrane concept held full sway, a modification was advanced by Sjöstrand (1969), based largely on electron microscope observations of biological membranes, particularly those from mitochondria. The standard bilayer arrangement of phospholipids was accepted in this concept, but the proteins, instead of coating each surface as then believed, were suggested to be within and through the phospholipid layers (Figure 1.9A). The proteins, of various size and location, are conceived as being rather closely packed and in orderly systems so that the components of multienzyme complexes could most readily interact with one another. Very little space is devoid of proteins, but in such gaps that do occur, the standard bilayer of phospholipids fills in. Some of the gaps remain open as pores, so that discontinuities exist that are surrounded by proteins. In addition, it is conceived that phospholipids interdigitate into some of the various proteins. Benson (1968) similarly suggested an interdigitation of lipids and proteins but did not detail the proposal thoroughly.

Proteinaceous-Membrane Concept. Recently this concept has been enlarged upon (Sjöstrand, 1978), mostly based on results obtained through new preparatory techniques that eliminated artifacts innate to standard procedures. Three types of membranes were found in mitochondria, on which organelle the study was largely founded. Hence, most of the details are more appropriate to the discussion on the mitochondrion (Chapter 9, Section 9.2.2) than to the

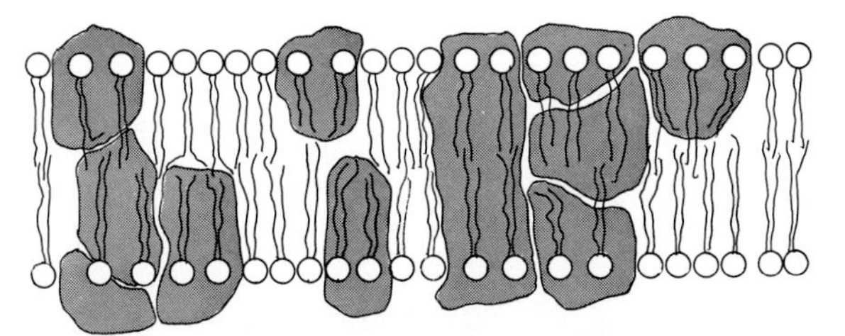

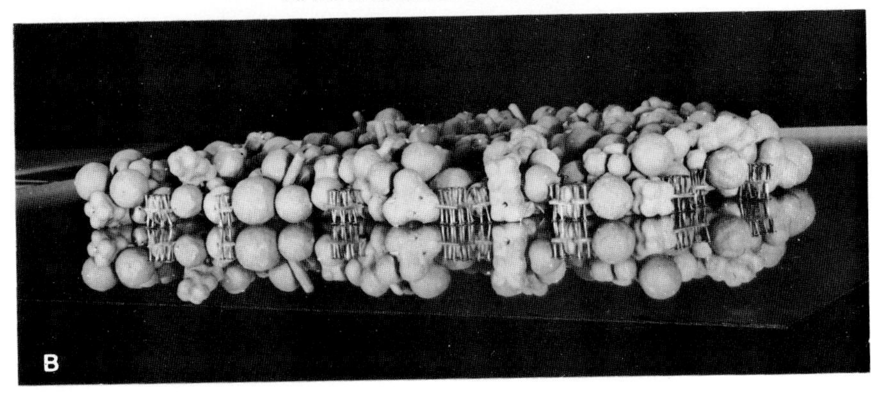

Figure 1.9. The interdigitated concept. (A) In this model the proteins not only are arranged into coordinated interacting systems, but the phospholipids interdigitate into the individual types. (B) Photogtaph of a model. (Courtesy of Sjöstrand, 1969.)

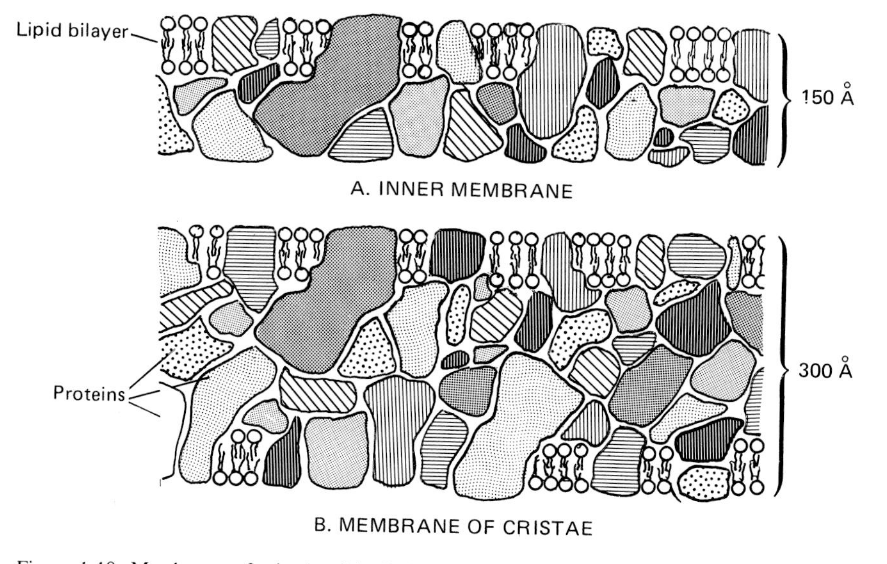

Figure 1.10. Membranes of mitochondria. In both the inner (A) and the outer (B) membranes of mitochondria, proteins are compactly and densely arranged in interacting systems, while phospholipids fill the interspaces.

present one. The conclusions important here are that in this one organelle, each membrane differed drastically from the others, as shown also by Crane and Hall (1972) and Dreyer *et al.* (1972). Consequently, it is evident that no single model of membrane structure can possibly apply to all membranes.

The outer membrane of the mitochondrion was in general similar to the fluid mosaic model of Singer and Nicolson (1972), except that the proteins were present in large masses, rather than as isolated molecules (Figure 1.10). In contrast, the inner one and that of the cristae were largely proteinaceous, and much thicker than usually measured—the inner one was found to be 150 Å thick and the cristal membrane 300 Å. In the former, the lipid bilayer was considered to be confined to one face of the membrane, while in the latter such a bimolecular film was found on each face (Figure 1.10B). The interior of each membrane was viewed as consisting of a nonaqueous portion made of polar and nonpolar phases mixed, whereas the surface was polar and aqueous.

One important feature of this concept is that it views membrane structure in terms of function rather than mere mechanical requirements. Hence, it is logical that the main functional molecules of biological subjects, the proteins, should provide both the bulk of each membrane as well as the chief distinctions between membranes of different types. The particulate nature of the membrane surface when viewed by freeze-dry techniques suggests a proteinaceous rather than lipid structure (Figure 1.11A,B; Demsey *et al.*, 1977, 1978). Still another

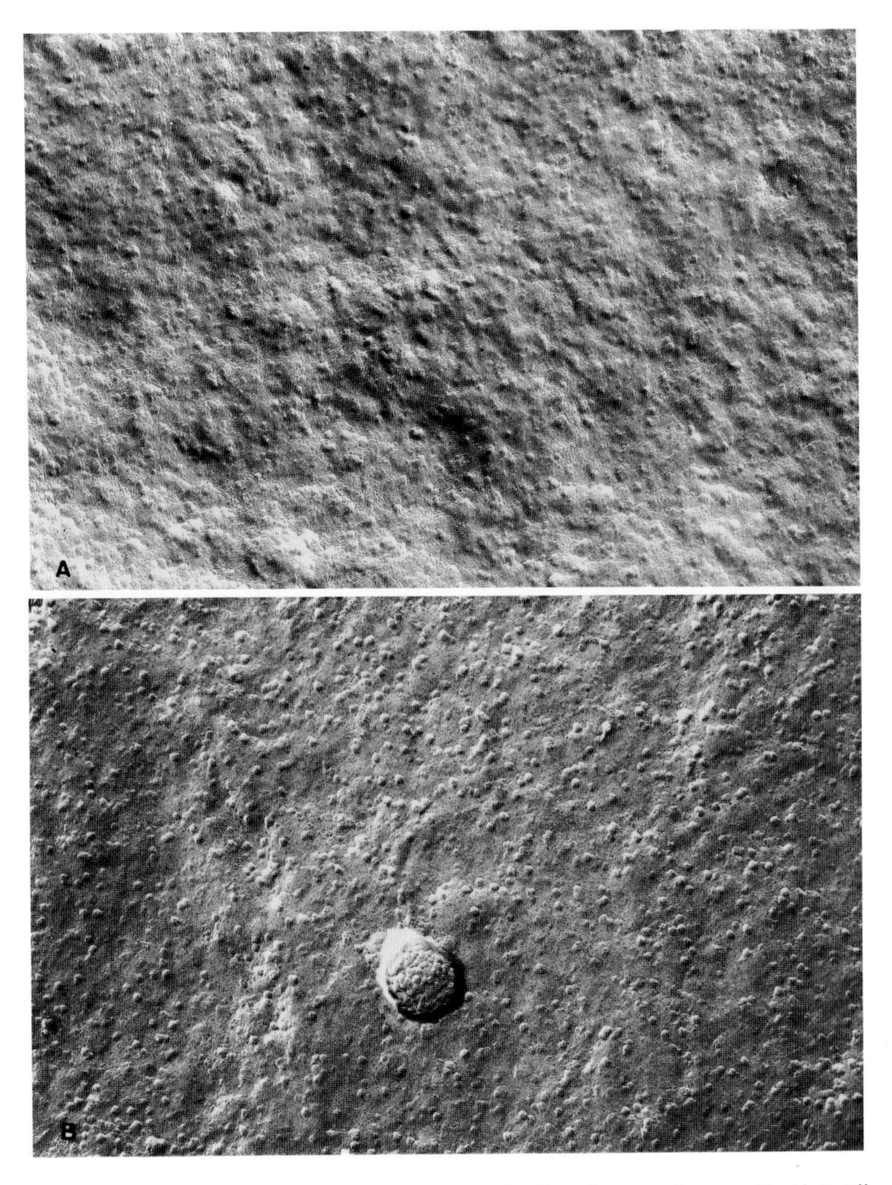

Figure 1.11. Outer surface of the plasmalemma. (A) Replica of a normal mouse fibroblast cell membrane shows particulate matter distributed in a fine matrix; the particles may be interpreted as proteins and the smooth matrix as the polar heads of phospholipids. (B) A similar cell membrane but infected with Rauscher leukemia virus; the projection in the membrane is a budding virus. Note that larger particles are found throughout the plasmalemma. (Courtesy of Demsey *et al.*, 1978, with permission of the *Journal of Ultrastructure Research*.)

reason for expecting membranes to consist largely of proteins is derived from evolutionary considerations. As the plasma membrane is the only such structure present in the simplest prokaryotes, that must be considered to be the oldest type. In turn, the capsid of virions has been shown to be the probable forerunner of the plasma membrane (Dillon, 1978). That structure consists predominantly of proteins in higher viral types and is exclusively of that type of substance in simpler forms, lipids being a later addition. As the first membranes were thus entirely proteinaceous, it is likely that their evolutionary derivatives are at least largely constructed in similar fashion.

Other additional desirable features of the present concept are at once evident. In the first place, it eliminates the problem pointed out earlier concerning the functionability of membranes at low temperatures. As almost all biological lipids solidify even at 10°C, it is difficult to perceive how the numerous living things of the ocean depths and frigid areas of the world could be active at 1 or 2°C, as they are, if the matrix is lipoidal. With a proteinaceous functional matrix, and interspersed lipids, the difficulty is alleviated. Second, the fluidity of the Singer and Nicolson model makes it impossible to conceive how membranes can be rigid, as many are in whole or part. The outer membrane of bacteria, for example, provides the shape to the organism, for when removed enzymatically, the resulting protoplasts immediately assume a spherical configuration. Moreover, the presence of complex structural features, such as grooves, raised lines, and pits on membranes, as described later (Section 1.2.3), are difficult to explain with a fluid concept of membranes. A predominance of proteins affords a ready explanation for these phenomena.

1.1.4. Functions of Biological Membranes

In view of the diversity of structure that exists among membranes, a comparable variety of function may well be expected, nor is one disappointed. The membranes of chloroplasts are involved in photosynthesis, those of the endoplasmic reticulum with protein synthesis, those of mitochondria with cell respiration, and so on. Because these specialized activities pertain more appropriately to the discussions of the individual types of organelles, here only those functions shared by all, or at least most, membranes regardless of type receive attention.

Permeability and Transport. Lying as they always do at an interface between one environment and another, such as between the watery external medium and the internal cytoplasm of a unicellular protozoan or between the cytoplasm proper (often called the cytosol) and the contents of an organelle, one obvious universal function is that of protection. Such protection is largely directed against loss of internal components and ingress from without of undesirable ions or chemicals.

Contrarily, if the cell is to feed or an organelle is to function, external ingredients must be able to enter as needed through the protective barrier. Too, the cell produces wastes and organelles form essential products that must pass outward through the membranous coat. This selective and bidirectional permeability is one of the most outstanding traits of biological membranes, and countless studies have been made on the particulars involved in the processes.

Ionophores. Undoubtedly the most thoroughly studied mechanisms of membrane transport are those concerned with the movement of such ions as K^+, Na^+, Cl^-, SO_4^{2-}, etc. Substances involved in these processes are referred to either as ionophores or as channel producers (Ovchinnikov, 1977), depending upon the mode of action. Among the better known of the ionophores is the antibiotic nonactin, a carrier of Na^+ ions whose crystallographic structure and possible mode of action have been described (Kilbourn *et al.*, 1967; Schmitt, 1971). The three-dimensional configuration of the molecule is not unlike a cage, or better, a crown (Figure 1.12A), the open end of which when empty faces the external milieu through the membrane. The cation is stripped of solvate water molecules as it enters the open cage, where it becomes bound coordinately with some of the eight oxygen residues that line the interior (Figure 1.12B,C). This bonding induces the ionophore to undergo a conformational change, so that the cage closes. When closed, the carrier passes through the lipid bilayer into the proximal side of the membrane, a process supposedly facilitated by the external side chains of the nonactin consisting largely of nonpolar amino acids. That the suggested movement of the carrier actually does occur is supported by the fact that transport ceases at temperatures below the point where lipids become gel crystalline, that is, "freeze" (Kotyk and Janáček, 1977). How the cage reopens to release the Na^+ ion into the cytoplasm is not clear, nor is it known how the nonactin regains its former position at the external surface.

This specificity of such carriers for particular ions, however, supplies one means by which cells are able to select Na^+, K^+, or others from among the numerous ions typically present in the environment for admittance through the plasma membrane. In other words, one might expect that each ion or molecule that is selectively taken into or removed from a cell would have a specialized peptide that carries it and only it.

The action of individual ionophores may be quite complex. For instance, the ionophore known as A23187, which primarily translocates Ca^{2+}, also stimulates the activity of lymphocytes and macrophages (Luckasen *et al.*, 1974; Hand *et al.*, 1977). In addition it has been found to enhance production of the eosinophil chemotactic factor and slow-reacting substance (Bach and Brashler, 1974; Czarnetzki *et al.*, 1976). That these reactions did not result from the Ca^{2+} brought in by way of the ionophores was demonstrated by their continuing in the absence of those ions (Resch *et al.*, 1978). The ionophore also induced changes in the membranes, such as enhancement in the rate of turnover of

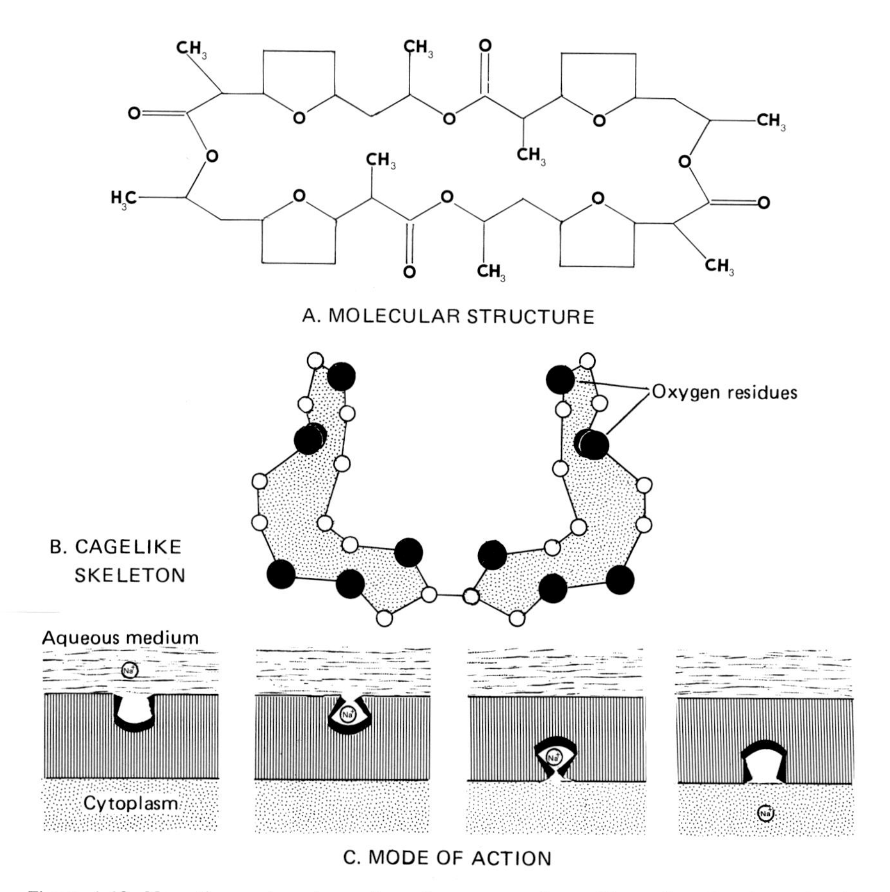

Figure 1.12. Nonactin, an ionophore. According to crystallographic studies (B), the molecular structure (A) is in the form of a crown, into which Na^+ cations can penetrate. (C) The crown is believed to close when it has received a Na^+ cation and then to rotate on its axis as it traverses the membrane to the inner surface, where the ion is released.

phosphatidylinositol or phospholipid fatty acids. More recently, it has been found that stimulation of macrophages with this substance was accompanied by release of prostaglandins, which in turn were believed to increase the levels of cAMP in the cells (Gemsa *et al.*, 1979). Quite different effects were noted when this and a related ionophore were used in the stick insect, *Carausius morosus* (Orchard *et al.*, 1979).

Channel Producers. The second type of ion conduction, channel formation, is best illustrated by the antibiotic gramicidin A; a number of issues about its mode of action still remain unresolved, however. It has been firmly established that two molecules of the substance are required to form one channel, but whether they do so by means of a $\pi_{L,D}$-helix or by a β double helix is not known (Veatch *et al.*, 1974). Both conformations apparently are capable of conducting cations through the pores they thus form, but are not selective in their action. Nonetheless, preferences are shown in the sequence

$$H^+ > NH_4^+ > K^+ > Na^+ > Li^+$$

as shown by Hladky and Haydon (1970). The probable existence of channels, rather than a carrier mechanism, is supported by that activity continuing at transition or actual gelation temperatures, with only minor reductions in rate. The presence in biological membranes of gramicidin A or any other antibiotic with similar action remains to be demonstrated, because these substances permit ions to leak out of the cells they attack. Thus the antibiotics are useful only as models of processes, especially for experimental approaches to the problem, but actual channel-forming proteins have been isolated from the mucosal membranes of the stomach (Sachs *et al.*, 1974). Several fractions were separated by polyacrylamide gel electrophoresis, each of which was tested individually to ascertain possible ion selectivity in a synthetic bilayer membrane. One such fraction of relatively low-molecular-weight members proved to display a selectivity for anions and exhibited a marked preference of Cl^- over SO_4^{2-}. In contrast, a second fraction, consisting of larger proteins, conducted cations of various kinds through the membrane; others, however, showed no selectivity, and still others, no conductance. Hence, it appears that only a few of the proteins of this membrane are channel formers active in ion conduction.

Active Transport. Transport of ions and other soluble substances, particularly of a variety of sugars and amino acids, typically occurs against both concentration and electrochemical-potential gradients and necessarily can be accomplished only with the expenditure of metabolic energy (French and Adelman, 1976). Hence, two outstanding features of active-transport systems are that they are unidirectional in flow across the membrane and that they require a continuous supply of energy. Consequently, any procedure that blocks cell respiration or other energy-storing activity also blocks active transport.

Several possibilities exist as to methods whereby chemical energy usage may be coupled to the transport of solutes. The simplest mechanism that can be envisioned is the movement of one particular ion or molecule from one side of the membrane to the other by direct employment of the energy, perhaps through changes in the charges on the particle (Garrahan and Garay, 1976; Caplan and Essig, 1977). But more complicated systems could involve the movement of a number of solutes simultaneously, two or more different types being linked to a common transport site before any can be translocated. Such linkage has been described in a system in which the electrochemical-potential gradient established by active transport of one solute provided the energy necessary for movement of a second substance against a gradient. Exemplifying this is the system in which the active transport of sodium is linked to that of potassium, which involves energy from ATP released by dehydrolysis of the membrane by ATPase (Skou, 1975, 1977a,b), which system is detailed in the following paragraph. Sometimes linkage seems to involve a third solute that moves as a by-product of the charges established by transport of the two others, and in some cases, an exchange across the membrane occurs, transport taking place in opposite directions.

The $Na^{+}+K^{+}$ ATPase System. The $Na^{+}+K^{+}$ ATPase system briefly mentioned in the preceding paragraph is perhaps the most important and widespread active-transport mechanism of the eukaryotic cells. Although by far the greater portion of studies have utilized the mechanisms in mammalian tissues as their subject, systems with similar properties are known from insects and other arthropods and mollusks (Bonting, 1970), as well as diatoms (Sullivan and Volcani, 1974). Among the shared characteristics are an absolute requirement of Na^+ and a need for a second monovalent cation, K^+, Rb^+, and $NH^{+}{}_4$ serving equally in this capacity (Albers, 1976). In addition all are inhibited by glycosides that are active on the heart, particularly ouabain, but these are active only when in the external medium (Schatzmann, 1953; Hoffman, 1978). The substrate is ATP in the presence of favorable divalent cations, Mg^{2+} yielding the highest rates of activity, followed in descending order by Mn^{2+}, Co^{2+}, Fe^{2+}, and Ni^{2+}. Ca^{2+} and many heavy metals have an inhibiting effect on interactions.

The system serves to pump sodium out of the cell and potassium in at apparently equal rates so that the sum of the sodium and potassium ions inside the cell remains constant; this can be expressed as $Na^+_i + K^+_i = 1$. Although the exchange of ions can occur in the absence of K^+_i and of Na^+_o (sodium outside) or both, K^+_o and Na^+_i are requisites (Hoffman, 1978). However, Na^+_o competes with K^+_o, and K^+_i does so with Na^+_i (Knight and Welt, 1974). The energy for the reactions comes from the breakdown of ATP into ADP and orthophosphate (P_i) in the presence of suitable divalent cations, all the products remaining within the cell. During active transport of these ions, antibody-inhibition studies have demonstrated that the ATPase undergoes conformational changes (Koepsell, 1979). It has been reported that the stoichiometry of the enzyme is such that for each ATP hydrolyzed, three Na^+_i are pumped out for each two K^+_o moved in (Post and Jolly, 1957; Whittam and Ager, 1965; Garrahan and Glynn, 1967). However, it is difficult to understand how the unity in $Na^+_i + K^+_i$ would be maintained if more Na^+_i leaves than K^+_o enters, unless an additional amount of K^+_o enters through others channels or Na^+_i leaks out. As shown in Figure 1.13, the enzyme is phosphorylated in the presence of Na^+ and dephosphorylated in that of K^+ (Hoffman, 1972).

The enzyme molecule is believed to undergo conformational changes as the various events occur, so that two states of the enzyme, E_1 and E_2, may be in the vectorial transport activities. The first of these is to be considered as hav-

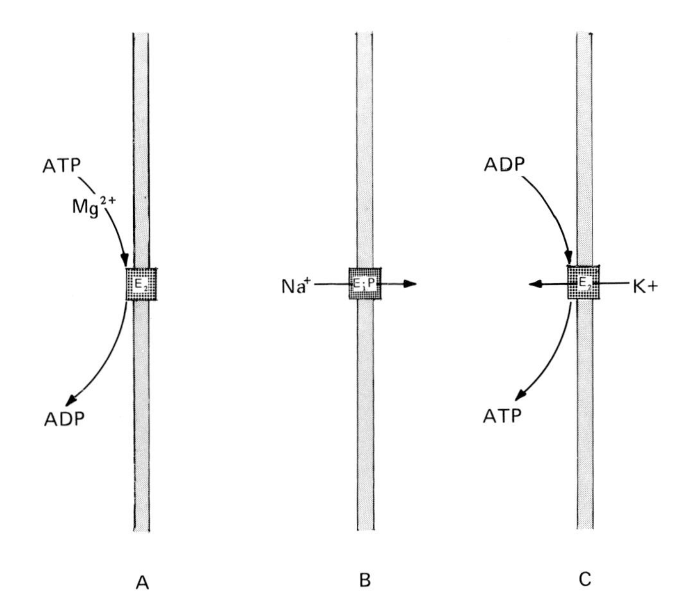

Figure 1.13. Steps in the transport of Na^+ and K^+ cations by the $Na^+ + K^+$ – ATPase pump.

ing its ionophore site oriented inwardly, the second one, outwardly (Albers, 1976). The interactions that occur may be expressed as occurring in the following six steps:

$$E_1 + ATP \rightleftharpoons E_1(ATP)$$

$$E_1(ATP) + 3Na_i^+ \underset{}{\overset{Mg^{2+}}{\rightleftharpoons}} E_1 - P : 3Na^+ + ADP$$

$$E_1 - P : 3Na^+ \rightleftharpoons E_2 - P + 3Na_o^+$$

$$E_2 - P + 2K_o^+ \rightleftharpoons E_2 - P : 2K^+$$

$$E_2 - P : 2K^+ \overset{Mg^{2+}}{\rightleftharpoons} E_1 - P : 2K^+$$

$$E_1 - P : 2K^+ \rightleftharpoons E_1 + 2K_i^+$$

ATPases in Membrane Transport. ATPases like that just mentioned are abundant in membranes of many types, not all of which are stimulated by such monovalent cations as those of the $Na^+ + K^+$ system. One type that appears widespread among eukaryotes, including the plasma membrane of *Neurospora* (Scarborough, 1977), is activated by divalent cations, especially Mg^{2+}. This is represented by two varieties, one of which is suppressed by ouabain and is independent of Ca^{2+}, whereas the second is just the opposite. The enzyme of

Neurospora is distinctive in that it is equally responsive to Mg^{2+} and Ca^{2+} and also does well with either Mn^{2+} or Zn^{2+}. Moreover, it requires the presence of a limited number of phospholipids, such as phosphatidylglycerol or phosphatidylserine, but is inoperative with phosphatidylcholine or phosphatidylethanolamine.

In bacteria and other prokaryotes ATPases do not appear to be present in the plasma membrane, in spite of the existence there of energy-requiring transport systems (Konings and Boonstra, 1977). Quite the opposite, ATP is synthesized in that membrane, both by aerobic and anaerobic processes such as nitrate and fumarate reduction. However, ATPase systems are present in the cell envelope as described in Section 1.2.4.

A vertebrate Na^+ pump that is driven by ATPase activity has received much attention and has been found in a number of tissues, including liver, skeletal muscle, kidney (Ismail-Beigi and Edelman, 1970, 1971), brain (Barker, 1951), and adipose tissue (Barker, 1964). In each of these cell types, the mechanism is stimulated by thyroid secretion, by way of receptors similar to those described later. The Na^+ pump is also limited to the production of ATP in the mitochondrial membrane, so that the two processes are tightly coupled. In resting cells, this ATP-utilizing transport accounts for 20 to 45% of the total energy expended (Ismail-Beigi, 1977).

Rate of Transport. The usual Michaelis–Menten formulation (Quinn, 1976) can be used to calculate the rate of active transport, regardless of the mechanism:

$$J = SV_{\max} / (K_m + S)$$

In this equation J is the rate of flow of a solute in one direction and S is the concentration of that substance at the membrane surface from which the flow takes place. $V_{\max}$ represents the maximum rate of flux and K_m is the concentration of the solute when $J = \frac{1}{2} V_{\max}$. In this case flux is assumed to be strictly unidirectional. If it is actually bidirectional, the net flux J_{net} can be calculated from

$$J_{\text{net}} = J_{\alpha-\beta} - J_{\beta-\alpha}$$

in which β and α represent the opposing sides of the membrane. The kinetics of various linkage groups, as well as those of carrier-mediated processes, are beyond the scope of this book, but they are provided in depth by Laprade *et al.* (1975) and Kotyk and Janáček (1977, p. 207 ff.).

Sugar and Amino Acid Transport. Other transport systems that have been actively explored are those involved in the movement of sugars across membranes, some of which also seem capable of transporting amino acids (Quinn, 1976). Many highly specific systems of this sort are found, not surprisingly, in

Figure 1.14. The brush-border region of a small-intestinal epithelial cell. A freeze-fracture preparation 90,000×. (Courtesy of Haase *et al.*, 1978, with permission of the *Biochemical Journal*.)

the brush-border membranes of the kidney and small intestine (Figure 1.14), areas particularly active in the movement of such substances. Silverman (1974) described one from the kidney that transports glucose and certain related sugar molecules only when they are in the chair configuration, while another used only D-mannose as substrate. In some instances, the sugar-transport systems have been suggested to be coupled to sodium movement (Kinne, 1976), but

Kimmich (1973) could find no evidence to support that supposition. In general in vertebrate tissues, glucose transport is related to cellular metabolic rates and often limits the rate of glucose use (Elbrink and Bihler, 1975). An additional sugar-transport system has come to light in *Chlorella* and in several bacteria and fungi, in which hexose movement is coupled to the influx of protons (Tanner, 1969; West and Mitchell, 1972; Seaston *et al.*, 1973; Tanner and Komor, 1975).

In a study of sugar transfer across erythrocyte membranes, a mechanism involving a tetrameric agent was proposed (Stein, 1972). According to this concept, the monomers are of two types, one having high affinity for the substrate, the other having low. In the tetramer, two representatives of each are arranged in an alternating fashion, so that one of each type lies side by side on the respective surfaces of the membrane (Figure 1.15A). At the center is an internal cavity the walls of which are formed by the monomers. All of these are assumed to be capable of rotating a quarter turn, but rotation is so synchronized that the active faces of one pair of opposing subunits are brought into the internal cavity while those of the second pair are moved to the respective surfaces (Figure 1.15B). A molecule of substrate that reaches the surface can be brought into the internal cavity by either a high- or a low-affinity monomer; however,

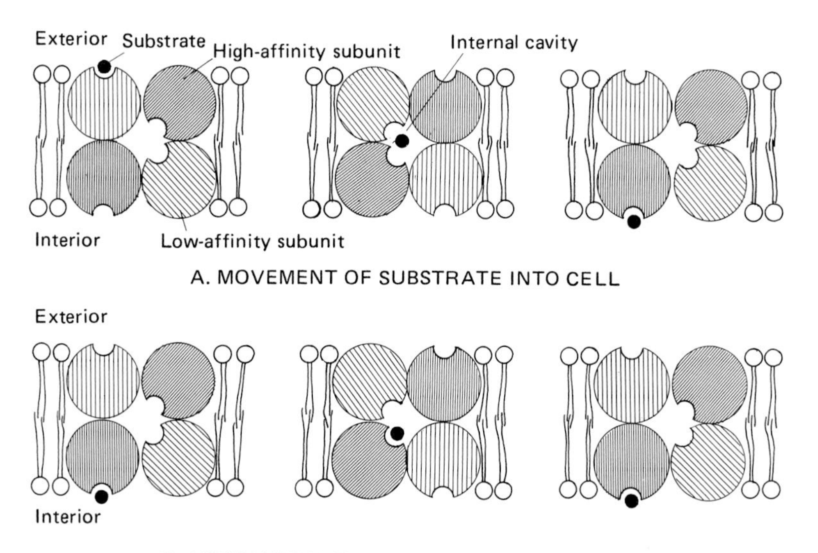

Figure 1.15. Tetrameric model of sugar transport. High- and low-substrate-affinity subunits are arranged in groups of four, one of each type being present on each face. Each is capable only of quarter turns, all movements being synchronous. If entry is by way of low-affinity subunits, movement across the membrane results (A), but no net movement ensues through entry by way of high-affinity subunits (B). (Based on Stein, 1972.)

since on reaching that location the active sites of the two types are available, it would certainly tend to react with the high-affinity type. Thus had arrival been by way of a low-affinity subunit, the molecule would leave by the high-affinity subunit and thus traverse the membrane. But if entrance had been made into the cavity via a high-affinity type, the molecule would leave only by the same route and be returned to the surface from which it entered.

Amino acid transport in the intestine has been reported to be by three groups of systems, each specialized for movement of certain categories of substrate (Quinn, 1976). Of these, one type interacts only with glycine, proline, and hydroxyproline, and another is specific for neutral amino acids that bear only one amino and one carboxyl radical. The last of the three is limited to the conduction of dibasic amino acids. In rat renal brush-border cells, a γ-glutamyl transpeptidase has been described that is a glycoprotein and an integral part of the membrane (Horiuchi *et al.*, 1978). Its active site has been shown to be oriented outwardly. Another system has been detected in Ehrlich ascites cells that is specific for those amino acids that have two carboxyl radicals, such as glutamic acid. Still another, in the plasma membrane vesicles of *E. coli*, is concerned solely with the transport of isoleucine (Yamato and Anraku, 1977). Finally, the kidney brush-border membrane has been shown to transport L-arginine (Busse, 1978).

1.2. THE PLASMA MEMBRANE (PLASMALEMMA)

So much of the experimentation performed on membranes reported in the foregoing section is based on plasma membranes that few additions can be made about structure or modes of functioning (Rothstein, 1978). Certainly it is far too early to attempt phylogenetic comparisons. Consequently, discussion is here confined to particular functions and characteristics unique to the plasma membrane, and to such topics that have been more thoroughly studied for this type of membrane than any other. Because of the intimate relationship that exists between the cell wall (outer membrane) and plasma membrane (inner membrane) of bacteria, the discussion includes the bacterial cell envelope in the closing section.

1.2.1. Endocytosis and Exocytosis

Activities that are largely, if not totally, confined to the plasma membrane of eukaryotes are those that involve the taking in of water, solutes, or large particles, known as endocytosis (Figure 1.16; Chapman-Andresen, 1977; Silverstein *et al.*, 1977), and the letting out of such substances, called exocytosis. The former activity is of two types, depending on whether the material admitted is liquid or particulate. Although the process of taking in particulate matter

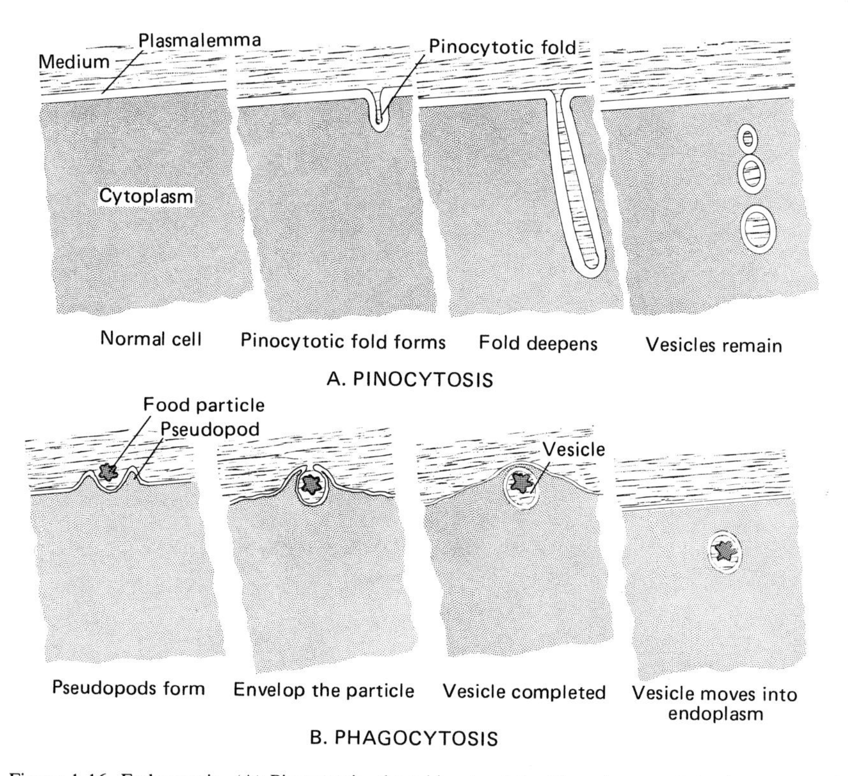

Figure 1.16. Endocytosis. (A) Pinocytosis, the taking in of liquids and suspended substances, and (B) phagocytosis, the taking in of larger particles, are the two aspects of endocytosis.

shares many steps of that in which water is imbibed, a few distinctive features are found, so that separate discussion is required. As the basic events are most clearly shown by liquid imbibition, it is examined first.

Pinocytosis. The taking in of liquids by a cell, including those containing dissolved substances, is referred to as pinocytosis (Lewis, 1931; Mast and Doyle, 1934; Holter, 1959). Viewed superficially, this act seems quite simple. At the surface, a shallow invagination appears in the plasma membrane (Andersen and Nilsson, 1960; Roth, 1960; Roth and Daniels, 1961; Andersen, 1962; Brandt and Pappas, 1962). As it deepens, it may also widen somewhat, but in many cells the dimensions remain relatively small. When a depth of sufficient proportions has been attained, the invagination is pinched off by the plasma membrane fusing over it, forming a vesicle filled with liquid (Figure 1.16A). This vesicle is then transported to wherever its contents may be needed.

Phagocytosis. Phagocytosis resembles the foregoing set of events in many details, but involves the taking in of particulate matter into the cell. When a cell encounters certain objects in the environment, this activity begins by the protrusion of a cup-shaped pseudopod around the particle (Figure 1.16B). The cup deepens until it surrounds the particle on all sides and extends beyond the object. Then the mouth of the cup closes and the plasma-membrane folds fuse as they make contact with one another, thus forming a vacuole containing both fluid and the particle. An enzyme-containing vesicle known as the lysosome (Chapter 7, Section 7.1) soon approaches and connects to such food vacuoles, or primary phagosomes, and hydrolysis of the substance it contains quickly begins.

Once the lysosome has become attached to the food vacuole, the combination is often referred to as a secondary phagosome (Saier and Stiles, 1975). As the digestive enzymes act on the included substance, they appear also to affect the structure of the former external surface of the plasma membrane, which now lines the interior of the phagosome. This lining becomes altered in appearance and, as degradation proceeds, it becomes increasingly permeable, first to molecular-sized particles, and then to larger fragmental by-products of the particle enclosed in the vacuole. These pass into the cytoplasm, where the larger pieces undergo further enzyme action until they too are broken down into molecules. In the meanwhile, the contents and lining also continue to be digested until the whole vesicle ultimately is consumed.

Exocytosis. During the breakdown of the phagosome, indigestible parts are usually present that must be disposed of. Usually these are collected in fine vesicles that form along the vacuolar wall; when full these are pinched off as microvesicles that are transported to the plasma membrane. After contact has been made, the membranous coat of the microvesicle fuses with the plasma membrane, a minute pore opens in the latter and enlarges, while the microvesicle expands, empties its contents, and then remains as a part of the plasma membrane (Figure 1.17).

This process of exocytosis is used not only to eliminate undesirable rem-

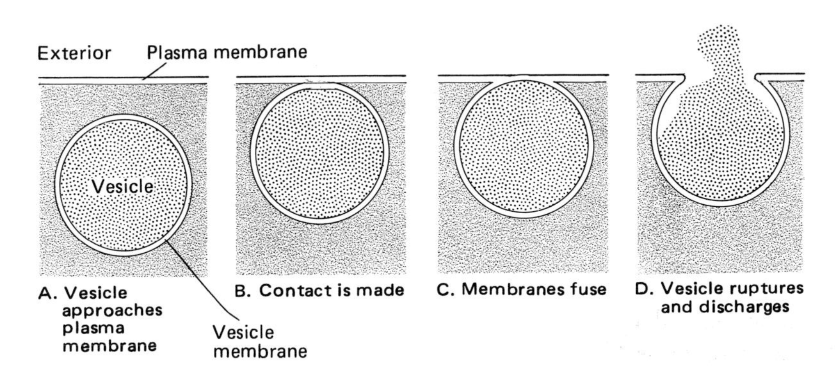

Figure 1.17. Exocytosis. Exocytosis, the discharge of matter from a cell, involves fusion of the membranes of a vesicle and the plasmalemma, followed by rupturing and discharge.

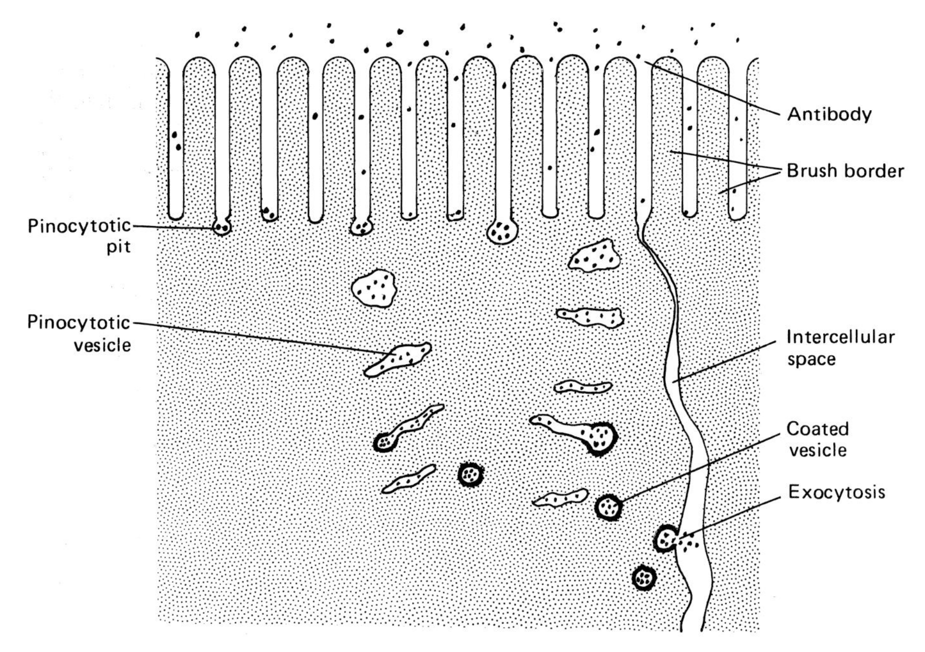

Figure 1.18. Transport and exocytosis of antibody in intestinal brushborder cells of rat. (Based in part on Rodewald, 1973.)

nants of digestion but also provides a means for secretory cells of glands to place their products into the bloodstream, for instance, or the lumen of the digestive tract, perhaps (Figure 1.18). The decidual cells of the human ovary behave in similar fashion (Herr and Heidger, 1978). One clear example of pinocytosis along with exocytosis is provided by Rodewald (1973), in which antibodies are taken into an intestinal cell of a rat and transported (Figure 1.18). The antibody molecules penetrate between the microvilli of the brush-border cells and appear to be bound selectively to the plasma membrane at specialized pits. These pits then deepen and split off to form pinocytotic vesicles (Figure 1.18), from which the molecules of the antibody are absorbed along with water. These nearly empty vesicles then enlarge and become coated with an unknown substance; eventually these spheroid coated vesicles reach the plasma membrane on the side of the cell, where exocytosis takes place (Figure 1.18). What product is contained in the coated vesicles is still undetermined.

Molecular Events in Endocytosis. Very little has been accomplished toward understanding endocytosis at the molecular level, although several studies have been made of phagocytic activities with the scanning electron microscope (Lockwood and Allison, 1963; Horn *et al.*, 1964; Goodall and Thompson, 1971; Lawson *et al.*, 1977). Perhaps the chemical requirements to trigger the engulfing reaction have been the most thoroughly explored area. In general only substances that carry a positive charge on their surface are capable of stimulating endocytotic reactions; neutral and negatively charged particles produce no reaction. Fresh bacteria artificially offered to such phagocytic cells as leukocytes fail to induce engulfing, but if the bacteria are first immersed in blood serum where a coat of a cationic substance known as opsinin is applied over them, the organisms are quickly phagocytized when offered to the leukocytes (Menkin, 1956).

On polymorphonuclear leukocytes, which have received much attention in the literature, two different kinds of surface receptors have been recognized, from both murine and human sources (Mantovani, 1975; Scribner and Fahrney, 1976). One of these, the C3b receptor, mediates the binding of particles to the cell, an action that is an essential prelude to phagocytosis. The second one, known as the Fc receptor, induces both particle binding and endocytosis (Wilton *et al.*, 1977). One interesting action was brought out by the latter investigators that bears indirectly on the influence of environmental factors on the genetic mechanism in cells in general; in this case, how one activity may influence subsequent ones. In man, they pointed out, the polymorphonuclear leukocytes reach the gingival crevice of the mouth by passing between the cells of the oral epithelium. After they have thus entered the cavity, they are less capable of phagocytosing the blastospores of such yeast as *Candida albicans* than they are when still in the bloodstream. The results of this study indicated that this loss of capability ensued from a decrease in the effectiveness of the C3b receptor. Because this capacity is an inherited characteristic, some alteration in gene function appears to be involved.

The scanning electron microscope studies mentioned at the beginning of this section showed that the sites at which bacteria became attached to the plasma membrane involved limited areas of the surface, only 0.01 to 0.25 nm^2 in extent. They also demonstrated that the attachment sites were located on extensions of the membrane that were present before, during, and after phagocytosis of the bacteria. Recently a study was made of phagocytosis in polymorphonuclear leukocytes using a combination of freeze-fracture scanning and thin-section transmission electron microscopy (Moore *et al.*, 1978). Because membranes during freeze-fracture explorations tend to separate between the two bimolecular lipid layers, the investigation largely concentrated on the relative densities of the intramembranous particles this technique revealed. Although obviously the formation of phagocytotic vesicles necessitates the formation of greater membrane area, no change in density of the particles could be noted during phagocytosis of bacteria by the white blood cells. Another transmission electron microscopic study of phagocytosis, this time on the ciliate *Nassula* sp., made it clear that the phagocytotic formation of food vacuoles at the cytostome involved a flow of highly gelated cytoplasm alongside rows of arm-bearing microtubules (Tucker, 1978). The invagination during the formation of phagosomes appeared to be covered by two membranes, one from the plasma membrane, the other from an unknown source.

1.2.2. Receptor Sites

Another important feature of the plasma membrane is, like endo- and exocytosis, confined to the eukaryotes as far as indicated by current knowledge. Many prokaryotes have been found to have specialized areas for the reception of specific substances, as already alluded to in the foregoing discussion, but the number of sophisticated mechanisms appears to be limited. As shown below, many of those of eukaryotes that have been studied in detail involve endocytosis as a significant feature.

Secretion of Thyroxine. In the processes of synthesizing the hormone thyroxine, a precursorial molecule known as thyroglobulin is first translated into existence on the ribosomes. From there, it is transported in the form of vesicles to the plasma membrane, where the tyrosine residues become iodinated (Figure 1.19). After this step is completed, exocytosis occurs and iodinated thyroglobulin becomes stored in the lumen of the follicle in the form of a colloidal suspension (Stanbury, 1972). The thyroglobulin is thus stored until a signal is received by way of the bloodstream. This signal, in the form of thyroid-stimulating hormone (TSH) from the pituitary, interacts with a special receptor site located in the plasma membrane; this consists of the enzyme adenyl cyclase, which converts ATP into cyclic AMP, releasing inorganic pyrophosphate (Onaya and Solomon, 1969).

How the cyclic AMP is transported across the cell has not been firmly established. Either it is conducted directly from the side of the cell adjacent to the blood vessel to the region facing the follicular lumen or it enters a lysosome (Chapter 7) and reaches its destination in that form. At any rate, the presence of the nucleotide induces the entrance of colloidal thyroglobulin into the cell from the follicular lumen by means of pinocytosis (Figure 1.19). As the endocytotic vesicles thus formed are being conducted across the cell toward the bloodstream, lysosomes attach to them; the enzymes thereby admitted into this secondary phagosome commence to hydrolyze the thyroglobulin, converting it into thyroxine. By the time a vesicle has reached the vascular surface, it contents have been fully converted into the hormone, ready to enter the blood vessels by means of exocytosis (Figure 1.19).

Adenylate cyclase appears to serve as a receptor for specific hormones and other chemicals in a variety of target tissues, as well as serving in a number of other processes. Recently its presence in the present capacity has been noted in renal cells of mammals, where it is activated by vasopressin (Jard *et al.*, 1977). Earlier its activation by glucagon had been described in rat liver (Pohl *et al.*, 1969), and it had been reported also to be activated by gonadotropins (Suther-

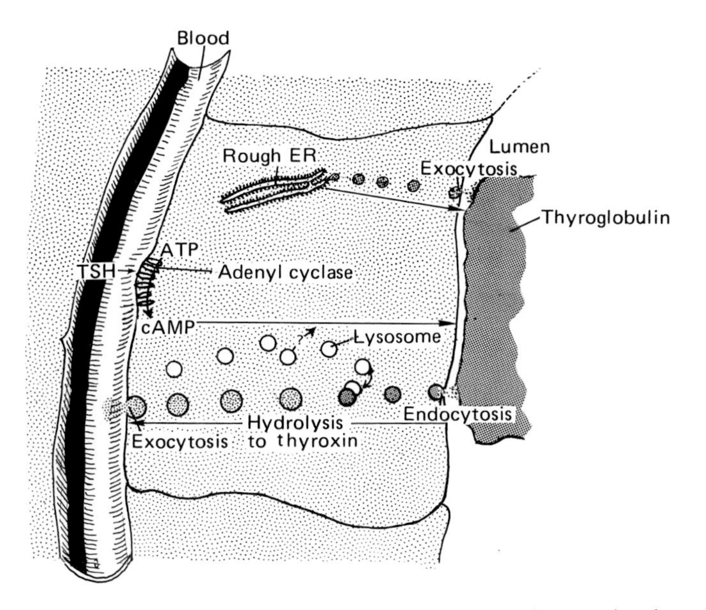

Figure 1.19. Endocytosis and exocytosis in cell functioning. In thyroxine secretion, known steps include, first, the biosynthesis of a precursor molecule (thyroglobulin) in the rough endoplasmic reticulum, followed by transport to the plasmalemma, where iodination occurs. The iodinated thyroglobulin is then discharged by exocytotic processes into the lumen for storage. Upon receipt of a signal at a receptor site in the form of thyroid-stimulating hormone (TSH), cAMP is transported to the appropriate region, stimulating the entrance of the thyroglobulin into the cell by pinocytosis. While being transported to the bloodstream, modification of the precursor into thyroxine occurs by lysosomal action; later the secretion is expelled from the cell, again by exocytosis.

land *et al.*, 1962). Other substrates known to be recognized include neurotransmitters (Robinson and Sutherland, 1971), histamines, serotonin (Kakiquchi and Rall, 1968), fluorides, and cardiac glucosides. Usually a diesterase is also present that acts as a control by converting cAMP to 5′AMP, thus inactivating it.

Cyclic AMP has also been implicated in the intercellular communication shown to exist in many of the cellular slime molds. As the mold amoebae (myxamoebae) begin to starve, some members of the colony emit rhythmic pulses of chemical signals, or acrasins. These are slowly relayed outwards from these initiation centers, inducing the movement of the myxamoebae toward those focal points in the form of complex bands (Newell, 1978). In the case of *Dictyostelium*, the acrasin is cAMP (Konijn, 1972), but in other forms, including *Polysphondylium*, it is a small peptide (Wurster *et al.*, 1976; Bonner, 1977).

Lipoprotein Receptor Mechanism. Another interesting receptor mechanism described in recent years (Brown and Goldstein, 1974, 1975, 1976; Goldstein and Brown, 1976, 1977) is associated with a number of human cell types, but was first discovered in the endothelial cells of the artery. The mechanism is concerned with the absorption of low-density lipoprotein molecules and their conversion into cholesterol; because this substance in too large quantities leads to atherosclerosis but at the same time is required for membrane biosynthesis, sensitive cellular controls are obviously involved. Although all details have not been fully elucidated, the main features appear to be as follows.

A receptor capable of binding the specific low-density lipoprotein is located in the plasma membrane; while the identity of the active protein remains unknown, the actual site has been observed under the electron microscope (Goldstein *et al.*, 1977). The receptor protein has an extremely high affinity for the lipoprotein, even when it is present in the milieu in very low concentration. After the receptor and lipoprotein have interacted, the latter is taken into the cell enclosed in pinocytotic vesicles. As in the case of thyroxine, lysosomes join these vesicles and after they fuse together, their contents hydrolyze the lipoid and protein moieties of the lipoprotein into free cholesterol and amino acids. The cholesterol can then be employed in membrane synthesis, but its presence has two other actions. First, it suppresses the cell's own biosynthesis of that lipid by reducing the activity of one of the essential enzymes (3-hydroxy-3-methylglutaryl coenzyme A reductsse) involved in those processes. And second, it activates an enzymatic complex (microsomal acyl-CoA:cholesterol acyltransferase), so that any excess cholesterol taken in by the low-density lipoprotein receptor system is reesterified for storage by the cell. In turn, when such stored cholesterol esters have reached an adequate level, the biosynthesis of the receptor protein is suppressed, thereby restricting further endocytosis of the lipoprotein.

Miscellaneous Receptor Systems. A number of other acceptor systems have been brought to light, but as these have not been thoroughly investigated, the nature of the acceptor usually remains undetermined. For instance, aldosterone and related steroid hormones interact with an acceptor in the plasma membrane of toad bladder cells that somehow activates the genetic mechanism to produce one or more proteins. In turn these proteins stimulate Na^+ ion transport across the plasma membrane (Sharp and Leaf, 1966; Cofré and Crabbé, 1967). Luteinizing hormone is known to interact at the plasma membrane with target cells, with about 5000 molecules binding to the receptor (DeKretser *et al.*, 1971). In a series of experimental studies along these lines, insulin has been bound to large Sepharose beads to prevent its entrance into the cells, thereby restricting the interactions to surface receptors (Monod *et al.*, 1965; Mueller and Rudin, 1968; Cuatrecasas, 1969, 1971; Miller *et al.*, 1971). In contrast, growth hormone (STH) appears to exercise its action upon the plasma membrane in general, rather than at a specific receptor site. For instance, in the presence of STH the membrane proteins of human erythrocytes were found to decrease in mean ellipticity by about 25% (Schneider *et al.*, 1970; Sonenberg, 1971).

1.2.3. Intercellular Communication

In addition to the chemical communications alluded to in the foregoing activator–receptor discussion, cells may communicate in a more direct way on a one-to-one basis by contact. Recognition or interaction may in turn involve chemical communicants, too, possibly in the form of antigen–antibody reactions. But often in tissues where two cells contact each other, special modified structures are developed, which seemingly are for communicative purposes. A small number of types of junctions have now been described, plus a wide diversity of lesser varieties within the major categories (Figure 1.20A). At the present time, this aspect of membrane biology is particularly active.

Gap Junctions. Perhaps the commonest type of communication mechanism known is the gap junction or nexus (Revel and Karnovsky, 1967; NcNutt and Weinstein, 1970, 1973; Merk *et al.*, 1973; Staehelin, 1974; Gilula, 1977; Kensler *et al.*, 1977) for they occur in many tissues (Figure 1.20B). Included in the list of tissues where they have been noted are cardiac and smooth muscle, epithelium from a diversity of organs and connective tissues, and at electrical synapses of the nervous system (Ribi, 1978). By use of fluorescent dye, actual transport has been observed to occur at such junctions but not elsewhere along the lines of contact between cells (Bennett *et al.*, 1978). On the surface, too, the nexus appears to be the simplest type, for sections show regions of the plasma membranes of two adjacent cells to lie close but with an apparent gap between their smoothly parallel surfaces. If the gap is filled with an electron-opaque substance, such as procion brown, lanthanum, or ruthenium red, a latticelike reticulum is seen to fill the space. When the cells are separated at such a gap junction and the corresponding surfaces examined, a large number of close-set polygonal particles 8–9 nm in diameter are found on the plasma (P) surface and a complementary series of pits on the external (E) face (Figure 1.21). This structural arrangement appears to characterize metazoan cells in general, except in arthropods (McNutt and Weinstein, 1970). In the latter animals, the particles are larger (10–30 nm) and more widely spaced (Figure 1.21B; Flower, 1972; Johnson *et al.*, 1973; Peracchia, 1973; Gilula, 1974).

Such gap junctions sometimes appear to be formed in large numbers between adjacent cells in normal tissues, even between different cell types in the same tissue (Johnson *et al.*, 1973; Simionescu *et al.*, 1978). McNutt and Weinstein (1971) reported that, whereas the row of young cells in the cervical epithelium had only 5 to 10 nexuses per cell, those of the intermediate zone had between 100 and 200. In contrast, the cells of invasive carcinoma of that same tissue had a mean number less than one. These junctions of normal tissue

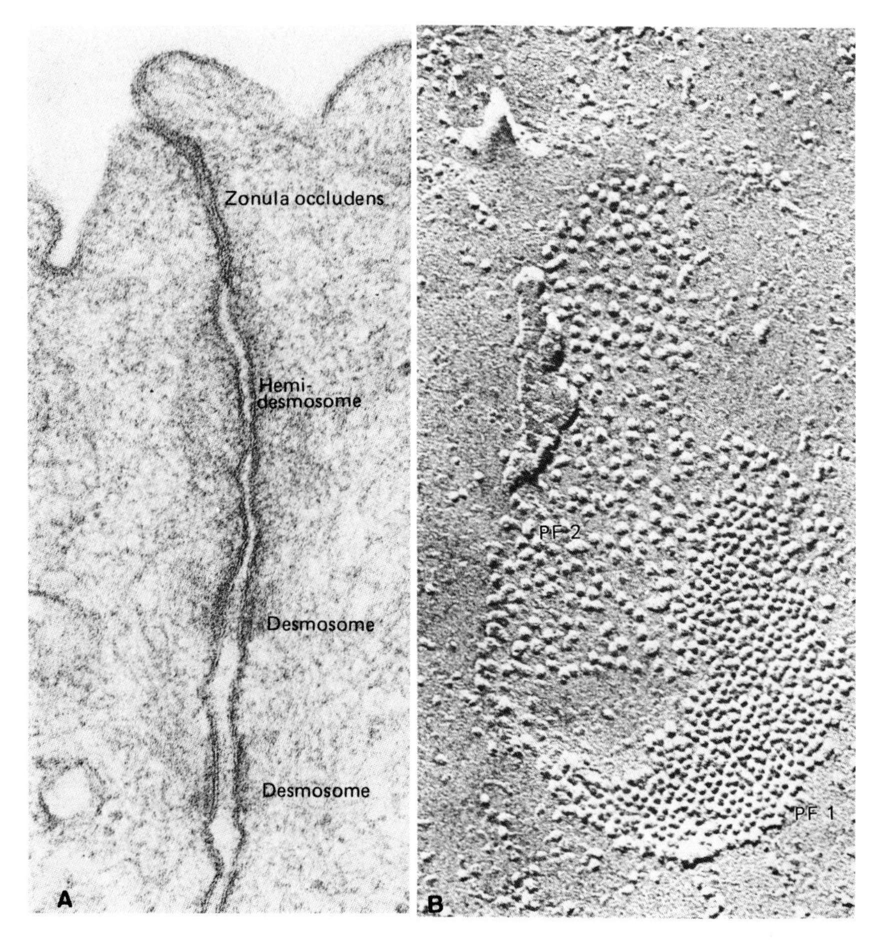

Figure 1.20. Types of intercellular junctions. (A) The region of contact between two epithelial cells of rat bladder shows several common types of junctions. 95,800×. (B) Freeze-fractured plasmalemma of a bladder epithelial cell shows two types of gap junctions on the plasma face. The PF-1 type has closely spaced small intramembranal particles 9 nm in diameter, whereas those of the PF-2 type are coarser and more widely spaced. 129,000×. (Courtesy of Pauli *et al.*, 1977, and *Laboratory Investigation.*)

→

Figure 1.21. Gap junction structure. (A) Plasmalemma of eipthelial cell of skin wound of rat. The P face shows the slightly spaced fine granules and the E face, the complementary depressions. 172,000×. (Courtesy of Gabbiani *et al.*, 1978, and *Journal of Cell Biology.*) (B) In the gap junction of arthropod tissues, such as this *Limulus* epithelium, the granules are much coarser than those of most metazoans. 112,000×. (Courtesy of Johnson *et al.*, 1973, and *Journal of Ultrastructure Research.*) (C) Diagram of one interpretation of gap junction structure.

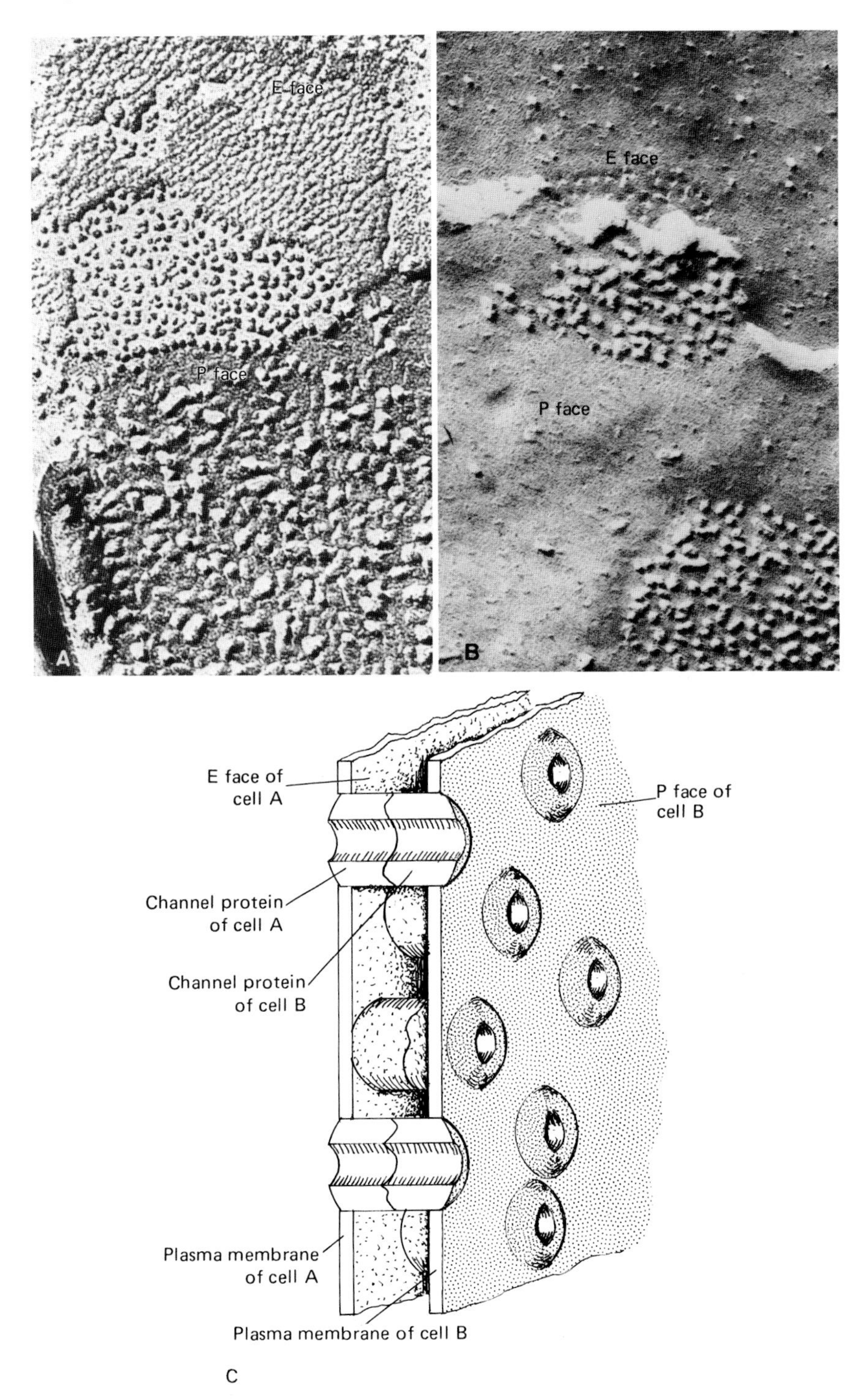
E face
P face
A
E face
P face
B
E face of
cell A
P face of
cell B
Channel protein
of cell A
Channel protein
of cell B
Plasma membrane
of cell A
Plasma membrane of cell B
C

require a longer time to develop in aging cells and are fewer in number (Kelley *et al.*, 1979). Another feature of at least certain gap junctions was originally described by Loewenstein (1966). At first detected by such simpler methods as employment of fluorescent and color tracers injected into large epithelial cells, the presence of channels subsequently has been confirmed by more sophisticated techniques (Goodenough and Revel, 1970; Loewenstein, 1975, 1977). The channels, which are estimated to be in the vicinity of 200 Å in length and to have a lumen about 10 Å in diameter, are each formed by two protein molecules, opposing one another in adjacent cell membranes, and are precisely aligned and interlocked (Figure 1.21C). It has been suggested that the particles that form the lattice on the surface of the gap junctions are parts of channel-forming proteins (Loewenstein, 1977); however, this view does not seem to explain the corresponding pits noted in the opposing side of the membrane.

An ontogenetic study makes it clear that gap junctions, and probably other types, are genetically controlled and undergo an orderly sequence of developmental steps (Gros *et al.*, 1978). This research was conducted on mouse myocardium at three fetal stages (at 10, 14, and 18 days postcoitum) and in the adult. Apparently the first indication of a gap junction was the appearance in the 10-day fetus of linear series of 9-nm particles on the P face of cardiac muscle cells. At one end of the series was a patch of similar particles that served as the nucleus for further growth of the junction. Enlargement appeared to be by accretion of the linear arrays and other particles around this center. The other types of junctions described in the following paragraphs seem similarly to undergo orderly series of developmental events.

The formation of channels at gap junctions has also been the subject of study recently (Loewenstein *et al.*, 1978), in which the conductance was measured during development. The channels that typify such junctions were assumed to be exactly alike. Because such pores do not exist in isolated cells but develop within a few minutes after two cells contact one another, high-resolution measurements of current conductance were made of newly united cells. The conductance was found to increase in quantum jumps as the membrane channels developed. Once formed they persisted for some time, in contrast to many ionic channels, including those of gramicidin, which have been found to be short-lived (Neher and Stevens, 1977; Bamberg *et al.*, 1978; Urban *et al.*, 1978).

Figure 1.22. Zonula occludens structure. (A) Section through zonula occludens of horny cells of rabbit skin. 150,000×. (Courtesy of Woo-Sam, 1977, and *British Journal of Dermatology*.) (B) Fractured plasmalemma of epithelial cells of toad urinary bladder. The height of the ridges (H_1) is equal to the height of the E face (H_2). 104,000×. (Courtesy of Wade and Karnovsky, 1974, and *Journal of Cell Biology*.) (C) In some cases, the elevated lines of the P face form an intricate network, as in this cell from the rat rete testis. 68,000×. (Courtesy of Suzuki and Nagano, 1978a, and *Anatomical Record*.)

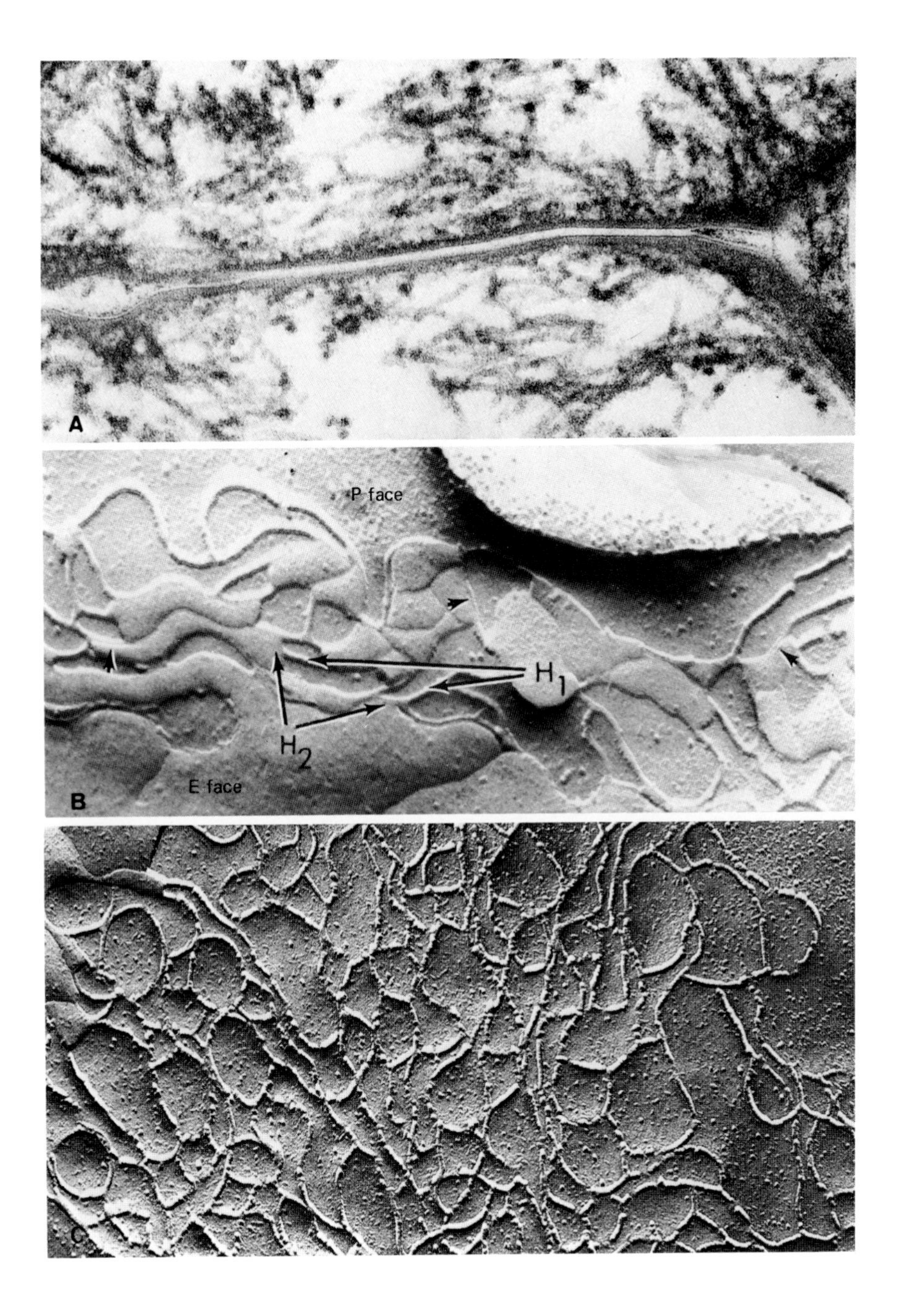
A
P face
H_1
H_2
E face
B
C

Zonula Occludens. A structure of less frequency than the gap junction is the zonula occludens or tight junction, which is confined principally to the apices of adjacent cells in columnar epithelia, such as those of the digestive tract, urinary bladder, and various glands (Pauli *et al.*, 1977). This feature consists of a narrow region (Figure 1.22A; Woo-Sam, 1977; Koga and Todo, 1978; Suzuki and Nagano, 1978a) where the two plasma membranes contact one another so closely that the combined thickness (140 Å) is less than the sum of the two separately (150 Å). It has been shown by freeze-fracture studies (Wade and Karnovsky, 1974) that on the P surface of the junction are series of ridges, opposite which on the E surface is a similar system of grooves (Figure

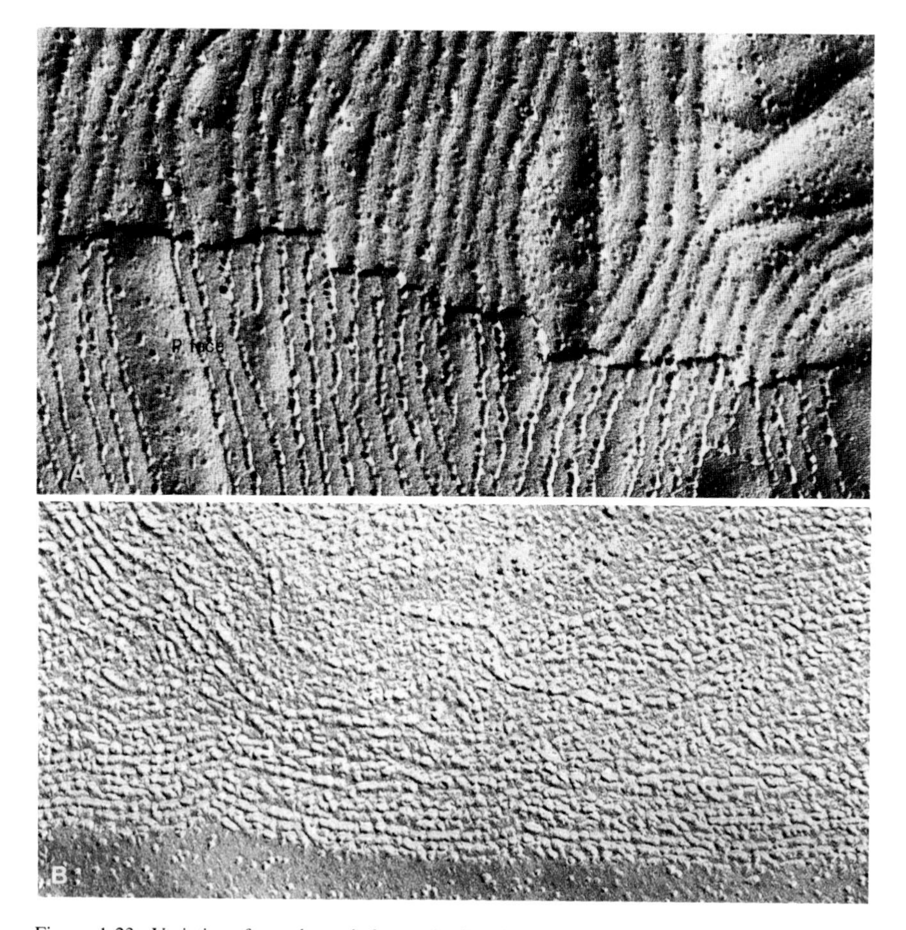

Figure 1.23. Varieties of zonula occludens. (A) Sertoli cells of rat have linear arrangements of elevations on the P face, with corresponding grooves on the E face. 71,000×. (Courtesy of Suzuki and Nagano, 1978a, and *Anatomical Record*.) (B) A septate junction from glandular vas deferens of *Ascaris*. 110,000×. (Courtesy of R. C. Burghardt, unpublished.)

1.22B,C; Suzuki and Nagano, 1978a,b), but a number of variations occur (Figure 1.23). Sometimes, as in the urinary and gallbladders, particulate substances in the junction form fibrils between the membranes, and in the salt gland cells of birds, the zonula is of markedly simple structure (Riddle and Ernst, 1979). As the name implies, the zonal connection occludes the substances of large molecular size from penetrating through the intercellular space. In a variant of this type of intercellular connection, the macula occludens, large molecules are able to penetrate throughout the intercellular space (Quinn, 1976).

The Desmosome. Another type of junction, most frequently found in epithelial tissues but also in muscle, is one of much greater complexity than those described above (Curtis, 1967; Shimono and Clementi, 1976; Woo-Sam, 1977). Rather than being a particular species of membrane organelle, the structure differs from one tissue to another. In muscle, for example, there is a pair of thick plaques of electron-dense material *within* the intercellular space where the plasma membranes of two cells more or less parallel one another (Curtis, 1967). Within each cell, slightly removed from the membranes, is another narrower and more diffuse plaque, which may or may not be fibrillar.

Another quite distinctive type has been reported from newt epithelial cells (Quinn, 1976). The adjacent membranes are parallel as in the foregoing example, but in the present case the electron-opaque plaques lie *beneath* the surfaces, while the intercellular space appears filled with transverse fibers (Figure 1.24A,B). From the plaques toward the interior of the cells extends a mass of tonofilaments, the function of which remains unknown. Near the plaques these filaments bear granules, or loops according to some laboratories (Quinn, 1976). Similar desmosomes are found in healing wounds in normal epithelial cells (Gabbiani *et al.*, 1978), but occur in smaller numbers in healing tissue (Figure 1.20A).

1.2.4. The Bacterial Cell Envelope

In bacteria the cell membrane structure varies with the type. Those forms that stain with Gram's reagent, that is, are gram-positive, have a plasma membrane over which is a simple wall, known as the dense layer (Figure 1.25A). This is made of peptidoglycan, also known as murein. Over this wall is laid an S layer, made of repeating subunits (Sleytr, 1978). In bacteria that do not stain with Gram's reagent, that is, are gram-negative, like *Escherichia coli*, an additional membrane is added between the dense and S layers (Figure 1.25B). Around this is still another covering, usually referred to as the capsule or slime. As the structure of the gram-negative coat really embraces that of the gram-positive organisms, it is necessary only to examine the former in detail.

The Cell Envelope of Gram-Negative Bacteria. To study the cell envelope of *E. coli* or other gram-negative bacteria, the organisms are first treated with lysozyme, an enzyme that hydrolyzes the peptidoglycan of the dense

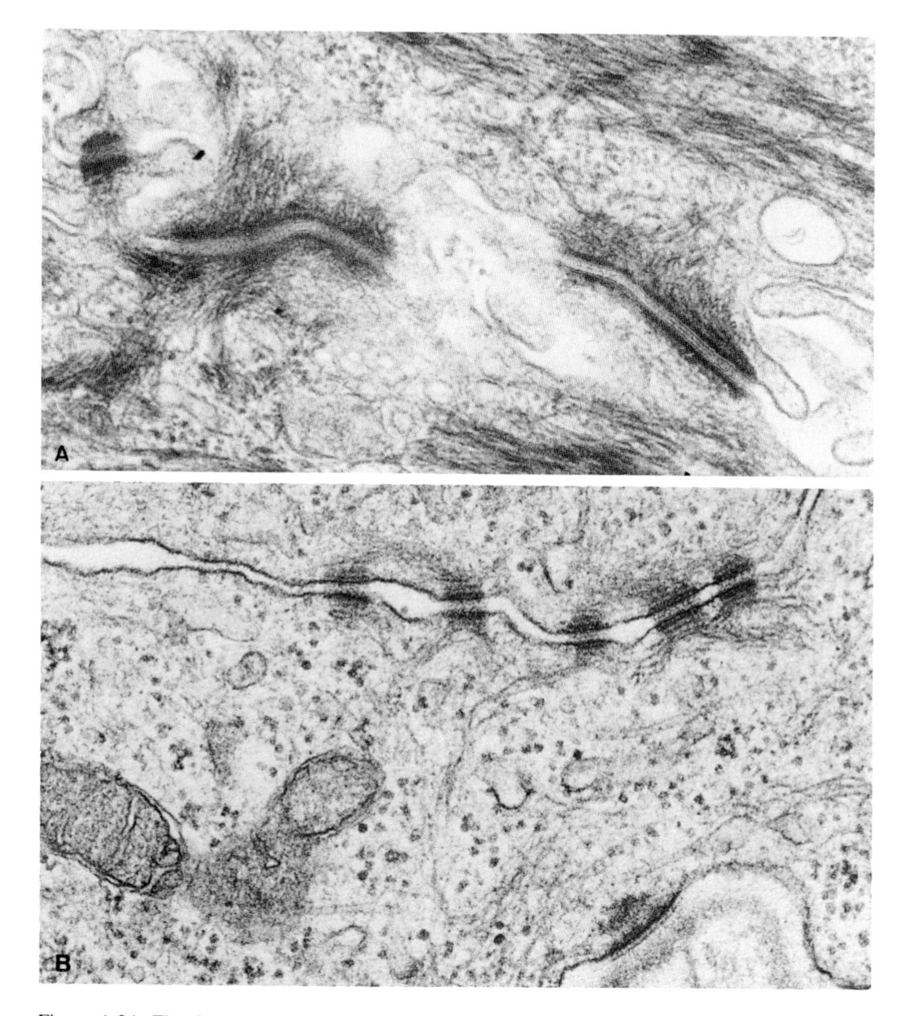

Figure 1.24. The desmosome. Plaques of electron-dense material occur variously on the exterior surface of the membrane, or as here, on the plasma surface. (A) Basal epidermis of rat. 51,400×. (Courtesy of Gabbiani *et al.*, 1978, and *Journal of Cell Biology*.) (B) This section of epithelium of the rat urinary bladder shows a series of five desmosomes, complete with interconnecting tonofilaments, and a hemidesmosome at lower right. 6,800×. (Courtesy of Pauli *et al.*, 1977, and *Laboratory Investigation*.)

layer. The result of this treatment is the detachment of the outer membrane from the inner; as the former is denser than the latter, the two are readily separable by sucrose density gradient centrifugation. When the membranes of another gram-negative form (*Salmonella typhimurium*) were separated in this fashion, the outer was found to contain three times as much protein as the inner, but the latter contained many more species (Lee and Inouye, 1974;

Gmeiner and Schlecht, 1979). The proportion of phospholipid to protein was 0.3–0.98 : 1, and of lipopolysaccharide to protein 0.53–0.61 (lipid): 0.09–0.14 (saccharide): 1 (protein) (Osborn *et al.*, 1972).

The Outer Membrane. The proteins of the outer membrane have received especially detailed attention (Inouye, 1975). Four main peaks have been found by gel electrophoresis, plus seven minor ones. That which produced peak 4, referred to as the matrix protein, had a strong affinity for peptidoglycan, to which substance it appeared to be bound in the intact organism (Rosenbusch, 1974). It consisted of a chain of 336 amino acid residues and had a molecular weight of 36,500; perhaps 40% of the amino acids were thought to be hydrophobic in nature. While the precise function of this protein is unknown, it is quite abundant, there being 150,000 molecules present per cell, a quantity sufficient to cover about 54% of the peptidoglycan layer. Thus this protein could assist in maintaining cell shape, along with the peptidoglycan.

The most abundant protein is that which formed peak 11, about 750,000 molecules being present per cell; it was determined to be quite small, having a molecular weight of only 7500 (Inouye, 1975). Like the foregoing matrix protein, it is bound to the peptidoglycan layer, but possible projects through the lipid bilayer to form channels. The two remaining proteins that were identified occur in much smaller proportions. That which formed peak 6, known as the Y protein, was shown to be related to DNA synthesis by means of thymine starvation. When DNA synthesis was thus inhibited, peak 6 decreased significantly in height. The last of the four proteins formed peak 7 and was named the *tol* G protein, as it may be involved in certain mutants known by that name. Because this protein, of molecular weight between 37,000 and 40,000, was rapidly degraded when the membranes were treated with protease, it was suspected to lie on the outer surface of the membrane where it would be readily accessible to the enzyme.

In addition there are an unknown number of other proteins present, among which are a variety of receptor types. One of these is the receptor that is sensitive to attack by bacteriophage T5 as well as to colicin M and has been reported

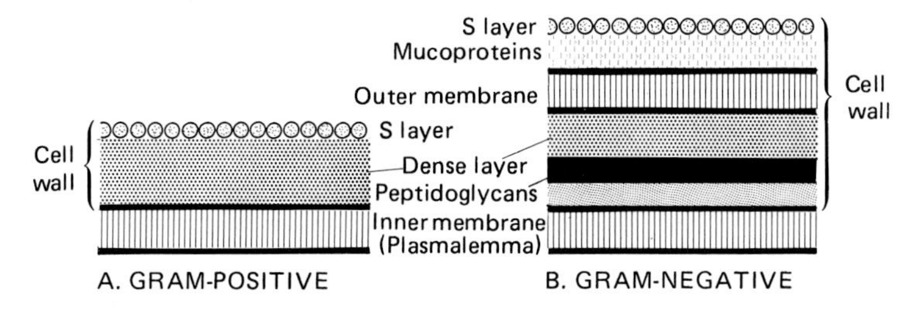

Figure 1.25. The cell envelope of bacteria. (Based in part on Sleytr, 1978.)

as having a molecular weight of 85,000 (Braun and Wolff, 1973; Braun *et al.*, 1973). So far it has not been established whether this and similar but less well-known receptors react only with phages and colicins or whether they have other functions (Tamaki *et al.*, 1971; Leive, 1974). Moreover, there is a phospholipase that appears to be covalently linked to the lipid A portion of an abundant lipopolysaccharide of the outer membrane. As the lipopolysaccharide is present in about 3.5×10^5 molecules per cell, the protein should be a major one; nevertheless, it has not been identified with any of the four protein peaks described in the foregoing paragraphs (Wu and Heath, 1973).

During recent years a number of proteins of gram-negative bacterial outer walls have been documented to serve as pores through which nutrients and other solutes may pass, including several proteins associated with the peptidoglycans (Fischman and Weinbaum, 1967; Pugsley and Schnaitman, 1978; van Alphen *et al.*, 1978b). Apparently all three proteins that have been thoroughly investigated were found to be associated with lipopolysaccharides and arranged across the membrane, forming aqueous pores. Moreover, they served as the protein portion of phage receptors (Lugtenberg *et al.*, 1978; van Alphen *et al.*, 1978a, 1979; Verhoef *et al.*, 1979).

The Dense Layer. The layer between the outer and inner membranes of gram-negative forms and covering the plasma membrane of gram-positive species is highly complex and widely variable from species to species. Basically the layer consists of peptidoglycans (Schmit *et al.*, 1974; Kotyk and Janáček, 1977), which in turn are built on a frame formed by essentially alternating molecules of two related substances, *N*-acetylmuramic acid and *N*-acetylglucosamine (Figure 1.26A,B). The proportions of these two constituents are rarely 1:1. For instance, in *Micrococcus lysodeikticus* and *Bacillus subtilis,* two gram-positive forms, the ratios are 1.7:1 and 1:2, respectively (Perkins and Rogers, 1959; Salton and Marshall, 1959). These chains of polymers are held in parallel series by short peptides of three to ten amino acid residues in length. In these oligopeptides, any glutamic acid, alanine, or asparagine that may be present usually is in the dextrorotatory (D) rather than the levorotatory (L) configuration that typifies those in proteins. In addition, in the dense layer there are diverse quantities of various polysaccharides, and polymers of ribitol or glycerol phosphate, collectively called teichoic acids (Figure 1.26C; Rogers and Perkins, 1968; Salton, 1973) and teichuronic acids (Ivatt and Gilvarg, 1978).

The S Layer. On the outside of the outer membrane, lying upon the polysaccharide layer or capsule when present, is another covering discovered some time ago but characterized only recently (Houwink, 1953; Costerton *et al.*, 1961; Sleytr, 1978). This is the S layer, a term derived from its being comprised of repeating subunits (Figures 1.25, 1.27). In some gram-negative species, there may be two or even more S layers superimposed upon one another. Usually the subunits are subspherical, but in certain forms, including

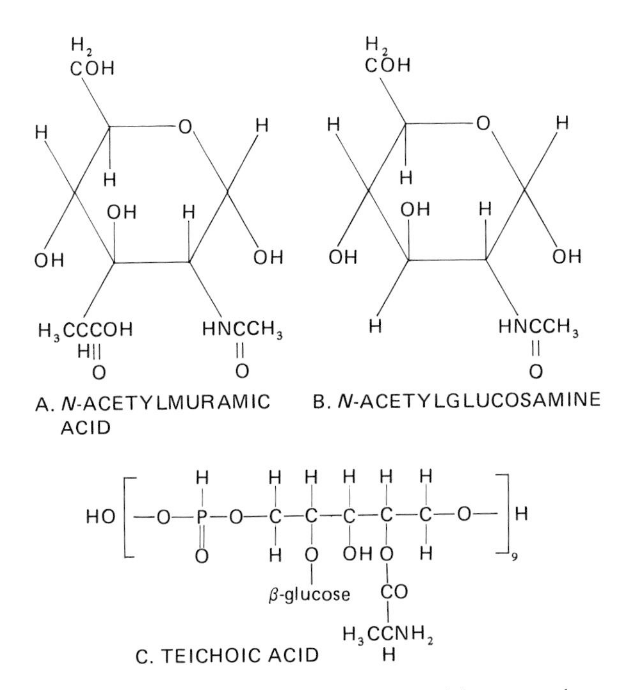

Figure 1.26. Important constituents of the bacterial outer membrane.

Flexibacter polymorphus, they are seen to be goblet shaped when viewed in sections; the stems of the goblets appear sufficiently long to penetrate through the outer membrane and may possibly reach even the inner one (Ridgway *et al.*, 1975).

The arrangement of the subunits in the layer also varies with the species. They may be arranged linearly, as in *Clostridium botulinum* (Takumi and Kawata, 1974), tetragonally, as in *Bacillus stearothermophilus* (Sleytr, 1978), or hexagonally, as shown by *Bacillus* sp. (Leduc *et al.*, 1977). They have been demonstrated to be largely proteinaceous. Those of *Bacillus* sp. were analyzed and found to consist of a single polypeptide having a molecular weight of about 140,000 (Brinton *et al.*, 1969). In addition, four hexose molecules are apparently associated with every subunit. Each of these may be attached to the underlying layers by hydrogen bonds, but not by covalent ones (Sleytr and Thorne, 1976).

Various functions have been suggested for the S layer. Several investigators have presented evidence that it serves in controlling the release of internal factors (Wecke *et al.*, 1974). They found that those species of *Clostridium,* such as *C. botulinum* and *C. tetani*, that accumulate toxins intracellularly during rapid growth possess an S layer, whereas those that do not do so, lack that

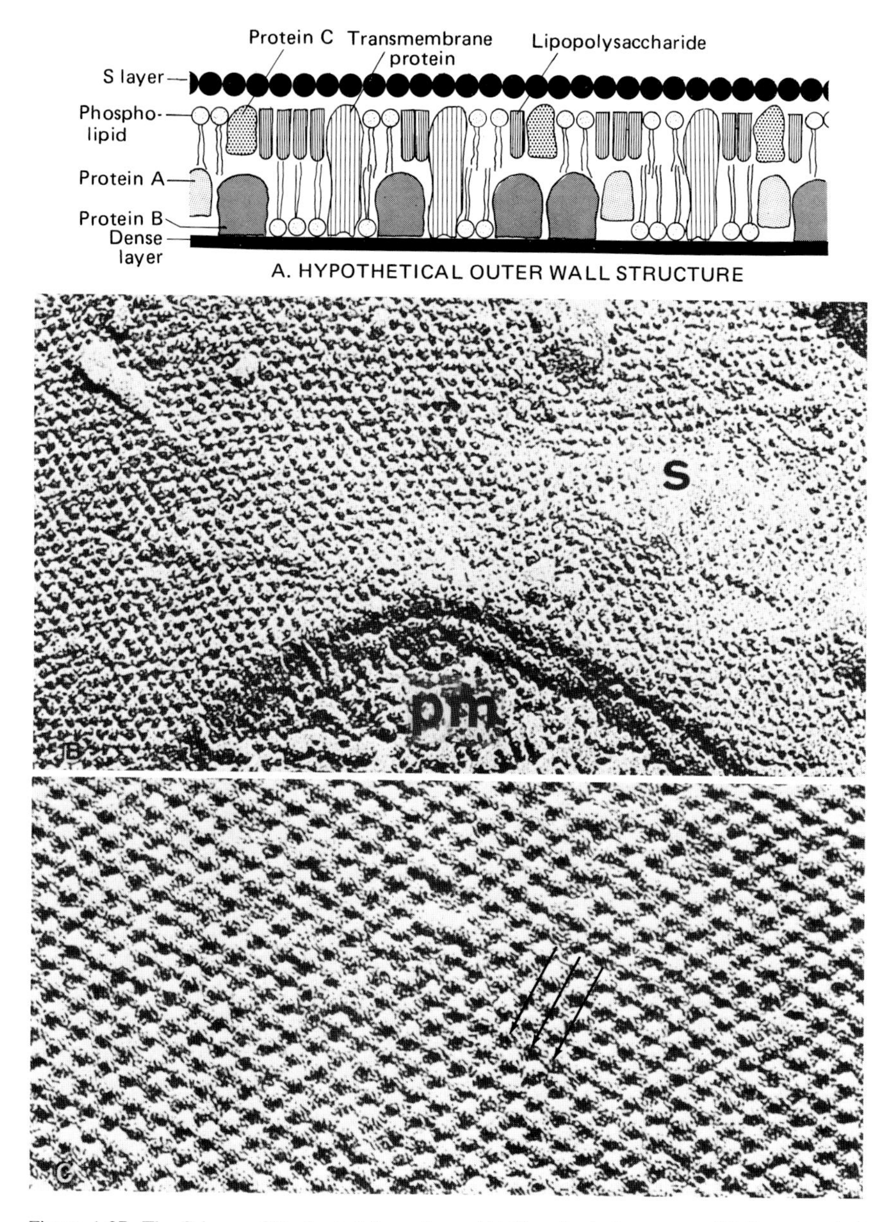

Figure 1.27. The S layer of the bacterial envelope. (A) Hypothetical structure. (B) Freeze-etched preparation of *Acinetobacter* sp. 180,000×. *s,* S layer, *pm,* exposed plasmalemma. (Courtesy of Sleytr, 1978, and the American Society of Microbiologists.) (C) Platinum-shadowed replica of *Spirillum* envelope, showing the S subunits (triple arrows). 200,000×. (Courtesy of Beveridge and Murray, 1976, with permission of the *Canadian Journal of Microbiology*.)

covering. Other workers in this area have demonstrated enzymatic activities on the part of the subunits (Thorne *et al.*, 1976); in *Acinetobacter* sp., by way of illustration, the subunits had phospholipase-A_2 activity. Still others have suggested that the goblet-shaped subunits discussed earlier played a role in the peculiar gliding locomotion found in *Flexibacter polymorphus* (Ridgway *et al.*, 1975).

Other Membrane Components. The structure of the inner membrane of both gram-positive and gram-negative bacteria has been elucidated less extensively than those of the outer parts, but what is known indicates that it is much more complex than the latter. In addition to the phospholipids that comprise about 30% of the total, there is a large variety of proteins (about 60%) and, at least in gram-positive forms, between 0 and 10% carbohydrates, including teichoic acid (Osborn *et al.*, 1974; Schmit *et al.*, 1974). Here again, then, is evidence that membranes are predominantly proteinaceous.

Among the proteins that have been identified are various cytochromes, including the cytochrome *c* oxidase discussed earlier (Section 1.1.3), and others engaged in electron-transfer activities. Then there are whole series, mostly unidentified, that carry out permease functions, such as channel formers and ionophores. The remainder are enzymes of a highly diversified nature, many of those that have been isolated being sensitive to various antibiotics. For example, in *Staphylococcus aureus*, four major proteins were found to bind penicillin, one with a molecular weight of 115,000, two of 100,000, and a fourth of 46,000 (Kazarich and Strominger, 1978). The smallest one demonstrated transpeptidase and carboxypeptidase activities, in which it used various oligopeptides as substrates and glycine and hydroxylamine as acceptors. In addition, a number of enzymes mediate the biosynthesis of the constituents of both the inner and the outer portions of the envelope, especially the lipoproteins of both regions and the lipopolysaccharides of the outer membrane.

The envelope of *E. coli* contains four transport systems that can accumulate K^+ within the bacterium, one of which involves an ATPase in the outer membrane and three structural proteins in the inner membrane (Epstein *et al.*, 1978). Other ATPase systems, including one that transports Ca^{2+} and Mg^{2+}, have a soluble binding-protein localized in the periplasmic space, as well as structural proteins in the inner envelope. In an alkalophilic bacterium (*Bacillus alcolophilus*), an ATP-requiring system was reported to be involved in the transmembrane transport of β-galactoside (Griffanti *et al.*, 1979).

One of the more significant specialized sites in the bacterial membrane is an association between that structure and the DNA molecule of the nucleoid (Ganesan and Lederberg, 1965). The attachment of the molecule is of a permanent nature (Fielding and Fox, 1970; O'Sullivan and Sueoka, 1972) and is located at the point where replication has its origin (Harmon and Taber, 1977; Nicolaidis and Holland, 1978). Thus in addition to protein synthesis, RNA polymerase activity, and DNA polymerase (Dillon, 1978), replication of the nucleoid requires the presence of this membrane–DNA complex. The complex formed between the membrane and the origin of replication has unique chemical properties that suggest that the binding is mediated by a protein (Craine and Rupert, 1978). It is interesting to note in connection with this requisite association that DNA replication in many viruses similarly requires attachment of a specific point to the cell membrane (Dillon, 1978).

1.3. THE NUCLEAR ENVELOPE

One of the chief features that distinguishes eukaryotes from prokaryotes is the partitioning off of the chromatin from the cytoplasm by means of a nuclear envelope. Yet the pair of membranes that comprise that structure is not indispensable for survival in these organisms, for in many taxa it breaks down completely during mitosis and meiosis. Moreover, the nuclear envelope is absent in the sperm of coccids (Moses and Wilson, 1970), as it is also in other sperm cells after it has fertilized the egg, where it remains fragmented or absent until the male pronucleus forms (Longo and Anderson, 1968; Zamboni, 1971).

Real interest in the nuclear envelope could only become manifest when the advent of the electron microscope made such studies feasible, but since that time research activities into its properties have increased rapidly. So many investigations have been conducted on it that a number of reviews have been published (Wischnitzer, 1960; David, 1964; Gall, 1964; Franke, 1970, 1974; Feldherr, 1972; Franke and Scheer, 1974; Kasper, 1974; Zbarsky, 1978), but these pertain almost exclusively to the metazoan envelope. Because this structure has been most thoroughly explored in these multicellular animals, the envelope from that source is examined before the comparative aspects are discussed.

1.3.1. The Nuclear Envelope of Advanced Eukaryotes

Characteristics. In addition to being made of two membranes, separated by a perinuclear space, the nuclear envelope is outstandingly distinct from all other membrane systems in being covered with pores (Figure 1.28). While these openings are present in the nuclear covering in all eukaryotic cells that have been examined, the number and density vary with the cellular activity. Both membranes of the envelope are perforated in unison; however, far from being simple openings, they are so complex that separate attention is given them later.

In living cells, the envelope apparently develops intimate associations with other organelles, particularly the endoplasmic reticulum (ER) (Watson, 1955; Whaley *et al.*, 1960; Sagara *et al.*, 1978). In part, unions that occur between them are responsible for the now defunct unitary-membrane concept (Section 1.1.1). The connections with the ER are quite frequent, in fact multiple much of the time in most vertebrate cells. The only cells in which this condition is

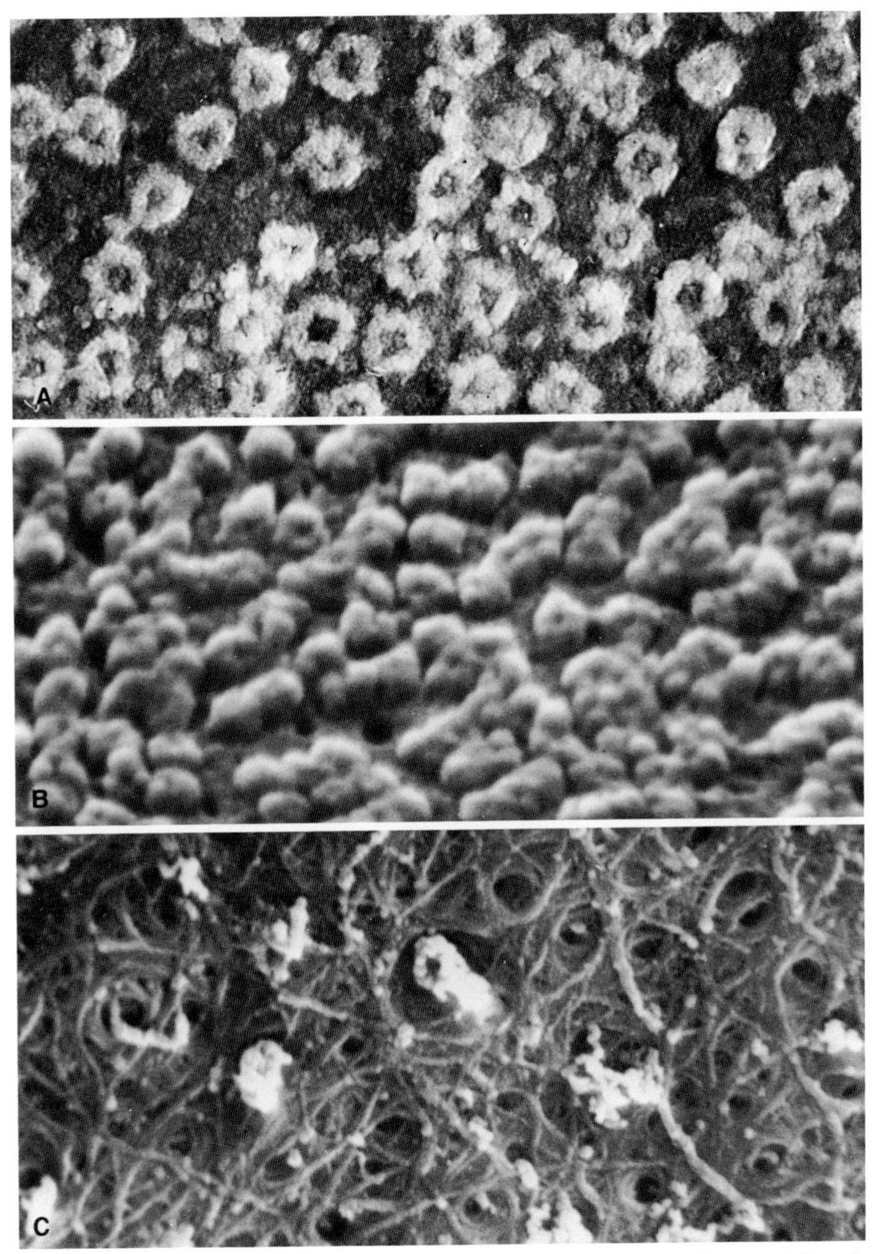

Figure 1.28. Nuclear envelope structure in two amphibians. (A) *Triturus viridescens,* shadowed with chrome. 76,000×. (Courtesy of Gall, 1964.) (B) *Xenopus laevis* oocyte. The densely packed pore complexes are often grouped to form clusters that share common walls. 65,000×. (C) The matrix associated with the inner surface of the nuclear envelope in *Xenopus* appears as a series of funnels circumscribed by fibers. Chains or masses of particles are inserted into many of the pores. 13,500×. (B and C, courtesy of Schatten and Thoman, 1978, with permission of the Rockefeller University Press.)

rare are those lacking large quantities of ER, such as sperm cells in the late stages of maturation and the erythrocytes of amphibians and birds. Often the P face of the outer membrane shows a resemblance to the ER structurally in bearing numerous ribosomes.

Occasionally the nuclear envelope is associated with those microtubules and microfilaments that are described in the next chapter (Manton *et al.*, 1969; Fawcett *et al.*, 1971; Fuge, 1971; Yamada *et al.*, 1971; Rattner, 1972). One such instance has been reported in the mature spermatozoon of a fern (*Pteridium* sp.), where the microtubules around the nucleus form a complex with the envelope in such a way that the perinuclear space is eliminated (Bell, 1978). Moreover, it is often joined to chloroplasts (Franke and Scheer, 1974) and to dictyosomes (Golgi material; Bouck, 1965; Weston *et al.*, 1972; Franke, 1974). If not actually connected to it, mitochondria at least may be intimately in contact with the nuclear envelope (Meyer, 1963; Hsu, 1967; Aikawa *et al.*, 1970; Rowley *et al.*, 1971). These organelles at times adhere closely to the outer membrane when nuclei are isolated, but it is still not clear whether this condition results from preparatory techniques or not. Finally, the nuclear envelope's relationship with the centriole is an observation of many decades' standing; in some cases the association is so intimate that the centrioles lie in a separate pocket in the membrane system.

The Pore Complex. In vertebrate nuclei, the structure of pores has been found to be unexpectedly complex. Around the perimeter of the impression, both on the external and on the inner surfaces of the envelope, is a series of light granules (Figure 1.29). Those on the cytoplasmic side appear larger than those on the nuclear surface, but both sets seemingly are fibrillar, with two loose fibers projecting into the surrounding substance (Franke and Scheer, 1974). On the nuclear side, the granules frequently seem to bear an associated thin sheet (Figure 1.29A), which has been suggested to be ribonucleoprotein. The lumen is not entirely open, but is crossed by eight vertically arranged septa, composed of fibrous materials. These taper as they converge upon a vertical elongate cylinder of fibers, to which they are attached near the midpoint of the latter's length. In scanning electron micrographs, the granules around the rim appear much broader than the pore itself (Figure 1.28B; Schatten and Thoman, 1978).

The pores are not always randomly distributed over the surface of the envelope. Sometimes dense clusters have been noted, and hexagonal and square packets have been reported (Merrian, 1962; Pitelka, 1963; Flickinger, 1970; Folliot and Picheral, 1971; Thair and Wardrop, 1971). At other times, observations have been made of the pores being confined to certain regions, so that other areas of the envelope are totally devoid of these structures (LaCour and Wells, 1972); this condition is especially common in spermatocytes and in sperm heads. Changes in number are correlated with the activity of the cell. During germination of cucurbits the number of pores per square nanometer

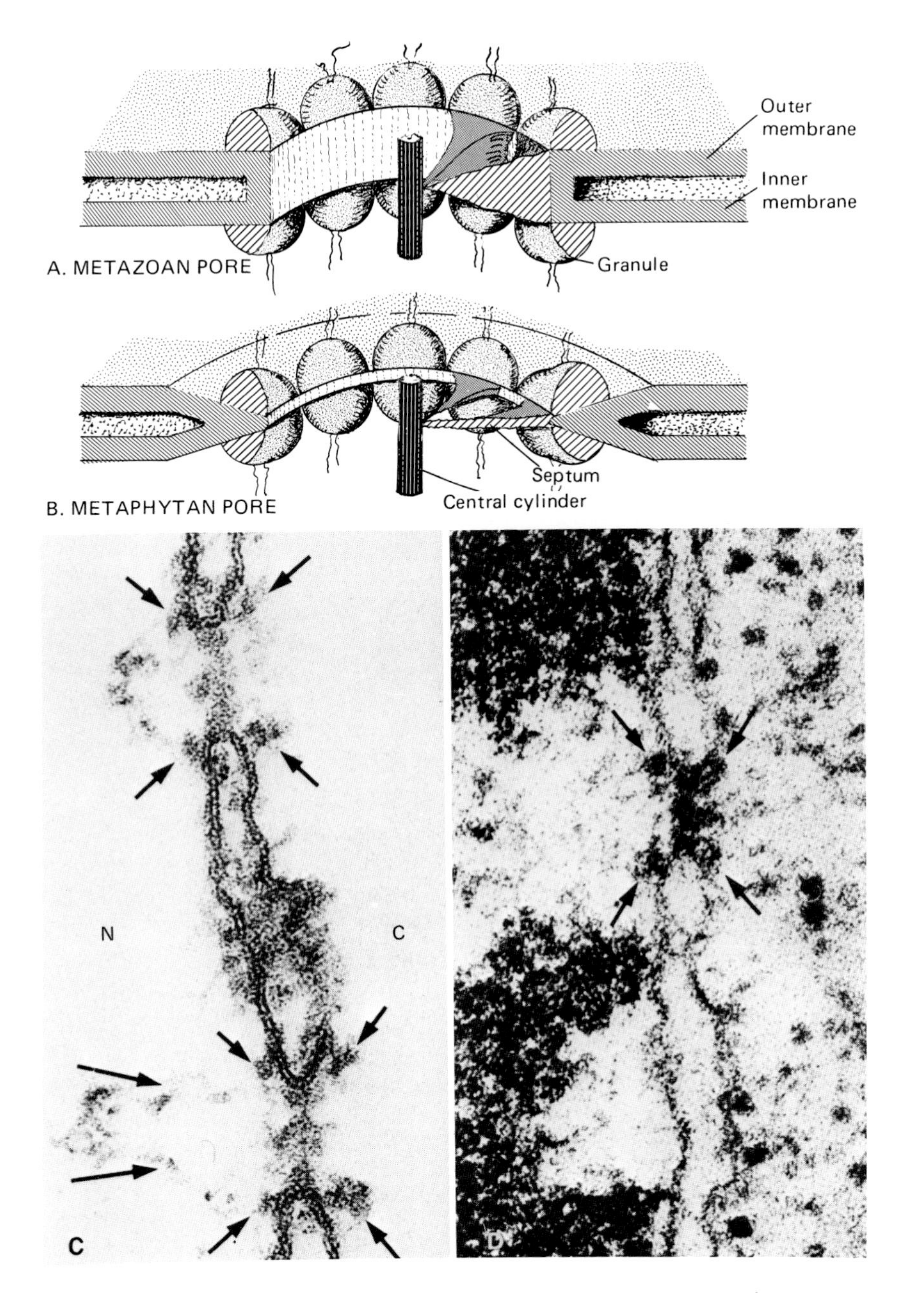

Figure 1.29. Nuclear pore structure. A difference not previously noted between the metazoan (A, C) and metaphytan (B, D) pore structure is the presence in the latter of a depression in which the granules lie. (C and D, courtesy of Franke, 1970.)

increased as the cell changed from a storage function to an actively photosynthesizing structure and decreased as senescence set in (Lott *et al.*, 1972; Lott and Vollmer, 1975). Moreover, pore number has been shown to be related to the environmental temperature at which growth occurred (Figure 1.30; Lott *et al.*, 1977).

Functions of the Envelope. While the actual function of the pores remains unestablished, it is not unlikely that the products of transcription—the various RNAs—pass through them en route into the cytoplasm to participate in translation and their other roles. Moreover, some of the proteins that result from translation must certainly pass into the nucleus through the pores from the cytoplasm, for the various enzymes involved in transcription, processing the transcripts and replication of DNA, are synthesized on cytoplasmic ribosomes. One study has found support for using the pore number per unit surface area of the nucleus as a valid method for estimating transport capacity (Severs, 1977).

On the other hand, rather than passing through these openings, some substances could equally be passed through the membranes by various transport mechanisms. In passing through, they would enter the perinuclear spaces, where processing might conceivably occur. Or from those cavities, they could enter the ER for further processing. Although the exact routes of the various materials have not been determined, the permeability had been established (Paine *et al.*, 1975). Molecules greater than 95 Å in diameter, such as ferritin and colloidal gold particles, do not enter the nucleus, and below that limit, rapidity of permeation is related inversely to size. Proteins of molecular weights of 12,000 to 67,000, with radii between 15 and 35 Å, pass through slowly; substances lighter than 4200 were not slowed in transit.

Some doubts have been raised as to the complete validity of this conclusion by use of *Xenopus* oocytes in which the nuclear membrane was disrupted artificially (Feldherr and Pomerantz, 1978). In such cells, it was shown that puncturing the membrane had no apparent effect either quantitatively or qualitatively on the uptake of endogenous polypeptides by the nuclei. Consequently, it was concluded that the endonuclear accumulation of specific nuclear polypeptides was not controlled by the envelope but by selective binding within the nucleoplasm.

Lowered temperatures have been reported to decrease abruptly the transport of materials through the nuclear membrane. For example, nuclei treated with 0.3% Triton X-100 have been demonstrated to show a linear decrease in RNA release with lowering of the temperature from 28°C (Herlan *et al.*, 1979). Electron spin resonance detected a corresponding decrease in the lipid fluidity. Thus it was proposed that chilling induced a change in the nuclear pore complexes from an open to a relatively closed state.

Enzymes of the Nuclear Envelope. Numerous enzymes of a diversity of types have been reported from the nuclear envelope of vertebrate cells, as well as a smaller number from metaphytan tissues. For a detailed listing of

Figure 1.30. Temperature-induced changes in nuclear envelope structure. (A) Nuclear pores (NP) are sparse and often widely spaced in this freeze-etched *Euglena gracilis* nuclear envelope, from a culture grown at 30°C. 64,000×. (B) In a similar preparation from a culture grown at 15°C, nuclear pores are relatively close set. 64,000×. (Courtesy of Lott *et al.*, 1977, with permission of the *Journal of Ultrastructure Research*.)

these, reference should be made to Zbarsky (1978). By far the majority of the enzymes of rat liver are phosphatases of various sorts, including alkaline, acid, and pyrophosphatases (Kartenbeck *et al.*, 1973). In addition, there are a number of enzymes that mediate the breakdown of phosphorylated sugars, such as glucose-6-phosphate, glucose-1-phosphate, and mannose-6-phosphate (Sikstrom *et al.,* 1976). Others catalyze the transfer of phosphate from tri- or diphosphorylated nucleotides to sugars, including ATP:glucose phosphotransferase, ADP:glucose phosphotransferase, GTP: and GDP:glucose phosphotransferase, and the like (Kartenbeck *et al.*, 1973). Still others like inosine diphosphatase, adenosine diphosphatase, and adenosine triphosphatase hydrolyze those phosphorylated nucleotides (Koen *et al.*, 1976). Acetylesterases, arylsulfatases, and proteinases round out the principal categories of the list of 34 nonoxidative metazoan enzymes listed (Zbarsky, 1978). In the nuclear envelope of onion cells, acid phosphatase, glucose-6-phosphatase, 5′-nucleotidase, inosine and adenosine diphosphatases, and two adenosine triphosphatases have been identified (Philipp *et al.*, 1976) belonging to this class.

In addition, 20 enzymes involved in oxidation–reduction reactions have become known in the nuclear envelopes of vertebrate tissues. Oxidases include those of cytochrome *c* (Franke *et al.*, 1976), tetrachlorohydroquinone, succinate, monoamines (Ono *et al.*, 1976), and NADH. Reductases are even more varied. Among their numbers may be listed those of succinate-cytochrome *c* succinate-PMS, NADH-cytochrome *c* and three others involving NADH or NADPH, and glutamic acid dehydrogenase. Only a few from nuclear envelopes or onion root cells have as yet been identified (Philipp *et al.*, 1976). Among them are cytochrome *c* oxidase, NADH oxidase, and NADH- and NADPH-cytochrome *c* reductases. Over and above these in mammalian tissues six cytochromes (b_5, P-450, a, a_3, c, and c_1) have been found. Doubtlessly, more will come to light in future years as this organelle receives the attention it deserves.

1.3.2. Comparative Aspects of Envelope Structure

Many of the studies of the nuclear membrane cited below have been conducted incidentally to investigations of the whole organism, and others may not prove to be altogether reliable, as they were completed before the advent of modern techniques. However, they are presented to provide a few samples of differences that have been noted in several representative species of protozoans and other unicellular forms. As these results are largely derived from studies on a single species, they may not necessarily pertain to all members of the respective taxa.

The Metaphytan Nuclear Envelope. In the seed plants, including such monocots as the onion and corn (Whaley *et al.*, 1959; Franke and Scheer, 1974) and dicots such as the pea (Roland *et al.*, 1977), the structure of the nuclear envelope closely resembles that of metazoans, consisting of two membranes with an enclosed perinuclear space. Pore structure also seems to be closely similar, in that each is surrounded on both exposed surfaces by eight granules and with a comparable eight-parted diaphragm and a central fibrous cylinder. Perhaps the only distinction to be perceived is detectable both in cross sections and in surface view. Instead of being at the same level as the bare areas of the membrane, the surface of the pore is depressed into a pit, in which the granules lie, as shown in Figure 1.29B. The pore complexes, too, appear to be larger in diameter and somewhat less densely distributed, but these points may not prove to be universally true.

Amoeboid Protozoan Envelopes. The nuclear envelope of amoebae varies extensively from one genus to another and thus each type requires separate discussion. In the giant, multinucleate (that is, syncytial) amoeba, *Pelomyxa palustris*, the two membranes are seen in section to lie parallel throughout their entire circumferences and to have the perinuclear space filled with a porous material (Daniels *et al.*, 1966; Daniels and Breyer, 1967). Although no pore complexes can be noted in any of the electron micrographs, their structural features are not discernible (Figure 1.31). Just outside the envelope is a lamellated covering, which at lower magnifications seems to be part of that envelope itself, making the structure appear quite thick. At higher power, however, the

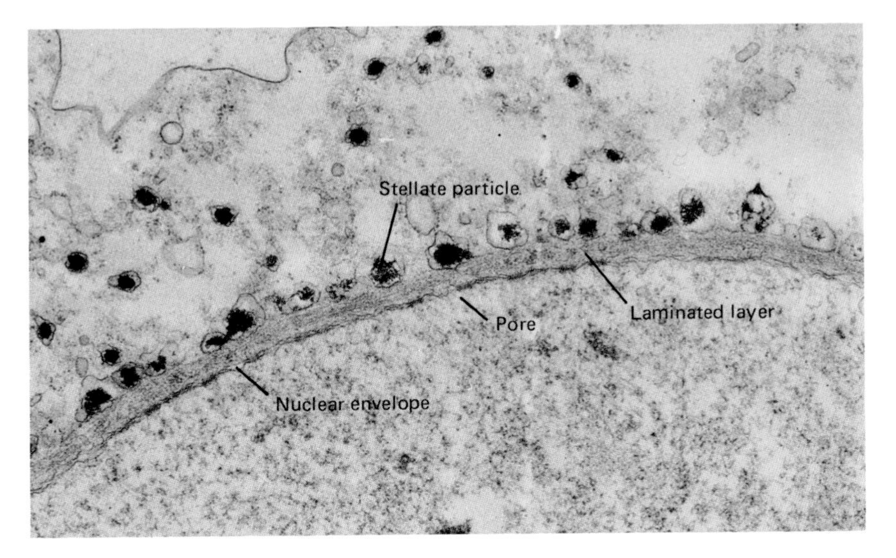

Figure 1.31. Nuclear envelope of the giant amoeba, *Pelomyxa*. The nuclear membrane is surrounded first by a thick layer of laminated material and then by a loose covering of irregular membrane-coated particles. The endosymbiontic bacteria usually present in the region do not appear in this view. 25,000×. (Courtesy of Daniels *et al.*, 1966.)

lamellated covering is seen to be vesicular and to lack the typical trilamellar pattern of dehydrated membranes. Outside of this is a broken ring of stellate particles and often a number of live bacteria that seem to be a nearly universal characteristic of this type of amoeba.

In the familiar species, *Amoeba proteus*, the envelope is also greatly thickened, but in this case the thickening is internal and results from the peculiar pore complex. The pores are close set over the entire surface of the envelope, so each forms a hexagon with its neighbors (Pappas, 1956). On the inner surface are walls that follow the outline of the complex, so when sectioned tangentially, a honeycomb appearance is created about four hexagons in width. When sectioned vertically, the walls appear parallel, and are about five or six times as thick as the dual-membrane system.

In *Endamoeba blattae*, the envelope is also extensively thickened, but the extra thickness results from a heavy covering of microtubules (Beams *et al.*, 1959; Driml, 1961). The pores in this species are not so close set and bear granules somewhat as in the higher plants and animals. Here, too, a perinuclear space appears much as in advanced forms. On the outer surface, the complex of microtubules provides a covering three or four times as thick as the envelope itself. Some of the tubules appear to run parallel to the nuclear covering, but many bend vertically to connect directly with the pores.

Other Unicellular Eukaryotes. For the most part among unicellular forms, the nuclear envelope shows no distinctive features. This holds true for such diverse examples as the zoospores of the estuarine fungus *Bryopsis plumosa* (Kazama, 1972), a number of yeasts, including *Rhodotorula glutinus, Saccharomyces cerevisiae, Hansenula wingei*, and *Candida parapsilosis* (Thyagarajan *et al.*, 1962; Conti and Brock, 1965; Clark-Walker and Linnane, 1967; Kellerman *et al.*, 1969), the dinoflagellate, *Anemonia sulcata* (Taylor, 1968), the diatom, *Nitzschia palea* (Drum, 1963), the flagellate *Trypanosoma conorhena* (Milder and Deane, 1967), and the ciliate, *Paramecium aurelia* (Jurand, 1976). In some of these reports, pore structure was not clearly shown, as the investigators were interested in aspects of ultrastructure other than that of the nuclear membrane, so some differences may yet be found in the taxa mentioned. For example, a freeze-etch study of the yeast shows the pores to be relatively large and few in number (Moor and Mühlethaler, 1963). Moreover, the pores appear as pits, without clear indication of the large granules that surround the structure in the advanced metazoans, nor are the septa and central cylinder in evidence. If these observations are not the result of artifacts, then the yeast nuclear membrane must certainly be considered primitive.

A number of more distinct varieties of nuclear envelopes have been reported. In one of the multiflagellated protozoans, *Lophomonas blattarum*, the envelope is densely surrounded by a maze of microtubules (Beams *et al.*, 1961), not unlike that described above in *Endamoeba*. Another protozoan, in this case a gregarine coccoid (*Gregarina melanopli*), shows a resemblance to

Amoeba proteus in that the pores are large and closely appressed to one another (Beams *et al.*, 1957). In this case, however, the pores are rounded, not hexagonal, and do not form a thick internal layer. In a pair of studies of *Euglena* sp. the pores are shown to be large and pitlike, much as in the yeast, but here they show a series of granules within the depressed area (Leedale, 1967; Lott *et al.*, 1977). These appear smaller and much more widely separated than those of metazoans and metaphytans, and without the central cylinder. At any rate, with what is known about pore structure, additional critical studies of the nuclear envelope in a sufficient number of simpler eukaryotes could reasonably be expected to shed much light on the evolutionary history of this organelle, and therefore on that of the organisms themselves.

1.3.3. Molecular Organization and Development

Here only those aspects of molecular organization need consideration that pertain solely, or at least predominantly, to the nuclear envelope. Related to this structural nature are the means by which the nuclear envelope undergoes replacement after its loss during mitosis.

Molecular Nature of the Outer Envelope. As reported in an earlier section, the outer leaflet of the envelope is frequently continuous with the endoplasmic reticulum, and similarly is often coated with ribosomes. Consequently, the presence of enzymes common to both these organelles is not especially surprising (Goldfischer *et al.*, 1964; Franke *et al.*, 1970; Kasper, 1971). That some of these may actually be identical structurally has been intimated by immunological reactions to antibodies by enzymes extracted respectively from microsomes and the outer envelope (Sagara *et al.*, 1978). Among those tested and shown to react identically were NADPH-cytochrome *c* oxidase, cytochrome P-450, NADH-cytochrome b_5 reductase, and cytochrome b_5.

Several other enzymes characteristic of the endoplasmic reticulum have been found in rat liver nuclear membrane and their relative activities compared (Mukhtar *et al.*, 1979). Among those assayed were aryl hydrocarbon hydroxylase, aminopyrine and benzphetamine *N*-demethylases, and epoxide hydrase. Nuclear membrane activities, while similar to those of the microsomal fraction, were only in the range of 2 to 12% of the latter. Comparisons of epoxide hydrase activity in nuclear envelope from several tissues showed that liver membranes were more active than kidney and the latter to be greater than lung, with spleen and heart envelopes even lower.

Consequently, it is evident that these two membrane systems do contain some of the same enzymatic properties. However, the presence of highly structured pores and the other distinctive features of the nuclear envelope strongly suggest the probable presence in the outer envelope of characteristic proteins, lipids, and other molecules, an observation supported also by the lower levels of activities that are displayed. Moreover, the comparative studies strongly intimate that the molecular organization varies from one tissue to another. Nothing appears to be known of the macromolecular constituents of the inner leaflet.

Development of the Nuclear Envelope. Few investigations have been made into the reconstitution of the nuclear envelope at the close of mitotic division, so that a recent paper on the topic in *Euplotes* is of particular interest (Ruffolo, 1978). In these euciliates as a whole, mitosis of the micronucleus takes place completely within the envelope, which remains intact throughout the processes. At telophase the DNA of the micronucleus begins to become diffuse and replication is initiated (Prescott *et al.*, 1962); at the same time, in the daughter nuclei a new nuclear envelope begins to form within the old one, extensive regions of double envelope being present. Often forks connecting the membranes of the new and old envelopes may be seen at the ends of these double regions, but at times small membranous vesicles are visible at the ends of the newly forming one. While pores remain in the outer envelope, they are not as abundant as in the inner. Shortly after telophase is completed, the DNA has been fully replicated, so that the cell is in the G_2 phase of the cell cycle, and the micronuclei have lost the outer envelope. It was proposed that the forks represent initiation sites of new membrane formation and that their arrangement suggests that both leaflets of the inner envelope are derived from the substance of the inner component of the outer one. This concept, however, does not appear in harmony with the outer leaflet's chemical composition being similar to that of the endoplasmic reticulum.

In vertebrates the breakdown and reconstruction of the nuclear envelope seems to be under the control of cytoplasmic factors, at least in part. One such substance called maturation-promoting factor from developing *Xenopus* oocytes has recently received attention (Wasserman and Smith, 1978); this chemical induces nuclear envelope breakdown when injected into amphibian oocytes. The peak of activity in newly fertilized eggs was found to be attained late in the G_2 stage just prior to mitosis, after which the activity disappeared. This cyclic appearance and disappearance of the factor's activity occurred even in enucleated eggs, so its cytoplasmic nature became evidenced.

1.4. BIOGENESIS OF MEMBRANES

At the present time, any discussion of how membranes are synthesized by the cell must necessarily be sketchy, because of the unsatisfactory state of knowledge. In the only symposium devoted to this topic in recent years (Tzagoloff, 1975a), the only membranes reviewed were those of mitochondria and chloroplasts and the cell wall (outer membranȇ) of bacteria. As Tzagoloff 1975b) points out, while the bacterial outer membrane has been studied rather widely, the inner (plasma) membrane has been largely overlooked. Because of the nature of current knowledge, here the assembly of membranes of prokaryotes is surveyed first, then bacterial cell wall structure is examined, followed by a summary of what is known of the processes in eukaryotes.

1.4.1. Biogenesis of Prokaryotic Membranes

Although considerably more has been learned about the biosynthesis and assembly of envelope constituents in bacteria than in eukaryotes, present knowledge is still in a preliminary state. Osborn *et al*. (1974) have presented evidence supporting the idea that the lipopolysaccharide of the outer membrane is synthesized in the inner membrane and then is translocated to its definitive position. Moreover, they have shown also that the phospholipids of the outer membrane have a similar biosynthetic pathway. However, in neither case is the mechanism for translocation known. In some organisms, including *Lactobacillus casei* and *L. plantarum*, some of the enzymes active in the formation of outer-membrane constituents have been reported to be present in the cytoplasm and the mesosome (Cota-Robles, 1966; Shockman *et al*., 1974).

The latter study makes it clear that envelope growth involves two separate processes, wall enlargement and wall synthesis. The first of these results in the increase in wall surface, as during growth of the organism; these activities are centered in a relatively small number of sites located on the cell surface near points where new septa are beginning to form. The second set of processes, in contrast, results in thickening of the wall and is carried out by a large number of sites densely distributed throughout the entire cell surface. If protein synthesis was inhibited experimentally, wall enlargement ceased, but thickening continued; hence, the organisms became covered with a thicker wall than normally.

1.4.2. Eukaryotic Membrane Synthesis

Several forceful reasons appear to underlie the general sparsity of information regarding membrane biosynthesis. In the first place, great complexity exists in what seemingly is a very simple structure. Second, probably as a consequence of this deceitful appearance, the conceptual basis for membrane structure has obstructed progress. Because both the unit-membrane concept and its first widespread replacement, the fluid mosaic membrane, are basically mechanical ideas of what membranes should be like, it has been difficult to reconcile actual observations to theoretical demands. Now that it is becoming evident (Sjöstrand, 1978) that membranes are largely proteinaceous organelles, like the remaining parts of the cell, perhaps progress can be more rapid. The discussion that follows may shed light on the reasons for conflict between observed facts and theory.

Distribution of Proteins. As already shown, each membrane type possesses proteins characteristic of itself alone. To a degree, later adjustments to the unit concept were made that permitted the presence of unique protein species in the several types, but within the framework of that theory, these had to be viewed as an addition to a set of species characteristic of all membranes. In actual closely related membranes, as for instance the inner and outer envelopes of mitochondria, completely different sets have been found. Schnaitman (1969) reported that whereas 12 proteins existed in the outer membranes of rat liver mitochondria, 23 occurred in the inner ones, and no single species appeared to be common to both.

In pointing out that this example and the innumerable others that are known cannot be explained in terms of ordinary metabolism, Wallach and Winzler (1974, p. 351) proposed two alternatives for the creation of membranes. First, they suggested that natural membranes were formed by assembling protein units that matched one another as far as possible and permitted optimal interaction between the lipids and the proteins. This, however, does not explain how 23 proteins match in the inner membrane of mitochondria and how 12 do so in the outer. In the suggested explanation, it would appear that no regard for functional differences has been paid, so that each membrane is treated as of structural importance only. In contrast, the earlier discussions of membrane junctions, recognition sites, ionophores, and other distinctive functional units should indicate clearly that such mechanical suppositions are without real foundation. Second, these investigators advocated that preexisting membranes might act as templates. In contradiction to this hypothesis, they brought out the fact that many viruses of eukaryotes develop their capsids at the host membrane but have compositions quite foreign to that of the eukaryote. Even such simple organisms as *Mycoplasma capricolum*, which requires sterols in their environment, do not incorporate cholesterol directly into their envelopes but require active metabolic growth (Clejan *et al.,* 1978; Rottem *et al.*, 1978).

Moreover, cyclic changes in membrane composition in *Dictyostelium* were inhibited when development was interfered with (Hoffman and McMahon, 1978). Nor would a requirement for template explain how membranes are formed *de novo*, as those of Golgi bodies always are (Chapter 6, Section 6.2.3), and how others, such as nuclear and mitochondrial membranes, occasionally are.

Briefly stated, the problem is this: While some mitochondrial and chloroplast membrane constituents are synthesized within the individual organelles, the remainder are synthesized on the ribosomes of the cytoplasm, as are those of all the other organelles, including the plasma membrane. Hence, there must be a system whereby the proteins synthesized for the plasma membrane are transported and inserted into it where needed, and those for the nuclear membrane and each other cell part are produced and carried in corresponding fashion. No model mechanism that has the capability of recognizing the individual proteins has as yet been formulated, nor has any been discovered in the cell.

A Second Type of Problem. A problem associated with membrane biosynthesis has been especially stressed by Tzagoloff (1975b). On the supposition that many proteins are hydrophobic, he raised the question as to how such constituents that would tend to aggregate when in water might be conducted through the cytosol, which is essentially aqueous. Although he proposed such solutions to the problem as the existence of a special class of ribosomes in association with each kind of membrane or special classes of mRNAs with specific binding sites for membranes, the problem is really academic. As shown in an earlier discussion, no protein is completely hydrophobic, nor even largely so; all may have hydrophobic regions of *limited* extent, but none have been demonstrated to have more than around six or eight nonpolar amino acids in sequence. It is interesting to note that Tanford (1978) states that hydrophobic forces are responsible for the assembling of membranes and intracellular organelles, rather than a difficulty to be overcome.

The problem is really more fundamental than Tzagoloff states, for it actually concerns moderate-sized or even small molecules more than it does macromolecular items like proteins. In the cell strictly nonpolar chemicals like fatty acid chains and certain amino acids must be transported from their point of origin to where they are to be used, including the various membranes. If hydrophobic substances like those can be transported readily through the aqueous interior milieu of the cell, then proteins or lipoproteins that contain nonpolar parts along with many polar ones should not present any more serious difficulties. It would be most interesting to see a theoretical explanation of the molecular interactions that exist at an interface between an oil deposit and its bimolecular membrane covering, such as those found in yolk cells for instance.

The Ultimate Problem. Much of the lack of knowledge about membrane assembly results from the persistent viewing of the processes as being mechanical rather than cellular. In turn, this approach to researches in this area results from adherence to mechanical models of membrane structure. It may prove beneficial here to outline the whole problem in light of what has been shown about membrane structure in the preceding pages:

1. The membrane is not primarily a bimolecular sheet of lipids, bearing isolated proteins. Rather it is a sheet of proteins bearing bimolecularly arranged patches of lipids.
2. The proteins in these sheets are not entirely uniformly distributed within a given kind of membrane, but are arranged (a) asymmetrically in relation to the two faces of the membrane, and (b) in accordance with functional needs. The location of receptors on particular sides of a given cell type, as in the thyroid follicle cells for one example, indicates this asymmetry. Intercellular junctions perhaps are even better illustrations of the location at small, separated areas of protein structures of sometimes amazing complexity—the porelike proteins found at gap junctions certainly are not to be found elsewhere.
3. Protein and lipid components of membranes are constantly being recy-

cled, in company with all the other molecules of the cell. Such processes occur throughout a given membrane, so that a desmosome for instance continually has its molecules replaced, as do receptors, nexuses, other specialized regions, and even the plain unspecialized surface areas.

4. Membranes undergo growth and maturational changes. Even specialized portions do not come into existence suddenly, but develop characteristic traits with time, presumably as they gain their distinctive proteins. This has been made especially clear through such studies as the ontogenetic changes shown in gap junctions (Gros *et al.*, 1978).

Thus the proteins are synthesized on ribosomes under the direction of the genetic mechanism, the given gene activity being governed in turn by the age

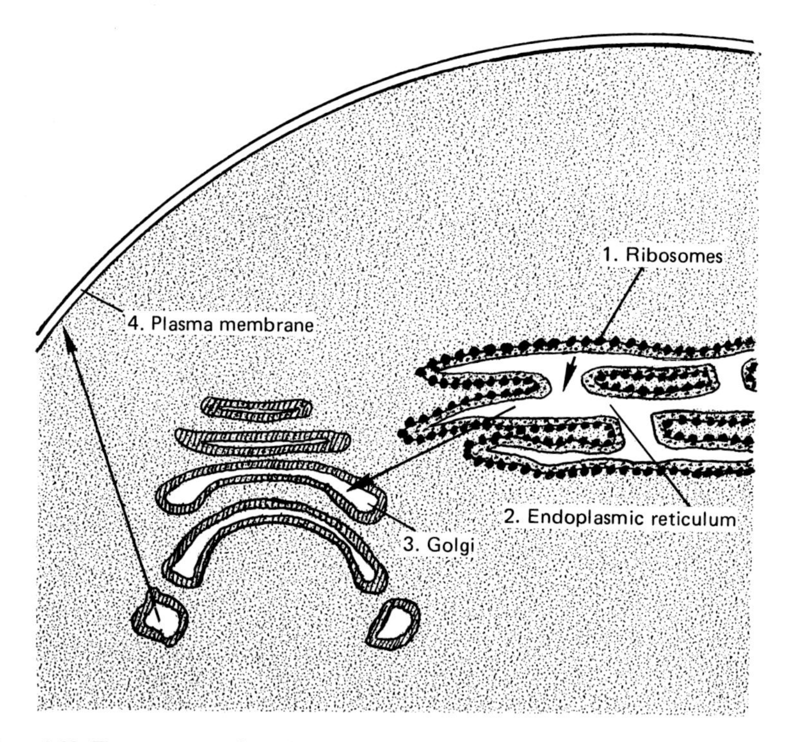

Figure 1.32. The movement of membrane proteins through the cell. Most of the proteins probably are formed on ribosomes lying free in the cytoplasm, rather than on those of the endoplasmic reticulum. Steps in protein formation: (1) Formed in ribosomes; (2) passed into ER, where it is modified; (3) passed next into Golgi body, where it is further modified; (4) then to plasma membrane. The proteins (enzymes) in each of the other membranes that carry out the modifications had gone through similar sets of events.

and functional activity of the cell at that specific point in time. Concurrently, lipids are being synthesized, probably in the endoplasmic reticulum, by the enzymes of those membranes. The proteins of these membranes, too, would be undergoing continual recycling, or sometimes replacement by a different species, in accordance with the needs of the cell at that moment, under the regulation of the genetic mechanism.

After synthesis each type is conducted through the cytosol, sometimes enclosed in vesicles but more frequently not, for further modification by other enzymes, first in the ER and then in the Golgi material; the enzymes in these membrane systems are equally subject to constant replacement (Figure 1.32). Finally, each is conducted to the exact site in the membrane where that molecule is needed, sometimes to replace a like component that is to be recycled, sometimes to assist in the growth of the structure as the cell itself enlarges, and sometimes, as during maturation of the cell, to replace a forerunner or less definitive molecule. The sites for each are often highly specific, as at junctions and receptors. When the various mechanisms involved have been elucidated, then and only then will we understand how living things really function.

2

Microtubules and Microfilaments

Next to membranes, there is probably no other cellular element as ubiquitous as the microtubule and related fibrillar structures. Although their presence was first suggested by Sigmund Freud (1882), then a young cell biologist working on live crayfish neurons, their actual existence could only be asserted with some assurance even as late as the mid-1950s (Porter, 1955). It remained for the advent of glutaraldehyde fixation (Sabatini *et al.*, 1963; Karnovsky, 1965), combined with other refinements in electron microscopic techniques (Palay *et al.*, 1962), to confirm their reality convincingly. Subsequently, researches on these organelles have reached a relatively high level of frequency, and a number of review articles have appeared in print (Dustin, 1972, 1978; Bardele, 1973; Margulis, 1973; Olmstead and Borisy, 1973; Shelanski, 1973; Bryan, 1974; Mazia, 1975; Snyder and McIntosh, 1976; Stephens and Edds, 1976; Kirschner, 1978). Generally three categories of fibrillar elements are recognized, distinguished for the present on the basis of their diameters: microtubules, ranging from 180 to 250 Å in thickness, intermediate filaments, between 80 and 100 Å, and nanofilaments,* from 40 to 70 Å (Burnside, 1975). Among others, the second category includes neurofilaments and tonofilaments, the latter of which will be recalled as occurring in intercellular junctions (Chapter 1, Section 1.2.3). Because microtubules are by far the most extensively investigated of these structures, they are reported in detail before other types are given attention.

* The term nanofilament is here proposed for this class, formerly known as microfilaments, because the latter name is generally applied to cellular filaments of all types collectively, exclusive only of microtubules.

2.1. MICROTUBULES

Cytoplasmic Microtubules. Despite the ubiquity of microtubules and the diversity of tissues in which they occur, their functions fall into just three categories: support, transport, and secretion. A role in support is exemplified by their frequent association with the plasmalemma, as in *Euglena* (Leedale, 1967), and during the development of cells (Overton, 1966) and tissues (Gibbins *et al.*, 1969; Tilney and Gibbins, 1969; Fellous *et al.*, 1975; Hardham and Gunning, 1978). A comparable activity may also be indicated by their proximity to the nuclear envelope in metaphytans (Figure 2.1B,C; Bell, 1978) and metazoans (Kessel, 1966) and, as already noted (Chapter 1, Section 1.3.2), in several unicellular forms (Marchant, 1979). Their involvement with the movement of chromosomes during mitosis and meiosis is a long-established observation (Figure 2.1A), whereas their concern with the movement of such particles as melanin granules in fish and amphibian pigment cells (melanophores) and in the axonal transport in neurons are more recent developments (Dahlström, 1971; Rebhun, 1972; Murphy, 1975; Schliwa, 1978). Their occurrence in such organelles of locomotion as flagella and cilia has been known since the 1950s (Chapter 4), but their apparent universality in pseudopods remained for the ensuing decade. However, they are not confined to particular regions of the cytoplasm but ramify through the entire cell (Figure 2.2; Osborn *et al.*, 1978; Weber *et al.*, 1978). Finally the secretory-related activities of microtubules were discovered, processes that now have been described in connection with the secretion of vasopressin (Taylor *et al.*, 1973, 1975), scale cells in insects (Overton, 1966), postphagocytotic enzymes (Malawista, 1975), thyroxine (Wolff and Bhattacharyya, 1975), and liver enzymes (Redman *et al.*, 1975). They have also been reported in the healing processes of wounds in mammals, along with microfilaments (Gabbiani and Montandon, 1977). Attention to these activities, however, must await a synopsis of the structure, chemistry, and synthesis, for reasons that become quite obvious as the discussion proceeds.

Although the foregoing remarks clearly intimate that these organelles are confined strictly to eukaryotic organisms, as appeared for many years to be the actual case, their occurrence in several types of bacteria has been reported in recent years (Figure 2.3; Pope and Jurtshuk, 1967; van Iterson *et al.*, 1967; Corfield and Smith, 1968; Eda *et al.*, 1976, 1977; Margulis *et al.*, 1978). They have also been described in several blue-green algae (Bisalputra *et al.*, 1975;

→

Figure 2.1. Some occurrences of microtubules. (A) Macronucleus of *Blepharisma* undergoing division; here the microtubules assist in the elongation of the nucleus. 45,000×. (Courtesy of Jenkins, 1977, with permission of the Society of Protozoologists.) (B and C) The microtubules that are associated with the nuclear envelope of the spermatozoid of the fern *Pteridium* sp. play an unknown role, possibly in support or conduction. (B) 60,000×; (C) 360,000×. (Courtesy of Bell, 1978.)

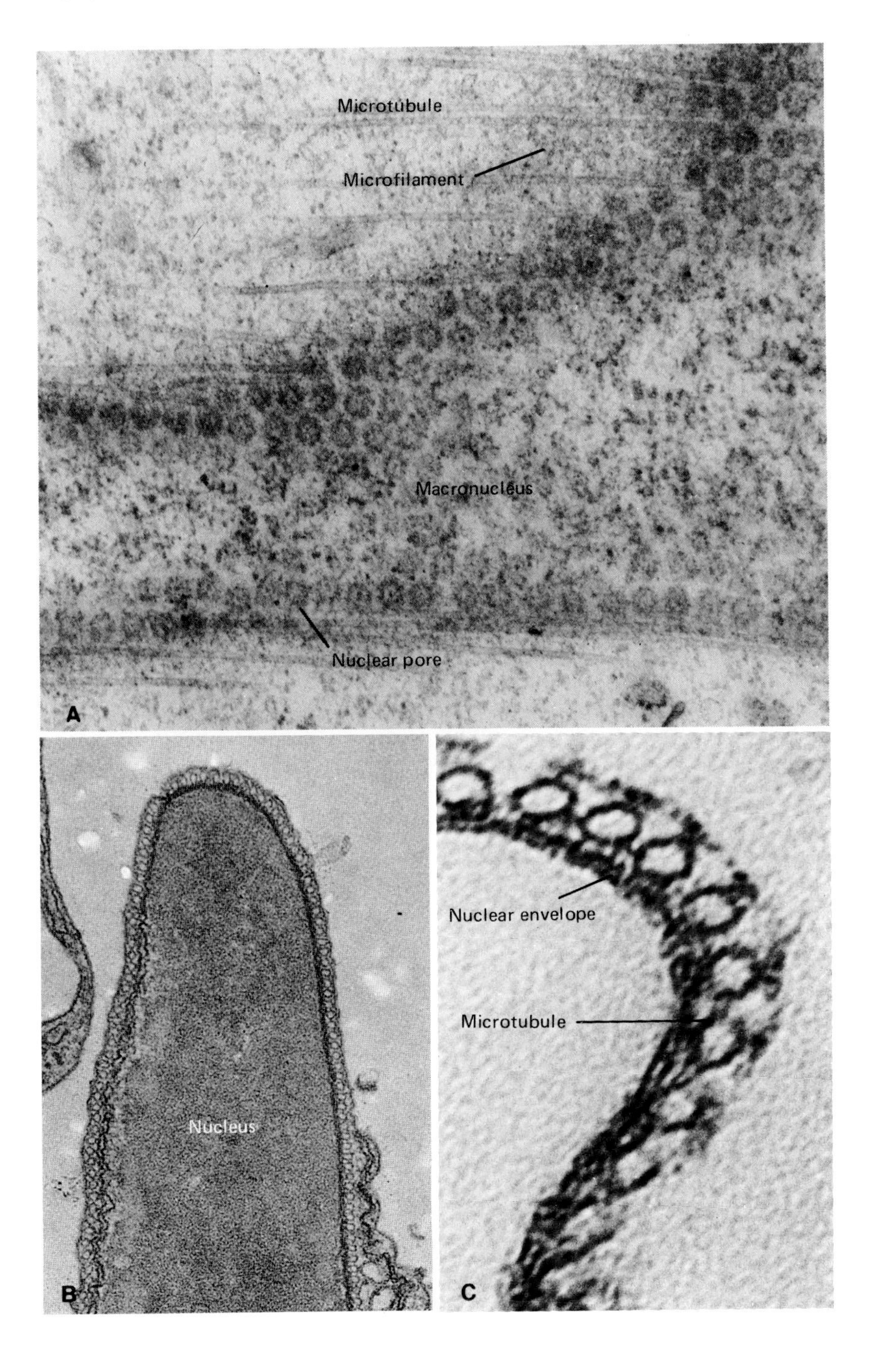
Microtubule
Microfilament
Macronucleus
Nuclear pore
A
Nucleus
B
Nuclear envelope
Microtubule
C

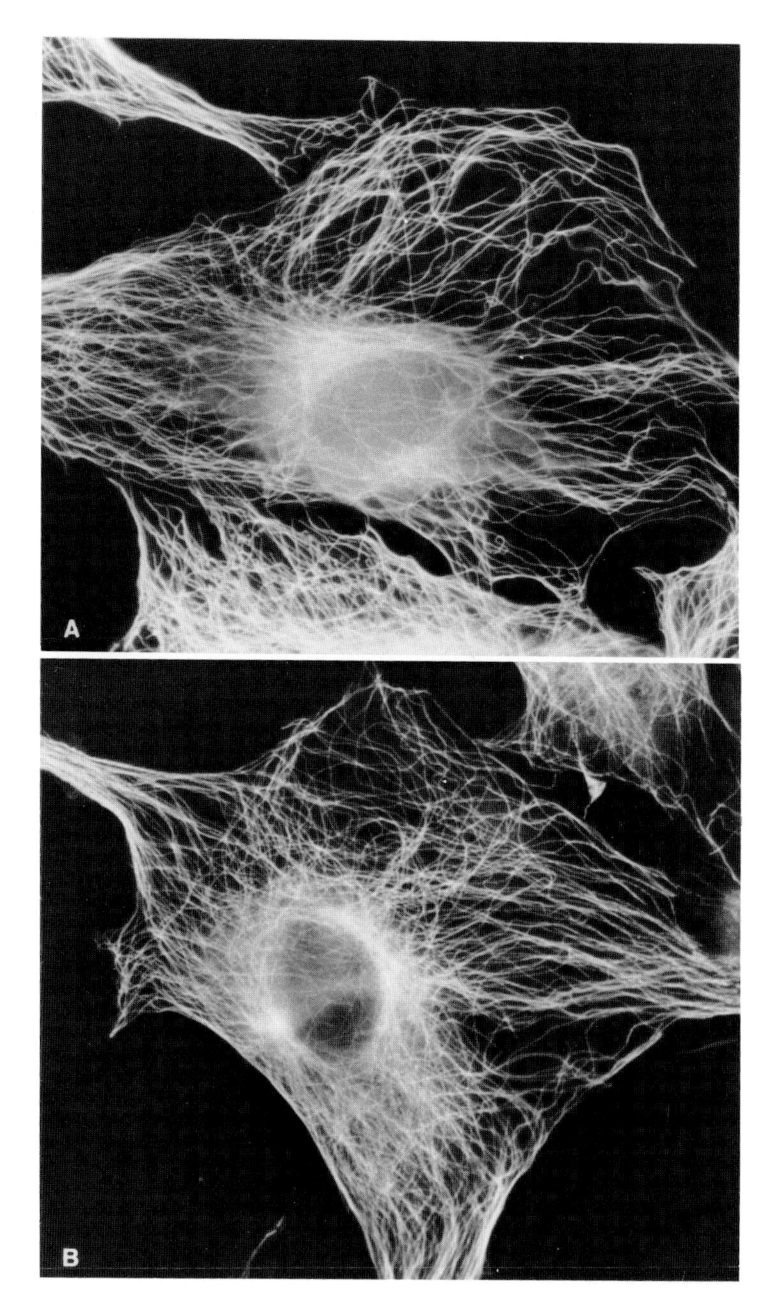

Figure 2.2. Microtubule distribution in mouse 3T3 cells. When visualized by immunofluorescent techniques that employ monospecific antitubulin antibodies, the microtubules are perceived to extend frequently from the plasmalemma toward the nucleus, which appears as a partly obscured oval area. (A) Cell fixed with glutaraldehyde; (B) cell fixed with formaldehyde. Both 750×. (Courtesy of Weber *et al.*, 1978.)

Khan and Godward, 1977). Because almost nothing more than their presence, not even their chemistry, has become known in subsequent years, little can be said about the function or assembly in prokaryotic source materials.

Nuclear Microtubules. In the majority of protozoan and algal cells, light microscopic studies showed that mitosis, complete with the characteristic microtubular spindle, takes place entirely within the intact nucleus, a finding that later received confirmation by electron microscopic investigations (Carasso and Favard, 1965; Sommer and Blum, 1965; Vivier, 1965; Wise, 1965; Tamura *et al.*, 1969; Jurand and Selman, 1970). Other ultrastructural researches have even found microtubules present in intact, dividing nuclei of such metazoan source materials as insect epidermal cells and crane fly spermatocytes (Smith and Smith, 1965; Behnke and Forer, 1966). This occurrence within dividing nuclei is treated in detail in the chapter on mitosis (Chapter 11); however, microtubules have also been reported from the nuclei of nondividing cells. For instance, DeBrabander and Borgers (1975) found them in abundance in mast cells and less frequently in eosinophils of rat after the tissue had been placed for 1 or 2 hr in a culture medium.

2.1.1. Fine Structure of Microtubules

Molecular Organization. As a general rule, microtubules consist of 13 so-called protofilaments arranged side by side around a central lumen, parallel to the long axis of the structure and in staggered sequence (Figure 2.4). Each protofilament consists of the protein tubulin, which is described in a subsequent section. Because the structural units run linearly along the length, the assemblage of macromolecules exhibits no periodicity such as occurs with DNA, collagen, and others having spirally arranged components. Although corresponding points of protofilaments are offset somewhat from those of their neighbors, the repeating units are dimers of two peptides (α- and β-tubulin) of similar size and shape, as shall be seen shortly (Fujiwara and Tilney, 1975); hence, the resulting spiral arrangement of the monomers is not visible in the microtubule itself.

While almost all microtubules consist of 13 protofilaments, regardless of whether the source material is from bacteria, protozoans, algae, plants, or metazoans, there are occasional exceptions. The most frequent exception is the B component of the doublets of cilia and flagella (Figure 2.5), which contains only nine protofilaments, in contrast to the A component, which has the standard number. More recently, microtubules with only 12 filaments have been described from several arthropod sources, including the axons of neurons from crayfish and epidermal cells from cockroaches (Hinkley and Burton, 1974; Nagaro and Suzuki, 1975).

Bridges and Arms. In addition to the basic structure just described, microtubules bear a number of appendages called bridges and arms. Often these

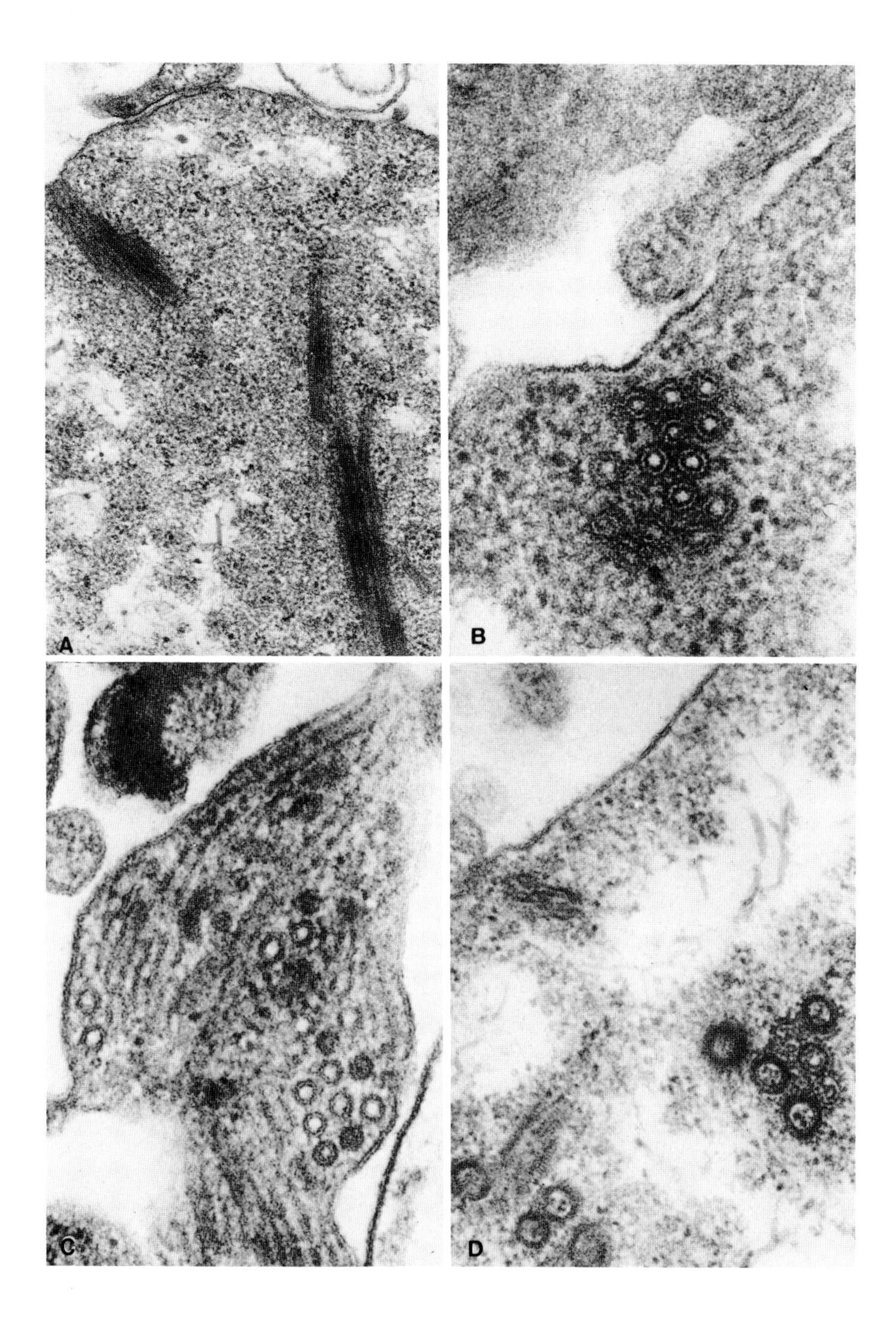
A
B
C
D

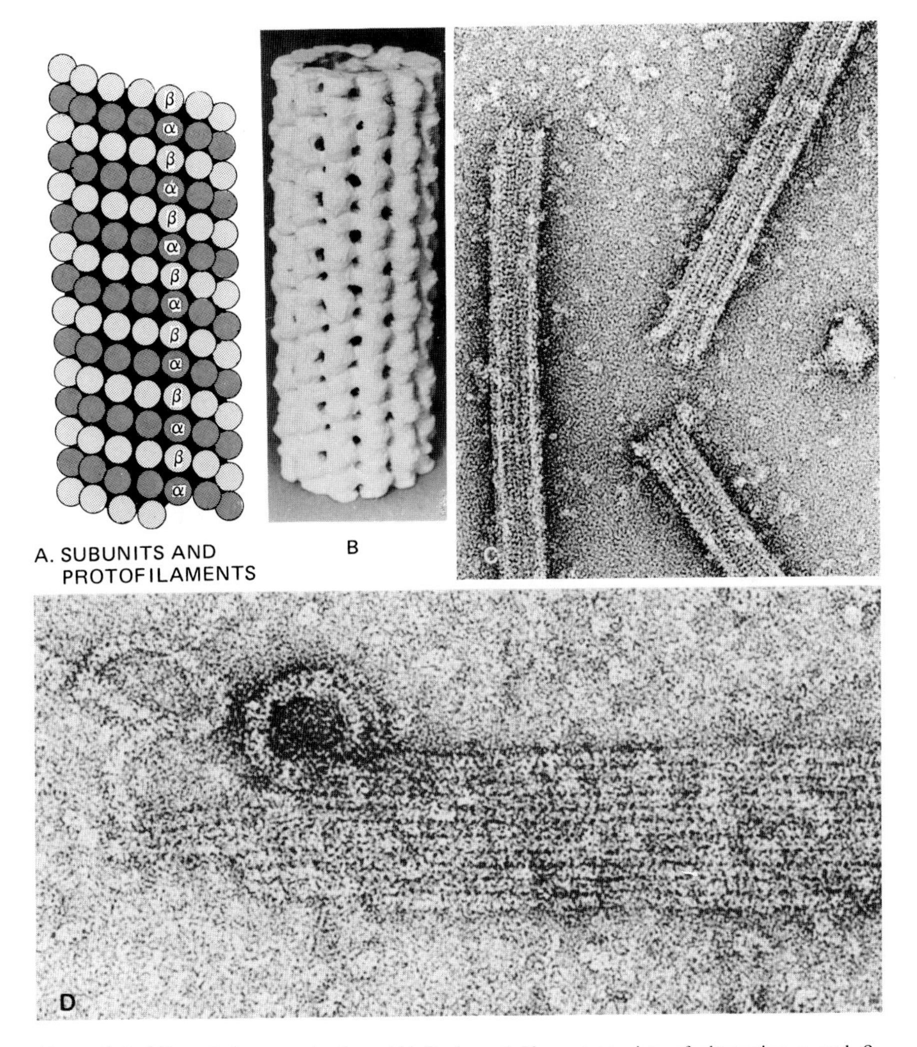

Figure 2.4. Microtubular organization. (A) Each protofilament consists of alternating α and β subunits, which are aligned side by side in intact microtubules. (B) Photograph of a model of a microtubule. (Courtesy of Amos and Klug, 1974.) (C) Microtubules from pig brain show the linear arrangement of the subunits especially clearly. 205,000×. (Courtesy of R. C. Williams.) (D) Part of a flattened microtubular wall, showing some of the 13 protofilaments that comprise these structures. No periodicity is in evidence. 350,000×. (Courtesy of Erickson, 1975.)

←

Figure 2.3. Microtubular structures in L-form bacteria. (A) Longitudinal section, 50,000×, and (B) transverse section, 160,000×, of microtubular structures in L-form *Staphylococcus aureus*. (C) Similar structures in L-form *E. coli*. 100,000×. (D) Much coarser tubules from L-form *Streptococcus*, having diameters approaching 1000 Å. 80,000×. (Courtesy of Eda *et al*., 1977, with permission of the *Journal of Bacteriology*.)

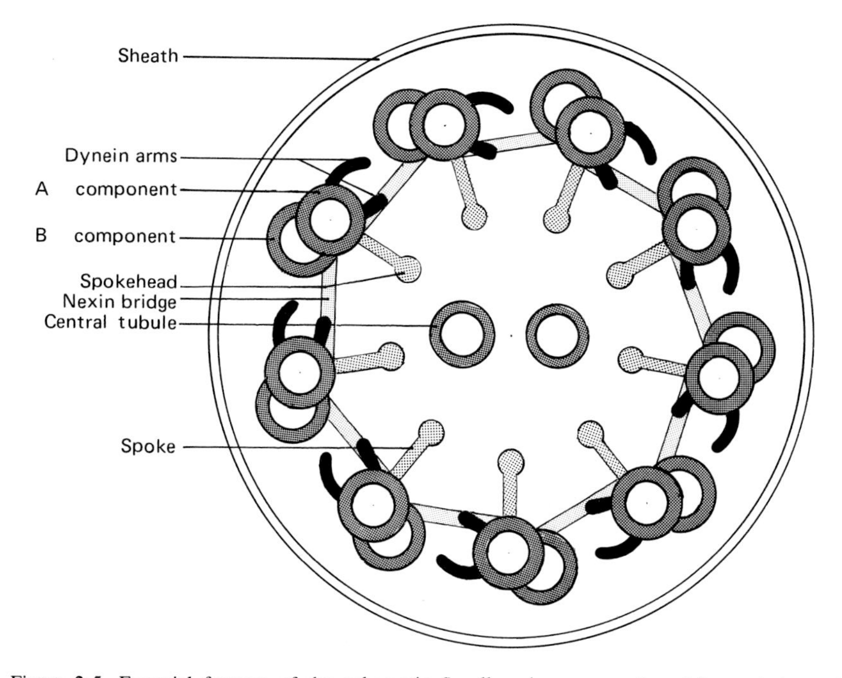

Figure 2.5. Essential features of the eukaryotic flagellum in cross section. Many tubular and filamentous structures occur within the flagellum, or cilium, of eukaryotes, which receives closer attention in Chapter 4. The spoke head is also referred to as the intermediate fiber. Only the central singlets and outer doublets are microtubular elements.

are delicate filaments that connect one microtubule to another or to membranes and the like, as in the nexin bridges between doublets (Figure 2.5). In other cases, they may be broad arms that merely project outwards into the cytosol, like those do on the A component of flagellar doublets.

The arrangement of these parts is remarkably precise, as indeed is that of the other microtubular components themselves. As shown in Figure 2.5, if, during the preparation of a print, the negative of a flagellar section is repeatedly rotated at precise intervals, a procedure called a Markham rotation, nine interval exposures of 40° each reinforce the images of every whole doublet, but more than that, the images of each arm and all of the protofilaments are found to coincide exactly. In other words, not only are the doublets themselves revealed as being oriented at precise distances from the center of the flagellum and from one another, but all their corresponding molecular parts are shown to be of equal dimensions and identically spaced, so that the nine exposures can be superimposed in minutest detail (Fujiwara and Tilney, 1975).

Bridges and arms functionally represent two contrasting types. The former are largely static linkages that connect microtubules to other organelles or to one another, whereas the latter are enzymatic structures that are chemically active. Bridges are by far the more widespread, for they have been reported in numerous tissues from a diversity of source organisms (Tilney and Porter, 1965; MacDonald and Kitching, 1967; Harris, 1970; Huang and Pitelka, 1973; McIntosh, 1973). Contrastingly, the only well-known representatives of the enzymatic type are the dynein arms that occur at frequent intervals along the long axis of the A component of flagellar doublets (Gibbons, 1968). These arms have been shown to possess ATPase activity that has been postulated to mediate sliding between the A and B components of the doublets and thus induce beating in the flagellum or cilium (Gibbons and Gibbons, 1972, 1973; Summers, 1974). However, certain bridgelike parts have been found to be chemically active, as becomes apparent in the chapter on the flagellum. Other organelles supported by microtubules also have bridgelike interconnections, such as the cytopharyngeal basket of certain ciliates (Figure 3.8).

Other Structural Properties. In descriptions of certain properties of microtubules, *in vitro* and *in vivo* observations are sometimes contradictory. Their cylindrical nature is supposed to confer a high level of rigidity, a property quite apparent when isolated, for they are usually straight or feebly curved, and they break, rather than bend sharply. Yet red blood cells, whose lateral walls are supported by a series of encircling microtubules, fold strongly when passing through capillaries, and flagella containing 20 microtubules, such as those of *Euglena* and other protozoans, bend freely into tight loops. Thus the rigidity seen *in vitro* may be an artifact of preparation.

Whether the lumen plays an active role in the functioning of a microtubule is still a moot point, although currently few cell biologists believe that it does (Burnside, 1975). Nevertheless, as the cavity is sufficiently large to accept negative stains, it would be equally capable of carrying many biologically important molecules. Indeed in a few instances in which exceptionally rigid microtubules have been reported, the lumen is found to be filled with electron-opaque substances (Huang and Pitelka, 1973).

X-Ray Structural Features. X-ray diffraction studies of flagellar microtubules (Cohen *et al.*, 1971, 1975) have largely tended to confirm the results obtained by electron microscopy. The diameter of the tubules was found to be similar (220 Å) with comparably thick walls (40 Å), as were also the arrangement of protofilaments in half-staggered sequences and the number of protofilaments per microtubule (12 or 13). However, the subunits appeared to be not so spherical as visualized in ultrastructural studies. This nonspherical construction has been made especially clear by the computer-reconstructed images of Erickson (1974a,b, 1975). His images of flattened microtubules could be interpreted in two ways, as Erickson pointed out. First, the α and β subunits could be considered as elongate oblongs placed parallel to one another but oriented obliquely in the filaments. In the second interpretation, the subunits could be viewed as semilunular molecules arranged in staggered fashion, with their outer curvatures adjacent. Radial sections appear to confirm the first of these suppositions, and in addition suggest that the α particles project distally beyond the β subunits and that the latter behave similarly proximally (Figure 2.4A).

2.1.2. Chemical Nature of Tubulin

The Nature of Tubulin. In many eukaryotic tissues, tubulin ranks among the most abundant proteins. For instance, in rat brain it comprises about 10% of the total protein present (Eipper, 1975) and about 20% of the soluble protein (Morgan and Seeds, 1975). This abundance and its ability to form complexes with such reagents as colchicine, podophyllotoxin, and vinblastine (Figure 2.6) have contributed greatly to researches into its nature (Renaud *et al.*, 1968; Shelanski and Taylor, 1968; Stephens, 1968; Weisenberg *et al.*, 1968; Jacobs and Mc Vittie, 1970; Bryan and Wilson, 1971; Eipper, 1972; Meza *et al.*, 1972; Witman *et al.*, 1972). Because the first two of those reagents compete with one

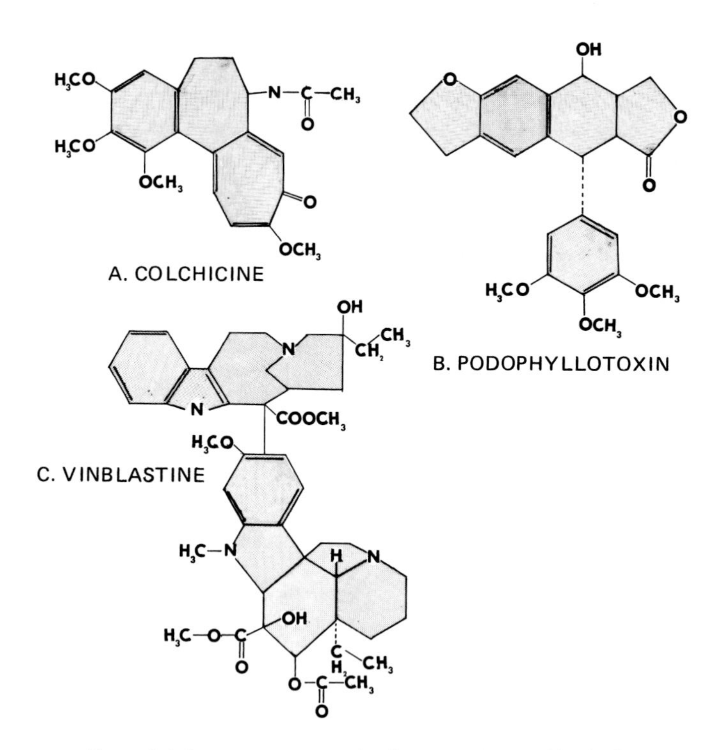

Figure 2.6. Important reagents that form complexes with tubulin.

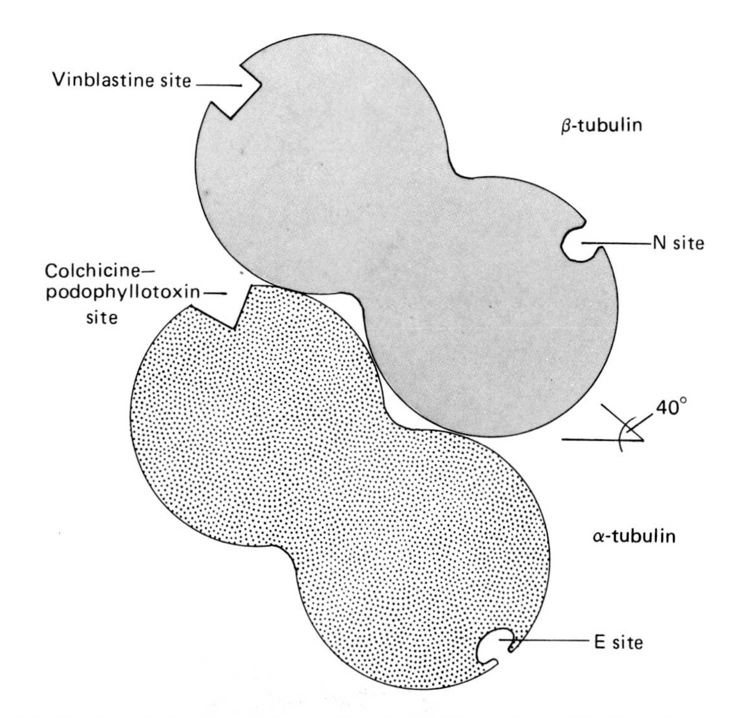

Figure 2.7. Possible relationships of the tubulin subunits. The exchangeable (E) and nonexchangeable (N) sites represent possible binding points for GTP; in the diagram they are arbitrarily assigned to β and α subunits, their actual association being unestablished. The combinations of sites on a given subunit have been determined, however. (Based in part on Bryan, 1974, and Erickson, 1974a,b.)

another for binding, they seem to be recognized by the same site on one of the two peptides that make up the dimeric subunit of tubulin, while the third compound apparently reacts with a site on the second member, in addition to a probable second but undetermined site (Bryan, 1975).

The two monomers that make up tubulin are known to differ chemically in other ways. Perhaps the greatest distinction is that β-tubulin is phosphorylated, whereas α-tubulin is not (Eipper, 1972, 1975). Furthermore, while both monomers carry bound guanosine triphosphate (GTP), that substance on the colchicine-accepting component is exchangeable with other substances in the medium, but that of the vinblastine-binding one is not (Figure 2.7; Stephens *et al.*, 1967; Yanagisawa *et al.*, 1968; Berry and Shelanski, 1972; Bryan, 1974). What the function of these energy-active nucleosides is has not been made clear as yet, although the obvious suggestions have been made that they may regulate assembly of the monomers or may be the energy source of polymerization.

They could equally be involved in intertubule reactions to form assemblages or even in the functioning of the organelle as a whole.

Another difference between the two types of tubulin involves a unique posttranslational modification of the α chain. In the presence of ATP after translation has been completed, an enzyme known as tubulin tyrosine ligase adds a tyrosine residue by means of a peptide linkage to the α-carboxyl of terminal glutamic acid residue (Barra *et al.*, 1974; Agaraña *et al.*, 1977; Raybin and Flavin, 1977a,b). Through the use of the highly purified enzyme, it has been found that every one of the more than 20 tubulin preparations from brain cells that were tested were composed of three species of α-tubulins (Nath and Flavin, 1978). One class already bore the tyrosine residue, the second could have such a residue added, and the third could not be acted upon by the ligase (Kobayashi and Flavin, 1977). By refining the techniques to separate the membrane and cytoplasmic fractions of each source, it was consistently found that the membrane-bound α-tubulin could accept tyrosine but normally was devoid of that residue, while the cytoplasmic α chains characteristically bore the terminal tyrosine. Thus these observations provide evidence that the microtubules associated with membranes are not completely identical chemically to those that lie free in the cytosol.

Tubulin Primary Structure. As yet the complete amino acid sequences of the two components of tubulin have not been established, but 25-site-long series from the NH_2 end of both α- and β-tubulins from sea urchin flagella and chick embryo brains have been reported (Luduena and Woodward, 1973, 1975). These studies indicated that both monomers have molecular weights between 55,000 and 60,000 and that the dimer ranges between 110,000 and 120,000. Although the β monomers have different amino acids from those of the α monomers in around 50% of the corresponding sites, little variation between like monomers from different sources was found, as shown in Table 2.1. The α-tubulins definitely differed from one another only at site 25, where that

Table 2.1
Partial Amino Acid Sequences of Tubulin

α-Tubulin	
Chick brain	MRZSISIHVTQATVQITXAS?ZLYS
Sea urchin	MRESISIHVTZATVZITXAS?ZLYA
β-Tubulin	
Chick brain	MRGIVHIQATQSTXQIT-AFWZVIS
Sea urchin	MRGIVHMZATZSTXZIT-AF??VIS

monomer from sea urchin had alanine, in contrast to the chick brain material, which had serine. Similarly the β-tubulins differed from each other only at site 7, the echinoderm subunit having methionine instead of the isoleucine of the vertebrate, and possibly at site 21, where the tryptophan found in the vertebrate material may be absent. While these results were claimed to indicate the evolutionary conservativeness of these tubulins, the preliminary nature of the analyses scarcely warrants the drawing of any conclusions. Whether or not the monomeric forms have been derived from a common gene, as proposed in the articles cited, should be more clearly indicated by analyses of tubulins from yeast or other protistan microtubules, or better yet, of those from prokaryotic sources.

A point detailed in a section on flagellar structure (Chapter 4, Section 4.3) clearly indicates that several varieties of tubulin exist, not only in the same organism, but even in the same cell. Moreover, the distinction between tubulins of different cell types has been made especially lucid by a current investigation of that substance from blood platelets (Ikeda and Steiner, 1979). Although tubulin from brain tissue has been demonstrated to possess protein kinase activity, its phosphorylating property is known to be due to an associated protein. In contrast, tubulin from platelets appeared to possess kinase activity itself, for the latter could not be separated from the colchicine-binding property common to all tubulins. Cytoplasmic tubulin of brain tissue consists of two groups of α subunits and at least a like number of β subunits, and that from synaptosomes has a third type of α subunit (Marotta *et al.*, 1979).

Tubulin as a Complex. In addition to the enzymatic and structural bridges already discussed, tubulin appears to be complexed with macromolecules of several types. Small quantities of proteins are present, at least one of which seems to be essential in assembling the tubulin dimers to form microtubules (Weingarten *et al.*, 1975), as described in the section that follows. Moreover, the presence of some variety of lipid has been reported, on the basis of irradiation experiments (Bryan, 1975). The results suggested a distinct likelihood that the phosphorylation mentioned in a preceding paragraph may involve this lipid, for treatment with enzymes that break down phospholipids, including phospholipase A from bee venom and phospholipase C from *Clostridium,* inhibited assembly of tubulin into microtubules.

2.1.3. Assembly into Microtubules

Historically speaking, studies on the assembly of tubulin into the definitive microtubules have followed the pattern that typifies most investigations into biological processes at the molecular level. First, as in many other instances, microtubule formation was considered a relatively simple synthesis, being a self-assembly process, involving merely dynamic equilibrium between the soluble subunits (dimers) and the tubules (Inoué, 1953; Inoué and Sato, 1967;

Borisy and Olmsted, 1972; Weisenberg, 1972; Olmsted and Borisy, 1973; Borisy *et al.*, 1974; Burnside, 1975). When the equilibrium was disturbed by addition (or biosynthesis) of subunits, it was restored by their polymerization into microtubules; conversely, a decrease in concentration of subunits was restored by dissociation of the organelles. But further data have subsequently shown, as in most features of cell metabolism, that that simplistic view is untenable, for a whole series of steps appears to be involved in the combining of the definitive microtubules into the paired, doublet, or multiunit assemblages that frequently characterize their presence in the living cell. Furthermore, again as in other living processes, enzymes and other factors are required for tubulin polymerization to occur. However, this facet has received little attention in the literature. while the basic topic has received much; hence, there is a corresponding inequality in the treatment of these subjects in the present synopsis.

Direct Incorporation of Subunits. A number of diverse observations have led to the decreased acceptability of the dynamic-equilibrium concept. First, temperature plays an important role in the formation and disintegration of microtubules, a feature especially shown by the mitotic spindle and *in vitro* studies. At elevated temperatures, such as 23°C with the oocytes of the annelid *Chaetopterus* (Inoué *et al.*, 1974) and 37°C with sea urchin sperm tail and mammalian brain tissue (Weisenberg, 1972; Olmsted and Borisy, 1973; Farrell and Wilson, 1978), tubulin subunits become arranged into microtubules, and, as they do so, the birefringence of the system increases (Inoué, 1952, 1953). In contrast, at temperatures of 6°C or lower, microtubules disintegrate rapidly into the subunits. Obviously, therefore, the relationship between monomer and polymer is not simply dependent on relative concentrations. Furthermore, certain classes of microtubules, such as those of sperm flagella, are not labile but remain stable, regardless of extreme temperature changes (Farrell *et al.*, 1979), even though their composition appears to be the same as that of the others.

Nevertheless, a number of observations that have been made indicate that tubulin dimers may enter into the growth of microtubules, although they seem unable to initiate the formation of the organelles (Erickson, 1975; Erickson and Voter, 1976). The subunits that entered into microtubule elongation were dimeric and sedimented with a coefficient of 6 S. About 90% were heterodimers ($\alpha\beta$), while the remainder consisted of $\alpha\alpha$ or $\beta\beta$ particles (Kirschner, 1978). Erickson (1975) presented a scheme that seems to explain how the subunits contribute to longitudinal growth of an existing microtubule. This postulated the presence of two longitudinal bonds on each of the ends of the 6 S subunits, while only one bond occurred on each side. As the exposed ends of the growing tubule thus offered two bonds, while the sides had only one, the resulting greater reactivity at the terminals favored elongation rather than widening in sheet fashion (Figure 2.8). However, this point of view fails to explain why free subunits do not unite with one another to form chains.

Depletion of the reserve of α- and β-tubulins from cells stimulates the syn-

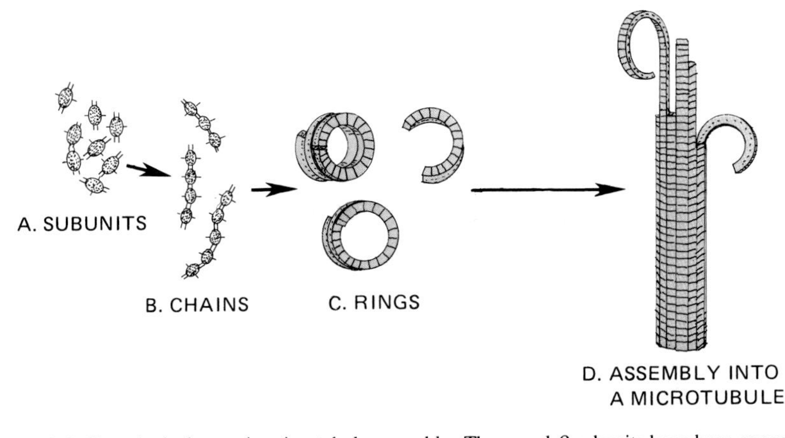

Figure 2.8. Hypothetical steps in microtubule assembly. The α and β subunits have been suggested to have more chemically active sites at their ends than at the sides (A); however, chains, as in (B) are not actually known to exist. (C) Ring forms are produced by subunits, which in turn assemble into microtubules.

thesis of a new supply of the proteins, according to experimental evidence from sea urchin embryos (Merlino *et al.*, 1978). The cells of these embryos carried a large cytoplasmic pool of microtubular proteins, sufficient to regenerate cilia several times without the synthesis of new material. When the blastulas were artificially deciliated three times, labeled methionine was incorporated into α- and β-tubulins at a rate two- or three-fold as great as previously. This ceased when mRNA synthesis was inhibited by actinomyosin D, as did the growth of new cilia; thus multiple deciliation induced an enhanced rate of transcription of the tubulin genes.

Requirement for Ring Structures. What has proven essential to the formation of microtubules in addition to the presence of 6 S subunits is that of ring structures (Figure 2.8). These rings, which sediment at 30 or 36 S, uniformly have the appearance of consisting of continuous filaments 50 to 60 Å thick that form circles 420 to 490 Å in outside diameter (Figure 2.9; Erickson, 1975; Kirschner *et al.*, 1975a,b). Under elevated pH conditions, 20 S components are formed instead of the 30 or 36 S ones (Dönges *et al.*, 1976). As a rule, the latter types of circles consist of 24 subunits, and when the rings are double, as is occasionally the case, the inner circle contains 18 subunits. The dimensions of these circles obviously preclude them from representing cross sections of definitive microtubules, yet they do become arranged into long stacks under certain conditions (Borisy *et al.*, 1975). More typically, however, the filaments of the rings may continue beyond the simple ring, forming coils of varying lengths.

Apparently the rings, or better coils, represent protofilaments of consider-

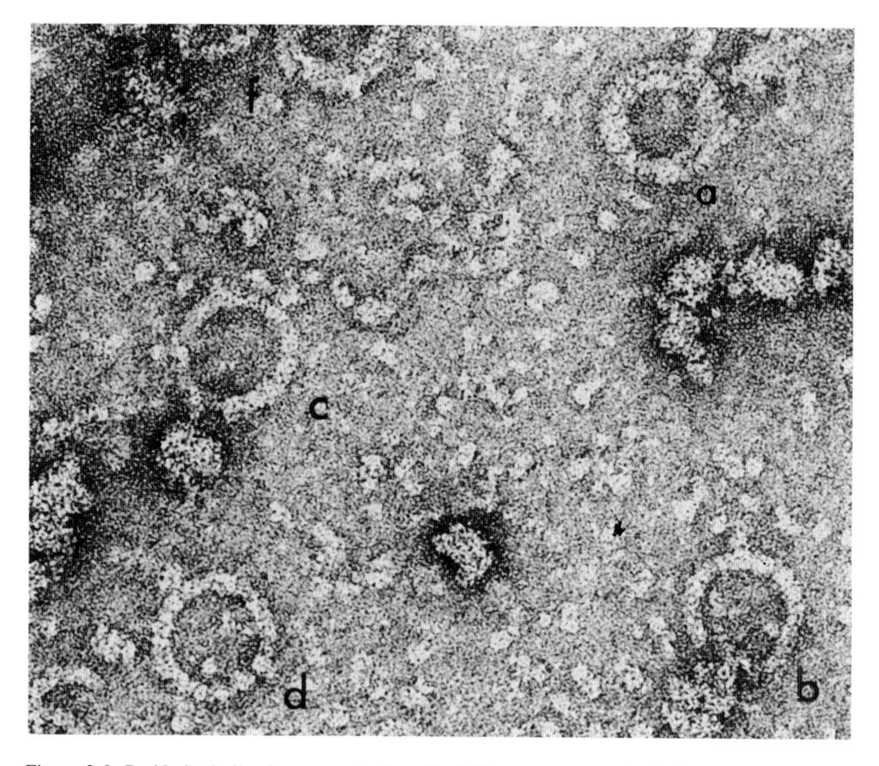

Figure 2.9. Purified tubulin rings, negatively stained. The structure marked *a* is a double coil while *b* is an incomplete ring; *c* and *d* suggest that the complete rings are actually coils. 350,000×. (Courtesy of Erickson, 1974b.)

able length, for under appropriate conditions they unroll and attach side by side to one another to produce sheets. In turn, the latter appear to roll into cylinders to form the actual microtubules (Figure 2.10). Not all 6 S subunits are capable of forming rings. When separated on an agarose column, those 6 S subunits contained in the lowermost of the two bands that were formed on the column were found to be incapable of uniting into rings, whereas the small amount that was associated with the ring structures, which made up the bulk of the upper band, readily did so (Kirschner *et al.*, 1975a,b; Kirschner, 1978). The former was referred to as X-tubulin and the latter as Y-tubulin.

In addition to the microtubular and sheet forms of tubulin polymers, a new variety, known as hoops, has been described (Mandelkow *et al.*, 1977). Produced in *in vitro* systems, these hoops are around 1000 to 1400 nm in circumference and may be comprised of 40 to 80 protofilaments in width (Figure 2.11). These appeared to result from the winding up of shallow spirals of sheets in such a way that the side that would normally be the inside one within the lumen in a microtubule served as the outside face of the hoop.

Polarity of Growth. While the unfolding rings in Figure 2.10 may be noted to be located at one end of the growing sheets, in some cases they occurred at both termini in this *in vitro* system. The existence of this predominantly unipolar growth was first demonstrated *in vivo* in the regeneration of the flagella of such protists as *Chlamydomonas* and *Ochromonas* (Rosenbaum and Child, 1967), for most of the labeled amino acids were incorporated in the distal region. However, because some was incorporated proximally in this intact system, it could not be definitely ascertained whether the basal radioactivity that also was observed represented growth of the tubules or that of the membrane or cytosol of the flagella. More recently, use of radioisotopes and electron microscopy have made it clear that the flagellar microtubules are assembled primarily by distal addition of tubulin (Witman, 1975), while proximal additions are made at about half the rate of those at the distal end (but see Chapter 4, Section 4.3, for a fuller discussion of these processes). Similar results have been obtained by *in vitro* studies on brain microtubules, where assembly was made on "nucleation centers," such as pieces of microtubules (Snell *et al.*, 1974), centrioles (McGill and Brinkley, 1975), chromosomal kinetochores (Gould and Borisy, 1978), and basal bodies (Allen and Borisy, 1974; Rosenbaum *et al.*, 1975; Stearns *et al.*, 1976). The Rosenbaum report showed that unidirectional growth was strongly marked when the concentration of tubulin was low and that bipolar synthesis increased with increments in the concentration.

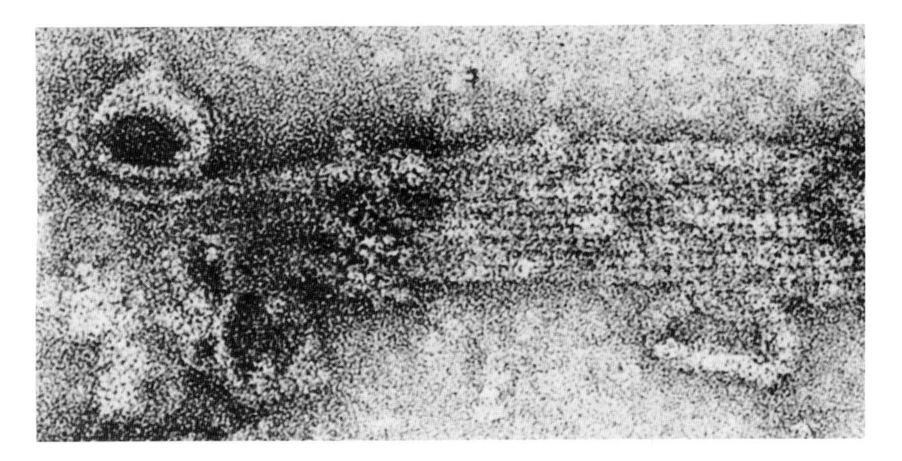

Figure 2.10. Assembly of rings into microtubules. The continuity of the protofilaments and rings is clearly shown in this flattened section of a forming microtubule wall. 350,000×. (Courtesy of Erickson, 1974b.)

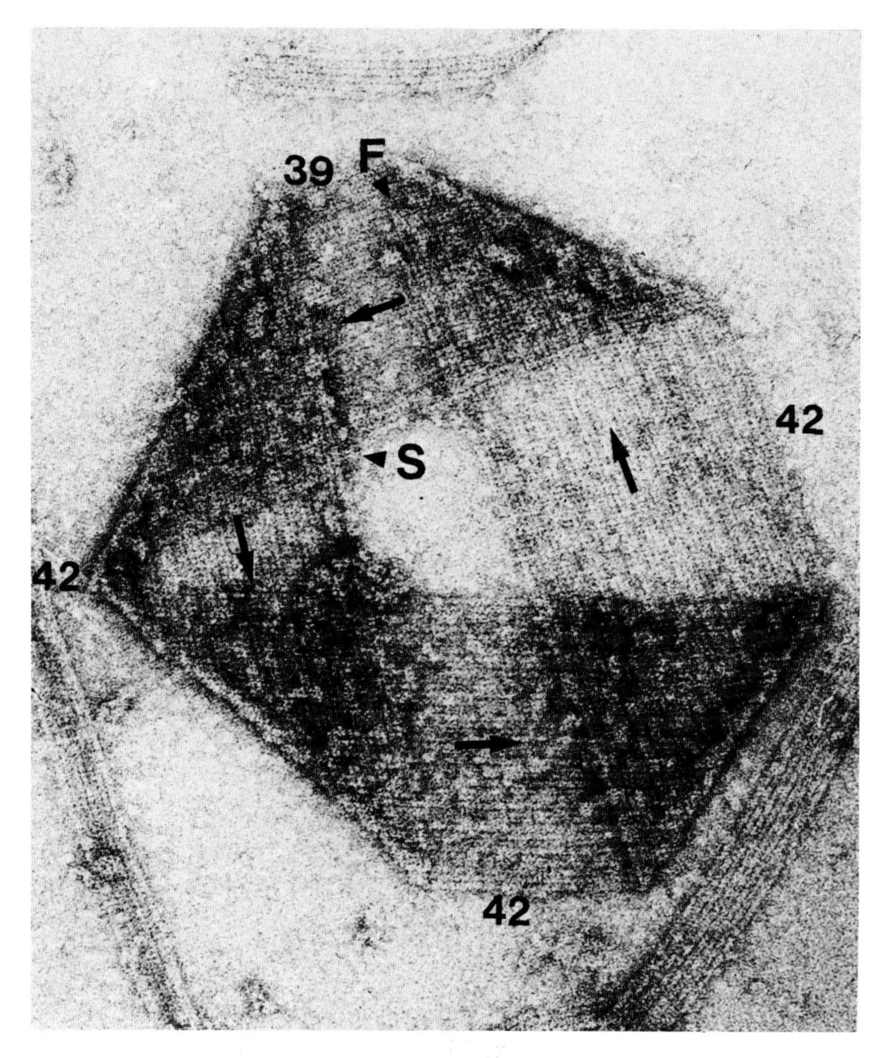

Figure 2.11. A tubulin hoop. An opened hoop, with sides flattened on the grid. The sides contain 42 protofilaments except that including the start (S) of the coil. The growing ends (F) are nearby. 250,000×. (Courtesy of Mandelkow and Mandelkow, 1979.)

An unusual mutation in *Tetrahymena* has made possible *in vivo* observations on growth of microtubules. At elevated temperatures the mutant homozygous for the recessive gene *mol*^{b} fails to divide properly, resulting in the two daughter organisms remaining attached. In swimming, the pair of cells bend so strongly at the point of attachment that some of the longitudinal microtubules located beneath the plasmalemma become severed, making the broken ends available for observations on regeneration (Ng, 1978). In a series of studies of that sort, the ends of microtubules toward the original posterior end of the mother cell become elongated as a rule, whereas the opposite ends only rarely are added to.

Nucleotides in Assembly. Purified tubulin was demonstrated to contain GTPase activity and to react specifically with 2 moles of GTP or GDP (Weisenberg *et al.*, 1968). One mole of the nucleotide that became attached at the "N site" was nonexchangeable, whereas that which bound at the "E site" was freely exchangeable with the medium. Later, after Weisenberg (1972) showed that either GTP or ATP was requisite for tubulin polymerization, it was assumed that these nucleotides might serve directly in microtubule assembly. At first this supposition was strengthened by investigations that indicated that nonhydrolyzable analogs of GTP were ineffective in microtubule assembly (Gaskin *et al.*, 1974; Olmsted and Borisy, 1975). Later, however, Penningroth *et al.* (1976) found that this inactivity resulted from the inhibition of binding by GDP and that, in fact, nonhydrolyzable analogs did support microtubule formation (Arai and Kaziro, 1976, 1977). The GTP at the N site was found not to be hydrolyzed during assembly *in vitro* for it remained present after three cycles of assembly with analogs (Kobayashi, 1975; Kobayashi and Simizu, 1976). Thus the analogs clearly replace only the GTP at the E site.

The interchangeability of ATP and GTP in polymerization for a time offered a difficult problem. Rather recently ATP was shown not to enter directly into these processes by combining with the tubulin but served in phosphorylating GDP at the E site of tubulin (Kobayashi and Simizu, 1976; Penningroth *et al.*, 1976; Penningroth and Kirschner, 1977). If the GDP was completely removed from the E site, ATP and its analogs could serve in its stead, however (Penningroth and Kirschner, 1978). A number of enzymes, including pyruvate kinase and phosphoenolpyruvate, were capable of phosphorylating GDP at this site, a step necessary for polymerization. Consequently, one step in assembly appears to be the phosphorylation of GDP at the E site, but GTP hydrolysis does not seem to be essential (Kirschner, 1978). Because the GTP at the N site remains intact throughout assembly, its role is still unknown at present. Even *in vivo* it has been found to turn over slowly, at a rate of only once in 24 hr in Chinese hamster ovary cells (Spiegelman *et al.*, 1977).

The Tau Factor. By means of ion-exchange chromatography on phosphocellulose, Weingarten and co-workers (1975) separated a protein that they designated as the tau (τ) factor. In the absence of this protein, tubulin existed solely as the 6 S dimer, which would not polymerize *in vitro*. Addition of τ to the system resulted in restoration of microtubule formation, and under low temperatures or other nonpolymerizing conditions, 36 S rings were formed. Later it was demonstrated that this factor also was required for tubulin polymerization on such microtubule nucleation sites as flagellar microtubules (Witman *et al.*, 1976).

When purified, τ was found to migrate during electrophoresis on acrylamide gels as four narrowly separated bands, approximating molecular weights between 55,000 and 62,000 (Cleveland *et al.*, 1977a; Herzog and Weber, 1978). The native protein sedimented with a coefficient of 2.6 S and had a molecular weight of 57,000. A neutral or basic protein, the factor was found to be phosphorylated by a protein kinase that became purified together with microtubules (Cleveland *et al.*, 1977b). It was suggested that each τ molecule promoted assembly by binding to several 6 S tubulin molecules, thus increasing the local concentration of tubulin. These molecules are distributed continuously along the length of microtubules (Connolly *et al.*, 1978; Sherline and Schiavone, 1978), but as they occur in a ratio of only 1 per every 30 tubulin dimers (Kirschner, 1978), it is not likely that they bring about assembly in the manner just suggested.

Other Factors. After calf brain microtubule preparations had been purified by repeated cycles of assembly and disintegration, electrophoretic analyses revealed the presence of many proteins in addition to α- and β-tubulin (Berkowitz *et al.*, 1977). This assortment, constituting about 17% of the total material, could be divided into two groups, one of which was removed during purification and the other was not. Of the latter group, the bulk appeared to be constituents of neurofilaments, intermediate filaments with diameters of 100 Å that are discussed later. Although τ had originally been reported from hog brain, it was not detected in this calf brain preparation. High-molecular-weight microtubule-associated proteins (MAPs) (Figure 2.12; Mandelkow and Mandelkow, 1979), with apparent molecular weights of 250 to 350 $\times$ 10^3, were present in a ratio of about 1:8 or 1:9 with tubulin, while corresponding proteins with molecular weights between 30 and 35 $\times$ 10^3 were in ratios of around 1:80. That type known as MAP-2 has now been purified and shown to be active in the assembly of tubulin (Kim *et al.*, 1979; Stearns and Brown, 1979). Studies on the A tubule of flagellar doublets similarly showed a wide diversity of proteins to be present. Interestingly, those of cilia from scallops differed in many electrophoretic bands from those of sea urchins (Linck, 1976). Mostly high-molecular-weight proteins were found associated with the mitotic spindle by immunofluorescent procedures (Sherline and Schiavone, 1978), while one of the 68,000-dalton proteins was found in purified microtubules of HeLa cells (Weatherbee *et al.*, 1978). Others from the latter source had molecular weights of 210,000 and 120,000 (Bulinski and Borisy, 1979). MAPs from mammalian brain microtubules decorated the newly formed microtubules with fine filamentous projections as they stimulated outer doublet tubulin assembly (Binder and Rosenbaum, 1978).

Few of the proteins have as yet received detailed attention. Among those that have received partial characterization is a group of six enzymes that me-

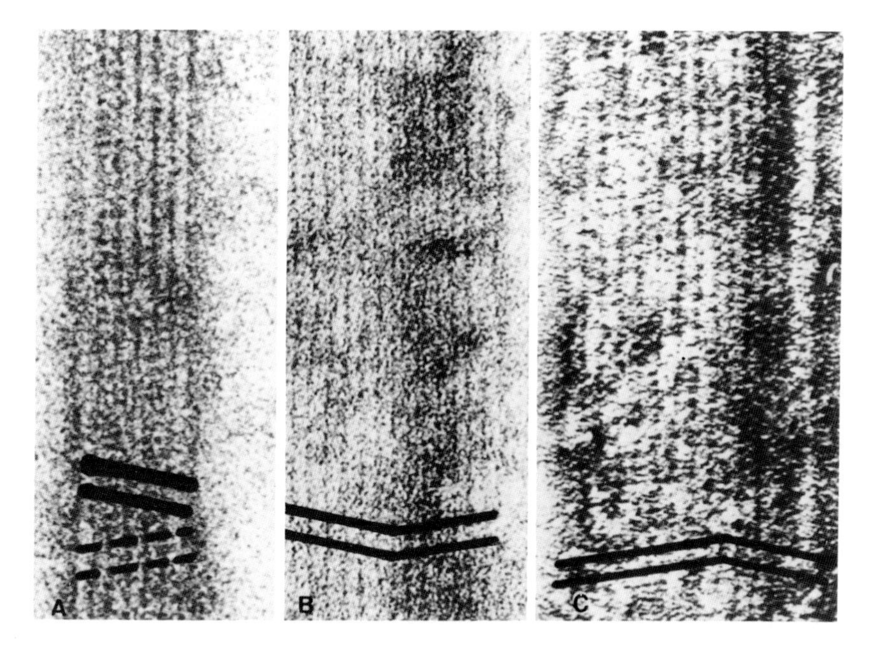

Figure 2.12. The influence of microtubule-associated proteins (MAPs). (A) A negatively stained microtubule; the heavy bars indicate the slope of the helix on the visible wall, while the broken bars indicate that of the concealed wall. 330,000×. (B and C) Sheets of protofilaments, one of which (B) has been assembled in the presence of MAPs, the other (C) in their absence. (Courtesy of Mandelkow and Mandelkow, 1979.)

tabolized nucleotides, but these seem to be confined to *Chlamydomonas* (Watanabe and Flavin, 1976)—at least many of them were not detected in cilia from *Tetrahymena*. To date, the dynein that forms arrays of bridges on the A tubule of flagellar doublets is the only protein that has been thoroughly studied and that solely on a preliminary basis (Warner *et al.*, 1977; Warner and Mitchell, 1978). Extraction of mollusk or *Tetrahymena* cilia with 0.5 M KCl solubilized the outer rows of dynein arms from the A tubules of doublets. Examination of this fraction under the electron microscope using nagative contrast techniques revealed a homogeneous array of molecules 93 Å in diameter, which were considered to represent a monomeric form of dynein, having a molecular weight of 360,000. This particle was believed to be only a portion of the outer dynein arms, for individual arms could be seen that were composed of three or four subunits. Dynein possesses ATPase activity and is arranged regularly along the length of the A tubule, where it is supposed to be involved in the bending of the organelle. The details, however, are more appropriate to the discussion of flagella and cilia (Chapter 4, Section 4.2).

Some conflicting results have been obtained with these microtubule-associated proteins insofar as their polymerizing abilities are concerned. Whereas some laboratories reported that τ effected microtubule formation, as reported in a preceding section, other laboratories found that only two of the high-molecular-weight proteins, referred to as MAP-1 and MAP-2, were active (Murphy and Borisy, 1975; Sloboda *et al.*, 1976; Murphy *et al.*, 1977). More recently the differences were resolved by Herzog and Weber (1978), who demonstrated that either MAP-2 or τ was able to induce tubulin polymerization in the absence of the other. These several varieties of MAPs have been found to be distributed along the lengths of microtubules, where they form a filamentous coat (Connolly *et al.*, 1978; Sherline and Schiawone, 1978; Kim *et al.*, 1979).

Calcium and Other Ions in Assembly. The role of divalent ions in the assembly of microtubules is scarcely more clear-cut than are those of the various protein factors. Weisenberg (1972) first showed that successful repolymerization of microtubules from brain tissue by any given buffer depended on the ability of the latter to chelate calcium and thus decrease the Ca^{2+} level. Addition of substances that chelate calcium preferentially to magnesium also enhanced microtubule assembly. Only fractional millimolar levels of Mg^{2+} ions may be present to permit tubulin polymerization. For instance, no inhibition was found with 0.1 mM, but assembly ceased with 2 mM (Olmsted and Borisy, 1975; Olmsted, 1976). Moreover, the presence of Mg^{2+} ions increased the inhibiting effects of calcium. With 0.5 mM $MgCl_2$, even 35 μmoles $CaCl_2$ failed to inhibit microtubule formation, whereas with 5 mM $MgCl_2$, only 5 μmole quantities of $CaCl_2$ reduced assembly to 50% (Rosenfeld *et al.*, 1976). Nevertheless, $MgCl_2$ was necessary along with GTP to bring about the polymerization of tubulin extracted from flagellar microtubules (Pfeffer *et al.*, 1978). However, the levels used experimentally, such as those cited, are higher than those found in living cells, so it is not likely that changes in Ca^{2+} ion concentrations *in vivo* are involved in the cellular control of the processes, as has been proposed (Weisenberg, 1972; Massini and Lüscher, 1976). The pH value of the system also is important in assembly. At pH 6.0, only branched protofilaments are found, normal microtubules forming only at pH levels approaching 7.7 (Burton and Himes, 1978)

2.2. MICROFILAMENTOUS STRUCTURES

Besides the neurofilaments mentioned in the preceding section, a number of nontubular fibrous structures and fiber-producing proteins have been described during the past decade. Indeed, the two interacting fibers of muscle tissue, actin and myosin, have been recognized for nearly half a century; their discovery in nonmuscle cells, however, has been a more recent event. No fewer than 12 contractile proteins have been cataloged (Pollard, 1976, 1977), but the majority of these are confined to particular taxa or are too poorly known to merit discussion.

2.2.1. The Actin–Myosin System

Two interacting proteins, actin and myosin, appear to be the chief, if not sole, force-generating substances in cells in general; in other words, this pair is often viewed as *the* proteins that bring about movement in cells. Because certain data also implicate microtubules directly in the movement of parts, at one time it was advocated that tubulin was identical to actin. That this supposition is erroneous was clearly shown by Mohri and Shimomura (1973), for the two molecules differ in dimensions and weight, amino acid composition, and peptide maps. Further, tubulin fails to interact with myosin to the same extent that actin does. The actin–myosin system appears to be universally distributed in cells of all kinds, for both major proteins have now been extracted from *E. coli* (Nakamura *et al.*, 1978) as well as from a diversity of eukaryotes.

Thin Filaments and Actin. Much of the information about the contractile fibers has been derived from studies of vertebrate muscle tissue; only recently have probings been made into the details of related molecules from invertebrate and other eukaryotic sources (Figure 2.13). Thus discussion here is principally centered on vertebrate material, most of which in turn is limited to mammalian tissues (Pollard and Weihing, 1974; Mannherz and Goody, 1976).

The Actin Molecule. Actin is a globular protein. In rabbit skeletal muscle it has a molecular weight of 41,785, a diameter of 45 to 55 Å, and consists of a single peptide chain 376 amino acid residues in length. Because of the number of resemblances in this substance from various sources, a topic discussed in detail later, it was concluded that actin has been highly conserved evolutionarily (Squire, 1975; Pollard, 1977). However, the actins from human heart and blood platelets have been described as encoded by different genes (Elzinga *et al.*, 1976), and more than one species has been reported from a single tissue, including calf muscle cells (Whalen *et al.*, 1976), and three in *Drosophila* (Horwitz *et al.*, 1979). Comparisons of the primary structure of actin from bovine brain and muscle also show differences (Lu and Elzinga, 1976). Hence, the supposed stability seems to have been based on too limited observations. Indeed, considerable differences in properties between this protein from vertebrate muscle and from *Acanthamoeba* have been evidenced (Gordon *et al.*, 1976). In cells, the native actin usually is in the form referred to as G-actin, each molecule of which bears a number of ATP residues and a single calcium ion. This is converted to F-actin through hydrolysis of the ATP to ADP; it is this form that makes up the characteristic fibers discussed in a later paragraph.

These two forms of actin differ greatly in their reactions with various re-

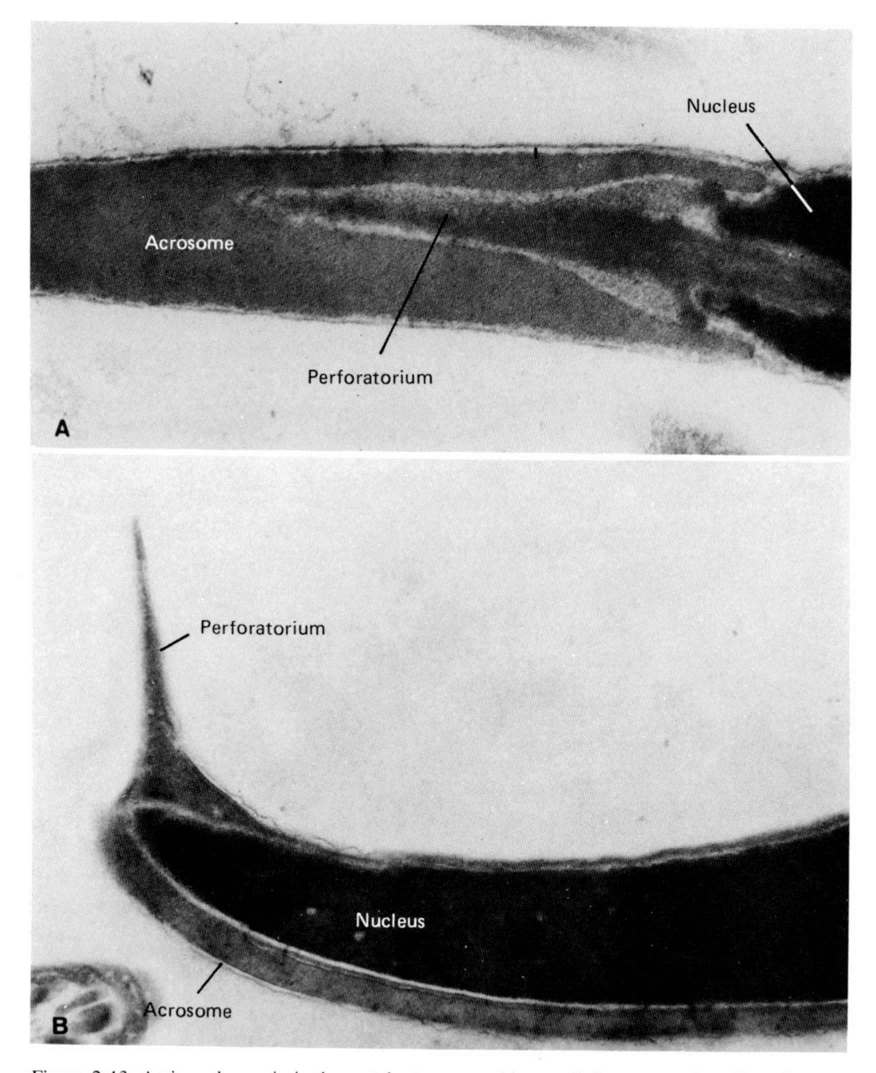

Figure 2.13. Actin and myosin in the vertebrate sperm. Almost all tissues contain actin and myosin fibers, as here in the sperm head where these and other proteins form the perforatorium. (A) The perforatorium of this turkey sperm head consists of closely packed microfilaments. 62,500×. Its arrangement here between the acrosome and nucleus is more typical of metazoan sperm in general than (B), the laterally placed structure in the rat spermatozoon. 23,000×. (Courtesy of C. Campanella and co-workers, 1979.)

agents. For example, the five cysteines of the actin molecule, located at sites 10, 217, 256, 283, and 373, display considerable distinctions. With 3,2′-dicarboxy-4-iodoacetamidoazobenzene, the residues at sites 10, 283, and 373 are labeled in G-actin without affecting the ability to polymerize, whereas in F-actin the cysteine at site 10 alone interacts (Lusty and Fasold, 1969). Similarly, diethylpyrocarbonate adds a carbethoxyl radical to the site 40 histidine in G-actin but not in F-actin (Hegyi *et al.*, 1974); in this case polymerization of the G-actin is inhibited. Because inhibition occurs also when the tyrosine of site 53 in G-actin has reacted with diazonium tetrazone (Bender *et al.*, 1974), it has been speculated that a region close to the NH_2 terminal is involved in conformational changes that accompany polymerization. In addition to these two states of actin, multiple species are now known to exist, at least three of which occur in rat brain (Palmer and Saborio, 1978).

Immunofluorescent microscopy has made it possible to locate actin filaments, microtubules, and other filaments in intact cells, unexpectedly including such secretory cells as chromaffin granules (Meyer and Burger, 1979). In these procedures antibodies to the particular protein under investigation are prepared and labeled with special dyes for examination with the fluorescent microscope. Through these means actin and myosin have been found to be nearly universal constituents of eukaryotes, being associated with motile parts in profusion, including those that move by means of pseudopods, such as *Dictyostelium* (Uyemura *et al.*, 1978). What was surprising was the recent discovery of actin in the nuclei of the oocytes of the clawed frog (*Xenopus laevis*). This protein proved to constitute about 6% of the total protein present in these nuclei (Clark and Merriam, 1977); 75% of this amount was diffusible, while 25% was closely associated with an insoluble nuclear gel. In this gel were embedded the chromosomes, nucleoli, and nuclear granules; actin was the most prominent chemical component, comprising 16% of the total protein present.

Primary Structure of Actin. In the absence of a number of established amino acid sequences, few comments on the primary structure of this protein can be made. So far only rabbit skeletal muscle actin has been sequenced completely (Elzinga and Collins, 1973, 1975; Elzinga *et al.*, 1973; Collins and Elzinga, 1975). Its primary structure, 319 residues in length, is presented in Table 2.2, along with four fragments of actin from trout muscle that center around cysteine residues (Bridgen, 1971). It is apparent that, at the region immediately following the amino terminus, about 25% of the 17 sites differ, as do 14% of sites 255 to 261, but at the carboxyl end only one occupant of 13 sites shows differences.

In addition, the sequence of actin from calf brain has been partly determined (Table 2.2; Lu and Elzinga, 1977). Of the 157 residues that have been firmly established, 14 differ from those of the corresponding sites in the rabbit muscle material, not 11 as stated by the article cited. Thus in the sequenced portions, the two sequences differ to an extent of 9%. Probably these should be

Table 2.2
Amino Acid Sequences of Actin Molecules[a,b]

	10 20 30 40 50 60
Rabbit muscle	ACDETEDTALVCDDGSGLVKAGFAGDDAPRAVFPSIVGRPRHQGVMVGMGQKDSYVGDEAQS
Trout muscle	ACDDEETTALVCDDGGSLVK( )
Calf brain	ACDEDEAAILVVDNGSGMCKAGFAGDDAPRAVF( )VGMGQKDSYVGDEAQS
	70 80 90 100 110 120
Rabbit muscle	KRGILTLKYPIEH[2]WGIITNDDMEKIWHHTFYNELRVAPEEHPTLLTEAPLNPKANREKMT
Trout muscle	( )
Calf brain	KRGILTLKYPIEH[2]( )
	130 140 150 160 170 180 190
Rabbit muscle	QIMFETFNVPAMYVAIQAVLSLYASGRTTGIVLNSGDGVTHNVPIYEGYALPHAIMRLDLAGRDLTDYLM
Trout muscle	( )
Calf brain	( )FETFNTPAM( )AILR( )DLTDYLM
	200 210 220 230 240 250 260
Rabbit muscle	KILTERGYSFVTTAEREIVRDIKEKLCYVALDFENEMATAASSSLEKSYELPDGQVITIGNERFRCPGTL
Trout muscle	( )RCPTSL
Calf brain	( )ATAASSSLEKSYELPDGQVITI( )
	270 280 290 300 310 320 330
Rabbit muscle	FQPSFIGMESAGIHETTYNSIMKCDIDIRKDLYANNVMSGGTTMYPGIADRMQKEITALAPSTMKIKIIA
Trout muscle	( )KKCDIDIRDL( )
Calf brain	( )ESCGIHETTFNSIM( )YANNVMSGGTTMYPGIADRMQKEITALAPSTM( )
	340 350 360 370 374
Rabbit muscle	PPERKYSVWIGGSILASLSTFQQMWITKQEYDEAGPSIVHRKCF$_{OH}$
Trout muscle	( )DEAGPSLVHRKCF$_{OH}$
Calf brain	( )ISK( )KCF$_{OH}$

[a] Based on Bridgen (1971), Elzinga and Collins (1973, 1975), Collins and Elzinga (1975), and Lu and Elzinga (1977).
[b] Segments that have not been sequenced are enclosed in parentheses.

viewed largely as tissue differences, rather than those of source species, for other related proteins from corresponding tissues that have been sequenced from various mammals show few variations. This receives stress from an investigation into pure actins from a diversity of metazoan muscles that showed that the cysteinyl residues were similarly located in the primary structure in each case (Nakamura *et al.*, 1979).

Tropomyosin. In living cells actin filaments appear to be consistently associated with two other types of proteins, tropomyosin and troponin. The tropomyosin molecule is a narrow rod about 410 Å long, with a molecular weight of around 70,000. It consists of two subunits, which are closely similar in composition; both are 100% α-helical polypeptides and are twisted about one another (Cohen and Holmes, 1963). Although the two subunits are claimed not to be present in identical proportions, the subunit composition of rabbit muscle tropomyosin being $\alpha_4\beta$ (Casper *et al.*, 1969; Hodges and Smillie, 1970), the reported molecular weights (α chain, 37,000; β chain, 33,000) do not permit such a ratio if the total molecular weight cited above is correct. Tropomyosin molecules from such nonmuscle tissues as calf blood platelets and pancreas, mouse fibroblasts, and rabbit brain are only 350 Å in length, have a molecular weight of 65,000, but similarly consist of two subunits (Fine and Blitz, 1975). Thus the latter reference suggested the existence of two distinct subclasses of this substance.

Primary Structure of Tropomyosin. The amino acid sequence of the α chain of tropomyosin from rabbit skeletal muscle has now been fully established (Table 2.3; Hodges *et al.*, 1972, 1973; Edwards and Sykes, 1978; Sodek *et al.*, 1978; Stone and Smillie, 1978). It was found to contain 284 residues and to be acetylated at the amino end. With the primary structure of this tissue now in hand and with the techniques that have been developed for separating the α and β chains in sufficient quantities, perhaps analyses of the compund from other source animals will be forthcoming. The two types of subunits have been found to be somewhat similar to one another and to be present in molar ratios of either 3 or 4 α chains to 1 β chain. In cells the two are arranged in a nonstaggered configuration, and each has 14 actin-binding sites. In addition one site binds troponin-T.

Current investigations on the functional organization of the α-tropomyosin molecule have indicated that two sites may bind troponin-T, rather than just one. By cleaving the molecule with cyanogen bromide, two large fragments were obtained: A, comprised of residues 11 to 127, and B, residues 142 to 281 (Ueno, 1978). Alone, neither of these displayed any ability to bind to troponin. Further cleavages of the intact molecule with trypsin and similar enzymes produced other segments, which when combined with A or B made it possible to deduce that two troponinbinding regions exist, one near site 150 and the other near the carboxyl end beyond the cysteine at site 190.

Troponin. The second of the actin-associated proteins is troponin, whose

Table 2.3
Amino Acid Sequence of Rabbit Skeletal Muscle Tropomyosin[a]

```
          10        20        30        40        50        60
AcMDAIKKKMQMLKLDKENALDRAEQAEADKKAAEDRSKQLEDELVSLQKKLKGTEDELDKY

          70        80        90       100       110       120
  SEALKDAQEKLELAEKKATDAEADVASLNRRIQLVEEELDRAOERLATALQKLEEAEKAA

         130       140       150       160       170       180
  DESERGMKVIESRAQKDEEKMEIQEIQLKEAKHIAEDADRKYEEVARKLVIIESDLERAE

         190       200       210       220       230       240
  ERAELSEGKCAELEEELKTVTNNLKSLEAQAEKYSQKEDKYEEEIKVLSDKLKEAETRAE

         250       260       270       280
  FAERSVTKLEKSIDDLEDELYAQKLKYKAISEELDHALNDMTSICOOH
```

[a] Based on Stone and Smillie (1978).

molecular weight of 76,000 is the total of three subunits that are named on the basis of their functions. Troponin-T, or Tn-T, binds to tropomyosin and has a molecular weight of 37,000 (Greaser *et al.*, 1972). The second subunit, troponin-C (Tn-C), with a molecular weight of 18,000, binds Ca^{2+} ions (Greaser and Gergely, 1971). The third subunit, troponin-I Tn-I), has a molecular weight of 20,850 and is an inhibitor, interfering with the interaction between actin and myosin that is described later (Perry *et al.*, 1972; Wilkinson *et al.*, 1972). The calcium-binding subunit has two Ca^{2+}-specific binding sites and two others that can bind either Ca^{2+} or Mg^{2+} (Collins *et al.*, 1977). It has not been clearly established that a troponin system is associated with smooth muscle contraction, but a troponin-C-like protein has now been shown to be present in mammalian smooth muscles (Grand *et al.*, 1979).

It should be noted that troponin-T, besides interacting with tropomyosin, binds to troponin-C more readily that it does to troponin-I (Ebashi *et al.*, 1974). Troponin-C bonds strongly with both of the other subunits, but not to actin or tropomyosin, whereas troponin-I binds to actin but not to tropomyosin (Hartshorne and Dreizen, 1972; Hitchcock *et al.,* 1973; Margossian and Cohen, 1973). At low Ca^{2+} concentrations, troponin-I and troponin-T compete for troponin-C, and at higher levels a ternary complex of all three subunits is produced (Ebashi 1974), but the three thus combine only when troponin-I is in a reduced state (Horwitz *et al.*, 1979).

Troponin-C Primary Structure. Amino acid sequences of each of the subunits have been established to a greater or lesser extent. That for troponin-C has been completed for four different sources of skeletal muscle (Collins *et al.*, 1973; Collins, 1974; Romera-Herrara *et al.*, 1976; Wilkinson, 1976) and one for cardiac muscle (van Eerd *et al.*, 1978), so that pertinent comparisons are now possible (Table 2.4). Examination of the four sequences of skeletal muscle troponin-C quickly reveals close kinships between them. That from human

Table 2.4
Amino Acid Sequences of Various Vertebrate Troponin-C's[a,b]

	Sequence
Skeletal muscle	
	10 20 30 40 50
Rabbit	$_{Ac}$DTQQAEARSYLSEEMIAEFKAAFDMFDA-DGGGDISVKELGTVMRMLGQTP
Human	DTQQAEARSYLSEEMIAEFKAAFDMFDA-DGGGDISVKELGTVMRMLGQTP
Chicken	ASMTDQQAEARAFLSEEMIAEFKAAFDMFDA-DGGGDISTKELGTVMRMLGQNP
Frog	$_{Ac}$AQPTDQQMDARSFLSEEMIAEFKAAFDMFDT-DGGGDISTKELGTVMRMLGQTP
"Primitive"	$_{Ac}$AQPTDQQAEARSFLSEEMIAEFKAAFDMFDA-DGGGDISTKELGTVMRMLGQTP
Cardiac muscle	
Cattle	$_{Ac}$MDDIYKAAVEQLTEEQKNEFKAAFDIFVLGAEDGCISTKELGKVMRMLGQNP
	60 70 80 90 100 110
Rabbit s.m.	TKEELDAIIEEVDEDGSGTIDFEEFLVMMVRQMKEDAKGKSEEELAECFRIFDRNADGYI
Human s.m.	TKEELDAIIEEVDEDGSGTIDFEEFLVMMVRQMKEDAKGKSEEELAECFRIFDRNADGYI
Chicken s.m.	TKEELDAIIEEVDEDGSGTIDFEEFLVMMVRQMKEDAKGKSEEELADCFRIFDKNADGFI
Frog s.m.	TKEELDAIIEEVDEDGSGTIDFEEFLVMMVRQMKEDAQGKSEEELAECFRIFDKNADGYI
"Primitive" s.m.	TKEELDAIIEEVDEDGSGTIDFEEFLVMMVRQMKEDAKGKSEEELAECFRIFDKNADGYI
Cattle c.m.	TPEELQEMIDEVDEDGSGTVDFDEFLVMMVRCMKDDSKGKSEEELSDLFRMFDKNADGYI
	120 130 140 150 160
Rabbit s.m.	DAEELAEIFRASGEHVTDEEIESLMKDGDKNNDGRIDFDEFLKMMEGVQ$_{OH}$
Human s.m.	DPEELAEIFRASGEHVTDEEIESLMKDGDKNNDGRIDFDEFLKMMEGVQ$_{OH}$
Chicken s.m.	DIEELGEILRATGEHVTEEDIEDLMKDSDKNNDGRIDFDEFLKMMEGVQ$_{OH}$
Frog s.m.	DSEELGEILRSSGESITDEIIEELMKDGDKNNDGKIDFDEFLKMMEGVQ$_{OH}$
"Primitive" s.m.	DSEELGEILRASGEHVTDEIIEELMKDGDKNNDGRIDFDEFLKMMEGVQ$_{OH}$
Cattle c.m.	DLEELKIMLQATGETITEDDIEELMKDGDKNNDGRIDYDEFLEFMKGVE$_{OH}$

[a] Based on Collins *et al.* (1973), Collins (1974), Romero-Herrara *et al.* (1976), Wilkinson (1976), and Van Eerd *et al.* (1978).
[b] Dark areas indicate identities between cardiac and skeletal muscle molecules; light areas, identities in all vertebrate skeletal muscle troponin-C's.

skeletal muscle, for a case in point (Romera-Herrera *et al.*, 1976), differs from the corresponding one of rabbit (Collins *et al.*, 1973, 1977; Collins, 1974) at only site 112, where the human material has proline instead of alanine. Thus at the level of the class, troponin-C appears evolutionarily stable. The frog and chicken skeletal muscle sequences (Wilkinson, 1976; van Eerd *et al.*, 1978) are similar in having three additional bases at the amino terminal (Table 2.4), the first residue of which is acetylated as it is in the mammalian type. Beyond that point the four are remarkably alike, especially in the premiddle region. From site 14 through 112 there are variations at only six points, sites 31, 41, 53, 92, 101, and 108. From site 113 through 142, variations are frequent, but from there to the carboxyl terminal only a single substitution occurs, the frog material having lysine at site 149, in contrast to the arginine of the other three. In many respects the chicken troponin-C is intermediate between the frog and mammal materials, sometimes at given points having amino acid residues identical to those of the amphibian, at others like those of the mammals.

Since birds are derivatives of the diapsid line of reptiles and mammals are derivatives of the synapsid line, at sites where correspondences exist only between the frog and the chicken compounds, the mammalian molecules may be considered to have diverged. In contrast, at points where the chicken troponin-C differs from the other three, it may be viewed as a mutant form, and where the frog material alone is distinct, it may be thought to have diverged from the primitive stock. Although the latter supposition may not always be valid, it provides a basis for establishing a sequence that possibly approaches that of ancestral reptiles, if not the amphibian stock that preceded the latter. A primitive sequence derived in this manner is given in Table 2.4.

What is surprising are the differences that exist between the troponin-C from heart muscle of cattle (Collins, 1974; van Eerd and Takahashi, 1975, 1976; van Eerd *et al.*, 1978), and that from mammalian skeletal muscle, as well as those of the chicken and frog (Table 2.4). In the first place, it differs from all the skeletal muscle sequences in having an addition in the form of a leucine residue at site 31, necessitating the insertion of hyphens in the other sequences to keep homologous sites aligned. Moreover, in comparison with the two mammalian series, it has an addition at the extreme amino terminus, agreeing with the chicken troponin-C in having methionine there but differing in having only one extra site, not three. From there on through site 64 many distinctions occur, so that the longest identical segments that can be observed are only seven residues in length. However, there are two of this length, sites 21–27 and 46–52. Then from site 65 to 73 there is a segment with nine residues identical to the others, which is followed by two series of eight homologous regions, sites 78–85 and 92–99. The rest shows mostly scattered similarities, but near the carboxyl end, from site 135 through 151, is the longest identical segment, consisting of 17 residues. Thus this cardiac muscle type, while undoubtedly sharing a common ancestry with that from skeletal muscle, has obviously had a separate evolutionary history for many eons. At what point in time the two molecules began to diverge cannot be established until the primary structures of troponin-C from the heart and the skeletal muscle of the lancelets, or at least a cyclostome, have been determined.

Comparisons with Fish Parvalbumins. Because troponin-C binds Ca^{2+}, comparisons have been made with other proteins having similar properties, especially the parvalbumins from fish (Collins *et al.*, 1973, 1977; Kretzinger and Barry, 1975; Stone *et al.*, 1975). Amino acid sequences of those from the skeletal muscle of three species of fish have been established (Capony *et al.*, 1973; Coffee and Bradshaw, 1973; Frankenne *et al.*, 1973), as shown in Table 2.5. These are small molecules, with molecular weights of approximately 12,000, formerly thought to occur in abundance only in the tissues of lower vertebrates but not in birds, mammals, or reptiles (Collins *et al.*, 1973). However, a parvalbumin has now been isolated from the muscles and numerous other tissues of chicken (Heizmann *et al.*, 1977; Heizmann and Strehler, 1979), so they probably are absent only from mammals. While their function is unknown, each molecule has been found capable of binding 4 Ca^{2+} ions, two of which can be occupied by Mg^{2+}. In Table 2.5 a "primitive" sequence has been

Table 2.5
Amino Acid Sequences of Various Calcium-Binding Proteins[a]

	10 20 30 40 50
Pike parvalbumin	$_{Ac}$AKDLLKADDIKLKALDAVKAEGSFNHKAFFAKVGLKAMSANDVKKVFKAI
Hake parvalbumin	$_{Ac}$AFAGILADADITAALAACKAEGSFKHGEFFTKFGLKGKSAADIKKVFGII
Carp parvalbumin	$_{Ac}$AFAGVLNDADIAAALEACKAADSFDHKAFFAKVGLTSKSADDVKKAFAII
Primitive parvalbumin	$_{Ac}$AFAGLLADADILAALDACKAEGSFNHKAFFAKVGLKAKSANDVKKVFKII
Primitive troponin-C	$_{Ac}$AQPTDQQAEARSFLSEEMIAEFKAAFDMF
CNP regulator	$_{Ac}$ADELTEEQIAEFKEAFSLF
	60 70 80 90 100 109
Pike parvalbumin	DADASGFIEEEELKFVLKSFAADGRDLTDAETKAFLKAADKDGDGKIGIDEFETLVHEA$_{OH}$
Hake parvalbumin	DQDKSDFVEEDELKLFLQNFSAGARALTDAETATFLKAGDSDGDGKIGVEEFAAMVKG$_{OH}$
Carp parvalbumin	DQDKSGFVEEDELKLVLQNFKADARALTDGETKTFLKAGDSDGDGKIGVDEFTALVKA$_{OH}$
Primitive parvalbumin	DQDKSGFVEEDELKLVLQNFAADARALTDAETKTFLKAGDSDGDGKIGVDEFEALVKEA$_{OK}$
Primitive troponin-C	DADGGGDISTKELGTVMRMLGQTPTKEELDAIIEEVDEDGSGTIDFEEEFLVMMVRQMK$_{OH}$
CNP regulator	DADGDGTITJKELGTVMRSLGQDPTZAZLZXMIDEVXXXGAGTIDFPEEELTMMARKMK$_{OH}$

[a]Based on Capony *et al.* (1973), Coffee and Bradshaw (1973), Frankenne *et al.* (1973), Collins (1974), and Dedman *et al.* (1978).

established, largely based on identical sites. Where only one difference occurs at a given set of sites, the residue in the two like ones has been employed, but where all three corresponding residues differ, that of the pike has been used on a strictly arbitrary basis.

If the primitive sequence of troponin-C is then aligned with this primitive parvalbumin following the method of Collins and his colleagues (1973), identical residues are found only at ten sites, shown boxed in the table. Thus there appears little basis for an ancestor–descendant relationship, so doubt must be cast on the validity of the calcium-binding sites that have been proposed (Dedman *et al.*, 1978; van Eerd *et al.*, 1978). Sites 52 to 84 have been shown by nuclear magnetic resonance studies to include one active calcium-binding region (Birnbaum and Sykes, 1978).

Amino Acid Sequences of Troponin-I. Recently the sequencing of troponin-I from two different source organisms has been completed, both from fast muscle. The first of the substances to have its primary structure established was from rabbit fast skeletal muscle (Wilkinson and Grand, 1975); in this study it was pointed out that genetically distinct polymorphic forms of this protein existed in at least three types of muscle, fast and slow skeletal and cardiac, such as those now confirmed for troponin-C. In the chicken, troponin-I's from major muscles of the breast were found to be identical to those of leg muscles, so these organs may be viewed as being of the same fast type (Wilkinson, 1978). The protein from this avian contains 182 residues, with a molecular weight of 21,136 (Wilkinson and Grand, 1978). This is slightly heavier than the protein from rabbit and is four residues longer (Table 2.6), but whether the differences

Table 2.6
Primary Structure of Fast Muscle Troponin-I[a]

```
                 10        20        30        40        50        60
Rabbit    AcGDEEKRNRAITARRQHLKSVMLQIAATELEKEEGRREAEKQNYLAEHCPPLSLPGSMAEV

Chicken   AcSDEEKKRRAATARRQHLKSAMLQLAVTEIEKEAAAKEVEKQNYLAEHCPPLSLPGSMQEL


                 70        80        90       100       110       120
Rabbit      QELCKQLHAKIDAAEEEKYDMEIKVQKSSKELEDMNQKLFDLRGKFKRPPLRRVRMSADA

Chicken     QELCKKLHAKIDSVDEERYDTEVKLQKTNKELEDLSQKLFDLRGKFKRPPLRRVRMSADA


                130       140       150       160       170       180
Rabbit      MLKALLGSKHKVCMDLRANLKQVKKEDTEKE---RDVGDWRNKIEEKSGMEGRKKMFES-ES COOH

Chicken     MLRALLGSKHKVNMDLRANLKQVKKEDTEKEKDLRDVGDWRNKIEEKSGMEGRKKMFEAGES COOH
```

[a] Based on Wilkinson and Grand (1978).

result from additions in the chicken protein molecule at sites 152–154 and 180 or from deletions in the mammalian material cannot be established until the primary structure of reptilian or amphibian troponin-I has been determined. Agreement between the sequences on a site-to-site basis is nearly as close as those of troponin-C, showing an 81% homology between the two in the present case.

Primary Structure of Troponin-T. To date the amino acid sequences of troponin-T from rabbit and chicken skeletal muscle have been only partially established (Jackson *et al.*, 1975; Wilkinson, 1978), so that meaningful comparisons cannot be made. It is of interest to note, however, that unlike the case with troponin-I, the troponin-T from the leg muscle of chicken differed from the breast muscle protein in having a molecular weight of 30,500 compared to 33,500 for the latter. This contrast is believed to result largely from the breast muscle troponin-T having an extra polypeptide chain about 24 sites long at the amino terminus (Wilkinson, 1978). As already noted troponin-T is phosphorylated. Now it has been confirmed that a specific protein kinase is involved that adds a phosphoryl radical to the serine at the amino terminal (Gusev *et al.*, 1978), an addition that takes place only with the intact substrate molecule.

α-Actinin. A final protein that is associated with actin filaments is α-actinin, which, like the previous ones, occurs with thin filaments in muscle as well as nonmuscle tissues (Schollmeyer *et al.*, 1974; Lazarides and Burridge, 1975; Lazarides, 1976a,b). For example, brain tissue and intestinal epithelia have been noted to contain this protein (Schook *et al.*, 1978; Geiger *et al.*, 1979). Regardless of the type of source tissue, at least those from mammalian stocks, the actinins appear to be chemically closely allied, for they respond to the same antisera. A molecular weight between 90 and 100×10^3 has been reported, which seems to result from dimerization of two subunits with molecular weights in the neighborhood of 45 to 50×10^3 (Hatano and Owaribe, 1976). In striated muscle tissue this globular protein apparently is associated with the Z lines, and in nonmuscle tissue it evidently is localized at the vertices of the intertwining actin filaments or in electron-opaque bodies like intercalated disks and desmosomes (Schollmeyer *et al.*, 1976).

Thin Filaments. In the presence of salt, G-actin polymerizes to form the so-called thin filaments, which have diameters around 110 Å. Each consists of a double helix of monomers, with a pitch of 59 Å and a periodicity of 760 Å. In negatively stained preparations, the filaments thus have an appearance of two strands of pearls entwined around one another (Figure 2.8; Squire, 1975; Mannherz and Goody, 1976). While actin contributes over 60% of the total molecular weight of these filaments, two other proteins make up the remainder, in molar proportions of 7 actin : 1 tropomyosin : 1 troponin (Ebashi and Endo, 1968; Potter, 1974).

The narrow rod-shaped molecule is about 410 Å long, a length that approximates seven times the 55-Å diameter of actin molecules. Because of this

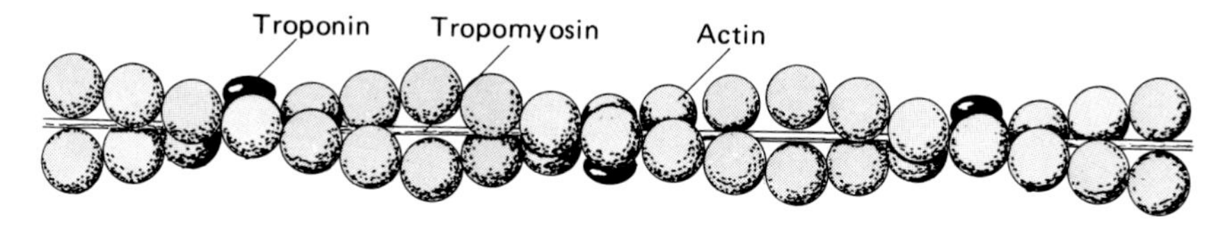

Figure 2.14. Molecular organization of a thin filament. (Based on Ebashi *et al.*, 1975.)

relationship, it was postulated that the tropomyosin molecules lay end to end in the grooves formed by the double-helical arrangement of the actin chains, providing the observed 7:1 molar proportions (Figure 2.14; Gregely, 1976). In addition each tropomyosin molecule binds one troponin complex near one of its termini and thereby completes the functional unit with actin. These relationships have been confirmed by X-ray diffraction pattern studies (Moore *et al.*, 1970; Haselgrove, 1972; Huxley, 1972).

Thick Filaments and Myosin. A second type of protein and filament of its polymer occurs in many types of cells other than vertebrate muscle (Figure 2.15, Zigmond *et al.*, 1979), but, as is the case with actin and its thin filaments, myosin and its thick ones have been most thoroughly investigated in mammalian skeletal muscle.

The Fast Muscle Myosin Molecule. Unlike actins, myosins from different tissues of the same organism vary greatly in size and composition; marked differences are found even between contrasting muscle types, such as white (fast) and red (slow) skeletal muscles of the same mammal or bird. Consequently, that which has been most thoroughly documented, the myosin of fast skeletal muscle tissue of rabbits, is described first. Here as in other skeletal muscle, it forms the A bands, whereas actin is localized in the I bands (Hanson and Huxley, 1953). The myosin molecule consists of six polypeptide subunits, referred to as two heavy and four light chains, and has a total molecular weight of 470,000 (Gershman *et al.*, 1969; Godfrey and Harrington, 1970; Mannherz and Goody, 1976). These parts are so arranged that they form a long rodlike molecule, at one end of which are two closely associated heads (Figure 2.16; Huxley, 1963; Lowey *et al.*, 1969; Arata *et al.*, 1977; Inoue *et al.*, 1977). The rodlike portion has a length of 1450 Å and is almost entirely α helical (Lowey *et al.*, 1969; Weeds and Frank, 1972); improved techniques, however, show the rod not to be rigid but to be quite flexible, especially at the center (Takehashi, 1978). The two heavy subunits have molecular weights approximating 200,000, whereas the light chains are of three size classes. LC-1 behaves as having a molecular weight of 25,000 by gel electrophoresis, but the amino acid sequence indicates the weight to be only 20,700 (Frank and Weeds, 1974). LC-2 is nearly as heavy, with a weight of 18,000 and LC-3 has a weight of 16,000.

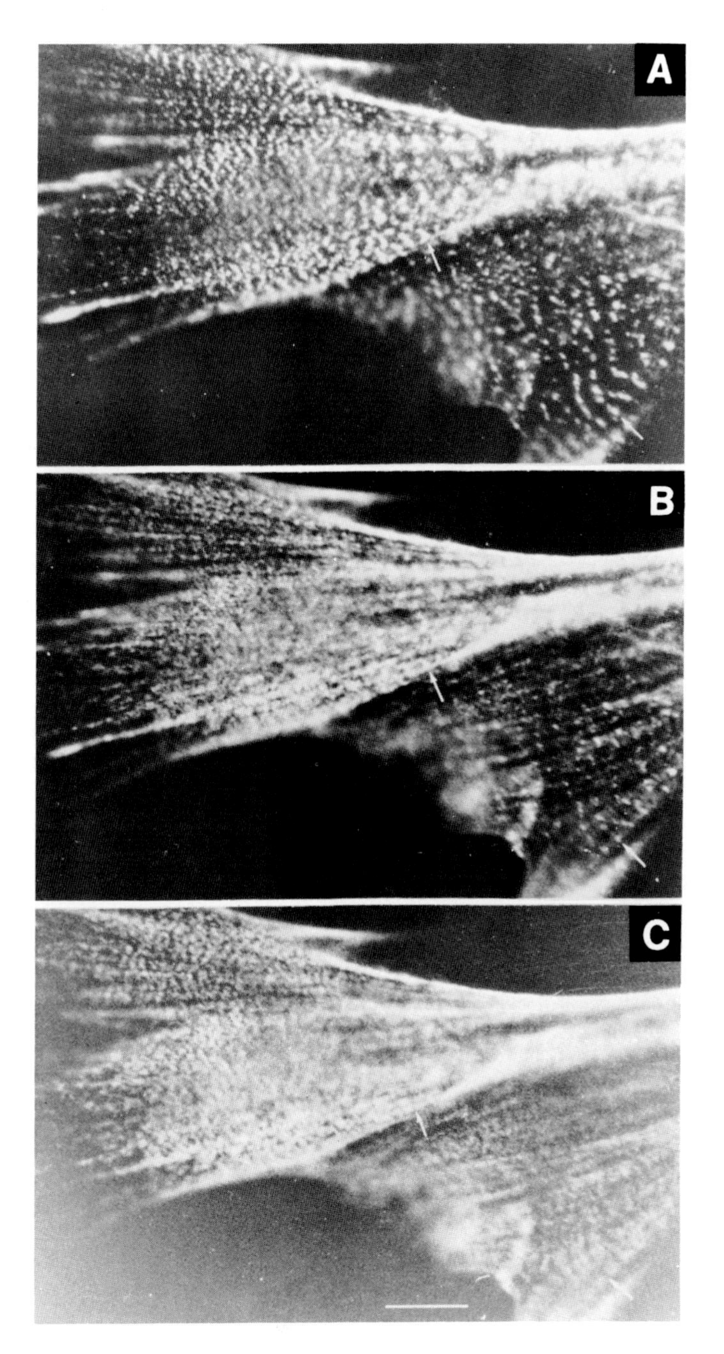

Figure 2.15. Interrelationships of fibrils in a human fibroblast. (A) α-Actinin visualized by means of fluorescein staining. (B) Both α-actinin and myosin fibrils are made visible, the former being somewhat brighter. (C) Myosin fibrils stained with rhodamine. All 13,000×. (Courtesy of Zigmond *et al.*, 1979.)

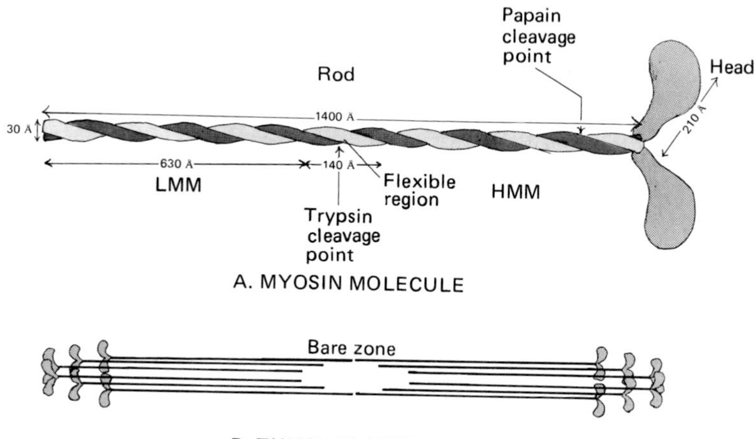

Figure 2.16. The molecular organization of myosin and thick filaments. HMM, heavy meromyosin; LMM, light meromyosin.

Two copies of LC-2 are present per complete molecule. Functionally, LC-1 and LC-3 are closely related. The amino acid sequence of the latter, although much shorter, is quite similar to that of LC-1, as shown later; these two are classed as essential to myosin function, while LC-2 is viewed as unessential (Mannherz and Goody, 1976). Because the two essential subunits are present in a ratio of 2:1, even in histochemically homogeneous samples, myosin molecules are thought to contain either two LC-1 or LC-3 subunits, and those bearing the former are twice as abundant as those bearing the latter (Sarkar, 1972).

Cleavage of the myosin molecule with trypsin yields two fragments, the soluble one of which is called heavy meromyosin or HMM and has a molecular weight of 350,000. It consists of two head subunits referred to as S-1A and S-1B, which have molecular weights of 115,000, plus a portion of the rod; each is approximately 100 to 150 Å long and is more or less banana-shaped, with a diameter of 30 to 40 Å (Moore *et al.,* 1970). The second fragment, S-2, is insoluble and consists of the remainder of the rod (Sutoh *et al.,* 1978); because its molecular weight is only around 150,000, it is referred to as light meromyosin (LMM). Two of the characteristic properties of myosin, ATPase activity and the ability to interact with actin, are present only in those fragments that include heads—in other words, they pertain to heavy meromyosin or more realistically to S-1 itself. Light chains, too, seem to be contained in the heads. Current investigations have found that the two essential thiol groups, Sh-1 and SH-2, which appear to be involved in ATP binding, are located in a short polypeptide of 20,000 daltons (Bálint *et al.,* 1978).

Before myosin can interact with thin filaments, as described shortly, it must be phosphorylated. Because the phosphate group is added to one of the light chains, the protein that catalyzes the transfer of the γ-phosphate of ATP to myosin is termed the myosin light-chain kinase. The specific light polypeptides that are phosphorylated in the myosin from various skeletal and heart muscles are those that have molecular weights between 18,000 and 20,000 and are designated the P light chains or LC-P. Two such LC-Ps per total myosin molecule are present (Scordilis and Adelstein, 1978). The kinase from skeletal muscle seems to differ from those of myoblasts.

Myosin of Other Muscle Tissues. Very little detailed information appears available pertaining to the myosin molecule from other types of muscle tissue. Many of the known distinctions stem from light-chain differences in the number of subunits present and in their molecular weights. For instance, the myosin from the slow skeletal muscles of rabbit contains only three light chains, instead of the four of the fast type, and these have molecular weights of 29,000, 27,000, and 19,000 (Sarkar *et al.,* 1971; Frearson and Perry, 1975). While chicken gizzard myosin has been shown to have a number of distinctive properties (Kelly and Rice, 1968; Suzuki *et al.,* 1978), its subunit structure has not been determined. Cardiac muscle was originally reported to contain only two classes of light chains, with molecular weights of 27,000 and 19,000, but that has proven true only of atrial myosin. Muscle of the ventricle has now been found to have three components (Hoh *et al.,* 1977, 1979). While smooth muscle of human uterus also has only two classes, they have molecular weights of 19,000 and 16,000 (Pollard and Weihing, 1974). The smallest of the light-chain species from slow and cardiac muscle share a distinctive property in that each is phosphorylated by a kinase specific for myosin light-chain polypeptides (Frearson and Perry, 1975). In contrast, myosin of aortic smooth muscle consists of three subunits having molecular weight of 192,000, 19,000, and 15,000 (Frederiksen, 1979).

The myosin from a number of invertebrate muscles (Chantler and Szent-Györgyi, 1978) and especially that from nematode muscles have also received thorough attention (Epstein *et al.,* 1976). Soil nematodes have muscle tissue of two contrasting types, body wall and pharyngeal. All the muscles of the body wall are arranged longitudinally and are used only in locomotion by providing an undulating motion to the body. In contrast to this relatively slow and intermittent use, the pharyngeal muscles contract rapidly and continually as the worm feeds on bacteria. Accordingly two differing types of myosins have been found. That from body wall muscle has two different heavy chains of 210,000 daltons each, whereas that from the pharyngeal muscle consists of one of 210,000 and one of 206,000. Both types have an undetermined number of light chains in the range of 15,000 to 20,000 daltons.

Primary Structure of Light Chains. Amino acid sequences of the two light chains of myosin from rabbit skeletal muscle have been established (Frank and

Table 2.7
Amino Acid Sequences of Myosin Light Chains

	Sequence
	10 20 30 40 50 60
Rabbit LC-1 (A1)	NH_2-PKKNVKKPAAAAAPAPKAPAPAPAPAPAPKEEKIDLSAIKIEFSKEQQDEFKEAFLLYDR
Rabbit LC-3 (A2)	NH_2-SFSAXZIAZFKEAFLLYDR
	70 80 90 100 110 120
Rabbit LC-1	TGDSKITLSQVGDVLRALGTNPTNAEVKKVLGNPSDEQMNAKKIEFEQFLPMLQAISNNK
Rabbit LC-3	TGDSKITLSQVGDVLRALGTNPTNAEVKKVLGNPSDEQMNAKKIEFEQFLPMLQAISNNK
	130 140 150 160 170 180
Rabbit LC-1	DQGTYEDFVEGLRVFDKEDGTVGMGAELRHVLATLGEKMKEEEVEALMAGEEDSNGCINY
Rabbit LC-3	DQGTYEDFVEGLRVFDKEDGTVGMGAELRHVLATLGEKMKEEEVEALMAGEEDSNGCINY
	190
Rabbit LC-1	EAFVKHIMSI-OH
Rabbit LC-3	EAFVKHIMSI-OH

Weeds, 1974), but those of no other organism are available as yet. These two, LC-1 and LC-3, of course differ in length, the former having a polypeptide chain of 41 residues at the amino end that is lacking in the latter (Table 2.7). From site 42 through 50 of LC-1 differences occur frequently, seven of the ten residues being different in the two substances. However, beyond that point to the end, the two proteins are identical throughout. The problem is made complex by the presence of isozymes, five of which have been reported for chicken skeletal muscle (Hoh, 1979).

Protein C. Several diverse proteins occur together with myosin, at least in vertebrate muscle cells. Among the more abundant of these is protein C, which is present to the extent of 5% of myosin and has a molecular weight of 140,000 (Moos *et al.*, 1972, 1975; Offer, 1973). In shape, it seems to be an ellipsoid rod, totaling around 350 Å in length, while in structure it appears completely devoid of α-helical regions. At low ionic levels, it binds very strongly to myosin and also binds to the deheaded myosin rod and to meromyosin. In myosin filaments one molecule of protein C is present to every 5.7 of myosin (Offer, 1973).

Thick Filaments. In vertebrate skeletal muscle, the myosin molecules polymerize to form thick filaments about 16,000 Å long and 140 Å in diameter. X-Ray and electron microscope studies have indicated that the heads of the molecules project from this filament and are helically arranged, with an axial spacing of 145 Å (Huxley, 1963; Huxley and Brown, 1967; Ullrick *et al.*,

1977). Although much of the molecule displays a rough appearance because of the projecting heads, there is a smooth region, or bare zone, at the middle of the filament about 1500 to 2000 Å long. Consequently, it has been proposed that the projecting heads provide the cross bridges that interact with actin and that the molecules are arranged in an end-to-end, or antiparallel, fashion medially to form the bare zone (Figure 2.16B). This bare zone corresponds to the M region of vertebrate muscle (Luther and Squire, 1978). About 2000 of the molecules are found per filament and are so arranged that three heads form a period of 430 Å in length. On this backbone, the protein C molecules are associated, so that one (not two as shown by Offer, 1973) occurs per period, that is, two per full pitch of 12 myosin molecules to provide the observed 1:6 ratio of protein C to myosin.

However, this theoretical arrangement needs to be adjusted to other observations. For instance, Pepe (1973) showed that the subunit structure basically consists of three subunits at the center, around which are nine peripheral ones (Figure 2.16). This proposal, however, raises difficult questions as to how the heads of the central subunits reach the surface to serve as bridges, unless exposed bridges are of different lengths.

Myosins from Other Tissues. By no means are myosins confined to muscle, for their presence in a diversity of tissues has now been thoroughly documented (Pollard and Weihing, 1974). Indeed, a heavy meromyosin has been reported from bacterial sources (Someya and Tanaka, 1979). Among those found in vertebrate materials are guinea pig granulocytes, human blood platelets, mouse fibroblasts, and rat brains (Booyse *et al.,* 1971; Adelstein *et al.,* 1971, 1972; Berl *et al.,* 1973; Stossel and Pollard, 1974; Bray, 1977). Several physical properties of certain of these myosins have been described. The molecular weight of human platelet myosin is around 540,000, considerably greater than that from muscle (Booyse *et al.,* 1971). Yet this report is in conflict with studies on the subunit composition, which found two heavy subunits of 200,000, and four light ones, two with molecular weights of 19,000 and two of 16,000 (Adelstein *et al.,* 1971). In astrocytes of mice the heavy chains have molecular weights of 200,000; two light chains are present, having molecular weights of 20,000 and 15,000. The larger of these is phosphorylated by an endogenous myosin light-chain kinase (Scordilis *et al.,* 1977). Myosin from chick brain was found to be composed of a 200,000-dalton heavy chain and three light ones. While capable of self-assembly, this myosin did not interact with that protein of chicken skeletal muscle to form hybrid filaments (Kuczmarski and Rosenbaum, 1979). Furthermore, antibody recognition studies showed that myosins from chicken brain, skeletal muscle, and smooth muscle were quite distinct.

Myosins from invertebrate and protozoan sources have also received attention. That reported from the slime mold *Physarum* resembled the vertebrate material in molecular weight (458,000), but its subunits have been found to include two of 240,000 and an unknown number of 14,000. Other light forms appeared to be absent (Adelman and Taylor, 1969; Nachmias, 1973). Recently myosin has been extracted from squid muscle and partly characterized after purification (Tsuchiya *et al.*, 1978). The heavy chain was found to have a molecular weight between 180,000 and 200,000. With it were associated two light chains, one of which had a molecular weight of 12,000, the other one, 15,000. Arthropod myosins also have had a share of attention, the structure of the molecule from lobster muscle being particularly thoroughly investigated. The molecule has a molecular weight totaling 492,000 and resembles most vertebrate material in consisting of six chains, including two heavy meromyosin chains of 210,000 each. Unlike the vertebrate protein, the light chains fall into just two weight classes, 20,000 and 16,000, with two of each chain per complete molecule (Siemankowski and Zobel, 1976). Thick filaments are 1559 Å long, the tail having a length of 1335 Å and the two heads being 225 × 45 Å in dimensions.

As might be expected, the principal species, myosin-I, from *Acanthamoeba castellani* was considerably simpler than others that have been observed, for it showed a molecular weight of only 180,000 and consisted of one heavy subunit of 140,000, a light one of 16,000, and another of 14,000 (Pollard and Korn, 1973a,b,c). The rodlike portion of the heavy subunit appears to be lacking in this species. In addition, filaments resembling those of myosin in size and shape have been detected in electron micrographs of several other amoebae, including *Amoeba proteus* (Pollard and Ito, 1970), *Chaos carolinensis* (Nachmias, 1968), and *Trichamoeba villosa* (Bhowmick, 1967).

More recently a second species of myosin from *Acanthamoeba* has been characterized (Pollard *et al.*, 1978). This myosin-II was more than twice as heavy as myosin-I, having a total molecular weight of 400,000. The molecule had the usual two heads at one end and a rodlike tail 90 nm long. When polymerized, the resulting thick filament was 205 × 66 nm and had a bare zone 97 nm in length. Proteinase treatment resulted in either five or six subparticles, two heavy ones with a molecular weight of 175,000, two light chains of 16,500, and either one or two of 17,500. While the precise chemical structure of this species awaits further elucidation, the molecule was found to be rich in acidic residues and to have no fewer than 32 cysteine residues.

Interactions between Myosin and Actin. *In vivo,* as well as *in vitro,* the filaments of actin and myosin interact in specific fashions (Figure 2.17). Moreover, those of actin interact with heavy-meromyosin subunits, a reaction that has been employed experimentally in detecting the presence of these filaments in various organisms and tissues.

"Decoration" of Actin Filaments. When heavy meromyosin is added to a system containing actin filaments, those molecules bind to the latter in a specific manner. The resulting meromyosin complexes project laterally at angles from the filament, giving an appearance under the electron microscope of a

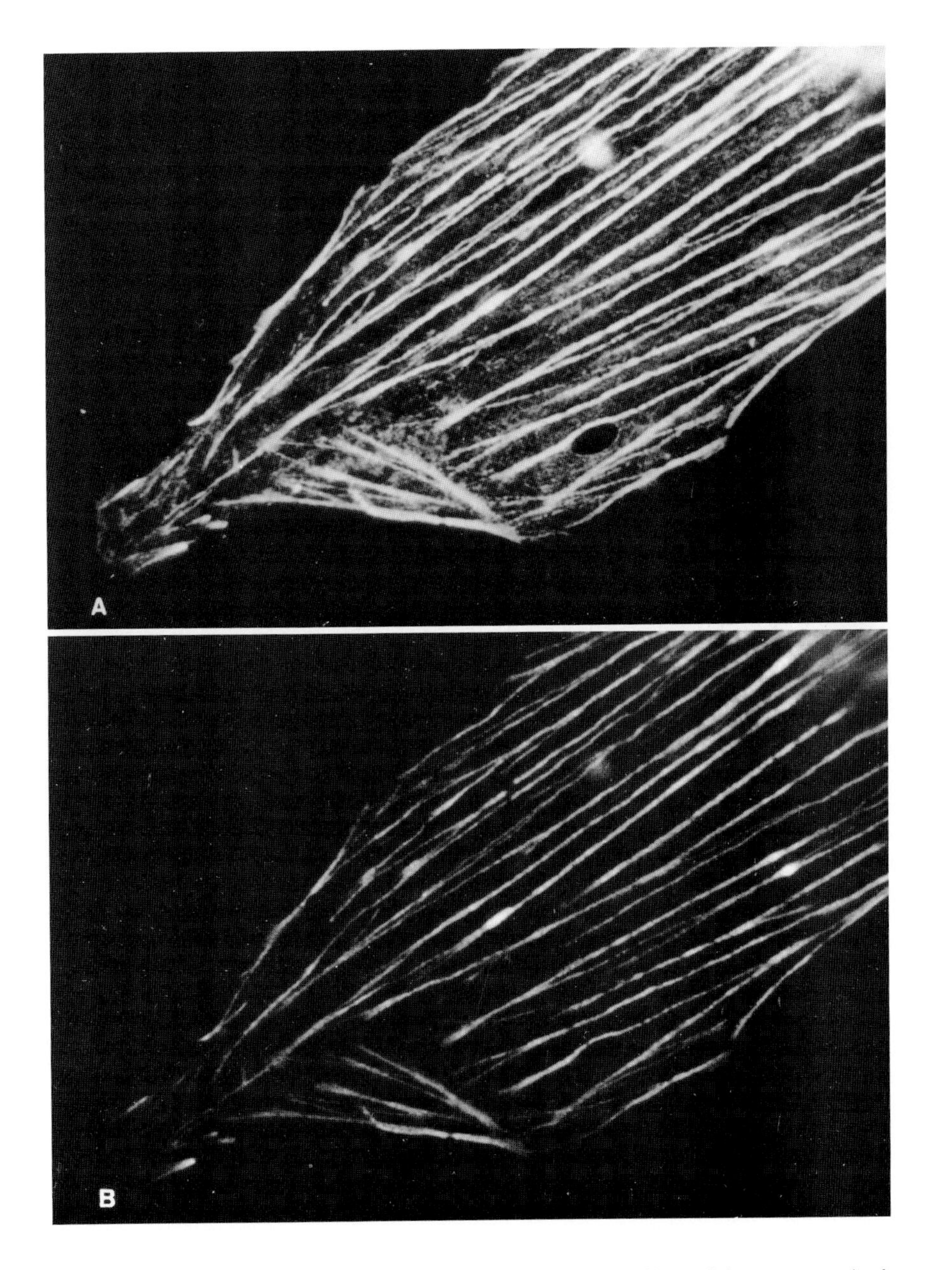

Figure 2.17. Fibers of actin and myosin in combination. This fibroblast of the rat was stained sequentially with rabbit antiactin antibody (A) and then with human antimyosin antibody (B). As the identical microfilaments were stained in each case, the close association of the two proteins in these microfilaments is obvious. 2000×. (Both courtesy of Toh *et al.*, 1979.)

series of arrowheads. This "decoration" of actin, as it is generally called, has enabled the periodicity of the filaments to be ascertained at 36 nm with remarkable consistency.

Decoration of actin filaments with the myosin component S-1, combined with fixation in tannic acid (Ishikawa *et al.,* 1969), makes it possible to observe the existence of polarity in their orientation (Begg *et al.,* 1978). In each cell type examined in which the actin filaments were attached to the plasma membrane, the arrowheads consistently pointed away from the membranes. This orientation was especially clearly shown in microvilli of sea urchin eggs and in brush-border cells. When Chinese hamster ovary cells were treated so as to induce their elongation, stress fibers composed of numbers of actin filaments were observed, which ran parallel to the long axis of the cell. In such bundles the filaments were oriented in both directions. With the latter were associated myosin, tropomyosin, and α-actinin, so it is possible that such stress fibers may be able to induce contractility.

Thick and Thin Filaments in Muscle Contraction. During muscle contraction the thick and thin filaments interact in a complex series of events that still are only partly understood to produce a complex often referred to as actomyosin. Activation of contraction apparently begins with the plasma membrane where a number of reactions occur, including the release of Ca^{2+} ions. The propagation of this excitation involves another series of steps (see Caputo, 1978, for a review), while the actual contraction centers in the interplay between actin and myosin filaments, plus other factors, known and unknown (Offer and Elliott, 1978; Ando and Asai, 1979; von Olenhusen and Wohlfarth-Bottermann, 1979). Because muscle tissues have developed contractile activities much further than has any other, most attention has been devoted to this aspect of the subject.

Ultrastructurally, one myosin filament is situated within a possibly cylindrical series of actin filaments, the length ratios between the two types perhaps providing the basis for the differing abilities of contraction (Pollard, 1977). At rest, the actin filaments are in the "off" condition, in which tropomyosin blocks the myosin attachment sites on the actin (Squire, 1975), but an unknown factor located on the thick filaments may also be involved. In this state, the myosin heads may be retracted against the rods, but X-ray and ultrastructural studies indicate the presence of bridges, that is, projecting heads. When excitation of the tissue occurs, the thin filaments become activated to the "on" state by movement of the tropomyosin away from the myosin-attachment sites, and the myosin filament itself may undergo some sort of change. One such alteration possibly centers on movement of the heads outward toward the thin filaments, and another, a change in molecular organization, with a shortening of the helical repeat to an extent of 1%. The movement between the two filaments during contraction is generally taken to result from the bridges changing their orientation from an angle of 90° to one of 45°. At the same time as these angles are changed, ADP and inorganic phosphate are released.

Other Contraction-Related Proteins. A number of other proteins extracted from various muscle tissues have been found to act upon actin; these have been described as being able to produce gels with that protein, similar to those produced by other cytoplasmic extracts (Stossel, 1978; Ishiura and Okada, 1979). Because of this discovery of gel–sol transitions *in vitro,* speculation has been revived that such transformations might be involved in cell motility and, hence, in muscle contraction (Brotschi *et al.,* 1978). Two proteins of high molecular weight from muscle have been partially characterized, filamin from chicken gizzards and another referred to as actin-binding protein. These resemble one another in molecular weight and amino acid composition (Shizuta *et al.,* 1976; Wang and Singer, 1977), and in being able, like myosin, to produce gels with F-actin. In contrast, four light proteins of molecular weights of 38,000, 32,000, 28,000, and 23,000 have been noted in *Acanthamoeba* extracts that formed actin gel, while almost no reaction occurred with high-molecular-weight fractions (Maruta and Korn, 1977).

2.2.2. Other Filamentous Systems

If knowledge of the vitally important actin–myosin filamentous system remains incompletely comprehended after many decades of intensive investigation, it is to be expected that miscellaneous fibers, largely of unknown function and chemical composition, have been scantily explored indeed. This lack of knowledge is further confounded by confusion resulting from actin and myosin filaments often being referred to in the literature simply as microfilaments, distinguished only from microtubules. Consequently, in spite of their involvement in such important processes as wound healing (Rudolph *et al.,* 1977; Gruber, 1978), little more than mere mention of the several more common types can be provided in the following discussion.

As pointed out in the introduction to this chapter, two different size groups occur, intermediate filaments approximately 100 Å in diameter and nanofilaments with diameters in the range of 50 to 75 Å. In the literature nanofilaments largely receive only passing mention, mostly in studies of actin and myosin filaments, but otherwise they appear to have been neglected. In the intermediate class is a miscellany of types including presumed microtubule derivatives of melanocytes, neurofilaments of nerve cells, glial filaments of glial cells and astrocytes, tonofilaments of dermal gaps, and "100 Å" or "10 nm" filaments of various other tissues, as well as skeletin filaments in heart Purkinje fibers (Benitz *et al.,* 1976; Davison *et al.,* 1977; Eriksson and Thornell, 1979; Franke *et al.,* 1979). Immunological and biochemical comparisons, while still in the elementary stages, indicate that these differ in subunit makeup and that the tonofilaments of epidermis are completely unrelated to the others in consisting of keratin (Steinert *et al.,* 1976; Brysk *et al.,* 1977; Franke *et al.,* 1979).

Microtubule Derivatives. At least in vertebrate tissues, several sets of observations indicate that microtubules may be broken down to form groups of filaments that are consistently of smaller diameter. In melanocytes of frogs, for example, darkened by melanocyte-stimulating hormone (MSH) or cyclic AMP, the projecting extensions or processes contain an abundance of filaments approximately 10 nm in diameter and only a few microtubules. After the pigment has been permitted to move from the projections to become concentrated within the cell, processes thus emptied possess a number of microtubules but no microfilaments (Moellmann and McGuire, 1975). Apparent intermediate forms, such as microtubules dissociating into filaments, have been encountered, so it may be that the latter are merely alternative states of tubulin. However, the actual chemical nature of the filaments has not been investigated, and a body of evidence suggests that microtubules are not convertible into microfilaments (DeBrabandor *et al.,* 1975).

Neurofilaments. During the development of neural cells, microtubules and microfilaments are present in varying numbers, but their interrelationships are less clear than in the case of melanocytes. For a time it appeared that these neurofilaments of 10 nm diameter might result from microtubule breakdown (Peters and Vaughn, 1967; Yen *et al.,* 1975), but in contrast to the melanocyte system, protein synthesis has been found to be essential to their formation (Roisen *et al.,* 1975). That the neurofilaments are totally distinct from microtubules has now been clearly demonstrated by studies on the transport of the involved proteins from the cell body through the axon (Lasek and Hoffman, 1976). Tubulin polypeptides and three others, with molecular weights of 68,000, 145,000, and 200,000, were found together in the slow-moving components of transport, the latter three seemingly being associated with neurofilament formation. Similar results have also been obtained by electrophoretic analysis (Micko and Schlaepfer, 1978). Indeed, there is some evidence that myosin is associated with these filaments and that side arms are present on them, not unlike the projecting heads on thick filaments (Wuerker, 1970; Hoffman and Lasek, 1975).

Further evidence of the distinctiveness of neurofilaments is derived from structural aspects. Whereas microtubules are comprised of globular subunits, as are thin filaments, the subunits of neurofilaments appear to be fibrous and twisted together in rope fashion (Krishnan and Lasek, 1975). Thus these filaments are distinct also from the thick ones comprised of myosin. Because the structural observations just cited were made on the giant axon of a polychaete annelid, neurofilaments might be suspected to occur generally among organisms, at least among invertebrates. The polychaete neurofilaments measured between 85 and 110 Å and consisted of two intertwined strands 40–55 Å in diameter, having a pitch of 1000 to 1300 Å in length. In turn, each strand consisted of protofilaments twisted together, having diameters of 20–25 Å and a pitch 120–130 Å long. Considerable α-helical content was noted.

Currently it is not clear whether or not neurofilaments from vertebrate brain tissue differ from those of nerve fibers. Recently it has been shown that rabbit brain neurofilaments share three polypeptides with those from sciatic nerve, having molecular weights close to 200,000, 150,000, and 70,000 (Anderton *et al.,* 1978). In addition, however, a 50,000-dalton major polypeptide was always present in the brain tissue and consistently absent from nerve. This light species may have been the result of the proteolysis employed during the preparatory steps, nor could the possibility of contamination from glia (Shelanski *et al.,* 1976) be completely ruled out. Nevertheless, it could equally be possible that the neurofilaments vary with the tissue, as has been so thoroughly documented for myosins and other filament-forming proteins.

Desmin Filaments. When the actomyosin has been extracted at high ionic strength from smooth muscle tissues, such as the mammalian stomach and chicken gizzard, an insoluble residue is left that is rich in intermediate filaments (Cooke and Chase, 1971). Two proteins have been found to predominate in this residue, actin and one referred to as desmin (Lazarides and Hubbard, 1976; Lazarides and Balzar, 1978; Hubbard and Lazarides, 1979). The latter is a protein of molecular weight around 50,000 that previously was found in skeletal and cardiac muscle at the Z lines and where those lines come into apposition with the plasma membrane (Hikida, 1978). Thus desmin has been viewed as forming a network that links the actomyosin filaments together and connects them to membranes. In a current study, it was shown that actin and desmin purify together from extracts of chicken gizzard and that they interact to form a copolymer in the form of 100-Å filaments (Hubbard and Lazarides, 1979). However, it was also demonstrated that desmin could polymerize into 100-Å filaments in the absence of actin.

3

CELL MOTILITY: I

Cytoplasmic Movements

One important cell function with which microtubules and microfilaments are closely associated is the movement of cytoplasm. Such movement can take a number of different forms, in addition to those endo- and exocytotic processes already described (Chapter 1, Section 1.2.1); however, these activities reappear here in various guises. Furthermore, there are three other major types of this basic function—amoeboid movement, microvillus production, and cytoplasmic streaming—which, in that same sequence, provide the main topics for discussion.

3.1. AMOEBOID MOVEMENT

Undoubtedly the most familiar and most thoroughly investigated type of cytoplasmic motility is that categorized by the term amoeboid movement. In general terms such movement involves the production of the temporary extensions of the cell body known as pseudopodia that typically result in the cell's progress along the substrate. However, there are a number of varieties of these projections and thus of amoeboid movement, in at least one of which no actual pseudopods are produced and others in which those structures result in no forward progress of the cell. Consequently, it is evident that no strict definition (Allen, 1961) can be provided for either the term pseudopodium or amoeboid movement, but these facets become clearer as attention is given to the various types that have been named.

3.1.1. Pseudopodial Structure

On the basis of their shape, pseudopodia fall into five major categories: (1) lobopodia, (2) filopodia, (3) rhizopodia or reticulopodia, (4) axopodia, and (5)

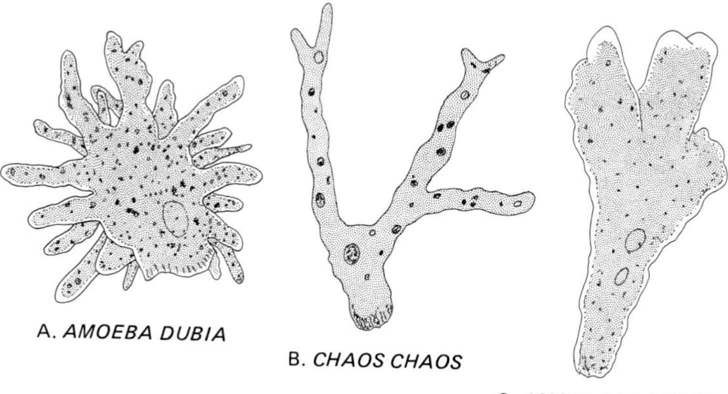

Figure 3.1. Variety of form among lobopodia.

lamellipodia (Ward and Becker, 1977). These differ not only in shape as their names imply but also in their mode of production and use.

The Types of Pseudopodia. The most thoroughly studied of the five types is the first one listed, the lobopodium, for it occurs in two common laboratory species, *Amoeba proteus* and *Chaos carolinensis*. This kind is broad and lobelike as viewed from above; rarely branched, it is formed relatively quickly and may be just as rapidly retracted. As shown in the illustration (Figure 3.1), this type is not always lobeform, but varies in shape from species to species, each having its own characteristic form, as pointed out by Allen (1961), but nearly all examples appear to involve the ectoplasmic as well as the endoplasmic layers of the cytoplasm. In profile (Figure 3.2) these structures can be

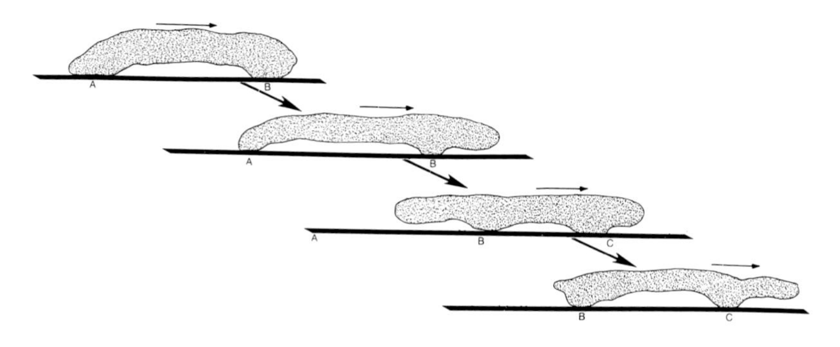

Figure 3.2. Pseudopodial action in profile. When viewed from one side, lobopodial locomotion can be observed to involve the production of ventral as well as forward projections.

noted to have leglike extensions that hold them above the surface on which the amoeba is moving (Dellinger, 1906; Jahn *et al.*, 1960; Nowakowska and Grebecki, 1978; Preston and King, 1978).

Axopodia are the extreme opposite of the foregoing type, for they are elongate and slender to the point of being rodlike and involve only the ectoplasmic layer. Confined to the Heliozoa and Radiolaria, the organelles are semipermanent and consist of an axial rod covered by a cytoplasmic envelope. Although arising from the interface between the endo- and ectoplasmic layers, the axial rods in some species penetrate more deeply into the central body and may even reach the nucleus; composed of fibrils of an undetermined nature, the rods are often resorbed and then redeveloped. Axopodia are employed exclusively in food-getting and apparently contain a toxic substance, because small protozoans that happen to touch one of these structures quickly become paralyzed. Following contact of this sort, the supportive rod is resorbed in the immediate area; then after the pseudopodial cytoplasm has increased in mass around the prey, the latter is transported to the cell body by flowing movements.

Like the axopodium, filopodia are also slender and comprised solely of ectoplasm, but they differ in being threadlike and of uniform thickness throughout their length. Sometimes this type of pseudopod branches, but if it does so, the branches do not anastomose. Although confined largely to testate amoeboids, such as *Euglypha,* they do occur among a few other representatives, including *Naegleria* (Preston and King, 1978) and *Entamoeba* (Figure 3.3B; Lushbaugh and Pittman, 1979).

Another filamentous type is the rhizopodium found in certain testate amoebae and in numerous Foraminifera; as slender and threadlike as the filopodium, this variety differs in being branched and in having the branches anastomose. Thus these pseudopods often form a large network about the organism that serves as a net for food-getting rather than for locomotion. Not only the capturing of prey occurs here, however, for *Lieberkuhnin* has been observed actually to digest at the point of contact any small ciliate that was caught by the rhizopodia.

The final kind of pseudopod, the lamellipodium, may be viewed as a modification of the lobopodium, for like that type it is a broad protrusion involving both the ecto- and endoplasm. Present knowledge indicates it to be confined to the neutrophils of vertebrates. It is distinguished from the lobopodium of amoeba in being thin and flat and probably contacts the substrate directly, rather than by leglike extensions.

This last type makes clear that the several categories of pseudopodia are not always sharply demarked, but may grade into one another, or two or more varieties may occur in the same organism. For example, the pseudopods of *Amoeba radiosa* are lobopodia when first formed, but they soon send out filopodial extensions. Moreover, as shown in the scanning electron micrographs (Figure 3.3A), two varieties are often present simultaneously on the same spec-

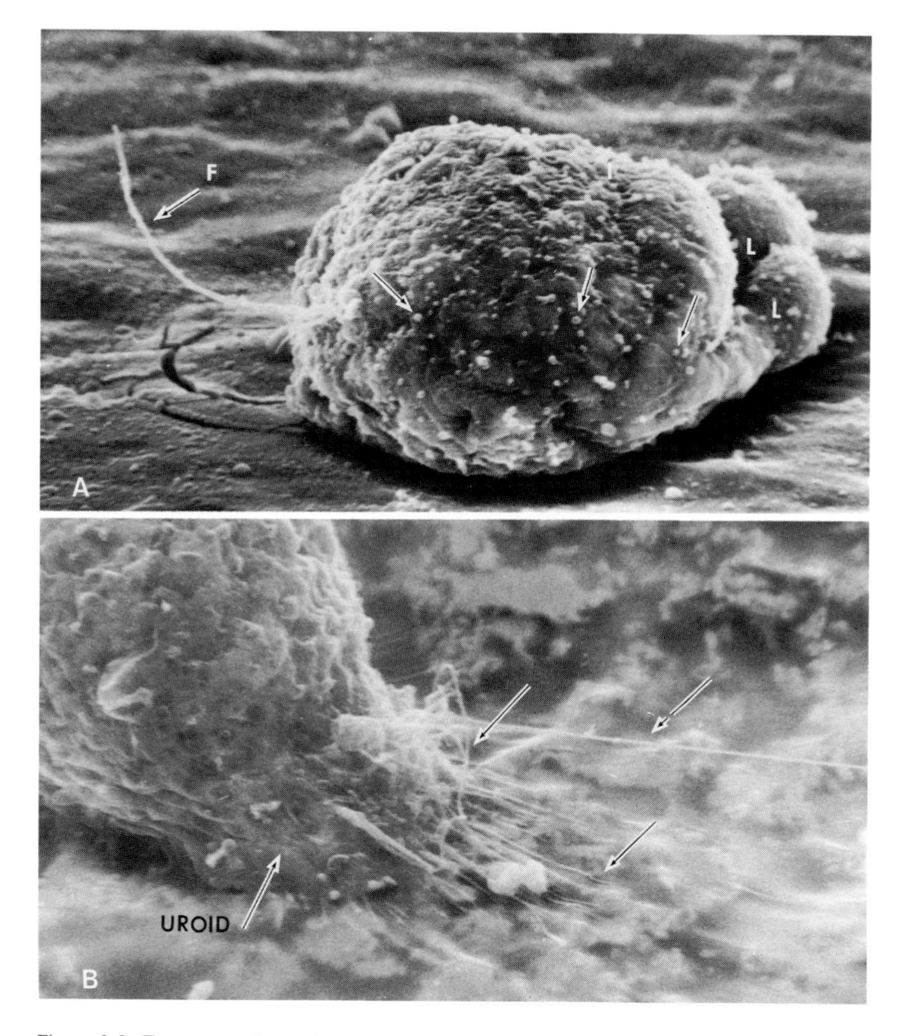

Figure 3.3. Two types of pseudopods concurrently on an amoeba. (A) These scanning electron micrographs of *Entamoeba histolytica* show the presence of lobopodia (L) at the forward side and a filopodium (F) on the uroid. 3000×. (B) The enlarged view of another specimen reveals the presence of numerous filopodia on the uroid. 4000×. (Both courtesy of Lushbaugh and Pittman, 1979.)

imen. Another instance of intergradation is provided by the common members of the test-bearing genus *Arcella,* in which the pseudopods are lobopodia, except that they consist entirely of ectoplasm. Moreover, changes in pH, ionic content, or temperature in the medium may induce alterations in pseudopod structure and form.

Amoeboid Movement. Ever since the early nineteenth century, microscopists have been fascinated by the movement of amoeboid organisms, but Dujardin (1835, 1838) seemingly was the first to suggest a mechanism for its production. Since then, nearly as many theories have been presented to explain pseudopodial locomotion as there have been observers. Consequently, no historical review of the concepts can be included here, for a full account of which reference should be made to Allen (1961). It can be stated, however, that the early proposed mechanisms involved surface tension, contractility, and ultimately the sol–gel concepts in which either contractility (Pantin, 1923) or elastic recoil of the ectoplasm (Mast, 1926) provided the driving force. But before the latter theories and recent ones are summarized, the changes that can be observed to occur in an amoeba moving over a surface by lobopod formation need attention.

During the formation of a lobopodium in a typical amoeboid, the inner layer of the cytoplasm, the axial endoplasm, can be seen to flow forward as in Figure 3.4. Then upon attaining a region somewhat just behind the tip of the growing organelle, the granules that the endoplasm contains may be perceived to move toward the sides at this fountain zone, as it is called, while the very tip (the hyaline cap) remains clear (Marsot and Couillard, 1978). The deposits

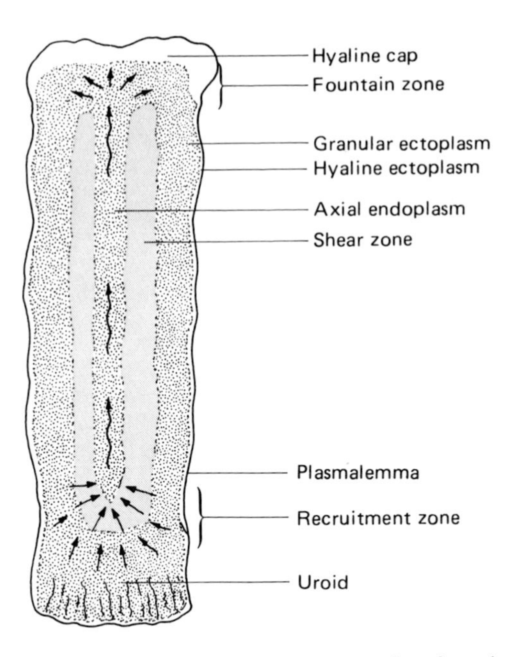

Figure 3.4. Cytoplasmic flow during lobopodium formation.

formed by these processes thus come to form the granular ectoplasmic layer, leaving a more or less immobile zone (the shear zone) between the motile axial endoplasmic and granular layers. At the end of the cell opposite the pseudopod is a rather wrinkled area, referred to as the uroid; it is in this region that the forward streaming of axial endoplasm is initiated. There is thus a one-way flow of endoplasm forward through the entire length of the organism, which results in the movement of the organism across the substrate.

3.1.2. Theories of Amoeboid Movement

Sol–Gel Concepts. Since almost no conflict in the basic observations exists, differences of opinions expressed in the literature have been confined to the proposed explanations of the mechanisms that propel the axial endoplasm and thereby provide amoeboid movement. Because the sol–gel theory mentioned earlier hypothesized that the endoplasm was a fluid sol that became gelated into granulated ectoplasm upon reaching the growing pseudopod, it was readily shown to be untenable by a simple set of experiments. First, the amoebae were fed a number of particles of such heavy metals as gold or iron, the former having the advantages of being heavier and nontoxic (Griffin and Allen, 1960). The organisms were then individually enclosed in tubes so that each could be tilted and rotated on its axis under a microscope as the movements of the ingested particles were observed. Obviously, if the interior had a low viscosity, the heavy metal bits would drop freely through the cytoplasm, whereas if it were highly viscous or gelated, the metallic fragments would move slowly or not at all. In amoebae that were actively sending out pseudopods, such experiments demonstrated clearly that, although the shear zone around the axial endoplasm was a sol, the remainder of the interior was too viscous for the particles to drop through the various subdivisions. Contrastingly, in amoebae that had been shocked into immobility, the entire interior was found to be fluid, as evidenced by the metallic fragments moving freely through that region. Furthermore, just before pseudopod formation was initiated again, the viscous state reappeared. Consequently, the fluid interior hypothesized by the sol–gel concepts was clearly demonstrated not to exist.

Contraction Concepts. Since the 1950s, models of amoeboid movement predominantly have been based on ectoplasmic contraction, a principle that had been invoked frequently in earlier years, too (Mast, 1926; Meyer, 1929; Seifriz, 1929; Monné, 1948). The earlier of those that appeared since the mid-century tended to view ectoplasmic contraction as occuring in the uroid region. For instance, Goldacre and Lorsch (1950) and Goldacre (1952a,b) proposed that contraction was produced by the folding of interlinked proteins of the ectoplasmic gel; then in becoming a sol in the endoplasm in accordance with the sol–gel concepts still prevalent at that time, the proteins were thought to become superfolded and to lose the linkages. Upon reaching the anterior of the amoeba, these substances unfolded completely and reformed linkages, thereby producing the gel of the ectoplasm once more.

More recently, the anterior region (fountain zone), at the point where the endoplasm flows into the hyaline cap and turns to the sides of the cell to become ectoplasm (Figure 3.4), has been suggested to be the actual zone of contraction (Allen *et al.*, 1960; Allen, 1961). Because in this region the endoplasm undergoes syneresis, that is, shortens and exudes water, there is little reason to doubt that it condenses somewhat in becoming ectoplasm, but because the exudate enters the hyaline cap, there must also be reabsorption of that fluid, otherwise the cap would continually increase in size during amoeboid movement.

Many experiments that applied counterpressure or negative pressures externally have been performed in attempts to test the pressure-gradient hypotheses, as these concepts eventually were called (Mast, 1931). A number of the investigations involved the use of capillary tubes of various types and sizes in which different species of amoebae were enclosed while external pressures were applied. Kamiya (1959, 1964), for instance, found that the flowing cytoplasm of that portion of an amoeba contained within a short tube responded appropriately to small changes in pressure applied to the ends of the tube, while the exposed portions did not cease flowing. By and large, however, the experimental results failed to support the pressure-gradient concept, for application of pressure in a direction that would be expected to accelerate streaming failed actually to do so until pressures of around 10×10^4 dyn/cm^3 were attained. Similar nonsupporting results were reported from experiments employing the external application of vacuum (Allen *et al.*, 1971; Jahn and Votta, 1972). As extension of the other pseudopods was not prevented and no reversal of direction of flow occurred, it was concluded that cytoplasmic streaming could not result from a positive pressure gradient generated along the length of the endoplasmic stream (N.S. Allen, 1974; Allen and Allen, 1978). Nor is it clear why these same experiments would not also negate the concept of contraction at the fountain zone, as proposed by the last citation. Still more difficult to comprehend is how the axial endoplasm could possibly flow forward into a pseudopod if contraction is occurring in that zone as theorized.

A Filamentous Basis for Pseudopod Formation. Although much has been established as to the presence of myosin, actin, and other microfilamentous substances in amoebae, as cited in the preceding chapter, little use has been made of that information in theories of amoeboid movement (Wohlman and Allen, 1968). Taylor and Wang (1978) are among the few who have employed fluroescein-labeled actin in exploring this problem. These investigators injected that dye into *Chaos chaos,* where its fluorescence was uniformly distributed throughout the organism; on the other hand, in the amoeboids of *Physarum,* it became associated with actin bundles. Obvious experiments that need to be conducted involve the use of fluorescein-labeled antibodies to actin, myosin, tubulin, and other fiber-associated proteins, which could make the fibers visible in intact active amoeboids. When this technique is employed, such fibers probably will be seen to form and dissociate as pseudopods form and movement continues.

That this suggestion holds a high degree of probability is supported by recent investigations into motile extracts from *Dictyostelium* (Eckert *et al.*, 1977). After being gelled by heating to 25°C, the extracts contracted in response to micromolar amounts of Ca^{2+} and to pH higher than 7.0. Moreover, any condition that induced solation of the gel elicited contractive responses in extracts containing myosin (Condeelis and Taylor, 1977). Actin, myosin, and a 95,000-dalton polypeptide were found to be concentrated in the contracted extract; furthermore, F-actin filaments could be demonstrated by electron microscopy, and actomyosin ATPase activity was shown to increase up to tenfold. No contraction occurred if the myosin was removed. On the basis of these observations, it was concluded that gelation in this slime mold involves an interaction between actin and several other components in the presence of Ca^{2+} and that contraction depended on the presence of myosin. Actin and myosin could not interact when in the gelled state, but solation resulted in the release of F-actin capable of interacting with myosin, thereby producing contraction.

Similar results have been obtained with echinoderm coelomocytes, except that a different protein was found to be active with the actin (Otto *et al.*, 1979). When these cells attached to a substrate, they formed large lamellipodia around the entire circumference, but when they were shocked with a hypotonic medium, this type of pseudopodium was replaced by filopodia. During this transformation, dense cores of actin filaments developed at the periphery and grew in thickness and length deep into the cytoplasm. After the cores of the filopodia had been completed, the plasmalemma covering them retracted to expose the pseudopodial structure completely (Figure 3.5). In the experiments, an antibody was used to reveal the distribution of the 58,000-dalton protein, fascin, that was found to organize the actin into the cores, apparently by cross-linking adjacent thin filaments.

Macrophages from mouse peritoneum have been shown to differ considerably from the foregoing in their pseudopodial structure. Under the electron microscope, the thick pseudopodia have been observed to contain a number of microtubules that were closely associated with endoplasmic reticulum and cytoplasmic bodies (Katsumoto *et al.*, 1978), and in addition to contain an ill-defined filamentous network at the tip. Similarly fibroblasts, which have a lamellipodial type of pseudopod, appear to have microtubules involved in their amoeboid movement, for antimicrotubule agents greatly reduced motility in these cells (Armstrong and Armstrong, 1979). Moreover, pseudopods of greatly contrasting behavior in thyroid epithelial cells were found to contain a meshwork of microfilaments. These pseudopods play a role in the early stages of secretion of thyroid hormones, rolling into balls and thereby forming colloi-

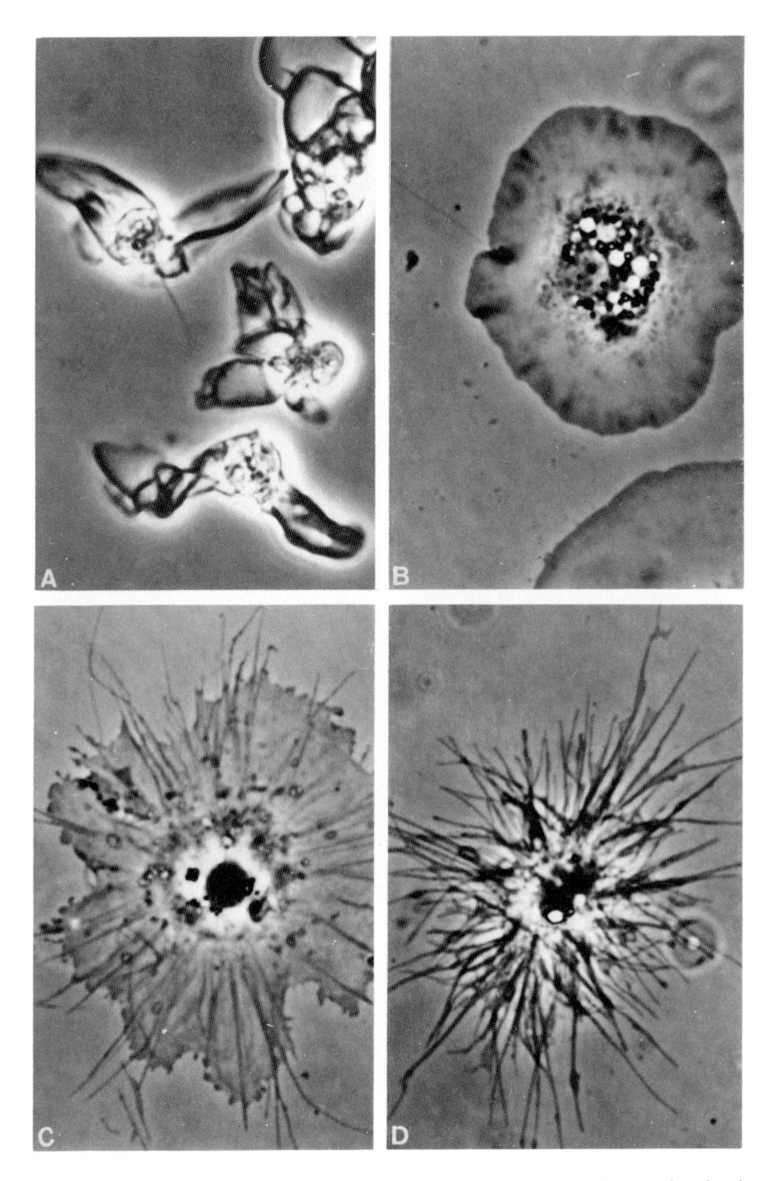

Figure 3.5. Transformation of a sand dollar coelomocyte. (A) The petaloid stage, bearing irregular bladderlike appendages. (B) After attaching to the substrate, the petals are replaced by a broad lamellipodium around the circumference. (C) Actin filaments form within the lamellipodium to provide the central rods of axopodia (D). After the latter are completed, the lamellipodium is retracted into the cell. All 2000×. (Courtesy of Otto *et al.*, 1979.)

Figure 3.6. A filopodium of a marine proteomyxid. The presence of microtubules (Mt) throughout the organelle is an obvious feature, along with mitochondria (M), vesicles (V), and various granules (PG). 9800×. (Courtesy of Anderson and Hoeffler, 1979.)

dal droplets (Zeligs and Wollman, 1977), an activity that is interrupted by thyroid-stimulating hormone. Electron microscopic examination revealed the tips of the pseudopods to contain networks of microfilaments so dense that other organelles, such as mitochondria, were excluded from that region. The presence of microtubules is not confined to metazoan pseudopods but have been demonstrated in the filopodia of protistans, including a marine proteomyxid (Figure 3.6; Anderson and Hoeffler, 1979).

The Universality of Actin in Pseudopods. Thus it appears that actin filaments are a universal feature of pseudopods and that longer or thicker types of those projections also contain microtubules. However, the several varieties of proteins that have been reported from different sources as being active in cross-linking the microfilaments imply that pseudopods have had a number of separate origins, at least in different metazoan tissues, all of which employ the ubiquitous substance actin as their basis.

However, it should be clearly noted that the existence of microfilaments within the organelles themselves does not solve the problem of how pseudopods elongate or thicken. Obviously contractile filaments that are located within such structures can serve only in constricting and shortening them; thus increase in diameter and elongation can arise solely from the activities of the cytoplasmic mass. Similarly the microtubules within pseudopods can be elongated as those projections grow and thereby contribute support to them, but there is no evidence that indicates them to be actively involved in that growth. The distinct possibility remains, however, that as the cytosol adds tubulin molecules to the growing ends of microtubules or actin to fine filaments, those acts themselves may serve to extend the tip of the pseudopod a short distance, a suggestion made earlier by McGee-Russell (1974). These small gains, repeated hundreds of times, could be the basis for pseudopodial growth, a point made clearer by the following discussion of microvilli, structures obviously closely akin to the present ones. But the filaments and/or tubules are needed along with the cytosol, for cytoplasm isolated from *Nitella* was found not to be excitable by itself (Koppenhöfer *et al.*, 1977), although that isolated from *Chaos chaos* was reported earlier to show streaming (Allen *et al.*, 1960). In addition, fibrillar contractions at constantly changing sites within the cell body are probably also essential to pseudopod growth and amoeboid movement (Taylor *et al.*, 1976).

3.2. MICROVILLI AND RELATED CELLULAR PROJECTIONS

In eukaryotic cells that have been isolated by explantation and in cells undergoing mitosis, numerous fine extensions often form rapidly on the surface and as quickly are retracted. These are usually referred to as microvilli (Lewis and Lewis, 1924; Gey, 1956), a name also employed for the related but slow-growing projections on intestinal and kidney epithelial cells (Chapter 1). Sometimes certain varieties of these structures are given the name microspikes (Weiss, 1961; Taylor and Robbins, 1963; Taylor, 1966), but as no structural differences occur, all are treated together under the term microvilli, along with several other less specialized cell projections of comparable nature.

Microvilli of Brush-Border Cells. Microvilli, particularly those of the intestinal brush-border cells, have received far more attention in recent years than have any of the related structures (Tilney and Cardell, 1970; Tilney and Mooseker, 1971, 1976; Tilney, 1977; Bretscher and Weber, 1979). As a whole each projection in this tissue appears to be supported by a compact core of about 20 fine filaments made of actin (Hiramoto, 1955), arranged parallelly in a highly ordered array (Mooseker and Tilney, 1975; Mooseker, 1976). At the proximal end, just below the base of the microvillus, the core originates on a structure called the terminal web (Brunser and Luft, 1970; Mukherjee and Staehelin, 1971; Rodewald *et al.,* 1976). The filamentous components then are attached individually to the plasmalemma covering the projections at their apical termini with the tip of the microvillus by means not understood at present. Laterally they are connected to the inner surface of the membrane by a long series of cross-links arranged in a polarized manner (Nachmias and Asch, 1976). On the basis of their molecular properties, these cross-links have been considered to consist of α-actin (Chapter 2; Section 2.2.1; Tilney and Cardell, 1970; Podlubnaya *et al.,* 1975), an identification based on immunofluorescent microscopy observations and on the molecular dimensions and weight of around 100,000 (Podlubnaya *et al.,* 1975). However, quite recent techniques have permitted the cores of the microvilli to be isolated intact, and studies employing the extracted structures have yielded contrasting results (Bretscher and Weber, 1979). One set of studies that employed antibodies to α-actin and immunofluorescent microscopy found that the microfilaments of the core did not react, while the terminal web did. When a sodium dodecyl sulfate polyacrylamide gel electrophoretic analysis was made, the presence of two polypeptides of similar molecular weight was revealed, α-actinin (from the web) and a new substance (villin), having a weight of 95,000 that constituted the cross-linkages.

Microvilli of Developing Eggs. Another group of subjects popular for investigations into microvillus structure and growth consists of developing eggs, particularly those of amphibians and sea urchins. The sea urchin egg microvilli increased rapidly from short projections less than 1 nm in length to 5 or 10 nm within 1 hr after fertilization had occurred (Eddy and Shapiro, 1976; Burgess and Schroeder, 1977). Ultrastructural studies, such as those cited, have reported the existence of bundles of fine filaments arranged in polarized fashion within the microvilli that extend just into the cell proper, but the presence there of a terminal web has not been confirmed as yet (Kidd *et al.,* 1976; Mann *et al.,* 1976). Although the polarized condition was attributed to meromyosin, in view of the recent finding of villin in intestinal brush-border cells, its identity needs to be reexamined. Other differences in structure do occur, however. For instance, the transverse stripes on each bundle have a periodicity of 330 Å in the intestinal microvilli (Mooseker and Tilney, 1975), whereas those of the sea urchin projections have intervals of 120 Å (Burgess and Schroeder, 1977). Similar stripes, including those with repeating units of 160 Å that have been described for microvilli of blood platelets (Nachmias *et al.,* 1977), have now been demonstrated to reflect the locations of bridges that unite the actin filaments into a bundle (Spudich and Amos, 1979). In addition, the latter report showed that, *in vivo,* the microvillar bundles were supercoiled and gave rise to axial repeat units 1500–2000 Å in length. On the average each of the bridges interacted with four actin monomers, linking two on one thin filament to two on an adjacent one; they were not precisely spaced, however, but were arranged quasiequivalently along the longitudinal axes of the bundles in steps of either four or five actin subunits in length.

As a whole, what has been reported about the microvilli of amphibian and rat oocytes (Franke *et al.,* 1976) shows them to be comparable in structure to those just described, for the actin filaments are similarly arranged into bundles that run the length of the projection. Nevertheless, broad generalizations are not to be made, because the microextensions of human conjunctival cells and chick embryonic kidney cells showed the presence of distinct microtubules (Taylor, 1966). These extended outwards through the projection, in association with unidentified filaments, and also were in abundance in the cell proper, just beneath the plasmalemma.

Tentacles of Suctorians. The peculiar microprojections that characterize those modified ciliates known as Suctoria have attracted a modest number of investigations into their mode of operation. These tentacles, as they are called, are used in feeding upon such protozoans as *Tetrahymena,* and a large number of solutions to the problem of how the prey cytoplasm is ingested have been offered over the years. In substance all these concepts propose a suction mechanism (reviewed by Canella, 1957; Hull, 1961; Bardele, 1974); however, the activity involves efforts of the entire organism, not those of the tentacles alone.

Moreover, ultrastructural studies have provided significant evidence against a role of suction in the feeding activities of suctorians (Rudzinska, 1970; Bardele, 1972, 1974). Each tentacle consists of a long shaft that terminates distally in a broader rounded knob and basally extends deeply into the cytoplasm (Figure 3.7). When sectioned transversely and viewed under the elec-

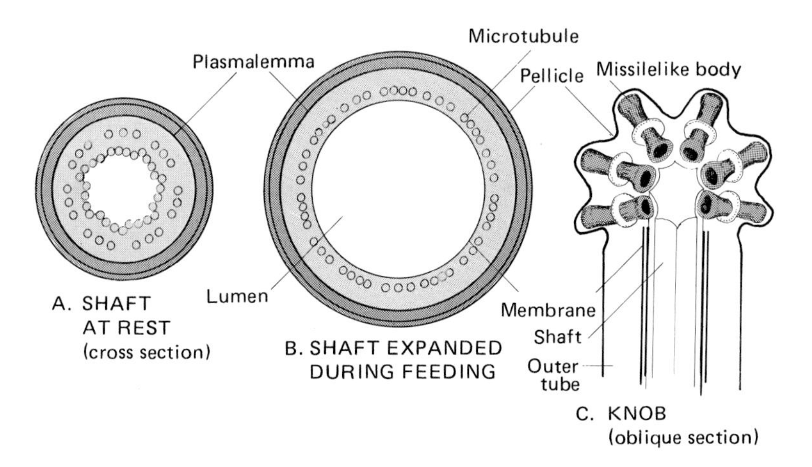

Figure 3.7. Tentacles of a suctorian. The tentacles of suctorians are not simple sucking structures but take in the cell contents of the prey by cytoplasmic streaming of a complex nature. (Based on Rudzinska, 1973.)

tron microscope, seven arcuate groups of microtubules are seen to be spaced around the narrow lumen that extends the length of the tentacles (Figure 3.7). These groups vary in number depending on the organism; in *Tokophrya,* the form illustrated, there are four close-set proximal microtubules associated with three more widely spaced ones that form a circle distad (Jurand and Bomford, 1965; Bardele and Grell, 1967; Hauser, 1970; Rudzinska, 1973; Bardele, 1974). While the microtubules attain the cell proper, they do not extend into the peculiar knob (Figure 3.7C). During the processes of feeding, the tentacle shortens and thickens, the lumen becoming many times wider than previously (Figure 3.7B). Whether the shortening results from contraction is still an unresolved question, but as shown in the next chapter dealing with the flagellum, this could readily be the case.

Contact with suitable prey triggers a number of events, beginning with the release into the prey of missile-shaped bodies, or toxicysts, stored in the knob (Figure 3.7C). As the prey thus becomes paralyzed, the shaft of the tentacle shortens and spreads, and dense bodies and vesicles move from the cytoplasm of the suctorian into the lumen of the shaft. When the shaft widens, the microtubules enter the knob and diverge laterally in all directions, perhaps aiding in penetration of the knob into the victim. After penetration has occurred, the prey's ruptured plasmalemma unites with that of the suctorian so that the two organisms are tightly fused. Once inside the cell of the prey, the membrane of the knob moves into the lumen of the shaft, carrying the cytoplasm of the prey with it (Rudzinska, 1973). Thus invaginating processes are involved, just as in other types of phagocytosis.

Certain gymnostomate ciliates, particularly the members of the genus *Actinobolina,* show affinities with the suctorians in having comparable tentacles (Holt and Corliss, 1973; Holt *et al.,* 1973), but in these organisms the tentacles are distributed among the cilia that cover the cell surface. When the ciliate is at rest, the long slender tentacles are extended, but when swimming, they are completely retracted. Each tentacle contains a toxicyst that is discharged into the prey upon contact, the paralyzed cell of which is then moved to the cytostome by means of ciliary action. Hence, the tentacles of these ciliates are capturing mechanisms, not feeding devices. However, ultrastructural studies reveal a somewhat comparable, if less complex, structure, in that circles of microtubules extend from the cytoplasm to the tip of the tentacle (Holt *et al.,* 1973).

3.3. CYTOPLASMIC STREAMING

The first recorded observations of cytoplasmic streaming were those of the circular movement now variously known as cyclosis and rotational streaming observed in the cells of the stonewort *Chara* two centuries ago (Corti, 1774). By and large, this and other forms of cytoplasmic flowing are more characteristic of cells of green plants than of animals or fungi, and many reviews have been published on the subject from a strictly botanical point of view (Seifriz, 1943, 1952; Kamiya, 1959, 1962; Allen and Allen, 1978), while only a few include observations from a broader perspective (Hepler and Palevitz, 1974; Pollard and Weihing, 1974).

Kamiya (1959, 1962) has recognized a number of types of cytoplasmic streaming, including saltation, cyclosis, circulation, fountain, reverse fountain, multistriate, shuttle, and sleevelike. As some of these are confined to a limited number of taxa or tissues, they receive little or no further attention. For instance, sleevelike streaming is found solely in the hyphae of fungi, shuttle streaming is restricted to slime molds, and fountain streaming is known primarily to occur in pollen tubes of *Plantago* and *Lilium* (Kumagaya, 1950; Iwanami, 1956). Somewhat more widespread, reverse-fountain streaming is found in pollen tubes of many flowering plants and in the root hairs of *Tradescantia* (Condeelis, 1974). Hence, attention here centers principally on the commonest and most thoroughly investigated of these phenomena, cyclosis, after definitions of the other types have been provided.

3.3.1. Miscellaneous Types of Streaming

Saltation. As its name implies, saltation is a type of streaming that involves a more or less linear excursion of a particle for a distance greater than that contributable to Brownian movement. Because of the presence in the cytoplasm of endoplasmic reticulum, other membrane systems, and especially microfilaments, very long saltations exceeding 100 nm in length occur frequently only along such structures, whereas between them only Brownian movement can be observed (Kamiya and Kuroda, 1973; R. D. Allen, 1974). This association with cytoplasmic membranes or fibers is also suggested by the tendency of saltating particles to follow straight or slightly arcuate courses through the cell. According to Rebhun (1964) actin- or myosinlike filaments seem most probably to be involved. In *Nitella,* the filaments associated with saltation are bundles of polarized thin filaments (Palevitz, 1976; Allen, 1977). Thus it has been proposed that saltation is produced by myosin-bearing particles interacting with such filament bundles (Huxley, 1969).

Circulation. In the cortex (ectoplasm) of hair cells of a number of seed plants, including *Tradescantia, Gloxinia, Saxifraga,* and *Cucurbita,* and in parenchymal cells of many monocotyledons, short-term, usually linear flowing of particles may be noted. Such movements, referred to as circulation, are characteristically along transvacuolar strands, and thus the streams may branch and rejoin. In *Caulerpa* the streaming was correlated to bundles of microtubules, and accordingly the latter was suspected to play a role in the activity, a proposal that receives emphasis in view of the observed absence of microfilaments (Sabnis and Jacobs, 1967; Szamier *et al.,* 1975).

Transport by Cytoplasmic Movement. Perhaps the most thoroughly explored aspect of transport of substances by cytoplasmic movement is that of the phloem. It has long been thought that the conduction of substances through this tissue was related to cytoplasmic movement, and more recently it has been considered to be produced by cytoplasmic filaments (Thaine, 1961, 1969; Johnson, 1968; Aikman and Anderson, 1971; MacRobbie, 1971; Robidaux *et al.*, 1973; Fensom and Williams, 1974). Rather than the microfilaments being comprised of actin or other substance of catholic occurrence, they appear to consist of P-protein (Cronshaw and Esau, 1967, 1968; Hepler and Palevitz, 1974). Because little actual information is as yet available on this unusual substance, the process of phloem transport, important as it is, does not merit further attention here.

Another important aspect of transport is that observed in neurons (Dahlström *et al.,* 1974; Heslop, 1974). Because the organelles and most of the macromolecules of these nerve cells develop in the cell bodies, they need to be transported thence through the axons and dendrites, a process that in longer types may require several days (Rebhun, 1972). In these cells two classes of movement are recognized, fast and slow. The former is independent of the cell body and continues in sections removed from the rest of the neuron or after a ligature is tied behind an advancing zone of labeled substance (Heslop and Howes, 1972). Transport rates, measured as about 16 or 17 mm/hr, were not correlated to the diameter of the axon (Ochs, 1972). In this type of movement, microtubules appear definitely to be involved, because colchicine and other inhibitors of tubulin polymerization blocked such transport (Kristensson, 1970).

3.3.2. Rotational Streaming (Cyclosis)

General Characteristics. Undoubtedly the most remarkable of the cytoplasmic streaming phenomena is rotational streaming, often referred to as cyclosis, but equally commonly classed simply as cytoplasmic streaming. Apparently this rapid type of movement, having speeds between 50 and 100 nm/sec, is confined to the stoneworts and the higher green plants (Kamiya, 1962). Among the latter taxa, *Anacharis* leaves have been especially well studied. In these structures the endoplasm streams in truly circular fashion, being almost exclusively confined to paths adjacent to those faces of a cell that are in contact with others. Under laboratory conditions of constant temperature, light, and ionic content of the medium, flow rate is unchanging day and night, but seasonal effects have been noted in wild plants (Allen and Allen, 1978).

The favored organism for laboratory observation is the stonewort *Nitella,* a common plant of shallow temperate lakes and ponds. Among its outstanding attributes for such studies are the absence of differentiated chloroplasts and the presence of transparent walls on the rhizoid cells. Internodal cells are also of frequent use, because of the lengths of 1 to 5 cm and ability to remain viable for weeks after isolation from the remainder of the plant. Because the cells are only 1 mm in diameter despite their great length, the rotational streaming follows a somewhat different path than in *Anacharis,* best described as quadrate, flowing along one long side, abruptly turning at the end, and returning along the opposite face. Between the counterflowing streams are two irregularly shaped so-called indifferent zones in which the endoplasm is absent or virtually so, separated from one another by the large central vacuole (Pickard, 1974; Allen and Allen, 1978). Chloroplasts are largely arranged in spiral rows within the ectoplasmic layer, but some are free in the endoplasm, where they may be noted to rotate on one axis in concert with some of the nuclei in these multinucleated cells.

Effect of Light. Various wavelengths of light have a direct effect on protoplasmic streaming in root hairs of barley. White light stimulates streaming if the grains have been germinated in its presence, but if germination has occurred in the dark, it generally inhibits the flow (Augsten and Finke, 1978), the precise effect varying with the length of time of exposure to the radiation. After germination in the dark, red light, however, promotes streaming, which continues after the tissue is removed to the dark once more. While illumination with far-red light has an inhibiting effect, the presence of shorter waves of red light overcomes that influence. Because of these reactions to light, the existence of a photoreceptor has been suggested, possibly the phytochrome system.

Possible Role in Transport. One aspect of transport of substances through the plant body is concerned with the mechanism of lateral movement of organic ions from the soil across the cortex of the root to the xylem vessels. Two major pathways have been proposed, one by way of the extracellular apocytoplasm of the tissues, the other through the cytoplasm of the individual cells (Anderson, 1976; Pitman, 1977). Because apocytoplasm is absent from the endodermis, however, ions ultimately must enter the cytoplasm to cross that barrier to the stele. To provide a mechanism for this transport, cytoplasmic streaming has long been considered an active factor, and in *Chara* a strong correlation has been demonstrated between the rates of streaming and that of the transfer of Cl^- ions between adjacent cells (Bostrom and Walker, 1976).

Contrastingly, other investigations have concluded that lateral transport is not correlated to rotational streaming in plants that, unlike *Chara,* have small cells (Tyree, 1970; Tyree *et al.,* 1974; Anderson, 1976). In a recent study on the problem, translocation of K^+ labeled with ^{86}Rb was used, along with cytochalasin *b,* a potent inhibitor of cytoplasmic streaming (Glass and Perley, 1979). After the tops had been removed from various species of cultivated plants, radioactivity in the exudate from the xylem was measured before and after addition of cytochalsin *b* to the medium in which the roots were immersed. Although this reagent induced complete cessation of cytoplasmic streaming in root cells within 15 min, no effect was noted on total ^{86}Rb uptake or on the rate of exudation. Hence, cytoplasmic streaming was ruled out as a factor in movement of K^+ across the root cortex.

Proposed Mechanisms. Early data had seemed to indicate that the shear zone between the ectoplasm and circulating endoplasm was quite narrow, so the motive forces propelling the latter were conceived to lie close to the proximal edge of the cortical layer (Linsbauer, 1929; Breckheimer-Beyrich, 1954; Kamiya and Kuroda, 1956). This concept, however, was shown to be invalid by the subsequent discovery that the fibrils of the cortex sent branches into the stream (R. D. Allen, 1974). By use of laser illumination with differential interference microscopy, the latter investigation was able to observe undulating filaments less than 100 Å in diameter, producing sinusoidal waves 25 μm in length and 5 μm in amplitude. Calculations demonstrated that these fibrils could generate force more than sufficient to drive the flowing endoplasm in its circuit around the cell's interior. When cells were shocked mechanically or electrically, and streaming was halted momentarily, the filaments remained straight. Upon resumption of activity, saltation of particles first was observed along the subcortical as well as endoplasmic filaments, followed by rotational streaming as the undulations of those fibrils increased (Allen, 1976).

Ultrastructural investigations of pertinent cell regions have disclosed bundles of microfilaments beneath the ectoplasm in the two genera of stoneworts (Nagai and Rebhun, 1966; Pickett-Heaps, 1967; Nagai and Hayama, 1979). Although microtubules also were present, they were arranged adjacent to the cell wall in a less orderly fashion than in seed plants; thus they were considered not to be in a position to be effective in cyclosis. Chloroplasts were held together in long files by fibrils lying below the ectoplasm (Palevitz and Hepler, 1975; Palevitz, 1976). Actually these and the fibrils that sent undulating branches into the flowing endoplasm proved to be nanofilaments with diameters between 50 and 80 Å. Because they became decorated by rabbit skeletal muscle heavy meromyosin in the form of the characteristic arrowhead arrays, they undoubtedly were F-actin (Ishikawa *et al.*, 1969; Williamson, 1974). Moreover, the arrowheads were aligned in the same direction (opposite to the direction of flow), indicating the fibrils to be highly polarized (Palevitz *et al.*, 1974; Kersey *et al.*, 1976). It also has been demonstrated that actomyosin solutions derived either from plasmodia of *Physarum* or rabbit striated muscle can undergo a type of cytoplasmic streaming when introduced into microcapillary tubules in the presence of ATP (Oplatka and Tirosh, 1973).

Now that a mechanism for propelling the endoplasm has been pinpointed, the next question to be resolved is how these thin filaments produce the undulations that provide the drive. Once this problem has been fully answered, then light will certainly be thrown on such other pressing riddles as muscle contraction and perhaps also on flagellar undulations, as described in the succeeding chapter.

3.3.3. Unclassified Types of Streaming

Cytoplasmic streaming is not confined to metaphytans and algae but occurs even in chlorophyll-less organisms, too, including various protozoans as well as metazoans. In a number of instances, movement of cytoplasmic particles occurs adjacent to series of microtubules; for instance, ribosomes (Macgregor and Stebbings, 1970), various types of vesicles (Hepler and Jackson, 1968; R. D. Allen, 1974; Gray, 1975), and pigment granules (Murphy and Tilney, 1974; Schliwa, 1975) have been reported to be translocated in this man-

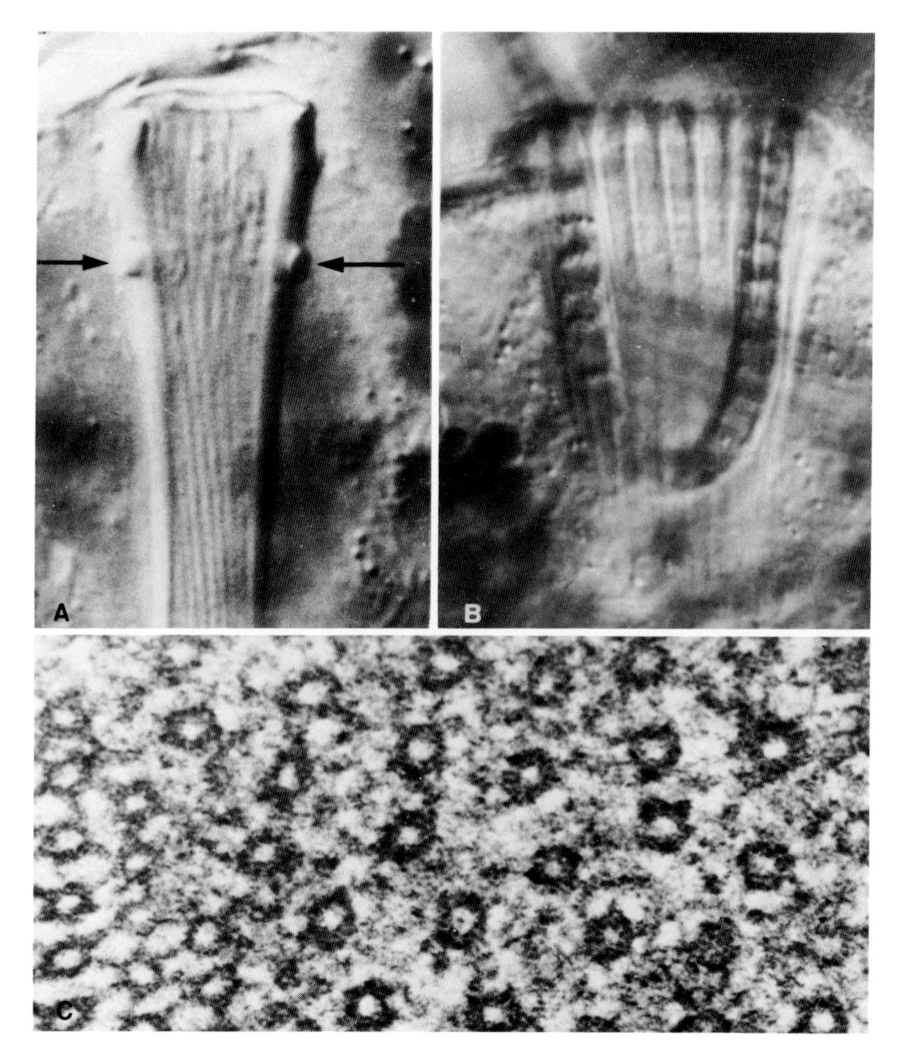

Figure 3.8. An expandable sheath in the ciliate *Nassula*. The cytopharyngeal sheath is normally contracted (A), but during feeding on filamentous algae, it can be expanded as in (B). 3350×. (C) A section of an unexpanded sheath shows the microtubules to have loose, interconnecting bridges between them. To the left, in the section of a supporting rod, the intertubular bridges are stout and compact. 230,000×. (Courtesy of Wellings and Tucker, 1979.)

ner. In most cases, it appears likely that such movements are the result of cytoplasmic streaming.

Streaming in a Ciliate. One example of this type of transport has come to light in the ciliates of the genus *Nassula* (Tucker, 1978), in which there is a pharyngeal basket through which filaments of blue-green algae are ingested. This basket is comprised in part of a bundle of microtubules, interconnected by numerous bridges (Figure 3.8). During ingestion, prolonged and highly oriented streaming of gelated cytoplasm occurs, which carries the algal filaments through the basket to the forming food vacuole. How the propulsion of the cytoplasm was achieved could not be determined, but it appeared to involve the microtubules and the microfilaments attached to those structures.

Cytoplasmic Flow in Metazoan Cells. Other types of movement have been described in cultured BHK-21 cells that have been treated with trypsin and then replated; cells subjected to this treatment change shape from a round to a fibroblastic form (Wang and Goldman, 1978). Time-lapse cinematomicrography of such spreading cells revealed that their organelles became redistributed by saltatory movements from a location near the nucleus outwards into the cytoplasm. Spreading continued in both directions along the long axis of the cell, proceeding at average speeds of 1.7 μm/sec during the initial period that increased to 2.3 μm/sec in fully spread cells. Concurrent electron microscopic studies showed that these movements were correlated to the changing pattern of microtubules and 10-nm filaments. If microtubule formation was inhibited with colchicine, the saltatory movements ceased, but later pseudopodial projections formed around the cell periphery, accompanied by streaming of the cytoplasm into the extensions. The pseudopods were shown to contain actin microfilaments that bound heavy meromyosin from skeletal muscle in the fashion typical of actin filaments.

Similar associations of microtubules and microfilaments have also been demonstrated in metazoan structures, one common example of which is provided by the nutritive tubes that are found in the ovarioles of the aquatic hemipteran, *Notonecta glauca* (Macgregor and Stebbings, 1970; Hyams and Stebbings, 1977). These tubes are membrane-limited channels, up to several millimeters in length, that assist in the transport of ribosomes from the apical trophic end of the ovariole to the developing oocytes. Although microfilaments are absent from these structures, microtubules are abundant, over 30,000 being present per tube (Macgregor and Stebbings, 1970). During movement, the ribosomes, however, could not be noted to come into contact with the microtubules, nor were any bridge structures present. Hence, it could not be ascertained how, or even whether, the microtubules were active in the transport of the particles. However, it was later determined that microtubule-associated proteins (MAPs) were present (Hyams and Stebbings, 1979), which possibly could play a role in the processes.

4

CELL MOTILITY: II

The Flagellum

In many ways, this second chapter on cell motility is a continuation of the discussion of microtubules pursued in the two that precede it, because flagella and cilia, like pseudopods, contain microfibrillar components in abundance. Actually, these structures could very well be viewed as microtubular organelles, but only in the eukaryotes, for in the prokaryotes they are of completely different structure, displaying an absence of homology, which is discussed more fully later.

4.1. THE FLAGELLUM

Because cilia are present in only a few taxa of protozoans and in certain metazoan tissues, the term flagellum here embraces both types of organelles, for it is characteristic of a wide diversity of unicellular organisms, as well as cells of a number of tissues from metaphytan and metazoan sources. While thus sufficiently similar to merit simplification of this sort, cilia may differ from flagella in other characteristics than relative length and number per cell, but the few differences that have been ascertained can readily be pointed out wherever they occur. For simplicity of discussion a summary of the prokaryotic organelle is presented first, followed in turn by descriptions of the structure and the development of that of eukaryotes.

4.1.1. The Prokaryote Flagellum

In general the flagella of bacteria are simple filaments, about 10 to 20 μm long and between 12 and 15 nm in diameter (Brock, 1974), but there is considerable variation, especially among those that are encased in a sheath. Certain of the latter type attain thicknesses of around 20 nm (Chalcroft *et al.*, 1973), and

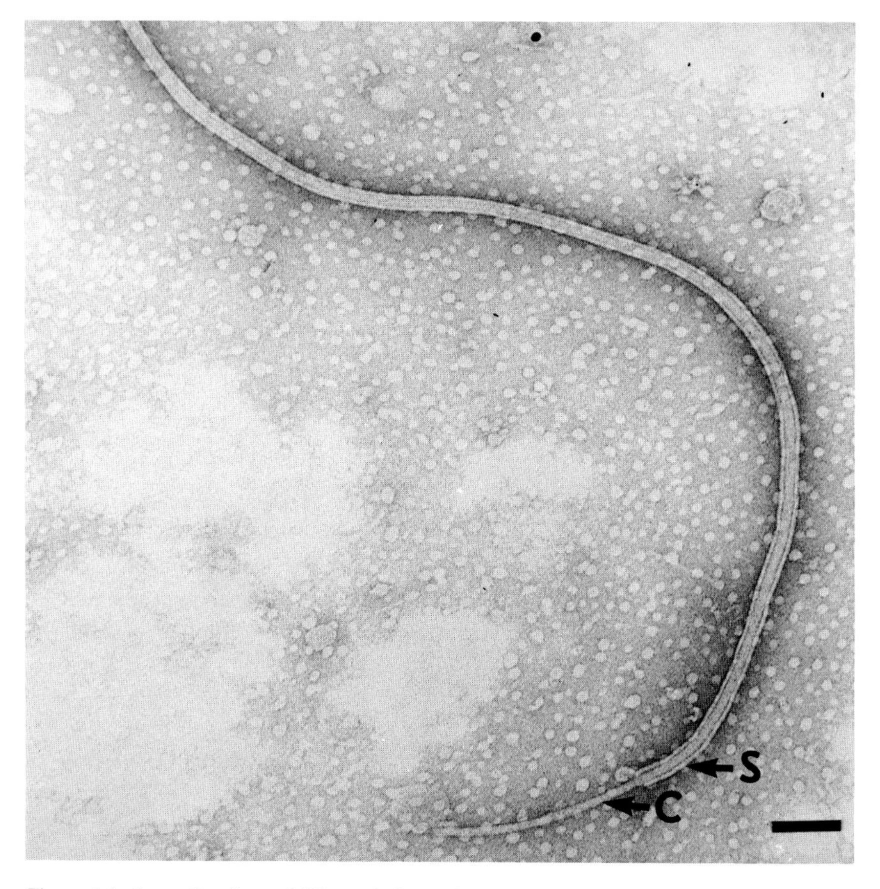

Figure 4.1. Intact flagellum of *Vibrio cholerae*. In this species the tip of the core (C) projects beyond the sheath (S). 90,000×. (Courtesy of Yang *et al.*, 1977.)

that of an unidentified bacterium reported by To and Margulis (1978) reaches 32 nm. Commonly, the flagellum consists of a slightly helical, threadlike part called the filament that is sometimes enclosed in a sheath (Figure 4.1); at the proximal end is the so-called root or basal granule (Tawara, 1965; Hoeniger *et al.,* 1966; van Iterson *et al.,* 1966), which is far more complexly structured than its name implies. This is intimately associated with the cell wall and plasma membrane. Between the basal granule and the filament is a curved part referred to as the hook structure, which is inappropriately named, as is pointed out in a following section.

Flagella are variously located on bacteria, according to the general type. Many, like *Lactobacillus* and most cocci, lack flagella throughout life; others have them in young cultures but lose them in aging ones. The polar types may have either a single flagellum (the monotrichous group) or two or more at the same end (the cephalotrichous group). Lophotrichous bacteria possess flagella at both ends, whereas peritrichous ones may lack polar flagella but bear a number along the sides. The latter category includes important gram-negative species along with *Proteus mirabilis,* which organism is of particular importance later in the discussion of the mechanism of flagellar movement.

The Filament. In view of bacterial motility's having been implicated in the pathogenesis of at least such diseases as cholera (Williams *et al.,* 1973; Guentzel and Berry, 1975; Schrank and Verway, 1976), it is surprising that proportionately few investigations have been conducted into the structure of the bacterial flagellum. Generally speaking, the filament consists entirely of a protein called flagellin, the molecules of which are about 45 Å in diameter and in *E. coli* have a molecular weight of around 54,000 (Hilmen *et al.,* 1974; Hilmen and Simon, 1976). These molecules are arranged in parallel, longitudinal rows, where they are offset sufficiently so that those of one row fit into the interspaces of the adjacent ones. Thus the filament, while resembling microtubules in consisting of parallel microfibers, differs in the fibrillar arrangement and the slope of the helix (Iino, 1974). Moreover, while the resulting complete assembly is cylindrical, it is solid rather than hollow.

How many microfibrils are present in a given filament does not seem to have been established, nor is much documented information available on the presence and composition of a sheath. However, electron microscopic studies have been made on several sheathed filaments, including those of *Pseudomonas* (Fuerst and Hayward, 1969) and several species of *Vibrio* (Follett and Gordon, 1963; Glauert *et al.,* 1963; Yang *et al.,* 1977). In *Vibrio cholerae,* the diameter of the entire filament was 23.2 nm, and that of the core, as the filamin portion is called, was 10.5 nm, leaving a thickness of 6.35 nm for the sheath. The latter does not cover the core quite completely, a short tip being left exposed (Figure 4.1). Because ferritin-labeled antibody against the core reacted with the entire filament of older organisms, including the sheath portion (Figure 4.2), it was concluded that the sheath is tightly structured in very young bacteria but loosens with age, an observation that correlates to the increase in diameter in older specimens (Follett and Gordon, 1963).

Recently in certain strains of a soil bacterium, *Pseudomonas rhodos,* a new variety of flagellar structure was described, called the complex type, to distinguish it from the ordinary, or plain, type of other strains of the same species (Schmitt *et al.*, 1974a). It is characterized by its being enclosed by three helical bands. Similar sheath organization has been reported also from a second species of soil bacterium, *Rhizobium lupini* H13-3 (Schmitt *et al.*, 1974b; Maruyama *et al.*, 1978). In addition, the filaments were found to be more fragile, suggesting a greater rigidity, and the hook was distinctive, as described in a following paragraph.

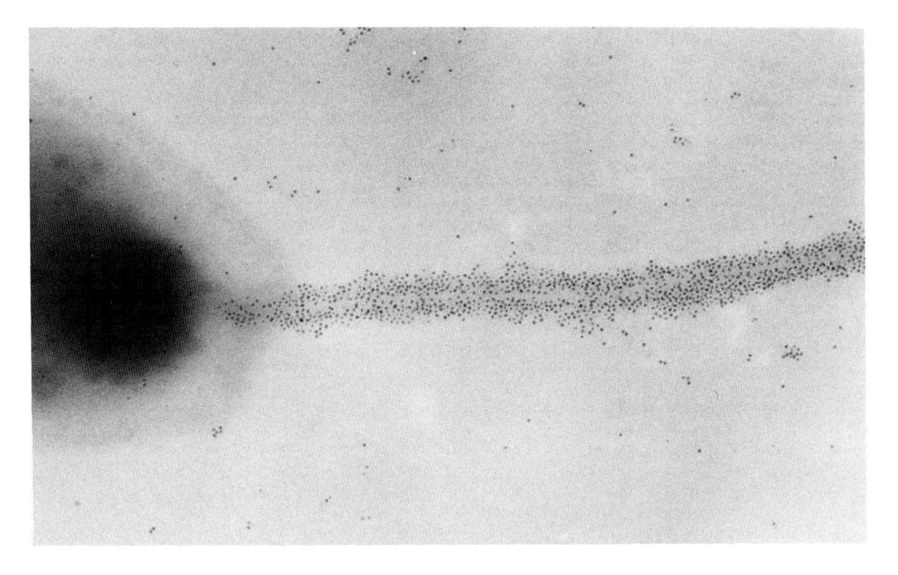

Figure 4.2. Distribution of flagellar-core antibody on a portion of a *Vibrio* cell. The antibody, labeled with ferritin, may be observed to be distributed evenly along the sheath but not on the cell surface. 98,000×. (Courtesy of Yang *et al.*, 1977.)

Flagellin. Although the primary structures of several flagellins have been incompletely sequenced, including those of two species of *Salmonella* (Davidson, 1971; Joys and Rankis, 1972), and *Proteus mirabilis* (Glossmann and Bode, 1972), the only one that seems to be completely determined is that from *B. subtilis* (DeLange *et al.,* 1973, 1976; Chang *et al.,* 1976; Shaper *et al.,* 1976), as shown in Table 4.1. When more have been entirely sequenced, considerable variation may be expected, for 16 strains of the same organism and several from *Salmonella* showed a wide range of variation when tested serologically (Kauffmann, 1964; Emerson and Simon, 1971). The variants of the first of these two bacteria fell into at least two sharply marked classes (Simon *et al.,* 1977). Two distinct species of flagellin have been reported in the flagellum of *Caulobacter crescentus,* each of which was capable of forming microfibers both independently and in combination with the other (Fukuda *et al.,* 1978).

The Basal Granule. The basal granule is a complex organ that differs somewhat in structure between gram-positive and such gram-negative forms as *E. coli* (De Pamphilis and Adler, 1971a,b; Dimmit and Simon, 1971; Adler, 1976). In the latter group it has been disclosed by the electron microscope to consist of a moderate-sized rod, around which are four rings or disks, arranged in two pairs (Figure 4.3). The members of the most proximal pair are believed, on theoretical grounds, not to be attached to one another. Of these the inner-

Table 4.1
Primary Structure of the Flagellin Molecule of Bacillus subtilis 168[a]

```
                 10         20         30         40         50         60
H2N-MPINHNIAALNTLNRLSSNNSASQKNMEKLSSGLRINRAGDDAAGLAISEKMRGQIRGLE

                 70         80         90        100        110        120
    MASKNSQDGISLIQTAEGALTETHAILQRVRELVVQAGNTTGQDKATDLQSIQDEISALT

                130        140        150        160        170        180
    DEIDGISNRTEFNGKKLLDGTYKVDTATPANQKNLVF(QIGANATQQISVNIED)MGADALG

                190        200        210        220        230        240
    IKEADGSIAALHSVNDLDVTKFADNAADTADIGFDAQLKVVDEAINQVSSQRAKLGAVQN

                250        260        270        280        290        300
    RLEHTINNLSASGENLTAAESRIRDVDMAKEMSEFTKNNILSQASQAMLAQANQQPQNVL

       304
    QLLR-OH
```

[a] Based on Chang *et al.* (1976).

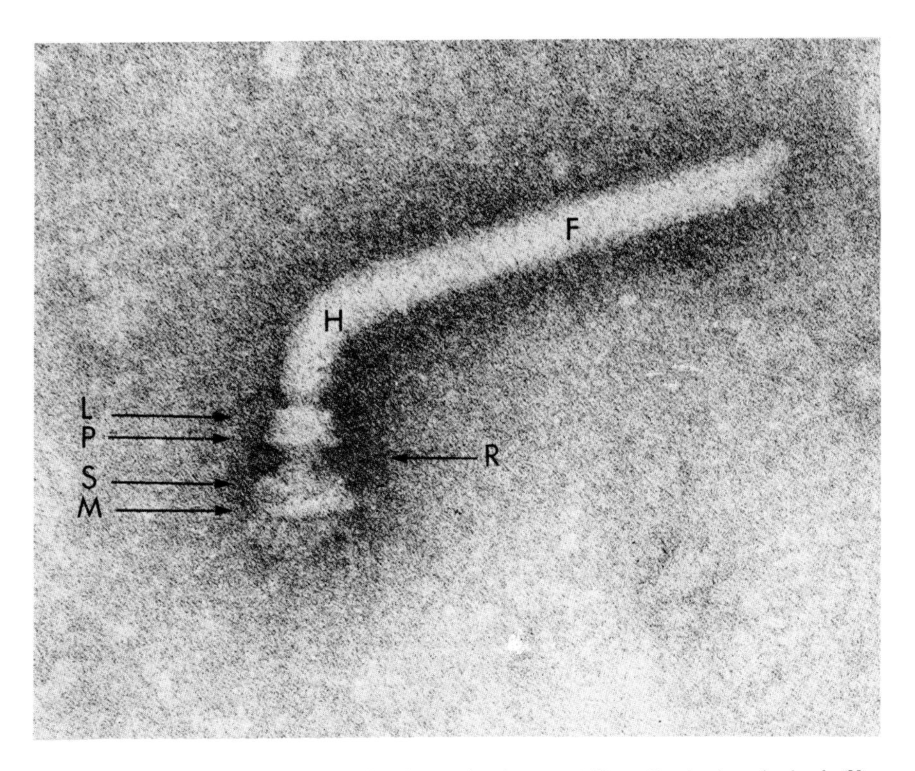

Figure 4.3. Electron micrograph of a basal granule of *Aquaspirillum*. Proximal to the hook (H) below the flagellum (F) is the complex basal granule. This structure consists of four rings, referred to as L, P, S, and M rings, grouped in two pairs on a central rod (R). 145,000×. (Courtesy of Coulton and Murray, 1978.)

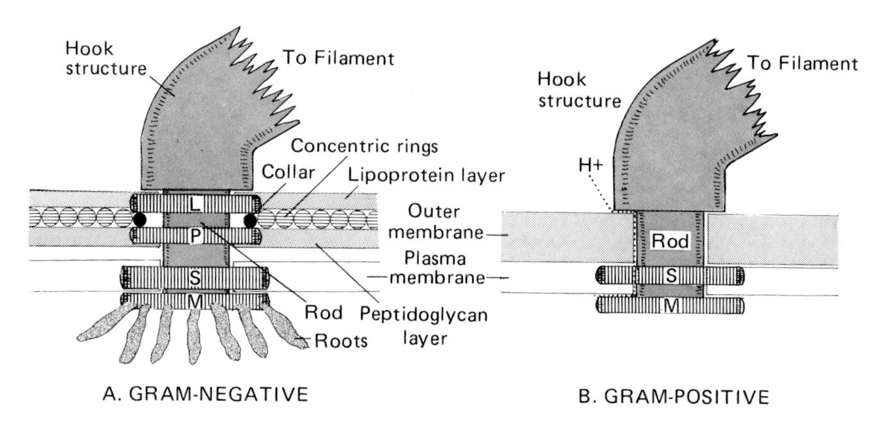

Figure 4.4. Basal granule structure. The basal granule of gram-negative bacteria (A) is far more complex than that of gram-positive forms (B). (Based in part on Coulton and Murray, 1977, 1978, and Läuger, 1977.)

most is the M ring, for it was originally believed to be embedded in the plasma membrane, whereas its distal mate is the S ring, for its supposed attachment to a supramembrane. However, more recent investigations place these parts as arranged in Figure 4.4 (Coulton and Murray, 1977, 1978). In contrast, the members of the distal pair appear broadly united, because negative stain fails to penetrate the space between them. Of these the outer component is referred to as the L ring, from its being fixed to the outer lipopolyprotein membrane, while its inner counterpart, fastened to the peptidoglycan layer, has been designated the P ring (De Pamphilis and Adler, 1971c). In some forms, such as *Caulobacter,* five rings are present (Johnson *et al.*, 1979). Among gram-positive forms the chief difference in structure is the absence of the distal pair of rings, the S ring being considered to be attached to the peptidoglycan layer. In some cases an area free of granules seems to be associated with the basal apparatus, as in the sheathed form of *Vibrio metchnikovii* (Glauert *et al.*, 1963).

It now appears that the basal granule disks may not be directly attached to the cell wall. Use of techniques that left the membranes intact around the flagellar mechanism, combined with electron microscopy, has recently led to the observation in *Aquaspirillum serpens* of concentric rings around the L disk (Coulton and Murray, 1977). The innermost of these is of distinctive structure and is known as the collar (Figure 4.4); it seems to lie immediately beneath the outer membrane and supports the L disk. In addition, seven concentric rings were observed around this structure, also located beneath the cell wall. In combination the eight rings comprised a plate about 90 nm in diameter.

Studies on the molecular aspects of the basal apparatus are still at an elementary level, but what has been determined reveals a complex biochemical organization. At least nine different proteins are involved in the structuring of the apparatus, ranging in molecular weight from 9000 to 60,000 (Hilmen and Simon, 1976).

The Hook Structure. Between the basal complex and the filament proper is a bent region known as the hook structure, but the term elbow would be more descriptive of its actual shape, for it is a curved sector that forms an obtuse angle between the filament and the basal granule. At least in insolated hooks, the angle varies from about 30° to close to 90° (Hilmen and Simon, 1976). Both in *E. coli* and in plain-flagellum-bearing *Pseudomonas rhodos,* this structure becomes slightly wider distally, so that it is a bent cone, but in the complex flagellar strains of the second species, it is cylindrical (Raska *et al.,* 1976). As a rule, the hook consists of a single protein with a molecular weight of around 43,000 in *E. coli* and *P. rhodos;* in molecular arrangement, it contrasts to the molecular organization of the filament in that the subunits are arranged in helices rather than in straight lines.

The Movement of Flagella. Unlike the flagella of eukaryotes, those of bacteria rotate to provide motility, an action first described in lophotrichous forms many years ago (Metzner, 1920). In these forms, the posterior flagella stand out from the cell body, while those at the anterior pole are curved back over it. During swimming the cell body also rotates, but in the opposite direction and at about one-third the rate of the filament. Through use of flashing lights, these rates were then determined to be 40 per second for the filament and 13 per second for the cell body (Metzner, 1920), but lower rates have been determined by more advanced procedures.

The rotation of the filament has recently been reconfirmed by the simple expedient of attaching the filaments to the glass slide by means of an antibody of flagellin. In bacteria (*E. coli*) anchored in this fashion, the cell body could be observed to rotate, usually counterclockwise but sometimes clockwise (Silverman and Simon, 1974a); in free bacteria, of course, the rotation of the filament would be in the opposite direction in each case. However, the rate of rotation of the cell observed was considerably slower than that cited earlier, being 1.5 revolutions per second or less (Berg, 1976). The direction of rotation was strongly influenced by the nature of the environment. If additions to the medium were made of substances that acted as attractants to the bacteria, the clockwise rotation of the flagella was maintained, whereas the addition of repellents induced rotation in the opposite direction (Larsen *et al.,* 1974). Such characteristic responses to repellents depended on the continuous presence of methionine in the milieu as a methyl donor (Adler and Dahl, 1967; Koshland, 1977). More recently, a possible receptor site has been identified in the plasma membrane of *E. coli,* in the form of a protein that contains a metabolically labile methyl radical donated by methionine (Kort *et al.,* 1975).

Although the evidence is clear that bacterial flagella propel by rotation, explanations of the mechanism are undoubtedly oversimplified. For instance, a current report described flagellation and swimming in *Thermoplasma acidophilum,* a bacterium that lacks the cell wall, a structure supposedly essential to support the distal bearinglike ring of the basal apparatus (Black *et al.,* 1979). Moreover, it is difficult to apply the concepts to multiflagellated forms, especially to such peritrichous types as *Proteus vulgaris,* which have the flagella located on the sides in rows. Moreover, partly rotating filaments have been observed, stationary portions being united end to end with rotating parts (Hotani, 1979). Nor is the nature of the driving mechanism clear. In gram-positive species, it has been proposed that the basal apparatus (M ring) is driven by the translocation of hydrogen ions down an electrochemical gradient through an ionophore in the cell membrane and channeled between the M and S rings (Figure 4.4B; Berg, 1974, 1975; Läuger, 1977). In contrast, Adam (1977) suggested that cytoplasmic streaming is involved.

Flagella and Spirochete Motility. Spirochetes are among the most motile of bacteria (To and Margulis, 1978), being able to elongate, shorten, bend, or loop the cell body, and also to produce planar and three-dimensional waves along its long axis; moreover, they can travel, even through agar gels, without aid of external flagella. Instead of being exposed, the flagella in these organisms are helically wound about the cell body, enclosed within the so-called periplasmic space between the plasma and outer membranes (Jahn and Bovee, 1965). About 1 to 100 flagella are present, depending on the species, and are attached at one pole or sometimes both. As in other bacteria, a hook and basal apparatus are found on each flagellum. The precise chemical nature of the flagellar apparatus remains unknown, but it probably is similar to that of the externally flagellated types, because antibodies to flagellin interact in a comparable fashion (Doetsch and Hageage, 1968).

Assembly of the Flagellum. Genetic studies on the assembly of the flagellum in bacteria show that at least 20 cistrons are essential to the processes (Iino, 1969; Yamaguchi *et al.,* 1972; Silverman and Simon, 1973; Vary and Stocker, 1973; Suzuki *et al.,* 1978). Flagellin has been found to be encoded in *E. coli* by the *hag* gene and by *H1* and *H2* in *Salmonella* (Horiguchi *et al.,* 1975). Although the specific roles for most of the remaining cistrons, usually designated as *fla,* remain unknown, it has been determined that *flaR* controls the length of the hook structure in both species (Silverman and Simon, 1972; Patterson-Delafield *et al.,* 1973) and that *flaT* is necessary at a step involving cyclic AMP (Silverman and Simon, 1974b; Komeda *et al.,* 1975). In addition, the protein encoded by *flaK* of *E. coli* has been found to be a constituent of the basal apparatus; in contrast, those of the other genes analyzed appeared to play only indirect roles in flagellar assembly (Komeda *et al.,* 1977; Matsumura *et al.,* 1977). For instance, mRNA for flagellin is absent in *flaL* mutants, so the transcription of the flagellin cistrons (Suzuki and Iino, 1975) appears dependent on the product of that gene. This implies that either a suppressor or related protein, or perhaps even a special RNA polymerase, is needed for the transcription of the cistrons involved. According to Suzuki and his co-workers (1978), the sequence of assembly events begins with the innermost parts of the basal apparatus and continues outwards, first through addition of the remaining disks, followed by the hook structure, and finally the filament. Thus the later parts are added through the cell envelope.

4.1.2. The Eukaryote Flagellum

By the mid-1950s, after the glass knife and suitable embedding plastics had been sufficiently developed to permit the routine securing of ultrathin sections, it had become clear that cilia and flagella of all eukaryotes were constructed on the same basic pattern (Houwink, 1953; Fawcett, 1954; Sager and Palade, 1957; Fauré-Fremiet, 1958; Afzelius, 1959). This has become known as the 9+2 pattern and has proven virtually universal, except for the few instances pointed out later. Two major components are considered to exist, the sheath, which is continuous with the cell membrane, and the remaining structures including the tubular framework, collectively known as the axoneme.

Structural Features

Standard Structure of the Axoneme. Basically the axoneme consists of a pair of spaced microtubules at the center and nine double microtubules around the periphery (Figure 4.5), with which are associated a ground substance and numerous small parts. The doublets are comprised of one complete microtubule, known as the A component or tubule A, that contains the standard 13 protofilaments. Contrastingly, the second member, the B component or tubule B, is incomplete and contains only nine protofilaments. Between each pair of doublets is a slender bridge made, at least predominantly, of the protein nexin (Gibbons, 1977; Witman *et al.*, 1978), and on each A component is a pair of projections made largely of dynein, referred to as the outer and inner arms. Both arms are very irregular in form, especially the outer one, but each is generally curved slightly toward the center of the flagellum. Bending of the organelle is considered to occur along a line, the axis, that passes between the two central microtubules. Because of the odd number of doublets, the even number of central microtubules, and a constancy of organization, one doublet is located on this axis, whereas an interspace lies at the corresponding point on the opposing side. To distinguish the individual doublets, numbering begins with the one on the axis and proceeds clockwise (Figure 4.5). The diagram follows standard procedures in representing the organelle as viewed from the base toward the tip.

Some new discoveries about the internal structure of the apical region of the flagellum of *Chlamydomonas* and *Tetrahymena* have been made, but are too recent for similar researches into other organisms to have been carried out.

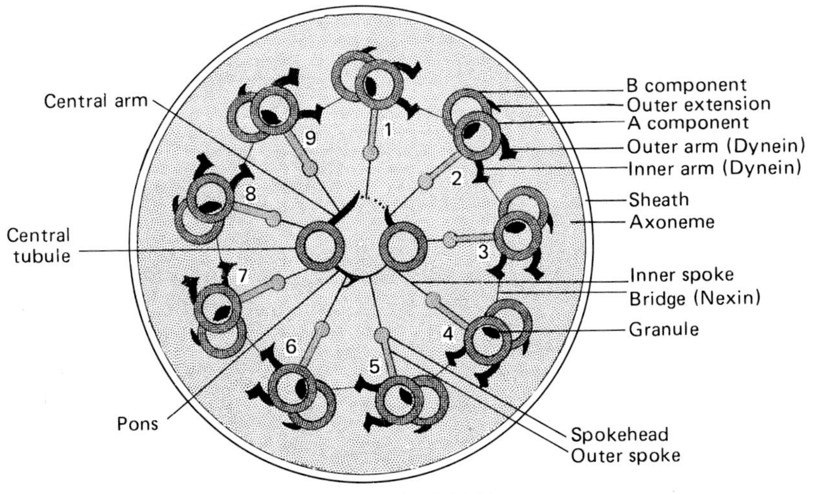

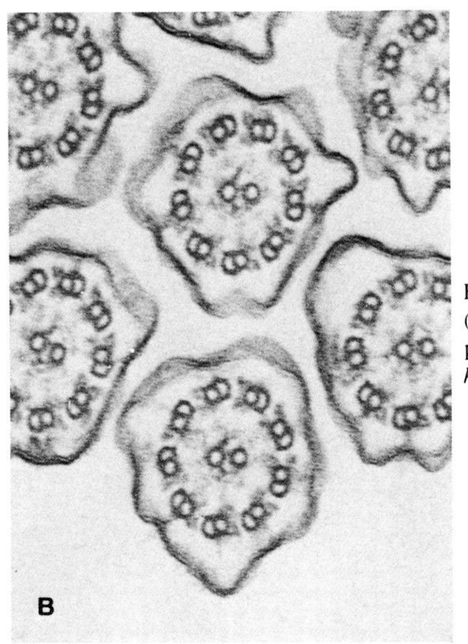

Figure 4.5. The internal structure of a flagellum. (A) Diagram showing the major structural components. (B) Transverse section of cilia of *Tetrahymena*. 100,000×. (Courtesy of Allen, 1968.)

Perhaps the most important of these is the observation that a cap covers the central pair of microtubules and attaches their distal ends to the tip of the flagellar sheath (Dentler and Rosenbaum, 1977). This cap is of complex organization and appears to consist of a ring from which a pair of extensions project that in some views can be seen to run proximally on the sides of the microtubules (Figure 4.6A). Distad to this can be noted a slightly smaller ring, upon which rests an apically indented bead. In this same region the doublets, too, are slightly modified, for each such set of microtubules at their ends bears several distal filaments (Figure 4.6B). These are attached to a spherical plug that fits into the distal lumen; in addition several filaments extend proximally from the sphere (Figure 4.7; Dentler, 1977). The presence of these several features should be borne in mind in connection with the discussions of flagellar movement and growth, especially the latter.

Less Constant Axoneme Features. A number of additional features occur repeatedly in the flagella of the eukaryotes but vary in form or presence among the various taxa. Unfortunately some of these have been included in published figures purportedly showing the ''standard'' structure of this organelle. One of these is the bridge between doublets 5 and 6, in which the dynein arms are longer than usual so as to form a connection between the A component of doublet 5 and the B component of the succeeding one (Figure 4.5). This feature,

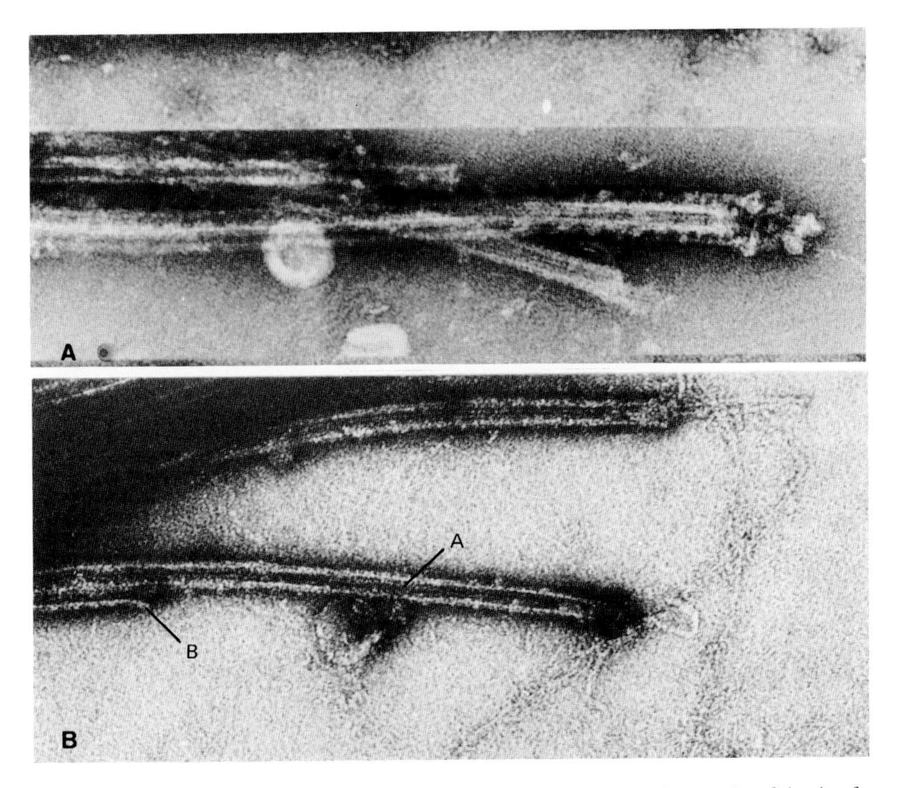

Figure 4.6. Structures at the distal end of the flagellum. These electron micrographs of the tip of a *Chamydomonas* flagellum reveal some remarkable structures. (A) The pair of central tubules bear a complex cap at their apex; nearer the base at the extreme left may be seen fragments of the central arms and pons. 80,000×. (B) On the A component of the doublet may be noted the distal filaments that terminate the microtubule; the slightly shorter B component lacks any special apical features. 120,000×. (Courtesy of Dentler and Rosenbaum, 1977.)

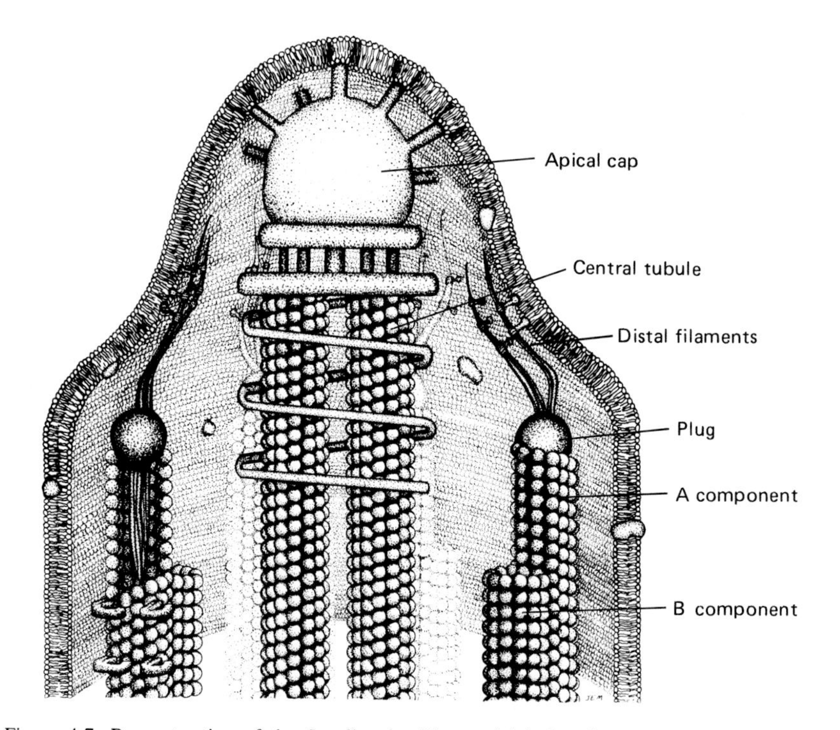

Figure 4.7. Reconstruction of the flagellar tip. The model is based on negative-stain and thin-section preparations of *Chlamydomonas* and *Tetrahymena* flagella. (Courtesy of Dentler, 1977.)

however, is of nearly constant occurrence only in the gill cilia of pelecypod molluscs, and even in those organisms, the inner dynein bar is frequently reduced to the usual arm (Gibbons, 1961; Satir, 1963, 1965, 1974). Sometimes a bridge is found in the same location in the spermatozoans of insects, such as that of the house cricket (Kaye, 1964); however, it is not in evidence in the electron micrographs of the sperm of a lepidopterous insect (André, 1961) and therefore does not characterize the insects as a group. In the instance cited, it appears to involve the inner arm rather than the outer. Recently, the cross-bridges made of these arms have been shown to be temporary structures that seem to form and break during flagellar beating (Warner, 1978).

Another structure that varies more than has been previously noted is the so-called inner sheath. In the first place, the name is a misnomer, for it neither encircles the entire central apparatus nor is it a continuous sheet as that term implies. Its longitudinal structure is more correctly visualized by Satir (1974), where it is shown as a series of rings, but more typically, rather than being a ring, it is a low arch that spans only one side of the gap between the central pair of microtubules. Because it is thus a bridgelike structure, it should be termed the inner bridge, or better, the pons, to avoid confusion; when present, it is on the side toward doublets 5 and 6. The presence of this pons is clearly shown in the sections of flagella of *Trichonympha* (Grimstone, 1963) and *Pseudotrichonympha* flagella (Gibbons and Grimstone, 1960), of the cilia of lamellibranch gills (Satir, 1965), in sperm flagella of a lepidopteran (André, 1961), leafhoppers (Phillips, 1969), and *Zamia* (Norstag, 1967). It does not show in the electron micrographs of *Giardia* flagella (Friend, 1966) and is modified in *Tetrahymena* cilia (Allen, 1968) by the presence of an outwardly projecting bar that probably corresponds to the midfiber of molluscan cilia (Gibbons, 1961). In the cilia of the swimming plates of ctenophores, it bears a prominent thickening at its center (Afzelius, 1961).

On the opposite side, where in diagrams the inner sheath is usually indicated to continue, are structures that are even more variable than the pons. In the cilia of pelecypods, a stout arm on each central tubule is placed parallel to the axis. The apices of the appendages of this pair are sometimes connected to one another by a high arch, as shown in electron micrographs (Satir, 1965, 1974), but most usually each is a separate entity. As this arch is only occasionally indicated in electron micrographs, while the arms are a consistent feature, it follows that the latter are nearly continuous structures along the length of the inner microtubule and the arch is formed of a fine thread placed at intervals.

One more characteristic of the axoneme, the spokes, similarly varies from taxon to taxon. In electron micrographs of insect sperm and mollusc cilia, rather broad streaks may be noted to extend from the A component of the doublets inward to end in a spoke head, formerly known as the intermediate filament (André, 1961; Kaye, 1964; Satir, 1965, 1974; Phillips, 1969). Rarely, however, one or two in a section may be seen to continue inwards to a central tubule, the pons, or an arm. Quite the opposite condition appears to prevail among protozoans, including *Pseudotrichonympha* (Gibbons and Grimstone, 1960), *Ochromonas* (Chen and Haines, 1976), and euciliates (Pitelka, 1961), and in the sperm of higher plants like *Zamia* (Norstag, 1967), for the inner segment of the spokes is almost always more distinct than the outer portion. In longitudinal sections of flagella, these spokes are sometimes seen to be arranged in pairs or triplets connected at the heads by fine fibrils (Figure 4.8). In *Chlamydomonas* the arrangement is by pairs, but in *Tetrahymena* and metazoans, the spokes are arranged in triplets (Hopkins, 1970; Witman *et al.*, 1978).

The Flagellar Sheath. Although relatively little attention has been devoted to the sheath of the eukaryotic flagellum, what has been learned clearly demonstrates that it is not merely a continuation of the cell membrane. Furthermore, it is not identical in construction throughout its length. Freeze-fracture investigations in which the flagellar membrane near the proximal end is split horizontally between its inner and outer lamellae show patterns, or plaques, of elevated

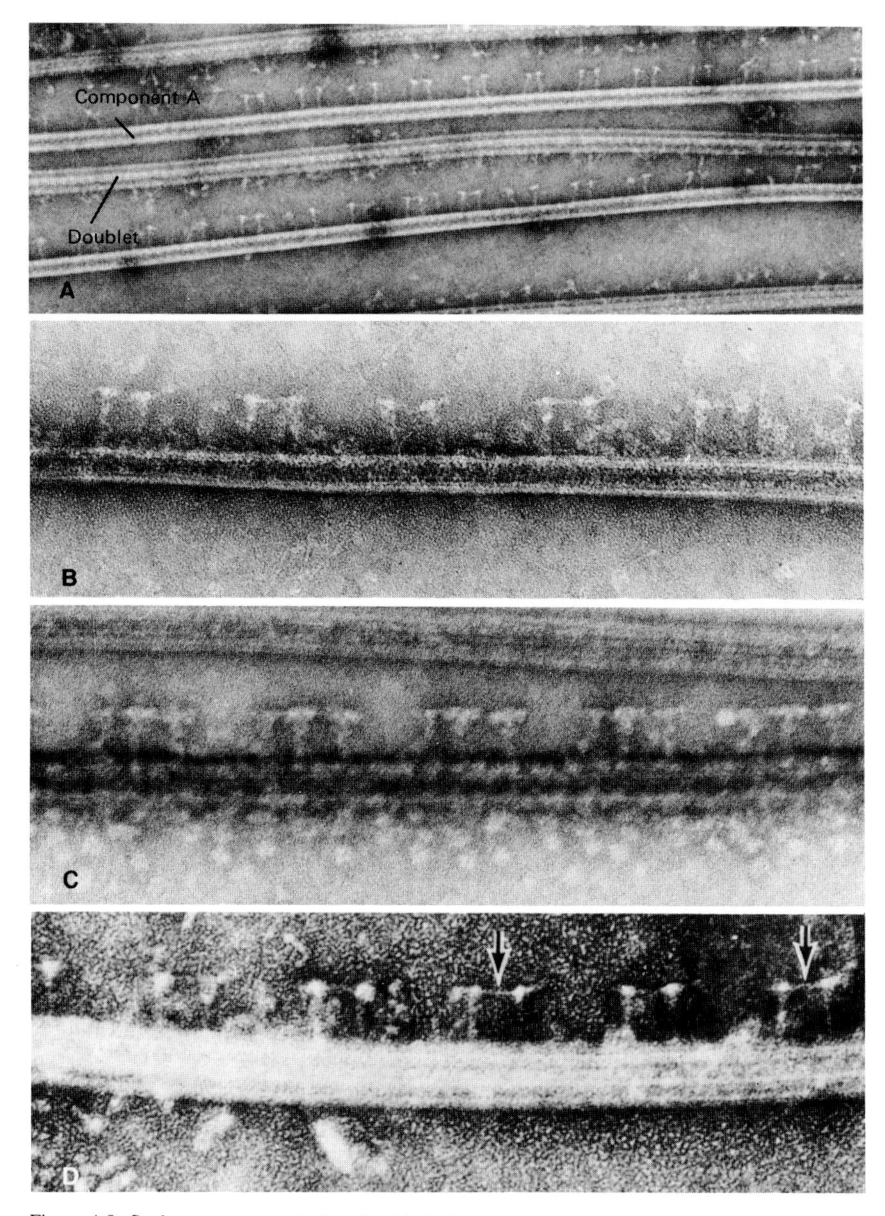

Figure 4.8. Spoke arrangement in flagella. (A) Isolated A components and intact doublets from the flagellum of *Chlamydomonas*. The spokes and spoke heads are arranged in pairs, interconnected by a fine filament. 79,000×. (B) Same but 210,000×. (A and B, courtesy of Hopkins, 1970.) (C) Isolated doublet of *Tetrahymena,* with spokes toward top of micrograph arranged in triplets, the left-hand member of which appears to be constructed differently than the other two. Dynein arms are seen on the lower surface. 225,000×. (Courtesy of Warner, 1978.) (D) Occasional pairs of spokes in *Chlamydomonas* may show apparent remnants of a former third member. 200,000×. (Courtesy of Witman *et al.*, 1978.)

particles on the inner component and corresponding pits on the outer, much like those described in Chapter 1 (Section 1.2.3). This plaque region is in the form of quadrate patches in *Paramecium* (Dute and Kung, 1978) and of a series of three undulating lines in mollusc gill cilia (Satir, 1974). Then distal to the basal area is a constricted region where these structures are lacking; this is followed in turn by an extensive apical region, where the granules and their matching pits are very unevenly distributed and the inner surface has a rough or crenulated appearance.

Studies on the chemical composition of the sheath have yielded some unexpected results. In the flagellar membrane of *Ochromonas,* neither phosphatidylcholine nor phosphatidylethanolamine were detected, while chlorosulfolipids made up the dominant fraction, occurring to an extent of 71 mole % (Chen *et al.,* 1976). These substances are unsuited for forming typical "bilayer" membranes, because, being highly polar, they are readily water soluble. Thus one ionic group of the molecules would be too deeply situated in the hydrophobic region of a bilayer membrane to provide stability, nor would the molecules be able to loop back to permit both polar groups to lie on a polar surface of the membrane.

In *Ochromonas danica* sheath, which has a diameter of 80 Å, five major proteins were detected, having molecular weights of 54, 47, 35, 31, and 28×10^3 (Chen and Haines, 1976). In addition, there was another major protein too heavy to be isolated by the procedures employed. Great uniformity of structure appeared to prevail in the flagellar sheath, whereas the cell membrane varied widely; thus the former was assumed to play an important role in flagellar activities. Electrophoretic analyses of the flagellar membrane of *Chlamydomonas* revealed a major heavy glycoprotein having a molecular weight greatly in excess of 170,000 (Witman *et al.,* 1972), which may correspond to the unidentified one of *Ochromonas*.

In a recent investigation, the sheaths of the gill cilia and those of the sperm flagella from the scallop *Aequipecten irradians* were analyzed comparatively by gel electrophoresis (Stephens, 1977). The flagellar sheath contained one major protein that accounted for nearly 70% of the total, a periodic acid–Schiff positive component that had a molecular weight of 250,000. In contrast, the major component of the ciliary sheath comprised around 65% of the total protein present, was weakly periodic acid–Schiff positive, and had a molecular weight of only 55,000. This was resolved into two subunits that comigrated with the α and β chains of doublet tubulin; further tests confirmed that this major component was tubulin, while that of the flagellar sheath was a glycoprotein. Thus it was proposed that cilia differ from flagella in their chemical composition; however, it could equally be that these contrasts reflect tissue differences. Whatever the source of the distinctions may actually be, it is clear that they strongly merit further investigation.

Mastigonemes. Often along the outer surface of the flagellar sheath are one or two longitudinal rows of bristles, formerly referred to as flimmer, but now called mastigonemes. In *Ochromonas* and other flagellates two types were observed, one being tubular and the other fibrous (Bouck, 1971; Chen and Haines, 1976). The tubular mastigonemes, with diameters near 200 Å and lengths under 200 nm, are themselves coated with hairs, called extramastigoneme filaments, while the fibrous ones, which vary from 1 to 3 μm in length but have diameters like those of the tubular type, are devoid of secondary hairs. In *Ochromonas*, five different glycoproteins were obtained from the tubular mastigonemes by electrophoretic analyses on urea-containing acrylamide gels, whereas those of *Chlamydomonas* yielded only one, having a molecular weight close to 170,000 (Witman *et al.*, 1972). Members of the latter genus have only one variety of mastigoneme, and this resembled the fibrous type in lacking extramastigoneme filaments. The mastigonemes have been observed to be formed in the Golgi vesicles, at first without the accessory hairs, the latter being added in the same organelle after the main shaft had been completed. During regeneration of flagella after experimental deciliation, the growing shaft bore mastigonemes from its first appearance as a mere stub (Leedale *et al.*, 1970).

Their precise arrangement is characteristic of the various taxa, as described in detail in a later section, as is also the method of their attachment to the flagellar shaft. In *Euglena* the long flexuous mastigonemes appear to be attached in a single row to the so-called paraflagellar rod on conical projections that reach the surface of the sheath (Bouck *et al.*, 1978). Contrastingly, the short hairs that accompany them are arranged in two helices that encircle the entire organelle, borne on minute elevations of the membrane that have fine extensions to the doublets of the axoneme. The mastigonemes of the flagella of *Ochromonas* have been reported to have attachments to the axoneme fibers by way of amorphous material (Markey and Bouck, 1977).

Atypical Internal Structure. Occasionally investigations into flagellar structure on a comparative basis have encountered atypical morphology in scattered taxa, most of which involved modifications of the central tubules and associated parts, and in some cases mutant individuals have been found. A notable case of the latter sort involved *Chlamydomonas* mutants that completely lacked the central microtubules and the radial spokes (Witman *et al.*, 1978) and were incapable of moving their flagella. Here, however, attention is confined to instances in which functional flagella deviate from the standard structure throughout a large taxon.

Cilia of Ctenophore Combs. In ctenophores the swimming plates are comprised of numerous rows of cilia, tightly packed in hexagonal fashion and all oriented in the same direction. Thin sections of such combs, as the series of plates are called, show several interesting deviant features, one of which, a thickening of the axoneme pons, has already been mentioned. An even more extraordinary trait occurs in the form of two longitudinal membranes that divide the peripheral portion of the axoneme into compartments. Both membranes extend from B components to the inner surface of the sheath, one from that of doublet 3 and the other from that of doublet 8 (Afzelius, 1961). Because of the uniform orientation of the cilia within a comb plate, the membranes of one cilium tend to be aligned with those of adjacent organelles. However, the role of these unique structures in the functioning of the plate has not been surmised as yet.

The Flagellum of Platyhelminth Sperm. Among the first deviant types of flagella discovered were those of sperm of various flatworms, in which a 9+1 pattern prevailed (von Bonsdorff and Telkkä, 1965; Hershenov *et al.*, 1966; Tulloch and Hershenov, 1967). However, the name given to this pattern is somewhat misleading, for no microtubule of any sort can be detected in the central region. In contrast, a study of the details of the anatomy showed that a unique complex structure occupied the entire central region (Silveira, 1969), being a cylinder 700 Å in width and consisting of three zones (Figure 4.9A,B). On the center was an electron-opaque axial zone, with a diameter of 300 Å, that consisted of a bundle of longitudinally oriented elements. Around this was the electron-transparent intermediate circular region that appeared to be comprised of globular particles. Enclosing the whole cylinder was an 80-Å-wide cortical sheath, to which the usual spokes from the doublets were attached. In the electron micrographs the spokes appear more robust than typical (Figure 4.9A), especially near the cylinder, often being broken toward the doublets. Similar modifications have been found throughout the Turbellaria (Silveira, 1973, 1974), in several polyclads (Hendelberg, 1965), and in at least one acoel and one fluke (Burton and Silveira, 1971).

Sperm Flagella of Acanthocephala. The sperm of the several species of Acanthocephala that have been investigated also have proven to have atypical structure, as well as a most unusual developmental history. During spermiogenesis in *Illiosentis furcatus africanus,* the flagellum first appears at the posterior end of the sperm, as is usual among the metazoans (Marchand and Mattei, 1976a). As the spermatozoan matures, the centriole and base of the flagella penetrate into the cytoplasm and begin to migrate anteriorly, a process that continues until that end is attained. After the centriole has disappeared, the flagellum continues to migrate anteriorly without increasing in length. When the processes are completed, the morphological extreme apex of the flagellum is located internally at the cell's posterior end, while the actual base of the flagellum now is at the tip of the anteriorly directed organelle. In other words, the flagellum slides entirely through the cell, so that the tip becomes the base and vice versa.

In these worms the internal structure of the flagellum retains the standard 9+2 construction, but the spokes have undergone modification (Marchand and Mattei, 1976b). Instead of being slender, barely visible lines, they are

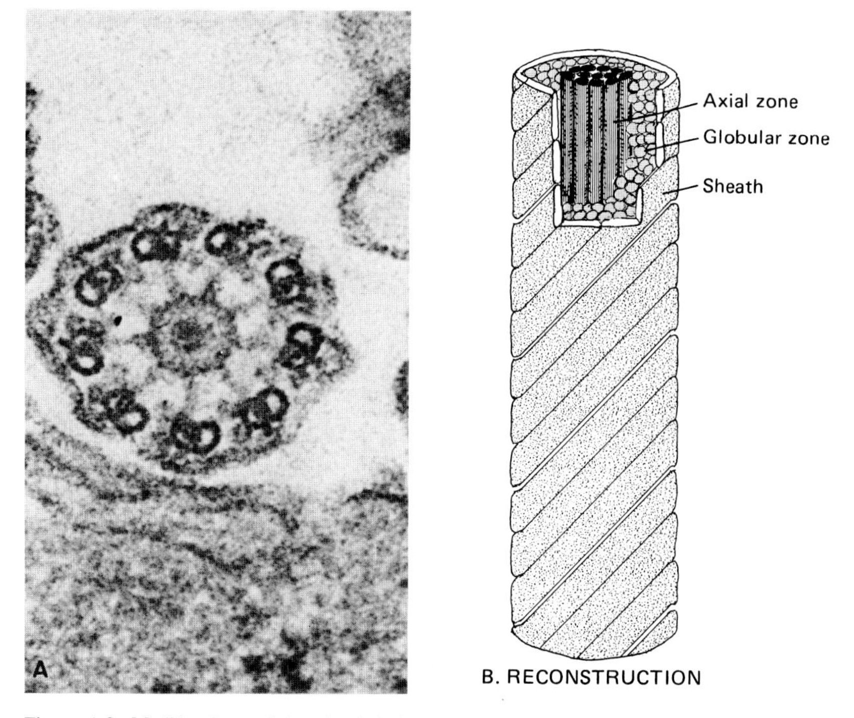

Figure 4.9. Modifications of the platyhelminth sperm flagellum. (A) Electron micrograph of a section through a sperm flagellum of *Dugesia tigrinum*. 160,000×. Instead of the usual two microtubules, an electron-dense cylinder fills much of the central region. (Courtesy of Silveira, 1969.) (B) Reconstruction showing the three regions of the central cylinder. (Based in part on Silveira, 1969.)

greatly enlarged and globular, so as to make almost the entire zone between the central and doublet microtubules quite dark. Neither pons nor arms can be noted on the central tubules. Instead there is a short, electron-opaque bar on each side of that pair of microtubules, which interconnects to two short threads in the interspace between them (Figure 4.10). In another species of these parasites, *Acanthosentis tilapiae,* the spokes are similarly dark and globular, but the central microtubules are also modified (Marchand and Mattei, 1976c). Modifications include a varying number of those structures from 0 to 5 and, furthermore, when present they are thick walled and more electron opaque than the doublets (Marchand and Mattei, 1977, 1978). Throughout this taxon the sperm appear to be so highly diversified in structure that they might very well provide clues to relationships among these highly interesting worms.

Modifications of Insect Sperm Flagella. Although the axonemes of insect sperm flagella as a whole have the general 9 + 2 pattern of other organisms, a number of exceptions are known to exist. Furthermore, an additional feature is common to the majority of the species, in the form of an extra set of nine single microtubules arranged around the periphery (Phillips, 1969). As a rule these are located opposite the interspace between doublets but are not centered there, tending to be closer to the A component of one doublet than to the B component of the next. In the development of the house cricket spermatozoon, they arose in the early spermatid as a long arm extending obliquely outwards from the A components to the sheath. Then after a tubule had been formed in each arm, the peripheral, doublet, and central microtubules gradually became filled with numerous filaments (Kaye, 1964); in the lumen of the peripheral microtubules about 14 were present, and in the inner ones, around 10. During spermatogenesis of Lepidoptera, the peripheral and central elements became elec-

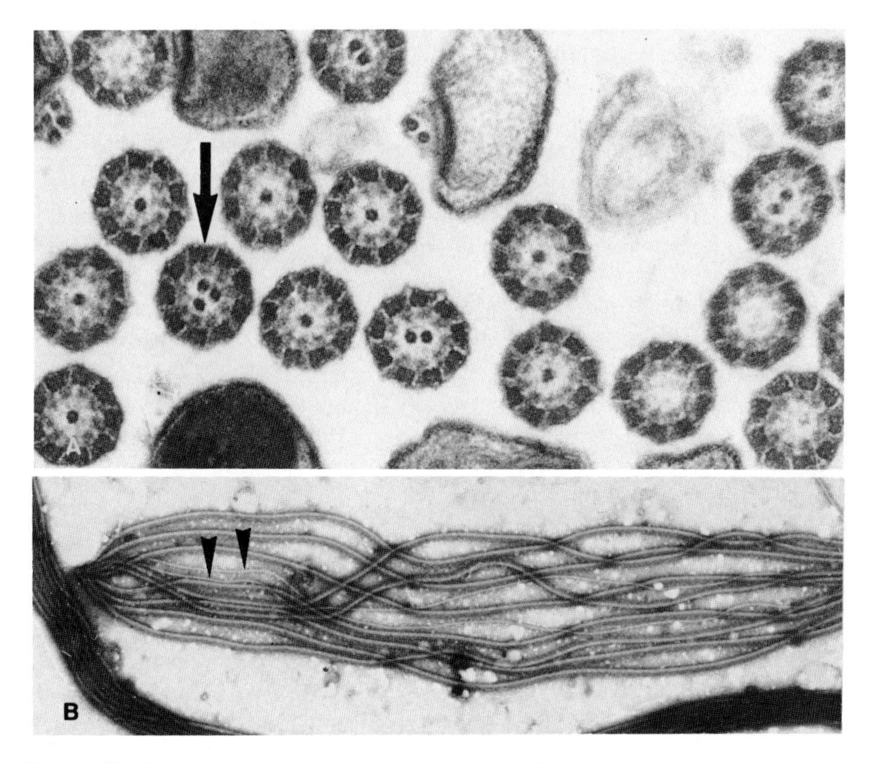

Figure 4.10. Flagellar structure in Acanthocephala sperm. (A) In these sections of the sperm of *Acanthosentis*, the central microtubules as well as the spoke heads are electron opaque. The arrow points to one containing three central tubules. 50,000×. (B) Negatively stained frayed flagellum; arrows indicate the two central microtubules surrounded by doublets. 12,000×. (Courtesy of Marchand and Mattei, 1976c.)

tron opaque, seemingly as a result of similar packing with fine filaments, but the doublets remained tubular (André, 1961).

The central region is subject to other modifications. In the mayflies, for example, the central tubules and associated structures have been shown to be absent, being replaced by a large cylinder (Phillips, 1969). While this recalls the structure found in the turbellarians to a degree, in the present case, only the wall of the cylinder was electron opaque and then only feebly so. Mosquito sperm, too, have lost the central microtubules, but here they are replaced with a single central tubular element that is of somewhat larger diameter than a microtubule and quite electron dense. Contrastingly, caddis fly sperm flagella contain seven dense microtubular elements at the center in place of the usual two.

Vertebrate Modifications. Few modifications of vertebrate flagella or cilia seem to have been reported, aside from those of mammalian sperm tails (Fawcett, 1957), but these appear to be the most widely divergent of all. Internally the structure is similar to that of other metazoans, especially in the central region, but a row of nine single filaments are placed peripherally to the doublets, as in insect sperm flagella. However, the entire peripheral region is greatly widened, so that the total flagellar diameter is about twice that enclosed by the doublets (Figure 4.11). Within this area is a series of eight irregular, coarse bands that spiral around the flagellum. In addition, two smaller inside ribs lie in the plane of the two central filaments and appear to run vertically along the flagellum. Outside the sheath proper is an additional coating that is thickened to form two large vertically oriented ribs (Figure 4.11).

Other Modifications of Flagella. In addition to the changes in the internal morphology of the axoneme, flagella have undergone a number of modifications in both unicellular and multicellular organisms. Many of the specializations possibly have use in locomotion, even though the actual value may be too subtle to be perceived as yet, while a number of others reveal a second function of these organelles, that of sensation.

External Modifications. In the earlier discussion of sheath structure, the presence and chemical composition of the mastigonemes that form two rows on the surface were noted, as was the dense mat of short hairs that covers the remainder of the sheath. As briefly pointed out there, although that pattern is widespread in the living world, a number of deviations exist, correlated to the systematic position of the organism concerned. First of all, the number of rows of mastigonemes may vary; the usual two are reduced to a single row in the members of just two known taxa, the euglenoids and dinoflagellates (Manton, 1952; Brown and Cox, 1954; Pitelka and Schooley, 1955; Pitelka, 1963; Leedale, 1967). Many green flagellates have flagella that entirely lack mastigonemes as well as the mat of short hairs; thus they are completely naked and smooth. The same bare condition prevails in flagellated sperm of higher plants, such as those of mosses and ferns. More primitive forms, like the members of

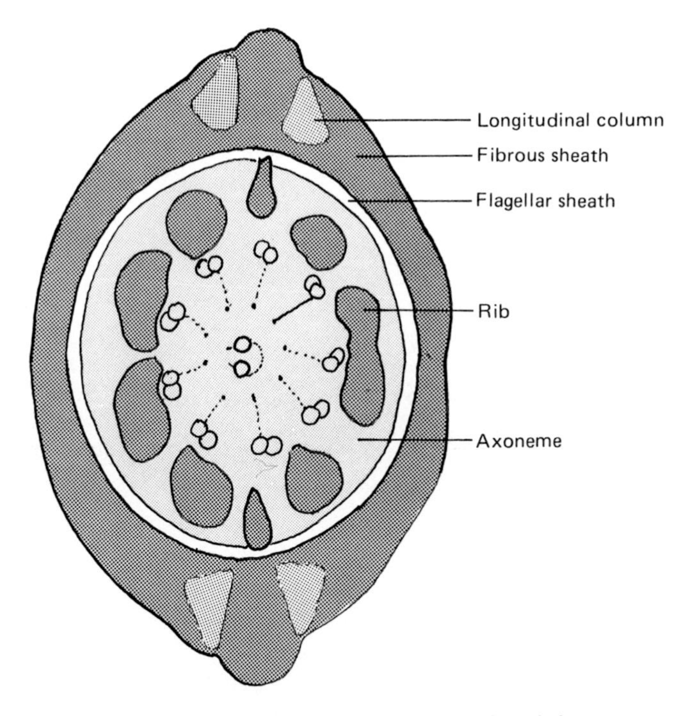

Figure 4.11. The flagellum of mammalian sperm. The usual microtubular axoneme and sheath provide only the core of the highly modified tail of mammalian spermatozoa. The diagram is of a mouse sperm tail.

the genus *Platymonas* that have pectinaceous walls rather than cellulose ones, bear the typical two rows of mastigonemes (Pitelka and Schooley, 1955; Lewin, 1958). Among the chrysomonads, brown algae, and the sperm of aquatic fungi, the fringed and naked conditions occur together. The forward-projecting flagellum of such forms bears mastigonemes, while the more laterally placed one is denuded. Metazoan cilia and sperm flagella also are bare.

The absence of the dense, short hairs in a number of organisms has already been mentioned. In some cases, however, their place is taken by a covering of scales, such as those on the flagella of *Mallomonopsis* (Bradley, 1966) and *Prasinocladus* (Parke and Manton, 1965). Both of these forms have mastigonemes, but in some genera of green flagellates, including *Heteromastix,* those hairlike structures are lacking (Manton *et al.,* 1965). In the chrysophyte *Sphaleromantis* there is a very heavy coating of scales (Harris, 1963), and the sperm of stone worts are similarly covered (Figure 4.12; Turner, 1968).

Besides the variations in occurrence, mastigonemes show other modifica-

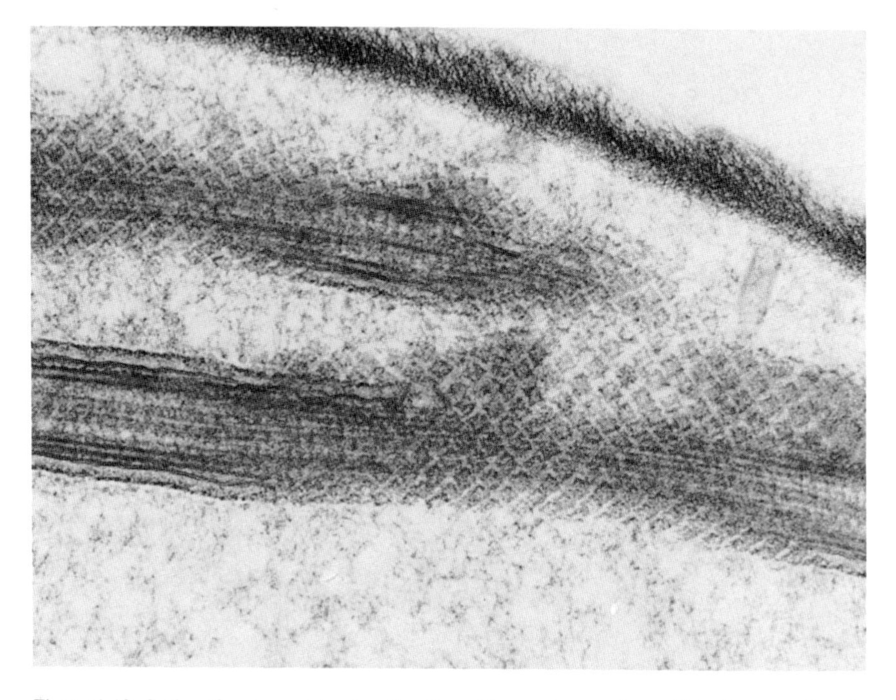

Figure 4.12. Scales of a stonewort sperm flagellum. The grazing sections of *Nitella* sperm flagella reveal their dense covering of fine scales. 57,000×. (Courtesy of Turner, 1968.)

tions that unfortunately have received too little close attention to permit full discussion. In the first place, great differences in relative lengths and shapes can be noted. For instance, in electron micrographs of intact *Euglena* flagella (Leedale, 1967), these fringing hairs are seven times as long as the diameter of the shaft, including the basal mat, are flexuous, and taper in thickness to a fine point distally. In electron micrographs of the *Mallomonopsis* flagellum (Bradley, 1966), on the other hand, the mastigonemes are only two and a half times as long as the shaft diameter and are thick and more or less rigid. Bradley described them as consisting of a tube about 80 Å in diameter around which was a spirally wound layer that brought the total diameter to 400 Å. At the tips were five fine filaments, each about 40 Å thick. Bouck (1969) reported in detail on the mastigonemes of the spermatozoid flagella of the brown seaweeds *Ascophyllum* and *Fucus,* showing them to be complexly structured. The basal region was found to be soft and flexuous, beyond which was a stiff microtubular portion, followed by a long, fine terminal filament. The thicker portion showed a length two and a half times the diameter of the flagellar shaft, a total length made double by the fine filament (Figure 4.13). In thickness, the microtubule measured 200 Å, well within the range of similar structures of the cell body cytoplasm and flagellar axoneme. A similar structure of the mastigonemes may be noted on the flagellum of the diatom *Lithodesmium* (Manton and von Stosch, 1966), the zoospore of the aquatic fungus *Saprolegnia* (Fawcett, 1961), and in the chrysophytan *Sphaleromantis* (Harris, 1963; Manton and Harris, 1966), although the flagellar structure is not discussed. In xanthophyceans the microtubular mastigonemes each terminate in two rather short fibers (Leedale *et al.*, 1970).

Modification of the Tip. In most organisms, including euglenoids, green algae, and metaphytans, the flagellar apex ends smoothly, being simply rounded without prior reduction in diameter. Contrastingly, the flagella of metazoan sperm and motile stages of xanthophytes, chrysophytes, and others that lack chlorophyll *b* are more or less drawn out into a slender process that has

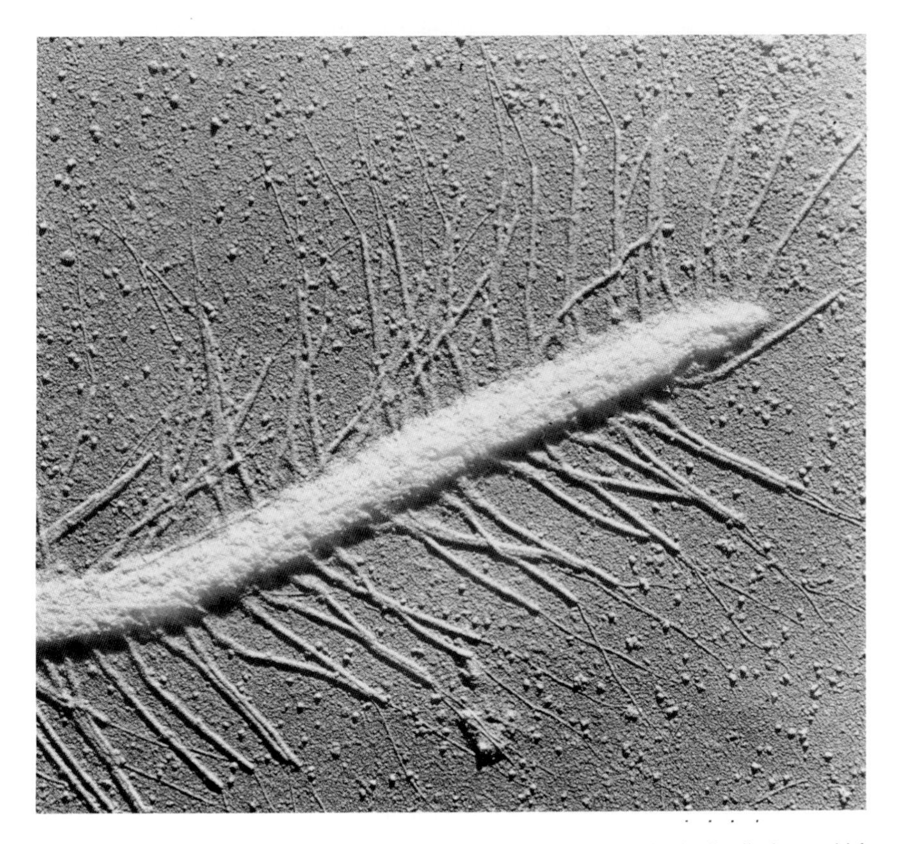

Figure 4.13. Mastigonemes. Each mastigoneme of this flagellum of *Polyedriella* displays a thick microtubular portion above a flexuous base and a single fine filament at the apex. 25,000×. (Courtesy of Hibberd and Leedale, 1972.)

been called a whiplash. The length of the whiplash varies. Among some chrysophyceans, such as *Sphaleromantis* (Manton and Harris, 1966), the apex merely tapers, rather than being prolonged, perhaps representing an elementary stage in the development of this peculiarity. Others, like *Prymnesium* (Manton and Leedale, 1963a), have a very long apical process, constituting more than 20% of the total flagellar length. Sometimes, for example in the xanthophycean *Tribonema* (Massalski and Leedale, 1969), the naked flagellum has an elongate whiplash, whereas that of the fringed one is scarcely indicated. Internally, the doublets are missing from the whiplash, but the two central filaments continue for a distance. In successive sections from the proximal to distal ends, the number of single filaments, at first 11, decreased one at a time until only 2 or 3 remained.

Sensory neurons of vertebrates often bear extremely long flagella, much of the length of which stems from the elongate whiplash. For a case in point, the olfactory sinus neurons of frogs may be 200 nm in length, two-thirds of which is contributed by the whiplash (Reese, 1965). The main shaft has a typical metazoan ultrastructure, but in the narrowed apical process, only singlets occur. Distally the total number present here becomes reduced from 11 to 7 near the end. It is believed that this narrow region beyond the shaft proper is the actual detector of olfactory stimuli.

It should not be supposed that this cited instance represents an exceptional modification of the flagellum; actually all known sensory structures of invertebrates and nearly all those of vertebrates are modified flagella (Brown and Bertke, 1974). Among those that have been demonstrated to be modified cilia are the lamellae of the eyes of scallops (Miller, 1958) and the rods and cones of the vertebrate retina (Willmer, 1955; deRobertis, 1956; Fawcett, 1961). Another unusual modification for sensory purposes has been described from the milkweed bug, *Lygaeus kalmii,* in which modified flagellar shafts were found in certain sense organs located on the antennae (Slifer and Sekhon, 1963). In several cases, the cilium is known to have become adapted for a secretory role. Within a vascular sac in fish brains, the epithelial lining bears crown cells, short club-shaped processes that project into the third ventricle. Originally suspected of being sensory in function, these later were demonstrated to engage in secretion and to have the pattern of internal structure typical of flagella (Bargmann, 1954; Bargmann and Knoop, 1955). Similar secretory modified cilia have been reported in cells of mouse hypophysis (Barnes, 1961); here as in most sensory cilia, the central fibers were absent.

4.2. THE BASIS OF FLAGELLAR ACTIVITY

Along with cellular movement in general, the basis for ciliary and flagellar activity has attracted the attention of biologists for many years, and currently the problem ranks among the more active fields of investigation. Because of the minute dimensions of the organelles and the rapidity of movement—the cilia of gills of mussels complete between 500 and 1000 cycles of beating per minute (Fawcett, 1961)—little real progress was possible until the advent and refinement of the electron microscope and the development of sophisticated preparatory and biochemical techniques clearly revealed the details of flagellar structure. But despite the existence of several concepts of flagellar movement, one of which is widely accepted, much still remains to be learned.

A Contractility Model. Soon after the "universal" pattern of flagellar structure had been ascertained, a theoretical model of the induction of flagellar motility based on that information was advanced (Bradfield, 1955). This made a number of assumptions about the various components that have not yet received substantiation. Among other things, it proposed that the nine outer doublets were contractile and that the central pair was not; moreover, it suggested that the cytosol filling the interspaces between the microtubules was more or less elastic and imparted some degree of rigidity to the entire organelle. Finally, it advocated the idea that the centriole or basal body initiated the beats and that this impulse was conducted by the central pair and doublet microtubules.

According to this model, a beat was initiated by the centriole or basal body under doublet 1 (Figure 4.5). The impulse induced a contraction of this fiber and propagated a wave of stimuli to each adjacent doublet in turn. That is, the impulse spread first from doublet 1 to the two that border it, numbers 2 and 9; from the latter, the signal next spread to 8 and 3. However, the stimulus would also be picked up by the central pair of microtubules and carried by them throughout the length of the flagellum, inducing limited localized contractions along all the doublets in sequences similar to the original one. These contractions thus would produce a wave of bending along the entire flagellum. In the meanwhile, the impulse would continue radially to doublets 7 and 4 and finally to 6 and 5 to produce the recovery stroke and thereby straighten the flagellum. Although today microtubules are not considered to be contractile and probably not conductive, the concept, nevertheless, has considerable value, as intimated later.

The Rotary-Motor Concept. A concept that flagellar movement involves rotation like that described for bacteria is not designed to explain flagellar action among eukaryotes in general but applies only to a certain taxon. The members of this group, the devescovinid flagellates that inhabit the hindgut of termites, have three curved, anterolaterally directed flagella attached to a lateral papilla, plus a single trailing flagellum that arises nearby. The papilla and its associated flagella are in turn located on a "head," marked off on the cell anteriorly by a constriction, at the base of which internally lies the nucleus, partly enclosed in a complex Golgi body. From the latter organelle to the posterior end of the cell extends a thick rod called the axostyle. In swimming the entire head, including the plasma membrane, papilla, and flagella, rotates clockwise as viewed from the anterior end, the axostyle rotating in unison with it (Kirby, 1941–1949; Tamm and Tamm, 1974, 1976; Tamm, 1976, 1978, 1979). Breaking the axostyle by means of a laser beam results in that organ continuing to rotate as before, while the head becomes stationary or rotates at a slower rate. Hence, it appears that the axostyle serves as the motor that rotates the head along with the flagellary structures (Langford and Inoué, 1979). Internally this organ consists of a spiral of parallel microtubules extending the length of the rod (Figure 4.14), but what induces the rotation remains unknown.

The Sliding-Filament Concept. A concept designed to explain eukaryotic flagellar movement in general, currently widely accepted, is known as the sliding-filament model (Brokaw, 1972). In part, the theory was originally based on the proposal that muscle contraction resulted from the thick and thin filaments sliding over one another (Hanson and Huxley, 1955), but a number of other observations entered into its formulation. That the basal apparatus of the cell did not seem to be involved was ascertained by separating intact flagella from the cell body. Such flagella swam independently with nearly normal movements until their energy supply was exhausted (Brokaw, 1961, 1963, 1965); thus it was taken as evidence that the movement resulted from endogenous mechanisms within the organelle and not from structures within the cell body. Furthermore, to maintain the waves of bending that characterize flagellar activity at constant amplitude from base to apex, the propelling processes must be distributed along the entire length of the flagellum.

That ATP was involved in movement was shown by those flagella whose sheath had been damaged or removed, for such organelles can be made to beat again, that is, can be reactivated, by the addition of the nucleotide to the medium (Hoffmann-Berling, 1955). According to the concept, the energy from ATP is converted to movement by a mechanochemical mechanism that includes the sliding of one doublet of the flagellum along an adjacent one or, in some versions, of an A component along the B component of the same doublet (Satir, 1968, 1974). Before an analysis of this concept is presented, it is first essential to review some details of structural changes that occur within active flagella and to compare the movements of these organelles to those of cilia. Then it will be shown that bending might produce sliding between the filaments, but sliding cannot produce bending. In the meanwhile three essential postulates of this model should be borne in mind: (1) the filaments are firmly attached at the base, (2) they are completely resistant to changes in their lengths, and (3) they are separated by a constant distance.

The Waveform of Flagellar Movements. Many observations by cinematomicrography or by multiple-flash exposures have been made on spermatozoons and flagellated eukaryotes to observe the speed and form of their undulations. In the sperm of a sea urchin, the waves traveled at a velocity close to a millimeter per second. In form the undulations were sinusoidal, consisting of a series of nearly semicircular arcs separated by short, straight lines (Ma-

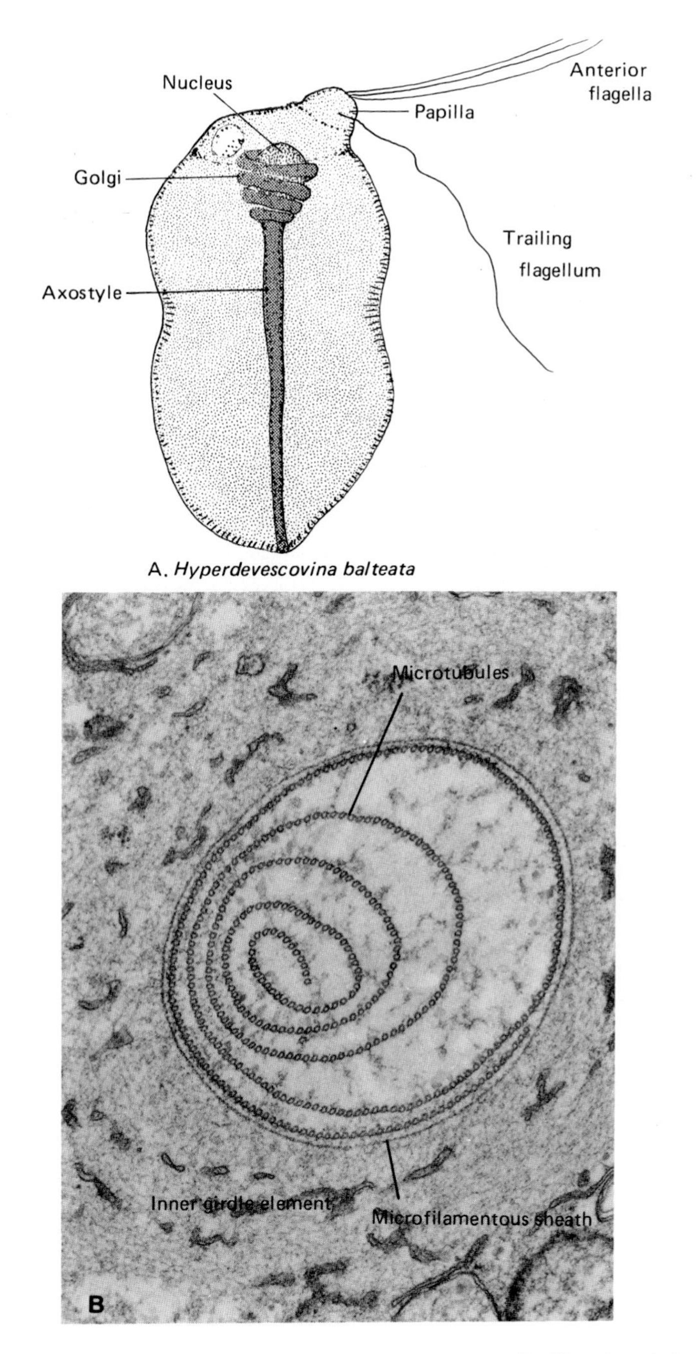

Figure 4.14. Axostyle structure. (B) Transverse section through axostyle. The microtubular bundle is surrounded by a sheath of microfilaments and by two membranous girdles, only the inner one of which appears here. 37,300×. (Courtesy of Tamm, 1978.)

chin, 1958; Brokaw and Wright, 1963; Brokaw, 1972; Brokaw and Simonick, 1976; Pommerville, 1978). Among a number of protists, the flagellum is held forward and pulls the organism along. To accomplish this, sinusoidal waves may be propagated along the organelle beginning at the tip and progressing posteriorly, as in *Mastigamoeba* and the trypanosomes (Bovee *et al.,* 1963; Jahn and Fonseca, 1963). However, in those forms like *Ochromonas* and innumerable others in which the forward-projecting flagella bear mastigonemes, waves are propagated from the proximal end anteriorly to the tip, against the direction of travel. In these forms the mastigonemes provide the necessary force as the sinusoidal waves produce changes in their orientation (Figure 4.15; Jahn *et al.,* 1964). What has not been taken into consideration is the recovery stroke of the mastigonemes, which is of equal importance in the net propulsive forces of the apparatus. As shown in the illustration, recovery of orientation occurs while the forward-moving mastigonemes lie within the inside bend of the waves, with their tips in contact and where they can exercise little backward movement to the system. In other words, there is a radial effect, producing greater movement and thus greater propulsive force on the outside of the bend than on the inside.

Before leaving this subject, it should be noted that those organisms like

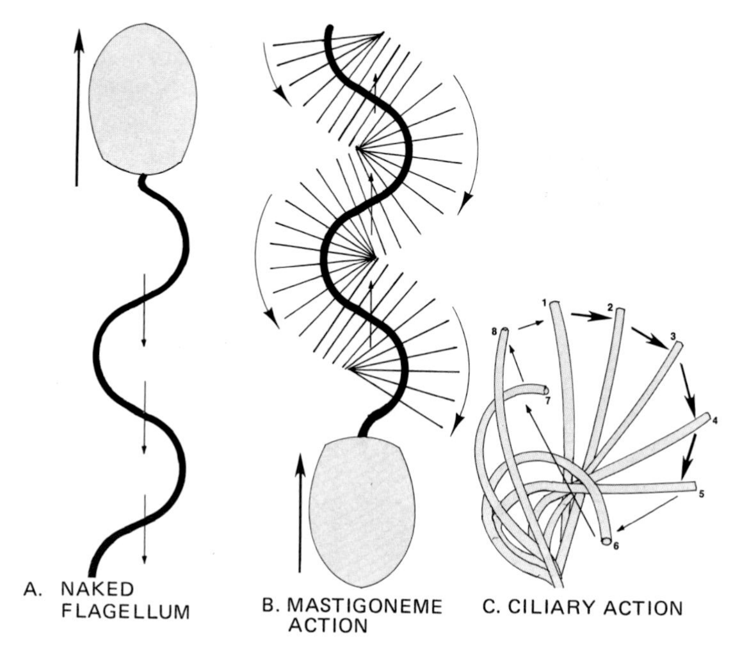

Figure 4.15. Action of flagella and cilia.

chrysomonads and brown seaweed zoospores that have one flagellum equipped with mastigonemes and one naked have had to undergo almost no evolutionary changes in the wave-propagating mechanism in deriving their present flagellation from the ancestral condition. In the primitive arrangement of two forward flagella bearing mastigonemes, waves would have been propagated distally, which direction of movement is conserved in those of these taxa. On the contrary, those green algae whose flagella lack the mastigonemes and are forward-directed had to acquire a wave-producing mechanism synchronized in the opposite sense. Thus, much more is involved in changes of flagellar structure than mere loss or gain of parts.

The Beating Movements of Cilia. Nor are cilia merely shortened flagella. As pointed out in a preceding section, the major protein of the ciliary sheath is a special type of tubulin, while that of flagella is a glycoprotein (Stephens, 1977). Moreover, although they retain an identical internal morphology, their pattern of beating is entirely distinct, being much more complex, especially in metazoans where they have been studied more closely. The effective stroke is direct and simple, involving the entire, more or less stiffened organelle, and is produced by a single abrupt bend close to the base (Figure 4.15C). Recovery, however involves a rolling motion starting at the base; a deep bend then progresses toward the distal end until finally the cilium is fully erect, with the shaft straightened, ready for another effective stroke (Figure 4.15C; Bradfield, 1955; Satir, 1974). Typically many cilia are present on cell surfaces where they occur, as in *Paramecium,* and on metazoan tissues, including the gills of molluscs. In each case, the beating is not synchronous with all effective strokes in unison, but is metachronal so that the effective strokes follow one another in orderly series (Satir, 1963). This provides a continual current or movement, rather than an intermittent jerky one. Whether the interconnecting system of fibers between cilia of the euciliates coordinates their beating is still an unsettled question (Sleigh, 1962), and even less is known about the coordination of the metazoan ciliary beating (Eckert, 1972).

Roles of Specific Axonemic Components. In investigations of the roles in beating of the several component parts of axonemes, it is common practice to sever the flagella from the cell bodies and then to remove the sheath by treatment with 0.05% Nonidet P-40 (Witman *et al.,* 1972; Allen and Borisy, 1974). When such denuded axonemes are treated with trypsin and ATP and observed with dark-field illumination, they may be seen to beat for a while and then suddenly to become greatly elongated or to disintegrate into a number of thin strands connected together at one end. If such thin strands and elongated axonemes are negatively stained and viewed under the electron microscope, they are found to consist of individual doublets or small groups that had separated presumably by sliding forces between adjacent outer doublets. This sliding apart can often be watched by light microscopy (Summers and Gibbons, 1971, 1973; Witman *et al.,* 1976). If the prepared axonemes are placed in a solution containing ATP but no trypsin, they begin to beat and are said to be reactivated, as pointed out earlier.

Several mutant forms of *Chlamydomonas* that lacked the central tubules and associated structures and one that had no radial spokes have been experimentally studied (Witman *et al.*, 1976). In such forms trypsin plus ATP treatment resulted in the normal sliding reactions, but axonemes of none of these were able to be reactivated. Thus these results seemed to support Warner and Satir's (1974) suggestion that radial spoke and central sheath (the pons) interactions play an important part in converting the sliding of doublets into local bending. However, it could equally be that some essential enzyme, possibly an ATPase that has been found located in the central area of wild-type organisms (Gordon and Barnett, 1967; Burton, 1973), might be missing along with the structural parts in these mutants. In these same mutant forms, the nexin links between doublets were found to be intact. Because sliding in disintegrating flagella did not occur until after trypsin had digested the nexin, those bridges were considered to be responsible for limiting the amount of sliding that could occur in normal intact organelles. Thus it is apparent that much still remains to be learned of this difficult subject of the specific functions of the individual structural parts.

The Dynein Arms. Much progress has been made in studies on the molecular nature of the dynein arms, but as the results are often contradictory, no clear picture of structural organization or mode of functioning has as yet emerged. As already pointed out (Section 4.12), dynein was originally described from sea urchin sperm flagella as an ATPase having a molecular weight of 600,000 (Gibbons and Rowe, 1965; Claybrook and Nelson, 1968) and was found to reconstitute the arms of doublets from those flagella and those of *Tetrahymena* (Shimizu, 1975). At first the protein was thought to consist of two subunits, variously designated as A- and B-, or α- and β-dynein (Linck, 1973; Burns and Pollard, 1974), but later Gibbons and co-workers (1976) distinguished five bands electrophoretically and the Warner laboratory (1977) separated six bands in the dynein fractions from molluscan and protozoan sources (Figure 4.16).

Most of these appear to be subunits of the native protein, however, for one species, dynein-1, from sea urchin sperm sedimented at 21 S, corresponding to a molecular weight of 1,250,000 (Gibbons and Fronk, 1979), whereas those of the above-mentioned multibanded separations ranged from either 600,000 or 360,000 to 300,000. The native dynein-1 was reported to consist of three subunits of approximate molecular weight 330,000, plus one each of subunits of 136,000, 95,000, and 77,000. This dynein was demonstrated to be the functionally active form of the enzyme in this organism (Gibbons and Gibbons, 1979). Quite a different approach to the problem was taken by Baccetti and his colleagues (1979), who extracted the dynein from organisms that naturally lack either the outer or inner set of arms, and are also devoid of radial spokes and

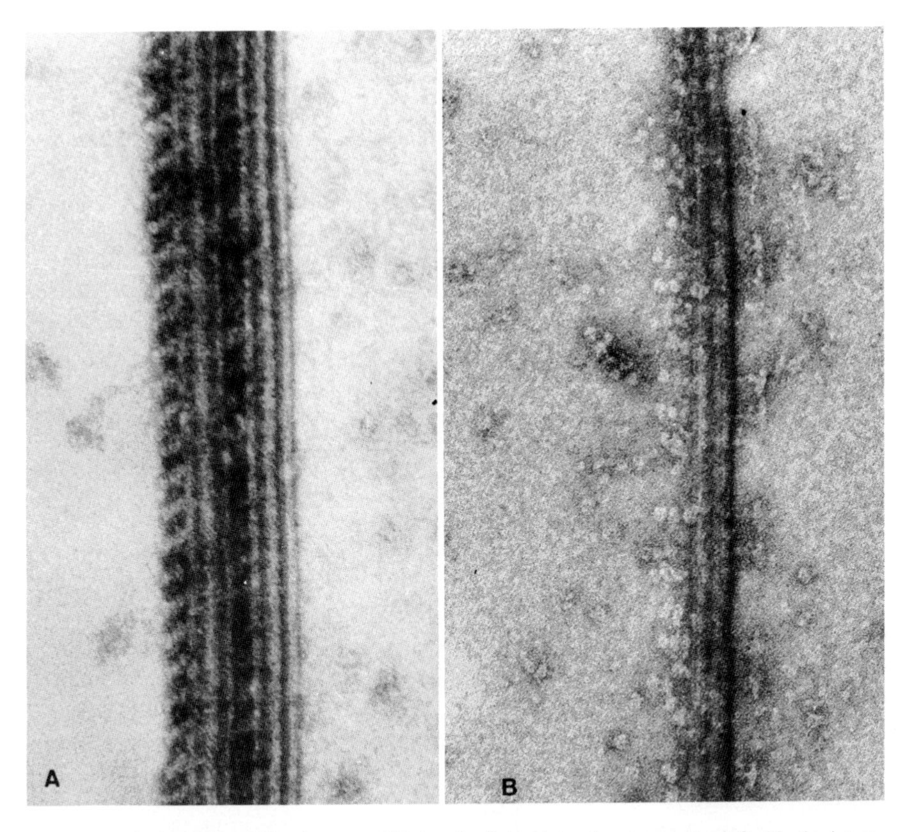

Figure 4.16. Structure of dynein arms. (A) A pair of doublets, showing on the left side the intact series of dynein arms, each of which slopes toward the base of the flagellum. 225,000×. (B) The subunit structure of the dynein arms on the left side of this doublet is clearly visible; taken from a *Tetrahymena* cilium. 203,000×. (Courtesy of Warner *et al.*, 1977.)

the central pair of tubules. In the gall midges of the genus *Diplolaboncus,* the flagella have outer arms only, whereas in the eels of the genus *Anguilla,* the inner arms alone are found (Baccetti and Dallai, 1976). Axonemic dynein from the insect comigrated with band A of the sea urchin protein, whereas that of the eel comigrated with band B, but in addition had a band D. Thus it was concluded that the outer arms contained A-dynein and the inner ones B-dynein, and that D was a constituent of the γ links. All types showed ATPase activity.

Although usually the dynein arms are accepted as binding only to the A components except during bridge formation, contrary observations have been made with dynein from *Tetrahymena*. In the first place, the molecule is much heavier than that of the sea urchin, sedimenting with a coefficient of 30 S

(Takahashi and Tonomura, 1978). When purified and added to doublets deficient in dynein arms, this protein bound almost equally to the A and B components, forming arms 22 nm long on the former and 24 nm long on the latter, in each case spaced about 22 nm apart along the axes. If ATP was added, only the B-component arms became dissociated from the filaments, and these became reassociated with the B component when the ATP was hydrolyzed. On arranging side by side several doublets that had been complexed with arms on both components and then treating them with ATP, the space between doublets increased from 14 to 17 nm. Another investigation that bears on the question of dynein arm attachment showed that Mg^{2+} ions induced the dynein arms of A components to attach to the adjacent B component (Warner, 1978; Zanetti *et al.*, 1979). This bridge formation reduced the axoneme diameter by 14% (Warner and Mitchell, 1978).

The Sliding-Filament Model: A Reassessment. It must first be pointed out that the sliding-filament model does fit certain solid data, such as the sliding motions of microtubules from disintegrating axonemes, and has provided a valuable framework for investigations for a number of years. However, it is in conflict with too many other observations to remain acceptable. Even at the time of its original formulation, it was in direct opposition to some facts that, strangely enough, were offered in its support. In that presentation, it was pointed out that the doublets were united at their bases (Brokaw, 1972), but the question has never previously been raised as to how two adjacent rodlike parts secured at one end can slide past one another to produce a bend, especially if they cannot change their lengths and must remain at constant distances, as postulated (p. 160).

What seems to have obscured this obstacle to the model's functioning possibly was the diagram that provided the basis for the mathematical equations. In this diagram, similar to Figure 4.16B, the flagellum is shown bent, making it possible to show shear forces between opposite microtubules. But it is one thing to bend two adjacent, basally secured rods and thereby induce one to slide past the other, providing that the resulting respective arcs are unequal, and another to try to slide an immovable rod to produce a bend. The relationships are made clear by Figure 4.17. In Figure 4.17A, that of a flagellum at rest, the opposing microtubules are shown attached at the base and with corresponding parts connected by dotted lines. Now if some exogenous force bends the flagellum as in B, shear forces are generated and the right-hand microtubule slides past its left-hand counterpart, because the latter bends on a longer outside arc. Why Brokaw also introduced an elongating movement into the right-hand unit is not clear, because by his definition no elongation is possible. In the present figure that movement is indicated by an arrow, whose length is somewhat exaggerated for the sake of clarity.

It should also be noted that the resulting shear strains, instead of being relieved by the bend of the flagellum, are actually increased, being much greater

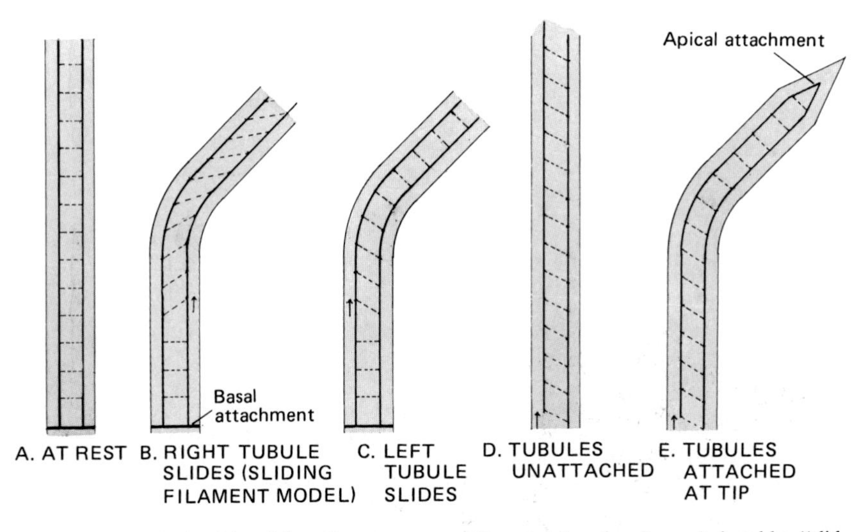

Figure 4.17. Analysis of the sliding-filament concept. Note that the stress forces induced by "sliding" are increased by the typical model of the concept shown by (B). In (C) the bending relieves those stresses. However, in view of the attachment at the base in (A)–(C), no sliding actually is possible. To induce bending by sliding, filaments must both be free at the base and attached at the apex as in (E); however, *in vivo* they are not free at the base.

beyond the bend than proximal to it—quite opposite to what should occur. However, if the left instead of the right microtubule slides, the strains are alleviated by the bend (Figure 4.17C), quite as in any similar natural system. Nevertheless, sliding of any type remains an impossibility so long as the tubules are bound together at their bases, as postulated by the theory; consequently, none of the conditions illustrated by the first three models would actually be capable of producing a bend in a flagellum. Nor would a fourth model (Figure 4.17D), even though the microtubules are shown not to be immobilized at the base and therefore free to slide. For while movement of the tubule is possible, there is no resistance at any point upon which the force can act to induce bending—all such movement can do is to engender shear strains as indicated. Nor could a rigid mechanical device in place of the shear lines be effective, not even series of inflexible transverse rods, for then the former immobile condition would be restored, except that immobilization would be produced throughout the length of the system, not just at the base as previously. Consequently, if bends by moving stiff rods are to be producible throughout the length of the shaft, the opposing microtubules must be secured at or close to the tip of the organelle, as indicated in Figure 4.17E. This, then, is the only model based on sliding filaments that is capable of inducing a bend in a flagellum.

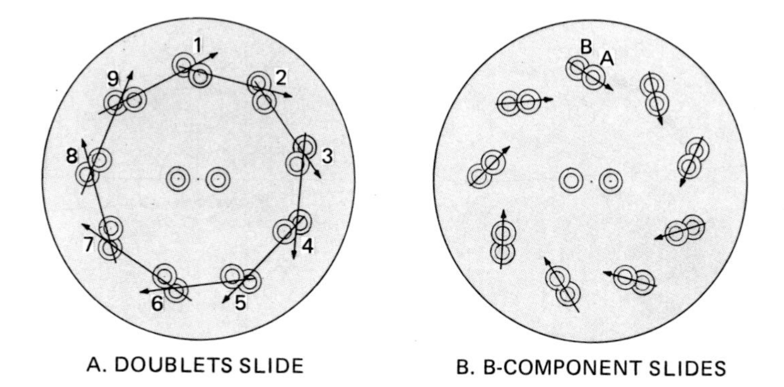

Figure 4.18. Cross-sectional view of generated forces. (A) If a doublet such as 1 slides past either of its neighbors, a rotational force is induced. (B) Similarly, if a B component slides on an A, a like force is generated.

The movements produced by sliding filaments need also to be considered from a cross-sectional view (Figure 4.18). There it may be seen that if filament 1 were to slide past filament 2 or 9, forces could be engendered only in a direction at an angle to the axis, not across it. Were such movements propagated around the entire axoneme, each would be produced at a similar angle to a radius, so the axoneme would rotate around its center, but would not bend. This rotation would ensue from the circular arrangement of the doublets. Similarly, if the A component were to slide against the B component of the same doublet (Figure 4.18), or vice versa, a comparable angular force would result, again producing rotation of the axoneme, but not bending of the shaft. But as pointed out above, sliding filaments are impossible under the conditions stated by the hypothesis.

A Realistic Model. The model presented in the above discussion, too, fails to meet the actual requirements of a beating flagellum. Whereas that proposed system can induce one bend in an organelle of that type, as a rule four bends in two opposing directions are present during flagellar propulsion (Figure 4.19). Moreover, a given sinusoidal bend is not just produced and then simply straightened out, but each bend represents a point in time and space of a traveling wave. Thus two additional reasons for the invalidity of the sliding-filament model become apparent. (1) No single or multiple series of sliding filaments can produce multiple curves in opposite senses simultaneously, and (2) no such system can engender traveling waves of bending and relaxation. Figure 4.19 provides two superimposed points in time and space of a set of waves moving along a flagellum. On it a number of pertinent corresponding segments are suitably labeled so that the changes in shape that occur may be followed. For clarity only a half-wave is shown, but similar changes in the opposite sense would occur during the passage of a complete wave.

If a segment between shear lines located in a straight region is examined first, such as segment 5, the two tubules (X and Z) it contains can be noted to be of identical lengths. However, when that segment is involved in a bend, as segment 5′, obviously the microtubule Z becomes longer than X. It also should be noted that distal to the bend the strain lines are directly transverse, so that no sliding has occurred beyond, nor before, the curvature. Similarly, if a segment is examined that already is in a curve such as segment 20 or 31, the X tubule is seen, of course, to be shorter than Z in the first case and vice versa in the second. But after the bend has traveled beyond that point as at 20′ and 31′, the opposite condition arises, X being longer than Z in 20′ and shorter than it in 31′.

Only two mechanical changes in the microtubules can be visualized to permit waveforms of this sort to be produced and propagated in the observed fashion. First, the microtubules could retain a constant length so that the flagellum assumes a strongly distorted ovate cross-sectional configuration wherever a bend occurs. This would be a consequence of the tubules on the shorter inside arcs of the wave moving adjacent to those on the outer edge in maintaining identical radii. No such distortion has been noted—cross sections of bending flagella are as circular as those of straight ones.

The second alternative thus remains the only one capable of fitting the observed facts—the microtubules themselves must change their length—either the outer ones must elongate to compensate or the inside rods must contract. Or perhaps both elongation and contraction occur, depending on the location of the tubule in the system at that point. Such a system, whether elongating, contracting, or both, would be capable of producing sinusoidal waves, but other local agents could induce the formation of the bend. Nevertheless, regardless of whether the tubules or a surrounding part actually supplies the necessary force,

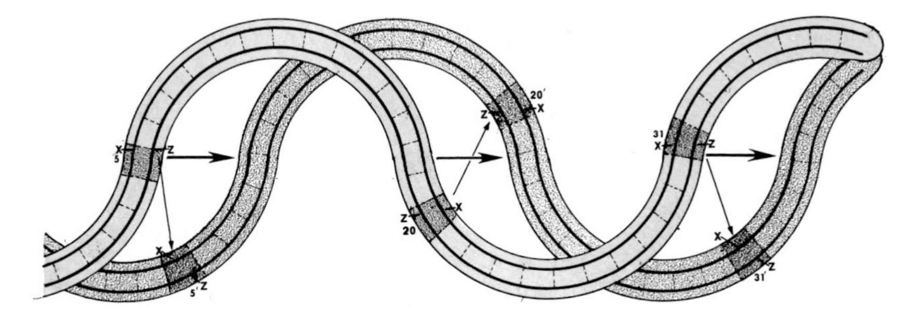

Figure 4.19. Changes in microtubule length during wave propagation. As a wave is propagated along a flagellum, changes in length are necessary in the microtubules on the outside and inside faces of the curves. Compare 20 and 20′, for example.

the microtubules have to elongate or contract rhythmically to maintain undistorted cross sections as a wave passes along a flagellum.* Such elongation or contraction would not necessarily produce a change in the total length of a given doublet for the existence of bends in opposite directions would compensate. Even if the microtubules were spiraled so that a given one would be located at either the outside or the inside of all four bends, it could be fully elongated or contracted for only a fraction of a second, because the wave travels along the flagellum rapidly. However, the microtubules do not spiral in sperm tails nor in mollusc gill cilia, nor do they do so in flagellated organisms. Microtubules had not been thoroughly tested for contractile or elongating properties, and therefore were not known to be able to carry out either such activity; however, they must be able to do one or the other in order to permit flagella and cilia to beat without change in cross-sectional shape of the shaft. Now such contractile properties have actually been demonstrated (Miki-Noumura and Kamiya, 1979). Moreover, although detached flagella are able to be reactivated by media containing ATP, they are not autonomous organelles, for the pattern of wave production in detached organelles is strongly atypical, although even reversal of direction can be effected by means of Ca^{2+} cations (Hyams and Borisy, 1978). Further, when attached, the wave patterns can be reversed and otherwise altered, as becomes evident as the nature of the basal apparatus is examined in the following chapter.

4.3. DEVELOPMENT OF FLAGELLA

Two opposite but related phenomena are involved in the overall picture of flagellar development, the formation of a new organelle and the gradual shortening of an existing one. Preceding cell division, most flagellates lose the flagella, not by shedding the organelles nor by retraction of the intact structure, as once thought, but by resorption of the flagellar constituents into the cell body (Tamm, 1967; Rosenbaum *et al.,* 1969). Shedding, however, can be induced mechanically, as in a homogenizer, or chemically. In contrast, use of such chemicals as pargyline can stimulate flagellar production (Milhaud and Pappas, 1968); this procedure, like synthetic shedding, receives wide use in investigations in this field.

Regeneration Studies. In cells in which the flagella have been resorbed or artifically removed (Kennedy and Brittingham, 1968), electron microscopic investigations show that a short section remains above the basal apparatus, a point of some importance to a later discussion (Figure 4.20A,B). Because this remaining portion retains a complex morphology, regeneration studies do not

*It is also necessary that the doublets on the sides of the bends change proportionately less than those on the outside or inside, or respond appropriately if both elongation and contraction occur.

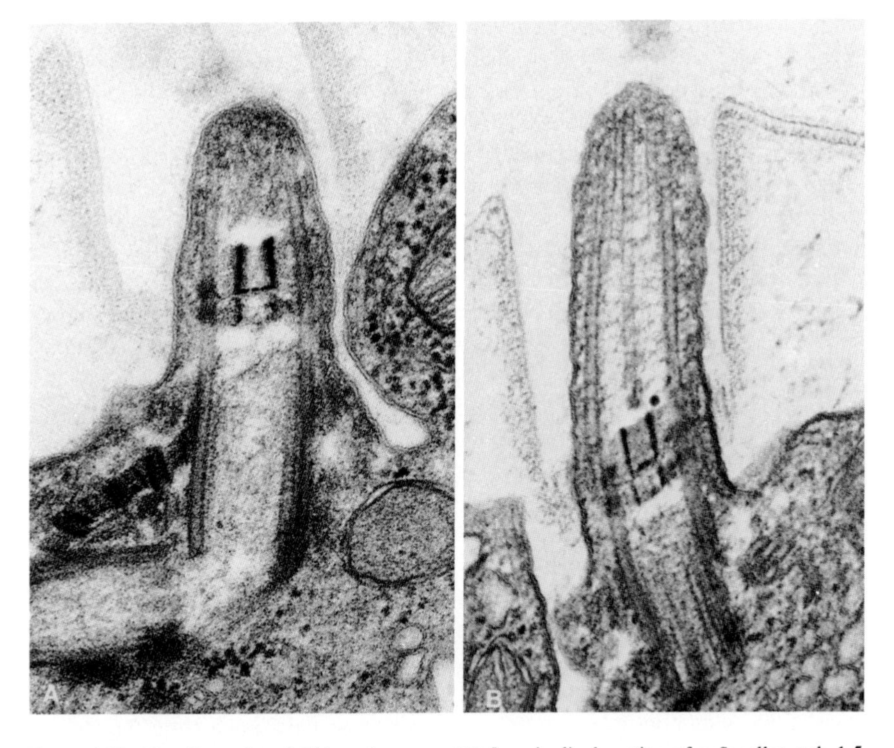

Figure 4.20. Flagellar stubs of *Chlamydomonas*. (A) Longitudinal section of a flagellar stub 1.5 min after amputation. (B) Same, after 4 min. Both 70,000×. (Courtesy of Rosenbaum *et al.*, 1969.)

describe the developmental processes fully, but only those that occur beyond this stub. An especially interesting result of one such investigation (Rosenbaum *et al.*, 1969) was an observation made on specimens that had been induced to lose one flagellum while the other remained intact. As the lost organelle was regenerated, the intact one was resorbed until the two were of equal length, when both began to elongate in unison. The materials that had been resorbed were shown to be employed in the regenerating organelle, after which protein synthesis was essential for further growth. Similar results have been reported from pressure-induced internal axoneme-bearing cells (Brown and Rogers, 1978).

As shown in the discussion of microtubules (Chapter 2, Section 2.1), growth has been found by radioautography to be localized largely at the tips in intact flagella (Rosenbaum and Child, 1967), so that flagellar microtubules have generally been considered to become elongated by additions to their distal ends (Witman *et al.*, 1976). Yet in each investigation about 20 to 35% of the total growth had occurred at the proximal end. Now recently described details of flagellar tip morphology have thrown further light on this topic. It will be recalled that in *Chlamydomonas* and *Tetrahymena,* the two central tubules were shown to bear a cap at their extreme apex, the bead of which bound them to the tip of the flagellar membrane, and that the A components of the doublets terminated in one or more filaments (Dentler and Rosenbaum, 1977). During flagellar growth, these filaments apparently had little effect on the elongating processes, an observation corroborated by *in vitro* assembly of neural tubulin on flagellar doublets. In contrast, the cap on the central singlets was found to inhibit completely the *in vitro* assembly of neural tubulin onto those microtubules. Hence, it was proposed that *in vivo* the doublets became elongated by distal additions, whereas the central microtubules grew by proximal additions, thus also accounting for the radioautographic results.

Artificial amputation of flagella has made possible the revealing of some details of the chain of molecular events that accompany growth. Deflagellation sets into motion the transcription of DNA into mRNAs (Guttmann, 1978) and the translation of the messengers into tubulins, dyneins, sheath protein, and at least 20 other structural proteins (Lefebvre *et al.,* 1978). If the cells were treated with cyclohexamide to block protein synthesis, flagella were regenerated to only half their normal length, seemingly indicating that a pool of flagellar proteins existed in normal cells (Auclair and Siegel, 1966). This pool was viewed as being drawn upon when deflagellation occurred to form new organelles quickly, after which new synthesis carried flagellar growth to completion and then continued to replace the material withdrawn from the steady-state pool. However, in those cases where only one flagellum has been removed from a biflagellated form, it is difficult to perceive why the one left intact was resorbed and its components used in regenerating the second flagellum, as pointed out above. In a following paragraph, moreover, it will be seen that both of these sets of observations are in conflict with results of recent, more precise studies.

Development of Cilia. In fibroblasts and smooth muscle cells from a variety of avian and mammalian sources, development of the cilia is known to proceed within the cytoplasm of the cell. In observations made of these tissues as they underwent differentiation, many of the cells have been noted to possess a short cilium on one of the mitotic centrioles, which later became modified, possibly for service in a sensory function. Growth of the cilium has now been followed by electron microscopic procedures and shown to be divisible into three phases (Sorokin, 1962). (1) In the earliest stage, a vesicle became attached at one end of the centriole, after which a flagellar bud appeared on the same end of that organelle; as the bud increased in length, it penetrated the vesicle, the membrane of which then served as a preliminary sheath. (2) The second stage is occupied by the elongation of the bud into a mature shaft; as this grew the sheath kept pace with it, probably by additions of membranes from secondary vesicles that appeared in the cytoplasm of the cell. (3) The new cilium then reached the surface of the cell and the sheath became joined to the plasmalemma. Later, these stages were followed by alterations, including loss of microtubules and the like, that fitted the cilium for service in its specialized role, but these are without significance here.

A similar investigation of ciliogenesis in embryonic epidermal and tracheal tissues of *Xenopus* yielded comparable results (Steinman, 1968), except that the entire development of the organelle took place on the surface of the cells. After the centrioles had been formed, as described elsewhere (Chapter 5), and had aligned themselves beneath the plasmalemma, a vesicle that contained microtubular elements was soon found over each of them. This vesicular bud then slowly grew into the mature shaft, sheath, and axonemic components, but no details were noted pertaining to where the additions were made.

Development of Modified Flagella. One further aspect needs to be included to provide a more complete view of the total picture of flagellar development, the formation of haptonema, those seemingly deficient flagella, which probably serve some important but unknown specialized function. The fine structure of this organelle has been investigated in the three related genera, *Chrysochromulina, Coccolithus,* and *Prymnesium* (Manton and Leedale, 1961, 1963a,b; Parke *et al.,* 1962; Manton, 1964), and similar structures called axopods have been examined in insects (Bradley and Satir, 1979). Often referred to as a short third flagellum, it differs from typical flagella in many details, yet it displays a sufficient number of resemblances to indicate possible kinship to them. Externally the only visible modification is the presence of a claw or stout papilla at the very tip. In cross sections, however, the axoneme is found to be enclosed by a thick wall that surrounds a cavity, both sides of the wall being covered by membrane (Figure 4.21). The inner side of the cavity is lined by the usual sheath over the axoneme, which is highly modified. In the latter only

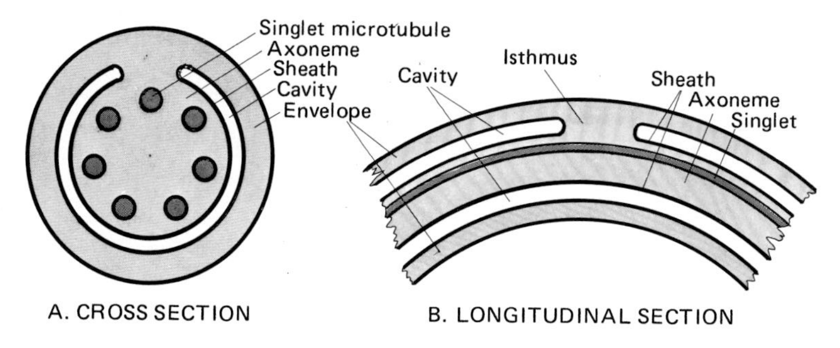

Figure 4.21. Haptonemal structure of *Prymnesium*. Haptonemes are often thought to be short flagella; however, they lack the typical internal structure (A) and are enclosed in an envelope (B). (Based on Manton, 1964.)

singlets are found instead of the usual doublets, and all these are peripheral, the central microtubules being absent. At the very base of the short shaft are nine of these singlets, but apically they gradually diminish to seven. Longitudinal sections through the proximal region reveal internal morphology similar to that of typical flagella, attached to a comparable basal apparatus, which is slightly modified, as described in the following chapter. Thus despite the basal apparatus being largely identical to the usual flagellar centrioles and apparently having been derived from them during its development, it nonetheless gave rise to a strongly aberrant structure. Consequently, formation of doublet microtubules in true flagella may be perceived not to be a mechanical consequence of the mere presence of centriolar triplets; much more than that is obviously needed for the initiation and growth of the doublets, singlets, and other complexities of flagellar organization. This statement receives further substantiation from the development of cilia in the euciliates as a whole, in which one member of a pair of centrioles produces a normal cilium and the other grows only a stub that later degenerates (Hufnagel and Torch, 1967; Sonneborn, 1970; Grimes and Adler, 1976, 1978). Moreover, some of the cilia become modified into cirri or other surface organelles, undergoing such alterations in structure that the existence of cortical genetic influences has been suggested to explain the ciliation patterns characteristic of individual species, genera, and families.

The Multiplicity of Tubulins. As reported earlier, many researches into flagellar growth have assumed that for the development of the microtubular elements, the necessary tubulin was withdrawn from a common pool and subsequently replaced. That this is not the actual case has been clearly demonstrated by investigations on *Naegleria,* an amoeba that undergoes transformation into a biflagellated form. In these organisms whole populations undergo transformation from the amoeboid phenotype to the flagellated one within an hour after the processes have been induced (Fulton, 1970). Instead of cells drawing upon the existing pool of tubulin that constitutes 12% of the total protein in the organism, new tubulin was reported to be synthesized. Thus to form the two flagella that when mature together will contain only 2% of the total tubulin in the pool, transcription and translation are requisite. By use of antiserum and ^{125}I-labeled tubulin from the flagella, the flagellar tubulin was shown to be distinct from that of the cell pool (Fulton and Simpson, 1976), appearing only during transformation to be employed solely in the formation of the axoneme doublets.

Furthermore, a number of other pertinent observations have been made. Among the more outstanding results was the demonstration that the tubulin of the A component of sea urchin sperm flagella differed from that of the B component (Stephens, 1970, 1975; Linck, 1976). Analysis of the subunits disclosed that the α-tubulins of the two components differed in at least 22 residues and the two β-tubulins in 16 or more. No studies have been conducted on the two central microtubules, but in view of the distinctive behavior of their molecules, they too possibly are constructed of a unique tubulin. In addition, the tubulin of ciliary sheaths has been demonstrated to differ from others (Stephens, 1977), as pointed out earlier. And, finally, several distinctive types of tubulin have been described from the cell body, one associated with the membrane and two within the cytosol (Chapter 2, Section 2.1.2). Thus in flagellated cells at least six contrasting sets of cistrons may be necessary to encode the various α- and β-tubulin subunits.

The Genetic Mechanism at the Cellular Level. Perhaps better than any other organelle of eukaryotes, the flagellum brings out the need for studies on the genetic mechanism at the cellular level. While the chapter on the membrane made clear the problem of conducting different proteins and other constituents to the specific membrane and site that required them, each macromolecule involved could be visualized as having marked chemical traits that set it apart from others. Thus the problem of recognition was somewhat mitigated. But the multiplicity of tubulin species, all of which share numerous similar molecular traits, including weight and mode of forming polymers, makes especially vivid the existence of a mechanism for the conduction and assembly of the various cell parts. The genetic processes at the molecular level, concerned with the synthesis of polypeptides and RNAs, while not completely elucidated by any means, have been explored thoroughly, as shown earlier (Dillon, 1978). However, the production of these raw materials, and even their refinement by posttranscriptional processing, important as all those processes are, becomes nearly devoid of meaning as long as knowledge is absent as to how each resulting product reaches its destination and is assembled into functional units.

The various tubulins certainly appear promising candidates as the experimental subjects needed in the required studies. They show marked contrasts in heat lability and other behavior patterns, can be readily localized in cells, and occur in sufficient quantities to make biochemical analyses feasible. Moreover, their assembly into polymers has already received considerable attention. But new researches in this area are essential, in which sweeping generalizations like dynamic equilibrium and other suggestions of autonomy of assembly are laid aside as no longer tenable. Spontaneity of polymerization simply cannot explain how two different α-tubulins, for instance, can each unite with the correct β-tubulin subunits to produce two distinctive fibrous structures that lie in intimate contact, as the A and B components of the axoneme doublets do. Nor will those assumptions explain the attachment at regular intervals of the nexin bridges, spokes, and dynein arms, as required during development of a flagellum, most of which parts are affixed only to one of the components. When it has been shown how the A-component subunits are conducted from the ribosomes through the cytosol and kept isolated from the other types of tubulins that exist there, and transported thence through the centriole to the growing tip of the doublet for assembly, and how the several types of dynein are placed correctly into arms, the nexin bridges added, and the spokes spaced in pairs as they are, then we shall understand flagellar development. And with that knowledge, we shall begin to comprehend the nature of the genetic mechanism at the cellular level.

5

CELL MOTILITY: III

The Basal Apparatus

As Fawcett (1961) has pointed out, the body at the base of the flagellum or cilium has been given many different names, not all of which are strictly homologous to the more general term centriole,* as shown later. Certain ones, like blepharoplast, apply to other types of bodies associated with this apparatus in one or more particular taxa of organisms, and still others, such as kinetosome, which are synonymous with centriole in some groups—in this case euciliates—have other uses elsewhere. Here this organelle is referred to as the centriole, while the term basal apparatus is employed to denote it plus any other structure that may be associated with it at the proximal end of the flagellum.

5.1. THE CENTRIOLE

As indicated in the footnote, the centriole commonly serves during mitosis in many organisms as a nucleation center for spindle fibers. Because the special aspects of the organelle during cell division are more appropriate to the discussion of the latter topic, here only general properties and those pertaining to the locomotory organelle are considered. For a simple-appearing, minute granule, the centriole is surprisingly complex in structure, some portions displaying considerable variation from one group of organisms to another. Unfortunately, no thoroughgoing studies of centriolar structure seem to have been made previously on a broad comparative basis.

* Fawcett employed ''basal body'' as the general term for the granule at the proximal end of the flagellar structure, but as this organelle is essentially identical to that which serves as the polar center for spindle fibers, the name centriole should be applied to both (Henneguy, 1897; Lenhossek, 1898).

5.1.1. Structure of the Centriole

In general, the centriole may be described as an elongate cylinder, closed or nearly so at the distal end, but open proximally. As a rule, flagellar centrioles are three or more times their diameter in length, whereas mitotic ones tend to be only twice as long as wide, a distinction not previously noted (Figure 5.1A,B; also see Roland *et al.,* 1977, for convenient comparative materials). For discussion, it may be viewed as consisting of three regions, the distal transitional zone, the intermediate cylinder, and the proximal dense region (Figure 5.2). The wall of the organelle consists largely of nine sets of triplet microtubules, one member of each (the innermost) being a complete microtubule like the A component of flagellar doublets, the other two resembling the B component in being partial. The number of protofilaments present in the respective components of the triplets has not been determined as yet, but likely will prove to be 13 in the A component and 9 in the B and C components, as in the flagellar doublets.

The Distal Transitional Zone. The distal transitional zone has typically

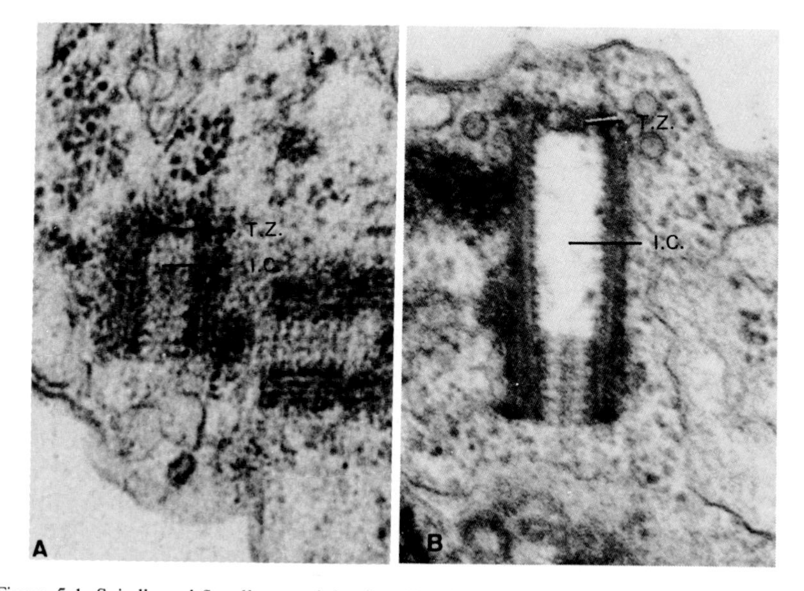

Figure 5.1. Spindle and flagellar centrioles from *Nitella*. (A) In an anaphase cell of this stonewort, these spindle-associated centrioles have just completed division. The body to the right is the daughter centriole. 62,000×. (B) Centriole of the sperm flagellum of the same organism. The greatly enhanced length of the organelle, attained mostly through the elongation of the intermediate cylinder (I.C.), distinguishes flagellar from spindle centrioles. The presence of a transitional zone (T.Z.) also should be noted. 84,000×. (Both courtesy of Turner, 1968.)

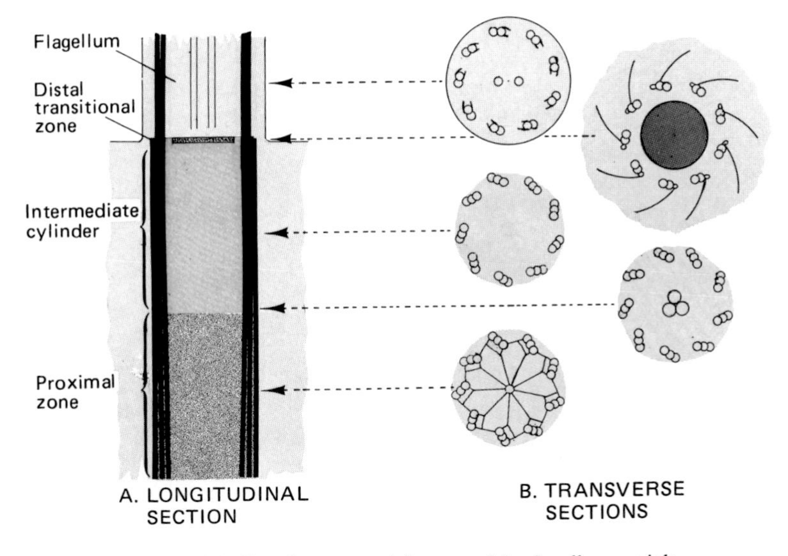

Figure 5.2. The ultrastructural features of the flagellar centriole.

been treated as a part of the flagellum, but as it remains intact on the centriole after the flagella have been resorbed or artificially removed, and is present on newly formed organelles before flagella develop on them, it seems more realistic to view it as here. Moreover, the walls in this region are in great part like those of the remainder of the centriole. Doubtlessly it is the most variable portion of the entire organelle, the morphology being very strongly correlated to systematic relationships. Perhaps the simplest condition is shown by such forms as *Prymnesium* (Manton, 1964), *Bodo* (Pitelka, 1961), *Trichomonas* (Anderson, 1955), *Aulacomonas* (Swale and Belcher, 1973), and trypanosomatids (Clark and Wallace, 1960), in which a distinct plate terminates the entire central region of the flagellar axoneme. This structure, usually referred to as the basal plate of the flagellum (Gibbons and Grimstone, 1960), actually marks the site of transition from that organelle to the centriole, for it is here that the doublets of the former become the triplets of the latter, the two central tubules having terminated just distad to the plate (Figure 5.3). Peripheral to the plate, there is sometimes a thickened rim just inside the sheath, so that the former part appears to extend completely across the shaft, as in *Aulacomonas* (Swale and Belcher, 1973) and *Dinobryon* (Kristiansen and Walne, 1977). In all cases this structure is aligned with the basal constriction of the flagellum referred to in the discussion of structural details of the sheath and is sometimes placed even with the surface of the cell but may be well elevated above it. Rarely is this structure located below the surface as in *Lophomonas* (Beams and Sekhon, 1969), except in such groups as the euglenoids, where the entire centriole is situated in a deeply invaginated continuation of the surface (Leedale, 1967).

Among the several other modifications of the transitional zone is the virtual or complete absence of the plate just discussed, as in the euglenoids and such flagellated amoeboids as *Naegleria*. In the latter, a central electron-dense granule occupies this position instead, from which extends a highly electron-transparent cylinder, enclosed distally by a thin arched membrane (Figure 5.3). This structure extends upwards into the flagellum well past the origin of the two central microtubules (Dingle and Fulton, 1966). Another variation of an opposite type is the presence of two plates here, as exemplified by the euciliate centriole (Favard *et al.*, 1963; Stewart and Muir, 1963; Bradbury and Pitelka, 1965), the thicker being the plate proper. Still another specialization is found in hypermastiginous flagellates like *Pseudotrichonympha,* in which a tall crescentic granule occurs distad to the plate on one side (Gibbons and Grimstone, 1960).

The most striking modification of the transitional zone is found among all members of the green algae and metaphytans that have been examined, a feature first pointed out as part of the flagellum in *Polytoma* by Lang (1963). In transverse sections of the flagellar base, nine-pointed starlike configurations are seen, one point of which extends to each A component of the doublets (Figure

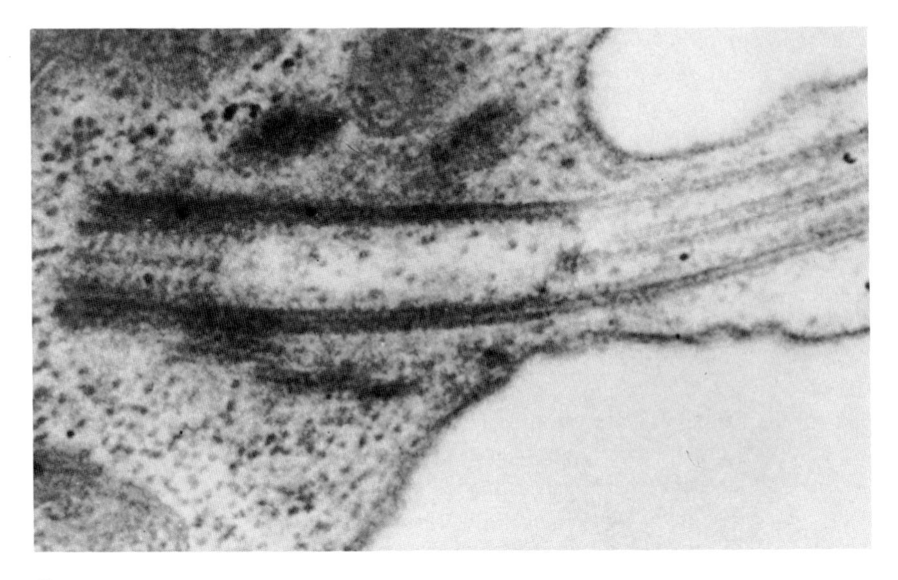

Figure 5.3. Centriole and base of flagellum. This longitudinal section of a *Naegleria* flagellum shows its two central microtubules to end just distad to the basal granule, whereas the peripheral doublets are continuous with the wall of the centriole. 97,000×. (Courtesy of Dingle and Fulton, 1966.)

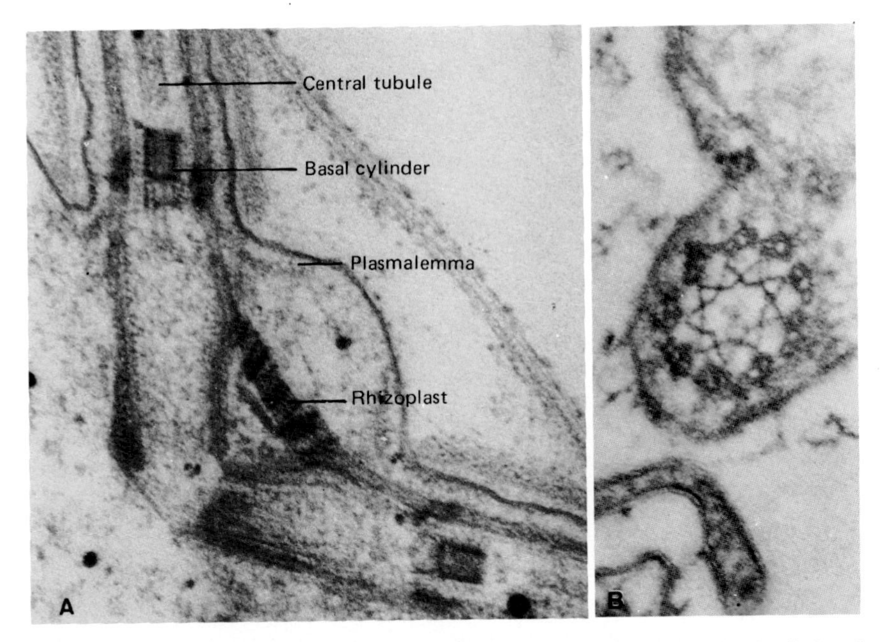

Figure 5.4. Transitional zone of the green plant flagellum. (A) The characteristic basal cylinder of green algae and metaphytans appears in each of the paired centrioles of *Chamydomonas*. A rhizoplast serves as a strut between the two organelles. 100,000×. (Courtesy of Friedmann *et al.*, 1968.) (B) Transverse section of the transitional zone of a *Marsilea* sperm flagellum. 140,000×. (Courtesy of Mizukami and Gall, 1966.)

5.4B). At the middle of the cross-pieces that reach across the base of the points is a distinct thickening, corresponding in position to the intermediate fibrils of the axoneme. In vertical sections a thick-walled cylinder is seen, closed proximally by the usual plate (Figure 5.4A); thus wherever present, the star-shaped cylinder may be readily identified in either transverse or longitudinal sections. Flagellated green protists in which the existence of this modification has been reported include *Pedinomonas* (Manton, 1964), *Nitella* (Turner, 1968), *Polytomella* (Brown *et al.*, 1976), *Chlamydomonas* (Lembi and Lang, 1965; Friedmann *et al.*, 1968), *Schizomeris* (Birbeck *et al.*, 1974), and *Tetraspora* (Lembi and Herndon, 1966), while among metaphytans in which the same feature has been noted are included the fern *Marsilea,* the cycad *Zamia* (Mizukami and Gall, 1966), and the hepatic, *Marchantia* (Carothers and Kreitner, 1968).

In developing the triplet condition from the doublets, no abrupt change occurs, but the third, or C component gradually forms. This transition may be effected by means of an intermediate condition, such as the formation of arcs of microtubular protofilaments, as shown by the ciliates (Figure 5.5) and *Naegleria* (Favard *et al.,* 1963; Dingle and Fulton, 1966). In other instances, for ex-

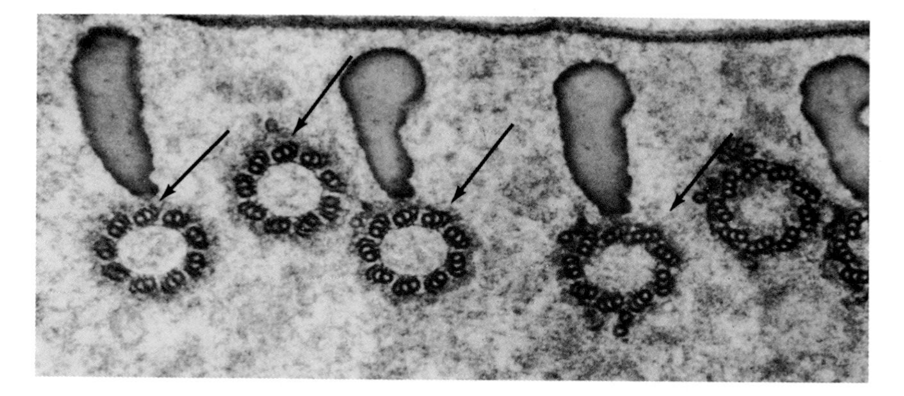

Figure 5.5. Development of the triplets from doublets. The four arrows point to what appear to be four stages in the development of the third microtubule of the triplets, beginning at the left. 76,000×. (Courtesy of Favard *et al.*, 1963.)

ample *Marsilea* (Hepler, 1976), the new component first appears as an extremely fine, incomplete capillary that gradually increases in diameter proximally. In other words, in these organisms the C components are tapered to points at their distal ends.

The Intermediate Cylinder. This is probably the most invariable portion of the centriole, undoubtedly because of its uniform simplicity of structure. Only two parts are present, the wall made of triplets and a homogeneous electron-transparent interior. At their outset, the triplets that comprise the wall lie at a sharp angle (about 80°) to the radii, so that a pinwheel appearance typifies cross sections of this region (Figure 5.2). Each triplet twists a bit on its axis as it passes proximally, so that at the extreme inner end they lie at an obtuse angle to the radii (40°). At certain undertermined levels of this cylinder, the triplets appear to be bound to one another by dense, irregular bridges that run from the B component of one triplet to the A of the next (Roth and Shigenaka, 1964; Turner, 1968). A few instances have been illustrated in which two and three tubular-appearing structures are seen to be present in the center of the cross sections (Gibbons and Grimstone, 1960). However, these are much larger in diameter than microtubules and are not aligned with the central pair of the flagellar axoneme. Rather than tubules, the parts are more than likely short cylinders, for they appear in very few electron micrographs. In longitudinal sections, this region displays extensive variation in length, from absence in mitotic centrioles to more than three times the length of the proximal region, as in the flagellar centriole of rabbit spermatid (Pedersen, 1969). In any one species, however, the intermediate zone is fairly constant in relative length.

The Proximal Region. Longitudinal sections of the proximal region of the centriole often appear more electron dense in electron micrographs than does the intermediate cylinder and, in addition, vague threads or tubules may be perceived in the interior region. The latter result from the "cartwheel" pattern of parts that characterizes this portion of the organelle in transverse sections. While the cartwheel configuration seems to be universally present, the details vary to some extent from taxon to taxon, but unfortunately relatively few clear micrographs of transverse sections of this region have been published. In *Pseudotrichonympha* (Gibbons and Grimstone, 1960), there is a short arm on the proximal edge of the A components of the triplets from which a feebly arcuate membrane extends spokelike to a central "hub," which appears to be a tubule somewhat larger in diameter than the usual microtubule (Figure 5.2). At each angle formed by the junction of an arm and spoke, another fine membrane extends parallel to the long axis of the triplet to join the distal surface of the A component, rather than to the bridge located here from the C component to the A (Figure 5.2). Among fungi a similar configuration occurs, except that the spokes tend to be straight (Lessie and Lovett, 1968). Ciliates (Favard *et al.,* 1963; Roth and Shigenaka, 1964), the flagellate amoeboids *Naegleria* (Dingle and Fulton, 1966) and *Tetramitus* (Outka and Kluss, 1967), insects (Berry and Johnson, 1975), and green plants, including *Nitella* (Turner, 1968), also have the spokes straight; consequently, the curved membranes of hypermastiginous flagellates may be considered to be exceptional to the general condition.

On the external surface of either this region or that just distal to it may frequently be found a variety of appended parts. Often these include an additional microtubule adjacent to every C component of the triplets, a condition exemplified by euciliates (Favard *et al.,* 1963); several such pericentriolar fibers occur in hepatics, ferns, and cycads (Mizukami and Gall, 1966; Carothers and Kreitner, 1968; Turner, 1968), along with armlike deposits of electron-dense material that provide a resemblance to paddle wheels. Comparable arms found on the molluscan centriole (Gibbons, 1961) are more complex, as they are also in jellyfish sperm tails (Szöllösi, 1964) and Chinese hamster ovary cells (Blackburn *et al.,* 1978), and probably in flagellated cells of all metazoans. In contrast, those of the euglenoid *Colacium* (Leedale, 1967) are simpler, more strongly arcuate, and located farther distad.

5.1.2. Macromolecular Properties and Replication

Partly as a consequence of the difficulties involved and partly as a result of the current lack of appreciation of its function, any discussion of the macromolecular organization of the centriole must necessarily be brief. Fortunately the related topic, replication of this small organelle, has been thoroughly investigated, but even there, certain aspects remain controversial at the moment.

Proteinaceous Constituents. As might be expected from the structure of the walls, the chief constituent of the centriole is tubulin (Fulton, 1971;

Heidemann *et al.,* 1977). However, the precise nature of this protein is unknown and, in view of the diversity of molecular structure shown in previous pages to exist among the several parts of the flagellum and cytoplasm, its mere presence is without real significance. For instance, the tubulin of the C component of the triplets is evidently different from that of the other two constituents, for in the centriolar fragments isolated by certain treatments, only doublets remained in the walls (Hoffman, 1965; Satir and Rosenbaum, 1965). Moreover, all of the centriolar microtubules show a greater affinity for osmic tetroxide than do most others of the cell, including those of the flagellum, and hence they may contain more lipid or other osmiophilic substances. The only additional protein that has been partially characterized from this organelle is an ATPase (Hartmann, 1964; Abel, 1969). This is probably a type of dynein and may be suspected to be localized in the arms that project from the A components. Because these differ markedly in form from the corresponding parts in the flagellum, here too the dynein of the centriole probably is distinct also in molecular structure.

Nucleic Acids. During the 1960s when DNA was first found to be present in mitochondria and chloroplasts, a search for that nucleic acid was made in other cell organelles, including the centriole, to determine whether all were capable of autonomy in replication. In early analyses, such as that of Seaman (1960), quantities as large as 6 parts DNA to 100 parts protein were reported, but as techniques for isolating centrioles were improved, the proportion of DNA reported was decreased to 0.6% (Argetsinger, 1965) and close to 0.0% (Hoffman, 1965). Other studies using cytochemical procedures, however, did find small quantities of DNA present (Randall and Disbrey, 1965; Smith-Sonneborn and Plaut, 1967), whereas many using autoradiographic procedures failed to detect any (Suyama and Preer, 1965; Dirksen and Crocker, 1966) or, contrastingly, found small quantities (Smith-Sonneborn and Plaut, 1967). Thus the evidence is contradictory to so large a degree that no conclusion can be firm, but the trend today is definitely toward considering DNA to be absent from these organelles (Fulton, 1971; Hartman, 1975; Dippell, 1976; Heidemann *et al.,* 1977).

The evidence for the presence of RNA shows greater consistency. When centrioles were treated with RNase, the short arms (the so-called feet) on the proximal side of the A components of the triplets disappeared (Stubblefield and Brinkley, 1967; Brinkley and Stubblefield, 1970), and the electron-opaque material that often obfuscates the proximal end of the centriole disappeared when treated similarly (Dippell, 1968, 1976). On the other hand, Heidemann *et al.* (1977) found no distinctions between untreated flagellar centrioles and those treated with RNase, but they did find dissimilarities in activity. For their experiments, results were assayed by using an observation made previously that flagellar centrioles, but not mitotic ones, from either *Chlamydomonas* or *Tetrahymena* induced the formation of asters when injected into unfertilized eggs of the clawed frog, *Xenopus laevis* (Heidemann and Kirschner, 1975). Flagellar centrioles treated with such low concentrations of ribonuclease A as 0.5 μmole/ml for 30 min lost the ability to induce aster formation, whereas controls could do so. Thus, rather than being concerned with replication of the organelle, the nucleic acids present, whatever they might prove to be, may be essential to carrying out the centriolar functions. In contrast, as RNA rather than DNA seems more likely to be present, Went (1977) speculated that a reverse transcriptase may be involved in the synthesis of DNA and thereby induce autonomous replication, but there is no evidence that such is the case.

Replication of Centrioles. The highly distinctive manner in which centrioles are replicated certainly does not imply the existence of any degree of autonomy for their formation. During the developmental processes, the forming centriole, referred to as a procentriole, lies at some distance—about 70 nm—from the original one, lying at a right angle to the latter and on one side of the proximal region. Hence, there is no clear-cut parent–daughter relationship between the two, nor is it evident that the existing organelle even serves as a template for the formation of the new one (Frisch, 1967). Moreover, procentriole formation is dependent upon protein synthesis by the cell, as shown by use of cycloheximide (Phillips and Rattner, 1976), and as many as eight of the bodies are formed simultaneously in mammalian lung epithelium (Dirksen and Crocker, 1966; Sorokin, 1968).

The first appearance of a procentriole is an electron-opaque disk, lying parallel to and opposite the equally dark-appearing proximal end of the mature centriole. In diameter, this disk is usually equal to that of the original, a size relationship that continues throughout replication (Figure 5.6). The procentriole then is increased in length, the parts being built up to definitive form from proximal to distal end, so that shortly after the apical tip has been added, the two centrioles are indistinguishable. Evidently the same features of centriolar reproduction prevail throughout the eukaryotic world, for they occur among protozoans (Pitelka, 1961; Gould, 1975) as well as metazoans (Harris and Mazia, 1962), but there is some evidence that the procentriolar structural parts, such as the triplets or plate, may not achieve definitive form until after the whole body has been laid down in a precursorial state (Outka and Kluss, 1967).

As a nearly catholic rule, the newly formed centriole is identical to the mature one, at least after maturation has been completed. One frequent exceptional type involves the formation of a flagellar centriole from a mitotic one, as in spermiogenesis of the snails of the genus *Viviparus* (Gall, 1961) and in that of the water mold *Allomyces* (Renaud and Swift, 1964). In both kinds of organisms the respective mitotic and flagellar centrioles have their individual typical form, the former variety being short, the flagellar product elongate. Another, more radical authenticated case of divergence is that which occurs during spermatogenesis in the fungus gnats of the genus *Sciara* (Phillips, 1966, 1967, 1970). In these insects, even in somatic tissues, the centrioles depart from the

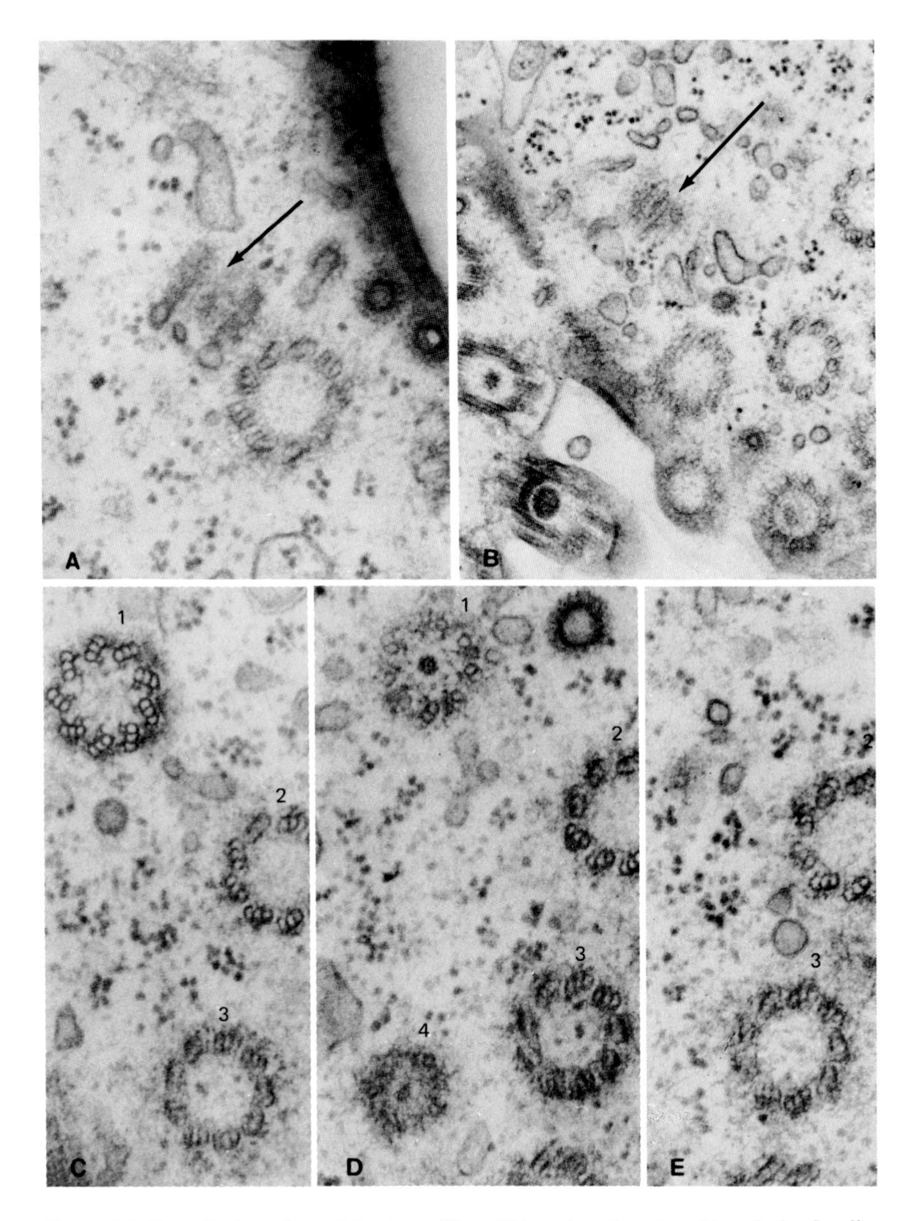

Figure 5.6. Reproduction of centrioles in a ciliate. This series of sections through the flagellar centrioles of *Oxytricha fallax* reveals some of the events in their formation. (A) A new centriole (arrow) is developing adjacent and at right angles to a mature one, as is typical of the replication of this organelle. 70,000×. (B) Sometimes new centrioles form at some distance from mature ones, as in the nascent plaque indicated by the arrow. 45,000×. (C–E) Serial sections taken parallel to the cell surface. A juvenile centriole (1) shows some of the developmental changes at two levels and a nascent centriole (4) has just been laid down. 75,000×. (All courtesy of Grimes, 1973a.)

general pattern in having nine doublets, rather than triplets, comprising their walls. When the primary spermatogonia divide, these atypical centrioles separate as usual, and then during replication, giant centrioles* containing 30 to 50 singlet microtubules are produced. These persist throughout the remaining stages of meiosis and even into the sperm, where giants of this type are associated with the flagellum. Occasional doublets are often included among the singlets of the walls, but no triplets have been noted. Thus in *Sciara* the 9-doublet centriole gives rise indirectly to multisinglet bodies, which are perpetuated into the embryo. At what point in the life history the 9-doublet type is regained has not been determined.

De Novo Production of Centrioles. The *de novo* origin of centrioles has now been thoroughly documented on a wide basis, the earliest confirmed instance being in the flagellate amoeboid *Naegleria*. In the members of this genus, centrioles were found to be absent during the amoeboid stage, but were readily detected when the organism prepared to enter the flagellated condition (Dingle and Fulton, 1966; Fulton and Dingle, 1971). These organelles first appeared just 10 min before the flagella began to develop. A similar condition was later reported from another flagellate amoeboid, *Tetramitus rostratus* (Outka and Kluss, 1967). Here, too, centrioles were absent from the amoeboid stages and were present in the flagellate, but developmental bodies, that is, amorphous preliminary stages, were found in association with the nucleus and dense bodies. In the euciliate *Oxytricha fallax,* some centrioles were noted to arise adjacent to mature ones, while many others formed at some distance from any other body (Figure 5.6B; Grimes, 1973a). A related study on the cysts of this same organism showed centrioles to be completely absent during this resting stage (Grimes, 1973b). Among green plants, particularly in the sperm of the fern *Marsilea,* flagellar centrioles arise from a peculiar mitotic body originally referred to as the blepharoplast, which itself has a clear-cut *de novo* origin (Hepler, 1976).

Among vertebrates, several studies have confirmed that centrioles arise in the absence of the mature organelle. During embryogenesis of *Xenopus laevis,* in the epidermis and trachea, the development of cilia began with the *de novo* formation of clusters of dense, amorphous procentriole-precursor bodies, which later were transformed successively into procentrioles and mature centrioles (Steinman, 1968). Comparable formation of cilia from amorphous granular material has also been described from mammalian lungs (Sorokin, 1968). Hence, there appears to be little reason to doubt that new centrioles can arise either *de novo* or adjacent to mature ones, depending upon cellular conditions and requirements.

* In neuropteran meiosis the term giant centriole has been applied to quite a different body (Friedländer and Wahrman, 1966), which is of typical centriolar morphology except in its greatly elongate form, being about 12 times as long as wide.

Functions of Centrioles. The functions of the centriole are not readily perceived and have often been greatly misunderstood. At one time it was proposed that this organelle served primarily in regulating the syntheis of the protein monomers needed for microtubules, a concept based on its association with the microtubule-containing flagellum and with the spindle during mitosis (deHarven and Bernhard, 1956; Seaman, 1962; DuPraw, 1968). Because centrioles do not occur at all in the cells of conifers and flowering plants, only sperm nuclei being involved in fertilization and mitotic centrioles being absent, their essentiality in such processes consequently is subject to great doubt.

The only function clearly and totally dependent upon this organelle is in the formation of the flagellum or cilium (Fulton, 1971), where it serves as a "crystallization center" upon which the microtubules are organized. This service as a nucleation center has already received mention in Chapter 2, and receives some support by the continuity that exists between the doublets and the first two components of the centriolar triplets. That much is clear. However, that point of view is based on the old notion of the self-assembly of microtubules and does not take into consideration the long-known observations that *in vitro* only singlet microtubules are formed on centrioles, not doublets as *in vivo*. Nor does it explain why triplets do not occur in flagella as they do in centrioles. As pointed out earlier, moreover, the A component of flagellar doublets is constructed of one type of tubulin, the B component of another, and the inner microtubules of a third type. The A component bears two arms of dynein and other proteins, which differ in composition, as well as radial spokes of complex structure. Moreover, the central tubules are joined at the apical end by a cap and bear a series of complexly arranged arms, sheaths, and other parts. Furthermore, at the base of the shaft is the transition zone, where plates, cylinders, 9-pointed star configurations, or other specializations occur. To suppose that such a highly ordered and complicated organelle merely grows by spontaneous combinations of molecules appears unrealistic indeed. Thus it seems self-evident that the centriole must play an important role in the development of flagella.

Possibly because flagella removed from the cells beat in the presence of ATP, the centriole is not currently even suspected of controlling those organelles *in vivo*. However, while the beating of detached flagella is fairly similar to the beating of those on cells, it is not identical and is always of one simple undulatory type. Contrastingly, the movements of the flagella of living protists are rarely constant but undergo changes in movement in accordance with the requirements of the organism. Changes in the rhythm and direction of the flagellar beating can alter the direction of swimming to the right or left or upwards or downwards, to say nothing of reversal. Many years ago Krijgsman (1925) observed and described a whole series of patterns of flagellar activity employed in swimming by members of the genus *Monas*, and numerous other reports exist that describe similar behavior in various protists, ciliates as well as flagellates. As it does not seem likely that the cell is subsidiary to the flagellum and has to go wherever that organelle takes it, the most plausible alternative seems to be that the cell exercises control over this organelle, probably through the mediation of the centriole. That flagellated organisms do exercise such control is amply substantiated by well-documented reports of phototactic and phototropic responses, too familiar to require detailing here, in which the photoreceptor clearly resides within the cell, not the flagellum.

Such control over flagellar activity does not expressly signify a neural function of a primitive sort for the centriole, but that role certainly remains a distinct possibility. It is pertinent in this connection to recall two sets of data. (1) All metazoan sensory detectors are modified cilia (p. 158), and (2) the centriole of protozoans is the actual light-receptor organ, not the stigma, as convincingly demonstrated about a century ago by Engelman (1882).

5.2. CENTRIOLE-ASSOCIATED ORGANELLES

The continuing activity of the centriole throughout the cell cycle is reflected in the variety of supplementary organelles that are associated with it. In general form these are often fairly constant among large groups of organisms, but the particulars frequently vary in the smaller component taxa. Some of these, like the blepharoplast of green plants, have attracted much attention in the literature and therefore are well known, while others, including that of vertebrates, are incompletely understood. Many, if not most, of these are constructed of microtubules to a greater or lesser extent. However, some centriole-associated parts are derivatives of other major organelles—for example, the kinetoplast of *Crithidia* and relatives is derived from the mitochondrion (Hill and Anderson, 1969)—and accordingly are described with the parental body.

The "Blepharoplast" of Green Plants. During the era of the light microscope, the term blepharoplast was applied by biologists to designate the centriole itself or various bodies loosely associated with it, either at the flagellum or during mitosis. Even in current literature of green-plant cells, the word has been applied to two entirely distinct organelles in separate but closely related taxa. Hence, it is proposed that the term be abandoned entirely and that other more appropriate and consistent terms be employed for each major type of fine structure exhibited by such organelles, as disclosed by electron microscopy.

Several terms, including blepharoplast, have been applied to a peculiar apparatus that lies adjacent to the flagellar centriole that has been most thoroughly described in such liverworts as *Marchantia*. Originally called the Dreiergruppe, that name was changed to the Vierergruppe* (Heitz, 1959; Carothers and Kreitner, 1967, 1968), because it is comprised of four layers (Figure 5.7). As Carothers and Kreitner (1967) show in detail, the distalmost layer consists of

* Perhaps a term like tetriole would be in more acceptable style and still be equally descriptive.

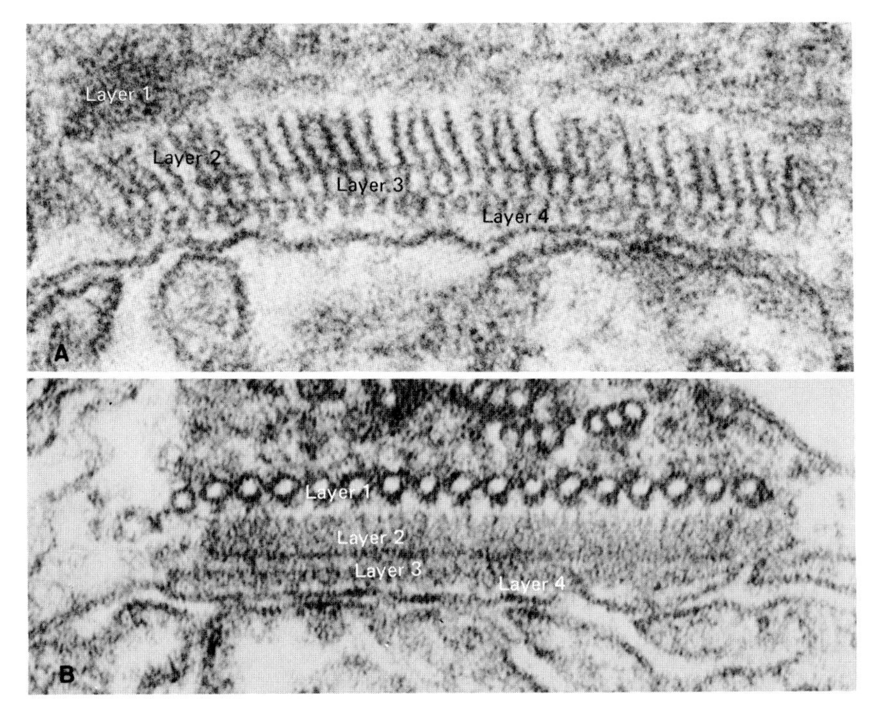

Figure 5.7. Sections through the tetriole of *Marchantia*. (A) Diagonal section through the organelle reveals the lamellar nature of layers 2 and 3 and the nanofilaments of layer 4. 199,500×. (B) A transverse section discloses the microtubular nature of layer 1. 166,000×. (Both courtesy of Carothers and Kreitner, 1967.)

about 17 microtubules bound together by numerous thick crosslinks. At an angle to this is the second layer, consisting of upright plates, the spaces between which form vertical oblongs; the third is similar, except that the plates are short so that the interspaces are quadrate in section. Proximal to this is the fourth and final lamella, made of intermediate filaments (Figures 5.7, 5.8). This tetriole occurs also in mosses (Paolillo, 1965), in which only 13 microtubules seem to be present. In each of these taxa, it continues into an elongate band that partially encircles the cell. This band, given the name spline (Carothers and Kreitner, 1968), gradually loses microtubules and other components as it leaves the centriolar region, only 6 of the original 17 extending through the entire length of the structure. The width of the spline varies even within a taxon, ranging to 38 microtubules in certain mosses and to 56 in the liverwort *Haplomitrium* (Carothers and Duckett, 1977). In the similar biflagellated sperm of the stoneworts, such as *Nitella,* the tetriole (under the name microtubular sheath) is comparable in structure but smaller, containing a maximum number of nine microtubules (Turner, 1968). It also occurs in ferns and related forms under the name multilayered body, as shown in a following paragraph.

The "Blepharoplast" of Ferns and Allies. A second, quite different body has been referred to as the blepharoplast in higher green plants, especially by investigators concerned with reproduction among ferns and horsetail rushes (Duckett and Bell, 1977). Because this organelle is totally unrelated to others known by the same name and because it is confined to ferns and related groups, the term pteridoplast might be preferable as a substitute. This is active in mitosis as well as in flagellation, at least in indirect ways. Here the particularly lucid account of its behavior during spermatogenesis in a water fern, *Marsilea vestita* (Hepler, 1976; Hepler and Myles, 1977), provides the main basis for the present discussion. It is best viewed as a spherical structure consisting of numerous cylinders, each of which matures into a procentriole first and later into a definitive centriole.

No centriole or other body could be observed during most of the first seven mitotic divisions of the microspore of *Marsilea* despite the examination of numerous sections of the polar regions of dividing cells; then at telophase of the seventh division the pteridoplast is formed in an indentation of each nucleus. Later as the prophase of the eighth division commences, this organelle becomes completely degenerated, only to reappear at telophase. During the ninth, or final, division, it remains near the polar region after dividing during prophase and migrating to opposite sides, reflecting centriolar behavior clearly.

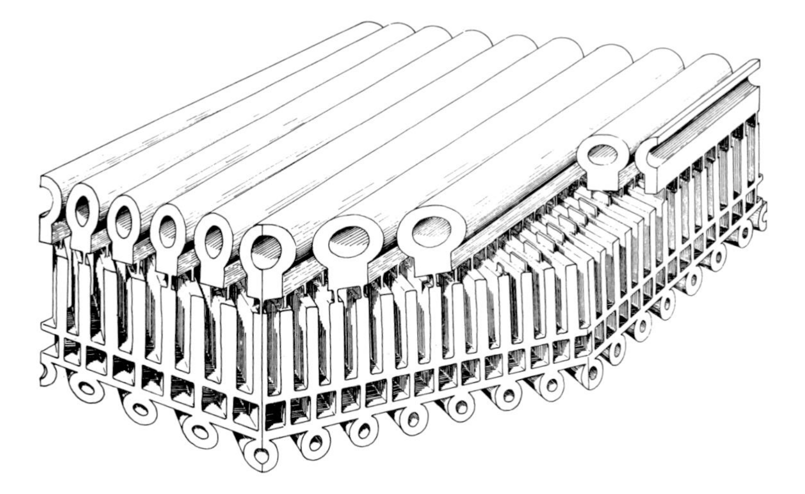

Figure 5.8. Reconstruction of the tetriole of *Marchantia*. (Courtesy of Carothers and Kreitner, 1967.)

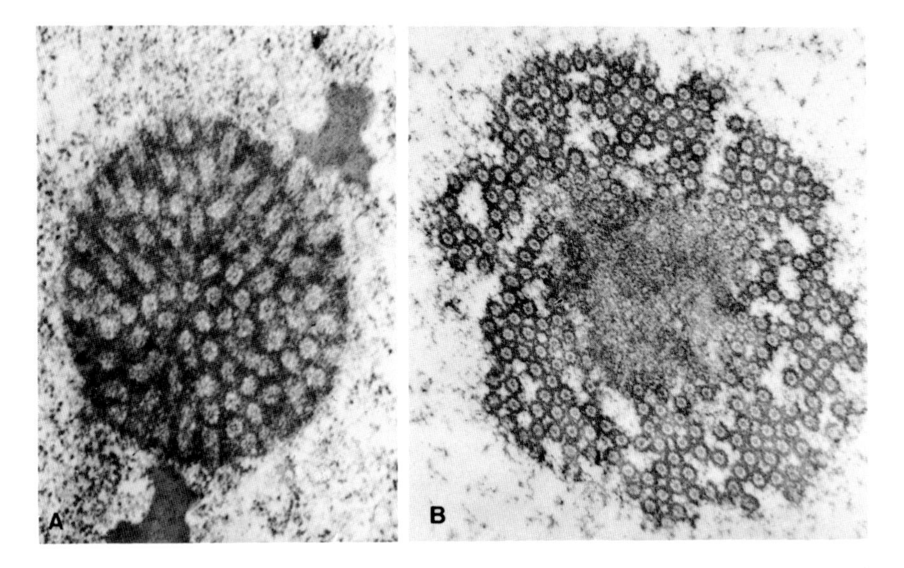

Figure 5.9. The pteridoplast of cryptogams. (A) A transverse section below tne surface of the pteridoplast of the fern *Marsilea*, showing the numerous procentrioles that make up this body. The characteristic starlike patterns are especially clear in the lower right sector. 45,000×. (B) A pteridoplast of *Zamia* reveals a similar construction. 21,000×. (Both courtesy of Mizukami and Gall, 1966.)

However, the two resulting pteridoplasts remain as focal points for the spindle only during prophase, moving off to one side of the poles at metaphase. Then late in anaphase within each of these organelles, the component parts (Figure 5.9A), which might be considered to be precursors of procentrioles, undergo maturation until they have acquired the ninefold radial symmetry of typical procentrioles (Mizukami and Gall, 1966). These later separate and grow into centrioles before they migrate to take positions along the spline that has been developing in the meanwhile. After giving rise to flagella, they serve as flagellar centrioles throughout the life of the spermatozoan. Similar structures were reported also from *Zamia* (Figure 5.9B).

As in the mosses and club mosses, the spline is an outgrowth of the four-layered tetriole, an organelle that in the present organisms has been known as the multilayered body (Duckett and Bell, 1969; Bell *et al.*, 1971; Bell, 1974a,b). Structurally and functionally this is similar to that already described, in consisting of one layer of microtubules, two layers having vertical-walled compartments, and one of intermediate filaments. However, it is much larger, the distal layer consisting of at least 80 microtubules and in later stages of development, of many more. In these more advanced forms, it has the obvious function of serving as a base for the numerous centrioles and their flagella, as pointed out in the preceding paragraph.

The Rhizoplast. Like the tetriole, the rhizoplast is a microtubular organelle, but unlike that structure, it is found among a diversity of organisms. Basically this cell part is an elongate bundle of microtubules associated with flagellar centrioles, frequently being rootlike in appearance; however, there are many variations on this general pattern. Often this structure has been described under the term root or rootlet, for it frequently appears like an outgrowth of the proximal end of a flagellar centriole. But it also is commonly found on the distal surface of that organelle, extending like the cable of a suspension bridge between the two members of a pair—sometimes the regularity of the cylindrical form of the centrioles is disturbed when viewed in vertical section (Figure 5.4A), suggesting the possible existence of stress as in *Chlamydomonas* (Friedmann *et al.,* 1968).

Unfortunately, the rhizoplast has received little detailed attention at the hands of cell biologists and none on a broad comparative basis; hence, much remains to be learned of the structural diversity that undoubtedly exists. Accordingly only a fraction of the total picture can be shown here, and even that little must be confined to differences seen in longitudinal sections of the organelle. Among such simple flagellates as *Euglena,* the rhizoplast appears to be short, as it is also in dinoflagellates (Dodge and Crawford, 1968), so the first known occurrence in its elongate form is in the flagellated amoeboid *Naegleria* (Figure 5.10A–C). In the latter organisms longitudinal sections show a banded pattern superimposed on a background of long fibers (Dingle and Fulton, 1966). The electron-opaque bands are about four times as high as broad, but that breadth is at least twice that of the pale bands that intervene; however, the periodicity has been found to be variable (Simpson and Dingle, 1971). In certain views, the rhizoplast can be seen to be connected to the flagellar centrioles by means of a band of microtubules* (Figure 5.10B,C).

In euciliates, such as the peritrichs, a similar heavily banded appearance is shown by the rhizoplast (Favard *et al.,* 1963; Anderson and Dumont, 1966; Lom and Corliss, 1971; Grim, 1966, 1972), but the electron-dense and -transparent lines are more equal in width (Figure 5.10D). A comparable pattern of alternating dark and light lines is to be seen in fungi like *Blastocladiella* (Lessie and Lovett, 1968), the flagellate amoeboid *Tetramitis* (Outka and Kluss, 1968), and advanced flagellates including *Trichomonas* (Joyon, 1963). Reduction of the dark bands is carried to the extreme in the green flagellates, *Chlamydomonas, Platymonas, Polytomella,* and *Prasinocladus,* in which the electron-transparent regions are broad bands only slightly less than twice as high as wide (Figure 5.11A), while the opaque parts are mere lines, each of which is double (Parke and Manton, 1965; Friedmann *et al.,* 1968; Stewart *et al.,* 1974; Brown *et al.,* 1976). Among metazoans ciliated epidermis is of widespread occurrence; wherever it is found, the associated centrioles consis-

*In a number of forms, such as *Tetramitis,* the band of microtubules is associated with several layers of electron-dense material, so it may be that this represents an ancestral form to the tetriole described in preceding pages. (See Outka and Kluss, 1967, Figure 8.)

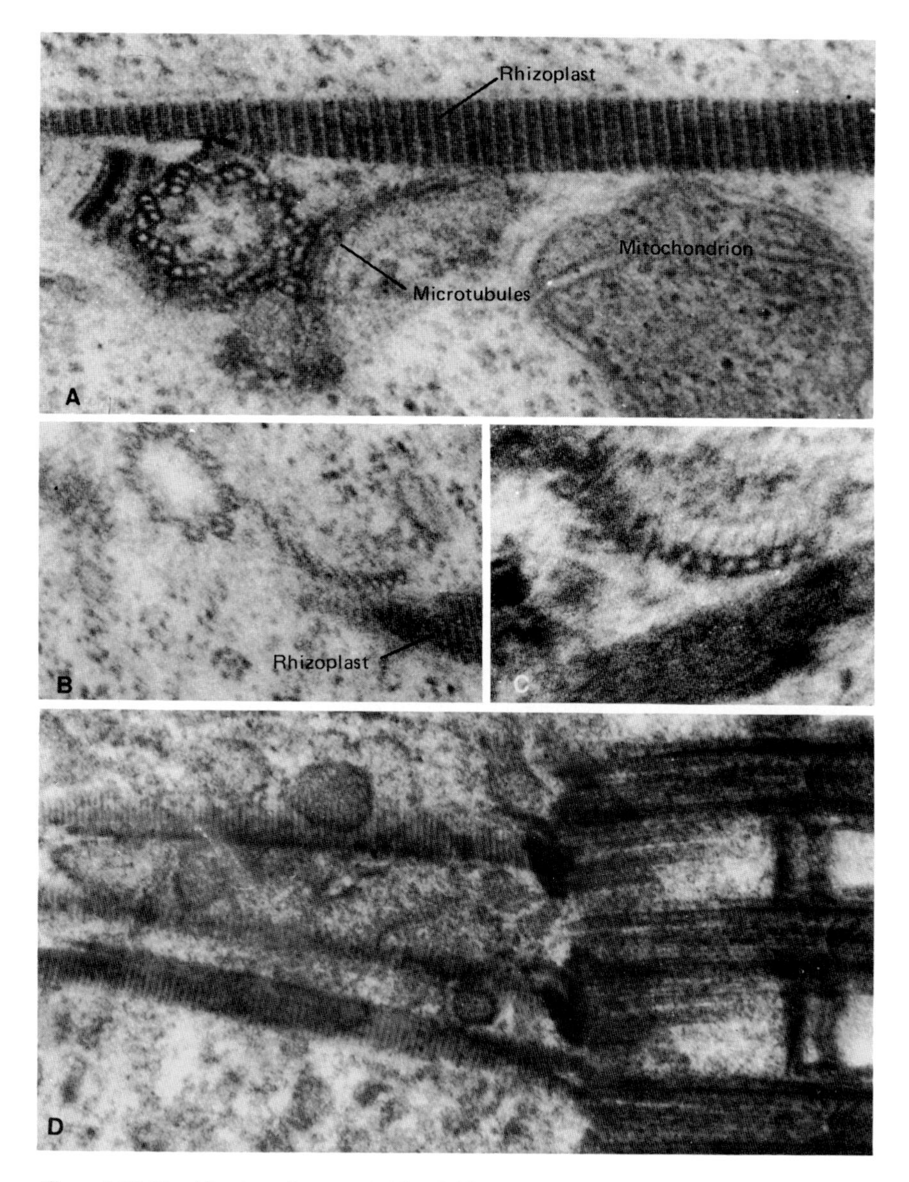

Figure 5.10. The rhizoplast of an amoeboid and ciliate. (A–C) The rhizoplast of *Naegleria gruberi*. (A) The electron-opaque structure of the rhizoplast in this flagellate amoeba is interrupted only by narrow transparent lines. 113,000×. (B) An arcuate row of microtubules extends between the rhizoplast and centriole. 88,000×. (C) The same as B but 107,000×. (A–C, courtesy of Dingle and Fulton, 1966.) (D) The electron-opaque and -transparent lines on this rhizoplast of the ciliate *Trichodina* are nearly equally wide. 52,000×. (Courtesy of Favard *et al.*, 1963.)

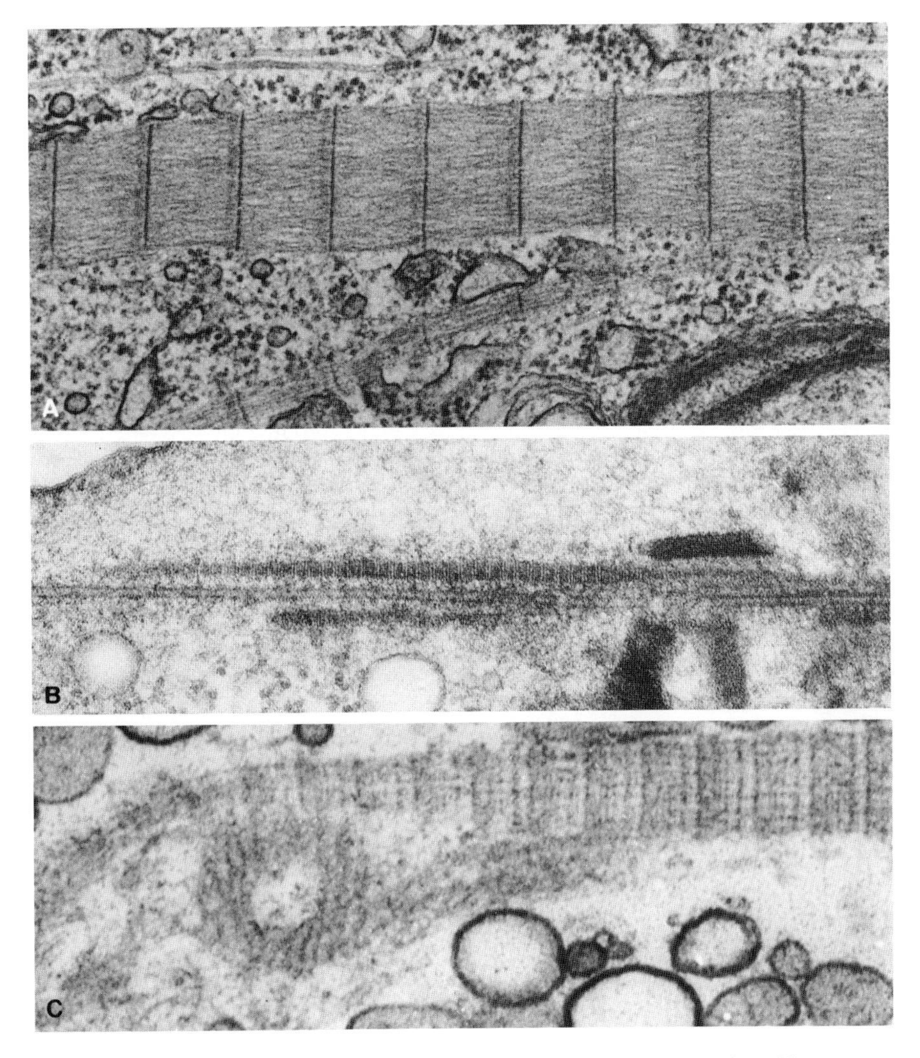

Figure 5.11. Phylogenetic variation in rhizoplast structure. (A) The rhizoplast from *Platymonas*, with a small branch. 100,000×. (Courtesy of Stewart *et al.*, 1974.) (B) The corresponding structure from *Polytomella*. 64,000×. (Courtesy of Brown *et al.*, 1976.) (C) As in the other electron micrographs, this rhizoplast of the jellyfish *Phialidium* displays a close relationship to a centriole, as well as an arcuate microtubular band. The periodic pattern, however, is far more complex. 85,000×. (Courtesy of Szöllösi, 1964.)

tently are accompanied by rhizoplasts showing great complexity of the banded pattern (Wheatley, 1967). However, too few clear electron micrographs are available to permit further phylogenetic comparisons; some variations that have been illustrated include those derived from jellyfish (Figure 5.11C; Szöllösi, 1964), molluscs (Gibbons, 1961), and rotifers (Lansing and Lamy, 1961).

5.3. A PHYLOGENY OF FLAGELLARY STRUCTURES

In spite of missing details, the flagellum and its associated structures in general are sufficiently well known to permit a phylogeny of its development to be proposed on fairly secure grounds. While the organelles of relatively few types of organisms have been sufficiently explored for inclusion, those that have been investigated are diversified evolutionarily to such an extent that a framework at least can be constructed. Nevertheless, some important points still remain completely unresolvable.

Origins of the Flagellar Organelles. Although there may be a possibility that the flagellum of the bacteria eventually provided the basis for that of eukaryotes, no supportive evidence can be found. A few advanced bacteria have flagella that perhaps are comprised of a group of the simple filaments found in the more primitive types and are sheathed by a membrane, but none have been reported of more complex structure. Hence, the main body of evidence is opposed to this point of view. In the first place, the filament is made of flagellin, whose molecular structure bears no resemblance to that of tubulin (compare Tables 2.1 and 4.1), nor does that organelle contain tubular structures. Hence, it would be necessary to replace the prokaryotic filament with microtubules to obtain any resemblance at all to the eukaryotic organelle. Furthermore, the hook and other basal structures in bacteria cannot readily be conceived to have become the nine-parted symmetrical centriole made largely of triplet microtubules. Consequently, the differences between the eukaryotic and prokaryotic basal parts are too great to be considered as having an ancestor–descendant relationship.

As yeasts possess many features that mark them as being among the most primitive of eukaryotes, examination of these organisms is instructive. These simple unicellular organisms are, of course, without flagella and even lack a definitive centriole during mitosis. Instead of the latter organelle, a plaque serves in nuclear division along with a cluster of microtubules. This cell part has a number of features that favor an interpretation of its being a possible forerunner of the centriole. Thus it can be deduced that the bacterial flagellum was lost by early eukaryotes and that an entirely new organelle arose at later levels of advancement. The sequence of events (Figure 5.12), then, might be viewed as (1) the acquisition of microtubules in higher bacteria, (2) the formation of a centriole precursor, the plaque, in yeast that is associated with microtubules during mitosis, (3) the further elaboration of the plaque into a centriole that served in nuclear division, and (4) a number of steps through which the microtubular flagellum was formed, all of which remained unknown.

The Elementary Level of Structure. After its origin, the flagellum, while containing the major elements of its definitive structure, appears to have lacked several of the refinements, if the morphology of the euglenoid organelle is to be believed. In cross sections the axoneme of those organisms lacks

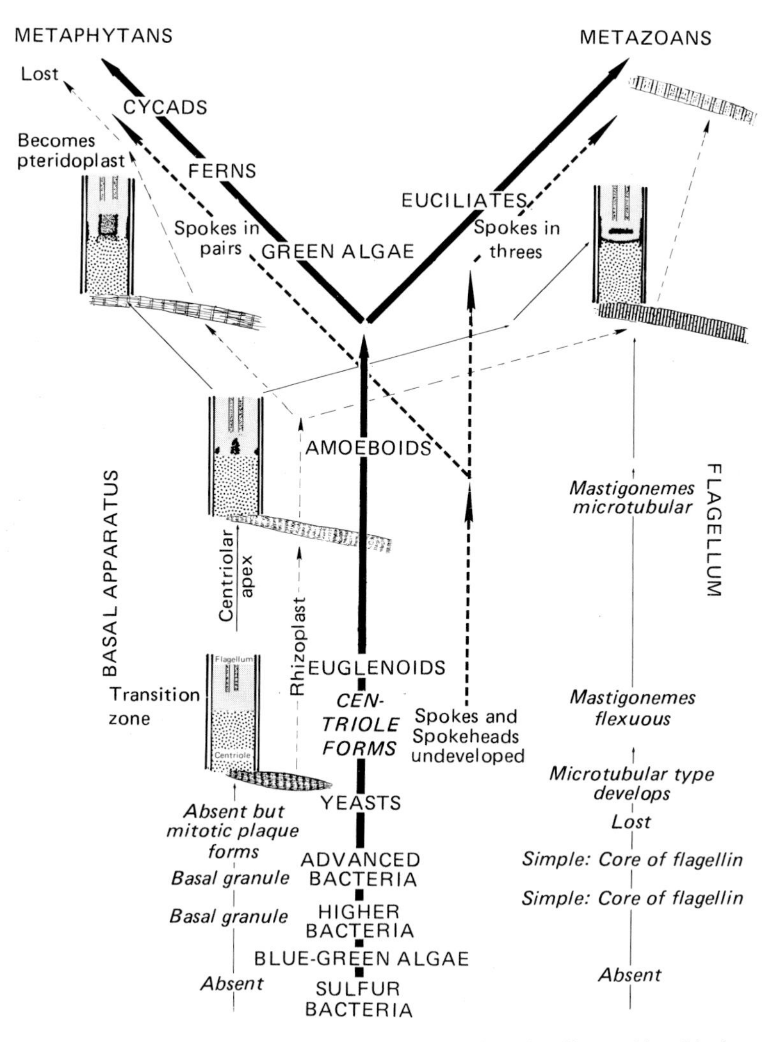

Figure 5.12. Sequence of events in the phylogeny of the flagellum and basal body.

spokes and spoke heads; in longitudinal sections these parts, while present, are slender and so irregularly arranged that they do not often become visible in transverse sections. Nor are any longitudinal interconnecting fibers in evidence. Moreover, the mastigonemes, located on only one side of the flagellum, are soft, flexuous, and hairlike, lacking the rigidity of structure that characterizes higher levels of evolutionary advancement. Finally, the centriole lacks complexity at the transitional zone, the rhizoplast is short, and other structures associated with the centriole in more advanced organisms are absent here. As dinoflagellates have similar flagella, mastigonemes, and basal apparatus, they must be placed nearby the euglenoids.

The Advanced Levels. The evidence of flagellar and centriolar structure suggests that two major lines of development eventually diverged near the apex of the evolution of these organelles. But because the two lines of specialization share such characteristics as microtubular mastigonemes and elongate rhizoplasts, considerable evolution must have occurred above the level represented by *Euglena* and the point of their divergence, during which period these two features were acquired (Figure 5.12). Internally, too, macromolecular adaptations were gained, especially noteworthy among which are the spokes and spoke heads, the latter bearing interconnecting fibers.

The Green Plant Branch. The members of the two diverging lines are respectively marked by distinctive developments, a statement that is especially true of the green plant branch. Among the latter organisms, the nine-pointed starlike cylinder of the flagellar base must have arisen early, for even those genera like *Polytoma* and *Polytomella* that lack several traits of the majority of members have this feature. Representatives of the genera mentioned have well-developed mastigonemes of the microtubule type, whereas the flagella of more advanced forms lack these structures. In this connection, it should be recalled that the acquisition of bare flagella involves not merely the shedding of hairlike appendages but also entails the acquisition of new methods of locomotion concurrently Chapter 4, Section 4.2.

When the flagellum gained another of its structural traits is not clear, for its discovery is too recent to have permitted comparative investigations. Apparently it was acquired fairly early, because the bipartite arrangement of the spokes has been detected in *Chlamydomonas* (Figure 4.8A,B). At this same level, the rhizoplast has become widely divergent from the simpler type found in *Naegleria*. In still more advanced forms, beginning with the stoneworts, the rhizoplast spurs have undergone great modification, first having become modified into the 8-microtubular tetriole of the stoneworts. Above that level, its changes follow the usual sequence of representative forms of plant morphologists. In mosses, it contains around 13 fibrils, then approximately 18 in liverworts, and finally as many as 80 microtubules in the horsetails and ferns, before being lost altogether in the conifers and flowering plants. Among those higher plants in which flagella and all associated parts have been lost with the flagellated spermatozoon, no cell fertilizes the egg, only a sperm nucleus.

The Metazoan-Related Branch. The second advanced line of ascent also is marked by flagellar traits, less striking perhaps than the starlike pattern of green plants, but distinctive nonetheless. One of these is the whiplash into which the tips of the paired flagella are drawn, contrasting with the simple rounded ends of other forms. Second, the forward-projecting flagellum always bears two rows of microtubular mastigonemes, while a second one that trails is always devoid of such structures. In some members, like certain fungi and most metazoan spermatozoans, only the trailing, denuded flagellum persists. Internally, the axoneme is complexly structured, having well-marked spokes and spoke heads, which in longitudinal sections are perceived to be arranged in groups of threes (Figure 4.8C). Because only ciliates and metazoan flagella have been suitably examined for this trait, however, it cannot be determined whether this arrangement characterizes all members of the line or not. Although it could be that this is a primitive feature and that the paired spokes of green plants represent the advanced condition, the nature of the triplets pointed out in that figure suggests otherwise.

Despite the transition zone between the flagellum and centriole displaying a number of specializations from taxon to taxon, no evolutionary pattern can be plainly detected. The rhizoplast, however, is more helpful. It is immediately clear that the euciliates, while more advanced than *Naegleria,* have a simpler rhizoplast than the hypermastiginous flagellates and that these in turn are simpler than are the metazoans insofar as this organelle is concerned. The remaining pattern of changes that are known to occur in the flagellum and centriole are best brought out by the diagram (Figure 5.12).

6

SECRETORY ORGANELLES: I

The Endomembrane System

Within many active cells there exists an extensive system of organelles, which, while consisting of membranes of various forms, are not enclosed by such structures. The two generally recognized types of this endomembrane system, as it has been called (Morré *et al.*, 1971; Morré and Ovtracht, 1977), may be closely associated or more or less isolated from one another and, being unrestrained by an enclosure, vary widely in size, form, and location between tissue types as well as with the activity of a particular cell at a given moment. The two kinds of organelles, currently known as the endoplasmic reticulum and the Golgi apparatus, are alike involved in secretion, a term that will be found to cover a great diversity of cellular functions. In some cases, these two organelles are so closely associated that they are difficult to distinguish except by cytochemical techniques. Such intimate associations are currently referred to as GERL, the letters of which acronym respectively refer to Golgi, endoplasmic reticulum, and lysosomes, the latter a secretory vesicle that is given attention in the following chapter along with others of a similar nature. When used correctly as a collective term, the acronym is of occasional value, especially in discussions of cellular conformations of these membranous structures whose proper identification has not been firmly determined.

6.1. THE ENDORETICULUM (ENDOPLASMIC RETICULUM)

The term endoplasmic reticulum should really be employed in the plural, for it actually embraces several organelles, which, although not infrequently interconnected, are quite different in structure and function. In biochemical investigations, these organelles are necessarily removed as microsomes, vesicular remains of the whole membranes that exist in cells. Unfortunately, too com-

Table 6.1
Comparison of Various Types of Endoreticulum

Class	Type	Source	Characteristics	Enzymes	Functions
Rough			With ribosomes		
	Cisternal	Vertebrates	Stacks of flat parallel cisternae; 12–60 per stack	Exoribonuclease Glucose-6-phosphatase	Synthesizes glucose-6-phosphate
		Metaphytans	Low stacks of irregular cisternae; 8–20 per stack	NADH-ferricyanide reductase	Processes precursorial polypeptides
		Protistans	Single irregular cisternae, often perinuclear	NADH- and NADPH-cytochrome *c* reductase	
	Granular vesicular	Protistans	Very irregular, scattered vesicles	β-Thioglucosidase	Certain elementary steps in biosynthesis of glycoproteins
	Granular tubular	Vertebrates (mammals only?)	Highly ordered series of tubules, with 12 rows of ribosomes	α-Amylase	

Smooth			without ribosomes		
	Vesicular	Vertebrates	Extensive regions of vesicles interconnected by irregular tubules	Glucose-6-phosphatase	Carbohydrate metabolism; catabolism of drugs
		Metaphytans		NADPH-cytochrome *c* reductase	
		Fungi			
		Unicellular forms		Oxidative dimethylase	
	Tubular	Vertebrate retina	Highly ordered or densely packed regions of uniformly diametered tubules	TPHN-cytochrome *c* reductase, and other drug-metabolizing enzymes	Synthesis of phospholipids; early steps of glycoprotein assembly
		Annelid luminous organs			
		Red algae			
	Cisternal smooth	Protistans	Cisternae associated with organelles		
		Metaphytans			
	Dictyosomal (GERL)	Many eukaryotes	Associated with dictyosomes	Acid phosphatase, alkaline phosphatase	

monly no distinction is made in the literature between the several types. One of the two major classes, the rough, or granular, endoplasmic reticulum, derives its name from the presence of ribosomes on its membranes, and thus is readily distinguished from the members of the smooth class that lacks those particles in whole or in large measure. One minor and two major kinds of the agranular reticula occur, tubular, cisternal, and vesicular, but although it has not been finally established that these represent distinct organelles, they are so considered here, rather than mere structural variations of one (Table 6.1). Moreover, in the present account, the term "endoplasmic reticulum" is reduced to "endoreticulum" or "endoreticular membranes," which names are more compact and less awkward than that in current use.

6.1.1. The Rough Endoreticulum

In eukaryotic cells as a whole, the rough, or granular, endoreticulum is by far the commoner class. In reality its components are combinations of two separate and distinct organelles, one membranous and the other granular. As the granule is the ribosome that forms part of the genetic mechanism at the macromolecular level described elsewhere (Dillon, 1978), only those features pertinent to present purposes receive attention, leaving most of the discussion to center on the membranous portions (Parry, 1978).

The Membranous Component. Because ribosomes occur free in the cytoplasm and nucleoplasm, as well as on the outer nuclear envelope and plasmalemma, in addition to their location on the present membranes, the latter therefore may validly be viewed as an organelle in its own right, even as the nuclear envelope is. However, the most characteristic form of this structure is not known to exist free of ribosomes, so closely are the two parts related functionally.

Structural Characteristics. In electron micrographs the typical configuration of the rough endoreticulum is a series of more or less parallel, flattened saccules called cisternae. These vary in shape and extent to some degree, being uniformly flattened in many cases and distended to variable amounts in others (Figure 6.1). Commonly their lumina are around 15 to 20 nm in thickness, while the membranes are around 60 Å in diameter, exclusive of the ribosomes.

→

Figure 6.1. Two varieties of metazoan rough endoreticulum. (A) The usual stacks of flat cisternae are located near the periphery of the cell at the lower edge of the print, while above it is an unusual whorl surrounded by glycogen (Gly) in this *Xenopus* liver cell. 15,000×. (B) At higher magnification, the whorl (as well as flat cisternae) may be seen to have numerous points of contact that form the fenestrae of freeze-fracture preparations (See Figure 6.2). 41,000×. (A and B, courtesy of Brown, 1978). (C and D) Similar stacks and occasional whorls occur in the seed plants, such as this pine cell. 18,750×. (Courtesy of Kupila-Ahvenniemi *et al.*, 1978.)

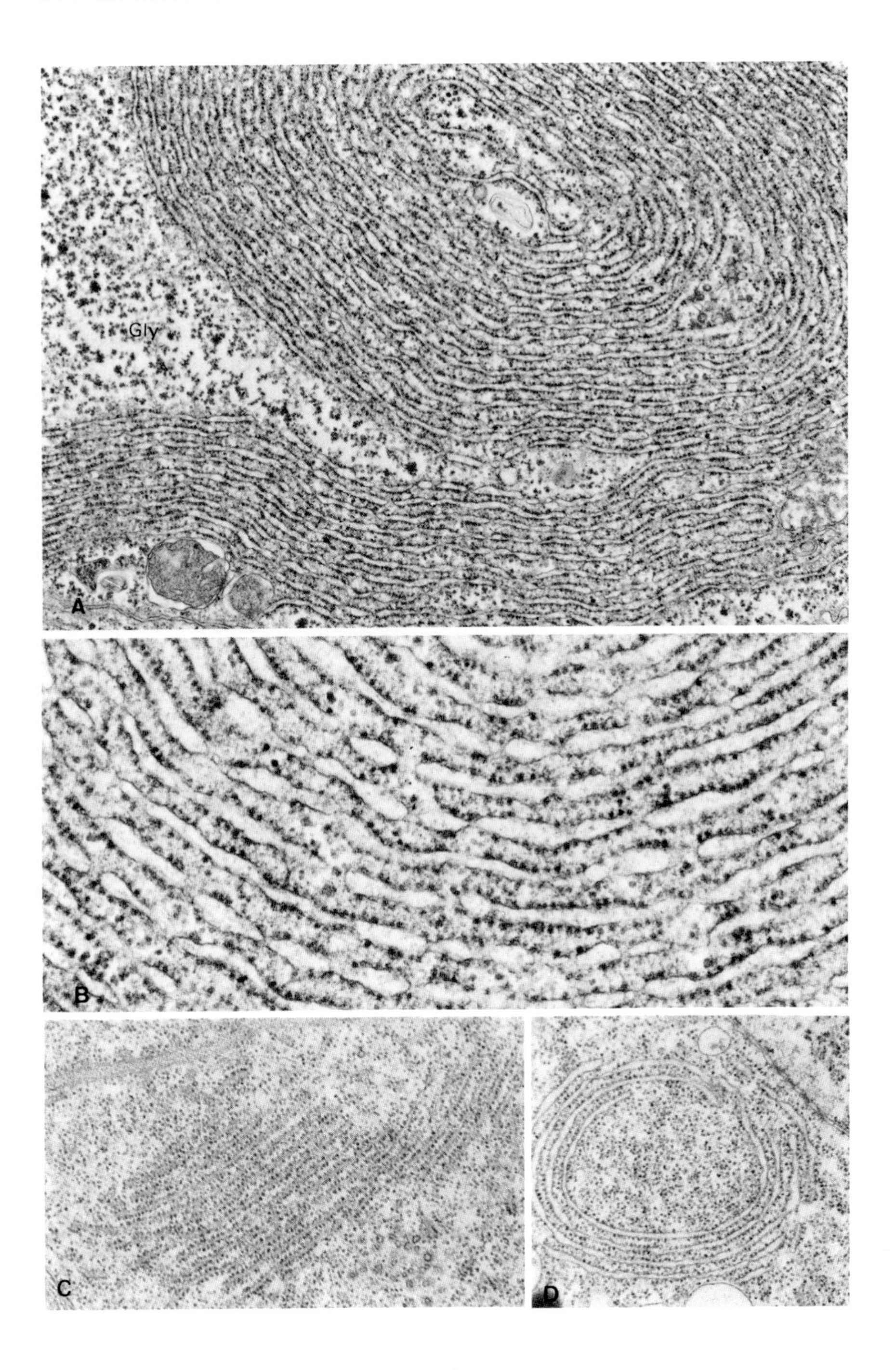
Gly
A
B
C
D

At times the cisternae may be swollen at their periphery to form vesicles, and here and there gaps may be noted to interrupt their continuity. Characteristically these gaps are considered to be pores, sometimes called fenestrae; they are best seen in face views, especially in freeze-etch preparations (Figure 6.2; Branton and Moor, 1964). Interconnections from one cisterna to an adjacent one, too, are generally to be noted (Figure 6.1B).

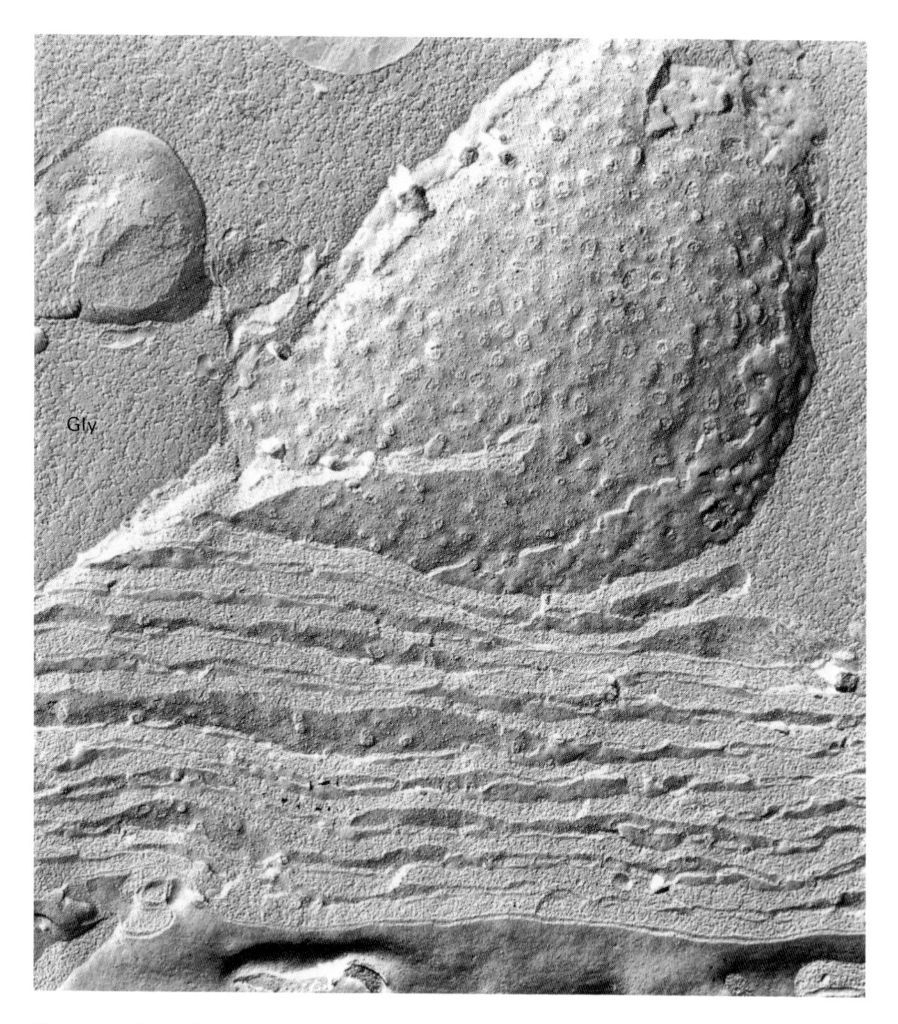

Figure 6.2. Freeze-fracture preparation of rough endoreticulum. In the whorl (above) and flat cisternae (below) in *Xenopus* liver cell prepared by freeze-fracture techniques, the interconnections that had existed between cisternae are now represented by the porelike fenestrae. 27,000×. (Courtesy of Brown, 1978.)

The breadth of the cisternae and their number per stack vary with the nature of the cell, being greatest in vertebrate tissues that are actively engaged in secretion of materials for export to other parts of the organism. In such cells, 30 to 50 or even more cisternae may be present in a single stack. However, in less active tissues, 12 to 15 is a more general rule, but in other instances only a few, or even a single one, may be present. In the liver cells of *Xenopus laevis*, stacks of cisternae are abundant along the periphery of the cells, but in addition large whorls of cisternae are found in most electron microscopic sections (Figure 6.1A; Brown, 1978).

Comparative Aspects. Such extensive parallel arrays of planar cisternae, however, are characteristic only of vertebrate cells. In lower metazoans, no stacks are present, only isolated irregular cisternae, whose lumina widen and narrow frequently, as in epitheliomuscular and nerve cells of *Hydra* (Lentz, 1965; Lentz and Barrnett, 1965). At most, three or four cisternae may be associated together, typically anastomosed at one edge, and never arranged in strictly parallel fashion, as shown by gland cells and cnidoblasts of *Hydra* (Lentz, 1965; Mattern *et al*., 1965). Generally, all the lumina are more dilated than are those of the vertebrates. In more advanced types, such as insects and echinoderms (Jacob and Jurand, 1963; Fabre *et al*., 1969), more complex and better ordered arrangements are to be found, but they rarely exceed five components per stack and the cisternae are always irregular in diameter.

Among higher plants, cells that secrete proteinaceous products, such as aleurone cells, display stacks of rough endoreticulum, the cisternae of which approach those of vertebrates in breadth; usually, however, these do not exceed 15 or 20 cisternal components per stack (Buvat, 1961; Chrispeels, 1977). Moreover, even in such stacks, the cisternae are never completely parallel and regular in diameter as they are in the vertebrate organelle, dilations and constrictions being frequent throughout their lengths. Interconnections between adjacent membranes also tend to be more frequent than in vertebrate cells, and pores too are more numerous (Jones and Chen, 1976). An unusual stack has been shown in the lacticifers of *Papaver*, consisting of a single continuous cisterna folded into about 15 layers, with ribosomes densely sandwiched in between (Nessler and Mahlberg, 1977). As a whole, however, parallel arrays are not characteristic of metaphytan cells, even where endoreticular membranes are relatively abundant. The more typical arrangement is a series of widely separated cisternae loosely associated in a broad region, with very little parallelism to be noted (Chrispeels, 1977).

Among protistans the limited amount of rough endoreticulum that is present occurs in two forms. The first type is like the flat cisternae found in vertebrates and metaphytans, but as in lower metazoans, it is confined to single units. In its most typical form, the single cisterna surrounds the nucleus to a greater or lesser extent, with frequent gaps interrupting the continuity, a condition not infrequent in metazoans (Figure 6.3; Peek and McMillan, 1979). This

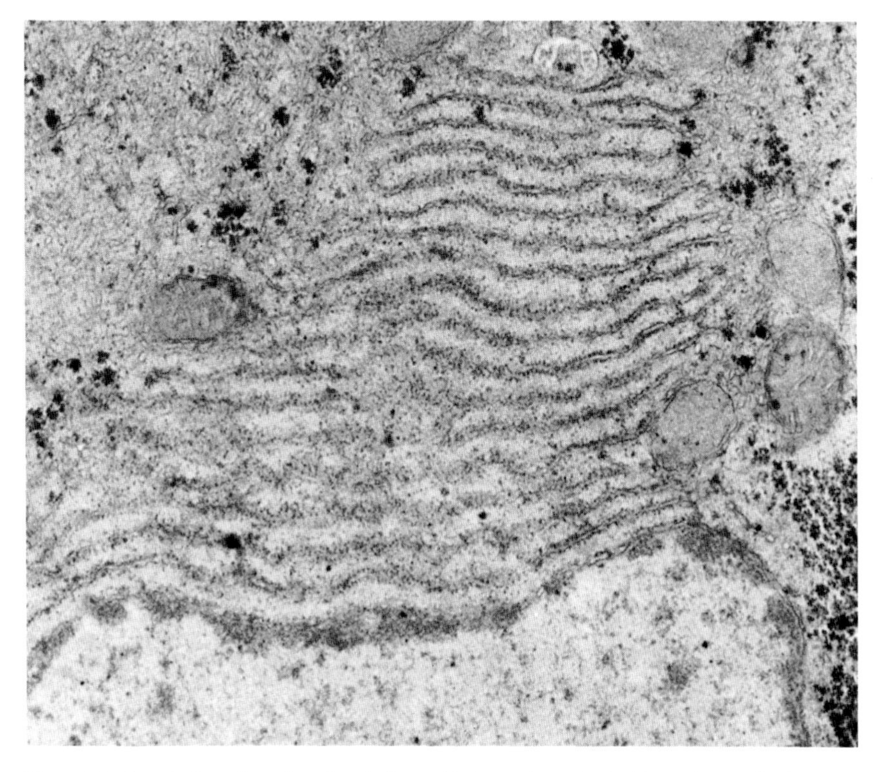

Figure 6.3. Association of rough endoreticulum with the nucleus. As in this nephron cell of a garter snake, rough endoreticulum frequently is found adjacent to the nuclear membrane in metazoans and many other types of eukaryotes. 30,000×. (Courtesy of Peek and McMillan, 1979.)

sort of arrangement has been reported for various algae, including *Mallomonas* (Figure 6.24) and *Schizomeris* (Mattox *et al.*, 1974; Wujek and Kristiansen, 1978). More frequently, even in many genera an endoreticular membrane is associated with the chloroplast that is usually categorized as of the rough type, some examples being the endosymbiotic alga of the dinoflagellate *Amphistegina* (Figure 6.4; Berthold, 1978), various xanthophyceans (including *Polyedriella*),

Figure 6.4. Associations of endoreticulum with various organelles. Generally, organelle-associated endoreticulum (ER) has been considered to be of the rough type, because of the presence of ribosomes. The latter are so sparse, however, as that shown near the chloroplast of the endosymbiotic diatom (A), that here it is considered to be of the smooth variety. 50,000×. (Courtesy of Berthold, 1978.) (B) This interpretation is supported by the total absence of ribosomes on the chloroplast-associated endoreticulum of the alga *Polyedriella*. 25,000×. (Courtesy of Hibberd and Leedale, 1972.) (C) Ribosomes do occur on those cisternae that are associated with microbodies of kidney cells from the garter snake. 25,000×. (Courtesy of Peek and McMillan, 1979.)

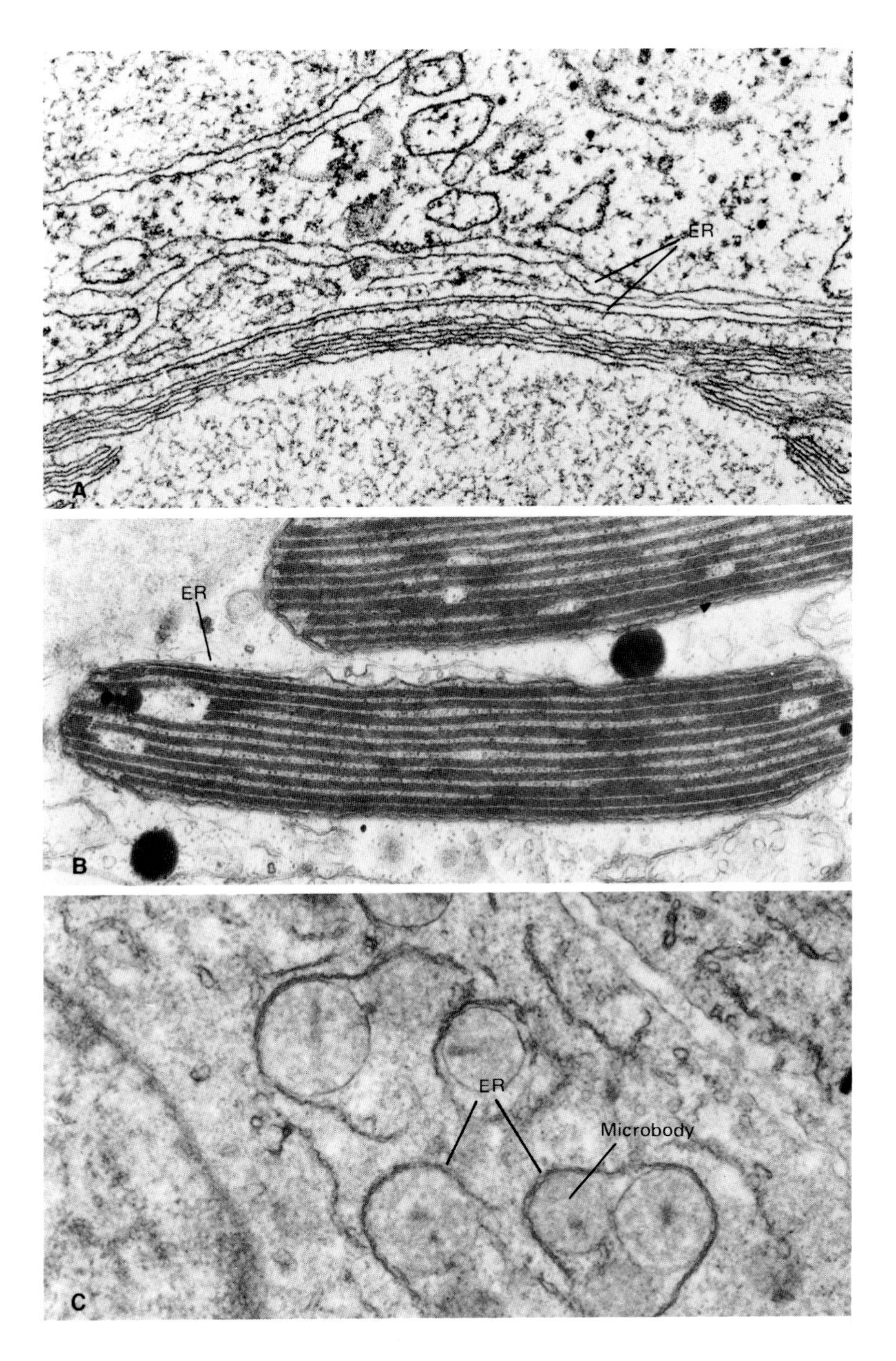
ER
A
ER
B
ER
Microbody
C

and eustigmatophyceans (Hibberd and Leedale, 1972). This association with the chloroplast also has been shown in electron micrographs of brown seaweeds (Bouck, 1965) and seed plants like *Lycopersicum* and *Solanum* (Abreu and Santos, 1977). Occasionally it may surround a mitochondrion, as in the fungus *Neobulgaria* (Moore and McAlear, 1963), or some other body, such as an invading bacterium, as in *Mixotricha* (Cleveland and Grimstone, 1964), or the oral canal as in *Euglena* (Leedale, 1966). Here this chloroplast- or mitochondrion-associated type is considered to be a member of the smooth group, because while scattered ribosomes are occasionally present, they are widely separated as a rule, with broad stretches of agranular membrane between. Perhaps both types are actually present, but until thorough cytochemical investigations have been conducted, it seems more realistic to class it as smooth (Section 6.1.2). Endoreticular membranes clearly of the granular type do occur around particles known as peroxisomes (Figure 6.4; Peek and McMillan, 1979).

In *Tritrichomonas*, a series of cisternae has been observed partly enclosing a mass of smooth reticulum (Nichols, 1968), but in many organisms scattered single cisternae lie free in the cytoplasm, as shown by the brown flagellate *Prymnesium* (Manton, 1964), the red alga *Porphyridium* (Gantt and Conti, 1965), members of the green algal family Volvoceae (Lang, 1963), and such varied fungi as *Aspergillus* (Tsukuhara and Yamada, 1965), *Phlyctochytrium* (Lange and Olson, 1977), and *Synchytrium* (Figure 6.5; Lange and Olson, 1978).

At the most elementary levels of the eukaryotic world, rough endoreticulum appears either to be absent or quite atypical in structure, as shown by a number of ultrastructural studies on yeasts (Conti and Brock, 1965; Avers, 1967; Clark-Walker and Linnane, 1967). In part the uncertainty arises from the lack of studies directed toward this particular organelle in the organisms, but it also stems in part from the generally poor contrast yeast membranes provide for ultrastructural studies. Moreover, in the yeast cell, such organelles as mitochondria exist as single unfolded membranes under certain conditions, adding further to the confusion. What appear to be rough endoreticular membranes in these organisms occur singly, never in stacks, mostly just below the plasmalemma, and may or may not be in the form of cisternae. Never are they abundant, most sections under the electron microscope showing only one to three membranes (Avers, 1967; Clark-Walker and Linnane, 1967). When freeze-etched preparations (Moor and Mühlethaler, 1963) reveal the surface of rough endoreticular membranes, the ribosomes may be noted to be densely placed, except for frequent circular areas that are free of those granules.

A second type of rough endoreticulum in protistans consists of numbers of vesicles that are highly irregular, both in size and form. Whether these are interconnected by tubular portions has not been clearly established, but in general they give the impression of being clusters of independent bodies. This vesicular

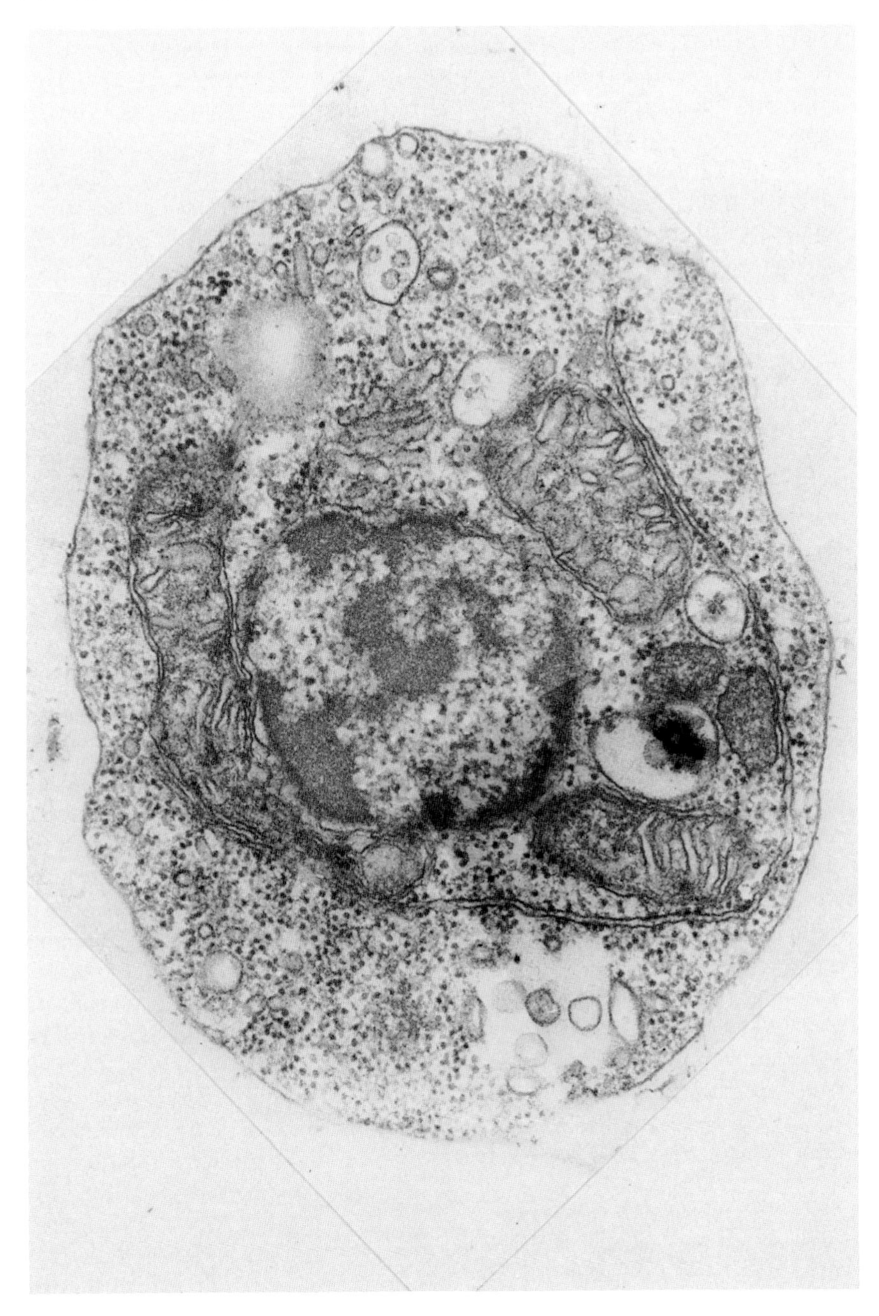

Figure 6.5. Cisternal rough endoreticulum. The zoospore of the fungus *Synchytrium* shows the typical form of the rough endoreticulum in unicellular eukaryotes in having a single cisterna extending widely through the cell. 42,500×. (Courtesy of Lange and Olson, 1978.)

type has been shown to occur in *Euglena* (Leedale, 1966), *Porphyridium* (Gantt and Conti, 1965), *Prymnesium* (Manton, 1964), *Tritrichomonas* (Nichols, 1968), and the myxomycete *Stemonitis* (Mims and Rogers, 1973), to name but a few of numerous examples. While widespread, it is neither as frequent nor as readily identified as the cisternal variety, for ribosomes are not so densely applied to either type of these structures among the protistans as they are in metazoans and metaphytans. Such irregular vesicles occasionally occur in metaphytan tissues, as in tapetal cells during pollen formation (Figure 6.6A; Santos *et al.*, 1979).

Tubular Rough Endoreticulum. An additional form of rough endoreticulum is occasionally encountered in vertebrate tissue, in which series of tubules, rather than flattened saccules, make up the organelle (Figure 6.6B; Taira, 1979). Typically, these are much larger in diameter than the cisternae, averaging close to 100 nm when viewed in cross section. One noted example of this type was originally reported from bat stomach, where the tubules occurred in large series, regularly packed in hexagonal fashion (Ito and Winchester, 1963). On the surface of each tubule were 12 ribosomes, arranged so systematically that adjacent granules often contacted one another to form linear arrays. Obviously, then, the ribosomes are not in the form of polysomes, as is so frequently the case with the usual cisternal stacks.

The Ribosome. Ribosomes, the final link in the direct chain of molecular events in the genetic processes (Tissières, 1974; Kolata, 1975; Kurland, 1977; Stöffler and Wittmann, 1977), also form part of the membranous system that constitutes a fraction of these activities at the cellular level. These are the particles in which translation of the messenger into the polypeptide chain occurs, but there are far greater numbers of ribosomes free in the cytoplasm carrying out this same function than are attached to these membranes, a point worthy of being borne in mind. It is also evident that, in a sense, the ribosomes functionally are intermediate between these two major facets of the genetic mechanism, because they terminate the molecular processes and commence the cellular. In addition to the RNA-containing body, DNA has also been reported associated with rough endoreticulum (Bond *et al.*, 1969).

Ultrastructural Aspects. The intermediacy of position also becomes evident from an ultrastructural point of view, for whereas the units of interaction at the molecular level are not resolvable by electron microscopy, the ribosome

Figure 6.6. Other forms of rough endoreticulum. (A) Sometimes the rough endoreticulum occurs as large irregular vesicles, as here in the cucumber tapetum. 30,000×. (Courtesy of Santos *et al.*, 1979.) (B) Acinar cell, with extensive rough endoreticulum in the form of large tubules. 43,000×. (Courtesy of Taira, 1979.) (C and D) Follicle cells of molluscs, displaying the abundant tubular rough endoreticulum. (C) From *Ilyanassa*, 22,000×; (D) from *Cratena*, 30,000×. (Both courtesy of Huebner and Anderson, 1976.)

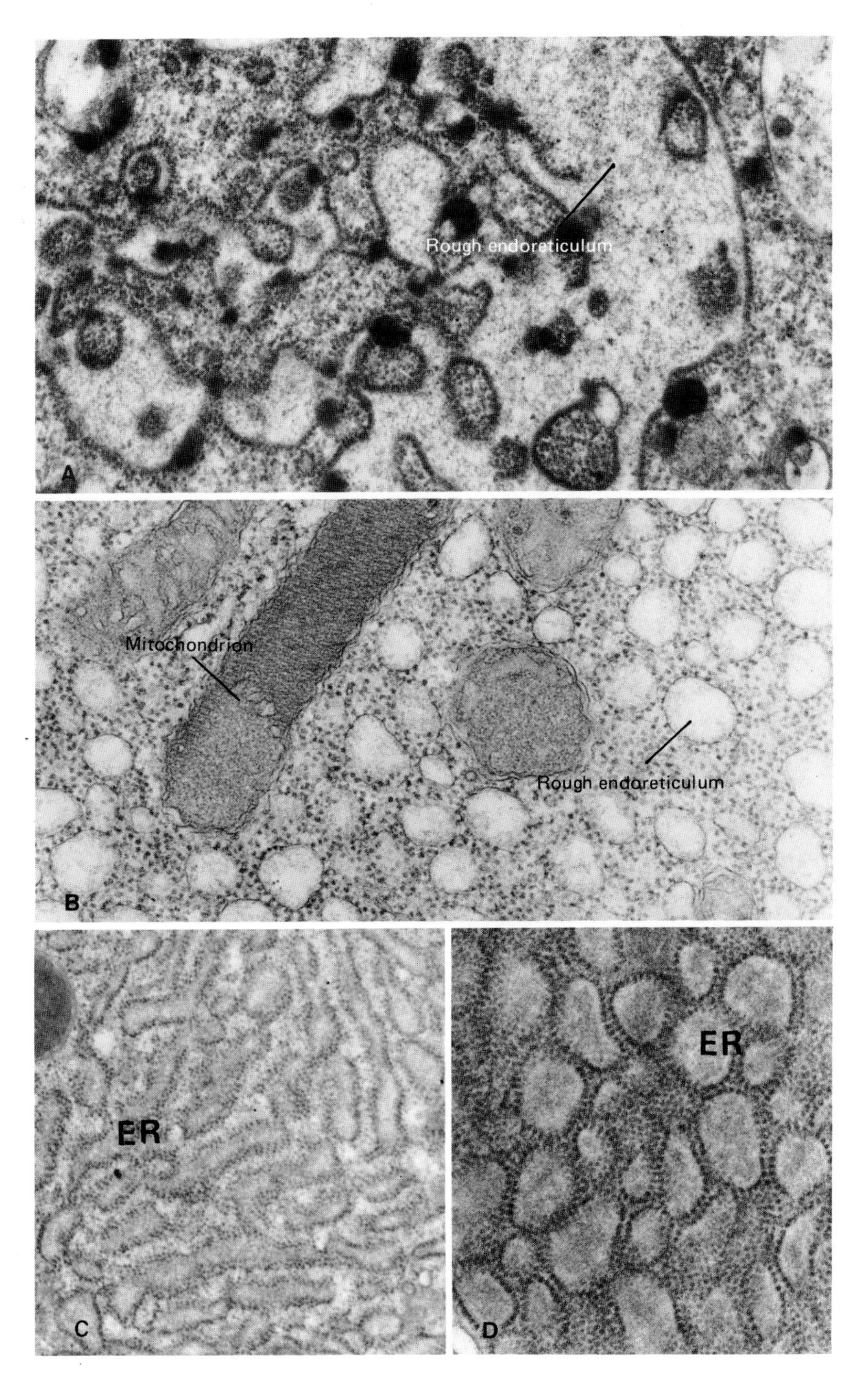
Rough endoreticulum
A
Mitochondrion
Rough endoreticulum
B
ER
C
ER
D

is. While it is true that the DNA molecule, the nucleoid of bacteria, and the chromosome of eukaryotes can be viewed readily by ultrastructural techniques, the individual genes are not discernible as such, nor are the polymerases or even the transfer RNAs. As the center of translation, ribosomes are indispensable in all cellular organisms and even occur in more advanced viruses (Dillon, 1978), but in prokaryotes, all are either free in the cytoplasm or on the plasmalemma, as endoreticulum of any type does not occur.

All known ribosomes consist of two unequal parts. In *E. coli,* the smaller subunit sediments at 30 S and consists of 21 proteins attached to a single RNA molecule that sediments at 16 S. The larger subunit has a sedimentary coefficient of 50 S and is comprised of about 32 proteins embedded on two RNA chains, one sedimenting at 5 S, the other at 23 S. The complete, more or less spherical ribosome (the monosome) containing one of each subunit sediments at 70 S and has a diameter of 225 Å (Van Holde and Hill, 1974; Boublik and Hellmann, 1978). Both subunits are quite irregular in shape, the small one being essentially a vertical, stout cylinder, with a broad projection on one side (Figure 6.7A). In contrast, the larger one is somewhat transverse and has two projections that give it the appearance of a teapot. Uncertainty still prevails as to how these two configurations can join to form a nearly spherical monosome

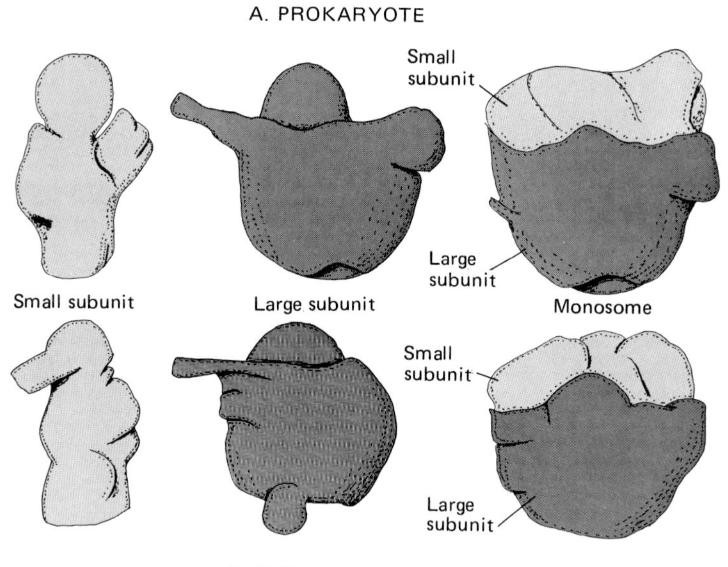

Figure 6.7. Ribosomal structure of prokaryotes and eukaryotes. In both types of organisms the complete ribosome (monosome) consists of a small and a large subunit.

(Lake *et al.,* 1974); the illustration provides one of several combinations that have been proposed (Lake, 1976; Cornick and Kretsinger, 1977; Stöffler and Wittmann, 1977; Grunberg-Manago *et al.*, 1978).

Among eukaryotes as a whole, the subunits are larger and more complex than those of prokaryotes. The smaller subunit sediments with a coefficient around 40 S and consists of approximately 30 proteins attached to a single RNA molecule that sediments at 18 S (Wool and Stöffler, 1974). The larger one has a sedimentation coefficient of 60 S and consists of about 40 proteins arranged on three RNA molecules that sediment with coefficients of 28, 5.8, and 5 S (Pene *et al.*, 1968; Weinberg and Penman, 1970; Rubin, 1973; Dillon, 1978). Even in thorough-going reviews of ribosomal structure, the presence of three RNA molecules in the large subunits is occasionally overlooked (Wool and Stöffler, 1974) or the 5.8 S is confused with the 5 S molecule (Boublik and Hellmann, 1978). The two are quite distinct and have completely different nucleic acid sequences, as shown elsewhere (Dillon, 1978, Table 4.4).

Various interpretations have been presented of the ultrastructure of the monosomic metazoan ribosome and its subunits (Kiselev *et al.*, 1974; Meyer *et al.*, 1974; Wittmann, 1976; Boublik and Hellmann, 1978) but high-magnification images suggest that the small subunit is an elongate body, as is the prokaryotic one. However, one end is strongly recurved and the lateral process is broad and widened distally (Figure 6.7B). The large subunit is less teapot-shaped than its prokaryotic counterpart, lacking a handlelike elevation and having a slender tubular process instead of a spout. In addition, there is a small, rounded appendage at the lower end, which is not to be noted in the prokaryotic structure. It is tempting to suggest that this possibly is the 5.8 S RNA molecule and its adhering proteins, but no evidence has as yet been advanced as to its actual identity. As in prokaryotes, these two irregular bodies are somehow able to unite to form a subspherical, monosomic body, as indicated in Figure 6.7B.

Phylogenetic Origins. Much speculation on the evolution of ribosomes has appeared in print, based particularly on similarities and differences in molecular weight of the respective monosomic body and subunits. For instance, it has been pointed out that there is a gradual increase in size of the organelle and its parts from prokaryotes to vertebrates (Noll, 1970). Even in the vertebrate series, beginning with amphibians and continuing to mammals, a trend in this same direction has been described (Loening, 1968). But in this latter sequence, the birds were indicated as being intermediate between reptiles and mammals, rather than as a separate lineage as in current views. Moreover, the overall trend from bacteria to higher eukaryotes was more sharply defined when relatively little information was available than it is now that data from a broader spectrum of organisms have been accumulated. To cite one case in point, *Euglena,* shown by flagellar and numerous other traits to hold a primitive position among eukaryotes, has been found to have the largest cytoplasmic ribosomes thus far investigated, with a sedimentation coefficient of 86 S compared to the 80 S of mammals (Rawson and Stutz, 1969). Because size comparisons necessarily involve the ribosomes of two cell organelles still to be examined, this subject is better reserved for consideration in connection with those discussions (Chapter 9, Section 9.3.3).

While at present the state of knowledge of this complex body necessarily limits comparative studies to sedimentation characteristics, it should be realized that the monosomic ribosomal molecular traits arise from a combination of many variables. Hence, the ribosomes from two types of organisms may have identical sedimentation coefficients and yet be nearly totally different in details of construction. By way of illustration, insect ribosomes proved to have sedimentation characteristics like those of metaphytans, the monosome sedimenting at 78 S while the RNAs of the small and large particles had coefficients of 17 S and 25 S (Loening, 1968; Noll, 1970). On the basis of these characteristics alone, insects would have to be considered more closely related to the seed plants than the fungi are, for the *Neurospora* monosome sediments at 77 S (Stutz and Noll, 1967).

In a closer view of some details, superficial as they are, the lack of significance of such comparisons becomes more apparent. Table 6.2 presents a summary of the numbers of proteins found in the ribosomes and their subunits from various tissues of eukaryotes. In the first place, it is immediately apparent that the numbers reported in some instances, as in rat and rabbit liver, vary over a broad range, depending on the laboratory that investigated them. These differ-

Table 6.2
Number of Proteins in Various Eukaryotic Ribosomes and Their Subunits

Source tissue	Protein numbers			References[a]
	Monomer	Large subunit	Small subunit	
Rat muscle	69	38	31	5
Rat liver	61-70	34-40	27-31	4,5,7,8
Mouse liver	67	38	29	9
Rabbit liver	68-75	38	30	2,9
Rabbit reticulocytes	70-73	38-40	32-33	1,3,9
Chicken liver	66	36	30	9
Yeast	80	50	30	6

[a]1, Chatterjee *et al.* (1973); 2, Huynh-van Tan *et al.* (1971); 3, Martini and Gould (1971); 4, Sherton and Wool (1972); 5, Sherton and Wool (1974); 6, Warner (1971); 7, Welfle *et al.* (1971); 8, Welfle *et al.* (1972); 9, Wool and Stöffler (1974).

ing results arose in spite of the use of the same general analytical procedures. Second, it is evident that, although all the intact ribosomes fall into the 80 S category on the basis of centrifugation studies, the finer points, that is, their true morphological traits at the molecular level, are quite disparate from one another. Even at still finer levels of organization, doubts must exist as to the significance of sedimenting characteristics in phylogenetic studies. The largest protein of the small unit Sl from *E. coli* occurs in only 10 to 30% of purified ribosomes (Wittmann, 1974), and *in vitro* systems with ribosomes reconstituted without this protein function quite normally (Held *et al.,* 1973). Therefore, this could be a protein of the cytosol rather than of the ribosome and might sediment with the ribosomal fraction in one prokaryote and with the supernatant in another, changing sedimentation characteristics accordingly.

Primary Structure of 5 S RNAs. More reliable characteristics exist in this particle and when the primary structures of many of the varied proteins and RNAs from a suitable series of organisms have been established the real sequence of evolutionary events may then be determined with a high degree of reliability. One type of RNA has now been investigated to this end in a number of organisms to at least provide insight into the nature of relationships that may be expected. This RNA is the 5 S variety found in the large ribosomal subunit of all organisms, prokaryotic and eukaryotic alike, over 50 sequences of which have now been reported. As the complete sequences are listed in detail elsewhere (Dillon, 1978; Erdmann, 1980), here, in order to make comparisons more meaningful, the primary structures have been grouped by selecting the most frequent base at each site (Table 6.3).

Counts of corresponding sites that differ between any given two sequences, while not to be taken overseriously by themselves, do justify the drawing of certain conclusions and reveal several general trends. Also they disclose a sufficient number of disparate results to cast doubt on the validity of phylogenetic trees based on numbers of base or amino acid changes. For instance, it appears sound to suggest that the single blue-green alga (*Anacystis*) probably represents the prokaryotic group most remote from the eukaryotes, for it differs from yeasts at 75 sites, whereas the *E. coli* and the *Bacillus* groups differ from that same taxon at only 66 and 61 sites, respectively. However, the blue-green alga cannot really be considered closely related to the two groups of bacteria, for 47 of its sites differ from corresponding ones of the *E. coli* group and 55 from those of the *Bacillus* type. Since the 5 S rRNA of the two bacterial groups are nonhomologous at 35 sites, three levels of advancement are to be noted among the prokaryotes, the bluegreen alga at the lowest position, *E. coli* and relatives above it at some distance, and the *Bacillus* group at the most advanced level. The comparisons of the two bacterial categories with the yeast sequence stated above give similar results, confirming the proposed three-level arrangement.

Among the eukaryotes, count comparisons result in confusion. The base

Table 6.3
Averaged Primary Structures of 5 S RNA from Major Taxa

Blue-green alga[a]	-UCCUGGUGUCU AUGGC GGUAUG GAACCACU CUGA CCCCAUCCCGAAC-UCAG UUGUGAAA CAU
E. coli group[b]	UGCCUGGCGGCC GUAGC GCGGUG GUCCCAC- CUGA CCCCAUGCCGAAC-UCAG AAGUGAAA CGC
Bacillus group[c]	--UCUGGUGGCG AUAGC GAAGAG GUCACAC- CCGU UCCCAUACCGAAC-ACGG AAGUUAAG CUC
Yeast group[d]	-GGUU-GCGGCC AUAUC UACCAG AAAGCAC- CGUU UCCCGU-CCGAUCAACUG ψAGUUAAG CUG
Green alga[e]	AUGCU-ACGUUC AUA-C ACCACG AAAGCAC- CCGA UCCCAU-CAGAAC-UCGG AAGUUAAA CGU
Metaphyta[f]	-GGAU-GCGAUA CCAUC AGCACU AAUGCAC- CGGA UC-CAU-CAGAAC-UCCG CAGUUAAG CGU
Metazoa[g]	-GCCU-ACGGCC AUACC ACCCUG AAAACAC- CGGA UCUCGU-CCGAUC-ACGG AAGUUAAG CAG
Vertebrata[h]	-GCCU-ACGGCC AUACC ACCCUG AAAGUGC- CCGA UCUCGU-CUGAUC-UCGG AAGCUAAG CAG
Blue-green alga	ACC UGCGGC-AACGAUAGCU ------CCCGGGUAG CCGGUGGCU AAAAUAGCU CGACGCCAGGUC-
E. coli group	CGU AGCGCC-GAUGGUAGUG -------UGGGGUCU CCCCAUGCG AGAGUAGGG AACUGCCAGGCAU
Bacillus group	UUC AGCGCC-GAUGGUAGGU -------GGGGGGUU GCCCCUGUG AGAGUAGGA CGUUGCCAGGC--
Yeast group	GUA AGAGCCUGACC-GAGUA GUGUAG--UGGGUGA CCAUACGCG AAACUCAGG UGCUGC-AAUCU-
Green alga	GGU UGGGCUCGAC--UAGUA CUGGGU--UGGAGGA UUACCUGAG UGGGAACCC CGACGU-AGUGU-
Metaphyta	GCU UGGGCGAGAG--UAGUA CUAGGA--UGGGUGA CCCCCUGGG AAGUCCUCG UGUUGC-ACCUU-
Metazoa	CGU CGGGCCUGGU--UAGUA CUUGGA--UGGGGGA CCGCCUGGG AAUACCAGG UGCUGU-AAGCUU
Vertebrata	GGU CGGGCCUGGU--UAGUA CUUGGA--UGGGAGA CCGCCUGGG AAUACCAGG UGCUGU-AGGCUU

[a] *Anacystis*.
[b] Mean of 16 sequences.
[c] Mean of 14 sequences.
[d] Mean of 6 sequences.
[e] *Chlorella*.
[f] Mean of 5 sequences.
[g] Proposed primitive sequence.
[h] Mean of 8 sequences.

sequence of yeast 5 S rRNA differs from that of the green alga (*Chlorella*) at 59 sites but from that of the seed plants at only 53, intimating that the yeasts are more closely related to the end products of that line of ascent than to its beginnings! This rRNA of the green alga obviously has undergone many alterations since the point of departure from the green plant main line, for with 43 differences it diverges from the metaphytan sequence almost as strongly as the 44 differences that exist between the seed plants and the metazoans. Further it can be observed that the yeasts are distinguished from the metazoans at only 41 sites, suggesting a closer kinship between these two taxa than between the former and the seed plants, where 53 sites are at variance.

Time considerations (chemical clocks) also most be viewed with caution. From this standpoint the nine nonhomologous sites between the vertebrate and proposed ancestral metazoan sequence suggest a rate of change of about 1 site per 50,000,000 years, which by itself appears reasonable in this highly conservative biochemical. However, the 55-site difference between the blue-green alga and *Bacillus* group (2,750,000,000 years) is difficult to reconcile with the age given by that of the 75-site difference between the former and yeasts (3,750,000,000 years). Moreover, the metaphytans would have to be considered to have diverged from the metazoan line 2,250,000,000 years ago and from the green alga at about the same time. Actually the beginnings of the metaphytan line can be traced back in the geologic record only to about 350,000,000 years ago. Moreover, the remote relationships indicated by altered sites in this rRNA between the green alga and seed plant are highly suggestive of the absence of uniformity in the rate of mutation in this substance throughout the living world. Until long series of such sequences have been established from a wide diversity of unicellular forms, counts of differences are of speculative interest only.

Direct Comparisons. Direct comparisons by scanning, however, do make some pertinent conclusions clear. First, the prokaryotic sequences, in spite of the 30 to 45% differences that exist between corresponding sites, show a high degree of interrelationships among themselves and, at the same time, an absence of close kinship to the eukaryotes, and similar statements are true for the latter. Sites 24 to 31 provide a case in point, as do also sites 85 to 90—the absence of this latter six-site sequence on the one hand and its presence on the other is particularly striking. Contrastingly, site-by-site comparisons of pairs of adjacent sequences strongly indicate relationships between the organisms concerned. This is true even for the line of contact between prokaryotic and eukaryotic sequences provided by those of the *Bacillus* and yeast groups. While a 50% difference does exist, the correspondences are of such a nature that a distinct flavor of common descent emerges. Interesting as such comparisons may be, at best present knowledge of 5 S rRNA primary structure merely confirms what was already well established. Only similar knowledge of this RNA from fungi, euglenoids, amoebae, ciliates, flagellates, and other unicellular organisms will really enhance present understanding of the evolutionary history of the living world.

6.1.2. Smooth Endoreticulum

The type of endoreticulum lacking the coating of ribosomes that characterizes the rough endoreticular system has been far less thoroughly investigated than the latter, for several reasons. First, the absence of a ribosomal covering makes it much less visible in thin sections under the electron microscope, and for another, its irregular configuration and vesicular construction render it difficult to describe. Third, it lacks any outstanding characteristic whereby it can be positively identified, and finally, it is less readily isolated from the rest of the cell for study. However, a definite trend toward investigating this organelle more fully can be noted in current literature.

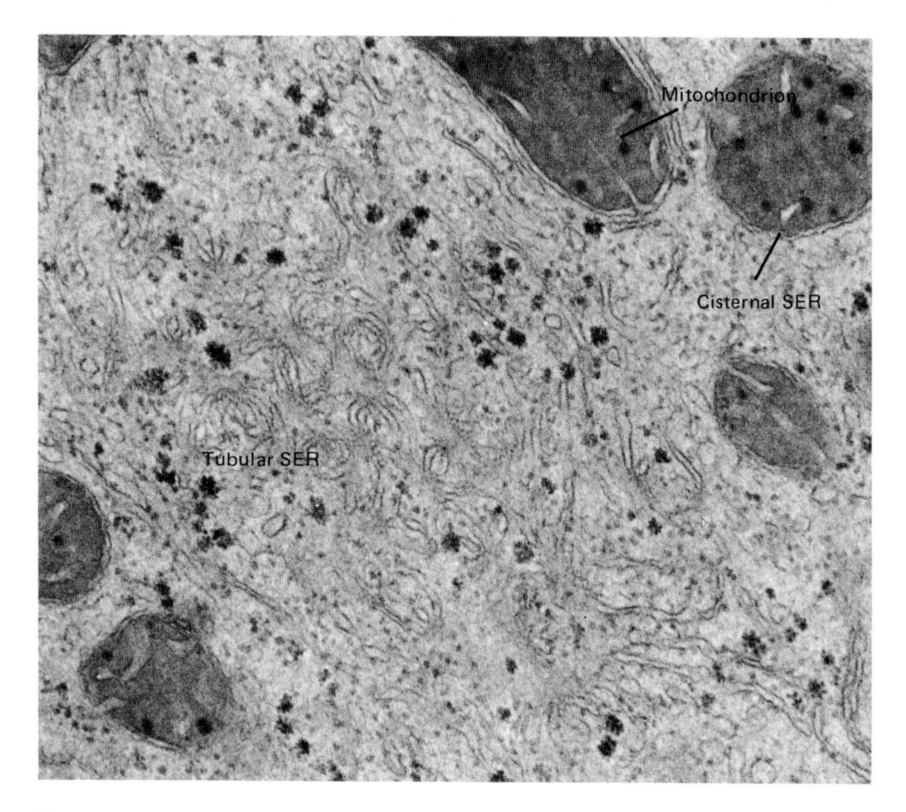

Figure 6.8. Smooth endoreticulum. Only infrequently do the two major types of smooth endoreticulum (SER) occur together as they do in this electron micrograph of the tubular nephron of a garter snake. The cisternal type frequently is associated with other organelles, such as a mitochondrion. 50,000×. (Courtesy of Peek and McMillan, 1979.)

Types of Smooth Endoreticulum. Two types of smooth (or agranular) endoreticulum are usually recognized, tubular and vesicular (Figure 6.8). Actually both varieties appear to be made of tubules, but in the vesicular kind, the tubules are irregular in form, being distended frequently and highly contorted throughout their length. In contrast, the second type is more visibly constructed of tubules, which in cross section are more regularly spaced and more nearly circular in outline. The latter type tends to lie closer to the surface of the cell, whereas the former is characteristically in the endoplasm (Porter, 1961a). Because transverse or, especially, oblique sections through the vesicular type tend to appear like isolated vesicles, there is much opportunity for misidentification of this organelle as Golgi vesicles or perhaps lysosomes. As the two types undoubtedly represent two entirely different organelles, not merely variations in the form of a single one, they are treated as such in the following discussions.

Vesicular Endoreticulum in the Mammalian Liver. Vesicular endoreticulum has been investigated more thoroughly in rat liver (see Figure 6.12A,B) than in any other tissue or cell. This is where they were first discovered and described (Fawcett, 1955) and where Porter and his co-workers conducted early investigations into their structure (Porter and Bruni, 1959; Porter, 1961a,b). In this tissue, vesicular endoreticulum is closely associated with glycogen particles, which are scattered throughout the cytoplasm of the hepatocytes bordering the sinusoids of the organ. The organelle is especially abundant in the glycogen regions of cells from animals that have fasted moderately (Cardell, 1977), where numerous vesicles and short sections of tubules can often be observed along with an occasional elongate portion. As the tubules have diameters averaging 50 nm and are quite distorted in outline, they cannot readily be mistaken for microtubules. Except for the abundant mitochondria, the entire glycogenous region is occupied by vesicular endoreticulum.

At the periphery of such regions, some of the tubules can be perceived to be united to rough endoreticular membranes, with the respective lumina continuous. Thus there is a distinct possibility that enzymes synthesized in the rough endoreticulum are passed into the vesicular tubules for conduction to the site of utilization. However, this conclusion should not be considered the sole means by which the smooth vesicles function, for experimental evidence shows otherwise. For instance, this organelle, but not the rough type, is stimulated by phenobarbital treatment (Higgins and Barrnett, 1972; Higgins, 1974). Such treatment resulted in an increase within the smooth endoreticulum of the enzymes that are involved in phospholipid synthesis, but these proteins were found to be consistently absent from the rough system.

Occurrence in Other Secretory Tissues. Vesicular endoreticulum as a whole is especially characteristic of secretory cells. One of the more striking examples of its occurrence is provided by the venom glands of a rattle snake (*Crotalus atrox*), in which nearly the entire cytoplasm is ramified by this organelle (Brown and Bertke, 1974). Another type of secretory tissue in which vesicular endoreticulum provides a prominent feature is that comprised of the interstitial cells of vertebrate testes (Christensen and Fawcett, 1960, 1961); unlike the liver, here only smooth vesicular reticulum is present. A further distinction in these cells is in the diameters of the vesicles, ranging from 30 to 45 nm, in contrast to the 10 nm of those of liver.

Vesicular Endoreticulum in Muscle Tissue. A second major type of tissue in which vesicular endoreticulum usually abounds is the contractile variety, particularly the skeletal and cardiac muscles of vertebrates, where it is known as the sarcoplasmic reticulum (Porter and Palade, 1957; Porter, 1961c). Although the precise configuration varies from one species and one muscle to another, in general the reticulum consists of series of irregular tubules and vesicles in the form of lacey sleeves around each bundle of muscle fibers. Each such sleeve terminates at a Z line, being interrupted there by desmosomes and transverse tubules, or T system, formed from the plasmalemma of the muscle cell (Peachey, 1965). Usually at each end and at the middle, there is an especially large vesicle, while the intermediate parts are often associated with mitochondria or clusters of glycogen granules. The frequent presence of glycogen in these vesicles is reminiscent of the association of this substance and vesicular endoreticulum found in liver cells. Two adjacent large terminal vesicles, which contain a diffuse granular material, are considered to form a triad with the transverse tubule between them, a region important in coupling the surface stimulus for contraction to the individual fibrillar bundles. It would seem that ionic currents might flow from the surface through the T tubules, exciting the adjacent smooth endoreticulum to secrete Ca^{2+} cations into the endoplasm and thereby initiate contraction (Freygang, 1965). Recently catalase has been found within the endoreticulum of hamster cardiac and skeletal muscle, as well as in peroxisomes (Figure 6.9; Christie and Stoward, 1979).

Vesicular Endoreticulum in Other Organisms. Vesicular endoreticulum appears to be widespread among eukaryotes in general, being perhaps a little less frequent in occurrence than the rough cisternal type but by no means rare. A particularly striking instance was described in the nucellus of cotton, in some cells of which the vesicles were large spheroids (Jensen, 1965). These were interconnected by fine tubules or flat cisternae, which occasionally showed unions with the outer nuclear envelope (Figure 6.10). In the micropylar cells of the same organism, vesicular endoreticulum similar to that of vertebrates was found. In the polypodiaceous fern, *Davallia fijiensis*, endoreticular vesicles were illustrated that recalled the condition frequently noted in rat liver cells, for the smooth reticulum occasionally seemed to be joined to the periphery of rough cisternae (Fisher and Evert, 1979). These were often associated with the dense, membrane-bound proteinaceous bodies called refractive spherules that are peculiar to vascular cryptogams. In the species named, these bodies were reported to be formed by either this organelle or the Golgi apparatus, but in lower metaphytans they have been shown to be derived solely from vesicular endoreticulum (Perry and Evert, 1975; Dute and Evert, 1977). Rhizoid cells of germinating ostrich fern spores displayed extensive regions of vesicular endoreticulum, intermingled with which were numerous single cisternae of the rough type (Gantt and Arnott, 1965). In *Chara,* associations of this organelle with spherosomes were sometimes detected (Pickett-Heaps, 1967).

Extensive regions of vesicular endoreticulum have been revealed by ultrastructural investigations of germinating fungal spores, as in the myxomycete *Stemonitis virginiensis* (Mims and Rogers, 1973). In fact, the entire cytoplasm

→

Figure 6.9. Catalase activity in hamster smooth endoreticulum. (A) Longitudinal section of cardiac muscle, showing catalase activity in the endoreticulum. (B) Transverse section of dystrophic skeletal muscle, with catalase activity showing in the endoreticulum. Both 560×. (Both courtesy of Christie and Stoward, 1979.)

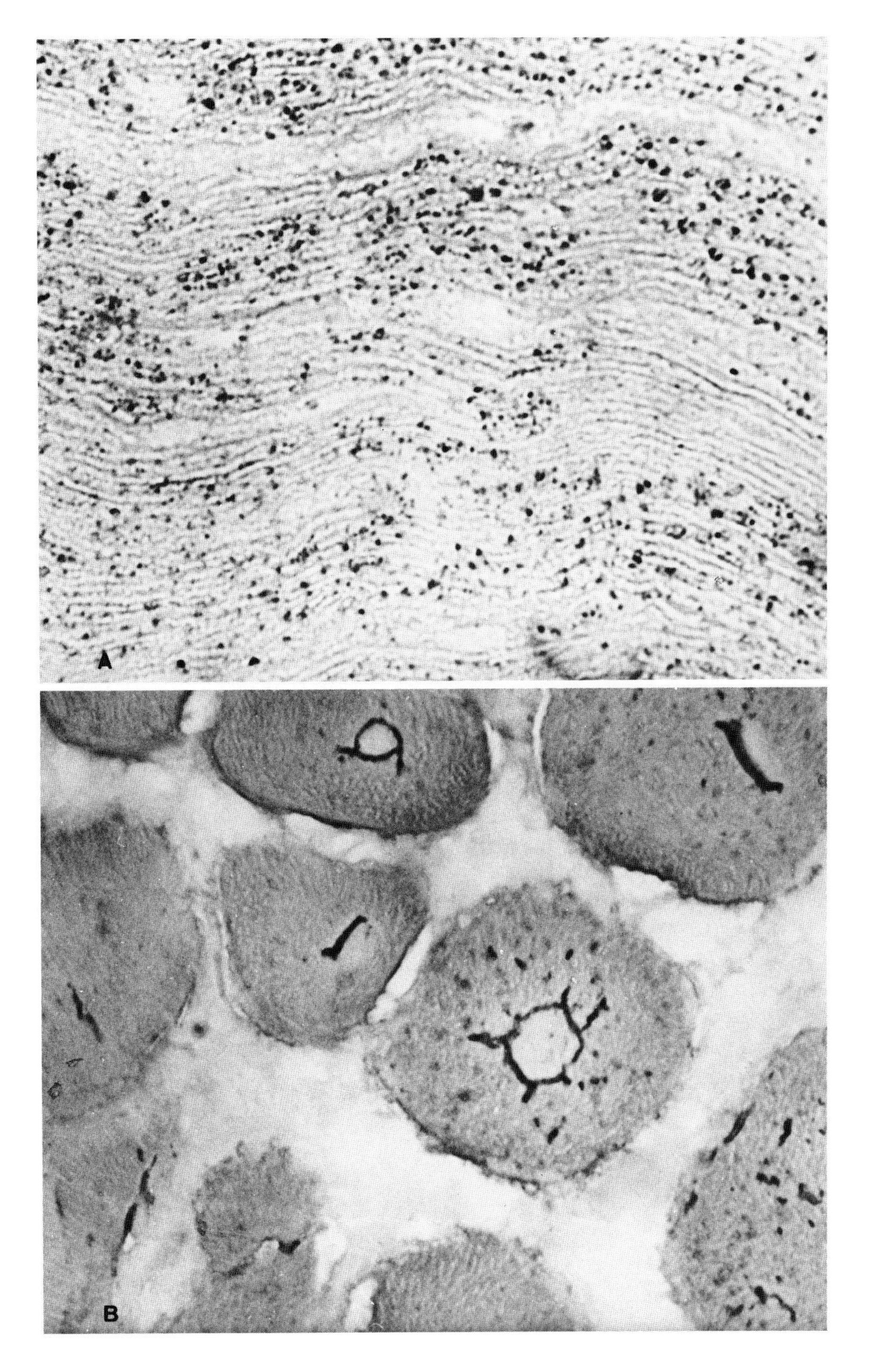
A
B

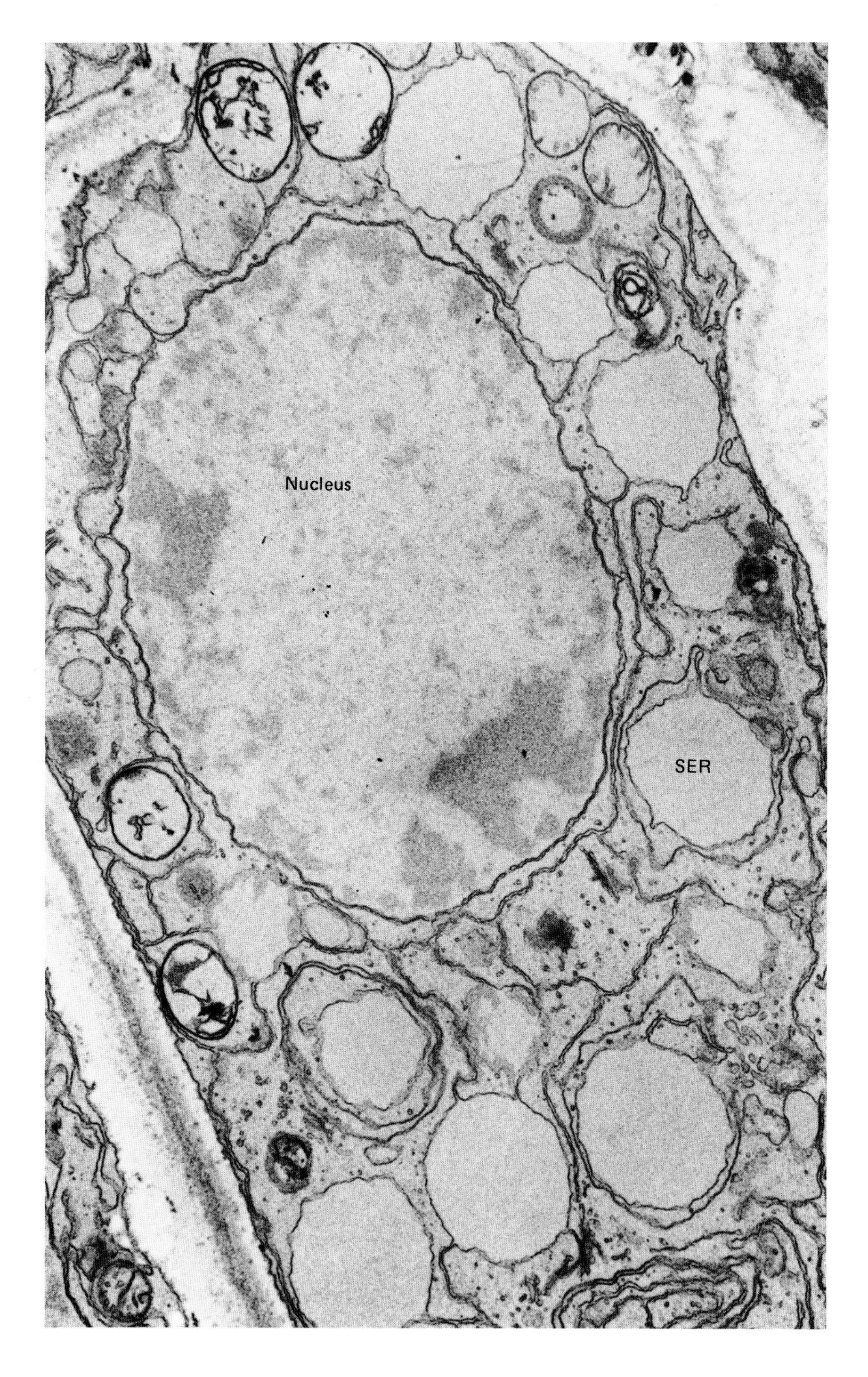
Nucleus
SER

of the mature meiospore of the aquatic phycomycete *Catenaria anguillulae* was nearly filled with this membranous organelle (Olson *et al.*, 1978). In many other protistans similar conditions have been observed. Electron micrographs of spores and mature cells of the sporozoan, *Schizoplasmodiopsis,* alike showed an abundance of vesicular endoreticulum (Dykstra, 1978), as did the trophozoite of a second member of the Sporozoa, *Didymorphyes* (Hildebrand, 1978). As in glycogen-containing cells of metazoans, floridean starch deposits in the red alga *Lomentaria baileyana* were found to have regions of this type of organelle associated with them, sometimes interspersed with which were single smooth cisternae (Bouck, 1962). In these same organisms, gland cells were literally packed with a mixture of swollen Golgi saccules and vesicles of smooth endoreticulum. Fusion cells in another genus of red algae (*Erythrocystis*) proved to have a most unusual arrangement of this organelle (Tripodi and DeMasi, 1978). While this structure itself was of typical form, it was enclosed along with numerous ribosomes within a cylinder of rolled membranes (Figure 6.11B,C).

Diurnal Rhythm in Smooth Endoreticulum. The vesicular type of endoreticulum has been found to undergo extensive changes in quantity and distribution in the liver of both normal and phenobarbital-treated rats. Cells taken from the pericentral region of untreated rat liver at 2 PM showed very little smooth endoreticulum, whereas those from the same region taken at 10 PM (Figure 6.12A,B) had far more smooth endoreticulum than rough, due to the proliferation of the former organelle, and the same change has been reported for phenobarbital-treated rats (Chedid and Nair, 1972). Cells from the periportal region displayed little or no change, as smooth endoreticulum never became extensive. This diurnal-rhythmic alteration in extent of the organelle was demonstrated to be accompanied by comparable rhythms in the microsomal enzyme content, levels of drug-metabolizing enzymes, such as hexobarbital oxidase, becoming enhanced as the smooth reticular membranes increased in abundance (Nair and Casper, 1969; Nair *et al.*, 1970). Similar increases have been described in pineal cells of baboon (Theron *et al.*, 1979).

Tubular Smooth Endoreticulum. The tubular type of smooth endoreticulum is distinguished from the vesicular type in having the tubules more or less constant in diameter, not fluctuating along their entire length nor expanded into chambers as is the case with the latter. When present, the tubular variety typically occurs in dense masses, one notable example of which is in the pigmented epithelial cells that lie just beneath the retina (Porter and Yamada, 1960). About 50% of the cytoplasm of this single layer of cells is occupied by a

←

Figure 6.10. Vesicular smooth endoreticulum. A collar cell of cotton nucellus with spherical enlargements of the smooth endoreticulum. Frequent connections to the nuclear membranes can be noted. 8000×. (Courtesy of Jensen, 1965.)

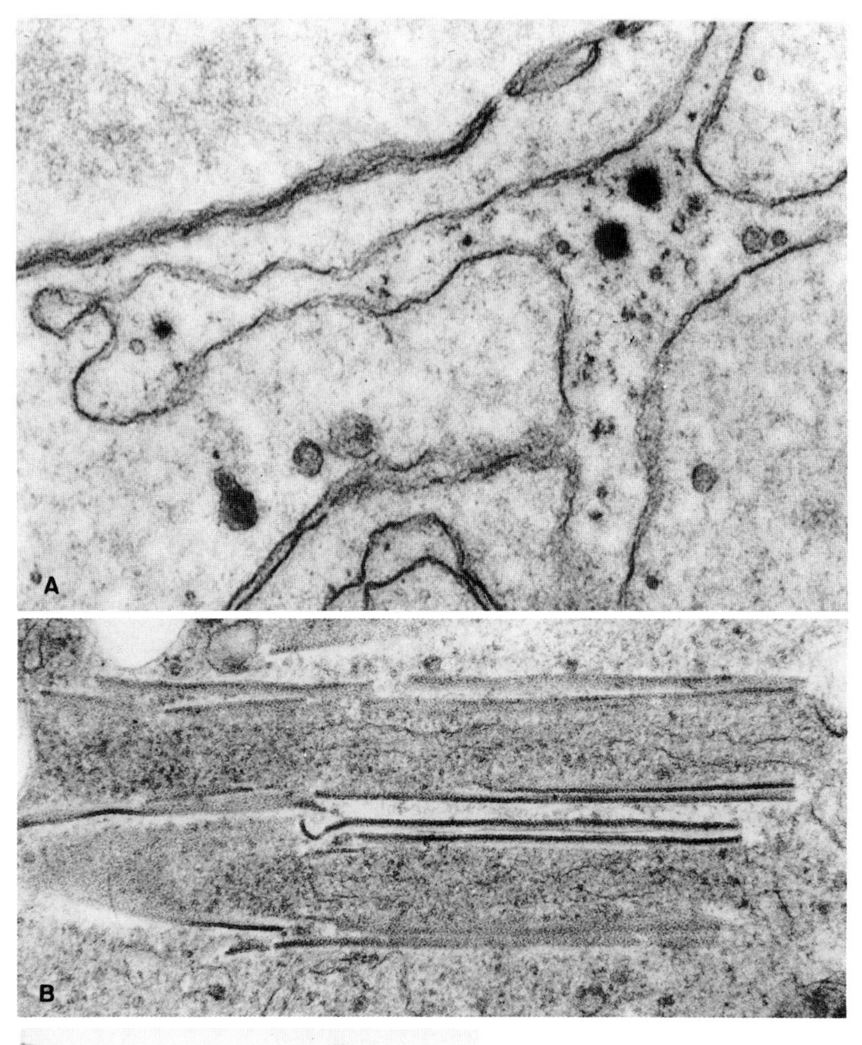

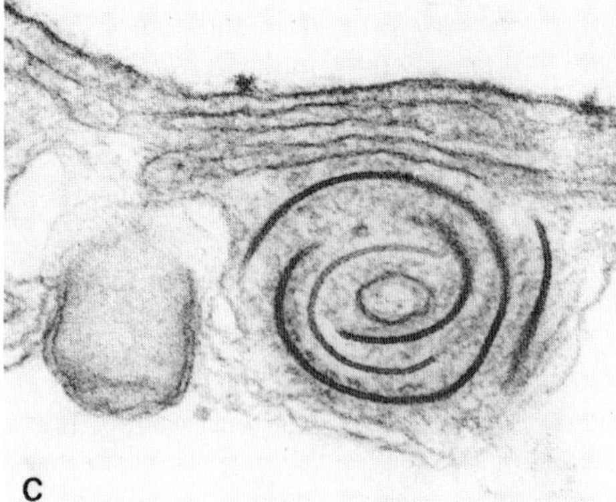

Figure 6.11. Unusual smooth endoreticulum. (A) This micropylar cell of cotton contains smooth vesicular endoreticulum that encloses granular material. 36,000×. (Courtesy of Jensen, 1965.) (B and C) In the fusion cells of the red alga *Erythrocystis,* a peculiar variant of smooth endoreticulum may be noted, consisting of long rolls, or perhaps tubules, enclosed by spiral membranes. (B) Longitudinal section, 50,000×; (C) transverse section, 58,000×. (Courtesy of Tripodi and DeMasi, 1978.)

loose network of tubules having a mean diameter of 75 nm. Here and there in the region of occurrence, a group of tubules expand into flat cisternae at their tips and form parallel arrays, known as myeloid bodies. Thus this modification represents an exception to the rule that the tubular type is of uniform diameter. The function of the myeloid bodies is unknown, but because of their structural similarity to known photoreceptor apparatuses, it has been speculated that they serve as photoreceptors for activating the cells of the pigment layer.

A second striking example of this organelle is provided by luminous tissues of the annelid *Annelides polynoinae* (Threadgold, 1976). In these cells, large areas of cytoplasm are occupied by a mat of highly ordered tubules, each of which is folded into a series of loops, arranged linearly. In turn, these looped tubules are placed in sequences on single planes, a series of which superimposed one above the other to form a stack comprise the tubular endoreticular complex. Frequently interconnections have been noted between the tubular elements and the outer nuclear membrane.

Still another unusual arrangement has received recent attention, in this case in the red alga *Erythrocystis* (Tripodi and DeMasi, 1977). In this organism, the fusion cells often contained rather densely placed patches of precisely oriented tubules (Figure 6.13). When sectioned parallel to the tubules, such patches appeared crystalline, except that numerous single tubules emanated in all directions from the entire surface.

Cisternal Smooth Endoreticulum. A more widespread variety of smooth endoreticulum consists of single cisternae. Typically each of these is closely associated with a particular organelle, which it partly encloses. So frequently noted is an association between the chloroplast and a cisterna that the latter is generally referred to as the chloroplast endoreticulum. It has been recorded in seed plants, and in *Volvox* (Bisalputra and Stein, 1966), *Platymonas* (Manton and Parke, 1965), and *Trichosarcina* (Mattox and Stewart, 1974) among the green algae, besides various brown seaweeds, including *Fucus* (Bouck, 1965) and *Egregia* (Bisalputra, 1966), and the chrysophycean *Prymnesium* (Manton, 1964). In vertebrates, similar cisternae have been observed around secretion granules (Figure 6.14), as in the luteal cells of several mammals (Gemmell and Stacy, 1979). At least in the metaphytan cells, the association commences early in the development of the proplastid (Diers, 1966; Abreu and Santos, 1977). A similar association with the Golgi apparatus and lysosomes has given rise to the concept of GERL mentioned in the introduction to this chapter. Often this cisternal variety bears a scattering of ribosomes, as pointed out in the preceding section, so that it sometimes is considered to be rough endoreticulum.

6.2. THE GOLGI APPARATUS

Of all the topics disputed by light microscopists, none attracted more prolonged controversy than the Golgi complex (Dalton, 1961). In the original

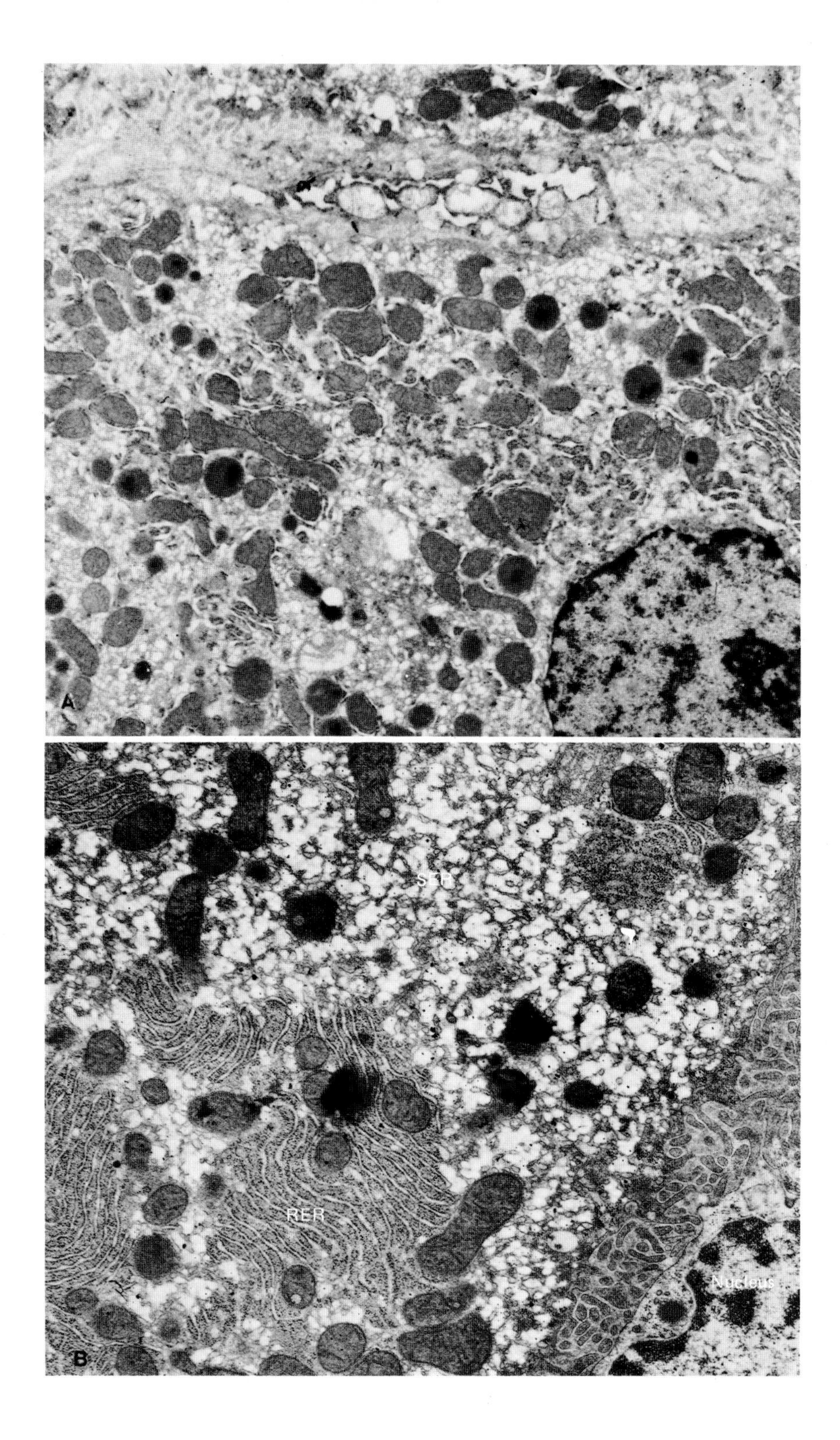
A
SER
RER
Nucleus
B

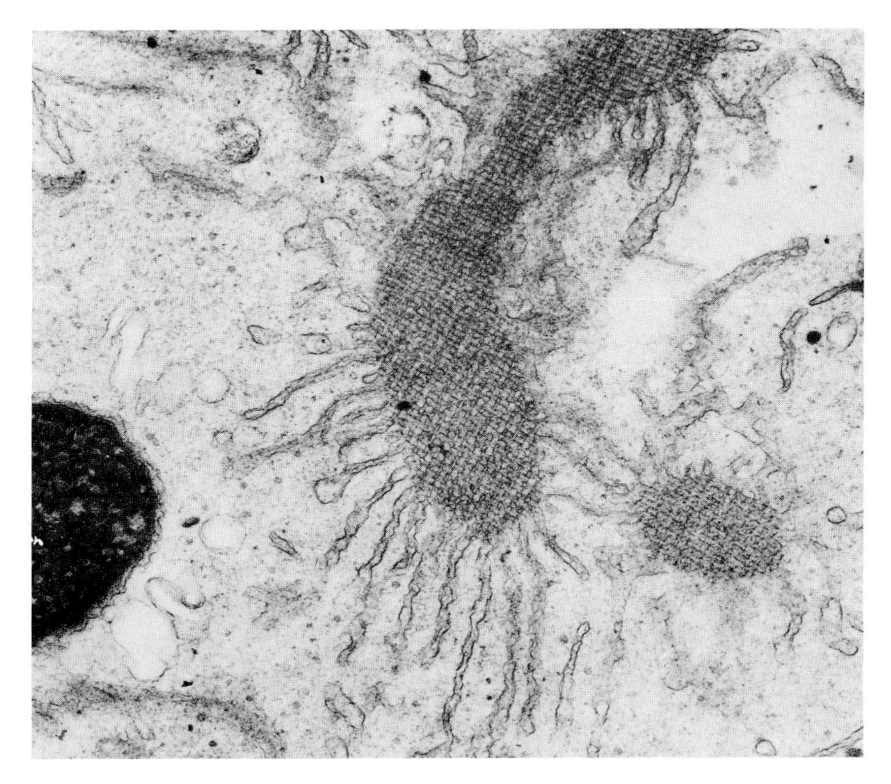

Figure 6.13. Paracrystallinelike endoreticulum. Tubular smooth endoreticulum occasionally has a paracrystallinelike structure, as in this fusion cell of the red alga *Erythrocystis*. 23,000×. (Courtesy of Tripodi and DeMasi, 1977.)

sense, the complex, or apparatus, included all those cell parts that reduced osmium tetroxide and silver nitrate or, as modified later, vitally stained with methylene blue. Many cytologists treated these observable traits as characterizing a distinct organelle or group of organelles, others considered them to be particular aspects of some familiar cell part such as the mitochondrion, whereas still others were convinced that they were mere artifacts. The electron microscope now has made it clear that the complex actually exists and consists of a single type of membranous structure, called the dictyosome, and its associated derivatives, sometimes called Golgi equivalents. Today either the two terms, Golgi complex and dictyosome, are treated as synonyms, or the former term is

←

Figure 6.12. Diurnal rhythmic changes in smooth endoreticulum. (A) Normal liver cell of rat at 2 PM. 3750×. (B) In the same type of cell at 10 PM, the smooth endoreticulum has become very extensive. 4000×. (Courtesy of Chedid and Nair, 1972.)

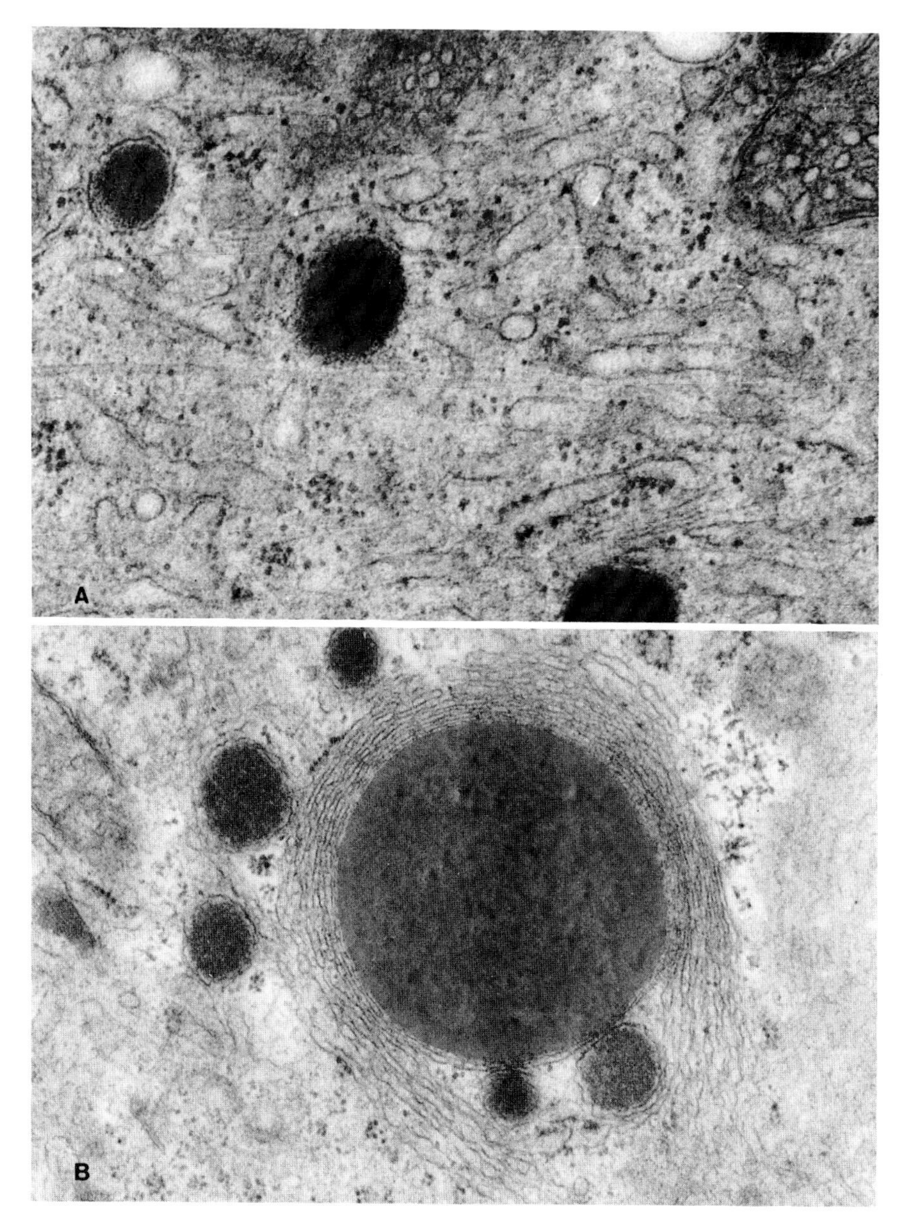

Figure 6.14. Cisternal smooth endoreticulum. (A) Smooth cisternal endoreticulum in mammalian corpora luteal cells. 22,000×. (B) Similar but more flattened cisternae may at times be associated with secretory granules as in this corpora luteal cell. 13,000×. (Both courtesy of Gemmell and Stacy, 1979.)

employed in a collective sense and the latter as the name for the individual organelle. Several digests of the subject exist, including one by Mollenhauer and Morré (1966) that reviews the plant organelle and another by Beams and Kessel (1968) that covers the entire field but accents the metazoan aspects.

6.2.1. Structural Traits

Briefly stated, a dictyosome is a stack of flattened saccules, the walls of which consist of single membranes of trilamellar appearance (Figure 6.15). These saccules are also referred to as sacs or lamellae, but the term cisternae, first proposed by Whaley and co-workers (1959), is the most widely employed. Usually so depressed as to have their opposite walls closely approximate, the individual cisternae often are not planar but may occasionally be arched and, at times, nearly bowl-shaped. Within their narrow lumina, material is either collected or secreted and then moved laterally toward the periphery, where the accumulating product distends the membranes to form vesicles; farther distally

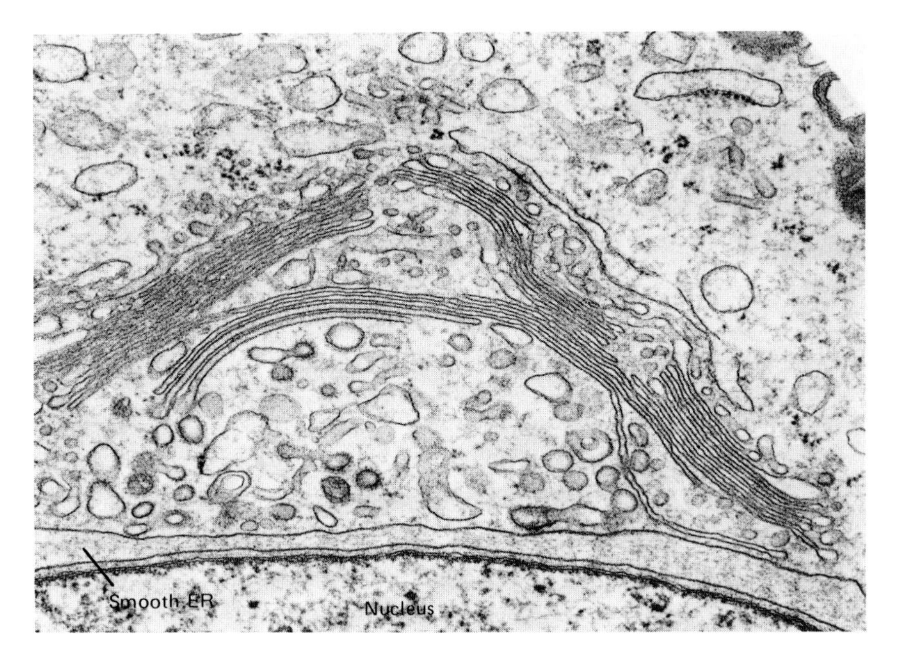

Figure 6.15. Typical structure of a Golgi body. The dictyosome of the developing spermatocyte in rat testis shown is more complex than that of most cells, but may be noted to have the small number of uniformly spaced cisternae characteristic of metazoan dictyosomes in general. No fewer than four interconnected dictyosomes are visible. 40,000×. (Courtesy of Hilton H. Mollenhauer, unpublished)

the vesicles gradually merge into a complex of vacuoles and tubules (Figure 6.15). The complexity of the outer reaches of the cisternae is best seen in surface view, in which a single component from the distal pole of a stack appears smooth centrally and becomes fenestrated and distorted toward the periphery (Figure 6.15; Cunningham *et al.,* 1966; Flickinger, 1969b).

Functional Organization. The cisternae are not permanent structures but undergo continual turnover, being destroyed in the course of carrying out their activities. According to a number of investigations (Grassé, 1957; Policard *et al.*, 1958; Schnepf, 1961, 1968a,b; Zeigel and Dalton, 1962; Mollenhauer and Whaley, 1963; Bruni and Porter, 1965; Bainton and Farquhar, 1966; Brown, 1969; Morré *et al.,* 1971), each active dictyosome follows a definite pattern of development and replacement. New cisternae appear to be formed one at a time at the proximal pole of the dictyosome. In at least some instances, the membrane of the new cisterna is not so thick as those of mature ones and is less electron opaque (Grove *et al.*, 1968; Morré and Ovtracht, 1977). With time the cisternae thereafter gradually increase in extent as their membranes slowly thicken and mature, and with their maturation, the enzyme systems they contain develop, permitting the cisternae to become actively functional. The resulting products then may be noted to move within the cisternae toward the periphery where they cause distensions in the form of the vesicles just described. As additional maturation is attained and the distensions have been enhanced, the vesicles begin to pinch free of the cisternae to move into the cytoplasm (Figure 6.15). When a given cisterna has arrived at the distal end of the dictyosome, its vesicles are then liberated in such quantities that its substance is exhausted. Because of this orderly sequence of events, it is evident that the position of a given cisterna within a stack indicates its age relative to the others.

In at least certain types of mammalian tissues, a somewhat more complex view of dictyosome structure and interrelationships has been emerging from recent use of either serial thin sections or thick sections and high-voltage electron microscopy. Employing either of these techniques, which permit the three-dimensional configuration to be reconstructed, a number of investigators have reported the cisterna at the forming face of a dictyosome to be interconnected by tubules to that of other stacks throughout such cell types as neurons, Leydig cells, and Sertoli cells (Rambourg and Chretien, 1970; Carasso *et al.,* 1971; Rambourg *et al.,* 1974). Two distinct regions of this interconnected network have been described, a "saccular region," corresponding to the familiar dictyosomes, and an "intersaccular connecting region," consisting of an osmiophilic series of tubules between the dictyosomes (Rambourg *et al.*, 1979). In addition, a close-knit mesh of anastomosing osmiophilic tubules has been found, underlain by three to seven fenestrated, closely apposed saccules (Figure 6.16). On the opposite face, a network of tubules connected to the most mature cisterna in each stack has been described. Thus many of the so-called vesicles around dictyosomes may prove to be transverse sections of tubules, and these in turn may some day be considered as a type of endoreticulum.

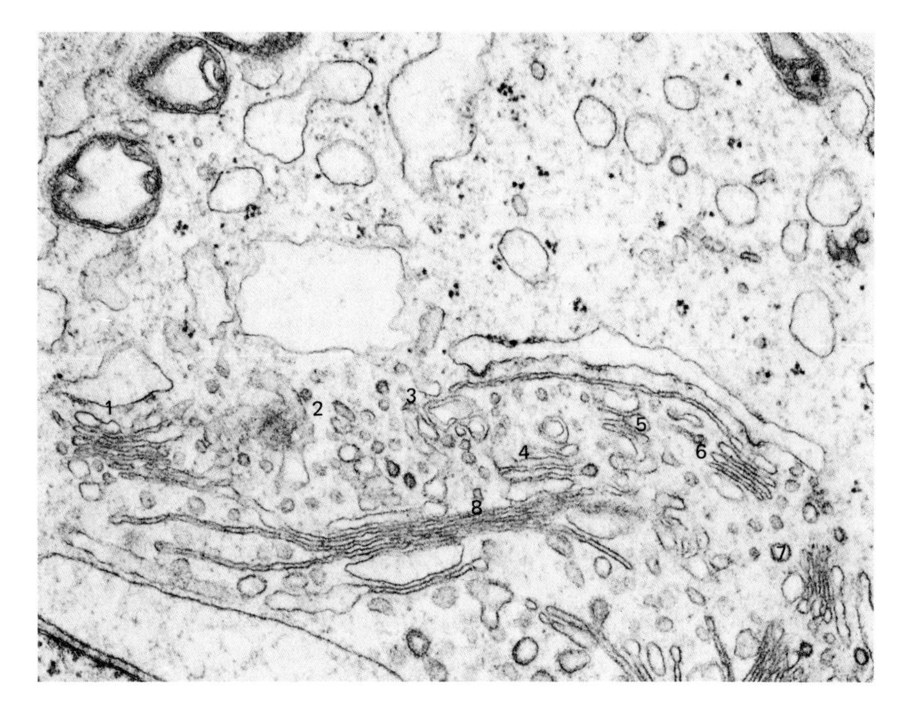

Figure 6.16. Golgi bodies as an interconnected system. Dictyosomes numbered 3, 5, and 6 in this rat testicle cell are clearly interconnected. 40,000×. (Courtesy of Hilton H. Mollenhauer, unpublished.)

Generally the loss of cisternae at the distal pole is compensated for by the formation of new ones at the opposite face, so that the total number present remains constant. However, if secretory functions are experimentally stimulated, the vesicles become filled and break free more rapidly, resulting in faster breakdown of the cisternae, accompanied by a reduction in their number (Grimstone, 1959; Weisblum *et al.*, 1962; Flickinger, 1968a,b, 1969a; Morré *et al.*, 1971). If vesicle production is inhibited, the dictyosomes become enlarged as a result of an increase in cisternal number (Hall and Witkus, 1964; Whaley *et al.*, 1964; Coombs *et al.*, 1968). During active secretion, one cisterna breaks down every 2 to 4 min, so that a dictyosome of eight cisternae would undergo complete turnover in about half an hour (Neutra and Leblond, 1966a).

The dimensions of the dictyosome and its parts also vary with species and cell type. In width the range is from 150 to 400 nm, the membrane thickness varies from 5 to 10 nm, and the narrowest diameter of the lumina averages between 5 and 10 nm, while vesicles may attain a diameter as great as 600 nm.

As the vesicles form and commence maturing, they become provided with tubules, but whether the latter structures have their source in the vesicles or in the cytoplasmic ground substance has not been determined. In longitudinal section these tubules appear similar to smooth endoreticulum and may extend for some distance into the cell, as just described; however, they are not universally present in every species nor in all types of metazoan cells. Sometimes, as in the processes of secreting milk in the mouse (Threadgold, 1976), the vesicles, after separating from the saccules, seemed to undergo a steady movement toward the plasmalemma, through which their contents ultimately left the cell. They appeared to bear no tubules, but the milk droplets they contained underwent a series of changes in physical properties during transit toward the exterior.

Zones of Exclusion. Very little is actually known of the cytosol, aside from its being a sol that is capable of undergoing sol-to-gel transformation (Wolman, 1955) and its containing the numerous types of filaments, microtubules, ribosomes, and other minute particles. Beyond these few points the electron microscope fails to provide any information. However, the areas beneath the plasmalemma and around centrioles, microtubules, and especially the Golgi material are zones of exclusion (Figure 6.17), as they are now called (Mollenhauer and Morré, 1978a), a more inclusive term than that under which they were first described, Golgi ground substance (Sjöstrand and Hanzon, 1954). In these areas, ribosomes and large organelles, like chloroplasts and mitochondria, are scarce or, more typically, absent. At times, especially in metaphytan and metazoan tissues, a zone of exclusion may envelop the entire dictyosome or centriole, but at others, it may be restricted to the mature face, a condition especially prevalent among more primitive eukaryotes like euglenoids (Mignot, 1965).

The zones are not characterized solely by their appearance but display other distinctive features. For example, they have been reported as having electrical properties different from those of the remainder of the cytosol (Giulian and Diacumakos, 1976) and as being rich in ribonucleoproteins (Ward and Ward, 1968). Moreover, there is a trend for them to be more electron dense than the remainder of the cytosol, and sometimes they show the presence of microfilaments (Mollenhauer and Morré, 1978a). Undoubtedly other distinctive features will come to light when techniques for their isolation are developed. Nevertheless, what little is known makes clear that the cytoplasm is highly structured, not only into discrete organelles and filamentous parts, but even in its seemingly free-flowing fluid portions.

Location and Number. In general in the cells of metazoans and metaphytans alike, the Golgi complex consists of only a few, small dictyosomes, except in cells that are actively secreting, where their number and size are greatly enhanced. The location of the organelle within the cell varies with the type from polar, as in epididymal tissue, to perinuclear, as in neurones. Typically among flagellates the single large dictyosome lies close to the flagellar base, whereas in ciliates it may be entirely absent (or periodic), or several may be found around the contractile vacuoles. Diatoms are distinct in having two dictyosomes (Figure 6.18), both situated close to the nucleus (Berthold, 1978). But it is not unusual for changes in abundance and location to

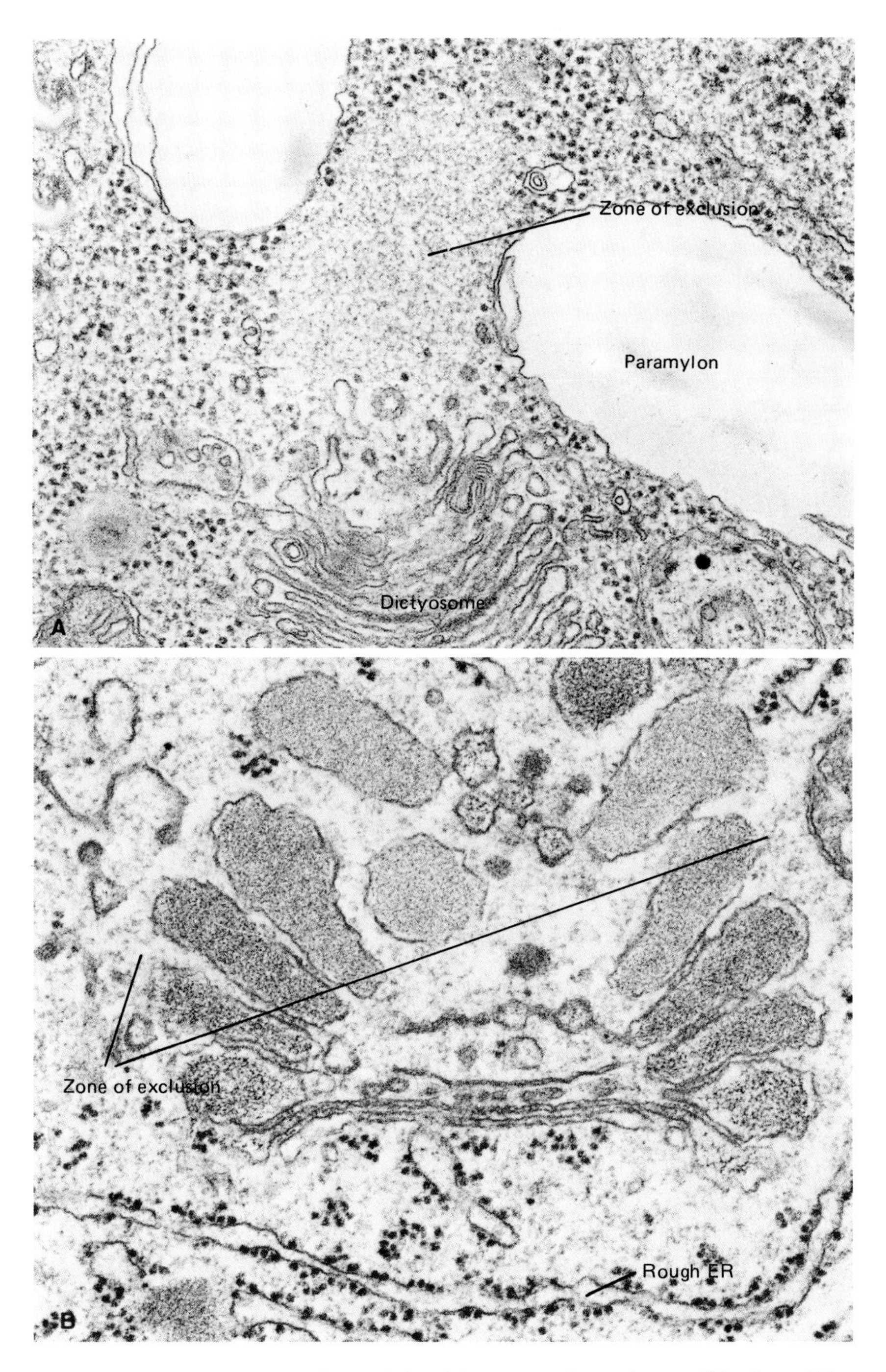

Figure 6.17. Zones of exclusion. Two varieties of the zone are shown, the consolidated type (A) as displayed by this *Euglena* dictyosome (27,000×), and a dispersed type (B) represented by maize root tip Golgi (45,000×). The several types occur in all taxa, however. (Both courtesy of Hilton H. Mollenhauer, unpublished.)

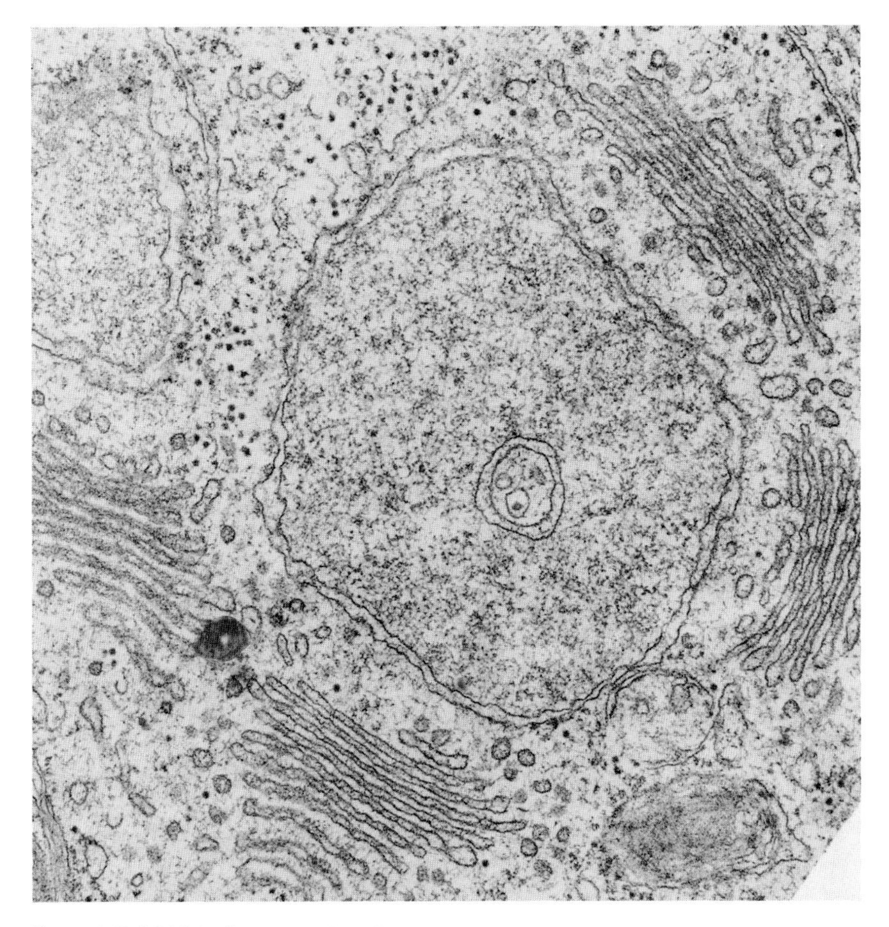

Figure 6.18. Multiple dictyosomes in a diatom. Most unicellular eukaryotes possess only a single dictyosome, but diatoms have two, located near the nucleus. In this electron micrograph of a recently divided specimen, four Golgi bodies are visible. 46,000×. (Courtesy of Berthold, 1978.)

accompany alterations in cellular activity. In dividing metaphytan cells, for instance, dictyosomes become abundant on each side of the incipient cell plates, in a region where they are absent during growth phases.

In hibernating mammals, radical changes in number and distribution of various cell organelles, including Golgi, have been recorded during the annual cycle of activity in various tissues, especially the chief cells of the parathyroid. Among those that have been described are the laboratory hamster (Kayser *et al.*, 1961), wild hamster (Stoeckel and Porte, 1973), bats (Nunez *et al.*, 1972), and the woodchuck (Frink *et al.*, 1978). Interruption of lactation by colchicine treatments likewise has been shown to induce a reduction in size of the dictyosomes (Thyberg *et al.*, 1977; Knudson *et al.*, 1978).

That the distribution of these bodies is closely correlated to cellular activity is lucidly shown by two experimental studies on primary root tissue of maize. Root cap cells exposed to light showed a far greater abundance of dictyosomes at the proximal end of cells than in the distal, whereas in the dark-maintained controls their distribution was random (McNitt and Shen-Miller, 1978). Comparable but opposite rearrangements of numbers were noted in geotropically stimulated tissues. Unlike control material, in which elongating cells had Golgi apparatus concentrated proximally and immature cells had larger numbers distally, cells in zones curving in response to gravity effects had the dictyosomes randomly distributed (Shen-Miller and McNitt, 1978).

6.2.2. Comparative Aspects

Variations among the Metazoa. One dictyosomal trait that varies from taxon to taxon, as well from one tissue type to another, is the average number of saccules constituting the organelle. As a rule it contains fewer among vertebrates than it does among invertebrates, but there is wide variation. For example, in the acinar cells of Brunner's gland in the mouse, the saccules ranged from 5 to 7 per dictyosome, a number found also in spermatocytes of guinea pig and epidermal cells of larval *Ambystoma*. On the other hand, in spermatids of the cat approximately 9–12 saccules were prevalent, paralleling the quantity found also in rabbit and mouse epidydimal cells.

Golgi in Protozoans. Although dictyosomes do not appear to be abundant in protozoans, they are present and of typical form (Figure 6.19; Pitelka, 1963). Among such amoeboids as *Gromia oviformis*, for example, the organelle consists of about 12 flattened cisternae (Hedley and Bertaud, 1962), but that of *Pelomyxa illinoisensis* contains only 5 or 6 (Daniels and Roth, 1961). Many flagellates seem to possess only a single dictyosome, but the cisternae are more numerous per organelle. In the polymastigote *Mixotricha paradoxa*, the dictyosome includes an average of 17 to 19 cisternae (Cleveland and Grimstone, 1964), while that of *Cryptomonas* contains around 12 and that of *Hydrurus* and *Chilomonas* 9–11 (Joyon, 1963). The Golgi body of Ochromonas is also similar (Kahan *et al.*, 1978). Dictyosomes of *Euglena* are distinctive in several ways. In the first place, they are especially abundantly supplied with cisternae, each consisting of approximately 15 to 30 (Leedale, 1966, 1967; Mollenhauer *et al.*, 1968). Second, the organelle is peculiar in that it often becomes bowl shaped toward both the distal and the proximal sides, as a consequence of the numerous vesicles along its periphery (Figure 6.20). The most notable peculiarity of the Golgi in this genus is the frequent thickening near the center of many of the cisternae. This thickening results from a deposit of fine electron-opaque material, through the midst of which may be seen a fine dense

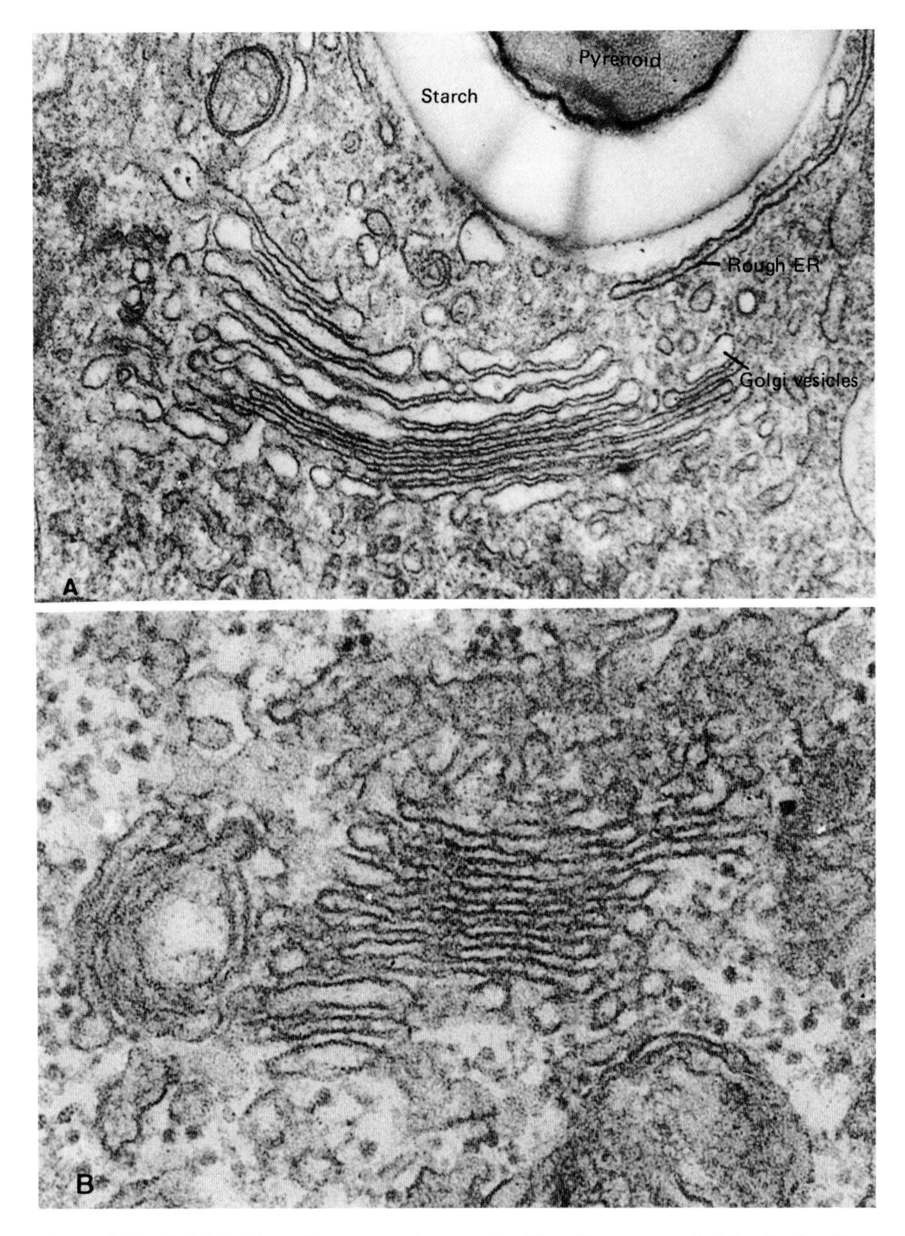

Figure 6.19. Golgi bodies of some protistans. (A) The dictyosome of this dinoflagellate *Amphidinium* has about seven or eight cisternae per stack; here it lies nearby the pyrenoid, a body associated with the chloroplast. 72,000×. (Courtesy of Dodge and Crawford, 1968.) (B) The single dictyosome of the zoospore of the parasitic fungus, *Synchytrium endobioticum*, contains nine unequal-sized cisternae. 90,000×. (Courtesy of Lange and Olson, 1978.)

line paralleling the membranous wall (Mollenhauer *et al.*, 1968; Mollenhauer and Morré, 1978b). As a rule, these short platelike formations appear in the third or fourth cisterna on the developing side and disappear just before the last stages are attained. Although the dictyosomes of other forms, such as the diatom *Amphipleura*, may occasionally show the presence of dense bodies (Stoermer *et al.*, 1965), these compact superimposed series of thickened regions appear unique to the euglenoids.

The euciliates are exceptional among eukaryotes in that the Golgi apparatus is infrequently encountered in ultrastructural studies. For example, Elliott and Zieg (1968) report that no dictyosome was encountered in several thousand electron micrographs taken of *Tetrahymena pyriformis* (see also Elliott and Bak, 1964), during the course of a thorough study of the occurrence of this organelle during the life cycle. However, sexually active strains, when starved to induce conjugation, were found to contain typical stacks of dictyosomal cisternae in the cytosomal region. When such individuals of opposite mating types were placed together in the same culture medium, the saccules became swollen at the periphery and formed vesicles in ordinary fashion. Some separated from the cisternae before conjugation and others toward its termination, so that some relationship of the secreted product with mating appeared evident.

Dictyosomes in the Metaphyta. Metaphytans, with five saccules, apparently have the lowest count per organelle of any eukaryote; it is their dictyosome that most clearly depicts the sequence of replacement outlined in a foregoing section. Isolated dictyosomal cisternae of these higher plants show two types of vesicles attached to the outlying tubule system, smooth and rough (Cunningham *et al.*, 1966). The significance of these vesicles is not known, but their presence demonstrates greater structural complexity for the organelle than might be suspected from thin sections alone (Figure 6.21). The spaces between the cisternae in a stack have received almost no attention, but the important differences between those of metazoans and metaphytans have recently been pointed out by Mollenhauer and Morré (1978b). Those of metazoans were shown to be of nearly uniform width throughout a given stack. In contrast, those of metaphytans were reported to increase from an average of around 8 nm between cisternae near the forming face to nearly 14 nm close to the mature face.

That these intercisternal areas are not simply unmodified regions of the cytosol has been clearly shown by an investigation of dictyosome structure in *Nitella* and *Chara* (Turner and Whaley, 1965; Pickett-Heaps, 1968a). In the first place, it was pointed out that the cisternae of a dictyosome remained associated as *in vivo* when that organelle was removed from a cell, intimating that they were bound together in some fashion. Thin vertical sections through Golgi stacks of vegetative and reproductive cells of those stoneworts disclosed the presence of an electron-dense line in each intercisternal space. As other sections showed the line as fine circles, it was deduced that the elements were

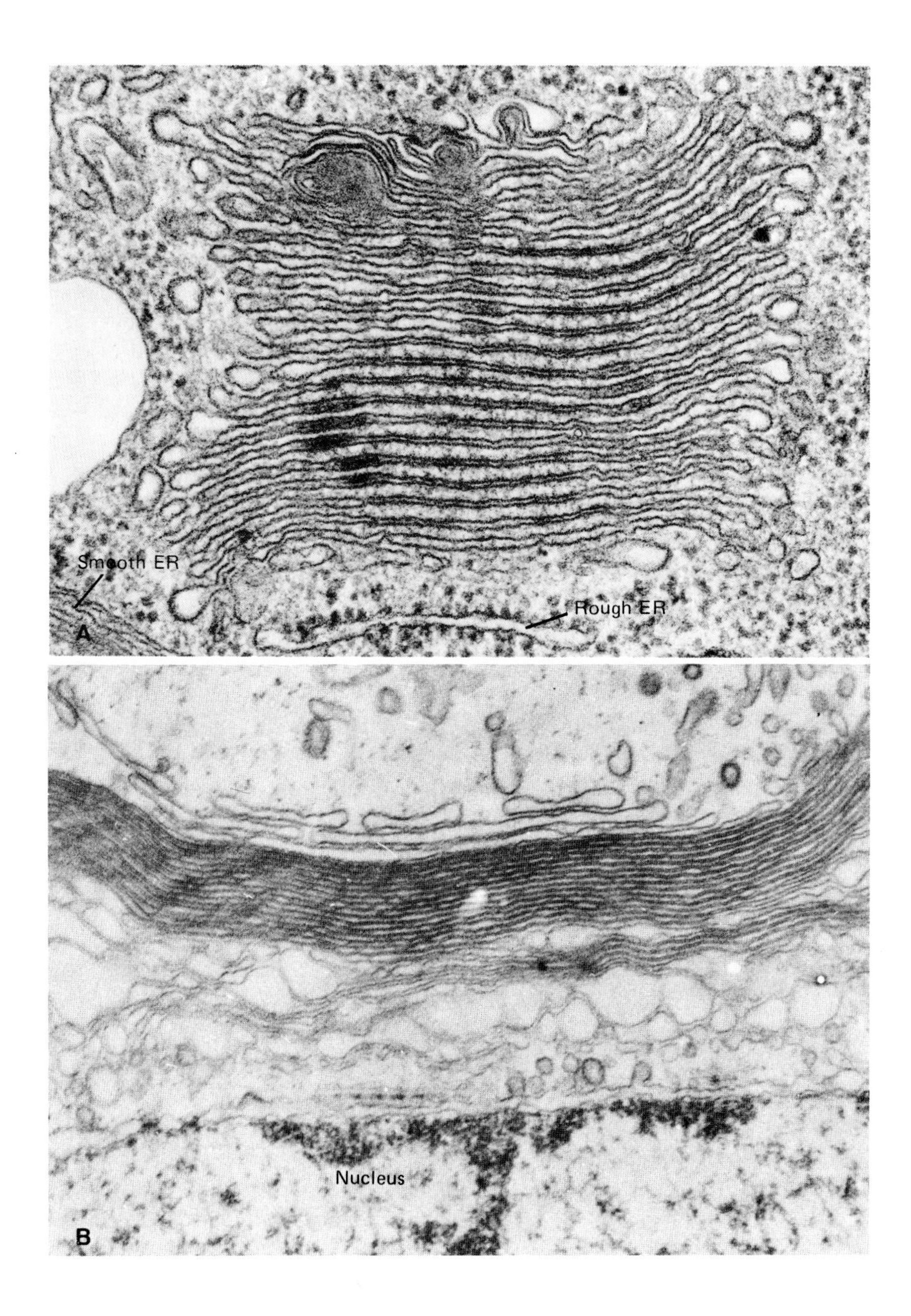
Smooth ER
Rough ER
A
Nucleus
B

Figure 6.20. The dictyosomes of several algae. (A) The dictyosome of *Euglena* contains a greater number of cisternae than almost any other organism; the distal face is toward the top of the micrograph. 72,000×. (Courtesy of Hilton H. Mollenhauer.) (B) In *Mallomonas,* the cisternae are so closely applied to their neighbors that no intercisternal spaces can be noted. 43,000×. (Courtesy of Wujek and Kristiansen, 1978.) (C) The dictyosome of this brown seaweed (*Cutleria*) has the cisternae compactly arranged proximally but more widely spaced distally; the numerous vesicles are especially striking. 26,000×. (Courtesy of LaClaire and West, 1978.)

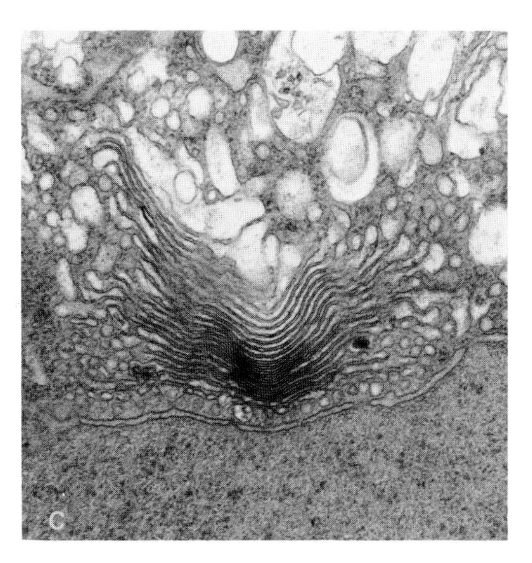

nanofilaments, with diameters of between 70 and 80 Å, arranged in rather loose parallel fashion. This interpretation was confirmed by sections cut parallel to the face of the cisternae. Thus far such elements appear to be confined to the stoneworts and higher green plants and are consistently absent from metazoans (Mollenhauer and Morré, 1978b). In the succulent plant *Aptenia cordifolia*,

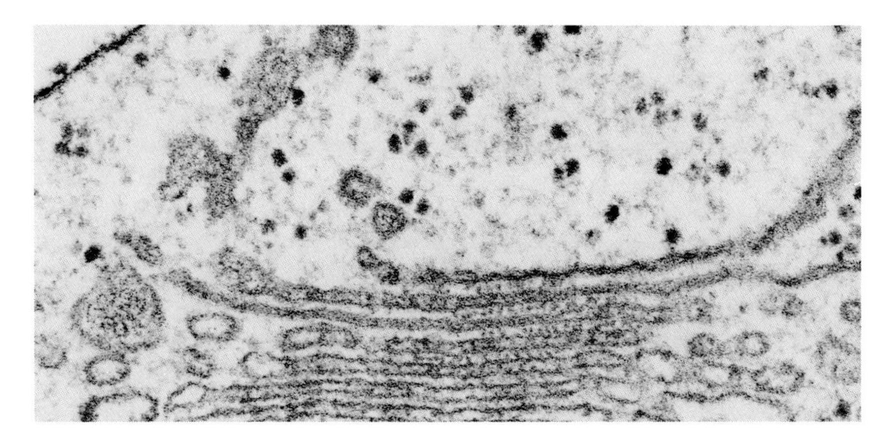

Figure 6.21. The dictyosome of a metaphytan. The microtubules between the cisternae that appear in this onion root dictyosome, along with the electron-opaque deposits of the membranes, are characteristic of metaphytan Golgi bodies as a whole. 75,000×. (Courtesy of Hilton H. Mollenhauer, unpublished.)

these elements were described as resembling strings of pearls, rather than being fibrillar (Kristen, 1978).

Dictyosomes of Algae. Among the most distinctive of all Golgi complexes is that of red seaweeds, including members of the genera *Polysiphonia* and *Callithamnion*. The dictyosomes have been especially well investigated in the sporangia of these plants, where they are extremely active in the secretion of materials for the rapidly growing walls. One peculiarity that was described pertains to the fast-expanding vesicles, which quickly grow in length as well as width, so that the outer cisternae are often beaker shaped, with relatively flat bottoms and nearly vertical sides (Hawkins, 1974a,b; Alley and Scott, 1977; Wetherbee and West, 1977). Even more striking, however, is the arrangement of the cisternae; these structures are so closely apposed to one another centrally that the intercisternal region is totally eliminated (Figure 6.20B). In contrast, the central lumina are quite broad except those of the outermost members. The vesicles too are unique, both in appearance and functionally (Wetherbee and West, 1976). In early sporangia, their activities seemed to be primarily with the formation of cell wall materials, as growth of that plant organ is very rapid. During that period the vesicles had a striated appearance, but as the sporangia approached definitive size, that type disappeared, being replaced by vesicles whose contents were crystalline.

That the curvature of the mature cisternae does not result entirely from the presence of swollen vesicles along the periphery is made evident by a recent analysis of the endosymbionts of the foraminiferan *Amphistegina lessonii* (Berthold, 1978). These organisms proved to be diatoms, most of which lacked a test. Usually all cisternae within a stack in a given endosymbiont were uniform in thickness from one edge to the other, the periphery being almost, if not entirely, devoid of vesicles. Nevertheless, in many cases the mature members of a dictyosome became strongly bowl shaped (Figure 6.22).

6.2.3. Development and Origins

Ontogenetic Origins. As has been customary for membranous organelles in general, various ontogenetic origins from other membranes have been suggested for Golgi bodies. Among sources that have been proposed are included the nuclear membrane (Moore and McAlear, 1963) and the plasmalemma (Frey-Wyssling *et al.*, 1964), with pinocytotic folds a possible mechanism of derivation in amoebae (Daniels, 1964). Manton (1966b), however, seems to be the first to advance some evidence that the Golgi complex may undergo division in flagellates, so that one other mode of origin is from other dictyosomes, as is the case with the mitochondrion and the chloroplast. Such duplication by actual division has been substantiated by studies largely of algae, such as desmids (Drawert and Mix, 1963) and diatoms (Berthold, 1978), and the higher plants (Buvat, 1958a,b; Diers, 1966; Clowes and Juniper, 1968;

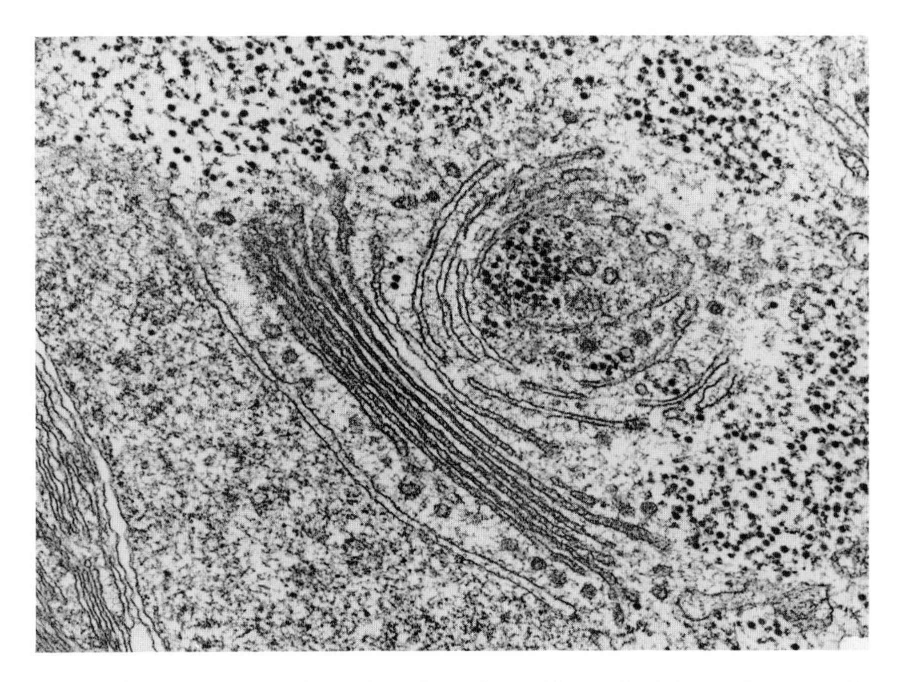

Figure 6.22. The Golgi body of an endosymbiotic diatom. The proximal cisternae become nearly spherical and enclose a mass of ribosomes and vesicles. Note that the distal components are nearly contiguous but become more widely spaced proximally. 60,000×. (Courtesy of Berthold, 1978.)

Di Orio and Millington, 1978). Moreover, similar processes have been uncovered in several invertebrates (Gatenby, 1960; Dalton, 1961; Kiermayer, 1970) and in a few vertebrates (Wischnitzer, 1962). In multiplying by fission, the entire stack in a dictyosome becomes severed perpendicularly to the plane of the individual cisternae (Figure 6.23) by the simple expedient of first forming two separated cisternae at the forming face. As these mature and new cisternae are added proximad to them, two stacks thus result. At least in more advanced taxa, the cisternae may become elongated before division takes place.

Daniels' suggested origin of these organelles from pinocytotic folds was based upon the resemblance of the contents of dictyosomal vesicles to the coating over the amoeba's surface. But Flickinger (1969a), working with amoebae enucleated for 5 days, offered a different interpretation. He believed the presence of the coat material actually represented its being secreted by the dictyosome, rather than the latter being derived from the plasmalemma. His concept was supported by the observation that when amoebae were induced to pinocytose ferritin, that electron-opaque substance appeared in some vesicles as well as the food vacuoles, but never in dictyosomes. Further, he proposed that

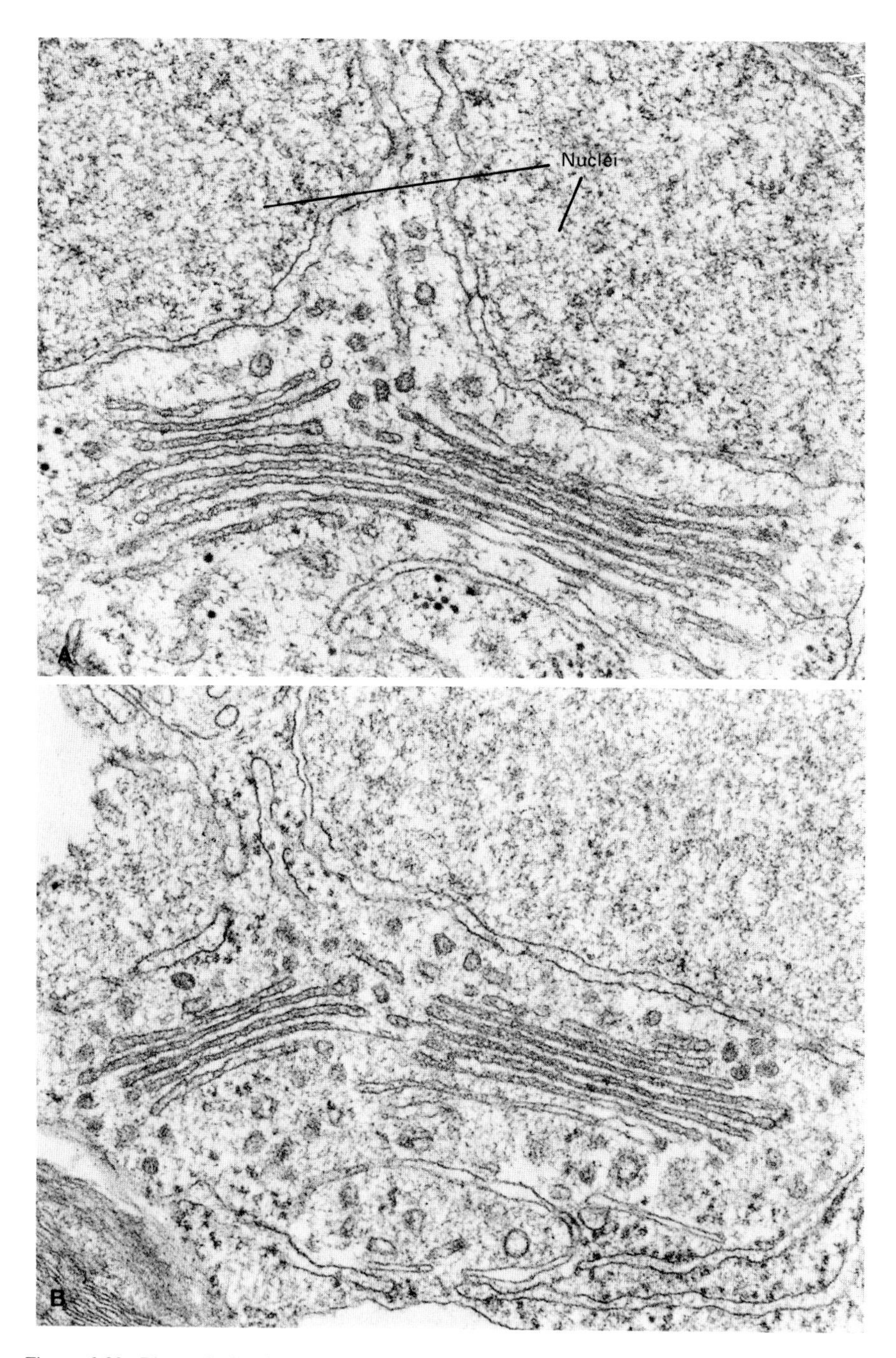

Figure 6.23. Binary fission in a dictyosome. (A) In this endosymbiotic diatom, several divided cisternae are to be seen lying near the nuclei that have already completed mitosis. 53,000×. (B) The dictyosome is now completely divided. 52,300×. (Both courtesy of Berthold, 1978.)

the dictyosome was derived from endoreticulum on the supposed resemblance of the luminal contents of the two organelles. No clear-cut similarity in appearance is evidenced in his electron micrographs, however, for the contents of the dictyosomes seem to be vesicular and slightly more electron opaque than the fine granular, or sometimes striated, contents of the endoreticular membranes.

One important aspect of the endomembrane concept mentioned in the introduction to this chapter is a supposed developmental, as well as structural, continuity between the various membranous components (Mollenhauer and Morré, 1966; Beams and Kessel, 1968; Morré *et al.*, 1971; Palade, 1975; Whaley, 1975; Morré and Ovtracht, 1977). Within this system are included the nuclear envelope, rough and smooth endoreticular membranes, the Golgi apparatus, and various vesicles of the cytoplasm, whereas the plasmalemma, lysosomes, and membranes of vacuoles are regarded as end products of the system. In part this point of view is based on the progressive change in the diameter of the respective membranes from thin to thick in going from the nuclear membrane to the endoreticula and secretory vesicles through the Golgi and finally to the plasma membrane. In thickness and staining properties, forming faces of dictyosomes always resemble endoreticular membranes and the mature faces are similar to the plasmalemma. Formation of the various membranes involves differentiation as progress is made from one organelle to another. Briefly stated, the concept implies that all membranes of the cell, except those of the mitochondrion and chloroplast, are actually nuclear membrane that has undergone progressive increase in thickness and differentiation. Thus rough endoreticulum is only slightly altered nuclear membrane, smooth a somewhat more modified derivative of rough, vesicles still more modified, and dictyosomal membranes represent transitional series of steps between the latter and the plasmalemma.

One would suspect that if this is an established pathway for any of the membranous organelles to develop and function, it would be fairly rigidly followed. If viewed closely, the concept is perceived to intimate that the enzymes needed to form rough endoreticulum are contained within the nuclear membrane so that it can give rise to the former. Then the rough endomembranes would necessarily have embedded in them those enzymes needed to modify its own membranes into those needed for forming the smooth endoreticular tubules and vesicles and that these in turn would consist in part of the enzymes involved in producing vesicles that are capable of forming the dictyosomes, and so on finally to the plasmalemma. Thus the plasmalemma basically is highly modified nuclear membrane. Because each step is thus essential to the subsequent one, there should be no short cuts, but a rather rigid, lock-step sequence of events. Yet Morré and Ovtracht (1977), in supporting this concept, reproduce an electron micrograph of a fungal cell (their Figure 8) in which blebs from the nuclear envelope are stated to enter into the formation of an adjacent dictyosome; similar blebbing toward a dictyosome has also been illus-

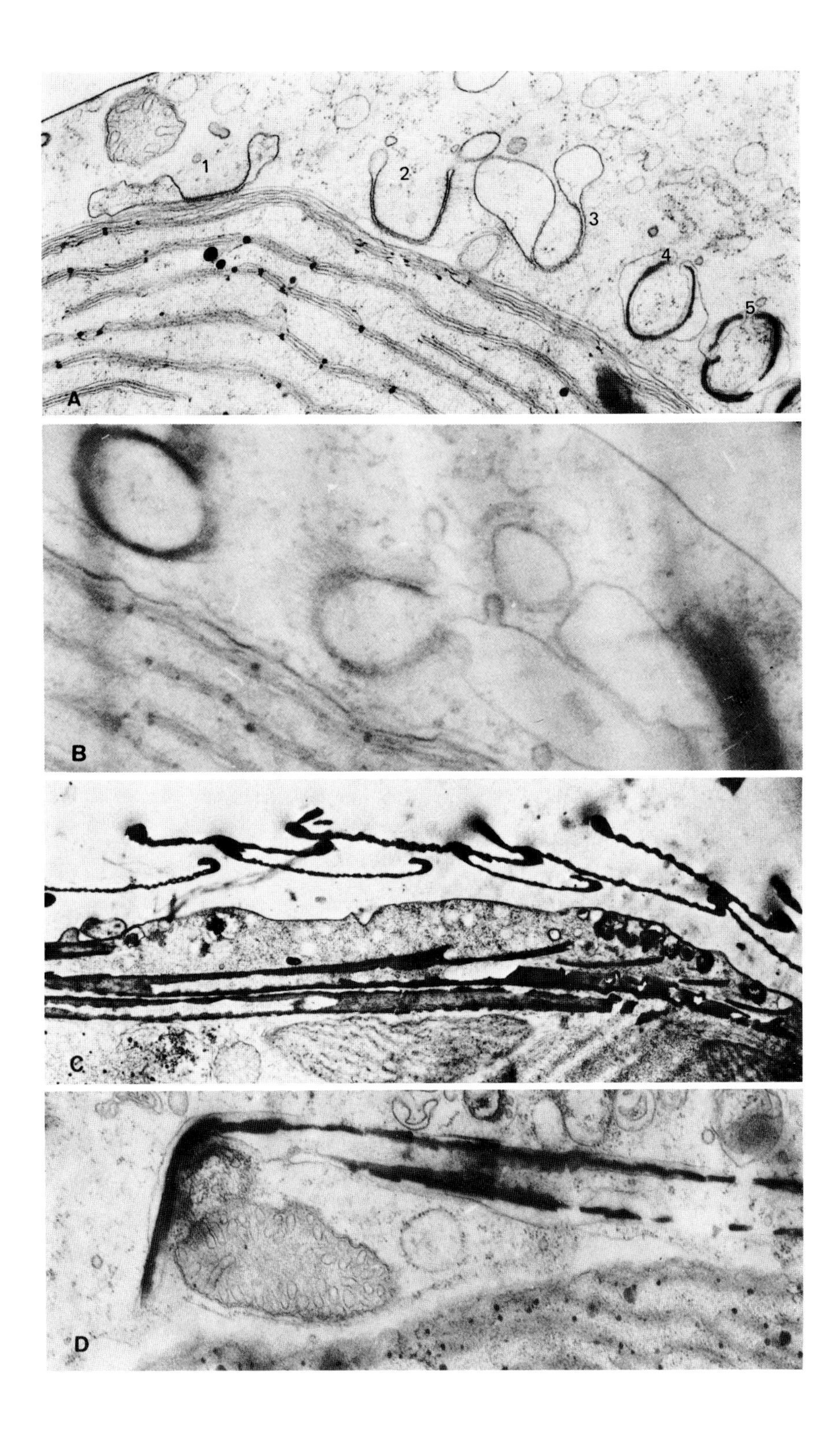

1
2
3
4
5
A
B
C
D

trated from the alga *Tribonema vulgare* (Massalski and Leedale, 1969). Thus the remainder of the usual system would seem unessential. Moreover, continual replacement of the nuclear membrane would be necessary just to compensate for the continual overturn of the dictyosomes (Parry, 1978). And as the necessary proteins are made on the ribosomes largely concentrated within the cytoplasm, there seems to be no need to send each and every one of the innumerable types that must comprise the several complex membranous organelles through a long, intricate route. Finally, in most organisms, there is rarely a close association between the endoreticulum and the Golgi (see Figure 6.29). It would seem far more reasonable and closer to most reported biochemical facts to suppose that each of the various organelles is essential in its own way to the cell, including some responsibilities of contributing to the growth and maintenance of others.

This idea receives support in part both from functional and enzymatic investigations. Scales and cellulosic and other materials for use in cell wall deposits certainly could be thought of as substances that could advantageously be secreted by the plasmalemma. Yet the enzymes for the production of these enveloping materials are usually confined to the Golgi apparatus, denying the notion that plasmalemma is merely modified dictyosomal membrane. Furthermore, investigations of the metabolic fate of ^{14}C-labeled glucose in corn root tips showed that only 10% of the radioactivity was contained in the Golgi-rich fractions, whereas 90% was in the microsomal (Bowles and Northcote, 1972, 1974). Hence, the researchers concluded that the endorecticulum played a major role in the synthesis of the pectin–hemicellulose portion of the cell wall. Indeed, in *Mallomonas* scales and bristles that cover the cell (Figure 6.24) are made in smooth cisternal endoreticulum (Wujek and Kristiansen, 1978). Obviously then, the route taken is not an assembly-line set of processes in a strict sense. Moreover, studies on enzyme distribution during formation of dictyosomes in amoebae indicate that a few enzymes, such as GDPase, are present in the nuclear envelope and endoreticulum (rough ER only, however), as well as the forming faces of new Golgi bodies (Flickinger, 1978). Thiamine pyrophosphatase (TTPase), in contrast, was found only in Golgi, not in the organelles whose membrane supposedly gives rise to that body directly. The closer view of the differences of functions displayed by the various membranous

←

Figure 6.24. Scale formation in *Mallomonas*. (A) Several stages in scale formation are shown in sequence in this unusual micrograph. At the extreme left (1) is the early cisterna formed by outpocketing from the smooth endoreticulum adjacent to the chloroplast. This then folds (2) and doubles over on itself (3, 4) to form a nearly circular vesicle (5). 16,000×. (B) The nearly circular vesicle (center) then becomes attached to a second vesicle in which the bristle portion of the scale is developed. 36,000×. (C) Apical portions of two almost completed bristles and sections of the scales covering the cell. 12,500×. (D) Bristle base in process of formation. 20,000×. (All courtesy of Wujek and Kristiansen, 1978.)

organelles presented later should also give pause to any concept of membrane continuity between them.

Thus it is self-evident that the cytosol is capable of secreting the membranes *de novo*, just as it does the centriole. Although no investigations seem to have been conducted into the association with the dictyosome of DNA, special ribosomes, or other items indispensable in current concepts for the synthesis of the particular proteins needed, the results of the study of enucleated amoebae mentioned above (Flickinger, 1969a) intimated that the necessary proteins were formed under the direction of the nucleus. Dictyosomes, absent during prolonged enucleation, began to reappear within 30 min after renucleation, became normal in size after 6 hr, and increased in abundance during a 3-day period. Hence, in these organisms, the organelles do not come from preexisting ones.

Life Cycle Alterations. During the ordinary course of its life cycle, the cell undergoes changes in its requirements, some of which have been noted in embryonic tissues and others have been documented in the processes of encystment. The general trend of the changes at first appears to go against statements made earlier to the effect that rapidly-secreting dictyosomes undergo reduction in size because of the accelerated rate of vesicle formation and accompanying enhancement of cisternal degeneration. That trend was the result of experimental stimulation of secretion; under the natural course of events quite the opposite results prevail.

Active examples of the quadriflagellated green organism *Polytomella agilis* are devoid of cell walls and usually possess a number of dictyosomes (Brown *et al.*, 1976a). The latter organelles present no unusual features, being rather small and consisting of about six cisternae; the cisternal lumina are quite narrow and at the periphery have relatively few vesicles. When the cells undergo encystment, however, a number of changes are observed, most of which stem from the development of a thick cell wall. During the early events of cyst formation, when the outermost of the cyst wall layers are laid down, the cisternae increase in length, their lumina widen, and they become filled with granular material. No increase in the number of vesicles occurs at that time (Brown *et al.*, 1976b), but later as the second layer of the cyst wall is deposited, the dictyosomes increase in depth to around eight cisternae and become doubly concave as a consequence of the numerous vesicles that form around their rims. Moreover, the cisternae are distended, and numerous vesicles containing fibrillar or laminar material are seen to surround them. Then as the third wall layer, consisting of an especially resistant and electron-dense material, is laid down, the Golgi bodies cease functioning and undergo reduction in size and vesicles become scarce.

Growth of the Golgi complex has been followed in the postnatal development of the epididymis of rats. In prenatal and newborn young, only from one to three dictyosomes can be noted per cell, each consisting of only four to six planar cisternae, nearly devoid of vesicles (Flickinger, 1969c). During the first

10 days after birth, the dictyosomes became more numerous and possessed distended and greatly fenestrated cisternae. The next 10-day period was marked by the cisternae increasing in breadth and number per dictyosome and by the appearance of large vesicles. After 3 weeks, growth of the Golgi proceeded at a slower rate but continued until the animals were 6 weeks old.

In the duodenal epithelial cells of fetal rats, the development of the Golgi apparatus followed a comparable course but was initiated earlier. Up to the 19th day of the mean 22-day intrauterine period, the intestinal lining showed almost no signs of engaging in pinocytosis. In such cells, as in the epididymis, the Golgi was limited to only one or so dictyosomes per cell, having few cisternae to a stack (Hayward, 1967). However, the vesicles were more numerous and larger than in that organ. At 20 days, pinocytosis became more active in the cells that covered the intestinal microvilli; in this layer the dictyosomes were larger and consisted mostly of vesicles, whereas elsewhere no change could be noted. During the last prepartum day, pinocytosis became more intensified as the Golgi bodies increased greatly in number and breadth with an enhanced number of cisternae per stack. Consequently, it seems apparent that, as cells undergo normal increases in secretory activities, the Golgi apparatus also becomes enlarged through the multiplication and overall enlargement of the dictyosomes. This appears to be true also of metaphytans, for in the antheridial filaments of *Chara* similar changes have been observed during the cell cycle (Kwiatkowska and Maszewski, 1979).

Phylogenetic Origins. Evidently the Golgi complex either came into existence suddenly with the advent of the eukaryotic nucleus or, more probably, its precursorial stages either occurred in now extinct taxa or are unrecognized among advanced extant prokaryotes. If application to higher eubacteria were made of some of the electron-staining techniques that have been developed (Friend, 1969; Rambourg *et al.*, 1969), perhaps some sort of particle might be detected that corresponds functionally to the dictyosomal material, but until that time, the Golgi complex has to be considered a strictly eukaryotic peculiarity. As the organelle is particularly active in the secretion of scales or cell wall material in several algal groups and metaphytans, it may be hypothesized that this activity, originally a general function of the plasmalemma in prokaryotes, became centered in a single region of that membrane as specialization increased. Perhaps this region later invaginated, and still later this infolded pocket pinched off to form a primitive Golgi body.

Moreover, because of the structural constancy between eukaryotic taxa pointed out before, no clear-cut sequence of evolutionary development is observable. Two reasons might be proposed for the morphological uniformity. On one hand, the very nature of the dictyosomal structure may preclude elaboration of pattern—it is difficult to conceive how a simple stack of discoidal membranous cisternae can be varied in form to any high degree. Perhaps the recently discovered network of tubules and cisternae that surrounds the dictyosome will eventually provide clues. Similarly, differences, in macromolecular constitution may very well exist, but this aspect remains virtually unexplored. On the other hand, the seeming lack of variety may reflect real absence of thorough comparative study of the organelle, for to date stress has been placed, logically enough, more upon the presence and similarities of the complex in various taxa than on differences.

Nevertheless, it can be suggested that possible early stages in the evolution of the dictyosome exist in such groups as the yeasts and euglenoids. Among such yeasts as *Saccharomyces cerevisiae* (Marquardt, 1962; Moor and Mühlethaler, 1963), both ordinary and freeze-etch preparations for the electron microscope showed the organelle to be notably simple and to consist of about three cisternae, so closely applied to one another medially that intercisternal spaces were lacking. In addition, the cisternae terminated abruptly in the cytoplasm, seemingly being devoid of the usual peripheral outlying vesicles. Both in *S. cerevisiae* (Moor and Mühlethaler, 1963) and in *Torulopsis utilis* (Linnane *et al.*, 1962), the vesicles that are derived from the mature cisternae contained lipid material.

If the euglenoids do represent a somewhat more advanced stage of phylogeny, then a great increase in number of cisternae can be deduced to have followed those simple beginnings, for in the single dictyosome of *Euglena* and *Distigma*, between 20 and 30 of these sacs have been counted (Migot, 1965; Leedale, 1967; Mollenhauer, 1974). Vesicles, mostly of conservative dimensions, may be noted at the periphery of most members of a stack, but few are to be seen in the zone of exclusion. Beyond this level, no sequence of events can be convincingly postulated, except that a trend in reduction of cisternal numbers exists. In *Amoeba* there are close to 10 per stack (Flickinger, 1968a), while most algae, protozoans, and fungi (Moore and McAlear, 1963; Hohl and Hamamoto, 1967; Hohl *et al.*, 1968) have between 6 and 9. In the metazoans the number ranges around 5 or 6, and in the metaphytans remains close to 5. Near the upper levels, the vesicles frequently distend into cavernous vacuoles, especially in the red (Gantt and Conti, 1965) and brown seaweeds (Bouck, 1965), the green algae, and advanced plants and metazoans. These trends in concert appear to indicate an increase in efficiency of functioning for the individual cisternae, so that fewer per stack are needed to produce large objects or quantities of any given required product.

6.3. FUNCTIONS AND ENZYMES OF THE ENDOMEMBRANE SYSTEM

Because of the close relationships of structure and function that exist among the several organelles comprising the endomembrane system, likenesses and differences can be better brought out by presenting their functions and enzymes together. Similarities between the two major classes of endoreticulum begin with the demonstration that the electrophoretic properties of their protein constituents are nearly identical (Hinman and Phillips, 1970). In addition to those activities discussed at this point, the endoreticular system contains two different electron-transport systems that can be examined more appropriately with others of the same nature (Chapter 10).

6.3.1. Functions of Rough Endoreticulum

In a list of the functions and enzymes of the rough endoreticulum, a broad spectrum of both becomes evident. This diversity should be considered to apply only to the class of organelle collectively, rather than to that from any given source organism or tissue. Vertebrates, complex invertebrates, and metaphytans of necessity carry out numerous specialized activities of no value to unicellular forms of life, so the enzymes and functions largely pertain only to the tissue named. Out of these diverse details, however, can be drawn an indication of the general nature of the organelle.

Ribonucleic Acid. Rough endoreticulum, in the form of microsomes that have been freed of ribosomes, has been shown to contain RNA, at least part of which included segments of polyadenylic acid [poly(A)] (DePierre and Dallner, 1975; Kreibich *et al.*, 1978a,b). Now evidence has been advanced that suggests that messenger RNA (mRNA) might interact directly with rough endoreticular membranes by means of these poly(A) segments or perhaps through nonpoly(A) moieties (Lande *et al.*, 1975; Cardelli *et al.*, 1976). This suggestion in turn implies the possible existence of a mechanism within these membranes that could engage in translation, thus bypassing the usual ribosomal activities. Analysis of washed rough microsomal membranes for the presence of mRNA has now shown none is to be found (Cardelli *et al.,* 1978). Consequently, the proteins processed by this organelle are not produced by its membranes but are products of translation on ribosomes, probably including those of the cytoplasm as well as those embedded on them. Related to the nucleic acid contents are the binding sites to which the ribosomes are attached. At least in large measure these sites appear to consist of two proteins called ribophorins that are unique to rough endoreticulum (Kreibich *et al.,* 1978a,b).

Important Enzymes. Also pertinent to the role of RNA in this organelle is the recent isolation of an exoribonuclease from rough microsomes of rat liver (Kumagai *et al.*, 1979). With a molecular weight between 80,000 and 83,000, this enzyme degraded poly(A) entirely into 5′-AMP by a processive mechanism proceeding in the 3′ to 5′ direction that needed both Mg^{2+} and K^{+} ions for maximum activity. It was suggested that this enzyme may be implicated in the cellular breakdown of mRNA in rat liver.

Among other important enzymes that appear to be unique to the present organelle is nucleoside diphosphatase (Novikoff *et al.*, 1962; Novikoff and

Heus, 1963; Novikoff, 1976). However, it may be characteristic of this endoreticulum only in vertebrate tissues, for in some cases this enzyme was absent. In these instances peroxidase seemed to be characteristic (Novikoff *et al.*, 1971a, 1974).

Examples of Activities in Metazoans. Rat liver rough endoreticulum, as well as that from rat kidney, has the capacity for synthesizing glucose-6-phosphate from glucose and inorganic pyrophosphate, a reaction catalyzed by glucose-6-phosphatase. This same enzyme also breaks down this product (Arion *et al.*, 1972), but only when artificially released into the cytoplasm by rupturing the reticular membranes. Extensive heterogeneity is exhibited both by the enzyme and by the rough endoreticulum, even within the liver, for the centrolobular zone reacted quite differently than the perilobular zone when the animal was treated with phenobarbital (Ménard *et al.*, 1974a,b). That treatment has been reported to induce a proliferation of smooth endoreticulum and decrease glucose-6-phosphatase activity (Ericsson, 1966; Garg *et al.*, 1971; Higgins and Barrnett, 1972). However, while those results characterized the reaction in the centrolobular zone, the cells of the perilobular region showed no changes from the untreated tissue (Figure 6.25). Because the enzyme is found largely in rough endoreticulum, the proliferation of the smooth type was stated to have led concomitantly to the observed decrease of reactivity. These results are in conflict with those of another study that showed that the newly formed smooth endoreticulum had an abundance of this enzyme within 24 hr postpartum (Leskes *et al.*, 1971).

The capacity for synthesizing glucose-6-phosphate has been described as increasing rapidly from low prenatal levels to a maximum between the second and fifth days postpartum, followed by a gradual decrease to adult levels (Goldsmith and Stetten, 1979). Latencies of the synthesizing and hydrolyzing functions of the enzyme displayed different age-related changes. While the synthesizing function reached latency levels of 60–80% shortly after birth, where it remained throughout life, the hydrolase function latency decreased with age, being more than twice as latent in the neonatal animal than in the adult.

Separation of microsomes from rat liver into size classes on a continuous sucrose gradient suggested that the enzymes of rough endoreticulum may not be distributed nonrandomly (Dallman *et al.*, 1969). Those classes consisting of the

→

Figure 6.25. Tissue differences in smooth endoreticular behavior. (A) A portion of a rat liver cell from the centrolobular region shows extensive proliferation of vesicular smooth endoreticulum among the glycogen deposits; these are sparingly filled with the reaction product indicating glucose-6-phosphatase activity. 7000×. (B) Several rat liver cells from the perilobular region with similar treatment. The smooth endoreticulum is not abundant, and the reaction product is more highly concentrated in the few patches where it occurs. 4200×. (Both courtesy of Ménard *et al.*, 1974a.)

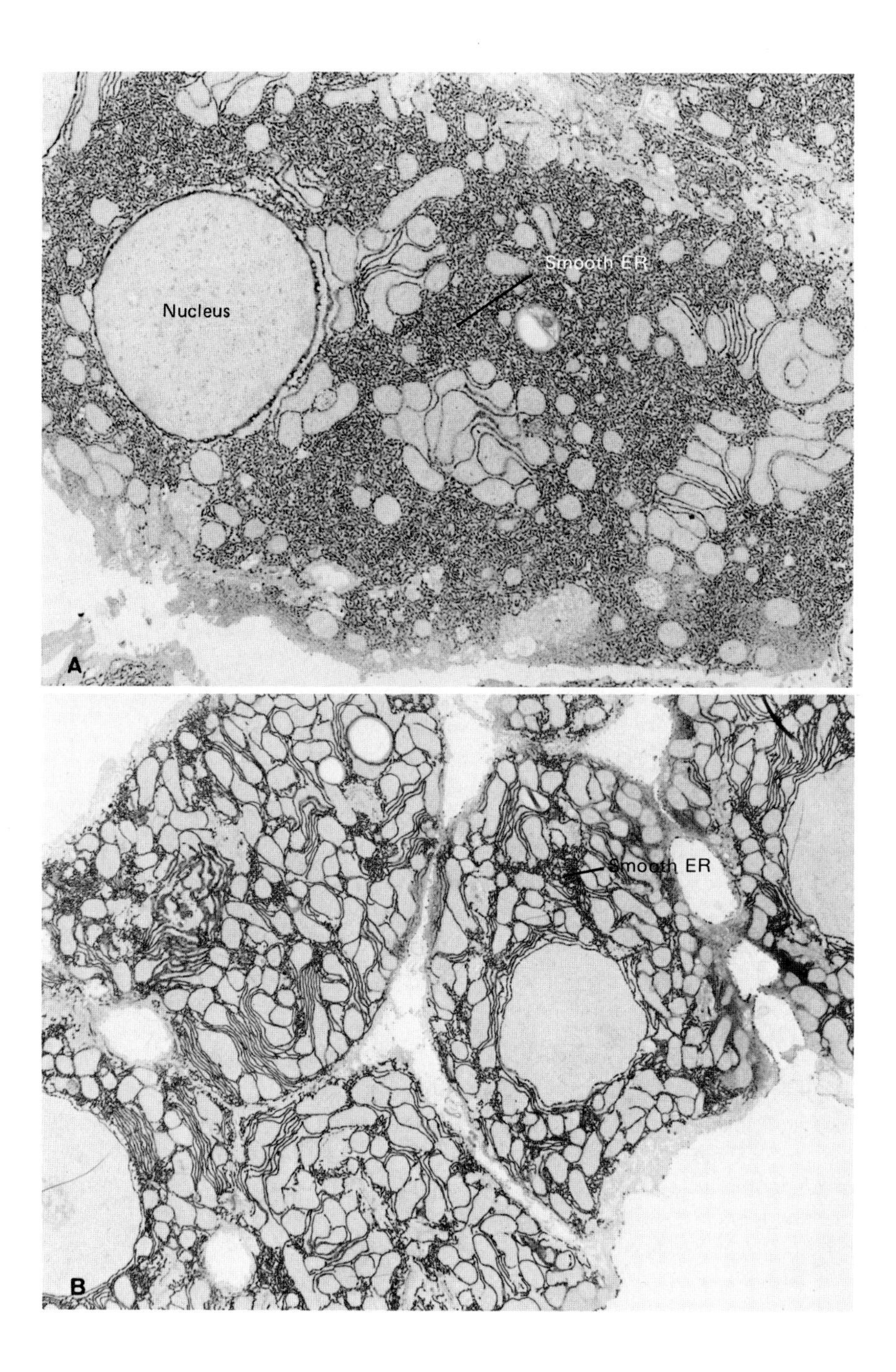
Smooth ER
Nucleus
A
Smooth ER
B

largest vesicles were found to contain the greatest concentration of NADH-ferricyanide reductase, NADH-cytochrome *c* reductase, and cytochrome b_5, whereas the NADPH-cytochrome *c* reductase and cytochrome P-450 were concentrated in the fractions of small vesicles.

Some of the events involved in the synthesis of substances exported by the cell have now been elucidated by series of *in vitro* studies on protein secretion in lactating mammary glands. The current hypothesis proposes that the precursorial molecules (or preproteins) from ribosomes that are to be processed by the rough endoreticulum (Blobel and Sabatini, 1971; Milstein *et al.*, 1972; Blobel and Dobberstein, 1975) contain terminal extensions of up to 30 residues of predominantly hydrophobic amino acids (Campbell and Blobel, 1976; Mercier *et al.*, 1978). Such segments serve as signals that interact with the rough endoreticular membranes, setting up the conditions for conduction through the membranes into the lumina of the cisternae. Once inside, the signals are removed during the processing of the preproteins into their definitive forms (Szczesna and Boime, 1976; Birken *et al.*, 1977; Maurer and McKean, 1977; Strauss *et al.*, 1978). In other studies, six major proteins of milk were synthesized in a cell-free system. Upon analysis of their molecular structures, three casines (α_s1, α_s2, and β) were found to bear terminal chains 15 amino acid residues in length, whereas κ-casein, β-lactoglobulin, and α-lactalbumin had chains of 21, 18, and 19 residues, respectively (Mercier *et al.*, 1978). These prelactoproteins currently have been demonstrated to be taken into the reticular cisternae and actually processed there into the definitive lactoproteins (Gaye *et al.*, 1979). Furthermore, it was shown that the signals could be removed by the proteases that were extracted from rough microsomes of mammary glands.

Another series of studies that provides firm evidence of specific activities of the rough endoreticulum was conducted on tendon cells from chick embryos. The collagens that such cells secrete are complex substances containing two sugars, galactose and glucosylgalactose, which are connected to hydroxylysine residues by *o*-glycosidic linkages (Kivirikko and Risteli, 1976; Miller, 1976). In contrast, the precursor molecule, procollagen, contains other carbohydrate residues in the amino acid segments that are appended to each end of the molecule (Furthmayer *et al.*, 1972; Murphy *et al.*, 1975; Clark and Kefalides, 1976; Duksin and Bornstein, 1977; Olsen *et al.*, 1977), there being 9 to 13 mannose residues in the chain on the carboxy terminal. Previous studies had clearly demonstrated that the hydroxylysine-linked carbohydrates were synthesized within the rough endoreticulum (Harwood *et al.*, 1976; Oikarinen *et al.*, 1976a,b, 1977; Risteli *et al.*, 1976). Now by inhibiting movement of procollagen from the rough endoreticulum into the Golgi bodies, it was evidenced that the mannose could be added in the former, too (Anttinen *et al.*, 1978).

Rough Endoreticular Functions in Seed Plants. Researches in rough endoreticular functions in plant tissues have not as yet progressed to the point where steps in the processing of proteins and glycoproteins have been ascertained. What has been established of specific functions, however, does contribute greatly to an appreciation of their functions. One study on aleurone bodies of barley has presented evidence that cisternal rough endoreticulum differs functionally from the vesicular type; this experimental project utilized gibberellic acid and/or actinomysin D combined with electron microscopy. In these bodies, α-amylase is among the important enzymes of the endomembrane system. In germinating seeds, this investigation showed that α-amylase activity increased rapidly but not until the stacks of endoreticular membrane cisternae had been completed. Once the cisternae were fully developed, they gave rise to vesicular endoreticulum, at which time α-amylase increased in amount. Thus this enzyme appears to be produced (at least processed) within the vesicular rough endoreticulum but not within the cisternal type. Polysaccharides that enter into the production of slime by maize root cells have been reported to be formed within the rough endoreticulum, but the type of this organelle was not made clear (Bowles and Northcote, 1974). Results of an independent study of the same enzyme using immunochemical techniques gave results somewhat contrary to those just presented; unfortunately, that report was not cited by the later article, nor were comparisons made (Jones and Chen, 1976). In cells not treated with gibberellic acid, the fluorescence was diffuse and was particularly low in the perinuclear region. When cells were treated with gibberellic acid, the fluorescence was found to be associated with the nuclear membrane and with the newly formed endoreticulum that is greatly proliferated by such treatments. In another report on an undesignated type of endoreticulum in plants, but presumably of the rough class, mannosyl transferase, which used GDP-mannose as donor, was found to be localized primarily in these membranes, but related enzymes were centered in dictyosomal membranes (Lehle *et al.*, 1978).

A few other reports exist on the activities of enzymes of rough endoreticulum in plants, including some inclined toward the systematic relationships of enzyme occurrence. Dilated cisternae of rough endoreticulum are characteristic of several taxa of seed plants (Figure 6.26), among them the families Brassicaceae and Capparaceae (Behnke and Eschlbeck, 1979). Electron microscopic cytochemical tests showed that β-thioglucosidase activity was displayed by these dilated cisternae. Another investigation reported some very unusual activities of rough endoreticulum, several aspects of which led into behavior more reminiscent of lysosomes (Chapter 7, Section 7.1) than of the present organelle. In this study, pollen development was followed in five species of dicotyledonous plants, in all of which the rough endoreticulum was in the form of dilated vesicles (Santos *et al.*, 1979). During these microspore-producing processes in the lining (tapetum) of the microspore case, the membranes of the vesicles invaginated, thereby engulfing cytosol and cytoplasmic organelles (Figure 6.27). These vesicles, which thus were lined with ribosomes, acted similarly to the typical autosomal lysosomes described later, in that their contents were digested. However, the rough endoreticulum was not alone in carry-

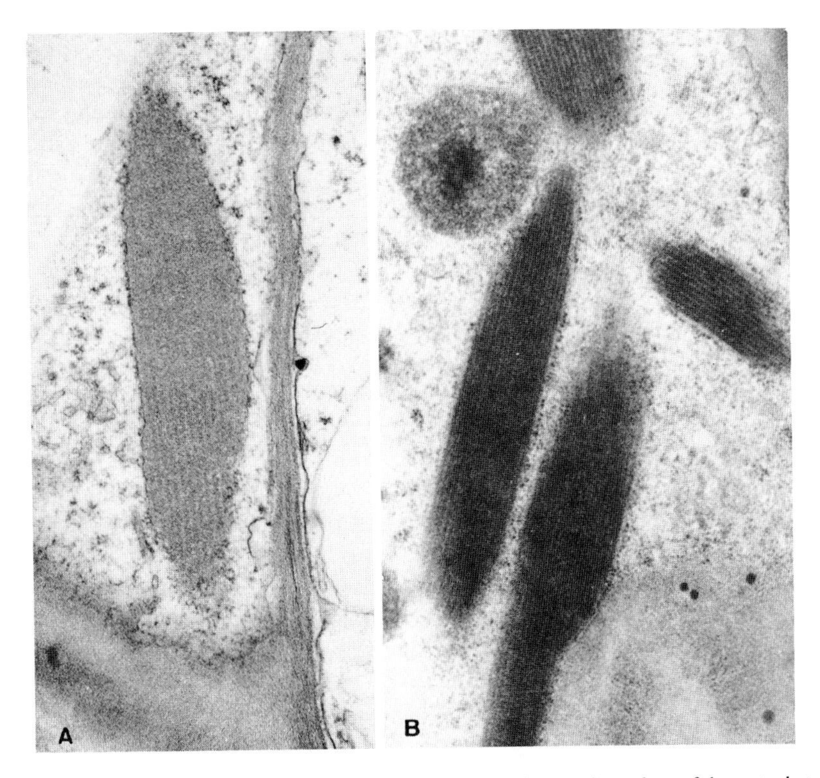

Figure 6.26. Endoreticular traits of systematic value. Various tissues of members of the metaphytan order Capparales have a distinctive type of rough endoreticulum. (A) Section of a stem vascular parenchymal cell with a cisterna of rough endoreticulum dilated by its contents of a tubular material, from *Bumais erucago,* a member of the Brassicaceae. (B) Similar tissue and endoreticulum from *Capparis badueca*, a member of the family Capparaceae. Both 20,000×. (Courtesy of Behnke and Eschlbeck, 1979.)

ing out these steps toward reduction of the cytoplasm; similar invaginations were formed on the outer nuclear envelope and on the plasmalemma.

6.3.2. Activities of Smooth Endoreticulum

Because smooth endoreticulum has not long been considered distinct from the rough type (Dallner *et al.*, 1966), older studies often fail to differentiate between the two, a practice not always avoided currently. As will be seen, the two are obviously related but each carries out specialized functions in their own right to such an extent that they should be viewed, as pointed out earlier, as being separate organelles—just as the flagellum and the flagellar centriole are viewed, in spite of the one always being attached to the other.

Functions in Vertebrate Tissues. A frequent point of view is that the enzymes are synthesized in the rough endoreticulum and conducted thence to the smooth (Dallner *et al.*, 1966). The study cited showed that glucose-6-phosphate and NADPH-cytochrome *c* reductase activities in developing rat liver cells appeared first in the former type and later in the latter as it was proliferated after birth (Section 6.1.2). Eventually these activities became uniformly distributed in the two organelles. However, these conclusions may need to be reexamined in light of the presence of RNA in the present organelle and the recognition of amino acid sequences described earlier. Although poly(A) is less abundant in smooth reticulum than in the rough, it is present to the extent of 20% of that of the latter (Cardelli *et al.*, 1978). In this organelle the poly(A) differed further from that of rough in being easily removed by high salt concentrations or by puromyocin.

The mere presence of the same enzyme in two organelles is not sufficient to demonstrate functional relationships, as shown by an article that seems to

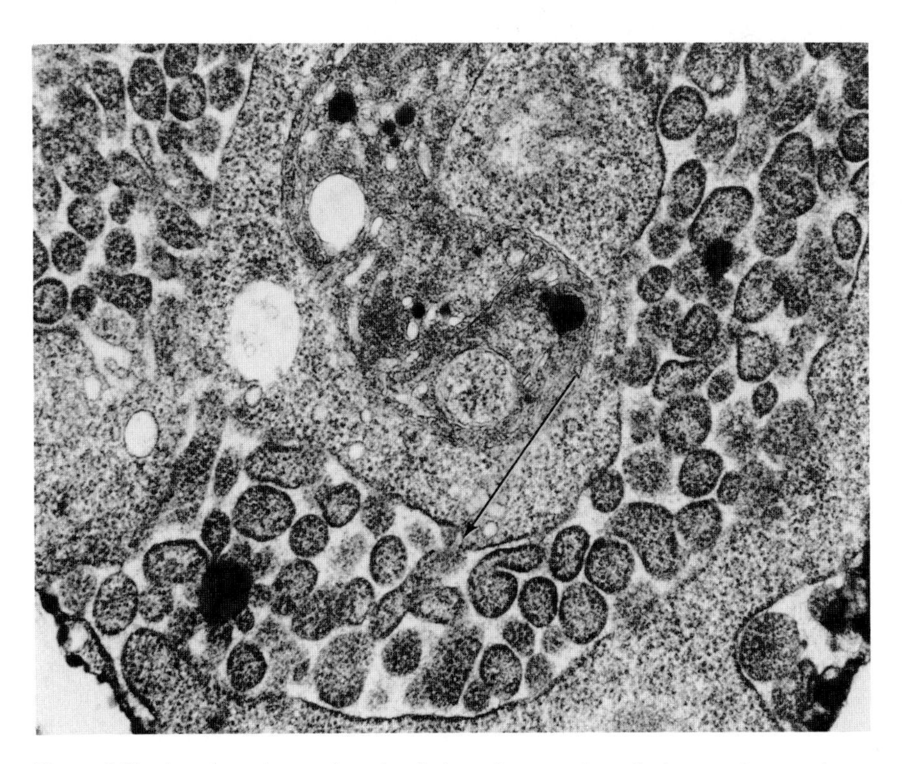

Figure 6.27. Autophagy by rough endoreticulum. In a number of plant species, rough endoreticulum is in the form of large vesicles, filled with smaller saccules containing ribosomes and other cytoplasmic materials. An invaginated saccule appears to be pinching off from the endoreticular membrane at the arrow. 27,600×. (Courtesy of Santos *et al.*, 1979.)

have been overlooked by recent researchers. This reported that the glucose-6-phosphatase of rough endoreticulum differed from that of the smooth in reactivity and in pH optima. In the former type of organelle, the enzyme was more reactive in the same preparations than that from the smooth, but showed greater changes in the latter with pretreatment with NH_4OH or deoxycholic acid. These patterns of differences persisted in starved, phenobarbital-treated, and alloxan-diabetic rats, just as in normal animals (Stelten and Ghosh, 1971). This lesser reactivity of smooth-membrane-derived glucose-6-phosphatase may underlie the decrease of enzyme reported in phenobarbital-treated rats described in an earlier paragraph (Ménard *et al.*, 1974a,b). In several investigations into phenobarbital-treated animals, alterations in enzymatic concentrations have been noted, especially in regard to an increase in several drug-metabolizing enzymes, including oxidative demethylase, and TPNH-cytochrome *c* reductase (Remmer and Merker, 1963; Orrenius, 1965a,b; Orrenius *et al.*, 1965; Ménard *et al.*, 1974a). Still another report on phenobarbital-treated rats demonstrated that the rapidly proliferated smooth endoreticulum of the liver was the site of phospholipid synthesis (Higgins, 1974). Greatly proliferated smooth endoreticulum is also characteristic of chloride cells of sea lamprey (Figure 6.28), in which case the tubular type rather than vesicular is present (Peek and Youson, 1979a,b).

Undesignated Endoreticular Enzymes. In a number of reports on enzymes of these organelles, no distinction was made between the rough and smooth varieties. Because one study concerns the localization of enzymes involved in the biosynthesis of phospholipids, its results may be tentatively assumed to appertain to the smooth in light of the report cited in the preceding paragraph. After the various membranous organelles had been isolated into separate fractions, each was assayed for activity of the various enzymes of phospholipid synthesis (Jelsema and Morré, 1978). The terminal enzymes of the cytidine diphosphate amine pathway, the phosphotransferases of choline and ethanolamine, were localized in the endoreticulum to the extent of 90%, the rest being confined to the Golgi apparatus. Both of these enzymes were absent from the nuclear membrane.

However, the enzymes of the second major pathway of phospholipid biosynthesis, the CDP-diglyceride sequence, were more widely distributed, with the exception of those that synthesize phosphatidylinositol. As before, these were confined largely to the endoreticulum of undesignated type and to a lesser degree the dictyosomes. Phosphatidylserine-synthesizing activity was primarily restricted to the endoreticulum but also was noted in the Golgi, plasmalemma, nuclear membrane, and cytosol fractions. Thus it was concluded that the [smooth?] endoplasmic reticulum is the principal site of phospholipid synthesis in the rat liver.

Functional Aspects in Metaphytans. Less attention has been devoted to enzyme distribution in plants than in metazoan tissues. The smooth endoreticulum of germinating castor bean endosperm has been found to be the exclu-

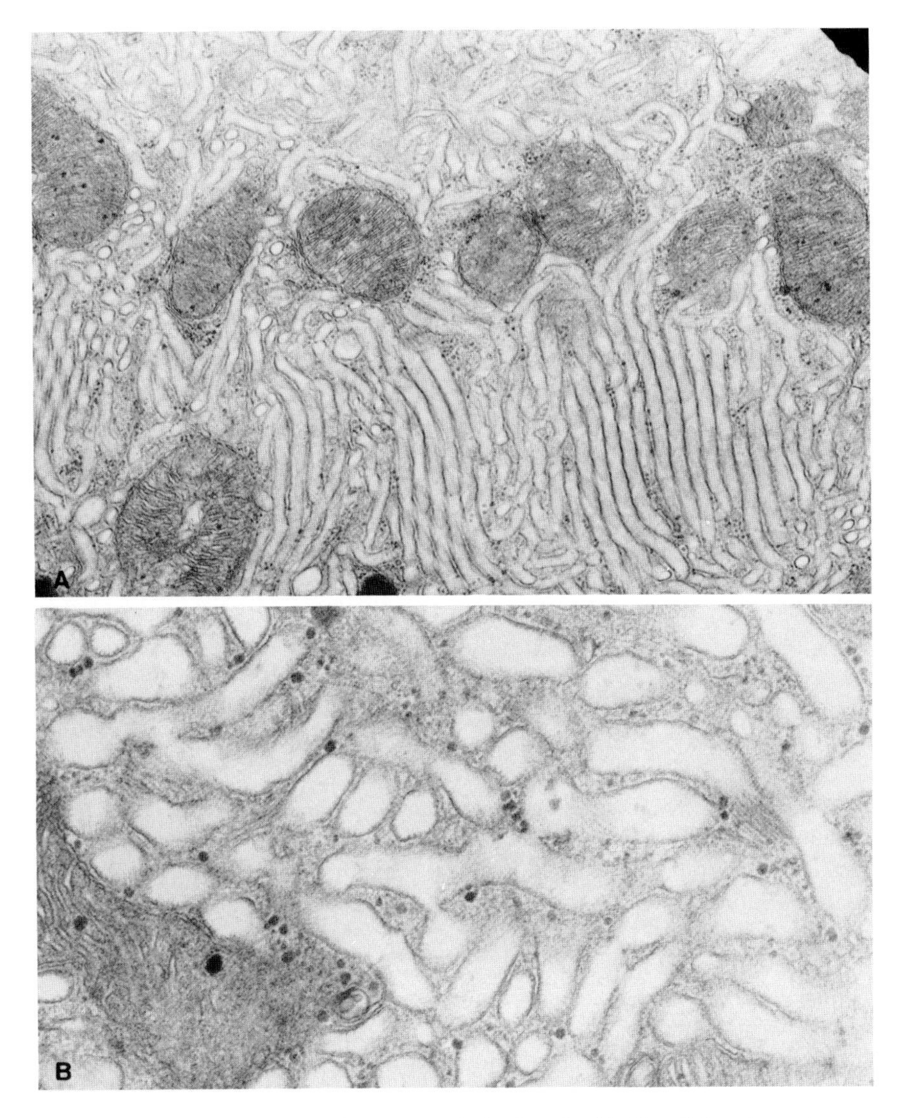

Figure 6.28. Proliferated tubular endoreticulum. (A) Section of a chloride cell from the anadromous sea lamprey in which smooth tubular endoreticulum pervades much of the cytoplasm. 22,000×. (B) The tubules at higher magnification. 40,000×. (Both courtesy Peek and Youson, 1979a.)

sive site of phospholipid synthesis (Moore *et al.*, 1973; Donaldson and Beevers, 1978), including such enzymes as phosphorylcholine-glyceride transferase. In addition it has been shown to contain an electron-transport system capable of mixed-function oxidase activity (Young and Beevers, 1976). This contrasts to a study of cytochromes in rat kidney brush-border cells, which reported only cytochrome *b* in the endoreticulum (García *et al.*, 1978).

No enzyme activities were investigated in an ultrastructural study of a parasitic flowering plant (*Tozzia alpina*), whose scale leaves lack chlorophyll (Renaudin and Capdepon, 1977), but possible functions were deduced from the intimate associations that were found between the tubular smooth endoreticulum and the starch-containing chloroplasts. Each such plastid is enveloped by four or five layers of tubular endoreticulum arranged in a compact, orderly fashion. Similar relationships have been noted in a number of other plants, including *Cryptomeria* (Camefort, 1970), *Acer* and *Pinus* (Wooding and Northcote, 1965), and *Calceolaria* (Schnepf, 1969). Consequently, it was suggested that storage of starch took place in the leucoplastid, whereas the endoreticulum served to conduct the glucose to the site and possibly contained enzymes for its conversion into starch. Thus these interrelationships between these two organelles parallel the functional relations between cisternal smooth endoreticulum and glycogen deposits in liver. This same type of organelle associated with the chloroplastid has been shown to produce vesicles in which the scales and bristles that coat the surface of *Mallomonas* were formed, together with vesicles from dictyosomes (Belcher, 1969; Schnepf and Deichgräber, 1969; Wujek and Kristiansen, 1978). In the members of this chrysophycean genus, the endoreticulum is so closely applied to the plastid that it was mistaken for the chloroplast envelope (Figure 6.24).

6.3.3. *Functions of GERL and Dictyosomes*

The discussion of smooth endoreticulum function necessarily continues into those of GERL and the dictyosome, for as already mentioned, tubular endoreticulum often is closely associated with Golgi material. This relationship in part is demonstrated not only by physical proximity but also by enzymatic activity. A sufficient number of differences also exists to indicate their respective uniquenesses.

Enzymes and Functions of GERL. Before the enzymes and functions of the cisternal endoreticulum–dictyosome–lysosome combination are discussed, it should be considered that the Golgi portion of that association is not necessarily identical enzymatically to other dictyosomes of the same cell. This is accented by the fact that the vesicles that were formed at the maturing face of GERL dictyosomes but not the tubular endoreticulum displayed acid phosphatase activity (Figure 6.29A), which is found elsewhere only in lysosomes (No-

Figure 6.29. GERL reacted with various reagents. (A) The dictyosome shows heavy reactivity for acid phosphatase activity, whereas the adjacent smooth endoreticulum (GERL) shows none. 49,000×. (B) The same but reacted for alkaline phosphatase; the deposition product appears in the dictyosomes and associated smooth endoreticulum. 25,000×. (C) Thiamine pyrophosphatase activity is indicated to occur only in the cisternae at the middle of the dictyosome. 20,000×. (All courtesy of Paavola, 1978a,b.)

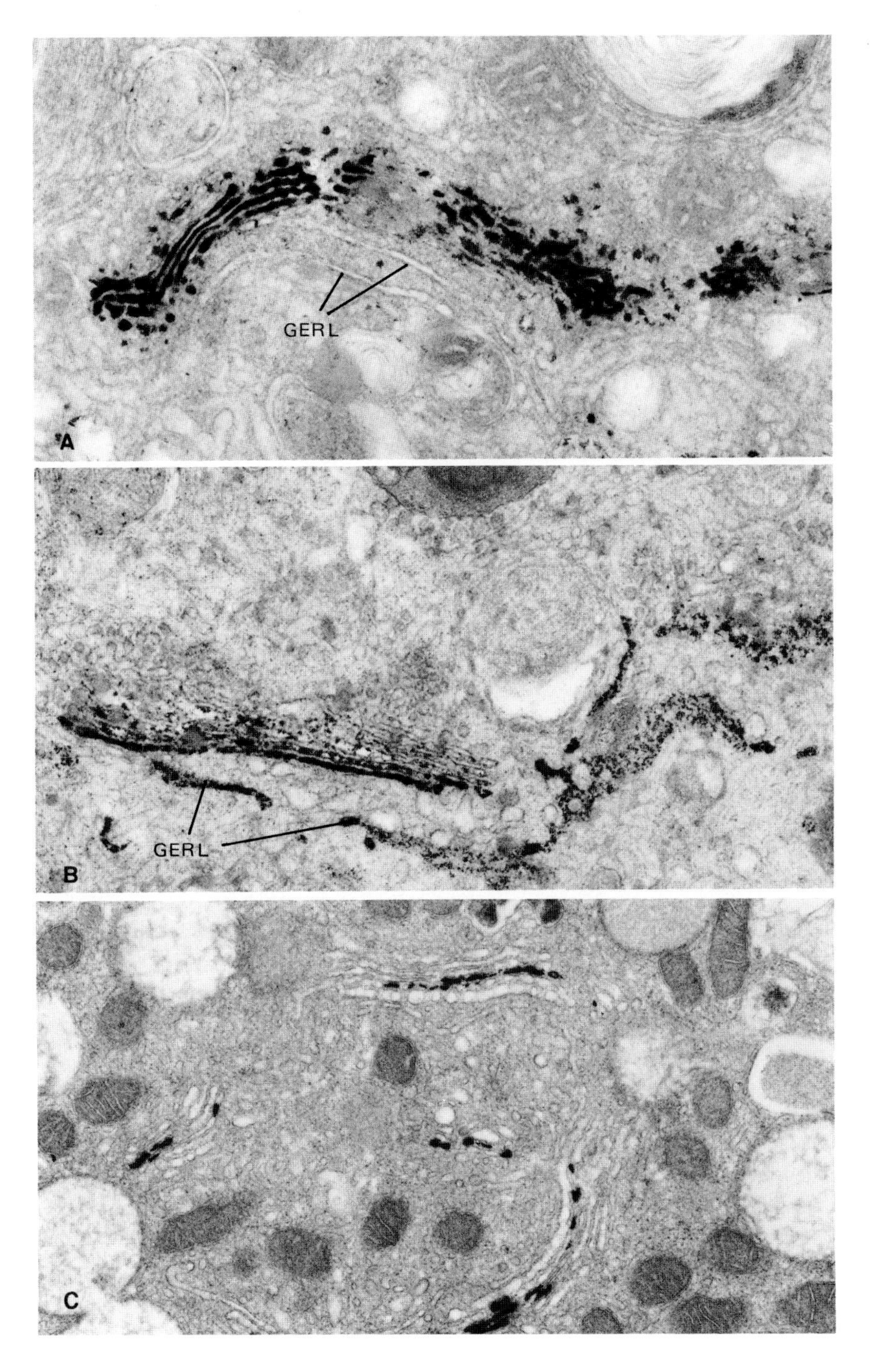
GERL
A
GERL
B
C

vikoff *et al.*, 1971b; Novikoff, 1976; Leblond and Bennett, 1977; Paavola, 1978a). Such vesicles are often referred to as presecretory granules or condensing vacuoles (Hand, 1971; Novikoff *et al.*, 1971a), but they could as validly be referred to as early primary lysosomes (Chapter 7, Section 7.1.1). Alkaline phosphatase activity, conducted on corpus luteal cells of the guinea pig, was evident throughout the mature saccules of dictyosomes and in the smooth endoreticular cisternae (GERL) associated with them, but occurred only to a limited extent on the periphery of some lysosomes (Figure 6.29B; Paavola, 1978b). Like dictyosomes in general, the central dictyosomal cisternae alone showed thymidine pyrophosphatase activity (Figure 6.29C; Novikoff and Goldfischer, 1961; Novikoff, 1976; Paavola, 1978b), and arylsulfatase activity was similarly distributed (Paavola, 1978b).

It should be noted that in those cells that were tested for alkaline and acid phosphatase activity, as well as arylsulfatase, the smooth cisternal was the only type of endoreticulum that reacted positively. Thus it is clear that it contains enzymes that are absent in the other reticular membranes. Too often, the term GERL is applied only to this endoreticular fraction of the system and therefore it is not really appropriate. The term dictyosomal endoreticulum would be far more suitable and precise in pinpointing discussions of the several elements in these complex associations.

Enzymes and Functions of Golgi Material. At least one general function of Golgi apparatus was evidenced in the preceding discussion. There it was shown that one kind of lysosome, if not all, was synthesized on the mature face of the dictyosome. Another common activity of this organelle is in the formation of glycoproteins, which are synthesized by a complicated series of reactions. After the polypeptide backbones of such molecules have been translated into existence by the usual genetic processes, carbohydrate side-chain formation is begun. Every such chain is formed by adding individual monosaccharide residues one by one, each monosaccharide being first phosphorylated and then complexed with a specific nucleoside before addition. These additive steps include the transfer of the carbohydrate residues from the nucleoside to the side chain or a branch in a reaction catalyzed by one of a number of glycosyltransferases, each highly specific for both the sugar to be added and the one to receive it.

When ^{3}H-labeled mannose was injected into young rats, the deposition of the sugar in thyroid tissue was found to occur solely in rough endoreticulum (Whur *et al.*, 1969). Thus simple side chains involving this sugar could be added to thyroglobulin in that organelle, because many begin with it. However, when labeled fucose, a carbohydrate that terminates many of the side chains of thyroglobulin, was injected, its presence was associated only with dictyosomes. Terminal residues consisting of galactose or sialic acid also have been shown to be added in Golgi bodies (Whur *et al.*, 1969; Leblond and Bennett, 1977). Hence, the terminal steps of glycosylation seem to be the exclusive property of dictyosomes. Because the sequence of events is complex, the several steps in the synthesis of a glycoprotein proceed from one organelle to another in assembly-line fashion, but this does not necessarily imply that all Golgi functions are secondary to those of the endoreticulum. The distribution of such enzymes as acid hydrolases, alkaline phosphatase, nucleoside diphosphatase, and thiamine pyrophosphatase lucidly demonstrates that many of the enzymes present in dictyosomes do not occur elsewhere (Sabatini *et al.*, 1963; Reid, 1967; Lane, 1968; Flickinger, 1978; Paavola, 1978a,b). Moreover, while specific details are as yet not forthcoming, it has been reported that many enzymes characterized in isolated Golgi fractions were not uniformly distributed in those membranes (Hino *et al.*, 1978a,b). Among the enzymes localized were acid phosphatase, 5′-nucleosidase, NADH- and NADPH-cytochrome *c* reductase, and galactosyltransferase, most of which behaved differently than corresponding enzymes of endoreticulum, lysosomes, and plasmalemma.

Other Secretory Activities. While the existence of enzymes within an organelle can only suggest its possible involvement in secretory activity, more direct evidence has been presented. DeRobertis and Sabatini (1960) examined the mechanism of secretion in chromaffin cells of the hamster adrenal gland, in which the origin and cellular pathway of the secretory products could be readily followed because their catecholamine contents actively reduced osmium tetroxide. The origin of the catecholamines within the dictyosomal vesicles was clearly indicated. Osmiophilic reaction, however, did not commence until the vesicles had been freed from the saccules; after liberation while moving toward the plasmalemma, the vesicles increased markedly in size, as did also the volume of the osmiophilic contents. A similar sequence of events has been shown for acinar cells of the rat pancreas in the formation of zymogen (Warshawsky *et al.*, 1963), but in this case the vesicles deposited their product within the cytoplasm for later release through the plasmalemma. In relation to such secretory activities, it is pertinent to note that receptor sites specific for insulin and human growth hormone labeled with ^{125}I have been found in Golgi fractions, comparable to those of the plasmalemma but somewhat lower in reactivity (Bergeron *et al.*, 1978; Posner *et al.*, 1978a,b).

In an investigation of the origin of granules in rabbit polymorphonuclear leukocytes (Bainton and Farquhar, 1966), the results conflict to some degree with Mollenhauer and Whaley's (1963) proposed developmental sequence of the dictyosome as outlined in a preceding section. First, the report showed that two distinct types of granules were produced, "azurophil" and "specific," and second, that both types were formed by the Golgi complex (Figure 6.30). The azurophil granules were secreted only during the progranulocyte stage of leukocyte development, whereas the less dense and smaller specific granules arose during the later myelocyte stage. The first type was secreted from the mature face of the dictyosome, whereas the second variety was produced and released from the cisternae at the forming face (Figure 6.30B).

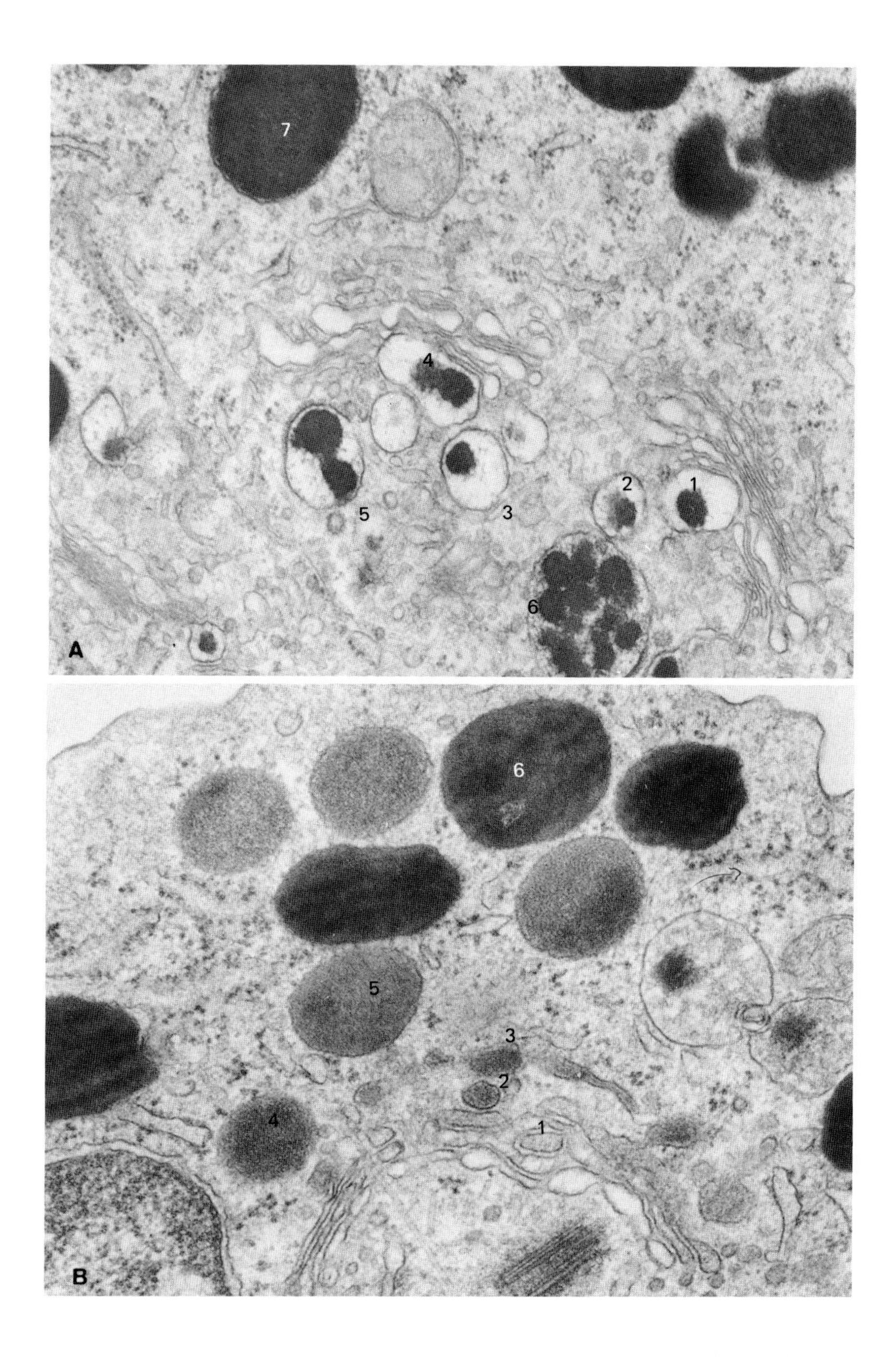
A
7
4
5
3
2
1
6
B
6
5
3
2
4
1

A more recent study of Golgi apparatus in rat cells of many types (Rambourg *et al.*, 1969) used two different electron stains for glycoproteins. Each procedure gave identical results and demonstrated that the carbohydrate was synthesized within the dictyosome, with the mature cisternae showing strong staining reactions, the immature ones little or none, and the intermediate ones a gradient of reactivity. Comparable results have been reported for the carbohydrate of mucus in goblet cells of the colon (Neutra and Leblond, 1966a,b). Thus, in summary, these studies on the vertebrate dictyosome suggest that the Golgi apparatus provides glycoproteins for cellular consumption in the lysosomal system, the cell coat, and basement membrane, and various secretions for organismal use (Whaley *et al.*, 1972).

Cell Wall Formation. In metaphytan cells secretion of carbohydrates has also been established as a Golgi function. Investigating formation of the cell plate by electron microscopy, Whaley and Mollenhauer (1963), as well as Frey-Wyssling and co-workers (1964), found that the dictyosome synthesized carbohydrates and perhaps other materials that served as precursors for the cell plate and cell wall (Mollenhauer and Mollenhauer, 1978). By use of labeled glucose, Northcote and Pickett-Heaps (1966) showed that polysaccharide synthesis was conducted in the Golgi complex of wheat root-cap cells. As in other organisms reported upon above, vesicles containing the labeled product migrated to the plasmalemma from the mature pole of the dictyosome for incorporation into the cell wall and slime layer. Analysis of the vesicular contents indicated that the major product was a pectic substance. A later study (Pickett-Heaps, 1968b), employing electron staining, indicated that polysaccharides were present in the vesicles and, to a lesser extent, in the cisternae too. It was noted that the farther a vesicle had migrated from the dictyosome, the greater its content of polysaccharide became. An enzyme involved in carbohydrate metabolism has now been isolated from the dictyosomes of the brown alga, *Fucus serratus* (Coughlan, 1977). This enzyme transferred L-fucose from a sugar nucleotide to free fucose to build dimers of that substance. More recently in the same organism, thiamine pyrophosphatase, inosine diphosphatase, and galactosyltransferase have been identified (Coughlan and Evans, 1978).

A series of remarkable demonstrations of Golgi function in two groups of flagellates has been conducted by Manton and her co-worker (Manton, 1966a,b, 1967; Manton and Harris, 1966). The experimental subjects included

←

Figure 6.30. Formation of granules in polymorphonuclear leukocytes. (A) A polymorphonuclear progranulocyte of rabbit, showing successive stages in the formation of azurophilic granules. These develop from the proximal face of the dictyosome. 80,000×. (B) The myelocyte of the same tissue, showing stages in the development of the specific granules from the distal face of the dictyosome. 50,000×. The Golgi bodies surround a centriole (bottom center) and represent the so-called centrosphere of metazoan cells. (Both courtesy of Bainton and Farquhar, 1966.)

members of three genera of brown flagellates (*Prymnesium, Sphaleromantis,* and *Chrysochromulina*) and a single one of green flagellates (*Pryamimonas*). All forms agreed in possessing only a single relatively large dictyosome, which was closely associated with the flagellar base. The dictyosome in each case was concerned with producing the scales that cover the organisms. Within the cisternae and vesicles, scale material was deposited until a mature scale had been formed. Then, still within the vesicles, the completed scales moved toward the plasmalemma and, after contacting it, seemed to exit by way of a small pit lying adjacent to the flagellum. Beyond that point it was impossible to trace how the scales moved into position on the external surface, nor could their chemical composition be established. Similar scale production in dictyosomes has been observed in the prasinophycean *Pyramimonas* (Moestrup and Walne, 1979). The lorica of the choanoflagellate *Stephanoeca diplocostata* has been demonstrated as being formed in comparable fashion (Leadbeater, 1979). In this instance costal strips were produced seemingly by Golgi bodies, then transported through the plasmalemma for incorporation into the lorica (Figure 6.31). In this connection, the instances reported earlier (Chapter 4, Section 4.1.2) of the secretion of mastigonemes and scales of flagella in the Golgi bodies should also be recalled (La Claire and West, 1978).

A related study on the coccoid marine chrysophycean *Pleurochrysis scherffelii* (Brown, 1969) similarly reported a single dictyosome to be present; this structure resembled those cases described in the foregoing paragraph in its large size and cavernous vesicles. It was proposed that in this organism the Golgi apparatus was involved in an orderly sequential deposition of cell wall material that became distributed throughout the inner surface of the wall by the continuous streaming of the cytoplasm. In its specific location in the cytosol, the dictyosome underwent periods of differential secretory activities in forming the wall units as it revolved with the rest of the protoplast. As might be expected, the organelle was most active immediately following cell division, when the new transverse wall was being formed. Because of the distinct stratification of the wall in this chrysophycean, it appeared of particular value in thus showing the origin of the new walls. The secretory products were in the form of wall fragments or scales, as shown in a later investigation (Brown *et al.,* 1969), and consisted in part of a cellulosic component together with galactose, ribose, arabinose, and glucose.

Protein Packaging. In cases where the cell's secretory product is largely proteinaceous, the role of the dictyosome is less clear-cut and appears to be largely that of concentrating enzymes secreted by the ribosomes, and possibly transported there by the rough endoreticulum, at least in part. One instance is provided by the series of studies on guinea pig pancreatic exocrine cells (Caro and Palade, 1964; Jamieson and Palade, 1967a,b). Using slices of pancreas pulse-labeled for 3 min with [^{14}C]leucine, coupled with cell fractionation and ultrastructural procedures, these studies reported the label first appeared in

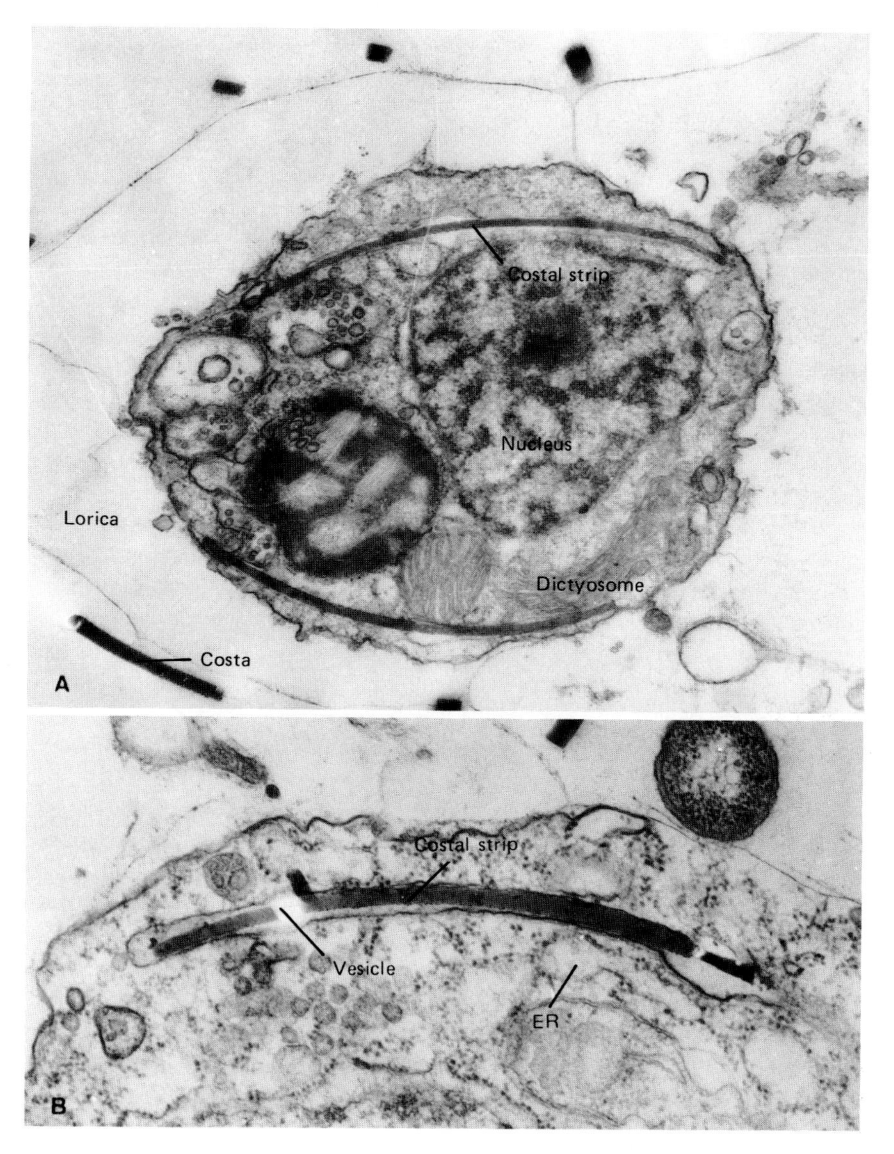

Figure 6.31. Costa production in a choanoflagellate. The choanoflagellate *Stephanoeca diplocostata* surrounds itself with a lorica supported by siliceous costae. (A) The costae are secreted in vesicles of unknown origin but always associated with a dictyosome, as shown in the lower portion of the cell. 20,000×. (B) At greater magnification, the vesicle can be noted to lie adjacent to a vesicle of rough endoreticulum. 40,000×. (Both courtesy of Leadbeater, 1979.)

the granular endoreticulum (after 7 min incubation), later in the Golgi vesicles (after an additional 17 min), and finally in the zymogen granules. Thus it appeared that the secreted proteins were conducted from the ribosomes by the endoreticulum toward the dictyosomes; within the latter they apparently became more concentrated in the peripheral portions as vesicles were produced. Then the latter were freed in typical fashion and developed into zymogen granules.

Packaging and modification of proteins by the Golgi apparatus occurs in such a way in testate amoeboids that this function is seen to be related to the cell wall formative activities of plants and scale production in flagellates. Proteins from the ribosomes and endoreticulum were modified in the Golgi apparatus and carried to the plasmalemma, where the test was gradually constructed. In *Centropyxis discoides* the shell consisted of up to four layers of hollow polyhedral compartments, between 200,000 and 500,000 of which were required for the shell (Netzel, 1975b). Similar but more granular compartments were employed in forming the test of *Arcella vulgaris* (Netzel, 1975a). All the components were considered to consist in large measure of keratinoid scleroprotein.

Summary of Functions. To summarize, the dictyosome appears functionally to be a most versatile organelle, whose activities, and therefore the enzymes it contains, vary with the needs of the organism and with the specializations of the cell. It may be actively secretory, but then as a whole its products appear to be carbohydrates; it sometimes condenses and packages, as in the case of pancreatic products; it occasionally combines carbohydrates (which it probably secretes) with proteins (which it possibly receives from ribosomes and endoreticulum), and it secretes or assembles complex scales and other body-covering materials.

In addition to these general functions, the Golgi complex at times is associated with the formation of several other specialized cell organelles, but in what capacity remains unknown. Among the associated structures are the acrosome of the metazoan spermatozoon, the metaphytan cell plate (in other ways than that mentioned previously), and the parabasal body. As each of these instances is of sufficient importance to merit full discussion individually, attention will be devoted to them in appropriate chapters. But one body, formerly considered a separate organelle, the centrosphere of metazoans and brown algae, appears to consist entirely of a spherical aggregate of dictyosomes (Figure 6.30B). Hence, it represents only a special configuration, not a true organelle, that in all likelihood has no function other than those that pertain to individual dictyosomes (Bainton and Farquhar, 1966).

7

SECRETORY ORGANELLES: II

The Vesicular System

In addition to the cisternae and tubules that comprise the endomembrane system just given attention, several major types of vesicles are involved in the secretory processes of cells. One of these, the lysosome, has already received mention as being part of GERL (Chapter 6) and still earlier in connection with endocytosis and receptor sites in the discussion of membranes (Chapter 1, Sections 1.2.1 and 1.2.2). Here even more clearly than in the membranous organelles, secretion is revealed as the underlying mechanism for most cellular functions, for these little bodies carry the secreted products to where the action is finally to take place. But much more is involved in the functioning of these vesicular particles than mere conduction of finished products, as becomes evident as the lysosome, peroxisome, spherosome, and others of lesser importance are discussed.

7.1. THE LYSOSOME

In the early days of differential centrifugation, one cell particle proved particularly vexing to investigators because it possessed a density intermediate to those of the mitochondrion and the microsome. Consequently, before the presence of a discrete organelle was suspected, certain enzymatic properties, including acid phosphatase activity, had been assigned variously to the mitochondrion or the microsome, depending on which fraction it happened to sediment (de Duve, 1964). By employing highly standardized and refined fractionation procedures, de Duve and his collaborators (1955) were able to separate a cell fraction that contained 40% of the acid phosphatase activity but only 10% of the protein. After investigating the properties of this fraction more completely, these biochemists decided a new cell organelle was involved, which they named the lysosome. When examined under the electron microscope, a distinctive granule was found that consisted of an electron-opaque matrix enclosed in a single membrane. After application of histochemical methods verified the localization of acid phosphatase in these particles, their identity with that called the lysosome was finally established. Since that time this organelle has acquired so great a popularity as a subject for biological investigations that the literature has become very extensive. For greater detail, reference may be made to the several reviews of the subject (Novikoff, 1961; Novikoff *et al.*, 1964; Tappel, 1968; de Duve, 1971; Dingle, 1972; Matile, 1975; Holtzman, 1976).

7.1.1. General Properties

Because most of the researches into the nature of the lysosome have been conducted with particles extracted from liver, kidney, spleen, and similar tissues of rats or other mammals, the general properties given here pertain largely to material from those sources. Then after these have been described, the fairly ample comparative materials now available can be examined.

Distinctive Features. Classically, two characteristics distinguish the lysosome from other organelles or particles: first and foremost, the presence of acid hydrolases (Tappel, 1968; de Duve, 1971; Novikoff, 1976), and second, a single enclosing membrane. The latter trait, shared with several other particles, including the peroxisome described later, may be identified by electron microscopic procedures or by the latency of reactivity under experimental conditions (de Duve, 1969). Here as elsewhere, latency refers to the observation that enzymes enclosed in membranes do not react when added to a medium containing suitable substrate until they have been ruptured mechanically or chemically. The importance in the cell of this membrane-induced latency of lysosome reactivity is indicated by an investigation on muscular dystrophy in the chicken pectoralis muscle. Apparently the disease involves a reduced efficiency of membrane formation throughout the organism but affects certain muscles more than other muscles and organs (Owen, 1979). As a result of the membrane defect, proteases are released prematurely from lysosomes; it has been reported that dystrophic muscle tissue grew normally when protease synthesis was inhibited (McGowan *et al.*, 1976). In addition, other reports demonstrate that greater quantities of lysosomal proteases are synthesized in diseased than in healthy muscles (Bird *et al.*, 1978; Pearson and Kar, 1979).

Aside from these two features, however, lysosomes lack consistent morphological and biochemical properties; the enzymes vary with the tissue and the origin of the particle within the cell. Indeed, as soon shall be seen, even the sole enzymatic type given in the definition of the particle (acid hydrolase) is occasionally absent. Furthermore, acid phosphatase reactivity may occur in several different organelles in the same cell (Figure 7.1), as recently reported in the latissimus dorsi muscle of the chicken (Trout *et al.*, 1979), and two dif-

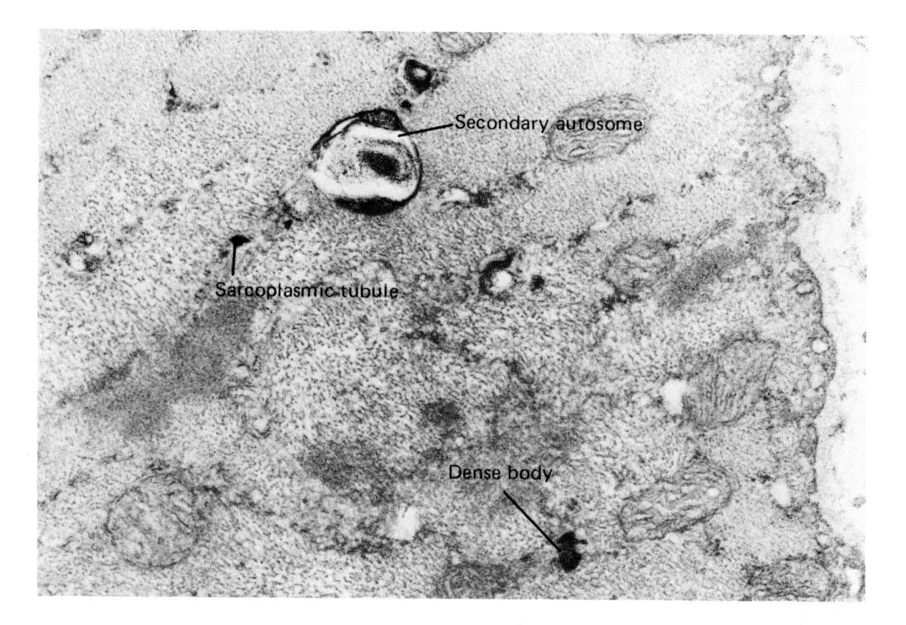

Figure 7.1. Sites of acid phosphatase activity in the anterior latissimus dorsi muscle of rabbit. All three of the particles labeled show the opaque reaction product of the acid phosphatase, the secondary autosome being the only body showing the activity that is classed as a lysosome. 40,000×. (Courtesy of Trout *et al.*, 1979.)

ferent species of lysosomes showing this activity may be present (Rome *et al.*, 1979). There, reactivity was detected in lysosomes, and in the sarcoplasmic reticulum near the I lines. Thus the name lysosome does not actually refer to a single organelle but an entire class, a few members of which have now been sufficiently characterized to name and discuss separately, as shown later.

Recognized Categories. Some of the categories that have been recognized and named refer to age classes of lysosomes, not individual types; for instance, primary and secondary lysosomes are stages in development of any given type, while a third, the residual body, is the final stage of many varieties.

As a class primary lysosomes are small vesicles, often formed from Golgi bodies or from endoreticulum, rough or smooth, and are difficult to recognize. At times, these small bodies are insufficiently developed to show much acid hydrolase activity and thus are indistinguishable from other vesicles in the region. In some instances, however, they may be "coated," that is, covered with an irregularly arranged substance that appears as radiating projections in transverse sections (Friend and Farquhar, 1967; Holtzman *et al.*, 1967; Holtzman, 1976). Even these coated particles do not represent a particular type, for it has been shown that such bodies in the vas deferens and other tissues have a mean diameter of 50 to 75 nm, while others that participate in endocytosis average 100 nm (Friend and Farquhar, 1967). However, many, if not most, primary lysosomes lack such coatings. Those discussed in the section on the Golgi apparatus in rabbit polymorphonuclear leukocytes (Chapter 6, Section 6.3.3; Figure 6.30) provide one example of primary lysosomes that are uncoated during their formative and subsequent stages, up to the time when they have interacted with their target substance or other vesicle.

Secondary lysosomes typically are those that have interacted with their respective targets, the several varieties of which are more readily recognizable than the primary. Among the small diversity of types that are currently recognized is the autophagic lysosome, which, as the name implies, consumes the substance of the cell within which it lies. Typically this type may be identified by the presence within them of mitochondria or other cytoplasmic organelles. Another example, already pointed out (Chapter 1, Section 1.2.1), consists in part of the vesicles formed by phagocytosis that contain prey or debris. These vesicles, often referred to as primary phagosomes, may be called secondary phagosomes after they have fused with one or more lysosomes. Additional names given to this combination of phagosome and particle (which is actually a type of secondary lysosome) include phagocytic digestive vacuoles (Holtzman, 1976) and heterolysosome (de Duve and Wattiaux, 1966; Mego, 1973). Pinocytosis, too, often is involved in the production of secondary lysosomes called pinocytotic digestive vacuoles, which can be recognized as such if the exogenous material contains electron-dense or otherwise identifiable particles. At times pinocytosed substances accumulate in vesicles that interact with primary lysosomes to form a structurally distinct secondary type known as the multivesicular body. These three frequently occurring examples, plus one other, are discussed in more detail in the following section.

Residual bodies, the final current term, is applied to the last stages of many types of secondary lysosomes. As a rule they have an electron-dense, amorphous matrix in which may be contained lipid droplets, opaque granules, lamellar substances, or other structured material. The actual appearance varies somewhat, depending upon the type of lysosome from which the specific residual body was derived.

7.1.2. Some Representative Types

Because of the diversification that lysosomes as a class have undergone in specializing for a wide variety of roles in cells, it is not feasible to attempt to describe all known cases in detail. Instead, it is believed more meaningful to concentrate attention on a few representative forms, during which discussion an insight is provided into the numerous functions these bodies perform. The five major types described here are given names to assist in distinguishing them. A preview of this quintet with their functions is provided in Table 7.1 to assist in clarification of this complex topic.

Heterophagic Lysosomes. Because the terms heterolysosome, pinocytotic digestive vacuole, and the like strictly refer only to secondary lysosomes,

Table 7.1
Summary of Lysosome Types and Their Functions

Class	Other names of secondary bodies	Known sources	Specific examples	Functions
Heterosome	Secondary phagosomes; heterolysosome; pinocytotic digestive vacuole	Eukaryotes in general	Azurophilic specific granules	Degrade exogenous organic matter taken in by endocytosis, including the plasma membrane folds
			Eosinophilic granules	
Exophagosome	---	Dinoflagellates, *Ochromonas*, fungi	---	Deposit digestive enzymes into milieu
Autosome	Autophagic lysosomes, cytolysomes	Eukaryotes in general	Multivesicular body	Degrade endogenous organic material, providing turnover of cellular substances; destroy excess secretions (crinophagy); break down stored foodstuffs
			Vacuole of metaphytans	
			Yolk platelets	
			Aleurone grains	
			Guard cell lysosomes	
			Isolation membrane	
Pungiosome	---	Multicellular organisms	Haustoria; pollen tube tips	Aid in penetration of exogenous cells or tissues
			Acrosome	
Histosome	---	Multicellular organisms	Modifiers of proinsulin into insulin	Modify endocellular products, carry intracellular chemical signals, etc.

the corresponding primary particles are left without distinctive varietal names. This lack of distinctive collective terminology persists despite a large body of literature pertaining to certain primary forms, particularly to the present type involved in phagocytosis (Novikoff, 1961; Hirsch and Cohn, 1964; Zucker-Franklin and Hirsch, 1964; Steinman *et al.*, 1976; Bowen *et al.*, 1979). There can be a little doubt that this is by far the most thoroughly documented variety of lysosome and is probably the only one well known in protistans. For convenience it is here proposed that the term heterolysosome be simplified to heterosome, a name that may apply to both primary and secondary bodies of the type, with the addition of suitable prefixes.

Heterosomes of Protistans. Information regarding the heterosomes of protistans remains at an elementary level, for far more attention has been devoted to the processes of phagocytosis in those organisms than to the particles associated with them. This is especially true for flagellated photosynthetic organisms (Aaronson, 1973). In two related genera, *Euglena* and *Astasia*, acid phosphatase was reported associated with these particles (Price, 1962; Sommer and Blum, 1965; Bennun and Blum, 1966); the last citation described three different species of this enzyme from *Euglena* that were particle-bound. Other than the identification of several hydrolases associated with particles in two species of *Ochromonas* (Lui *et al.*, 1968; Shio, 1971) and the digestion of yeast by lysosomes in *Acanthamoeba* (Bowen *et al.*, 1979), little information exists regarding heterosomes in this type of organism.

Zooflagellates have been a little more thoroughly explored for these particles than have their photosynthetic counterparts (Eeckhout, 1973). In *Crithidia,* acid proteinase and deoxyribonuclease were identified in lysosomes (Eeckhout, 1970, 1972). Biochemical analyses of *Trichomonas* homogenates disclosed the presence in particles of four acid hydrolases: a proteinase, a phosphatase, β-glucuronidase, and β-*N*-acetylglucosaminidase (Müller, 1972, 1973). Some of the enzymes reacted optimally at the rather high pH of between 6 and 6.5, whereas most others had optima between 4 and 5. A third reported instance of heterosomes was in the protociliate *Opalina*, in which acid phosphatase was demonstrated to be present in these particles (Sergeyeva, 1967, 1969). For clarity, the known occurrence of enzymes in the various classes of lysosomes is summarized in Table 7.2.

Tetrahymena has had its heterosomes characterized both biochemically and ultrastructurally, and in large measure they agree with those of more advanced organisms. Five hydrolases having acid optima have been found, acid phosphatase, ribonuclease, deoxyribonuclease, proteinase, and amylase (Müller *et al.*, 1966). In addition, permeability studies have been conducted on these heterosomes; the particles were found to be permeable to the cryoprotective nonelectrolytes glycerol and dimethylsulfoxide and to several hexoses (Lee, 1970). Among the latter were included glucose, galactose, and α-methylglucose. Furthermore, they were permeable to the hexitols (mannitol, sorbitol, and inositol) but not to such disaccharides as sucrose, maltose, and melibiose.

Heterosomes of Fungi. In the amoeboids (myxamoebae) of the cellular slime mold, *Dictyostelium discoideum*, heterosomes have been described biochemically to a considerable extent (Ashworth and Wiener, 1973). In these organisms a series of acid hydrolases has been detected, including acid phosphatase, ribonuclease, deoxyribonuclease, proteinase, amylase, β-*N*-acetylglucosaminidase, α-mannosidase, and β-glucosidase. The synthesis of each one present has been shown to be controlled individually and not by the same operon as is the case in bacteria (Watts and Ashworth, 1970). Most differed from corresponding metazoan lysosomal enzymes in having a lower pH for optimal activity, around 3 instead of the usual 4 to 5. Moreover the possible presence of isozymes was indicated by the existence in some cases of two pH optima and acid phosphatase showed three (Solomon *et al.*, 1964). In addition, secondary heterosomes of slime molds have been examined under the electron microscope (Hohl, 1965; Ashworth *et al.*, 1969). One type of secondary heterosome noted contained partially digested bacteria, which are the principal food of these fungi; in addition these vesicles contained concentric annular lamellae, probably consisting of phospholipids or other indigestible remnants of the bacteria (Rosen *et al.*, 1965). Upon isolation the particles were shown to contain the bulk of the acid hydrolase activities (Wiener and Ashworth, 1970), but more recently it has been reported that the particles were actually heterogeneous and also included lysosomes having functions other than phagocytosis (Ashworth and Quance, 1972).

Heterosomes in Plant Cells. Because of the heavy cell walls that envelop plant cells, pinocytotic and phagocytotic activities are relatively rare events, yet some evidence indicates heterosomes are occasionally present. Viruses and macromolecules can enter these cells by pinocytosis (Wheeler and Hanchey, 1971), and bacterial DNA in solution has been shown to be able to enter stem cells of *Lycopersicon* and *Vicia faba* by similar means (Stroun *et al.*, 1966; Gahan, 1973). After 12 hr of immersion in the DNA solution, DNase activity could be detected in pinocytotic vesicles in the stem cells, suggesting the possible involvement of heterosomes in the breakdown of their contents.

Extracellular Activities. The complexity of the lysosomal system of cellular particles is indicated by a related activity of many cell types, that of extracellular digestion. On the surface it would seem logical that if heterosomes digest particulate matter brought in by invaginations of the plasmalemma, they might be equally capable of depositing their enzymes through that membrane to digest prey in various guises in the environment. A number of instances of the deposition of specific enzymes into the medium have been documented among a variety of organisms, including metazoans, fungi, and higher plants. Among the examples that can be cited is *Ochromonas*, which has been found to secrete amylase, invertase, proteinases, and lipases extracellularly (Pringsheim, 1952).

Table 7.2
Known Enzymes of Various Types of Lysosomes

Class	Source	Phosphat-ases	Acid prote-ase	Acid nucle-ase	Amyl-ase	Gluco-sidase	Other phosphat-ases	Ester-ases	Perox-idase	β-Acetyl glucosamin-idase	Lyso-zyme	Others
Hetero-somes	*Trichomonas*	Acid	*							*		*[b]
	Tetrahymena	Acid	*	Both[a]	*							
	Fungi	Acid	*	Both	*	β				*		*[c]
	Azurophil (neutrophil PNM)	Acid	*						*			
	Specific granule (neutrophil PNM)	Alka-line							0			
	Eosinophil	Acid							*		0	
	Azurophil (monocyte)	Acid							*		*	*[d]

	Rat liver	Acid	Both	Both			*[g]	*[h]		
Auto-somes	Multivesicular body	Acid								
	Aleurone	Acid	*	RNase	*	α				
	Root cap cell	Acid					*[i,j]			
	Guard cell	Acid								
Pungio-somes	Mycoparasite	Acid					*[i,j]			
	Pollen tube	-		RNase	*		*[i,j]	*[k]		
	Acrosome (see Table 7.3)	Acid and Alkaline				α, β		*	*	*[c,d,l,m]
Histo-somes		Acid								

[a] RNase and DNase.
[b] β-glucuronidase.
[c] α-mannosidase.
[d] Aryl sulfatase.
[e] Histamine.
[f] Heparin.
[g] Phosphoprotein phosphatase.
[h] Acid phosphodiesterase.
[i] β-glycerophosphatase.
[j] Acid naphtholphosphatase.
[k] α-naphthol esterase.
[l] β-mannosidase.
[m] Hyaluronidase.

Another case pertains to a dinoflagellate (*Amphidinium carteri*), which was shown to impart alkaline phosphatase activity to the culture medium (Strickland and Solorzano, 1966). On the other hand, yeasts do not secrete proteases into the medium and are unable to utilize proteins extracellularly (Matile, 1968a,b), so such particles are probably absent.

Only a few investigations of these processes have examined the nature of the particle that was involved, but those few intimate that special lysosomal types may be responsible. *Neurospora* cells secrete two acid proteases into the medium so that this organism, like many other fungi, can make use of extracellular protein. By cytochemical means, those enzymes have been detected in particles that contained no other hydrolases (Matile, 1964, 1965) and similar specialized granules have been suggested to be present in *Aspergillus, Penicillium, Rhizopus,* and *Coprinus* (Matile, 1969). Consequently, lysosomes that deposit their enzymes through the plasmalemma appear to represent a class distinct from heterosomes and are tentatively assigned the name exophagosome. Supporting this treatment are the results of an investigation of *Tetrahymena pyriformis* as to the source of the hydrolases that organism secretes into the medium (Müller, 1972). During starvation conditions, the ciliate lost most of the enzymes normally secreted in this fashion, including the β-glucuronidase, β-*N*-acetylglucosaminidase, α-glucosidase, and amylase. When resuspended in a nutrient medium, there was a continuous increase of secretion of these enzymes. The source of those substances was attributed to a special population of lysosomes that disappeared from the cells during starvation, but reappeared and became quite abundant when the organism was placed in nutrient media.

Heterosomes in Metazoans. As knowledge of the lower metazoan heterosome is largely confined to inferences derived from studies on intracellular digestion rather than from direct observation (Tiffon *et al.*, 1973), attention is necessarily centered on vertebrate material. However, results from studies on mammalian and avian tissues are sufficiently diversified to provide a firm basis for the appreciation of this organelle and its activities. Indeed, the lysosomes of phagocytic leukocytes are in themselves nearly totally satisfactory for present purposes, so discussion is largely confined to those cell types.

The major lysosomes of neutrophils have already received mention in connection with the discussion of the Golgi apparatus, where the two types of granules that characterize this variety of polymorphonuclear blood cell were shown to have different origins, the azurophilic granule on the proximal pole and the specific granule later on the distal. The azurophilic particles contain acid hydrolases and are therefore clearly lysosomes (Seeman and Palade, 1967); in addition they also possess peroxidase (Figure 7.2; Archer and Hirsch, 1963; Bainton *et al.*, 1976) of the myeloperoxidase variety. In contrast, the specific granules lack both types of enzymes, but have alkaline phosphatase instead (Bainton and Farquhar, 1970; Bainton *et al.*, 1971; West *et al.*, 1974). Among amphibians the corresponding blood cells seem to possess only a single kind of granule, in the present case the azurophilic type (Curtis *et al.*, 1979).

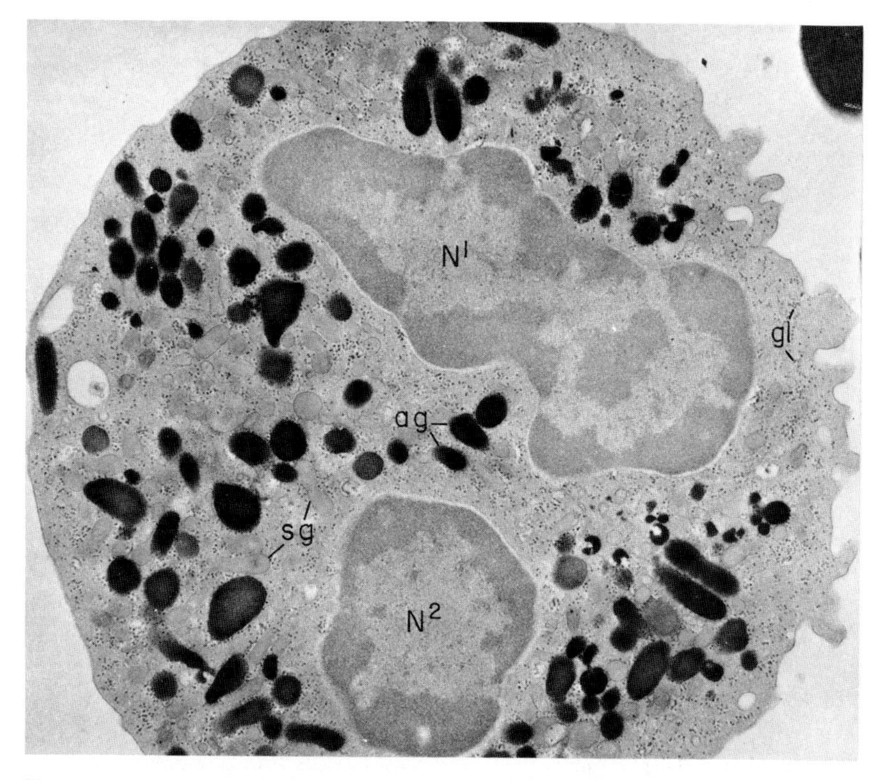

Figure 7.2. Peroxidase activity in the heterosomes of human neutrophils. The azurophilic granules (ag) in this mature human neutrophil show peroxidase activity (electron opacity), whereas the specific granules (sg) do not. Thus two distinct species of heterosomes are present. The Golgi bodies (gl) are quite small at this stage. 12,500×. (Courtesy of Bainton *et al.*, 1976.)

During the phagocytosis and digestion of microorganisms, which are the function of these blood cells, both types of particles play active roles (Robineaux and Frederic, 1955; Hirsch and Cohn, 1960; Cohn and Wiener, 1963). According to a detailed study of the sequence of activity on newly formed primary phagosomes (Bainton, 1972), the specific granules were found to unite with those vesicles first. Thus these particles emptied their enzymes into the vesicles (Hirsch, 1962; Lockwood and Allison, 1963), when the contents of the latter had a pH near 7, under which conditions the alkaline phosphatases of the specific granules react optimally. After several minutes had elapsed and the pH had become lowered somewhat, the azurophils attached to the vesicles in numbers. Ten minutes after phagocytosis, the secondary heterosomes thus formed had increased to large vacuoles and showed the presence of the enzymes of both types of granules. Thus in lacking acid hydrolases, the specific granules do not fit the definition of the lysosome, but as they clearly serve in the same capacity as typical representatives, they must nevertheless be recognized as one species of primary heterosome while the azurophils represent a second.

The primary heterosomes of eosinophils are quite distinct from the two of neutrophils, both in structure and in enzyme content. While much variation exists from one species of organism to another, typically the particles have an elongate core of thin needlelike crystals (Figure 7.3A,B; Miller *et al.,* 1966; Wetzel, 1970). Briefly stated the enzyme picture of eosinophilic particles differs in having a greater abundance of peroxidase and in lacking lysozyme, phagocytin, and alkaline phosphatases (Archer and Hirsch, 1963). However, the latter enzyme is present in the rat eosinophil (Figure 7.4A,B; Williams *et al.,* 1979). Furthermore, the peroxidase is distinct chemically, being a heme protein quite unlike the myeloperoxidase of the neutrophils (Archer *et al.*, 1965). Functionally these heterosomes behave much as those described earlier, fusing with primary phagosomes into which their enzymes become emptied. Although the cells in which they occur do engage in the phagocytosis of microorganisms, their chief function in the vertebrate body appears to be that of engulfing antigen–antibody complexes, particularly those involving immunoglobulin E (Ishikawa *et al.*, 1974). In contrast neutrophils take in particles containing immunoglobulin G. It is pertinent to note that only azurophilic granules in leukocytes of all varieties have been found to contain proteases (Bainton *et al.*, 1976).

In basophilic polymorphonuclear leukocytes (Figure 7.5A) the granules are not heterosomes, for there is no evidence that they contain acid hydrolases (Table 7.2) or even alkaline phosphatases (Ackerman, 1963; Wetzel *et al.*, 1967; Bainton *et al.*, 1976). Thus according to current definitions, they cannot even be classed as lysosomes, in spite of their having been produced by a Golgi apparatus that shares common ancestry with that of other polymorphonuclear leukocytes. As far as known, their chemical contents include peroxidase, histamine, and heparin, much like the granules of mast cells, except that no serotonin is present. When basophils carry out phagocytosis of red blood cells, as they occasionally do, the granules discharge their contents into the cytoplasm rather than into the phagosomes (Sampson and Archer, 1967). Thus they represent a distinct class of lysosomes that is left unnamed for the present; in their method of functioning they can be noted to resemble the autosomic lysosomes of certain plants described in the next section.

Monocytic leukocytes (Figure 7.5B), at least in those of man, appear to contain two types of azurophilic granules, one of which contains peroxidase, whereas the second does not. That which is peroxidase-positive also contains arylsulfatase, lysozyme, and acid phosphatase (Nichols *et al.*, 1971; Nichols, and Bainton, 1973), thus qualifying it as a lysosome; no enzymes are known

→

Figure 7.3. Heterosomes of eosinophils. (A) The heterosome (eosinophilic granule) of a rat eosinophil contains a highly structured core (C) in the midst of a granular matrix (M) and is enclosed in a typical trilamellar membrane (GM). 375,000×. (Courtesy of Miller *et al.*, 1966.) (B) In the human eosinophil (from bone marrow) heterosomes contain cores of irregular form and lack ordered structure. 33,000×. (Courtesy of Bainton *et al.*, 1976.)

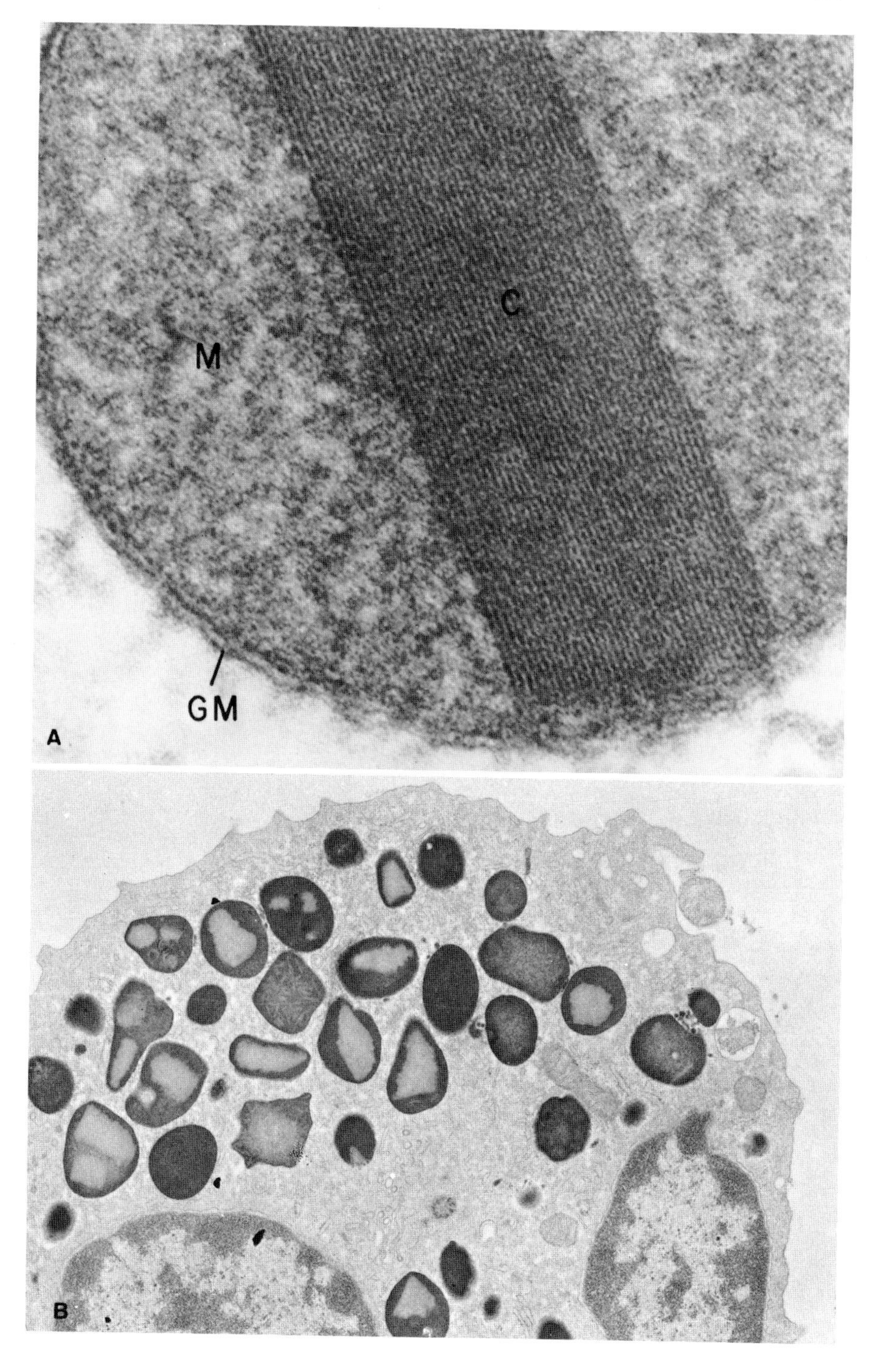
C
M
GM
A
B

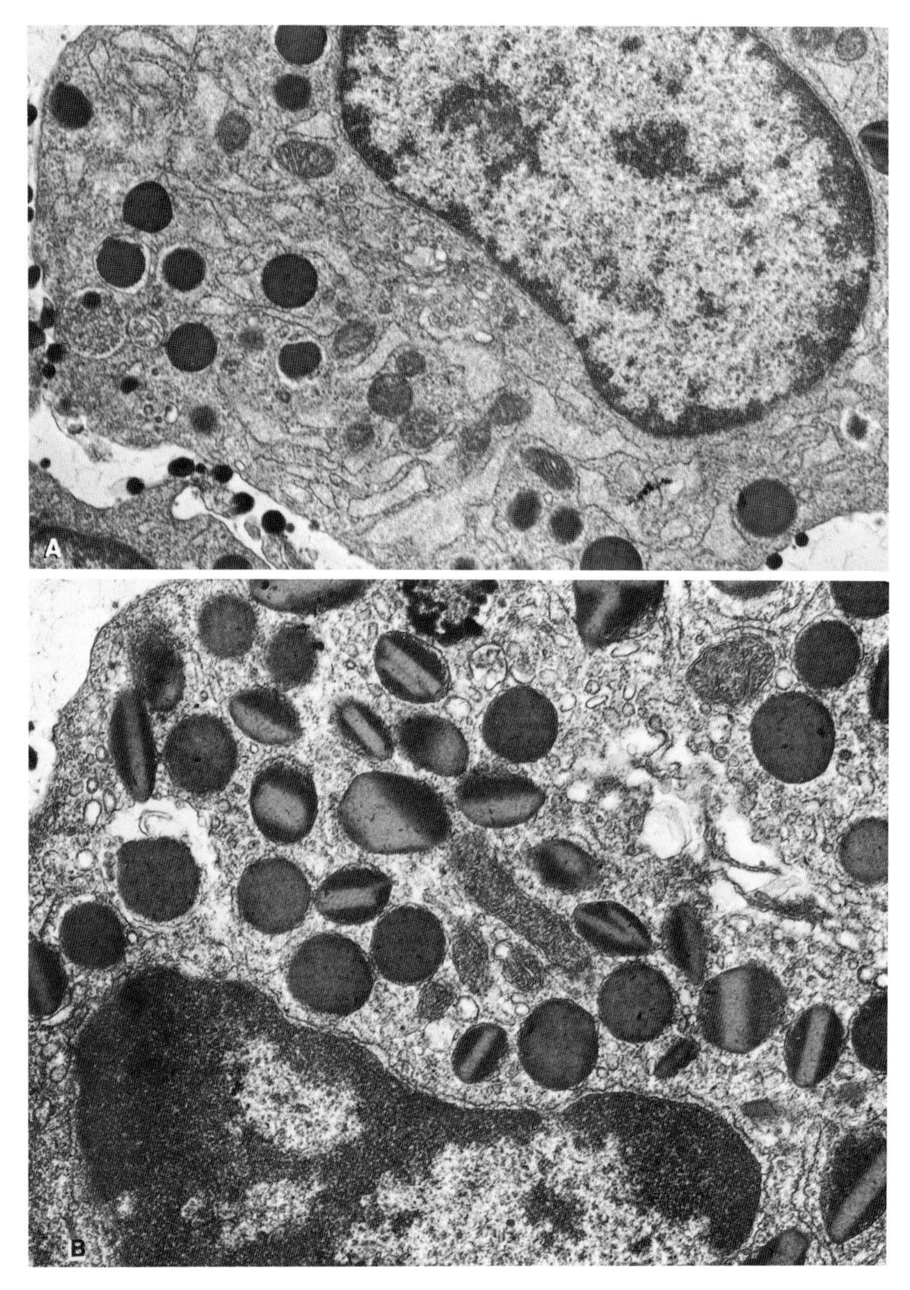

Figure 7.4. Development of alkaline phosphatase activity in eosinophils. (A) Late eosinophil promyelocyte from rat bone marrow, the first stage in which cores can be noted in the granules. The reaction product not only appears in the rather sparse heterosomes but also to a slight degree in association with the plasmalemma. 14,000×. (B) In a later stage (eosinophil late myelocyte), the granules are more abundant, cores are more numerous, and the spherical reaction deposits are denser. 11,500×. (Both courtesy of Williams *et al.*, 1979.)

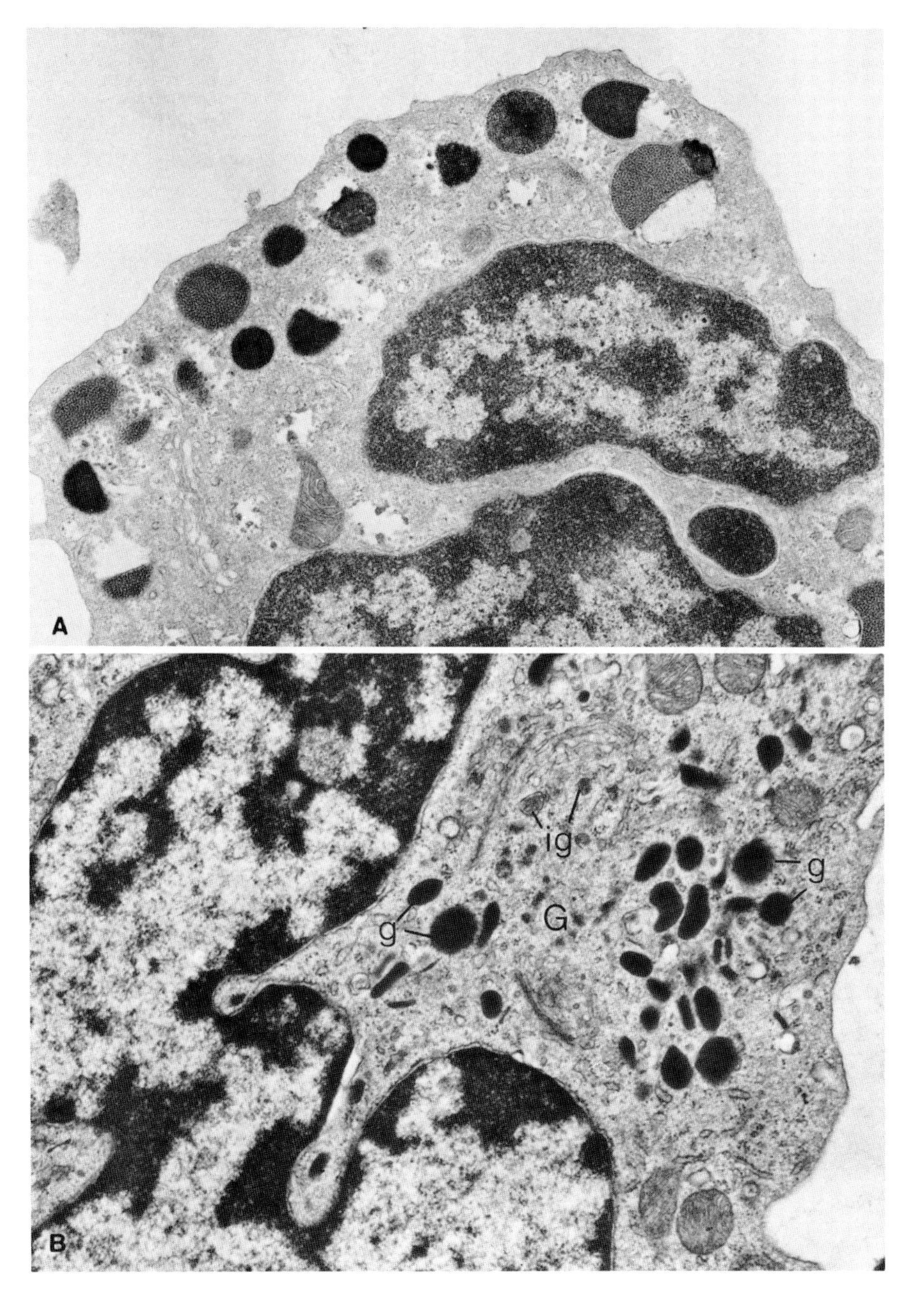

Figure 7.5. Lysosomes of basophils and monocytes. (A) This human basophil shows peroxidase activity in its lysosomes; no acid hydrolases or alkaline phosphatases are present, so these organelles cannot be classed as heterosomes. 15,000×. (Courtesy of Bainton, 1972.) (B) This monocyte from rabbit bone marrow is not fully mature, for it contains immature granules (ig) as well as mature ones (g). 18,600×. (Courtesy of Bainton *et al.*, 1976.)

for the second variety (Nichols and Bainton, 1973), so its classification remains an open question. Because the former type unites with phagosomes into which their contents are emptied, they are readily recognized as belonging to the heterosome class (Stossel, 1974; Klebanoff and Hamon, 1975). In contrast to the polymorphonuclear leukocytes, which soon perish after commencing their phagocytic activities in the tissues, monocytes develop into macrophages, becoming larger than previously and losing the azurophilic granules. These mature cells, which may live for 2 months or longer, contain a number of small vesicles, many of which are coated (van Furth and Cohn, 1968). These appear to transport arylsulfatase, acid phosphatase, and proteases (Cullen *et al.*, 1979), which are more abundant than previously, to the numerous phagosomes that are present. Thus a new type of heterosome develops after loss of the azurophilic granules.

A few examples of heterosomes known from other types of vertebrate tissues are also worthy of attention. During the latter half of development of the chick (stages 27–31, Hamburger and Hamilton, 1951), some of the muscle cells in the bulbus arteriosus of the embryonic heart undergo spontaneous death as the aorta and pulmonary artery begin to form (Hurle *et al.*, 1977). This cellular debris has now been shown to be taken into neighboring myocardial cells by phagocytosis, where it is digested within typical secondary heterosomes (Figure 7.6; Hurle *et al.*, 1978). Another example of ingestion of ad-

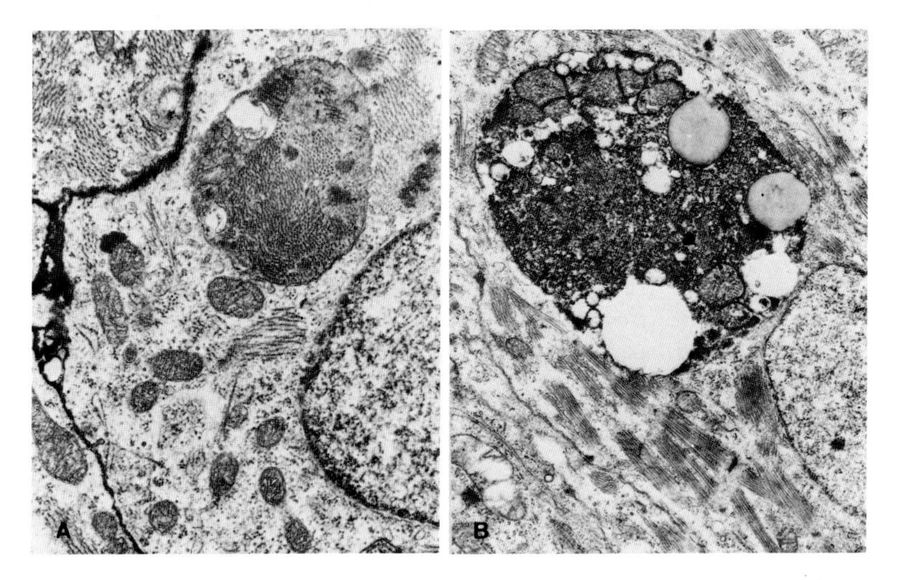

Figure 7.6. Heterosome activity during development of chick heart. (A) Myocardial cell showing a phagocytosed cell fragment that contains muscle filaments. Ruthenium red coats the cell surface. 15,500×. (B) A heterosome of a myocardial cell containing ingested mitochondria and lipid droplets. 7800×. (Both courtesy of Hurle *et al.*, 1978.)

jacent tissue by developing organs was reported from the growing placenta of sheep, in which the trophoblastic epithelial cells were studied (Figure 7.7; Mayagkaya *et al.*, 1979). These cells took in erythrocytes by endocytosis, the secondary heterosomes thus formed showing acid phosphatase activity around the engulfed blood cells but not within them. From the larger heterosomes, minute vesicles containing small numbers of erythrocytes were occasionally pinched off, giving rise to tertiary heterosomes, referred to in the report as erythrolysosomes. In both the secondary and the tertiary particles, the hemoglobin was digested before the plasma membranes of the erythrocytes were.

A number of other types of mammalian tissues engage similarly in endocytosis to a greater or lesser extent. Intestinal cells, fibroblasts, tumor and embryonic cells, lung tissue (Heath *et al.*, 1976), and thyroid epithelium (Zeligs and Wollman, 1977) are among those in which phagocytosis has been recorded and the associated heterosomes identified. Lysosomes, presumably of the heterosome type, in rat liver have been demonstrated to possess at least six major classes of enzymes that act on bonds involving phosphorus. These include acid phosphatase, phosphoprotein phosphatase, acid ATPase, acid phosphodiesterase, acid ribonuclease, and acid deoxyribonuclease (Nakabayashi and Ikezawa, 1978). Among some of these, isozymes were noted; for example, three were recorded for acid ribonuclease, but only one of acid deoxyribonuclease.

Autophagic Lysosomes. Although the presence of actual bodies that consumed the cell's own substance had long remained unsuspected by biologists, such autophagic lysosomes, or autosomes as they are named here, are now known to be quite widespread in normal tissues, including neuronal, adrenal medullar, and corpus luteal (Paavola, 1979a,b), and in Schwann cells (Ericsson, 1969; Holtzman, 1976), as well as pathological material (Novikoff and Essner, 1962; Kerr, 1973; Poole, 1973). Normal aspects of autophagy alone need to be considered, for the particles are quite abundant. Indeed, they are involved in one of the fundamental properties of cells in general, the continual replacement, or turnover, of proteins and other biochemicals of protoplasm (Segal *et al.*, 1978). Thus this class of lysosomes must be universally present in eukaryotes and a corresponding system must be located somewhere in the prokaryotic cell.

That this class of lysosome is distinct from heterosomes has been shown by results of a comparative study of the intracellular breakdown of exogenous and endogenous proteins (Poole *et al.*, 1978). Whereas chloroquine and ammonia inhibit the digestion of exogenous proteins by heterosomes, that by autosomes is not affected. The membrane of these particles has been shown to contain an ATPase-driven proton pump (Schneider and Cornell, 1978), but this probably is a feature not confined to this type of lysosome.

Unlike the situation with heterosomes, autosomes in vertebrates are often known in response to experimental treatments of various sorts, such as injection

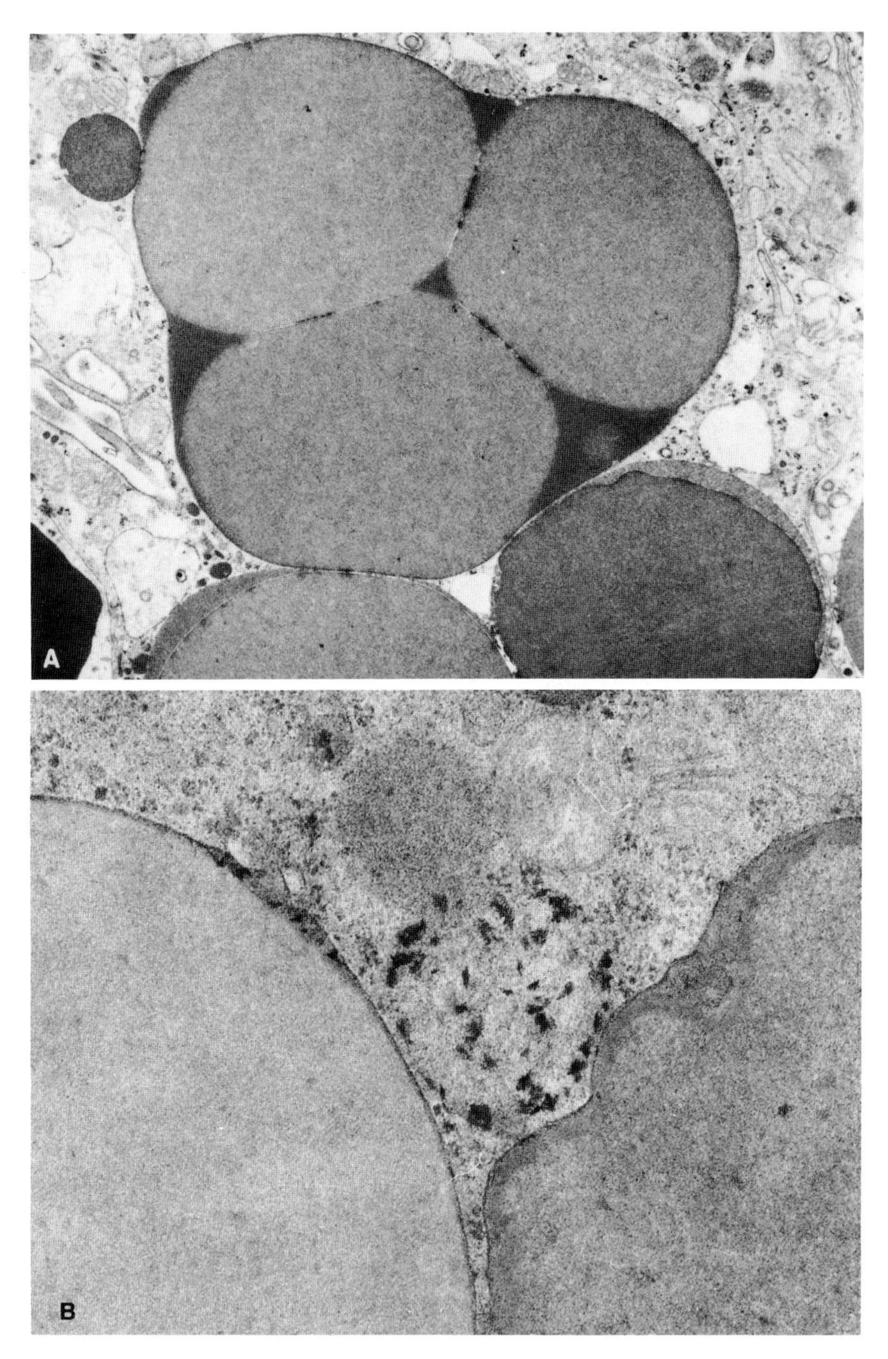

Figure 7.7 Activities of heterosomes in growing sheep placenta. (A) A heterosome containing three erythrocytes in approximately equal stages of digestion. 15,000×. (B) Two adjacent heterosomes, containing an erythrocyte in different stages of digestion; acid-phosphatase-reacted product is visible between the lysosomal and erythrocyte membranes. 104,000×. (Both courtesy of Myagkaya *et al.*, 1979.)

with glucagon, Jectofer (an iron complex), or other substances, and consequently their presence may be viewed as the result of a pathological condition. As a rule the secondary autosomes, or cytolysomes as they are also called (Gahan, 1973), may be recognized by their containing cell organelles, as already pointed out. But there are a number of instances where less radical autophagy occurs in which such structures could be absent. One such case has been reported in conjugating ciliates of the genus *Stylonychia* (Sapra and Kloetzel, 1975). When conjugation commences and feeding by the protozoans is interrupted, small typical autosomes do appear, in which mitochondria and other cell organelles may be observed, but after the pair has separated, large autosomes filled with electron-dense matrix can be noted, perhaps resulting from fusion of small autosomes, followed by condensation. Acid phosphatase activity was detected in such bodies but only to a slight extent.

Multivesicular Bodies. One of the more frequent, but by no means the only, type of secondary autosome is the multivesicular body. Usually these have a diameter in the range of 50 to 75 nm and are enclosed by a single membrane. The name of this type is derived from the presence of a number of vesicles within the saclike body. Not infrequently a flattened protrusion is present on one side, formed by a fold of the membranes; this handlelike projection is frequently coated externally (Holtzman, 1976). Quite often the particles appear to be formed from pinocytotic membranes, as for example, following the uptake of protein from the blood in fat body cells of skipper butterflies of the genus *Calpodes* (Locke and Collins, 1968). In this organ the pinocytotic vesicles fused in small clusters to become multivesicular bodies, to which primary autosomes later attached and discharged their acid phosphatase contents. If these observations are correct, the multivesicular bodies may be considered a means for disposal of some of the plasmalemma taken in by endocytosis.

Another role played by this type of particle is that known as crinophagy, the consuming of excess secretory products in cells. In the cells of the anterior pituitary of lactating rats, it was demonstrated that if the young were abruptly separated from the female, numerous secretion granules accumulated in the form of multivesicular bodies (Smith and Farquhar, 1966). Similar processes were later noted in other regions of the anterior pituitary (Farquhar, 1969, 1971; Smith, 1969). Later the secretory cells adjusted their activity, but at the time of the sudden change in physiological requirements, the multivesicular bodies aided in the control of output.

A third function of these vesicles is demonstrated by yolk platelets of the eggs of molluscs, echinoderms, and annelids, and to a lesser extent by those of amphibians and birds (Pasteels, 1973). Most of these particles have a moderately dense, granular stroma that reacts with Gomori medium, thereby showing acid phosphatase activity. They have been documented as showing a definite series of changes with time, which began as a dense mass of yolk enclosed

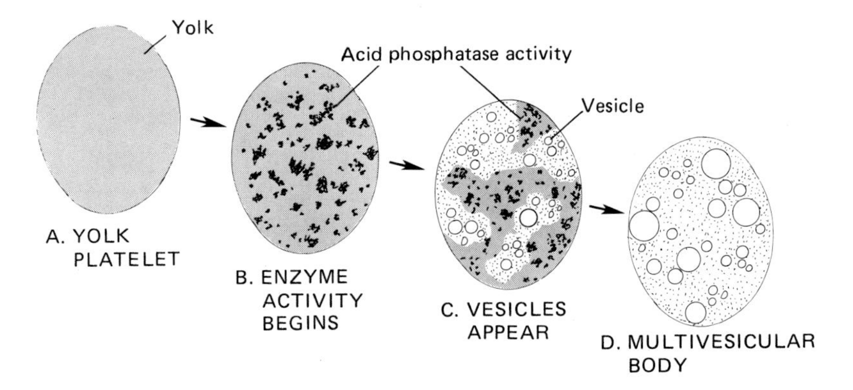

Figure 7.8. Changes in yolk platelets (autosomes) during development. (Based in part on Pasteels, 1976.)

within a membrane. At first this body was not reactive but soon showed an increasing amount of acid phosphatase activity, a stage that was followed shortly by a gradual decrease in yolk (Figure 7.8). As the yolk supply decreased, the structure showed a number of small vesicles, completing the conversion into a multivesicular body. This series of steps can be interpreted as showing the transformation of a yolk-storage particle into a vesicular body that digests that material and liberates the products into the cytosol.

Autosomes in Fungi. Changes in cell type in the slime mold *Dictyostelium* from the active myxamoeboid to the sessile vegetative stages are accompanied by loss of protein and RNA and by the appearance of new carbohydrates. While the latter phenomenon is of no importance to the present topic, the loss of substance has to a slight degree been indicated to arise from autosomal activity. During differentiation, acid phosphatase activity has been reported to triple (Gezelius, 1966; Wiener and Ashworth, 1970), while the proteinase activity remained constant. Throughout this period of change, synthesis of new proteins proceeded at a considerable rate, so it has been suggested that the autophagic processes (possibly by lysosomes) included some mechanism that distinguished those proteins that were to be degraded from those that were to be conserved (Sussman and Sussman, 1969; Ashworth and Wiener, 1973). Because there was similar biosynthesis and degradation of RNA during the same period, there may be a comparable mechanism for distinguishing obsolete from still useful RNA molecules.

Autosomes in Higher Plants. Autophagy occurs in higher plants in several ways, the first of which involves the formation of heterosomes much like those of metazoans. First it must be recalled that the characteristic vacuole of the plant cell contains acid hydrolases and is formed by fusion of numerous vesicles that are derived from the endoreticulum and dictyosomes. Thus it is really a large lysosome. During cell differentiation this vacuole increases in size and thus is distinctive of mature rather than undifferentiated cells. As it increases in volume, the tonoplast (vacuolar membrane) invaginates in such a fashion that cytoplasmic material is carried into the vacuole, thus forming vesicles within the latter (Matile and Moore, 1968; Thorton, 1968; Gahan, 1973). The first report cited suggested that the membrane enclosing these vesicles became labile after invagination was completed, thereby exposing the contents to the lysosomal enzymes of the vacuole.

Usually the proteinaceous food reserves of seeds are contained in vacuoles referred to as aleurone grains or protein bodies (Gibson and Paleg, 1972). Within these granules are a number of acid hydrolases, including phosphatase, proteases, ribonuclease, β-amylase, and α-glycosidase, so that, together with the protein derived from the cell, these bodies are obviously lysosomes of the autosomal variety that seemingly digest the stored protein and make it available to the cell (Matile, 1968a,b, 1969). During germination these vacuoles become enlarged and fuse to one another, gradually replacing the volume within the cell originally occupied by the cytosol and organelles (Hinklemann, 1966). This observation implies that much cytoplasmic material becomes incorporated into the vacuole as the latter grows, possibly by a mechanism like that described in the preceding paragraph. Two isozymes of acid phosphatase were noted in germinating peas, isozyme I associated with aleurone grains and isozyme II with the cell wall (Bowen and Bryant, 1978). The levels of the former remained constant during and immediately following germination, whereas those of the latter increased sevenfold in the first 12 days.

In part the root cap carries out its function of protecting the underlying meristematic layer by producing cells that can be lost by abrasion against soil particles as the root grows. To replace the cells that are lost, mitosis, maturation, and senescence thus must take place rapidly in this relatively thin tissue, making the whole sequence of events continually present in vertical sections. By electron microscopy, the development of mitochondria, dictyosomes, and a large endoreticular system was noted during the first 24 hr of growth of the root cap, and later lysosomes were observed in association with the endoreticulum (Berjak and Villiers, 1970). Development of senescent, loosely attached cells on the outer regions of the tissue occurred within 48 hr and involved active secretion by the Golgi apparatus and the rapid growth of lysosomes. In the oldest cells, the lysosomes finally ruptured and acid phosphatase activity was detected throughout much of the ground cytoplasm, thereby probably inducing the senescence of the cells.

One additional example should suffice to provide a full appreciation of the variety of uses to which these autophagous particles are put in normal cells. In the guard cells of the stomata of *Campanula persicifolia*, the acid phosphatase activities have been found to fluctuate in response to changes in light intensities (Sorokin and Sorokin, 1968). During bright illumination, when the guard cells are turgid, thereby opening the stomata fully, β-glycerophosphatase and acid naphthol phosphatase are inactive, but regain activity with darkness and the cessation of photosynthesis. The mechanism that produces these observations depends on changes in osmotic pressure and pH (Levitt, 1967; Gahan, 1973). Photosynthesis in these cells synthesizes hexose more rapidly than it can be converted to starch, the resulting accumulation leading to endosmosis and an increase in turgidity. Concomitantly, the pH of the cytoplasm becomes elevated to 7 or 8, at which value the enzymes of the lysosomes are inactive. With darkness, photosynthesis ceases, but the hexose continues to be converted to starch, decreasing the concentration of sugar in the cytoplasm and lowering the osmotic pressure. The oxidative decarboxylation that accompanies hexose conversion leads to a decrease in the pH to 5 or 6, at which level the enzymes of the autophagic lysosomes become activated. Thus indirectly the autosomes serve in regulating the guard cell operations.

Isolation Membranes. In a strict sense, isolation membranes are not lysosomes because they consist of a double membrane, not the single one stated in the definition of that particle. Yet they are clearly autosomes in behavior, as is shown in the fat body cells of *Calpodes* in which organ multivesicular bodies were just described. In these cells, peculiar bodies called peroxisomes, to be discussed later in this chapter, begin to undergo senescence after carrying out their particular functions (Locke and McMahon, 1971). When they have become slightly reduced in size, a flattened cisterna termed an isolation membrane appears on one side and commences to extend in all directions around a peroxisome until it has completely enveloped the latter. After envelopment is completed, acid hydrolases appear within the lumen of the internal membrane and digest the peroxisome enclosed therein. The *de novo* origins of the cisterna should be especially noted.

Penetrating Lysosomes. Lysosomes that function in penetration have been described from several sources, but as only a few types have been thoroughly documented, they alone can be discussed in detail. As no collective name exists for these particles, the term pungiosomes (piercing bodies) is proposed.

Pungiosomes in Plant Parasites. The involvement of pungiosomes in the penetration of host tissues by parasites has been investigated in a number of plants to greater or lesser extents. In the infective processes of the mycoparasite *Mycotypha microspora* in the fungus *Piptocephalia virginiana*, lysosomes were observed moving down the germ tube of the parasite into the haustorium (Armentrout and Wilson, 1969). At the tip of the latter, concentrations of acid phosphatase were detected by histochemical means. Similar results have been noted during the invasion of the red alga *Harveyella* into its host (Goff,

1979a,b) and in the infection of orchid protocorms by the endophytic fungi during the initiation of mycorrhiza formation. After the fungi had penetrated the protocorm to some extent, the hyphal membranes broke down, releasing quantities of acid naphthol phosphatase into the host cells (Gahan, 1973). At the same time β-glycerophosphatase was shown to be associated with lysosomes still within the hyphae. The lysis of the hyphae was suspected to result from activity of host lysosomes, rather than being a normal process of invasion.

Pungiosomes of Pollen Tubes. Penetration of tissue by foreign cells of a different nature has also been investigated on a preliminary basis and was found to involve pungiosomes. The extension of the pollen tube through the style of the pistil or into the megasporangium has been followed in 50 species of pines, horsetail rushes, and flowering plants (Knox and Heslop-Harrison, 1970). Beneath the inner cellular part of the wall of the pollen grains (the intine), acid hydrolases were found in all particles, uniformly congregated at the point of emergence. Among the enzymes identified were acid naphthol AS phosphatase, ribonuclease, acid protease, amylase, β-glycerophosphatase, and α-naphthyl esterase.

The Acrosome of Metazoan Sperm. The large electron-dense body located at the tip of the metazoan spermatozoon, known as the acrosome (Figure 7.9), is now generally viewed as a lysosome (Allison and Hartree, 1970; Hartree,

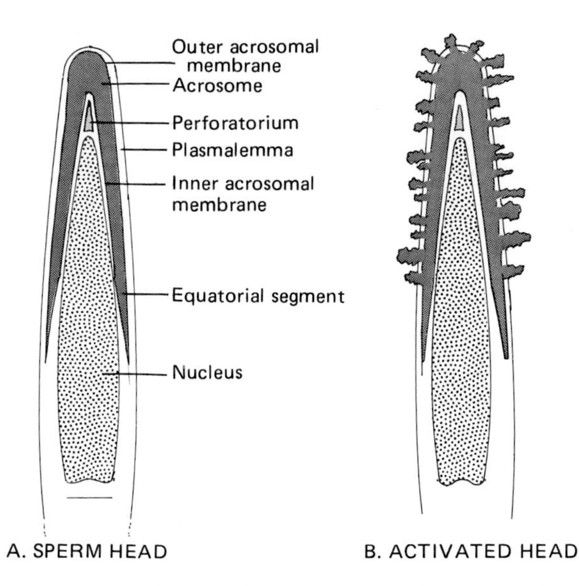

Figure 7.9. The acrosome (pungiosome) of a vertebrate sperm head. After release into the female reproductive tract, the acrosome frees its enzymes and thereby becomes activated, a process that may aid in penetration of the egg. (Based on Morton, 1976.)

1975; Morton, 1976). In a sense it is analogous to the vacuole of plant cells in that it is formed by the fusion of innumerable small vesicles derived from the Golgi apparatus (Clermont and LeBlond, 1955; Seiguer and Castro, 1972). Its lysosomal nature also is revealed by its staining a bright orange-red with acridine orange and by its reacting appropriately with various cytochemical reagents for specific lysosomal enzymes (Table 7.3). Like some of the leukocyte granules, the acrosome contains a few alkaline hydrolases, as well as the more varied acid-reactive types. These and the other major enzymes of the acrosome are listed in Table 7.3.

When present in the lower invertebrates, the acrosome of the spermatozoon is relatively small, being more or less discoidal and much narrower than the nucleus (Figure 7.10A; Jamieson *et al.*, 1978; Jamieson and Daddow, 1979). However, it is occasionally absent as in such remotely related phyla as the Entoprocta and Ectoprocta and also the arthropod class Pycnogonida (El-Hawawi and King, 1978; Franzén, 1979). With evolutionary advancement this body increases in relative size, becoming more or less spherical in certain forms (Afzelius, 1977; Franzén, 1977; Afzelius and Ferraguti, 1978), until at the most advanced stages it is a conical body, enveloping the entire anterior half of the nucleus. At its posterior rim is a broad equatorial segment or collar whose function is imperfectly understood (Figure 7.9). After sperm have entered the female reproductive tract, they have to undergo a period of maturation, termed capacitation, before they are able to fertilize an egg. During capacitation the outer membrane of the acrosome fuses with the plasmalemma of the spermatozoon to form whole series of vesicles, thereby releasing most of the enzymes of the acrosome. How this shedding of enzymes aids in penetrating the egg envelopes is not known, but apparently sufficient enzymes remain attached to the inner acrosomal membrane to assist in penetrating the coverings to the point where the sperm can be brought by egg microvilli into the cytoplasm for fertilization to be consummated.

Histosomes. At least in multicellular plants and animals, lysosomes are involved in the performance of a number of physiological activities, although their precise roles in such functions remain undiscovered. One or two examples of their participation in secretion in this fashion have already been cited in connection with receptor sites in the plasmalemma (Chapter 1, Section 1.2.2). This type of lysosome that is specialized for aiding the cell in carrying out nonlysosomal activities is here named the histosome.

Little is known of the ultrastructural and enzymatic properties of histosomes; indeed, their existence has only recently been suspected (Werb and Dingle, 1976). In some cases their enzymes may be implicated in the activation or modification of nonlysosomal substances (Holtzman, 1969; Smith and van Frank, 1975), such as the conversion of proinsulin into insulin (Kemmler *et al.*, 1973). Further, some evidence suggests that histosomal enzymes may influence

Table 7.3
Principal Enzymes of the Acrosome

Type	Enzymes	Source of spermatozoa		Remarks
		Present	Absent	
Carboxyl ester hydrolases	Esterases	Human, bull, rodents	—	Multiple species exist
Phosphoric monoester hydrolases	Alkaline phosphatase	Rabbit, ram	Human, rodents, shrew	—
	Acid phosphatase	Mammalian types	—	Multiple species
Sulfuric esterases	Arylsulfatases A and B	Ungulates, rodents, rabbits	—	—
Glycosidases	Hyaluronidase	Most mammals	Dog, donkey, birds	—
	α-Glucosidase	Primates, bull, ram, dog, rabbit	Horse, pig	—
	β-Glucosidase	Bull	—	—
	β-Galactosidase	Rat, ram	Mouse	—
	α-Mannosidase	Rat, pig, rabbit	Horse	Multiple species
	β-Mannosidase	Bull, ram, dog, man	Pig, horse	—
	β-Acetylglucosaminidase	Rat	—	—
Peptide hydrolases	Acrosin	Most mammals	—	—

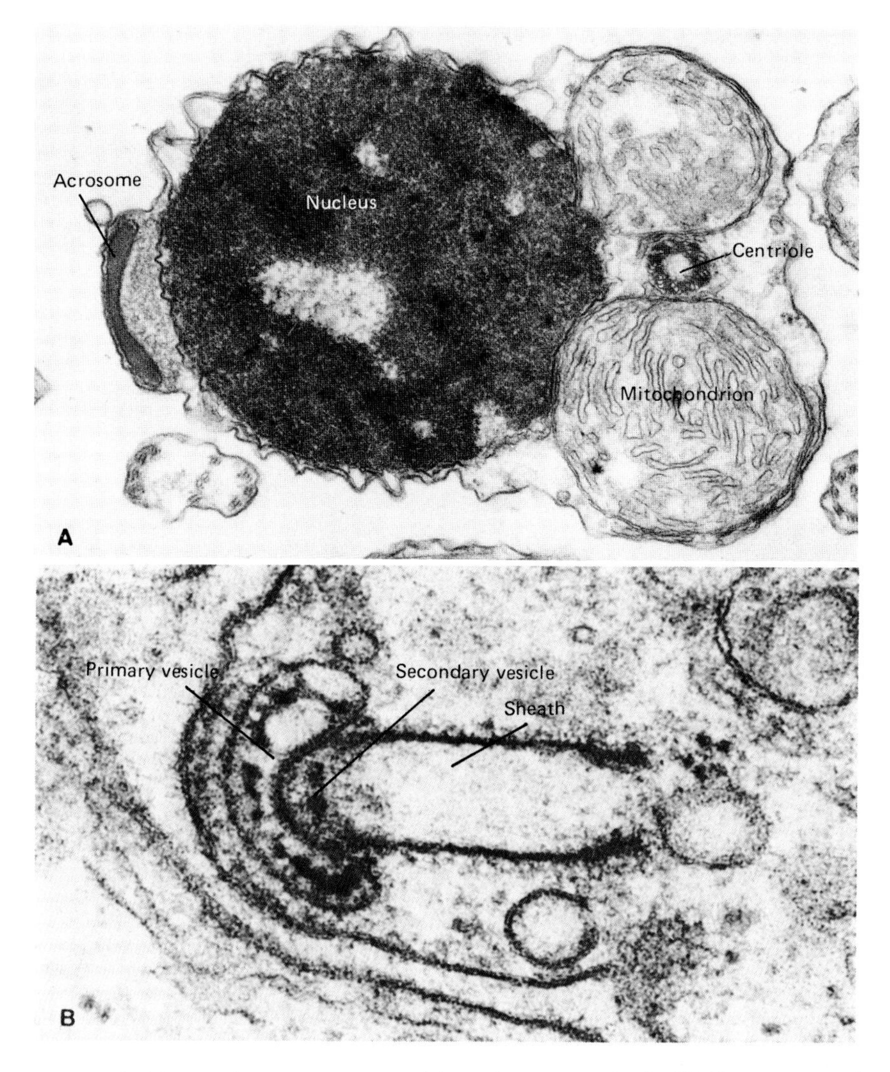

Figure 7.10. The pungiosome (acrosome) of invertebrate sperm head. (A) Spermatozoon of *Priapulus*; in many invertebrates the acrosome is quite small, as here. 28,000×. (Courtesy of Afzelius and Ferraguti, 1978.) (B) Developing acrosome in a spermatid of a tubificid worm. These vesicles appear at least in part to be derivatives of dictyosomes. 50,000×. (Courtesy of Jamieson and Daddow, 1979.)

nuclear function (Szego, 1974). Two very unusual granules recently described from the ejaculatory duct of a polychaete worm probably belong in this category (Westheide, 1978, 1979). Type I granules were electron dense, enclosing a transparent region, whereas type II were largely electron transparent. Around the periphery of the latter, which were shown to be products of dictyosomes, was a tubular element in the form of either a single or double spiral. During passage of the sperm, these granules entered the lumen of the duct and were assumed to serve in capacitation of the spermatozoa.

The complexity of the interrelationships between acceptor sites and pinocytosis on the part of the plasmalemma (Chapter 1, Section 1.2.2) and lysosomal activities is indicated by a current study of human fibroblast behavior (Sando *et al.*, 1979). Under experimental conditions, these cells take in certain acid hydrolases selectively by pinocytotic processes that involve recognition of a specific marker site on the enzyme by receptors on the plasmalemma of the fibroblast (Hieber *et al.*, 1976; Kaplan *et al.*, 1977, 1978; Sando and Neufeld, 1977; Ullrich *et al.*, 1978). It has been proposed that these activities might be of importance for packaging endogenous hydrolases into the fibroblast histosomes (Neufeld *et al.*, 1977). On a preliminary basis, existing evidence has indicated that a phosphorylated sugar acts as a major determinant of the recognition marker on the enzymes, because a number of such substances, including mannose-6-phosphate and fructose-1-phosphate, compete with one another in inhibiting the receptor-mediated uptake of the enzyme (Kaplan *et al.*, 1978; Ullrich *et al.*, 1978). In addition, certain amines act to inhibit these processes. Because this inhibition by amines of various kinds is noncompetitive, it is obviously different from that of the phosphorylated sugars (Sando *et al.*, 1979). As more is learned of the details of cell functionings, there exists a distinct likelihood that this type of lysosome will be found to be involved in many of the routine but complex activities in a diversity of metazoan tissues and in a number of metaphytan cells also.

7.2. THE PEROXISOME

Another particle more readily identified by biochemical than by ultrastructural methods has been variously named the peroxisome, glyoxysome, and microbody (de Duve *et al.,* 1960; Baudhuin *et al.*, 1965a,b; Breidenbach and Beevers, 1967). Its distinctive enzyme is catalase, which breaks down hydrogen peroxide, and thus provides the basis for the customary name.

7.2.1. Structural Traits

The particles extracted from rat liver cells, in which they were first discovered, have a mean diameter of 0.5 μm, and are bounded by a single membrane; those from eosinophilic leukocytes and other tissues of mammals, however, are quite small and have been termed microperoxisomes (Figure 7.11A,B; Novikoff *et al.*, 1973; Arnold and Holtzman, 1978; Mooradian and Cutler, 1978). Although the stroma of the granule viewed under the electron microscope appears finely granular and fairly electron transparent, a large, electron-dense area is usually present centrally, referred to as the core or nucleoid (Baudhuin,

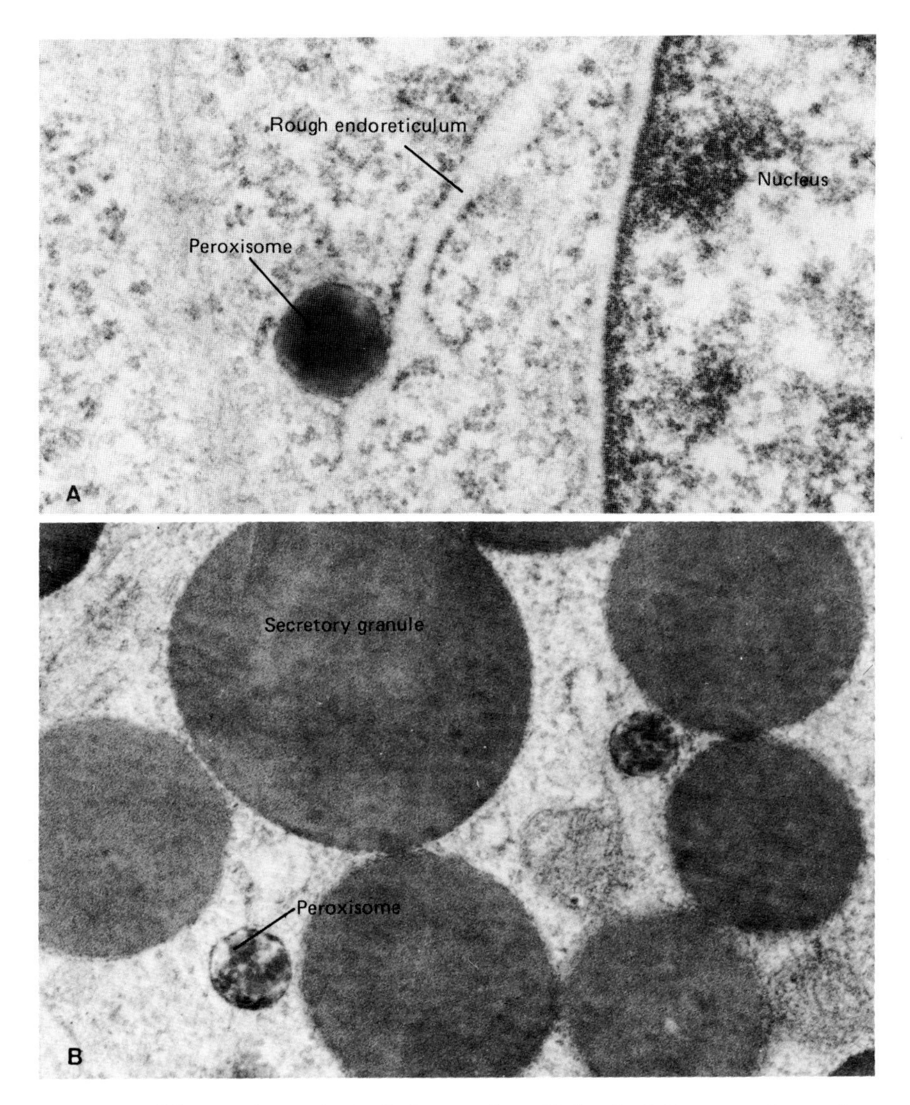

Figure 7.11. Microperoxisomes in developing rat submandibular gland. Some peroxisomes, here reacted for catalase activity, are quite small as shown by comparison with rough cisternal endoreticulum (A) and with secretory granules (B). (A) 32,500×; (B) 50,000×. (Courtesy of Mooradian and Cutler, 1978.)

1969; Hruban and Rechcigl, 1969). When present, the core may have a latticelike appearance, as in rat and guinea pig tissue, or it may lack a definite pattern, as in hamster and mouse cells (Hruban and Swift, 1964; de Duve and Baudhuin, 1966; Shnitka, 1966). This structure seems to result from the presence of tubules, not from uricase as once claimed.

In some vertebrates, including several birds (Shnitka, 1966), the core is absent. A number of plant tissues, including tobacco root cortex, possess similar tubular cores, but in other types, such as oat coleoptile parenchyma, this region is crystalloid and arranged in geometric configurations (Hruban and Rechcigl, 1969). Comparable particles have been detected in various protistans, but because they are strongly divergent in appearance from the foregoing examples, they are reserved for attention until the enzymatic properties have been described.

7.2.2. Enzymes of Vertebrate Peroxisomes

Catalase. Catalase, the major enzyme of this particle, breaks down hydrogen peroxide and is probably the only one universally present in peroxisomes (Figure 7.12A). This enzyme appears to be a constant feature of all forms of life, constituting up to 5% of the dry weight of many prokaryotes, such as *Micrococcus lysodeikticus* (Herbert and Pinsent, 1948; Clayton, 1959). Although it occurs in all tissues of the vertebrate body, it is especially abundant in liver and kidney cells and in leukocytes. Mammalian erythrocytes also are rich in the enzyme, whereas those cells of a number of birds, including ducks and geese, have only small quantities present (Silveira and Hadler, 1978).

The protein has a mean molecular weight between 225,000 and 250,000 and in both vertebrates and higher plants consists of two different subunits (Nicholls and Schonbaum, 1963; Dixon and Webb, 1964) combined into tetramers. That of bacteria is similar, except that the molecular weight is close to 232,000. While little specific information is available, catalase varies widely in molecular structure in different tissues and organisms. Moreover, there is much evidence attesting to its heterogeneity even in the same tissue (Patton and Nishimura, 1967).

The breakdown of H_2O_2 by catalase can occur in two ways. The first of these, known as the catalatic reaction, involves the decomposition of the peroxide, with vigorous liberation of oxygen:

$$2H_2O_2 \xrightarrow{\text{catalase}} 2H_2O + O_2$$

The second one, the peroxidatic reaction, involves a receptor substance that is oxidized, so no oxygen is given off:

$$H_2O_2 + RH_2 \xrightarrow{\text{catalase}} 2H_2O + R$$

The acceptor can be any number of substances—even nitrous acid will serve—but more typically it may be an alcohol, such as methanol (Mannering *et al.*, 1969), ethanol, or *N*-propanol, an organic acid, including formic acid, or a phenol of various types, like pyrogallol or *p*-cresol. While thus the *in vitro* ac-

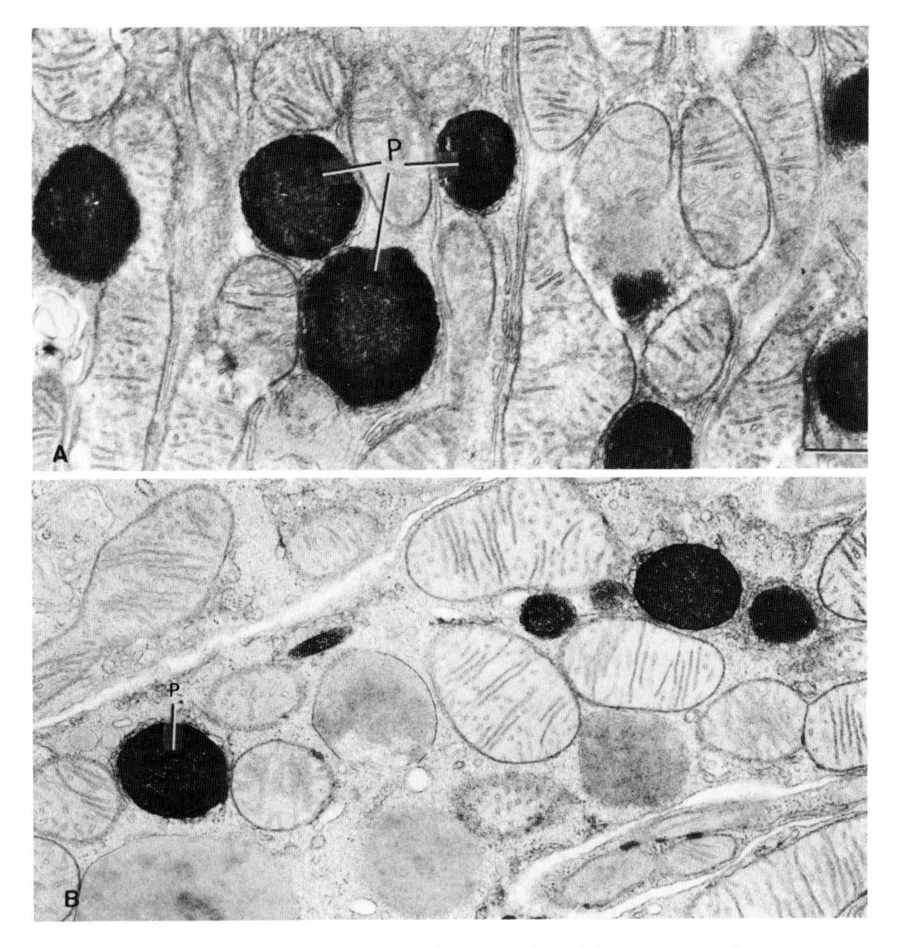

Figure 7.12. Peroxisomes (P) of kidney tested for enzymatic activity. These peroxisomes are more typical in size than those shown in Figure 7.11. Enzymatic activity for catalase is demonstrated in (A) and for D-amino acid oxidase in (B). Both 22,000×. (Courtesy of Arnold *et al.*, 1979.)

tivities of catalase are thoroughly established, its actual role *in vivo* remains unknown (Thurman and Chance, 1969).

Oxidases. In rat liver peroxisomes, three enzymes that produce hydrogen peroxide are also abundantly present, the most prominent of which is uric acid oxidase, also known as uricase (Noguchi *et al.*, 1979). This converts uric acid to allantoin and hydrogen peroxide, with the liberation of CO_2, as shown in Figure 7.13. Although most metazoans have this enzyme, it is absent from monkeys and higher primates, birds, and terrestrial reptiles. In the chicken, the enzyme is present for the first 7 days of embryogenesis and then disappears. In yeast it can be induced to be present by addition of uric acid to the culture medium, as it can also in bacteria by similar procedures (Kaltwasser, 1968).

A second common oxidizing enzyme acts on the dextrorotatory enantiomorphs of amino acids to provide a basis for their conversion into the levorotatory forms that are employed nearly exclusively in natural proteins. This D-amino acid oxidase (Figure 7.12B), a flavoprotein, oxidatively deaminates D-amino acids, which can then be reaminated asymmetrically to form the L enantiomorph (Berg, 1959; Arnold *et al.*, 1979). A second pathway provided by the enzyme to this end is that of its removing the amino radical in the same fashion, which is then transferred by a second enzyme system to an appropriate α-keto acid, several of which are available by way of the respiratory and other cycles described in Chapter 8. Nevertheless, in view of the lack of the natural abundance of D-amino acids, the actual value of the enzyme to organisms remains an enigma, for the only source of such substances is from microorganisms, of which a few types are employed in certain antibiotics. However, some evidence has been forthcoming that suggests that the enzyme also may act upon the neutral amino acid, glycine, as well as on certain L enantiomorphs like L-proline, with L-alanine a possible third additional substrate (Neims and Hellerman, 1962; Scannone *et al.*, 1964; Wellner and Scannone, 1964). Under its influence glycine is oxidized to glyoxalic acid, which then enters into a transamination reaction with glutamic acid, catalyzed by another enzyme usually present in these particles (Kisaki and Tolbert, 1969).

In addition to the two foregoing oxidizing enzymes, a third one exists in most vertebrate tissues that similarly produces hydrogen peroxide. This L-α-hydroxy acid oxidase is a flavin mononucleotide flavoprotein (Robinson *et al.*, 1962), which reacts according to the following equation:

$$R \cdot CHOH \cdot COOH + O_2 \rightarrow R \cdot CO \cdot COOH + H_2O_2$$

Whether or not the enzyme is identical in all tissues is still an open question, for in hog kidneys two species are present, one of which employs long-chain

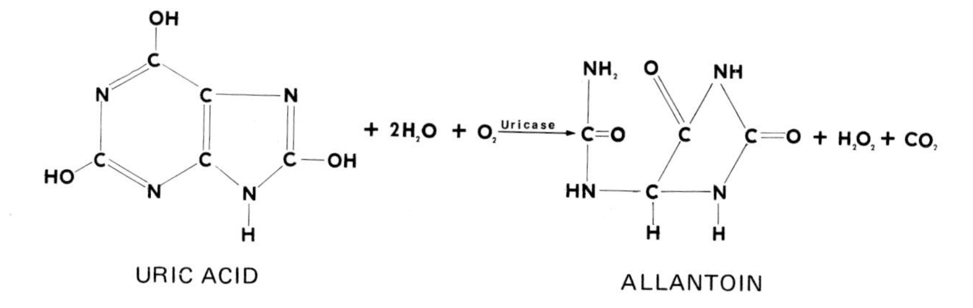

Figure 7.13. The action of uric acid oxidase of peroxisomes.

and the other short-chain hydroxy acids as substrate. From some sources, moreover, the enzyme acts also on L-lactic and L-amino acids, but that of carnivores, guinea pig, rabbit, ox, and sheep tissues is incapable of doing so (Blanchard *et al.*, 1944).

Turnover of Peroxisomal Enzymes. Several enzymes of peroxisomes exhibit marked and similar half-lives, suggesting that the particles may be degraded as entire units, possibly through autosomal activities (Poole, 1971; de Duve, 1973). Because radioactivity is lost from pulse-labeled particles with simple exponential kinetics and without any lag, it would appear that the mechanisms involved do not discriminate between new and mature organelles (Holtzman, 1976). Some evidence indicates that peroxisomes break down independently of the autosomes, their enzymes dispersing into the cytoplasm, as in hepatocytes of drug-treated animals (Svoboda and Reddy, 1972). Similar processes have also been observed in insect fat bodies (Figure 7.14) and in plant tissue (Locke and McMahon, 1971; Vigil, 1973). However, the precise rate of turnover is subject to variation; for instance, it varied greatly in different strains of mice (Rechcigl, 1971; Giranello and Axelrod, 1973).

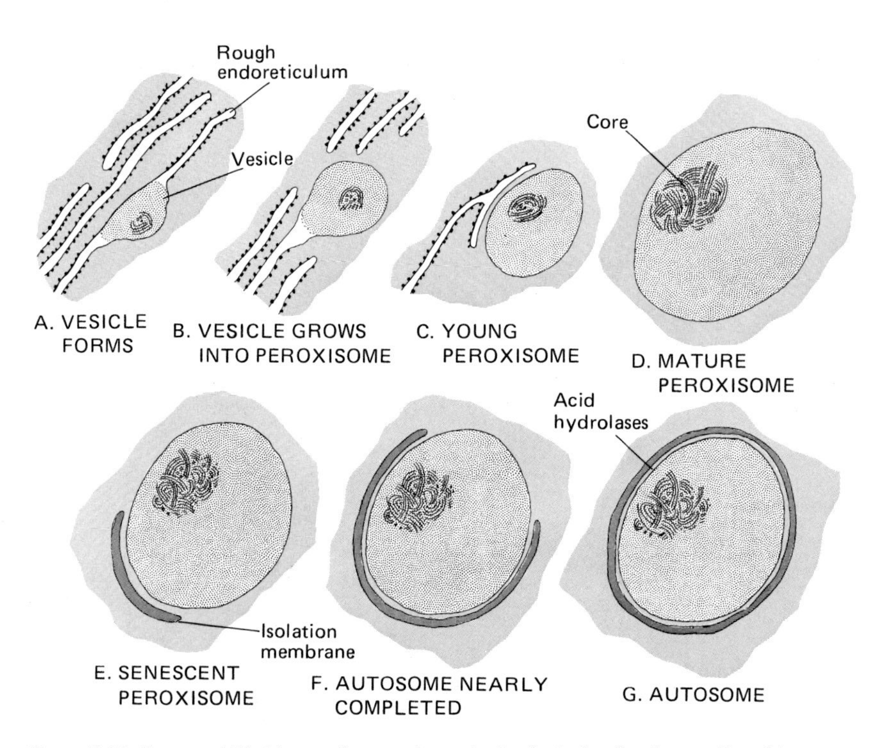

Figure 7.14. Suggested life history of a peroxisome in the fat body of an insect. (Based in part on Locke and McMahon, 1971.)

7.2.3. Comparative Aspects

During the time zoological biochemists were studying this particle as the peroxisome, plant scientists were actively pursuing investigations into the body they called the glyoxysome, now usually considered to be the same organelle. As pointed out earlier, these have a similar general type of morphology as the vertebrate particle and contain many of the same enzymes. For instance, catalase is present in large proportions and uric acid oxidase also is found in castor bean endosperm during germination; however, the latter is absent from the peroxisomes of spinach leaves (de Duve, 1969). Absent also from these plant particles are D-amino acid and L-α-hydroxy acid oxidases. However, many additional similarities become apparent as the discussion of the metaphytan particle progresses.

The Metaphytan Peroxisome. The investigations into the peroxisome in the leaves of higher plants necessitated reevaluation of much information obtained earlier on photosynthetic CO_2 fixation and subsequent carbon metabolism via the glycolic acid pathway (Tolbert, 1962; Tolbert and Yamazaki, 1969). Leaf peroxisomes, often closely associated with chloroplasts (Frederick and Newcomb, 1969), were found to be respiratory organelles, responsible for what had been known as photorespiration. Now that the same enzyme systems and processes have been found to exist in vertebrate kidney and liver tissues in the absence of light, the term peroxisomal respiration has been proposed for this activity (Tolbert and Yamazaki, 1969). Thus peroxisomes in a sense compete functionally with the mitochondrion in metazoans and with both that organelle and the chloroplast in plants. Because the processes enter into the photosynthetic pathways, the details of chemistry are reserved for discussion with those activities.

The main features of peroxisome respiration are pertinent, however. Among the chief steps are included a cyclic series that occurs only in light and differs distinctly from those that occur in the dark. In tobacco leaves, light-dependent respiration proceeds at a rate five times that of the dark (Zelich, 1968) and can be measured either by the O_2 consumed or the CO_2 produced. The substrate is glycolic acid, a major product of the CO_2 that is fixed during photosynthesis; thus peroxisomal respiration in plants reduces the net amount of sugar produced directly by the photosynthetic processes, but then supplements it, for one of its own end products is a hexose (Figure 7.15). Thus far in time only four electron carriers have been identified in peroxisomes, FMN, NAD, NADP, and the heme moiety of catalase (Tolbert and Yamazaki, 1969); the nature of electron carriers in general and of the substances cited is discussed where more appropriate (Chapter 8).

The Enzymes of Plant Peroxisomes. Because plant peroxisomes also engage in some of the photosynthetic processes, one might expect that their enzyme systems would diverge widely from those of vertebrates, but such is not

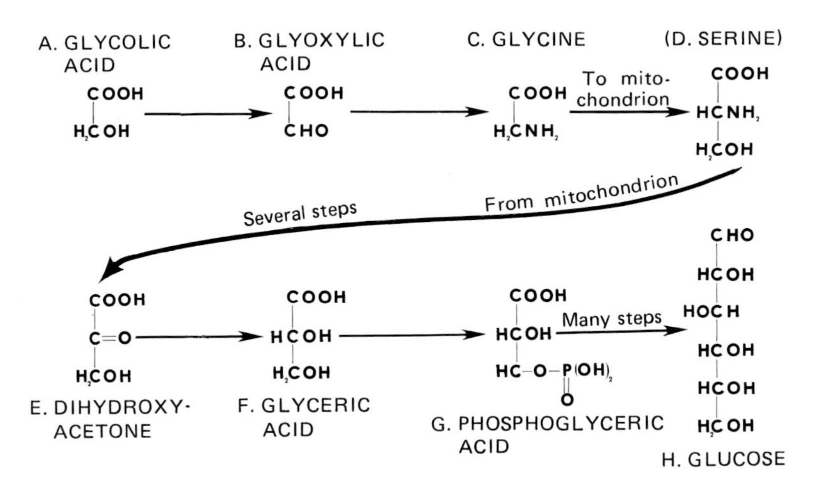

Figure 7.15. Enzymatic processes engaged in by plant peroxisomes. These steps receive attention again in Chapter 10.

the case, as already intimated. More than 15 different enzymes have been identified in these particles of plants (Breidenbach, 1969), 7 of which participate in peroxisome respiration in plants and presumably also in rat liver (de Duve, 1969). Two of these are glyoxylic acid reductases, one of which acts together with NAD, while the second needs NADP; the remaining four are dehydrogenases, two of glyceric acid (one with NAD, a second with NADP), and one each of malic (with NAD) and isocitric acids (with NADP). The final one in this series is an oxidase of glycolic acid.

A group that appears unique to plants is a series of transaminases, five of which are known in those organisms, whereas none have been found in metazoans. Four of these serve to transfer amino radicals to or from glutamic acid, whereas the last serves similarly with alanine. Germinating castor beans contain an additional group of enzymes, which so far have been identified in neither mature leaves nor metazoan peroxisomes. Among these are included isocitric acid lyase, malic and citric acid synthases, aconitase, and enoyl-CoA hydratase.

Peroxisomes are also active in plants during germination of the seed. In cucumber, to cite a specific case, two peroxisome enzymes, isocitric acid lyase and catalase, were investigated by immunoelectrophoresis (Lamb *et al.*, 1978). The first of the pair constituted 0.56% of the total protein, while the second was only one-sixth as abundant; determinations of their molecular weights also were made, which proved to be 325,000 and 225,000, respectively. The lyase apparently consisted of five subunits having mean weights of 63,500, and the catalase, as usual for the enzyme, had four units of 54,500. Dry seeds also have been shown to contain peroxisomal enzymes, including malic acid and citric acid synthases, malic acid dehydrogenase, isocitric acid lyase, catalase, and crotonase (Köller *et al.*, 1979).

Peroxisomes of Protistans. Although the hallmark enzyme of peroxisomes, catalase, had been reported from numerous protistans in earlier studies, only a few have been examined to determine its association with this particle. Among those that have been reinvestigated to this end are amoeboids of three genera, *Amoeba*, *Chaos*, and *Acanthamoeba* (Tomlinson, 1967; Müller and Møller, 1969a,b), flagellates of the genus *Ochromonas* (Lui *et al.*, 1968), and the ciliate *Tetrahymena pyriformis* (Müller *et al.*, 1968; Müller, 1969). In all cases except *Chaos*, 95% of the total catalase activity was found associated with peroxisomes; the exceptional protozoans had only 77% centered in those bodies. In general the particles possessed most of the ultrastructural traits of the organelle in metaphytans and metazoans. The mean diameter in *Tetrahymena* was found to be identical to those of higher forms (0.5 μm) and the appearance was also comparable, showing a granular stroma enclosed by a single membrane (Williams and Luft, 1968; Müller, 1969). No core was observed, however. In addition to catalase, uric and lactic acid oxidases, fumarase, citric acid synthase, aconitase, and dehydrogenases of isocitric, succinic, and malic acids have been identified in the particles from *Tetrahymena* (Müller, 1969).

The yeast, *Saccharomyces cerevisiae*, has received more extensive attention, and a number of reports have been made of the presence of peroxisomes (Baudhuin *et al.*, 1965b; Avers and Federman, 1968; Tolbert *et al.*, 1968; Szabo and Avers, 1969). In ultrastructure the yeast particles resemble those of other organisms, having a mean diameter around 0.5 μm, a moderately electron-dense, granular stroma, and a single membrane enclosing it. However, most of the particles may be noted to have a small number of diffuse cores scattered throughout the stroma (Szabo and Avers, 1969). The last citation also compared the content of certain enzymes of this particle to that of the mitochondria in the same organism. The two organelles were shown to be nearly equal in the activity of two enzymes, glyoxylic acid reductase (NADPH-associated) and isocitric acid lyase, but differed markedly in all others. In malic acid dehydrogenase activity, the mitochondrion accounted for two-thirds of the total and in cytochrome *c* oxidase, it exceeded the peroxisome about 3 : 1. In contrast, the peroxisome displayed about four times as much catalase activity as the mitochondrion, and about the same amount for malic acid synthases and glycolic acid oxidase.

The presence of peroxisomes has also been reported in the aquatic fungus *Entophlyctis* (Figure 7.16; Powell, 1979) and in the slime mold, *Dictyostelium discoideum* (Parish, 1975), and now a study of some of its enzymes has been published (Hayashi and Suga, 1978). The latter is of especial interest in that it described some of the changes in enzyme content as the organism changed under starvation conditions from an active feeding amoeboid to a stationary

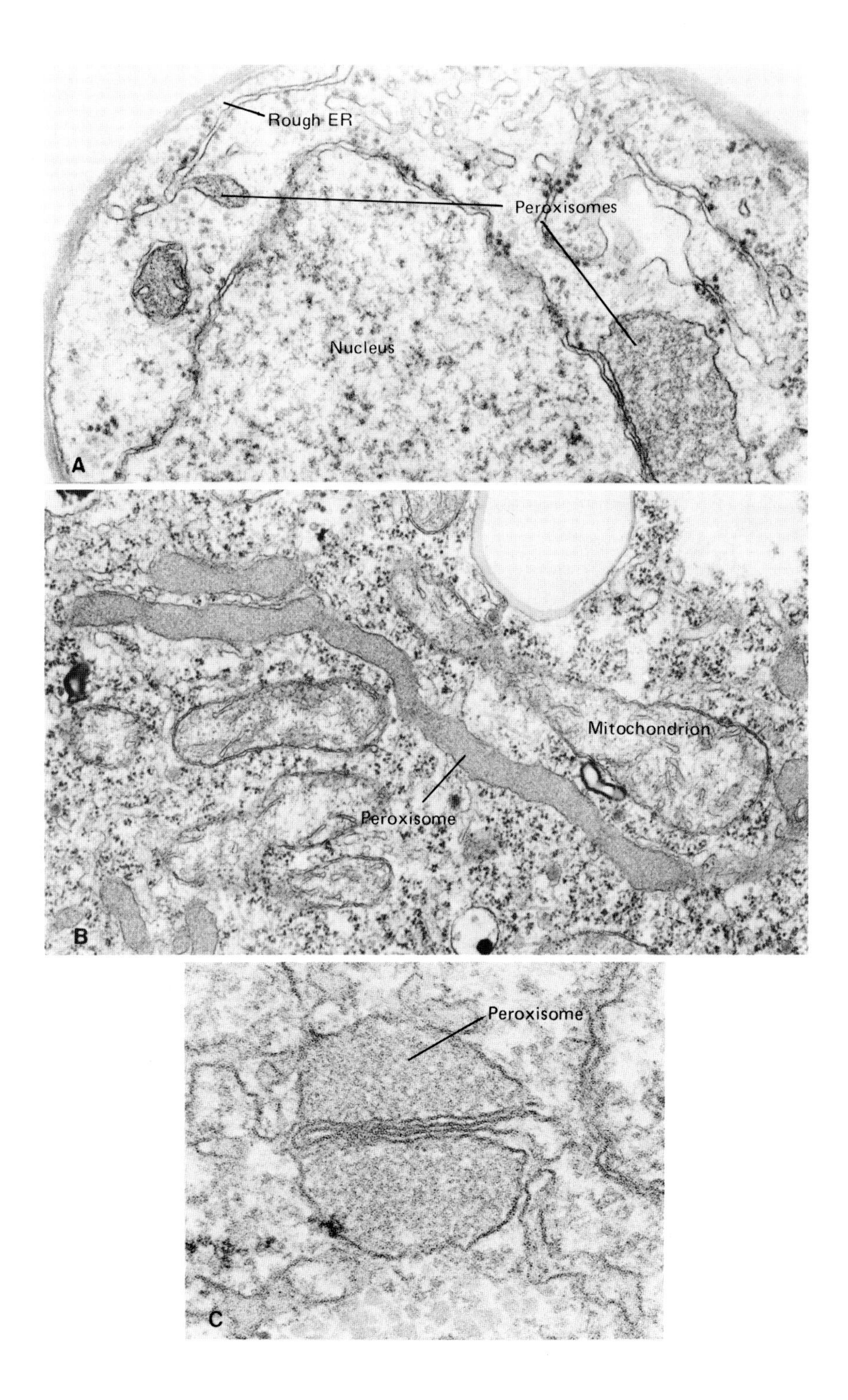
Rough ER
Peroxisomes
Nucleus
A
Mitochondrion
Peroxisome
B
Peroxisome
C

form. Uric acid oxidase and catalase, along with lesser amounts of acid phosphatase, were the principal enzymes of the particle, but no D-amino acid oxidase nor L-hydroxy acid oxidase could be detected. With the absence of food and the change in cell habit, the activity of both uric acid oxidase and catalase increased in the particle. On the other hand, acid phosphatase activity was enhanced only in the cytosol, whereas that of the peroxisomes was reduced to about one-half.

In the aquatic fungus mentioned in the foregoing paragraph and its relatives, many of the peroxisomes were of unusual form, being elongate, slender rods, rather than ovate or spherical particles (Powell, 1978). These were especially characteristic of the young zoosporangia, in which organs they were aligned along bundles of microtubules, and were often continuous with rough endoreticular membranes (Figure 7.17; Lange and Olson, 1977, 1978; Powell, 1979). In older zoosporangia, the peroxisomes tended to be ovate and were often arranged in pairs placed one on each side of an endoreticular cisterna. It was suggested that these particles arose in three different ways: (1) from blebs of tubular endoreticulum, (2) as vesicles of endoreticular cisternae, and (3) as buds of the elongate peroxisomes. Among the enzymes that were identified in the zoospore organelle were catalase, malic acid synthase, and isocitric acid lyase (Powell, 1976). In a parasitic fungus, *Sclerotium rolfsii*, catalase, glyoxylic acid dehydrogenase, and isocitric acid lyase were detected, but no malic acid synthase was found (Armentrout *et al.*, 1978).

7.2.4. Formation of Peroxisomes

The ontogenetic formation of peroxisomes still is not clear, but the involvement of endoreticulum appears to be a certainty. In metazoan tissues even the earliest reports indicated members of this system to be the source (Hruban and Swift, 1964), without designating the particular type of organelle—a practice too frequently followed to the present. However, Novikoff and Shin (1964) and Svoboda *et al.* (1967) reported continuities to exist between the membranes of smooth endoreticulum and those of peroxisomes, especially in regenerating or cancerous rat liver. These investigators proposed that the particles arose directly from vesicles that became separated from the reticular membranes, or in some cases, remained attached. Other studies, contrastingly, have implicated the rough endoreticulum. Catalase, the chief enzyme of the peroxisome, was described as being formed on rough endoreticular membranes, but in an inac-

Figure 7.16. Peroxisomes of the aquatic fungus *Entophlyctis*. The peroxisomes of this fungus are more irregular in shape than those of metazoans, and sometimes are associated with rough endoreticulum (A); occasionally the bodies may be elongate (B) or ovate (C). In the latter micrograph a cisterna of rough endoreticulum lies between the paired peroxisomes. (A) 42,100×; (B) 25,100×; (C) 77,900×. (All courtesy of Powell, 1979.)

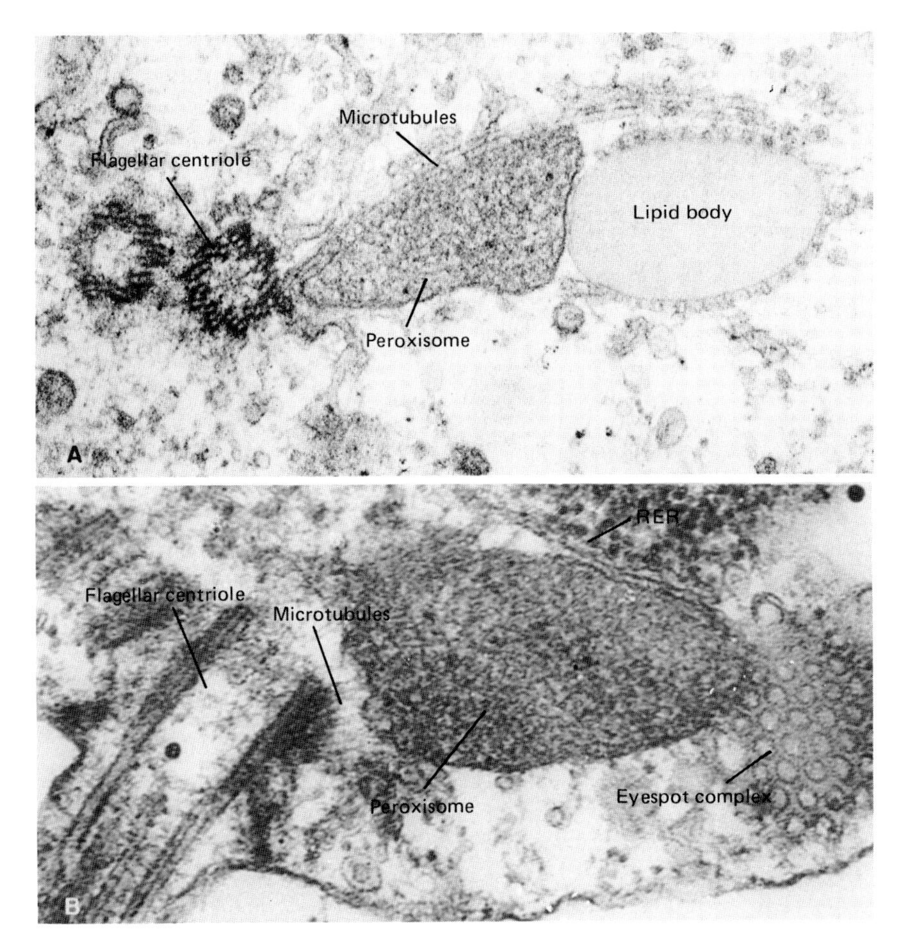

Figure 7.17. Peroxisomes in the zoospore of the fungus *Phlyctochytrium*. (A) In this view transecting the flagellar centrioles, the peroxisome is seen to form part of a complex, so-called eyespot system. 68,000×. (B) The same complex sectioned at right angles to the foregoing. 56,000×. (Both courtesy of Lange and Olson, 1977.)

tive state (Higashi and Peters, 1963a,b). Later the origins of the bodies themselves were suggested to have arisen as smooth vesicles at the periphery of rough cisternae (Essner, 1966, 1967). Within these, the nucleoids developed, after which the particle either broke free or remained attached. Proliferation of smooth endoreticulum does not result in an increase of peroxisome numbers and *vice versa* (Reddy and Kumar, 1977).

Among metaphytans, observations of the origins of peroxisomes have been little more precise. The phospholipids of the enclosing membrane of these organelles have been shown to be synthesized on endoreticulum (Kagawa *et al.*, 1973), as all such substances seem to be. Later, evidence was presented that the enzymes of the peroxisomes of castor bean might also be produced by the endoreticular membranes (Gonzalez and Beevers, 1976). During germination, according to the latter report, when peroxisomes were just beginning to form, about half of the malic and citric acid synthases were associated with the endomembrane system. Later, when the particles were fully developed, only 10% of the activity remained in the endoreticulum. No distinction was made between the rough and smooth types of membranes.

7.3. MISCELLANEOUS BODIES

Whether all miscellaneous particles that have been described should be considered specific cell organelles is doubtful, for there is good reason to believe that some are inert masses of fatty substances, glycogen, or other materials deposited in the cytoplasm, possibly for storage. Most such varieties are unnamed, but many others, often confined to a single taxon, have been given distinctive terms. Among this miscellany, several are of sufficient importance to deserve attention here.*

7.3.1. The Spherosome

One important body, now known as the spherosome, has received special attention because it is a characteristic feature of green plants in general. This organelle was actually first described under the name microsome (Hanstein, 1880), many years prior to the employment of the same term for sedimented fragments of endoreticulum (Claude, 1943). Because the later-named structure quickly became an extremely popular subject for investigation by animal cytologists and biochemists, botanical workers eventually were forced, for the sake of clarity, to adopt the current term for their particle (Perner, 1953).

Basic Characteristics. Unfortunately clarity was not fully attained by the change in name, for even in the current literature two quite different particles are referred to as spherosomes. The type to which that name should strictly appertain includes small spherical organelles that rarely are neither less than 0.8 μm in diameter nor more than 1 μm. As a rule, in metaphytans they are the most conspicuous bodies present in the cytoplasm, as they are abundant and have a high refractive index. Each is covered by a single membrane (Drawert and Mix, 1962; Grieshaber, 1964) enclosing a stroma that is largely pro-

*One body, which has received the name liposome, is a synthetic product widely used in biological and medicinal investigations. Several current reviews or symposia include Pagano and Weinstein (1978) and Papahadjopoulos (1978).

teinaceous, along with an ample supply of lipid material. The latter substance, however, does not generally surpass 40% of its total dry weight. At one time many investigators considered them to be similar to small fat droplets while others have suggested that they gave rise to such particles (Frey-Wyssling *et al.*, 1963). However, oil droplets never are membrane covered and often attain a size up to 150 μm (Sorokin, 1967).

Because of the liberal quantity of lipid present (Tu *et al.*, 1978), this body resembles other fatty particles in reacting strongly with osmic acid, Sudan black, and other lipid dyes used in histochemical investigations. From the others, spherosomes are distinct in displaying the Nadi reaction and in showing under the electron microscope a finely granulated stroma, not an amorphous one, when fixed with permanganate or osmic acid. Whereas other lipid bodies most likely are not actively secretory, the spherosome has been clearly indicated to produce enzymes. For example, in onion bud scales they have been demonstrated to liberate phosphate from glycerol phosphate (Walek-Czernecka, 1962) through the action of an acid phosphatase. Hence the transesterification of glycerol phosphate, the final step in fat synthesis, occurs within the spherosome and nowhere else in the cell, according to the histochemical tests employed. In *Crambe abyssinica,* the organelle was likewise shown to contain acid β-glycerophosphatase (Smith, 1974), an enzyme that may prove to characterize a whole class of spherosomes. Because that enzyme is also a distinctive feature of lysosomes, the spherosome at some date may need to be considered to represent a special category of that particle. As far as known, they are confined to higher plants, being absent even from their close allies, the green algae.

Ontogeny of the Spherosome. While the functional aspects of the spherosome thus remain obscure, its development has been rather thoroughly explored (Jarosch, 1961; Frey-Wyssling and Mühlethaler, 1965). Osmiophilic material commences to accumulate within a peripheral vesicle of the endoreticulum, which body eventually constricts at the cisternal connections and, while still quite minute, breaks off the parent organelle. The detached vesicle then increases in size, undergoing maturation as it does so. Upon reaching a diameter of around 100 nm, the granulated stroma characteristic of the mature organelle becomes readily perceptible, bringing into being a stage sometimes distinguished by the term prospherosome. Beyond this point no change other than size can be noted, the definitive stage being attained when a diameter of 500 nm or more has been achieved. Eventually the spherosome undergoes a second series of changes. First, as some of the stroma is lost, fat commences to accumulate centrally; these processes, fat accumulation and stroma loss, continue until all granulated stroma has been metabolized and a full-fledged lipid body of typical form eventually comes into existence. Thus some fat particles represent residual bodies of spherosomes.

Whether the suggested origin of the first minute vesicle from the endoreticulum is more apparent than real remains obscure. In view of the secretory powers of the definitive spherosomal membrane, a precursor incapable of secreting identical enzymes appears superfluous. In order to grow, the membrane as well as the stroma of the minute vesicle after it has broken free certainly requires additions of new proteins, which can only come from the surrounding cytoplasm. And if the cytoplasm is thus capable of providing the materials essential to growth, why can it not equally produce the initial vesicle?

7.3.2. The Oleosome and Other Lipid Bodies

The Oleosome. The second application of the name spherosome mentioned earlier is to a particle of higher lipoidal content that more correctly is termed the oleosome. The example recently described in detail from the carrot, *Daucus carota*, may be taken as typical (Kleinig *et al.*, 1978). The isolated oleosomes were found to be lipid bodies, consisting of 97% triacylglycerols and only 1 or 2% protein, thus contrasting strongly in composition with the spherosome proper. Moreover, in electron micrographs, the boundary between the particle and the cytosol was not a true phospholipid-containing trilamellar membrane, nor even a semimembrane (Figure 7.18).

Contrastingly, these particles from cotyledons of rape, sunflower, and watermelon seedlings have been reported to be enclosed by a semimembrane (Wanner and Theimer, 1978). This was extended into a flat handlelike projection on one side, not unlike that described earlier for certain primary lysosomes. Accordingly this extension was of trilamellar membrane construction. In developing cells, before any of the stored fats were used, the handle bore a number of ribosomes (Figure 7.19), but later, the latter were lost while the membrane became much thicker, reaching a diameter of 220 Å. The handlelike extensions were believed to represent lipase-carrying membranes.

Containing such a small fraction of protein as they do, these particles could not be expected to display much enzymatic activity, yet they have been demonstrated to contain some enzymes. Among those that have been reported are acyl-CoA hydrolase and acyl-CoA : 1,2-diacylglycerol acyltransferase (Kleinig *et al.*, 1978), the former being much more abundant than the latter.

Other Lipid Bodies. A survey of the major groups of organisms makes it apparent that under the term lipid body is embraced a number of unrelated structures whose features differ from taxon to taxon. In yeasts, freeze-etch studies show that the lipid deposits that the trilamellar membrane encloses to consist of bimolecular layers (Moor and Mühlethaler, 1963), but similar investigations on fungi (Hess, 1968) reveal lack of orderly arrangement in this lipoidal material. In the fungus *Aspergillus*, for example, lipid granules were large and so irregular in form as to suggest the possible absence of an enclosing membrane (Tsukahara and Yamada, 1965). But in another member of the same taxon, *Phytophthora parasitica*, the lipid granules were somewhat electron

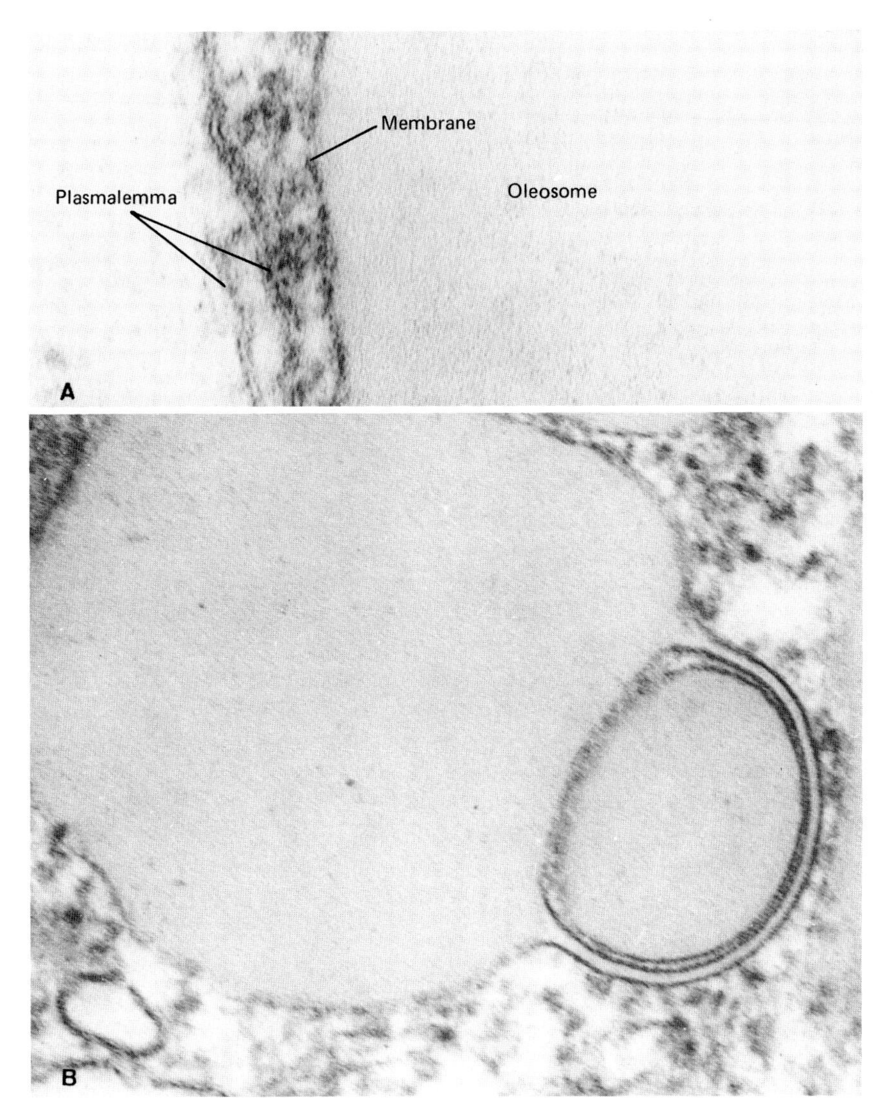

Figure 7.18. Structure of the membrane of an oleosome. (A) The lack of trilamellar construction is evident in contrast to the plasmalemma. 280,000×. (B) The oleosome of rape seed. 120,000×. (Both courtesy of Wanner and Theimer, 1978.)

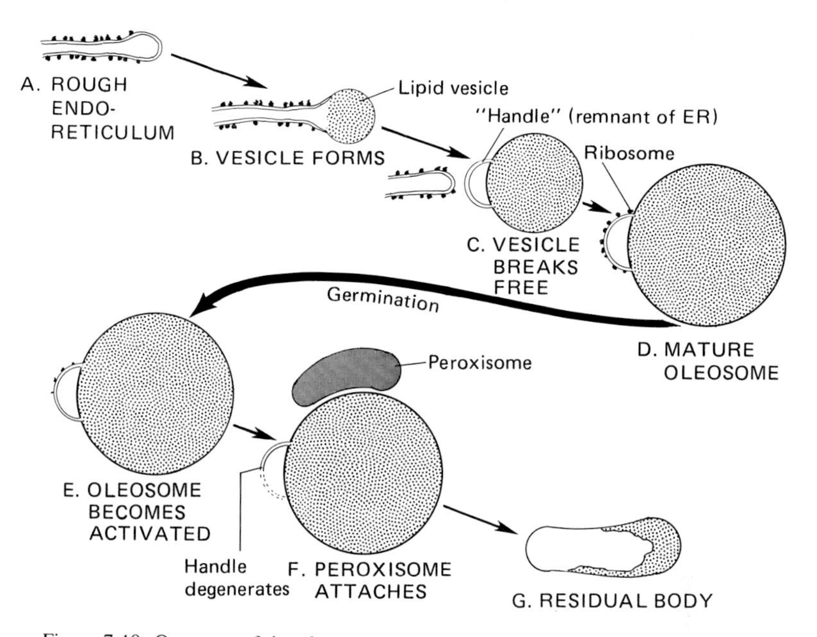

Figure 7.19. Ontogeny of the oleosome. (Based on Wanner and Theimer, 1978.)

transparent and enclosed within a distinct trilamellar membrane (Hohl and Hamamoto, 1967). Furthermore, they were scattered throughout the cytoplasm in the latter genus but were predominantly terminal in *Aspergillus*. Among the algae, few ultrastructural investigations identify lipid deposits within their subjects. One of the exceptions, dealing with the diatom *Nitzschia palea* (Drum, 1963), showed small oil droplets within the chloroplastid and large oil bodies distally, none of which were membrane enclosed.

Among the Vertebrata, lipid inclusions consisting of fatty acid triglycerides occur either with or without an enclosing trilamellar membrane. The latter condition is the more widespread, liver tissue and Sertoli cells from the testes providing typical examples (Fawcett, 1966). Frequently, such inclusions may appear to be surrounded by a membrane, but, because typical trilaminar construction is lacking, this appearance is usually explained as representing a layer of reduced osmium tetroxide. But now that membranes of atypical construction are known to surround the oleosome, this interpretation may need reexamination. During lipid absorption, intestinal epithelial cells accumulate large numbers of fat droplets, each enclosed in a membrane-limited vesicle. Unlike similar bodies in other organisms, these lipid deposits within their vesicles are surrounded by a narrow zone of stroma (Palay and Revel, 1964).

As yet it is not clearly established whether these fat-containing bodies are produced via the endoreticulum and dictyosome or are strictly products of pinocytosis. But because some biochemical evidence shows that only monoglycerides and fatty acids are absorbed from the intestinal lumen, the triglycerides appear to be newly synthesized. Ontogenetic origins of lipid bodies in other organisms and tissues have not received investigation to date.

7.3.3. The Lomasome

A peculiar membrane-coated body, originally described from fungi, has also been detected in the mesophyll cells of wheat (Manocha and Shaw, 1964). These lomasomes, as they are called (Moore and McAlear, 1961), are bodies continuous with the cell wall in the fungi, bounded on their proximal region by the plasmalemma. The nature of their contents varies from coarsely granular to vesicular. The walls of hyphae of Ascomycetes and Phycomycetes seem to be the locations where these structures are most frequently encountered (Girbardt, 1961), but they apparently do not occur in the walls of conidia, ascospores, conidiospores, nor the cells from which those parts arise (Brenner and Carroll, 1968). However, upon germination, basidiospores and conidiospores develop lomasomes in vacuoles adjacent to the cell wall (Kozar and Weijer, 1969). At one time an attempt was made to synonymize these particles to the mesosomes of bacteria (Zachariah and Fitz-James, 1967), but there is no basis for this action.

The latter study proposed a developmental sequence for the lomasome,

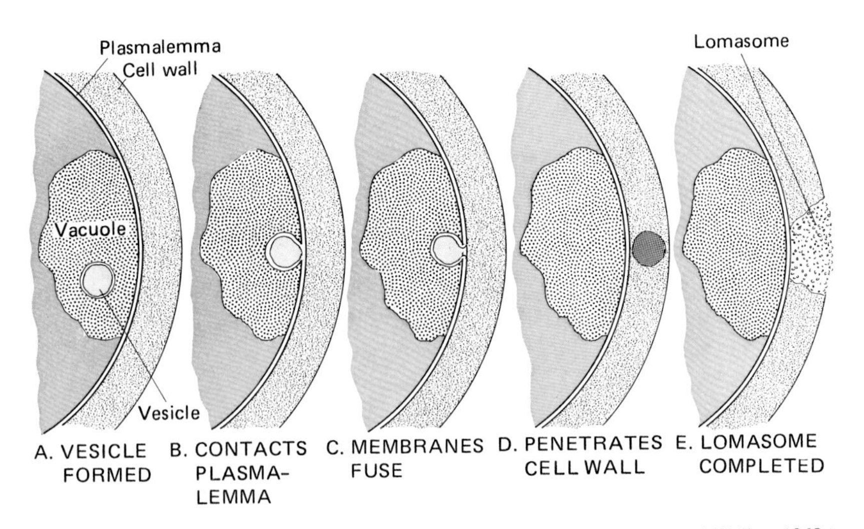

Figure 7.20. Development of a lomasome in *Neurospora*. (Based on Kozar and Weijer, 1969.)

beginning with their formation within large vacuoles. The newly formed, electron-transparent body was thought to move to the plasmalemma with which its membranous coat temporarily fused. Then the point of fusion disintegrated, permitting the body to pass through the plasmalemma into the cell wall (Figure 7.20), where it became more electron dense. Almost nothing is entirely clear as to their function. Because of their presence in basidia in which vacuoles were forming, it had been suggested at one time that lomasomes were involved in the breakdown of cytoplasm (Wells, 1965), but this presence now is considered to reflect the formation of these particles, as already stated. Others have proposed that they were more involved directly in the synthesis of the plasmalemma or to serve as a mitochondrion (Zachariah and Fitz-James, 1967), and still others believed them to be involved in cell wall formation (Wilsenach and Kessel, 1965). In addition, they have been viewed as being active in a secretory role (Moore and McAlear, 1961; Kozar and Weijer, 1969).

7.4. THE ENDOMEMBRANE AND VESICULAR SYSTEMS IN RETROSPECT

Here in the discussions of the endomembrane and particulate systems of organelles, a first glimpse has been caught into the genetic mechanism at the cellular level. Especially revealing was the brief view given in Chapter 6 of the modification of proteins within the lumina of rough endoreticular cisternae and the synthesis of phospholipids by the smooth. Thus a role for these structures was exposed in the refinement of the raw materials provided by the molecular genetic processes to produce ingredients suited for incorporation into cell organelles or for cellular functions. Further, some steps were suggested as to how certain simple organelles may be synthesized. But these chapters also have treated the first organelles that are sufficiently extensive to provide an insight into other workings of the cellular genetic mechanism.

Growth of the Endoreticulum. Unlike other cell organelles, the members of the endoreticular system are not permanent structures but increase and decrease in extent as the needs of the cell vary. This continual change was particularly indicated by the daily fluctuations in size, or even presence, of smooth endoreticular membranes, which increase in quantity from near absence at 2 PM to great abundance 8 hr later (Chapter 6, Section 6.1.2). This rhythmicity was not an automatic process throughout the liver but appeared only in those cells having the correct genetic constitution. Thus the circadian changes suggest that the endoreticular cisternae, tubules, and vesicles are formed by the cell in time of need and then broken down when the need passes. While the degradation of the organelles has not been observed, it must be carried out in a systematic manner, for there do not appear to be any reports describing their disintegration into vesicles or other remnants, nor have any observations of their being consumed by lysosomes been recorded, except perhaps in pathological tissues.

The waxing and waning of the endoreticular system is clear, but questions still remain as to how its growth and resorption are carried out. The endomembrane system concept claims that smooth endoreticulum is the product of rough, and the latter in turn is produced by the nuclear membrane. But if the nuclear envelope gives rise to all these components, what breaks them down? And in those tissues like interstitial cells of vertebrate testes that lack rough endoreticulum, what gives rise to the extensive smooth membrane system found there? How does either type arise in amoeboids and innumerable other cell types in which the nuclear membrane is devoid of the ribosomes needed for protein synthesis? If made on some central structure, how do enzymes reach the outer components of the network without ever appearing in the inner tubules through which they must pass, according to this view? Because no answers, only more questions, are aroused by these efforts, perhaps it might be fruitful to examine the Golgi apparatus as to its mechanisms of growth and renewal.

The Golgi Apparatus Growth and Turnover. As the regular progression of Golgi cisternae from the forming face of the dictyosome to the mature one has already been sufficiently detailed, some events associated with those processes need consideration. Existing concepts propose that the new cisternae result from fusion of vesicles derived from the associated endoreticulum, a view based on the occasional presence of blebs on the reticular membranes and small vesicles near the forming faces. However, the more recent investigations with thick sections and high-voltage electron microscopy have found these "vesicles" to be sections through tubules, thereby removing a substantial portion of the supposed supporting evidence. It will be recalled that these tubules not only partly enclose and interconnect the dictyosomes but also are attached to the newly developing and to the fully mature cisternae, but not to the intermediate ones. Thus the tubules appear to detach after a new cisterna has been formed to a certain degree, and later to reattach when a mature saccule is breaking down into vesicles. Consequently, the tubules must behave to some extent like endoreticulum but on a shorter time scale, forming attachments to a new cisterna, moving with it for a while, then breaking free and growing fresh extensions to the next new Golgi component that forms. At the same time, other tubular arms from the same system have to grow toward the most fully mature cisterna, attach to it, and move with its vesicles as they break free. When the latter have matured, the tubules detach and form other connections with the succeeding arrival at the mature end of the dictyosome. These complex movements, growth, and degradation on the part of the cisternae, together with the tubular components, provide a view of the organelle that is fully in keeping with the *dynamic nature of the cell*.

In a sense, dictyosomal cisternae are never really formed, for they are no more static in behavior than is the system as a whole. After its first framework has been laid down, each one passes into the intermediate region where its walls become thickened as new enzymes and structural components are added, processes that continue as the saclike compartment progresses toward the mature face. Once added, the enzymes do not necessarily remain where placed, but frequently are subject to removal or replacement. For instance, thymine pyrophosphatase and arylsulfatase are found only in the intermediate cisternae, while acid phosphatase is confined to mature ones. Thus these dictyosomal components are not mere mechanical sacs into which a prescribed mixture is poured; rather they are living parts of the living systems and experience continual gain, loss, and change throughout their existence.

The Cellular Genetic Mechanism. In a strictly mechanical view of dictyosomal growth and the function of the associated tubular network, it could be supposed that the latter supplied the enzymes and other biochemicals and thus played the major role in the formation and maturation of the cisternae. However, this proposal would supply only a superficial explanation of Golgi apparatus activities, for several reasons. First, the tubular network is itself a closed system, being connected only to all the dictyosomes. Consequently, it would have to receive at least the precursors of the essential ingredients to pass to the growing or maturing cisternae, for without ribosomes, it cannot itself engage in protein synthesis. Second, enzymes and other biochemicals have to be brought to the beginning cisternae that are obviously different from those brought to the component at the maturing face. Because both types of these saccules are continually forming, some intraluminal mechanism would have to exist that could guide each substance to its proper destination. This arrangement, while not unfeasible, would still not provide any means for supplying the intermediate cisternae with their particular and distinctive enzymes.

The last situation makes it clear that the only visible source of supply for the thymine pyrophosphatase and other distinctive chemical components of the saccules in this middle region is the cytosol. Then it could be the cytosol that adds new extensions to the tubular system of the dictyosomes and destroys old ones, just as it builds new centrioles and flagella. And if the cytosol is capable of building new centrioles and Golgi-associated tubules, why can it not also construct new cisternae and modify old ones? Or for that matter, why should it be incapable of building rough or smooth endoreticular membrane systems and destroying them when no longer needed? Perhaps some of the ingredients do arrive at the site enclosed in vesicles, but all need not be transported in that form.

It would seem far more feasible for new additions to be made by some mechanism such as a cytosolic one, located directly at the site of action than by a device located in a remote membrane of some sort, whether that of the nucleus or the endoreticulum. The proteins or precursors are, of course, synthesized on ribosomes under the directions of the macromolecular genetic apparatus. The processing of the precursors, the synthesis of carbohydrates and lipids, and the complexing of these various raw materials is one aspect of the cellular genetic mechanism about which some knowledge has been gained, as shown in these two related chapters. But the other, more difficult facets of the mechanism that involve the conduction of these finished ingredients to their respective specific destinations and their assembly there into the various organelles by the cytosol have not been explored even on a limited basis.

8

ENERGY-ORIENTED ORGANELLES AND ACTIVITIES: I

Cell Respiration

In eukaryotic cells two specialized organelles are involved in energy transfer, the chloroplast, which actively stores solar energy in the form of carbohydrates or lipids, and the mitochondrion, which provides a site for many respiratory processes that make the energy thus stored available to the cell. These processes share many common chemical components, so it is advantageous to discuss their ingredients without regard to function. Then after those have received attention, the respiratory processes become the center of focus. Next the energy-related organelles are reviewed in the following two chapters, the mitochondrion first, followed by the chloroplast and photosynthesis.

8.1. MOLECULAR ASPECTS OF CELL RESPIRATION

Because the mitochondrion serves as the center in which much of cell respiration is localized, a discussion of the respiratory reactions and their biochemicals appears a necessary prelude to any account of the organelle itself. Two major types of respiration are generally recognized, anaerobic and aerobic. Although differing in oxygen requirements and enzyme systems, the two have much in common, including an electron-transport system and the production of energy-transferal compounds. The anaerobic (fermentative) processes, however, are examined together with photosynthesis in Chapter 10, where the two activities will be found to share many identical steps.

8.1.1. The Energy-Transferal Substances

The ultimate energy-transferal substances in cells are limited exclusively to the triphosphorylated nucleosides and adenylated compounds. In certain

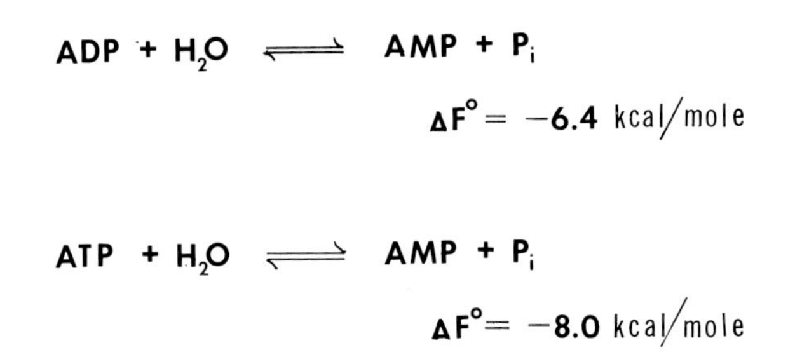

Figure 8.1. Inconstancy of bond energies. Although in each case the same bond is broken, the energy involved is different.

quarters, it is the vogue to attribute the energy-carrying capacities of these compounds to the presence of "high-energy phosphates" or "energy-rich bonds." But "it is impossible," to quote Gillespie and his co-workers (1953), "for any simple bond dissociation process to be energy-producing, and the concepts of 'energy-rich' bonds and the storage of energy in bonds are physically meaningless. Bond energies are always positive—energy is required to break a chemical bond."

Furthermore, as DuPraw (1968, p. 31) points out, the physical chemistry of the energy transfer associated with the phosphoryl radical differs from phosphate to phosphate and even from reaction to reaction of the same phosphate. For example, when adenosine diphosphate (ADP) is hydrolyzed to adenosine monophosphate, the measured free energy (ΔF°) is calculated at about -6.4 kcal/mole; when ATP is hydrolyzed to the same monophosphate, however, the ΔF° has been determined as -8.0 kcal/mole. Yet the bond concerned (Figure 8.1) is identical in both cases. As it thus appears that the structure of the entire molecule is involved in the release of free energy, it is not correct to speak of "high-energy phosphates" in the sense of radicals or residues, but only as compounds. And it is in this manner that the term is employed here.

Using quantum mechanics techniques, Pullman and Pullman (1960) calculated the electron structure of various phosphates. In their excellent report, four classes of high-energy phosphates are recognized, as follows: (1) carboxyl phosphates, such as 1,3-diphosphoglyceric acid of glycolysis, in which the phosphate radical replaces the hydrogen constituent of a terminal carboxyl radical; (2) enolphosphates, represented by 2-phosphoenolpyruvic acid of glycolysis; (3) organic pyrophosphates, including the very familiar ATP and ADP; and (4) amino phosphates (often called phosphogens), with creatine and arginine phosphates as familiar examples. In all the high-energy compounds, an unusual arrangement of π-electron charges is found, in that a chain of at least three adjacent atoms bears net positive charges as a result of a deficiency of π electrons. Sometimes the chain may include five or six positively charged atoms. Such a charge distribution leads to strong electronic repulsions, introducing thermodynamic instability in the system. The pyrophosphates are found to possess a net positive, while the rest show a net negative, repulsion energy. From a biological standpoint, it may be suggested here that perhaps this difference accounts for the adenosine phosphates alone being capable of serving as energy-transfer agents in many cellular processes, including contraction of muscle elements.

On the other hand, the carboxyl phosphates are shown in the same report to owe their high-energy property to the high value of the complementary opposing resonance that results from the greater structural changes undergone during their formation than is the case with other high-energy phosphates. Contributing also to the property is their free energy of ionization, not known to be a factor in the other types. The energy of the enolphosphates is attributed to the lowering of energies during tautomerization from the enol to the keto forms after removal of the phosphate group, but no explanation is advanced to account for the energy capacities of the amino phosphates.

Before leaving this brief summary of this important class of compounds, it needs to be made clear that pyrophosphates other than those of adenosine are also of biological importance, especially those of uridine and guanidine. What is more important, certain reactions normally involve a particular one of these phosphorylated nucleosides, rather than the more ubiquitous adenosine compounds, but the molecular basis for the specificity of the reaction is not immediately apparent. Furthermore, it should be pointed out that energy transfer is a complex process into which many outside factors may also enter. For example, in the hydrolysis of ATP to ADP, Burton (1959) has calculated the ΔF° to be -8.6 kcal/mole in the absence of magnesium ions but only -7.0 kcal/mole when those ions are present. The difference is explained as stemming from the fact that ADP does not bind magnesium so strongly as ATP; hence, the equilibrium is shifted toward the ATP side of the reaction. Fluctuations in the precise value of ΔF° also arise from differences in buffer systems and in the enzymes employed to catalyze the reaction.

The phosphoryl group is not the only one commonly associated with energy transfer; the adenyl group also can serve, as it does in the hydrolysis of ATP to AMP and pyrophosphate. Moreover, energy from both the phosphate and the adenosine residues can at times be interchanged with carboxyl esters of coenzyme A, the resulting products likewise showing high ΔF° values for hydrolysis. These high-energy compounds also transfer energy, usually in association not with phosphoryl, but with acetyl, succinyl, and other acyl esters; they are especially prominent in the entry into and within the citric acid cycle.

8.1.2. Major Features of Cell Respiration

The aerobic processes of cell respiration consist of a cyclic series of events, which can be considered to begin with the synthesis of the six-carbon compound citric acid. Hence, one of the names commonly applied to this

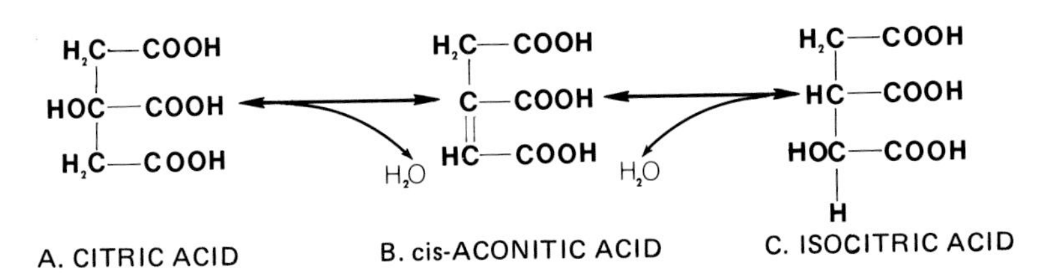

Figure 8.2. Early reactions in the tricarboxylic acid cycle.

sequence of chemical reactions is the citric acid cycle; however, it is also frequently referred to as the Krebs cycle, after the biochemist who first described its cyclic nature, and the tricarboxylic acid cycle, from the three carboxyl radicals that characterize citric acid. The last of these is the one in current use.

The Tricarboxylic Acid Cycle. In essence the tricarboxylic acid cycle consists of the reduction of a tricarboxylic to a dicarboxylic acid, a process that involves ten major stages and the formation of a number of energy-storing compounds. As shown in Figure 8.2, the first two steps involve the rearrangement of the citric acid molecule to form isocitric acid, by means first of dehydration and then by a rehydration reaction, *cis*-aconitic acid being the intermediate product.

Next, the resulting isocitric acid is dehydrogenated to form a short-lived α-keto acid, known as oxalosuccinic acid, in a reaction catalyzed by isocitric acid dehydrogenase. In eukaryotes, that enzyme can employ either the substance known as NAD or the related chemical NADP as the coenzyme; the protein that requires the latter substance in eukaryotes is located largely in the cytosol, whereas that which is specific for the former is wholly in the mitochondrion. In nearly all bacteria, however, only the NADP-requiring enzyme is present, a point worthy of particular note. Once the oxalosuccinic acid has been formed in this fashion, it apparently does not detach from the enzyme but undergoes spontaneous decomposition into α-ketoglutaric acid and CO_2 (Figure 8.3). This, the first production of CO_2, should be noted not to require free O_2 for its completion. The production of NADH·H, which as shown later is the usual first step in the production of energy-storage chemicals, should be observed to precede, rather than accompany, the release of CO_2. It is of interest that the NAD-requiring dehydrogenase does not appear able to catalyze the reverse reaction, whereas the scarcer mitochondrial NADP-dependent enzyme is capable of carrying out the reaction in either direction. The former had been reported to consist of four subunits, having molecular weights between 39,000 and 41,300, but under certain conditions of temperature and pH, the active enzyme contained four subunits, but under others it actually consisted of eight (Giorgio *et al.*, 1970; Fan *et al.*, 1975; Rushbrook and Harvey, 1978).

The next two major steps are far more complex than the others in the

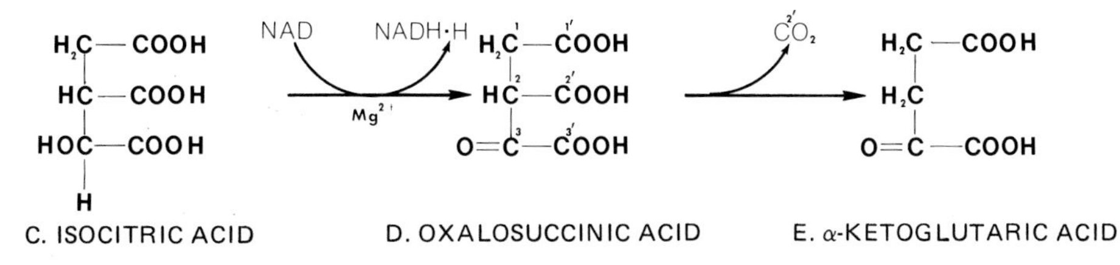

Figure 8.3. The formation of α-ketoglutaric acid in the tricarboxylic acid cycle. Note that the carbon of the (C2′) second carboxyl radical is lost as this chemical is formed (D, E).

cycle, all of which are similar to those already described; this point is pertinent to evolutionary aspects of these cyclic events discussed in a later section. In the first of these reactions, the α-ketoglutaric acid is oxidatively decarboxylated by a multienzyme complex consisting of three proteins and five coenzymes. Among the latter are coenzyme A, NAD, thiamine pyrophosphatase (TPP) mentioned in the discussion of Golgi bodies, and a flavoprotein. While all the details are not pertinent to the discussion, two intermediate stages are involved. The first reaction results in the production of succinate semialdehyde-TPP and the liberation of CO_2; it is evident that here also no free oxygen is consumed in its production. The succinyl residue is then transferred first to lipoic acid and finally to coenzyme A, resulting in the synthesis of succinyl-CoA as illustrated (Figure 8.4), the union of the two substances being by way of a thioester bond.

To conserve the energy of the above compound, the second of these complex steps is believed to consist of a series of three reactions, as follows:

$$\text{succinyl} \cdot \text{S-CoA} + \text{enzyme} \longleftrightarrow \text{CoA-S} \cdot \text{enzyme} + \text{succinic acid}$$

$$\text{CoA-S} \cdot \text{enzyme} + P_{\text{i}} \overset{Mg^{2+}}{\longleftrightarrow} \text{phosphoryl} \cdot \text{enzyme} + \text{CoA-SH}$$

$$\text{phosphoryl} \cdot \text{enzyme} + \text{GDP} \overset{Mg^{2+}}{\longleftrightarrow} \text{enzyme} + \text{GTP}$$

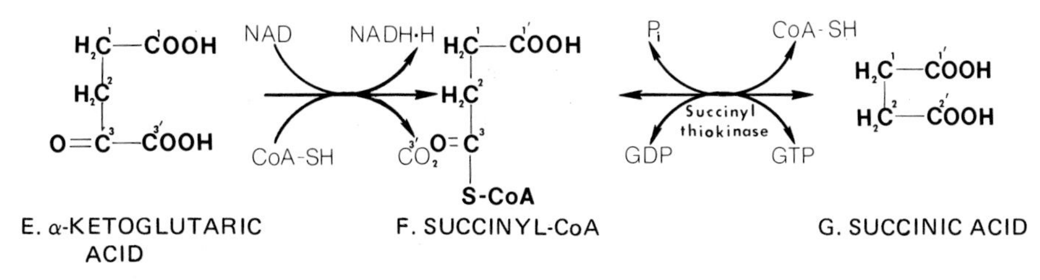

Figure 8.4. The steps leading to succinic acid synthesis. Note that the third carbon (C3) of the α-ketoglutaric acid (E) becomes the carbon of the second carboxyl radical (C2′) in succinic acid (G).

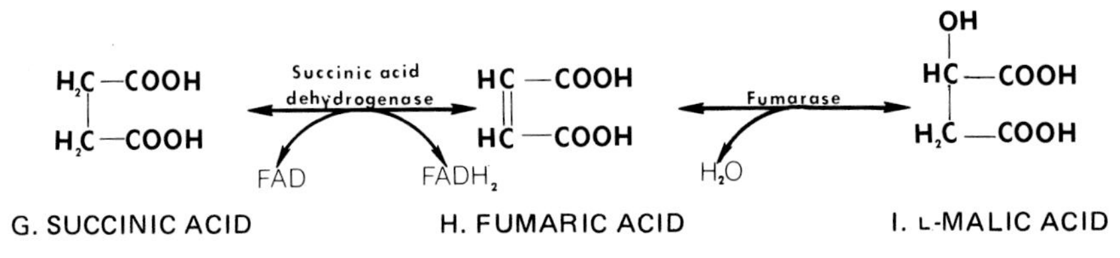

Figure 8.5. The synthesis of L-malic acid in the tricarboxylic acid cycle.

The enzyme is succinyl thiokinase and is specific for either GDP or IDP; for convenience, in equations the guanosine diphosphate is usually indicated. Now the hydrocarbon skeleton of the substrate molecule contains only two carbons, number 3′ having been lost in the CO_2, while number 3 becomes carbon 2′ (Figure 8.4).

As stated, the remaining steps of the cycle are relatively simple ones. Immediately beyond this point, the succinic acid is dehydrogenated by an enzyme called succinic acid dehydrogenase, a flavoprotein containing 1 molecule of FAD per molecule of enzyme. The product of dehydrogenation, fumaric acid (Figure 8.5H), is then hydrolyzed by fumarase to form L-malic acid, a reversible reaction (Figure 8.5I). Among the outstanding characteristics of this latter substance is an ability to penetrate the mitochondrial membranes, so that it enters or leaves freely. Thus it is able to play an active role in a shuttle mechanism between the cytosol and the mitochondrion.

The last of the routine reactions centers on the dehydrogenation of the L-malic acid by malic acid dehydrogenase to form oxaloacetic acid (Figure 8.6). Multiple species of malic acid dehydrogenase occur in many organisms, including amphibian embryos and oocytes (Petrucci *et al.*, 1977) and *Zea mays* (Yang *et al.*, 1977). Four NAD^+-requiring isozymes of mitochondria have now been identified in the latter organism, all of which were encoded by genes of the

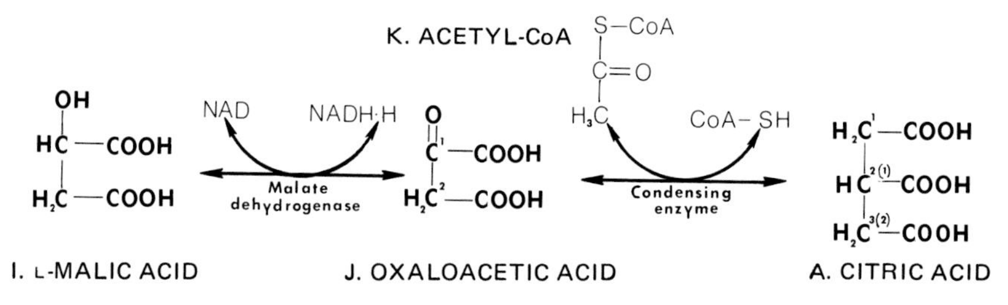

Figure 8.6. The recycling steps of the tricarboxylic acid cycle. The acetyl radical added here is considered to contain carbons 1 and 1′, whereas the ones formerly known as 1 and 2 of the oxaloacetic acid skeleton now become carbons 2 and 3. In the diagram of the citric acid that is thus formed (A), the former numbers are enclosed in parentheses.

nucleus. However, in yeast only two isozymes appear to be present, one in the cytosol, the second in the mitochondrion (Hägele *et al.*, 1978).

The oxaloacetic acid then receives an acetyl group, usually from pyruvic acid in the cytosol, carried by coenzyme A, resulting in the synthesis of citric acid, the substance that commences these cyclic events (Figure 8.6). It should be noted that the acetyl radical has been added to the 1 carbon of the old substrate molecule; hence, when the molecule is renumbered, the acetyl radical is designated as carbon 1 and the former 1 and 2 carbons now become 2 and 3. Because the last of these is the one that becomes the carboxyl radical, carbon 2′, during the production of the second CO_2, the cyclic events may be conceived to involve the peeling off of the terminal skeletal portion with each cycle. Thus it requires three cycles to disintegrate completely a given molecule of citric acid.

The Electron-Transport System. At several points in the foregoing cyclic series of events, substances such as NAD (nicotinamide adenine dinucleotide) and FAD (flavin-adenine dinucleotide) were shown in the diagrams as accepting hydrogens. These chemicals are among those that carry hydrogens or electrons through a series of oxidation–reduction activities that terminate in a reaction with atmospheric oxygen. Consequently, the observation made earlier—that although none of the events in the tricarboxylic acid cycle itself required oxygen, it was nonetheless considered to be aerobic—now becomes clear. These side reactions require oxygen; thus when all available NAD or FAD in the cell has become reduced to NADH·H or FADH·H, the tricarboxylic acid cycle cannot continue until those substances have become oxidized and again available for hydrogen residues.

Because the essential ingredients are examined in detail shortly, here only the sequence of events and related topics are of importance. In the intact chain the involved ingredients normally react in the sequence:

Substrate → NAD → FAD → CoQ → cytochrome b → cytochrome c_1 → cytochrome c → cytochrome c oxidase → O_2

These substances interact in the order of the E_0' values (Chance, 1977), proceeding from the one having the most negative value (NAD = − 320 mV) to that possessing the highest positive one (O_2 = + 820 mV). The first three of the reactants actually carry hydrogens; then, as shown later, the cytochrome b removes protons to the environment and passes electrons to the next cytochrome in the sequence. As the cytochromes thus transport electrons, this chain of substances is commonly regarded as an electron-transport chain. These proteins carry out this function largely by way of their heme prosthetic group, the active part of which is an iron atom. In the passage of electrons down the chain, the first ferric iron (Fe^{3+}) gains an electron to become ferrous iron (Fe^{2+}); when it loses the electron to the next cytochrome (c_1) in the line, it thereby becomes oxidized to the ferric state once more. The ferrous iron of the

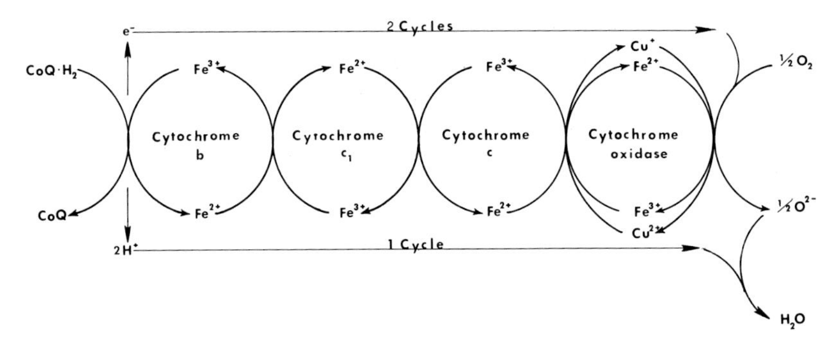

Figure 8.7. Events in electron transport in the metazoan system.

cytochrome c_1 then transfers the electron to the ferric iron of the cytochrome c heme, with similar consequences, as visualized in Figure 8.7. There it can be observed that the oxidase contains copper as well as iron, both of which substances must be reduced before the reaction with $\frac{1}{2}O_2$ occurs. Furthermore, it is evident that because coenzyme Q carries two hydrogens, two passages of electrons must take place before that reaction can occur.

Oxidative Phosphorylation. The energy that is released in the oxidation of the hydrogens is not dissipated as heat but is largely conserved in high-energy phosphate compounds, typically adenosine triphosphate (ATP). In the processes of the electron-transport chain and the rest of the system, there is no suggestion as to how ATP is produced; usually the synthesis of that substance is viewed as a separate process called oxidative phosphorylation (Slater, 1977; Wilson *et al.*, 1979). Typically the reaction is coupled to those of the transport system (Green and Blondin, 1975; Boyer, 1977), but it can become uncoupled by means of a number of various reagents. If uncoupled, the transport-chain processes continue, whereas the production of ATP ceases (Fritz and Beyer, 1969).

In the complete electron-transport system, the production of the ATP has been suggested to occur at those three points where the free energy change ($\Delta G^{\circ\prime}$) expressed in calories per mole exceeds that required to break the phosphate bond between the outer two phosphate residues of ATP (7500 cal/mole). Those points are at the interaction between NADH·H and FADH·H (12,450 cal/mole), cytochrome b and c (10,150 cal/mole), and cytochrome oxidase and $\frac{1}{2}O_2$ (24,450 cal/mole). Consequently, in the full eukaryotic system, three units of ATP per original hydrogen are synthesized, but because NAD carries two hydrogens, six actually result from each complete oxidation of NADH·H (Ochoa, 1943). However, this concept has proven contrary to data derived from actual measurements of the ATP synthesized per unit of oxygen consumed, reported more fully in connection with the discussion of the chemiosmotic concept that follows shortly. At the reaction in the tricarboxylic acid cycle where succinic acid reduces FAD, the first of these points is bypassed, so that only four units of ATP result. Now if the entire cycle is reviewed, it can be observed that NADH · H is produced at three points, between substances C and D, E and F, and I and J, resulting in the synthesis of 18 ATP molecules. In addition, 4 result in the G→H transformation, for a total of 22 ATP molecules plus 1 GTP, produced in the F→G reaction—the virtual equivalent of 23 ATPs, as each of the common triphosphorylated nucleosides can be converted to the others.

ATPase, the enzyme chiefly involved in the production of ATP in the mitochondrion, functionally is a complex of several components, including an ionophore, a membrane factor that directs H^+ cations to the catalytic site, an oligomycin-sensitizing protein, a soluble factor called F_1, and an ATPase inhibitor of unknown function (Gómez-Puijou *et al.*, 1976). The coupling factor F_1 has been shown to possess several ATP- and ADP-bonding sites, for labeled ADP and ATP can be bound concurrently (Slater *et al.*, 1979). Moreover, there appear to be at least two binding sites for ATP to one for ADP. In the ATPase of yeast a complex of F_1 and the oligomycin-sensitizing protein has been reported to have a molecular weight of 500,000 (Ryrie, 1977) and to consist of at least nine subunits. The amino acid sequence of subunit 9 has now been established (Wachter *et al.*, 1977), and more recently, the nucleotide sequence of the gene that encodes that portion has been determined (Hensgens *et al.*, 1979). Thus that section is definitely encoded by the mitochondrial genome. Some evidence has also been presented that, at least in certain bacteria, the ATPases may be glycoproteins (Andreu *et al.*, 1978).

Organizational Units of the Respiratory Chain. The respiratory chain can be separated into four (actually five) multienzyme complexes (Hatefi *et al.*, 1962; Hatefi, 1966; Chance, 1977; DePierre and Ernster, 1977), the second one of which might be viewed as a side reaction. Thus the three of the chain proper (Figure 8.8) include Complex I (NADH-CoQ reductase), Complex III (CoQ · H_2-cytochrome *c* reductase), and Complex IV (cytochrome *c* oxidase). The side chain is Complex II (succinic acid-CoQ reductase). Those of the main chain are energy-coupling sites, whereas Complex II is not, and as has already been pointed out, can be used for ATP production by way of an oligomycin-sensitive enzyme, ATPase (Senior, 1973).

Complex I, with a total molecular weight near 850,000, is by far the largest and most elaborate of these subdivisions and may prove to be the largest proteins of the mitochondrial membrane. Among its 16 subunits is one having an FMN prosthetic group and five or six others bearing a total of between 16 and 24 Fe-S prosthetic groups variously arranged in clusters. An integral part of the cristal and/or inner membrane, the NADH is considered to be located on the matrix (M) side, and the ubiquinone near its center (Copenhaver and Lardy,

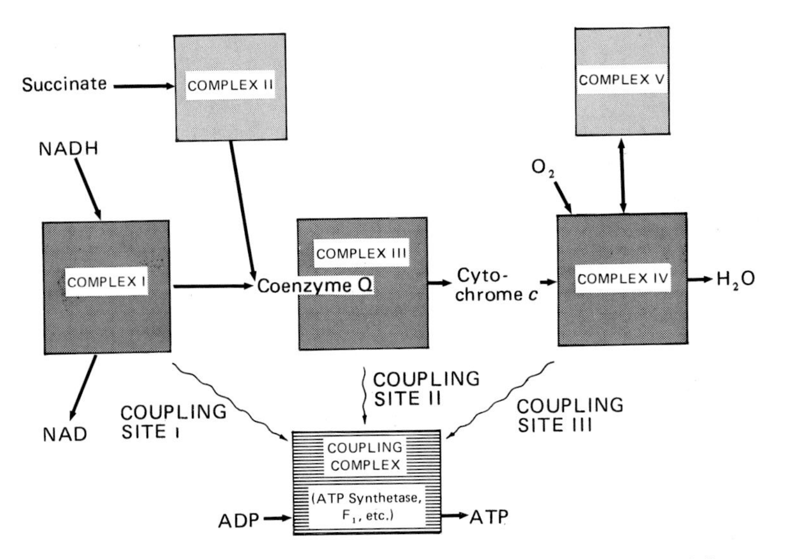

Figure 8.8. The relationships of the various components of the respiratory chain.

1952; Hatefi, 1966; Harmon *et al.*, 1974; Harmon and Crane, 1976; Hatefi and Stiggall, 1976; Ragan, 1976). This localization, based on the failure of ferricyanide to be reduced by NAD^+-linked substrates in intact mitochondria, militates against the chemiosmotic concept. With the NAD are associated several dehydrogenases, one of which from *Drosophila* larvae has been found to have a molecular weight of 79,000 (Hermans, 1979). Others in the same organism may be induced by heat shock treatment.

The side component, Complex II, molecular weight 97,000, has been found to consist of four polypeptides (Capaldi *et al.*, 1977). The largest of these, succinic acid dehydrogenase, having a molecular weight of 70,000, contained covalently bound FAD, while the next largest, with a weight of 24,000, contained two Fe-S clusters. The other two were much smaller proteins, with molecular weights of 13,500 and 7000; whether either of these contained a single cluster of Fe-S as the prosthetic group is still an unresolved question (Lee *et al.*, 1967; Quagliariello and Palmieri, 1968; Harmon *et al.*, 1974; Case *et al.*, 1976; Hatefi and Stiggal, 1976). This too is an integral part of the inner or cristal membrane, in which it is similarly oriented, with succinic acid dehydrogenase on the M side, coenzyme Q at the center of the membrane, and polypeptide CII-3 on the cytosol-surface (DePierre and Ernster, 1977; Merli *et al.*, 1979).

With a molecular weight of 280,000, Complex III (CoQ · H_2-cytochrome *c* reductase) is the second largest portion and consists of 6 to 12 subunits

(Gellerfors and Nelson, 1975; Leung and Hinkle, 1975; Bell and Capaldi, 1976). At least two, and possibly three, of these are isozymic cytochromes *b* (Davis *et al.*, 1972), another is cytochrome c_1, still another is an Fe-S protein, while the rest are known respectively as "antimycin-binding protein" and two species of "core protein." They are present in the molar ratio of 2:1:2:1:2, in the same sequence (Yu *et al.*, 1974). As an integral constituent of the membrane, this complex is considered as having the coenzyme Q site situated at the middle of the membrane and the cytochrome c_1 site on the cytosol (C) surface, but the arrangement of the numerous constituents is still being investigated (Hatefi, 1966; Baum *et al.*, 1967; Wikström, 1973; Guerrieri and Nelson, 1975; Papa, 1976; Rieske, 1976; Trumpower, 1976; Gellerfors and Nelson, 1977; Mendel-Hartvig and Nelson, 1978; Chiang and King, 1979).

Between the foregoing and the final complexes is inserted another unit consisting entirely of cytochrome *c*. Because of its reactivity in intact mitochondria and the ease with which it is extracted with salt solutions (Jacobs and Sanadi, 1960), it is considered to lie on the C surface of the inner membrane. There it is bound stoichiometrically to cytochrome *c* oxidase, which reaction is believed to secure it in its superficial location (Nicholls, 1964, 1974). Interestingly, the cytochrome *c* is bound to the membrane more securely when oxidized than when reduced, possibly due to a regulatory effect of ATP (Vanderkooi *et al.*, 1973a,b; Ferguson-Miller *et al.*, 1976).

Because the final component, Complex IV, contains only cytochrome *c* oxidase, which consists of cytochromes *a* and a_3, it is better described in detail in the section that follows. This combination, with a molecular weight of about 200,000, is considered to span the mitochondrial membrane, with the cytochrome *c* site at the C surface and the O_2 site possibly at the M surface, but the latter is not firmly established. In the collective complexes, the cytochromes are present in two molar ratios, with the exception of cytochrome c_1, which is represented by half that quantity (DePierre and Ernster, 1977).

Recently a fifth complex (Complex V) has been isolated, characterized by very high ATP–P_i exchange rate and ATPase (Griffiths, 1976; Stiggall *et al.*, 1978). Thirteen major and minor polypeptide bands were revealed by dodecyl sulfate–acrylamide gel electrophoresis, among which were five factor F_1 subunits of ATPase, the oligomycin sensitivity-inferring protein, and the uncoupler-binding polypeptide. But very little more seems to have been learned about its nature.

The Chemiosmotic Mechanism of Coupling. The strictly biochemical concept of coupling between respiration and phosphorylation in mitochondria, in which the chain just described is considered to be an "electron-transport chain," is not universally accepted. A second school of thought (Mitchell, 1961) believes a different combination of the chain, transferal substances, and mitochondrial membrane exists in what is known as a chemiosmotic mechanism. In its hypothesis, protons rather than electrons receive emphasis. Basically four sets of factors enter into the chemiosmotic system: (1) a proton-translocating reversible ATPase, (2) a proton-translocating respiratory chain, (3) exchange–diffusion systems that couple translocation of protons to that of anions and cations, and (4) an ion-impermeable coupling membrane (Mitchell and Moyle, 1967; Ernster, 1977; Mitchell, 1977; Hill, 1979). As visualized in Figure 8.9, NAD, FMN, and coenzyme Q are each postulated as translocating two hydrogen equivalents across the membrane and then removing two electrons. After removal, the electrons are translocated back across the membrane successively by the iron sulfide system, cytochrome *b*, and cytochrome c_1, the latter closely followed by the remainder of the oxidase system (Baird and Hammes, 1979).

Perhaps the most radical departure from the strictly chemical view of the reactions lies in the energy relationships. In the proposed system, it is evident that ATP is formed by the proton transport, the proton-motive force being one of the outstanding aspects of the concept (Moore *et al.*, 1978a,b; Moore and Wilson, 1978). Perhaps the strongest evidence favoring this concept is derived from recent measurements of the amount of ATP actually produced per unit of oxygen consumed, the so-called P/O ratio (Hinkle and Yu, 1979). As reported earlier, in the chemical view of these processes, one ATP is produced by one electron at each of three points, so with the passage of an electron a P/O ratio of 3 was formulated (Ochoa, 1943). More accurate measurements yielded a value close to 2 with NADH-linked substrates and only 1.3 instead of the 2 expected with succinic acid (Hinkle and Yu, 1979). While in strong conflict with the chemical concepts, these are the results predicted with the proton-motive forces of the chemiosmotic view.

Other data presently do not favor this hypothesis, however. For instance, the present hypothesis assigns an electron-carrying role to cytochrome *b*. However, as pointed out above, the effect of pH on the oxidation–reduction potential of this substance shows that, between pH 6.4 and 8.2, one proton is released during oxidation. Thus reduced cytochrome *b* carries a proton as well as an electron (Straub and Colpa-Boonstra, 1962; Slater, 1967), which is virtually the equivalent of stating that it carries a hydrogen atom. These and other problems make it difficult to reconcile present knowledge of respiratory-chain properties with the chemiosmotic theory. At the same time the concept receives support from other data that appear in conflict with the strictly chemical coupling hypothesis. A related concept avoids use of the membrane arrangement but postulates similar use of protons for the synthesis of ATP (see Williams, 1978, for a review).

8.2. THE RESPIRATORY CHAIN

In general, the respiratory chain may be said to consist of pyridine nucleotides, flavoproteins, a series of iron–porphyrin- (i.e., heme or hemin) contain-

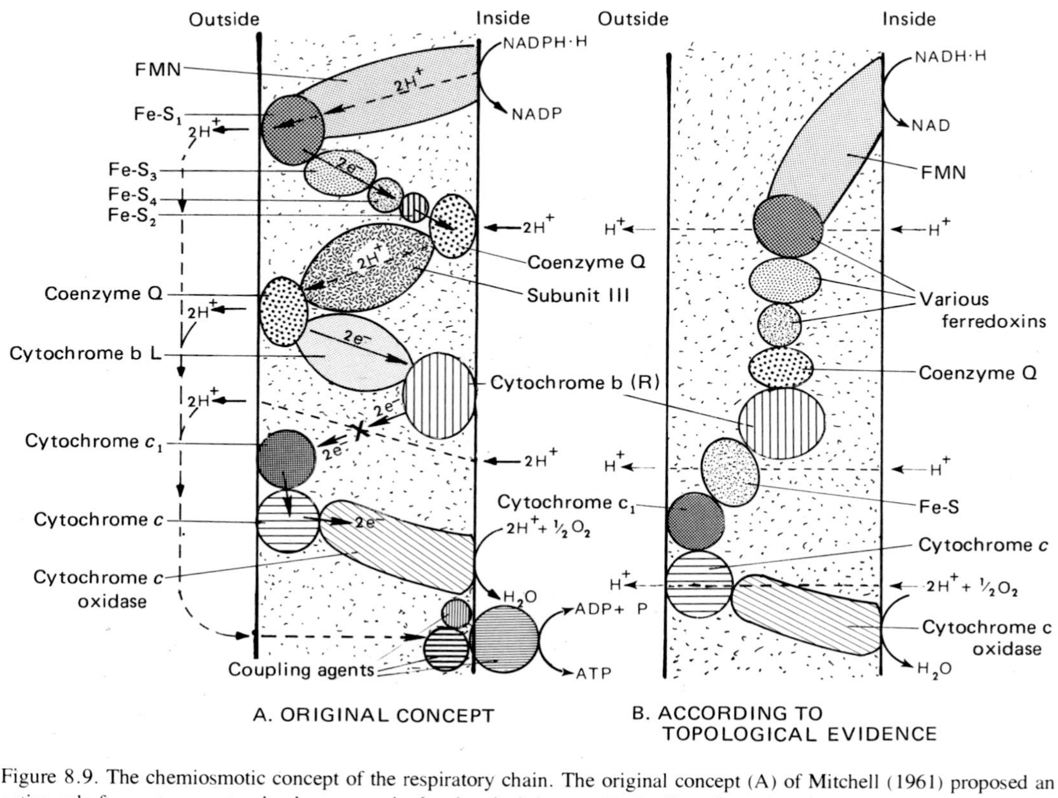

Figure 8.9. The chemiosmotic concept of the respiratory chain. The original concept (A) of Mitchell (1961) proposed an active role for protons, not only electrons as in the chemical theory. (B) Studies of membrane topography suggest a number of changes in orientation for the several components. (Based in part on Mitchell, 1961, and DePierre and Ernster, 1977.)

ing proteins known as cytochromes, nonheme iron-containing proteins, and copper. As the name of this combination of substances implies, the system is capable of transferring hydrogen equivalents or protons and electrons from various substrates, or from such excited molecules as those of chlorophyll, to a receptor. Although the main focus is on the chain employed in cell respiration, some attention is given here to comparable systems in other organelles, for transport of hydrogens or electrons is an essential feature of a wide spectrum of cellular activities.

8.2.1. The Cytochromes c

The cytochrome portion, with five or more distinct varieties occurring in the respiratory chain in many cells, constitutes the actual electron-transporting series. Although these five work closely together, they do not form a single structural entity, as shown by the preceding discussion of the several multienzyme complexes. In each cytochrome, an attached heme group* is the center of activity, the iron component becoming oxidized or reduced as one electron is lost or gained, explaining why the cytochrome molecule is capable only of transporting single electrons. However, cytochrome *b* is exceptional in this regard, for as shown by the effect of pH on its oxidation–reduction potential, this compound in the reduced state carries a proton as well as an electron (Straub and Colpa-Boonstra, 1962). As already indicated, the five members of the eukaryotic series are currently arranged functionally in the sequence b, c_1, c, a, and a_3 on the basis of the electric potentials carried in the oxidized state, expressed as E_0', a cytochrome bearing a higher value being capable of donating an electron to the next lower member in the series (Kagawa, 1976). Although the essential features are uniform throughout the higher eukaryotes, variations in details abound, as for example in the trypanosomatids (Martin and Mukkada, 1979).

Being involved in so many essential cellular processes, the cytochromes of all sorts have received extensive attention from molecular biologists, so that now they must rank among the most thoroughly studied of proteins. This is especially true of the cytochromes *c*, a component more readily extractable than the others—a full analysis of the amino acid sequences of cytochromes *c* that have been established could readily fill many pages (for sequences see Dayhoff, 1972, 1973, 1976; Boulter, 1973; Lemberg and Barrett, 1973; McLaughlin and Dayhoff, 1973; Brown and Boulter, 1974; Carlson *et al.*, 1977; Niece *et al.*, 1977). Hence, extensive consolidation of the available information has been essential in this presentation.

* Such attached active centers as the heme of hemoglobin are referred to in chemistry as prosthetic groups; the protein deprived of its prosthetic group is termed the apoprotein or apoenzyme, and the apoprotein plus the prosthetic group is the holoenzyme.

Distinctive Characteristics. Currently the cytochromes of type *c* (Vanderkooi and Erecińska, 1976) are defined as proteins having one or more protohemes IX, the prosthetic group being covalently bound to the polypeptide chain by thioether linkages (Figure 8.10C; Marks, 1969). These bonds result from condensation of vinyl groups of the heme with the sulfhydryl groups of cysteine in the protein chain (Lemberg and Barrett, 1973; Bartsch, 1977; Salemme, 1977). However, the primary structures that have been established show that several subclasses are embraced by this definition (Dickerson and Timkovich, 1975; Ambler *et al.*, 1976; Dickerson *et al.*, 1976), but the subclass on which attention first centers consists of polypeptide chains 85 to 135 residues in length that carry a single heme group near the amino terminus. In addition, the members have histidine and methionine serving as axial ligands of the heme iron (Figure 8.10) and high oxidoreduction potentials between +150 and +380 mV. Among the chief members are included cytochrome *c* proper, cytochrome *f*, and various others currently bearing such subscript designations as cytochromes c_2, c_{551}, and c_{555}. Some of the distinctive traits of a number of variants are given in Table 8.1.

Cytochromes c Proper. Table 8.2 summarizes in comparative form representative amino acid sequences of cytochromes *c* and related types from a broad spectrum of organisms. When examined, it is immediately obvious that such subscript designations as c_2 and c_{550} are not only unnecessary but undesirable, for they suggest distinctions that are no more real than those of other instances of ancestor–descendant relationships. That of purple bacteria referred to as c_2, for a case in point, shows homologies with that of yeast and other eukaryotes throughout the molecule—a count indicates that correspondences between this sequence and that of yeast or eukaryotes in general occur at 72 sites, including almost all the eukaryotic invariable sites (Yockey, 1977). A similar statement can be made for *Paracoccus* cytochrome c_{550} (Timkovich and Dickerson, 1976) and for the cytochromes c_2 from two genera of nonsulfur purple bacteria (Ambler *et al.*, 1976), but to a slightly lesser degree. Even in the sequence of *Pseudomonas* cytochrome *c*, where resemblances are still fewer in number, correspondences are nevertheless sufficient to indicate homology. Comparable conclusions have been drawn from studies of structural homologies of bacterial cytochromes c_2 and eukaryotic *c* (Dickerson *et al.*, 1976; Cookson *et al.*, 1978), but the prokaryotic type appears to function in photosynthesis as well as respiration (Baccarini-Melandri *et al.*, 1978).

Actually a better case for a distinctive subscript designation could be made for *Euglena* and *Crithidia* cytochromes *c*, for those molecules differ remarkably from all others of the class in having the heme moiety attached by only one cysteine, instead of the two otherwise universally present (Pettigrew *et al.*, 1975). Because such designation could interfere with its being viewed as homologous with others of eukaryotic origin, no such distinction is recommended. Unlike that of *Euglena*, the *Crithidia* sequence shows many points of similarity

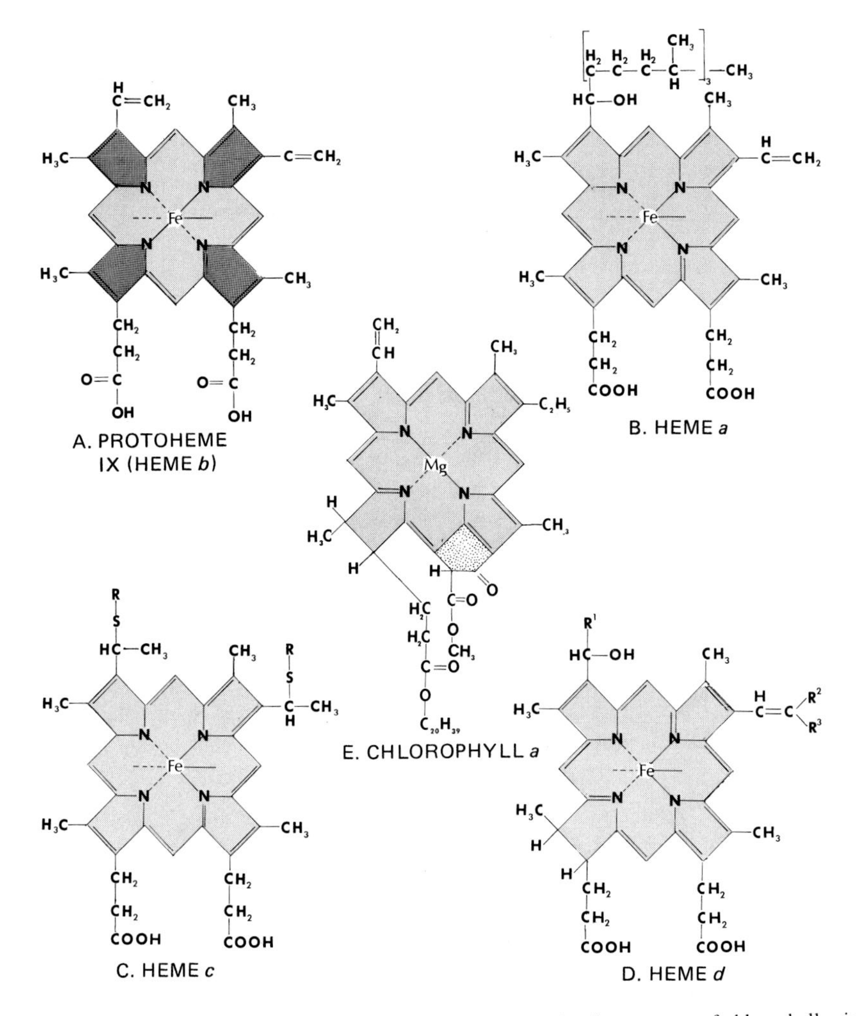

Figure 8.10. The important hemes of the cytochromes. The molecular structure of chlorophyll *a* is also provided to show the similarities of structure.

with that from the green alga, *Enteromorpha* (Meatyard and Boulter, 1974). It is pertinent to note that little intraspecific heterogeneity has been observed in the cytochromes *c*, one known exception being the protein from mammalian spermatocytes (Goldberg *et al.*, 1977), another being that from the hippopotamus (Thompson *et al.*, 1978). Mules and hinnies likewise carry equal proportions of horse and donkey cytochromes *c*, but that condition is a consequence of their hybrid origins (Walasek and Margoliash, 1977).

Table 8.1
Characteristics of Cytochromes c and Related Molecules

Cytochrome	Spectral characteristics: Reduced (nm) α	Reduced (nm) β	Reduced (nm) γ	Oxidized (nm) γ	Molecular weight	E'_0 (V) (at pH 7)	Occurrence	Characteristic traits
c	550	521	415	407	13,000	+0.254	Virtually universal	Readily extracted
c_1	553	523	418	410	37,000	+0.22	Metazoan mitochondria	Tightly bound
c_2	550-552	521-525	415-418	410-416	14,000-32,000	+0.25-+0.32	Photosynthetic and denitrifying bacteria	—
c_3	552-553	521-525	418	410	13,000	-0.205	Sulfate-reducing bacteria	Autoxidizable; here named cytochrome *s*
c_4	551	522	414	414	—	+0.30	*Azotobacter* and other aerobic bacteria	—
c_5	555	524-526	418-420	414	—	+0.32	*Azotobacter* and other aerobic bacteria	—
c_{556}	556	526	422	409	18,500	—	Snail hepatopancreas	—
f	555	525	423	413	110,000	+0.365	Chloroplasts	Serves in photosynthesis

Table 8.2
Cytochromes of the c Type[a]

Cytochrome	Sequence (10 20 30 40 50)
Pseudomonas c	EDPEVLFKNKGCVACHAIDTKMVGPAYKDVAAKFAGQAGAEAG
Paracoccus c_{550}	NGNAAKGEKEFNK-CKACHMIQAPDGTDIKGGKTGPNLYGVVGRKI
Purple bacteria c_2	AGDAVKGEQVFKQ-CKICHQVGPGAKNKVGPELNGVFGRKAGARPG
Saccharomyces c	TEFKAGSAKKGATLFKTRCELCHTVEKGGPHKVGPNLHGIFGRHSGQAQG
Euglena c	+GDAERGKKLFESRAAQCHSAQKGV-NSTGPSLWGVYGRTSGSVPG
Neurospora c	GFSAGDSKKGANLFKTRCALCHGEGGNLTQKIGPALHGLFGRKTGSVDG
Crithidia c	PKAREPLPPGDAAKGEKIFKGRAAQCHTGAKGGANGVGPNLFGIVNRHSGTVEG
Moth *(Samia) c*	GVPAGNAENGKKIFVERCAQCHTVEAGGKHKVGPNLMGFYGRKTGQAPG
Vertebrate *c*	+GDVEKGKKIFVQKCAQCHTVEKGGKHKTGPNLHGLFGRKTGQAPG
Enteromorpha c	+STFANAPPGNPAKGAKIFKAKCAQCHTVDAGAGHKQGPNLNGAFGRTSGTAAG
Wheat *c*	+ASFSEAPPGNPDAGAKIFKTKCAQCHTVEAGAGHKQGPNLHGLFGRQSGTTAG
Generalized *f*	GDDANGEKVFSANCAACHAGGGNVIMAGKTLKKSDAEEYLANGMD
Euglena f	GGADVFADNCSTCHVNGGNVISAGKVLSKTAIGGYLDGGY-
Blue-green alga *f*	ADAAAGGKVFNANCAACHASGGGQINGAKTLKKNALT---ANGKD

[a] +, Acetyl residue; underscores in the generalized *f* sequence indicate invariable sites. A heme-attachment group is bracketed.

Although it has been established that cytochrome *c* is synthesized in the cytoplasm in eukaryotes, its precise site of formation has remained in doubt. Recently it has been shown that, contrary to earlier reports, this substance is not biosynthesized on the ribosomes of the rough endoreticulum (Robbi *et al.*, 1978a). Further, it was demonstrated that cytochrome *c* is either produced adjacent to mitochondria and transferred immediately to those organelles, or it is generated by cytoplasmic ribosomes attached to the outer mitochondrial membrane (Robbi *et al.*, 1978b). Similar rapid transport into the mitochondrion from the cytosol site of synthesis also was detected in *Neurospora*, but in this case the apocytochrome was clearly confined to the cytosol and the holoenzyme to the mitochondrion (Korb and Neupert, 1978).

Cytochromes f. The only true *c* type that is obviously distinct from the others is that known as cytochrome *f*, a chemical active in photosynthesis, in the discussion of which topic it receives further attention (Chapter 10). Relationships to the cytochromes *c* of respiration are indicated by the manner of attachment of the prosthetic group and by the location of that point in the molecule. Moreover, a number of its invariant sites, italicized in the generalized *f* sequence in Table 8.2, correspond to those of the cytochromes *c* that are cited. However, the few identities are confined to the amino terminal region of the polypeptide (Ambler and Bartsch, 1975; Laycock, 1975), and although more could be suggested by inserting a deletion here or there, this action would add little to the general absence of homology that exists in the greater portion of the molecules. On both structural and functional bases, then, cytochrome *f* warrants being recognized as a separate subclass of this group of substances (Table 8.2). This statement is reinforced by further comparisons of the primary structures of cytochromes *c* and *f* from *Euglena*, both of which sequences are provided in the table. The shortening of the sequence of *Euglena* cytochrome *f* at the amino end and its prolongation at the carboxyl region are especially puzzling. This is particularly so in light of the similarity that exists between the blue-green algal cytochrome *f* and generalized sequences (Aitken, 1977).

Cytochrome c_{553} ($_{550}$) from *Desulfovibrio desulfuricans* may warrant the erection of a new subclass to receive it and related species, but the partial amino acid sequence that has been established is too short (only 14 sites) to permit the drawing of any firm conclusions (Fauque *et al.*, 1979). Another distinctive type seems to be represented by the cytochrome c_{553} of *Chlorobium*, which has its heme attached by a single cysteine residue (Yamanaka, 1976). A still more important distinction in this molecule is the presence of two subunits, one of molecular weight 11,000 being a polypeptide that bears the heme, the other of 47,000 being a flavoprotein. As cytochrome c_{552} of *Chromatium* also consists of two subunits, one of which is a flavoprotein with a molecular weight of 46,000, it may prove to be related to that of *Chlorobium* (Fukumori and Yamanaka, 1979).

Cytochrome c_1. Another cytochrome of the *c* type that appears to be a valid member of the group is cytochrome c_1, a strictly eukaryotic variety, as far as known. Knowledge of this enzyme has not increased rapidly, largely because it has proven difficult to extract intact from its close association with cytochrome *b* and the other components of the respiratory chain. The molecule, with a molecular weight of 31,000 (Ross and Schatz, 1976a), has a single heme attached to the polypeptide chain by thioether linkages, but consists of two subunits arranged as an $\alpha\beta$ combination; the α subunit, which is about twice as heavy as the β, bears the heme (Dickerson and Timkovich, 1975). Furthermore, its amino acid composition bears no resemblance to that of vertebrate cytochrome *c*; proline and leucine are especially abundantly represented and 12 cysteine and 14 histidine residues are present compared to the two of each in cytochrome *c*. However, one could scarcely expect this pair to be too similar in composition, for c_1 is the reductase of *c* and necessarily precedes it in the respiratory chain. In yeast, as is the case with cytochrome *c*, the present one is synthesized on the ribosomes of the cytosol (Ross and Schatz, 1976b).

Modified Bases. Modified bases are few in the cytochromes as a whole, including cytochrome *c*. One that has proven to be rather common is ϵ-*N*-

trimethyl lysine. This base, first detected in wheat germ and *Neurospora* cytochrome *c* (DeLange *et al.*, 1969), has subsequently been found in the same substance from a number of fungal and higher plant sources (Boulter *et al.*, 1970; DeLange *et al.*, 1970). The cytochrome-*c*-specific methylase III responsible for the modification has been identified in *Neurospora* and was found to use *S*-adenoxyl-L-methionine as the methyl donor (Nochumson *et al.*, 1977; DiMaria *et al.*, 1979).

A Proposed New Class of Cytochromes. Another type of cytochrome, currently passing under the designation $\mathbf{c}_3$, is entirely distinct from those enumerated above. The only relationship displayed with the foregoing is that the heme group (protoheme IX) is bonded to the polypeptide by way of thioester linkages, supplemented by histidine. Thus it fits the definition of members of this class given on a preceding page. Here it is proposed that that definition is too broad, embracing nonhomologous substances; accordingly it should be redefined to include only those cytochromes that have a single heme group. On the other hand, those like cytochrome c_3 that contain three hemes would be placed in a new category and those like cytochrome c_4 of *Azotobacter* probably merit similar treatment (Campbell *et al.*, 1973), when they become better known.

This new category containing these former cytochromes c_3 having three hemes, presumably attached by way of thioester linkages, should be referred to as cytochromes *s*, the latter referring to sulfate reducing, for all known representatives of this class occur in sulfate-reducing bacteria. Several genera of such organisms are presently known, including *Desulfatomaculum, Desulfovibrio,* and *Desulfuromonas* (Fauque *et al.*, 1979). The latter two genera appear to include related species, for neither forms spores as the first does, and both contain cytochrome *s*, a protein lacking in the other. Amino acid sequences of cytochromes *s* from three species of *Desulfovibrio* have now been determined, those from *D. desulfuricans* (Ambler *et al.*, 1971), *D. gigas* (Ambler *et al.*, 1969), and *D. vulgaris* (Ambler, 1968; Trousil and Campbell, 1974), all of which are reproduced in Table 8.3. The homology that exists between these three is immediately apparent, for all contain four possible sites for the binding of prosthetic groups, although only three hemes actually are present. Usually in two of the putative heme-binding sites, the cysteine residues are separated only by two sites, as in typical cytochromes *c*, but four intervene between the cysteines of the other two, except in *desulfuricans* in which three binding sites are typical and only one atypical (Ambler *et al.*, 1971). The alignment of the trio of sequences is modified from that of the latter reference (Table 8.3).

Examination of those sequences discloses, despite their obvious homology, that they display an amazing number of distinctions from one another, considering that all are from species assigned to the same genus (Postgate and Campbell, 1966). This points out again the absence of close kinship among

Table 8.3
Primary Structures of Cytochromes s[a]

	10 20 30 40 49
D. desulfuricans	VDAPADMVIKAPAGAKVTKPVAFSHKGHASMDCKTCHHKWDGAGAIQP
D. gigas	VDVPADGAKIDFIAGGEKNLVVFNHSTXKDVKCXDCHHZPGXKG-YRK
D. vulgaris	APKAPADG----LKMEATKQPVVFNHSTHKSVKCGDCHHKWDGAGAIQP
	60 70 80 90 100
D. desulfuricans	CQASGCHANTESK-KGDDSFYMAFHERKSE-KS-CVGCHKSMKKGPTK-----
D. gigas	CTTAGCHNILDKADKSVNSWYKVVHDAKGGAKP-CISCHKDKAGDDKELKKK-
D. vulgaris	CGTAGCHDSMDKKDKSAKGYYHVMHDKNTKFKSTCVGCHVEVAGADAA-KKKD
	110
D. desulfuricans	---CTE--CHPKN
D. gigas	LTGCKGSACHPS
D. vulgaris	LTGCKKSKCHE

[a]Based on Ambler (1968), Ambler *et al*. (1969, 1971), and Trousil and Campbell (1974). Heme-attachment groups are bracketed.

many bacteria that has already been shown in earlier discussions. In the present case, this lack is substantiated by the marked differences that exist also in the DNA base compositions (Sigal *et al.*, 1963; Saunders *et al.*, 1964). Further, it is nearly needless to point out that almost no homologies can be found between these three species of cytochromes *s* and the cytochromes *c* and *f*.

8.2.2. Cytochrome c Oxidase

The substance variously referred to as cytochrome oxidase, cytochrome *c* oxidase, and cytochrome *c*:oxygen oxidoreductase is the terminal member of the eukaryotic electron-transport chain (Wikström *et al.*, 1977). It forms an integral part of the mitochondrial inner (cristal?) membrane, where it may represent as much as 20% of the total protein (Capaldi and Briggs, 1976). In its position at the end of the chain, it catalyzes the reaction:

$$4H^+ + 4e^- + O_2 \rightleftharpoons 2H_2O + \Delta$$

Under normal coupled conditions, much of the energy that results is not released as heat but is employed by the ATP synthetase complex to form ATP, the interaction with cytochrome *c* being carried out in several steps (Petersen and Andréasson, 1976).

Cytochrome c Oxidase as a Complex. Cytochrome *c* oxidase is a complex in more than one sense. First, it is a duplex that can be resolved into an electron-transfer and an ionophore-transfer complex (Kessler *et al.*, 1977; Wikström, 1977, 1978; Fry and Green, 1979). Coupling requires an interaction between the moving electron in the first of these components and a moving, positive-charged ionophore–cation adduct in the second, more briefly called the ion-transfer complex (Fry and Green, 1979). Thus the energy-coupling systems are resolvable into exergonic (energy-driving) and endergonic (energy-driven) units, the two necessarily being in close juxtaposition for functioning (Green and Blondin, 1975). The ion-transfer complex has been demonstrated to consist of three of the six or seven subunits that comprise the whole molecule, namely subunits I, II, and III that are encoded by the DNA of the mitochondrion (Mason and Schatz, 1973; Rubin and Tzagoloff, 1973); however, only subunit I is the actual ionophore (Fry and Green, 1979). In addition to these subunits, the complex is embedded in a matrix of phospholipids, some of which are cardiolipins (Awasthi *et al.*, 1971).

The second portion of the duplex molecule, the electron-transfer complex that consists of the three or four remaining subunits, contains equimolar amounts of heme and copper (Palmer *et al.*, 1976). In the minimum functional unit, it is probable that 2 moles each of heme and copper are present to accommodate the four electrons cited in the equation in a preceding paragraph (Rosén *et al.*, 1977; Wikström and Saari, 1977). Moreover, four equivalents of the reducing agent are required to reduce the oxidized cytochrome *c* oxidase completely (Mackey *et al.*, 1973), so both metal ions are believed to be oxidized, presumably as Fe^{3+} for the heme iron and Cu^{2+} for the copper. One of the two hemes present reacts with carbon monoxide and is referred to as heme a_3, whereas the other does not and is distinguished as heme *a* (Figure 8.10B; Capaldi and Briggs, 1976). Thus two distinct cytochromes, known as cytochromes *a* and a_3 (Table 8.4) are considered in some quarters to be present, whereas other researchers in the area do not—even the purification and elucidation of subunit structure have not completely resolved the question (Briggs *et al.*, 1975).

Subunit Structure. As already has been intimated, this enzyme consists of seven to ten subunits, depending on the taxon, which in the aggregate have a molecular weight between 200,000 and 250,000. Beef heart cytochrome *c* oxidase usually had been considered to be constructed of six subunits, the individual weight of which varied with the author. The largest one had been reported to have a weight close to 30,000, followed in turn by polypeptides weighing 22,300–25,000, 15,000–19,000, 11,200–14,000, 9,000–10,000, and 5300–8000 (Kuboyama *et al.*, 1972; Rubin, 1972; Rubin and Tzagoloff, 1973; Yamamoto and Orii, 1974; Briggs *et al.*, 1975). More recently ten subunits

Table 8.4
Properties of Cytochromes a

Cytochrome	Spectral characteristics: Reduced (nm) α	β	γ	Oxidized (nm) γ	Molecular weight	E'_0 (V) (at pH 7.0)	Occurrence	Characteristic traits
a	603–604	—	450	407	—	+0.29	Nearly universal; absent in lower prokaryotes	Not sensitive to CO or cyanide
a_1	590	—	—	—	—	—	Bacteria	Resembles a in ractivity
a_2	652	629	460	412	90,000	—	*Pseudomonas*	Contains a dihydroporphyrin; serves in place of $a + a_3$
a_3	600	—	445	—	—	—	Eukaryotes and higher prokaryotes	Sensitive to CO and cyanide; acts in conjunction with a
$a + a_3$	—	—	—	—	200,000–250,000	—	Eukaryotes and higher prokaryotes	Contains 1 atom copper per molecule

have proven to be present, plus 7% phospholipids (Ludwig *et al.*, 1979; Steffens and Buse, 1979). In mammals, the combination of enzymes spans the cristal membrane (Ludwig *et al.*, 1979), where it is immobile (Künze and Junge, 1977); further, electron paramagnetic resonance investigations have demonstrated that the hemes are oriented perpendicularly to the surface of the membrane (Erecińska *et al.*, 1977).

In yeast cytochrome *c* oxidase seven subunits have been found, with weights of 40, 33, 22, 14.5, 12.7, 12.7, and 4.6 × 10^3 (Poyton and Schatz, 1975; Phan and Mahler, 1976a,b). The smaller of these components have been shown to be exposed on the intact complex, while the 33,000 one is partly, and the largest one completely, buried (Eytan and Schatz, 1975). The three largest are synthesized in the mitochondrion and the rest in the cytosol as in mammals (Mason and Schatz, 1973; Saltzgaber-Müller and Schatz, 1978), and a similar arrangement has been found in *Neurospora* (Sebald *et al.*, 1972). Correspondingly, in the mammalian enzyme the two largest subunits have been demonstrated to be synthesized in the mitochondrion, but it has not proven possible as yet to show the points of origin for the remainder (Jeffreys and Craig, 1977; Yatscoff *et al.*, 1977).

The Primary Structure of Subunits. In view of the wide variations reported in the molecular weights of the subunits by different laboratories, it comes as no surprise that few of their amino acid sequences have as yet been determined. The only complete primary structure is of one subunit from beef heart; however, this is an especially important component, as it contains the heme *a* group (Tanaka *et al.*, 1977; Yasunobo *et al.*, 1979); it has a molecular weight of 11,600 and is 109 residues in length, as shown in Table 8.5. Unfortunately the prosthetic group was removed by the preparatory steps, so its exact site in the molecule remains unknown. A number of potential heme ligands

Table 8.5
The Primary Structure of the Heme-a-Containing Subunit of Beef Heart Cytochrome Oxidase[a]

```
         10              20              30
SHGSHETDEEFDARWVTYFNKPDIDAWELR

         40              50              60
KGMNTLVGYDLVPEPKIIDAALRACRRLND

         70              80              90
FASAVRILEVVKDKAGPHKEIYPYVIQELR

        100
PTLNELGISTPEELGLDKV
                   COOH
```

[a] Based on Tanaka *et al.* (1977).

were found, including cysteine (at site 55), methionine (site 33), three histidines (sites 2, 5, and 78), and seven lysines (sites 21, 31, 46, 72, 74, 79, and 108). Although 20% homology was claimed, it actually showed only 10% with the hemoglobin β chain with which comparisons were made. Another small polypeptide, known as VIIIa, has been analyzed, but its sequence has not been established (Buse and Steffens, 1978).

Bacterial Terminal Oxidases. Bacterial equivalents of the eukaryotic cytochrome *c* oxidase are quite numerous and diversified, but as a whole are lamentably poorly known. One of the better explored enzymes is the cytochrome *c* oxidase of *Pseudomonas*, whose properties approach those of the eukaryotic one at many points. Like it, this cytochrome has two heme prosthetic groups, but in this case, one each of heme *c* and *d*, the latter being autoxidizable like a_3. Also as in that enzyme, it is sensitive to CO and cyanide (Horio, 1958a,b). However, no copper nor metal other than iron is present, and the molecule is somewhat smaller, having a molecular weight of 120,000 (Parr *et al.*, 1974). Moreover, only two subunits exist, one bearing the heme *c*, the other the heme *d* (Kuronen and Ellfolk, 1972).

This oxidase has been shown to receive electrons from two other oxidation–reduction components of the bacterial chain, a small cytochrome *c* and the blue copper-containing protein called azurin (Horio *et al.*, 1961; Parr *et al.*, 1976). Either of these appear to function with it in catalyzing the four-electron reduction of O_2 to H_2O and the one-electron reduction of NO_2 to NO (Yamanaka *et al.*, 1961; Barber *et al.*, 1976). The latter activity, its more important function *in vivo,* was reported to be inhibited by cyanide but was unaffected by CO; because of this role, the enzyme is also known as nitrite reductase (Yamanaka and Fukumori, 1977). Eukaryotic cytochromes *c* have been shown to react very poorly with this prokaryotic oxidase, whereas corresponding bacterial enzymes in general reacted rapidly with it (Yamanaka, 1972, 1975). Similar reactions were obtained with cytochrome *c* oxidase from *Thiobacillus novellus*, although it has heme *a* as the prosthetic group (Yamanaka and Fukumori, 1977). Still another type has been isolated recently, in this case from *Photobacterium phosphoreum*; this was found to possess both a protoheme and a heme *d* as prosthetic groups (Watanabe *et al.*, 1979).

One other important terminal cytochrome, cytochrome *o*, is classed as a *b* type, and is accordingly described in the following section.

8.2.3. The Cytochromes b

The cytochromes of type *b* form a large and complex group, having diverse functions alike in prokaryotes and eukaryotes (Table 8.6). Some of the members possibly will prove sufficiently distinct when their primary structures have been fully determined to warrant their placement in separate groups, as proved to be the case with certain former members of type *c*. At present the group is defined as electron-carrying proteins that have protoheme IX as the prosthetic group and the location of the α band of the absorption spectra in the ferrous state ranging from 554 to 560 nm (Hagihara *et al.*, 1975). As a whole the nature of the ligands bonding the heme group is unknown, except that thioether linkages are not involved. In the variety known as cytochrome b_5, two widely separated histidine residues serve as ligands, as shown later. Table 8.6 summarizes certain of the physicochemical properties of the principal members of this type.

Respiratory Cytochromes b. Because the main focus of the present discussion is on cell respiration, it appears appropriate to describe the several cytochromes *b* on a functional basis, at least insofar as present knowledge permits. This approach facilitates comparisons of structure, about which almost no details have been firmly established. However, it has been determined that at least two species of cytochromes *b* are present in the intact mitochondria of metazoans, metaphytans, and yeasts, which differ kinetically (Chance *et al.*, 1970), potentiometrically (Dutton *et al.*, 1970, 1971), and spectroscopically (Slater *et al.*, 1970; Sato *et al.*, 1971; Wikström, 1971).

Metazoan Respiratory Members. The heterogeneity of type *b* cytochromes in mammalian mitochondria was first demonstrated by Chance (1958), but even today the number actually present has not become firm. Three clearly exist. Classical cytochrome *b*, sometimes referred to as cytochrome b_K, is reduced with succinic acid even in the absence of ATP, whereas cytochrome b_T is reduced with that substance only when ATP is present. Both of these are associated with Complex III (CoQ · H_2-cytochrome *c* reductase) of the respiratory chain (Davis *et al.*, 1973), and neither combines with CO. The third confirmed species, cytochrome b_5, is not a member of the respiratory chain but is located in the outer membrane of the mitochondrion; hence, it is considered more fully in a subsequent section. A fourth one, unfortunately referred to as sulfite oxidase, is located between the outer and inner mitochondrial membranes but is a member of this class of cytochromes (Ito, 1971). Still another species may occur, a component of Complex II (succinate-CoQ reductase) of the respiratory chain, most conveniently designated as cytochrome b_{558} (Lindsay *et al.*, 1972; Berden and Opperdoes, 1972).

Other Respiratory Cytochromes b. Three kinds of cytochromes *b* also occur in the higher plants (Chance *et al.*, 1968; Bonner and Slater, 1970), cytochromes b_{565}, b_{560}, and b_{556}, also called b_{562}, b_{557}, and b_{553}, respectively, depending on the criterion employed. The second of these, cytochrome b_{560}, corresponds functionally to cytochrome *b* of the metazoans and shows similar spectral and potential properties, while the first is probably homologous to cytochrome b_T. In contrast, the third one has no clear-cut counterpart in animals.

The cytochromes *b* of bacteria are a highly diverse and complex group, the molecular properties of which are largely unexplored. One exception to that

Table 8.6
Characteristics of the Various Cytochromes b

Cytochrome	Spectral characteristics: Reduced (nm) α	β	γ	Oxidized (nm) γ	Molecular weight	E'_0 (V) (at pH 7.0)	Occurrence	Characteristic traits
b	562	532	429	418	28,000	+0.05	Nearly universal	—
b_1	559-560	528-530	426-428	415	—	+0.25	Bacteria	The common *b*-type cytochrome of bacteria
b_2	557	528	424	413	170,000	—	Yeast	Functions as a lactic dehydrogenase, with a low-molecular-weight DNA (10,000-20,000)
b_3	559	529	425	—	40,000	+0.04	Metaphytan microsomes	Readily extracted; autoxidizable
b_4	554	521	415	—	—	+0.19	Salt-tolerant bacteria	Often considered the *c*-type cytochrome, c_{554}
b_5 (=*m*)	556-560	525-526	423-424	413-415	16,900-100,000	+0.02	Metazoan microsomes and liver mitochondria	—
b_6	563	—	—	—	—	-0.06	Chloroplasts	Functions in photosynthesis
b_7	560	529	—	—	—	-0.03	Widespread in metaphytons	Autoxidizable
o	568	535-537	416-418	430?	—	—	*Acetobacter*, *E. coli*	Reduced by *c*, oxidized by O_2
P-450	450	—	411	417	45,000	-0.24	Endoreticulum and mitochondria; *Pseudomonas*	A monooxygenase; binds CO

condition is cytochrome b_{556} of *E. coli*, which now has been obtained in a highly purified state (Kita *et al.*, 1978). This oligomer of identical polypeptides having molecular weights of 17,500 had a total weight of approximately 65,000 and carried one heme group per holoenzyme. Its oxidation–reduction potential of − 45 mV suggests that it serves as the first cytochrome of the electron-transport chain and thus parallels the cytochrome *b* of metazoans functionally.

Cytochrome o. As pointed out earlier, a *b*-type cytochrome serves as the terminal oxidase of the respiratory chain in a number of prokaryotes as well as in a few protistans (Webster and Hackett, 1966; Perlish and Eichel, 1971; Srivastava, 1971). This cytochrome *o*, however, has been far more thoroughly characterized from bacterial sources, to which discussion is necessarily confined. For most studies, more specifically, the substance from *Vitreoscilla* has provided the basis and has been shown to exist in three chemical states in living cells. The reduced state occurred under anaerobic conditions and was replaced by an oxygenated state in respiring cells. In turn, the latter was converted predominantly into the oxidized state in starved cells (Webster and Orii, 1977), during which reaction hydrogen peroxide was produced (Tyree and Webster, 1979a,b). This is the first known instance of the existence of an oxygenated state in a cytochrome. *Vitreoscilla* cytochrome *o* consisted of two identical polypeptide chains of 13,000 daltons, each bearing a molecule of protoheme IX that interacted with cyanide (Tyree and Webster, 1978a,b, 1979a,b).

The dual polypeptide chains and hemes of this cytochrome possibly indicate an ancestor–descendant role with cytochrome *c* oxidase of the eukaryotes, albeit a remote one. To make the transformation, the polypeptide would have had to increase in size by evolutionary means, the protoheme IX would have had to evolve to provide the distinctions noted in the eukaryotic heme *a*, and two copper atoms would have had to be acquired. Thus it could appear that so few real differences at the molecular structural level exist between cytochromes *a* and *o* that they may need eventually to be combined as type *a*. But this would depend upon the existence of a large degree of homology in their primary structures.

Membrane Cytochromes b.

A protoheme-bearing protein exists abundantly in the endoreticulum of mammalian liver cells that was first referred to as cytochrome b_1 (Keilin and Hartree, 1940). Now, however, it is called cytochrome b_5 and has been found also in Golgi and in the outer membrane of the mitochondrion (Hagihara *et al.*, 1975).

Properties of Cytochrome b_5. With a molecular weight of 11,000, cytochrome b_5 compares with classical cytochromes *b* and *c* in size (Rashid *et al.*, 1973; Mathews and Czerwinski, 1976). *In vivo* it is tightly bound to the membrane, where it becomes reduced by a flavoprotein, cytochrome b_5 reductase, which is similarly firmly bound to the membrane. This electron-transporting substance appears to vary widely from one source to another, for rat liver material shows an asymmetrical α-absorption band having a peak at 556 nm, whereas that from metaphytan sources (mung bean) is at 555 nm (Shichi and Hackett, 1962). A corresponding species from *Aspergillus* has the α band at 554, but that from *Neurospora* is at 556. Tissue differences have also been reported (Ichikawa and Yamano, 1965). Moreover, the enzyme from *Ascaris* differs quite strongly, especially in the molecular weight of 100,000 (Matuda, 1979a,b), compared to 60,000 for the mammalian cytochrome (Nisimoto *et al.*, 1977).

The heme group is attached to the molecule by way of two histidine residues, in which imidazole linkages are involved (Ozols and Strittmatter, 1968; Nóbrega *et al.*, 1969). Because the two residues are located similarly to those linking the heme in the β chain of hemoglobin, it has been suggested that the latter is probably a derivative of the former (Strittmatter and Huntley, 1970). However, the present heme protein is also a glycoprotein (Elhammer *et al.*, 1978). It seems to be synthesized only on the ribosomes of rough endoreticulum (Harano and Omura, 1977), where at least one moiety each of glucosamine and galactose are added to the hydrophilic region of the molecule. Where the remaining part of the oligosaccharide chain is completed has not been determined, but it seems to be in the endoreticular membranes. In the rough endoreticulum of mammary gland epithelial cells, cytochrome b_5 was found to be complexed with another cytochrome called P-420 (Bruder *et al.*, 1978).

The roles of this cytochrome in the membranes have not been fully established. One activity that seems clear is its participation in cholesterol biosynthesis, mainly in introducing the double bond on carbon 5 (Reddy *et al.*, 1977). In these processes it serves as part of an electron-transport chain that includes NADH, a flavoprotein, itself, and cytochrome *c*, usually with a cyanide-sensitive factor and O_2.

Primary Structure of Cytochrome b_5. The primary structures of cytochrome b_5 from five species of mammals and that of chicken have been determined (Ozols and Strittmatter, 1969; Tsugita *et al.*, 1970; Nóbrega and Ozols, 1971; Hagihara *et al.*, 1975). A generalized mammalian species, deduced from the five by averaging and by employing the bovine type in regions where the others were lacking, is shown along with the avian species in Table 8.7, examination of which at once discloses the extensive homology that exists—69 sites have identical residues. With these is included the sequence of the so-called cytochrome b_2 of yeast (Guiard *et al.*, 1974). That this species deserves to be considered homologous to the others is clearly indicated by the occurrence of 31 identical occupants of corresponding points. Experimental shortening of the carboxyl end of the mammalian molecule has indicated that the last 20 or so sites at that end are involved in binding the molecule to the membrane (Dailey and Strittmatter, 1978).

Cytochrome b_5 of Yeast. Because of the obvious relationships of the heme-containing portion of the lactate dehydrogenase of yeast, formerly called cy-

Table 8.7
Primary Structures of Cytochromes b_5 and b_2[a]

	Residues 1–50 (positions 10, 20, 30, 40, 50 marked)
Mammalian b_5	NH_2 Q A A S D K A V K Y Y T L E **E** I E **K H N** N S K S T **W** L I L H H K **V Y D L T** K **F L** E E **H P G G** E E **V** L
Avian b_5	NH_2 G R Y Y R L E **E V** Q **K H N** N S Z S T **W** I I V H H R I **Y D** I **T** K **F L** D E **H P G G** E E **V** L
Yeast b_2	NH_2 K Q K I S P A **E V** A **K H N** K P D D C **W** V V I N G Y **V Y D L T** R **F L** P N **H P G G** Q D **V** I

	Residues 51–100 (positions 60, 70, 80, 90, 100 marked)
Mammalian b_5	R E Q **A G** G **D** A **T** E D **F E** D V G **H** S T **D** A R E L S K T F - I I **G** E **L** H P D D R **P** K **L** T K **P** S E S_{COOH}
Avian b_5	R E Q **A G** G **D** A **T** E D **F E** D V G **H** S T **D** A R A L S E T F - I I **G** E **L** H P D D R **P** K **L** R_{COOH}
Yeast b_2	K F N **A G** K **D** V **T** A I **F E** P L - **H** A P **D** V I X K Y I A P Q K L **G** P **L** E G S M P **P** E **L** V C **P** P Y A G E T K_{COOH}

[a] Based on Ozols and Strittmatter (1969), Tsugita *et al*. (1970), Nóbrega and Ozols (1971), Guiard *et al*. (1974), Hagichara *et al*. (1975). The shading points out homologous sites of b_5 and so-called b_2.

tochrome b_2, it is referred to as cytochrome b_5 in the remainder of the discussion (Guiard *et al.*, 1974). This treatment is further supported by the report that the heme ligand is either one or two histidine residues (Groudinsky, 1971), as in the remainder of this class. As in the other species of the cytochrome, the present one in part occurs in the mitochondria, where its chief role is the mediation of the electron transfer from L-lactic acid to the cytochrome *c* of the respiratory chain (Ohnishi *et al.*, 1966). Furthermore, it is complexed with flavin mononucleotide (FMN) and perhaps DNA, but the latter does not appear essential to its normal functioning (Hagihara *et al.*, 1975).

The nature of the complex has not been fully resolved, but the present trend is toward considering the protein to consist of two subunits, one of molecular weight of 34,500, the other of 22,000 (Pajot and Groudinsky, 1970; Mével-Ninio, 1972). Each subunit is a globulin, joined to the other in the intact enzyme by a bridge that is preferentially cleaved by many proteases. Data have now been presented that indicate that the larger α subunit contains the heme, that is, it consists in part of the cytochrome b_5 (Mével-Ninio *et al.*, 1977). Apparently in the intact enzyme, it is located at the amino terminus of the α chain (Guiard *et al.*, 1975). Thus it is probable that the β subunit contains the flavin moiety. In functioning, the electrons follow a path from L-lactic acid to FMN and then to the heme (Morton and Sturtevant, 1964; Thusius *et al.*, 1976), and finally to cytochrome *c* (Pajot and Claisse, 1974), the lactic acid being oxidized to pyruvic acid (Blazy *et al.*, 1976).

Cytochrome P-450. Another heme protein that occurs in membranes is known merely as cytochrome P-450, because of the uncertain nature of its heme (Mathews and Czerwinski, 1976; Wiseman *et al.*, 1978). This has been identified as a protoheme and accordingly this substance should be classed as a type *b* cytochrome (Omura and Sato, 1964; Yu *et al.*, 1974b). However, more recently the prosthetic group has been shown to be linked to the polypeptide chain by cysteine residues (Jefcoate and Gaylor, 1969; Stern and Peisach, 1976), so that it may need to be considered a cytochrome *c*. Thus it may be perceived that, while the present classification of the cytochromes has been satisfactory for many years, newer, more detailed knowledge seems to be necessitating a revision in the near future. In short, the scheme has been fairly adequate for strictly chemical needs, but it fails to provide the basis for grouping homologous and structurally related types required in biochemical and evolutionary biology.

The existence of multiple forms within the same tissues has hampered studies on the nature and occurrence of this cytochrome (Haugen *et al.*, 1975; Johnson and Muller-Eberhard, 1977a; Johnson *et al.*, 1979), most of which have been based on liver microsomes from mammalian sources, especially rat and rabbit. As the principal species is synthesized in abundance in phenobarbital-treated animals, it should be suspected of being an enzyme of the vesicular smooth endoreticulum (Chapter 6, Section 6.1.2). This isozyme has been variously designated as P-450$_{LM_2}$ and P-450*d* (Johnson and Muller-Eberhard, 1977a; Peterson *et al.*, 1977), while the major type induced by 2,3,7,8-tetrachlorodibenzo-*p*-dioxin has been referred to as P-450$_{LM_4}$ or P-450*c* (Johnson and Muller-Eberhard, 1977b,c; Peterson *et al.*, 1977; Chiang and Coon, 1979). Two minor forms, P-450*a* and P-450*b*, are induced also by the same reagent. A molecular weight of 60,000 has been cited for P-450*b*, 50,000 for P-450*a*, 54,500 for P-450*c*, and 51,000 for P-450*d* (Johnson and Muller-Eberhard, 1977a; Botelho *et al.*, 1979). Apparently all forms are one-electron acceptors (Peterson *et al.*, 1977), not two- as had been reported earlier, and *in vivo* appear to be clustered in groups around a single NADPH · cytochrome P-450 reductase within the endoreticular membrane (Peterson *et al.*, 1976). A fifth form, P-448, has been isolated from nuclei as well as from microsomes of mammalian livers treated with a number of polycyclic aromatic hydrocarbons (Thomas *et al.*, 1979).

These terminal components in a monooxygenase system that oxidizes drugs and other reagents introduced into the body are uniquely structured. Because the molecule consists of 16 subunits but only 8 hemes, it evidently is comprised of two different subunits, only one of which bears a protoheme (Tilley *et al.*, 1976). Apparently the cytochromes P-450 from such bacterial sources as *Pseudomonas* and *Bacillus megaterium* (Berg *et al.*, 1979) are similarly structured, for their chemical properties compare closely to those from mammalian tissues.

Photosynthetic Cytochromes b. In addition to cytochrome *f*, discussed in connection with cytochromes *c*, chloroplasts contain at least two cytochromes *b*.

Cytochrome b_6. The principal cytochrome *b* of chloroplasts is that known as b_6, discovered by Davenport (1952) and Hill (1954). In the metaphytans, its oxidation–reduction potential ranges between 0.0 and −60 mV, that of the blue-green alga *Nostoc muscorum* similarly being 0.0 mV (Knaff, 1977). This substance functions as an electron carrier in the noncyclic electron-transport pathway that is associated with that part of the photosynthetic processes called photosystem I, described in Chapter 10. Within that system it can be both oxidized and reduced in light-requiring reactions (Hind and Olson, 1966; Böhme, 1976), where it seems to occupy a position between cytochrome *f* and a substance known as ferredoxin (Knaff, 1977), to be discussed shortly. This autoxidizable enzyme with a molecular weight of around 40,000 is not reduced by ascorbic acid and does not combine with CO. Similar substances have been identified from a number of other algal sources, including *Euglena*, red algae, and green algae (Lemberg and Barrett, 1973).

Cytochrome b_{559}. The second major cytochrome *b* of chloroplasts is that first described as b_3 (Lundegårdh, 1964), but because another substance known by that same name is a soluble cytochrome of the endoreticulum, and this is membrane bound, the present one has been renamed as b_{559} (Boardman and Anderson, 1967). Its oxidation–reduction potential has been variously reported but appears to be around + 350 mV, except in *Euglena*, where it was found to be between + 300 and + 320 mV (Ikegami *et al.*, 1968). This autoxidizable enzyme has a molecular weight close to 30,000 and is completely reducible with ascorbid acid (Hagihara *et al.*, 1975). In chloroplasts, three forms of cytochrome b_{559} occur, which differ in their redox potentials; thus high-, middle-, and low-potential types are present (Wada and Arnon, 1971).

8.2.4. Other Components of Electron-Transport Chains

In addition to the cytochromes, electron-transport chains contain several substances of other chemical types, some of which are well known, whereas others of equal or even greater importance have been neglected, usually because of difficulties of purification. None of these types is represented by a wide diversity of species, as was the case with the cytochromes, and therefore they can be treated together in this one major section. Among the ones involved in respiratory-function chains are pyridine nucleotides, flavin-containing substances, quinones, and the ferredoxins mentioned earlier.

The Pyridine Nucleotides. The pyridines of the respiratory chain include two closely related chemicals that differ from one another only in the number of phosphate residues present. The basic member of this group is NAD (nicotinamide adenine dinucleotide), formerly also known as DPN and coenzyme I. In molecular structure this substance could be considered to consist of ADP to which a ribosylate nicotinamide residue is attached (Figure 8.11A). Doubtlessly this is one of the most ubiquitous of all hydrogen acceptors, occurring in almost all prokaryotes and every eukaryote that has been investigated; moreover, it is known to serve as an acceptor with no fewer than 250 dehydrogenases of a wide diversity of types (Dalziel, 1975). In accepting an electron or hydrogen, this substance is reduced to NADH, the amide radical of the nicotinamide moiety being the active receptor. However, it is also capable of carrying a second hydrogen, so it characteristically acquires two electrons or hydrogens to become NADH·H.

The closely related NADP, formerly also called coenzyme II, is identical to the former in composition except that it bears an additional phosphate group on the ribose moiety of the adenosine (Figure 8.11A). It too functions with innumerable dehydrogenases as an acceptor, in which reactions the corresponding radical does the actual accepting. Although these molecules differ relatively slightly, many dehydrogenases are highly specific as to which of the pair they react with, as has been intimated in relation to the peroxisome in the preceding chapter.

In vivo, NADP is synthesized from NAD, the two substances being readily interconvertible (Apps, 1970; Griffiths and Bernofsky, 1972; Kaplan, 1972).

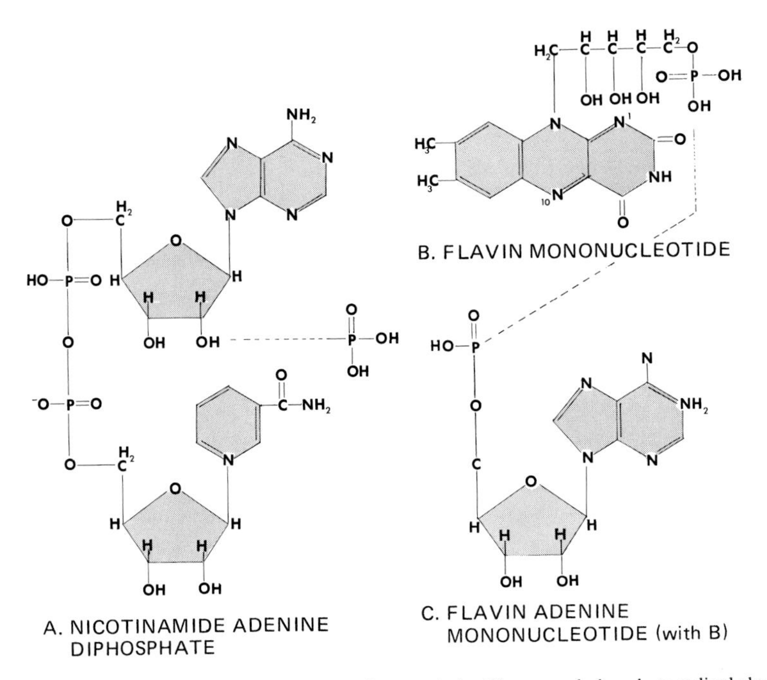

Figure 8.11. Important constituents of the respiratory chain. The second phosphate radical shown in (A) can unite with the NAD to form NADP, replacing the H of the hydroxyl group indicated.

Nevertheless, the former consistently occurs in lower concentrations in cells, usually being present at levels of 1 to 30% of the NAD pool (Kaplan, 1972). The higher concentrations are correlated to greater biosynthetic activities, but the metabolic factors involved in regulating the respective levels are not understood in any detail (Ting *et al.*, 1977). In early mouse embryos, about one-third of the *de novo*-formed NAD is converted into NADP (Kuwahara and Chaykin, 1973). Moreover, NADH proved to be only a minor source of NADPH in *E. coli*, the main sources being hydrogens from positions 1 and 6 of glucose (Csonka and Fraenkel, 1977).

One of the interesting sidelights that aids in illuminating the nature of these respiratory-chain constituents occurs among such bioluminescent bacteria as *Photobacterium fischeri*. These organisms contain an oxidoreductase that catalyzes the reduction of FMN by either NADH·H or NADPH·H, the reduced FMN being employed by an enzyme named luciferase (RCHO) to produce light (Jablonski and DeLuca, 1977), as follows:

$$\text{NAD(P)H}\cdot\text{H} + \text{FMN} \xrightarrow{\text{enzyme}} \text{NAD(P)} + \text{FMNH}_2$$

$$\text{FMNH}_2 + \text{O}_2 + \text{RCHO} \longrightarrow \text{FMN} + \text{RCOOH} + \text{H}_2\text{O} + \text{light}$$

Flavin Nucleotides. Two flavin nucleotides are associated with electron transport, either free or bound to proteins as a prosthetic group. One of these, FMN (flavin mononucleotide or riboflavin 5′-phosphate), consists of a dimethylisoalloxazine ring structure, plus a phosphorylated pentose carbohydrate (Figure 8.11B). Note that the pentose derivative is a ribitol residue, the sugar alcohol of ribose, and thus it lacks the ring structure of that pentose sugar (Figure 8.11B). In the electron-transport systems, FMN accepts two hydrogen atoms, one each on the nitrogens indicated at positions 1 and 10.

The second flavin nucleotide is similar in construction to the foregoing, even in regards to the nature of the pentose derivative, which likewise is a ribitol residue. It differs in having an AMP residue joined to the ribitol phosphate and hence is known as flavin-adenine dinucleotide or FAD (Figure 8.11C). Oxidation–reduction reactions involve the same two nitrogens as in the mononucleotide. Like the following, FAD often serves bound in a number of enzymes, including β-D-glucose oxidase, in which free and bound forms are readily interconverted (Okuda *et al.*, 1979).

Flavoproteins. A number of NAD- or NADP-containing proteins active in electron transfer have become known from mammalian tissue as well as from bacterial sources (Beinert, 1963; Mayhew and Ludwig, 1975), some of which carry out specialized functions in the cell. Most of these activities are not directly related to present concerns (Bright and Porter, 1975; Massey and Hemmerich, 1975), so attention is restricted to two classes. The first of these, however, that containing ETF (electron-transferring flavoprotein) from mammals, is difficult to obtain in high purity, so it has not been investigated in detail. Moreover, it is active mainly in a side reaction, that of transferring electrons from acylated coenzyme A and sarcosine dehydrogenases to the mitochondrial cytochrome chain or other electron acceptors. Hence, it can be passed by in favor of the more thoroughly explored second class, the flavodoxins of prokaryotes and protistans.

These flavodoxins, containing one equivalent of FMN, have been isolated from a diversity of bacteria, including *Azotobacter*, *E. coli*, *Clostridium pasteurianum*, and *Rhodospirillum* (Cusanovich and Edmundson, 1971; Vetter and Knappe, 1971; van Lin and Bothe, 1972), two genera of blue-green algae, *Anacystis* and *Synechococcus* (Smillie, 1965; Crespi *et al.*, 1972), and one of green algae, *Chlorella* (Zumft and Spiller, 1971).

The flavodoxins are relatively light in molecular weight, ranging between 15,000 and 23,000. Two subclasses of these substances have been proposed, one of which (such as that of *Desulfovibrio*) exhibits a red shift of the 450-nm absorption band relative to unbound FMN. In addition, in this subclass, the apoenzyme binds FMN at a high rate and riboflavin less rapidly, but the binding of the prosthetic group effects little change in the circular dichroism spectrum (D'Anna and Tollin, 1972; Dubourdieu and Fox, 1977). In the second subclass, represented by the *Clostridium* substance, no red shift is exhibited, the apoenzyme cannot bind riboflavin, and large changes in circular dichroism

Table 8.8
Comparisons of the Primary Structure of Bacterial Flavodoxins[a]

	10 20 30 40 50 60
Clostridium pasteurianum	M-KVNIIYWSGTGNTEAMAKLIAEGAQEKGAEVKLLNVSDAKEDDVKEA-DVVAFGSPSM
Clostridium MP	M-K--IVYWSGTGNTEKMAELIAKGIIESGKDVNTINVSDVNIDELLNE-DILILGCSAM
Peptostreptococcus	M--VEIVYWSGTGNTEAMANEIEAAVKAAGADVESVRFEDTNVDDVASK-DVILLGCPAM
Desulfovibrio	MPKALIVYGSTTGNTEYTAETIARELANAGYEVDSRDAASVEAGGLFEGFDLVLLGCSTW
Azotobacter	-AKIGLPPGSNTGKTRKVAKSIKKRFDDETMS-DALNVNRVSAEEDFAQYQFLILGTPTL
	70 80 90 100 110 120
Clostridium pasteurianum	GSEVVE----ES--EM---FLDVVS-SIVTGKKVALFGSY---G-------WMRNWVSEM
Clostridium MP	GDEVLE----ESEFEP---FIEEIS-TKISGKKVALFGSY---GWED--GKWMRDFEERM
Peptostreptococcus	GSEELE----DSVVEP---FFTDLA-PKLKGKKVGLFGSY---GWGS--GEWMDAWKQRT
Desulfovibrio	GDDSIE-L--QDDFIPL--FDSLEE-TGAQGRKVACFGCGDS-SYEYF CGAVDAIEEKL
Azotobacter	GEGELPGLSSDCENESWEEFLPKIEGLDFSGKTVALFGLGDQVGYPDNYLDALGELYSFF
	130 140 150 160 170 180
Clostridium pasteurianum	NLGANVVNDG----------------------LIVQE----APE----LGKNLGRELV
Clostridium MP	GYGCVVVETP----------------------LIVQNEPDEAEQDCIEFGKKI--ANI
Peptostreptococcus	DTGATVIGTA-----------------------IVNEMPDNAPE-CKELGEAA--AKA
Desulfovibrio	NLGAEIVQDG----------------------LRIDGDPRAARDDIVGWAHDV-RGAI
Azotobacter	KDRGAKIVGSWSTDGYDFDSSDAVVDGKFVGLALDLDNESGKTGERVAAWLAQIAPEFGLSL

[a]Based on Fox *et al.* (1972), Tanaka *et al.* (1974, 1975a,b), and MacKnight *et al.* (1974).

spectra result when it binds the prosthetic group. However, as shown in the chart of representative primary structures (Table 8.8), no obvious structural basis in the molecule can be detected for those functional subdivisions.

Several conclusions can be drawn from those sequences, which repeat and confirm observations made in earlier chapters on the primary structures of other substances, the most evident of which is that bacteria do not represent a single close-knit taxon. For example, the sequences from two species of *Clostridium* and that of *Peptostreptococcus* obviously are more closely related to each other than any of them is to the *Desulfovibrio* substance. Further, the great length of the *Azotobacter* sequence removes it widely from all the others. Even the two species of *Clostridium* are shown not to be really congeneric, for if they were, the respective falvodoxin sequences should exhibit a degree of homology comparable to that shown by the mammalian cytochromes *c*. In contrast, the present protein molecules differ at 71 of 130 sites, indicating a homology of only about 45%. In fact, *C. pasteurianum* exhibits virtually as close a kinship to *Peptostreptococcus* as it does to its nominal congenor, for comparison reveals that only 73 sites differ.

The Quinones. One of the components of the respiratory chain is a quinone, which usually is referred to as coenzyme Q (CoQ) or ubiquinone, the latter term referring to its ubiquitous occurrence throughout the living world. The quinone bears an isoprene side chain the length of which varies with the source. In mammalian mitochondria the chain contains the maximum number of ten repeating units (Figure 8.12), and the same is true for the substance from *Rhizobium japonicum* (Daniel, 1979). Others, such as that from yeast mitochondria (Goewert *et al.*, 1977; Brown and Beattie, 1977), have as few as six, while diatoms have nine (Shimazaki *et al.*, 1978a,b). Not only does CoQ occur in mitochondria, where it is more abundant than the cytochromes, but also in other organelles, including the nucleus. During oxidation–reduction processes, the two double-bonded oxygen residues are reduced to hydroxyls to form a quinol compound, the hydrogens involved in this reaction being readily lost to the next component in the chain to restore the quinone. When succinate-cytochrome *c* reductase was depleted of coenzyme Q and phospholipids and then reconstituted, both the latter substances proved to be necessary to restore activity. Furthermore, restoration of full function required the addition of the coenzyme prior to the phospholipids (Yu *et al.*, 1978). Besides its role in the respiratory chain, coenzyme Q seems to be active in oxidative phosphorylation (Bertoli *et al.*, 1978). In recent years a related substance, coenzyme Q semiquinone, has been suspected of playing a role in the respiratory chain, too, but its function has not been firmed to date (Konstantinov and Ruuge, 1977; Tikhonov *et al.*, 1977).

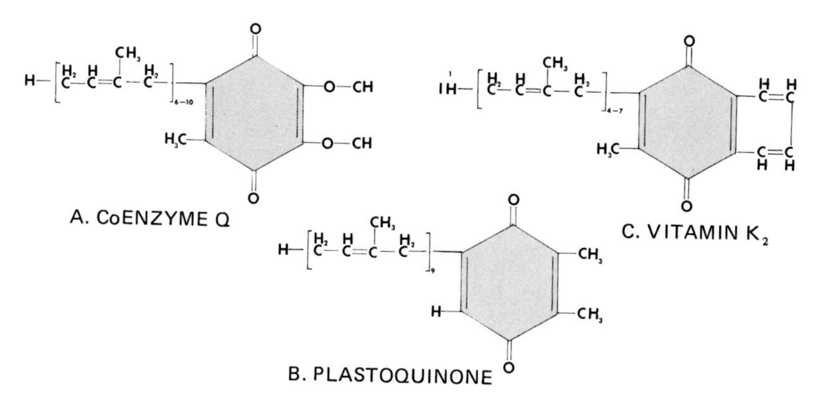

Figure 8.12. Structure of several quinones.

CoQ occurs also in chloroplasts and may participate in photosynthesis among eukaryotes as well as in certain bacteria (Clayton and Sistrom, 1964). In addition, two other related quinone-containing classes of substances occur in the chloroplasts of higher plants, plastoquinone and phylloquinone (vitamin K_1). The first of these is quite similar in structure to CoQ in having an isoprenoid chain of nine repeating units, but it differs in the nature of the other side chain (Figure 8.12); however, at least four subclasses are recognized, respectively designated as A–D. The vitamins K occur also in bacteria in the form of several vitamins K_2; these recall CoQ itself in having a diversity of isoprenoid chains, varying in the number of repeating units from four to seven (DuPraw, 1968).

The Ferredoxins. One final class of substances of the respiratory chain needs brief attention, the ferredoxins. These are nonheme iron- and sulfur-containing proteins that occur in the metaphytans, green algae, and photosynthetic and anaerobic prokaryotes, blue-green algae as well as bacteria (Wakabayashi *et al.*, 1978). In all cases the iron is bound to inorganic sulfur and to the polypeptide, the latter by means of the sulfur of cysteine (Dayhoff, 1972). Recently it has been shown that at least two different isozymes may occur in the same organism, known instances of which include blue-green algae as well as higher plants (Wakabayashi *et al.*, 1978).

As a considerable number of amino acid sequences of ferredoxins have been determined, comparisons and phylogenetic trees based on those data are abundant (Dayhoff, 1972, 1973, 1976; Tanaka *et al.*, 1974, 1975a,b; Wada *et al.*, 1975; Hase *et al.*, 1978a,c,d; Wakabayashi *et al.*, 1978). Two major classes occur, the members of one of which contain two clusters of $4Fe,4S^{2-}$ groups, probably chelated to a total of 8 cysteine residues (Adman *et al.*, 1973). That type, here designated as ferredoxins$_8$, appears to be confined to sulfur and other strictly anaerobic bacteria (Glass *et al.*, 1977). In contrast, the second class possesses only a single $2Fe,2S^{2-}$ cluster, with four cysteine residues serving as the ligands (Hase *et al.*, 1977a, 1978d); these ferredoxins$_2$ appear to be confined to advanced bacteria, such as *Halobacterium*, and to photosynthetic organisms, especially the eukaryotic forms.

It has been claimed that if the bacterial sequences are split into two beyond site 29 and the resulting halves aligned, clear evidence of gene duplication is revealed (Dayhoff, 1973; Matsubara *et al.*, 1978). While it is true that a number of resemblances occur, more careful analysis casts doubt upon this interpretation. To show the relationships, sequences of ferredoxins of two bacteria are shown in Table 8.9, one of which is that of a primitive anaerobe, *Clostridium pasteurianum*, and the second is from a more advanced type, *Micrococcus aerogenes*. In the first instance, 13 of the 29 sites in each half are homologous and in the second one, 12 of the 29; these 45 and 41% correspondences could appear to intimate a possible gene duplication. However, if the nature of the identical sites is examined, it becomes obvious that a large

Table 8.9
Comparisons of the Two Halves of Bacterial Ferredoxins

Clostridium NH_2 half	AYKI-ADSCVSCGACASECPVNAISQGDS
Clostridium OH half	IFVIDADTCIDCGNCANVCPVGAPVQE--
Micrococcus NH_2 half	AYVI-NDSCIACGACKPECPVN-IQQG-S
Micrococcus OH half	IYAIDADSCIDCGSCASVCPVGAPNPED-

portion of them are cysteines and adjacent amino acids involved in bonding the iron and sulfur; these are shown in dark shading in Table 8.9, while non-ligand-related homologies are lightly shaded. Hence, the majority are structural necessities and reflect the presence of two groups of four iron and four sulfur atoms, not ancestor–descendant relationships (Hase *et al.*, 1978d). When the nonstructural homologous sites are counted, only three are found in either sequence, making the proposed gene duplication quite doubtful.

What could have been noted instead of that proposal is that the ligands for each $4Fe,4S^{2-}$ cluster are identically spaced, paralleling the arrangement in the cytochromes *c*, as pointed out earlier. In this case the first three cysteine residues (at sites 8, 11, and 14, and 37, 40, and 49) are each separated by two sites, while the fourth (at sites 18 and 53) is separated by three. Second, the second cysteine in each cluster is followed by a glycine, and the fourth one of each group is followed by a proline and preceded by a glutamic acid or valine residue. In the chart of known $8Fe,8S^{2-}$ ferredoxin sequences (Table 8.10), the constancy of these sites is clear, except between the second and third cysteines in the carboxyl half of the molecules from advanced bacteria. These larger sequences indicate the cysteines to function in pairs. Pairs one and three are C--C, and pairs two and four are C---CP. Further, Table 8.10 shows that increases in gene length for a given protein do not necessarily derive from duplications, for although the molecules of the three *Chlorobium* species are somewhat longer than the primitive forms, and the *Chromatium* ferredoxin (Hase *et al.*, 1977a,d) is about one-third longer than all the others, the additional segments show no likeness to other portions.

Two-Iron Ferredoxins (Ferredoxins₂). The ferredoxins$_2$ are involved in photosynthesis and perhaps in other functions in multicellular organisms. Aside from having an iron and sulfur combination as the prosthetic group, this class of substances shows no relationship to those of the other class, either in structure or function. In addition to the $2Fe,2S^{2-}$ active center, the members are alike in having molecular weights of about 11,000, prokaryotic, algal, and higher plant molecules being of similar length, about 100 sites *in toto*.

Because Wakabayashi *et al.* (1978) supply a comparison of 16 amino acid

Table 8.10
Comparisons of Ferredoxins$_8$[a]

	10 20 30 40 50 60
Cl. butyricum	AFVINDSCVSCGACAGECPVSAITQGDTQFVIDADTCIDCGN------CANVCPVGAPNQE
Cl. pasteurianum	AYKIADSCVSCGACASECPVNAISQGDSIFVIDADTCIDCGN------CANVCPVGAPVQE
Cl. acidi-urici	AYVINEACISCGACDPECPVDAISQGDSRYVIDADTCIDCGA------CAGVCPVDAPVQA
Micrococcus	AYVINDSCIACGACKPECPVN-IQQG-SIYAIDADSCIDCGS------CASVCPVGAPNPED
Chl. thiosulfatophilum	ALYITEECTYCGACEPECPTNAISAGSEIYVIDAAGCTECVGFADAPACAAVCPAECIVQG
Chl. limacola$_1$	ALYITEECTYCGACEPECPVTAISAGDDIYVIDANTCNECAGLDEQ-ACVAVCPAECIVQG
Chl. limacola$_2$	AHRITEECTYCAACEPECPVNAISAGDEIYIVDESVCTDCEGYYDEPACVAVCPVDCIIKV
Chr. vinosum	ALMITDQCINCNVCQPECPNGAISQGDETYVIEPSLCTECVGHYETSQCVEVCPVDCIIKDP
	70 80
Chr. vinosum	SHEETEDELRAKYERITGEG

[a]Based in part on Fox *et al.* (1972), Tanaka *et al.* (1974, 1975a,b), Dayhoff (1976), and Hase *et al.* (1978a). The organisms, listed in abbreviated form, are *Clostridium butyricum, pasteurianum,* and *acidi-urici; Micrococcus aerogenes; Chlorobium thiosulfatophilum* and *limicola;* and *Chromatium vinosum*. Shading indicates constant regions.

sequences, including those of both isozymes mentioned earlier from four different organisms, it is not necessary to repeat that information. Instead, comparisons are provided of generalized sequences derived by averaging methods of higher plant (Hase *et al.*, 1977b,c; Masaki *et al.*, 1977; Takruri and Boulter, 1979) and green, red (Takruri *et al.*, 1978), and blue-green (Hase *et al.*, 1976a,b, 1978b; Hutson *et al.*, 1978) algal material, along with one from an aerobic bacterium (*Halobacterium*) capable of photophosphorylation (Kerscher and Oesterhelt, 1977; Hase *et al.*, 1977a, 1978a). The latter shows about a 40% level of homology with the blue-green ferredoxin, 40 sites being identical, and around 34% with the metaphytan protein. This observation, together with the 22-residue-long appendage at the amino end and another 5-residue-long section at the carboxyl end, strongly suggests that the bacterial enzyme has other functions than in photophosphorylation. Indeed, it has been demonstrated to serve as the physiological electron acceptor in the enzymatic oxidation of α-ketoglutaric, pyruvic, and α-ketobutryic acids (Kerscher and Oesterhelt, 1977). The ferredoxin from *Cyanidium*, many of whose other characteristics approach those of yeast, also suggests a primitive position among the eukaryotes for that organism, for the amino acid sequence (Hase *et al.*, 1978e) is distinctly intermediate between those of the blue-green algae and the red seaweed (Table 8.11).

In metazoans, ferredoxins have been identified especially in mitochrondrial systems, in which they serve as an electron-transfer intermediate between cytochrome P-450 and an NADPH-dependent ferredoxin reductase. They have also been shown to be involved in the hydroxylation of cholesterol, along with cytochrome P-450 (Kapke *et al.*, 1979).

8.3. COMPARATIVE ASPECTS OF CELLULAR RESPIRATION

Although occasional reference to processes and to enzymes and diverse substances of other than metazoan origin was made in preceding pages, several topics of special interest of a comparative nature remain. The first of these pertains to the electron-transport chains of bacteria and various eukaryotes, the second concerns differences that exist in the tricarboxylic acid cycle in various phyla, and the third provides a phylogenetic view of the respiratory processes.

8.3.1. Comparative Aspects of the Respiratory Chain

Among the macromolecular constituents that enter into the respiratory processes, much variation, as well as general constancy, is encountered. For in-

Table 8.11
Ferredoxin$_2$ Sequences from Major Sources[a]

Source	Residues 1–60 (numbered 10, 20, 30, 40, 50, 60)
Halobacterium	PTVEYLNYETLNNQGWDMDDDDLFEKAANAGLDGEDYGTMEVAEGEYILEAAEAQGYDWP
Blue-green alga	ATYKVTLINEAEGINETIEVPDDEYILDAAEEAGYDLP
Cyanidium	ASTKIHLVNKDQGIDETIECPDDQYILDAAELQGLDLP
Red alga	ADYKIHLVSKEEGIDVTFDCSEDTYILDAAEEEGIELP
Green alga	ATYKVTLKT-PSG-DQTIECPDDTYILDAAEEAGLDLP
Metaphytan	AAYKVTLVT-PSG-TQTFDCPDDVYILDAAEEAGLDLP

Source	Residues 61–end (numbered 70, 80, 90, 100, 110, 120)
Halobacterium	FSCRAGACANCASIVKEGEIDMDMQQILSDEEVEEKDVRLTCIGSPAANEVKIVYNAKHLDYLQNRVI
Blue-green alga	FSCRAGACSTCAGKLVSGTIDQSDQSFLDDDQIEAGYV-LTCVAYPTSDCVIQTHQEEGL-Y
Cyanidium	YSCRAGACSTCAGKLLEGEVDQSDQSFLDDDQVKAGFV-LTCVAYPTSNATILTHQEESL-Y
Red alga	YSCRAGACSTCAGKVTEGTVAQSDQSFLDDEQMLKGYV-LTCIAYPESDCTILTHVEQEL-Y
Green alga	YSCRAGACSSCAGKVEAGTVDQSDQSFLDDSQMDGGFV-LTCVAYPTSDCTIATHKEEDL-F
Metaphytan	YSCRAGACSSCAGKVVSGTVDQEDGSFLDDEQIEEGWV-LTCVAYPESDVTIETHKEEEL-TA

[a]Sequences based on Sugeno and Matsubara (1969), Hase *et al.* (1978a,e), Takruri *et al.* (1978), and Wakabayashi *et al.* (1978). Shading indicates invariant regions.

stance, there is at once remarkable uniformity in the presence of the electron-transport chain and a striking variation in the specific types of cytochromes represented, as is best evidenced by an examination of several major taxa.

The Bacteria. Among such gram-positive bacteria as *Bacillus subtilis* and *Sarcina lutea*, the cytochromes included in the chain are the same as those of yeast and metazoans, a, a_3, b, c, and c_1, all with spectral peaks approximating those of the eukaryotes. Nevertheless, some variation is encountered within the group, especially in the a type, for quite frequently, as in *Micrococcus pyogenes*, cytochromes $a + a_3$ are absent, o serving as the end component of the chain in their stead.

Typical gram-negative bacteria like *E. coli* and *Proteus vulgaris* differ strikingly from the foregoing and all other types of organisms. Three so-called cytochromes c are known to be present, along with five b types, and three oxidases, a, a_2, and o.* In addition to NADH and NADPH, coenzyme Q_8 occurs under aerobic conditions, whereas a naphthoquinone called menaquinone$_8$ is abundant during anaerobiosis (Haddock and Jones, 1977). In several other gram-negative species, further variations in the respiratory chain are found, among which is an increase in the number of components. *Azotobacter vinelandii*, an obligate aerobe, is a nitrogen-fixing species and uses molecular oxygen as the terminal oxidant for respiration. Among the known constituents of its respiratory chain are included NADH, NADPH, Fe,S-containing proteins, FMN, coenzyme Q_8, flavoproteins, and six major cytochromes (b, c_4, c_5, a_1, o, and d) (Jones and Redfearn, 1966; Haddock and Jones, 1977). The chain in the gram-negative *Paracoccus denitrificans* also can use molecular oxygen as the terminal oxidant and closely approaches that of eukaryotes in composition (Lawford *et al.*, 1976). Present are both NADPH and NADH, FAD, Fe,S-containing proteins, coenzyme Q_{10}, cytochromes b_{562} and b_{565}, c_1 and c, and $a + a_3$. The last of these constituents, which lacks copper, is used during aerobic conditions, whereas NO_3^-, NO_2^-, and cytochrome o serve during anaerobiosis (Haddock and Jones, 1977).

One final major difference that has been reported among bacteria is the complete lack of cytochromes of all types. Among those in which this condition has been detected are included *Beggiatoa* and *Thiothrix* of the colorless sulfur bacteria and certain strains of *Streptococcus faecalis* (Smith, 1954; Ritchey and Seeley, 1976). Apparently just above these lie those sulfate-reducing bacteria like *Desulfovibrio,* which contain only a single cytochrome and that is of the peculiar s type (Section 8.2.1) (Thauer *et al.*, 1977).

Although in many cases the cytochromes of bacteria, as in most organisms to be discussed below, carry the same names, they are not always identical to those of mammals of corresponding designation, for, as already pointed out, the names merely signify that they show similar absorption spectra and other

*Blue-green algae appear to have only an o-type terminal oxidase (Webster and Hackett, 1966).

properties. Consequently, bacterial cytochromes $a + a_3$ do not oxidize mammalian cytochrome *c*, or will do so only very slowly, and bacterial *c*-type cytochromes are oxidized not at all by mammalian cytochromes $a + a_3$. Yamanaka and Okunuki (1968) made use of this species specificity to compare reactions of 47 kinds of cytochrome *c* derived from various microorganisms, higher plants, and metazoans with cytochrome oxidase from cattle and *Pseudomonas* to determine phyletic relationships. Because the phylogeny derived from these studies showed man to be closely akin to oyster, cattle to pigeons and toads, and the horse to stand midway between mackerel and squid, it is difficult to determine how the reactions may contribute to a better understanding of evolutionary kinships. Nevertheless, the results do make it clear that the bacteria are not a natural, close-knit group.

Further distinctions and similarities can be noted between bacterial and mammalian cytochrome oxidases. Although both *a* and a_3 occur together in gram-positive bacteria, they do not appear to form a complex like that of mammals. Nor in bacteria is a heavy-metal component present to correspond to the copper of mammals (Smith, 1968). Resemblances between the members of the two taxa are in the reactions with cyanide and carbon monoxide and in the prosthetic group being heme *a*. In fact, cytochrome a_1 seems also to have the same group. On the other hand, cytochrome a_2 has a quite different prosthetic group in the form of an iron complexed with an unusual chlorin (Barrett, 1956). Cytochrome *o*, in contrast, seems to have a protoheme, more comparable to that of the *b*-type compounds, as shown in a preceding discussion.

When more than one cytochrome oxidase is present in these unicellular forms, any single one of them appears capable of enabling most of the respiratory processes to continue (Castor and Chance, 1959). If cytochrome *a*, for instance, is lost from *Staphylococcus epidermidis*, no diminution of respiratory function is noted (Jacobs and Conti, 1965). Hence, it is difficult in many cases to comprehend what functional basis might exist for the multiple cytochrome oxidases. However, in certain species, one cytochrome *c* oxidase may be more efficient than another; one instance of the sort is found in *Rhodopseudomonas sphaeroides*, in which the *a* type appears more effective than the cytochrome *o* (Kikuchi and Motokawa, 1968).

The Yeasts. As in the vast majority of eukaryotic organisms, the electron-transport system of yeasts includes cytochromes *b*, *c*, c_1, and $a + a_3$. The statement made above regarding the absence of complete chemical identity to those of their mammalian counterparts receives emphasis through a study of yeast cytochrome *c*. Haneishi and Shirasaka (1968) compared this substance biochemically and serologically from 104 strains of 53 species representing 14 genera and found no less than four major types present. While types I and II differed only slightly, principally in motility on an Amberlite column, type III showed contrasting reactions with the antisera. Further, contrary to the strong reaction with *Saccharomyces* cytochrome *a* obtained with types I–III, type IV showed only slight reactivity, was highly inhibited by the antisera, and was markedly more motile on the Amberlite column. In the genus *Saccharomyces*, 19 of the 21 tested species had type I and two had type IV, whereas of the 11 species of *Candida* studied, three had type I, five type II, two type III, and one type IV. As a whole, types II and IV appeared to be the most widespread, each occurring in seven of the tested genera; type III was detected in only two genera, *Pichia* and *Candida*. Type IV, with its weak affinity for Amberlite CG-50 and slight reaction with the cytochromes *a*, most closely approached the cytochrome *c* of such bacteria as *Bacillus subtilis* and *Rhodospirillum rubrum*, as well as that of the red alga, *Porphyra tenera*. Type I of these authors, at least in *Saccharomyces*, appeared to correspond to the iso-1-cytochrome *c* of other workers (Sherman *et al.*, 1965; Slonimski *et al.*, 1965), but whether or not their type IV was identical to iso-2-cytochrome *c* of the latter remains unclear.

Functionally in mitochondria, the differences in the various cytochromes may be less striking. At least Mattoon and Sherman (1966) reported results from a cytochrome-*c*-deficient strain of yeast (*S. cerevisiae*) suggestive of this conclusion. They found that electron transport and oxidative phosphorylation in mitochondria isolated from the mutant strain could be reconstituted by the addition of purified cytochrome *c* from either wild-type yeast or horse heart. The efficiency level of oxidative phosphorylation in such reconstituted mitochondria equaled that of normal yeast mitochondria under the same conditions, regardless of the source of the cytochrome *c*.

Other Primitive Taxa. Comparative studies involving photosynthetic forms are still somewhat confused by difficulties caused by ambiguity as to whether derivation is from mitochondria or chloroplasts. Yamanaka and Okunuki (1963), for example, showed that *c*-type cytochromes from algal sources could not be oxidized by mammalian cytochrome oxidase and thus were probably not associated with respiration. Similar results on a broader basis, covering 31 species from three major taxa, have also been reported by Sugimura *et al.* (1968). Brown and red algae were stated to lack *b*-type cytochromes but to have typical algal *c*-type cytochromes (Katoh, 1959); unfortunately it was not made clear whether the latter were typical of respiratory or photosynthetic function. Green algae, quite to the contrary, were found to possess cytochromes *b* and *c*, with the latter showing wide range in characteristics among the various representatives. Some of these approximated the types found in photosynthetic bacteria, others were more typical of respiration, and still others approached the cytochrome *f* characteristic of the metaphytan chloroplast. No mention was made of cytochrome *a*.

In contrast to the yeasts, such aquatic fungi as *Leptomitus lacteus* and *Apodachlya punctata* (Gleason, 1968) appear to have two *b*-type cytochromes (564 and 557 nm) besides cytochromes *a*, a_3, and *c*, but seem to lack c_1. However, it is not clear as yet whether this is typical of the entire taxon or not.

The Higher Plants. In a paper by Lance and Bonner (1968), the respiratory-chain components of a number of higher plants have received detailed attention. The study was based on mitochondria from white potato, Jerusalem artichoke, cauliflower, mung beans, and skunk cabbage spadix. With one exception, all were found tightly coupled, the monocot skunk cabbage spadix mitochondria alone showing no appreciable coupling. Furthermore, all species proved similar in both the qualitative and the quantitative composition of their electron carriers. Basically, close resemblance was found to the cytochrome components of the metazoan chain, including a cytochrome oxidase optically resolvable into cytochromes a and a_3. Of the two c types, one resembling c_1 was firmly bound to the mitochondrial membrane. A distinctive feature that was pointed out earlier appeared among the b types, for three members of this class of cytochromes were uniformly present.

Still others have been observed, some of which may prove to be peroxidases (Lundegårdh, 1962), but to the present it has not been possible to position the several components relative to others in the chain. Furthermore, a second cytochrome oxidase has been demonstrated in mung bean mitochondria (Bonner, 1964; Hiroshi *et al.*, 1964) that differed from cytochrome a_3 in spectral and kinetic properties. Its γ band at 438 nm compared to that of cytochrome a_3 at 445 nm, and its complex with carbon monoxide had a band at 420 nm and that of cytochrome a_3, 430 nm. It has not as yet been established whether this cytochrome oxidase is sensitive to cyanide; if it should prove not to react with CN^-, it might provide the cyanide-insensitive path to oxygen utilization that characterizes many plant tissues.

8.3.2. Comparative Aspects of the Respiratory Processes

The major processes of cellular respiration, including the tricarboxylic acid cycle and oxidative phosphorylation, occur with striking consistency throughout the greater part of the living world. Because the second of these processes has been the subject of a particularly thorough review (Thauer *et al.*, 1977), that aspect receives mention only as it is especially pertinent. Several significant variations in the tricarboxylic acid cycle have been reported from among the more primitive taxa of organisms, which provide the center of attention as they hold more than usual biological interest.

The Bacteria. The tricarboxylic acid cycle appears to be a basic cellular requirement of typical bacteria in general, whether gram-negative or gram-positive, but oxidative phosphorylation does not seem to be so closely coupled in this taxon as it is among vertebrates. Consequently, the number of phosphate radicals gained per atom of oxygen used (the P/O ratio) is close to 1, whereas it is generally viewed as being 2 or 3 in mammals (Dolin, 1961; Hinkle and Yu, 1979).

In the respiratory chain of several bacteria, *Azotobacter vinelandii* for one (Brodie and Gray, 1956; Haddock and Jones, 1977), a soluble protein is present in addition to the usual components; other species, including *Alcaligenes faecalis*, have a requirement for a polynucleotide of the RNA type (Pinchot, 1957a,b). But in neither case is it clear as to the location of the additional substance relative to the chain components. Among still other bacteria, inorganic compounds may serve as electron acceptors during particular sets of reactions. For example, in the oxidation of formate in *E. coli*, nitrate serves as an electron acceptor (Takahashi *et al.*, 1956), and in the oxidation of sulfur compounds among the thiobacilli, oxygen serves in that capacity. On the other hand, sulfate-reducing bacteria, including *Desulfovibrio desulfuricans*, employ sulfate and other sulfur-containing substances as terminal electron acceptors. In the last named example, the electron-transport chain includes cytochrome *s*, an autoxidizable substance; ATP is also required, as the reactions appear to involve the production of adenosine-5-phosphosulfate (Peck, 1962; Peck *et al.*, 1965; Thauer *et al.*, 1977).

The Glyoxylate Bypass. An especially important variation of the tricarboxylic acid cycle, called the glyoxylic acid bypass, is found among many bacteria (Krampitz, 1961; Stadtman, 1968). As shown in Figure 8.13, the majority of the steps are identical to those of the tricarboxylic acid cycle, but a shortcut is provided between isocitric and malic acids. Isocitric acid is broken into 4-carbon and 2-carbon compounds (succinic and glyoxylic acids, respectively) by a reaction catalyzed by isocitratase. The glyoxylic acid then is condensed with acetyl-CoA through the action of the enzyme malate synthetase to form malic acid; the latter substance, as well as the succinic acid, can then continue through the cycle in regular fashion. In addition to shortening the usual cycle by two steps, the formation of CO_2 is avoided. Furthermore, two molecules of acetic acid are consumed during the course of a single cycle, and two molecules of oxaloacetic acid are produced, so that an increasing amount of acetic acid can be oxidized. As this cycle thus is adapted for utilization of substrate acetic acid, it is scarcely surprising that, as far as is known, one of its unique enzymes, malic acid synthetase, has been demonstrated only in microorganisms cultured in media in which either acetic or glycolic acid was the sole source of carbon (Wong and Ajl, 1956, 1957; Kornberg, 1961). Most interestingly, as shown by Falmange and co-workers (1965), each of these two substances induced the synthesis of a different type of malic acid synthetase. Although catalyzing the identical reaction, the two enzymes possessed different sensitivities to thermal denaturation and were separable by chromatographic procedures; moreover, acetic acid simultaneously induced the production of isocitratase, whereas glycolic acid did not.

Numerous other examples of modified steps in the respiratory activities may be found in standard textbooks and periodicals devoted to bacteriology, but as most pertain only to a limited number of species, it is not necessary to cite them here. The majority, moreover, involve the production of one or more peculiar enzymes that are synthesized only under the influence of a particular

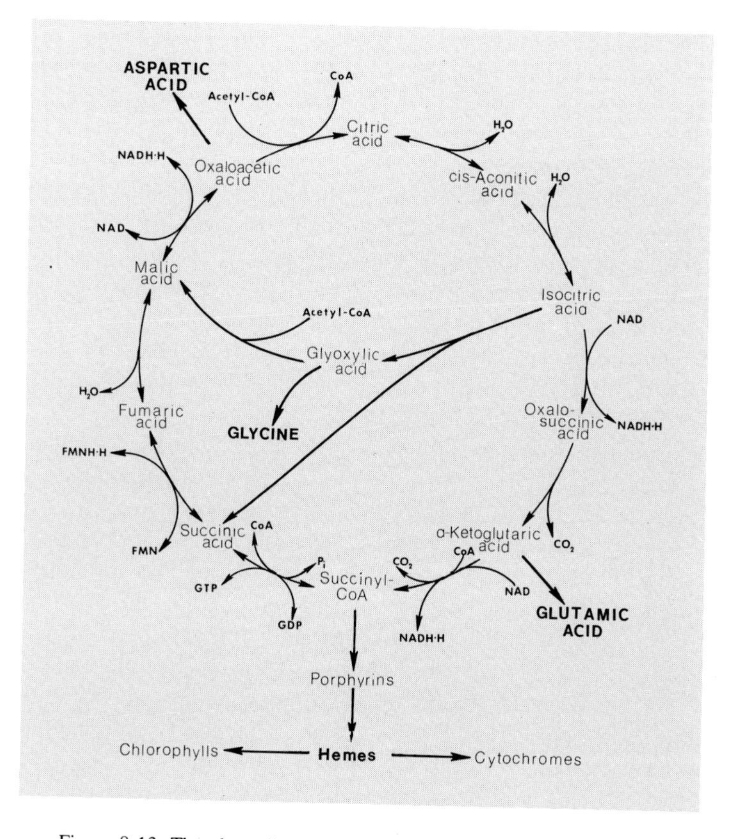

Figure 8.13. The glyoxylic acid bypass in the tricarboxylic acid cycle.

substance or condition of the substrate. The molecular basis of such environmental influences upon the genetic control of protein synthesis is perhaps of as much biological importance as the variations in the processes themselves, if not more, but remains unexplored as yet.

Variations in Other Taxa. Only a decade ago was the tricarboxylic acid cycle of any blue-green alga given detailed attention, in two studies on *Anabaena variabilis* (Hoare *et al.*, 1967; Pearce *et al.*, 1969). While most of the enzymes of the cycle were detected, both α-ketoglutarate dehydrogenase and succinyl-CoA synthetase were lacking; hence, the cycle was shown to be interrupted by the absence of the step between ketoglutaric and succinic acids (see Figure 8.19). Furthermore, succinic acid dehydrogenase was present in such small quantities as to be virtually wanting. This limitation may be responsible for the inability of the investigating teams to detect the actual employment of the glyoxylate bypass when the alga was grown on acetic acid, despite their finding the special enzymes of that system, isocitratase and malic synthetase. Instead of the succinic acid being oxidized, it appeared to be activated to succinyl-CoA by way of acetoacetate-CoA for entrance into porphyrin formation. Hence, this portion of the broken cycle may be adapted for production of photosynthetic pigments rather than for energy storage. As ultrastructural studies show the blue-green algal cell to be continually growing and to contain relatively large quantities of photosynthetic lamellae (Figures 10.19, 11.1, 11.3), it is logical to suspect that a constant supply of new chlorophyll might be a most essential feature of the organism's economy. In this connection it is pertinent to note that interrupted tricarboxylic acid cycles similarly lacking α-ketoglutarate dehydrogenase have been reported for two photosynthetic bacteria, *Chloropseudomonas ethylicum* (Calley *et al.*, 1968) and *Chromatium* D (Fuller *et al.*, 1961).

A biosynthetic role must also be assumed for the highly incomplete cycle found in *Beggiatoa* (Burton *et al.*, 1966), as a number of the typical enzymes were absent, in addition to all the cytochromes, as reported earlier. Among the lacking enzymes were included citrate-condensing enzyme, α-ketoglutarate dehydrogenase, fumarase, and aconitase, as well as the enzymes of glycolysis, so that substrate glucose could not be metabolized. During metabolism of substrate acetate, no CO_2 was liberated by growing cultures. Much of the system appears to be concerned with the production of various families of amino acids through the oxaloacetate–aspartate and α-ketoglutarate–glutamate pathways (See Figure 8.17), but no use of the succinic acid could be detected in this colorless sulfur bacterium.

Undoubtedly many other variations of the cycle will be disclosed as more types of prokaryotes are studied. Some presently unexplored groups that should prove especially fruitful are other members of the colorless sulfur bacteria (such as *Thioploca*), the spirochaetes, and genera of the blue-green algae other than *Anabaena* reported upon above. Of the latter taxon, *Oscillatoria* and its allies might be found to possess a more widely broken cycle than that of *Anabaena*, while one would suspect those members of the group shown to have a mesosomelike body, including *Anacystis*, might at least approximate the glyoxylate bypass dual system of the eubacteria.

8.4. EVOLUTION OF THE TRICARBOXYLIC ACID CYCLE

Obviously, nothing of such complexity as the tricarboxylic acid cycle, with its amazing continuous chain of events, could possibly have appeared, fully developed, in the primordial organisms from the primitive environment. But until quite recently no clue had existed as to where the primitive cycle may have begun nor as to how any single step may have functioned isolated from the others to make a useful starting point (Broda, 1971; Buvet and LePort,

1973; Yamanaka, 1973; Hartman, 1975). With the disclosure of the nature of the respiratory cycle in the blue-green algae by Bisalputra *et al.* (1969), coupled to the earlier analysis of events in *Beggiatoa* (Burton *et al.*, 1966), a preliminary proposal of the evolutionary events now becomes feasible. Briefly stated, the phylogenetic history presented here proceeds in the reverse of the discussion in preceding pages of the processes themselves. As in other biochemical evolutionary events examined earlier, each suggested step is completely functional, involves the addition of only a single set of enzymes at a given point, and results in the synthesis of a product of obvious survival value.

Earliest Stage. Because the biochemical events leading to the earliest living things and their subsequent evolution to a primitive cellular form have been elaborated upon elsewhere (Dillon, 1978), this account can be confined to the subsequent steps. Whereas the early atmosphere had been rich in carbon dioxide and ammonia, so that the protobionts and the chlamydidlike early cellular forms could synthesize their own amino acids, the carbon dioxide content probably began to become dilute. Thus the organisms drew increasingly upon the content of the primeval seas, until that source of those biochemicals ultimately became nearly exhausted. Consequently, the ability to convert other organic substrates into glycine and aspartic and glutamic acids, from which all the other amino acids could be derived (Dillon, 1978), became essential for survival. As shown in Figure 8.14, it appears that the necessary enzymes were acquired by evolutionary processes in the primitive cellular organisms to permit the formation of these three amino acids by aminating glyoxylic, oxaloacetic, and α-ketoglutaric acids, respectively.

The Second and Third Stages. Little further advance was made in the synthesis of glutamic acid until fairly late in early prokaryote evolution, so the next several series of advances center around glycine and aspartic acid synthases, particularly those of the latter. The second step in the phylogeny of the tricarboxylic acid cycle was the advent of the malic acid dehydrogenase, so that a fresh supply of oxaloacetic acid was thereby made available from malic acid. As NAD is the hydrogen acceptor in this reaction, it too must have come into existence in early organisms, as later evidence suggests. Malic acid itself can be derived from glyoxylic acid (Figure 8.15).

Perhaps the dual usage of glyoxylic acid for glycine and aspartic acid by way of malic acid made it evolutionarily advantageous for the acquisition of an enzyme system that converted isocitric to glyoxylic acid. Similarly the enzymes necessary for the deacylation of isocitric acid to form succinic acid seem to have provided a third major source of the synthesis of aspartic acid. Be that as it may, these two enzyme systems appear to have developed at this time (Figure 8.16).

The Fourth Major Stage. Although all the foregoing proposed stages lack solid evidence in the form of actual organisms possessing the respective combinations of enzyme systems, it is highly probable that appropriate inves-

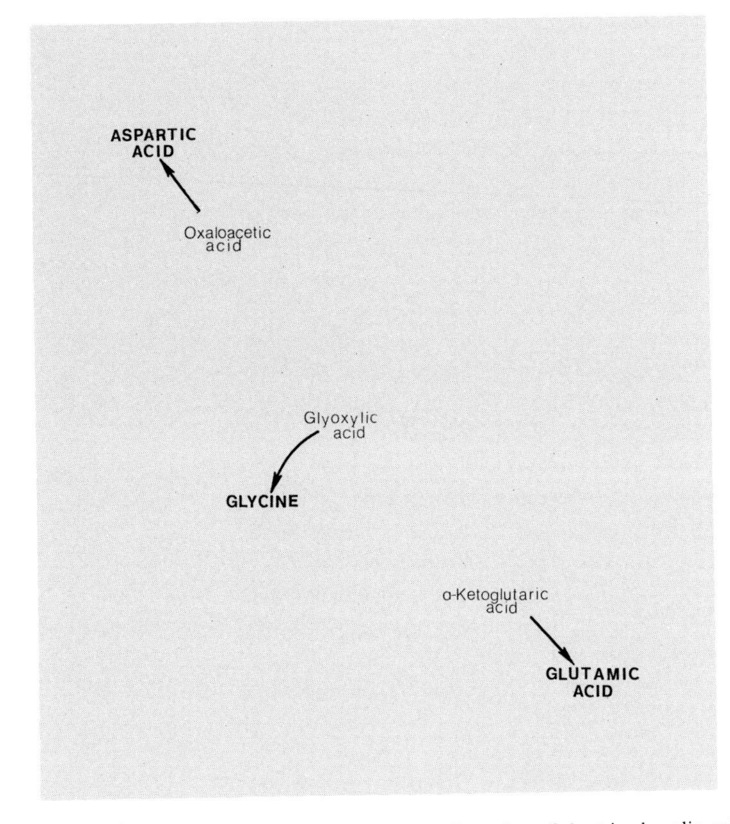

Figure 8.14. The earliest of the reactions that led to the formation of the tricarboxylic acid cycle were involved only in amino acid production.

tigations of various rickettsias and perhaps the chlamydids would disclose their actual existence. In contrast, here at the fourth major level, the ingredients that constitute the partial cycle that exists in the colorless sulfur bacteria, such as *Beggiatoa*, have been documented (Burton *et al.*, 1966). As visualized in Figure 8.17, several important steps became established at this level, the simplest of which hydrated fumaric acid to provide an additional source for aspartic acid by way of malic and oxaloacetic acids. Whether coenzyme A had existed in simple form prior to this time cannot be stated, as a result of the absence of suitable investigations. But that coenzyme has been established as being present in *Beggiatoa,* where it plays two roles in the cycle, one in the acetylation of glyoxylic acid to produce malic acid, the second in a new reaction involving succinic acid to result in succinyl-CoA (Figure 8.17). However, the usefulness of this latter product remains a mystery, for porphyrins and hemes, including

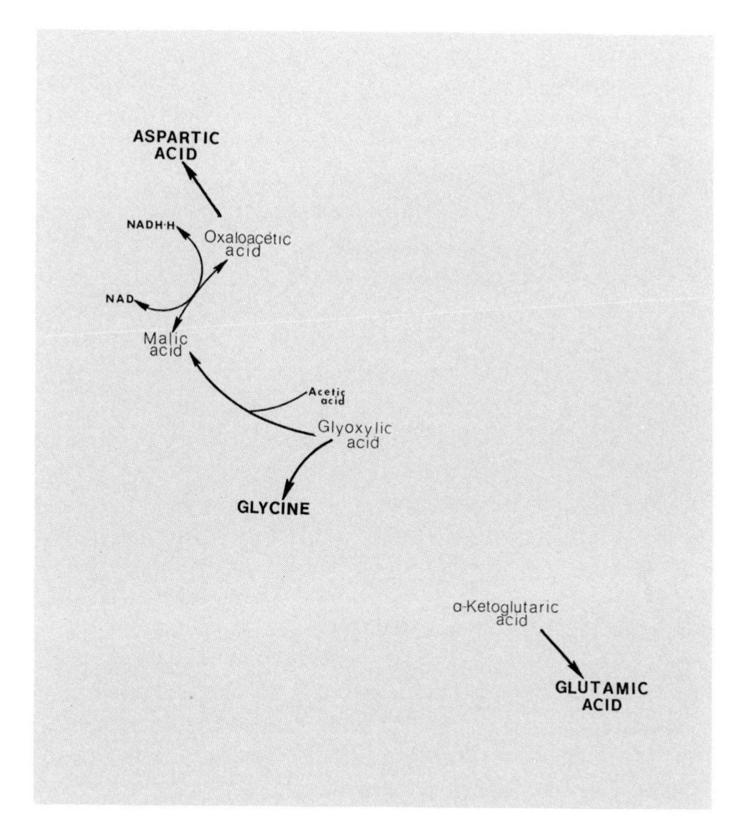

Figure 8.15. The derivation of malic acid from glyoxylic acid provided a supplementary source for aspartic acid.

cytochromes, are absent in this photosynthetic sulfur bacterium. Nor is it apparent in the absence of heme proteins whether the NADH · H synthesized in the malic to oxaloacetic acid reaction is actually involved in the production of energy-storing substances. Explorations into the ATP-forming mechanisms of this, the simplest of living cellular things, should be most instructive, especially in view of the sulfur-based photosynthetic processes and the absence of cytochromes.

The Fifth-Stage Events. Between the preceding and the next level of events supported with evidence from researches on living things, a number of developments occurred, whose sequence of appearance must await the uncovering of supportive data. While their order of establishment is thus undeterminable, their appearance during this interim is clear. One set of these is concerned with the biogenesis of isocitric acid, which, through its employment as a

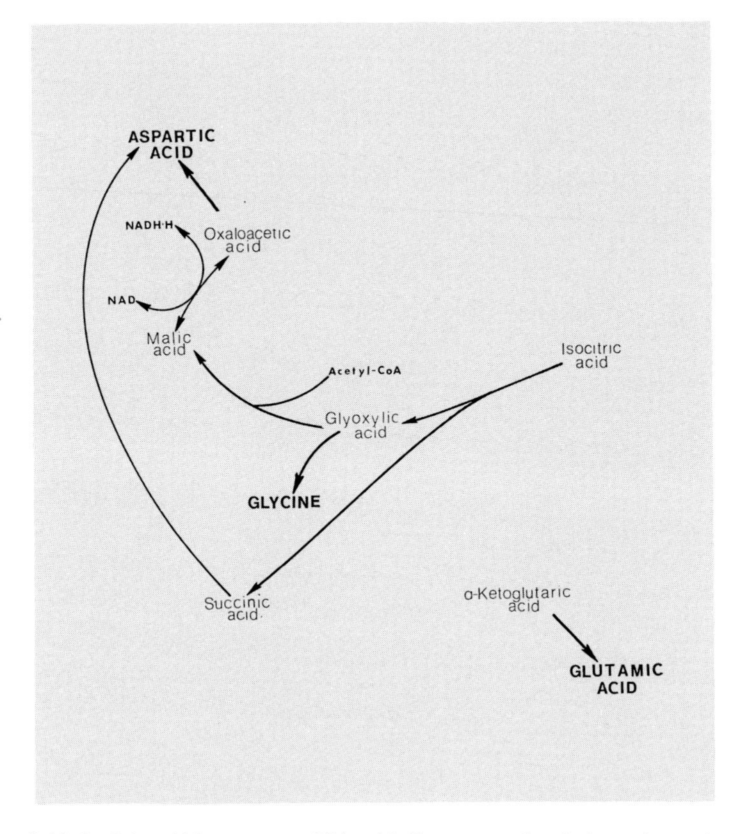

Figure 8.16. Isocitric acid becomes an additional indirect source for glycine and aspartic acid.

basis for glycine and succinic and malic acids, seems to have become depleted in the early seas at this stage. First, the enzymes for hydrating *cis*-aconitic acid were developed, followed by a second enzyme that converted citric acid into that substance by dehydration (Figure 8.18).

About this same time a second set of processes was developing whose final products have had inestimable impact on the evolution of simple cells into the complex forms of higher life. The succinyl-CoA, together with glycine, provided the springboard for a series of nine successive reactions that ultimately resulted in the formation of protoporphyrin and thus hemes (Figure 8-18).

Features of the Sixth Stage. The sixth stage, represented by the incomplete tricarboxylic acid cycle of such blue-green algae as *Anabaena* (Hoare *et al.*, 1967; Bisalpoutra *et al.*, 1969), thus corroborates that the proposed several additions did appear in the interval above the colorless sulfur bacteria. Of the features that occurred at this level, two are of outstanding importance and

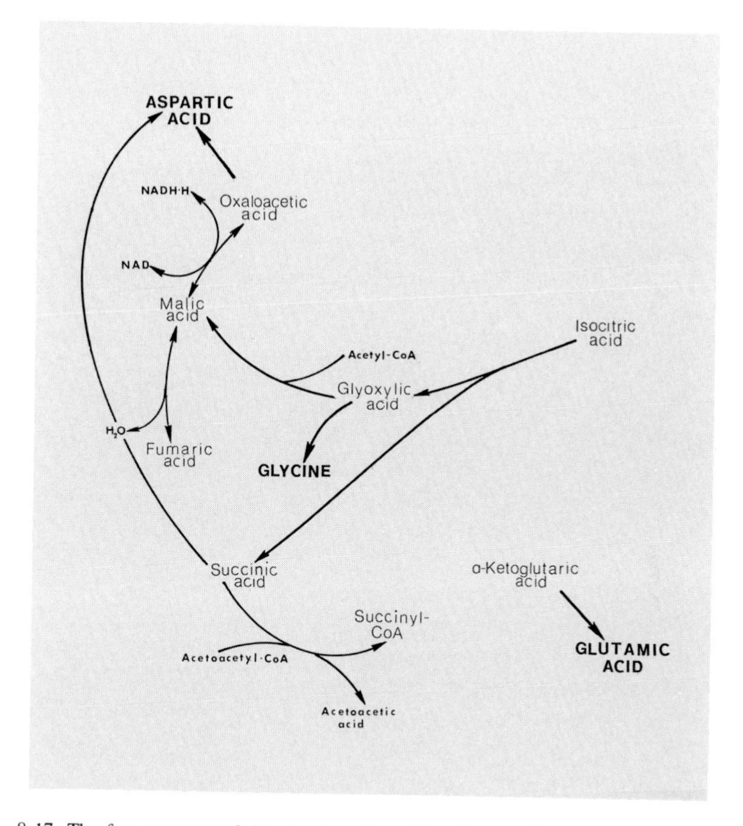

Figure 8.17. The few enzymes of the tricarboxylic acid cycle actually present in *Beggiatoa* clearly indicate its amino-acid-synthesizing functions.

two of lesser impact (Figure 8.19). Among the minor happenings is an enzyme system that catalyzes the dehydrogenation of isocitric into oxalosuccinic acid, utilizing NADH · H as hydrogen acceptor, the first appearance of this coenzyme. This new substance then spontaneously decomposes into α-ketoglutaric acid, liberating CO_2 in the process, and thereby enriches the basis for glutamic acid biosynthesis.

Among the principal events is the evolutionary origin of the condensing enzyme that unites a molecule of acetic acid from acetyl-CoA to oxaloacetic acid (Figure 8.19). As this late step in the evolution of the tricarboxylic acid cycle usually employs acetic acid derived from breakdown of a sugar, it is self-evident that the cycle was not originally a device for deriving energy from glucose or other sugars produced by the chlorophyll type of photosynthesis, but was mainly involved in the biogenesis of amino acids. By means of this step,

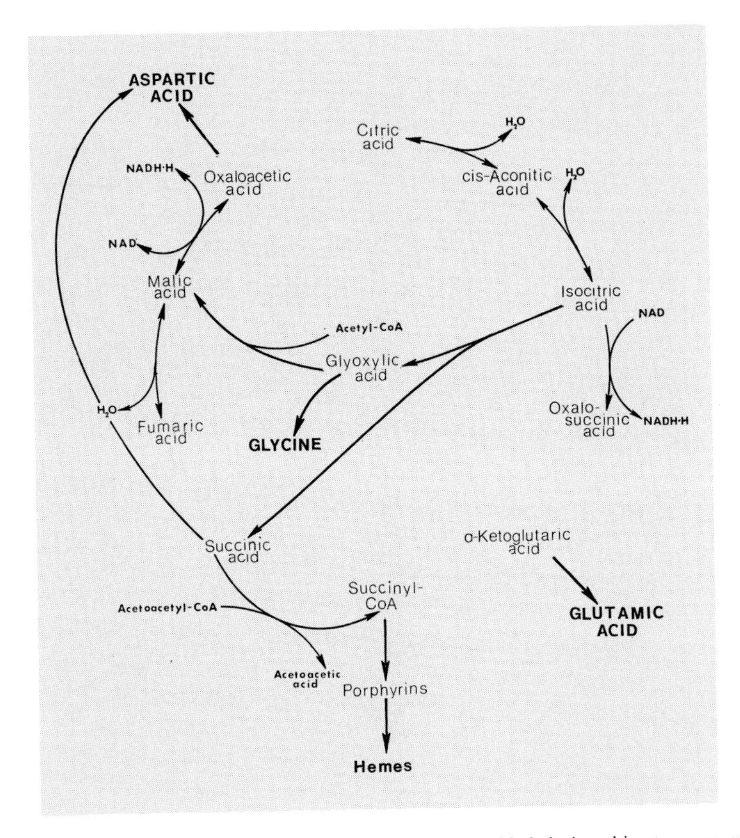

Figure 8.18. Supplementary sources for isocitric acid were added during this stage, as was the synthesis of porphyrins and hemes. These acquisitions probably were made one at a time in several successive steps.

these interrelated reactions finally became cyclic to a degree, in combination with the glyoxylic bypass. The second major event substantiates the observation concerning the original lack of correlation to photosynthesis, for chlorophyll is produced here for the first time. Doubtlessly there were—or still may be—simpler types of blue-green algae whose molecular structure of cytochromes and chlorophylls would throw much needed light on the evolutionary steps in the formation of both of these classes of biochemicals, which remain unknown now.

The Final Stages. Several reactions still need to be gained by evolutionary means to complete the cyclic set of steps. The simpler of these is the addition of the NAD-dependent decarboxylation of α-ketoglutaric acid to form succinyl-CoA (Figure 8.20), in which the second liberation of CO_2 occurs in the cycle. Thus it is evident that although succinyl-CoA is present in the in-

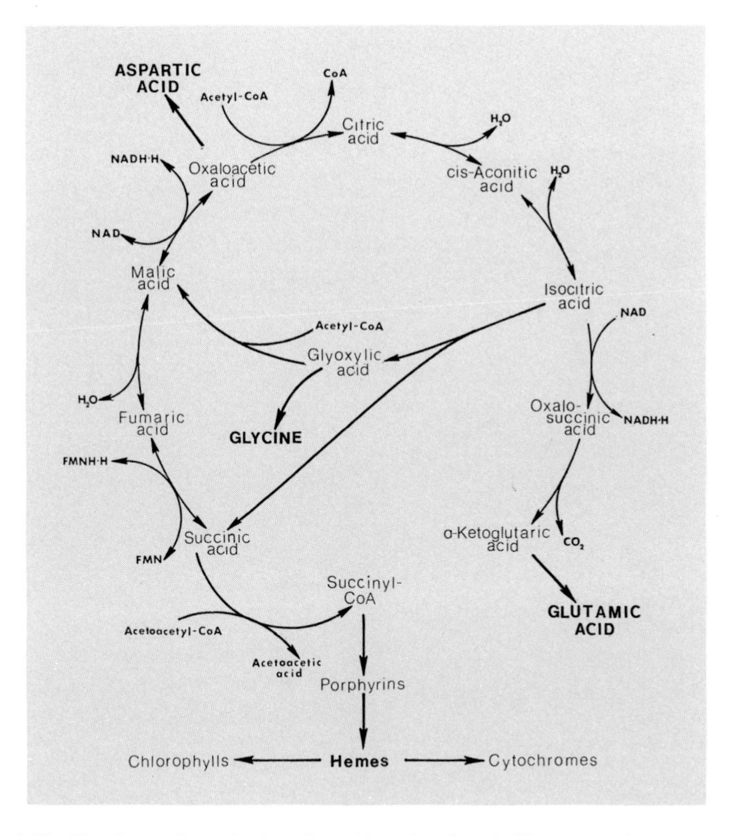

Figure 8.19. The incomplete tricarboxylic acid cycle of such blue-green algae as *Anabaena* provides a basis for cytochrome and chlorophyll synthesis as well as for amino acid and energy-transfer substances.

complete cycle of the blue-green algae, CO_2 is released at only one point, not two as in the complete series. This simple-appearing step is actually one of the two multienzyme-requiring reactions pointed out in the discussion of the cyclic reactions. Consequently, its origin late in the cycle's evolution as indicated by the evidence is quite understandable.

Whether any organisms will be found to contain the second additional step beyond those found in the blue-green algae remains for the future. But the final gap to be closed is the second of the multienzyme processes leading from succinyl-CoA to succinic acid. Available data do not disclose whether this last reaction was actually acquired in two steps or not, but possibly the first several reactions led to an intermediate product, succinyl phosphate being a likely candidate. This suggestion is supported by the uniqueness of the reaction that completes the cycle in which succinic acid is synthesized accompanied by the

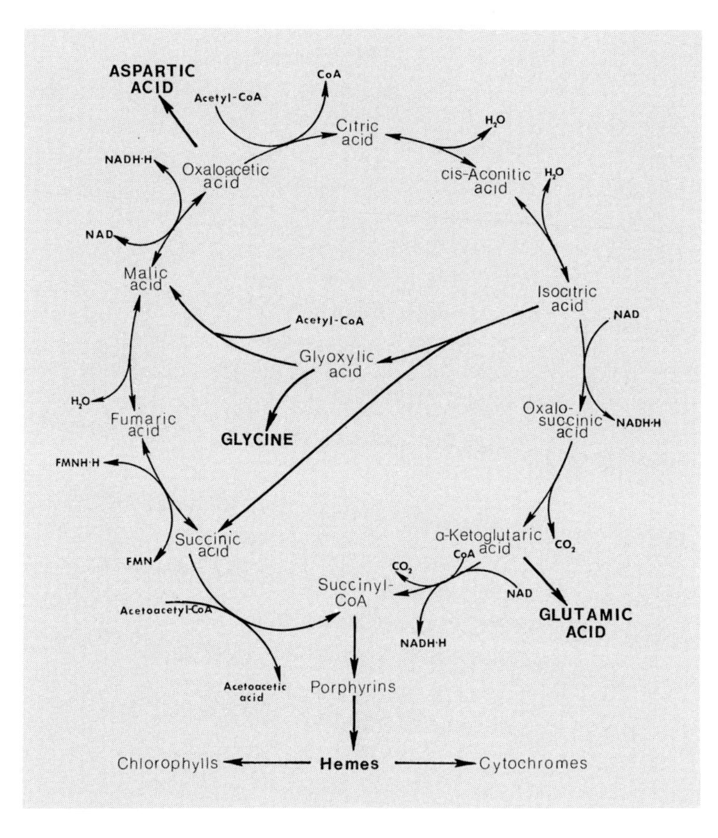

Figure 8.20. The nearly complete tricarboxylic acid cycle. The complete one is illustrated in Figure 8.13.

formation of one unit of GTP (or ITP), without intervention of the cytochrome chain. However the last gap may have been closed, it is known to be complete in all eubacteria and identical in its essential features from these organisms throughout the eukaryotes (Figure 8.13), except for the loss of the glyoxylate bypass in the metazoans.

Functional Evolution. It may be profitable to reexamine the foregoing hypothetical steps from the standpoint of function. That the first three steps suggested are primarily involved in the biosynthesis of amino acids is self-evident; only the last of these results in a by-product recognizable as a member of the energy-transferal compounds in extant organisms. Whether this NADH served in some such capacity or merely in dehydrogenating reactions is not clear now, because the metabolism of the simplest organisms still remains insufficiently explored. However, what is known of *Beggiatoa*, a form lacking cytochromes that represents an even higher level, suggests that the involvement of NADH in energy production is quite unlikely at this stage in evolutionary development.

Even through the fourth and fifth stages, respiratory functions cannot be ascribed to the evolving cycle on any available evidence, for molecular O_2 is still not employed as far as can be perceived, nor is it apparent how ATP is biosynthesized. In primordial organisms it was probably derived from the environment, where its presence was demanded for the origin of the nucleic acids. Or perhaps some ATP was engendered by the nonchlorophyll type of photosynthesis, such as is presently found among the colorless sulfur bacteria that represent the level under discussion. That the atmosphere was still of either the reducing or the oxidized type when these levels arose is indicated by the persistent absence of chlorophyll-dependent photosynthesis.

Only after the advent of the porphyrin- and the heme-protein-synthesizing systems found in the sixth and seventh stages that permitted chlorophyll-based photosynthesis to evolve could the atmosphere gradually be converted to an oxidizing type. Because the conversion would require a considerable span of geologic time, the first chlorophyllaceous organisms (blue-green algae) probably used oxygen from the breakdown of water that was trapped intracellularly for their respiratory needs. Furthermore, fermentative and other known anaerobic respiratory processes, including glycolysis, could not have existed until the familiar variety of photosynthesis arose, for only then would glucose and its degradation products, perhaps excluding pyruvic acid, assume positions of importance in the cell's economy, as discussed more completely in a succeeding chapter. On this basis anaerobic respiration of the types that have been investigated to date is a more recent event than aerobic, although other means doubtlessly existed. But until the metabolic processes of such extant organisms as *Beggiatoa*, *Clostridium,* and other very early forms have received thorough analysis, no hint as to the possible origins of these most fundamental activities of all living things can be made. Unfortunately, the eubacteria, which have been nearly universally treated as primitive, are actually far too advanced to throw much light upon the earliest developments.

In viewing the evolution of the cycle in retrospect, problems of even greater moment may be perceived. Throughout the developmental history, whether by the steps proposed here or by any other means, new sets of enzyme systems are brought into existence, one by one. But from where do the necessary catalyzing proteins come? And how does the sequence of nucleotides that codes each enzyme become incorporated into the nuclear DNA molecule? These are problems that have attracted much attention, but the proposed solutions remain unsupported by firm evidence. For example, one current theory suggests that the new substances are produced by modification of existing ones, but this explanation presents problems too. As the original proteins are likely to have been serving a useful function within the cell, modifications to them might be deleterious and even make them incapable of performing their original tasks.

Hence, it has also been hypothesized that the gene encoding the original enzyme underwent duplication and later one of the duplicate copies became mutated into the new protein. While existing evidence supports the idea that isozymes may have arisen in this fashion, for example, the ferredoxins I and II of chloroplasts (Hase *et al.*, 1977b; Matsubara *et al.*, 1978), no data support the concept that any protein having a new function has evolved in this fashion. In view of all the enzyme systems that had to develop to mediate the complete citric acid cycle, quite a number of fortuitous events need to have occurred. At any rate, the gaps in the cycle that exist between the levels respectively represented by *Beggiatoa* and the blue-green alga, *Anabaena*, and again between that of the latter and the true bacteria appear to be ideal subjects for investigations into these very basic problems.

9

ENERGY-ORIENTED ORGANELLES AND ACTIVITIES: II

The Mitochondrion

Among eukaryotes the cyclic respiratory processes just described occur for the greater part in an organelle of the cytoplasm known as the mitochondrion, while glycolysis and related activities are confined to the cytoplasm. In the prokaryotes mitochondria are absent, but many types of those organisms possess a membranous body in which the citric acid cycle may proceed. Because the eukaryotic organelle has been far more extensively explored, its structure receives attention prior to the simpler bodies found in bacteria and their relatives. Although the primary function of the mitochondrion is in cell respiration, it seems to be involved in numerous other aspects of the cell's economy, as attested by the differing enzyme systems found from tissue to tissue. Most of these roles remain unknown, but some are coming to light at the current time. For instance, in earthworm spermiogenesis it plays an evident part in the condensation of the chromatin in the nucleus (Figure 9.1; Martinucci and Felluga, 1979). Moreover, it has been found to be active in mediating the action of luteinizing hormone in the synthesis of steroids in Amphibia (Wiebe, 1972).

9.1. MITOCHONDRIAL STRUCTURE

As a consequence of both the relative ease with which its fine morphology is revealed and the innumerable experiments that have elucidated its major functions, the mitochondrion doubtlessly ranks among the best known organelles of the cell. Yet, well explored as it is, much about it still remains either entirely unknown or subject to strongly conflicting interpretations. Furthermore, the vast majority of electron microscopic and experimental investigations have been made upon the mitochondrion of but a single group of organisms,

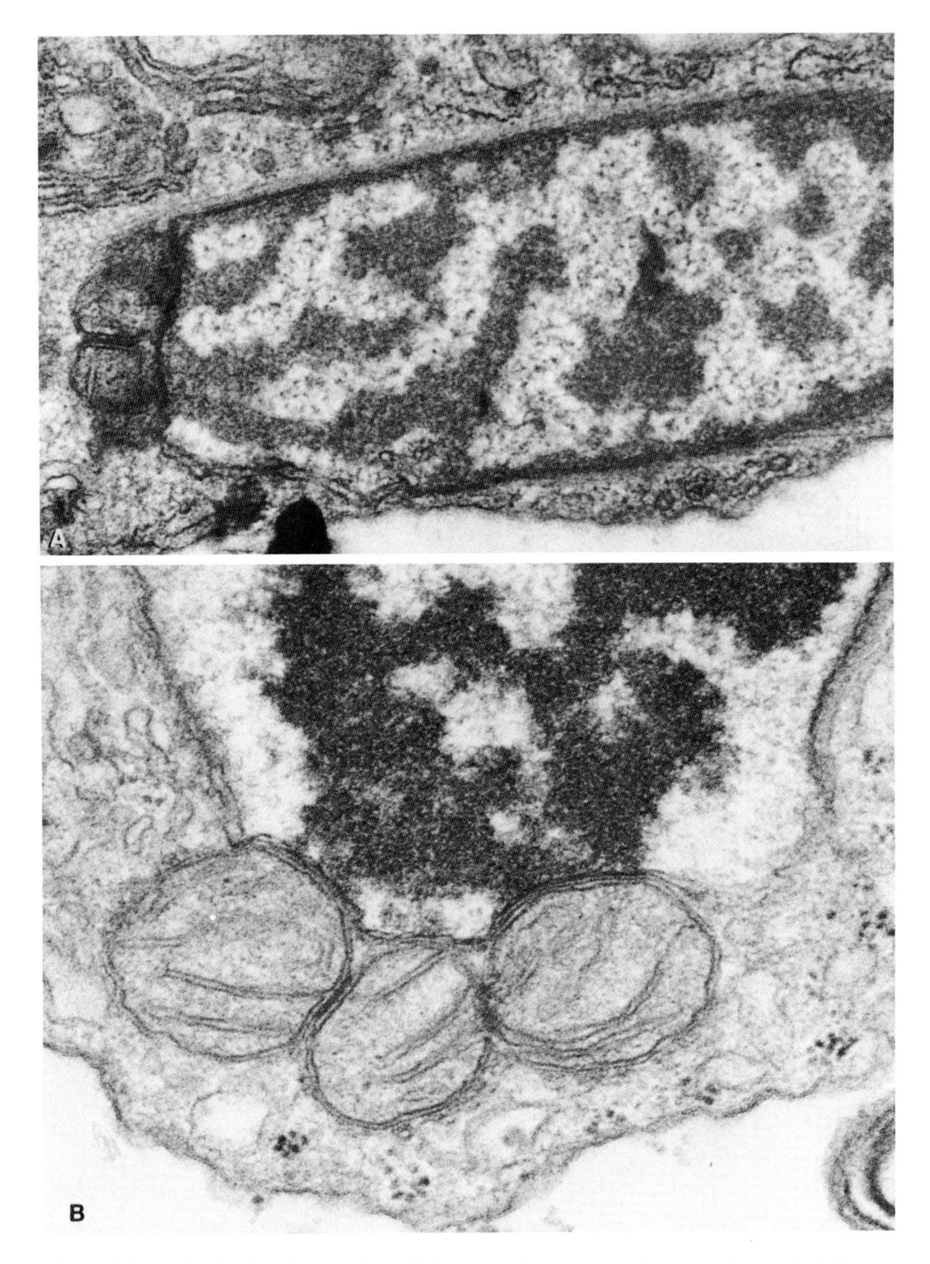

Figure 9.1. A role for the mitochondrion during spermiogenesis. During spermiogenesis in lumbricoid annelids, the mitochondria play a major role in condensing the chromatin into bodies called prochromosomes. (A) A number of electron-opaque prochromosomes appear in the nucleus. 20,000×. (B) In this enlarged view of the end of the foregoing, the relationship between condensing chromatin and three mitochondria is made clear. 63,000×. (Both courtesy of Martinucci and Felluga, 1979.)

the metazoans—or more specifically, the vertebrates. Because of this concentration of knowledge within a single area, it is advantageous first to discuss the organelle found among those animals, even though it be an advanced type, and then to make comparisons with the others later.

9.1.1. The Mitochondrion of the Metazoa

The Standard Structure. So familiar is the standard ultrastructure of the metazoan mitochondrion that it is almost superfluous to describe it here (Steinert, 1969). According to the familiar interpretations of the morphology, the organelle is enclosed by two parallel, adjacent membranes, the narrow space between being the outer compartment (Whittaker, 1966). The large inner compartment (Figure 9.2A) is filled with a more or less homogeneous ground substance, usually called the stroma or matrix. Across the stroma extends a number of cristae, membranous structures that typically are flattened sacs, usually thought to be formed by invagination of the inner enveloping membrane. Within the lumen of each crista (the intracristal space) is a limited quantity of material that typically is somewhat more electron transparent than the stroma itself (Figure 9.2A).

The familiar view of the organelle is currently undergoing a somewhat radical change in some of its features as a result of improved techniques. First, it was shown that in mitochondria from tissue that was frozen within a few sec-

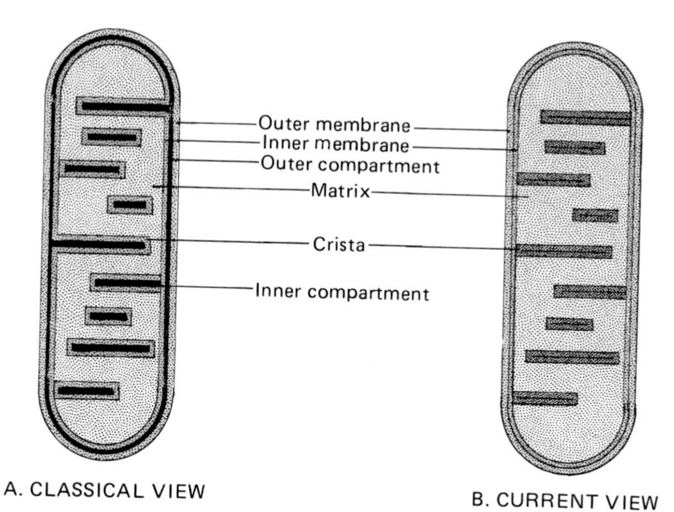

Figure 9.2. Two views of standard mitochondrial structure. Techniques currently being developed suggest that neither an outer compartment nor the intracristal space actually exists (B), as they do when prepared by standard techniques (A).

onds after death, the two sides of the cristae were appressed, so that they formed a single double membrane without an intracristal space separating them (Malhotra, 1966). Moreover, with this treatment, no outer compartment was present (Figure 9.2B). Avoiding use of osmium tetroxide, which has a deteriorating effect on membranes, and combining freeze-drying with low-temperature embedding in plastic, Sjöstrand and his co-workers confirmed and extended these observations (Sjöstrand and Kretzer, 1975; Sjöstrand and Bernhard, 1976; Sjöstrand, 1977, 1978; Sjöstrand and Cassell, 1978). Thus it appears that in living material, the mitochondrion consists simply of a dual-membrane enclosure containing a matrix-filled inner compartment that is traversed by a number of flat double-membraned structures called cristae (Figure 9.3B).

Although the major ultrastructural characteristics are remarkably consistent throughout the Metazoa, variations upon the general theme are often encountered. For example, the number of cristae per mitochondrion varies from one tissue to another, being high in striated muscle and low in kidney and liver cells. Sometimes, as in kidney cells of overwintering frogs, the cristae parallel the longitudinal axis of the mitochondrion, whereas those of active summer frogs are of typical construction. In the inactive frogs, the longitudinal orientation is associated with low cytochrome oxidase activity (Karnovsky, 1963). A further example of this variation is found in the mature eggs of *Tubifex*, a freshwater annelid, and still others have been reported from pineal cells of the rat (Lin, 1965) and human uterine mucosa cells during the midmenstrual period (Merker *et al.*, 1968). In addition to a frequent elongate condition, the mitochondria of the pineal cells were often filled with a number of fiber-containing cylinders that replaced most of the central cristae.

Modifications in Chloride Cells. In the gills of a number of marine organisms, devices for eliminating salt from the body, or conversely, absorbing it from the sea, have been developed that involve the mitochondrion, often highly modified for the function. In certain fish, including *Fundulus heteroclitus* (Philpott and Copeland, 1963), the mitochondria have the cristae oriented longitudinally, not unlike the arrangement just described in hibernating frogs. Moreover, a second more diversified type is present, but its discussion needs to be reserved for a later section. Still different mitochondrial modifications exist in the sea lamprey, in which this organelle is associated with a proliferation of microtubules (Figure 9.4A; Peek and Youson, 1979).

A greatly modified variant of the cristate mitochondrion, however, has been described from salt cells of the blue crab (*Callinectes sapidus*). In this arthropod these specialized cells are most frequently employed for absorbing salt from the medium when the animal is living under low saline conditions (Copeland and Fitzjarrell, 1968), but they can also serve in the opposite capacity and secrete salt during high salinity periods (Mantel, 1967). While the usual form of mitochondrion is also present, specialized ones are arranged singly within the numerous, close-set pinocytotic folds of the plasmalemma. This

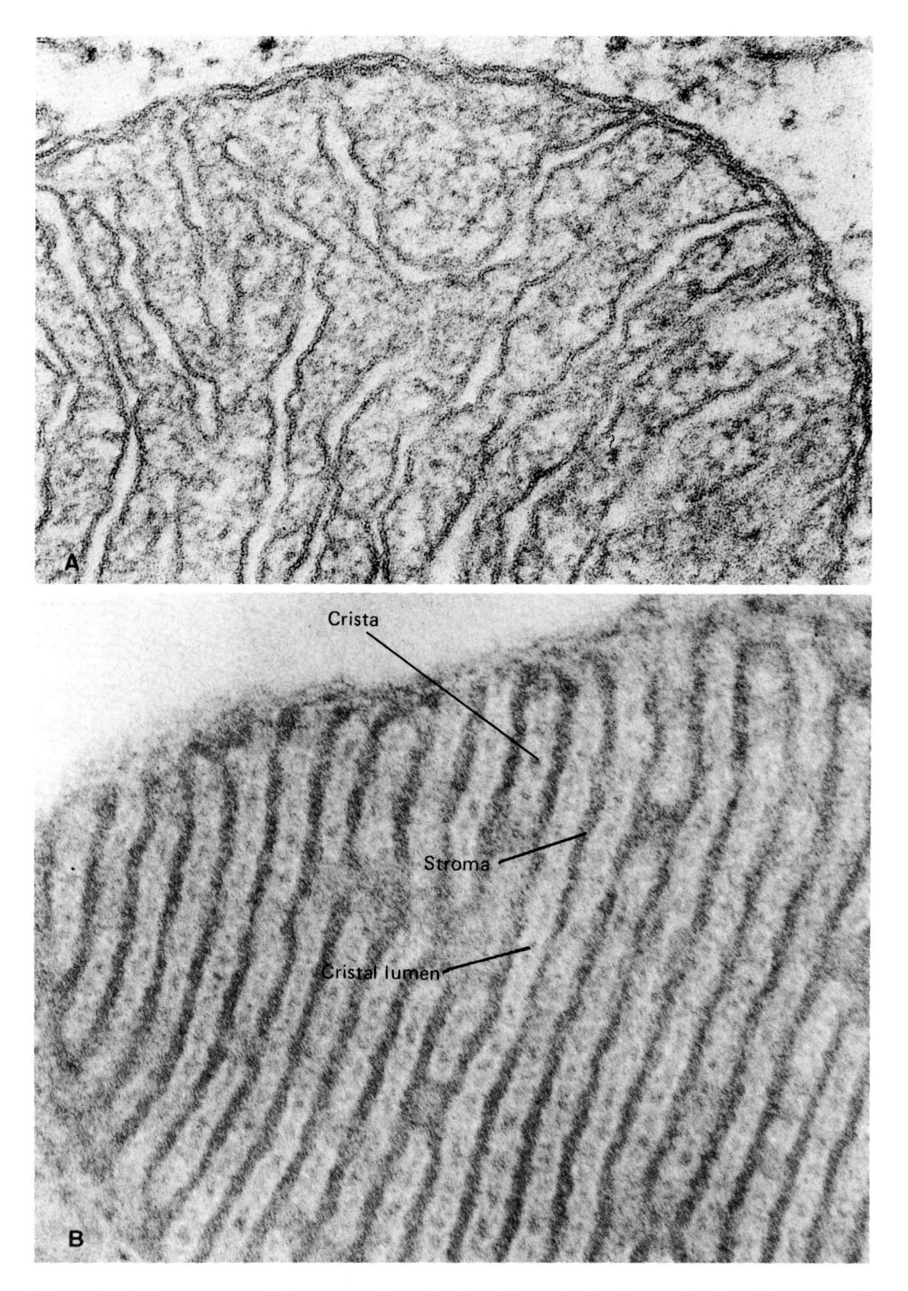

Figure 9.3. Ultrastructure of the mammalian mitochondrion. (A) Rat heart mitochondrion prepared by standard techniques shows much stroma and distinct intracristal lumina. 120,000×. (B) Same but prepared by modified techniques. The stroma is confined to narrow interspaces between the compressed cristae. 180,000×. (Both courtesy of Sjöstrand, 1977.)

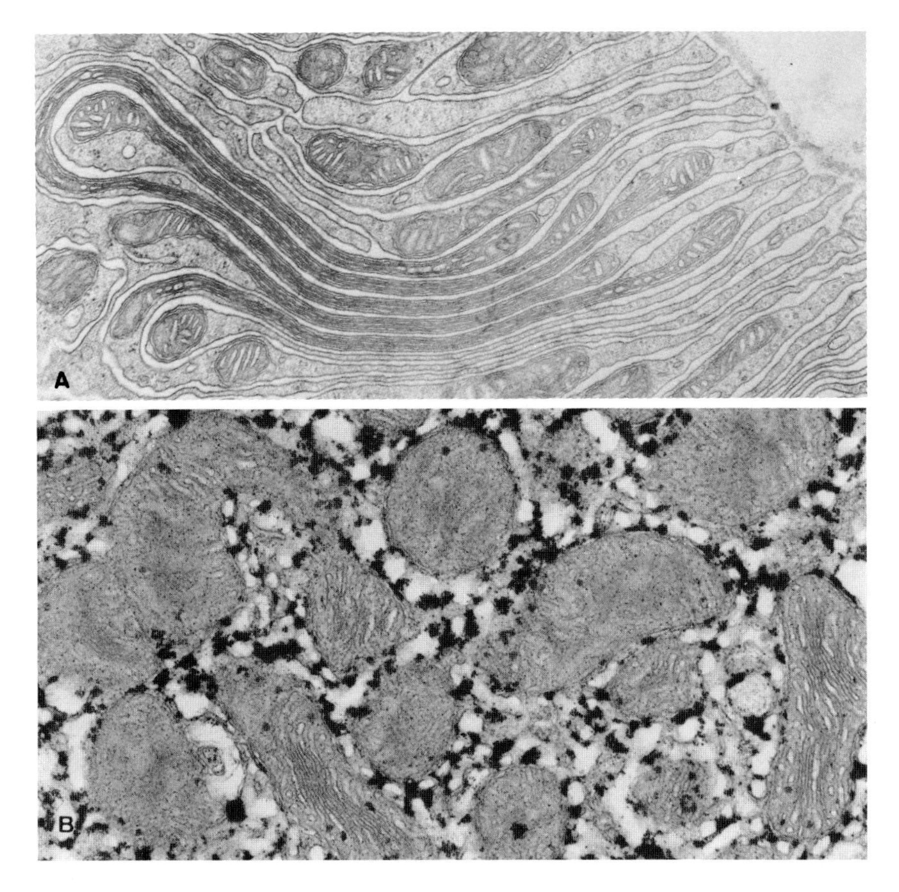

Figure 9.4. Modifications of the metazoan mitochondrion in salt cells. (A) Cup-shaped mitochondria in the salt cells of the blue crab. 17,000×. (Courtesy of Copeland and Fitzjarrell, 1968.) (B) Section of epithelium of the chloride cells of the pinfish, treated to show sites of Na^+ + K^+-ATPase activity, most of which is localized in the plasmalemma labyrinth rather than in the mitochondria. 26,000×. (Courtesy of Hootman and Philpott, 1979.)

modified type is greatly elongate and has the periphery swollen; often it is discoidal but frequently may be infolded into a cup-shaped configuration (Figure 9.4A). In the prothoracic gland of fourth-instar larval silkworms, the mitochondria are similarly elongate and cup shaped, with a single crista, but the enlarged peripheral portion is vesicular rather than cristate (Beaulaton, 1968). Since most of the Na^+ + K^+ ATPase activity has been shown to be located in plasmalemma folds (Figure 9.4B), the mitochondrion may simply supply the necessary energy (Hootman and Philpott, 1979).

The Microvillose Pattern. In several instances among the Metazoa, the organelle displays a morphology so strikingly different that it is precluded from being considered a mere variant of the familiar type. Instead of flat disks, numerous tubular cristae, often called microvilli, penetrate the stroma from all sides (Figures 9.5, 9.12B). Among reptiles, birds, and mammals, this microvillose variety appears mainly to characterize cells of the adrenal cortex, but it also has been reported from the tail muscle cells of larval *Xenopus*. Because most of the details of structure somewhat resemble those of the organelle in many ciliated and amoeboid protozoans and other protistans discussed later, it appears to be more primitive than the cristate form. Unfortunately, almost no attention has been given to the finer particulars of its structure, function, or molecular organization.

A modification of the foregoing, although occurring most frequently in such active cells as those of skeletal and cardiac muscle, has been observed also in numerous other tissues and in a great variety of metazoans. In this variant, the microvilli are bent sharply several times into a zigzag pattern. As a rule they are more highly ordered than in the commoner microvillose type, forming several rows down the length of the organelle, with the denticles of each row dovetailing into the adjacent ones. Some chloride cells from gills of *Fundulus heteroclitus* have this pattern, not the longitudinal cristae noted above; the same cells from the skin of a very young sardine (*Sardinops caerulae*) have a similar structure (Lasker and Threadgold, 1968). Other examples of this type among metazoans include gastric mucosal gland cells of the frog and the right ventricular papillary muscle of the cat.

Another variation of the microvillose type has been reported from the vinegar eelworm (Zuckerman *et al.*, 1973). Although most of the mitochondria were of the normal flat cristate type, an occasional large one was encountered in which the cristae were tubular, with greatly thickened membranes (Figure 9.5). In addition, a variable number of paracrystalline arrays were present. Still another variant of the microvillose type occurs in a number of metazoans, including such diversified subjects as bats (pancreas and cricothyroid muscles), grass frog (gastric mucosa), salamander (liver), and the medicinal leech (nervous tissue) (Revel *et al.*, 1963). In addition, it is found with the other types mentioned in preceding paragraphs as occurring in the gill chloride cells of *Fundulus*. The outstanding characteristic of this variety is the prismatic form of the cristae, which are triangular in cross section. As a rule, the prismatic tubules are located amid others of the usual type, often forming a central bundle as illustrated (Figure 9.6).

Differences between metazoan mitochondria are not confined to structural features but also are being demonstrated at the functional level, as exemplified by a study on the rat liver cell (van Berkel and Kruijt, 1977). In this mammal, about 65% of the liver mass consisted of parenchymal cells (hepatocytes) and about 10% nonparenchymal. Isolation of pure fractions of intact cells of each type permitted the demonstration of distinct differences in the activities of the

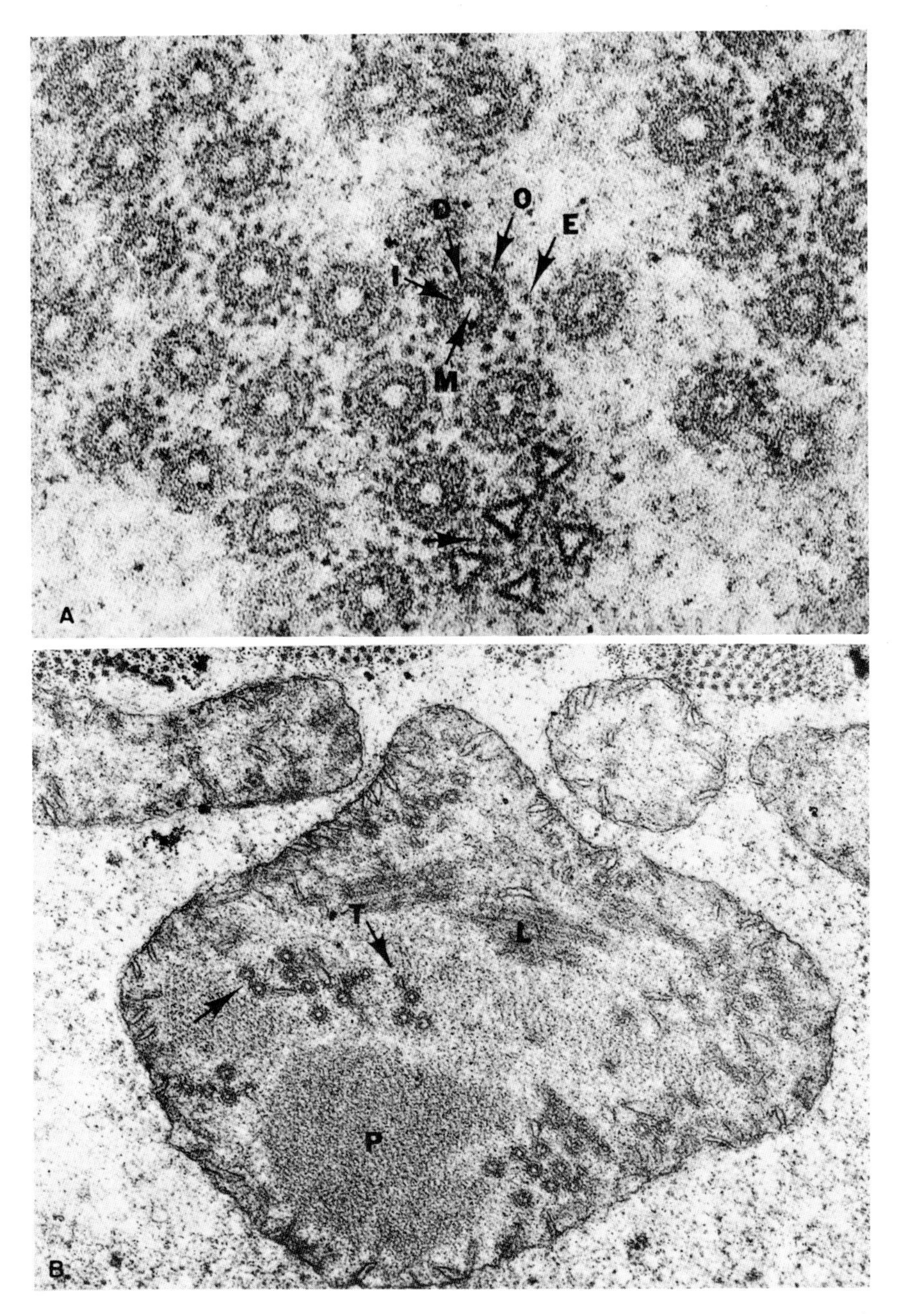

Figure 9.5. Unusual cristae in a microvillose mitochondrion. (A) The mitochondrion of the vinegar eel (*Turbatrix aceti*) consists of short microvilli around the periphery, longer modified ones centrally, and paracrystalline arrays. E, capitate particles; D, matrix; O, outer membrane; I, inner membrane; M, lumen. 33,000×. (B) The modified cristae are doughnut shaped and bear capitate particles over their surface; a few that are triangular in cross section are also evident. P, paracrystalline arrays; L, longitudinal section; T, transverse section. 150,000×. (Both courtesy of Zuckerman *et al.*, 1973.)

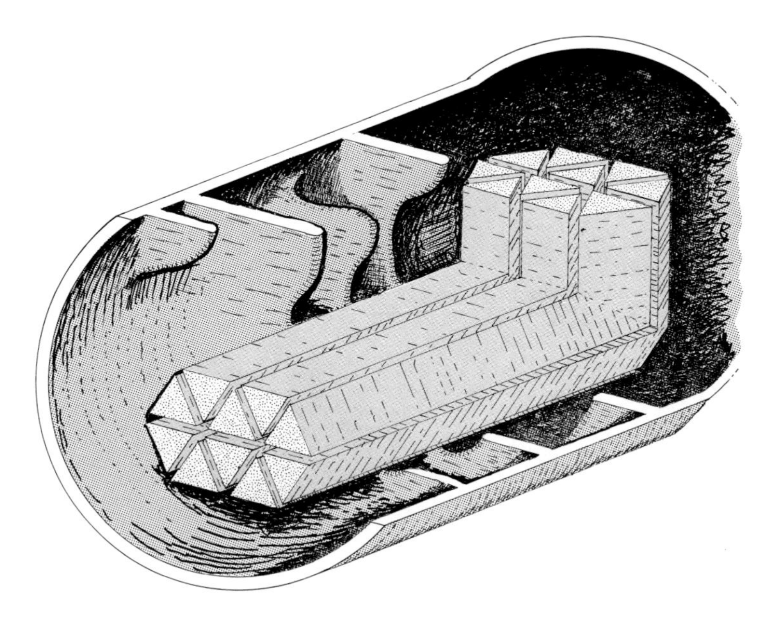

Figure 9.6. A prismatic variation of the microvillose mitochondrion. (Based on Revel *et al.*, 1963.)

mitochondrial enzymes. Except for monoamine oxidase, which was equally active in both cell types, the specific activities were generally lower in nonparenchymal cells. Pyruvic acid carboxylase activity was only 2% that of the parenchyma, glutamic acid dehydrogenase 4.3%, and cytochrome *c* oxidase 79.4%. Moreover, only one form of the pyruvic acid carboxylase was present, instead of the two of the hepatocytes.

Physiological-State Effects. Recent studies indicate that the actual morphological state of any given mitochondrion is associated in large measure with physiological conditions (Green *et al.*, 1968; Penniston *et al.*, 1968; Harris *et al.*, 1969; Cieciura *et al.*, 1978, 1979). A "nonenergized" (orthodox) pattern, in which the cristae are flat disks as in the typical metazoan organelle, is seen in electron micrographs of rat heart mitochondria prepared by the usual techniques. Similar configurations are obtained by use of 2,4-dinitrophenol, a reagent that discharges the energized state of the membranes. Under energizing conditions, such as those induced by the presence of a 10 mM concentration of sodium succinate, the cristae become swollen (condensed or energized) or swollen and zigzagged (energized-twisted) (Hackenbrock, 1966, 1968). Mitochondria from other tissues (Figure 9.7), including rat liver and kidney, ox retina, and canary flight muscle, react in identical fashion (Vail and Riley, 1972; Grimwood and Wagner, 1976). Similar changes have been noted to occur in the eggs of sea urchins. In the unfertilized egg of these animals, the

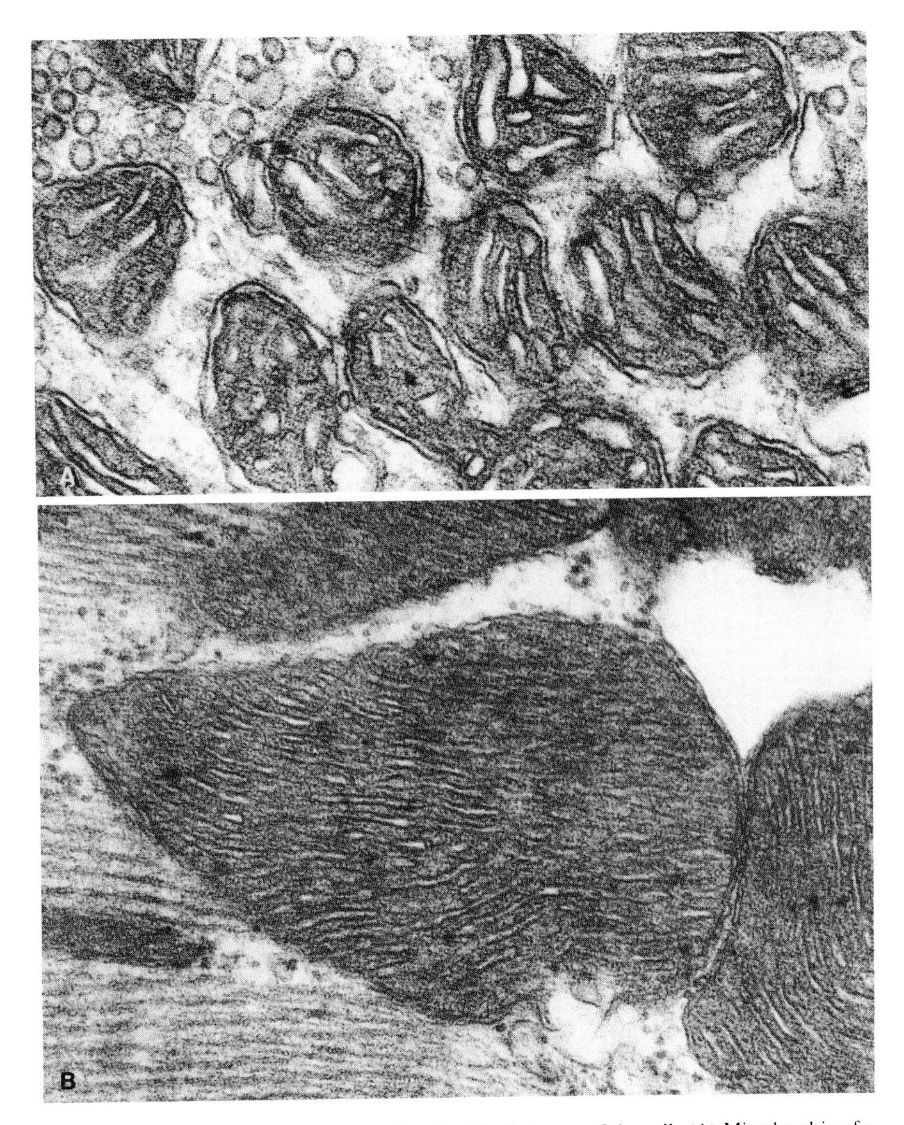

Figure 9.7. Changes in morphology with the physiological state of the cell. (A) Mitochondria of a mossy fiber of the granular layer of the rat cerebellum approach the condensed, or energized, state. 67,500×. (B) Mitochondria of rat heart muscle cell. 67,500×. (Both courtesy of Cieciura *et al.*, 1979.)

mitochondria were found to be in the condensed (energized) configuration, but all were in the orthodox form after fertilization had occurred (Innis *et al.*, 1976).

Effects of the physiological state of the organism as a whole have also been reported on numerous occasions, several of which, such as overwintering frogs, have already received mention. Hibernation in rodents, such as the dormouse, induced alterations in mitochondria of the kidney tubule cells (Amon *et al.*, 1967). These organelles accumulated ferritin as the cristae degenerated and the matrix became more electron opaque. Near-freezing temperatures in the environment altered both the size and the number of cristae in rats, the organelle of brown adipose tissue becoming considerably enlarged and the cristae more numerous (Suter, 1969). These changes were reversed when the animals were returned to warm temperatures for a week or more. Moreover, prolonged starvation in the Japanese newt led to the development of three types of inclusions (Figure 9.8; Taira, 1979).

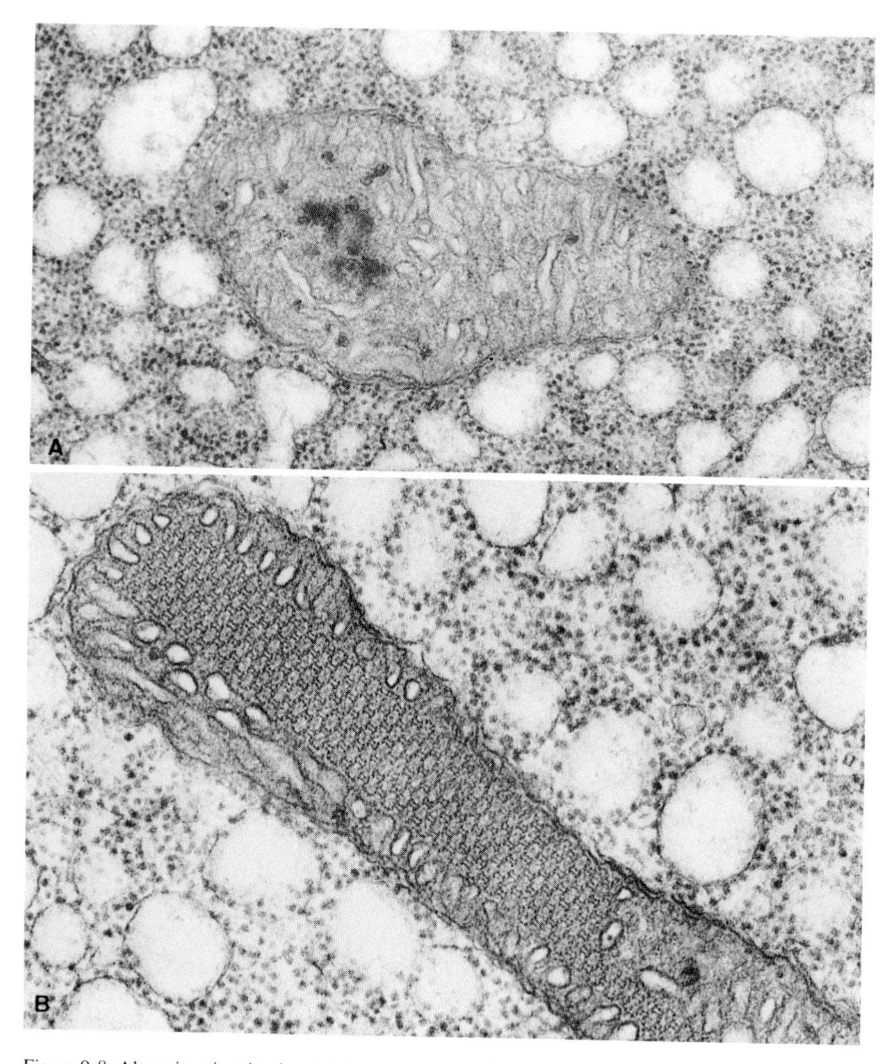

Figure 9.8. Alterations in mitochondrial structure induced by starvation. Three types of inclusions were found in mitochondria of Japanese newt cells after 1 week of starvation, including electron-opaque bodies (A) and crystalloids (B). (A) 43,000×; (B) 52,000×. (Both courtesy of Taira, 1979.)

One case involving reproductive physiological events has already been enumerated, but many others are known. For instance, in the females of the genus *Locusta*, the mitochondria of the corpora allata became greatly enlarged as sexual maturation proceeded. With this enlargement, the cristae elongated and became hooplike, while the matrix increased in electron density (Fain-Maurel and Cassier, 1969). When the ovaries commenced to ripen, the second type of this organelle developed, having a dumbbell shape and longitudinal cristae. Moreover, morphological and physiological changes have been noted in the germ cell mitochondria of rats (DeMartino *et al.*, 1979). In secondary spermatocytes and again in spermatids, those organelles assumed a rounded configuration and the condensed arrangement of the cristae. At the same time, their respiratory control ratio was very high, suggesting that they were functionally very active.

9.1.2. The Mitochondrion of Higher Plants

Although the mitochondrion of the Metaphyta has received considerable attention (Hackett, 1959; Diers and Schötz, 1965; Parsons *et al.*, 1965), it has not been studied nearly so intensively as that of the metazoans. Perhaps some of the relative neglect of the green plant organelle results from the greater difficulties associated with its isolation and preparation, or perhaps it stems in measure from the greater attention focused by plant cytologists upon that unique organelle of their subjects, the chloroplast. Nevertheless, enough has been learned that the major features of its structure and biochemistry may be outlined.

Structure. The mitochondrion of higher plants is generally considered similar in structure to the metazoan organelle (DeRobertis *et al.*, 1970); close comparison, however, discloses a number of small but persistent differences. By and large, in the plant there is a far higher percentage of stroma present, with the cristae few in number and tending to be irregular, both in form and disposition (Figure 9.9). Frequently, the plane in which the expanded portion of the cristae is oriented varies widely, so that in a single surface exposed by sectioning, some of these structures are cut longitudinally and others transversely. As microvillose cristae often are interspersed among flattened discoidal ones, electron micrographs of the higher plant mitochondrion reveal a strange mixture in the stroma, consisting of cylinders sectioned at various angles, ghostlike images of membranes cut parallel to their surfaces, and the familiar partitionlike formations of metazoans. Moreover, the outer membrane appears to be completely covered by minute pits; by actual measurement in white potato mitochondria, the pits were found to be from 25 to 30 Å in diameter and spaced at about 45-Å intervals, center to center (Parsons *et al.*, 1965).

In mesophyll cells and others actively engaged in photosynthesis, the cristae of the mitochondria are more close set and tend to be relatively short;

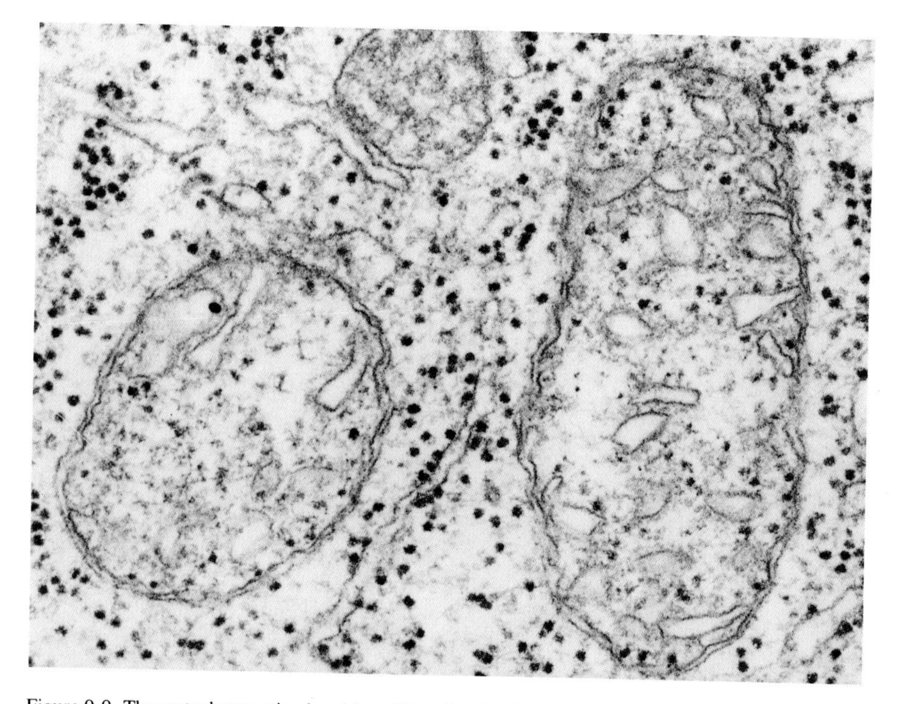

Figure 9.9. The metaphytan mitochondrion. The mitochondrion of higher plants contains a mixture of inflated saclike and tubular cristae, sparsely placed, as in this root tip cell of corn. 40,000×. (Courtesy of Hilton H. Mollenhauer, unpublished.)

frequently, they merely fringe the periphery, leaving a central area free. The few studies by freeze-etch techniques (e.g., Branton and Moor, 1964) have thus far contributed little to further understanding of the mitochondrion's morphology but confirm the observations made by standard procedures.

9.1.3. The Mitochondrion of Other Eukaryotes

Among the remainder of the eukaryotes, mitochondrial features distinctive for the respective taxa are disappointingly meager. In broad terms, the amoeboid and ciliated protozoans possess mitochondria with microvillose cristae, a condition that is also catholic among most of the eukaryotic algal groups, including both chlorophyllaceous and colorless flagellates. Only the green algae are exceptional to some extent, in that the organelle, not surprisingly, more closely approximates that of the metaphytans (Lang, 1963; Lembi and Lang, 1965; Lloyd and Venables, 1967). Among important variants known to exist are the several that follow.

Mitochondria of Fungi. In general, the mitochondria of fungi are like those of other protistans in being of the microvillose type (Figure 9.10), the microvilli being rather coarse and greatly distended, as in *Synchytrium* (Lange and Olson, 1978). Moreover, they are relatively long, usually half the width of the inner chamber and sometimes nearly fully as long. How many of these traits would alter with the use of more recent techniques remains for the future to disclose. In some cases the overall appearance differs from the preceding description, the microvilli being fine in cross section, relatively sparse, and shorter than half the width of the inner chamber. In addition, there is a Feulgen-positive central body (Kuroiwa *et al.*, 1976a,b), which can be stained with thionine for viewing by light microscopy. Under the electron microscope, it is seen as an electron-dense cylinder, oriented with the long axis of the mitochondrion (Kuroiwa *et al.*, 1977). This body has been shown to contain a large fraction of the DNA present in the organelle and has been observed to undergo division in unison with the rest of the structure (Kuroiwa, 1974). During these processes, the central cylinder, called the nucleoid, elongates as the entire organelle does likewise; then after a cleavage furrow has formed around the equatorial region of the mitochondrion and has eventually deepened, it becomes divided into two, much as in the bacterial nucleoid (Chapter 11). The close affinity to the other fungi of the so-called protozoan taxon Mycetozoa is clearly indicated by the presence of this unique feature in their mitochondria, as shown by *Didymium* (Schuster, 1965). Among botanists this genus is already classed as a slime mold.

In addition, bodies somewhat resembling the mesosome of bacteria, described later in this chapter, have been reported for various lower fungi (Zachariah and Fitz-James, 1967; Malhotra, 1968; Kozar and Weijer, 1969). However, examination of the published electron micrographs discloses that most of these structures are located within vacuoles and in some cases appear to be degenerating mitochondria. Others possibly are artifacts, as has been suggested (Curgy, 1968).

The Euglenoid Mitochondrion. At first glance the mitochondrion of *Euglena* and allies appears so similar to that of the Metazoa that it might be passed over as not distinctive; however, closer scrutiny reveals small, but consistent, differences. In the first place, the organelle is highly irregular in outline, almost as though it had become partially shriveled during preparation (Figure 9.11). While the cristae are undoubtedly flattened sacs, they are uniformly short, never extending more than halfway across the stroma, and are so broadly attached to the inner membrane that continuities between cristal and inner membranes are frequently observed. While most cristae lie with their broad surfaces parallel, a few may be set at 45 or 90° angles to the rest; consequently, face views of the sacs are commonplace in electron micrographs. In such cases, they appear irregular in outline or, more typically, bilobed.

In colorless species of euglenoids, the mitochondria are relatively more

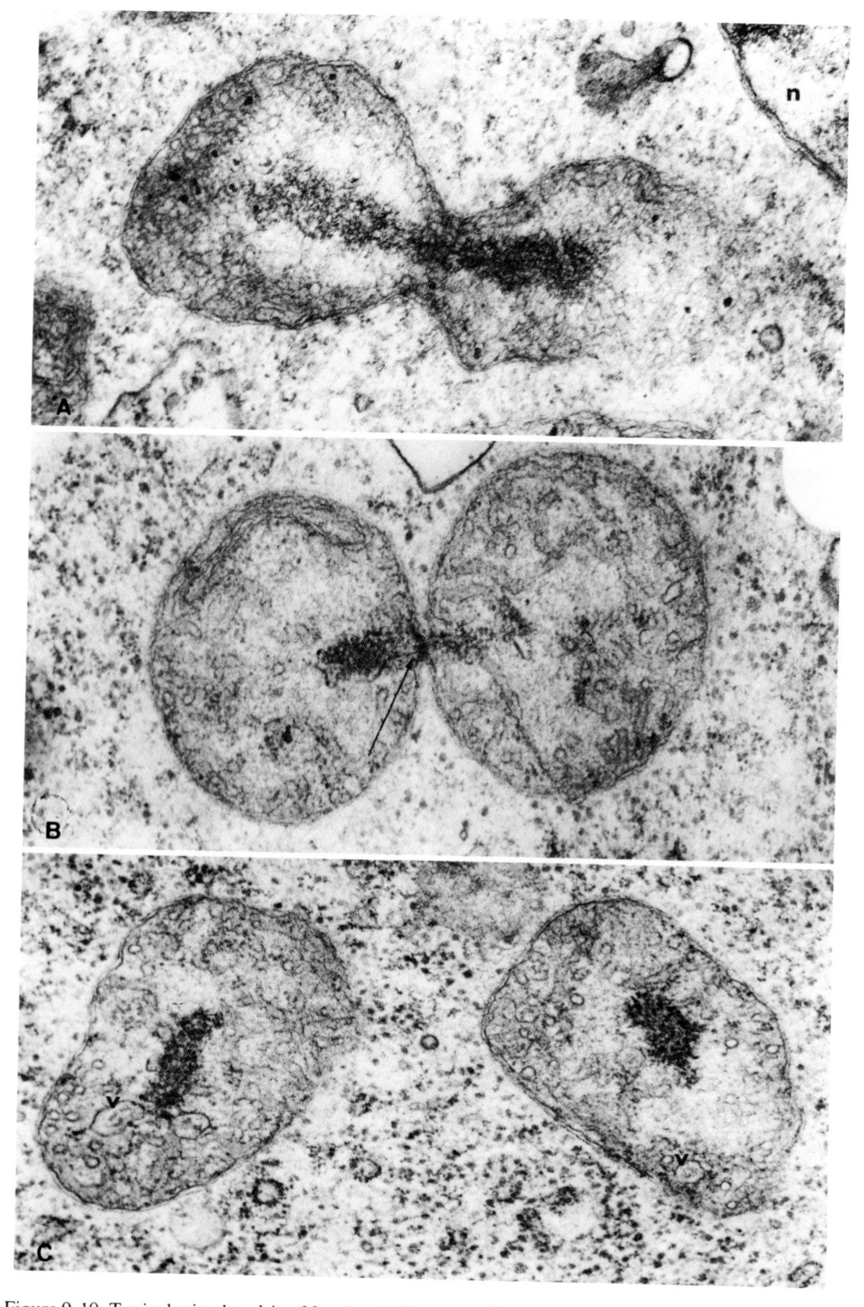

Figure 9.10. Typical mitochondria of fungi. (A) The microvillose mitochondrion of *Physarum* contains an electron-dense bar consisting in large measure of DNA, which divides as the organelle itself undergoes cleavage (B,C). All 32,000×. (Courtesy of Kuroiwa *et al.*, 1977.)

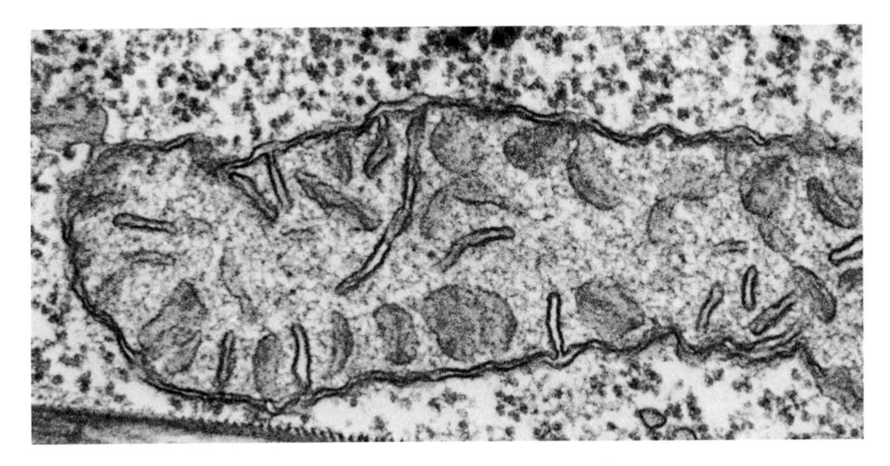

Figure 9.11. Mitochondrion of *Euglena*. This section of a mitochondrion shows the short, saclike cristae that typify this flagellate both in cross section and face view. In some species the sacs are bilobed. 72,000×. (Courtesy of Hilton H. Mollenhauer, unpublished.)

abundant within a single cell than in green forms; similarly after a green form has been treated with heat or streptomycin to inactivate the chloroplasts, the decolorized cells show a several-fold increase in the volume of mitochondria present. This increase is usually explained as reflecting the change from a phototrophic to a heterotrophic type of nutrition.

The Organelle in Other Protistans. If the organelle of *Amphidinium* may be taken as representative, the mitochondrion of dinoflagellates is relatively small, with short, rather swollen microvilli that usually are clustered close to the periphery (Dodge and Crawford, 1968). Consequently, there are broad stretches of matrix in the center of the inner compartment that are uninterrupted by cristae (Figure 9.12A). They offer a sharp contrast to the organelle of such euciliates as *Colpidium* (Foissner and Simonsberger, 1975), in which the inner compartment is nearly filled with elongate, slender microvilli (Figure 9.12B). Among radiolarians, as represented by *Aulacantha* (Ruthmann and Grell, 1964), and testaceans, including *Arcella* (Netzel, 1975), the microvilli are as long and slender as in the ciliates, but exhibit a marked tendency toward being branched. Moreover, they are much less abundant in a given mitochondrion, so that the interior is largely matrix but without the broad uninterrupted areas of the dinoflagellates. Other amoeboids, like *Gromia* (Hedley and Bertaud, 1962), have microvilli that are nearly as short and swollen as those in the dinoflagellates, but here they are slightly more elongate and much more closely set.

Among the green flagellates and filamentous algae, the mitochondrion exhibits a variety of forms. At the more primitive levels, such as *Volvulina*

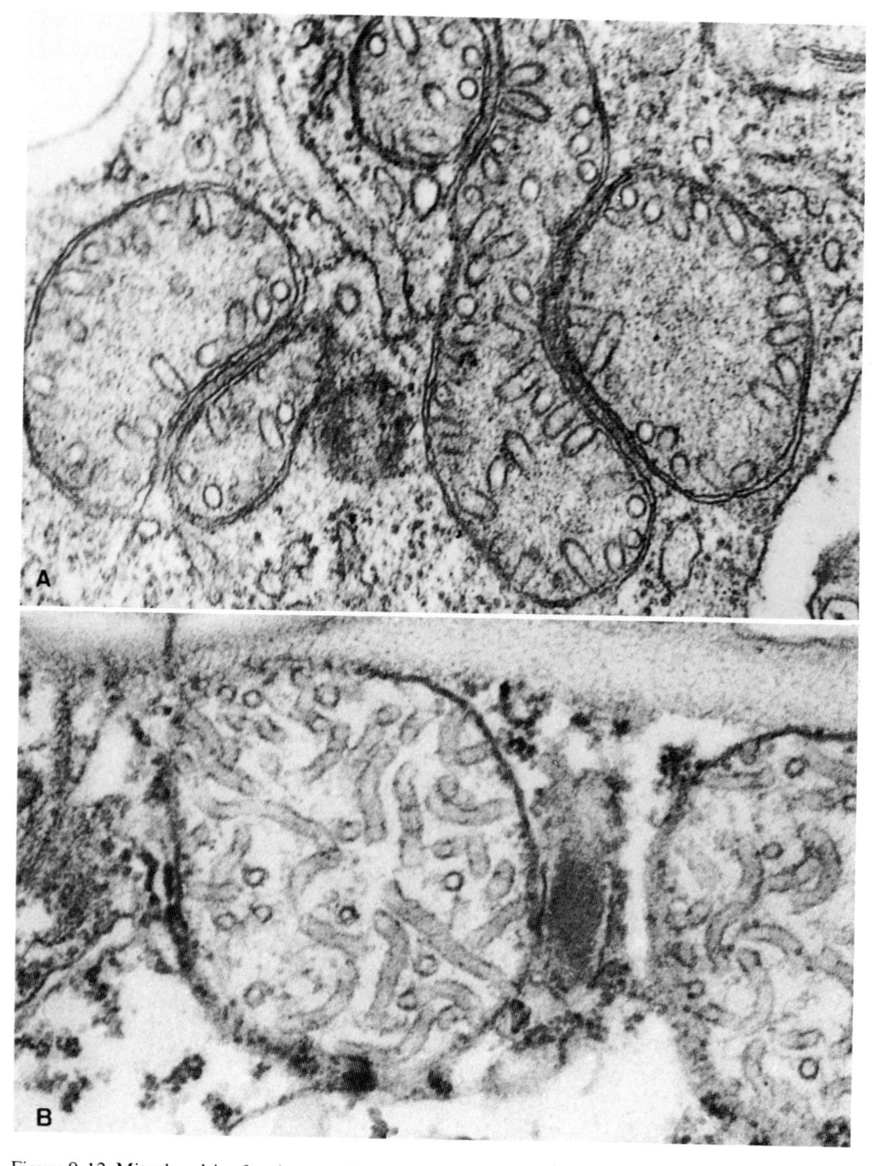

Figure 9.12. Mitochondria of various protistans. (A) The mitochondria of dinoflagellates such as that of *Amphidinium* shown here contain microvillose cristae that are exceptionally short, so that they protrude only slightly into the stroma. 50,000×. (Courtesy of Dodge and Crawford, 1968.) (B) The organelle of euciliates in general, like this one from *Colpidium,* displays the long microvilli that typify this variety in many higher eukaryotes. 56,000×. (Courtesy of Foissner and Simonsberger, 1975.)

(Lang, 1963), the microvillose type prevails, the tubules being slender and long, often reaching nearly across the inner chamber. As a rule, these are oriented directly outwards from the sides, but a few in each organelle may run longitudinally. In the less advanced of the filamentous species, the microvilli are largely replaced by paddle-shaped cristae, the blade of which in such forms as *Ulothrix* and *Klebsormidium* (Marchant and Fowke, 1977) varies in size and shape. These are often intermingled with saclike cristae and linear constituents as in metaphytans, except that these inner structures are more numerous. Quite in contrast, the mitochondria of the brown seaweeds, including *Fucus* (McCully, 1968) and *Egregia* (Bisalputra, 1966; Bisalputra and Bisalputra, 1967), approach those of certain metazoan tissues in being of the microvillose type. Usually the organelles are very densely packed with somewhat swollen, moderately long cristae. However, in *Egregia,* the tubules may be quite robust and less abundant than in *Fucus*. In the ripe sperm of these seaweeds, filamentous or tubular components extend through the entire length of the intracristal compartment (Pollock and Cassell, 1977).

The Mitochondria of Yeasts. The mitochondrion of yeasts is of particular interest, for its structure varies extensively in response to the availability of free oxygen. Under aerobic conditions it appears not too unlike that of the metazoans, with the usual double-membrance jacket and flattened cristae, but several differences may be perceived. In the first place, among the mitochondrial population of a given cell, a decided absence of uniformity is a prominent feature, whether in size, shape, number, or disposition of cristae (Thyagarajan *et al.*, 1961; Marquardt, 1962; Takagi and Nagata, 1962; Clark-Walker and Linnane, 1967; Stevens, 1974). The size, usually proportionately small for the organelle, is rarely consistent, and may at times even approach the cell in length, while the outline varies from circular to ovoid or irregular. Frequently, the cristae may be oriented both longitudinally and transversely, not to mention obliquely also, within the confines of a single organelle, and occasionally a circular arrangement can be observed (Federman and Avers, 1967). When a cristal membrane is exposed in face view, it appears as a transversely ovoid sac, often somewhat bilobed as in the euglenoids. If a number of mitochondria are examined, a discontinuity in the investing membranes is usually to be found in one or two organelles, and sometimes a set of membranes is observed to continue into the stroma as a scroll-like fold, scarcely differing from the cristae in appearance. The inconsistencies of structure and number are especially marked in the mutant forms known as petite strains (Smith *et al.*, 1969).

The presence of the discontinuities and prolongations is more readily understood after anaerobically grown yeast mitochondria are examined, especially if an obligate aerobic species is considered, such as *Candida parapsilosis* (Kellerman *et al.*, 1969). This species will not grow in the complete absence of oxygen, even in media supplemented by yeast extract, unsaturated fatty acids, and ergosterol; low oxygen tension, just sufficiently high to support growth, thus was used to induce anaerobic effects. Under these conditions, mitochondrial sections were rare, and those few organelles present contained only a few, poorly defined cristae (Fig. 9.13A). Instead, there was an abundance of elongate, detached membranes stretched through the cell, four elongate and several short examples of which appear in the electron micrograph (Figure 9.13A). Several of these seem to be continuous with the plasmalemma, nucleus, or vacuole. Like the definitive mitochondrion, the membranes have been revealed to be centers of cytochrome *c* peroxidase activity (Avers, 1967) and, consequently, appear to have a respiratory function. When grown in the presence of chloramphenicol, mitochondria and a few cytoplasmic membranes develop in the organisms, but cristae remain few and poorly defined in the former (Figure 9.13B).

Several interpretations of the events have been advanced. One group (Linnane *et al.*, 1962; Linnane, 1965; Clark-Walker and Linnane, 1967; Watson *et al.*, 1970) presents evidence that indicates the possibility that the mitochondria were formed *de novo* by the action of a population of vesicles. These saccules, containing only an electron-transparent stroma during anaerobiosis, became filled with an electron-opaque substance following the restoration of aerobic conditions. Because mitochondria in various stages of development were later found attached to the vesicles, the latter were assumed to have produced the organelles. Other workers in the area (Marquardt, 1962; Schatz, 1963; Plattner and Schatz, 1969; Plattner *et al.*, 1970), however, show "premitochondria" in their electron micrographs; these are minute bodies structurally similar to mitochondria except they lack cristae. The premitochondria then gradually form cristal folds from the inner membrane as they increase in size to become mature organelles.

If a number of ultrastructural studies of the yeast cell are examined (including some not especially devoted to the mitochondrion so as to avoid prejudice), a somewhat different concept of the origin of the organelle in aerobic yeast unfolds (Kawakami, 1961; Thyagarajan *et al.*, 1961, 1962; Takagi and Nagata, 1962; Conti and Brock, 1965; Kellerman *et al.*, 1969; Nagata *et al.*, 1975). In these, the electron micrographs reveal a series of configurations that suggests that the open membranes characteristic of low-oxygen-tension conditions may fold over near their ends to form loose loops. As the loops elongate, open folds are then produced initially, and later as the processes continue, open vesicles result, containing few or no cristae. Finally, the vesicles mature into typical mitochondria, with more numerous cristae and completely fused membranes. A number of partly open mitochondria appear in Figure 9.13B. If this interpretation is valid, explanation is provided for the openings, gaps, or pores and other features frequently observed even in the mitochondria of aerobically grown yeast (Moor, 1964). Stepwise increases in enzymatic activities and cytochrome concentrations have been shown to accompany respiratory readaptation (Nejedly and Greksák, 1977).

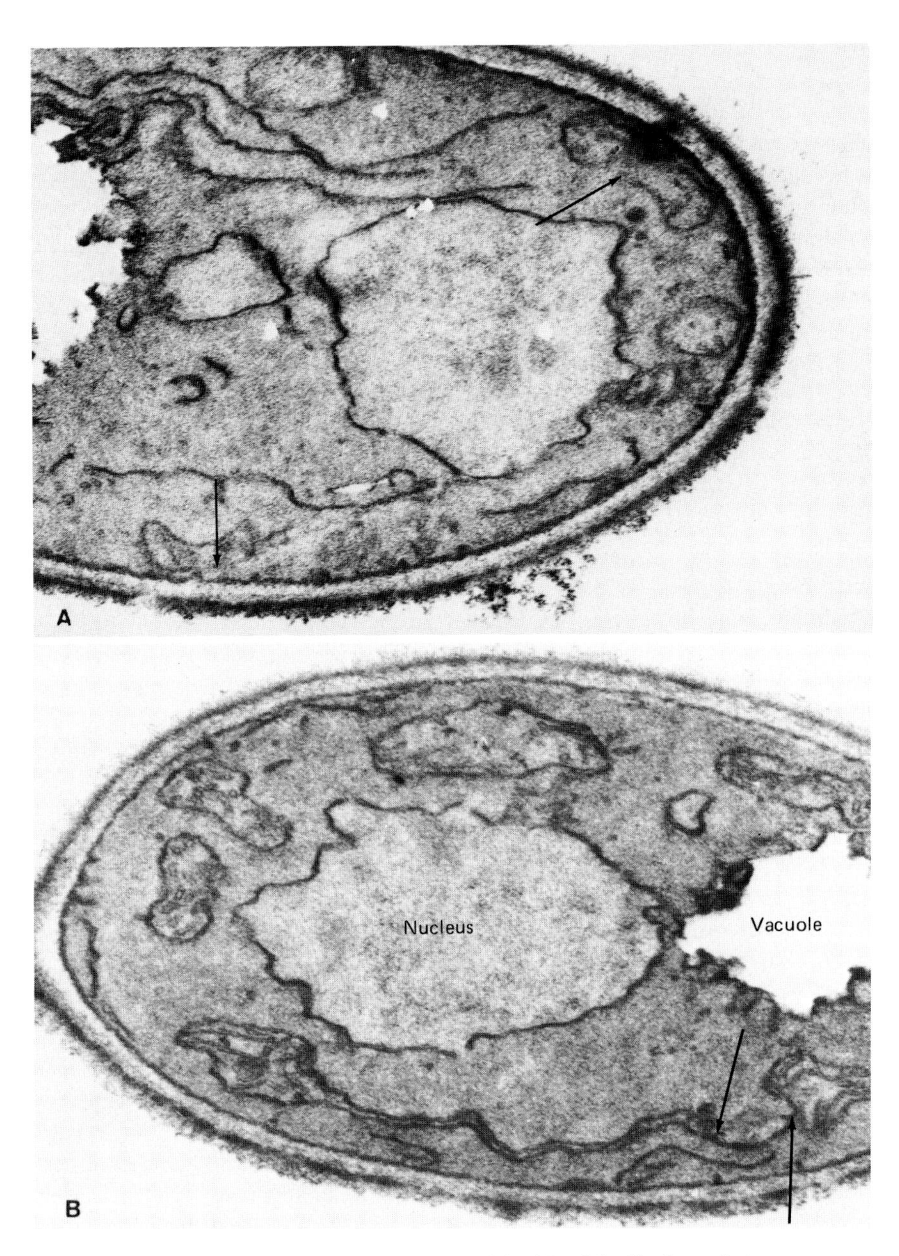

Figure 9.13. Mitochondria of yeasts. (A) The mitochondria of *Candida*, like those of other true yeasts, contain few cristae and those are often vague in outline. Moreover, many of these organelles show broad openings in the investing membranes (arrows). In this anaerobically grown specimen, mitochondria are few, whereas elongate single membranes are abundant. (B) When transferred to aerated cultures, the mitochondria increase in number while the membranes become scarce. Elongate interconnecting regions can often be noted, some of which contain openings (arrows). Both 42,500×. (Courtesy of Kellerman *et al.*, 1969.)

In electron micrographs of frozen-etched yeast cells (Moor and Mühlethaler, 1963), the surface of the outer mitochondrial membrane is seen to be irregularly covered with minute pits and a few fine furrows; its simplicity of sculpturing contrasts sharply with a comparable view in similarly prepared fungus, *Basidiobolus ranarum* (Bauer and Tanaka, 1968). Both surfaces of the outer membrane in that organism are described as showing a ''fingerprint''-like structure, because of the pattern of broad ridges and furrows that constitute the membrane. It is also of interest to note here that in *Cyanidium caldarum,* a thermophilic, acidophilic alga of uncertain affinities (Rosen and Siegesmund, 1961), the mitochondria apparently develop in the fashion suggested above for yeast; it is evident that in cell shape and nuclear characteristics, this organism also closely resembles the yeasts, except that simple chloroplasts are present.

9.1.4 Behavioral and Other Properties of Mitochondria

As has undoubtedly been noted when the various figures of mitochondria were being examined, an extreme variability in size and shape characterizes this organelle. Nowhere are any two of these similar, not even where multitudes exist in the same cell. While some of the variation stems from the differing angles and areas passed through by the sectioning knife, and which therefore are more apparent distinctions than real, other contrasts in configuration or extent are actual. The basis for the variability becomes more comprehensible as behavioral and certain organizational traits are discussed.

The Motility of Mitochondria. Many organelles have a more or less fixed location in the cell, such as dictyosomes adjacent to the flagellar centriole, and chloroplasts laterally, but not so the mitochondrion. Cinematomicrography and other techniques have revealed it to be a most actively motile structure, as first shown many years ago by Lewis and Lewis (1914). In *Euglena* continuous activity was observed in the chondriome as the various parts elongated, branched, fragmented, and fused (Leedale and Buetow, 1970). As acid phosphatase had been demonstrated to be present in the organelles, it was suggested that this enzyme might be involved in the fragmentation and fusion processes, for the mitochondria of cells starved for more than 7 days lacked both these abilities and the enzymes.

The mitochondria of *Xenopus laevis* tadpole heart epithelial cells have been studied by cinematomicrography (Bereiter-Hahn and Morawe, 1972) and found similarly to move saltatorially. Although cytosol movement was enhanced by HCN, the movements of these organelles ceased, nor were they increased when ATP was added as were those of the cytoplasm. Consequently, their motility is independent of that of the rest of the cell. Mitochondria in cells influenced by 2,4-dinitrophenol displayed three different maxima in their velocity distribution curves, suggesting the existence of heterogeneity among these organelles, a topic discussed more fully in a later section. Through use of phase-contrast microscopy with the same type of cells, mitochondria were noted to travel at speeds up to 100 μm/min, their movements falling into four categories (Bereiter-Hahn, 1978): (1) alternating extension and contraction; (2) formation of lateral branches; (3) peristaltic wavelike action along the length of the organelle; and (4) contraction and expansion of individual regions.

The Mitochondrion as a System. Some of the apparent variation in size and appearance also may be attributable to a condition that has thus far been described in only a few organisms, but which eventually may prove to be a nearly universal characteristic of cells in general, except perhaps those of multicellular plants and animals. Like the tubular interdictyosomal connections recently found to unite the Golgi bodies of a cell into a single system, mitochondria also have proven to be interconnected. However, in the present case, the connections are actually continuities between expanded and contracted portions of the same body. Three-dimensional models derived from 80 to 150 consecutive serial sections of entire yeast cells showed that all the mitochondrial profiles were sections through a single, highly branched, very irregular tubular structure that traversed the entire cell just beneath the plasmalemma and continued into the forming bud (Figure 9.14; Hoffman and Avers, 1973).

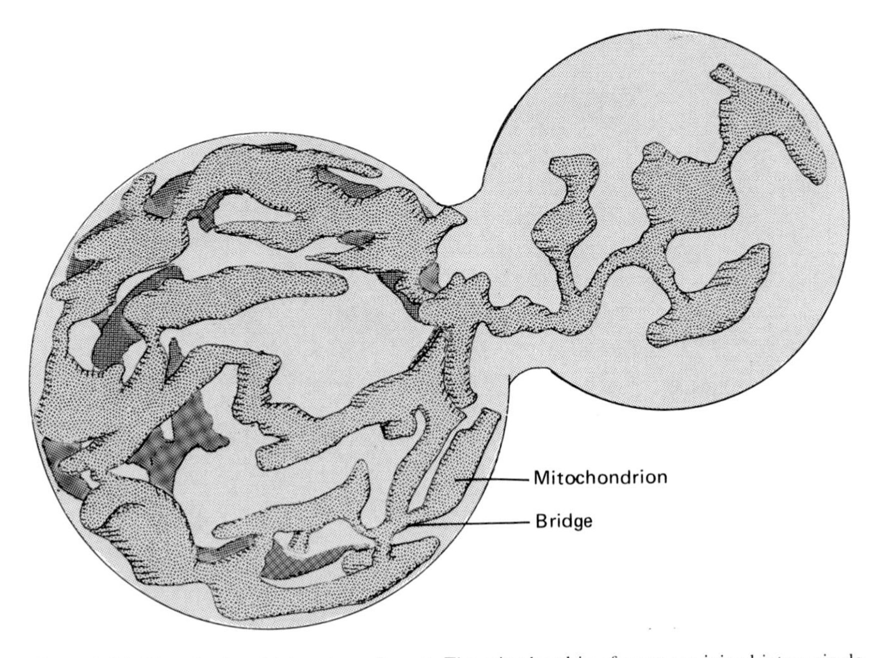

Figure 9.14. The mitochondrial system of yeast. The mitochondria of yeast are joined into a single continuous system. (Based on Hoffmann and Avers, 1973.)

Apparently cristae extend throughout the length of the mitochondrion; thus small mitochondria are merely sections through a neck that interconnects two expanded regions. Because in life the entire body as a unit and its separate parts individually expand and contract, elongate and shorten, bend and straighten, no two cells can ever be similar in regards to the size, shape, or number of this organelle.

The condition just described for the typical yeast cell may not be repeated in precise fashion throughout the biological world, but the general pattern probably is. In *Chlamydomonas,* reconstruction of a cell model from numerous sequential serial sections demonstrated the existence of eight mitochondrial bodies (Arnold *et al.*, 1972; Schötz *et al.*, 1972). As in the yeast, these were found to be contorted, branched, elongated tubular structures, most of which were concentrated below the plasmalemma, while the rest seemed to be largely associated with the chloroplast. Although the mitochondrial system, or chondriome, of *Euglena* has been studied only in living cells (Leedale and Buetow, 1970), that investigation demonstrated similar continuities between the respiratory organelles. Often the interconnections, and even the mitochondria themselves, were threadlike—the ''threads'' and ''mitochondria'' are actually the same thing, the diameters and form being temporary states common to the entire system.

9.1.5. Highly Modified Mitochondria

In addition to the various morphological modifications in the crista that have been presented, the mitochondrion as an entity has undergone specializations in structure. Moreover, at least in one general type, the organelle has given rise to a distinct part that is so highly modified that it lacks all the characteristics of its parent body.

The Mitochondrial Helix of Spermatozoa. In the midpiece of the metazoan spermatozoon, that is, the region just posterior to the head, is a series of greatly elongated mitochondria, usually twisted into a compact helix around the flagellar basal region. Among mammals a variable number of rows of mitochondria commonly comprise the helix, and the organelles themselves may show slight internal modifications as well (Threadgold, 1976). In bull sperm, three or four rows are present that together form 64 gyres around the flagellum, whereas in rabbit four or five rows encircle the flagellum 41 times (Figure 9.15; Phillips, 1977). Contrastingly, rhesus monkey spermatozoa have only a double helix, as do mouse sperm. In some instances, fusion may take place between adjacent mitochondria, but such is not the case in rat sperm (André, 1962).

Among invertebrates, mitochondria occasionally remain within the sperm head at the posterior of the nucleus rather than spiraling along the flagellum, but several striking variations on this theme have been described. In a pulmonate snail, *Testacella haliotidea,* the mitochondria are very numerous during

A
B
C

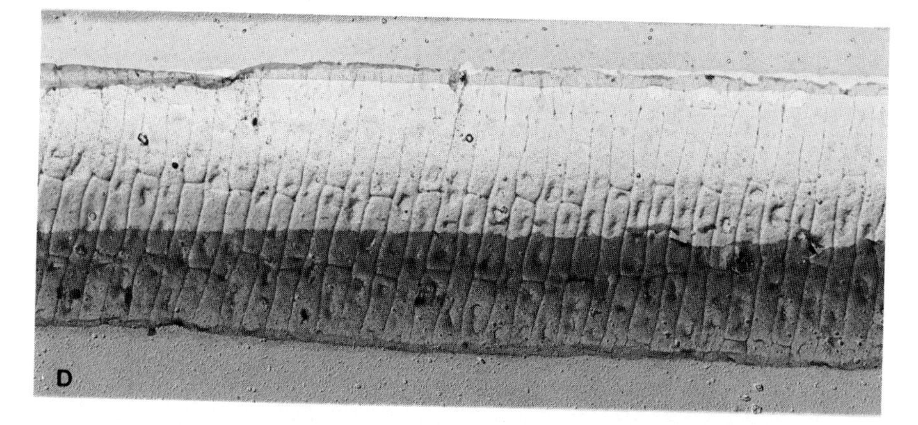

Figure 9.15. Modified mitochondria of mammalian sperm. At the midpiece of mammalian spermatozoa, mitochondria are arranged in helices around the flagellar sheath, as shown in these electron micrographs of platinum-carbon replicas. Shown are the midpieces of bull sperm (A, 15,000×), mouse (B, 10,000×), and rhesus (C, 19,000×). (D) At higher magnification, the mitochondria of Chinese hamster are seen to be precisely shaped and arranged. 41,000×. (All courtesy of Phillips, 1977.)

spermatogenesis; late in the spermatid stage, still within the head, they fuse into a single mass, out of which later is produced one gigantic derivative. Still later this itself undergoes metamorphosis, the final product being a paracrystalline rod that surrounds the flagellar base; most of the substance of the rod is protein, arranged in numerous parallel tubules about 90 Å in diameter (André, 1963). A second example occurs among the insects, in which taxon the mitochondria of the sperm of a sphinx moth (*Macroglossum stellatarum*) undergo a comparable metamorphosis. However, the end product is quite different, for it consists of a thick-walled, elongate rod that encloses a structured electron-transparent matrix surrounding the flagellum.

The Kinetoplast. Among the trypanosomatids and their relatives, the bodonids, a peculiar structure is present near the flagellar centriole, one of several organelles formerly referred to as the blepharoplast by light microscopists. Now it has been revealed by electron microscopy usually to be a rodlike mass (Figure 9.16), circular or ovate in cross section, consisting of densely packed transverse fibers (Pitelka, 1961; Rudzinska *et al.*, 1964). Also it has been shown to be part of a single mitochondrion and to consist largely of DNA, the combination of organelle and rod now being known as the kinetoplast (Trager, 1964; Simpson, 1972). The Feulgen-positive rod of DNA has proven to be fairly consistent in width for a given species; in *Crithidia fasciculata,* for example, its diameter is between 0.23 and 0.29 μm (Anderson and Hill, 1969).

The kinetoplast contains the largest known deposit of DNA outside of the

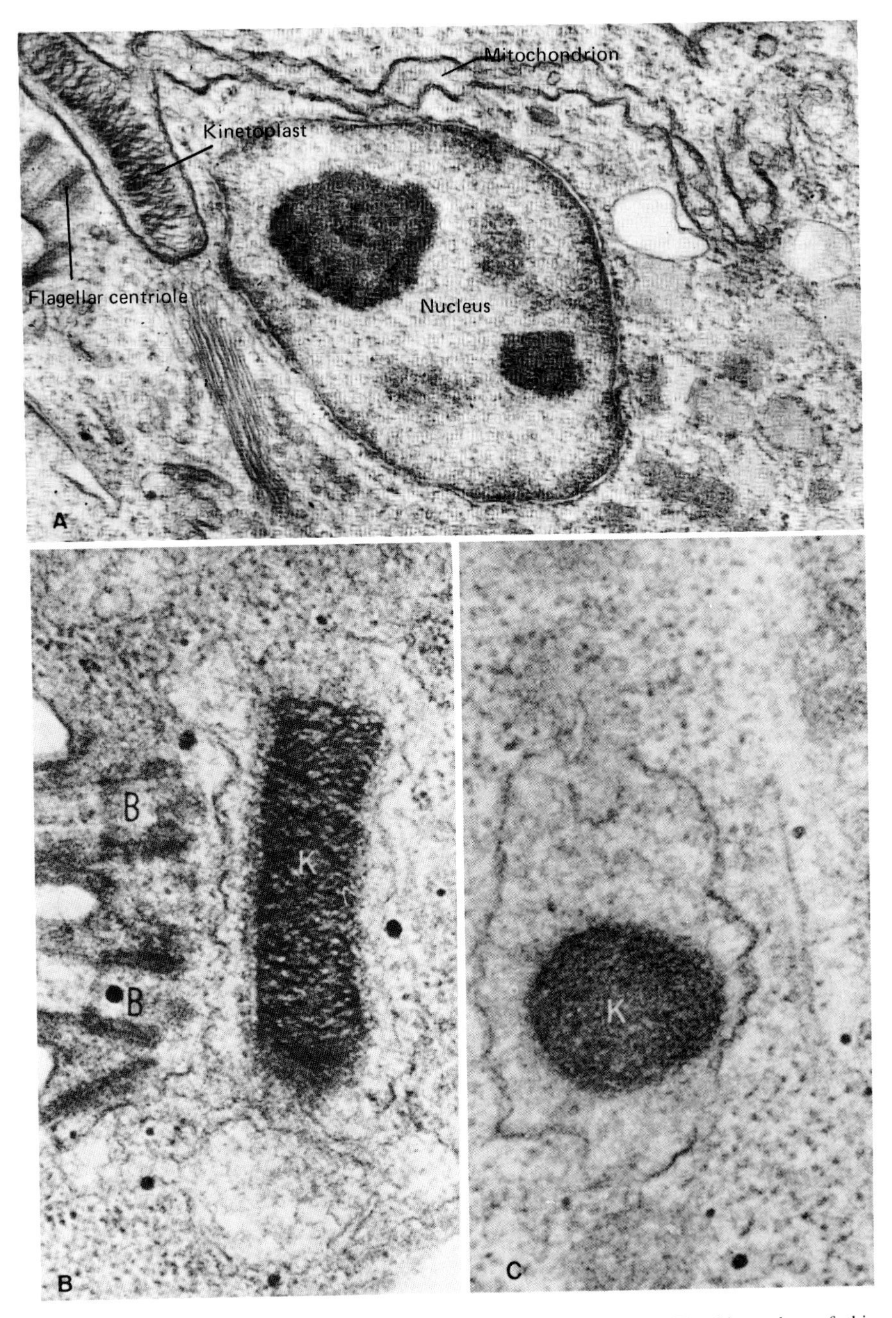

Figure 9.16. The kinetoplast, a derivative of the mitochondrion. (A) The kinetoplast of this *Trypanosoma conochina* cell is seen to be an expanded portion of the mitochondrion enclosing an electron-opaque body. 35,000×. (Courtesy of Milder and Deane, 1967.) Longitudinal (B) and transverse (C) sections of the kinetoplast (K) of *Crithidia* reveal similar structure and orientation close to the flagellar centrioles (B). Both 40,000×. (B and C, courtesy of Hill and Anderson, 1969.)

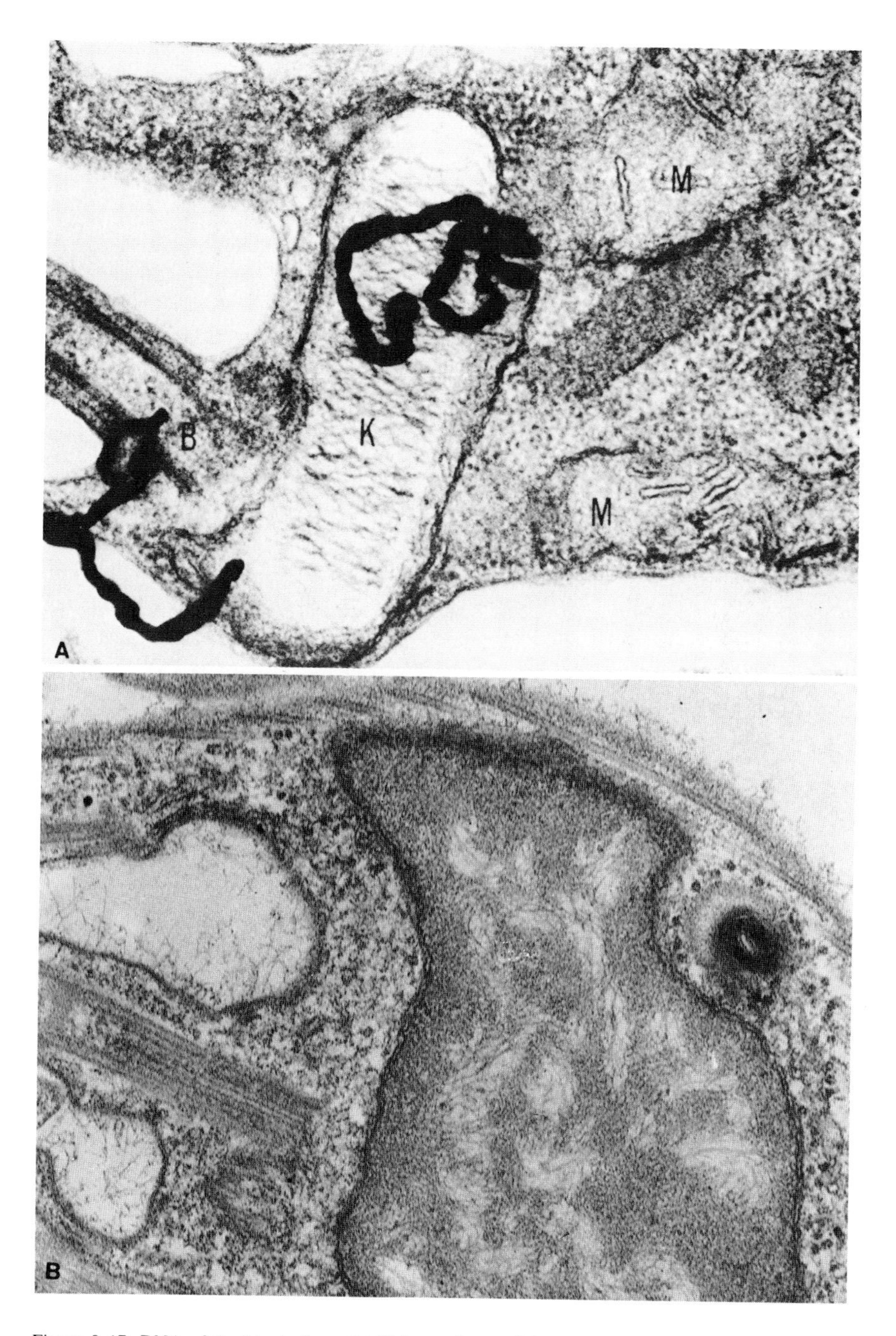

Figure 9.17. DNA of the kinetoplast. (A) This specimen of *Crithidia* has been pulsed with labeled thymidine, the activity being confined to the kinetoplast (K). 80,000×. (Courtesy of Hill and Anderson, 1969.) (B) In *Cryptobia* the DNA is distributed widely throughout the kinetoplast, but especially near the flagellar base. 70,000×. (Courtesy of Vickerman, 1977.)

nucleus (Figure 9.17), but its role in the cell has not been deciphered as yet. Nor has it been determined why it is so closely associated with the flagellar base (Vickerman, 1977). That it is not entirely essential is clearly indicated by its normal absence in *Trypanosoma equinuum* and in some strains of *T. equiperdum* and *T. evansi* (Simpson, 1972). Although part of its DNA is in the form typical of many other mitochondria, as described later, the bulk of the molecules are atypical to such an extent that the ability to serve a genetic function has been questioned (Borst and Fairlamb, 1976).

The kinetoplast of the bodonids differs somewhat in form from the corresponding organelle of the trypanosomatids. In *Bodo* perhaps the greatest distinction is that the kinetoplast is continuous with the so-called "cord" of mitochondria that encircles the cell, and not merely a rod surrounded by a mitochondrial remnant (Hill and Anderson, 1969). Instead of the short cristae found in *Trypanosoma,* opaque granules fill the space between the DNA fibrils and the limiting membrane (Pitelka, 1961). Another member of the group has quite a different arrangement but varies between the dimorphic forms that occur in this species, the common "thin" and the rare "broad" forms. The latter have the kinetoplast DNA concentrated in a mass near the flagellar centriole (Vickerman, 1977), but in the more frequent arrangement the DNA is dispersed throughout the mitochondrial network.

9.2 MOLECULAR ORGANIZATION OF MITOCHONDRIA

As already brought out, the foremost enzymes of the metazoan mitochondrion are those involved in oxidative phosphorylation, including the electron-transport chain and many enzymes of the citric acid cycle, but excluding those of glycolysis. Hence, the mitochondrion stands out as the ATP-producing center *par excellence* of the cell.

9.2.1. Enzymes of the Mitochondrion

Energy-Requiring Activities. Within this organelle a number of reactions that utilize ATP also are known to occur, including the synthesis of protein discussed in a later section. After their biosynthesis, certain proteins are phosphorylated by a protein kinase located on the stroma side of the mitochondrial inner membrane of mammals (Vardanis, 1977), as well as those of yeast (Rogobello *et al.*, 1978). Among other synthetic activities may be listed the formation of phosphatides, hippuric and *p*-aminohippuric acids (Kielly and Schneider, 1950; Leuthardt and Nielson, 1951), and citrulline (Siekevitz and Potter, 1953). Carboxylations and phosphorylations of nucleoside diphosphates in general are likewise localized here (Herbert and Potter, 1956). In addition, amino acid metabolism may be a particular function of the mitochondrion, because transaminases (Hird and Rowsell, 1950), glutaminase I (Shepherd and Kalnitsky, 1951), and glutamic dehydrogenase are present in abundance (Hogeboom and Schneider, 1953; Schneider, 1959). Among specific amino acids actually known to be oxidized are thyroxine and triiodothyronine.

The metabolism of glutamine in avian liver provides an interesting insight into some mitochondrial processes that are not energy related. The synthetase of this amino acid was found to be located within the matrix of this organelle (Vorhaben and Campbell, 1977), together with dehydrogenases of malic and glutamic acids and that of NADP-dependent isocitric acid. Glutamine synthetase parallels carbamyl phosphate synthetase I functionally but occurs in the liver of uric-acid-secreting (uricotelic) species instead of urea-secreting (ureotelic) ones as the latter does (Vorhaben and Campbell, 1972; Campbell, 1973; Campbell and Vorhaben, 1976). Both enzymes mediate the initial steps in the detoxification of ammonia through the synthesis of either uric acid or urea. When glutamic acid is catabolized by glutamic acid dehydrogenase, the resulting ammonia is used in the synthesis of citrulline by way of carbamyl phosphate or, in uricotelic forms, it is employed in the production of glutamine through the action of glutamine synthetase. In both cases the products then leave the mitochondria for final conversion into the excretory compounds. In contrast to the glutamine synthetase, which is localized in the matrix, phosphate-dependent glutaminase is confined to the mitochondrial outer membrane in avian liver; in rat liver, however, it is found in the interior of the organelle (Curthoys and Weiss, 1974). The enzyme paralleling glutamine synthetase functionally in rat liver, carbamyl phosphate synthetase, has been reported to be contained completely within the inner membrane, where it makes up about 15% of the total mitochondrial protein (Clarke, 1976a). Its molecular weight has proven to be in the neighborhood of 165,000.

Schneider (1959) lists also a number of oxidases that are localized in mitochondria. Among the more important are those of mesotartrate and sarcosine, cholestrol and other steroids, all the fatty acids, xylitol and other polyols, itaconate, and kynureinine. The inorganic pyrophosphate produced by the β oxidation of fats or activation of amino acids is known to be hydrolyzed by a pyrophosphatase located on the inside surface of the inner membrane (Schick and Butler, 1969).

Heterogeneity of Enzymes. In multicellular organisms, tissue differences in mitochondrial enzymes occur, just as has been the case of the other organelles that have already received attention. One investigation that compared several dehydrogenase activities in mitochondria of four rat tissues (Ohkawa *et al.*, 1969) showed that the liver and kidney organelles were nearly equal in being more highly reactive for β-hydroxybutyric and glutamic acid dehydrogenases than either white or brown adipose tissue, and that kidney ranked highest in succinic acid dehydrogenase. However, mitochondria of brown adipose tissue were at least ten times as active in α-glycerophosphate dehydrogenase as the others, white adipose tissue being the least reactive for the enzyme. This latter difference is all the more remarkable, because the substrate is known to be a precursor for glycerophosphatides and triglycerides, substances important in lipid metabolism. In contrast, the enzymes for another set of reactions associated with fat metabolism, the production of ketone bodies, were found to be located in liver cell mitochondria (Chapman *et al.*, 1973).

Even in such unicellular organisms as bakers' yeast, the mitochondria are not all identical, in spite of their being united into a single system. Schatz (1963) was the first to distinguish two distinct populations of these organelles in *Saccharomyces*. One of these carried a high degree of succinate-cytochrome *c* oxidoreductase activity, whereas the second group was highly active in succinate dehydrogenase. This presence of at least two types was later confirmed by means of biochemical and quantitative electron microscopic procedures (Matile and Bahr, 1968); in addition, the two types were reported to differ noticeably in relative weight. HeLa cell mitochondria also have been demonstrated to be heterogeneous in function (Storrie and Attardi, 1973).

9.2.2. Mitochondrial Membrane Activities

Mitochondrial Membrane Structure. As the numerous concepts regarding the structure of membranes, including those of the mitochondrion, have already been discussed (Chapter 1), here only the one that appears to apply particularly well to mitochondrial properties receives attention, that of Sjöstrand and his co-workers (Sjöstrand, 1977, 1978; Sjöstrand and Cassell, 1978). In brief, the most pertinent characteristics are (1) the outer, inner, and cristal membranes are each distinctively structured (Figure 9.18). Hence, the cristae are not to be viewed as mere outpocketings of the inner membrane. (2) The membranes are largely (75%) protein. (3) The inner membrane has lipid on only one surface, that bordering the matrix. (4) The cristal membranes are 150 Å thick, but so closely appressed that a single crista is 300 Å in thickness. (5) The outer membrane readily permits the passage of water, whereas the inner one limits water penetration, possibly through lipids sealing the interstices between protein molecules. (6) In these proteinaceous membranes, both polar and nonpolar regions alike can be exposed to the surface or just as readily embedded within the interior, wherever their specific properties may be needed. (7) No outer compartment nor intracristal spaces are present. This view of the enclosing membranes being closely applied to one another affords an explanation of the earlier observation that freeze-fracture preparations do not disclose the outer face of the inner membranes (Ruska and Ruska, 1969).

Substructure of the Membranes. Using negative staining techniques on isolated beef heart mitochondria, Fernández-Morán (1961) found a macromolecular repeating unit in both the cristae and the inner membrane. Later, these units, called subunits or elementary particles, were shown by Fernández-

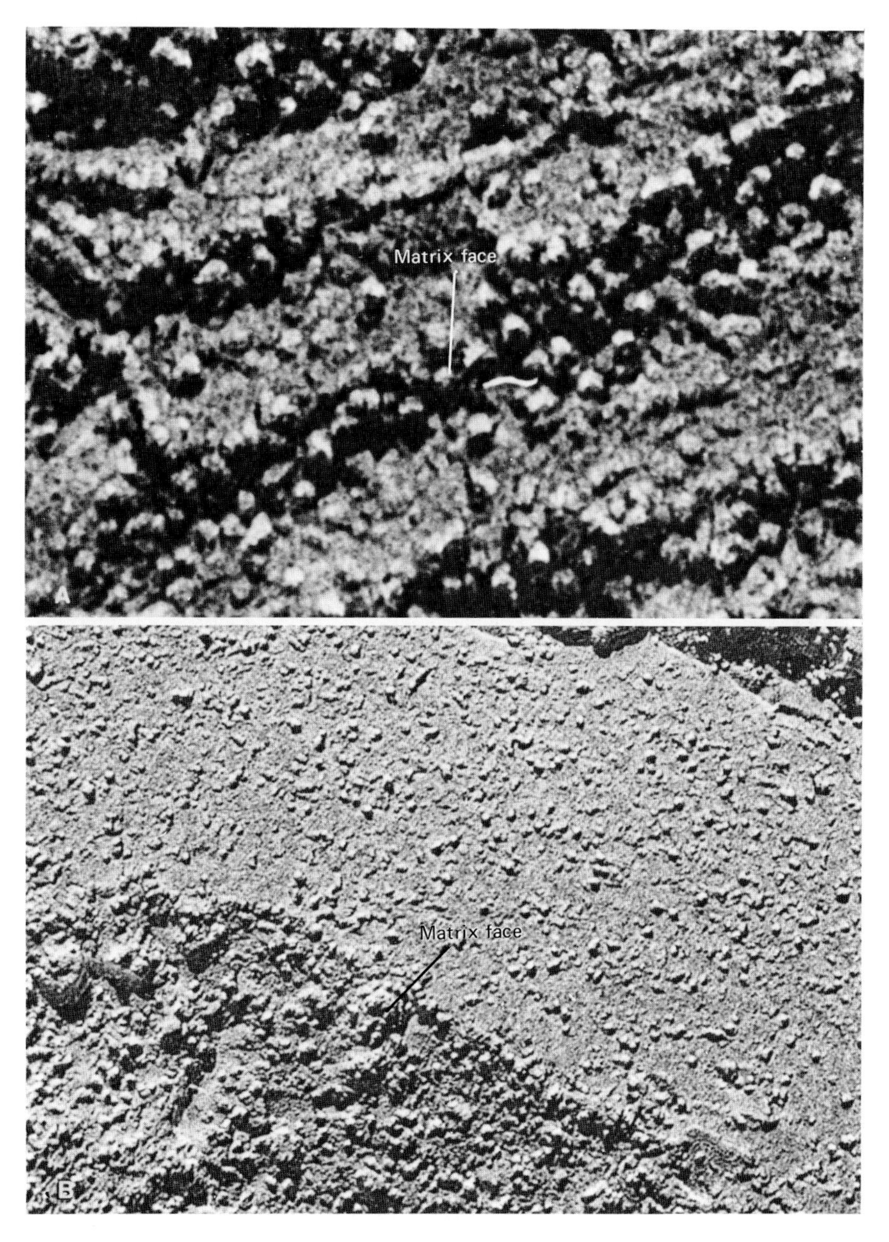

Figure 9.18. Mitochondrial membrane substructure. (A) Freeze-fracture preparation cut obliquely through the cristae; the matrix surface is densely covered with coarse particles. 380,000×. (B) Same but exposing the matrix surface of the inner membrane of the envelope; large convexities and irregular particles are present. 130,000×. (Both courtesy of Sjöstrand and Cassell, 1978.)

Morán *et al.* (1964) to consist of three parts: a more or less spherical or polyhedral head, 80 to 100 Å in diameter, a slender stalk, 30 to 40 Å wide and approximately 50 Å in length, and a base piece of 40 × 100 Å dimensions. The base pieces formed an integral part of the outer electron-dense lamella of the membrane, while the heads projected into the stroma in regular arrays. From the same source, subunits resembling these elementary particles in size later were isolated and found to contain complete electron-transport chains. Because cytochromes *a* and a_3 require dissolved O_2 in their reactions, these researchers believed the end of the chain to be localized in the head piece projecting into the stroma. On the other hand, because the dehydrogenases, some of the non-heme iron proteins, coenzyme Q, and cytochrome *b* interact with NADPH or succinic acid, they were considered to be located in the base piece and the remainder of the transport chain in the stalk.

Later Green and Perdue (1966) removed the head pieces of the subunits by sonication; then, by ultracentrifugation, they separated fractions containing inner and outer mitochondrial membranes, respectively. Upon testing, the entire electron-transport system proved to be localized in the base pieces of the cristae and inner membrane. The head pieces, these biochemists suggested, were not all alike but actually were different kinds of enzymes, including ATPase as demonstrated by Racker and Conover (1963). Later it was reported that the stalks consisted of fibrous proteins (Hall *et al.*, 1969).

On the other hand, in the mitochondrial outer membrane the subunits lacked both stalked head pieces and electron-transport activity (Green and Perdue, 1966; Parsons *et al.*, 1966). In this membrane the particles were about 90 Å in diameter and seemed to consist of various enzymes of the citric acid cycle, as well as those of fatty acid oxidation and elongation, and of hydroxybutyrate and amino acid oxidation (Allmann *et al.*, 1966).

The ATPase–ATP synthase complex of enzymes has been demonstrated to be asymmetrically oriented in the inner* mitochondrial membrane (Hootman and Philpott, 1979), the factor F_1 being located in the matrix side (Racker, 1970). In the processes of ATP metabolism, thiol groups of the several enzymes play important roles (Senior, 1973). Those involved in the mechanism of ATP synthesis have now also been localized in a nonpolar area on the matrix surface of the inner membrane (Blanchy *et al.,* 1978). It is of interest to note that the gene for the ATPase in yeast has now been sequenced (Macino and Tzagoloff, 1979).

In addition, similar studies have revealed an orientation of various components in the same membrane but in the opposite direction (Krebs *et al.*, 1979). For instance, a total of seven classes of polypeptides were separated by dodecyl sulfate gel electrophoresis (Clarke, 1976b), having molecular weights of

*It should be borne in mind that the term inner membrane in these reports includes the cristal membrane as well as the inner investing one.

130,000, 87,000, 73,000, 31,000, 26,000, 16,000, and 10,500, respectively. Although all of these were localized on the exterior surface of the inner membrane, their specific functions were not ascertained, but they were assumed to be involved in transmembrane transport. However, complex II of mitochondrial membrane, a fragment containing succinic acid-ubiquinone reductase activity and two polypeptides, has been reported as spanning the inner membrane (Merli *et al.*, 1979). The two polypeptides, called CII3 and CII4, had molecular weights of 13,500 and 7000, respectively; the first of these was found to be located on the outer surface of that membrane, while the succinic acid dehydrogenase flavoprotein was on the matrix side.

The sites of some enzyme activities have now been localized in submitochondrial fractions of plant tissues. Intact mitochondria from the endosperm of castor bean were isolated, ruptured, and separated into inner and outer membrane fractions, each of which was then examined for the presence of certain enzymes (Sparace and Moore, 1979). The results indicated that the synthesis of phosphatidylglycerol, CDP-diglyceride, and phosphatidylcholine occurred in the inner membrane, whereas that of phosphatidic acid took place in this as well as the outer membrane. Furthermore, the phospholipids themselves have been shown to be asymmetrically distributed within the inner membrane itself (Krebs *et al.*, 1979).

To refer back to a matter discussed in the first chapter, it is evident that the two membranes investing this single organelle present further evidence negating the concept of universality of structure among membranes. The protein content in the form of enzymes and coenzymes of one bears no resemblance to that of the other, any more than does the form of the subunits that compose each one. Furthermore, both membranes, like those of the chloroplast but unlike most others, appear in freeze-etch preparations to lack a central lipid layer (Staehelin, 1968).

Miscellaneous Membrane Enzymes. In addition to the enzymes enumerated above, the mitochondrial membranes contain a number of others that have been identified but not localized in further detail. One of the more important is an actinomyosinlike protein (Ohnishi and Ohnishi, 1962) that probably is responsible for much of the organelle's marked abilities of elongation and contraction. Moreover, a structural protein with a monomeric molecular weight of around 22,000 has been isolated (Criddle *et al.*, 1962); immunological tests show it to be similar in the mitochondria of such diverse organisms as yeast, *Neurospora*, and beef. *In situ,* the protein is polymeric and insoluble in water, but treatment with alkali or anionic detergents converts it to the soluble monomer. Upon restoration of the pH to neutrality or removal of the detergent, polymerization proceeds rapidly, during which processes the protein can combine with phospholipids to form a complex (Fleischer and Klouwen, 1960). With any purified cytochrome, whether *a, b,* or *c,* the structural protein similarly can unite to form water-insoluble complexes in a 1:1 molar ratio. Combination of the structural protein and cytochrome *b* is a slow process, requiring a minimum of 4 hr. When thus combined, cytochrome *b* undergoes a significant alteration in its oxidation–reduction potential (Goldberger *et al.*, 1962), changing from $E'_0 = 0.34$ V to one more positive than 0.0 V.

Long-chain fatty acids from dietary sources play a role of considerable importance in the composition of the mitochondrial membranes and even the functioning of the organelle (Awasthi *et al.*, 1971; Blomstrand and Svensson, 1974). To cite one instance, the erucic acid of rapeseed has been reported to be incorporated preferentially into cardiolipin, an integral constituent of cytochrome *c* oxidase, with resulting decrease in mitochondrial ATP synthesis (Houtsmuller *et al.*, 1970; Clandinin, 1976).

Transport across the Mitochondrial Membranes. Mitochondria possess marked abilities in accumulating cations, including monovalent ones like K^+ and divalent ones such as Ca^{2+}, Mn^{2+}, and Sr^{2+}, as well as anions, among them phosphate, arsenate, acetate, sulfate, and those of the citric acid cycle. These capabilities, together with the energy pathways present, make the mitochondrion an ideal system for studying movement across living membranes. Such studies have been rather numerous, especially since a number of agents whose effects upon specific sites along the pathways have become known.

While such investigations are not pertinent to the present account, it is not surprising that they have afforded some insight into mitochondrial function. For example, Siekevitz and Potter (1953) observed an increased rate of respiration in mitochondrial suspensions upon addition of small amounts of Ca^{2+} in the absence of ADP. In later experiments, Chance (1964, 1965) found that this elevation of the respiratory rate did not endure indefinitely but returned to the resting rate after a period of time. In a complete stoichiometric study of these Ca^{2+}-induced "respiratory jumps," Rossi and Lehninger (1963) showed that between 1.7 and 2 Ca^{2+} cations were required to activate each respiratory site and that virtually all the calcium was accumulated by the mitochondrion during the period of the respiratory jump. Lehninger (1966) demonstrated that this calcium was not irreversibly retained within the organelle but was maintained at a steady state by equal rates of Ca^{2+} ingress and egress through the membranes. Although the mitochondria most thoroughly studied were derived from rat liver, those of other cell types showed similar Ca^{2+} relations; sources included kidney, brain, heart, and skeletal muscle, maize and bean seedlings, the fungus *Neurospora crassa,* and yeasts. Chappell *et al.* (1963) and Carafoli (1979) reported comparable reactions for Sr^{2+}, and the former investigators for Mn^{2+} as well, but Mg^{2+} did not induce respiratory jumps nor was it accumulated within the mitochondrion.

Addition of larger quantities of Ca^{2+} to the medium induced massive accumulation of the cation by respiring organelles, a process blocked by 2,4-dinitrophenol or cyanide, but not by the inhibitor oligomycin (DeLuca and Engstrom, 1961; Vasington and Murphy, 1962; Brierley and Slautterback, 1964). Rossi and Lehninger (1963) established that Ca^{2+} accumulation was coupled to inorganic phosphate uptake in a ratio of 1.67 Ca^{2+}: 1.0 P_1. With these higher concentrations of Ca^{2+}, mitochondria did not phosphorylate ADP; consequently, massive Ca^{2+} loading may be viewed as a process alternative to oxidative phosphorylation of ADP—explaining why calcium ions have traditionally been considered an uncoupling agent. Ultrastructural investigations of treated mitochondria show the calcium phosphate to be deposited within the stroma; hence, it is clear that chemicals can be transported directly across both investing membranes. Strontium together with organic phosphate may be accumulated in the same manner as calcium, but massive loading of either substance leads to mitochondrial damage.

Molecules as large and complex as amino acids and proteins, including cytochrome *c*, are also apparently conducted into the mitochondrion from outside sources, as becomes apparent in later discussions, but the mechanisms involved to date seem to have been explored to only a limited extent (Chua and Schmidt, 1979; Schatz, 1979). Similarly, such large molescules as NADH · H are known to be employed in various reactions, including gluconeogenesis, within the cytosol, although the chemical named has been established as being synthesized solely in the mitochondrion (Krebs *et al.*, 1967). As mitochondria show low permeability for both NAD and NADH · H, some carrier mechanism may be present. The investigators cited suggest that the carrier, possibly the malate-oxaloacetate system, may transfer reducing equivalents of intramitochondrially generated NADH · H rather than the substance itself. If so, one concept or another seems destined for change—either that of the direct involvement of NADH · H in the cytoplasmic reactions or of its formation occurring solely in the mitochondria.

The transfer of the acetyl group of acetyl-CoA from the inner compartment through the inner membrane also is an essential step in the *de novo* synthesis of fatty acids from glycolytic precursors. Investigations into the mechanism for transport of such radicals have shown that most of the activity is by way of citric acid (Watson and Lowenstein, 1970; Lowenstein, 1971). However, more recently it has also been established that between 15 and 20% is transferred through the inner membrane by an additional pathway, probably involving free acetate (Walter and Söling, 1976).

Still another mechanism for transport through the mitochondrial membranes has been uncovered, one that pertains to proteins made in the cytoplasm and then imported into the mitochondrion. The proteins involved in this investigation were subunits of the F_1 ATPase of the yeast organelle, the three largest of which were found to be synthesized as still larger precursors (Maccecchini *et al.*, 1979). These then penetrated both enclosing membranes, apparently undergoing trimming as they did so, for only the mature subunits occurred within the inner compartment.

9.2.3. Other Molecular Aspects

Mitochondrial Aging. A number of investigations have been directed at the problem of aging in mitochondria (Waite *et al.*, 1969; Parce *et al.*, 1978), the results of many of which suggested that endogenous phospholipase A_2 might be involved in the observed deterioration of energy-linked functions in the organelle. The first report cited showed that a direct relationship existed between the activity of this enzyme and irreversible swelling of the mitochondrion, which change has been long associated with loss of energy-related processes. Moreover, Scarpa and Lindsay (1972) obtained evidence intimating that the enzyme was likewise responsible to some degree for the decay of respiratory control that accompanies aging. In the aging processes, two main stages have been recognized (Parce *et al.*, 1978), the first involving a reversible loss of energy-linked functions, including a decline in the level of ATP, a decreased ability to reacylate monoacylphospholipids, and a decline of fatty acid oxidation rates. Both of the last two effects are really derived from the diminution of the ATP content. The second phase is marked by the irreversible loss of energy-linked activities, including the ability of the inner membrane to produce the energized configuration. Apparently, this final phase is in part the consequence of phospholipase A_2 activity, which results in the loss of phospholipids from the membrane.

Composition of the Stroma. Because investigations into the macromolecular composition of mitochondria have centered attention upon the membranes, the matrix remains relatively poorly explored. It should not be supposed, however, that all the enzymes enumerated in the preceding section are thus implied to have been established as being part of the membranes—only those of ion and metabolite transport and the majority of those involved in protein synthesis and in respiration have been clearly localized as being there. But few of the remainder have actually been consigned to the matrix. Among them are the two motile constituents of the electron-transport chain, that is, coenzyme Q and cytochrome *c*. This pair, unlike the other members, has been suggested by Green and Wharton (1963) as not being fixed so that they migrate among the rest and react by random collision. Another set believed to be in the matrix, or perhaps only weakly attached to the membranes, are the dehydrogenases of the citric acid cycle, because they are generally lost when mitochondria are fragmented. The matrix also contains an extensive pool of free amino acids (Roodyn *et al.*, 1961; Truman and Korner, 1962; Roodyn, 1965), and amino-acid-activating enzymes, many, if not all, of which differ sharply from their counterparts in other portions of the cytoplasm (Barnett *et al.*, 1967). Although evidently different in molecular structure, the system can serve to replace cytosol in ribosomal systems of protein synthesis (Truman and Korner, 1962).

Ultrastructural studies have clearly established that the matrix includes another component, one whose presence in the organelle holds several important lessons for biologists and biochemists alike. During the past several decades, it had been the practice of investigators in general to insist that all the DNA of the cell was contained in the nucleus. When on occasion DNA was well authenticated as being in the cytoplasm (for references see Gahan and Chayen, 1965), the results were generally passed off as being in atypical cells. As Roodyn (1967, p. 145) has pointed out, the history of the discovery of this substance within the mitochondrion "is an interesting example of how an oversimplified view can easily develop into a false dogma that can then retard the development of further research." Roodyn continues by stating "that of the thousands of studies on mitochondria in the fifties and early sixties, there were extremely few reports of DNA assays in the mitochondrial fraction." The remaining studies merely assumed that the 90 to 95% recovered in the nuclear fraction represented the total recoverable by the procedures. Unfortunately for the progress of science, false dogmas and oversimplifications are more readily perceived in retrospect than through the eyes of the present. They can only be ferreted out by continual reexamination of concepts.

The first to break with tradition were Chèvremont and his co-workers (1959) who, by employing autoradiographic and cytochemical techniques upon fibroblasts in tissue culture, demonstrated the presence of DNA in the mitochondrion. However, all doubt on the subject was removed only by the electron microscopic studies by Nass and Nass (1962), whose micrographs clearly showed DNA fibrils in the mitochondrial stroma (see Figure 9.20). Later Nass *et al.* (1965) demonstrated a similar condition to exist in mitochondria from a broad spectrum of biological material. Further details of this nucleic acid and the rest of the mitochondrial genetic system are given in connection with replication of the organelle (Section 9.3.2.).

Dark Granules. Frequently, in mitochondria relatively large, dense granules are present, averaging about three per organelle. Their major constituents had long been reputed to be calcium, inorganic phosphate, magnesium, and carbonate, along with organic material, until they were reinvestigated by Pasquali-Ronchetti and colleagues (1969), who employed 2,4-dinitrophenol and electron microscopy. In treated cells the number of dense granules per mitochondrion was found to be reduced to two within 12.5 min and to one by 30 min exposure to the chemical. Because 80% of the calcium present was eliminated from the mitochondrion within 2 min without a corresponding reduction in the number of granules, the investigators concluded that this element can no longer be considered a major constituent. On the other hand, magnesium was found to be associated with the particles, but the relationship remains obscure.

Furthermore, the extensive participation of calcium in these dense granules was negated by use of high-temperature microincineration (Thomas and Greenawalt, 1968), for they were found, unlike calcium phosphate, to be completely volatilized. Moreover, they have been shown to possess osmiophilic properties (André and Marinozzi, 1965; Ashworth *et al.*, 1966), suggesting a lipid constituent to be present. However, the numerous minute granules that are also present, having diameters less than 0.1 μm, have been demonstrated to contain both calcium and phosphorus (Sutfin *et al.*, 1971).

Macromolecular Organization of Metaphytan Mitochondria. In general, macromolecular organization of the higher plant mitochondrion likewise approximates that of the metazoans. Among the similarities are the repeating units of the membrane, complete with stalked, projecting head pieces (Cunningham *et al.*, 1967) and cytochromes (Baker, 1962). Undoubtedly, many of the same enzymes and enzyme systems of animal mitochondria are present in the plant organelle also, but too few studies have been made to establish their actual existence there.

That it is not safe to make too sweeping generalizations is indicated by the widespread lack of transhydrogenases in plant mitochondria (Hackett, 1963). While good exchange of hydrogen occurs between NAD^+ and NADPH, only one-tenth as much occurs between NAD^+ and NADPH (Ragland and Hackett, 1961). Other, more striking differences are known to exist. For example, although an electron-transport chain replete with cytochromes is present, the series, as pointed out earlier, is by no means identical, for cytochrome c_1 is absent, and several of the *b* type have been found.

One other feature is distinctive of plant mitochondria, an ability for NADH to pass through the membranes (Wiskich and Bonner, 1963). It has been suggested (Lieberman and Baker, 1965) that this property either stems from small crevices present in the membranes of isolated mitochondria despite their seeming intactness, or is perhaps a real capability arising from actual differences in physical or biochemical composition. It might eventually prove to derive from the porosity of the outer membrane described in an earlier section.

9.3. DNA AND MITOCHONDRIAL REPLICATION

Since the discovery of DNA in the mitochondrion mentioned earlier, that substance from this organelle has been investigated extensively as to its physicochemical properties, role in replication, and possible genetic functions. Consequently, the body of literature on this subject has increased to voluminous proportions, including several reviews (Nass *et al.*, 1965; Nass, 1969; Borst, 1972, 1974; Lloyd, 1974; Birky, 1976; Milner, 1976; Borst *et al.*, 1977). As a rule, several DNA molecules are present per mitochondrion, the primary structures of all of which are reportedly similar or nearly identical (Borst, 1972). They are double-stranded, covalently closed molecules, often circular and in most cases collectively comprising about 1% of the total DNA of the Cell.

9.3.1. General Properties of Mitochondrial DNA

Because the physical properties of mitochondrial DNA (mtDNA) have been employed in support of a theory regarding the phylogenetic origins of the mitochondrion, it is necessary to devote more attention to that biochemical than might otherwise be desirable. While much has been accomplished on the comparative facets of mtDNA structural characteristics, some of its organizational properties have decreased in clarity as knowledge has increased, as becomes apparent almost immediately.

mtDNA in Yeast. Among the protistans, the mtDNA of yeast has been investigated especially thoroughly, so that a number of physical maps of the genome have been published (Groot-Obbink *et al.*, 1976; Heyting and Sanders, 1976; Nagley *et al.*, 1976; Rabinowitz *et al.*, 1976; Sanders *et al.*, 1976a; Schweyen *et al.*, 1976, 1977; Tzagoloff *et al.*, 1976; Borst *et al.*, 1977; Grivell and Moorman, 1977; Grosch *et al.*, 1977; Jacq *et al.*, 1977; Lewin *et al.*, 1977; Van Ommen and Groot, 1977; Linnane and Nagley, 1978). In diploid specimens, the molecule was usually considered to be a circular duplex about 25 μm in circumference, having a molecular weight of 4.5×10^7, equivalent to around 70,000 base pairs (Hollenberg *et al.*, 1970). A study of the mtDNA at various phases of the growth curve, however, has presented a different picture of the structure. Far from being uniformly circular, four Gaussian subpopulations of linear molecules were found and quantified, having mean lengths of 2.2, 4.0, 6.0, and 10.0 μm (Guérineau *et al.*, 1975). With them were found open circular molecules whose circumferences ranged from 0.5 to 10.0 μm. During the exponential growth phase, only 5% of the total DNA molecules were circular and their mean length was 2.2 μm, with a range to 15 μm. With the advent of the transitional phase, circularity increased to 15% and the mean length of the linear forms became 4.4 μm, with a range to 20 μm. Later in the transitional phase, these trends continued, until circular molecules comprised 20% of the total and the greatest length of linear units reached 25 μm. However, in the stationary phase, the circular mtDNA disappeared entirely and the length of the linear forms decreased once more to a mean of 2.2 μm. Furthermore, application of other techniques revealed that the mtDNA became rearranged during the exponential phase, as the mitochondria were differentiated, sometimes forming molecules 40 μm in length (Guérineau and Paoletti, 1975). The reported linearity is in direct conflict with the circularity depicted in all genetical maps of the yeast DNA. It is pertinent to note that in mutants that lack mtDNA, the mitochondria are devoid of cristae (Figure 9.19A–C; Montisano and James, 1979).

When cleaved with restriction endonucleases, the molecule was broken into about 100 fragments, the size distribution of which varied greatly from one strain of *S. cerevisiae* to another. Much of this variability has been attributed to the presence of numerous insertions and/or deletions, many of which were short sections 25 to 50 base pairs in length (Sanders *et al.*, 1977). In addition, in some strains several large insertions between 900 and 1500 base pairs, or even as much as 2600 to 3000 base pairs, in length were present (Sanders *et al.*, 1976b, 1977). For instance, the single gene for the large rRNA has been shown to contain one insert (Bos *et al.*, 1978), while that for cytochrome *b* includes at

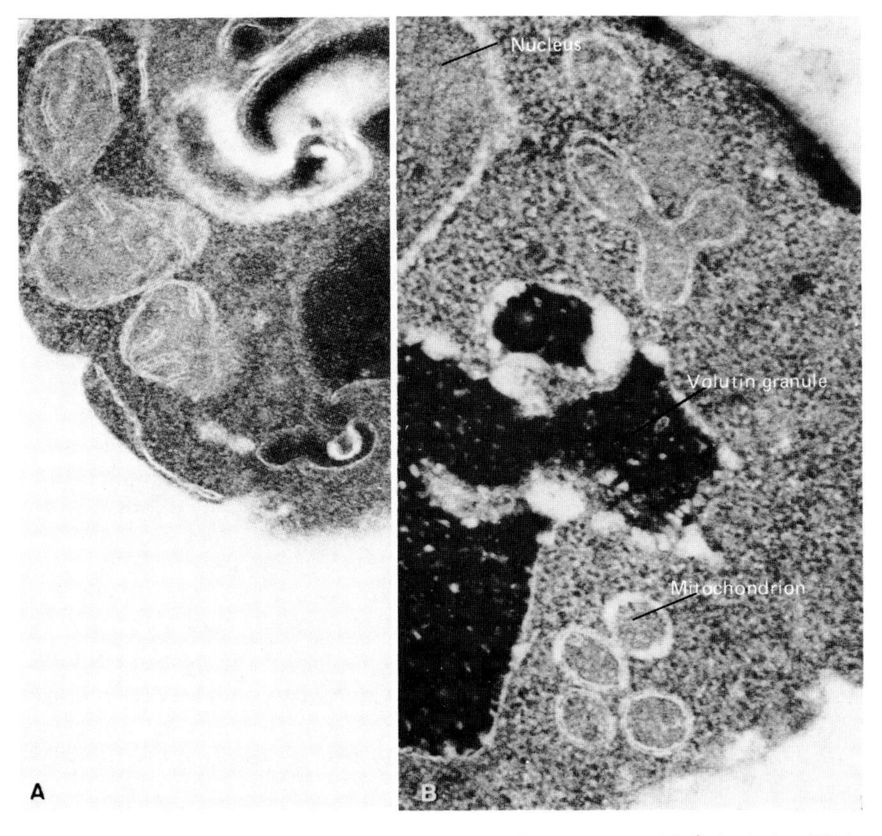

Figure 9.19. Effects of mtDNA on mitochondrial structure. In yeast mutants (*rho*°) that lack mtDNA, the mitochondria (B) lack the cristae found in those of the wild-type specimens (A). (A) 70,000×; (B) 66,000×. (Both courtesy of Montisano and James, 1979.)

least four intervening sections (Borst and Grivell, 1978; Grivell *et al.*, 1979). This condition contrasts strongly with prokaryotes in which no interrupted genes appear to occur.

Comparisons have been made of the sequence homologies of nuclear and mtDNAs of four different species of yeast, *S. cerevisiae, S. carlsbergensis, Candida utilis,* and *Kluyveromyces lactis*. Both nuclear and mitochondrial DNA showed around 90% homology to exist between the two species of *Saccharomyces,* but homologies between other combinations of the four were low, being mostly in the range of 10% (Groot *et al.*, 1975). Moreover, it was demonstrated that the conserved regions of the mtDNAs had acquired several times as many mutated sites than had the nuclear DNA. As a whole the regions that formed DNA · DNA hybrids represented areas that encoded rRNAs and perhaps tRNAs, those of the nuclear genome being more stable than those of the mitochondria.

With a G + C content of 18 mole %, yeast mtDNA has one of the most A + T-rich genomes known to occur in nature (Bernardi *et al.*, 1970, 1976). The latter reference presented some evidence that was in harmony with considering the A + T-rich segments as spacers, the 70 copies present totaling nearly 50% of the DNA. Some of these segments appeared to be palindromes about 47 nucleotide residues in length and proved to be identical in two strains (Cosson and Tzagoloff, 1979). Of the latter, 24 coded for tRNAs, 2 for rRNAs, and 7 for polypeptides, while the rest remained undetermined (Casey *et al.*, 1972; Reijnders and Borst, 1972; Reijnders *et al.*, 1972; Schatz and Mason, 1974).

One of the strange genetic properties of *S. cerevisiae* is that it has proven relatively easy to isolate mutant forms involving changes in the mtDNA (Williamson *et al.*, 1977). Ease of mutation usually implies the existence of a low ploidy, that is, a relatively small number of genetically active copies of the genome. However, each mitochondrion contains several mtDNA molecules, and every cell encloses perhaps 20 or more mitochondria. Although these organelles seem to be bound together into a single system, around 50 or more DNA molecules would appear to be the minimum number in any given yeast cell. Whereas Williamson *et al.* (1977) give a figure of 44 to 67 for the haploid cell, Grimes *et al.* (1974) found only 10 or 11 in haploid and 22 in diploid specimens. As each molecule apparently is equally capable of being replicated, it would seem virtually impossible for a mutation to segregate out and result in a homoplastic clone, yet, as stated before, segregation occurs rapidly.

Although no adequate explanation has been forthcoming to date, the eventual explanation may involve an organizational unit of the genome that has recently been uncovered (Williamson and Fennell, 1975; Williamson *et al.*, 1977). Through use of a fluorescent dye specific for DNA, it was found that the DNA of yeast mitochondria rarely existed separately but was clustered into aggregates called chondriolites, each containing from one to eight mtDNA molecules (Williamson, 1976). On this basis, each bud might inherit just three to five chondriolites, thus greatly enhancing the possibility of segregating mutants. The actual unit of inheritance in these cells, however, has not been established, nor is it known whether all the DNA molecules in a chondriolite are identical in genetic makeup.

mtDNA in Unicellular Protistans. The mtDNAs from a small number of unicellular protistans have been at least partially characterized. Like some of the yeast types, the double-stranded molecule from *Euglena gracilis* is linear and quite short, having a mean length of between 0.9 and 1.0 μm and a molecular weight of only 1.9×10^6 (Edelman *et al.*, 1966; Krawiec and Eisenstadt, 1970a,b; Manning *et al.*, 1971; Nass *et al.*, 1974). The last report cited also described the presence of a small number of equally small but circular molecules. This smallest of all known eukaryotic genomes has a G + C content of 25% (Fonty *et al.*, 1975). A + T-rich spacers like those of yeast were also reported.

A somewhat greater depth of information has been garnered for *Tetrahymena pyriformis,* whose mitochondria also contain linear DNA molecules (Suyama and Muira, 1968). The latter were found to have a mean length of 17.6 μm corresponding to a molecular weight of 34×10^6 (Suyama and Preer, 1965; Brunk and Hanawalt, 1969). About eight molecules are contained in each mitochondrion (Suyama and Muira, 1968; Flavell and Jones, 1970), all of which have extremely similar primary structures and consist of around 26% G + C (Flavell and Jones, 1971a). Like adenovirus DNA, the linear molecules have been shown to have a terminal duplication, so that extracted molecules that were not renatured formed hairpin structures (Garon *et al.*, 1972; Wolfson and Dressler, 1972; Goldbach *et al.*, 1976), but in contrast to that virus, the duplicated-inversion sector in this ciliate was extremely long, containing about 3500 sites and constituting about 8% of the entire molecule (Figure 9.20). Comparisons of the mtDNAs from five different strains of the organism indicated close kinships to exist only between strains T and ST, which showed about 90% homology (Goldbach *et al.*, 1976). In contrast, homologies between other strains ranged from only 4% to a maximum of 16%. However, all had the terminal duplication-inversion and ragged duplex ends (Arnberg *et al.*, 1977; Goldbach *et al.*, 1977). *Paramecium aurelia* differs strongly from the preceding ciliate in the character of this molecule. Its linear duplex molecule, about 14 μm in length, weighed around 35×10^6 daltons and consisted of 40% G + C (Suyama and Preer, 1965; Flavell and Jones, 1971b; Goddard and Cummings, 1975; Cummings *et al.* 1976). The structure of the mtDNA from *Acanthamoeba castellanii* has been investigated independently by two laboratories (Bohnert, 1973; Bohnert and Herrmann, 1974; Hettiarachchy and Jones, 1974). Approximately 80% of the molecules were circular, with a circumference of 12.7 μm and a molecular weight of 25.7×10^6. Close to 30% of the base composition was attributed to G + C.

An unusual DNase-sensitive structure has been found in the mitochondria of those cells of red algae that give rise to carpospores (Tripodi *et al.*, 1972). Instead of the fibrils that usually characterize the DNA of this organelle, many mitochondria of *Polysiphonia* were in the form of a twisted, complex body (Figure 9.21). This had the appearance of consisting of two series of parallel tubules, the opposing sets being irregularly connected by long fibrils. Similar bodies have been reported from rat hepatocytes and corpus striatum cells and slime mold (Mugnaini, 1964; Behnke, 1965; Schuster, 1965; Blecher, 1967).

Kinetoplast DNA. From other unicellular organisms, little further information regarding this molecule of the mitochondrion seems to have been garnered, except that of the kinetoplast of hemoflagellates, which has been relatively thoroughly explored (Borst *et al.*, 1976; Vickerman, 1977). It has been

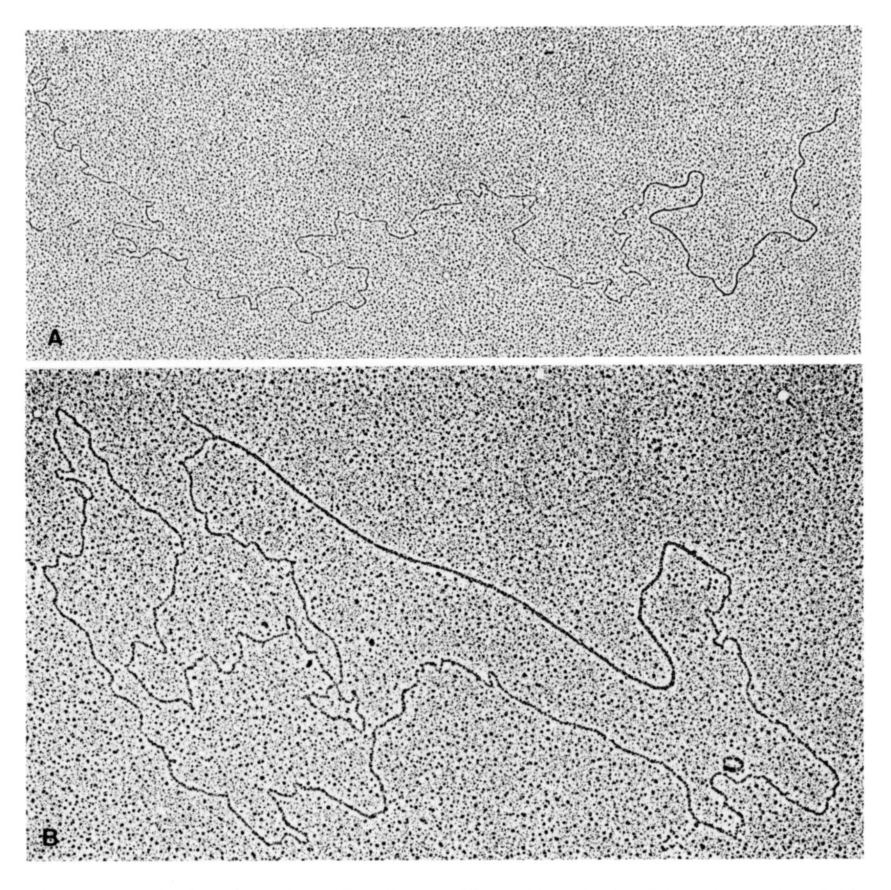

Figure 9.20. Mitochondrial DNA of *Tetrahymena*. The DNA molecule of the mitochondria of this ciliate occurs both as linear (A) and circular (B) forms. (A) 30,000×; (B) 42,000×. (Courtesy of Arnberg *et al.*, 1977.)

found to be quite a complex mixture, four types usually being present (Riou and Delain, 1969; Renger and Wolstenholme, 1970, 1971; Simpson and da Silva, 1971). The greatest quantity was concentrated in the first class (1) consisting of large, apparently linear molecules, whose weight of 43×10^6 resembled that of other unicellular forms (Laurent and Steinert, 1970). Another fairly abundant class (2) consisted of small figure-of-eight molecules plus minicircles catenated with one another. In *Leishmania tarentolae, Trypanosoma cruzi,* and *Crithidia fasciculata,* these had molecular weights of 0.56, 0.94, and 1.49×10^6, respectively (Newton, 1967, 1968; Riou and Paoletti, 1967; Riou and Delain, 1969; Riou and Pautrizel, 1969; Simpson and da Silva, 1971). The

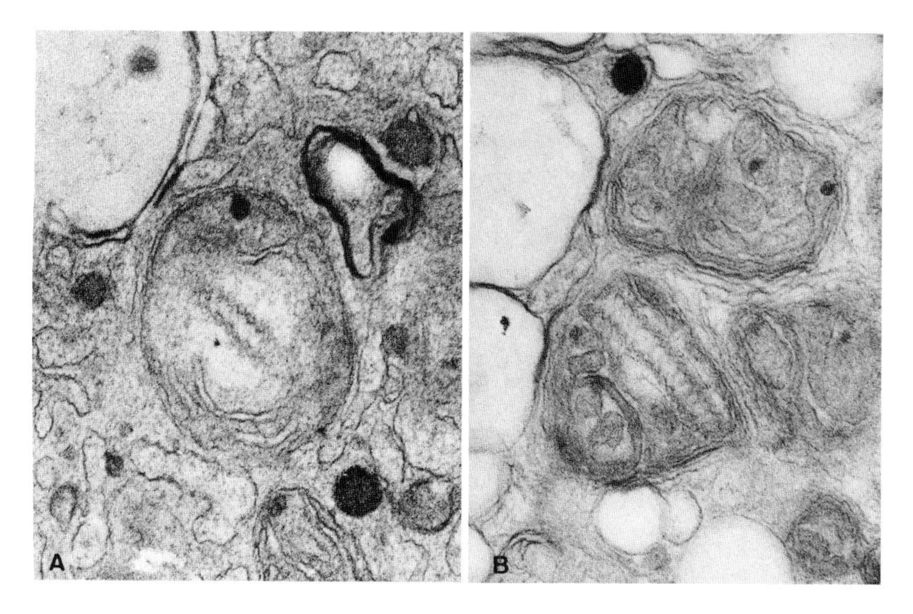

Figure 9.21. DNase-sensitive bodies in *Polysiphonia* mitochondria. (A and B) Such complexes are destroyed by treatment with DNase. (A) 60,000×; (B) 54,000×. (Courtesy of Tripodi *et al.*, 1972.)

minicircles have proven to be highly heterogeneous in base sequence (Kleisen *et al.*, 1976a; Price *et al.*, 1976; Riou and Yot, 1977) and in many forms are linked by thousands in an intricate network of two types (Englund *et al.*, 1977). In form I all the minicircles are covalently closed while in form II they are open. The final two classes were comprised of (3) short, catenated minicircles and (4) individual minicircular molecules. Hybridization investigations uncovered only 2 to 4% homology between the pooled mtDNAs of *L. tarentolae* and *T. cruzi* (Steinert *et al.*, 1973; Lloyd, 1974). More recent analyses have tended to show that a large circular form probably is the true DNA of intact kinetoplast mitochondria (Kleisen *et al.*, 1976b). In addition, two RNA components have been found associated with the DNA in the organelle (Nichols and Cross, 1977).

mtDNA of Fungi. The characteristics of the mtDNA of a small number of fungal types have now been described. That substance from *Allomyces macrogynus* behaved as a single species and proved to have a molecular weight of 52×10^6 and a Gp + pCp content of 42% (Dizikes and Burke, 1978); the configuration of the molecule was not determined, however. *Neurospora crassa* originally was reported to have linear mtDNA molecules 25 μm long and a molecular weight of 40×10^6 (Wood and Luck, 1969; Schäfer *et al.*, 1971; Clayton and Brambl, 1972), with a G + C content close to 40% (Richter and Lipmann, 1970). Because both the reported linearity and the homogeneity of the native molecule appeared questionable, further investigations were conducted using improved techniques; these obtained large closed circular molecules (Clayton and Brambl, 1972) and minicircles were also detected (Agsteribbe *et al.*, 1972). The closed circular configuration of the molecule is now well established, as is also a circumference of 19 μm (Terpstra *et al.*, 1977). Moreover, several genetic maps of the mtDNA have appeared in print (Bernard and Küntzel, 1976; Terpstra *et al.*, 1976).

Data available from other fungal sources similarly are incomplete or less than satisfactory. The mtDNAs of *Aspergillus nidulans* and *Kluyveromyces lactis* had molecular weights in the range of 20 to 26×10^6 (Sanders *et al.*, 1974; Lopez Perez and Turner, 1975). This substance from the mitochondria of *Physarum polycephalum* consisted of heterogeneous linear molecules with molecular weights between 40 and 50×10^6 and lengths up to 28 μm (Sonenshein and Holt, 1968; Kessler, 1969); however, more extensive investigations possibly would show that the molecule is actually circular as in *Neurospora*. A base composition including 26% G + C has been reported (Guttes *et al.*, 1967; Evans and Suskind, 1971), while that of the mtDNA from *Dictyostelium* was determined as being 28% G + C (Sussman and Rayner, 1971).

At least a large amount, but not all, of the mtDNA of fungi is concentrated into the so-called mitochondrial nucleus, or better, nucleoid, which is Feulgen positive and undergoes binary fission in a manner reminiscent of the prokaryotic body of the same name (Figure 9.10; Schuster, 1965; Guttes *et al.*, 1966, 1969; Kuroiwa, 1973, 1974; Kuroiwa *et al.*, 1976b). In addition to DNA, these bodies contain RNA, which they synthesize, and proteins, at least one of which is basic (Kuroiwa *et al.*, 1976a). During its synthesis the RNA is distributed nonrandomly on the nucleoid. A somewhat similar body of DNA has been reported in *Polysiphonia,* a red seaweed (Tripodi *et al.*, 1972); this differs strikingly from the fungal nucleoid in having a coiled construction (Figure 9.21).

mtDNA in Higher Plants and Metazoans. In the metaphytans the mtDNA, with lengths up to 62 μm, appears to be the largest reported (Wolstenholme and Gross, 1968; Borst and Kroon, 1969). Quite to the contrary, the molecules from metazoans of all sorts are the very smallest, except for the euglenoids, the circular supercoiled molecule having a mean circumference of 5 μm and a molecular weight of 10×10^6, which corresponds to between 15,000 and 18,000 base pairs (Borst, 1972; Dawid *et al.*, 1976). The actual known range in length is from 4.45 μm in a sea urchin to 5.85 μm in the echiuroid worm, *Urechis caupo* (Dawid and Brown, 1970), and 6.2 μm in *Drosophila melanogaster* (Fauron and Wolstenholme, 1976). Included in the products coded by the molecule are two rRNAs, a set of tRNAs, and a few poly(A)-containing RNAs that appear to be messengers for mitochondrial proteins (Hirsch *et al.*, 1974; Ojala and Attardi, 1974). Apparently the mtDNA molecules of at least mouse and HeLa cells are not pure DNA but contain a fraction of RNA, for they were nicked when treated with a ribonuclease (Grossman *et al.*, 1973). The major fraction seemingly contained perhaps ten ribonucleotide residues.

Genetic maps of the mtDNA from several vertebrates have been published, including those of two species of *Xenopus* (Dawid *et al.*, 1976), rat liver (Saccone *et al.*, 1976, 1977), and mouse LA 9 cells (Moore *et al.*, 1977; Parker and Watson, 1977). Some years ago, it was reported that the genome coded for only nine peptides, supposedly of the respiratory enzyme complex (Ashwell and Work, 1970; Borst, 1972; Schatz and Mason, 1974), but the latter supposition may have been premature. Unusual heterogeneity in primary structure has been brought to light by a recent investigation that compared different tissues from the same animal and also like tissues from different individuals of the same species, the ox (Coote *et al.*, 1979). Significant differences were found both between liver and brain cell mtDNA and between different specimens. Consequently, the employment of this substance to establish evolutionary histories of a given taxon, as has been done with parthenogenetic lizards of the genus *Cnemidophorus* (Schoen *et al.*, 1979), can scarcely carry great conviction.

The molecular form of the mtDNA of metazoans is also proving to be more heterogeneous than originally suspected (Hayashi *et al.*, 1978a,b). For a period, forms other than monomeric circles were believed to characterize genetically or physiologically abnormal cells, such as cultured tissues, virus-transformed cells, and pathological tissue (Matsumoto *et al.*, 1976). But now, in apparently normal tissues from larvae of three species of *Drosophila,* dimeric and oligomeric circular forms have been found in high proportions, which proved to be head-to-tail concatemers (Shah and Langley, 1977). The number of mtDNA molecules per cell also are variable. Mouse L cells were demonstrated to have 1100 copies per cell of a 1×10^7-dalton molecule, whereas others had 900 copies of a 2×10^7-dalton molecule, and HeLa cells contained about four times as many molecules per cell (Bogenhagen and Clayton, 1974). Moreover, while individuals possessed only one type, two kinds of mtDNA were shown to exist in populations of horses, rats, and man; one of these in the rat proved to have a small core complexed to it (Francisco *et al.*, 1977). Perhaps this is a remnant of a former larger core similar to that of fungi.

9.3.2. Replication and Transcription

The activities of the mitochondrial genetic system at the molecular level are just beginning to be understood to some degree, so any account of the processes must necessarily be brief and greatly incomplete (Young and Hunter, 1979). However, what has been learned reveals differences and similarities to those of both prokaryotes and eukaryotes.

Replication of mtDNA. That DNA was synthesized within the mitochondrion was first established by Parsons and Simpson (1967) who showed that it required the presence of all four deoxyribonucleotide triphosphates and Mg^{2+}, as in the nuclear processes. However, it was not clear whether this was a replicatory or repair activity until Karol and Simpson (1968) demonstrated that long segments of DNA are synthesized within the organelle, strongly suggestive of replication. This has been further confirmed by use of hybrid mouse–human cells, analyses of which provided evidence that both parental types of mtDNA were synthesized (Coon *et al.*, 1973); such cells have also displayed recombination in producing differently labeled RNAs (Horak *et al.*, 1974). As in viral, prokaryotic, and nuclear DNA of eukaryotes, mtDNA replication involves a membrane–DNA complex (Shearman and Kalf, 1977) and is a discontinuous process (Koike and Wolstenholme, 1974).

The replicative intermediate molecules have been indicated to fall into two contrasting categories. The first class was comprised of branched circles (Figure 9.22), the two branches of which were of equal length (Kirschner *et al.*, 1968; Tobler and Gut, 1974), while the second consisted of closed circular duplexes to each of which was attached a short, single branch, called the E strand (Kasamatsu *et al.*, 1971; Ter Schegget and Borst, 1971a,b). This branch was hydrogen-bonded to the duplex, displacing one of the strands to form a D loop (displacement loop), and could be melted off without breaking the circular portion. As it hybridized exclusively to the L strand of the mtDNA, it probably represented a fragment of the H strand. Both in chick liver mtDNA and that of thyroid, the length of the D loop was 3.5% of the total molecular length (Arnberg *et al.*, 1971). In mouse cells about 50% of the total mtDNA (Kasamatsu *et al.*, 1971; Berk and Clayton, 1976), and in chick liver about 30% (Arnberg *et al.*, 1971), consisted of D loops.

DNA polymerases are present in mitochondria (Soriano *et al.*, 1974; Philippe and Chevaillier, 1978), but little else is known about them. One in yeast mitochondria was found to differ from that of the nucleus in molecular size (Iwashima and Rabinowitz, 1969; Cottrell *et al.*, 1973). At least one of these enzymes is active in repair synthesis, as shown by mitochondria of *Tetrahymena* damaged by ultraviolet light (Westergaard and Pearlman, 1969; Westergaard, 1970; Keiding and Westergaard, 1971). The data of these investigations suggested that repair required concurrent RNA synthesis and that the polymerase was synthesized on cytoplasmic ribosomes with mRNA transcribed from nuclear DNA. However, at least one DNA polymerase has been found to be localized in the mitochondrion (Tanaka and Koike, 1978), one that has proven to be a high-molecular-weight protein, sedimenting at 9.2 S and using poly(A) · oligo(dT) as a template primer (Tanaka and Koike, 1977). Thus in some schemes it may be classified as a DNA polymerase-γ (Weissbach *et al.*, 1975), but the terminology for this class of enzymes has still not become stabilized (Dillon, 1978).

Replication of the mtDNA in the two ciliates that have been investigated shows a number of differences, not only from other eukaryotes but also from

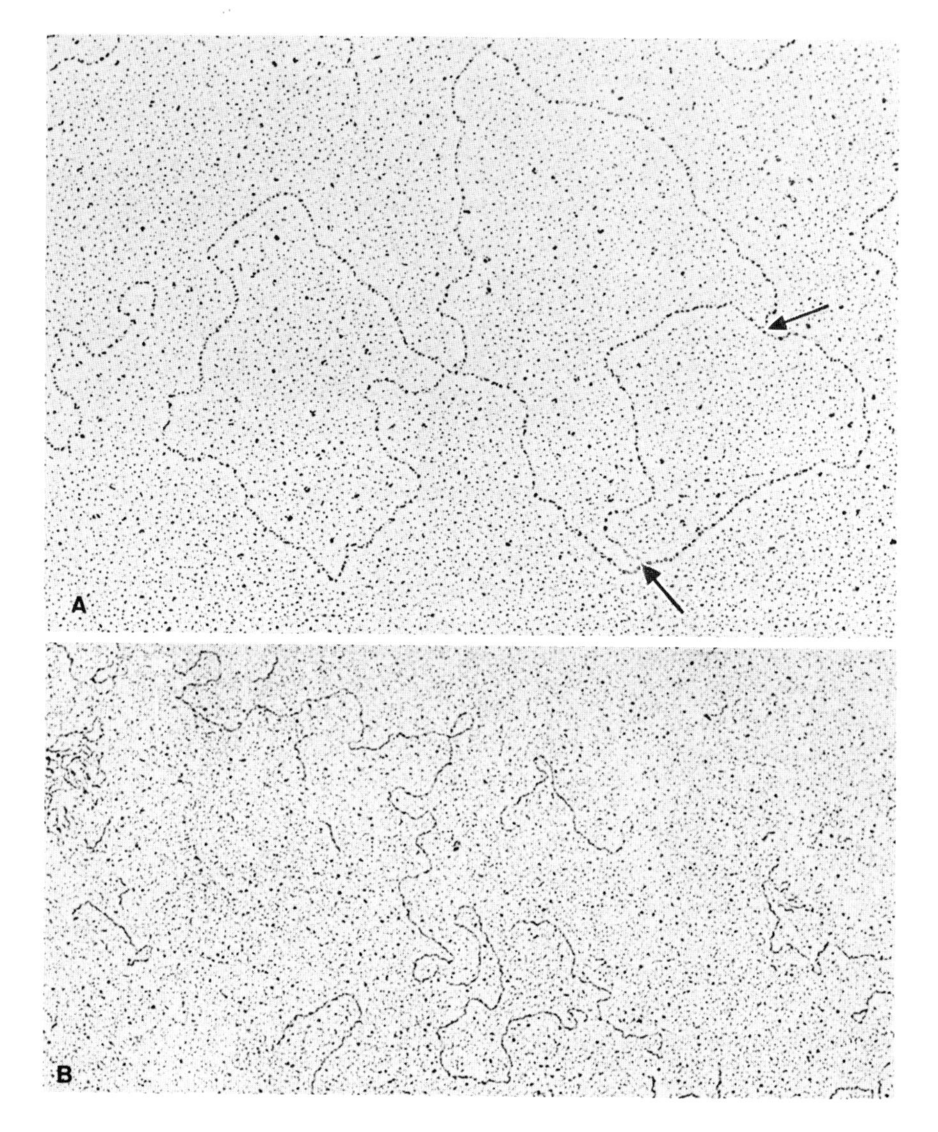

one another. Synthesis of the 15-μm linear duplex molecule of *Tetrahymena* was apparently initiated at the center of the molecule and proceeded bidirectionally from that point toward both ends (Figure 9.23; Arnberg *et al.*, 1974; Clegg *et al.*, 1974; Upholt and Borst, 1974). In the end these processes resulted in the formation of "eye molecules" that later were processed directly into two molecules like the original one. Although the mtDNA of *Paramecium aurelia* was similarly linear and nearly equal in length (14 μm), its replication was shown to proceed unidirectionally from one end, resulting in lariat molecules and linear dimers (Goddard and Cummings, 1975). In the lariat forms, it was

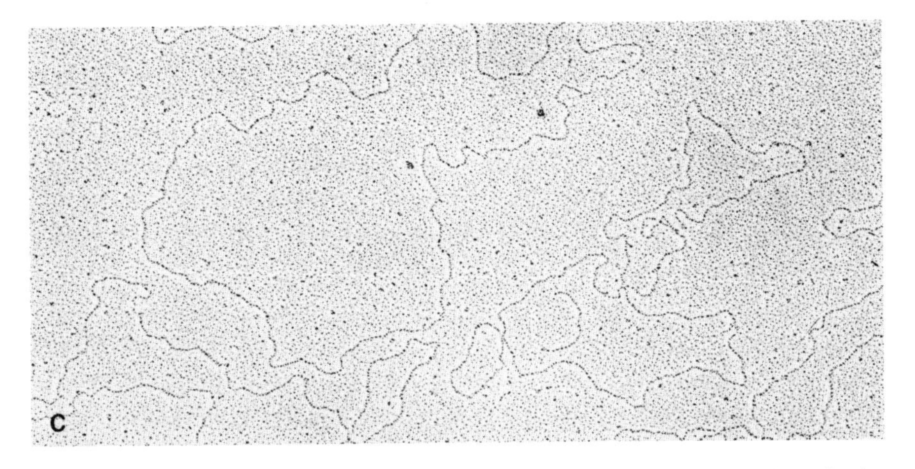

Figure 9.22. Replicative forms of mtDNA from *Ascaris* embryos. (A) Double-forked circular molecule from four-cell-stage embryo; the two segments between the fork (arrows) are of equal length. 96,000×. (B) The heavy band mtDNA from the same source contains linear, unbranched molecules of varied lengths. 20,000×. (C) The light satellite molecules are circular. 60,000×. (All courtesy of Tobler and Gut, 1974.)

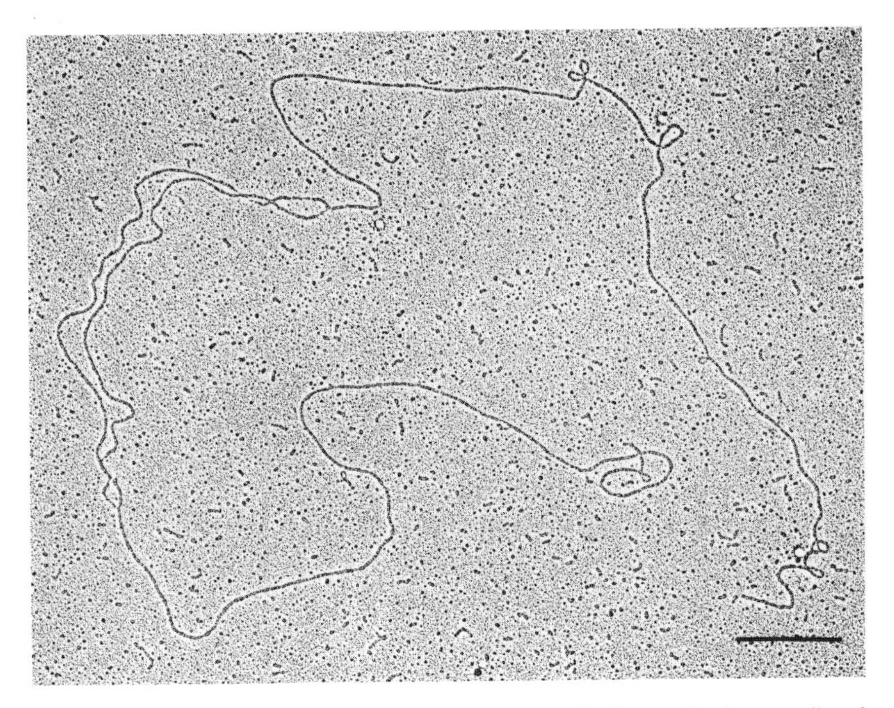

Figure 9.23. A replicating mtDNA molecule from *Tetrahymena*. The linear molecules are replicated bidirectionally from a central point. 48,000×. (Courtesy of Goldbach *et al.*, 1977.)

found that one-half the circumference of the loop plus the length of the tail was consistently equal to 14 μm. It was later demonstrated that the lariats and dimers were true replicative intermediates, the latter being formed directly from the former (Goddard and Cummings, 1977). The dimers were in a head-to-head configuration, each monomer containing a covalent crosslink at the initiation end of the molecule that was retained throughout the processes of replication.

Timing of Replication. Investigations into the correlation of mtDNA synthesis with specific phases of the cell cycle have met with varying degrees of success. In yeasts, mtDNA replication appeared to follow nuclear DNA synthesis and was completed before bud formation began (Smith *et al.,* 1968; Cottrell and Avers, 1970). The second paper cited also detected evidence of synchrony of mitochondrial DNA replication in synchronized cell cultures. Among protistans, *Tetrahymena pyriformis* and *Physarum polycephalum* have been found to replicate mtDNA throughout the cell cycle (Parsons, 1965; Guttes *et al.,* 1967; Charret and André, 1968; Braun and Evans, 1969). In mammals, results are somewhat conflicting, depending on how the cells were synchronized. The Chinese hamster ovary cells synchronized by mitotic selection showed evidence of mtDNA replication after 4 hr, which processes continued for 9 hr, thus resembling nuclear DNA (Ley and Murphy, 1973). Contrastingly, when the cells were synchronized by isoleucine deprivation, mtDNA production did not begin until 9 to 12 hrs after addition of isoleucine but was completed within 3 hr.

Transcription of mtDNA. The DNA of mitochondria is transcribed into various RNAs by specific DNA-directed RNA polymerases (Tzagoloff, 1977) that are distinct from those of both the eukaryotic nucleus and the prokaryotes. In the first place, these enzymes differ in consisting of only a single subunit, and second, unlike the high-molecular-weight complexes of the bacterial and nuclear enzymes, they are relatively small. That of yeast has been reported to be in the range of 59,000 to 67,000 (Rogall and Wintersberger, 1974; Scragg, 1974), that of *Neurospora,* 63,000 (Küntzel and Schäfer, 1971), *Xenopus,* 46,000 to 50,000 (Wu and Dawid, 1972), and rat liver, 65,000 to 70,000 (Reid and Parsons, 1971; Mukerjee and Goldfeder, 1973; Gallerani and Saccone, 1974). In contrast, the RNA polymerase of *Lactobacillus* has been determined as having a molecular weight of 425,000 (Stelter and Zillig, 1974) and that of *E. coli,* 495,000 (Matsura, 1973). The enzyme from both these latter sources had a subunit composition of $\beta\beta'\alpha_2\sigma$ (Travers and Burgess, 1967). Third, while some of the mitochondrial polymerases resembled those of prokaryotes in being sensitive to rifampicin (Reid and Parsons, 1971; Gallerani and Saccone, 1974; Scragg, 1974), those of yeast and *Xenopus* mitochondria were not (Wu and Dawid, 1972, 1974; Rogall and Wintersberger, 1974). At least rat mtDNA-dependent RNA polymerase, quite unexpectedly, has proven to be as sensitive to cycloheximide, a chemical frequently employed to suppress cytoplasmic protein synthesis in eukaryotes (Saccone and Quagliariello, 1975). That of yeast has recently been shown to be transcribed from nuclear DNA and translated on cytoplasmic ribosomes (Scragg, 1976).

One of the most unusual features of transcription of the mtDNA is that both strands are transcribed completely, producing RNA that hybridizes along the entire lengths of both the H and L chains of the DNA (Saccone and Quagliariello, 1975). Consequently, the immediate product of transcription must be processed extensively to bring the functional molecules into being. Further, this method of transcription precludes the existence of a control mechanism, for one unit of every ultimate product is produced by each cycle of transcription.

Processing the Transcripts. Processing of the single transcript is poorly understood, but it is known to differ from those activities of the cytoplasm and prokaryotes. First, the precursor molecule appears to be cleaved into several large segments, many of which are then polyadenylated on their 3′ termini (Saccone and Quagliariello, 1975). Some of the bases are modified (Borst, 1972; Mahler and Raff, 1975), but whether this occurs before or after polyadenylation remains unestablished. The most distinctive features of processing concern the addition of the poly(A) segments. Like hnRNA, polyadenylation of the nascent mtRNA proceeds rapidly, but here large sectors are recipients. For instance, large chains have been identified that include rRNA in hamster mitochondria; one polyadenylated RNA molecule that sedimented at 20 S has been identified as the precursor for the 17 S RNA of the mitochondrial ribosomes (Attardi *et al.*, 1976; Cleaves *et al.*, 1976; O'Brien, 1977).

Not all the large cleavage products are polyadenylated, however; Attardi *et al.* (1976) separated 18 size classes of poly(A)-containing transcripts and 14 lacking such appendages. In *Neurospora,* to a limited extent a more precise picture of cleavage has been provided, in that it has been shown that a 32 S precursor was cleaved into two smaller precursors. When further processed, the latter ultimately resulted in the mature 25 S and 19 S RNAs of the mitochondrial ribosomes (Saccone and Quagliariello, 1975). Moreover, yeast transcripts as a whole appear to be deficient in poly(A) segments.

In vertebrates, the poly(A) segments have a gel electrophoretic mobility around 4 S, corresponding to a length of between 50 to 80 nucleotides (Perlman *et al.*, 1973). Mouse ascites cell transcripts contained poly(A) sequences only 35 to 55 nucleotides in length (Avadhani, 1979). On the basis of the decay rate, these fell into two classes, one having a half-life of 45 min, the other, 210. In the germinating conidia of the fungus *Trichoderma viride,* the transcripts of the cytoplasmic and mitochondrial fractions were compared directly (Rosen and Edelman, 1976), and it was found that about 10% of the total mitochondrial fraction was polyadenylated. The poly(A)-bearing RNAs of the mitochondria showed only two peaks on polyacrylamide gels, one at 22 S and the other at 29 S, whereas those of the cytoplasm were heterogeneously distributed along the gel. The poly(A) segments of the mitochondrial fraction were only 20 to 25 nucleotides long, while those of the cytoplasm were 50 to 60.

9.3.3. Mitochondrial Ribosomes and Transfer RNAs

The Mitochondrial Ribosomes. The ribosomes of mitochondria have proven to be far less consistent along phylogenetic lines than have their counterparts in the cytoplasm (Table 9.1); hence, it is nearly impossible to make broad generalizations. In the cytoplasm, the monosomic bodies fall with remarkable uniformity into two classes, 70 S in prokaryotes and 80 S in eukaryotes. Quite contrastingly, the monosomic ribosomes of mitochondria range from 55 S in *Paramecium* and rat liver to 78 S in maize and turnip root, and 80 S in *Tetrahymena*. Equally dissimilar results have been noted with the large and small subunits, which show a fair degree of constancy in being 50 S and 30 S in prokaryotes and 60 S and 40 S, respectively, among eukaryotes (Table 9.1). Mitochondrial large subunits, however, range from 39 S in rat liver to 60 S among higher plants, and the small subunit of mitochondria ranges from a low of 30 S in rat liver to a high of 55 S in *Tetrahymena,* in which ciliate the large and small subunits are identical in size (Suyama, 1967; Chi and Suyama, 1970). In addition to these types within the stroma, cytoplasmic ribosomes are bound to the outer surface of the organelle, at least in yeast (Kellems and Buetow, 1974).

The preceding comparisons, disparate as the results are, still do not show the true lack of consistency of the mitochondrial ribosomes. For instance, the RNAs of their subunits differ strongly from one taxon to another, that of the large subunit showing a range from 16 S in rat liver to 23 S in *Neurospora* (Table 9.1), whereas the RNA of the large subunit of the cytoplasmic ribosomes only varies between 25 S and 29 S. Where they have been established, protein numbers likewise show a broad base. This statement is also true of the cytoplasmic particle, which has totals ranging from 25 to 71, but to a lesser degree than the mitochondrial ribosome, where only 20 proteins have been detected in the yeast and as many as 107 in *Neurospora*. Thus it is obvious that the ribosomes of mitochondria have little in common with either the prokaryotic or the eukaryotic cytoplasmic granule.

This distinctiveness is accentuated by the observation that all the proteins of the mitochondrial ribosomes except one (S-5) in *Neurospora* are synthesized by the cytoplasmic organelle (Küntzel, 1969; Neupert *et al.,* 1969; Lizardi and Luck, 1972; Lambowitz *et al.,* 1976, 1979; Buetow and Wood, 1978), and all seem to be thus formed in yeast (Davey *et al.,* 1970; Schmitt, 1970, 1971). In *Tetrahymena,* a few of the very small proteins of the mitochondrial particle are synthesized within that organelle, but even the synthesis of these is under the control of the nuclear system (Millis and Suyama, 1972). A similar statement applies to the mitochondrial ribosome proteins of *Paramecium* (Tait *et al.,* 1976). That the situation is complex has been shown by some investigations on interspecific hybrids of *Xenopus laevis* and *X. mulleri* (Leister and Dawid, 1975). The F_1 hybrids received nuclear genes from both parents, but the mi-

Table 9.1
Comparison of Ribosomal Structures

Source	Monosome	Large subunit		Small subunit		5 S rRNA	5.8 S rRNA	Number of proteins		
		Total	rRNA	Total	rRNA			Total	Large subunit	Small subunit
Prokaryote										
E. coli	70 S	50 S	23 S	30 S	16 S	Present	Absent	56	34	21
Eukaryotes										
Yeast	80 S	60 S	25 S	40 S	16 S	Present	Present	---	36–55	30–40
Tetrahymena	80 S	60 S	26 S	40 S	17.5 S	Present	Present	25	---	---
Neurospora	77 S	---	25 S	---	17 S	Present	Present	---	---	---
Rat liver	83 S	47 S	32 S	32 S	16 S	Present	Present	---	---	---
Pea	78 S	55 S	---	35 S	---	Present	Present	---	---	---
Turnip	80 S	60 S	25 S	40 S	18 S	Present	Present	---	---	---
Mitochondria										
Yeast	72 S	---	22 S	---	18 S	Present	(13S)	20	---	---
Tetrahymena	80 S	55 S	21 S	55 S	14 S	Present	Absent	---	---	---
Neurospora	73 S	52 S	24 S	39 S	17 S	Absent	Absent	107	---	---
Rat liver	55 S	39 S	16 S	30 S	12 S	Absent	Absent	84(107)	40	44
Turnip	78 S	60 S	---	46 S	---	Present	Absent	---	---	---
Chloroplasts										
Euglena	---	---	22 S	---	16 S	---	---	---	---	---
Chlamydomonas	70 S	60 S	---	45 S	---	---	---	---	---	---
Maize	78 S	60 S	---	44 S	---	Present	Absent	---	---	---

tochondrial genes came solely from the egg. When the proteins of the large subunit of the mitochondrial ribosomes were compared, one species was observed to be absent from each F_1 progeny. The presence of certain others was related to the sex and species of the parent, one being absent if that species of clawed frog had been the mother, and four others if the corresponding species had been the father. One explanation offered was that, although the hybrid nuclear genomes contained genes from both parents, only maternal proteins were actually incorporated into the mitochondrial ribosome. However, this does not explain the presence in one case of the paternal protein.

Unlike the proteins of the mitochondrial ribosomes, the rRNAs are encoded by mtDNA and transcribed and processed within the mitochondrion (Reboul and Vignais, 1974). Little appears to have been ascertained regarding processing, but in hamster cells it includes a low degree of methylation (Dubin, 1974). The 17 S nucleic acid of the large subunit was demonstrated to contain 0.13 methyl group per 100 nucleosides; further analysis showed that one methylated ribose moiety and one unidentified methylated residue were present per molecule. The 13 S RNA of the small subunit was slightly more heavily methylated, having 0.37 methyl group per 100 nucleosides; the modified bases appeared to include one thymidine, one N^6-dimethyladenosine, and one methylated cytidine residue per molecule.

Mitochondrial tRNAs. The mitochondrial genome codes for a variety of tRNA species (Casey *et al.*, 1972, 1974a,b; Cohen and Rabinowitz, 1972), but the precise number encoded is highly diversified among eukaryotes. In *Tetrahymena,* only two types were found to be coded by the mtDNA, those for phenylalanine and leucine; however, three isoaccepting $tRNAs^{Leu}$ were later identified (Chiu *et al.,* 1974). These four that were formed in mtDNA were referred to as "native" tRNA, while others that did not hybridize to mtDNA were considered "imported" from cytoplasmic sources (Chiu *et al.,* 1975). *Neurospora crassa* has been shown to lie at the opposite end of the scale, for its mitochondrion-encoded tRNAs numbered 18 (Barnett and Brown, 1967); this is exceeded by yeast, whose mtDNA coded for 20 tRNAs (Martin *et al.,* 1977), including two $tRNAs^{Met}$. Previously elsewhere, three valine tRNA-isoaccepting species had been reported to be mtDNA products (C. Schneller *et al.,* 1975). The location of the genes for many of these on the mitochondrial genome has been determined (Fukuhara *et al.,* 1976; Martin *et al.,* 1977); some isoaccepting species have been found to be transcribed from different regions of the yeast mitochondrial genome (N. C. Martin *et al.,* 1976). Three distinct $tRNAs^{Ser}$ were shown to be encoded by the mtDNA, while others from the cytoplasm were imported selectively (Baldacci *et al.*, 1976, 1977).

Among mammalian sources, the number of tRNAs varies greatly from species to species (Aujame and Freeman, 1979). In HeLa cells, Lynch and Attardi (1976) detected 17 tRNAs that could hybridize with mtDNA; these specified 16 different amino acids, the four missing species being for asparagine, glutamine, histidine, and proline. Thus this is the sole source from which a mitochondrial $tRNA^{Cys}$ has been reported. Only six amino acids have been found provided with mitochondrial tRNAs in rat liver (Buck and Nass, 1969; Nass and Buck, 1970; Jakovcic *et al.,* 1975), and an identical number have been described from calf brain (Charezinski and Borkowski, 1973; Borkowski and Brzuszkiewicz-Zarnowski, 1975). Contrastingly, four isoaccepting species of $tRNA^{Met}$ have been identified in mouse liver mitochondria (Wallace and Freeman, 1974b). That the relationship among tRNAs of mammals is limited largely to intraordinal boundaries was indicated by the finding that the mitochondrial $tRNA^{Leu}$ from rat liver would hybridize under stringent conditions only to that mtDNA of mouse and guinea pig, in addition to its own (Jakovcic *et al.,* 1975).

Molecular Traits of Mitochondrial tRNAs. Some years ago molecular weights of mitochondrial tRNAs were reported to be slightly lighter in some cases than those of the corresponding cytosolic types (Dubin and Friend, 1972), but so far this condition has actually been demonstrated only for the $tRNA^{Leu}$ of the hamster (Aujame *et al.,* 1978). In that rodent, two isoaccepting species of leucine from the mitochondria were found to have molecular weights of 23,000 and 24,000, respectively, in contrast to the 27,000 of the same type from the cytosol. Moreover, at least mammalian sources have a lower level of unusual bases. In particular, 7-methylguanosine has been noted to be rare in mitochondrial varieties (Dubin and Friend, 1974; Chia *et al.,* 1976; Wallace *et al.,* 1978), which nucleoside in prokaryotic and cytosolic tRNAs occurs only in site C of the short type arm IV (Dillon, 1978). However, while present there in most species that have been sequenced, it is absent with a high degree of variability from one source to another. This lack of homology between tRNAs of the mitochondria from diverse organisms continues at the gene level, for hybridization experiments with tRNAs and mtDNA from different sources show only low degrees of interaction (Jakovcic et al., 1975).

Primary Structure of $tRNA^{Met}$. The base sequence of one mitochondrial tRNA has been determined, that of the initiator $tRNA^{Met}$ of *Neurospora crassa* (Heckman *et al.,* 1978). Comparison of its primary structure with others of the same tRNA type from other sources (Table 9.2) suggests close relationships with neither those of prokaryotes nor eukaryotes. Although it is formylated for use in initiation as in most prokaryotic but no known eukaryotic tRNAs, not all of the former types are so modified. For instance, *Halobacterium cutirubrum* does not employ formylated $tRNA^{Met}$ for initiation (Heckman *et al.,* 1978). On the other hand, it resembles those of eukaryotic sources in having the bases in sites 1 and 1′ of arm I form a standard Watson–Crick pair, while the initiator species of prokaryotes cannot, *Halobacterium* again being exceptional (Heckman *et al.,* 1978).

In the sequence, arms homologous in composition to those of other sources are scarce. Arm IA (Table 9.2) shows resemblances to eukaryotes as

Table 9.2
Comparison of the Initiator RNA of Neurospora[a]

	Arm IA	Bend 1	Arm II	Bend 2	Arm IIIA	Anticodon
E. coli$_{F1}$	p CGCGGGG	U_tG	GAGC AGCCUGGD--A GCUC	G	$UCGGGC'_2U$	CAU
N. crassa$_{mt}$	p UGCGGAU	UA	UUGU AA-DAG-D--A ACAU	A	UUUGGCU	CAU
S. cer$_F$	p AGCCGCG	UG_1	G_2CGC AG-D-GG--AA GCGC	G_2^2	CAGGGCU	CAU
Mus$_4$	p GCCUCGU	UA	G_2CGC AG-DAGGD--A GCGC	G_2^2	ΨCAGΨCU	C_2AU

	Arm IIIB	Arm IV	Arm V	Arm IB	Preterm
E. coli$_{F1}$	AACCCGA	AGG_7UG	GUCGG TΨCAAAU CCGGC	CCCCGCA	A
N. crassa$_{mt}$	G_1ΨCCGAA	-UGAC	AUAGG UGCAAAU CCUGU	AUCCGCA	U
S. cer$_F$	$A_tACCCUG$	AUG_7DC_5	CUCGG $AVCGA_1AA$ $CCGA_mG_m$	CGCGGCU	A
Mus$_4$	A_tAΨCUGA	AGG_7DC_5	GUGAG TΨCGA_1UC CUCAC	ACGGGGC	A

[a] Based on Dillon (1978) and Heckman *et al*. (1978).

well as prokaryotes, while bend 1 is strictly eukaryotic in composition. In arm II, the double-stranded sector displays kinships to neither group, but the single-stranded loop is predominantly eukaryotic in nature, especially in having two dihydrouridine residues present. Bend 2 is totally unique, as is also the double-stranded region of arm III. Also in the latter arm, sites A and B of the unpaired loop correspond to those of eukaryotic species, whereas those of sites C and D are unrelated to any other type. A similar lack of relationships in primary structure is found in the remainder of the molecule, including the preterminal site.

Taken as a whole, the molecule is unusually rich in uridine residues. If the modified species of this nucleoside, such as pseudouridine and dihydrouridine, are included, the count of such residues totals 23, which compares to 11 and 10 for the *E. coli* and yeast initiator tRNA, respectively, and to 15 for the mouse $tRNA_4^{Met}$. Adenosine residues are also more abundant than elsewhere but less strikingly so than uridine, 19 being present here compared to 13 for the bacterial, and 15 for the eukaryotic cytoplasmic types. Finally, a paucity of modified bases can be noted (Wallace *et al*., 1978), there being only 4 present, contrasting to 6 in the prokaryote, 11 in the yeast, and 15 in the mouse species. Because each modified site requires a separate modifying enzyme, this lack of modification could well be correlated to the reduction of the genome that characterizes the mitochondrion of all eukaryotes.

This loss in turn could also apply to the noted absence (J. Schneller *et al*.,

1975a,b) of the hypermodified base referred to as Y that characterizes the tR-NA^{Phe} of many eukaryotes (Dillon, 1978). The sparsity of modified bases and the high frequency of uridine and adenosine have been recorded from such mammalian sources as the mitochondria of rat liver and Morris hepatomas 5123D and 7777 (Chia *et al.*, 1976). Studies on the methylases of yeast showed that, while a large variety of types occurred in the cytoplasm, only those of 1-methyl- and 2-methyl-guanosine, 2-methyl-N_2, N_2-dimethylguanosine, and ribothymidine occurred in the mitochondria (C. Schneller *et al.*, 1975, 1976). Other studies have shown that in addition to the products of these methylases, inosine, pseudouridine, and t^6A also occurred in the mitochondrial tRNAs of yeast (R. Martin *et al.*, 1976).

Mitochondrial tRNA Ligases. The tRNA ligases (or synthetases as they are too frequently called) of mitochondria have been investigated to only a limited extent, but that little has sufficed to reveal a complicated picture, the complexity being derived in large measure from the presence in the mitochondria of native and imported tRNAs. Frequently, the mitochondrial ligases form a class distinct from those of the cytoplasm, as reported for yeast (Boguslawski *et al.*, 1974; Baldacci *et al.*, 1975; J. Schneller *et al.*, 1975c), *Neurospora* (Epler *et al.*, 1970), *Tetrahymena* (Suyama and Eyer, 1967; Chiu and Suyama, 1975), and metazoans (Barnett *et al.*, 1967; Buck and Nass, 1969; Chia *et al,* 1976). For a case in point, all three isoaccepting mitochondrial tRNAs^{Leu} from *Tetrahymena* were reported to be native species that were aminoacylated by a valyl-tRNA ligase that showed no structural relationship to the corresponding enzyme of the cytoplasm (Chiu and Suyama, 1975; Suyama and Hamada, 1976). In contrast, the two isoaccepting tRNAs^{Val} are imported and are aminoacylated in the mitochondrion by a ligase that is indistinguishable from that of the cytoplasm (Suyama and Hamada, 1978). The suggestion has been made that perhaps the ligase plays a key role in the transport of the tRNA across the mitochondrial membrane, but no evidence exists either in support of or in opposition to that proposal (Suyama and Hamada, 1976, 1978). Not all the ligases of specific imported mitochondrial tRNAs are identical to the corresponding enzyme of the cytoplasm, however; those of the four aminoacyl types tested in addition to valine showed chemical properties markedly different from those of the counterparts in the cytoplasm (Suyama and Hamada, 1978, Table 3).

9.3.4. Protein Synthesis in Mitochondria

Mitochondrial Protein Biosynthesis. Mitochondria are capable of synthesizing proteins (Scragg *et al.*, 1971; Stegeman and Hoober, 1974; Innis and Craig, 1978), but maximum rates are more sensitive to ionic composition of the medium than are the cytoplasmic systems. Relatively high concentrations of Mg^{2+} and K^+ cations are especially important (Roodyn *et al.*, 1961; Beattie,

1971; Mockel, 1972); Na^+ or NH_3^+ may be substituted for the K^+, but they do not yield as rapid rates of amino acid incorporation (Grivell *et al.*, 1971). In addition, an energy source, such as ATP or an ATP-generating system, also is essential, along with the other common nucleoside triphosphates. Furthermore, the mitochondrial system has been clearly demonstrated to be completely uncoupled from the nuclear system (Rouslin, 1977).

Initiation of protein synthesis involves an interaction between the small subunit of the mitochondrial ribosomes, mRNA bearing the initiation triplet (AUG), formylated methionyl-$tRNA_F$, GTP, and initiation factor IF-2 as in cytoplasmic systems (Jay and Kaempfer, 1975; Dillon, 1978, p. 163 ff). The requirement for formylated initiator $tRNA^{Met}$ still has some problem areas. First, the source of the formyl radical has not been clearly established, but the most probable donor is N^{10}-formyltetrahydrofolate (Galper and Darnell, 1969; Galper, 1974). When cultured in the presence of methotrexate (a drug that depletes the C_1-folate pools) and metabolites whose synthesis is mediated by such pools, mouse L cells and KB cells grew normally, with no apparent effect on the formylation of the initiator tRNA (Wallace and Freeman, 1974a). Hence, the mechanism through which the initiator tRNA was formylated under these conditions remained obscure, nor could it be discerned whether or not protein synthesis had proceeded normally in the possible absence of formylation. In all systems, the initiation factors are located on the ribosomes, but those of mitochondrial ribosomes are not interchangeable with those of the cytoplasm (Avadhani and Buetow, 1974). Another problem, which has received no attention, is how initiation occurs in *Tetrahymena* in the mitochondria of which ciliate no $tRNA^{Met}$ of any sort, native or imported, formylated or unformylated, has as yet been detected (Suyama and Eyer, 1967; Chiu *et al.*, 1974, 1975; Suyama and Hamada, 1976).

Peptide-chain elongation and termination processes in the mitochondrial system have received little or no attention to date. The elongation factors EF-G and EF-T are known to be present, however, and differ from those of the cytoplasm (Avadhani and Buetow, 1974; Lewis *et al.*, 1976). In contrast, those of *Euglena* are at least partially interchangeable with the corresponding cytoplasmic enzymes (Avadhani and Buetow, 1972a,b).

The quantity of protein synthesis that takes place in mitochondria has been investigated to some degree. In *Chlamydomonas,* the rate and extent of incorporation of labeled leucine into mitochondrial proteins were such that it appeared that all proteins encoded by the mitochondrial genome were probably synthesized in that organelle (Stegeman and Hoober, 1974). In a trypanosomatid, *Crithidia luciliae,* a series of studies presented evidence that perhaps as much as 50% of the protein biosynthesis of the entire cell took place in the single large mitochondrion (Laub-Kupersztejn and Thirion, 1969, 1972, 1974), but other laboratories using the same organisms found that only about 3% of the total actually occurred there (Kleisen and Borst, 1975). Actually more than

90% of the proteins employed by mitochondria are encoded by nuclear DNA and synthesized on cytoplasmic ribosomes (Ashwell and Work 1970; Linnane *et al.*, 1972), but the interrelationships of subunits of large proteins are proving to be complicated. For example, four of the subunits of cytochrome *c* oxidase are synthesized in the cytoplasm and then assembled into the holoenzyme in the mitochondria with three subunits synthesized there (Groot and Poyton, 1975; Poyton and Groot, 1975; Eggitt and Scragg, 1975; Eggitt, 1976). It is possible that subunits of this oxidase are the sole products of mitochondrial protein synthesis in *Neurospora* (Rowe *et al.*, 1974). A number of subunits similar to that of cytochrome *c* oxidase and with corresponding sites of production (three in the mitochondria, four in the cytoplasm) have been reported for the cytochrome bc_1 complex (Katan *et al.*, 1976; Marjana and Ryrie, 1976; Buetow and Wood, 1978). While the details have not become firmed as yet, mitochondrial ATPase appears to have a comparable dual origin (Tzagoloff *et al.*, 1973, 1974; Jackl and Sebald, 1975).

Other Products Synthesized in Mitochondria. Several other classes of substances in addition to proteins and RNAs are produced within the mitochondrion. Some of the minor products are guanosine polyphosphates, such as 5′-diphospho-guanosine-3′-diphosphate (ppGpp) and 5′-triphosphoguanosine-3′-diphosphate (ppGpp) (Horváth *et al.*, 1975). These substances characteristically are placed at one end of RNA molecules in prokaryotes as well as eukaryotes, where they play an unknown role.

Of more importance to the economy of the organelle are the origins of pyridine nucleotides, such as nicotinamide mononucleotide (NMN) and nicotinamide adenine dinucleotide (NAD), to which the inner mitochondrial membrane is nearly impermeable (von Jagow and Klingenberg, 1970). However, nicotinic acid and nicotinamide can penetrate the membrane readily. When labeled, either of these substances could be found incorporated into NMN or NAD within isolated intact or sonically disrupted mitochondria (Grunicke *et al.*, 1975; Lange and Jacobson, 1977). An absolute requirement of ATP, 5-phosphoribosyl-1-pyrophosphate, and $MgCl_2$ was demonstrated.

9.3.5. Mitochondrial Replication

Two major aspects of the biogenesis of mitochondria merit attention, one at the cellular level and the other at the molecular. Viewed as the whole organelle, there are three possible sources for the origin of new mitochondria: (1) other membranous portions of the cell, (2) division of existing mitochondria, and (3) *de novo* synthesis (Berger, 1964; Baxter, 1971; Afzelius, 1972).

Replication at the Cellular Level. As has appeared on more than several occasions in preceding pages, earlier workers occasionally advocated an origin for the mitochondrion from various other cell organelles, including the nuclear envelope, endoreticulum, and dictyosomes (Novikoff, 1961) and even the plasmalemma (Robertson, 1959). In view of the specialized nature of each of the membranes and the diverse natures of their enzymatic content, there appears to be little justification for such simplistic concepts. Perhaps an organelle of one sort or another could serve as a focal point on which a mitochondrion might be assembled as flagella are, but that is quite another matter and amounts to *de novo* synthesis.

Little doubt persists that the most frequent method of producing more copies of this organelle is from preexisting mitochondria by growth and division. The details of the processes are elusive, however. Indirect evidence of fission was first advanced by Luck (1963), who used labeled choline and a choline-requiring mutant of *Neurospora*. Radioautograms from samples taken at timed intervals following exposure to the radioactive substance revealed a random pattern of labeling in the mitochondria of daughter cells, rather than the nonrandom pattern that would have resulted either by *de novo* synthesis or from preexisting membranous organelles. The fine series of electron micrographs on division of the organelle and cylinder of DNA in fungi reproduced in earlier pages (Figure 9.10) and others from cinematomicrographic investigations cited in relation with mitochondrial motility (Section 9.1.4) leave little room for doubt that one means of reproduction in this organelle is by binary fission (Tandler *et al.*, 1969).

One source of evidence for *de novo* production of mitochondria is from anaerobically grown bakers' yeast. Provided that no ergosterol or unsaturated fatty acids have been added to the culture media, *S. cerevisiae* cells grown in the absence of air contain no trace of mitochondria (Linnane, 1965). When aeration of such organisms continues for about 3 hr, vesicles and cytoplasmic membranes appear, followed in another hour by incomplete mitochondria (Section 9.1.3.), and after 8 hr by complete ones. It should be noted that the anaerobically grown cells did not contain cytochromes a, a_3, or b, although small amounts of cytochrome c were present. However, all the cytochromes were detected after aeration.

In a second set of evidence indicative of *de novo* origins, the events parallel those of anaerobic yeast. The source material, the inner segment of the photoreceptor cells of fish retina, was claimed to be ideal for investigations of developing mitochondria (Figure 9.24), because those organelles become extremely abundant, are contained in a well-defined space, and their genesis occurs in a gradient (Berger, 1964). The steps in development, deduced from serial-sectional analysis of embryonic guppy eyes, began with double-membrane sheets and small vesicles in the cytoplasm. Then the membranes folded around the vesicles, until the ends met and subsequently fused. In the meanwhile the vesicles elongated and compressed to form cristae. After the latter had become more abundant and the entire organelle had increased in size, the definitive mitochondrion was thereby brought into being.

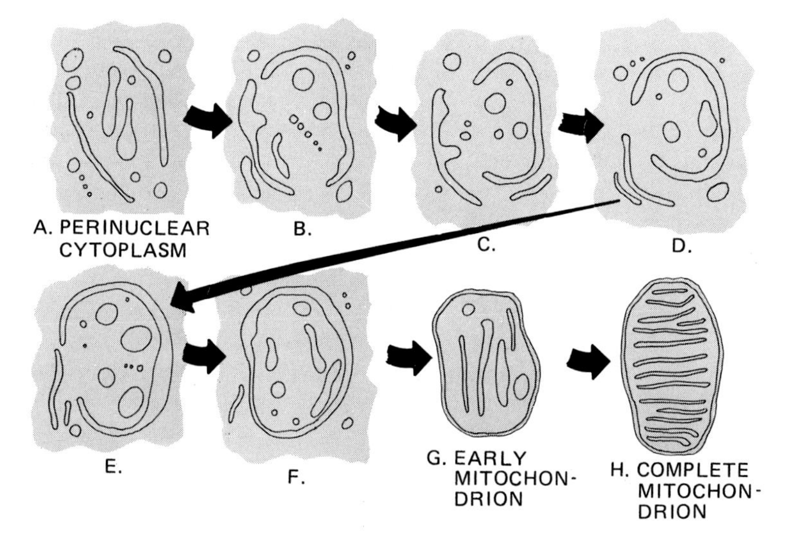

Figure 9.24. Stages in the *de novo* origins of mitochondria in photoreceptor cells. (Based on Berger, 1964.)

Regeneration of Mitochondria. Events in the regeneration of mitochondria in organisms emerging from quiescent stages such as cysts parallel those of the yeast to some extent, except in not being entirely *de novo*. One notable investigation of these processes followed the changes in the organelle subsequent to emergence of *Artemia salina* (the brine shrimp) from the encysted stage (Schmitt *et al.,* 1973, 1974) as follows. When induced to form the cyst by dehydration, embryonic growth ceased at the gastrula stage of development. After being rehydrated, it underwent development to the prenauplius stage, which process took place in 15 hr without cell division. Further incubation then led to hatching, followed by the resumption of cell division and development into the free-swimming nauplius. If oxygen consumption is expressed as $Q_{O_2} = \mu$moles O_2 per hour per gram of dried cysts, an increase from $Q_{O_2} =$ 11 to $Q_{O_2} = 27$ occurred within an hour after incubation was initiated at 30°C. This quantity reached a plateau at 32 in the prenauplius and rose to 75 after hatching. These increases were paralleled by increments in cytochrome oxidase activity and also in quantities of cytochromes *b* and *c* present. Concurrent studies of the changes of the mitochondria were also conducted. In the cysts, mitochondria were present but lacked cristae; however, within an hour after incubation was commenced, a few such structures could be noted along the periphery. As development continued, cristae became abundant as the mitochondria attained mature size in the nauplius.

The main event, then, in regeneration seems thus to be the formation of the cristae, four avenues for their origins being available: They could form (1) as mere outpocketings of the inner membrane or (2) *de novo;* (3) they might represent remnants of the inactive inner membrane, while a new inner investing membrane formed; or (4) both the inner membrane and the cristae could be combinations of new membrane with old remnants. The only known evidence, the differences in enzymatic content of cristal and inner membranes, militates against the first alternative, but no data are available that either support or conflict with the remainder. However, it is not unlikely that the fourth possibility will eventually prove to be the route actually followed.

9.4. MESOSOMES: PRIMITIVE RESPIRATORY ORGANELLES?

A structure found in bacteria nearly three decades ago, while thoroughly explored morphologically, is still controversial in regard to its physiology (Burdett and Rogers, 1972; Greenawalt and Whiteside, 1975). Part of the lack of understanding comes from conflicting results through use of differing experimental procedures, part from the confusion of too many proposals based solely on limited ultrastructural observations, and part from the difficult nature of the organelle itself. As shown by the discussion that follows, discordant reports have characterized the body throughout its known history.

9.4.1. The Bacterial Mesosome

Many years prior to its identification by electron microscopy of the intact organism, a membranous organelle of bacteria, the mesosome (Figure 9.25), had been identified *in situ* by selective staining as a respiratory structure, and later it was isolated by means of cell fractionation and ultracentrifugation. While Mudd and his colleagues (1951a,b, 1956; Mudd, 1954, 1956) were seemingly pinpointing the respiratory organelle by cytochemical methods, Kellenberger and Huber (1953) separated a layer of high respiratory activity from bacteria and showed it to consist of structureless particles. Still later, a number of electron microscopic investigations, such as those by Niklowitz (1958) and Giesbrecht (1960), correlated this particular with a peculiar body in these organisms, often attached to the nucleoid and cytoplasmic membrane.

More recently, combinations of biochemical and ultrastructural procedures comparable to those employed with metazoan cells have revealed active respiratory sites in intact bacteria. The first attempts along these lines were not conclusive. Using 2,3,5-triphenyl tetrazolium, Vanderwinkel and Murray (1962) and Takagi and his associates (1965) demonstrated sites of oxidation–reduction activity but were unable to relate them to regularly occurring, clearly defined organelles. Subsequently, van Iterson and Leene (1964) employed tellurite, and Sedar and Burde (1965) made use of a tetrazolium salt, trinitro blue

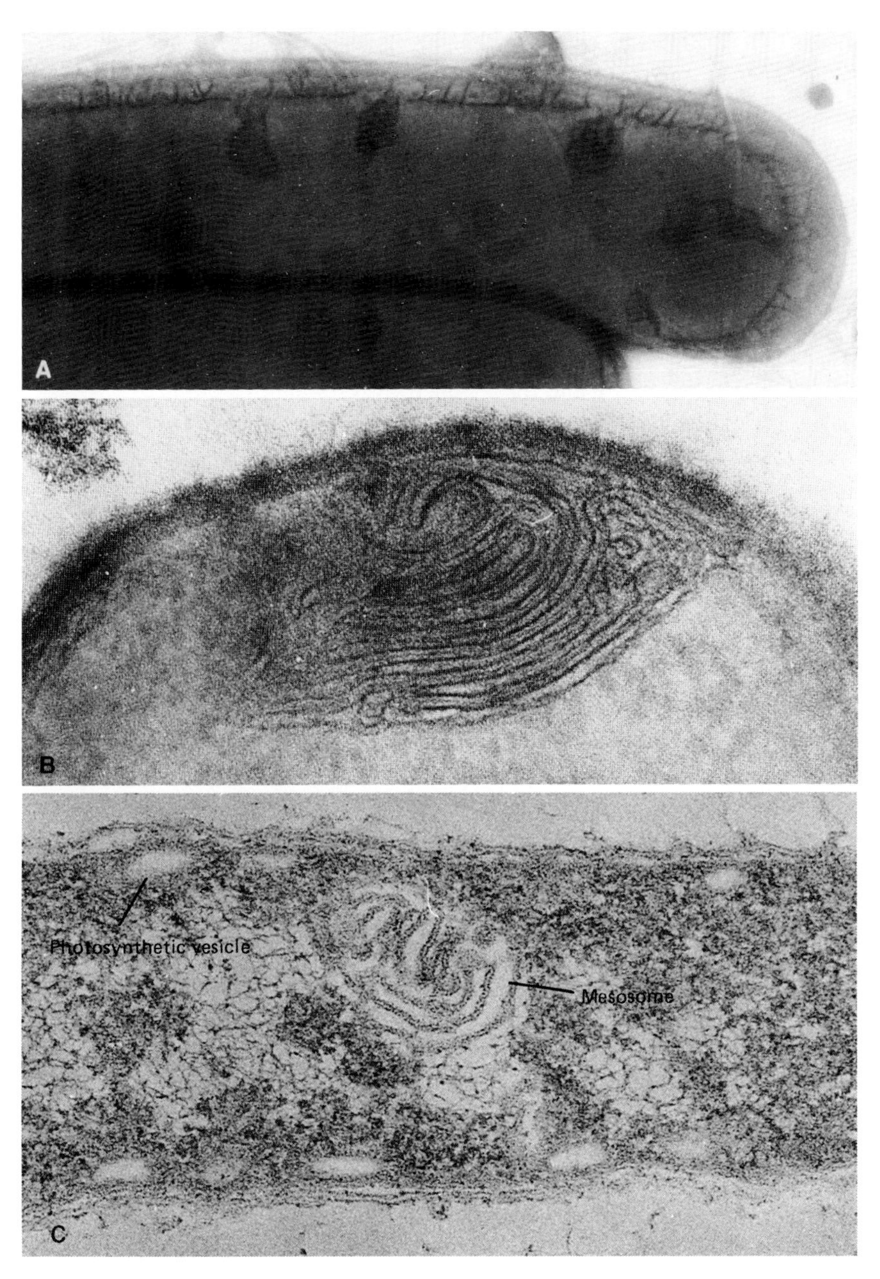

Figure 9.25. Mesosomes of two bacteria. (A) Mesosomes may be distributed throughout the bacterial cell, as in this specimen of *Bacillus licheniformis*. 41,000×. (B) The mesosome of this species when sectioned is seen to display an unusually high degree of complexity. 196,300×. (Both courtesy of Burdett and Rogers, 1972.) (C) The corresponding organelle of *Chlorobium thiosulfatophilum* appears to be less complex and more loosely folded than the foregoing. 120,000×. (Courtesy of Cohen-Bazire *et al.*, 1964.)

tetrazolium, whose reduction product is insoluble in the reagents utilized during tissue preparation. Both teams of investigators were able to demonstrate clearly that the principle center of succinic acid reduction lies in the organelle known as the mesosome. Yet when isolated from the cell, in some laboratories succinic hydrogenase activity was so low as to suggest its total absence (Reaveley and Rogers, 1969), and the same statement is true of other components of the respiratory cycle (Salton and Owen, 1976; Salton, 1978). In contrast, other laboratories reported a full spectrum of the enzymes to be present (Pangborn *et al.*, 1962).

The mesosome, a name first applied by Fitz-James (1960), is variously called also the onion-shaped body, chondrioid (Kellenberger *et al.,* 1958; van Iterson, 1965), and plasmalemmasome (Edwards and Stevens, 1963). By both ordinary and freeze-etch techniques (Remsen, 1966, 1968), the organelle has been shown to consist of a spiral or much-folded membrane, closely associated usually with the plasmalemma and frequently with the nucleoid, as mentioned above (Ryter, 1968). More recent work, with improved techniques, has demonstrated that the membranes are heavily associated with numerous tubules and moreover are often themselves folded into tubulelike structures (Burdett and Rogers, 1972). In many electron micrographs the organelle actually appears to be derived from the cell membrane; indeed, the study on *Diplococcus* by Tomasz and co-workers (1965) leaves little room for doubt as to this derivation. Although the mesosome quite frequently is located toward one end of the bacterium in proximity to, or even upon, the septum, this location is by no means constant, for as reported below, the body actually migrates during the cell's cycle of growth and, at times, numerous examples are distributed throughout the cell (Figure 9.25A; Higgins and Daneo-Moore, 1974).

The papers cited above show also that not all the aerobic respiratory reactions occur within the mesosome. On the electron micrographs, poorly defined, but large, areas of the nucleoid, too, typically appear to be active in these processes. In several of the studies, no mention is made of these electron-dense regions, but Sedar and Burde (1965) state their inability to account for the dark deposits, for the regions of activity are too consistently present in the nucleoid to be explained as mere artifacts of preparation.

Unfortunately, these methods have as yet been applied to just a few bacteria, so that now it is impossible to state what body serves as the respiratory site among other major types of bacteria. Particularly needed are investigations of such forms as *Beggiatoa* and *Thiothrix* among the colorless sulfur bacteria, the Myxobacteria as a whole, and the spirochaetes. Because the mesosome appears to be a derivative of the plasmalemma, it might be anticipated that in at least some of these types the cell membrane could prove to be a site of respiration, as no mesosome is currently known to exist in any of the taxa mentioned.

Nor is it clear what the significance is of the structural differences in the mesosome found among various typical bacteria. In *Bacillus subtilis* (Vanderwinkel and Murray, 1962) and *Chlorobium thiosulfatophilum* (Cohen-Bazire *et al.*, 1964), the membranes are arranged rather spirally and upon them are located scattered, vague, round bodies, perhaps three times the diameter of the membranes in size. Although the membranes are of trilamellar construction, frequently they seem to consist of series of smaller particles, a condition that is especially clear in the *Chlorobium* species mentioned (Figure 9.25C). On the other hand, in a second species of the same genus (*C. limicola*) described in the latter article and in *Diplococcus* (Tomasz *et al.*, 1965), the membrane becomes folded to form several smaller vesicles arranged around a larger central one. At least during the early stages of formation, the membrane here, too, is composed of series of small particles.

Furthermore, while mesosomes are frequent in gram-positive bacteria, they are often absent or poorly developed in gram-negative forms. In *E. coli* and *Proteus vulgaris* they are not present under ordinary conditions of culturing but are developed under elevated temperatures, such as 40°C. In contrast, the mesosome is as abundant in the gram-negative stalked bacteria that comprise the family Caulobacteriaceae as in the gram-positive forms (Poindexter *et al.*, 1964). Ryter and Jacob (1966) explain the difficulty in observing the mesosome in *E. coli* as resulting from the fragility of the membranes that form it; they believe that only when the plane of section falls perpendicular to the mesosomal folds are the latter visible. More recently Pontefract *et al.* (1969) confirmed these observations with refined techniques, but even with their special preparations, the organelle in micrographs was not so clear-cut as in gram-positive forms.

As will be recalled, the last named authors also hypothesized that the mesosome was essential to nuclear division, a suggestion that receives further attention in a later chapter. While interesting, their hypothesis does not receive support from a study of mesosomal behavior in *Bacillus licheniformis* (Highton, 1969). During the growth cycle of the organism, the mesosome was found to move from the end where it had been formed to the middle of the elongating cell. At this point it divided as the new cell septum grew, one product of division going to each daughter cell before cytoplasmic division was completed. Because nuclear division in exponentially growing bacteria occurs when the mesosome is still moving toward the end where division will occur, its seeming association with nuclear division is possibly more apparent than real.

9.4.2. Blue-Green Algae

The location of the actual respiratory site in the Cyanophyceae remains obscure in spite of the appearance in print of several excellent studies on the subject. In great part the present confusion results from two major factors, the first of which is the relative paucity of investigations so far conducted on the respiration of these organisms. While this situation currently seems to be undergoing improvement, the second may prove difficult to correct, for revision of taxonomic concepts may be involved. In short, the Cyanophyceae, like the bacteria discussed in several preceding chapters, may need to be viewed as representing a number of phyletic lines, rather than a single "natural" or monophyletic group—at least ultrastructural studies on the site of respiration appear to suggest this possibility.

On one hand, Echlin (1964) reported an organelle in *Anacystis nidulans* that closely resembled the mesosome described above in the gram-negative bacteria. As in that group, the organelle consisted of tightly spiraled membranes extending into the nuclear region from the plasmalemma. However, because its actual involvement in cell respiration had not been explored, Echlin proposed that the term "lamellasome" be employed until homology of function was established. Besides possessing this similar-appearing structure, the present organism resembled the Eubacteriales in having a more or less continuous nucleoid and in lacking the polyhedral bodies and structured granules so characteristic of more typical, filamentous blue-green algae (Figures 11.1, 11.3). *Anacystis nidulans,* currently considered a unicellular type in the Chroococcales, may according to Drouet (1962) actually represent a pseudounicellular type consisting of short but multicellular filaments and should probably be assigned to the Nostocales, perhaps in the genus *Phormidium*. Similar structures have been reported from *Synechocystis aquatilis*, *Anabaena variabilis* (Avakyan *et al.*, 1978), and *Chlorogloea fritschii* (Baulina *et al.*, 1978).

On the other hand, filamentous members of the division possess no specific organelle of respiratory function. In an investigation of possible respiratory sites in *Nostoc sphaericum,* Bisalputra and colleagues (1969) reported that reduction of tellurite and trinito blue tetrazolium occurred upon the photosynthetic lamellae, clearly indicative of the location there of respiratory function. Similar results have been obtained in the author's laboratory (Dillon and Dillon, unpublished).

9.5. PHYLOGENETIC ORIGINS OF THE MITOCHONDRION

The problem of the phylogenetic origins of the mitochondrion has received widespread attention, based on a number of diverse observations. One proposal has suggested that blue-green algae developed the ability to engage in phagocytosis, a process claimed to be somewhat parallel to that of cell division (Cavalier-Smith, 1975). Through modification of the membranes brought into the interior by this activity, endoreticulum, lysosomes, dictyosomes, mitochondria, and the other cell organelles were thought to have developed, thereby giving rise to the eukaryotic cell. However, as no prokaryote is known to be capable of phagocytosis, the concept lacks a firm base of factual support. The same weakness is shared by a second concept, which conceived the mitochondrion to have been derived from the peroxisome (de Duve, 1973).

9.5.1. The Endosymbiontic Concept of Mitochondrial Origins

Undoubtedly the most widely accepted concept of the origins of mitochondria, as well as those of two other organelles, has been the endosymbiontic theory. This idea was originally advanced many years ago (Wallin, 1927) but attracted little support until it was readvocated on a more sophisticated basis by Margulis (formerly Sagan) in 1967, 1968, and 1970 (Nass, 1969; Raven, 1970; Schnepf and Brown, 1971; King, 1977). Simply stated, the theory holds that the mitochondrion is a type of prokaryote that penetrated the ancestral eukaryotic cell, originally as a parasite but later acquiring mutualistic qualities. This interesting view was forwarded by the discovery that DNA was present in the organelle, thereby suggesting the existence of autonomy. At that time it was believed that the mitochondrion's entire genetic apparatus was encoded by its genome, but as shown in an earlier section, present knowledge has revealed how few of the essential enzymes, tRNAs, and ribosomal proteins are actually synthesized within that organelle.

Much additional evidence against the theory has come to light during the past decade, some of the main points being as follows:

1. The physical properties of mitochondrial ribosomes have little in common with prokaryotic ones (Table 9.1; Pace, 1973).
2. The DNA resembles that of viruses more closely than it does that of bacteria, often being circular. In many organisms, however, it is linear, thus differing even from the viral types.
3. Replication and transcription similarly involve distinctive features and are largely carried out by cytoplasmic enzymes; initiation and elongation are particularly unique.
4. The single nucleotide sequence of a mitochondrial tRNA ($tRNA^{Met}$) that has been established is neither prokaryotic nor eukaryotic.
5. Many of the elements of the respiratory cycle themselves, which supposedly enabled the invading prokaryote to become evolutionarily advantageous to the ancient host, differ strikingly from those of any known prokaryote. For example, the enzyme that catalyzes the oxidation of isocitric acid in prokaryotes requires NADP, whereas that of mitochondria utilizes only NAD.
6. The cytochromes of bacteria differ greatly from those of mitochondria, particularly cytochrome *c* and its oxidase. The cytochromes $a + a_3$ of bacteria lack the two copper atoms present in the eukaryotic enzyme and either do not oxidize mammalian cytochromes *c* or do so very slowly and the same system of mammals does not act

at all on bacterial cytochrome *c*. Consequently, any prokaryotic invader could not have aided cellular respiration as proposed.

7. Mitochondria possess an actin–myosin system that enables them to move actively about the cell, but all bacteria and blue-green algae lack such a system and comparable capabilities.
8. Mitochondria are of a number of varied types, including microvillose and septate, not to mention the numerous configurations found in chloride cells and other specializations described in the preceding pages. Different types often occur in the varied tissues of the same organism and several distinct populations may occur together. No prokaryote is extant that can assume so many diverse ultrastructural traits.
9. Mitochondria give rise to highly specialized cell parts such as the kinetoplast of hemoflagellates, the spiral at the base of the metazoan sperm tail, and the inner segment of vertebrate retinal cells, adaptations impossible to reconcile with their being a prokaryotic invader.
10. These organelles often actively participate in such cell activities as mitotic and meiotic division and the condensation of chromatin in sperm nuclei, functions much more amenable to the organelle's being a specialized adaptable part of a cell rather than an endosymbiont.
11. In yeasts, the organelle can arise from open sheets of membranes, and in other cells *de novo* origins have not been completely ruled out—observations not at all compatible with the endosymbiontic concept. Consequently, it appears no longer possible to consider this organelle other than an evolutionarily derived, specialized compartment of the evolving eukaryotic cell itself.

9.5.2. The Episome Theory

Dissatisfaction with the weaknesses of the endosymbiont concept stimulated the proposal of other views (Cohen, 1970), the most outstanding of which was the episome theory (Raff and Mahler, 1972, 1973). Episomes are plasmids that are attached to the nucleoid of bacteria, plasmids being covalently closed-circular DNA molecules that replicate independently of the nucleoid DNA. Hence, the nucleoid of the mitochondrion may be loosely viewed as a eukaryotic plasmid. In brief, the hypothesis suggests that the organelle developed from the ancestral eukaryotic plasmalemma, combining with an episome (plasmid) that carried the genes for mitochondrial ribosomes, tRNAs, and some of the respiratory enzyme subunits. Unfortunately, the data employed in support of the proposed derivation are drawn mainly from the ribosomes and genome of the mitochondrion, a source already shown to be too controversial to provide substance to phylogenetic considerations (Uzzell and Spolsky, 1973). The same statement holds true also for a closely related proposal (Reijnders, 1975), as well as for the analyses of sequence differences of various nucleic acids and proteins presented by Schwartz and Dayhoff (1978). The composite evolutionary tree of the latter clearly indicates the unreliability of difference counts, as mammals come off the tree at two widely separated points and "plants" (metaphytans) at three. A related concept has postulated that a part of a prokaryotic nucleoid was transferred into a eukaryotic cell by a virus (Ostroumov, 1977).

9.5.3. A Biological Concept

Actually the episome concept agrees well with many factual observations of which its authors appear to have been unaware; in short, its main flaws are a lack of sufficient depth of supportive data and a restricted view of the nature of mtDNA. However, the genome of the mitochondrion does not necessarily represent a prokaryotic episome or plasmid; its frequent linearity demonstrates that point clearly. In trying to decipher the probable past events in the development of this organelle, some facts disclosed by the analyses of the respiratory cycle, the cytochromes, and ribosomal 5 S RNA need to be recalled. These largely pertain to the existence of several evolutionary levels among the prokaryotes over and above those of bacteria and blue-green algae. Rather, the colorless sulfur bacteria *(Beggiatoa, Thioploca,* and related genera) need to be considered a separate phylogenetic stage, apparently at the simplest known cellular level above the rickettsiae; *Clostridium* (at least in part) is shown to be somewhat more advanced. Then there exist two, or perhaps more, successively higher levels among the blue-green algae, followed by a series of stages among the bacteria proper. Thus *E. coli* and *Bacillus megaterium* do not represent the bacteria as a whole; they can be considered only as representatives of separate lines from among the numerous ones that exist within that group. That the true yeasts are the most primitive of the existing eukaryotes has previously been indicated here by several sets of evidence. The euglenoids then come off the main line at some distance above the yeasts, followed by a series of steps leading to the remainder of the protistans, and finally to the branches ending in the metaphytans and metazoans.

Accordingly a logical sequence of events in mitochondrial phylogeny could be as follows:

1. Because the earliest functional respiratory cycle occurs in the blue-green algae in which it is located on the photosynthetic thylakoids, themselves derivations of the plasmalemma, the respiratory functions might be considered to have been situated originally on the plasmalemma. Further discussion of the relations that exist between respiration and photosynthesis is provided in the chapter on the chloroplast that follows.

2. In the second (higher) branch of blue-green algae, these two sets of processes became separated as the mesosome formed. This organelle also is a derivative of the plasmalemma; that is, it is derived from a particular region of the plasma membrane that is specialized for the tricarboxylic acid cycle. Thus the groundwork was set for the eventual formation of the two separate specialized organelles, the mitochondrion and chloroplast.
3. Through the several lines of eubacteria that followed, cell respiration and photosynthesis essentially remained properties of separate regions of the plasmalemma; mesosomes typically were present, but in cases where they have failed to differentiate from that membrane, no serious problem is presented—the ability to produce intracellular membranes specialized for respiratory functions has still persisted.
4. In ancestral eukaryotes, these intracellular membranes seem to have undergone radical modification, as intimated by those present in oxygen-deficient cultures of yeast. Such organisms have the membranes present as compressed vesicles, each consisting of double membranes as shown in Figure 9.13A.
5. The membranes later became increasingly folded over on themselves, thereby gradually enclosing a matrix. At first, these folded membranes remained slightly open. It is during this stage, represented among yeast transferred into an oxygen-rich environment, that a genome was taken in along with the ground substance. The source of the DNA could have been a plasmid (episome), as Raff and Mahler (1972) have proposed, but the frequent occurrence of linear molecules militates against that concept. Its actual origin could readily be investigated through the use of yeast recovering from anaerobic conditions as the experimental subject. At the same time, ribosomes also were taken into the enclosed matrix, probably then being identical to those of the cytoplasm. As with the DNA, it should also be possible to investigate the actual origins of these particles in the anaerobic yeast system.
6. At the level of aerobically grown yeast, this premitochondrion closed entirely and developed cristae, to become a complete but primitive mitochondrion.
7. Following its origin in this manner, the organelle underwent a series of changes, as indicated by the diagram. Among the alterations not shown seems to have been a gradual loss of much of its genome, especially in the line leading to the metazoans. Such losses appear to have occurred independently along several side branches, accounting for the inconsistencies in the encoded tRNAs that have been demonstrated among the various major taxa.
8. It should be especially noted that the acquisition of a genome and ribosomes occurred in a stock prior to the earliest actually known

eukaryote. As this stock would have been evolutionarily between the most advanced prokaryote and the lowest living eukaryote, the likenesses to and dissimilarities from the DNA and ribosomes of both types of organisms are accounted for.

10

ENERGY-ORIENTED ORGANELLES AND ACTIVITIES: III

The Chloroplast and Photosynthesis

The second type of energy-related organelle, the chloroplast, is so closely interwoven with its chief function, photosynthesis, that it is difficult to discuss one without the other. Furthermore, less obvious, but just as real, interrelations exist between this light-dependent production of sugars and other complex series of metabolic events such as glycolysis and photorespiration. Consequently, this organelle and these processes together provide that basis for the bulk of this chapter, along with less thoroughly explored topics that throw light on the ancestry of the modern organelle and its activities.

10.1. ENERGY-RELATED PROCESSES

As indicated above, the photosynthetic production of glucose is in large measure closely akin to the metabolism of the same substance. The latter processes, however, are more straightforward, because they lack the complexities that derive from the light-driven breakdown of water and the capture of energy from the sun that characterize photosynthesis. Accordingly, many advantages are gained by viewing the catabolic steps before the synthesizing ones.

10.1.1. Glycolysis

Glycolysis, or the Embden–Meyerhof pathway, is the anaerobic process of breaking down glucose or other sugars to pyruvic or lactic acids. Ten steps are involved, the majority of which are by way of phosphorylated intermediates, and all of which can proceed in the absence of oxygen. At several points ATP is produced, while at others that substance is catabolized, so a net of only 2 moles/mole of sugar results. Although the chloroplast contains the enzymes for the majority of the steplike series of events, two important ones are lacking, as is pointed out where appropriate.

Consequently, the reactions largely occur within the cytosol or various other organelles in metaphytans and metazoans alike. In the latter organisms, mitochondria of sperm, for instance, metabolize the end product, pyruvic acid (Hutson *et al.,* 1977), and the nuclei of many tissues have been demonstrated to contain all the glycolytic enzymes, as well as the intermediates of glycolysis (Siebert and Humphrey, 1965; Allfrey, 1974). That the nuclear content is not merely a reflection of that of the cytosol is indicated by the differences that exist in the respective concentrations of the various proteins; for instance, the nuclear concentrations were observed to increase as much as 400% during liver regeneration, while those of the cytoplasm decreased (Ottaway and Mowbray, 1977). Moreover, at least the chemical properties of some of the nuclear enzymes differ from those of the cytosol; lactic acid dehydrogenase, a substance actually synthesized within the nucleus, provides an example of an isozyme that differs in kinetic properties from that of the cytosol (Kuehl, 1967). In addition thymus cell nuclei contain the enzymes of the tricarboxylic acid cycle and can produce $^{14}CO_2$ from labeled glucose (McEwen *et al.,* 1964; Giddings and Young, 1974). Other cell organelles that have been reported to contain the glycolytic enzymes include the glycogen granules of many tissues and the sarcoplasmic reticulum of muscle cells (Ottaway and Mowbray, 1977). In phylogenetic and other comparative studies, it is important to realize that, although the steps of the glycolytic processes are nearly identical wherever they occur, the respective enzymes that mediate the reactions often differ widely from organism to organism and even from tissue to tissue in many cases, despite their bearing identical names. This has been especially effectively demonstrated for the enzyme known as glyceraldehyde-3-phosphate dehydrogenase (Lebherz and Rutter, 1967; Steinbach *et al.,* 1978).

Three major sequences of events may be recognized in the glycolytic pathway: (1) phosphorylation of the carbohydrate to be degraded; (2) breakdown of that phosphorylated compound to phosphoglyceraldehyde and oxidation of the latter to phosphoglyceric acid; and (3) dehydration and dephosphorylation to produce pyruvic acid or derivatives.

Phosphorylation—The Preparatory Step. As a rule, the processes begin with the phosphorylation of glucose, but it is not unusual for other preparatory reactions involving such storage types of complex carbohydrates as glycogen and starch to precede this step. Usually the preliminary steps simply hydrate and phosphorylate the stored substance, often by the enzymes of the smooth endoreticulum, but these activities soon lead to a common point with glucose (Figure 10.1A–D). Glucose itself is first phosphorylated with a phos-

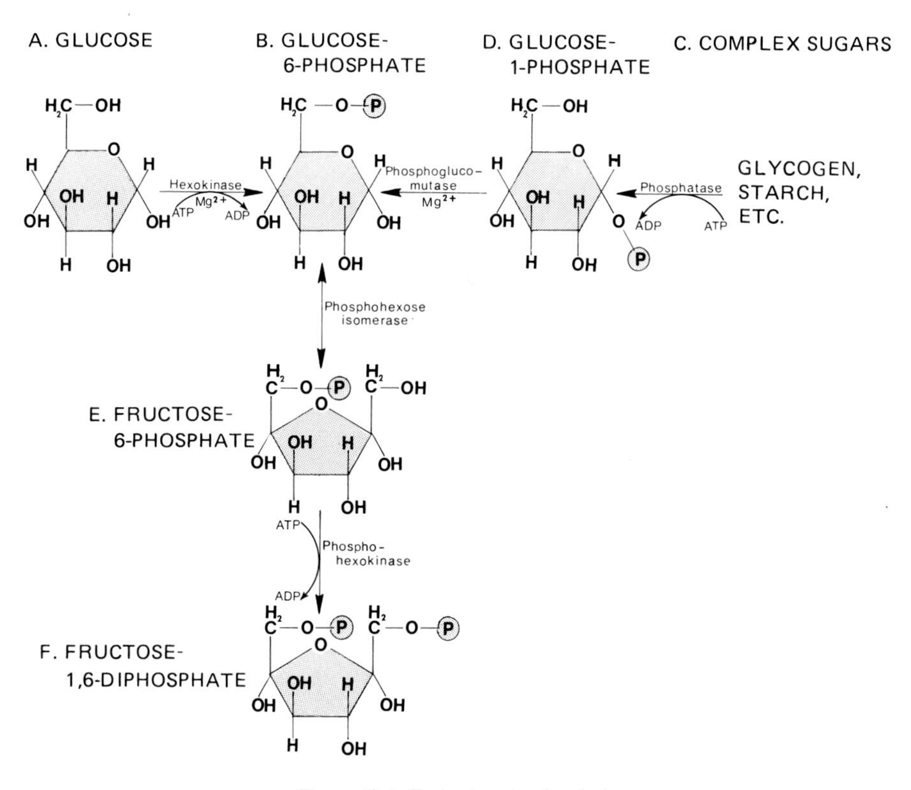

Figure 10.1. Early steps in glycolysis.

phate radical from ATP in the presence of Mg^{2+} in a reaction catalyzed by hexokinase, an enzyme that can use a number of different D-hexoses as substrate; sometimes, however, as in mammalian liver cells, an additional enzyme, glucokinase, is present that is specific only for D-glucose. The product of this step, glucose-6-phosphate, is produced also from glycogen after the latter has been first converted to glucose-1-phosphate with inorganic phosphate in a reaction mediated by phosphorylase; a second enzyme, phosphoglycomutase, then transforms the substrate to the glucose-6-phosphate. Consequently, the reactions starting with glucose require energy from ATP, whereas those beginning with glycogen do not. In addition to these two processes, glucose-6-phosphate serves as a central point of departure for several other metabolic pathways.

In glycolysis, this substance is next converted to fructose-6-phosphate (Figure 10.1E), in which reaction phosphohexose isomerase serves as the enzyme. Although these two substances are freely interconvertible, equilibrium is reached when the concentration of glucose-6-phosphate is nearly double that of fructose-6-phosphate. Additionally D-fructose can enter this chain of events at this point by having a phosphate residue from ATP added to it by means of the enzyme hexokinase.

The Intermediate Steps. Because the diversity of substrates and entrance points ceases with production of fructose-6-phosphate, the preparatory steps may be considered to close there, ready for the actual glycolytic processes. The first step in the latter again requires energy from ATP, in a reaction catalyzed by phosphofructokinase that terminates in the production of D-fructose-1,6-diphosphate (Figure 10.1F). This reaction, being highly energy requiring, is physiologically irreversible, and accordingly represents the first unique stage in glycolysis; as there is no returning by this route, it is often referred to as the "committed" step. However, a second enzyme, known as fructose diphosphatase, mediates the reverse reaction, requiring water and yielding fructose-6-phosphate and inorganic phosphate.

A second sequence may be considered to commence with the splitting of the fructose-1,6-diphosphate, a process catalyzed by aldolase, that actually results in two different products, dihydroxyacetone phosphate and glyceraldehyde 3-phosphate (Figure 10.2A–C). The former compound may be immediately isomerized to the latter by the enzyme triosephosphate isomerase, so that 2 moles of glyceraldehyde 3-phosphate in theory could be formed from each mole of glucose, only one of which is followed beyond this point in the diagrams. At the next step, the formation of 1,3-diphosphoglyceric acid (Figure 10.2D), inorganic phosphate provides the phosphoryl radical, the oxidation–*reduction* reaction being catalyzed by the enzyme phosphoglyceraldehyde dehydrogenase. This protein, which has been shown to exist in several isozymic states in certain plants (Cerff and Chambers, 1978), is complexed with 4 molecules of its coenzyme NAD in metazoans and NADP in many plants (Cerff, 1978a,b, 1979). In the latter case, after the NADP has accepted the hydrogen released in the reaction, it is reoxidized by 4 units of NAD from the cytoplasm. This free NADH may be utilized by the cell in many diverse fashions, depending upon the organism and the chemical conditions.

The next step, catalyzed by phosphokinase, is closely correlated to the above and has an absolute requirement of ADP and any one of three divalent cations, Mg^{2+}, Mn^{2+}, or Ca^{2+}. In this reaction, the ADP accepts the phosphate radical to become ATP; as 1 mole of this compound is produced from each of the 2 moles of glyceraldehyde 3-phosphate, the 2 molecules of ATP employed during the preparatory steps are thus restored to the cytoplasm at this time. The immediate product of the present stage, 3-phosphoglyceric acid (Figure 10.2E), is next mutated to 2-phosphoglyceric acid, in a reaction mediated by phosphoglyceromutase, for which the presence of 2,3-diphosphoglyceric acid is demanded. The need for the latter becomes clear in Figure 10.3, where it may be perceived that the original diphosphorylated compound becomes the 2-phosphoglyceric acid and the original monophosphate becomes the diphosphate.

The Final Steps. The final sequence in glycolysis may be considered to

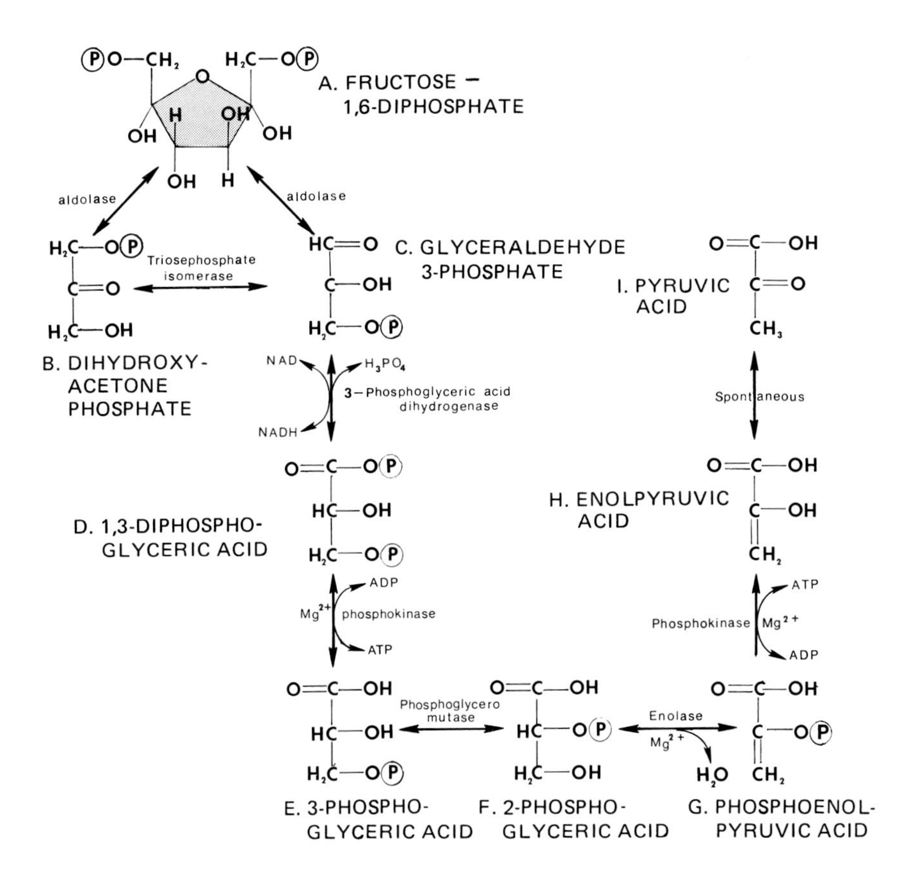

Figure 10.2. Intermediate and later steps in glycolysis.

begin with the dehydration of the 2-phosphoglyceric acid in the presence of enolase and Mg^{2+} to form the enol compound, phosphoenolpyruvic acid (Figure 10.2G). When the latter yields its phosphate radical to ADP, and only then, the processes of glycolysis prove profitable to the cell by netting 2 molecules of ATP/molecule of glucose. Under aerobic conditions the pyruvic acid can later enter the tricarboxylic acid cycle detailed in Chapter 8, after becoming converted to acetic acid.

The chain of events in this strictly anaerobic degradation of glucose to pyruvic acid is not an efficient set of reactions. When the substances involved are burned, the ΔF^0 for glucose is found to be -688 kcal/mole and for 2 moles of pyruvic acid a total of -546 kcal. Hence, 142 kcal/mole of glucose are actually released during glycolysis, of which 14 kcal are stored in the 2 moles of ATP netted—an efficiency level of only 10%. To this, however, must be added

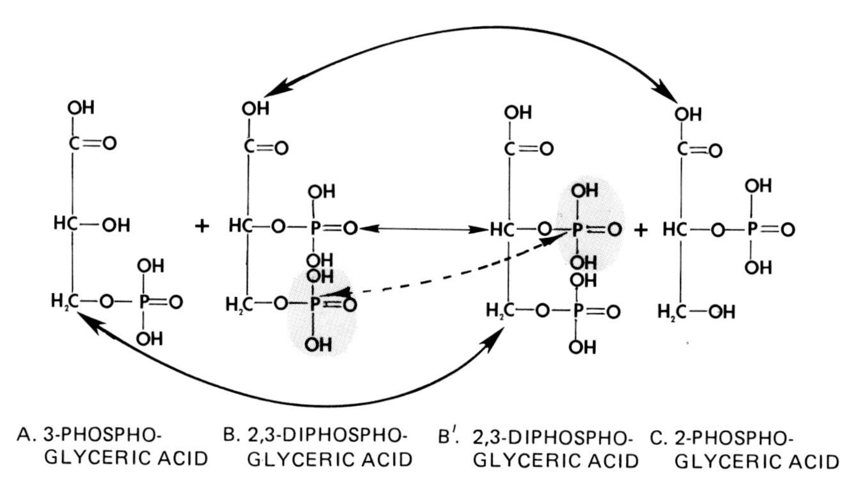

Figure 10.3. The need for 2,3-diphosphoglyceric acid. This compound (B) becomes 2-phosphoglyceric acid (C), while the 3-phosphoglyceric acid (A) receives the phosphate to become 2,3-diphosphoglyceric acid (B′).

the energy stored in the 2 moles of NADH that also are produced; the ΔF° of this compound amounts to 104 kcal for the 2 moles. While the total energy utilization at this point reaches the impressive *potential* level of over 80% efficiency, the *actual* level varies widely, depending upon the chemical pathway followed by the NADH. Under aerobic conditions, most eukaryotic cells appear

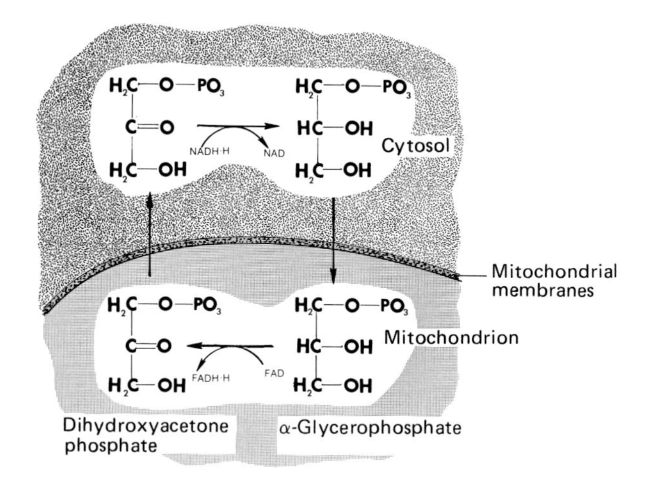

Figure 10.4. The α-glycerophosphate shuttle.

to oxidize the NADH extramitochondrially by the reduction of dihydroxyacetone phosphate to α-glycerophosphate (Figure 10.4). Then after its transport to the mitochondria, the latter substance is oxidized back to the former, with FAD serving as the electron acceptor. From a phosphorylating reaction that ensues, 2 moles of ATP are formed per mole of NADH. Hence, of the 104 kcal contained in the 2 moles of NADH, only 28 kcal are actually made available to the cell in the form of 4 moles of ATP.

Under anaerobic conditions various fermentative processes can be followed, which do not serve in all instances to produce more energy-transferal substances; some merely reoxidize the NAD so that the glycolytic processes may continue. However, interesting though these anaerobic steps may be, they do not pertain even indirectly to the present concerns of the photosynthetic production of sugars and the chief metabolic pathways of those products.

10.1.2. Dark Reactions of Photosynthesis

A number of important reactions of photosynthesis do not require energy from light, and, because they can proceed in the dark as well as the light, they are often referred to as the dark reactions. One of the most important of these begins with the fixation of CO_2 in a series of events that parallels those described in the preceding discussion of glycolysis. In broad terms the processes of photosynthesis are similar in prokaryotes and eukaryotes, including cyclic and noncyclic events, but somewhat different sets of pigments are often involved, especially among the bacteria (Frenkel, 1970).

The Principal Chain of Events

The Fixation of CO_2. Use of labeled CO_2 permitted the pathway of photosynthetic fixation of that substance to be elucidated by Calvin and Benson (1948), researches complicated by the fact that even mammalian liver cells can carry out a similar activity. One example of this is the carboxylation of phosphoenolpyruvic acid (PEP):

$$\text{PEP} + CO_2 + \text{GDP} \rightleftharpoons \text{oxaloacetic acid} + \text{GTP}$$

In the experiments on plants, 3-phosphoglyceric acid was the primary product found to contain the radioactive CO_2, but clues to the real source of this material depended on two key observations. (1) When photosynthesizing organisms were deprived of light, the amount of phosphoglyceric acid increased while the pentose sugar, ribulose diphosphate, decreased. (2) When the plants were deprived of CO_2, the ribulose diphosphate level was raised and that of phosphoglyceric acid was depressed. Hence, it became clear that the latter was formed at the expense of the former. Now it is known that the reaction is catalyzed by a system of enzymes called the carboxydismutase enzyme system and proceeds

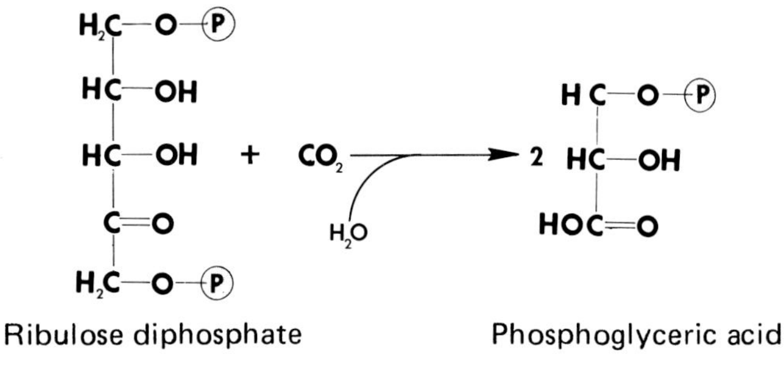

Figure 10.5. The primary reaction in CO_2 fixation.

through several stages (Figure 10.5). First, the keto form of the pentose is converted to the enol. When CO_2 is added, an unstable intermediate is formed, which breaks down in the presence of water to form 2 moles of 3-phosphoglyceric acid. Thus fixation involves the combination of CO_2 with a five-carbon substance to form a six-carbon compound that breaks into two copies of a three-carbon substance. One of the important enzymes in this series of reactions is ribulose-1,5-diphosphate carboxylase, the genetic and structural properties of which are currently the subjects of active investigations (Kelly *et al.,* 1976).

Intermediate Steps. In the next step of this cyclic series of events known as the Calvin cycle, the 3-phosphoglyceric acid is converted into 1,3-diphosphoglyceric acid by a kinase, with energy and a phosphate from ATP (Figure 10.6). This product next receives hydrogen from NADPH, with one of the glycolytic enzymes, glyceraldehyde-3-phosphate dehydrogenase, proceeding in the reverse direction. Thus the end products include glyceraldehyde-3-phosphate, inorganic phosphate, and $NADP^+$.

Undoubtedly some of the glyceraldehyde-3-phosphate is converted by isomerase to dihydroxyacetone phosphate, because equilibrium favors the latter substance. These together then are used in another reversible reaction of glycolysis to produce fructose-1,6-diphosphate, using aldolase as the enzyme (Figure 10.6). As described earlier, a nonglycolytic enzyme, α-phosphatase, is required at the next step, for the corresponding reaction of glycolysis is not reversible. This results in fructose-6-phosphate, which is readily converted to glucose-6-phosphate, bringing these processes to a close. However, as only one new carbon (from carbon dioxide) was acquired in the formation of this six-carbon sugar, six cycles are required to produce the equivalent of one complete molecule of that substance.

The Regeneration of Ribulose. Because CO_2 fixation, the very first step in the main chain of dark reactions, depends on the presence of the five-carbon sugar ribulose diphosphate, it is self-evident that some mechanism must exist constantly to generate a new supply of that phosphorylated pentose.

Because the processes are complex, and the same substances enter into the reaction at two or even three widely separated points, a flow sheet is used to make the process clearer.

The Early Reactions. The early reactions begin with the triose 3-phosphoglyceraldehyde (Figure 10.7A), which plays important roles at no fewer than three separate points in the total processes. Of these, the first is with the six-carbon sugar fructose-6-phosphate (Figure 10.7B); through the mediation of transketolase, these two substances interact to form two phosphorylated sugars, one a four-carbon sugar, erythrose-4-phosphate, and the other a pentose, xylulose-5-phosphate (Figure 10.7C,D). This five-carbon sugar is also synthesized in a second reaction described later. In the meanwhile a second product derived from 3-phosphoglyceraldehyde, the substance phosphodihydroxyacetone, is synthesized in a reaction catalyzed by phosphotriose isomerase (Figure 10.7E). This combines with the erythrose-4-phosphate in the presence of aldolase to produce a seven-carbon sugar, sedoheptulose-1,7-diphosphate

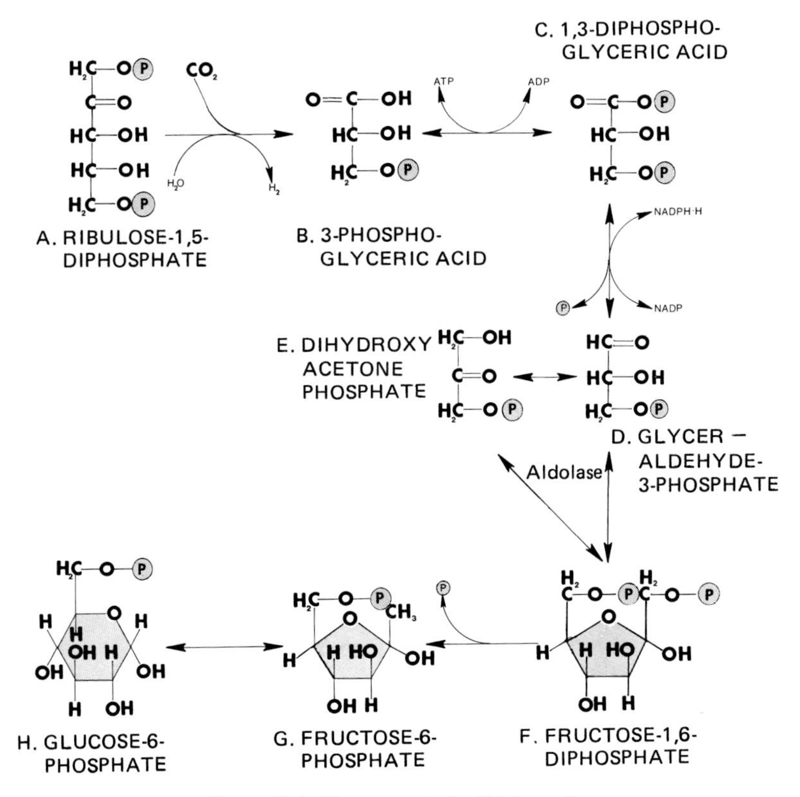

Figure 10.6. The events in the Calvin cycle.

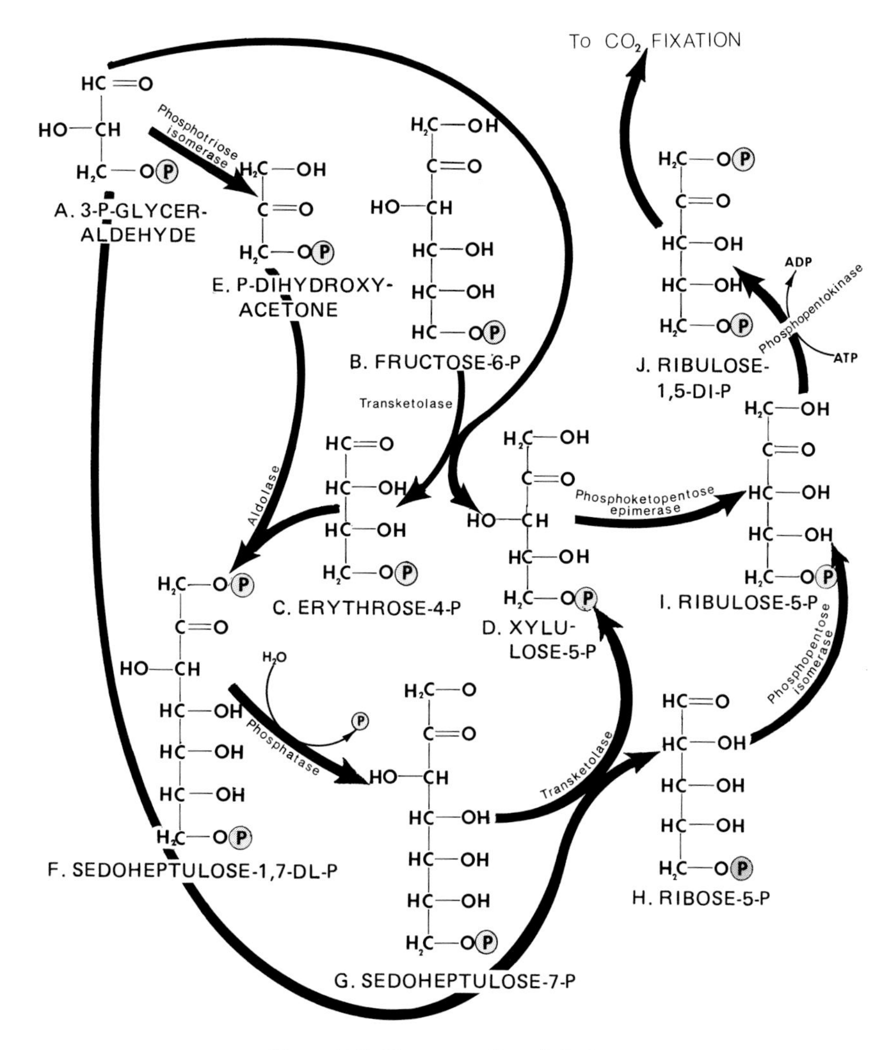

Figure 10.7. The regeneration of ribulose.

(Figure 10.7F). Although not shown on the diagram, 3-phosphoglyceraldehyde and phosphodihydroxyacetone can similarly combine through the mediation of that same enzyme, aldolase, to produce fructose-1,6-diphosphate, a reaction corresponding to those already discussed under the topics of glycolysis and CO_2 fixation.

The Terminal Reactions. Although likewise including combination and cleavage of different sugar molecules, the terminal set of steps is somewhat more straightforward than the preceding. First, the sedoheptulose-1,7-diphosphate loses one of its phosphate radicals through the mediation of phosphatase in the presence of water to produce sedoheptulose-7-phosphate (Figure 10.7G). Next, a transketolase transfers a two-carbon length of this substance to 3-phosphoglyceraldehyde, rearranging the molecules to form xylulose-5-phosphate (the second formation of this sugar in these processes) and ribose-5-phosphate, respectively (Figure 10.7A,D,H). Each of these pentose molecules is then rearranged to form ribulose-5-phosphate, the first by the enzyme phosphoketopentose epimerase and the second by phosphopentose isomerase. After this new pentose has received a second phosphate radical from ATP in a reaction catalyzed by phosphopentokinase to form ribulose-1,5-diphosphate, that product is ready to enter into the CO_2-fixing processes.

The Dark Reactions in C_4 Plants. Unlike spinach, pea, or similar plants, when sugarcane, maize, grasses, sedges, and many woody plants were supplied with $^{14}CO_2$, the first stable compounds to display radioactivity proved to be such four-carbon (C_4) chemicals as malic or aspartic acids (Kortschak *et al.*, 1965; Hatch and Slack, 1966; Gibbs, 1971). Now it is known that all higher plants whose leaves contain bundle sheaths possess this modified pathway and that the cells of that structure carry out some of the steps (Black, 1973; Kelly *et al.*, 1976; Rathnam and Edwards, 1977); moreover, several grass species have also been described in which intermediate structure and activity were found (Brown and Simmons, 1979; Morgan and Brown, 1979).

The Basic Reactions. The C_4 plants, as the organisms that carry out this type of CO_2 fixation are called, preferentially synthesize either malic or aspartic acid, although both substances are formed, albeit in unequal amounts. Only the major one produced serves as a transport metabolite and donor of a carboxyl radical to the usual Calvin cycle, which is centered in bundle sheath cells (Chapman and Hatch, 1979); in contrast, the minor product remains as a pool in the mesophyll cells (Hatch, 1971; Hatch and Kagawa, 1976). Thus the major reaction of the C_4 pathway is carboxylation of suitable monocarboxyl substances in mesophyll cells to form either malic or aspartic acid, which is then transported to the bundle sheath. There decarboxylation occurs, the released CO_2 then being refixed via the Calvin cycle (Kagawa and Hatch, 1977).

Currently C_4 plants are classified into three groups, based on the type of major enzyme present, which may be either NADP-malic enzyme, NAD-malic enzyme, or phosphoenolpyruvic carboxykinase (Hatch *et al.*, 1975). Actually the second of these occurs in the mitochondria of all species, but in smaller quantities than in those where it is the major type; each of the others is confined to one group of plants, in which it is found solely in the chloroplasts (Rathnam and Edwards, 1975). It now appears that in maize and other plants possessing both the NAD- and the NADP-malic enzymes, only the latter localized in the chloroplast plays a part in photosynthesis (Hatch and Mau, 1977). However, even in plants containing only the mitochondrial enzyme, a role in photosynthesis has been found for the protein, although all the steps have not been established (Chapman and Hatch, 1977). This NAD-malic enzyme was found to be activated several fold by coenzyme A and acetyl coenzyme A (Hatch and Kagawa, 1974). In the aggregate the steps in the C_4 processes appear to be as follows. Aspartic acid is deaminated by aspartic acid aminotransferase to form oxaloacetic acid; this chemical then receives a hydrogen from NADH to form malic acid and NAD in a reaction mediated by NAD-malic dehydrogenase. Decarboxylation by the NAD-malic enzyme then occurs, with the liberation of CO_2 and the hydrogenation of NAD (Chapman and Hatch, 1977).

10.1.3. The Light Reactions of Photosynthesis

As the actual synthesis of glucose from CO_2 proceeds in the dark, the light reactions are essential only in supplying the energy. This energy, in the form of ATP and NADPH, was shown in a preceding discussion to be requisite at two points, the first of these substances in the conversion of 3-phosphoglyceric acid to 1,3-diposphoglyceric acid, and the second in the hydrogenation of the latter compound to form glyceraldehyde-3-phosphate. As ATP and NAD(P)H are used widely in cells of all types, their synthesis by light energy would possess survival value even in cells that do not create sugar. Thus the light-gathering mechanism is of primary value (Crill, 1977), unrelated to any secondary use, a point of value in evolutionary considerations.

General Features. Characteristically the events of the light reactions of photosynthesis are visualized in an N pattern, the legs of which are provided by the two major light-harvesting systems that form an essential feature of the molecular apparatus. Customarily photosystem II is placed as the left leg and photosystem I as the right-hand member (Figure 10.8).

The General Pattern. The two photosystems play the key role in the transfer of electrons from water to NADP, but the route followed is long and highly complicated (Avron, 1971; Renger, 1975). Each system in metaphytans has chlorophyll a as the principal light receptor, but two different forms exist, known as chlorophyll a_I and a_{II}, respectively. Chlorophyll a_I, also referred to as chlorophyll a_{P700}, is more reactive to far-red light, whereas chlorophyll a_{II} (or chlorophyll $a_{P680-690}$) responds to light of shorter wavelength (Fenton *et al.*, 1979).

In broad terms photosystem I results in the production of weak oxidants but also strong reductants that enter into the reduction of CO_2 during its fixation. On the other hand, photosystem II produces weak reductants but strong oxidants that enter into the breakdown of water and the evolution of O_2 (Clayton, 1965; Levine, 1968). Moreover, this latter system generates thermoluminescence, whereas the other does not (Demeter *et al.*, 1979).

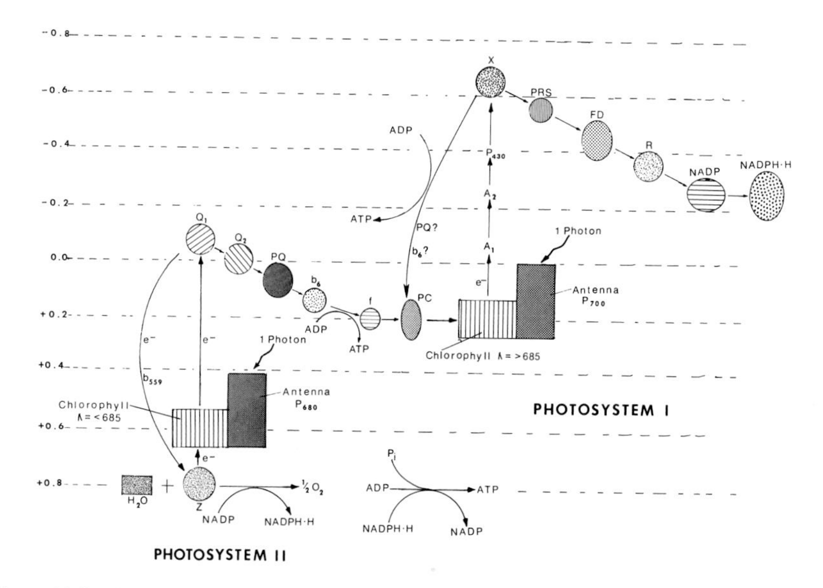

Figure 10.8. The light reactions of photosynthesis. The positions of such new elements as A_1, A_2, and Q_2 should be considered tentative.

Photosystem II. In addition to the major reaction center, called P_{680}, which consists of chlorophyll and other light-accepting ingredients described later, photosystem II contains an electron-transport chain. This begins with an electron-acceptor designated as Q, which symbolizes its ability to quench the fluorescence induced in chlorophyll by exposure to light (Rijersberg *et al.*, 1979). Sometimes, however, that letter is employed in the literature as an abbreviation for quinone, resulting in a degree of confusion. There now appears to be two species of this electron acceptor present in the system, designated as Q_1 and Q_2 (Golbeck and Kok, 1978, 1979; Bowes *et al.*, 1979). In a typical chain, the Q_1 and Q_2 are followed in turn by a plastoquinone, cytochrome *b*, cytochrome *f*, and plastocyanin; finally the electron is passed to a complex of substances called P_{700}, the major constituent of the reaction center of photosystem I (Avron, 1971). One quantum of light excites the reaction center of photosystem II to reduce Q (Figure 10.8), which then can be reoxidized by the passage of an electron through the chain to P_{700}. At the same time an unknown substance called Z is oxidized, which in turn oxidizes water. However, it should be noted that the actuality of this H_2O breakdown during photosynthesis has been questioned (Metzner, 1975).

When the kinetics of photosystem II photochemistry are studied, they are found to be biphasic (Melis and Homann, 1975, 1976). These results have been interpreted as a reflection of the presence of two different reaction centers, α and β, which appear to be special varieties of plastoquinones capable of being reduced to their respective plastosemiquinones upon continuous illumination (Melis, 1978). Moreover, it has been shown that the primary electron acceptor, plastoquinone, of the β center has a midpoint oxidation–reduction potential considerably higher than that of the α center. What relationship these dual centers bear to the two physiological donors D_1 and D'_1 that also have been proposed remains unresolved (Conjeaud *et al.,* 1979). One molecule of ATP appears to be normally coupled to the noncyclic flow of an electron at the step from cytochrome b_6 to cytochrome f (Avron and Chance, 1966). In addition, there is a cyclic flow of electrons from Q to Z, which seemingly is mediated, at least in part, by cytochrome b_{559} (Cramer and Whitmarsh, 1977; Hind *et al.,* 1977; Heber *et al.,* 1979). Whether or not photophosphorylation occurs on this pathway has not been established.

However, another noncylic photophosphorylative event occurs with the oxidized product Z. This unidentified substance brings about the breakdown of water, in an unknown number of steps as stated earlier, resulting in the release of $\frac{1}{2}O_2$ and reducing NADP to NADPH · H. In turn this induces the addition of P_i to ADP to form ATP (Figure 10.8; Bassham, 1973).

Photosystem I. The electrons from the plastocyanin of the noncyclic electron-transport chain of photosystem II finally reach the reaction center of photosystem I. Light energy absorbed by this latter center (P_{700}) then changes the E'_0 of the electron from +0.2 V to −0.6 V when it reduces the substance designated as X (Figure 10.8). As in photosystem II, the receptor to carrier sequence is more complex than originally believed and now is considered first to involve the series

$$P_{700} \longrightarrow A_1 \longrightarrow A_2 \longrightarrow P_{430}$$

(Shuvalov *et al.,* 1979a,b). From this point, the electron can follow either of two alternative pathways. The first is noncylic and leads initially to the reduction of ferredoxin, which is reoxidized by the transfer of its electron to NADP (Davis and San Pietro, 1977). The second is cyclic (Slovacek *et al.,* 1978, 1979; Crowther *et al.,* 1979) and results in the production of 1 mole of ATP, but the components of the transferal chain are unknown, except that the plastocyanin of the photosystem II noncylic chain seems to serve as the terminal member. More recently, cytochromes b_6 and f along with plastoquinone have also been implicated in this system (Böhme, 1977, 1979; Slovacek *et al.,* 1979). Thus ATP can be produced at three points (McCarty, 1978; McKinney *et al.,* 1979), or perhaps four, if the recently reported cyclic system of photosystem II proves to be productive, while NADPH · H is produced at two points. These substances then are the net products of the light reactions of photosynthesis and, as already seen, are largely employed in the fixation of CO_2. Here in photosynthesis, two schools of thought regarding coupling exist, the chemical and chemiosmotic, just as in oxidative phosphorylation, and here too no immediate solution of the problem is available (Baird and Hammes, 1979). It is pertinent to note, also, that a reverse flow of electrons from NADPH to Q may occur, resulting in the reduction of the latter; this pathway has been viewed as a mechanism for control of the NADPH level in the cell (Mills *et al.*, 1979).

In *Euglena* a circadian rhythmic effect has been noted in the evolution of O_2, the control mechanism for which appeared to be operative between the two photosystems (Lonergan and Sargent, 1979).

Photosynthetic Pigments

The Chlorophylls. The principal photosynthetic pigments, the chlorophylls, carry a prosthetic group that, like the hemes, is a porphyrin, but here it contains an Mg atom instead of an Fe. Furthermore, an additional ring derived from a 4-ethyl-6-β-ketopropionic acid chain and usually an esterifying alcohol called phytol are present (Figure 10.9). Four major varieties are recognized, chlorophylls *a, b, c,* and *d,* the first two of which are found in the metaphytans as a whole. Three of them are obviously closely related. As may be seen in Figure 10.9 *a, b,* and *d* differ from one another at only one or two points, *b* having the CHO^- radical on carbon 3 in place of the methyl of chlorophyll *a,* and *d* having the CHO^- on carbon 1, instead of the vinyl of *a* and *b*. Type *d* is found only among certain red seaweeds. In contrast, chlorophyll *c* that occurs among a number of plant groups, including the brown seaweeds, diatoms, and chrysophytes, differs markedly from the others in lacking the phytol chain and the hydrogens on carbons 7 and 8. Two varieties are known, one bearing an ethyl, the other a vinyl radical on carbon 4 (Figure 10.9). Treatment of these as mere varieties does not appear equivalent to that given the other major types, for they differ as greatly from one another as either *b* or *d* does from *a*.

It is also obvious that chlorophyll *c* diverges more widely from the others than does bacteriochlorophyll *a* of the purple bacteria, which differs in only two minor points: the modification of the vinyl group of position 1 to an acetyl and one less double bond in ring II, so that hydrogens occur on carbons 3 and 4 (Jones, 1973). Unfortunately, the name given to the latter is misleading in that it suggests fundamental differences from the others, thereby concealing possible ancestor–descendant relationships—cholorphyll *e,* for example, would appear to be far more appropriate. Two sets of chlorophylls related to chlorophyll *a* and the others show a somewhat simpler molecular structure. Each of these chlorobium chlorophylls consists of mixtures of at least six pigments, which differ in the combination of radicals of certain side arms (Figure 10.10). Chlorobium chlorophyll 660 has been found in all species of that family of green sulfur bacteria except *Chlorobium thiosulfatophilum,* which has that known as chlorobium chlorophyll 650. Both sets are unique in possessing one additional radical (between rings I and IV), in lacking the acetyl group on ring V, and in having the usual phytol ($C_{20}H_{40}O$) replaced by farnesol ($C_{15}H_{26}O$) or

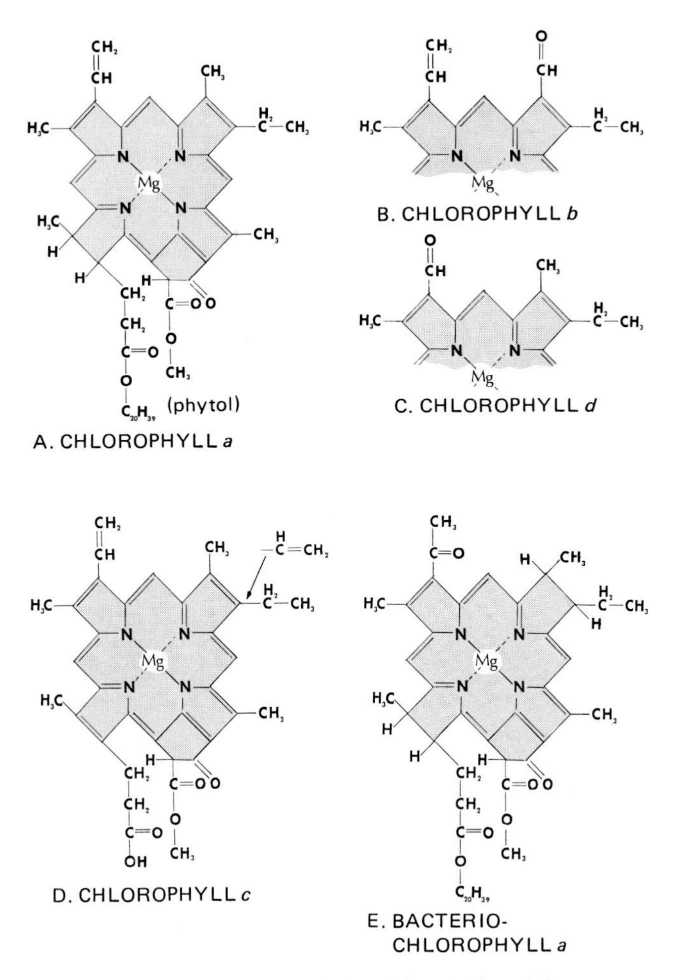

Figure 10.9. The principal varieties of chlorophyll.

other alcohol. Thus these types, being somewhat simpler in structure, could possibly be intermediates in an evolutionary chain leading to the definitive types. It is of interest to note that all species of *Chlorobium* also have a small quantity of bacteriochlorophyll *a* present. The obfuscation of relationships among the various types will be worsened if a suggestion is followed, which proposes that chlorobium chlorophylls 660 and 650 be referred to as bacteriochlorophylls *c* and *d* (Gloe *et al.*, 1975; Caple *et al.*, 1978). Actually it would appear much more advisable to rename all the chlorophylls in light of modern knowledge, using Roman numerals or other designations to prevent confusion and to allow for additions as they arise.

Carotenoids. Carotenoids are a diversified group of pigments found throughout the living world, although they are not synthesized by the metazoans and certain protozoans. A number of varieties are found in chloroplasts of the higher plants in association with chlorophyll, but, as they vary widely from one taxon to another, here only two major types, β-carotene and the xanthophyll called zeaxanthin (Goodwin, 1973), are diagramed (Figure 10.11). Both carotenoids and xanthophylls have long, unsaturated hydrocarbon chains, typically ending in an ionone residue (Figure 10.11A), the xanthophylls being distinguished by the presence of oxygen in the form of a hydroxyl or other radical, which is lacking in the carotenoids.

While the functions of these pigments have not been fully established at present, they are of utmost importance, for mutations that prevent their synthesis are lethal. According to experiments with mutants of the photosynthetic bacterium *Rhodopseudomonas spheroides,* they play a role in protecting chlorophylls from being photosensitized. Under anaerobic conditions or in the dark, the organisms that lacked these pigments grew normally, but when exposed to light in the presence of oxygen, they were killed within 12 to 14 hr (Stanier and

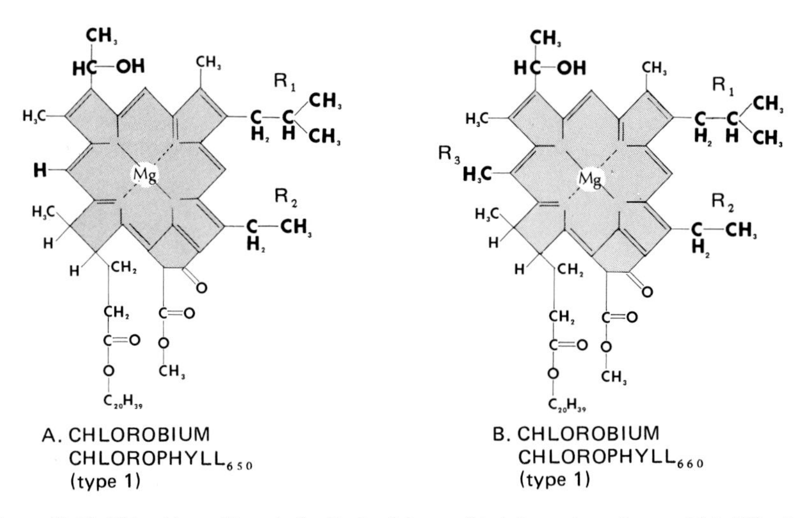

Figure 10.10. Chlorobium chlorophylls. Each of the two kinds have six variants, which differ from those illustrated as follows: (A) At R_1 type 3 has isobutyl as shown, types 2 and 5 have *N*-propyl, while 4 and 6 have ethyl; at R_2 types 2 and 4 have ethyl and 3, 5, and 6 have methyl. (B) R_1 in type 2 is isobutyl as shown, 3 and 4 have *N*-propyl, and 5 and 6 have ethyl; R_2 is an ethyl group in all variants except 6, which has a methyl; and R_3 is methyl in all except 2 and 4, which seem to have an ethyl radical there.

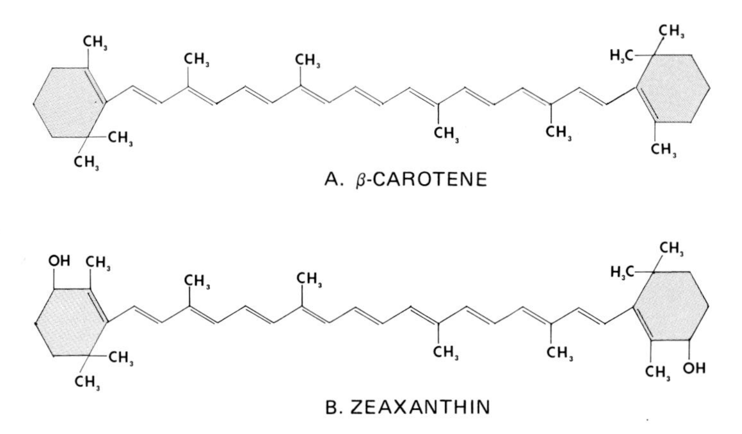

Figure 10.11. Structure of carotenoids.

Cohen-Bazire, 1957; Krinsky, 1971). In addition, the pigments normally also are able to participate in the absorption of light for photosynthesis, at least in certain algae (Goodwin, 1973). Flavonoids also are an important type of pigment (see Giannasi, 1978, for a review).

The Reaction Centers, or Antennae. In photosynthetic organisms in general, including bacteria, the chlorophylls do not occur alone in the active apparatus but are complexed with other pigments and various proteins. For example, *Rhodospirillum* was found to have a photosynthetic complex consisting of 4 moles of bacteriochlorophyll, 2 of bacteriopheophytin, and 1 of spirilloxanthin (van der Rest and Gindras, 1974). Moreover, iron-sulfur (ferredoxin-related) centers have been identified in the members of this genus as well as those of green- and purple-sulfur bacteria (Malkin and Bearden, 1978). Similar complexes in eukaryotes are organized into groups, often referred to as reaction centers or antennae. Early indications of this arrangement were secured by sonicating chloroplasts in the presence of such surfactants as Triton X-100 or digitonin (Wolken, 1973), which yielded micelles ranging in diameter from 100 to 1000 Å. Particles 200 Å in diameter were calculated to contain 1 cytochrome, 1 ferredoxin, 225 chlorophyll, and 55 carotenoid molecules. Gentler techniques, including gel electrophoretic analyses, have now indicated that all the chlorophyll of the organelle exists in similar complexes with protein (Markwell *et al.*, 1979).

The reaction center of photosystem I appears to have been more thoroughly characterized than that of photosystem II; it consists of chlorophyll *a*, proteins, β-carotene, and a trace of lutein (Bailey *et al.*, 1966). The chlorophyll and P_{700} have been reported to be associated with six polypeptides, named subunits I to VI in descending order of molecular weights of 70, 25, 20, 18, 16, and 8×10^3 (Bengis and Nelson, 1975, 1977; Golbeck *et al.*, 1977). P_{700} was established as being associated with the heaviest of these, subunit I, whereas bound ferredoxins were suggested to be associated with subunits IV, V, and VI. Two subunits I, each containing 20 chlorophyll *a* molecules, were required for each P_{700} unit. In a preliminary model of the arrangement of the various particles within the thylakoid membrane (Figure 10.12), at least the relative positions of P_{700} and the ferredoxins were in part corroborated by the report that proton translocation is localized on the reducing side of both these substances (Tyszkiewicz *et al.*, 1977).

Although the photosystem II reaction center has not been as thoroughly analyzed as that of its counterpart, it probably consists of a number of subunits, too; in the green algae and higher plants it contains chlorophylls *a* and *b*, proteins, xanthophylls, and a small amount of β-carotene (Bailey *et al.*, 1966; Tevini and Lichtenthaler, 1970). The primary electron donor, P_{680}, appears to have at least two physiological electron donors associated with it. That known as D_1 reduced P_{680} very rapidly, in less than 1 μsec with flashed light, while D'_2 reacted more slowly (6 and 22 μsec after flashes) and was operative under deficient O_2 conditions (Conjeaud *et al.*, 1979). In thylakoid membranes studied by freeze-fracture techniques, this center has been shown to appear as a large particle, as will be seen later (Miller *et al.*, 1976).

The Antennae of Protistans. As mentioned earlier, *Cyanidium* is a thermophilic, acidophilic alga of uncertain relationships that lacks chlorophyll *b* but

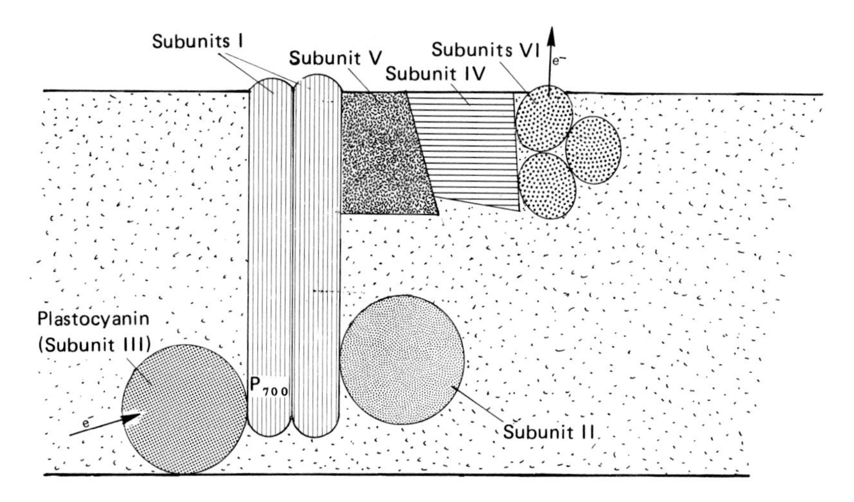

Figure 10.12. Model of membrane arrangement of photosystem I. (Based on Bengis and Nelson, 1977.)

has an abundance of phycocyanin C. Normally the latter pigment appears to be associated with chlorophyll *a* in the antenna of photosystem II in this organism. Recently a mutant lacking the pigment made possible an analysis of the reaction center, which showed that only 20 molecules of chlorophyll *a* occurred per center (Diner and Wollman, 1979), whereas *Chlorella* has been reported to contain 240 chlorophyll *a* + *b* units per center.

The Antennae of Blue-Green Algae. Using gentle procedures, three chlorophyll–protein complexes have been isolated from such blue-green algae as *Phormidium* and *Anabaena* (Reinman and Thornber, 1979). Complexes I and II had red-absorbance maxima of 676 nm, indicating that they were photosystem I components, whereas complex III had a maximum of 671 nm. Molecular weights of 250,000 were reported for complex I, 118,000 for II, and 58,000 for III. The intermediate-weight complex was reported to resemble the P_{700}–chlorophyll *a*–protein complex of seed plants and green algae, whereas complex III was believed to represent the photosystem II antenna of these blue-green forms. As the subunit report of seed plant antenna I implies a total molecular weight of about 235,000 (Bengis and Nelson, 1977), complex I in those organisms would appear to correspond more closely. However, correspondences, or even similarities, between cyanophycean and higher plant antennal structures are difficult to perceive, especially in view of there being three unrelated subunits in the former.

Some aspects of the relationships between the antennae and the substructure of the thylakoids have been made possible by the nitrogen-fixing specialized cells of these blue-green algae, known as heterocysts. These have been demonstrated to lack photosystem II activity, along with the major accessory pigments. Whereas normal vegetative cells have moderate-sized densely placed particles on the protoplasmic fracture face (P face), with diameters of 70 Å (Figure 10.13), heterocysts have slightly smaller ones. However, greater differences are observed on the exoplasmic fracture face (E face) on which, in the vegetative cells, particles of two sizes occur, one 75 Å, the other 100 Å in diameter (Figure 10.13B; Giddings and Staehelin, 1978, 1979). Heterocysts, however, have only the 75-Å particle.

The Antennae of Bacteria. Some progress has been achieved in analyses of bacterial reaction centers, but the complexity provided by the multiple phylogenetic levels of these organisms acts as a retardant (Vernon, 1968; Frenkel, 1970). The ones that resembled the antennae of higher plants most closely were those of *Rhodopseudomonas* and *Rhodospirillum* (Sauer and Austin, 1978). As a rule, the bacteriochlorophyll, about 95% of which is unbound, plays an essential but rather passive role in light harvesting, absorbing light and then transferring the resulting electronic excitation to the 5% contained in the reaction centers. These centers have been found to consist of complexes containing 4 bacteriochlorophylls, 2 bacteriopheophytins, 2 CoQ, and 1 iron atom, together with 3 polypeptides having molecular weights of 28, 23, and 21×10^3, respectively (Straley *et al.*, 1973; Feher and Okamura, 1977). However, the cytoplas-

Figure 10.13. Thylakoid structure of a vegetative blue-green alga. These freeze-fracture electron micrographs show the arrangement of particles on the protoplasmic face (P face) and exoplasmic face (E face) of the thylakoids. (A) In addition to a general view of the thylakoids, sections of cross walls and the P face of the plasmalemma are shown. 40,000×. (B) Parallel thylakoids at higher magnification; the particles of the P face are densely placed and uniform in size, whereas those of the E face are sparse and of two sizes. The extremely narrow lumen is indicated. 100,000×. (Both courtesy of Giddings and Staehelin, 1979.)

mic pigments are not totally free, for the larger portion appears to be linked to a polypeptide of 10,000 daltons and a phospholipid (Sauer and Austin, 1978).

Associated with the bacteriopheophytin of the reaction center is the unknown real primary electron acceptor referred to as I in these organisms (Dutton *et al.*, 1978). Furthermore, the bacteriochlorophyll in these centers is in the form of dimers (Dutton *et al.*, 1975), and the CoQ is complexed with Fe. In isolated complexes, the primary charge separation occurred within 10 psec after exposure to a light flash, followed by an electron transfer within 200 psec (Fenton *et al.*, 1979).

The complexes of the green sulfur bacterium, *Chlorobium limicola,* are far more varied than previously believed. Instead of the 6 varieties of chlorobium chlorophyll$_{660}$ reported earlier, 14 have now been demonstrated (Caple *et al.*, 1978), 4 of which contained farnesol as the esterifying alcohol, two had phytol, and the rest any one of four other alcohols. Each of the 14 was combined respectively with a distinctive type of bacteriopheophytin and one of four types of bacteriomethylpheophorbide.

Halobacterium halobium was shown to have a radically different reaction center, in that the light activated bacteriorhodopsin; when thus photoactivated, that pigment transferred protons—not electrons as usually is the case—from the interior to the exterior of the cell membrane (Oesterhelt and Stoecknius, 1973). Now this organism has been shown to fix CO_2 in a reaction possibly involving the abundant ferredoxin that is present (Danon and Caplan, 1977).

Peroxisomal Respiration. As mentioned in connection with the discussion of the peroxisome (Chapter 7, Section 7.2), peroxisomal respiration has until recently been referred to as photorespiration (Andrews and Lorimer, 1978), because it has been investigated chiefly in the leaves of higher green plants. But because the same processes occur in livers and kidneys of vertebrates, the more appropriate term peroxisomal respiration has been proposed (Tolbert and Yamazaki, 1969) in reference to the site of this activity. In that chapter, too, it was pointed out that further attention would be given the topic in relation to photosynthesis, because the two sets of processes share a number of ingredients.

The Basic Ingredient. Outstanding among the shared ingredients is glycolic acid (Krause *et al.*, 1977), which is the basis for the entire chain of events in peroxisomal respiration. Among the higher plants this two-carbon substance is biosynthesized in the chloroplast in a reaction that requires light and O_2, and similarly is an abundant product of algal metabolism (Lewin, 1962). Yet in spite of numerous investigations, the enzymatic pathway for its formation remains unknown in plants as well as metazoans. One distinct possibility is by direct condensation of two single-carbon units (Warburg and Krippahl, 1960; Zelitch, 1965), as proposed elsewhere for the primitive amino acid biosynthesizing processes (Dillon, 1972, 1978). In the higher plants glycolic acid is generated only in the light and therefore is viewed as a product of photosynthesis alone (Tolbert and Yamazaki, 1969)—in the dark it appears to be converted to malic acid or other constituents of the tricarboxylic acid cycle.

The Early Reactions. Using O_2 from the environment, the first reaction to occur is the oxidation of the glycolic acid to glyoxylic acid and H_2O_2, with the aid of the enzyme glycolic acid oxidase. As the immediate destruction of the peroxide by the abundant catalase present in the peroxisome releases $\frac{1}{2}O_2$, the net consumption of oxygen is an identical amount, accounting for the uptake of this element by these organelles during the processes.

The glyoxylic acid thus formed then receives an amino radical through the mediation of a peroxisome-specific enzyme glyoxylate-glutamate transaminase (Kisaki and Tolbert, 1969) to form glycine (Figure 7.15). Thus glutamic acid from the cytosol needs to enter the organelle. Moreover, the next reaction in this pathway does not take place in the peroxisome, for the necessary enzyme is confined to the mitochondrion (Figure 10.14; Tolbert and Yamazaki, 1969). In this organelle the enzyme serine hydroxymethyltransferase converts 2 moles of glycine to 1 each of serine and CO_2, but what disposition is made of the excess amino radical is not clear.

The Later Reactions. Following its synthesis in the mitochondrion, the serine evidently is transported back to the peroxisomes, for the next two en-

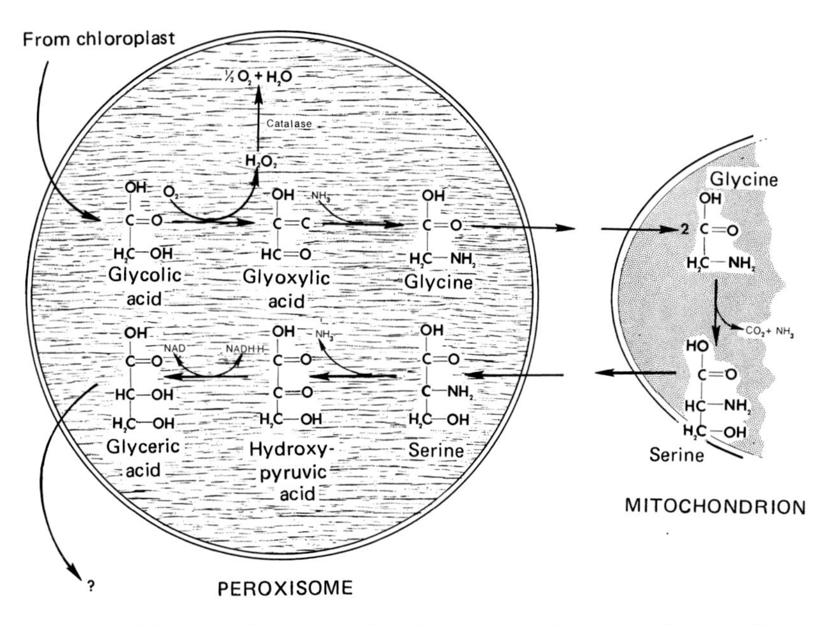

Figure 10.14. Interdependency of peroxisomal, mitochondrial, and chloroplastic reactions.

zymes of these processes are found only in the latter organelle. In these particles the serine is deaminated by serine-pyruvate aminotransferase to result in dihydroxyacetone, which substance will be recognized as one of the ingredients of glycolysis as well as of the dark reactions of photosynthesis. Another equally familiar substance, glyceric acid, is next produced from the dihyroxyacetone by a very active enzyme, NAD-D-glyceric acid dehydrogenase, in a reaction deriving a hydrogen equivalent from NADH and perhaps energy from the same source.

Beyond this point the glyceric acid is phosphorylated to become 3-phosphoglyceric acid, but the location of the kinase involved in this reaction remains unknown—it has not been found in the peroxisome of spinach leaves (Tolbert and Yamazaki, 1969). Also it has not been determined how this substance finally enters into the processes of sugar synthesis, nor where that conversion takes place. However, it is clear that in leaves, peroxisomal consumption of O_2 is actually involved in photosynthesis, whereas in metazoans it is associated with cell respiration. Hence, the close kinship pointed out elsewhere that exists between these two fundamental processes receives emphatic confirmation. Moreover, if the speculation by Tolbert and Yamazaki (1969) proves to be justified, then the ample supply of serine produced by photosynthesizing chloroplasts may similarly be sent to the mitochondria for conversion into dihydroxyacetone, making additional ties between the two processes evident.

10.2. THE CHLOROPLAST

As might be expected, the chloroplast of higher plants has been far more extensively investigated than has that of algae and bacteria. Consequently, treatment here corresponds to that given the mitochondrion, where the metazoan organelle was described in detail before attention was given to comparative aspects. In that case, however, the structure of the definitive type was less strongly divergent from that of lower forms than is the case with the chloroplast.

10.2.1. Metaphytan Chloroplast Structure

Also in contrast to the mitochondrion, the present organelle does not appear to undergo morphological changes in response to various physiological states, nor have preparatory techniques been demonstrated to affect its morphology. Consequently, its structure has been found remarkably uniform throughout the metaphytans and in many of the higher green algae. In all these organisms, the chloroplasts, which vary in shape from nearly spherical to elongate-oblong and crescentic, are enclosed by two membranes, the inner one of which appears to be somewhat thinner. Often in sections, these membranes may be invaginated irregularly or may form vesicles.

The Stroma. Enclosed within the body is a granular-appearing stroma, the extent of which varies with the tissue. Its biochemical content is largely protein, the main catalytic activity of which, collectively known as fraction I, is associated with ribulose-1,5-bisphosphate carboxylase, an enzyme whose biochemistry and genetics currently are receiving much attention (Chua and Schmidt, 1978; Link *et al.,* 1978). In addition, there may be a number of starch grains lying free in the matrix and, with osmic acid fixation, small electron-opaque bodies referred to as plastoglobuli appear in nearly all chloroplasts. Generally the latter particles are considered to consist of such lipids as plastoquinone and tocophorylquinones (Leggett-Bailey and Whyborn, 1963; Lichtenthaler, 1969), but in older leaves they may also accumulate carotenoids. Also located here are the ribosomes, which receive attention in a following section.

The Lamellar System. In all chloroplasts, metaphytan and algal alike, the lamellae (Granick, 1963) are in the form of closed flat sacs, called thylakoids (Menke, 1962), the lateral ones of which in longitudinal section parallel the arching enclosing membranes, while those medially are planate. However, in cross-sectional views, the membranes appear to be spiral (Coombs and Greenwood, 1976). Here and there the thylakoids are stacked together with smaller ones to form grana, to which structures photosynthesis was at one time believed to be confined. Although this arrangement of membranes and grana is constant, much variation in the number of individual parts exists from tissue to tissue and even within the same cell. Those thylakoids extending between grana are perforated or subdivided to such an extent in the metaphytans that they are usually referred to as frets (Weier and Thomson, 1962; Heslop-Harrison, 1963). In forming a granum the fretwork is arranged obliquely while it sends out extensions that expand into the small flat disks that comprise those bodies (Figure 10.15). As a rule these multiple extensions that form the successive layers are arranged as a right-handed helix, several parallel frets often being involved in the structuring of a given granum (Paolillo and Falk, 1966; Paolillo *et al.,* 1969; Paolillo, 1970). Present interpretation of the membrane system proposes that all the frets, grana, and thylakoids, large and small, are actually folds and extensions of a single continuous double sheet (Heslop-Harrison, 1966). Thus the interior of a chloroplast may be viewed as containing only two compartments, a large one occupied by the stroma and the other enclosed within the continuum that comprises the lamellar system (Coombs and Greenwood, 1976).

The thylakoids that make up the individual grana do not actually contact one another but are narrowly separated by a deposit of material called the partition. It is within these partitions where much of the diformazon reaction product is deposited that is formed from TNBT when it is reduced by light, suggesting the localization of photosynthetic activity to those areas (Weier *et al.,*

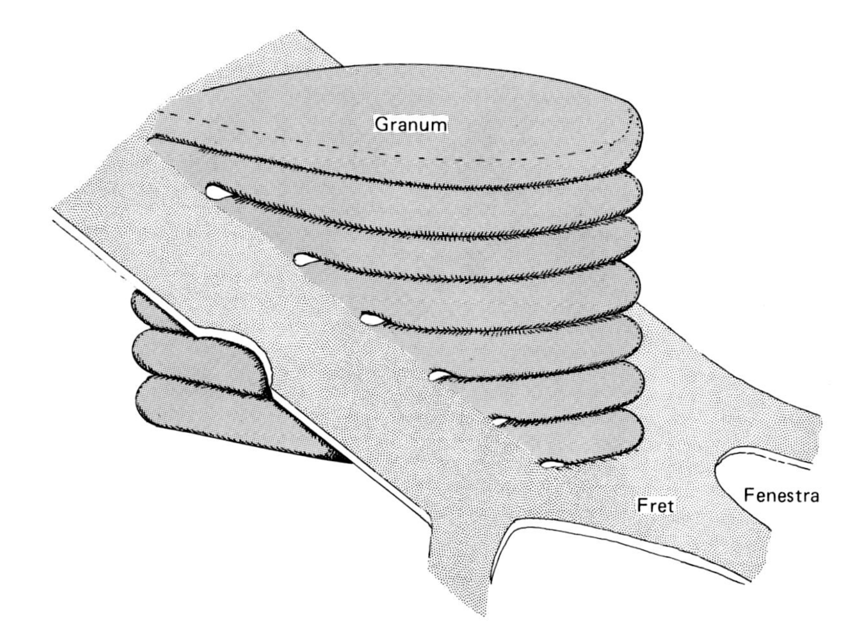

Figure 10.15. The structural relationships between a fret and granum. (Based on Heslop-Harrison, 1963.)

1966a). However, in the electron micrographs, the intergranal frets also show some deposition. More recently these partitions have been demonstrated to have a tripartite structure, consisting of two electron-transparent lines sandwiching an electron-opaque band (Casadora and Rascio, 1978a,b).

Molecular Structure of the Membrane System. Here in the chloroplast, the unit-membrane concept of thylakoid structure at the molecular level soon yielded to a particulate view as techniques were improved and higher resolution electron micrographs became feasible. Among the first of these were investigations of granal structure that clearly revealed the membranes to consist of globular subunits (Weier *et al.*, 1965, 1966b). Freeze-etch studies corroborated the presence of such globules and demonstrated that three different membrane surfaces existed. The A face, the surface of the membrane exposed to the stroma, showed low irregularities and indistinct particles; the B face, one of the internal surfaces revealed by splitting the membranes, had sparse large grains, whereas on the C face, the granules were smaller and more densely placed (Branton and Park, 1967). At magnifications of 500,000×, the larger particles were found to be oriented in precise fashion and to be structurally identical throughout the exposed surface; in diameter they ranged from 13 to 17 nm and about 8 to 9 nm in height. However, in the fret regions of the thylakoids these particles of the B face were few in number or even completely absent, being mostly located close to partitions of the grana. In contrast, the granules of the C face occurred throughout the membranes, regardless of location. Further investigations and technique refinements have now permitted the proposal of a model of structural organization of this membrane (Miller, 1976, 1978).

The presence of the particles and the concentration on granal surfaces revitalized a concept originated by Emerson and Arnold (1932), which proposed that individual particles, or subunits, were capable of responding to a quantum of light repeatedly in carrying out the photosynthetic processes. Thus the term quantasome was applied to those large particles of the B face exposed by freeze-etching (Park, 1965). However, as the so-called quantasomes are scarce or absent on fret membranes, which nevertheless contain chlorophyll and participate in photosynthesis, the concept has fallen into disuse once more.

Other methods of investigating the substructure of the membranes have been employed, with some degree of success. For instance, dodecyl sulfate polyacrylamide gel electrophoresis yielded three relatively sharp bands of membrane proteins (Menke and Schölzel, 1971). These had apparent molecular weights of 66,000, 62,000, and 59,000, the two lighter ones being components of the factor that couples ATP synthesis to photophosphorylation (McEvoy and Lynn, 1973; Nelson *et al.*, 1973). Consequently, they correspond in part to the indistinct particles of the A face that had been found to be correlated to ATPase activity (Coombs and Greenwood, 1976). More recently, the 60,000-dalton band was shown to consist of several components, three of which have been isolated and identified. Those called PS-I_{96} and PS-I_{88} appeared to be associated with photosystem I, while that named PS-II_{42} was active with photosystem II (Koenig *et al.*, 1977).

Other studies have isolated fractions of uniform weight from thylakoid membranes and localized them *in situ* by means of specific antisera. In one report, four polypeptide fractions having molecular weights of 11,000, 17,000, 40,000, and 48,000 were examined (Koenig *et al.*, 1976). Of these the smallest was found by inhibition reactions to be located on the H_2-degrading side of photosystem II, while the antiserum to the second largest polypeptide (40,000) demonstrated that the polypeptide was normally active in photosystem I, probably as the γ component of the coupling factor. The largest protein was similarly revealed to be related to the electron chain between the two systems, seemingly just prior to the plastocyanin.

Another investigation along comparable lines is of importance in relation to membrane structure theory as well as to the morphology of this organelle (Radunz, 1977). Antibodies to individual lipids never were found to cover more than 17% of a membrane surface, and those that were reactive with many lipids collectively did not coat more than 34% of the area. In contrast, the sera that contained antibodies to proteins covered the entire accessible area. Consequently, it became evident that these membranes consisted largely of proteins, lipids serving only to fill in interstices between those molecules, as in the proposal by Sjöstrand reported earlier (Chapter 9, Section 9.22). Thus neither the unit-membrane nor fluid mosaic concept is applicable to the thylakoid membrane.

10.2.2. Algal Chloroplast Structure

Because current terminology on chloroplast structure developed largely through investigations of the metaphytan organelle, it is appropriate first to review its morphology in the closely allied group, the green algae, before the other types of chlorophyll-bearing plants are examined. In this way, the relative complexity of the chloroplast in each individual taxon can be better evaluated.

The Higher Green Algae. Chloroplasts replete with grana and other hallmarks of the metaphytan organelle are found in the stoneworts, *Chara* and *Nitella* (Dodge, 1973), and in such advanced filamentous algal types as *Schizomeris* (Mattox *et al.,* 1974). But in the latter, the grana are low, containing only three or four loosely associated thylakoids. For the greater part, however, such specialized structures are absent, although pseudograna are present in such genera as *Carteria* (Joyon and Fott, 1964). In general the lamellae are continuous from one end of the organelle to the other and characteristically are comprised of three thylakoids each, with starch grains stored between the lamellae in the stroma. Nevertheless, there is considerable variation. In *Caulerpa verticellata,* for instance, many of the thylakoids were abbreviated to a greater or lesser extent (Calvert *et al.,* 1976), and the lamellae appeared more compact and electron dense, forming what was called thylakoid bands. One of the most outstanding features in this genus and other members of the Caulerpales and Dichotomosiphonales was the concentric lamellar system that was located at one end of the plastid (Hori and Ueda, 1967; Calvert *et al.,* 1976; Calvert and Dawes, 1976). These bodies, to which a number of diverse names have been applied (Descomps, 1965; Sabnis, 1969; Borowitzka and Larkum, 1974; Turner and Friedmann, 1974), consisted of 3 to 15 pairs of arcuate membranes, arranged around an amorphous center from which a number of microtubules extended (Figure 10.16). Although details differ among the several investigators, the consensus holds that these bodies represent developmental stages of the thylakoids.

As stated earlier, grana characterize the chloroplasts of the advanced green algae, but these structures occur in diverse shapes and sizes. Among the largest in relative size are those of the genera *Pediastrum* and *Hydrodictyon* (Hawkins and Leedale, 1971). In the first genus, the individual grana may approach half the length of the chloroplast and up to 40% of its width, and may include as many as 22 long thylakoids closely applied to one another in an irregular fashion. Usually four or five such bodies occur in a given organelle, along with several starch grains. In *Hydrodictyon* they are somewhat shorter and not so

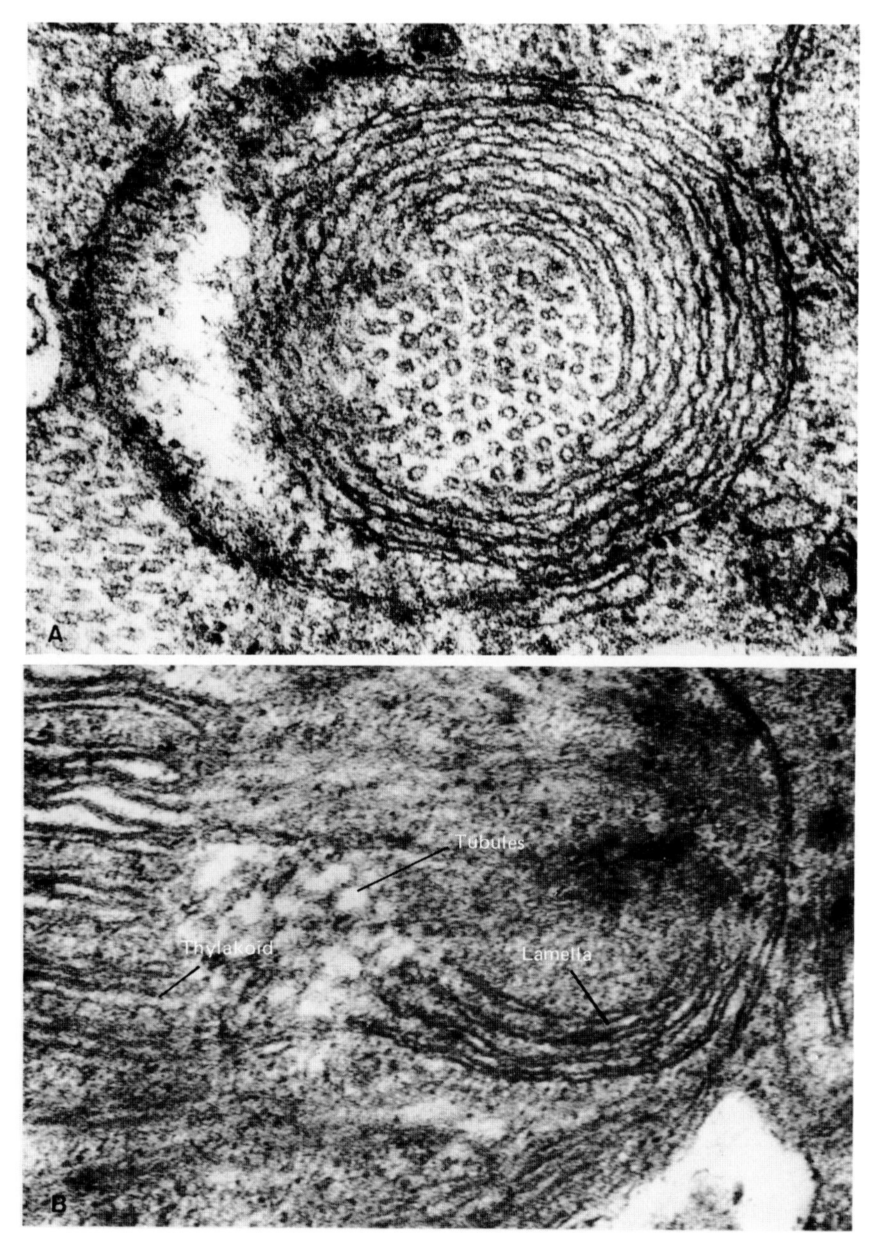

Figure 10.16. Concentric lamellar system of a chloroplast of *Caulerpa*. In this chlorophycean, the chloroplast contains a specialized region thought to be involved in the ontogenetic production of thylakoids. (A) In the center of the system is a set of membrane tubules arranged in a regular triangular pattern. 85,000×. (B) The tubules are considered intermediates in the transition from the lamellae to thylakoids. 55,000×. (Both courtesy of Calvert and Dawes, 1976.)

high, but are so numerous that, along with the large starch grains, little space for stroma remains. The rigidity and numbers of these enclosed bodies impart an angular configuration to the chloroplast that is not encountered elsewhere. *Acetabularia* chloroplasts also possess elongate grana, which often are one-third as long as the organelle itself (Crawley, 1963). In these, however, the thylakoids are less closely applied in stacks, leaving interstices between them in irregular fashion and having a more flexuous appearance than in the two foregoing genera (Crawley, 1963).

The grana of *Schizochlamys* and other members of the Tetrasporales approach those of the metaphytans in appearance, being short, and with more or less straight ends (Wujek and Gretz, 1977). The *Volvox* chloroplast also has grana that approach those of metaphytans, but in this case, they are more widely spaced, and the stroma lamellae are reduced in number, the net result being a more open appearance and a greater proportional amount of stroma (Bisalputra and Stein, 1966). Similar conditions prevail in other members of the Volvocales, including *Haematococcus* (Lang, 1968a).

In general, however, the chloroplast of these algae is devoid of grana, consisting simply of a variable number of elongate, continuous lamellae. Although each lamella is comprised of three thylakoids, as in most algal chloroplasts, in the present group the individual members are nearly lacking in lumina and are closely appressed to one another. Consequently, the lamellae appear thin and flexuous. Some examples of this characteristic structure are provided by *Pleurastrum* (Molnar *et al.,* 1975), *Platymonas* (Manton and Parke, 1965), and *Chlorella* (Graham and Graham, 1978).

Chloroplasts of Lower Eukaryotes. The thylakoids of the euglenoids show many of the traits that characterize the nonchlorophycean algae as a whole. Throughout these many groups, the lumina of the thylakoids exceed the enclosing membranes in diameter (Figure 10.17A), whereas in the green algae, the cavities are scarcely perceptible, as just described. The thylakoids are usually arranged as groups of threes to form the stromal lamellae, and grana are consistently absent (de Haller, 1959; Siegesmund *et al.,* 1962; Ben-Shaul *et al.,* 1965; Leedale, 1966, 1967). As a rule, a number of the membranes do not attain the ends of the chloroplast. Furthermore, individual thylakoids on the stromal side of a lamella occasionally are abbreviated, resulting in two-thylakoid regions of variable lengths. Branching, however, does not occur, the components of each lamella remaining in its own system and never contacting or coalescing with those of adjacent lamellae (Figure 10.17A). Cross-sectional views of chloroplasts in general rarely appear in print; one of the few exceptions, that of *Euglena spirogyra,* shows the lamellae to form concentric rings parallel to the envelope (Siegesmund *et al.,* 1962). Consequently, the central plate is quite narrow. Stroma is fairly abundant, near the middle of the length often being distinctly wider in transverse sections than the lamellae.

In the dinoflagellates, a group that shows relationship to the euglenoids in flagellar and nuclear characteristics, the chloroplast is similarly rather simple

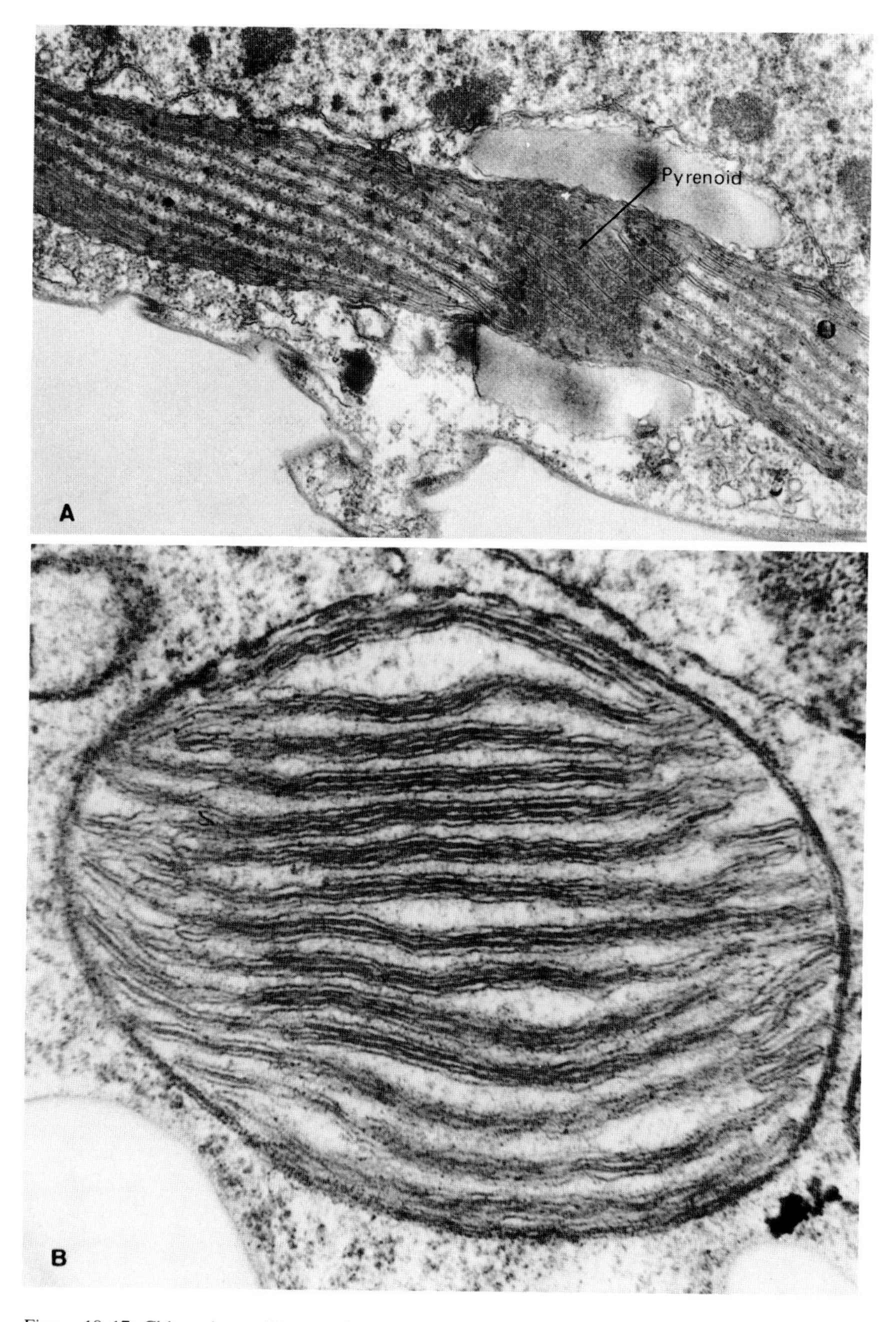

Figure 10.17. Chloroplasts of lower eukaryotes. (A) Chloroplast of *Euglena*. 33,600×. (Courtesy of Salvador *et al.*, 1971.) (B) As in the euglenoids, in the dinoflagellate *Amphidinium* the three thylakoids of a lamella have wide lumina, are loosely associated with one another, and are of variable lengths. 22,000×. (Courtesy of Dodge and Crawford, 1968.)

(Dodge, 1968). As in that taxon, the lamellae usually extend from end to end, but a few are exceptional in being somewhat abbreviated. They likewise consist of three thylakoids, with swollen lumina; a number are similar also in being short and creating regions in the lamellae that contain only two thylakoids. In *Prorocentrum* and to a lesser extent *Dissodinium,* however, some of the lamellae contain four or even more members (Dodge, 1973, 1975), but the thylakoids do not interconnect with those of other lamellae. Some contrast to the euglenoids is provided by the thylakoids being loosely arranged, so that gaps between adjacent members often can be detected, as in *Amphidinium* (Figure 10.17B). Cross-sectional views are strikingly different from the preceding taxon, for the membranes are not concentric; rather, at least in the pyrenoid region, they extend from the sides toward the center, many being in parallel arrays, and others irregularly oriented (Dodge, 1975).

Chloroplasts of Chlorophyll-c-Containing Taxa. A number of algal groups indicate their possible interrelationships by sharing the unusual chlorophyll termed *c* in addition to *a;* in contrast, most of the remainder contain chlorophyll *b* as well as *a,* while a few have only the latter. As the differences between *a* and *b* are so very slight (Figure 10.9), repeated separate phylogenetic origins for the second of these from the first remains a distinct possibility, whereas *c* is too widely divergent for that to be a likelihood. Therefore, it would appear to be no coincidence that in this assemblage of types the chloroplast also shares a trait found in no other taxon. Here the general form of the organelle and its internal structure are like those of the dinophyceans. That is to say, the lamellae extend to each end of the organelle and consist of three loosely associated thylakoids, distinct interthylakoid spaces being abundant, as shown in *Mallomonas* (Figure 10.18A; Wujek and Kristiansen, 1978). However, a consistent difference can be noted: the individual thylakoids frequently cross from one lamella to a neighbor. Thus the number of components in a given lamella is subject to rapid changes. Specific examples of this condition in the Chrysophyceae and *Chrysochromulina* and *Prymnesium* (Manton, 1964, 1966, 1967), the diatom *Nitzschia palea* (Drum, 1963), and the brown seaweeds, including *Giffordia, Chorda* (Bouck, 1965), and *Sphacelaria* (Figure 10.18B; Galatis *et al.,* 1977). At least in the latter organisms (Phaeophyceae), one set of thylakoids, called the girdling lamella, continues around each end of

→

Figure 10.18. Chloroplasts of chlorophyll-*c*-containing organisms. Those organisms that utilize chlorophyll *c* also have a unique structural trait in the chloroplast. Here and apparently nowhere else, a thylakoid of one lamella diverges to join another, to form a loose reticulum. Moreover the lamellae are quite thin relative to the stromal interspaces. (A) A chloroplast of the chrysophycean *Mallomonas*; arrows indicate several interconnecting thylakoids. 18,500×. (Courtesy of Wujek and Kristiansen, 1978.) (B) A chloroplast of the brown alga *Sphacelaria*, with interconnecting thylakoids similarly indicated. 23,000×. (Courtesy of Galatis *et al.*, 1977.) (C) This view of the chloroplast of the red alga *Chorda* shows only a few interconnecting thylakoids (arrows). The eyespot is also shown. 30,000×. (Courtesy of Toth, 1974.)

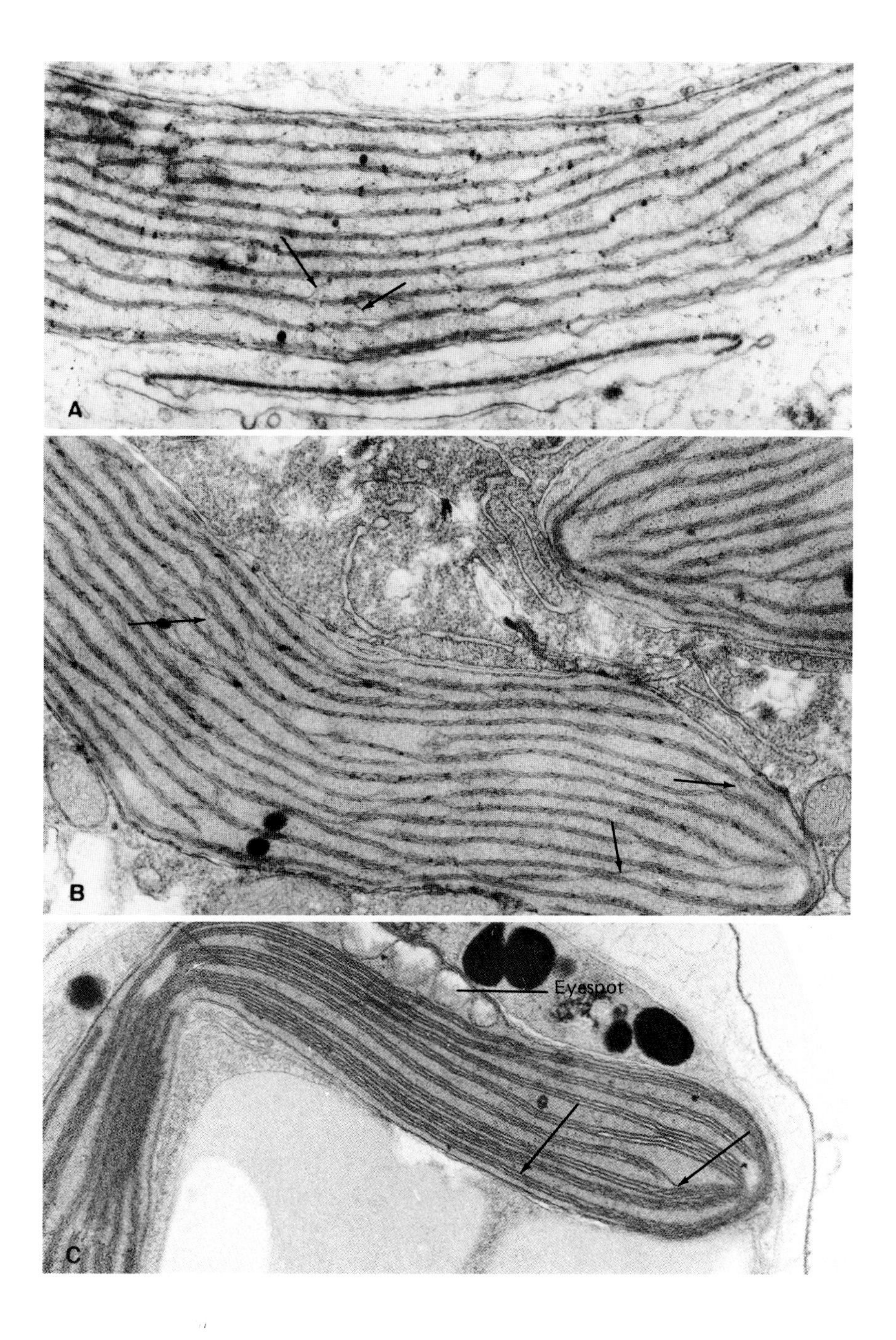
A
B
C
Eyespot

the organelle, so that none of the others can reach fully from one end to the other. Similar arrangements are sometimes present in the Chrysophyceae and usually occur in diatoms.

Chloroplasts of Red Seaweeds. The Rhodophyceae, or red seaweeds, probably possess the most distinctive of all chloroplasts. Each lamella, consisting of only a single thylakoid and extending from end to end of the chloroplast, bears on its two surfaces a number of large, evenly spaced particles that give them a serrate appearance. These features have been observed in *Porphyridium* (Gantt and Conti, 1965, 1966a,b; Nichols *et al.*, 1966; Dodge, 1973), *Laurencia* (Bisalputra and Bisalputra, 1967), and *Nitophyllum* (Honsell *et al.*, 1978), in each of which there is also a girdling lamella around the ends. While the thylakoids remain free of one another in longitudinal sections, in cross-sectional views they often can be seen to interconnect (Dodge, 1973). Thus the chloroplast provides some evidence supporting the relationship to the chlorophyll-*c*-bearing plants proposed on a different basis earlier (Dillon, 1962). The particles on the membrane surfaces appear at least to some degree to consist of phycobiliprotein, a substance believed to participate in harvesting the light for photosynthesis (Gantt and Conti, 1966a,b). These vary in size considerably, depending on growth conditions.

However, these traits characterize only advanced members of the taxon, for in the purportedly simplest in the order Laminariales, they are noticeably lacking. For example, the sporangium of *Chorda* contains chloroplasts having sinuous, trithylakoid membranes like those of the chrysophyceans and allies that interconnect quite frequently (Figure 10.18C; Toth, 1974, 1976). Moreover, the phycobilin particles are absent. Hence, there appears to be ample evidence to support the view that the single-thylakoid lamellae and phycobilin particles have been acquired secondarily by the higher members and are not suggestive of a primitive level of evolutionary advancement for the division as a whole.

Photosynthetic Membranes of Blue-Green Algae. Several features of the photosynthetic apparatus of blue-green algae seem to indicate kinship to the chloroplasts of red algae. In the first place, the lamellae consist of single thylakoids, and second, they bear regular arrays of phycobilin particles, as in *Nostoc* (Lang, 1968b; Gray *et al.*, 1973), *Fremyella,* and a genus of saltwater forms (Gantt and Conti, 1969, 1975). There, however, all resemblances cease. As a rule, in forms where the structure is clear, the thylakoids are short, perhaps half or less the length of the cell, as in the genera mentioned above and *Synechocystis* (Figure 10.19A; Schulz-Baldes and Lewin, 1976), *Calothrix* and *Porphyrosiphon* (Ris and Singh, 1961), and *Cylindrospermum* (Miller and Lang, 1968). In all these genera, the membranes are highly irregular in form, often being curled and forming tangled masses. Moreover, they frequently may be observed to emanate in groups from a common site on the plasmalemma, but the short length is the principal characteristic.

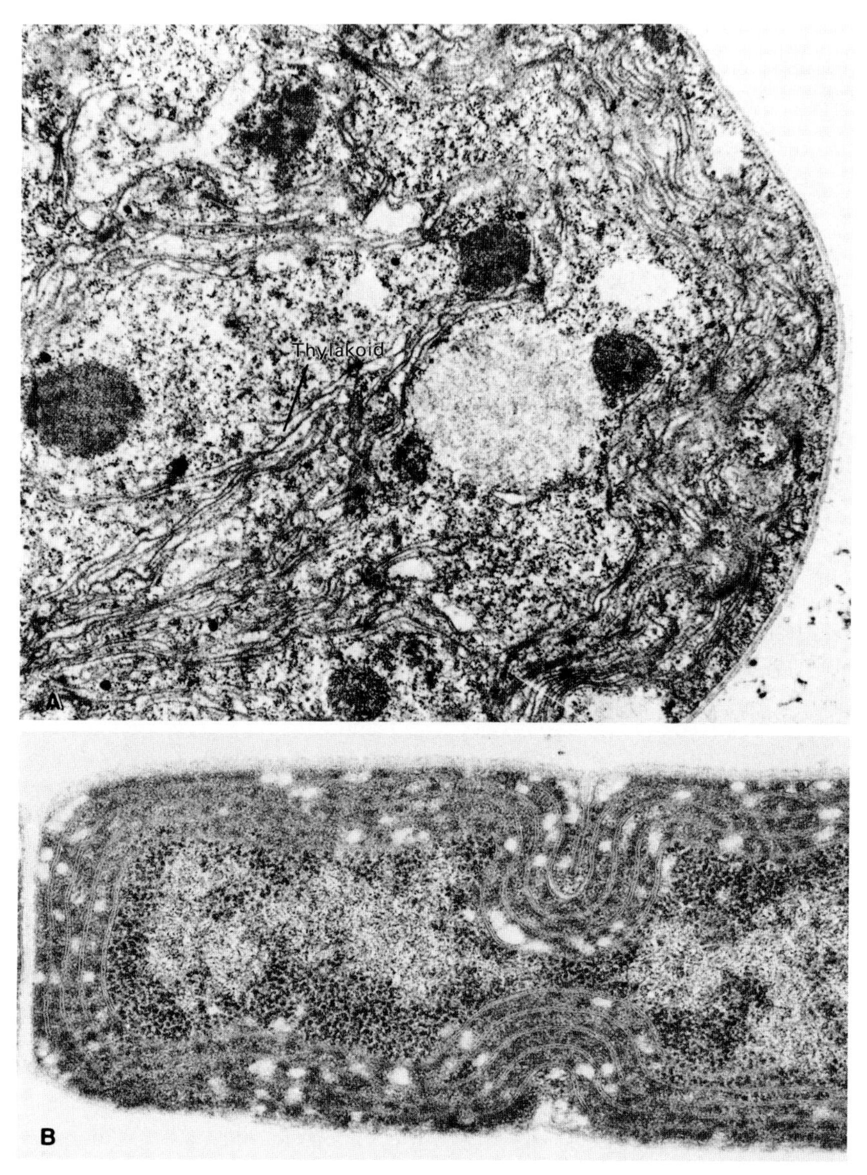

Figure 10.19. Photosynthetic lamellae of blue-green algae. (A) In such members of this group of algae as *Synechocystis,* the thylakoids are irregular in disposition and much shorter than the cell length. 20,300×. (Courtesy of Schulz-Baldes and Lewin, 1976.) (B) In many others, as in this specimen of *Calothrix paritina,* the thylakoids are long and parallel, frequently being continuous around the ends of the cell. 45,000×. (Courtesy of Mercedes R. Edwards, unpublished.) See also Figures 11.1 and 11.3.

A second group of genera, however, are obviously different, in that the membranes are arranged in parallel aggregates in which all the constituents are fully as long as the cell (Figures 10.19B, 11.1, and 11.3). This condition occurs in *Anabaena* (Lang and Rae, 1967; Lang, 1968b), *Agmenellum* (Brown and Van Baalen, 1970), and *Oscillatoria* (Jost, 1965). Because this morphological character is so contrasting from the others, it probably represents a fundamental difference between the two groups. Thus, as indicated also by the presence of the mesosome in some of these algae and its absence in others, two levels of evolution exist in these organisms; in other words, at least two divisions of blue-green algae need to be recognized.

Even in those forms that have long, parallel lamellae, the chloroplast shows few additional likenesses to the red seaweeds. In the present organisms, the lamellae are confined to the sides of the cell, leaving the central third of the sections free. In that region are found the polyhedral bodies, nucleoid, and ribosomes, along with polyphosphate granules and lipid droplets in some genera, but few, if any, chlorophyll-bearing lamellae. Furthermore, as pointed out in the preceding chapter, the lamellae in the present organisms are dual-purpose structures and engage equally in cell respiration as in photosynthesis, as shown by tellurite and TNBT reduction products (Bisalputra *et al.,* 1969).

Photosynthetic Structures of Bacteria. As in the blue-green algae, the photosynthetic pigments of bacteria may be associated with membranous thylakoids. Two outstanding examples with this type of structure were found in salt flats. One of them, identified only as strain SL-1, contained two stacks of unequal-sized thylakoids, each near a pole of the organism (Raymond and Sistrom, 1967) and often arranged at right angles to one another. Each stack contained 7 to 12 thylakoids. Whether or not the major chlorophyllous pigment was bacteriochlorophyll could not be established, but the chief carotenoid present was identified as spirilloxanthin. The second representative, belonging to the genus *Ectothiorhodospira,* whose members deposit sulfur outside the cell, was more amply supplied with membranes. While only two or, rarely, three stacks were present per cell, the thylakoids approached, or even exceeded, half the length of the organism in extent (Trüper, 1968). In number of components, the stacks varied widely, ranging from a minimum of 7 to 23 or more. The thylakoids were generally rather uniform in size; moreover, occasional ones often continued along the periphery and then doubled back through the stack at another level, thus nearly enclosing the whole stack in membranes. Bacteriochlorophyll *a* was found to be present, along with spirilloxanthin as in the other species, and capitate particles could likewise be noted on the surfaces of the thylakoids. Somewhat similar bundles have also been reported for *Rhodomicrobium vannielii* (Vatter and Wolfe, 1958; Boatman and Douglas, 1961).

A more unusual type of photosynthetic apparatus has been described in *Thiocapsa pfennigii* (Cohen-Bazire, 1971). In this organism, the cell was largely filled with elongate tubular membranes that usually ran longitudinally through the cell. The tubules were gently sinuous, with rather large lumina, so that about 22 of them arranged side by side attained the full width of the organism. Occasional branching could be noted in the electron micrographs, and capitate particles were reported to occur on the surface of the tubules. In this case, bacteriochlorophyll *b* was the chief photosynthetic pigment.

Instead of membranes or tubules, a number of photosynthetic bacteria contain vesicles. As a rule these are small, ovate bodies that typically fill much of the interior of the cell; unfortunately they have been given the name "chromatophores" already applied to several different structures of eukaryotes. In size they vary from 70 to 300 nm, depending on the age of the cell as well as the species (Schachman *et al.*, 1952; Löw and Afzelius, 1964; Gibson, 1965). When the membranes were fractured, capitate subunits were observed on the surfaces and compared with the mitochondrial membrane vesicles from beef heart (Löw and Afzelius, 1964). If vesicles are removed from the cell, centrifuged, and suitably sectioned for examination under the electron microscope, sometimes they are found not to be enclosed by membranes but to be comprised entirely of a mass of parallel fibrils (Figure 10.20A,B; Cohen-Bazire *et al.*, 1964; Cohen-Bazire, 1971). Many of the vesicle-bearing types that have been investigated are members of the genera *Chlorobium, Rhodopseudomonas,* and *Rhodospirillum,* but a number of others are known, including *Chloroflexus* (Madigan and Brock, 1977). The particles in the latter genus contained bacteriochlorophylls *a* and *c*.

10.2.3. Modifications of the Chloroplast

The specialized organelle within the chloroplast, the pyrenoid, which consists largely of proteins, is found in algae of all types except the blue-green, but does not occur in the higher plants. Because its supposed function, based on its close association with such substances as starch and paramylon granules, is that of the dark reactions and conversion of the glucose into the storage products, its absence from the seed plants is difficult to comprehend. Several reviews of this structure have been published that summarize the morphological features (Leyon, 1954; Gibbs, 1962a,b; Manton, 1966; Griffiths, 1970; Dodge, 1973). In general, the gist of the synopses is that few traits have developed on an overall phylogenetic basis between groups and that parallel evolution has occurred within the respective major taxa from similar ancestral types. Consequently, in the present discussion the evolutionary changes that have taken place within a single group of organisms are outlined first to serve as a model for the others, followed by notes on modifications that are characteristic of other taxa.

The Pyrenoid of the Dinophyceae. As a whole, this summary of the dinophycean organelle follows the excellent study of Dodge and Crawford

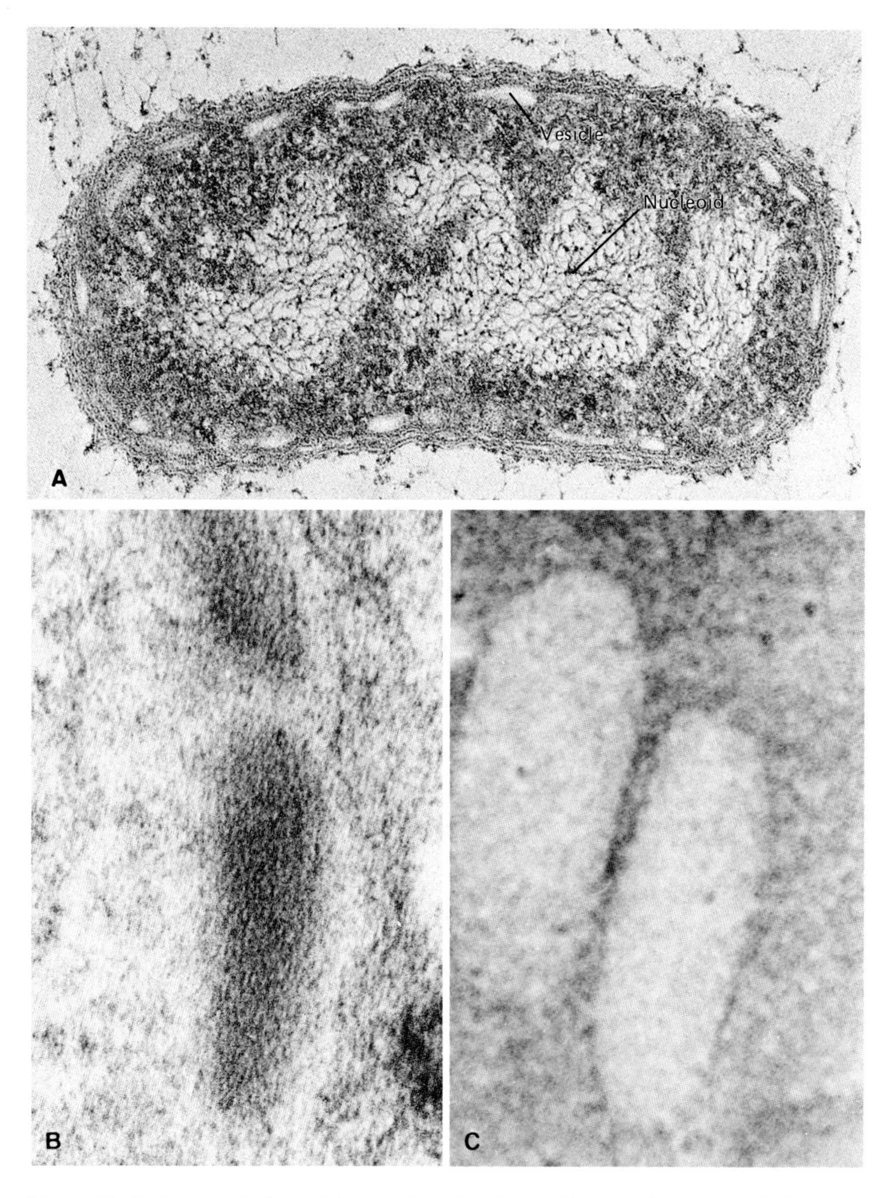

Figure 10.20. Photosynthetic vesicles of bacteria. (A) Such green bacteria as *Chlorobium thiosulfatophilum* have chlorophyllaceous vesicles arranged beneath the plasmalemma. 94,000×. (B) When greatly enlarged, the vesicles are found to be entirely fibrous and not covered by a membrane. 400,000×. (A and B courtesy of Cohen-Bazire *et al.*, 1964.) (C) Other forms, including this example of *Chloroflexus,* are packed throughout the cell with vesicles that are particulate rather than fibrous. 365,000×. (Courtesy of Madigan and Brock, 1977.)

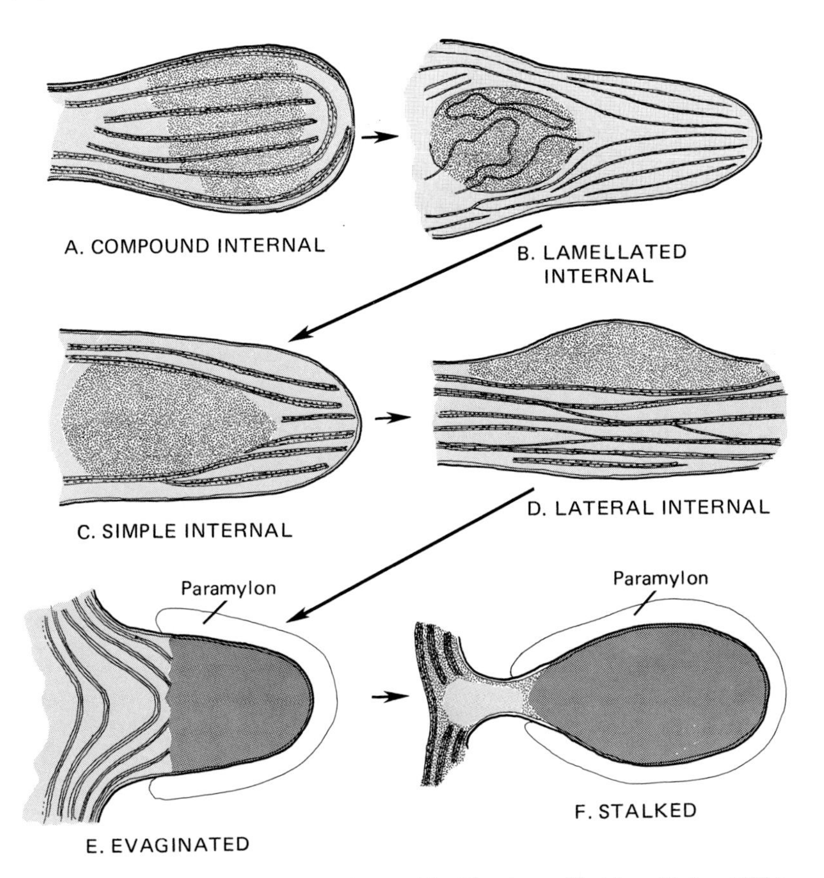

Figure 10.21. Types and evolution of pyrenoids. (Greatly modified from Dodge, 1973.)

(1971), except that their second stage is treated as being the earliest and one or two additional stages are inserted. The simplest pyrenoid, in the Dinophyceae as well as elsewhere, is probably that called the compound internal pyrenoid (Figures 10.17A, 10.21A). In this type, the organelle is represented merely by specialized areas of the chloroplast stroma that lie adjacent but are separated by the lamellae, as in *Prorocentrum*. Thus there are actually a number of pyrenoids per chloroplast at this stage A of phylogenetic development, rather than a single one.

What probably represents the next stage (B) is not known to occur among the dinoflagellates, so an example is drawn from the red seaweed, *Porphyridium* (Dodge, 1973). This variety, here named the lamellated internal type (Figures 10.21B, 10.22), is viewed as having been derived from the preceding

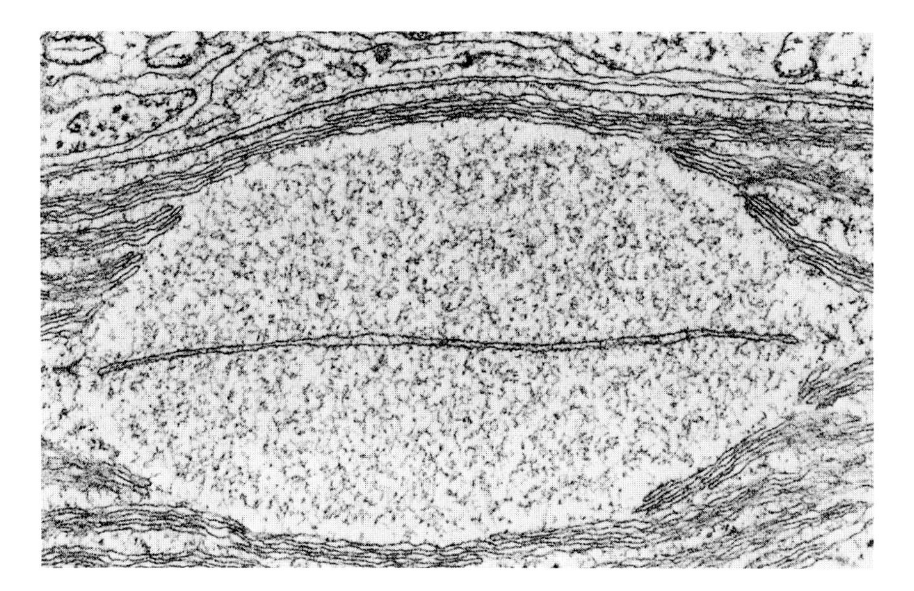

Figure 10.22. A lamellated internal pyrenoid. In this unidentified symbiont of *Amphistegina*, the pyrenoid contains so few thylakoids that it approaches the type C, the simple internal type. 32,000×. (Courtesy of Berthold, 1978.)

by fusion of the several separated structures, as suggested by the presence of a number of irregular lamellae within this organelle. Hence, at this stage the pyrenoid has nearly become a unified body, rather than a series of adjacent ones. Indeed, many supposed examples of the next higher level of advancement actually are at this intermediate phase. Although that third (C) stage, the simple internal type, is defined as including those pyrenoids that are devoid of internal lamellae and are confined to the space between two chloroplast lamellae (Figure 10.21C), it is clearly exemplified among the dinoflagellates only by the organelle of *Gymnodinium* (Dodge and Crawford, 1971), for the others cited in that reference (the diatom *Melosira* and the dinoflagellate *Glenodinium*) really represent stage B, because the lamellae more or less divide the stroma into sections.

At times representative organelles at either stage B or C of evolution may undergo a modification into what is termed the multiple-stalk pyrenoid (Dodge and Crawford, 1971). In this variety, the body, still centered within the chloroplast, becomes somewhat separated from the latter by means of vacuoles, stored particles, or a combination of both (Figure 10.23). A few typical lamellae from the plastid ramify this structure, clearly suggesting it to be still a part of that organelle, as before, but in this case running through the connections or stalks that unify the two types of bodies. *Amphidinium* of the dinoflagellates has been shown to possess this kind of pyrenoid.

The next major level (D) is here named the lateral internal type and apparently has not been noted among the dinophyceans, so an example is drawn from the chrysophyceans (Figure 10.21D). In this variety, instead of being surrounded by chloroplast lamellae on all sides, the pyrenoid, which when fully developed has no lamellae penetrating it, has become located at one side of that body, as in *Uroglena* (Wujek, 1976). In the diatoms endosymbiontic in *Amphistegina,* the pyrenoid is partially moved to the side but still is penetrated by a lamella (Berthold, 1978) and thus represents a transitional phase.

The last two evolutionary stages to be described are really continuations of the lateral movement, which may be interpreted as signifying increasing degrees of functional independence. In the first of this pair, the evaginated stage (E), the pyrenoid projects outwards from the chloroplast as a thick process (Figure 10.21E), and does not appear to be represented among the dinophy-

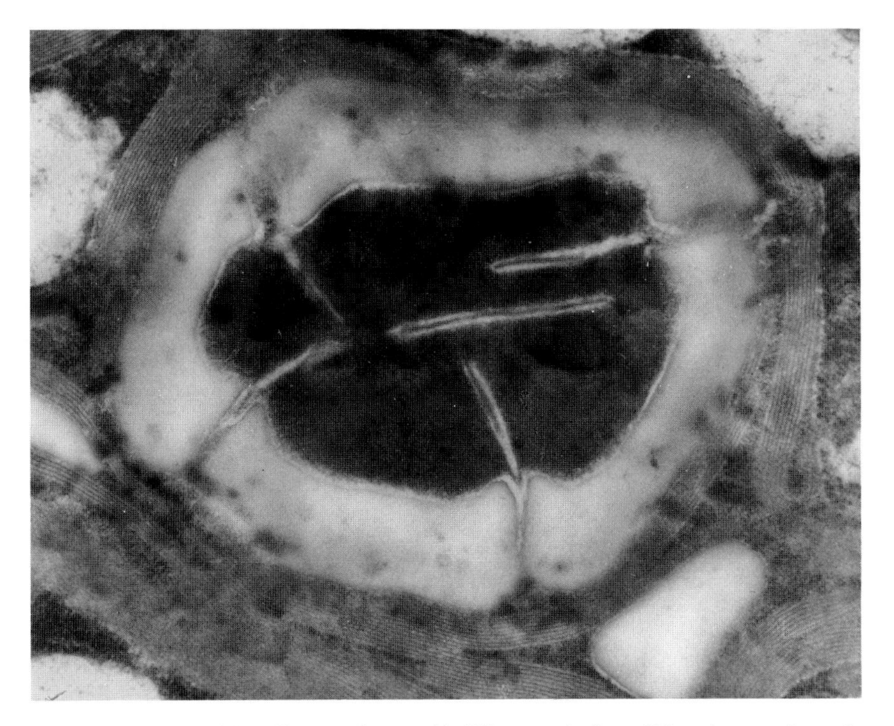

Figure 10.23. A multiple-stalk type of pyrenoid. This example from *Chlamydomonas* is nearly separated from the chloroplast by starch deposits, except that four stalks still connect it, through which the thylakoids extend. 32,000×.

ceans. However, members of two euglenoid genera, *Trachelomonas* (Dodge, 1973) and *Colacium* (Leedale, 1967), have pyrenoids at an advanced stage of this level of development, in both examples of which the projection is covered by a cap of paramylon. At the very most advanced stage (F), the broad region connecting the projecting type of pyrenoid to the chloroplast has become constricted into the narrow pedicel that provides the basis for its name, the stalked type (Figure 10.21F). Whereas in stage E, photosynthetic lamellae still contact the pyrenoid, no such structures penetrate through the stalk of the present type, the matrix of which is more electron transparent than that of the body proper, as in *Glenodinium* (Dodge and Crawford, 1971), and also in *Polyedriella* of the Eustigmatophyceae (Figure 10.24; Hibberd and Leedale, 1972). Thus the separation of this organelle from its parent body now appears to be complete.

One additional step is found among the Dinophyceae that may be confined

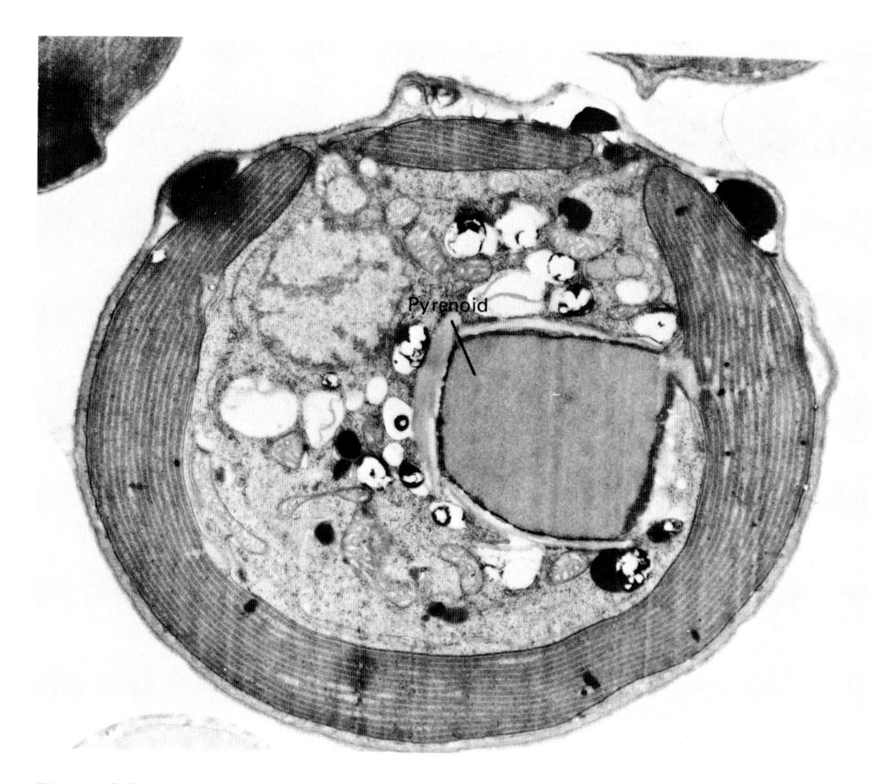

Figure 10.24. A stalked type of pyrenoid. In well-developed stalk-type pyrenoids, no lamellae penetrate the base as in this example from *Polyedriella*, a green alga. 10,000×. (Courtesy of Hibberd and Leedale, 1972.)

to that taxon—as shown by *Heterocapsa*, the surface of the stalked pyrenoids sometimes bears deep invaginations, which permit the cytosol to extend far into its interior (Dodge, 1973).

Variations in Other Taxa. Few variations actually exist that have not already received mention in the foregoing outline. The primitive nature of the euglenoids is indicated by two conditions that prevail: first, in that taxon alone do all the chloroplasts have pyrenoids; in contrast, in other algal groups only one such body is found per cell. Second, the pyrenoids of euglenoids are largely of the elementary compound internal type, although, as already discussed, *Trachelomonas* has the fairly advanced evaginated type. Red algae, too, are primitive in this regard, for the pyrenoid is at the second level of evolution, the lamellated internal type as shown in *Porphyridium* (Gantt and Conti, 1966a) and *Smithora* (McBride and Cole, 1972). In fact, this stage B has been reported to occur in seven different algal groups, including chrysophyceans, xanthophyceans, chlorophyceans, and diatoms, in addition to those already mentioned (Dodge, 1973; Chelune and Wujek, 1974). Most chlorophyceans have the pyrenoid confined to the leucoplast, the usually colorless, starch-containing modification of the organelle found only in that taxon. Perhaps its confinement to this modified chloroplast led eventually to its subsequent loss in the seed plants.

The lateral pyrenoid of the chrysophycean genera *Uroglena* and *Uroglenopsis* typically lies adjacent to the nucleus, to which cell part they frequently show close relationships (Wujek, 1976). At times these associations are displayed by a central projection of the pyrenoid extending deeply into the nucleus, but at other times, the nucleus sends extensions into the pyrenoid. However, the invaginations may exist without receiving nuclear processes. A similar extension of the nucleus penetrates the pyrenoid of the chlorophyceans of the genus *Prasinocladus;* in this case, the pyrenoid, while likewise of the lateral type, is enclosed within the strongly arcuate chloroplast (Parke and Manton, 1965).

A Second Major Type of Chloroplast Modification. A second major structural modification of the chloroplast had its origins in a cytoplasmic structure of lower eukaryotes but later became associated with the chloroplast in higher protistans. Primitively, as in the euglenoids, the cell part under discussion, the eyespot or stigma, lies within the cytoplasm quite remote from any chloroplast, being located near the base of the flagellum, which has a swollen extension (Dodge, 1969). Here as in other groups, the structure consists largely of osmiophilic globules that are brightly pigmented. As a number of dinoflagellates and xanthophyceans have a similarly constructed eyespot, central evolution appears to be lacking, as was seen to be the case with the pyrenoid.

Among many dinoflagellates, including certain species of *Wolozynskia* (Crawford *et al.*, 1971) and *Peridinium* (Messer and Ben-Shaul, 1969), the eyespot has become enclosed within one of the numerous chloroplasts, this modified one being located at one side of the organism just beneath the cell envelope (Figure 10.18). Except for the xanthophyceans mentioned above, in all other algal types, the stigma is situated within the chloroplast, in which organelle, however, it is variously positioned. In *Chlamydomonas,* for instance, it is arranged on the side closest to the cell surface, rather than near the flagellar base, and in the cryptophycean *Chroomonas,* it occurs near the end of a short spur of the chloroplast that projects beyond the base of the stalked pyrenoid (Dodge, 1973).

Although the function of this organelle still remains in doubt, a number of theories concerning its activities have been advanced (for a review, see Halldal, 1964). At the level of its first known appearance, the stage represented by the euglenoids, the eyespot has been demonstrated to serve as a shade for the flagellar base, as discussed in Chapter 5, Section 5.3. Hence, the latter was the real photoreceptor, a point emphasized by the occurrence of flagella and basal bodies in the photoreceptors of metazoans in general. As no central evolution seems to have occurred along the main line of ascent, there is no conflict between these isolated pieces of evidence and other functions that may have developed among members of intermediate branches. In *Chlamydomonas* the location of the stigma remote from the flagellar base but adjacent to the plasmalemma, where light receipt would be maximal, strongly intimates a light-receiving rather than a shading function in this green alga. In this site, the stigma might serve in detecting the intensity of the light and signal the cell into moving the chloroplasts appropriately, as described in the following paragraph.

Chloroplast Movement. Except in higher plants where chloroplasts may be circulated throughout the cell by cytoplasmic streaming (Chapter 3), this organelle displays none of the movements characteristic of the mitochondrion. Consequently, the chloroplast behavior has not been the subject of many investigations. Cells of a number of mosses, ferns, and seed plants, as well as green algae, have been found to orient the chloroplast or chloroplasts in response to the intensity of light. If a single plastid is present per cell, as in *Selaginella,* under low intensities of illumination that organelle becomes centered opposite the exposed surface, where it receives the greatest quantity of light (Figure 10.25; Senn, 1908; Haupt, 1963, 1973). When the intensity is greatly increased, the chloroplast is moved to one side, where as little exposure to the rays as possible is obtained. Comparably in *Vaucheria* and other multichloroplast genera, movement is toward the center of the field under dim light and onto the sides in bright light (Figure 10.25; Zurzycki, 1955). While the complete picture has not been established, indications presently are that the movement is provided by cytoplasmic flow and that the organelle itself contributes little if anything to its reorientation (Fischer-Arnold, 1963; Schönbohm, 1973). While the light-receiving mechanism is still under investigation, present results implicate the cytoplasm rather than the chloroplast as containing the receptor that initiates the response (Mayer, 1971).

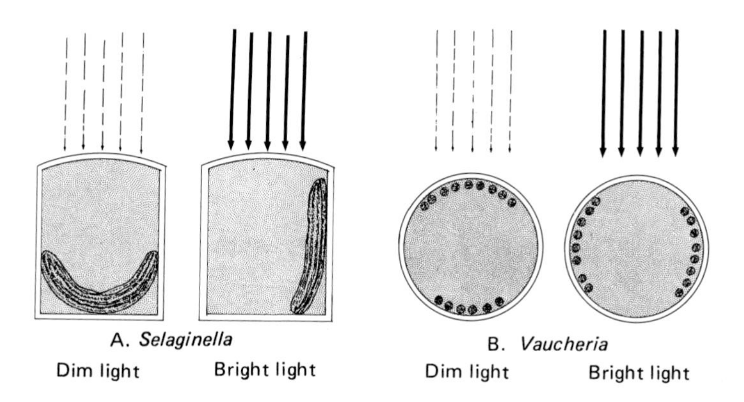

Figure 10.25. Movement of chloroplasts in response to light. Under dim light the chloroplasts are centered where the maximum light is received, whereas under bright illumination, they are moved to the sides. (Based on Haupt, 1973.)

10.3. REPLICATION OF CHLOROPLASTS

In general, the account of chloroplast replication parallels much of the discussion of reproduction of the mitochondrion given in the preceding chapter. However, while numerous similarities exist, the particulars are quite distinctive in each case.

10.3.1. The Chloroplast Genetic Mechanism

As was the case in the mitochondrion, biochemists were reluctant to accept the presence of DNA outside of the nucleus, so its existence in the chloroplast similarly was explained away as an artifact whenever detected. But much progress has been made since it was positively identified as being present in the organelle.

The Chloroplast Nucleoid. In metaphytans, considerable knowledge has been gained toward understanding the nature of the chloroplast nucleoid at both the ultrastructural and the molecular levels. As a rule the nucleoids are multiple and more extensive than in the mitochondria and typically are arranged as a series of patches mostly near the circumference of the organelle (Figure 10.26; Herrmann and Kowallik, 1970; Gibbs *et al.,* 1974a,b) or in close association with granal membranes (Rose, 1979). Apparently they represent multiple copies of a double-stranded circular molecule having a molecular weight in the range of 85 to 97 $\times$ 10^6 (Tewari and Wildman, 1966, 1970; Tewari *et al.,* 1977), and contour lengths between 45.2 nm *(Oenothera)* and 45.9 nm *(Beta).*

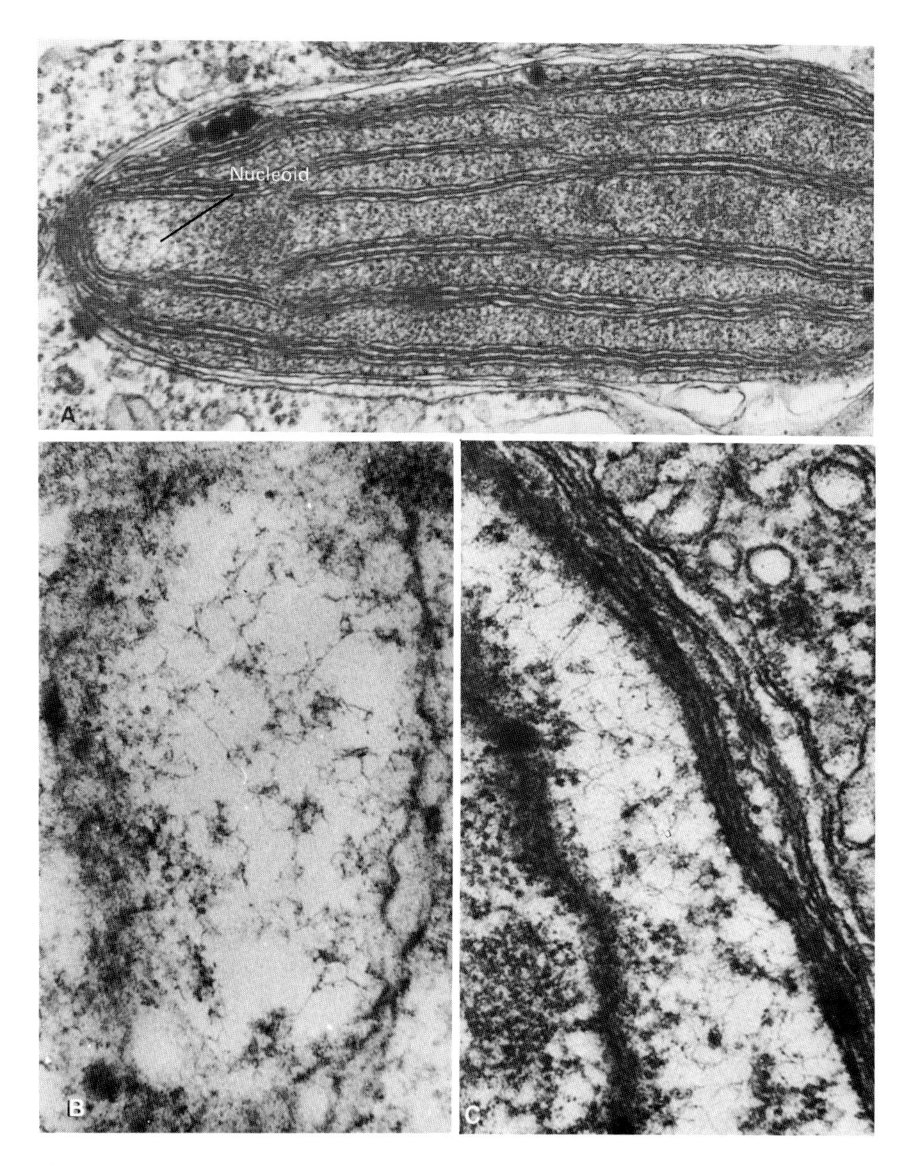

Figure 10.26. The nucleoid of the chloroplast of *Ochromonas*. (A) The nucleoid of the chloroplast in this genus of chrysophyceans is in the form of a continuous ring near the periphery. 47,400×. (Courtesy of Gibbs *et al.*, 1974a.) (B and C) Increase in DNA content during greening of the chloroplast. (B) The chloroplast nucleoid in this dark-grown cell has a relatively scant DNA content. (C) After 102 hr in the light the DNA fibrils have greatly increased in concentration. Both 78,400×. (B and C, courtesy of Gibbs *et al.*, 1974b.)

About 15% of the total population in spinach chloroplast CNA was in the form of circular dimers (Herrmann *et al.,* 1975), whereas only 3% of the pea chloroplast molecule was in that state (Tewari *et al.,* 1977). In monocots like corn as well as several dicots, the DNA molecule was found to include a long, inverted region of repeated sequences. The total amount of chloroplast DNA present in a cell varied with the age and type of tissue; in fully greened leaves of peas, it approached 12% of the total present or about 9500 copies per cell (Lamppa and Bendich, 1979).

This inverted region has been demonstrated to contain two copies of each of the rRNA genes, in the sequence 16 S–23 S–5 S (Table 9.1) in both maize and spinach (Bedbrook and Bogorad, 1976; Whitfeld *et al.,* 1976, 1978; Bedbrook *et al.,* 1977; Hartley and Head, 1979). In addition, the spacer between the 16 S and 23 S genes of spinach chloroplast DNA has been found to contain a tRNA gene (Bohnert *et al.,* 1979). While the RNAs of the chloroplast ribosomes may be produced endogenously, they may be synthesized in the cytoplasm, too, for the rRNAs also hybridize with nuclear DNA in some plants (Woodcock and Bogorad, 1971; Smillie *et al.,* 1973; Bogorad, 1975). The origins of a recently discovered rRNA, having only 80 to 103 nucleotides and sedimenting between 4 S and 4.5 S, have not been established as yet (Hartley, 1979). At least some of the ribosomal proteins are encoded by nuclear genes (Bourque and Wildman, 1973), thus paralleling the condition described for the mitochondrion. A higher number of these proteins are present than in the cytoplasmic counterparts, in the case of wheat totaling between 70 and 75 (Jones *et al.,* 1972; Chua and Luck, 1974). In peas the ribosomes have been shown to be closely associated with the thylakoid membranes in the light, but become detached in the dark or in the absence of O_2 (Alscher-Herman *et al.,* 1979).

As to the genes for the tRNAs of this organelle, little specific information appears to be available, but it has been shown by hybridization experiments that the chloroplast DNA possibly contains 30 to 40 tRNA genes (Tewari *et al.,* 1977); consequently, it is possible that the genome encodes for at least one species of tRNA for each of the 20 amino acids. However, cytoplasmic tRNAs also are present in chloroplasts (Weil *et al.,* 1977). According to the limited amount of information currently available, the tRNA ligases of the chloroplast seem to be specific for chloroplast DNA-encoded tRNAs, as they did not charge those from the cytosol. Isolated intact chloroplasts from pea were shown by SDS gel electrophoresis to synthesize at least six major proteins, one of which was a thylakoid constituent known as the large subunit of fraction I protein (Blair and Ellis, 1973; Sakano *et al.,* 1974; Ellis, 1977). Even when isolated, intact chloroplasts from pea leaves could incorporate [^{32}P]orthophosphate into several polypeptides that have molecular weights ranging from 7000 to 70,000 (Bennett, 1977) and could synthesize cytochrome *f* (Doherty and Gray, 1979). Among the most abundant of these phosphoproteins are three constituents of the thylakoid membrane, one having a molecular weight of 9000 and two of 26,000. Together the latter form a doublet associated with the chlorophyll *a* + *b*-binding protein complex (Bennett, 1979b). With this complex is also included a protein kinase that carries out the phosphorylation. Both the phosphorylation process and the synthesis of the kinase have been found to be light dependent (Bennett, 1979a).

Nevertheless, large numbers of proteins essential to chloroplast functioning are encoded by nuclear genes. Among those that have been thoroughly established are several involved in porphyrin synthesis particularly for steps between protoporphyrin IX and Mg–protoporphyrin IX (Von Wettstein *et al.*, 1971), and another between the latter compound and protochlorophyllide *a*. Other nuclear genes control the synthesis of chloroplast-specific lipids, and still others govern the insertion of chlorophyll into the thylakoids. Moreover, some of the proteins synthesized within chloroplasts were not processed there (Grebanier *et al.*, 1978), and in general, the biosynthesis of chloroplasts appeared to be tightly controlled by the cellular genetic apparatus. However, chloroplast mRNA has been found capable of being translated by an *in vitro* system from *E. coli* but not by one from wheat germ, whereas the reverse situation existed for cytoplasmic mRNA from the same organism (Bottomley *et al.*, 1976).

The Chloroplast Genome of Euglena. Among the more thoroughly explored algal chloroplast genomes is that of *Euglena,* but even in that flagellate, the details are still sketchy. Here as in all forms investigated, including *Chlamydomonas,* the DNA molecule is a double-stranded, closed circle, with a circumference of 44.5 or 62 nm (Vandrey and Stutz, 1973; Rochaix, 1978), 60–80 copies being present per organelle (Allet and Rochaix, 1979). In *Euglena,* however, the density was somewhat lower, suggesting a G + C content of 25%, whereas that of the denser *Chlamydomonas, Chlorella,* and seed plant molecules was around 38% (Kirk, 1971). All three were smaller than that of *Acetabularia,* which had a molecular weight of 1.52×10^9, compared to 0.26×10^9 in *Chlamydomonas,* but in this case only three genomes were present per organelle (Padmanabhan and Green, 1978).

The genes controlling the ribosomal constituents have been especially well explored. The ribosomes of this organelle sediment at 68 S, compared to 87 S for those of the cytoplasm (Rawson and Stutz, 1969; Heizmann, 1970), whereas the subunits sediment at 32 S and 49 S (Freyssinet, 1977), contrasting to 41 S and 62 S for those of the cytoplasm. As in the higher plants, chloroplast rRNA hybridizes with the chloroplast DNA (Scott, 1973; Gruol *et al.*, 1975), as well as with that of the nucleus. Furthermore, the genes are arranged in the same fashion as in the metaphytans, in the sequence 16 S–22 S–5 S, in three tandemly repeated sequences of 5600 base pairs each (Gray and Hallick, 1978, 1979; Jenni and Stutz, 1978, 1979). In addition, an apparently unique situation occurs, in the form of a supplementary 16 S rRNA cistron (Jenni and Stutz, 1979). As elsewhere, nuclear genes encode a number of the ribosomal proteins (Freyssinet *et al.*, 1976), of which the small subunit bears between 22 and 24 and the large, 30 to 34; on the other hand, the cytoplasmic ribosomal subunits have 33 to 36 and 37 to 43 in the same order (Freyssinet, 1977). Synthesis of at least 12 of the chloroplastic ribosomal proteins was reported to be encoded by nuclear genes and translated on cytoplasmic ribosomes (Freyssinet, 1978). The tRNA genes (at least 26 cistrons) are located in the chloroplast genome, for no hybridization of chloroplast tRNAs was noted with the nuclear genome (Schwartzbach *et al.,* 1976). In contrast, the tRNA ligases are nucleus encoded and synthesized on cytoplasmic ribosomes (Parthier, 1973; Parthier and Krauspe, 1973).

Few particulars concerning transcription or translation appear to be available, aside from the recent report that the chloroplast protein-elongating factor EF-G is encoded by a nuclear gene and synthesized on cytoplasmic ribosomes (Breitenberger *et al.,* 1979). The chloroplast mRNAs that bore poly(A) sequences have been found to undergo changes in nature during light-induced development of the organelle (Chelm and Hallick, 1976; Rawson and Boerma, 1976; Verdier, 1979a,b). After 1 hr of illumination, the high complexity of messenger sequences present suggested that nearly all had been transcribed from nuclear genes. After 24 hr exposure, 73% of the total messenger population were similarly highly diversified, intimating a like origin, while the remaining 27% probably were largely products of chloroplast genes.

The Chloroplast Genome of Chlamydomonas. In *Chlamydomonas* nuclear genes play as important a role in the biosynthesis of plastids as do those reported earlier in barley and contain two sets of rRNA genes within two inverted repeated sections (Rochaix, 1978). A list of proteins affected and unaffected by chloroplast-specific antibiotics demonstrated that five major components were under the control of the chloroplast genome. These included cytochrome 559, factor Q, chloroplast lamellar organization, and pyrenoid formation (Surzycki *et al.,* 1970). In contrast, about 20 substances or processes were unaffected by the antibiotics, clearly indicating nuclear control. Among these were included cytochromes 553 and 564, P700, ferredoxin, chlorophyll (to a large degree), carotenoids (about 50%), eyespot formation, starch synthesis, plastocyanin, membrane formation (about 50%), and around 8 to 10 enzymes important in photosynthesis. Although at least seven chloroplast ribosomal proteins were shown to be encoded by chloroplast genes (Chua and Luck, 1974), the vast majority were coded by nuclear genes and synthesized in cytoplasmic ribosomes (Honeycutt and Margulies, 1973). As phospholipids and plastoquinones A, B, and C were found to be biosynthesized in isolated chloroplasts from *Acetabularia* (Moore and Shepherd, 1977; Moore and Tschismadia, 1977; Tschismadia and Moore, 1979), it is not improbable that those substances, too, are produced in these organelles of *Chlamydomonas.*

In the chloroplast genome for the large rRNA of *Chlamydomonas* (Rochaix and Malnoe, 1978; Allet and Rochaix, 1979) and that of certain strains of *Tetrahymena* (Wild and Gall, 1979), intervening sequences are present that are not found in the mature product. Such intervening regions seem to be characteristic only of eukaryotic and viral genes—at least they have not been found among prokaryotes to date (Crick *et al.*, 1979). Such spacers have been found among cytosolic as well as mitochondrial and chloroplastic genes. Hence, processing the transcript must include breakage of the molecule, followed by splicing, to eliminate the inserted portion.

10.3.2. Replication of Chloroplasts

Because almost no investigations have been conducted into the genetic processes of biosynthesis of chloroplast proteins and RNAs, the macromolecular aspects of replication must remain for the future. Hence, only the cellular and ultrastructural features of the replicatory steps can be examined at this time. Two routes of biogenesis appear to be followed by this organelle, binary fission and ontogenetic development; *de novo* formation does not appear to have been reported.

Binary Fission. That chloroplasts undergo binary fission in conjunction with cell division in algae is an observation of long standing. The actual processes, however, have not been subjected to much study on a modern basis, so few details are available (Bouck, 1962; Gantt and Arnott, 1963; Gantt and Conti, 1965). In most cases, as in *Chlamydomonas,* the chloroplast divides by furrow formation into two approximately equal portions, usually just prior to cytokinesis in the cell as a whole. Thus the number of individual chloroplasts remains constant, not only in those types that possess just one or two but those also like *Euglena,* where 10 to 30 may be present (Schiff, 1970). Even when grown in the dark for as long as several years, so that the chlorophyll is lost and the structures regress by loss of lamellae, the remnants in *Euglena* undergo division in unison with the rest of the cell (Leedale, 1967). Exceptional cases, however, have been reported in which multiple divisions occur (Mitrakos, 1960; Honsell *et al.,* 1978). After division is completed, the chloroplast undergoes growth; for instance, in moss an increase from 2 to 7 μm was recorded (Kaja, 1955). At the same time, the number of lamellae per chloroplast went from 8 in young plants to 18 in older ones (Von Wettstein, 1959) and the number of grana increased from 5 up to as many as 80. If these figures are significant, it is clear that division involves regression in size of the organelle and internal structure. Otherwise mere doubling should be involved.

Ontogenetic Generation of Chloroplasts. The degenerate body of *Euglena* just mentioned is commonly referred to as the proplastid, but because it occurs in newly formed tissues of higher plants, it might also be considered a juvenile chloroplast. Actually in higher plants, all replication of plastids appears to be confined to the production of proplastids, probably by fission during the early stages of tissue differentiation. Generally speaking, proplastids appear as colorless vesicles, having diameters between 0.7 and 1.5 μm and being enclosed within two concentric membranes (Kirk and Tilney-Bassett, 1967). The processes are not always uniform, not even in the same species; for example, in the red alga *Nitophyllum punctatum,* two distinct types occur (Figure 10.27A,B; Honsell *et al.,* 1978).

The differentiation of the proplastids in the presence of light into mature chloroplasts has been followed by electron microscopic procedures. According to those studies, the processes commenced with the budding off of closed vesicles from the inner investing membrane (Von Wettstein, 1959, 1961; Epstein and Schiff, 1961; Ben-Shaul *et al.,* 1964); in turn these gave rise to small thylakoids, which became oriented into several layers. As the latter increased in dimensions, they contacted one another and fused, gradually resulting in the lamellae of the stroma and eventually in grana. However, it should also be recalled in this connection that evidence has been presented which suggests that, in a three-dimensional system, all the thylakoids and lamellae are derived from a single lamella. In *Ochromonas* the chloroplast increases fourfold in volume, and as it does so, its DNA content is multiplied by a like amount (Figure 10.26B,C; Gibbs *et al.,* 1974b).

In the dark, a somewhat different sequence of events occurs. Without light, synthesis of chlorophyll is blocked at the protochlorophyll stage, so although vesicles are budded off the inner investing membrane, they fail to become oriented into layers in the absence of that pigment. Instead, they accumulate into a mass in the stroma where they develop into a complex, orderly assembly of tubules, known as the prolamellar body (Bartels and Weier, 1967). In the seed plants in general, the chlorophyll production proceeds by way of Mg-2-vinylpheoporphyrin a_5 to chlorophyllide *a* in the light (Rebeiz and Castelfranco, 1973; Mattheis and Rebeiz, 1977a,b). Then that precursor is esterified to chlorophyll *a* by reactions that can occur in the dark. While chlorophyll *b* appears to be synthesized from chlorophyll *a,* it is not known how the several varieties of the latter become differentiated (Belanger and Rebeiz, 1979).

The sensitivity of chloroplast differentiation to the presence or absence of chlorophyll and other biochemicals has stimulated investigations into the effects of appropriate mutations on this aspect of ontogenetic development. For instance, chloroplast ontogeny was described in a mutant of tobacco that lacked photosystem II (Miller and Cushman, 1979). Analyses showed that the chlorophyll *a*/chlorophyll *b* level was almost normal, as was that of photosystem I activity. However, the chlorophyll/protein ratio was about 50% that of the wild-type plant, and CO_2 fixation was close to 15% of normal. The organelles nevertheless developed fully, with single-thylakoid lamellae of usual appearance, except that they were far more numerous than in the wild-type tobacco, and grana were almost wanting (Figure 10.28A,B). According to a report on a barley mu-

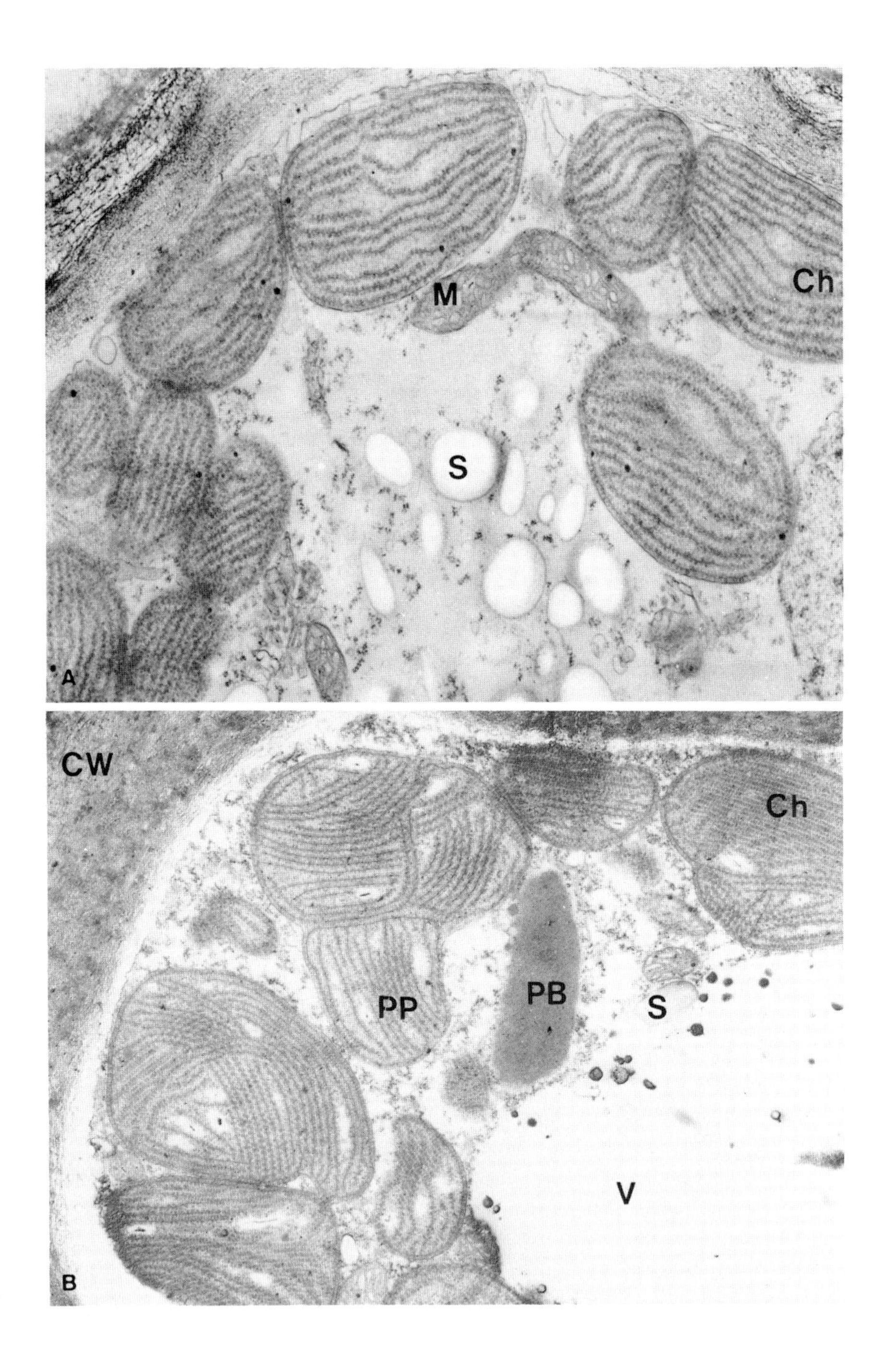

M
Ch
S
A
CW
Ch
PP
PB
S
V
B

tant that lacked chlorophyll *b,* that pigment was an absolute requirement for the appearance of the chlorophyll *a/b* protein in the thylakoid membranes (Apel and Kloppstech, 1978; Neilsen *et al.,* 1979).

Examination of the thylakoid membrane structure in the tobacco mutant and another of the same type in *Chlamydomonas* (Olive *et al.,* 1979) revealed further differences. Although only similarities could be noted between the P faces and the two types, the E face of the mutant lacked the large particles that characterize that surface of the wild type (Figure 10.29A,B), and its fine particles also seemed to be wanting. Hence, it was concluded that the large particles represent amalgams of many of the components that comprise photosystem II, and, because of the absence of one or two of the constituents in the mutant, the photosystem was unable to develop and function. This in turn likewise influenced the development of the thylakoids and absence of grana. Somewhat similar results were obtained by growing pea and bean seedlings in intermittent light, except that grana were completely absent (Argyroudi-Akoyunoglou *et al.,* 1976; Armond *et al.,* 1977).

Enzymatic Activities of Proplastids. At least in developing castor bean seeds, proplastids are not functionless embryonic chloroplasts but are essential in glycolysis and the pentose phosphate pathway, in combination with the cytosol (Simcox *et al.,* 1977). Apparently the processes begin in the latter region, where the necessary enzymes for the production of hexose phosphate from sucrose occur, as well as glucose-6-phosphate dehydrogenase that mediates the first step in the pentose phosphate pathway. The products of these activities, glucose-6-phosphate and 6-phosphogluconic acid, are transported to the proplastids where the remaining steps leading to pyruvic acid formation are found (Simcox *et al.,* 1977; Thompson *et al.,* 1977; Simcox and Dennis, 1978a,b). Comparable modified proplastids have been described from the lacticifers and latex of *Papaver,* which similarly seemed to engage in enzymatic activities (Nessler and Mahlbert, 1979a). Although identification was not firm, the enzymatic contents appeared to be peroxidase (Nessler and Mahlberg, 1979b).

Structural Varieties of Chloroplasts. In higher plant tissues too deep to be reached by light, proplastids often are converted to storage organelles possessing distinctive structural traits. Commonly the juvenile forms develop normally to the prolamellar stage before acquiring specialized activities; as all of the variant forms lack chlorophyll, the term leucoplast is applied to these bodies in general. Among the most frequent type of specialization is that of storing starch grains; deposits of these particles usually vary greatly in number and size from one of these so-called amyloplasts to another (Mühlethaler, 1971). An-

Figure 10.27. Two types of chloroplast development in a red alga. In the homogeneous type of development (A) in *Nitophyllum*, the thylakoids are approximately parallel and in a single system (17,000×), whereas in the heterogeneous type (B) the thylakoids are in irregular groups (14,700×). (Courtesy of Honsell *et al.*, 1978.)

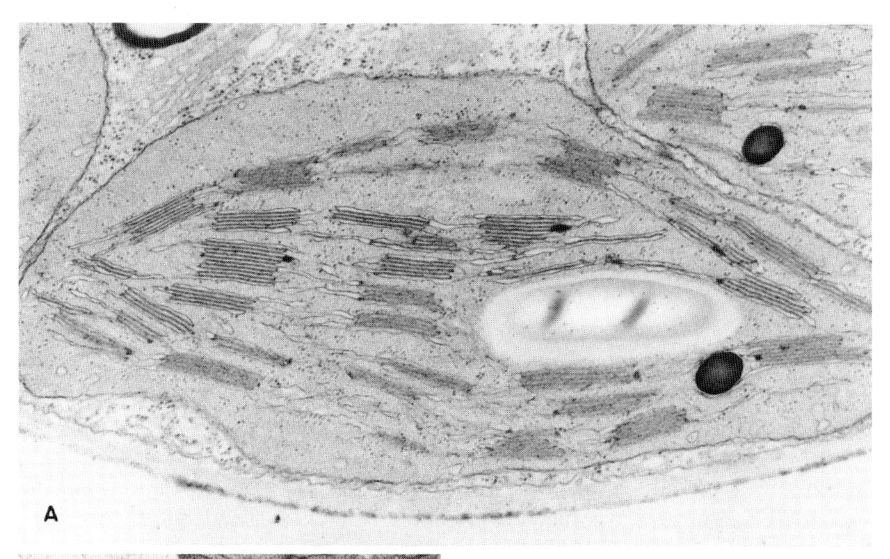

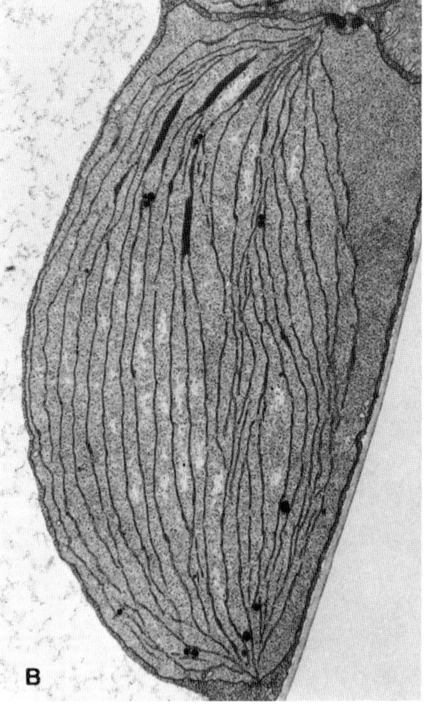

Figure 10.28. Wild-type and mutant tobacco chloroplasts. (A) In the wild-type plant, the chloroplast has the typical structure of those of metaphytans in general, with numerous stacks of thylakoids (grana) and large starch grains. 20,000×. (B) In the mutant lacking photosystem II, grana were nearly absent, while single thylakoids were more numerous. 15,000×. (Both courtesy of Miller and Cushman, 1979.)

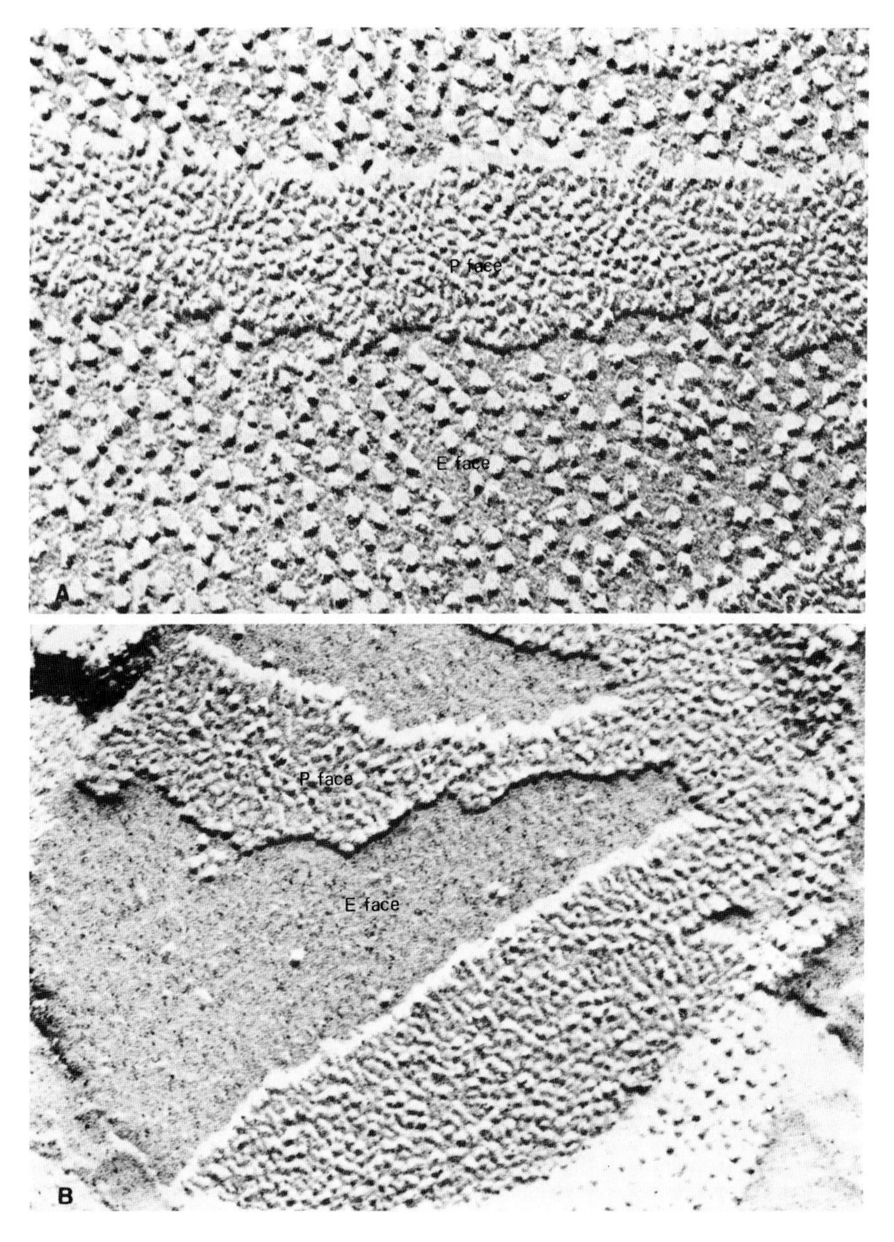

Figure 10.29. Grana membranes of wild-type and mutant tobacco chloroplasts. The P face in each case bears similar densely placed particles; on the E face the large particles of the wild type (A) are nearly wanting in the mutant (B) that lacks photosystem II. 150,000×. (Both courtesy of Miller and Cushman, 1979.)

other type, the lipoplast, or elaioplast, is similar, except that lipid material is stored instead of starch. In addition to the young leaves of many orchids, such lipid-bearing varieties also are known to occur in diatoms, but no information appears to be available at the ultrastructural level regarding their ontogeny. A third kind of storage organelle is the proteinoplast, in which protein crystals are accumulated; this type is especially characteristic of seeds and embryonic tissues, such as embryo sacs (Küster, 1951) and also the milk sap of *Cecropia*. Among its most outstanding distinctions is the presence of a membrane around the inclusion; as the deposits develop within invaginations of the inner investing membrane, this covering has been viewed as being identical to a thylakoid (Newcomb, 1967). Ribosomes appear to be more abundant in large proteinoplasts than in other varieties; whether or not these bodies develop into those known as aleurone grains seems to be an unresolved problem at present.

A final type, the chromoplast, may serve merely in storage or in providing color; this carotene-containing sort differs further from the preceding types in being derived from either proplastids or mature chloroplasts. Three subtypes are recognized, depending on the manner in which the carotene is stored; the first two of these, the globular and tubular, develop from mature chloroplasts, while the third, the crystalline, forms from proplastids (Mühlethaler, 1971). Globular chromoplasts, which characterize the color-imparting bodies of petals and perianths of many flowers, contain carotenes suspended in lipid droplets that are termed globuli (Steffen and Walter, 1958). During development the first globuli appear in chloroplasts just after the grana have been completed; then they increase in size and number, ultimately completely destroying the thylakoid system. Further maturation involves the movement of the granules to the periphery, where they become more or less arranged in rows.

Tubular chromoplasts typically occur in the outer covering of fruits and are distinguished by the pigment being arranged in long tubules. As fruit matures, the chloroplasts gradually lose their chlorophyll and starch granules and assume an elongate spindle shape. As those changes occur, tubules of a carotene–protein complex develop in increasing numbers and become aggregated into bundles. Like the globular variety, these changes result in the breakdown of the lamellar system and grana, but few details of the processes appear to be available (Mühlethaler, 1971). The third type, the crystalline chromoplast, occurs in such roots as carrots, but it also imparts color to the corona of *Narcissus poeticus*. As the name implies, the carotenoids are present in the form of microscopic "crystals" that vary in shape from filamentous to polygonal or even spiral (Straus, 1953, 1961), the polygonal being more frequent. Development of this type begins with starch-filled proplastids of early seedlings, the carotenes being deposited around the starch grains (Frey-Wyssling and Schwegler, 1965). As these deposits increase in size, the starch gradually disappears, and further increment leads to the formation of multilayered tubular sheets, the so-called crystals. As the sheets become more abundant, the stroma and other contents of the chloroplast break down into a watery fluid.

Chloroplast Development in Chlamydomonas. Among protistans, ontogenetic and assembly studies on chloroplasts have been largely concentrated on two familiar laboratory types, *Chlamydomonas* (Bogorad, 1975) and *Euglena*. Basically the results of such investigations closely parallel those just reported in showing that pertinent genes of the nucleus outnumber those of the chloroplast. However, it still is not firmly established in *Chlamydomonas* whether all of the mRNAs formed in the nucleus are translated on cytoplasmic ribosomes, or whether some enter the chloroplast for translation (Ohad, 1977). At least one case using the latter combination has been demonstrated, involving a component required for the formation of photosystem II. In a heat-sensitive mutant, known as T4, a polypeptide of 44,000 daltons was found to be encoded by a nuclear gene but was translated within the chloroplast (Kretzer *et al.*, 1977). The formation of chlorophyll–protein complex I required the presence of one or more polypeptides with molecular weights around 64,000 that were translated in the chloroplast, but at least one was proven to be encoded by a nuclear gene (Bar-Nun and Ohad, 1974; Chua *et al.*, 1975). A second one with origins similar to those of the latter was found essential for development of photosystem I activity, and one both encoded and translated within the nucleus has been reported necessary for the watersplitting activity of photosystem II (Kretzer *et al.*, 1977).

As in the development of higher plant plastids, light and other physical parameters play major roles. Two proteins, each transcribed from nuclear genes and translated on cytoplasmic ribosomes, required the presence of light having a wavelength of 632 nm for synthesis (Ohad, 1975); this pair in unison combined with 4 to 6 chlorophyll *a* and *b* molecules to form the chlorophyll–protein complex II. Once the latter had been formed in the presence of light, much of the synthesis of chloroplast membranes could proceed even in darkness (Ohad, 1977). Temperature also has been found strongly influential on development. If the heat-sensitive mutant T4 mentioned earlier is cultured at 37°C, in the resulting absence of photosystem II activity the chloroplast develops a dense network of lamellae in some regions of the organelle, whereas in the remainder of its compartment these are few and irregular. When the culture was removed to cooler growth conditions, such as 25°C, the dense stack of thylakoids dissociated and normal lamellae consisting of three thylakoids were formed throughout the organelle, as photosystem II activity was gradually restored (Ohad, 1977).

In growing cells of *C. reinhardii,* the chloroplast membranes were found to contain a large number of ribosomes, some of which had a polysomelike appearance (Margulies and Michaels, 1974). As some of the membrane proteins are synthesized within the organelle, it was suggested that these membrane-bound ribosomes synthesize membrane proteins, especially in view of the particles being attached to those structures by nascent polypeptide chains (Margulies *et al.*, 1975; Michaels and Margulies, 1975).

Chloroplast Formation in Euglena. Chloroplast formation in *Euglena* has recieved more attention from molecular biologists than have those processes in *Chlamydomonas,* and a detailed review has appeared in print (Nigon and Heizmann, 1978). The DNA molecules of the organelle, with a molecular weight of 93×10^6, had an overall G + C ratio of 25%, but the distribution of the bases was quite heterogeneous. Five segments were found, ranging in size from one (segment I) that comprised 40% of the molecule and had a G + C ratio of 22%, to segment V, which made up only 7% of the entire DNA and had a G + C content of 41% (Manning and Richards, 1972; Slavik and Hershberger, 1975, 1976). As in other species studied, the DNA coded for chloroplast tRNAs, but to a rather limited extent (McCrea and Hershberger, 1976)—only 7 species seemed to be encoded by the chloroplast genome alone, whereas 18 hybridized with both nuclear and chloroplast DNA. Also as elsewhere the greatest portion of the organelle's protein was encoded by nuclear genes and was synthesized on cytoplasmic ribosomes (Schiff, 1970). Here as in cytoplasmic ribosomes, the rRNAs are not transcribed into the mature form but into precursors that require trimming by enzymes. Two such precursors, pre-16 S and pre-23 S, have now been found in *Euglena* (Heizmann, 1974).

Because the effects of light on development of plastids of dark-grown specimens are similar in *Euglena* to others discussed (Figure 10.30A–D; Salvador *et al.*, 1971), the results of the opposite procedure are of interest. When mature cultures of *Euglena* were placed in the dark, the chloroplasts showed no sign of regression for 24 hr (Ben-Shaul *et al.,* 1965), but after 48 hr the organelle appeared swollen, with wide spaces of stroma separating many of the lamellae. The latter, too, appeared to be less electron opaque than usual and largely consisted of single thylakoids, rather than of pairs or triplets. After 3 days of darkness, the organelles had decreased in size, while the number of lamellae was nearly halved. These losses in size and lamellar content continued until by the 6th day, the chloroplasts had been reduced to proplastids, completely lacking any trace of the membranous structures.

10.4. PHYLOGENETIC ORIGINS

Like the mitochondrion, many origins have been proposed for chloroplasts, but also as in that organelle, the endosymbiontic concept proved to be the most widely accepted. In this case, a blue-green alga was supposed to be the mutualistic invader of the cell (Sagan, 1967; Ellis, 1969; Margulis, 1970; Taylor, 1970; Cohen, 1973; Stanier, 1974; Zablen *et al.*, 1975). However, Stanier could only find close resemblances between these algae and the chloroplasts of red seaweeds, because these two groups similarly had phycobilin par-

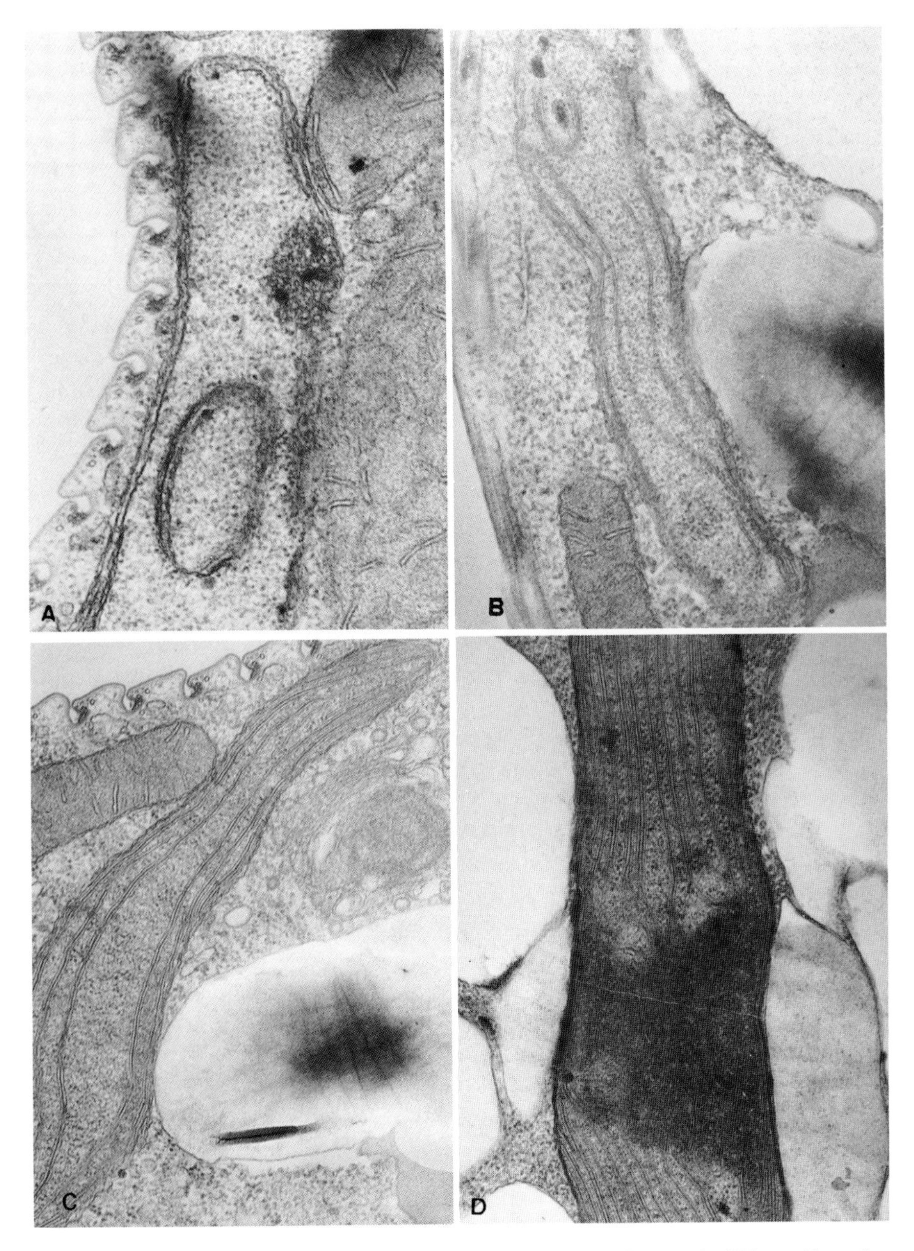

Figure 10.30. Developmental changes in the chloroplast of *Euglena*. (A) The proplastid formed in total darkness is nearly devoid of internal membranes. 39,000×. (B) Proplastid after 3 hr of illumination. 32,000×. (C) The same after 8 hr exposure to light. 32,800×. (D) Membranes nearing completion after 48 hr of light. 35,000×. For a mature chloroplast see Figure 10.17A. (All courtesy of Salvador *et al.*, 1971.)

ticles distributed along the thylakoids. For the remaining eukaryotes, he proposed separate origins from unknown sources, a proposal seconded by Bonen and Doolittle (1976). When one considers the complexity of the photosynthetic processes and the general uniformity of structure that exists among eukaryotic chloroplasts, it would seem far more reasonable to propose separate origins for the phycobilins than for the organelle itself.

Points in Opposition to an Endosymbiontic Origin. At the time when the endosymbiontic concept was readvanced, the presence of DNA, ribosomes, tRNAs, and other genetic components seemed to point to autonomy for the chloroplast, as they did also for the mitochondrion. But as in the latter, further investigations have shown that the organelle is not independent of nuclear control and cannot grow or function without nuclear-encoded products. Furthermore, other types of evidence clearly contraindicate the blue-green algae for a direct role in chloroplast origins, which may be summarized briefly in outline form:

1. The nucleoid of chloroplasts is arranged in patches and is located peripherally, not centrally as is the long, continuous nucleoid of blue-green algae. Moreover, the latter is far larger in proportion to the size of the organism than is that of the chloroplast.
2. The genome encodes only a fraction of chloroplast biochemical components, excluding some of utmost importance in replication, translation, and photosynthesis. Many cytochromes and ribosomal proteins, for instance, are products of nuclear genes.
3. The ribosomes have RNAs similar to those of prokaryotes but the resemblances cease there, for the sizes of the two subunits are quite disparate from those of bacteria.
4. The tRNAs are frequently of nuclear origin and are neither clearly prokaryotic nor eukaryotic.
5. The photosynthetic lamellae of blue-green algae also carry out cell respiration and therefore do not resemble chloroplast membranes.
6. The lamellae of those algae consist of single thylakoids, not multiples as in most lower eukaryotes (Figure 10.31), and grow constantly.
7. Chloroplasts usually divide by furrow formation, whereas blue-green algae undergo fission by ingrowth of a new transverse wall or septum.
8. The antennae of blue-green algae consist of three subunits totally different in molecular weights, components, and other properties from those of higher plants.
9. Chloroplasts also can revert into and develop from proplastids completely devoid of lamellae and photosynthetic properties. No proplastidlike body is known to exist in prokaryotes of any type.
10. Chloroplasts contain pyrenoids in unicellular forms, which are absent in blue-green algae.
11. The green plant chloroplast, including many advanced unicellular

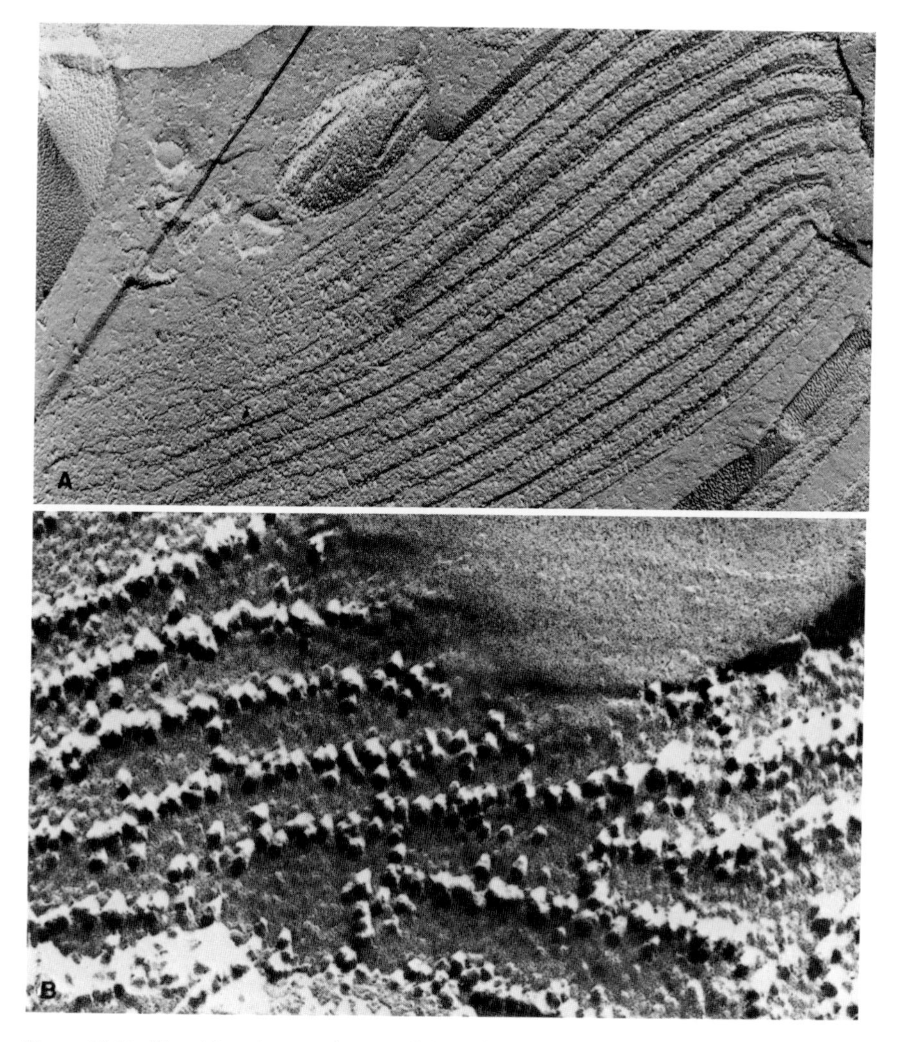

Figure 10.31. The chloroplast membranes of *Cyanidium caldarum*. (A) The chloroplast of this primitive eukaryote contains only single-thylakoid membranes, much like those of blue-green algae, but a comparison with Figure 10.13 reveals no other resemblances. 36,000×. (B) Further freeze-fracture preparations at higher magnification make clear that the particles of the thylakoid are quite distinct from those of the blue-green alga shown in Figure 10.16. 164,000×. (Both courtesy of Wollman, 1979.)

algal members, shows a remarkable series of evolutionary changes, in which the lamellae undergo regression while thylakoids become concentrated into grana. It seems improbable that an endosymbiont would behave in this fashion.

12. The chloroplast has developed other specializations, including various

kinds of leucoplasts and eyespots that had their origins in the cytoplasm. Thus rather than a foreigner invader, it seems far more reasonable to view the chloroplast as a cell part that has undergone evolutionary development along with the other organelles as the total cell became increasingly more complex and gave rise to the higher organisms.

A Biological Model of Chloroplast Origins. If the photosynthetic apparatus and processes are viewed as being what they appear to be, that is, organelles and functions that have evolved in cells along with those of other types, a biological model of phylogenetic origins can be proposed (Figure 10.32) slightly modified from that presented earlier (Dillon, 1962). In relation to the earliest levels, it is essential to recall those aspects of the respiratory processes that have been shown to share many features in common with photosynthesis. For instance, *Beggiatoa* and other colorless sulfur bacteria have only a highly incomplete tricarboxylic acid cycle (Section 8.3.2) and, in their inability to synthesize porphyrins, have neither cytochromes nor chlorophyll. Consequently, they must certainly be viewed as ranking among the most primitive of cellular organisms. In the H_2S-based photosynthesis that these bacteria utilize, that inorganic substance is employed as the light-energy receptor, the resulting sulfur being deposited as granules intracellularly. As these granules have not been found to be localized within vesicular invaginations of the plasmalemma (Maier and Murray, 1965), photosynthesis appears to be carried out directly within the cytoplasm. However, NADH has been found to be produced by these organisms, as well as ATP; accordingly, this, the most primitive form of photosynthesis known to exist, is involved only in providing energy-transferal substances, quite like that in all higher organisms, including the seed plants. Hence, the function of the processes has not changed; only the biochemical and structural mechanisms have evolved.

The utilization of the energy-transferal substances in carbohydrate synthesis came later, its earliest known occurrence being on photosynthetic infoldings of the plasmalemma after the cytochromes and chlorophylls had come into existence with the blue-green algae. Still later such thylakoids became more compactly arranged in various types of bacteria, including *Rhodomicrobium* and *Ectothiorhodospira,* but whether the tubules of *Thiocapsa* and the fibrous vesicles of *Chlorobium* and *Rhodospirillum* represent side branches or further advancements cannot be soundly conjectured, because of the absence of phototropic representatives among the most advanced prokaryotes and the most primitive eukaryotes.

What appears to be the earliest chloroplast among the eukaryotes occurs in the thermophilic, acidophilic alga *Cyanidium* (Rosen and Siegesmund, 1961), a form that shares many characteristics with yeast. These organisms possess membrane-enclosed chloroplasts, containing around eight to ten thylakoids ar-

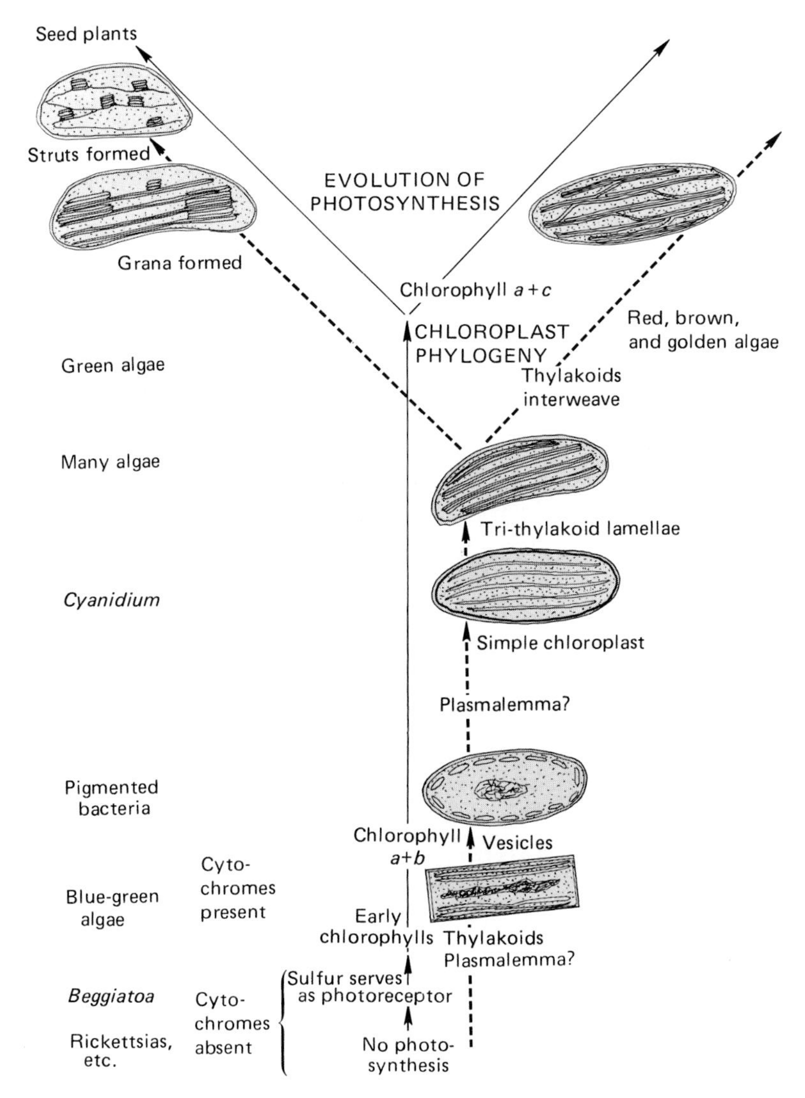

Figure 10.32. A phylogeny of chloroplast structure and photosynthesis.

ranged singly and not interconnecting; usually one to three of these continue around the ends of the organelle in the form of girdling lamellae, but this is a highly variable trait. Only a single chloroplast is present per cell, but because of its length and sinuous nature, it may appear two or three times in sections. However, its most distinctive trait is in the envelope consisting of only a single membrane, not two as in all other organisms.

Beyond this point the standard structure came into existence, with lamellae consisting of three thylakoids each and being enclosed in a dual-membrane envelope. Among the most advanced forms, two divergent lines are indicated. On that branch that includes the green algae, the lower members show few specializations in the structure of this organelle, whereas in the higher components, grana are developed, which become better organized and more numerous with phylogenetic advancement. Then at the highest levels, including the metaphytans, the trithylakoid membranes become modified into the frets that characterize the organelle in those organisms.

Although the branch just discussed continues a primitive trait in having a chlorophyll $a + b$ combination, the second one develops a new type of photosynthetic pigment, chlorophyll *c*, which occurs here along with *a*, but not *b*. Moreover, the storage product of the members of this line is never true starch as it is among the preceding ones. In addition their chloroplast shows a few distinctive traits, the most prominent of which is that individual thylakoids diverge to neighboring lamellae, interconnecting the internal membranes into a single system. It can be noted that the general configuration of the overall phylogenetic diagram (Figure 10.32) is Y shaped, quite similar to that provided in the chapter on the flagellum and its centriole (Chapter 5, Section 5.3).

11

Nuclear and Cell Division

Aside from an orderly net synthesis of proteins and other biochemicals that result in growth, the most important activity of cells of the vast majority of organisms is that of reproducing others of their own kind. These acts of reproduction involve division of the existing nucleus and cytoplasm into two, each resulting half resembling its counterpart as completely as possible. Because cytoplasmic division, or cytokinesis, is the simpler but less thoroughly known of this pair of activities, it is examined before nuclear division (mitosis or meiosis) receives attention. Then a closing section of this final chapter brings out salient points of the cell as a whole that need careful consideration.

11.1. CYTOKINESIS

Division of the cytoplasm is carried out by only two means, growth of a new transverse cell wall, or septum, and by constriction of the cell, typically by forming a furrow. However, in a few types of organisms a combination of these two processes occurs, and, in others, certain preliminary procedures are carried out before the actual divisionary steps are taken. Viruses are the only major groups of organisms that multiply without cell division, a distinctive act they accomplish by inserting the new copy of the genome and its accompanying proteins into a freshly synthesized capsule, a part that appears to be the remote ancestor of the cell proper (Dillon, 1978). Thus nothing resembling cellular or nuclear division occurs in these precellular forms.

11.1.1. Prokaryotic Cell Division

Division of the cell in prokaryotes appears to be such a simple process that virtually no investigations have been conducted to ascertain the actual mechanisms that must be involved. Unlike other activities among the blue-green algae, the present one offers more advantages for introducing the subject than does that of the bacteria; accordingly the processes of cytokinesis in members of the genus *Oscillatoria* provide the introduction to the subject.

Cytokinesis in Blue-Green Algae. Basically the division of the cytoplasm in many blue-green algae is by formation of a new septum, beginning at the periphery and gradually growing across the cell from all sides. At least in *Oscillatoria splendida,* the processes are continuous and have their inceptions two or more cell cycles in advance of completion (Figure 11.1). In the illustration, it can be noted that the largest of the forming septa, that at the center, extends about one-fifth of the diameter across the cell from the sides, its location and extent being made evident largely by the furls of the lamellae. Then about half-way between the center of the cell and each end, a second beginning septum can be noted, which is less than half the size of the first. Finally, midway between these second ones and the end septa, slight displacements of the lamellae mark the sites of minute incipient septa. In the electron micrograph, too, the ovate shape of the nucleoid makes it obvious that nuclear division had not begun as yet. In the enlarged view of the growing septum, it is apparent that that structure had grown through the outer lamellae (Figure 11.2) and was in the process of piercing another when the cell was fixed. Further, as indicated by the arrangement of the membranes, the mature septum had done likewise when it had formed. However, because in a strict sense the septa do not grow but are developed by additions of material upon the internal edge, the penetrations through the lamellae just noted were in reality acts of the cytosol as some of its enzymatic contents added the necessary biochemicals to the growing crosswalls.

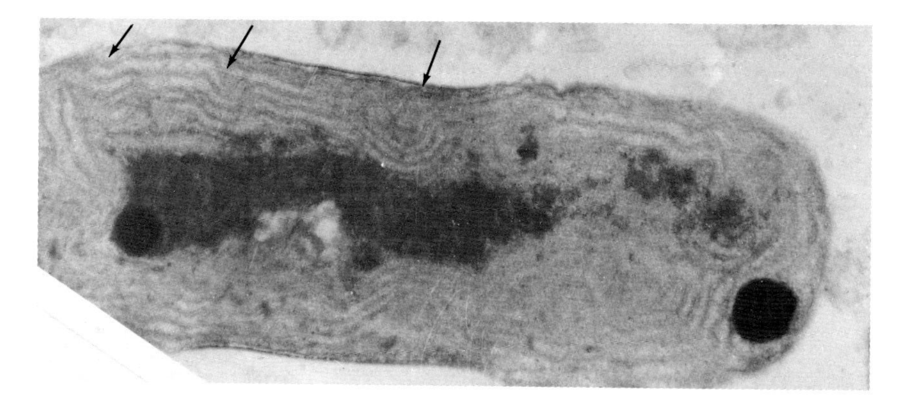

Figure 11.1. Cytokinesis in a blue-green alga. In this view of a dividing cell of *Oscillatoria splendida,* three sets of forming septa are apparent (arrows), the central one the longest, followed by successively shorter ones toward each side. 12,000×.

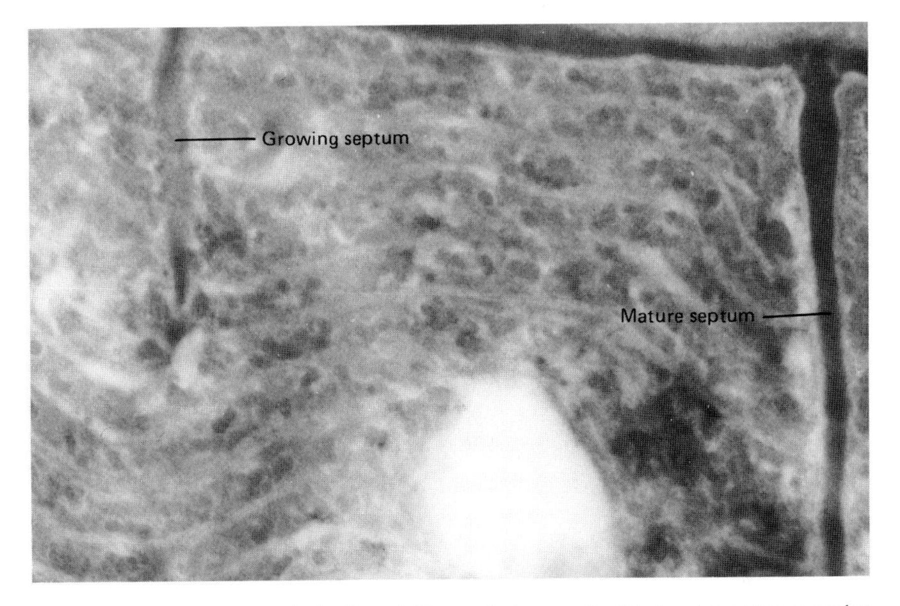

Figure 11.2. A growing septum in *Oscillatoria limosa*. During growth of the forming septum, a number of photosynthetic membranes have been severed, while others have become indented ahead of the advancing edge. 200,000×.

The presence of three sets of developing septa in blue-green algae or any other organism is exceptional and may be correlated in this instance to the elongate form of the species. Whether the correlation is one of cause or effect remains unknown, but in short-celled related genera, such as *Symploca muscorum* (Prankratz and Bowen, 1963) and *Lyngbya* (Lamont, 1969), only two sets are produced, the long medial one and those that are half as long situated at each quarter. In other forms with cells equally short as those of the latter species, only a single septum is produced (Figure 11.2), as in *Oscillatoria chalybea* (Lamont, 1969). Nevertheless all of the foregoing species share in a distinctive trait—the exterior of the cell provides no indication of the location of the septa, old or new, except for a narrow notch. In contrast, a number of genera like *Plectonema* and *Calothrix* (Kessel *et al.,* 1973) lack the surface notch opposite the developing septum but become constricted after a completed septum has undergone maturation. This constriction is shallow in the cases mentioned, but perhaps can deepen sufficiently to sever the filament, as in some of the more advanced stages described below.

Anabaena and *Anacystis* make clearer the nature of this constriction, because it deepens to a much greater extent than it is known to do in the preceding types. When the first stages of setpum development begin, which origins occur at the widest portion of the cell, no indication of its location can be noted

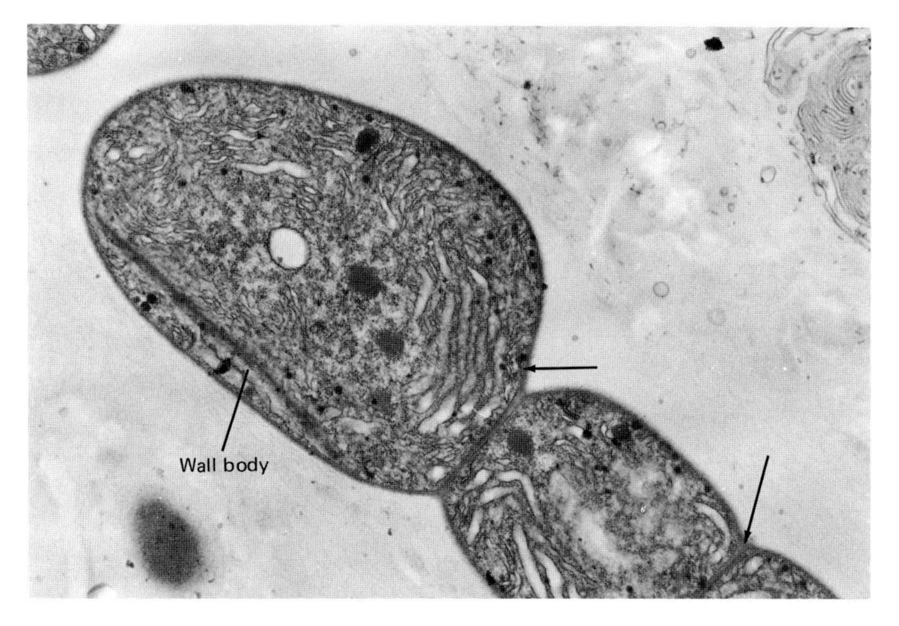

Figure 11.3. Cell division in *Anabaena*. During cell division among more advanced blue-green algae, the cell becomes slightly constricted (lower arrow) as a septum forms; the constriction then becomes deepened (upper arrow) as the septum is resorbed, ultimately liberating the individual cells. 16,000×. (Courtesy of Jensen and Ayala, 1977.)

on the exterior of the wall (Hopwood and Glauert, 1960; Echlin, 1964; Lang, 1965; Leak and Wilson, 1965; Jensen and Ayala, 1977), but as the septum matures, a constriction forms around the cell (Figure 11.3). At first shallow, this narrow belt deepens until it results ultimately in the separation of neighboring cells. In view of the rigidity of the walls, the now ovate form of the previously cylindrical cell implies that the entire cell contents and wall in the region have had to undergo changes during these pinching-off processes. Obviously disintegration and reconstruction of the outer wall have had to accompany the breakdown of the transverse structure. Although cell division has not been observed, the ovate form of the cell of *Synechocystis* (Schulz-Baldes and Lewin, 1976) makes it appear quite probable that cytokinesis in members of that genus is carried out by similar steps. It is pertinent to note that, despite the superficial similarity of this process and the familiar furrow formation of metazoans, the two activities have nothing in common. In the present instance, enzymatic dissolution of the septum and adjacent wall apparently occurs so that constricting really involves redistribution and growth of all the cell parts in the area, whereas among the Metazoa, furrow formation is actually by constriction, as shown later. This observation is also of significance in the discussion of cell division among bacteria that follows.

Cytokinesis in Bacteria. To a large degree, cytokinesis in bacteria parallels the several variations found in the blue-green algae, especially at the lower levels. Quite as in *Oscillatoria,* the cells of *Bacillus subtilis, Beggiatoa,* and *Clostridium pasteurianum* (Figure 11.4) carry out division of the cytoplasm by septum formation with little or no external indication of the site (Glauert *et al.,* 1961; Laishley *et al.,* 1973; Strohl and Larkin, 1978a,b). Only in *Beggiatoa,* however, is the septum permanent in form; in the others, it undergoes resorption and modification as the individual cells become separated. As *Beggiatoa* and *Clostridium* have been reported to have only primitive mechanisms for respiration, this process of cell division may be considered elementary, too.

Somewhat more advanced in this process are the members of such genera as *Mycobacterium* (Figure 11.5A,B; Barksdale and Kim, 1977), *Streptococcus* (McCandless *et al.,* 1968), *Achromobacter* (Colwell *et al.,* 1968), *Diplococcus* (Tomasz *et al.,* 1964), and *Leptotrichia* (Hofstad and Selvig, 1969), in all of which constriction of the cell accompanies septum formation. Thus the outer portion of the septum and adjacent cell wall are undergoing resorption or alteration concurrently with the addition of new material to the growing interior edge of the forming septum. All the cited examples of this latter level of cytokinesis have well-marked mesosomes. *Corynebacterium* (Hard, 1969) and

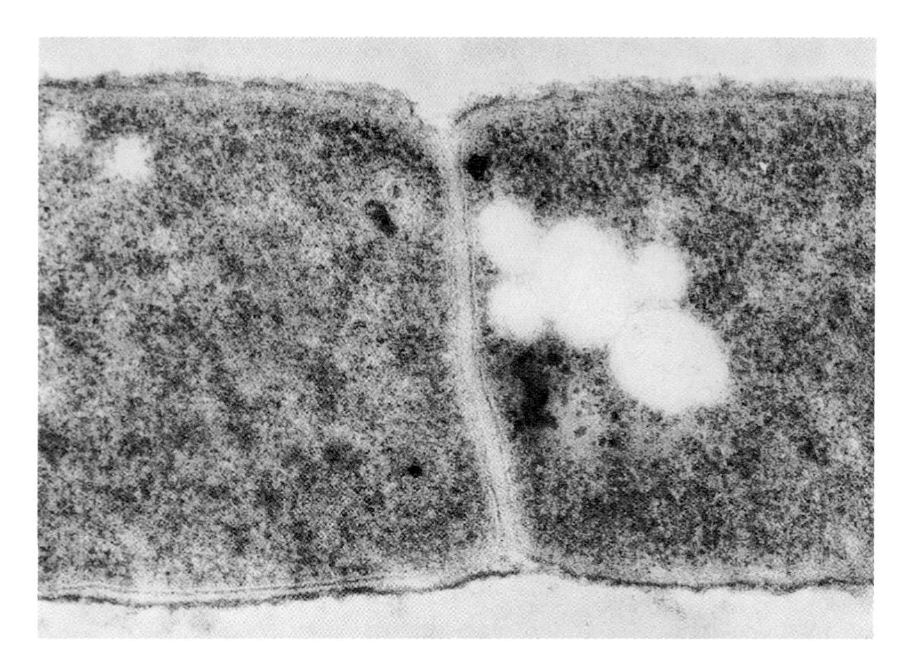

Figure 11.4. Cell division in a primitive bacterium. The septum of *Clostridium pasteurianum* resembles that of the more primitive blue-green algae like *Oscillatoria* in having only a slight impression externally. 40,000×. (Courtesy of Laishley *et al.,* 1973.)

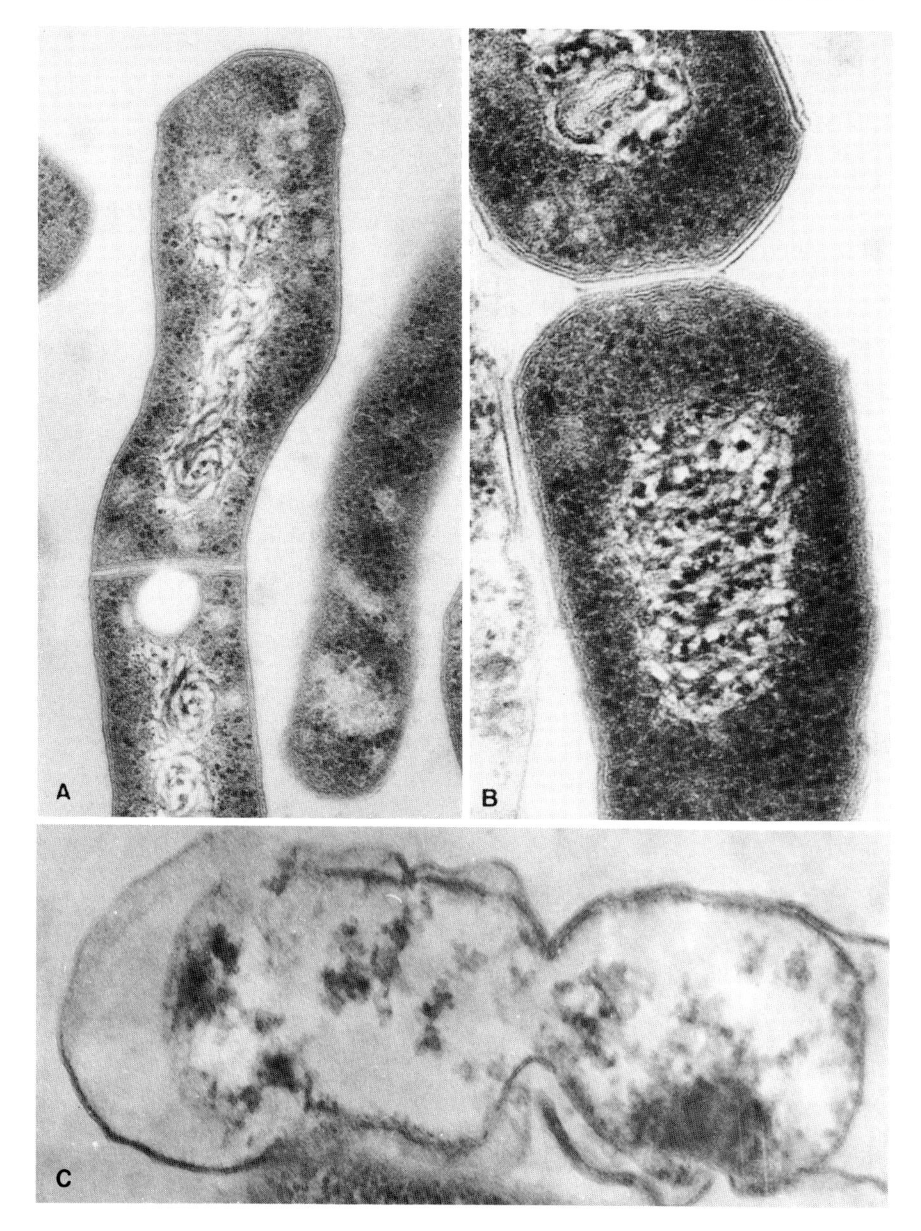

Figure 11.5. Cell division in more advanced bacteria. In such advanced bacteria as *Mycobacterium*, a septum is formed to divide the daughter cells (A); later, however, a deep furrow forms external to the septum (B) and deepens, ultimately freeing individual cells. (A) 65,000×; (B) 78,000×. (Both courtesy of Barksdale and Kim, 1977.) (C) Among still more advanced bacteria, no septum is formed, division being the result of an invagination of the cell envelope. In this partially autolyzed cell of *Erwinia*, the plasmalemma can be noted to maintain its invaginated condition even though the cell wall has been partly destroyed. 51,800×. (Courtesy of Grula and Smith, 1965.)

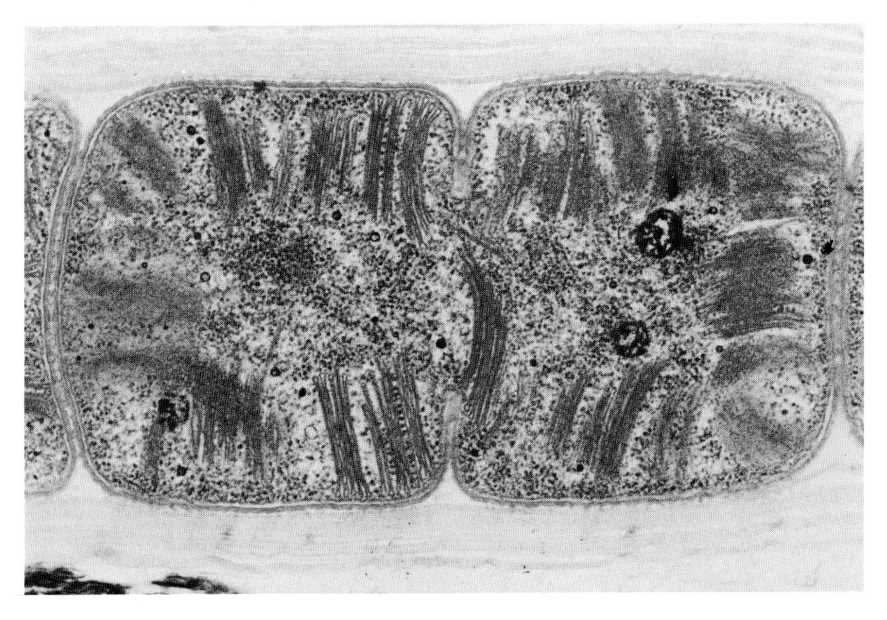

Figure 11.6. Cytokinesis in *Crenothrix*. This complex but nonphotosynthesizing cell bears a number of cisternae and fibrillar elements; cell division involves the invagination of the cell envelope, just inside of which is a short septum. Consequently, it may prove to represent a stage intermediate between the septum producers and those that divide by invaginating the cell envelope. 47,000×. (Courtesy of Völker *et al.*, 1977.)

Actinomyces (Duda and Slack, 1972) seem to be intermediate between this level of advancement and the preceding in that constriction does not accompany septum formation, but commences after the transverse partition has been completed.

Beyond this point in prokaryotic phylogeny, no septum is formed, cell division being carried out entirely by constriction. Representative types include such genera as *Wolbachia* (Burgdorfer *et al.*, 1973), *Mycoplasma* (Green and Hanson, 1973; Hirumi and Maramorosch, 1973), *E. coli* (Cota-Robles, 1966; Tudor *et al.*, 1969), and *Erwinia* (Figure 11.5C; Grula and Smith, 1965). The second genus named is somewhat distinct from the others in that its envelope is mucilaginous and often becomes drawn out into a thin tube after cell division has been completed, leaving two or more members interconnected in this fashion. Usually in the other forms mentioned the constriction is rather broad so that in longitudinal sections it appears to be V shaped as it deepens. Contrastingly in dividing cells of *Crenothrix* (Figure 11.6; Völker *et al.*, 1977), the constriction is narrow and furrowlike. Because the members of the latter genus possess a number of membranous cisternae and fibrillar elements, they and their mode of cell division may need to be considered to represent a separate level of bacterial development.

11.1.2. Cell Division of Eukaryotes

As in the prokaryotes, knowledge of the cytokinetic mechanisms remains largely at the cellular level, especially among unicellular types. But even among the metaphytans and metazoans, related subcellular phenomena have been explored very scantily, as becomes evident as the discussion proceeds.

Cytokinesis in Yeasts. In the true yeasts, such as *Saccharomyces* (Mundkur, 1960, 1961a,b; Robinow and Marak, 1966) and *Rhodotorula* (Thyagarajan *et al.*, 1962), cell division is carried out by processes not too dissimilar from those of *E. coli*, except for a prelude not found in that prokaryote. The early steps in yeast are involved with the formation of a bud, a bubblelike appendage that increases in size while the mother cell decreases. Because the organism is enclosed in a thick cell wall, any decrease in size of the parent obviously must involve resorption of wall material, as the necessary molecules are inserted into that organelle of the growing daughter. As the latter begins to approach about half the mass of the mother cell, the constriction deepens in the neck where the two are interconnected. As the lumen of the interconnection never is wide, constriction proceeds rapidly until a narrow double wall closes the opening. When this has been divided into two along the central plane by unknown enzymatic processes, the bud breaks free, leaving a cicatrix on the surface of both individuals.

Cell Division in Euglenoids. The pattern of cell cleavage among single-celled eukaryotes in general is well established in euglenoids, except for the presence there of a primitive trait. As an excellent account of division in this taxon is available (Leedale, 1967), the present description is confined to the major details. The nucleus migrates toward the flagella, and there, along with the anterior canals and flagellar centriole, undergoes division; when mitosis has progressed sufficiently, an infolding of the cell membrane and pellicle may be perceived anteriorly between the two daughter canals. This cleavage furrow develops posteriorly, spiraling as it does so in response to the helical construction of the pellicle. As it deepens, redistribution of cytoplasmic organelles can be noted just caudad of the advancing cleavage plane, which passes between the now divided nucleus, the progress of the furrow through the entire length of the organism requiring between 5 and 60 min. Division continues to the very tip of the cell, the two resulting cells remaining attached by their posterior ends for an extended period.

Here, as in the prokaryotes, dissolution and reconnection of the pellicle must be involved in furrow formation; secretion of new material is likewise requisite, for two small bodies have a greater surface area than a single one equal to their combined mass. In this protistan, cytoplasmic parts are not consistently distributed equally between the two daughters, for sometimes one may receive most, or even all, the chloroplasts or mitochondria, and at least one study has reported an instance in which both nuclei went to a single daughter (Leedale, 1959a). Among the chief features of orientation to be noted in these organisms is that the plane of nuclear movement is transverse, while that of cleavage is longitudinal.

Cytokinesis in Other Protistans. In nearly all protistans, cell cleavage follows the pattern just described, except for one marked modification. In the chrysophycean *Hydrurus,* chlorophytans *Pyramimonas* and *Platymonas,* and various fungi by way of examples, the chloroplast, where present, divides while the nucleus is in early prophase (Joyon, 1963; Stewart *et al.,* 1974; Pearson and Norris, 1975; Ashton and Moens, 1979). Then after the latter body has completed mitosis in a transverse direction, a cleavage furrow begins to form at the anterior end as in *Euglena,* similarly deepening and progressing posteriorly. But before it has passed completely between the daughter nuclei, a second furrow forms at the posterior end and grows forward between the chloroplasts until the two constrictions meet, the anterior one consistently remaining the deeper component.

Undoubtedly the commonest modification that occurs among protistans is in a change in direction of cleavage-furrow formation from longitudinal to transverse, the best known examples being provided by the euciliates such as *Blepharisma* (Jenkins, 1977; Sawyer and Jenkins, 1977), *Stentor* (Tartar, 1966), *Oxytricha* (Grimes, 1972), and *Paramecium* (Jurand and Selman, 1970). As shown in the light micrographs (Figure 11.7), a furrow develops around the middle of the length of the cell as the macronucleus prepares to undergo mitosis. Then as macronuclear division proceeds, the furrow deepens and finally appears to constrict the organism in two. In Figure 11.7C the off-center location of the cleavage furrow at the termination of cleavage makes clear that, just as in longitudinal division, the furrow formed on one side prior to the appearance of a counterpart on the opposite surface. A similar asymmetry of furrow formation in transverse fission is displayed by the fungus *Bullera* (Taylor and Wells, 1979).

The Sawyer and Jenkins and the Grimes references provide abundant evidence that the processes just described are deceptive in their simple appearance, making it quite lucid that mere pinching into halves does not really describe them. As those investigators show, in these ciliates a new oral groove develops, flagellar centrioles and associated basal apparatus multiply, and the rows of cilia undergo reorientation. Because the pellicle that covers the organism does not show wrinkling or other indications of being pursed together by the deepening cleavage furrow, it is obvious also that resorption of the covering and the underlying plasmalemma is a necessary accompaniment of cytokinesis. Oral groove formation is shown similarly to be a complex set of activities, in-

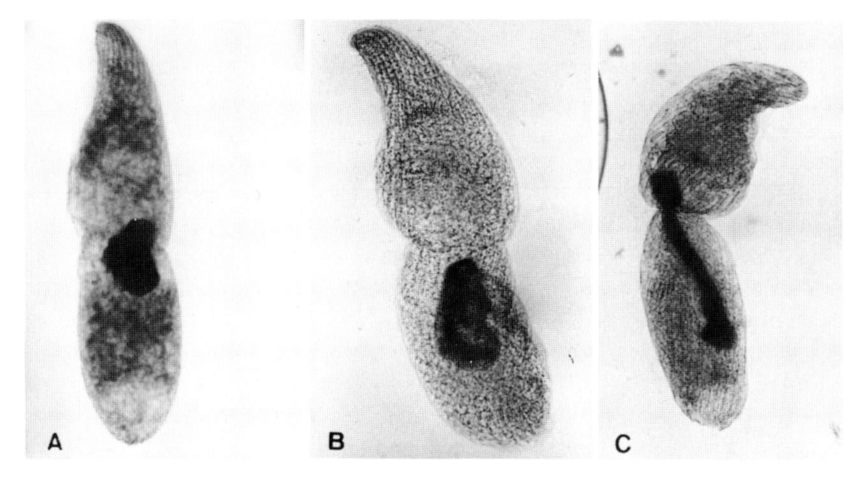

Figure 11.7. Cell division in the ciliate *Blepharisma*. In ciliates as a whole, cytokinesis takes place transversely by furrow formation. Where division is nearly complete (C), it is especially apparent that one side of the furrow is deeper than the other. (B) The macronucleus in this specimen failed to divide, so the anterior portion remained enucleate and died within 24 hr. 450×. (All courtesy of Jenkins, 1977.)

volving development of a dense set of modified cilia along one long side of the groove and the formation of an undulating membrane along the opposite edge, in addition to the creation of a gullet and associated parts. If the fact is kept in mind that these events occur 6 hr prior to nuclear division and cytokinesis (Tartar, 1966), the genetical phenomena involved in the control of these events can also be perceived to be highly complicated, a topic that receives further attention later. It should also be noted that nuclear division does not signal the occurrence of cytokinesis. Quite to the contrary, experimental removal of division lines and oral groove primordia has demonstrated that the cytoplasm provides the signal for nuclear division to commence (Tartar, 1966; Frankel, 1967a,b).

Fission among Chlorophytans. Although transverse fission prevails throughout all the filamentous types of green algae, clues as to the origins of this reorientation of longitudinal division are provided by a simple flagellated form. In the members of the genera *Chlamydomonas* and *Ulva* (Johnson and Porter, 1968; Løvlie and Bråten, 1968), a thick pectinaceous cell wall covers the flagellates throughout their life spans, structures far too resistant to undergo division by furrow formation. Normally a flagellate of this type would be expected to divide longitudinally like *Euglena,* and it actually does so, but only after the entire protoplast has made a quarter rotation within the cell wall. The onset of division is marked by the loss of the flagella, following which the cytoplasm rotates as a unit until the centrioles are located at the center of one side. After these organelles, the nucleus, and the chloroplast have divided, a cleavage furrow appears in the plasmalemma just beneath the cell wall. When it has deepened to some extent, a corresponding fissure appears beneath the opposite side of the cell wall and continues as described in other flagellated forms. This first division is followed by a second one, after the daughter cells have rotated an additional quarter turn, one clockwise, the other counterclockwise; consequently, both sets of flagellar centrioles are approximately at the very center of the original cell, this second cleavage otherwise being identical to the first. Hence, it is evident that cell division is longitudinal in these organisms, despite that of the first one happening to proceed in a transverse direction. In those filamentous types that possess only one chloroplast, the orientation of the various organelles in respect to the cleavage furrow also demonstrates that, insofar as the protoplast is concerned, cytokinesis still takes place longitudinally—it is only transverse relative to the cell wall. The same conclusion can therefore be extrapolated to the metaphytan and even to the metazoan cells, in all of which similar rotation of the cytoplasm or entire cell during meiotic division has been thoroughly documented. It should not be thought that the elongate form of the cell and attached ends in filamentous types necessitates the transverse fission, for the nuclei of hyphal cells of many fungi, including *Physarum,* carry out mitosis across the diameter (Aist and Wilson, 1965; Aldrich, 1969).

Evolution of Cell Plate Formation. The origins of another feature of the higher plants—cell plate formation—are similarly made clear by investigations of cell division in the green algae. Unfortunately, because of the lack of ultrastructural investigations of cytokinesis and mitosis in a large number of critical types, details of the processes must be held to the major steps and, in a few instances, based largely on light microscopic researches. That the latter are amazingly accurate has been confirmed repeatedly by the results of those studies utilizing the sophisticated instruments available today. Thus the sources cited are doubtlessly dependable; their findings merely need to have finer details made clearer by current techniques.

The first stage (I) in the phylogenetic history of cytokinesis among green plants is the primitive type that occurs in most flagellates—simple longitudinal division by furrow formation, as in *Euglena*. Among the numerous representatives known to undergo this type of division are *Platymonas* (Pitelka and Schooley, 1955), *Pleurastrum* (Molnar *et al.,* 1975), *Eudorina*, and *Volvox* (Zimmermann, 1921; Lewin and Meinhart, 1953). Following this is stage II, in which longitudinal division takes place after the cytoplasm has revolved one-quarter turn; as reported above in *Chlamydomonas,* morphological longitudinal fission is thus transverse relative only to the cell wall.

Between this and the next level of advancement, representatives appear to be unknown, for a radical departure in cleavage-furrow development marks the members of stage III. In such filamentous genera as *Zygnema* (Merriam, 1906; Cholnoky, 1932), *Mougeotia* (Peterschilka, 1922), *Spirogyra* (Czurda, 1922;

McAllister, 1931; Oura, 1953; Fowke and Pickett-Heaps, 1969a,b), *Trichosarcina* (Mattox and Stewart, 1974), and *Klebsormidium* (Floyd *et al.,* 1972), a cleft in the cytoplasm can be noted to form during late anaphase at the midpoint of the interzonal fibers, but no external furrow is in evidence. Deposition of granules on each side of the cleft then commences, first centrally and then spreading toward all sides, accompanied by the lengthening of the internal cleavage furrow. When the latter has completely severed the cytoplasm into two, a new septum grows centripetally from the entire cell wall. The final definable phase, stage IV, needs no additional description, for it is the familiar type represented by cytokinesis among all the higher plants, beginning at least with such filamentous types as *Ulothrix* and *Stigeoclonium* (Stewart *et al.,* 1973) and continuing thence through the stoneworts (Karling, 1926; Pickett-Heaps, 1967), mosses, ferns, and into the seed plants.

11.1.3. Fine Structural Events in Cytokinesis

Although a number of ultrastructural studies of cytokinesis have been made, relative to the immensity of the problem they have been too few to provide a clear picture of the subcellular aspects. Consequently, many of the mechanisms involved in cell division have not as yet been revealed.

Cell Plate Formation. That the Golgi material in metaphytans has long been known to be active in the secretion of cell plate material has been pointed out where appropriate (Chapter 6, Section 6.3.3), but the details were reserved for the present topic. The dictyosomes shown to be engaged in this activity do not arrange themselves directly in line with the forming plate, but remain at a slight distance (Whaley and Mollenhauer, 1963; Frey-Wyssling *et al.,* 1964). These organelles produce an abundance of small vesicles containing electron-dense material in their lumina, which increase in size and volume of contents as they move toward the midpoint between the divided nuclei (Bajer and Allen, 1966a). Upon arrival, the vesicles, now nearly doubled in diameter, at first accumulate over a broad belt, but a little later they become arranged in a single, somewhat wavering line. Thus aligned, many vesicles and their enclosed material commence to fuse with adjacent ones, terminating in a row of larger, elongate vesicles surrounding a series of tenuous filaments. As these processes continue and an increasing amount of fusion occurs, the alignment is straightened so that the component parts form a nearly flat plane, the end result being a continuous cell plate. Throughout the later stages, the products carried by the Golgi vesicles are added both to thicken the forming plate and to increase its diameter until the edges have attained the outer cell wall along the entire periphery. While the cell plate itself consists of pectic substances and hydrated hemicellulose, later additions that are made consist of cellulose.

Unfortunately the fused membranes of the vesicles have been viewed as becoming the plasmalemma, on the basis of their being located in a corresponding position (Frey-Wyssling *et al.,* 1964), but the current, much more detailed information regarding membrane structure necessitates a different interpretation. More than likely, after the fused vesicles had consolidated their contents to form the cell plate, their tasks being completed the membranes were replaced molecule by molecule with other substances while septal growth approached completion, until they were entirely converted into new plasmalemma. Similar Golgi involvement in septum formation has been documented in other organisms, including the marine chrysophycean genus *Pleurochrysis* (Brown, 1969). In this case the single dictyosome that is present becomes greatly enlarged, so as often to contain over 30 cisternae, the outer 12 or more of which enclose linear deposits of cell wall material. Among the brown seaweeds, septum formation corresponds closely to that of seed plants, involving Golgi vesicles as well as endoreticulum (Markey and Wilce, 1975).

In the Frey-Wyssling reference cited, the smooth cisternal endoreticulum that also was present in abundance received almost no attention, but the clear electron micrographs provided in that study reveal a number of additional details. These cisternae, too, can be noted to produce vesicles, in size much greater than those formed by the dictyosomes but lacking the electron-opaque material. They appear to intermingle with the Golgi vesicles, fusing with them and thereby playing an active, direct role in the formation of the cell plate. As a matter of fact, the numerous single endoreticular cisternae remain more closely associated with the plate throughout its growth in thickness and width than do the dictyosomes. In addition, fragments of microtubules, probably from the spindle (Hepler and Newcomb, 1967), and mitochondria also are present in large numbers adjacent to the plate, but whether the latter provide energy to the system remains for the future to disclose.

Furrow Formation in Metazoans. The processes of cleavage-furrow formation in vertebrate cells have been followed by a few investigators, including several at the ultrastructural level (Figure 11.8; for example, Zeligs and Wollman, 1979c). In some cases the researchers employed fluorescent heavy meromyosin to demonstrate the active participation of actin filaments (Schroeder, 1973; Sanger, 1975; Sanger and Sanger, 1976). During late anaphase, the thin filaments thus revealed became arranged in large numbers along a broad belt where the furrow would develop; then during telophase they contracted in coordinated fashion to engage in cleavage. Using similar techniques, other investigators have had contrasting results, indicating the actin not to be localized near the furrow during cleavage (Herman and Pollard, 1979). Moreover, because the contractility of a given fiber is extremely limited, reducing the cell from its region of greatest width to nil certainly cannot involve simple contraction only; besides becoming contracted, each fibril must undergo great reduction in length during these processes by activities on the part of the cytosol.

This involvement of the cytosol with the actin filaments continues throughout the cell cycle, as demonstrated in chick fibroblasts and other vertebrate cells

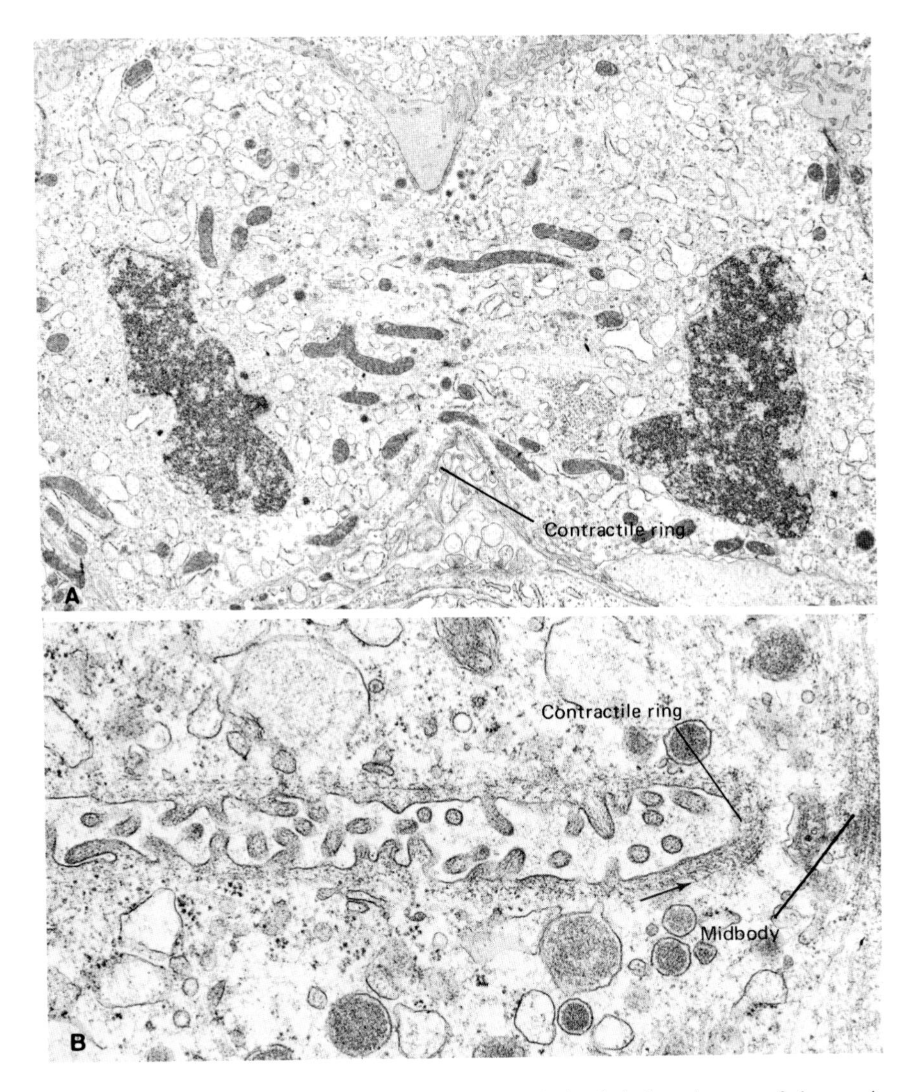

Figure 11.8. Cleavage-furrow formation in a metazoan. (A) A relatively early stage of cleavage in rat thyroid cells shows the furrow to be symmetrical; a contractile ring of fibers lines the innermost edge of the furrow. 5000×. (B) Apical portion of a late-stage furrow, with microvilli prominent within the invagination; the midbody developing centrally is shown in more detail in Figure 11.9. 25,000×. (Both courtesy of Zeligs and Wollman, 1979c.)

(Sanger and Sanger, 1976). At the onset of mitosis in these cells, the thin filaments were transferred from their interphase association with pseudopod formation to locations scattered beneath the plasmalemma, where they participated in microvillar expansion and contraction. Then when mitosis progressed, they were relocated to the equatorial belt to engage in cytokinesis, as just enumerated, after completion of which they were conducted to the poles of the daughter cells to take part in pseudopod formation once more. Similar cell-cycle-related changes in distribution of the fibrous protein fibronectin have been described (Stenman *et al.*, 1977).

11.2. NUCLEAR DIVISION

Because the molecular events that accompany nuclear division in organisms of all types have already received their full share of attention (Dillon, 1978), here discussion can be devoted to processes at the cellular and subcellular levels. First, the structural features of the metazoan cell during mitosis or meiosis are reviewed, including the associated parts such as the spindle and mitotic centriole. This then is followed by an appraisal of nuclear division on a comparative basis, including that of prokaryotic cells.

11.2.1. Nuclear Division in Metazoans

Throughout this chapter, no distinction is made between mitosis and meiosis, except those cases in which radical differences at the structural level exist. In other words, nuclear division is equated with both mitosis and meiosis, although the latter term is rarely employed.

Nuclear Events. The events that occur within the nucleus during its division on the surface appear to be so simple that it is easy to become deceived into believing them to be that way in actuality. In truth, they are amazingly complex and still are not fully known. Replication of the chromatin matter occurs chiefly during the S (synthetic) phase of the cell cycle, after a gap in the nuclear replicatory processes (G_1) subsequent to the last mitosis. This is followed by a second gap in nuclear division (G_2), following which the chromatin becomes condensed to form chromosome pairs, the members of each being held together at the kinetochores (centromeres).* Then as the nuclear membrane and nucleolus break down, the chromosome pairs become aligned on the equatorial plate (metaphase), each one being attached to a spindle thread. Just before the next stage, the kinetochores are thought to undergo division, thereby detaching the mates of each chromosome pair, but in view of kinetochore morphology, this concept appears to be simplistic. Thereafter the homologous chromosomes are moved apart (anaphase) to reach opposite poles of the cell, where they spread out to form the chromatin network, while the spindle breaks down and a new nuclear membrane is synthesized to bring the last phase (telophase) to a close. Typically the latter is accompanied by cell division.

* In metazoans ultrastructural examination of isolated intact chromosomes has demonstrated the kinetochore to be merely a region containing fewer DNA-protein fibers than the remainder of the chromosome (DuPraw, 1965, 1966).

The Spindle. The nature of the spindle apparatus has attracted a number of investigations during the past decade, and several reviews have appeared in print (Bajer and Molé-Bajer, 1972; McIntosh *et al.*, 1975, 1976). According to electron microscopic investigations, the fibrous elements of this apparatus are microtubules (Borisy, 1978; Oakley and Heath, 1978), three classes of which are recognized as forming the spindle. Kinetochore microtubules (1) are those that are attached to the kinetochores of chromosomes, and typically also terminate at a pole. Polar microtubules (2) terminate at a pole but are not attached to kinetochores; and free microtubules (3) are those that are unattached at both ends (Fuge, 1974; McIntosh *et al.*, 1975). Because the kinetochore microtubules are more stable chemically than are the others, they may prove to consist of a different type of tubulin (Brinkley *et al.*, 1967; Cande *et al.*, 1974; Brinkley and Cartwright, 1975). In vertebrates at least, spindle formation apparently is initiated largely at the kinetochores and to a lesser extent at the polar apparatus, according to studies on isolated chromosomes and other appropriate structures (Gould and Borisy, 1978). Since the discovery that microtubules were the major constituent of the spindle, a number of hypotheses have been advanced for the molecular mechanisms involved in the movement of chromosomes (Inoué and Sato, 1967; McIntosh *et al.*, 1969; Nicklas, 1971; Bajer, 1973; Oakley and Heath, 1978)—as was the case also with flagellar movement, sliding was one of the mechanisms proposed and contraction another.

However, the spindle does not consist of microtubules alone, for electrophoretic analyses have shown the presence of around 75 different substances (McIntosh *et al.*, 1976). Almost none of those constituents have been identified, but actin filaments have been demonstrated to occur in small numbers (Bajer and Molé-Bajer, 1969; McIntosh and Landis, 1971; Forer, 1976; Forer and Jackson, 1976; Sanger and Sanger, 1976; Herman and Pollard, 1979), and high-molecular-weight MAPs have been detected (Sherline and Schiavone, 1978). What has not been determined is how movement of the chromosomes is brought about by this array of fibers and proteins. In relation to the problem, several observations that have been reported are particularly pertinent. The first of especial note is the establishment of the spindle's lack of active participation in these processes, for the fibers have been shown neither to produce nor transmit force during movement of the chromosomes (Forer, 1966), a point consistently overlooked in discussions of this topic (Kubai, 1975). Moreover, it has been demonstrated with crane fly spindles that during anaphase the birefringence of the fibers is quenched only at their chromosomal ends (Forer, 1976), suggesting that depolymerization occurs at those points alone. Thus it may be suspected that no contractile system is engaged in this activity and that chromosomal movement during anaphase is brought about by the cytosol, in part by breaking down the microtubules (Novitski and Bajer, 1977).

The Mitotic Centriole. Associated with each pole of the spindle in the majority of cell types is a centriole, identical in structure as far as known to that of the flagellum except in having one region short or lacking (Figure 5.1; de Harven and Bernhard, 1956; Yamada, 1958; Bernhard and de Harven, 1960; de Harven, 1968). As a general rule, by anaphase the centrioles have already undergone replication in metazoan cells, ready for the next mitosis. Actually their reproduction has been commenced by the time mature sister centrioles have begun to separate in early prophase (Rattner and Berns, 1976); hence, these are the first organelles to undergo replication (Mazia *et al.,* 1960), a situation paralleled in lower eukaryotes also. However, growth of the daughter centrioles ceased after they had attained about two-thirds of their mature length (Zeligs and Wollman, 1979b), not to resume until late telophase or very early interphase. The precise role of the centrioles during nuclear division is difficult to appraise. In the first place, the axis of the centriole is unrelated to that of the spindle and its orientation varies greatly from one cell to another. Second, the spindle fibers do not really make contact with this organelle, although they do converge toward it. Hence, in metazoans it does not even serve as a nucleation center for the microtubules (de Harven, 1968).

Actually the centrioles appear to play no direct role in mitosis, as shown by several sets of evidence. One of these is provided by an experimental investigation in which the centriole was damaged by a laser microbeam during prophase (Berns and Richardson, 1977; Berns *et al.,* 1977). Although the treated cells completed mitosis and underwent cytokinesis, electron microscopic examination revealed that both centrioles had been severely damaged by the beam and remained to one side of the polar region during anaphase and subsequent events. The second set of evidence derives from several observations pertaining to the absence of centrioles from dividing mammalian cells. In one instance, cultured rat–kangaroo cells underwent meioticlike reduction division *in vitro* in the absence of centrioles (Brenner *et al.,* 1977); nevertheless, the microtubules of the spindle terminated at the poles in a pericentriolarlike cloud of material. An earlier study showed that although oogonia and oocytes of man and various rodents have centrioles during meiotic division up to the pachytene stage, these organelles then are lost and consequently are absent during later stages of the processes (Szöllösi *et al.,* 1972). Thus in vertebrates, as in higher plants, centrioles appear to play an insignificant role, if any, in nuclear division; furthermore, as shown in the descriptions of protistan processes, centrioles are often similarly inactive in unicellular organisms.

Contrastingly, the pericentriolar region seems to be of vital importance in such activities. Around each centriole of mitotic cells is a cloud of amorphous material, which varies extensively in appearance in different cells. Sometimes small electron-dense spherical particles have been observed here, but little more seems to be known about it. To judge from unicellular related organisms, these appear to be remnants of a cell part referred to as the centrosome. When this region, but not the centriole, after being sensitized with acridine orange, was damaged with an argon-ion laser microbeam (Berns *et al.*, 1977), the cells failed to undergo mitosis, although the nuclear membrane broke down, the chromosomes condensed and aligned on the metaphase plate, and cytokinesis occurred normally. Ultrastructural investigation demonstrated that kinetochore microtubules had formed in the usual fashion but no interzonal nor polar microtubules were present. Because the acridine orange sensitivity suggested the presence of RNA in this region, that nucleic acid was thought to play a role in spindle formation—the presence of ribonucleoprotein in the centrioles and kinetochores has actually now been confirmed in newt lung cell spindles (Rieder, 1979).

Other Cytoplasmic Events. Few investigations have been made at the ultrastructural level of other cytoplasmic events that accompany nuclear division, but a current one of rat thyroid epithelial cells indicates that changes in other organelles are manifold (Zeligs and Wollman, 1979a). From early metaphase to middle telophase, the usually flat cisternae of rough endoreticulum had been replaced by smaller vesicular types that bore less than half as many ribosomes as the cisternae had. At prophase the Golgi apparatus had moved from its usual polar location to surround the nucleus; then with the approach of metaphase, the dictyosomes gradually decreased in abundance until by the close of that stage, they were nearly absent. In their place appeared a number of uncoated vesicles, which became especially concentrated near the sides of the equatorial plate and continued to be associated with the periphery of the spindle through anaphase and early telophase. Later the usual dictyosomes could again be observed as the vesicles gradually disappeared. Even lysosomes underwent reorientation to some degree. After the onset of prophase, these organelles, although distributed throughout the cytoplasm as usual, tended to be grouped into clusters of 10 to 20 or more. Then as the cleavage furrow formed with telophase, they became especially abundant in that general region.

The Interzonal Fibers. One of the characteristic features of metazoan mitosis whose function remains inscrutable is the cluster of microtubules often called the interzonal fibers. According to the classification given earlier, this cluster comprises the type referred to as free microtubules, being unattached at the ends. Typically during anaphase and often in telophase, these increase in length as the regenerating daughter nuclei move farther apart (Roth *et al.*, 1966). Then as the cleavage furrow deepens during cytokinesis, the interzonal fibers receive an electron-opaque deposit that has been referred to as a midbody (Figure 11.9; McIntosh *et al.*, 1971; Mullins and Biesele, 1977; Zeligs and Wollman, 1979c). Deposition continues until by the end of cleavage the midbody is quite dense (Figure 11.9B) and lies at the surface, where it provides the only connection between the daughter cells (McIntosh *et al.*, 1979). Finally it undergoes either absorption or phagocytosis, thereby liberating the cells completely (Zeligs and Wollman, 1979c).

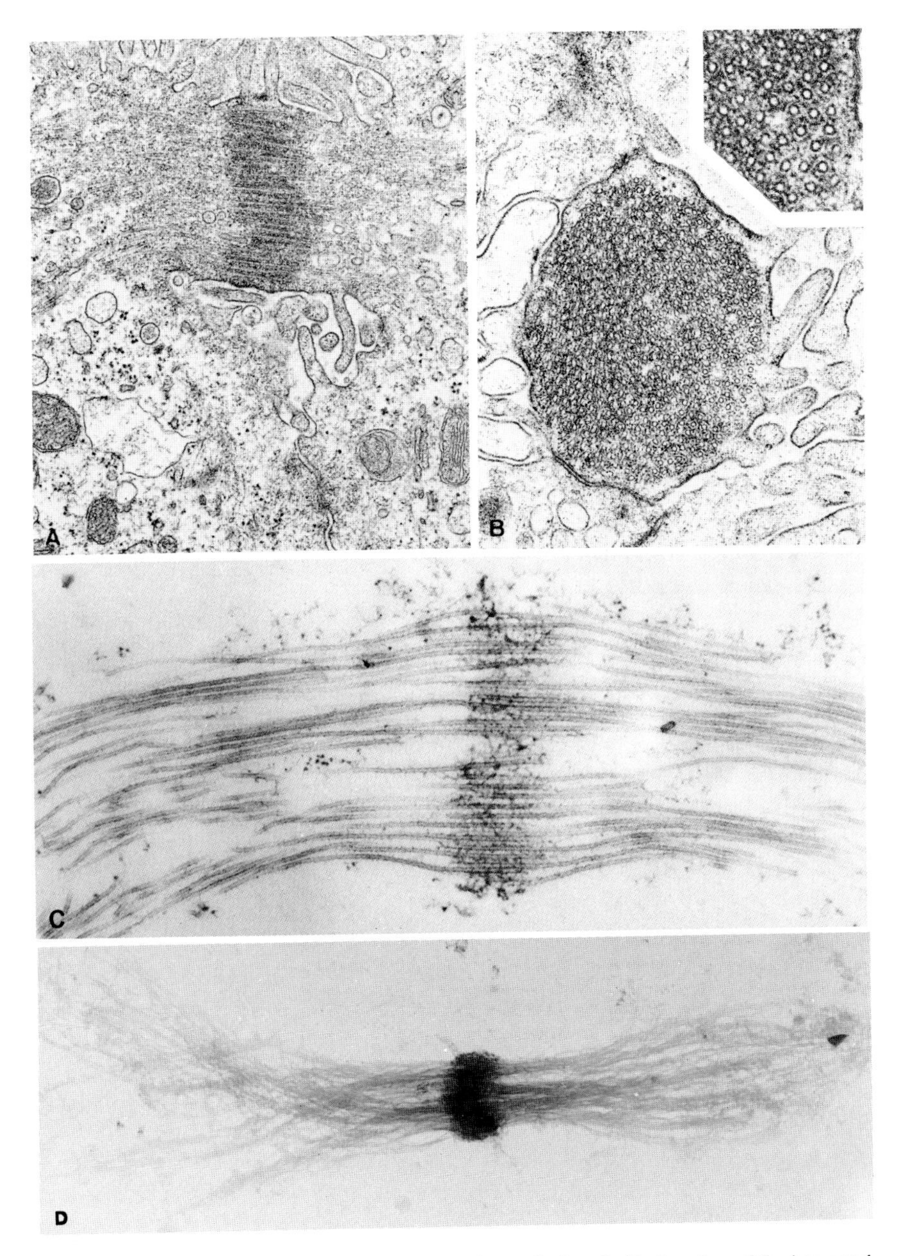

Figure 11.9. The midbody of metazoan interzonal fibers. (A) Longitudinal section of the interzonal fibers in a late-telophase rat epithelial cell, the microtubules of which have received deposits of electron-opaque particles to form a midbody. 20,000×. (B) In cross section, the midbody may be noted to contain around 460 microtubules, which are shown enlarged in the inset. 29,000×; inset, 56,000×. (A and B, courtesy of Zeligs and Wollman, 1979c.) (C) Longitudinal section of a telophase spindle from a cultured human cell. (D) A bundle isolated from a cultured human cell at telophase. C and D, 35,000×. (C and D, courtesy of McIntosh *et al.*, 1979.)

11.2.2. Nuclear Division in the Metaphyta

Now that the metazoan processes have been examined in some detail, it is possible to describe those of other eukaryotes more briefly and simultaneously devote greater attention to the differences that exist. Considering both the importance to the species of a precise method of nuclear division and the seeming simplicity of the events in the higher animals, an astonishing number of variations in details are encountered among eukaryotes.

Mitosis among Metaphytans. Few differences, however, are to be noted between the nuclear events of division in the metaphytans and those of metazoans (Bajer and Allen, 1966b; Fuseler, 1975; Laane *et al.*, 1977). Actually just two distinctions appear to exist, and of these, one is of only minor importance. That having little apparent significance pertains to nucleolar behavior; whereas the nucleolus of the Metazoa loses its staining properties early in prophase and disperses, that of the higher plants remains stainable and apparent until that stage nearly closes.

For many years the second distinctive feature of seed plant mitosis was considered particularly striking, but in view of current knowledge, it has lost some of its significance. The absence of centrioles in the present organisms is a consistent feature of mitosis, not just an occasional divergence, however, as it is among the higher animals. In this connection, the aberrant behavior of the centrioles in ferns and other cryptogams should be recalled, for even in those plants it was absent at times and at others it appeared *de novo* in greatly modified form (Chapter 5, Section 5.1.2). Thus in those forerunners of the flowering plants, there appears to be a phasing out of an active role in mitosis for this organelle.

One conclusion that might be derived from the number of traits during mitosis that are shared by the two great groups of multicellular organisms is that they must have had common ancestry relatively late in the nearly 4 billion span of years that life has existed on earth. These kinships are not direct ones, of course, but arise through their respective unicellular predecessors. In the case of the higher plants, these forebears were obviously the green algae, whose diversity of types represents any number of intervening steps along the path of ascent. In the case of the metazoans, however, the pathway is more obscure, not so much for lack of intermediate types, but largely because of obfuscations by dogma or other prejudices.

Other Aspects of Metaphytan Mitosis. One long-accepted feature of nuclear division shared by metazoans and metaphytans is the breakdown of the nuclear membrane in late prophase, the actuality of which has been questioned by evidence presented by Wada (1966, 1972, 1976). The truth of the matter probably lies with neither its continuing intact nor its breakdown, but with both. In all likelihood, in some cases it persists throughout much of mitosis to be replaced by a new one, as described in Chapter 1, Section 1.3.2, whereas in a number of other instances, possibly constituting a majority of the organisms, it breaks down at late prophase and is formed anew in telophase, with remnants persisting throughout division in certain species. That the breakdown does occur is clearly indicated by a recent study on *Marchantia,* in which the early-prophase nuclear membrane is clearly intact, whereas that of a late-prophase nucleus is broadly ruptured at one sector (Fowke and Pickett-Heaps, 1978). Through metaphase and anaphase the membrane is not present in electron micrographs, but at early telophase it reappears in interrupted fashion and becomes continuous later.

A structure whose existence has been established only relatively recently may or may not prove to occur in organisms other than seed plants. This organelle is the "preprophase band," first described in wheat meristem cells (Pickett-Heaps and Northcote, 1966); many subsequent reports have likewise been confined to monocotyledonous types (Kaufman *et al.,* 1970; Srivastava and Singh, 1972; Singh *et al.,* 1977), but it also has been shown to exist in *Phleum,* peas, and tobacco (Burgess and Northcote, 1967; Pickett-Heaps, 1969). It consists of a broad, flat bundle of microtubules that appear to have their inception close to the plasmalemma, whence they extend obliquely to the nuclear membrane. Originally they were believed to serve in orienting the nucleus prior to mitosis, but a study on the stomatal complexes of sugarcane cast doubt on that suggestion (Singh *et al.,* 1977). In this tissue, the subsidiary mother cells divided asymmetrically, while the guard mother cells did so symmetrically. The preprophase bands in the former type developed in the region where the future cell plate would contact the cell wall, but in the latter, they formed at polar regions, perpendicular to the plane of the cell plate, that is, in a location 90° away from that of the other type of cell.

11.2.3. Mitosis in Other Eukaryotes

Nuclear Division in Green Algae. As mitotic division in the higher plants differs so slightly from that of animals, the ancestral stock of the former groups likewise cannot be expected to show many important distinctions, and such has proven to be the actual case. Unfortunately, although excellent ultrastructural studies have been made of cell division in the green algae, the published electron micrographs have often tended to be devoted mostly to the details of cytokinesis. Nonetheless, the mitotic processes that have received sufficient attention establish clearly that the green algae represent an evolutionary series leading from relatively simple forms to those that merge into the embryophytes.

For example, in *Vacuolaria* the nuclear envelope breaks down and centrioles do not participate in mitosis (Heywood and Godward, 1974; Heywood, 1978), whereas in others the latter organelles do engage in those activities. One investigation into nuclear division of *Trichosarcina* and *Pseudoendoclonium* revealed an interesting pattern of behavior in centrioles (Mattox and Stewart, 1974). At the initiation of division, those organelles became positioned at the center of one side in line with the cleavage furrow that appeared precociously in these organisms. They then separated (after being replicated?) and commenced migrating toward the poles—by prometaphase they had attained a position on the persistent nuclear membrane equal to about 60% of the total distance of travel. By this time the spindle had developed within the nucleus and was directed toward one side of a gap at each end of the nuclear membrane. By metaphase, the spindle had become oriented to the terminal gap of the nuclear envelope, while the centrioles had moved to their definitive position in these algal cells lateral to the polar mass. The latter was a cloudy area, irregular in outline, made especially clear by the absence of ribosomes. Thus in these filamentous forms the centriole played no role in spindle formation nor in its orientation. In still other forms this organelle does play a direct role, as in many types of protistans. For instance, in *Chlamydomonas* the centriole of each of the two flagella becomes replicated at the initiation of mitosis, all four bodies then descending to the nuclear envelope (Johnson and Porter, 1968). There they give rise to four spindles that penetrate through fenestrae in the otherwise intact nuclear envelope and attach to the chromosomes.

One peculiarity, which seemingly is characteristic of protistan mitosis in general, has been revealed by ultrastructural studies on several types of green algae, too, particularly those belonging to advanced orders such as the Ulotrichales and Chaetophorales (Floyd *et al.*, 1972; Stewart *et al.*, 1973). In all of the members of the orders mentioned, as among algae of other types, kinetochores were absent from the chromosomes. Consequently, the latter structures lacked trailing arms and became arranged in flat plates at metaphase, and moved in the same form at anaphase. Because of their absence in these intermediate types, the kinetochores of metaphytan chromosomes cannot be homologous with those of metazoans, for they obviously must have had separate origins in each of these great groups of organisms. In the studies of the algal orders under discussion, the behavior of the centriole was not detailed; however, those organelles were directly involved in spindle formation, development being largely from the poles as the nuclear membrane became fragmented in early metaphase.

Mitosis in the Euglenoids. Because of the presence of chlorophylls *a* and *b* in the euglenoids as well as the green algae, close kinship has been supposed to exist between these two taxa (e.g., Leedale, 1967). On the basis of this suggested kinship, the mitotic processes of the present group should be expected to resemble those just described, but nothing could be further from the actual condition. Besides thus pointing out the dangers of proposing phylogenetic relationships on single characteristics, nuclear division in *Euglena* and its relatives offers both an extreme contrast to those just presented and an opportunity for introducing essential terms.

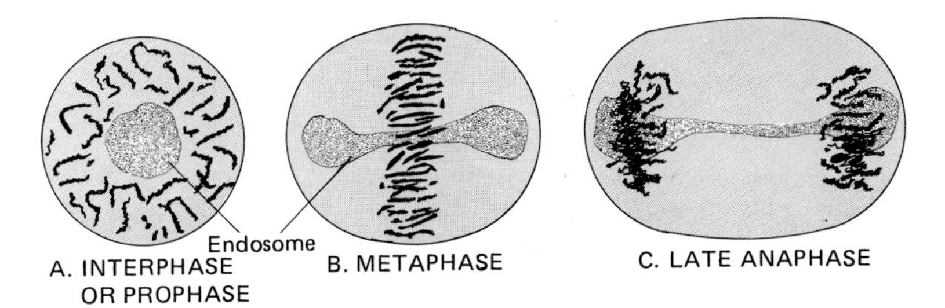

Figure 11.10. Mitosis among euglenoids. A number of unusual traits characterize nuclear division among euglenoids, including the presence of a central body known as the endosome and the absence of a spindle. Moreover, the chromosomes remain permanently condensed. (Based on Leedale, 1967.)

Few recent studies on these processes appear to be available, but what has been conducted at the ultrastructural level confirms those made earlier with the light microscope (see Leedale, 1958, for an extensive review). According to the micrographs in Leedale (1967), as in other accounts, the chromosomes in members of this group remain condensed throughout interphase as well as during mitosis (Figure 11.10). These surround a large electron-opaque body, called the endosome, which, as Leedale (1967) points out, is similar to nucleoli in containing RNA but not DNA, in its staining properties, and in the ability to fragment and recombine. It differs only in persisting and dividing during mitosis and in retaining RNA throughout nuclear division. In addition in the euglenoids, occasional transverse sections reveal the presence of a small body consisting of dual membranes that are spirally infolded, the folds being separated by a stroma that is continuous with the cytoplasm at one point.

At one time it was proposed (Dillon, 1962) that the endosome was a direct derivative of the bacterial nucleoid, but the data that have accumulated subsequently indicate that the suggestion is not valid. Because this strictly eukaryotic organelle plays an important role in evolutionary considerations, its behavior during mitosis in the euglenoids is especially worthy of note. In these organisms prophase actually can be recognized only by the changes that occur to this body, because of the persistence of the chromosomes in the condensed state during interphase. Thus the first sign of nuclear division is the slight elongation of the endosome from a spherical to an ovate conformation (Figure 11.10). As further elongation occurs, a constriction appears near the middle, which deepens as the bulk of the body becomes concentrated at opposite ends, eventually resulting in a dumbbell-shaped body. The chromosomes then become oriented in a circlet around the center of the dividing endosome, forming a typical metaphase plate of single or double chromosomes depending on the species. However, two outstanding traits need to be mentioned, the first being the persistence of the nuclear membrane throughout mitosis and the second and more remarkable, the complete absence of a spindle. Hence, as anaphase proceeds and the chromosomes separate (after replication in some cases), their movement poleward does not depend on the presence of microtubules. Whether actin filaments or others play a role in these processes has not been investigated fully (Blum *et al.,* 1965). Perhaps the absence of filaments and spindle accounts for the slowness of their movement, for anaphase occupies 60 to 70% of the total mitotic period, contrasting to 2.5 to 22% in all other known organisms and tissues. Telophase is marked only by the fracture of the central region of the endosome, the remnants of which flow rapidly to the main bodies of each daughter endosome; the latter then round out as the chromosomes become randomly distributed. These events are accompanied by the breakage of the elongated nuclear membrane at a midpoint and the resumption of the typical spherical configuration. Whether the nuclear membrane is replaced later as in other organisms does not seem to have been established. While absent in *Euglena,* a sparse spindle has been reported to develop in *Astasia* that is not associated with the centrioles, the latter remaining with the flagella (Blum *et al.,* 1965).

In view of the apparent absence of the spindle, the mechanisms of chromosomal movement in *Euglena* have been the subject of several hypotheses, none of which is entirely satisfactory. Although Leedale's (1958, 1967) observation that endosome division and chromosome movements are autonomous is firmly based, his intimation that chromosome behavior involves repulsion between sister chromatids is untenable, as no such repulsive forces are known to exist (Blum and Padilla, 1962; Edmunds, 1964). Moreover, sister chromosomes have already become separated in some species by metaphase and align singly on the plate. Nevertheless, his view that mitosis in these organisms is primitive has much merit, for it correlates well with a similar condition in such other organelles as the mitochondrion and dictyosome.

Nuclear Fission in Euglenoids. The primitive nature of the euglenoid nucleus is further substantiated by the occasional occurrence of a nucleus dividing directly by binary fission (Wenrich, 1924; Leedale, 1959a,b, 1967). Rather than being associated with cytokinesis, this behavior seems to be in response to unfavorable conditions of growth, for the two resulting "half-nuclei" remain within the cell. If such a cell subsequently undergoes division, as sometimes happens, the half-nuclei undergo mitosis in synchrony. Even cells that accidentally receive only one half-nucleus, as well as those that obtain three, are quite viable and are capable of producing similar progeny. Hence, although each whole nucleus seemed to divide merely by constricting into two to form half-nuclei, some degree of ordering of the chromosomes must have been an essential preliminary to fission. However, the chromosome number in half-nuclei varies between 40 and 45, whereas that of normal nuclei of *E. spirogyra* is consistently 86.

Nuclear Division of Yeast. If mitosis in euglenoids is to be considered primitive because of the absence of a spindle and the presence of an endosome, that of true yeasts must be archaic indeed, for identifiable chromosomes as well as spindles are nonexistent in this taxon proper* (Mundkur, 1961a,b). Feulgen-positive material (DNA) is present, however, but exists only in extremely fine granules dispersed through the nucleoplasm (Robinow and Marak, 1966). Hence, a logical interpretation appears to be that a fairly typical chromatin occurs in the yeast nucleus, which, although never condensing to form visible chromosomes, behaves genetically in chromosomal fashion. Consequently, genetical analyses have revealed the existence of linkage groups and many strains undergo meiosis; in other words, functional equivalents of chromosomes are present in these organisms, while actual structural units are not.

In addition, there is a large, crescent-shaped body of Feulgen-negative electron-opaque material, usually situated toward the periphery, often near the large vacuole that characterizes these cells. This crescentic body, variously referred to as the peripheral cluster, nucleolus, and crescent (Mundkur, 1961a; Thyagarajan *et al.,* 1962; Robinow and Marak, 1966), divides along with the nucleus in hour-glass fashion, and as it contains some RNA, it appears to correspond to the endosome of the euglenoids. However, in the present organisms, it is not so well developed as a rule, particularly in *Saccharomyces*. In *Rhodotorula,* it is not only more sharply defined than in bakers' yeast, but also is more centrally situated (Thyagarajan *et al.,* 1962), so that the correspondence to the typical endosome is even more marked.

Because the spindle and chromosomes are absent, nuclear division is extremely simple. As bud formation proceeds, the nucleus migrates partially into the growing daughter cell (Figure 11.11), followed by growth in size of the protrusion until the two ends of the hour-glass figure thus formed are equal, the nuclear envelope remaining intact until this point. The latter is then divided by constriction of the neck that joins the two cells, probably with the assistance of membrane-associated filaments (Byers and Goetsch, 1976). Inside the nucleus, the endosome has undergone similar stages and is divided concomitantly with the nucleus. Although no actual spindle is to be found in these organisms, a fibrous apparatus lying on the nuclear membrane has been described (Robinow and Marak, 1966), as pointed out earlier (Chapter 5, Section 5.1.2).

Mitosis in Euciliates. Nuclear division in the euciliates is complicated both by the presence of two types of nuclei, the macro- and micronucleus, and by the engagement of the organisms in conjugation and other forms of nuclear rearrangements. First, it should be recalled that the macronucleus, which includes more than 400 times as much DNA as the micronucleus (Woodard *et*

* It should be noted that chromosomes have been described in such genera as *Lipomyces* (Robinow, 1961) and *Wickerhamia* (Yuasa *et al.,* 1965); if further investigations confirm their presence, then the organisms need to be placed in some other systematic category, for none of the relatives of *Saccharomyces* possess visible chromosomes.

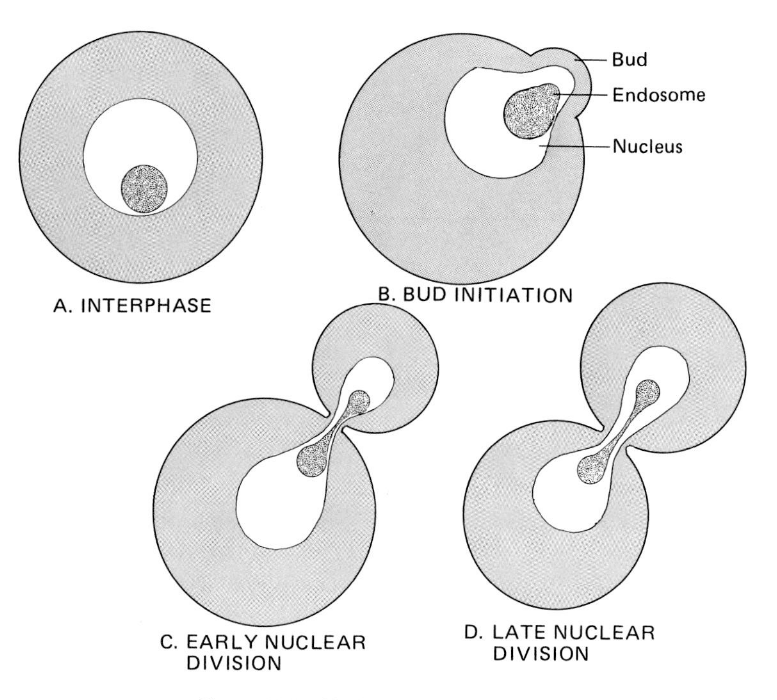

Figure 11.11. Nuclear division in a yeast.

al., 1961, 1966), contains the genes governing growth and fission—as Sonneborn showed many years ago (1938, 1946), ciliates can undergo fission indefinitely without a micronucleus but can do so only once or twice without the large type. In broad terms, the micronucleus lacks the ability to produce much RNA (Jurand and Bomford, 1965; Pasternak, 1967) and appears to serve during autogamy and conjugation in permitting genetic recombination and in giving rise to macronuclei after those processes are completed. The behavior of this type of nucleus following the initiation of conjugation when suitable mating types have made contact has been investigated to some extent with cycloheximide-treated *Blepharisma,* the results of which indicated that an unidentified protein had to be synthesized to a certain threshold before mitosis was triggered (Miyake *et al.,* 1979).

In *Paramecium,* during ordinary mitosis of the micronucleus, the basic features of the processes are essentially the same as those of metazoan cells (Jurand and Selman, 1970); however, a small number of distinctions have been observed. For instance, the nucleolus that is present during interphase stains poorly with methyl green and pyronin, suggesting that it contains little RNA. Moreover, a layer of microtubules convergent toward each pole remains throughout the cell cycle closely associated with the nucleoplasmic surface of the inner membrane of the nuclear envelope (Figure 11.12). This remains intact throughout division, as it does in all known members of the taxon (Carasso and Favard, 1965); at least in *Blepharisma,* during telophase a new one is formed within the nuclear substance (Jenkins, 1967; Inaba and Sotokawa, 1968; Giese, 1973). Although the chromatin condenses at prophase into compact chromosomes, the latter structures are not always rodlike in form but may appear as irregular balls; at metaphase, they align to form a typical equatorial plate, each attached to several spindle microtubules. Because of these attachments, kinetochores have often been assumed to be present, but electron micrographs fail to reveal any discrete structures, even in rodlike chromosomes (Rao, 1966). Rather, the spindle elements appear to attach randomly. These microtubules apparently are not essential to the processes, for the nucleus of *Tetrahymena* was able to divide normally when colchicine was present to prevent their growth (Jaeckel-Williams, 1978).

Following conjugation or autogamy, the micronucleus of *Paramecium* and *Kahlia* may be somewhat modified in configuration, often being crescentic or sickle shaped (Rao, 1966; Jurand, 1976). Usually these elongated nuclei have numerous, deep, longitudinal folds (Figure 11.12C), so that in transverse section they present an amoeboid appearance. Moreover, the nucleolus seemed to be lacking, as centrioles were also, but other details followed those of ordinary mitosis.

The macronucleus contains large quantities of DNA, proteins, and RNA. Although apparently becoming highly polyploid during its early formation from the pronucleus (Sonneborn, 1947; Kimball, 1953), radioautographic studies have shown that, after reaching maturity, this type of nucleus contains DNA-histone units that replicate only once each cell cycle (Gall, 1959; Ruffolo, 1978). Its polyploid nature is supported by genetic evidence, which has indicated that each segment of essential genetic information is represented by multiple copies (Doerder, 1979; Nanney and Preparata, 1979). In *Tetrahymena,* differentiation from the pronucleus has been established as including extensive alteration of the genome (Yao and Gall, 1979). Perhaps 10% of the micronuclear genome was lost or underrepresented in the macronucleus, some of the sequences hybridizing only with DNA from the micronucleus. Moreover, extensive amplification of the ribosomal DNA occurred, until around 200 copies were present per haploid genome, all of which were extrachromosomal and arranged as palindromes.

To understand its organization and the divisionary processes that occur, it is valuable first to review its formation from a pronucleus (or synkaryon) following conjugation, the discussion of which is based primarily on the detailed analysis by Kloetzel (1970). In the organisms that were studied, members of the genera *Stylonychia* and *Euplotes,* banded polytene chromosomes could be observed in thin sections of the early developmental stages of the forming

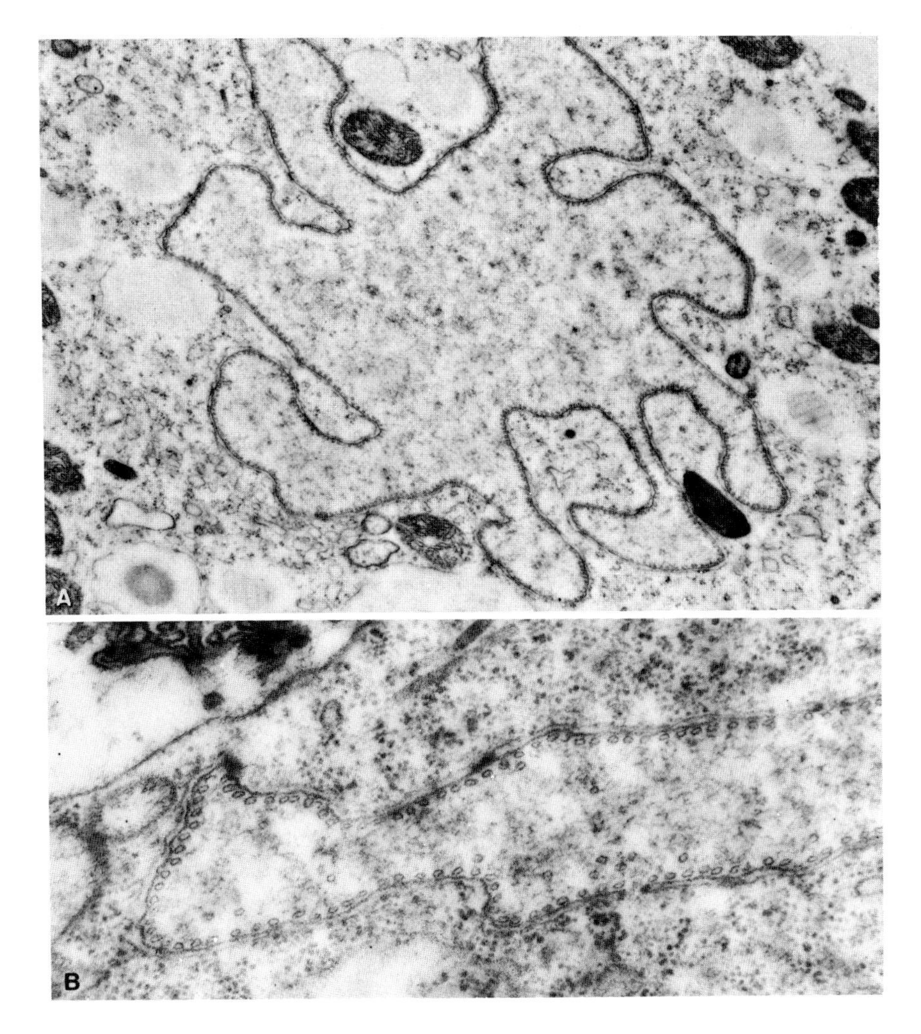

macronucleus. Both the electron-opaque and -transparent bands varied extensively in breadth and number, but nothing suggestive of a kinetochore could be detected in the published micrographs. Even in these early stages, fine membranes subdivided the entire nuclear contents into compartments, each of which enclosed one electron-opaque band of a chromosome along with the surrounding matrix of nucleoplasm. As development proceeded, the partitions increased in thickness, and additions of nucleoplasm gradually began to separate the compartments, while the chromosomal opaque bands slowly underwent relaxation and dispersal. By the time the chromatin had fully relaxed, the compartments enlarged, resulting in a nucleus literally filled with thousands of separate cham-

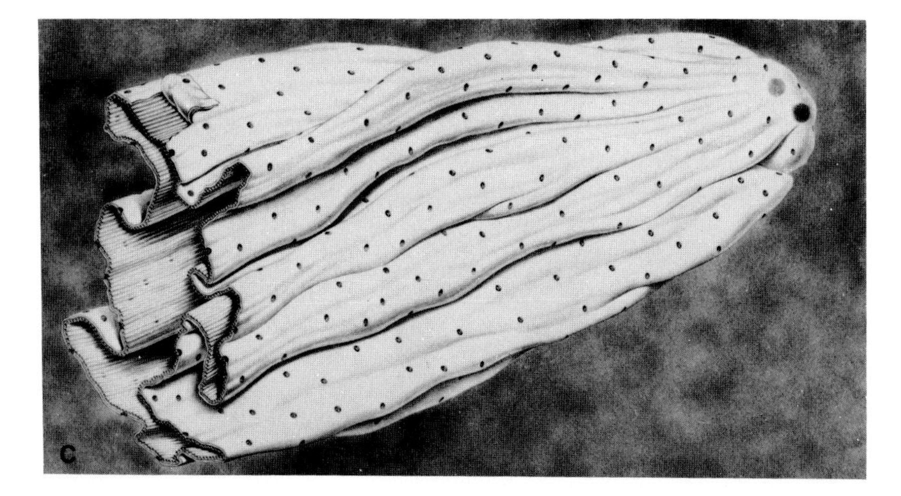

Figure 11.12. The postconjugation crescentic stage of the micronucleus in *Paramecium aurelia*. (A) Cross section, with a single layer of microtubules underlying the deeply folded nuclear envelope. 15,000×. (B) Enlarged view of a nuclear-envelope fold and microtubules. 56,000×. (C) Three-dimensional reconstruction of about one-quarter of a micronucleus at this stage. (All courtesy of Jurand, 1976.)

bers. Later, however, the partitions themselves underwent degradation and disappeared by the time the S phase of the cell cycle had been initiated and the so-called replication bands had appeared in the elongating macronucleus. Before their breakdown, the contents of each compartment usually were finely granular, but sometimes one or more clumps of electron-opaque material were present. As DNA synthesis proceeded, nucleolar (or endosomal) bodies appeared, so that the mature macronucleus contained a large number of chromatin granules and endosomal particles.

As chromosomes are absent in macronuclei, nuclear division is extremely simple and amitotic, resembling the processes in yeast in forming an hourglass structure within the persistent envelope. Here, however, the two ends were demonstrated to be of equal size throughout division. After the nucleus had greatly elongated, so that a long narrow stalk joined the two rounded ends, a furrow formed at a midpoint, deepened, and separated the daughter macronuclei (Jurand and Selman, 1970). Thus the electron-opaque chromatin granules, shown to be Feulgen positive by Wolfe (1967), became approximately equally distributed between the daughters, as was the case with the less dense endosomal particles, which had been shown to be pyronin positive (Nanney and Rudzinska, 1960). Recently, evidence gained by serial sectioning has been presented that intimated that the chromatin particles were not entirely discrete, but frequently were united into loose series of chromatin clumps (Ruffolo, 1978).

Nevertheless a mean difference in DNA content of 11.6% has been shown to exist between daughter cells (Witt, 1977; Berger, 1979; Doerder, 1979). Unlike the micronucleus, microtubules were present within the macronucleus only during division; a number of others lying on its exterior surface appeared to aid in its elongation, but not division (Jenkins, 1977).

Mitosis in the Fungi. During the past decade, many studies of fungal nuclear division have been conducted on a modern basis, many of which have centered on the spindle or peculiar spindle pole body (Lu, 1978; Littlefield and Heath, 1979), and thus have already received attention in appropriate sections of this chapter. A small fraction, however, have focused on the nuclear events, so that the latter have become quite well elucidated (see Girbardt, 1978, for a review). Basically, division of the nucleus closely resembles the processes of metaphytans and metazoans (Figure 11.13), a distinct nucleolus being present during interphase and early prophase but thereafter disperses until its recondensation in telophase (Ashton and Moens, 1979). Its similarity to metazoan mitosis is further accentuated in such primitive fungi as *Physarum* by the presence of a centriole that bears a few astral rays, as already pointed out (Aldrich, 1969; Kasama, 1974). In members of the genus mentioned, this and other details vary with the cell type, for in the myxamoebae haploid mitosis includes a well-marked extranuclear spindle associated with the centrioles, while the nuclear envelope breaks down during prophase. The chromosomes have distinct kinetochores (Moens, 1976; Heath, 1978), but these are quite different from those of the metazoans, being electron-dense caps that appear to clamp onto the chromosomes. Thus the term kinetochore has been loosely applied to a diversity of analogous structures (Wada, 1966), having a common function but different morphologies; consequently, a thoroughgoing study providing new terminology is a pressing need. Metaphase shows nothing distinctive, but anaphase toward its close is marked by a characteristic feature—the chromosomes undergo relaxation so as to form an apparently fused mass, which persists into telophase, during which period a new nuclear envelope develops. On the other hand, the fruiting plasmodia of this genus undergo diploid mitosis, which differs from the foregoing chiefly in the persistence of the nuclear envelope throughout division, the presence of an intranuclear spindle, and the absence of a centriole (Heath, 1978).

Although the basic features described above characterize the members of

→

Figure 11.13. Meiotic division in a hemiascomycete, *Cephaloascus fragrans*. (A) The nuclear processes are essentially those of metazoans, except that the nuclear membrane persists. A metaphase nucleus shows the central spindle fibers extending to a spindle-pole body, not a centriole; the chromosomes are not visible in these electron micrographs. 32,000×. (B) During anaphase, the chromosomes appear as vague electron-opaque clouds, while the spindle-pole body develops into a three-layered structure, and a centrosome becomes apparent. 35,000×. (Both courtesy of Ashton and Moens, 1979.)

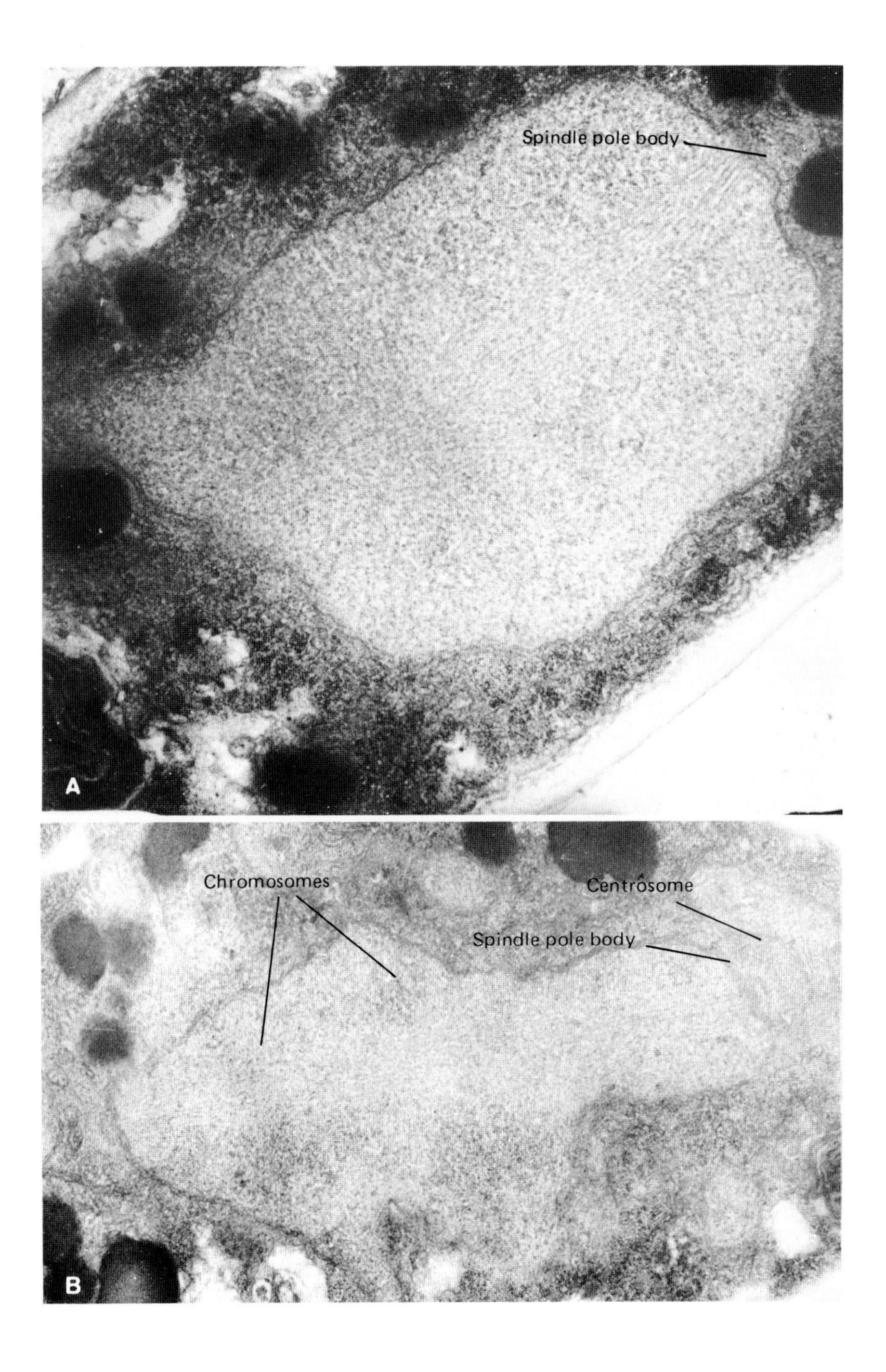
Spindle pole body
A
Chromosomes
Centrosome
Spindle pole body
B

the taxon as a whole, some striking specializations occur in some of the more advanced types. In the basidiomycete *Bullera alba,* for one example, the nuclear membrane remained intact throughout mitosis and the nucleolus persisted through anaphase (Figure 11.14; Taylor and Wells, 1979). The latter, however, stayed in that part of the nucleus that was to remain in the parent cell, while the spindle and chromosomes underwent their activities transversely in that portion that was to be passed to the daughter cell. At telophase the region containing the mitotic apparatus became greatly elongated in such a fashion that part of it returned to the parental body. How the nucleolus was subsequently reapportioned between the two regions could not be established, because this structure became diffuse in late anaphase. Unusual as such an eccentric spindle and nucleolus are, they are not unique to the fungi, but also occur in coccoid protozoans like *Eimeria magna* (Danforth and Hammond, 1972; Hammond, 1977). Thus the two groups may prove to be related. A second example of diversification among the fungi is shown by *Saprolegnia* (Heath, 1978), in which the kinetochores at prophase become closely associated with the nuclear envelope, adjacent to the spindle-pole body (Heath and Greenwood, 1970).

Nuclear Division in Dinoflagellates. Nuclear division in dinoflagellates has attracted a fair abundance of investigations at both the light and the electron microscopic levels (for example, Chatton, 1920; Hall, 1925b; Bělař, 1926; Leadbeater and Dodge, 1967; Kubai and Ris, 1969; Soyer and Haapala, 1973, 1974; Ris and Kubai, 1974; Oakley and Dodge, 1976; Spector and Triemer, 1979). Among the most outstanding of the distinctive features that these have brought to light is the morphology of the chromosomes, which have proven to be thick, banded rods, comprised of numerous supercoiled fibrils. Thus the intact chromosomes appear like coiled rope, a basis for a claim that they resemble the nucleoid of bacteria (Ryter and Kellenberger, 1958; Dodge, 1963a; Kubai and Ris, 1969; Soyer, 1971). However, other than a fibrous appearance, very little resemblance can be noted in thin sections, a coiled arrangement of the fibers being quite evident in dinoflagellate chromosomes that is absent in bacterial nucleoids (Soyer and Haapala, 1973, 1974). An additional distinction of the 20 to 300 chromosomes found in free-living members, as well as of the four of parasitic genera like *Syndinium* (Ris and Kubai, 1974), is in their being permanently attached to the nuclear envelope.

The spindle also is unique in that it consists of a bundle of extranuclearly formed microtubules, and becomes enclosed in a deep fold of the nuclear membrane. The individual spindle components do not become attached directly

→

Figure 11.14. Mitosis in the basidiomycete *Bullera alba*. (A) The nucleus partly migrates from the parent cell into the daughter that was produced by budding. As it does so the spindle-pole bodies begin to separate, while a spindle forms between them. 77,000×. (B) After the nucleus has partially penetrated the daughter cell, a metaphase configuration is formed across the cell. 40,000×. (Both courtesy of Taylor and Wells, 1979.)

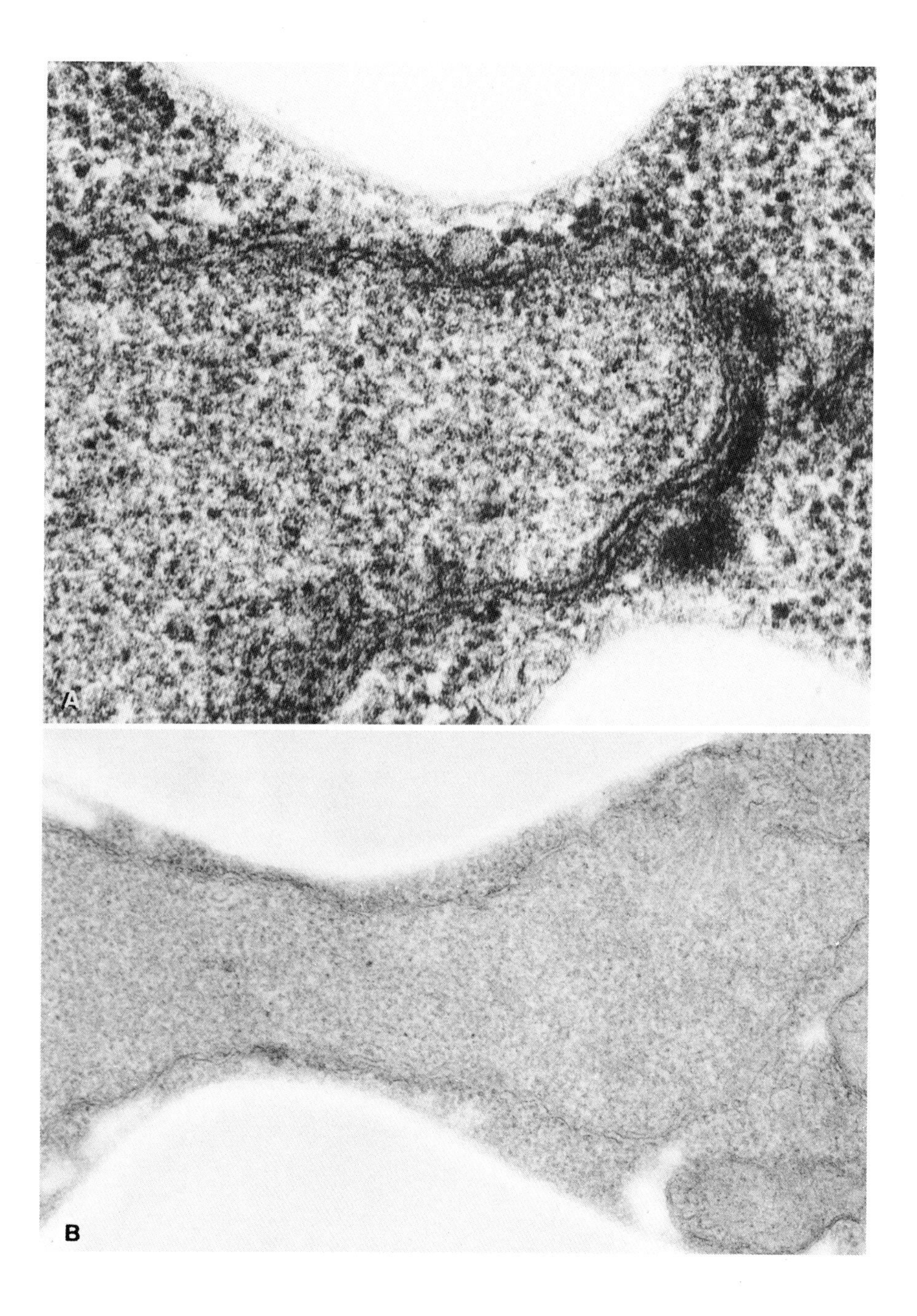
A
B

to the chromosomes but rather to specialized structures within the nuclear membranes referred to as kinetochores, even though they are not part of the chromosomes; however, the chromosomes are indirectly fastened to them (Kubai and Ris, 1969; Ris and Kubai, 1974; Spector and Triemer, 1979). In parasitic species the spindle forms in association with the flagellar centriole, but the latter does not engage in mitosis in free-living species (Ris and Kubai, 1974). The chromosomes, which are permanently in the condensed condition (Figure 11.15; Bouligand *et al.*, 1968), have been demonstrated to contain little or no acid-soluble proteins, such as histones (Dodge, 1964; Rizzo and Noodén, 1972, 1973). However, by use of gel electrophoresis on the small acid-soluble fraction, a protein has been found whose properties resembled those of histone H4, the smallest of the mammalian species (Dillon, 1978).

As Rizzo and Noodén (1972) pointed out, the several unusual traits displayed by the nucleus during division, together with the chemical and structural properties of the chromosome, attest to its representing a primitive stage in the evolution of that body. To this list must be added also the presence of an endosome in such primitive representatives as *Oxyrrhis* (Hall, 1925a; Dodge, 1963b), a feature often lost in more advanced forms (Hall, 1925b; Skoczylas, 1958). Because this structure behaves precisely as it does in euglenoids, at one time it was proposed that the two groups were members of a common branch (Dillon, 1962); subsequent studies having established the existence of an elementary spindle in the dinoflagellates, it now is evident that the two taxa represent separate branches, although closely situated ones. Consequently, the chromosomes of the euglenoids, which are the more primitive organisms, should be expected similarly to lack histones, a point worthy of investigation.

Mitosis in Amoeboids. Another large group of eukaryotes whose nuclear divisionary processes merit attention here has been the subject of almost no investigation into this activity at the ultrastructural level. Consequently, the meager accumulated data supplied by the few electron microscopic studies must be liberally supplemented from those by light microscopists. What makes the situation still worse is that these biologically and economically important organisms are presently lumped into a single class, the Sarcodina, whereas in actuality they are as diversified cellularly as the flagellates have proven to be. At least three major branches of living things have been shown to consist largely or entirely of amoeboid types (Dillon, 1962, 1963), and several others include groups of lesser systematic importance whose members are amoeboids. How in

Figure 11.15. Nuclear characteristics of dinoflagellates. (A) In this interphase nucleus of *Peridinium balticum*, the chromosomes are permanently condensed, as in other dinoflagellates and euglenoids. The large endosome is another similarity to the latter group. 2500×. (B) The banded chromosomes do not move en mass during anaphase but migrate individually. 6500×. (C) Even in midtelophase, as here, the chromosomes do not form a compact cluster. 3900×. (All courtesy of Tippit and Pickett-Heaps, 1976.)

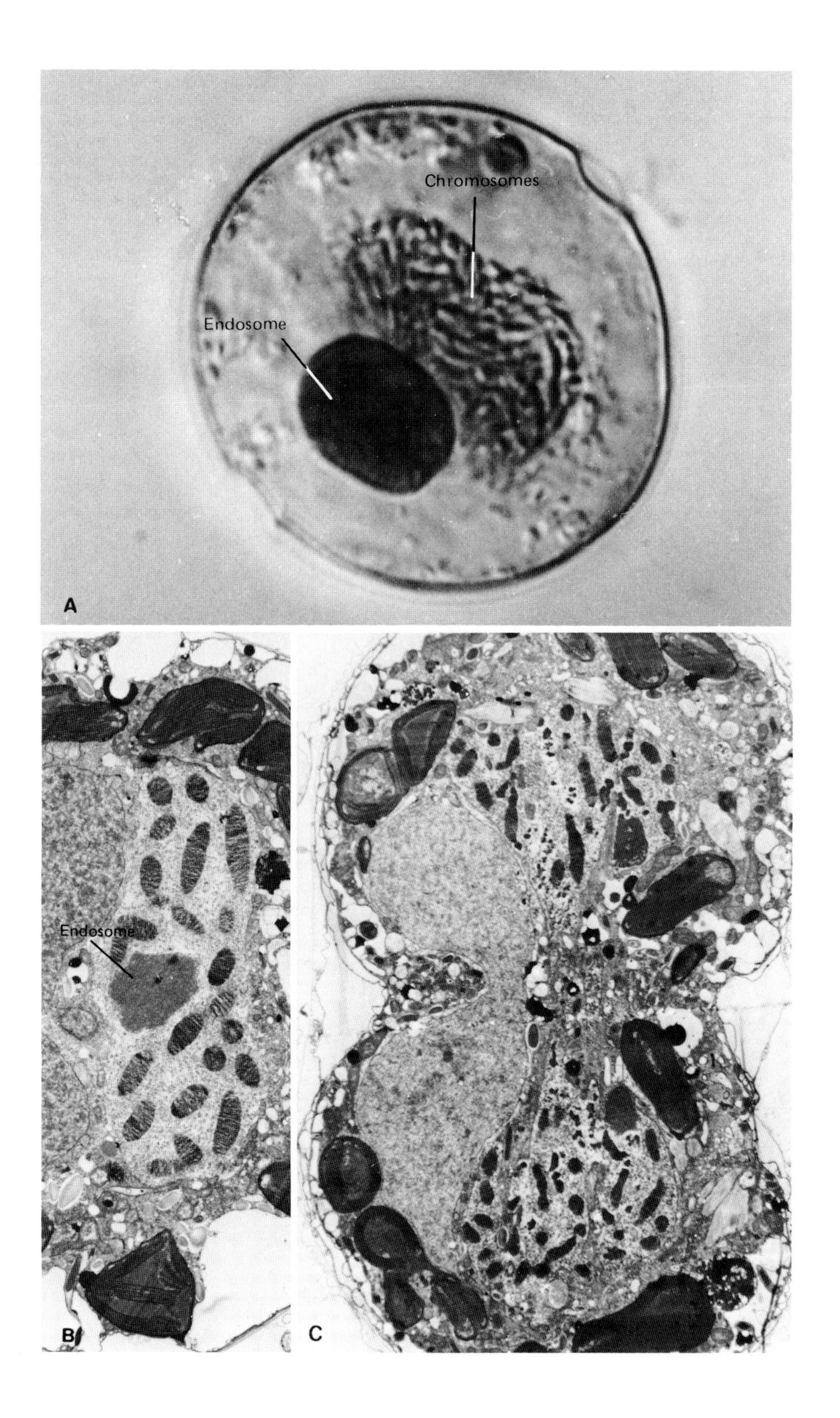
Chromosomes
Endosome
A
Endosome
B
C

protozoology the persistent practice can be justified of employing a single criterion (that of a major type of locomotory apparatus) as the sole basis for arranging the unicellular animal-like organisms into classes or subphyla is difficult to comprehend, especially in the face of the great masses of cytological, ultrastructural, and macromolecular evidence that clearly indicate the artificiality of this treatment. The arrangement given by algologists for the chlorophyllaceous pseudopod-bearing forms at least utilizes a few biochemical traits and, hence, is a far more natural scheme of classification.

The three major branches of amoeboid forms just mentioned were proposed on the differences in the behavior of the endosome during mitosis. Because they were found to be intermediate between the euglenoids and dinoflagellates on one hand and the most advanced organisms on the other, the differences that exist among them in this trait were taken as representing successive steps in nucleolar development. In *Naegleria* and other soil-dwelling amoeboid flagellates, like *Dimastigamoeba* and *Tetramitus,* the endosome is very large and, through metaphase, undergoes division much as that of *Euglena* (Figure 11.16), in that it remains intact as it elongates into a dumbbell-shaped body (Rafalko, 1947, 1951). At this point it is similarly encircled by the metaphase plate of chromosomes, but the latter are elongate, loose rods that are spaced in an orderly fashion and are associated with a spindle. Then as the

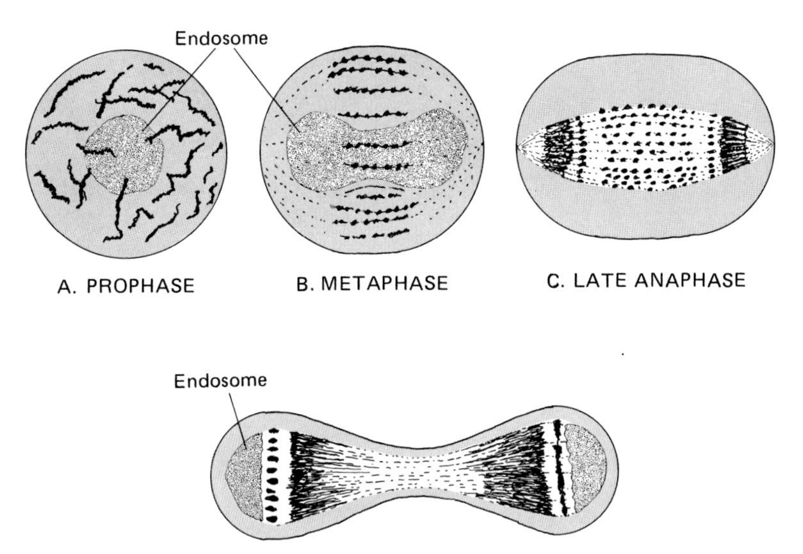

Figure 11.16. Mitosis in the flagellate amoeba *Naegleria*. (Based on Rafalko, 1947.)

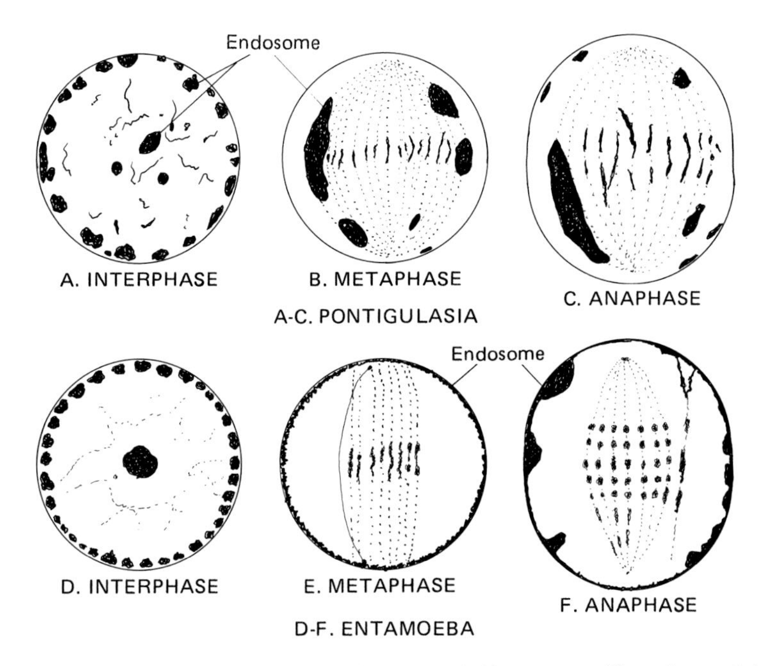

Figure 11.17. Stages in the nuclear divisions of some amoeboid protozoans. The endosome behaves increasingly as a nucleolus by becoming diminished in extent during the later stages of mitosis.

chromosomes move apart in anaphase, the endosome disintegrates into small particles, one cluster of which is located at each pole in advance of the chromosomes, and a third one medially. In all three locations, the particles appear to cling to the microtubules of the spindle and interzonal fibers. With telophase, the central cluster breaks into two, half of the particles moving into each daughter nucleus. The flagellar centriole of *Naegleria* does not participate in mitosis but disappears along with the flagella during the transformation from the flagellate into the amoeboid type (Fulton and Dingle, 1971).

The second, much more advanced group of amoeboids is represented by *Pontigulasia vas* (Stump, 1943), a free-living freshwater organism that encloses itself in a test of sand grains. As the nucleus of protociliates like *Opalina* shares many features in common with these forms, that taxon probably should be considered to be related to this group of amoeboid forms. In both types the endosome during interphase behaves much as a definitive nucleolus (Figure 11.17), except that it is broken into a number of particles (Ivanić, 1936; Stump, 1943). After a spindle has developed intranuclearly in prophase, large clumps of the nucleolus are found attached to it and persist through metaphase. Then in anaphase, most of the particles disappear, except for some small ones affixed to the interzonal fibers.

The third level resembles the second in having the endosome broken into a number of particles throughout interphase, as in *Entamoeba muris* (Kofoid and Swezey, 1925; Wenrich, 1940), *Lesquereusia* (Stump, 1959), and *Pelomyxa* (Roth and Daniels, 1962; Daniels and Roth, 1964; Daniels *et al.*, 1966). In the present case, these particles are largely concentrated into a layer beneath the nuclear envelope, where they often remain throughout mitosis (Figure 11.18). After anaphase has commenced, however, a number of small particles are to be

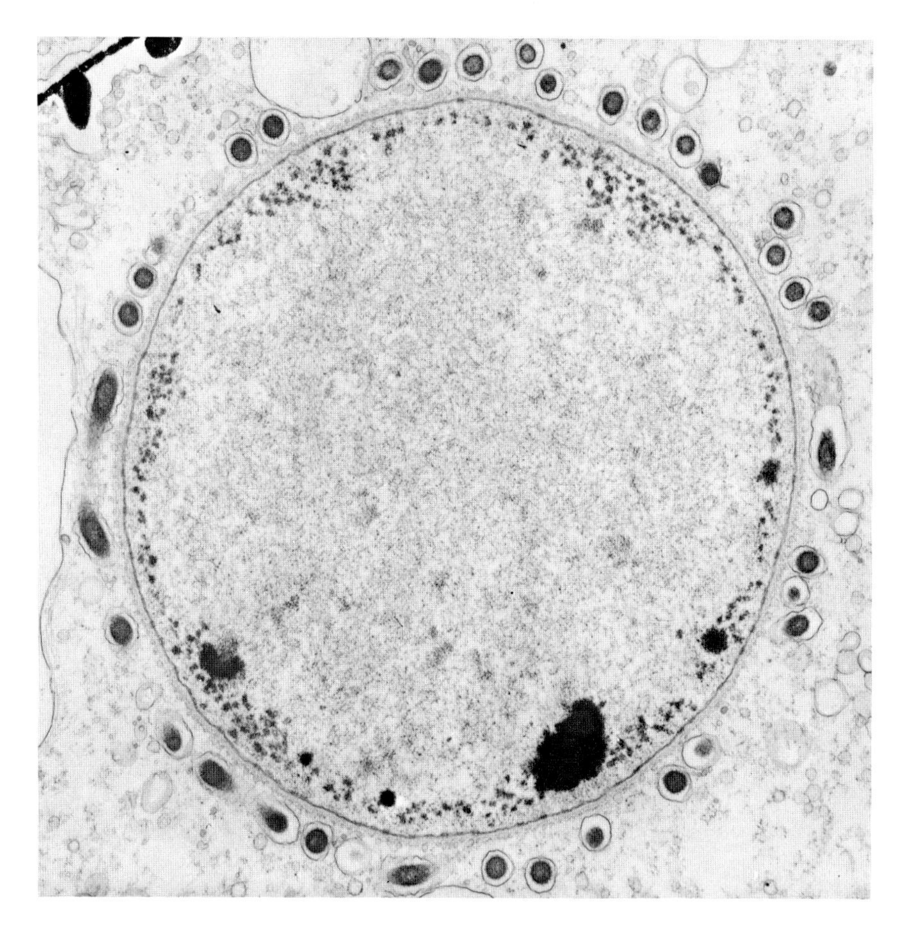

Figure 11.18. Interphase nucleus of *Pelomyxa palustris*. As in the other types of amoeboid forms, the endosome is broken into numerous particles deposited beneath the nuclear membrane. At the present level of evolution they remain there during mitosis, instead of aggregating into large clumps. 12,500×. (Courtesy of Daniels *et al.*, 1966.)

noted on the spindle and interzonal fibers, passing to each daughter nucleus in approximately equal numbers at telophase. In *Pelomyxa illinoisensis* they were considered to disappear in very late telophase, to be replaced by smaller, much more numerous nucleolar particles (Daniels and Roth, 1964). The nuclear membrane, at least in this latter species, disintegrates into a number of fragments that persist throughout the entire mitotic process. During early prophase after the 100 or so small spheroidal chromosomes have condensed, an irregular spindle is developed outwards from each of these bodies and subsequently grows toward each pole.

Mitosis in Hypermastiginous Flagellates. As the hypermastiginous flagellates have a type of nuclear division that is both striking and complex, these organisms must certainly represent one of the most advanced of the protistans. Fortunately their mitotic processes have received much attention at the ultrastructural as well as light microscopic levels, most observations being in general agreement. As *Barbulanympha* not only has been the most thoroughly investigated representative but also shows the essential features particularly clearly, it is especially suited for serving as the model here.

The first trace of mitosis is the appearance of two long rootlike structures, referred to as elongate centrioles (Cleveland, 1953, 1954, 1955, 1958, 1963; Inoué and Ritter, 1978; Ritter *et al.,* 1978). However, the structural features shown in the published electron micrographs are not those of that type of organelle; because of the morphological resemblances, they are here considered to be rhizoplasts (Chapter 6, Section 6.3.3). At the posterior end each is swollen into a knoblike part called the centrosome; strictly speaking this represents another misapplication of a term (see Chapter 5, Section 5.2), but because of the present insufficiency of data, that term is used nevertheless, leaving more precise terminology for a future investigator. Even before these centers contact the nuclear envelope, a rather sparse aster is developed from each rhizoplast, the fibers of which extend through each centrosome. These rays are rather flexible, the longer ones tending to drape over opposite sides of the nucleus as mitosis progresses (Hollande and Carruette-Valentin, 1971).

Later as the chromosomes condense, a few spindle fibers form between the two centrosomes, followed by a rapid increase in numbers. Along with this numerical growth, the spindle and interzonal threads become increasingly long, as the centers become more widely separated. At this point in time the chromosomes are fully condensed and are seen to be attached near their middles to a knoblike "kinetochore" embedded within the persistent nuclear membrane, recalling their behavior in dinoflagellates. The remainder of the processes also display resemblance to mitosis in those organisms, for some of the spindle fibers become attached to the kinetochores externally. A deep invagination then forms in the envelope into which the spindle fits, so that the nucleus ultimately becomes completely wrapped about the latter (Ritter *et al.,* 1978). Without forming a recognizable metaphase plate, sister chromatids become separated and after they have moved to their respective poles, the interzonal fibers become elongated, separating the polar bodies and chromosomes still farther. The lips of the cleft in the nuclear envelope then part as a furrow forms transversely, finally partitioning the nucleus into two (Inoué and Ritter, 1978). In *Trichonympha,* the events, as far as they have been described, appear to be comparable to those of *Barbulanympha* (Kubai, 1973).

Similar centrosomes have been reported in several groups of organisms. In foraminiferans, including *Allogromia laticollaris,* a corresponding structure has been noted that was membrane enclosed and contained a stroma consisting of granular and fibrillar material (Cesana, 1971; Schwab, 1972). In other members of that taxon, however, typical centrioles appeared to serve as a mitotic center (Schwab, 1968, 1969). To the contrary, the centrosome in the hypermastiginous flagellates was simply a fibrous mass, not covered by a membrane (Ritter *et al.,* 1978). The central body present in diatoms likewise was not protected by a membrane but consisted entirely of a dense granular substance (Drum and Pankratz, 1963; Tippit and Pickett-Heaps, 1977; Tippit *et al.,* 1978). In these organisms, too, loosely placed sinuous microtubules formed an irregular unorganized aster about each centrosome, which persisted through interphase. Whether these several variations on the common theme represent homologous organelles remains an open issue, for the existing data are far too scant to provide a firm answer to the problem. The three groups involved, nevertheless, do show other similarities that indicate their kinships, as shortly will be seen, suggesting the possibility of common descent for their centrosomes.

11.3. NUCLEOID DIVISION IN PROKARYOTES

Despite an abundance of ultrastructural and biochemical information concerning the events in division of the prokaryotic nucleoid, the mechanical details underlying the processes remain unknown. While it is well established that the DNA molecule has already undergone replication and that the two resulting sister duplex molecules become separated as the nucleoid is split into two, much remains for elucidation, as becomes evident in the discussion that follows immediately.

Nucleoid Division in True Bacteria. Division of the nucleoid in the true bacteria is most clearly shown by those forms in which the cell wall and nucleoid divide in concert, such as *Erwinia* (Grula and Smith, 1965; Grula *et al.,* 1968). In these and all other known prokaryotes, replication of the DNA is a continuous process, so that in the absence of discrete periods of synthesis, division, and gaps (Dillon, 1978), they possess nothing that resembles the cell cycle of eukaryotes. The nuclear body in a typical newly divided cell is a regular rod filling the greater portion of the interior of the cell, which in longitudinal sections may be observed to consist of long, intertwined fibrils lying in

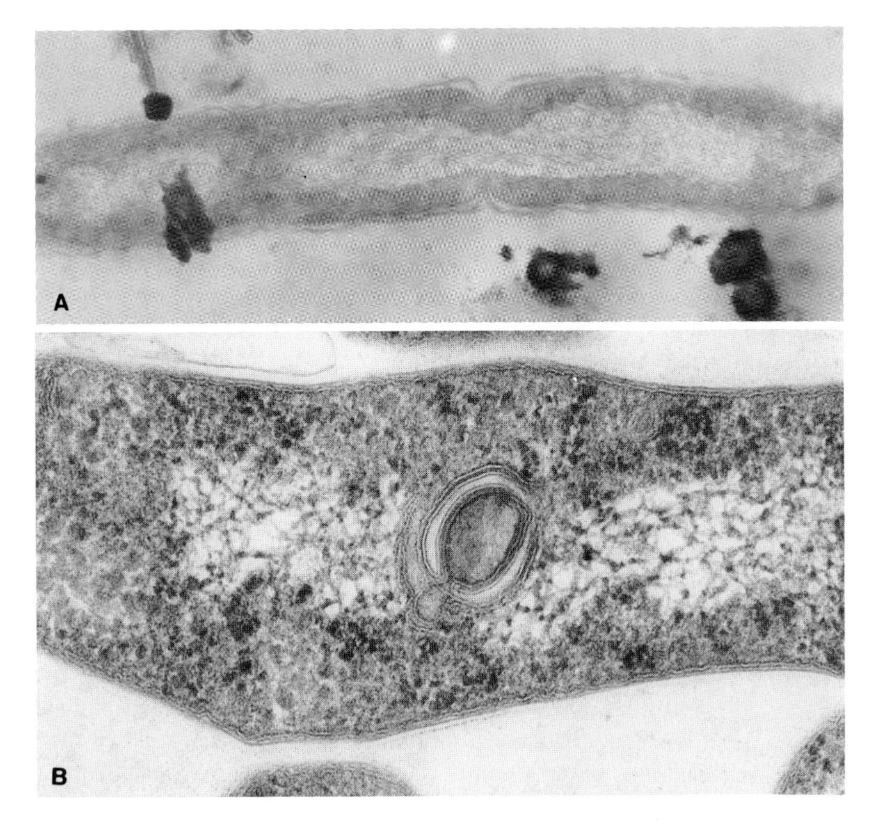

Figure 11.19. Nucleoid characteristics of higher bacteria. (A) In this example of a bacterium of the genus *Erwinia*, the elongate nucleoid is beginning to undergo division, cytokinesis being already well advanced. 46,500×. (Courtesy of Grula and Smith, 1965.) (B) Septum formation is nearly complete in this specimen of *Mycobacterium* before nucleoid division is completed; the fibrous nature of the DNA molecule should be especially noted. The mesosome has remained associated with the septum. 78,000×. (Courtesy of Barksdale and Kim, 1977.)

an electron-transparent stroma (Figure 11.19). During growth of the organism, this structure maintains its mass in such a fashion that the cytoplasm around it is about equally wide at maturity as it is immediately following division. When the mature dimensions of the cell have been attained, a constriction in the nucleoid becomes evident at the center, opposite to an invagination of the plasmalemma and cell wall. The constriction deepens until division into two nucleoids results, a step that always occurs slightly in advance of cytokinesis.

These gross aspects of nucleoid division are sufficiently clear; what is not comprehensible is (1) how two intimately intertwined molecules of great length can untwist and separate, a problem treated in detail at the molecular level

(Dillon, 1978). Although this difficulty is innate to all replicating DNA molecules, in the present organisms it is greatly intensified by the 200 or more folds required to pack the elongate circular molecule into a short space. Another problem (2) follows the unraveling of the replicated DNA, for the two intimately associated molecules must be moved to opposite ends of the nucleoid, but what mechanism thus separates them topographically has not even been suggested. Then (3) after this physical separation has been completed, constriction of the nuclear stroma is conducted by some still undetermined force. The ingrowing plasmalemma and wall cannot play a role in this process, for nucleoid division is completed well in advance of the ingrowing parts. Moreover, this ingrowth is not really a force but is a consequence of new material being deposited molecule by molecule by the cytoplasm. Nor can pressure generated by the constricting septal region be the causative agent, for that force would be distributed equally throughout the fluid cell contents. Whether actin or other contractile fibrous proteins participate in the division of the nuclear body has not been investigated.

In exponentially growing cultures of *E. coli* and many other species, nuclear division occurs nearly two full cycles in advance of cytokinesis, a particularly lucid exposition of which activity is provided by the time-lapse, phase-contrast sequence by Mason and Powelson (1956). As the original cell is seen to begin cytokinesis, two nucleoids commencing to divide are visible in each daughter (Figure 11.20). Thus it is evident that the daughter nucleoids not only had completed division, but their products had already begun to do likewise. Because cleavage-furrow formation is thus more than one and one-half laps behind fission of the nuclear matter and therefore quite remote from the narrowing site, the constricted regions that form in the nucleoid during those processes obviously cannot result from ingrowth of adjacent cell membrane.

Nuclear Division in Blue-Green Algae. The nucleoid of such blue-green

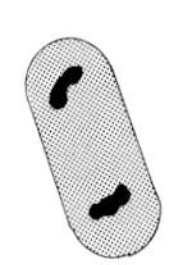
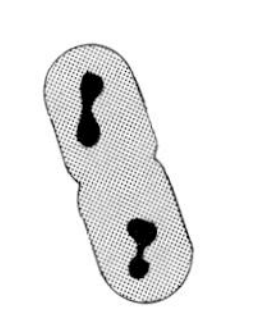

Figure 11.20. Cell and nucleoid division in exponentially growing bacteria. Under rapid-growth conditions, nucleoid division precedes cytokinesis by more than 1½ cell cycles. In (A), for example, the nucleoid has already divided once and is commencing another division before the first cleavage of the cytoplasm begins (C). (Based on Mason and Powelson, 1956.)

algae as *Oscillatoria splendida*, in which the photosynthetic thylakoids are more or less parallel to the cell wall, appears not too unlike that of the true bacteria in electron micrographs. Undoubtedly the most striking distinction is its irregularity (Figure 11.21A), a form suggestive of a poorer internal organization. Inspection of the illustration reveals a constriction commencing to form between the opposite sides of a developing septum, which is decidedly off-center in location. As a result, one end of the dividing nucleoid is distinctly larger than the other. Such eccentricity of nucleoid division and cytokinesis can also be noted in related genera, including *Plectonema* (Jensen and Sicko, 1974; Doonan and Jensen, 1977) and *Synechococcus* (Lawry and Jensen, 1979).

In blue-green algae like *Anabaena* and *Nostoc* (Jensen, 1979), in which thylakoids are in less than regular arrays, the nucleoid is so highly dispersed among the other organelles (Figure 11.21B) that early electron microscopists failed to find a nuclear body (Bradfield, 1956; Jensen and Ayala, 1976, 1977). Although visible in sections only as scattered patches, it is nevertheless not difficult to visualize those areas as representing diverticula of a highly irregular but continuous body. What is nearly inconceivable is how such a diffuse genome can possibly be replicated accurately and divided into identical halves. A beginning toward comprehension is provided by enlarged views of the nucleoid patches (Figure 11.21), in which the contents are disclosed as short fibrils; no known electron micrograph of any blue-green alga shows long, continuous fibers like those of bacteria (Figures 11.5, 11.19). Thus it may be supposed that the nucleoid of these blue-green algae exists as multiple copies of short DNA molecules that become randomly distributed to the daughter cells during division, much as in the macronucleus of ciliates described in a preceding section. In those forms having regular thylakoids, the nucleoid is largely granular or amorphous, so the DNA molecules may prove to be extremely small. As indicated by the processes of cytokinesis also, DNA structure strongly suggests that these represent the more primitive group of blue-green algae, the longer molecules present in the second type certainly seeming to be the more advanced condition.

Nucleoid Division in Primitive Forms. Several taxa representing differing levels of nucleoid development exist among bacteria and bacterialike organisms, all of which appear less advanced than the blue-green algae. The few ultrastructural studies on *Beggiatoa* and such other members of the colorless sulfur bacteria as *Thioploca* (Morita and Stave, 1963; Maier and Murray, 1965) show the nucleoid to consist of short fibrils and granules, the fibers being rarer and the entire nuclear material more highly dispersed than in the blue-green algae. In *Thioploca* fibrils evidently are entirely wanting, with only granules being present (Maier and Murray, 1965). Thus to judge from the published micrographs, the latter genus should be considered more primitive than *Beggiatoa*. While the latter's position relative to *Anabaena* has been confirmed by its

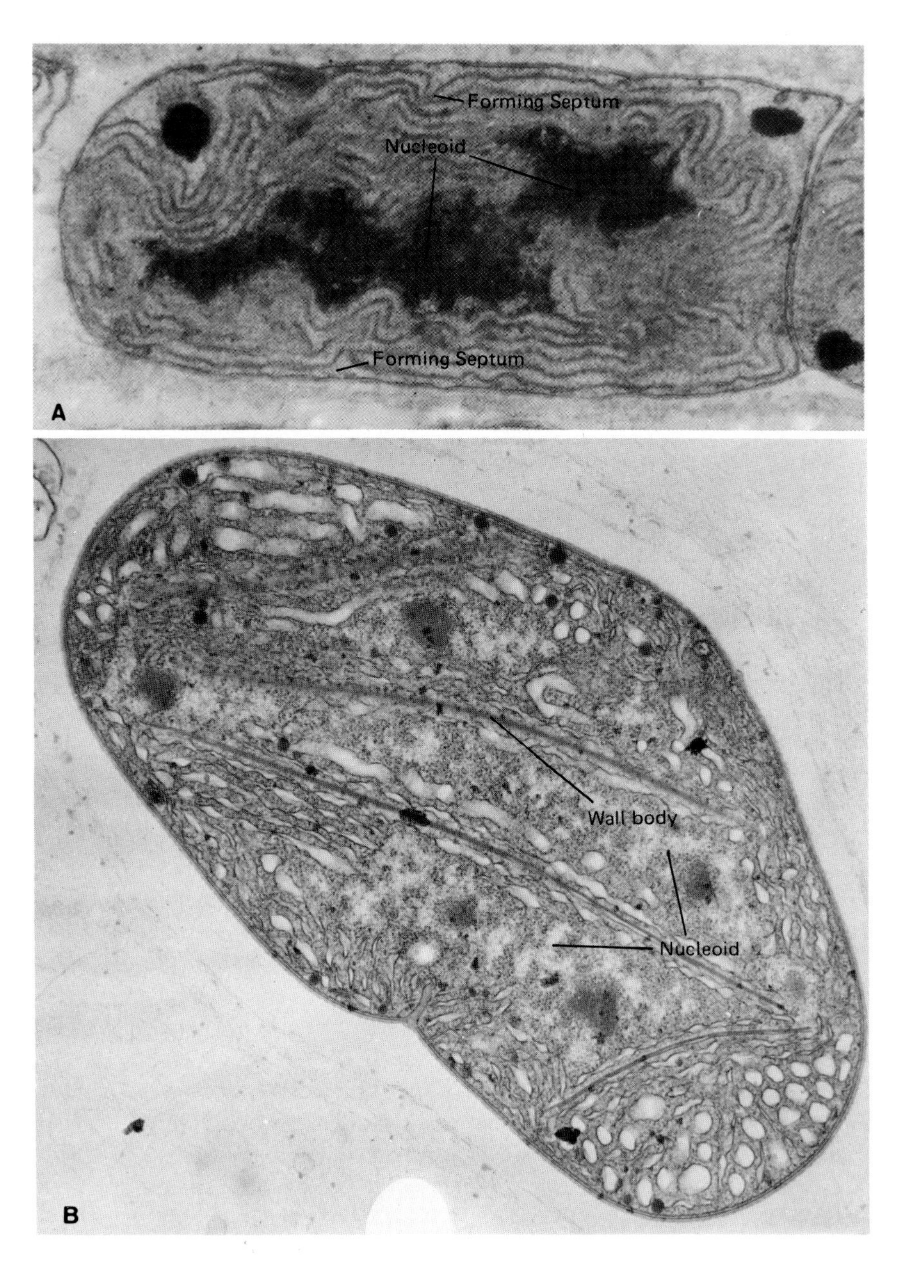

Figure 11.21. Nuclear division in the blue-green algae. (A) Unlike the higher bacteria in which the nucleoid is clearly fibrous, in blue-green algae related to this example of *Oscillatoria splendida*, it is amorphous or particulate. Note the asymmetry of cell and nuclear division as indicated by the location of the forming septa. 23,000×. (B) In *Anabaena* and others with irregular photosynthetic and respiratory lamellae, the nucleoid in section appears to be dispersed among the lamellae and other cell parts, but probably is a continuous, although highly irregular body. 13,500×. (B, courtesy of Jensen and Ayala, 1977.)

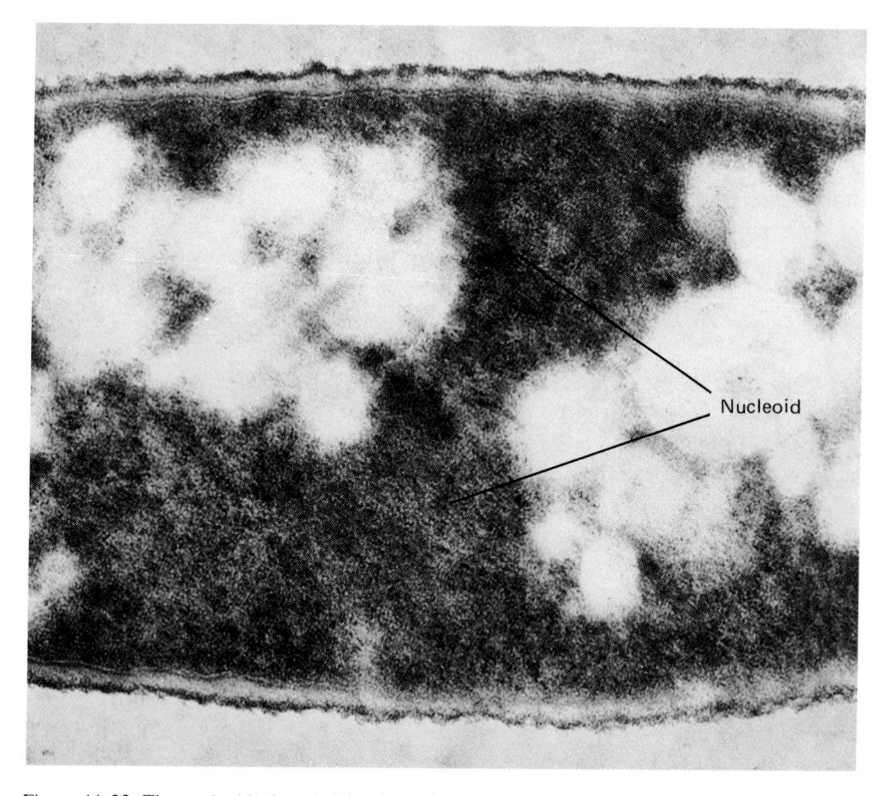

Figure 11.22. The nucleoid of a primitive bacterium. In *Clostridium pasteurianum* the nucleoid has a largely particulate or amorphous appearance, the continuous fine fibrils that characterize the genome of advanced bacteria being wanting. 51,000×. (Courtesy of Laishley *et al.*, 1973).

less complete tricarboxylic acid cycle (Section 8.3.2), the enzymes of this system have not received attention in *Thioploca;* however, one would suspect that they would be even fewer in number than in *Beggiatoa*. The position of *Clostridium pasteurianum* remains obscure, but electron micrographs show its nucleoid (Figure 11.22) to be as irregular in form as those of blue-green algae, with similar short fibrils dispersed among granules (Laishley *et al.*, 1973).

Finally, in *Rickettsia* the entirely amorphous DNA is dispersed throughout the cytoplasm (Hayes and Burgdorfer, 1979), both fibrils and patches of material being absent. These forms, then, are not to be viewed as degenerate bacteria, but as representatives of a group lying close to the point of origin of the cell, as advocated elsewhere (Dillon, 1978). It must be supposed that the nucleoid in all of these elementary taxa undergoes division by the cytosol as it extends the septum inwards during cytokinesis, as in *Beggiatoa*.

11.4. SOME PERTINENT CONCLUSIONS

Out of the maze of descriptive data that have been presented concerning the biochemical and ultrastructural properties of the cell, only one conclusion is at once obvious—it is a far more complex structure than biologists have dared to imagine. Still other generalizations can be drawn if the time is taken to sort the facts into related units and view the net results objectively. The several that are presented below appear both particularly firm and especially pertinent.

11.4.1. The Genetic Mechanism at the Cellular Level

Functions. In several chapters where material was unusually favorable, reference has been made to the existence of a genetic mechanism that operates at the level of the cell. In connection with the endoreticulum and dictyosome, it was seen that one function of this cellular mechanism was that of processing the products translated on the ribosomes. Such activities included trimming the nascent protein and, at times, piecing the parts together. Many instances involved the addition of various sugar or lipid moieties one at a time, some in the rough endoreticulum, others in the smooth, and still others in the Golgi apparatus, the growing glycoprotein or lipoprotein being shuttled back and forth between the respective organelles during those processes. Because only certain specific polypeptides are conjugated in this manner, whereas others undergo no such modification, recognition of some sort on the part of a particular substance or complex is an obvious necessity. After processing has been completed, the resulting product then needs to be conducted to its unique site and enzymatically inserted in correct relationships to the other molecules already situated there.

As discussed in part in the chapter on the flagellum, the protein tubulin makes it especially lucid that such recognition, pinpoint transport, and insertion require the utmost discrimination. In a flagellated cell, such as a component of ciliated epidermal tissue, at least eight different tubulins may be present, three in the flagellum, three during interphase in the cytoplasm, another in the chromosomal spindle, and possibly still another in the interzonal fibers. In addition, the triplets of the centriole may contain one more variety. As each type in turn consists of α and β components, certain ingredients of the cytosol are required to recognize and correctly conduct 16 or more different subunits.

Mention of the centriole leads to a third function of this cellular genetic mechanism, that of building the parts of the cell. As will be recalled, this organelle does not undergo division, but a sister body is constructed at a right angle some distance removed from the mature one. Moreover, as will be recalled further, replication of this organelle can take place even in the absence of a centriole to serve as a model. Comparable functions on a larger scale received passing mention in relation to cytokinesis among ciliates, preceding which processes new oral grooves and associated cilia and other structures were assembled prior to cell division. There can be little doubt that each and every organelle of the cell requires the same treatment by many specialized molecules of the cytosol in being constructed or replicated and maintained.

The Nature of the Cellular Apparatus. Briefly stated, the cellular genetic mechanism differs from the molecular one in that it orders into usable parts the macromolecules that the latter synthesizes. The well-known DNA–RNA–protein genetic apparatus may be viewed as the manufacturing part of a cell that produces the required raw materials, which are then refined, distributed, and carefully arranged in their definitive sites by the cellular genetic mechanism. In short, the molecular apparatus is the producer, the cellular one the finisher, distributor, and builder, and neither is completely useful to the cell in the absence of the other.

As the concept of a cellular genetic mechanism is new, nothing of its organization and almost nothing of its molecular nature are known. What little has been established pertains to the enzymes of the various endoreticula and Golgi material, which are, of course, proteins, and one would suspect that the rest of this apparatus consists similarly of proteins, possibly including phosphorylated ones and perhaps nucleoproteins. In cells even as simple as those of the blue-green algae and true bacteria, these proteins are needed, not only to process the raw molecules synthesized on the ribosomes, but also to conduct and arrange the finished products in orderly arrays. Indeed, these processing and assembling polypeptides have been reported in a number of advanced viruses like the T-even bacteriophages and bacteriophage λ, in which many more genes encode various assembling products than structural proteins (Dillon, 1978, pp. 361–372). Thus one can perceive that, as the cell came into existence and then increased in complexity, the DNA–RNA–protein system not only needed to produce more types of raw material to become arranged into new or more efficient organelles or enzymatic processes, but simultaneously had also to evolve new proteins for the cellular genetic apparatus that could recognize the new product and conduct it to the essential site. Certainly some of the proteins, for example, of a mitochondrion, in advanced cells are merely modifications of those of a more primitive organism, cytochrome *c* being an illustration. But to form the first mitochondrion out of the plasmalemma must have required additions of numerous novel proteins and other macromolecules to be incorporated into the new structure, along with the necessary subsidiary constructing types to do the assembling of each of those new products.

The Inadequacy of the Mechanistic Philosophy of Life. In the first member of this trilogy (Dillon, 1978), one of the conclusions drawn from the array of facts presented there concerning the molecular genetic mechanism was that a strict application of the mechanistic point of view of living things was an impossibility. The chemical organization of atoms and molecules of biological objects is indeed subject to the same physiochemical processes as those of the nonliving world; thus carbon, hydrogen, nitrogen, and other elements are united in biochemicals in precisely the same chemical ratios and by means of the same physical processes as in inorganic substances. Hence, to that extent a narrow mechanistic explanation of the biological world is adequate. But beyond that point, it is necessary to add to the concept an evolutionary ingredient, based on survival value to be sure, but in a complex and unpredictable manner. Self-ordering of macromolecules seemingly never occurs in the biological realm, as would be expected with strictly mechanical processes. Furthermore, as a general rule more efficient systems do not replace their forebears, but are built upon them, resulting in larger and larger complexes. One case in point pertains to the macromolecular genetic apparatus itself. Whereas 189 types of proteins are known to suffice in bacteria to synthesize a protein molecule by way of the DNA–RNA–protein processes, eukaryotes need 279 or more kinds to accomplish the same end (Dillon, 1978, pp. 206–207). Evolutionary processes, rather than simple physiochemical ones, have superimposed addition after addition upon the ancestral prokaryotic systems, as improvements were slowly gained, because those that already existed were vitally essential to the cell's being and their loss would therefore have been lethal.

Similar nonmechanical processes can be perceived at the cellular level. The assembly of α and β subunits of several varieties of tubulin is the most familiar example, those of one type being incorporated into the A component of the outer doublets of flagella, a second into the B components, a third into the central microtubules, and perhaps a fourth into the triplets of the centriole. Such subunits do not polymerize mechanically but are added enzymatically, after being duly recognized and transported to the correct site in each case. Were polymerization strictly a mechanical process, it is not unlikely that the subunits would interact near their points of synthesis on the cytoplasmic ribosomes, so that the resulting microtubules would ramify the cytoplasm in random fashion, rather than being oriented properly within the flagellum. A parallel instance is supplied by the fibrin filaments, which at one point of the life cycle are confined to the pseudopods, where the localization of a given fibril must be ephemeral indeed. For the pseudopod to provide movement to the cell, these fibers have to be polymerized, permitted to act, and then depolymerized and transported to a new site in an endlessly repetitious process. At other points of the cell cycle, the subunits are conducted below the plasmalemma as a whole to act in the production of microvilli, and at telophase relocated once more to participate in cytokinesis. Such behavior is never characteristic of molecules *per se*; it occurs only in living things. Formalization of the uniqueness of molecular behavior in the biological world is best provided by modifying the prevailing mechanistic point of view to what has been tentatively named the biomechanistic philosophy of life (Dillon, 1978, pp. 426–427).

11.4.2. Evolutionary Considerations

The exquisite organization of molecules into constantly growing, changing, and interacting organelles that characterizes metazoan and metaphytan cells must certainly have required many hundreds of thousands of millenia to be achieved through evolutionary means. Many of the intermediate stages in the development of these cells from the primitive ancestral type fortunately still are represented by descendants, which themselves have likewise undergone diversification since their origins. Thus the path followed by cells in evolving to their greatest complexity of form necessarily is obscure, a condition aggravated at the present by absence of suitable investigations into several critical types of organisms. Consequently, any attempt at tracing the phylogenetic history of the cell at this time has to be considered preliminary until more complete information is available. Tentative though it may thus be, the following outline perhaps will at least serve to guide investigators in exploring the gaps in present knowledge thereby exposed. To follow successfully any tracing based on an overall view of cell structure and function at subcellular and ultrastructural levels, it is first necessary to free the mind of preconceptions, including many terms of systematic biology. Words like bacteria, algae, protozoans, animals, and plants tend to tie thoughts to old accustomed paths and often preordain the rejection of new ones opened by current knowledge without suitable consideration.

The Procedures. The procedures are both obvious and straightforward—the information provided by all available aspects of cell structure and function are assembled in such a fashion that those organisms shown to be primitive have the simplest structures and metabolic processes, with the fewest possible deviations from that condition. Fortunately, a few guideposts exist at the lower levels of cellular evolution to assist in arranging the prokaryotes into a reasonable sequence, one of the chief guides being the tricarboxylic acid cycle and its precursors. Moreover, these processes are augmented by considerations of nucleoid structure and the methods of cell division. Among the eukaryotes, mitochondrial structure and, in some instances, chloroplast morphology provide clues, which, combined with the mitotic processes and nuclear organization, serve in suggesting a logical sequence of events. The major aspects of cytokinesis and mitosis are then assembled into a diagram (Figure 11.23); this diagram, and corresponding diagrams on the chloroplast and flagellum and the discussions in all the preceding chapters, form the principal bases for the phylogenetic tree.

The Phylogenetic Tree. As a whole the phylogenetic tree presented here (Figure 11.24) differs little from the one proposed some years ago (Dillon, 1962, 1963), the chief differences being at the earlier levels. The Chlamydiae are here viewed as the earliest existing type of cell (Dillon, 1978), followed by the Rickettsiae. Both of these taxa of organisms lack a nucleoid, cytochromes, and all the enzymes of the tricarboxylic acid cycle, the Chlamydiae being

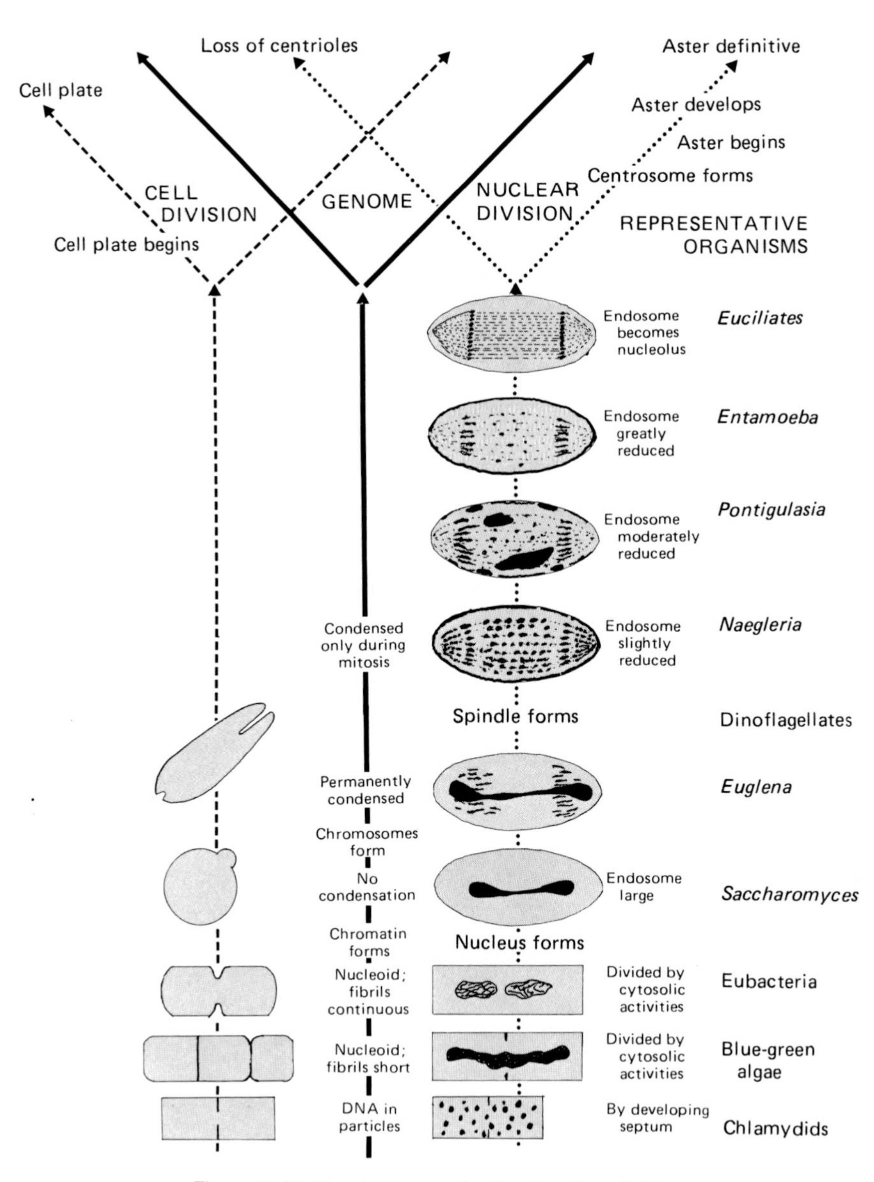

Figure 11.23. The phylogeny of cell and nuclear division.

simpler in having a viral-like nucleocapsid during part of their life history. Next above these two levels are *Beggiatoa* and *Thioploca*, the latter probably representing a separate, somewhat more elementary branch in that its ribosomes appear to be less complex than those of the former. For molecular biologists concerned with the functioning or structure of ribosomes, these four types of unicellular organisms appear far more desirable subjects for investigation than the much more advanced bacteria like *E. coli* and *B. subtilis* that are customarily used. Typical *Clostridium* species, those that cannot be induced to produce cytochromes, probably form a branch above the colorless sulfur bacteria, followed by at least two lines of blue-green algae. With the advent of these several groups, the nucleoid has become organized into a single body, instead of being scattered in clusters throughout the cell as in the earlier forms. Finally, cell wall and mesosome structure, the processes of cytokinesis, the electron-transport chain, and other metabolic activities indicate that several branches of bacteria complete the prokaryotic portion of the trunk of the tree (Figure 11.24).

The eukaryotic sector begins with such true yeasts as *Saccharomyces*, in which the chromatin is divided into a number of units that behave like chromosomes but do not condense into visible rodlike structures. In addition to an endosome persisting through nuclear division and the spindle being very primitive, such other characteristics as the Golgi apparatus, centrioles, and flagella being lacking and the endoreticular membranes and mitochondria poorly developed, in concert indicate undeniably an elementary level of development for the true yeasts. *Cyanidium*, with its extremely simple chloroplast, absence of Golgi apparatus, and mitochondria that behave like those of yeasts, probably represents the next level of advancement among the protistans, but its mitotic processes remain unknown. The euglenoids must then be assigned the third lowest position on the eukaryotic portion of the tree, because of their many primitive traits. Especially outstanding among them are the failure of the chromosomes to relax into chromatin, the absence of a spindle, and the presence of a well-developed endosome, in addition to the simple nature of its flagellum, dictyosomes, and other organelles pointed out in appropriate chapters. Dinoflagellates, which likewise have permanently condensed chromosomes and a persistent endosome in primitive types, seem to be just one stage higher than the euglenoids in possessing a spindle. The latter, however, is definitely primitive in behavior.

Beyond this point are several branches that represent three, or perhaps more, stages in which the endosome acquires the characteristics of the definitive nucleolus. Representatives include various types of amoeboids, the prociliates, and probably several of the lesser algal taxa like cryptomonads, whose mitotic processes are too incompletely known to assign a firm position on the phylogenetic tree. Once the nucleolus has acquired its definitive form, two major lines of cellular development are followed, one of which obviously

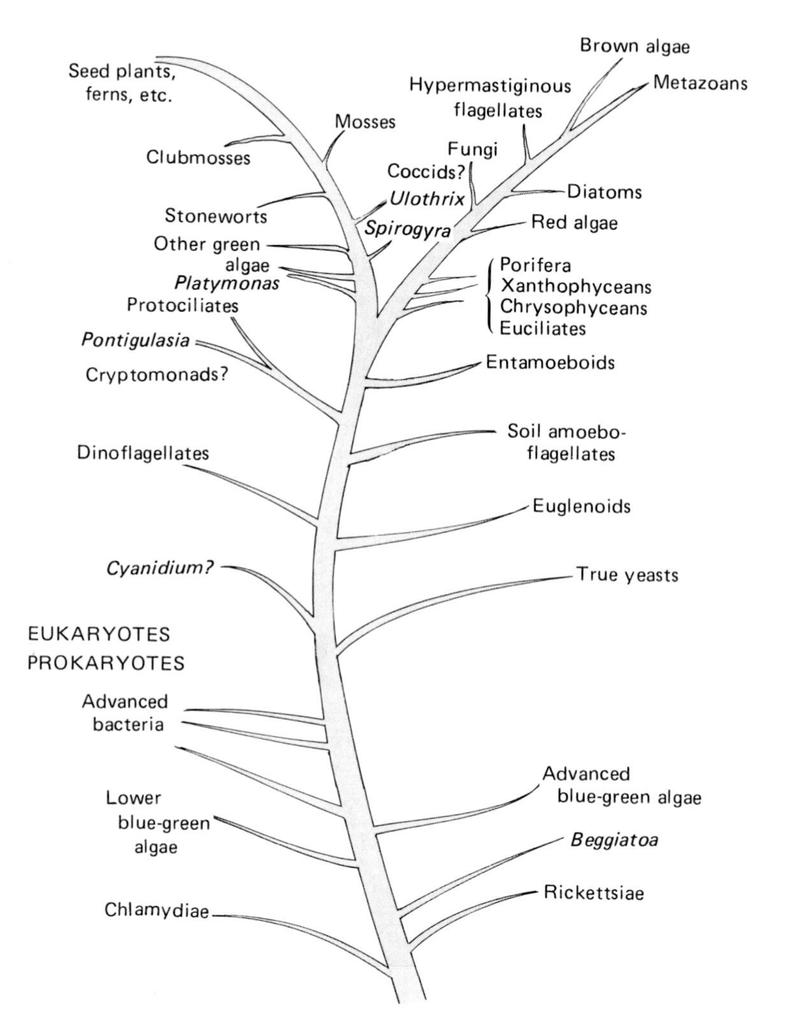

Figure 11.24. The phylogenetic stages of cell evolution. The diagram incorporates data from the discussions of preceding chapters as well as the present one, and thus is based on the evolutionary histories of numerous organelles.

leads through a number of levels of starch-producing green algae past the stoneworts, mosses, and fern allies into the seed plants. The several stages of altering the longitudinal type of fission first into a transverse variety and then of developing the cell plate provide one part of the basis for the pattern, while the reduction of centriolar activity during mitosis and the changing of the lamellate chloroplast into a grana-bearing type form the remainder.

The second major branch at the top of the diagram (Figure 11.24) lacks the cohesiveness of the first, because of both the diversification of its members and the insufficiency of knowledge concerning cell division of certain representatives. Nevertheless, a number of shared traits serve to unify them, outstanding among which are the storage products of metabolism being lipids rather than starch, peculiarities of the flagellar arrangement and fringing hairs, a chlorophyll *a* and *c* combination as a rule, and chloroplast structure. That these characteristics do in chorus provide a sufficiently firm basis for the proposed relationships is evidenced also by the fact that no argument against the arrangement has been raised since first proposed nearly two decades ago (Dillon, 1962). The sequence of its component taxa, however, cannot be fully firmed at present, again because of gaps in information at critical points. Probably, as indicated, the xanthophyceans, chrysophyceans, euciliates, and sponges and their proterospongid forebears are located near the base of the branch, as all these groups lack the hallmarks that distinguish more advanced members. Lower fungi show the advent of one advanced trait, an astral body that is so poorly developed in these organisms that it is secondarily lost among higher members of the taxon. In the diatoms the aster is also only slightly developed, but with their origin a second distinctive trait, the so-called centrosome, comes into existence and persists to the end of the branch. However, along the main line, the aster becomes of increasing importance, being fairly well developed in the hypermastiginous protozoans, and fully so in both the brown algae and the metazoans. As shown in Figure 11.25, the mitotic processes in the latter two taxa are identical, except that the nuclear envelope is degenerated at metaphase in the brown algae, rather than at late prophase as among metazoans.

Conclusions. In his masterly work on the classification of mammals, Simpson (1945) stated that taxonomy "eventually gathers together, utilizes, summarizes, and implements everything that is known about animals"—or better, in broader perspective, organisms. This statement, however, while certainly true, should be viewed as one corollary resulting from the information having been gathered together and assembled into phylogenetic format—a result, in other words, not a cause. And now that the information from all the parts of the cell have been collected, correlated, and summarized, its meaning in systematic terms becomes a pressing need.

Upon close examination of Figure 11.24, one observation becomes immediately clear: there is no point that divides plants from animals, nor protozoans from algae or any other grouping, for all share too many cellular properties to be sharply separated one from another into a few great kingdoms. One could conceivably divide prokaryotes from eukaryotes, as has frequently been advocated (e.g., Copeland, 1956; Whittaker, 1969; Margulis, 1970; Dodson, 1971), but that action is arbitrary and results in an artificial division reflecting grades of nuclear organization, rather than relationships. For instance, it would be equally valid to separate preemptively those organisms in which the chroma-

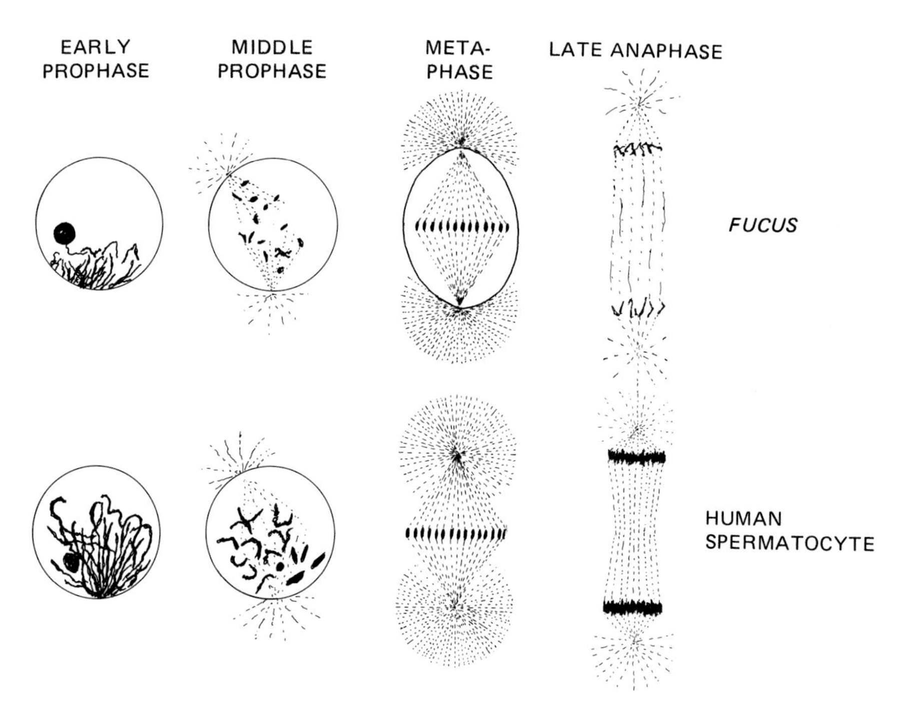

Figure 11.25. Meiotic division in metazoans and brown seaweeds. Comparisons of nuclear and astral structure reveal no differences, except in the nuclear envelope persisting through metaphase in these algae.

tin fails to condense or those in which the chromosomes are permanently condensed by cutting the main line of ascent above the yeasts in the first case and above the dinoflagellates in the other. Similar dogmatic separation into "kingdoms" could be made by combining the two topmost branches and separating them from all the others on the basis of their possessing a typical nucleolus, which all others lack. Thus no case can be made in support of a separate kingdom Monera to contain the nucleoid-bearing organisms, nor for the Protista to include all unicellular forms.

The truth of the matter is that the only natural subdivisions that can be recognized are represented by the branches. Whether such branches should be considered to be individual kingdoms is a question that should be decided by an international congress; the biochemical and cellular unity of life, however, is better expressed by treating them as subkingdoms of a single kingdom of living things as advocated elsewhere (Dillon, 1962, 1963). Another pressing topic for the agenda of such a congress is the fate of the present two codes of nomenclature, one for botanists and the other for zoologists, which depart radically from one another at a number of points. Because intimate relationships exist between many organisms consistently considered algae (and therefore plants) and those always treated as protozoans (and hence, animals), both codes currently apply equally to much of the entire tree of life. These and other related problems need prompt solutions if systematic biology is to achieve a scheme of classification that reflects the detailed information of the present century, rather than continuing to attempt justification of conclusions based on the much scantier knowledge of preceding times.

References

CHAPTER 1

Aikawa, M., Huff, C. G., and Strome, C. A. P. 1970. Morphological study of microgametogenesis of *Leucocytozoon simondi*. *J. Ultrastruct. Res.* **32**:43–68.

Albers, R. W. 1976. The (sodium plus potassium)-transport ATPase. *In*: Martonosi, A., ed., *The Enzymes of Biological Membranes*, New York, Plenum Press, Vol. 3, pp. 283–301.

Albrecht, O., Gruler, H., and Sackmann, E. 1978. Polymorphism of phospholipid monolayers. *J. Phys.* **39**:301–313.

Andersen, C. C. 1962. Studies on pinocytosis in amoebae. *C. R. Trav. Lab. Carlsberg* **33**:73–264.

Andersen, C. G., and Nilsson, J. R. 1960. Electron micrographs of pinocytosis channels in *Amoeba proteus*. *Exp. Cell Res.* **19**:631–633.

Bach, M. K., and Brashler, J. R. 1974. *In vivo* and *in vitro* production of a slow reacting substance in the rat upon treatment with calcium ionophores. *J. Immunol.* **113**:2040–2047.

Bamberg, E., Apell, H.-J., Alpes, H., Gross, E., Morell, J. L., Harbaugh, J. F., Janko, K., and Läuger, P. 1978. Ion channels formed by chemical analogs of gramicidin A. *Fed. Proc. Fed. Am. Soc. Exp. Biol.* **37**:2633–2638.

Barker, S. B. 1951. Mechanism of action of thyroid hormone. *Physiol. Rev.* **31**:205–243.

Barker, S. B. 1964. Physiological activity of thyroid hormone and analogues. *In*: Pitt-Rivers, R., and Trotter, W. R., eds., *The Thyroid Gland*, London, Butterworths, Vol. 1, pp. 199–236.

Beams, H. W., Tahmisian, T. N., Devine, R., and Anderson, E. 1957. Ultrastructure of the nuclear membrane of a gregarine parasitic in grasshoppers. *Exp. Cell Res.* **13**:200–204.

Beams, H. W., Tahmisian, T. N., Devine, R., and Anderson, E. 1959. The fine structure of the nuclear envelope of *Endamoeba blattae*. *Exp. Cell Res.* **18**:366–369.

Beams, H. W., Tahmisian, T. N., Anderson, E., and Wright, B. 1961. Studies on the fine structure of *Lophomonas blattarum* with special reference to the so-called parabasal apparatus. *J. Ultrastruct. Res.* **5**:166–183.

Bell, P. R. 1978. A microtubule–nuclear envelope complex in the spermatozoid of *Pteridium*. *J. Cell Sci.* **29**:189–195.

Bennett, M. V. L., Spira, M. E., and Spray, D. C. 1978. Permeability of gap junctions between embryonic cells of *Fundulus*: A reevaluation. *Dev. Biol.* **65**:114–128.

Benson, A. A. 1968. The cell membrane: A lipo-protein monolayer. *In*: Bolis, L., and Pethica, B. A., eds., *Membrane Models and the Formation of Biological Membranes*, Amsterdam, North-Holland, pp. 190–202.

Beveridge, T. J., and Murray, R. G. E. 1976. Superficial cell-wall layers on *Spirillum* "Ordal" and their in vitro reassembly. *Can. J. Microbiol.* **22**:567–582.

Bonner, J. T. 1977. Some aspects of chemotaxis using the cellular slime molds as an example. *Mycologia* **69**:443–459.

Bonting, S. L. 1970. Sodium · potassium activated ATPase and cation transport. *In*: Bittar, E., ed., *Membranes and Ion Transport*, New York, Wiley–Interscience, Vol. 1, pp. 286–392.

Bouck, G. B. 1965. Fine structure and organelle associations in brown algae. *J. Cell Biol.* **26**:523–537.

Brandt, P. W., and Freeman, A. R. 1967. Plasma membrane: Substructural changes correlated with electrical resistance and pinocytosis. *Science* **155**:582–585.

Brandt, P. W., and Pappas, G. D. 1962. An electron microscopic study of pinocytosis in ameba. II. The cytoplasmic uptake phase. *J. Cell Biol.* **15**:55–71.

Branton, D., and Moor, H. 1964. Fine structure in freeze-etched *Allium cepa* root tips. *J. Ultrastruct. Res.* **11**:401–411.

Braun, V., and Bosch, V. 1972. Repetitive sequences in the murein-lipoprotein of the cell wall of *E. coli*. *Proc. Natl. Acad. Sci. USA* **69**:970–974.

Braun, V., and Wolff, H. 1973. Characterization of the receptor protein for phage T5 and colicin M in the outer membrane of *E. coli* B. *FEBS Lett.* **34**:77–80.

Braun, V., Schaller, K., and Wolff, H. 1973. A common receptor protein for phage T5 and colicin M in the outer membrane of *E. coli* B. *Biochim. Biophys. Acta* **323**:87–97.

Bretscher, M. S. 1972. Asymmetrical lipid bilayer structure for biological membranes. *Nature New Biol.* **236**:11–12.

Bretscher, M. S. 1973. Membrane structure: Some general principles. *Science* **181**:622–629.

Brinton, C. C., McNary, J. C., and Carnahan, J. 1969. Purification and *in vitro* assembly of a curved network of identical protein subunits from the outer surface of a *Bacillus*. *Bacteriol. Proc.* **1969**:48.

Brown, M. S., and Goldstein, J. L. 1974. Familial hypercholesterolemia: Defective binding of lipoproteins to cultured fibroblasts associated with impaired regulation of 3-hydroxyl-3-methylglutaryl coenzyme A reductase activity. *Proc. Natl. Acad. Sci. USA* **71:**788–792.

Brown, M. S., and Goldstein, J. L. 1975. Regulation of the activity of the low density lipoprotein receptor in human fibroblasts. *Cell* **6**:307–316.

Brown, M. S., and Goldstein, J. L. 1976. Receptor-mediated control of cholesterol metabolism. *Science* **191**:150–154.

Busse, D. 1978. Transport of L-arginine in brush-border vesicles derived from rabbit kidney cortex. *Arch. Biochem. Biophys.* **191**:551–560.

Capaldi, R. A. 1974. A dynamic model of cell membranes. *Sci. Am.* **230**(3):26–33.

Caplan, S. R., and Essig, A. 1977. A thermodynamic treatment of active sodium transport. *Curr. Top. Membr. Transp.* **9**:145–175.

Chapman-Andresan, C. 1977. Endocytosis in freshwater amoebas. *Physiol. Rev.* **57**:371–385.

Christiansson, A., and Wieslander, Å. 1978. Membrane lipid metabolism in *Acholeplasma laidlawii* AEF22. *Eur. J. Biochem.* **85**:65–76.

Clark-Walker, G. D., and Linnane, A. W. 1967. The biogenesis of mitochondria in *Saccharomyces cerevisiae*. *J. Cell Biol.* **34**:1–14.

Clejan, S., Bittman, R., and Rottem, S. 1978. Uptake, transbilayer distribution, and movement of cholesterol in growing *Mycoplasma capricolum* cells. *Biochemistry* **17**:4579–4583.

Cofré, G., and Crabbé, J. L. 1967. Active sodium transport by the colon of *Bufo marinus*: Stimulation by aldosterone and antidiuretic hormone. *J. Physiol.* **188**:177–190.

Conti, S. F., and Brock, T. D. 1965. Electron microscopy of cell fusion in conjugating *Hansenula wingei*. *J. Bacteriol.* **90**:524–533.

Costerton, J. W. F., Murray, R. G. E., and Robinow, C. F. 1961. Observations on the motility and the structure of *Vitreoscilla*. *Can. J. Microbiol.* **7**:329–339.

Cota-Robles, E. H. 1966. Internal membranes in cells of *Escherichia coli*. *J. Ultrastruct. Res.* **16**:626–639.

Craine, B. L., and Rupert, C. S. 1978. Identification of a biochemically unique DNA–membrane interaction involving the *E. coli* origin of replication. *J. Bacteriol.* **134**:193–199.

Crane, F. L., and Hall, J. D. 1972. Diversity in membrane structure: From unit to binary. *Ann. N.Y. Acad. Sci.* **195**:24–34.

Cuatrecasas, P. 1969. Interaction of insulin with the cell membrane: The primary action of insulin. *Proc. Natl. Acad. Sci. USA* **63**:450–457.

Cuatrecasas, P. 1971. Insulin–receptor interaction in adipose tissue cells: Direct measurement and properties. *Proc. Natl. Acad. Sci. USA* **68**:1264–1268.

Curtis, A. S. G. 1967. *The Cell Surface: Its Molecular Role in Morphogenesis*, Woking, England, Logos Press.

Czarnetzki, B. M., König, W., and Lichtenstein, L. M. 1976. Eosinophil chemotactic factor (ECF). *J. Immunol.* **117**:229–234.

Dallner, G. 1977. Biosynthesis and transport of microsomal membrane glycoproteins. *In*: Abrahamsson, S., and Pascher, I., eds., *Structure of Biological Membranes*, New York, Plenum Press, pp. 95–106.

Dallner, G., Siekowitz, P., and Palade, G. E. 1965. Phospholipids in hepatic microsomal membranes during development. *Biochem. Biophys. Res. Commun.* **20**:142–148.

Danielli, J. F., and Davson, H. 1935. A contribution to the theory of permeability of thin films. *J. Cell. Comp. Physiol.* **5**:495–508.

Daniels, E. W. 1964. Origin of the Golgi system in amoeba. *Z. Zellforsch. Mikrosk. Anat.* **64**:38–51.

Daniels, E. W., and Breyer, E. P. 1967. Ultrastructure of the giant amoeba, *Pelomyxa palustris*. *J. Protozool.* **14:**167–179.

Daniels, E. W., Breyer, E. P., and Kudo, R. R. 1966. *Pelomyxa palustris* Greeff. II. Its ultrastructure. *Z. Zellforsch. Mikrosk. Anat.* **73**:367–383.

David, H. 1964. Physiologische und pathologische Modifikationen des submikroskopischen Kernstruktur. *Z. Mikrosk. Anat. Forsch.* **71**:412–456.

DeKretser, D. M., Catt, K. J., and Paulson, C. A. 1971. Studies on the *in vitro* testicular binding of iodinated luteinizing hormone in rats. *Endocrinology* **88**:332–337.

Demsey, A., Kawka, D., and Stackpole, C. W. 1977. Application of freeze-drying intact cells to studies of murine oncorna-virus morphogenesis. *J. Virol.* **21**:358–365.

Demsey, A., Kawka, D., and Stackpole, C. W. 1978. Cell surface membrane organization revealed by freeze-drying. *J. Ultrastruct. Res.* **62**:13–25.

Dillon, L. S. 1978. *The Genetic Mechanism and the Origin of Life*, New York, Plenum Press.

Dreyer, W. J., Papermaster, D. S., and Kün, H. 1972. On the absence of ubiquitous structural protein subunits in biological membranes. *Ann. N.Y. Acad. Sci.* **195**:61–74.

Driml, M. 1961. Electron and light microscope studies of *Endamoeba terrapinae*. *Proc. Iowa Acad. Sci.* **68**:581–585.

Drum, R. W. 1963. The cytoplasmic fine structure of the diatom, *Nitzschia palea*. *J. Cell Biol.* **18**:429–440.

Eisenberg, M., and McLaughlin, S. 1976. Lipid bilayers as models of biological membranes. *BioScience* **26**:436–443.

Elbrink, J., and Bihler, I. 1975. Membrane transport: Its relation to cellular metabolic rates. *Science* **188**:1177–1184.

Eletr, S., and Keith, A. D. 1972. Spin label studies of dynamics of lipid alkyl chains in biological membranes: Role of unsaturated sites. *Proc. Natl. Acad. Sci. USA* **69**:1353–1357.

Epstein, W., Whitelaw, V., and Hesse, J. 1978. A K^+-transport ATPase in *E. coli*. *J. Biol. Chem.* **253**:6666–6668.

Fahey, P. F., and Webb, W. W. 1978. Lateral diffusion in phospholipid bilayer membranes and multilamellar liquid crystals. *Biochemistry* **17**:3046–3053.

Fawcett, D. W., Anderson, W. A., and Phillips, D. M. 1971. Morphogenetic factors influencing the shape of the sperm head. *Dev. Biol.* **26**:220–251.

Feldherr, C. M. 1972 Structure and function of the nuclear envelope. *Adv. Cell Mol. Biol.* **2**:273–307.

Feldherr, C. M., and Pomerantz, J. 1978. Mechanism in the selection of nuclear polypeptides in *Xenopus* oocytes. *J. Cell Biol.* **78**:168–175.

Fielding, P., and Fox, C. F. 1970. Evidence for stable attachment of DNA to membrane at the replication origin of *E. coli. Biochem. Biophys. Res. Commun.* **41**:157–162.

Fischman, D. A., and Weinbaum, G. 1967. The formation of multiple layers of membrane-like structures in *Escherichia coli. J. Cell Biol.* **32**:524–528.

Fisher, R. W., and James, T. L. 1978. Lateral diffusion of the phospholipid molecule in dipalmitoylphosphatidylcholine bilayers. *Biochemistry* **17**:1177–1183.

Flickinger, C. J. 1970. The fine structure of the nuclear envelope in amebae: Alterations following nuclear transplantation. *Exp. Cell Res.* **60**:225–236.

Flower, N. E. 1972. A new junctional structure in the epithelia of insects of the order Dictyoptera. *J. Cell Sci.* **10**:683–691.

Folliot, R., and Picheral, B. 1971. Motif de répartition des pores nucléaires. *J. Microsc. (Paris)* **10**:219–224.

Fox, C. F. 1972. The structure of cell membranes. *Sci. Am.* **226**:30–38.

Fox, C. F., and Keith, A. D., eds. 1972. *Membrane Molecular Biology*, Stanford, Conn., Sinauer Associates.

Franke, W. W. 1970. On the universality of nuclear pore complex structure. *Z. Zellforsch. Mikrosk. Anat.* **105**:405–429.

Franke, W. W. 1974. Structure, biochemistry, and function of the nuclear envelope. *Int. Rev. Cytol. Suppl.* **4**:72–236.

Franke, W. W., and Scheer, U. 1974. Structures and functions of the nuclear envelope. *In*: Busch, H., ed., *The Cell Nucleus*, New York, Academic Press, Vol. 1, pp, 219–347.

Franke, W. W., Deumling, B., Ermen, B., Jarasch, E. D., and Kleinig, H. 1970. Nuclear membranes from mammalian liver. I. Isolation procedure and general characterization. *J. Cell Biol.* **46**:379–395.

Franke, W. W., Keenan, T. W., Stadler, J., Genz, R., Jarasch, E. D., and Kartenbeck, J. 1976. Characteristics of highly purified nuclear membranes in comparison with other membranes. *Cytobiologie Z. Exp. Zellforsch.* **13**:28–56.

French, R. J., and Adelman, W. J. 1976. Competition, saturation, and inhibition ionic interactions shown by ionic currents in nerve, muscle, and bilayer systems. *Curr. Top. Membr. Trans.* **8**:161–207.

Frey-Wyssling, A. 1955. Submikroskopische Struktur des Cytoplasmes. *Protoplasmologia* **2**(A):1–244.

Fuge, H. 1971. Spindelbau, Mikrotubuliverteilung und Chromosomenstruktur während der I. meiotischen Teilung der Spermatocyten von *Pales ferruginea. Z. Zellforsch. Mikrosk. Anat.* **120**:579–599.

Gabbiani, G., Chaponnier, C., and Hüttner, I. 1978. Cytoplasmic filaments and gap junctions in epithelial cells and myofibroblasts during wound healing. *J. Cell Biol.* **76**:561–568.

Gall, J. G. 1964. Electron microscopy of the nuclear envelope. *Protoplasmatologia* **5**:4–25.

Ganesan, A. T., and Lederberg, J. 1965. A cell-membrane bound fraction of bacterial DNA. *Biochem. Biophys. Res. Commun.* **18**:824–835.

Garrahan, P. J., and Garay, R. P. 1976. The distinction between sequential and simultaneous models for sodium and potassium transport. *Curr. Top. Membr. Transp.* **8**:29–97.

Garrahan, P. J., and Glynn, I. M. 1967. The stoichiometry of the sodium pump. *J. Physiol.* **192**:217–235.

Gebhardt, C., Gruler, H., and Sackmann, E. 1977. On domain structure and local curvature in lipid bilayers and biological membranes. *Z. Naturforsch.* **32c**:581–596.

Gemsa, D., Seitz, M., Kramer, W., Grimm, W., Till, G., and Resch, K. 1979. Ionophore A23187 raises cyclic AMP levels in macrophages by stimulating prostaglandin E formation. *Exp. Cell Res.* **118**:55–62.

Gilmore, R., Cohn, N., and Glaser, M. 1979a. Fluidity of LM cell membranes with modified lipid compositions as determined with 1,6-diphenyl-1,3,5-hexatriene. *Biochemistry* **18:**1042–1049.

Gilmore, R., Cohn, N., and Glaser, M. 1979b. Rotational relaxation times of 1,6-diphenyl-1,3,5-hexatriene in phospholipids isolated from LM cell membranes. Effects of phospholipid polar head-group and fatty acid composition. *Biochemistry* **18**:1050–1056.

Gilula, N. B. 1974. Junctions between cells. *In*: Cox, R. P., ed., *Cell Communications*, New York, Wiley, pp. 1–29.

Gilula, N. B. 1977. Gap junctions and cell communication. *In*: Brinkley, B. R., and Porter, K. R. eds., *International Cell Biology, 1976–1977*, New York, Rockefeller University Press, pp. 61–69.

Gmeiner, J., and Schlecht, S. 1979. Molecular organization of the outer membrane of *Salmonella typhimurium*. *Eur. J. Biochem.* **93**:609–620.

Goldfischer, S., Essner, E., and Novikoff, A. B. 1964. The localization of phosphatase activities at the level of ultrastructure. *J. Histochem. Cytochem.* **12**:72–95.

Goldstein, J. L., and Brown, M. S. 1976. The LDL pathway in human fibroblasts: A receptor mediated mechanism for the regulation of cholesterol metabolism. *Curr. Top. Cell Regul.* **11**:147–181.

Goldstein, J. L., and Brown, M. S. 1977. The low density lipoprotein pathway and its relation to atherosclerosis. *Annu. Rev. Biochem.* **46**:897–930.

Goldstein, J. L., Brown, M. S., and Anderson, R. G. W. 1977. The low-density lipoprotein pathway in human fibroblasts: Biochemical and ultrastructural correlations. *In*: Brinkley, B. R., and Porter, K. R. eds., *International Cell Biology, 1976–1977*, New York, Rockefeller University Press, pp. 639–648.

Goodal, R. J., and Thompson, J. E. 1971. A scanning electron microscope study of phagocytosis. *Exp. Cell Res.* **64**:1–8.

Goodenough, D. A., and Revel, J. P. 1970. A fine structural analysis of intercellular junctions in the mouse liver. *J. Cell Biol.* **45**:272–288.

Green, D. E. 1972. Membrane proteins: A perspective. *Ann. N.Y. Acad. Sci.* **195**:150–172.

Green, D. E., and Brucker, R. F. 1972. The molecular principles of biological membrane construction and function. *BioScience* **22:**13–19.

Green, D. E., and Perdue, J. F. 1966. Membranes as expressions of repeating units. *Proc. Natl. Acad. Sci. USA* **55**:1295–1302.

Griffanti, A. A., Blanco, R., and Krulwich, T. A. 1979. A requirement for ATP for β-galactoside transport by *Bacillus alcalophilus*. *J. Biol. Chem.* **254**:1033–1037.

Gros, D., Mocquard, J. P., Challice, C. E., and Schrevel, J. 1978. Formation and growth of gap junctions in mouse myocardium during ontogenesis: A freeze-cleave study. *J. Cell Sci.* **30**:45–61.

Haase, W., Schäfer, A., Murer, H., and Kinne, R. 1978. Studies on the orientation of brush-border membrane vesicles. *Biochem. J.* **172**:57–62.

Hand, W. L., King, N. L., Johnson, J. D., and Lowe, D. A. 1977. Requirement for magnesium influx in activation of alveolar macrophages mediated by ionophore A23187. *Nature* (*London*) **265**:543–544.

Harmon, J. M., and Taber, H. W. 1977. Some properties of a membrane–DNA complex isolated from *B. subtilis*. *J. Bacteriol.* **129**:789–795.

Herlan, G., Giese, G., and Wunderlich, F. 1979. Influence of nuclear membrane lipid fluidity on nuclear RNA release. *Exp. Cell Res.* **118**:305–309.

Herr, J. C., and Heidger, P. M. 1978. A freeze-fracture study of exocytosis and reflexive gap junctions in human ovarian decidual cells. *Am. J. Anat.* **152**:29–44.

Hidalgo, C., and Ikemoto, N. 1977. Disposition of proteins and aminophospholipids in the sarcoplasmic reticulum membrane. *J. Biol. Chem.* **252**:8446–8454.

Hladky, S. B., and Haydon, D. A. 1970. Discreteness of conductance change in bimolecular lipid membranes in the presence of certain antibiotics. *Nature (London)* **225**:451–453.

Hoffman, J. F. 1972. Sidedness of the red cell Na:K pump. *In*: Bolis, L., Keynes, R. D., and Wilbrandt, W., eds., *Role of Membranes in Secretory Processes*, Amsterdam, North-Holland, pp. 203–214.

Hoffman, J. F. 1978. Asymmetry and the mechanism of the red cell Na–K pump, determined by ouabain binding. *In*: Solomon, A. K., and Karnovsky, M., eds., *Molecular Specialization and Symmetry in Membrane Function*, Cambridge, Mass., Harvard University Press, pp. 191–211.

Hoffman, S., and McMahon, D. 1978. The effects of inhibition of development in *Dictyostelium discoideum* on changes in plasma membrane composition and topography. *Arch. Biochem. Biophys.* **187**:12–24.

Holter, H. 1959. Pinocytosis. *Int. Rev. Cytol.* **8**:481–504.

Horiuchi, S., Inoue, M., and Morino, Y. 1978. γ-Glutamyl transpeptidase: Sidedness of its active site on renal brush-border membrane. *Eur. J. Biochem.* **87**:429–437.

Horn, R. G., Spicer, S. S., and Wetzel, B. 1964. Phagocytosis of bacteria by heterophil leukocytes. Acid and alkaline phosphatase cytochemistry. *Am. J. Pathol.* **45**:327–335.

Houwink, A. L. 1953. A macromolecular mono-layer in the cell wall of *Spirillum* spec. *Biochim. Biophys. Acta* **10**:360–366.

Hsu, W. S. 1967. The origin of annulate lamellae in the oocyte of the ascidian, *Boltenia villosa* Stimpson. *Z. Zellforsch. Mikrosk. Anat.* **82**:376–390.

Inouye, M. 1975. Biosynthesis and assembly of the outer membrane proteins of *E. coli*. *In*: Tzagoloff, A., ed., *Membrane Biosynthesis: Mitochondria, Chloroplasts, and Bacteria*, New York, Plenum Press, pp. 351–391.

Inouye, S., Wang, S., Sekezawa, J., Halegowa, S., and Inouye, M. 1977. Amino acid sequence for the peptide extension on the prolipoprotein of the *E. coli* outer membrane. *Proc. Natl. Acad. Sci. USA* **74**:1004–1008.

Ismail-Beigi, F. 1977. Thyroidal regulation of active sodium transport. *Curr. Top. Membr. Transp.* **9**:367–388.

Ismail-Beigi, F., and Edelman, I. S. 1970. The mechanism of thyroid calorigenesis: Role of active sodium transport. *Proc. Natl. Acad. Sci. USA* **67**:1071–1078.

Ismail-Beigi, F., and Edelman, I. S. 1971. The mechanism of the calorigenic action of thyroid hormone: Stimulation of the Na^+, K^+ activated ATP activity. *J. Gen. Physiol.* **57**:710–722.

Israelachvili, J. N., and Mitchell, D. J. 1975. A model for the packing of lipids in bilayer membranes. *Biochim. Biophys. Acta* **389**:13–19.

Ivatt, R. J., and Gilvarg, C. 1978. Molecular structure of the teichuronic acid of *Bacillus megaterium*. *Biochemistry* **17**:3997–4003.

Jard, S., Roy, C., Rajerison, R., Butlen, D., and Guillon, G. 1977. Vasopressin-sensitive adenylate cyclase from the mammalian kidney: Mechanisms of activation. *In*: Nicolau, C., and Paraf, A., eds., *Structural and Kinetic Approach to Plasma Membrane Functions*, New York, Springer-Verlag, pp. 173–187.

Johnson, R. G., Herman, W. S., and Preuss, D. M. 1973. Homocellular and heterocellular gap junctions in *Limulus*: A thin section and freeze-fracture study. *J. Ultrastruct. Res.* **43**:298–312.

Johnston, L. W., Hughes, M. E., and Zilbersmit, D. B. 1975. Use of phospholipid exchange protein to measure inside–outside transposition in phosphatidylcholine liposomes. *Biochim. Biophys. Acta* **375**:176–185.

Jurand, A. 1976. Some ultrastructural features of micronuclei during conjugation and autogamy in *Paramecium aurelia*. *J. Gen. Microbiol.* **94**:193–203.

Kahlenberg, A., Walker, C., and Rohrlick, R. 1974. Evidence for an asymmetric distribution of phospholipids in the human erythrocyte membrane. *Can. J. Biochem.* **52**:803–806.

Kakiquchi, S., and Rall, T. W. 1968. The influence of chemical agents on the accumulation of adenosine 3′, 5′-phosphate in slices of rabbit cerebellum. *Mol. Pharmacol.* **4**:367–378.

Kaneshiro, T., and Marr, A. G. 1965. Phospholipids of *Azotobacter agilis*, *Agrobacterium tumefaciens*, and *E. coli*. *J. Lipid Res.* **3**:184–189.

Kartenbeck, J., Jarasch, E.-D., and Franke, W. W. 1973. Nuclear membranes from mammalian liver. VI. Glucose-6-phosphatase in rat liver, a cytochemical and biochemical study. *Exp. Cell Res.* **81**:175–194.

Kasper, C. B. 1971. Biochemical distinctions between the nuclear and microsomal membranes from rat hepatocytes. *J. Biol. Chem.* **246**:577–581.

Kasper, C. B. 1974. Chemical and biochemical properties of the nuclear envelope. *In*: Busch, H., ed., *The Cell Nucleus*, New York, Academic Press, Vol. 1, pp. 349–384.

Kavanau, J. L. 1965. *Structure and Function in Biological Membranes*, San Francisco, Holden–Day.

Kazama, F. Y. 1972. Ultrastructure and phototaxis of the zoospores of *Phlyctochytrium* sp., an estuarine chytrid. *J. Gen. Microbiol.* **71**:555–566.

Kellerman, G. M., Biggs, D. R., and Linnane, A. W. 1969. Biogenesis of mitochondria. XI. A comparison of the effects of growth-limiting oxygen tension, intercalating agents, and antibiotics in the obligate aerobe, *Candida parapsilosis*. *J. Cell Biol.* **42**:378–391.

Kelley, R. O., Vogel, K. G., Crissman, H. A., Lujan, C. J., and Skipper, B. E. 1979. Development of the aging cell surface. *Exp. Cell Res.* **119**:127–143.

Kensler, R. W., Brink, P., and Dewey, M. M. 1977. Nexus of frog ventricle. *J. Cell Biol.* **73**:768–781.

Kilbourn, B. T., Dunitz, J. D., Pioda, L. A. R., and Simon, W. 1967. Structure of the K^+ complex with nonactin, a macrotetrolide antibiotic possessing highly specific K^+ transport properties. *J. Mol. Biol.* **30**:559–563.

Kimmich, G. A. 1973. Coupling between Na^+ and sugar transport in small intestine. *Biochim. Biophys. Acta* **300**:31–78.

Kinne, R. 1976. Properties of the glucose transport system in the renal brush border membrane. *Curr. Top. Membr. Transp.* **8**:209–267.

Knight, A. B., and Welt, L. G. 1974. Intracellular potassium: A determinant of the sodium–potassium pump rate. *J. Gen. Physiol.* **63**:351–373.

Koen, Y. M., Perevoshchikova, K. A., and Zbarsky, I. B. 1976. Some enzymes of isolated rat liver and hepatoma 27 nuclear membranes and cell nuclei. *Biokhimiya* **41**:982–988.

Koepsell, H. 1979. Conformational changes of membrane-bound (Na^+–K^+)-APTase as revealed by antibody inhibition. *J. Membr. Biol.* **45**:1–20.

Koga, A., and Todo, S. 1978. Morphological and functional changes in the tight junctions of the bile canaliculi induced by bile duct ligation. *Cell Tiss. Res.* **195**:267–276.

Konijn, T. M. 1972. Cyclic AMP as a first messenger. *Adv. Cyclic Nucleotide Res.* **1**:17–31.

Konings, W. M., and Boonstra, J. 1977. Anaerobic electron transfer and active transport in bacteria. *Curr. Top. Membr. Transp.* **9**:177–231.

Korn, E. D. 1966. Structure of biological membranes. *Science* **153**:1491–1498.

Kornberg, R. D., and McConnell, H. M. 1971. Inside–outside transitions of phospholipids in vesicle membranes. *Biochemistry* **10**:1111–1120.

Kotyk, A., and Janáček, K. 1977. Membrane transport—An interdisciplinary approach. *In*: Manson, L. A., ed., *Biomembranes*, New York, Plenum Press, Vol. 9, pp. 1–348.

Kozarich, J. W., and Strominger, J. L. 1978. A membrane enzyme from *Staphylococcus aureus* which catalyzes transpeptidase, carboxypeptidase, and penicillinase activities. *J. Biol. Chem.* **253**:1272–1278.

Krab, K., and Wikström, M. K. F. 1978. Proton-translocating cytochrome *c* oxidase in artificial phospholipid vesicles. *Biochim. Biophys. Acta* **504**:200–214.

Krasne, S., Eisenman, G., and Szabo, G. 1971. Freezing and melting of lipid bilayers and the mode of action of nonactin, valinomycin, and gramicidin. *Science* **174**:412–415.

LaCour, L. F., and Wells, B. 1972. The nuclear pores of early meiotic prophase nuclei of plants. *Z. Zellforsch. Mikrosk. Anat.* **123**:178–194.

Laprade, R., Ciani, S., Eisenman, G., and Szabo, G. 1975. The kinetics of carrier-mediated ion permeation in lipid bilayers and its theoretical interpretation. *In*: Eisenman, G., ed., *Membranes: A Series of Advances*, New York, Dekker, Vol. 3, pp. 127–214.

Lawson, D., Raff, M. C., Gomperts, B., Fewtrell, C., and Gilula, N. B. 1977. Molecular events during membrane fusion. A study of exocytosis in rat peritoneal mast cells. *J. Cell Biol.* **72**:242–259.

Leduc, M., Rousseau, M., and Heijenoort, J. 1977. Structure of the cell wall of *Bacillus* species C.I.P. *Eur. J. Biochem.* **80**:153–163.

Lee, A. G. 1975. Functional properties of biological membranes: A physical-chemical approach. *Prog. Biophys. Mol. Biol.* **28**:3–56.

Lee, N., and Inouye, M. 1974. Outer membrane proteins of *E. coli*: Biosynthesis and assembly. *FEBS Lett.* **39**:167–170.

Leedale, G. F. 1967. *Euglenoid Flagellates*, Englewood Cliffs, N.J., Prentice–Hall.

Leive, L. 1974. The barrier function of the Gram-negative envelope. *Ann. N.Y. Acad. Sci.* **235**:109–127.

Lewis, W. H. 1931. Pinocytosis. *Bull. Johns Hopkins Hosp.* **49:**17–28.

Litman, B. J., and Smith, H. G. 1974. The determination of molecular asymmetry in mixed phospholipid vesicles and bovine retinal rod outer segment disc membranes. *Fed. Proc. Fed. Am. Soc. Exp. Biol.* **33**:1575 (abstr.).

Lockwood, W. R., and Allison, F. 1963. Electronmicrographic studies of phagocytic cells. I. Morphological changes of the cytoplasm, and granules of rabbit granulocytes associated with ingestion of rough pneumococcus. *Br. J. Exp. Pathol.* **44**:593–600.

Loewenstein, W. R. 1966. Permeability of membrane junctions. *Ann. N.Y. Acad. Sci.* **137**:441–472.

Loewenstein, W. R. 1975. Permeable junctions. *Cold Spring Harbor Symp. Quant. Biol.* **40**:49–63.

Loewenstein, W. R. 1977. Permeability of the junctional membrane channel. *In*: Brinkley, B. R., and Porter, K. R., eds., *International Cell Biology, 1976–1977,* New York, Rockefeller University Press, pp. 70–82.

Loewenstein, W. R., Kanno, Y., and Socolar, S. J. 1978. Quantum jumps of conductance during formation of membrane channels at cell–cell junction. *Nature* (*London*) **274**:133–136.

Longo, F. J., and Anderson, E. 1968. The fine structure of pronuclear development and fusion in the sea urchin, *Arbacia punctulata*. *J. Cell Biol.* **39**:339–368.

Lott, J. N. A., and Vollmer, C. M. 1975. Changes in the cotyledons of *Cucurbita maxima* during germination. *J. Ultrastruct. Res.* **52**:156–166.

Lott, J. N. A., Larsen, P. L., and Whittington, C. M. 1972. Frequency distribution of nuclear pores in *Cucurbita maxima* cotyledons as revealed by freeze-etching. *Can. J. Bot.* **50**:1785–1787.

Lott, J. N. A., Wilson, J. J., and Vollmer, C. M. 1977. Temperature-induced changes in the nuclear envelop of *Euglena* and *Scenedesmus*. *J. Ultrastruct. Res.* **61**:1–9.

Luckasen, J. R., White, J. G., and Kersey, J. H. 1974. Mitogenic properties of a calcium ionophore, A23187. *Proc. Natl. Acad. Sci. USA* **71**:5088–5090.

Lucy, J. A., and Glauert, A. M. 1964. Structure and assembly of macromolecular lipid complexes composed of globular micelles. *J. Mol. Biol.* **8**:727–748.

Lugtenberg, B., van Boxtel, R., Verhoff, C., and van Alphen, W. 1978. Preprotein *e* of the outer membrane of *E. coli* K12. *FEBS Lett.* **96**:99–105.

McLaughlin, S., and Harary, H. 1974. Phospholipid flip-flop and the distribution of surface charges in excitable membranes. *Biophys. J.* **14**:200–208.

McNutt, N. S., and Weinstein, R. S. 1970. The ultrastructure of the nexus: A correlated thin section and freeze-cleave study. *J. Cell Biol.* **47**:666–687.

McNutt, N. S., and Weinstein, R. S. 1973. Membrane ultrastructure at mammalian intercellular junctions. *Prog. Biophys. Mol. Biol.* **26**:45–101.

McNutt, N. S., Hershberg, B. H., and Weinstein, R. S. 1971. Further observations on the occurrence of nexuses in benign and malignant human cervical epithelium. *J. Cell Biol.* **51**:805–825.

Maddy, A. H., and Malcolm, B. R. 1965. Protein conformations in the plasma membrane. *Science* **150**:1616–1618.

Maier, S., and Murray, R. G. E. 1965. The fine structure of *Thioploca ingrica* and a comparison with *Beggiatoa*. *Can. J. Microbiol.* **11**:645–655.

Manton, I., Kowallik, K., and von Stosch, H. A. 1969. Observations on the fine structure and development of the spindle at mitosis and meiosis in a marine centric diatom (*Lithodesmium undulatum*). *J. Cell Sci.* **5**:271–298.

Mantovani, B. 1975. Different roles of IgG and complement receptors in phagocytosis by polymorphonuclear leukocytes. *J. Immunol.* **115**:15–17.

Mast, S. O., and Doyle, W. L. 1934. Ingestion of fluid by amoeba. *Protoplasma* **20**:555–567.

Menkin, V. 1956. *Biochemical Mechanisms in Inflammation*, Springfield, Ill., Charles C. Thomas.

Merk, F. B., Albright, J. T., and Botticelli, C. R. 1973. The fine structure of granulosa cell nexuses in rat ovarian follicles. *Anat. Rec.* **175**:107–125.

Merriam, R. W. 1962. Some dynamic aspects of the nuclear envelope. *J. Cell Biol.* **12**:79–90.

Metcalfe, J. C., Birdsall, N. J. M., and Lee, A. G. 1972. ^{13}C NMR spectra of *Acholeplasma* membranes containing ^{13}C labelled phospholipids. *FEBS Lett.* **21**:335–340.

Meyer, G. F. 1963. Die Funktions Struckturen des Y-chromosoms in den Spermatocytenkernen von *Drosophila hydei, D. neohydei, D. repleta,* and einigen anderen *Drosophila*-arten. *Chromosoma* **14**:207–255.

Milder, R. V., and Deane, M. P. 1967. Ultrastructure of *Trypanosoma conorhini* in the crithidial phase. *J. Protozool.* **14**:65–72.

Miller, J. H., Swartzwelder, J. C., and Deas, J. E. 1971. An electron microscopic study of *Entamoeba histolytica*. *J. Parasitol.* **47**:577–587.

Miller, J. V., Cuatrecasas, P., and Thompson, E. P. 1971. Partial purification by affinity chromatography of tyrosine aminotransferase-synthesizing ribosomes from hepatoma tissue culture cells. *Proc. Natl. Acad. Sci. USA* **68**:1014–1018.

Monod, J., Wyman, J., and Changeux, J.-P. 1965. On the nature of allosteric transitions: A plausible model. *J. Mol. Biol.* **12**:88–118.

Moor, H., and Mühlethaler, K. 1963. Fine structure in frozen-etched yeast cells. *J. Cell Biol.* **17**:609–628.

Moore, P. L., Bank, H. L., Brissie, N. T., and Spicer, S. S. 1978. Phagocytosis of bacteria by polymorphonuclear leukocytes. *J. Cell Biol.* **76**:158–174.

Moses, M. J., and Wilson, M. H. 1970. Spermiogenesis in an iceryine coccid *Steatococcus tuberculatus* Morrison. *Chromosoma* **30**:373–429.

Moyle, J., and Mitchell, P. 1978. Cytochrome *c* oxidase is not a proton pump. *FEBS Lett.* **88**:268–272.

Mueller, P., and Rudin, D. O. 1968. Resting and action potentials in experimental bimolecular lipid membranes. *J. Theor. Biol.* **18**:222–258.

Mukhtar, H., Elmamlouk, T. H., and Bend, J. R. 1979. Epoxide hydrase and mixed-function oxidase activities of rat liver nuclear membranes. *Arch. Biochem. Biophys.* **192**:10–21.

Neher, E., and Stevens, C. F. 1977. Conductance fluctuations and ionic pores in membranes. *Annu. Rev. Biophys. Bioeng.* **6**:345–381.

Newell, P. C. 1978. Cellular communication during aggregation of *Dictyostelium*. *J. Gen. Microbiol.* **104**:1–13.

Nicolaidis, A. A., and Holland, I. B. 1978. Evidence for the specific association of the chromosomal origin with the outer membrane fractions isolated from *E. coli*. *J. Bacteriol.* **135**:178–189.

O'Brien, J. S., and Sampson, E. L. 1965. Lipid composition of the normal human brain: Gray matter, white matter, and myelin. *J. Lipid Res.* **6**:537–544.

Oldfield, E. 1973. Are cell membranes fluid? *Science* **180**:982–983.

Oldfield, E., Chapman, D., and Derbyshire, W. 1972. Lipid mobility in *Acholeplasma* membranes using deuteron magnetic resonance. *Chem. Phys. Lipids* **9**:69–81.

Onaya, T., and Solomon, D. H. 1969. Effects of chlorpromazine and propanolol on *in vitro* thyroid activation by thyrotropin, long-acting thyroid stimulator, and dibutyryl cyclic-AMP. *Endocrinology* **85**:1010–1017.

Ono, H., Ono. T., and Wada, O. 1976. Amino acid incorporation by nuclear membrane fraction of rat liver. *Life Sci.* **18**:215–222.

Op den Kamp, J. A. F., Houtsmuller, U. M. T., and Van Deenen, L. L. M. 1965. On the phospholipids of *Bacillus megaterium*. *Biochim. Biophys. Acta* **106**:438–441.

Orchard, I., Osborne, M. P., and Finlayson, L. H. 1979. Ultrastructural changes in neurohaemal tissue of the stick insect, *Carausius morosus*, induced by the ionophores Br-X-537A and A-23187. *Cell Tiss. Res.* **197**:227–234.

Osborn, M. J., Gander, J. E., and Parisi, E. 1972. Mechanism of assembly of the outer membrane of *Salmonella typhimurium*. *J. Biol. Chem.* **247**:3973–3986.

Osborn, M. J., Rick, P. D., Lehmann, V., Rupprecht, E., and Singh, M. 1974. Structure and biogenesis of the cell envelope of Gram-negative bacteria. *Ann. N.Y. Acad. Sci.* **235**:52–65.

Osuga, D. T., and Feeney, R. E. 1978. Antifreeze glycoproteins from Arctic fish. *J. Biol. Chem.* **253**:5338–5343.

Osuga, D. T., Ward, F. C., Yeh, Y., and Feeney, R. E. 1978. Cooperative functioning between antifreeze glycoproteins. *J. Biol. Chem.* **253**:6669–6672.

O'Sullivan, M. A., and Sueoka, N. 1972. Multifork attachment of the replicative origins of a multifork (dichotomous) chromosome in *B. subtilis*. *J. Mol. Biol.* **69**:237–248.

Ovchinnikov, Y. A. 1977. Recent findings in the structural and functional aspects of the peptide ionophores. *In*: Abrahamsson, S., and Pascher, I., eds., *Structure of Biological Membranes*, New York, Plenum Press, pp. 345–372.

Paine, P. L., Moore, L. C., and Horowitz, S. B. 1975. Nuclear envelope permeability. *Nature (London)* **254**:109–114.

Pappas, G. D. 1956. The fine structure of the nuclear envelope of *Amoeba proteus*. *J. Biophys. Biochem. Cytol. Suppl.* **2**:431–434.

Pauli, B. U., Weinstein, R. S., Alroy, J., and Arai, M. 1977. Ultrastructure of cell junctions in FANFT-induced urothelial tumors in urinary bladder of Fischer rats. *Lab. Invest.* **37**:609–621.

Peracchia, C. 1973. Low resistance junctions in crayfish. II. Structural details and further evidence for intercellular channels by freeze-fracture and negative staining. *J. Cell Biol.* **57**:66–76.

Perkins, H. R., and Rogers, H. J. 1959. The products of the partial acid hydrolysis of the mucopeptide from cell walls of *Micrococcus lysodeikticus*. *Biochem. J.* **72:**647–654.

Philipp, E. I., Franke, W. W., Keenan, T. E., Stadler, J., and Jarasch, E. D. 1976. Characterization of nuclear membranes and endoplasmic reticulum isolated from plant tissue. *J. Cell Biol.* **68**:11–29.

Pitelka, D. R. 1963. *Electron-Microscopic Structure of Protozoa*, Oxford, Pergamon Press.

Pohl, S. L., Birnbaumer, L., and Rodbell, M. 1969. Glucagon-sensitive adenyl cyclase in plasma membrane of hepatic parenchymal cells. *Science* **164**:566–567.

Post, R. L., and Jolly, P. C. 1957. The linkage of sodium, potassium and ammonia active transport across the human erythrocyte membrane. *Biochim. Biophys. Acta* **25**:118–128.

Prescott, D. M., Kimball, R. F., and Carrier, R. F. 1962. Comparison between the timing of micronuclear and macronuclear DNA synthesis in *Euplotes eurystomus*. *J. Cell Biol.* **13**:175–176.

Pugsley, A. P., and Schnaitman, C. A. 1978. Outer membrane proteins of *E. coli*. VII. Evidence that bacteriophage-directed protein 2 functions as a pore. *J. Bacteriol.* **133**:1181–1189.

Quinn, P. J. 1976. *The Molecular Biology of Cell Membranes,* Baltimore, University Park Press.

Rattner, J. B. 1972. Nuclear shaping in marsupial spermatids. *J. Ultrastruct. Res.* **40**:498–512.

Resch, K., Bouillon, D., and Gemsa, D. 1978. The activation of lymphocytes by the ionophore A23187. *J. Immunol.* **120**:1514–1520.

Revel, J. P., and Karnovsky, M. J. 1967. Hexagonal array of subunits in intercellular junctions of mouse heart and liver. *J. Cell Biol.* **33**:C7–C12.

Ribi, W. A. 1978. Gap junctions coupling photoreceptor axons in the first optic ganglion of the fly. *Cell Tissue Res.* **195**:299–308.

Riddle, C. V., and Ernst, S. A. 1979. Structural simplicity of the *zonula occludens* in the electrolyte secreting epithelium of the avian salt gland. *J. Membr. Biol.* **45**:21–35.

Ridgway, H. F., Wagner, R. M., Dawsey, W. T., and Lewin, R. A. 1975. Fine structure of the cell envelope layers of *Flexibacter polymorphus*. *Can. J. Microbiol.* **21**:1733–1750.

Robertson, J. D. 1959. The ultrastructure of cell membranes and their derivatives. *Symp. Biochem. Soc.* **16**:3–43.

Robertson, J. D. 1960. The molecular structure and contact relationship of cell membranes. *Prog. Biophys.* **10**:343–418.

Robertson, J. D. 1967. The organization of cellular membranes. *In*: Allen, J. M., ed., *Molecular Organization and Biological Function*, New York, Harper & Row, pp. 65–106.

Robinson, G. A., and Sutherland, E. 1971. Cyclic AMP and cell function. *Ann. N.Y. Acad. Sci.* **185**:5–9.

Rodewald, R. 1973. Intestinal transport of antibodies in the newborn rat. *J. Cell Biol.* **58**:189–211.

Rogers, H. J., and Perkins, H. R. 1968. The mucopeptides. *In*: Rogers, H. J., and Perkins, H. R., eds., *Cell Walls and Membranes*, London, Spon, pp. 231–258.

Roland, J. D., Szöllösi, A., and Szöllösi, D. 1977. *Atlas of Cell Biology*, Boston, Little, Brown.

Rosenbusch, J. 1974. Characterization of the major envelope protein from *E. coli*. *J. Biol. Chem.* **249**:8019–8029.

Roth, L. E. 1960. Electromicroscopy of pinocytosis and food vacuoles in *Pelomyxa*. *J. Protozool.* **7**:176–185.

Roth, L. E., and Daniels, E. W. 1961. Infective organisms in the cytoplasm of *Amoeba proteus*. *J. Biophys. Biochem. Cytol.* **9**:317–323.

Rothstein, A. 1978. The cell membrane—A short historical perspective. *Curr. Top. Membr. Transp.* **11**:1–13.

Rottem, S., Slutzky, G. M., and Bittman, R. 1978. Cholesterol distribution and movement in the *Mycoplasma galliseptum* cell membrane. *Biochemistry* **17**:2723–2726.

Rowley, M. J., Berlin, J. D., and Heller, C. G. 1971. The ultrastructure of four types of human spermatogonia. *Z. Zellforsch. Mikrosk. Anat.* **112**:139—157.

Ruffolo, J. J. 1978. Micronuclear envelope formation after telophase of mitosis in the ciliate *Euplotes eurystomus*. *Trans. Am. Microsc. Soc.* **97**:259–263.

Sacks, G., Spenney, J. G., Saccomani, G., and Goodall, M. C. 1974. Characterization of gastric mucosal membranes. VI. The presence of channel forming substances. *Biochim. Biophys. Acta* **332**:233–247.

Sagara, Y., Harano, T., and Omura, T. 1978. Characterization of electron transport enzymes in the envelope of rat liver nuclei. *J. Biochem.* **83**:807–812.

Saier, M. H., and Stiles, C. D. 1975. *Molecular Dynamics in Biological Membranes*, New York, Springer-Verlag.

Salton, M. R. J. 1973. Structure and function of the bacterial cell wall. *In*: Markham, R., Bancroft, J. B., Davies, D. R., Hopwood, D. A., and Horne, R. W., eds., *The Generation of Subcellular Structures,* Amsterdam, North-Holland, pp. 235–262.

Salton, M. R. J., and Freer, J. H. 1965. Composition of the membranes isolated from several Gram-positive bacteria. *Biochim. Biophys. Acta* **107:**531–538.

Salton, M. R. J., and Marshall, B. 1959. The composition of the spore wall and the wall of vegetative cells of *B. subtilis*. *J. Gen. Microbiol.* **21**:415–420.

Scarborough, G. A. 1977. Properties of the *Neurospora crassa* plasma membrane ATPase. *Arch. Biochem. Biophys.* **180**:384–393.

Schafer, R., Hinnen, R., and Franklin, R. M. 1974. Structure and synthesis of a lipid-containing bacteriophage. *Eur. J. Biochem.* **50**:15–27.

Schatten, G., and Thoman, M. 1978. Nuclear surface complex as observed with the high resolution scanning electron microscope. *J. Cell Biol.* **77**:517–535.

Schatzmann, H. J. 1953. Herzglykoside als Hemmstoffe für aktiven Kalium und Natrium Transport durch die Erythrocyten membran. *Biochim. Biophys. Acta* **94**:89–96.

Schmit, A. S. Pless, D. D., and Lennarz, W. J. 1974. Some aspects of the chemistry and biochemistry of membranes of Gram-positive bacteria. *Ann. N.Y. Acad. Sci.* **235**:91–104.

Schmitt, F. O. 1971. Molecular membranology. *In*: Wallach, D. F. H., and Fischer, H., eds., *The Dynamic Structure of Cell Membranes*, Berlin, Springer-Verlag, pp. 5–36.

Schnaitman, C. A. S. 1969. Comparison of rat liver mitochondrial and microsomal membrane proteins. *Proc. Natl. Acad. Sci. USA* **63**:412–419.

Schneider, A. S., Schneider, M. J., and Rosenheck, K. 1970. Optical activity of biological membranes: Scattering effects and protein conformation. *Proc. Natl. Acad. Sci. USA* **66**:793–798.

Scribner, D. J. and Fahrney, D. 1976. Neutrophil receptors for IgG and complement: Their role in the attachment and ingestion phases of phagocytosis. *J. Immunol.* **116**:892–897.

Seaston, A., Inkson, C., and Eddy, A. A. 1973. The absorption of protons with specific amino acids and carbohydrates in yeast. *Biochem. J.* **134**:1031–1043.

Severs, N. J. 1977. Nuclear envelope transport capacity and the cell cycle in yeast (*S. cerevisiae*). *Cytobios* **18**:51–67.

Sharp, G. W. G., and Leaf, A. 1966. Mechanism of action of aldosterone. *Physiol. Rev.* **46**:593–633.

Shimono, M., and Clementi, F. 1976. Intercellular junctions of oral epithelium. I. Studies with freeze-fracture and tracing methods of normal rat keratinized oral epithelium. *J. Ultrastruct. Res.* **56**:121–136.

Shockman, G. D., Daneo-Moore, L., and Higgins, M. L. 1974. Problems of cell wall and membrane growth, enlargement, and division. *Ann. N.Y. Acad. Sci.* **235**:161–196.

Shockman, G. D., Kolb, J. J., Bakay, B., Conover, M., and Toennies, G. 1963. Protoplast membrane of *Streptococcus faecalis*. *J. Bacteriol.* **85**:168–176.

Sikstrom, R., Lanoix, J., and Bergeron, J. J. M. 1976. An enzymatic analysis of a nuclear envelope fraction. *Biochim. Biophys. Acta* **448**:88–102.

Silverman, M. 1974. The chemical and steric determinants governing sugar interactions with renal tubular membranes. *Biochim. Biophys. Acta* **332**:246–262.

Silverstein, S. C., Steinman, R. M., and Cohn, Z. A. 1977. Endocytosis. *Annu. Rev. Biochem.* **46**:669–722.

Simionescu, N., Simionescu, M., and Palade, G. E. 1978. Open junctions in the endothelium of the post capillary venules of the diaphragm. *J. Cell Biol.* **79**:27–44.

Singer, S. J. 1972. A fluid lipid-globular protein mosaic model of membrane structure. *Ann. N.Y. Acad. Sci.* **195**:16–23.

Singer, S. J. 1977. The fluid mosaic model of membrane structure. *In:* Abrahamsson, S., and Pascher, I., eds., *Structure of Biological Membranes,* New York, Plenum Press, pp. 443–461.

Singer, S. J., and Nicolson, G. L. 1972. The fluid mosaic model of the structure of cell membranes. *Science* **175**:720–731.

Sjöstrand, F. S. 1963a. A comparison of plasma membrane, cytomembranes, and mitochondrial membrane elements with respect to ultrastructural features. *J. Ultrastruct. Res.* **9**:561–580.

Sjöstrand, F. S. 1963b. A new ultrastructural element of the membranes in mitochondria and of some cytoplasmic membranes. *J. Ultrastruct. Res.* **9**:340–361.

Sjöstrand, F. S. 1969. Morphological aspects of lipoprotein structures. *In*: Tria, E., and Scanu, A. M., eds., *Structural and Functional Aspects of Lipoproteins in Living Systems*, New York, Academic Press, pp. 73–128.

Sjöstrand, F. S. 1978. The structure of mitochondrial membranes: A new concept. *J. Ultrastruct. Res.* **64**:217—245.

Skou, J. C. 1975. The (Na^+ + K^+) activated enzyme system and its relationship to transport of sodium and potassium. *Q. Rev. Biophys.* **7**:401.

Skou, J. C. 1977a. Coupling of chemical reaction to transport of sodium and potassium. *In*: Abrahamsson, S., and Pascher, I., eds., *Structure of Biological Membranes*, New York. Plenum Press, pp. 463–478.

Skou, J. C. 1977b. Coupling between ATP hydrolysis and sodium and potassium transport. *In*: Nicolau, C., and Parof, A., eds., *Structural and Kinetic Approach to Plasma Membrane Functions*, New York, Springer-Verlag, pp. 145–151.

Sleytr, U. B. 1978. Regular arrays of macromolecules on bacterial cell walls: structure, chemistry, assembly, and function. *Int. Rev. Cytol.* **53:**1–64.

Sleytr, U. B., and Thorne, K. J. I. 1976. Chemical characterization of the regularly arranged layers of *Clostridium thermosaccharolyticum* and *Clostridium thermohydrosulfuricum*. *J. Bacteriol.* **126**:377–383.

Sonenberg, M. 1971. Interaction of human growth hormone and human erythrocyte membranes studied by intrinsic fluorescence. *Proc. Natl. Acad. Sci. USA* **68**:1051–1055.

Staehelin, L. A. 1974. Structure and function of intercellular junctions. *Int. Rev. Cytol.* **39**:191–283.

Stanbury, J. B. 1972. Some recent developments in the physiology of the thyroid gland. *Rev. Physiol.* **65**:94–125.

Stein, W. D. 1972. The mechanism of sugar transfer across erythrocyte membranes. *Ann. N.Y. Acad. Sci.* **195**:412–427.

Sturgess, J., Moscarello, M., and Schachter, H. 1978. The structure and biosynthesis of membrane glycoproteins. *Curr. Top. Membr. Transp.* **11**:16–105.

Sullivan, C. W., and Volcani, B. E. 1974. Synergistically stimulated (Na^+, K^+)-ATPase from plasma membrane of a marine diatom. *Proc. Natl. Acad. Sci. USA* **71**:4376–4380.

Sundler, R., Sarcione, S. L., Alberts, A. W., and Vagelos, P. R. 1977. Evidence against phospholipid asymmetry in intracellular membranes from liver. *Proc. Natl. Acad. Sci. USA* **74**:3350–3354.

Sutherland, E. W., Rall, T. W., and Menon, T. 1962. Adenylcyclase. I. Distribution, preparations, and properties. *J. Biol. Chem.* **237**:1220–1227.

Suzuki, F., and Nagano, T. 1978a. Regional differentiation of cell junctions in the excurrent duct of the rat testis as revealed by freeze fracture. *Anat. Rec.* **191**:503–519.

Suzuki, F., and Nagano, T. 1978b. Development of tight junctions in the caput epididymal epithelium of the mouse. *Dev. Biol.* **63**:321–324.

Takesue, Y., and Nishi, Y. 1978. Topographical studies on intestinal microvillous leucine β-naphthylamidase on the outer membrane surface. *J. Membr. Biol.* **39**:285–296.

Takumi, K., and Kawata, T. 1974. Isolation of a common cell wall antigen from the proteolytic strains of *Clostridium botulinum*. *Jpn. J. Microbiol.* **18**:85–90.

Tamaki, S., Sato, T., and Matsuhashi, M. 1971. Role of lipopolysaccharides in antibiotic resistance and bacteriophage adsorption of *E. coli* K-12. *J. Bacteriol.* **105:**968–975.

Tamm, S. L. 1979. Membrane movements and fluidity during rotational motility of a termite flagellate. *J. Cell Biol.* **80**:141–149.

Tanaka, M., Haniu, M., Yasunobu, K. T., Yu, C. A., Yu, L., Weit, Y. H., and King, T. E. 1977. Amino acid sequence of the heme *a* subunit of bovine heart cytochrome oxidase and sequence homology with hemoglobin. *Biochem. Biophys. Res. Commun.* **76**:1014–1019.

Tanford, C. 1978. The hydrophobic effect and the organization of living matter. *Science* **200**:1012–1018.

Tanner, W. 1969. Light-driven active uptake of 3-*O*-methylglucose via an inducible hexose uptake system of *Chlorella*. *Biochem. Biophys. Res. Commun.* **36:**278–283.

Tanner, W., and Komor, E. 1975. Hexose–proton cotransport of *Chlorella vulgaris*. *Biomembranes* **35**:145–154.

Taylor, D. L. 1968. *In situ* studies on the cytochemistry and ultrastructure of a symbiotic marine dinoflagellate. *J. Mar. Biol. Assoc. U.K.* **48**:349–366.

Thair, B. W., and Wardrop, A. B. 1971. The structure and arrangement of nuclear pores in plant cells. *Planta* **100**:1–17.

Thorne, K. J. I., Oliver, R. C., and Heath, M. F. 1976. Phospholipase A_2 activity of the regularly arranged surface protein of *Acinetobacter* sp. 199A. *Biochim. Biophys. Acta* **450:**335–341.

Thyagarajan, T. R., Conti, S. F., and Naylor, H. B. 1962. Electron microscopy of *Rhodotorula glutinis*. *J. Bacteriol.* **83**:381–394.

Tucker, J. B. 1978. Endocytosis and streaming of highly gelated cytoplasm alongside rows of arm-bearing microtubules in the ciliate *Nassula*. *J. Cell Sci.* **29**:213–232.

Tzagoloff, A., ed. 1975a. *Membrane Biogenesis: Mitochondria, Chloroplasts, and Bacteria*. New York, Plenum Press.

Tzagoloff, A. 1975b. Current problems in membrane biogenesis. *In*: Tzagoloff, A., ed., *Membrane Biogenesis: Mitochondria, Chloroplasts, and Bacteria*, New York, Plenum Press, pp. 1–14.

Urban, B. W., Hladky, S. B., and Haydon, D. A. 1978. The kinetics of ion movements in the gramicidin channel. *Fed. Proc. Fed. Am. Soc. Exp. Biol.* **37**:2628–2632.

van Alphen, L., Van Selm, N., and Lugtenberg, B. 1978a. Pores in the outer membrane of *E. coli* K-12. Involvement of proteins b and c in the permeation functioning of pores for nucleotides. *Mol. Gen. Genet.* **159**:75–83.

van Alphen, L., Verkleij, A., Leunissen-Bijvelt, J., and Lugtenberg, B. 1978b. Architecture of the outer membrane of *E. coli* III. Protein–lipopolysaccharide complexes in intramembraneous particles. *J. Bacteriol.* **134**:1089–1098.

van Alphen, L., Lugtenberg, B., van Boxtel, R., Hack, A. M., Verhoef, C., and Havekes, L. 1979. *meo A* is the structural gene for outer membrane protein c of *E. coli* K-12. *Mol. Gen. Genet.* **169**:147–155.

van Deenan, L. L. M., de Gier, J., van Golde, L. M. G., Nauta, I. L. D., Renooy, W., Verkleij, A. J., and Zwaal, R. F. A. 1977. Some topological and dynamic aspects of lipids in the erythrocyte membrane. *In*: Abrahamsson, S., and Pascher, I., eds., *Structure of Biological Membranes*, New York, Plenum Press, pp. 107–118.

Vanderkooi, G., and Capaldi, R. A. 1972. A comparative study of the amino acid compositions of membrane proteins and other proteins. *Ann. N.Y. Acad. Sci.* **195**:135–138.

Vanderkooi, G., and Green, D. E. 1970. New insights into biological membrane structure. *BioScience* **21**:409–415.

Veatch, W. R., Fossel, E. T., and Blout, E. R. 1974. The conformation of gramicidin A. *Biochemistry* **13**:5249–5256.

Verhoef, C., Lugtenberg, B., van Boxtel, R., deGraaff, P., and Verheij, H. 1979. Genetics and biochemistry of the peptidoglycan-associated proteins b and c of *E. coli* K-12. *Mol. Gen. Genet.* **169**:137–146.

Wade, J. B., and Karnovsky, M. J. 1974. The structure of the *zona occludens*. A single fibril model based on freeze-fracture. *J. Cell Biol.* **60**:168–180.

Wallach, D. F. H. 1972. *The Plasma Membrane: Dynamic Perspectives, Genetics and Pathology*, London, English Universities Press.

Wallach, D. F. H., and Winzler, R. J. 1974. *Evolving Strategies and Tactics in Membrane Research*, New York, Springer-Verlag.

Wallach, D. F. H., and Zahler, H. P. 1966. Protein conformations in cellular membranes. *Proc. Natl. Acad. Sci. USA* **56**:1552–1559.

Warren, G. B., and Metcalfe, J. C. 1977. The molecular architecture of a reconstituted calcium pump. *In*: Nicolau, C., and Paraf, A., eds., *Structural and Kinetic Approach to Plasma Membrane Functions*, New York, Springer-Verlag, pp. 188–200.

Wasserman, W. J., and Smith, L. D. 1978. The cyclic behavior of a cytoplasmic factor controlling nuclear membrane breakdown. *J. Cell Biol.* **78**:R15–R22.

Watson, M. L. 1955. The nuclear envelope. Its structure and relation to cytoplasmic membranes. *J. Biophys. Biochem. Cytol.* **1**:257—270.

Ways, P., and Hanahan, D. J. 1964. Characterization and quantification of red cell lipids in normal man. *J. Lipid Res.* **5:**318–328.

Wecker, J., Reinicke, B., and Schallehn, G. 1974. Remarkable differences in the ultrastructure of the cell wall of toxigenic *Clostridia*. *Electron Microscopy, Proceedings of the 8th International Congress,* Vol. 2; p. 644.

Weibull, C., and Bergström, L. 1958. The chemical nature of the cytoplasmic membrane and cell wall of *Bacillus megaterium*, strain M. *Biochim. Biophys. Acta* **30**:340–351.

West, J. C., and Mitchell, P. 1972. Proton-coupled β-galactoside translocation in non-metabolizing *E. coli*. *J. Bioenerg.* **3**:445–462.

Weston, J. C., Greider, M. H., Ackerman, G. A., and Nikolewski, R. F. 1972. Nuclear membrane contributions to the Golgi complex. *Z. Zellforsch. Mikrosk. Anat.* **123:**153–160.

Whaley, W. G., Mollenhauer, H. H., and Kephart, J. 1959. The endoplasmic reticulum and the Golgi structure in maize root cells. *J. Biophys. Biochem. Cytol.* **5**:501–506.

Whaley, W. G., Mollenhauer, H. H., and Leech, J. H. 1960. Some observations on the nuclear envelope. *J. Biophys. Biochem. Cytol.* **8**:233–245.

Whittam, R., and Ager, M. E. 1965. The connexion between active cation transport and metabolism in erythrocytes. *Biochem. J.* **97**:214–227.

Wikström, M. K. F. 1977. Proton pump coupled to cytochrome *c* oxidase in mitochondria. *Nature (London)* **266**:271–273.

Wikström, M. K. F., and Krab, K. 1978. Cytochrome *c* oxidase is a proton pump. *FEBS Lett.* **91**:8–14.

Williams, R. J. P. 1978. Energy states of proteins, enzymes, and membranes. *Proc. R. Soc. London. Ser. B.* **200**:353–389.

Wilton, J. M. A., Renggli, H. H., and Lehner, T. 1977. The role of Fc and C3b receptors in phagocytosis by inflammatory polymorphonuclear leucocytes in man. *Immunology* **32**:955–961.

Wischnitzer, S. 1960. The ultrastructure of the nucleus and nucleocytoplasmic relations. *Int. Rev. Cytol.* **10**:137–162.

Woo-Sam, P. C. 1977. Cohesion of horny cells during comedo formation. *Br. J. Dermatol.* **97**:609–615.

Wu, M., and Heath, E. C. 1973. Isolation and characterization of lipopolysaccharide protein from *E. coli*. *Proc. Natl. Acad. Sci. USA* **70**:2572–2576.

Wurster, B., Pan, P., Tyan, G. G., and Bonner, J. T. 1976. Preliminary characterization of the acrasin of the cellular slime mold *Polysphondylium violaceum*. *Proc. Natl. Acad. Sci. USA* **73**:795–799.

Yamada, K. M., Spooner, B. S., and Wessells, N. K. 1971. Ultrastructure and function of growth cones and axons of cultured nerve cells. *J. Cell Biol.* **49**:614–635.

Yamato, I., and Anraku, Y. 1977. Transport of sugars and amino acids in bacteria. XVIII. Properties of an isoleucine carrier in the cytoplasmic membrane vesicles of *E. coli*. *J. Biochem.* **81**:1517–1523.

Zamboni, L. 1971. *Fine Morphology of Mammalian Fertilization*, New York, Harper & Row.

Zbarsky, I. B. 1978. An enzyme profile of the nuclear envelope. *Int. Rev. Cytol.* **54**:295–360.

Zwaal, R. F. A., Roelofsen, B., and Colley, C. M. 1973. Localization of red cell membrane constituents. *Biochim. Biophys. Acta* **300**:159–182.

CHAPTER 2

Adelman, M. R., and Taylor, E. W. 1969. Isolation of an actomyosin-like protein complex plasmodium and the separation of the complex into actin- and myosin-like fractions. *Biochemistry* **8**:4964—4975.

Adelstein, R. S., Pollard, T. D., and Kuehl, W. M. 1971. Isolation and characterization of myosin and two myosin fragments from human blood platelets. *Proc. Natl. Acad. Sci. USA* **68**:2703–2707.

Adelstein, R. S., Conti, M. A., Johnson, G., Pastan, I., and Pollard, T. D. 1972. Isolation and characterization of myosin fragments from cloned mouse fibroblasts. *Proc. Natl. Acad. Sci. USA* **69**:3693–3697.

Agaraña, C. E., Arce, C. A., Barra, H. C., and Caputto, R. 1977. *In vivo* incorporation of (^{14}C)tyrosine into the C-terminal position of the α subunit of tubulin. *Arch. Biochem. Biophys.* **180**:264–268.

Allen, C., and Borisy, G. G. 1974. Structural polarity and directional growth of microtubules of *Chlamydomonas* flagella. *J. Mol. Biol.* **90**:381–402.

Anderton, B. H., Ayers, M., and Thorpe, R. 1978. Neurofilaments from mammalian central and peripheral nerves share certain polypeptides. *FEBS Lett.* **96:**159–163.

Amos, L. A., and Klug, A. 1974. Arrangement of subunits in flagellar microtubules. *J. Cell Sci.* **14**:523–549.

Ando, T., and Asai, H. 1979. Conformational change in actin filament induced by the interaction with heavy meromyosin: Effects of *p*H, tropomyosin and deoxy-ATP. *J. Mol. Biol.* **129**:265–277.

Arai, T., and Kazior, Y. 1976. Effect guanine quanine nucleotides on the assembly of brain microtubules: Ability of 5′-guanylylamidodiphosphate to replace GTP in promoting the polymerization of microtubules *in vitro*. *Biochem. Biophys. Res. Commun.* **69**:369–376.

Arai, T., and Kaziro, Y. 1977. Role of GTP in the assembly of microtubules. *J. Biochem.* **82**:1063–1071.

Arata, T., Mukohata, Y., and Tonomura, Y. 1977. Structure and function of the two heads of the myosin molecule. VI. ATP hydrolysis, shortening, and tension development of myofilaments. *J. Biochem.* **82**:801–812.

Bálint, M., Wolf, I., Tarcsafalvi, A., Gergely, J. and Sréter, F. A. 1978. Location of SH-1 and SH-2 in the heavy chain segment of heavy meromyosin. *Arch. Biochem. Biophys.* **190**:793–799.

Bardele, C. F. 1973. Struktur, Biochemie, und Funktion der Mikrotubuli. *Cytobiologie Z. Exp. Zellforsch.* **7**:442–487.

Barra, H. S., Arce, C. A., Rodriguez, J. A., and Caputto, R. 1974. Some common properties of the protein that incorporates tyrosine as a single unit and its microtubule proteins. *Biochem. Biophys. Res. Commun.* **60**:1384–1390.

Begg, D. A., Rodewald, R., and Rebhun, L. I. 1978. The visualization of actin filament polarity in thin sections. *J. Cell. Biol.* **79**:846–852.

Behnke, O., and Forer, A. 1966. Intranuclear microtubules. *Science* **153**:1536–1537.

Bell, P. R. 1978. A microtubule-nuclear envelope complex in the spermatozoid of *Pteridium*. *J. Cell Sci.* **29**:189–195.

Bender, N., Fasold, H., and Rack, M. 1974. Interaction of rabbit muscle actin and chemically modified actin with ATP, ADP, and protein reactive analogues: Role of the nucleotide. *FEBS Lett.* **44**:209–212.

Benitz, W. E., Dahl, D., Williams, K. W., and Bignami, A. 1976. The protein composition of glial and nerve fibers. *FEBS Lett.* **66**:285–289.

Berkowitz, S. A., Katagiri, J., Binder, H. K., and Williams, R. C. 1977. Separation and characterization of microtubule proteins from calf brains. *Biochemistry* **16**:5610–5617.

Berl, S., Puszkin, S., and Nicklas, W. J. 1973. Actomyosin-like protein in brain. *Science* **179**:441–446.

Berry, R. W., and Shelanski, M. L. 1972. Interactions of tubulin with vinblastine and GTP. *J. Mol. Biol.* **71**:71–80.

Bhowmick, D. K. 1967. Electron microscopy of *Trichamoeba villosa* and amoeboid movement. *Exp. Cell Res.* **45**:570–589.

Binder, L. I., and Rosenbaum, J. L. 1978. The *in vitro* assembly of flagellar outer doublet tubulin. *J. Cell Biol.* **79**:500–515.

Birnbaum, E. R., and Sykes, B. D. 1978. NMR studies of a Ca^{2+}-binding fragment of troponin C. *Biochemistry* **17**:4965–4970.

Bisalputra, T., Oakley, B. R., Walker, D. C., and Shields, C. M. 1975. Microtubular complexes in blue-green algae. *Protoplasma* **86**:19–28.

Booyse, F. M., Hoveke, T. P., Zschocke, D., and Rafelson, M. E. 1971. Human platelet myosin: Isolation and properties. *J. Biol. Chem.* **246**:4291–4297.

Borisy, G. G., and Olmsted, J. D. B. 1972. Nucleated assembly of microtubules in porcine brain extracts. *Science* **177**:1196–1197.

Borisy, G. G., Olmsted, J. D. B., Marcum, J. M., and Allen, C. 1974. Microtubule assembly *in vitro*. *Fed. Proc. Fed. Am. Soc. Exp. Biol.* **33**:167–174.

Borisy, G. G., Marcum, J. M., Olmsted, J. D. B., Murphy, D. B., and Johnson, K. A. 1975. Purification of tubulin and associated high molecular weight proteins from porcine brain and characterization of microtubule assembly *in vitro*. *Ann. N.Y. Acad. Sci.* **253**:107–132.

Bray, D. 1977. Actin and myosin in neurones: A first review. *Biochimie* **59**:1–6.

Bridgen, J. 1971. The amino acid sequence around four cysteine residues in trout actin. *Biochem. J.* **123**:591–600.

Brotschi, E. A., Hartwig, J. H., and Stossel, T. P. 1978. The gelation of actin by actin-binding protein. *J. Biol. Chem.* **253**:8988–8993.

Bryan, J. 1974. Microtubules. *BioScience* **24**:701–711.

Bryan, J. 1975. Preliminary studies on affinity-labeling of the tubulin–colchicine binding site. *Ann. N.Y. Acad. Sci.* **253**:247–259.

Bryan, J., and Wilson, L. 1971. Are cytoplasmic microtubules heteropolymers? *Proc. Natl. Acad. Sci. USA* **68**:1762–1766.

Brysk, M. M., Gray, R. H., and Bernstein, I. A. 1977. Tonofilament protein from newborn rat epidermis. *J. Biol. Cell.* **252**:2127–2133.

Bulinski, J. C., and Borisy, G. G. 1979. Self-assembly of microtubules in extracts of cultured HeLa cells and the identification of HeLa microtubule associated proteins. *Proc. Natl. Acad. Sci. USA* **76**:293–297.

Burnside, B. 1975. The form and arrangement of microtubules: An historical, primarily morphological, review. *Ann. N.Y. Acad. Sci.* **253**:14–26.

Burton, P. R., and Himes, R. H. 1978. Electron microscope studies of the *p*H effects on assembly of tubulin free of associated proteins. *J. Cell Biol.* **77**:120–133.

Campanella, C., Gabbiani, G., Baccetti, B., Burrini, A. G., and Pallini, V. 1979. Actin and myosin in the vertebrate acrosomal region. *J. Submicrosc. Cytol.* **11**:53–71.

Capony, J. P., Ryden, L., Demaille, J., and Pechere, J. F. 1973. The primary structure of the major parvalbumin from hake muscle. *Eur. J. Biochem.* **32**:97–108.

Caputo, C. 1978. Excitation and contraction processes in muscle. *Annu. Rev. Biophys. Bioeng.* **7**:63–83.

Carasso, N., and Favard, P. 1965. Microtubules fusoriaux dans les micro- et macronucleus de ciliés péritriches en division. *J. Microsc. (Paris)* **4**:395–402.

Caspar, D. L. D., Cohen, C., and Longley, W. 1969. Tropomyosin: Crystal structure, polymorphism, and molecular interactions. *J. Mol. Biol.* **41**:87–107.

Chantler, P. D., and Szent-Györgyi, A. G. 1978. Spectroscopic studies on invertebrate myosins and light chains. *Biochemistry* **17**:5440–5448.

Clark, T. G., and Merriam, R. W. 1977. Diffusible and bound actin in nuclei of *Xenopus laevis* oocytes. *Cell* **12**:883–892.

Cleveland, D. W., Hwo, S. Y., and Kirschner, M. W. 1977a. Purification of tau, a microtubule-associated protein that induces assembly of microtubules from purified tubulin. *J. Mol. Biol.* **116**:207–225.

Cleveland, D. W., Hwo, S. Y., and Kirschner, M. W. 1977b. Physical and chemical properties of purified tau factor and the role of tau in microtubule assembly. *J. Mol. Biol.* **116:**227–247.

Coffee, C. J., and Bradshaw, R. A. 1973. Carp muscle calcium-binding protein. I. Characterization of the tryptic peptides and the complete amino acid sequence of component B. *J. Biol. Chem.* **248**:3305–3312.

Cohen, C., and Holmes, K. C. 1963. X-ray diffraction evidence for α-helical coiled coils in native muscle. *J. Mol. Biol.* **6**:423–432.

Cohen, C., Harrison, S. C., and Stephens, R. E. 1971. X-ray diffraction from microtubules. *J. Mol. Biol.* **59**:375–380.

Cohen, C., DeRosier, D., Harrison, S. C., Stephens, R. E., and Thomas, J. 1975. X-ray patterns from microtubules. *Ann. N.Y. Acad. Sci.* **253**:53–59.

Collins, J. H. 1974. Homology of myosin light chains, troponin-C, and parvalbumins. *Biochem. Biophys. Res. Commun.* **58**:301–308.

Collins, J. H., and Elzinga, M. 1975. The primary structure of actin from rabbit skeletal muscle. Completion and analysis of the amino acid sequence. *J. Biol. Chem.* **250**:5915–5920.

Collins, J. H., Potter, J. D., Horn, M. J., Wilshire, G., and Jackman, N. 1973. The amino acid sequence of rabbit skeletal muscle troponin-C: Gene replication and homology with calcium-binding proteins from carp and hake muscle. *FEBS Lett.* **36**:268–272.

Collins, J. H., Greaser, N. L., Potter, J. D., and Horn, M. J. 1977. Determination of the amino acid sequence of troponin-C from rabbit skeletal muscle. *J. Biol. Chem.* **252**:6356–6362.

Connolly, J. A., Kalnins, V. I., Cleveland, D. W., and Kirschner, M. W. 1978. Intracellular localization of the high molecular weight microtubule accessory protein by indirect immunofluorescence. *J. Cell Biol.* **76**:R781–R786.

Cooke, P. H., and Chase, R. H. 1971. KCl-insoluble myofilaments in vertebrate smooth muscle cells. *Exp. Cell Res.* **66**:417–425.

Corfield, P. S., and Smith, D. G. 1968. Microtubular structure in group D streptococcal L-forms. *Arch. Mikrobiol.* **63**:356–361.

Dahlström, A. 1971. Axoplasmic transport (with particular respect to adrenergic neurons). *Philos. Trans. R. Soc. London Ser. B* **261:**325–358.

Davison, P. F., Hong, B. S., and Cooke, P. 1977. Classes of distinguishable 10 nm cytoplasmic filaments. *Exp. Cell Res.* **109:**471–474.

DeBrabander, M., and Borgers, M. 1975. Intranuclear microtubules in mast cells. *Pathol. Eur.* **10**:17–20.

DeBrabander, M., Aerts, F., Van de Veire, R., and Borgers, M. 1975. Evidence against interconversion of microtubules and filaments. *Nature (London)* **253**:119–120.

Dedman, J. R., Jackson, R. L., Schreiber, W. E., and Means, A. R. 1978. Sequence homology of the Ca^{2+}-dependent regulator of cyclic nucleotide phosphodiesterase from rat testes with other Ca^{2+}-binding proteins. *J. Biol. Chem.* **253**:343–346.

Dönges, K. H., Biedert, S., and Paweletz, N. 1976. Characterization of 20 S component in tubulin from mammalian brain. *Biochemistry* **15**:2995–2999.

Dustin, P. 1972. Microtubules et microfilaments: Leur rôle dans la dynamique cellulaire. *Arch. Biol.* **83**:419–480.

Dustin, P. 1978. *Microtubules*, Berlin, Springer-Verlag.

Ebashi, S. 1974. Regulatory mechanism of muscle contraction, with special reference to the calcium–troponin–tropomyosin system. *Essays Biochem.* **10**:1–36.

Ebashi, S., and Endo, M. 1968. Calcium ion and muscle contraction. *Prog. Biophys. Mol. Biol.* **18**:123–183.

Ebashi, S., Onishi, S., Abe, S., and Maruyama, K. 1974. A spin-label study on calcium-induced conformational changes of troponin components. *J. Biochem.* **75**:211–213.

Eda, T., Kanda, Y., and Kimura, S. 1976. Membrane structures in stable L-forms of *E. coli*. *J. Bacteriol.* **27**:657–664.

Eda, T., Kanda, Y., Mori, C., and Kimura, S. 1977. Microtubular structures in a stable staphylococcal L-form. *J. Bacteriol.* **132**:1024–1026.

Edwards, B. F. P., and Sykes, B. D. 1978. Assignment and characterization of the histidine resonances in ^1H nuclear magnetic resonance spectra of rabbit tropomyosins. *Biochemistry* **17**:684–689.

Eipper, B. A. 1972. Rat brain microtubule protein: Purification and determination of covalently bound phosphate and carbohydrate. *Proc. Natl. Acad. Sci. USA* **69**:2283–2287.

Eipper, B. A. 1975. Purification of rat brain tubulin. *Ann. N.Y. Acad. Sci.* **253**:239–246.

Elzinga, M., and Collins, J. H. 1973. The amino acid sequence of rabbit skeletal muscle actin. *Cold Spring Harbor Symp. Quant. Biol.* **37**:1–7.

Elzinga, M., and Collins, J. H. 1975. The primary structure of actin from rabbit skeletal muscle. Five cyanogen bromide peptides, including the NH_2 and COOH termini. *J. Biol. Chem.* **250**:5897–5905.

Elzinga, M., Collins, J. H., Kuehl, W. M., and Adelstein, R. S. 1973. Complete amino acid sequence of actin of rabbit skeletal muscle. *Proc. Natl. Acad. Sci. USA* **70**:2687–2691.

Elzinga, M., Moran, B. J., and Adelstein, R. S. 1976. Human heart and platelet actins are products of different genes. *Science* **191**:94–95.

Epstein, H. F., Harris, H. E., Schachat, F. H., Suddleson, E. A., and Wolff, J. A. 1976. Genetic and molecular studies of nematode myosin. *In*: Goldman, R., Pollard, T., and Rosenbaum, J., eds., *Cell Motility*, Cold Spring Harbor, N.Y., Cold Spring Harbor Laboratory, Book A, pp. 203–214.

Erickson, H. P. 1974a. Microtubule surface lattice and subunit structure and observations on reassembly. *J. Cell Biol.* **60**:153–167.

Erickson, H. P. 1974b. Assembly of microtubules from preformed, ring-shaped protofilaments and 6-S tubulin. *J. Supramol. Struct.* **2**:393–411.

Erickson, H. P. 1975. The structure and assembly of microtubules. *Ann. N.Y. Acad. Sci.* **253**:60–77.

Erickson, H. P., and Voter, W. A. 1976. Polycation-induced assembly of purified tubulin. *Proc. Natl. Acad. Sci. USA* **73**:2813–2817.

Eriksson, A., and Thornell, L. E. 1979. Intermediate (skeletin) filaments of heart Purkinje fibers. *J. Cell Biol.* **80**:231–247.

Farrell, K. W., and Wilson, L. 1978. Microtubule reassembly *in vitro* of *Strongylocentrotus purpuratus* sperm tail outer doublet tubulin. *J. Mol. Biol.* **121**:393–410.

Farrell, K. W., Morse, A., and Wilson, L. 1979. Characterization of the *in vitro* reassembly of tubulin derived from stable *Strongylocentrotus purpuratus* outer doublet microtubules. *Biochemistry* **18**:905–911.

Fellows, A., Francon, J., Virion, A., and Nunez, J. 1975. Microtubules and brain development. *FEBS Lett.* **57**:5–9.

Fine, R. E. and Blitz, A. L. 1975. A chemical comparison of tropomyosins from muscle and non-muscle tissues. *J. Mol. Biol.* **95**:447–454.

Frank, G., and Weeds, A. G. 1974. The amino acid sequence of the alkali light chains of rabbit skeletal-muscle myosin. *Eur. J. Biochem.* **44**:317–334.

Franke, W. W., Schmid, E., Weber, K., and Osborn, M. 1979. HeLa cells contain intermediate-sized filaments of the prekeratin type. *Exp. Cell Res.* **118**:95–109.

Frankenne, F., Joassin, L., and Gerday, C. 1973. The amino acid sequence of the pike (*Esox lucius*) parvalbumin III. *FEBS Lett.* **35**:145–147.

Frearson, N., and Perry, S. V. 1975. Phosphorylation of the light-chain components of myosin from cardiac and red skeletal muscles. *Biochem. J.* **151**:99–107.

Frederiksen, D. W. 1979. Physical properties of myosin from aortic smooth muscle. *Biochemistry* **18**:1651–1656.

Freud, S. 1882. *Sitzungsber. Akad. Wiss. Wien Math. Naturwiss.* **85**:9.

Fujiwara, K., and Tilney, L. G. 1975. Substructural analysis of the microtubule and its polymorphic forms. *Ann. N.Y. Acad. Sci.* **253**:27–50.

Gabbiani, G., and Montandon, D. 1977. Reparative processes in mammalian wound healing: The role of contractible phenomena. *Int. Rev. Cytol.* **48**:187–219.

Gaskin, F., Cantor, C. R., and Shelanski, M. L. 1974. Turbidimetric studies of the *in vitro* assembly and disassembly of porcine neurotubules. *J. Mol. Biol.* **89**:737–758.

Geiger, B., Tokuyosu, K. T., and Singer, S. J. 1979. Immunocytochemical localization of α-actinin in intestinal epithelial cells. *Proc. Natl. Acad. Sci. USA* **76**:2833–2837.

Gergely, J. 1976. Troponin–tropomyosin-dependent regulation of muscle contraction by calcium. *In*: Goldman, R., Pollard, T., and Rosenbaum, J., eds., *Cell Motility*, Cold Spring Harbor, N.Y., Cold Spring Harbor Laboratory, Book A, pp. 137–149.

Gershman, L. C., Stracher, A., and Driesen, P. 1969. Subunit structure of myosin. III. A proposed model for rabbit skeletal muscle myosin. *J. Biol. Chem.* **244**:2726–2736.

Gibbins, J. R., Tilney, L. G., and Porter, K. R. 1969. Microtubules in the formation and development of the primary mesenchyme in *Arbacia punctulata*. I. The distribution of microtubules. *J. Cell Biol.* **41**:201–227.

Gibbons, B. H., and Gibbons, I. R. 1972. Flagellar movement and ATPase activity in sea urchin sperm extracted with Triton X-100. *J. Cell Biol.* **54**:75–97.

Gibbons, B. H., and Gibbons, I. R. 1973. The effect of partial extraction of dynein arms on the movement of reactivated sea urchin sperm. *J. Cell Sci.* **13**:337–357.

Gibbons, I. R. 1968. The biochemistry of motility. *Annu. Rev. Biochem.* **37**:521–546.

Godfrey, J. E., and Harrington, W. F. 1970. Self-association in the myosin system at high ionic strength. II. Evidence for the presence of a monomer–dimer equilibrium. *Biochemistry* **9**:894–908.

Gordon, D., Yang, Y. Z., Eisenberg, E., and Korn, E. D. 1976. Properties of actin isolated in high yield from *Acanthamoeba castellanii*. *In*: Goldman, R., Pollard, T., and Rosenbaum, J. eds., *Cell Motility*, Cold Spring Harbor, N.Y., Cold Spring Harbor Laboratory, Book B, pp. 493–498.

Gould, R. R., and Borisy, G. G. 1978. Quantitative initiation of microtubule assembly by chromosomes from Chinese hamster ovary cells. *Exp. Cell Res.* **113**:369–374.

Grand, R. J. A., Perry, S. V., and Weeks, R. A. 1979. Troponin C-like proteins (calmodulins) from mammalian smooth muscle and other tissues. *Biochem. J.* **177**:521–529.

Greaser, M. L., and Gergely, J. 1971. Reconstitution of troponin activity from three protein components. *J. Biol. Chem.* **246**:4226–4233.

Greaser, M. L., Yamaguchi, M., Brekke, C., Potter, J., and Gergely, J. 1972. Troponin subunits and their interactions. *Cold Spring Harbor Symp. Quant. Biol.* **37**:235–244.

Gruber, S. 1978. The myofibroblast. *Surg. Gynecol. Obstet.* **146**:641–649.

Gusev, N. B., Dobrovol'skii, A. B., and Severin, S. E. 1978. Skeletal muscle troponin and phosphorylation: Site of troponin T phosphorylation by specific protein kinase. *Biokhimiya* **43**:365–372.

Hanson, J., and Huxley, H. E. 1953. Structural basis of the cross-striations in muscle. *Nature (London)* **172**:530–532.

Hardham, A. R., and Gunning, B. E. S. 1978. Structure of cortical microtubule arrays in plant cells. *J. Cell Biol.* **77**:14–34.

Harris, W. F. 1970. The arrangement of the axonemal microtubules and links of *Echinosphaerium nucleofilum*. *J. Cell Biol.* **46:** 183–187.

Hartshorne, D. J., and Dreizen, P. 1972. Studies on the subunit composition of troponin. *Cold Spring Harbor Symp. Quant. Biol.* **37**:225–234.

Haselgrove, J. C. 1972. X-ray evidence for a conformational change in the actin-containing filaments of vertebrate striated muscle. *Cold Spring Harbor Symp. Quant. Biol.* **37**:341–352.

Hatano, S., and Owaribe, K. 1976. Actin and actinin from myxomycete plasmodia. *In*: Goldman, R., Pollard, T., and Rosenbaum, J., eds., *Cell Motility*, Cold Spring Harbor, N.Y., Cold Spring Harbor Laboratory, Book B, pp. 499–511.

Hegyi, G., Premecz, G., Sain, B., and Mühlrád, A. 1974. Selective carbethoxylation of the histidine residues of actin by diethyl-pyrocarbonate. *Eur. J. Biochem.* **44**:7–12.

Heizmann, C. W., and Strehler, E. E. 1979. Chicken parvalbumin. *J. Biol. Chem.* **254**:4296–4303.

Heizmann, C. W., Häuptte, M. T., and Eppenberger, H. M. 1977. The purification, characterization and localization of a parvalbumin-like protein from chicken-leg muscle. *Eur. J. Biochem.* **80**:433–441.

Herzog, W., and Weber, K. 1978. Fractionation of brain microtubule-associated proteins. *Eur. J. Biochem.* **92**:1–8.

Hikida, R. S. 1978. Z-line extraction: Comparative effects in avian skeletal muscle fiber types. *J. Ultrastruct. Res.* **65**:266–278.

Hinkley, R. E., and Burton, P. R. 1974. Tannic acid staining of axonal microtubules. *J. Cell Biol.* **63**:139a.

Hitchcock, S., Huxley, H. E., and Szent-Györgi, A. G. 1973. Calcium sensitive binding of troponin to actin–tropomyosin: A two-site model for troponin action. *J. Mol. Biol.* **80**:825–836.

Hodges, R. S., and Smillie, L. B. 1970. Chemical evidence for chain heterogeneity in rabbit muscle tropomyosin. *Biochem. Biophys. Res. Commun.* **41**:987–994.

Hodges, R. S., Sodek, J., and Smillie, L. B. 1972. The cyanogen bromide fragments of rabbit skeletal muscle tropomyosin. *Biochem. J.* **128**:102p.

Hodges, R. S., Sodek, J., Smillie, L. B., and Jurasek, L. 1973. Tropomyosin: Amino-acid sequence and coiled-coil structure. *Cold Spring Harbor Symp. Quant. Biol.* **37**:299–310.

Hoffman, R. N., and Lasek, R. J. 1975. The slow component of axonal transport. Identification of major structural polypeptides of the axon and their generality among mammalian neurons. *J. Cell Biol.* **66**:351–366.

Hoh, J. F. Y. 1979. Developmental changes in chicken skeletal myosin isoenzymes. *FEBS Lett.* **98**:267–275.

Horwitz, J., Bullard, B., and Mercola, D. 1979. Interaction of troponin subunits. *J. Biol. Chem.* **254**:350–355.

Huang, B., and Pitelka, D. R. 1973. The contractile process in the ciliate *Stentor coeruleus*. I. The role of microtubules and filaments. *J. Cell Biol.* **57**:704–728.

Hubbard, B. D., and Lazarides, E. 1979. Copurification of actin and desmin from chicken smooth muscle and their copolymerization *in vitro* to intermediate filaments. *J. Cell Biol.* **80**:166–182.

Huxley, H. E. 1963. Electron microscope studies on the structure of natural and synthetic protein filaments from striated muscle. *J. Mol. Biol.* **7**:281–308.

Huxley, H. E. 1972. Structural changes in the actin and myosin-containing filaments during contraction. *Cold Spring Harbor Symp. Quant. Biol.* **37**:361–376.

Huxley, H. E., and Brown, W. 1967. The low-angle X-ray diagram of vertebrate striated muscle and its behavior during contraction and rigor. *J. Mol. Biol.* **30**:383–434.

Ikeda, Y., and Steiner, M. 1979. Phosphorylation and protein kinase activity of platelet tubulin. *J. Biol. Chem.* **254**:66–74.

Inoue, A., Kikuchi, K., and Tonomura, Y. 1977. Structure and function of the two heads of the myosin molecule. V. Enzymatic properties of heads B and A. *J. Biochem.* **82**:783–800.

Inoué, S. 1952. The effect of colchicine on the microscopic and sub-microscopic structure of the mitotic spindle. *Exp. Cell Res. Suppl.* **2**:305–318.

Inoué, S. 1953. Polarization optical studies of the mitotic spindle. *Chromosoma* **5**:487–500.

Inoué, S., and Sato, H. 1967. Cell motility by labile association of molecules. *J. Gen. Physiol.* **50**:259–292.

Inoué, S., Borisy, G. G., and Kiekart, D. P. 1974. Growth and lability of *Chaetopterus* oocyte mitotic spindles isolated in the presence of porcine brain tubulin. *J. Cell Biol.* **62**:175–184.

Ishikawa, H., Bischoff, R., and Holtzer, H. 1969. The formation of arrowhead complexes with heavy meromyosin in a variety of cell types. *J. Cell Biol.* **43**:312–328.

Ishiura, M., and Okada, Y. 1979. The role of actin in temperature-dependent gel–sol transformation of extracts of Ehrlich ascites tumor cells. *J. Cell Biol.* **80**:465–480.

Jackson, P., Amphlett, G. W., and Perry, S. V. 1975. The primary structure of troponin T and the interaction with tropomyosin. *Biochem. J.* **151**:85–97.

Jacobs, M., and McVittie, A. 1970. Identification of the flagellar proteins of *Chlamydomonas reinhardii*. *Exp. Cell Res.* **63**:53–61.

Jenkins, R. A. 1977. The role of microtubules in macronuclear division of *Blepharisma*. *J. Protozool.* **24**:264–275.

Jurand, A., and Selman, G. G. 1970. Ultrastructure of the nuclei and intranuclear microtubules of *Paramecium aurelia*. *J. Chem. Microbiol.* **60**:357–364.

Karnovsky, M. J. 1965. A formaldehyde–glutaraldehyde fixative of high osmolality for use in electron microscopy. *J. Cell Biol.* **27**:137A–138A.

Kelly, R. E., and Rice, R. V. 1968. Localization of myosin filaments in smooth muscle. *J. Cell Biol.* **37**:105–116.

Kessel, R. G. 1966. The association between microtubule and nuclei during spermiogenesis in the dragonfly. *J. Ultrastruct. Res.* **16**:293–304.

Khan, Z. N. T., and Godward, M. B. E. 1977. Tubular elements—A new structure in blue-green algal cells. *J. Cell Sci.* **28**:303–308.

Kim, H., Binder, L. I., and Rosenbaum, J. L. 1979. The periodic association of MAP_2 with brain microtubules *in vitro*. *J. Cell Biol.* **80**:266–276.

Kirschner, M. W. 1978. Microtubule assembly and nucleation. *Int. Rev. Cytol.* **54**:1–71.

Kirschner, M. W., Suter, M., Weingarten, M., and Littman, D. 1975a. The role of rings in the assembly of microtubules *in vitro*. *Ann. N.Y. Acad. Sci.* **253**:90–106.

Kirschner, M. W., Honig, L. S., and Williams, R. C. 1975b. Quantitative electron microscopy of microtubule assembly *in vitro*. *J. Mol. Biol.* **99**:263–276.

Kobayashi, T. 1975. Dephosphorylation of tubulin-bound GTP during microtubule assembly. *J. Biochem.* **77**:1193–1197.

Kobayashi, T., and Flavin, M. 1977. Tubulin-tryosine ligase: Purification and application to studies of tubulin structure and assembly. *J. Cell Biol.* **75**:277a.

Kobayashi, T., and Simizu, T. 1976. Roles of nucleoside triphosphates in microtubule assembly. *J. Biochem.* **79**:1357–1364.

Kretsinger, R. H., and Barry, C. D. 1975. The predicted structure of the calcium-binding component of troponin. *Biochim. Biophys. Acta* **405**:40–52.

Krishman, N., and Lasek, R. J. 1975. The fine structure of neurofilaments in the giant axon of *Myxicola infundibulum* (Annelida, Polychaeta). *J. Cell Biol.* **67**:226a.

Kuczmarski, E. R., and Rosenbaum, J. L. 1979. Chick brain actin and myosin. *J. Cell Biol.* **80**:341–355.

Lasek, R. J., and Hoffman, P. N. 1976. The neuronal cytoskeleton, axonal transport, and axonal growth. *In*: Goldman, R., Pollard, T., and Rosenbaum, J., eds., *Cell Motility*, Cold Spring Harbor, N.Y., Cold Spring Harbor Laboratory, Book C, pp. 1021–1049.

Lazarides, E. 1976a. Actin, α-actinin, and tropomyosin interaction in the structural organization of actin filaments in nonmuscle cells. *J. Cell Biol.* **68**:202–219.

Lazarides, E. 1976b. Aspects of the structural organization of actin filaments in tissue culture cells. *In*: Goldman, R., Pollard, T., and Rosenbaum, J., eds., *Cell Motility*, Cold Spring Harbor, N.Y., Cold Spring Harbor Laboratory, Book A, pp. 347–360.

Lazarides, E., and Balzer, D. R. 1978. Specificity of desmin to avian and mammalian muscle cells. *Cell* **14**:429–438.

Lazarides, E., and Burridge, K. 1975. Alpha-actinin: Immunofluorescent localization of muscle structural protein in nonmuscle cells. *Proc. Natl. Acad. Sci. USA* **71**:2268–2272.

Lazarides, E., and Hubbard, B. D. 1976. Immunological characterization of the subunit of the 100 Å filaments from muscle cells. *Proc. Natl. Acad. Sci. USA* **73**:4344–4348.

Leedale, G. F. 1967. *Euglenoid Flagellates*, Englewood Cliffs, N.J., Prentice–Hall.

Linck, R. W. 1976. Flagellar doublet microtubules: Fractionation of minor components and α-tubulin from specific regions of the A-tubule. *J. Cell Sci.* **20**:405–439.

Lowey, S., Slayter, H. S., Weeds, A., and Baker, H. 1969. Substructure of the myosin molecule. I. Subfragments of myosin by enzymatic digestion. *J. Mol. Biol.* **42**:1–29.

Lu, R. C., and Elzinga, M. 1976. Comparison of amino acid sequences of actins from bovine brain and muscles. *In*: Goldman, R., Pollard, T., and Rosenbaum, J., eds., *Cell Motility*, Cold Spring Harbor, N.Y., Cold Spring Harbor Laboratory, Book B, pp. 487–492.

Lu, R. C., and Elzinga, M. 1977. Partial amino acid sequence of brain actin and its homology with muscle actin. *Biochemistry* **16**:5801–5806.

Luduena, R. F., and Woodward, D. O. 1973. Isolation and partial characterization of α- and β-tubulin from outer doublets of sea urchin sperm and microtubules of chick-embryo brain. *Proc. Natl. Acad. Sci. USA* **70**:3594–3598.

Luduena, R. F., and Woodward, D. O. 1975. α- and β-tubulin: Separation and partial sequence analysis. *Ann. N.Y. Acad. Sci.* **253**:272–283.

Lusty, C. J., and Fasold, H. 1969. Characterization of sulf-hydryl groups of actin. *Biochemistry* **8**:2933–2939.

Luther, P., and Squire, J. 1978. Three-dimensional structure of the vertebrate muscle M-region. *J. Mol. Biol.* **125**:313–324.

MacDonald, A. C. and Kitching, J. A. 1967. Axopodial filaments of Heliozoa. *Nature (London)* **215**:99–100.

McGill, M., and Brinkley, B. R. 1975. Human chromosomes and centrioles as nucleating sites for the *in vitro* assembly of microtubules from bovine brain tubulin. *J. Cell Biol.* **67**:189–199.

McIntosh, J. R. 1973. The axostyle of *Saccinobaculus*. II. *J. Cell Biol.* **56**:324–339.

Malawista, S. E. 1975. Microtubules and the mobilization of lysosomes in phagocytizing human leukocytes. *Ann. N.Y. Acad. Sci.* **253**:738–749.

Mandelkow, E. M., and Mandelkow, E. 1979. Junctions between microtubule walls. *J. Mol. Biol.* **129**:135–148.

Mandelkow, E. M., Mandelkow, E., Unwin, N., and Cohen, C. 1977. Tubulin hoops. *Nature (London)* **265**:655–657.

Mannherz, H. G., and Goody, R. S. 1976. Proteins of contractile systems. *Annu. Rev. Biochem.* **45**:427–465.

Marchant, H. J. 1979. Microtubular determination of cell shape during colony formation by the alga *Pediastrum*. *Protoplasma* **98**:1–14.

Margossian, S. S., and Cohen, C. 1973. Troponin subunit interactions. *J. Mol. Biol.* **81**:409–413.

Margulis, L. 1973. Colchicine-sensitive microtubules. *Int. Rev. Cytol.* **34**:333–361.

Margulis, L., To, L., and Chase, D. 1978. Microtubules in prokaryotes. *Science* **200**:1118–1124.

Marotta, C. A., Strocchi, P., and Gilbert, J. M. 1979. Subunit structure of a synaptosomal tubulin. *Brain Res.* **167**:93–106.

Maruta, H., and Korn, E. D. 1977. Purification from *Acanthamoeba castellanii* of proteins that induce gelation and syneresis of F-actin. *J. Biol. Chem.* **252**:399–402.

Massini, P., and Lüscher, E. F. 1976. On the significance of the influx of calcium ions into stimulated human blood platelets. *Biochim. Biophys. Acta* **436**:652–663.

Mazia, D. 1975. Microtubule research in perspective. *Ann. N.Y. Acad. Sci.* **253**:7–13.

Merlino, G. T., Chamberlain, J. P., and Kleinsmith, L. J. 1978. Effects of deciliation on tubulin mRNA activity in sea urchin embryos. *J. Biol. Chem.* **253:**7078–7085.

Meyer, D. F., and Burger, M. M. 1979. The chromaffin granule surface: The presence of actin and the nature of its interaction with the membrane. *FEBS Lett.* **101**:129–132.

Meza, I., Huang, B., and Bryan, J. 1972. Chemical heterogeneity of protofilaments forming the outer doublets from sea urchin flagella. *Exp. Cell Res.* **74**:535–540.

Micko, S., and Schlaepfer, W. W. 1978. Protein composition of axons and myelin from rat and human peripheral nerves. *J. Neurochem.* **30**:1041–1049.

Moellmann, G., and McGuire, J. 1975. Correlation of cytoplasmic microtubules and 10 nm filaments with the movement of pigment granules in cutaneous melanocytes of *Rana pipiens*. *Ann. N.Y. Acad. Sci.* **253**:711–722.

Mohri, H., and Shimomura, M. 1973. Comparison of tubulin and actin. *J. Biochem.* (*Tokyo*) **74**:209–220.

Moore, P. B., Huxley, H. E., and DeRosier, D. J. 1970. Three-dimensional reconstruction of F-actin, thin filaments and decorated thin filaments. *J. Mol. Biol.* **50**:279–295.

Moos, C., Offer, G. W., and Starr, R. L. 1972. A new muscle protein which affects myosin filament structure. *Abstr. Am. Biophys. Soc.* **1972**:208a.

Moos, C., Offer, G. W., Starr, R. L., and Bennett, P. 1975. Interaction of C-protein with myosin, myosin rod, and light meromyosin. *J. Mol. Biol.* **97**:1–9.

Morgan, J. L., and Seeds, N. W. 1975. Properties of tubulin prepared by affinity chromatography. *Ann. N.Y. Acad. Sci.* **253**:260–271.

Murphy, D. B. 1975. The mechanism of microtubule-dependent movement of pigment granules in teleost chromatophores. *Ann. N.Y. Acad. Sci.* **253**:692–701.

Murphy, D. B., and Borisy, G. G. 1975. Association of high-molecular-weight proteins with microtubules and their role in microtubule assembly *in vitro*. *Proc. Natl. Acad. Sci. USA* **72**:2696–2700.

Murphy, D. B., Valley, R. B., and Borisy, G. G. 1977. Identity and polymerization-stimulatory activity of the nontubulin proteins associated with microtubules. *Biochemistry* **16**:2598–2605.

Nachmias, V. T. 1968. Further electron microscopic studies on fibrillar organization of the ground cytoplasm of *Chaos chaos*. *J. Cell Biol.* **38**:40–50.

Nachmias, V. T. 1973. *Physarum* myosin: Two new properties. *Cold Spring Harbor Symp. Quant. Biol.* **37**:607–612.

Nagaro, T., and Suzuki, F. 1975. Microtubules with 15 subunits in cockroach epidermal cells. *J. Cell Biol.* **64**:242–245.

Nakamura, T., Takahashi, K., and Watanabe, S. 1978. Myosin and actin from *E. coli* K12 C600. *J. Biochem.* **84**:1453–1458.

Nakamura, T., Yamaguchi, M., and Yanagisawa, T. 1979. Comparative studies on actins from various sources. *J. Biochem.* **85**:627–631.

Nath, J., and Flavin, M. 1978. A structural difference between cytoplasmic and membrane-bound tubulin of brain. *FEBS Lett.* **95**:335–338.

Ng, S. F. 1978. Directionality of microtubule assembly: An *in vivo* study with the ciliate *Tetrahymena*. *J. Cell Sci.* **33**:227–234.

Offer, G. 1973. C-protein and the periodicity in the thick filament of vertebrate skeletal muscle. *Cold Spring Harbor Symp. Quant. Biol.* **37**:87–93.

Offer, G., and Elliott, A. 1978. Can a myosin molecule bind to two actin filaments? *Nature* (*London*) **271**:325–329.

Olmsted, J. B. 1976. The role of divalent cations and nucleotides in microtubule assembly *in vitro*.

In: Goldman, R., Pollard, T., and Rosenbaum, J., eds., *Cell Motility*, Cold Spring Harbor, N.Y., Cold Spring Harbor Laboratory, Book C, pp. 1081–1092.

Olmsted, J. B., and Borisy, G. G. 1973. Microtubules. *Annu. Rev. Biochem.* **42**:507–540.

Olmsted, J. B., and Borisy, G. G. 1975. Ionic and nucleotide requirements for microtubule polymerization *in vitro*. *Biochemistry* **14**:2996–3005.

Osborn, M., Webster, R. E., and Weber, K. 1978. Individual microtubules viewed by immunofluorescence and electron microscopy in the same PtK2 cell. *J. Cell Biol.* **21**:R27–R34.

Overton, J. 1966. Microtubules and microfibrils in morphogenesis of the scale cells of *Ephestia kühniella*. *J. Cell Biol.* **29**:293–305.

Palay, S. L., McGee-Russell, S. M., Gordon, S., and Grillo, M. A. 1962. Fixation of neural tissues for electron microscopy by perfusion with solutions of OsO_4. *J. Cell Biol.* **12**:385–410.

Palmer, E., and Saborio, J. L. 1978. *In vivo* and *in vitro* synthesis of multiple forms of rat brain actin. *J. Biol. Chem.* **253**:7482–7489.

Penningroth, S. M., and Kirschner, M. W. 1977. Nucleotide binding and phosphorylation in microtubule assembly *in vitro*. *J. Mol. Biol.* **115**:643–673.

Penningroth, S. M., and Kirschner, M. W. 1978. Nucleotide specificity in microtubule assembly *in vitro*. *Biochemistry* **17**:734–740.

Penningroth, S. M., Cleveland, D. W., and Kirschner, M. W. 1976. *In vitro* studies of the regulation of microtubule assembly. *In*: Goldman, R., Pollard, T., and Rosenbaum, J., eds., *Cell Motility*, Cold Spring Harbor, N.Y., Cold Spring Harbor Laboratory, Book C, pp. 1233–1257.

Pepe, F. A. 1973. The myosin filament: Immuno-chemical and ultrastructural approaches to molecular organization. *Cold Spring Harbor Symp. Quant. Biol.* **37**:97–108.

Perry, S. V., Cole, H. A., Head, J. F., and Wilson, F. J. 1972. Localization and mode of action of the inhibitory protein component of the troponin complex. *Cold Spring Harbor Symp. Quant. Biol.* **37**:251–262.

Peters, A., and Vaughn, J. E. 1967. Microtubules and filaments in the axons and astrocytes of early postnatal rat optic nerves. *J. Cell Biol.* **32**:113–119.

Pfeffer, T. A., Asnes, C. F., and Wilson, L. 1978. Polymerization and colchicine-binding properties of outer doublet microtubules solubilized by the French pressure cell. *Cytobiologie* **16**:367–373.

Pollard, T. D. 1976. Contractile proteins in nonmuscle cells. *In*: Fasman, G. D., ed., *Handbook of Biochemistry and Molecular Biology: Proteins Section*, Cleveland, CRC Press, Vol. 2, pp. 307–324.

Pollard, T. D. 1977. Cytoplasmic contractile proteins. *In*: Brinkley, B. R., and Porter, K. R., eds., *International Cell Biology, 1976–1977*, New York, Rockefeller University Press, pp. 378–387.

Pollard, T. D., and Ito, S. 1970. Cytoplasmic filaments of *Amoeba proteus* I. The role of filaments in consistency changes and movement. *J. Cell Biol.* **46**:267–289.

Pollard, T. D., and Korn, E. D. 1973a. The "contractile" proteins of *Acanthamoeba castellanii*. *Cold Spring Harbor Symp. Quant. Biol.* **37**:573–583.

Pollard, T. D., and Korn, E. D. 1973b. *Acanthamoeba* myosin. I. Isolation of an enzyme similar to muscle myosin. *J. Biol. Chem.* **248**:4682–4690.

Pollard, T. D., and Korn, E. D. 1973c. *Acanthamoeba* myosin. II. Interaction with actin and with a new cofactor protein required for actin activation of Mg ATPase activity. *J. Biol. Chem.* **248**:4691–4697.

Pollard, T. D., and Weihing, R. R. 1974. Actin and myosin and cell movement. *Crit. Rev. Biochem.* **2**:1–65.

Pollard, T. D., Stafford, W. F., and Porter, M. E. 1978. Characterization of a second myosin from *Acanthamoeba castellanii*. *J. Biol. Chem.* **253**:4798–4808.

Pope, L. M., and Jurtshuk, P. 1967. Microtubule in *Azotobacter veinlandii strain O*. *J. Bacteriol.* **94**:2062–2064.

Porter, K. R. 1955. The fine structure of cells. *Fed. Proc.* **14**:673–682.

Potter, J. D. 1976. The content of troponin, tropomyosin, actin, and myosin in rabbit skeletal myofibrils. *Arch. Biochem. Biophys.* **162**:436–441.

Raybin, D., and Flavin, M. 1977a. Enzyme which specifically adds tyrosine to the α chain of tubulin. *Biochemistry* **16**:2189–2194.

Raybin, D., and Flavin, M. 1977b. Modification of tubulin by tyrosylation in cells and extracts and its effect on assembly *in vitro*. *J. Cell Biol.* **73**:492–504.

Rebhun, L. I. 1972. Polarized intracellular particle transport. Saltatory movements and cytoplasmic streaming. *Int. Rev. Cytol.* **32**:93–139.

Redman, C. M., Banerjee, D., Howell, K., and Palade, G. E. 1975. The step at which colchicine blocks the secretion of plasma protein by rat liver. *Ann. N.Y. Acad. Sci.* **253**:780–788.

Renaud, F. L., Rowe, A. J., and Gibbons, I. R. 1968. Some properties of the protein forming the outer fibers of cilia. *J. Cell Biol.* **36**:79–90.

Roisen, F. J., Braden, W. G., and Friedman, J. 1975. Neurite development *in vitro*. III. The effects of several derivatives of cAMP, colchicine, and colcemid. *Ann. N.Y. Acad. Sci.* **253**:545–561.

Romera-Herrara, A. E., Castillo, O., and Lehmann, H. 1976. Human skeletal muscle proteins: The primary structure of troponin C. *J. Mol. Evol.* **8**:251–270.

Rosenbaum, J. L., and Child, F. M. 1967. Flagellar regeneration in protozoan flagellates. *J. Cell Biol.* **34**:345–364.

Rosenbaum, J. L., Binder, L. I., Granett, S., Dentler, W. L., Snell, W. S., Sloboda, R., and Haimo, L. 1975. Directionality and rate of assembly of chick brain tubulin onto pieces of chick brain neurotubules, flagellar axonemes, and basal bodies. *Ann. N.Y. Acad. Sci.* **253**:147–177.

Rosenfeld, A. C., Zackroff, R. V., and Weisenberg, R. C. 1976. Magnesium stimulation of calcium binding to tubulin and calcium induced depolymerization of microtubules. *FEBS Lett.* **65**:144–147.

Rudolph, R., Gruber, S., Suzuki, M., and Woodward, M. 1977. The life cycle of the myofibroblast. *Surg. Gynecol. Obstet.* **145**:389–394.

Sabatini, D. D., Bensch, K., and Barrnett, R. J. 1963. Cytochemistry and electron microscopy. The preservation of cellular ultrastructure and enzymatic activity by aldehyde fixation. *J. Cell Biol.* **17**:19–58.

Sarkar, S. 1972. Stoichiometry and sequential removal of light chains of myosin. *Cold Spring Harbor Symp. Quant. Biol.* **37**:14–17.

Sarkar, S., Stréter, F. A., and Gergely, J. 1971. Light chains of myosins from white, red, and cardiac muscles. *Proc. Natl. Acad. Sci. USA* **68**:946–950.

Schliwa, M. 1978. Microtubular apparatus of melanophores. *J. Cell Biol.* **76**:605–614.

Schollmeyer, J., Goll, D., Tilney, L., Mooseker, M., Robson, R., and Stromer, M. 1974. Localization of α-actinin in non-muscle material. *J. Cell Biol.* **63**:304a.

Schollmeyer, J. E., Furcht, L. T., Goll, D. E., Robson, R. M., and Stromer, M. H. 1976. Localization of contractile proteins in smooth muscle cells and in normal and transformed fibroblasts. *In*: Goldman, R., Pollard, T., and Rosenbaum, J., eds., *Cell Motility*, Cold Spring Harbor, N.Y., Cold Spring Harbor Laboratory, Book A, pp. 361–388.

Schook, W., Ores, C., and Puszkin, S. 1978. Isolation and properties of brain α-actinin. *Biochem. J.* **175**:63–72.

Scordilis, S. P., and Adelstein, R. S. 1978. A comparative study of the myosin light chain kinases from myoblast and muscle sources. *J. Biol. Chem.* **253**:9041–9048.

Scordilis, S. P., Anderson, J. L., Pollock, R., and Adelstein, R. S. 1977. Characterization of the myosin-phosphorylating system in normal murine astrocytes and derivative SV40 wild-type and A-mutant transformations. *J. Cell Biol.* **74**:940–949.

Shelanski, M. L. 1973. Microtubules. *In*: Schneider, D. J., Angeletti, R. H., Bradshax, R. A., Grasso, A., and Moore, B. W., eds., *Proteins of the Nervous System*, New York, Raven Press, pp. 227–242.

Shelanski, M. L., and Taylor, E. W. 1968. Properties of the protein subunit of central-pair and outer-doublet microtubules of sea urchin flagella. *J. Cell Biol.* **38**:304–315.

Shelanski, M. L., Yen, S. H., and Lee, V. M. 1976. Neurofilaments and glial filaments. *In*: Goldman, R., Pollard, T., and Rosenbaum, J., *Cell Motility*, Cold Spring Harbor, N.Y., Cold Spring Harbor Laboratory, Book C, pp. 1007–1020.

Sherline, P., and Schiavone, K. 1978. High molecular weight MAPs are part of the mitotic spindle. *J. Cell Biol.* **77**:R9–R12.

Shizuta, Y., Shizuta, H., Gallo, M., Davies, P., Pastan, I., and Lewis, M. S. 1976. Purification properties of filamin, an active binding protein from chicken gizzard. *J. Biol. Chem.* **251**:6562–6567.

Siemankowski, R. F., and Zobel, C. R. 1976. Comparative studies on the structure and aggregative properties of the myosin molecule. I. The structure of the lobster myosin molecule. *J. Mechanochem. Cell Motil.* **3**:171–184.

Sloboda, R. D., Dentler, W. L., Bloodgood, R. A., Telzer, B. R., Granett, S., and Rosenbaum, J. L. 1976. Microtubule-associated proteins (MAPs) and the assembly of microtubules *in vitro*. *In*: Goldman, R., Pollard, T., and Rosenbaum, J., eds., *Cell Motility*, Cold Spring Harbor, N.Y., Cold Spring Harbor Laboratory, Book C, pp. 1171–1212.

Smith, U., and Smith, D. S. 1965. A microtubular complex in the epidermal nucleus of an insect, *Carausius morosus*. *J. Cell Biol.* **26**:961–964.

Snell, W. J., Dentler, W. L., Haimo, L. T., Binder, L. I., and Rosenbaum, J. L. 1974. Assembly of chick brain tubulin onto isolated basal bodies of *Chlamydomonas reinhardii*. *Science*. **185**:357–360.

Snyder, J. A., and McIntosh, J. R. 1976. Biochemistry and physiology of microtubules. *Annu. Rev. Biochem.* **45**:699–720.

Sodek, J., Hodges, R. S., and Smillie, L. B. 1978. Amino acid sequence of rabbit skeletal muscle α-tropomyosin. The COOH-terminal half (residues 142–284). *J. Biol. Chem.* **253:**1129–1136.

Someya, A., and Tanaka, N. 1979. Heavy meromyosin- and ATP-binding protein from *Escherichia coli*. *FEBS Lett.* **101**:166–169.

Sommer, J. R., and Blum, J. J. 1965. Cell division in *Astasia longa*. *Exp. Cell Res.* **39**:504–527.

Spiegelman, B. M., Penningroth, S. M., and Kirschner, M. W. 1977. Turnover of tubulin and the N site GTP in Chinese hamster ovary cells. *Cell* **12**:587–600.

Squire, J. M. 1975. Muscle filament structure and muscle contraction. *Annu. Rev. Biophys. Bioeng.* **4**:137–163.

Stearns, M. E., and Brown, D. L. 1979. Purification of a microtubule-associated protein based on its preferential association with tubulin during microtubule initiation. *FEBS Lett.* **101**:15–19.

Stearns, M. E., Connolly, J. A., and Brown, D. L. 1976. Cytoplasmic microtubule organizing centers isolated from *Polytomella agilis*. *Science* **191**:188–191.

Steinert, P., Idler, W. W., and Zimmerman, S. B. 1976. Self-assembly of bovine epidermal keratin filaments *in vitro*. *J. Mol. Biol.* **108**:547–567.

Stephens, R. E. 1968. On the structural protein of flagellar outer fibers. *J. Mol. Biol.* **32**:277–283.

Stephens, R. E., and Edds, K. T. 1976. Microtubules: Structure, chemistry, and function. *Physiol. Rev.* **56**:709–777.

Stephens, R. E., Renaud, F. L., and Gibbons, I. R. 1967. Guanine nucleotide associated with the protein of the outer fibers of flagella and cilia. *Science* **156**:1606–1608.

Stone, D., and Smillie, L. B. 1978. The amino acid sequence of rabbit skeletal α-tropomyosin: The NH_2-terminal half and complete sequence. *J. Biol. Chem.* **253**:1137–1148.

Stone, D., Sodek, J., Johnson, P., and Smillie, L. B. 1975. Tropomyosin: Correlation of amino acid sequence and structure. *Proc. 9th Meet. FEBS* **31**:125–136.

Stossel, T. P. 1978. Contractile proteins in cell structure and function. *Annu. Rev. Med.* **29**:427–457.

Stossel, T. P., and Pollard, T. D. 1974. Myosin in polymorphonuclear leukocytes. *J. Biol. Chem.* **248**:8288–8294.

Summers, K. E. 1974. ATP-induced sliding of microtubules in bull sperm flagella. *J. Cell Biol.* **60**:321–324.

Sutoh, Ka., Sutoh, Ke., Karr, T., and Harrington, W. F. 1978. Isolation and physico-chemical properties of a high-molecular-weight subfragment-2 of myosin. *J. Mol. Biol.* **126**:1–22.

Suzuki, H., Onishi, H., Takahashi, K., and Watanabe, S. 1978. Structure and function of chicken gizzard myosin. *J. Biochem.* **84**:1529–1542.

Takahashi, K. 1978. Topography of the myosin molecule as visualized by an improved negative staining method. *J. Biochem.* **83**:905–908.

Tamura, S., Tsuruhara, T., and Watanabe, Y. 1969. Functioning nuclear microtubules in macronuclear division of *Tetrahymena pyriformis*. *Exp. Cell Res.* **55**:351–358.

Taylor, A., Mamelak, M., Reaven, E., and Maffly, R. 1973. Vasopressin: Possible role of microtubules and microfilaments in its action. *Science* **181**:247–250.

Taylor, A., Maffly, R., Wilson, L., and Reaven, E. 1975. Evidence for involvement of microtubules in the action of vasopressin. *Ann. N.Y. Acad. Sci.* **253**:723–737.

Tilney, L. G., and Gibbins, J. R. 1969. Microtubules in the formation and development of the primary mesenchyme in *Arbacia punctulata*. II. An experimental analysis of their role in development and maintenance of cell shape. *J. Cell Biol.* **41**:227–250.

Tilney, L. G., and Porter, K. R. 1965. Studies on the microtubules in Heliozoa. I. Fine structure of *Actinosphaerium* with particular reference to axial rod structure. *Protoplasma* **60**:317–344.

Toh, B. H., Yildiz, A., Sotelo, J., Osung, O., Holborow, E. J., and Fairfax, A. 1979. Distributions of actin and myosin in muscle and non-muscle cells. *Cell Tissue Res.* **199**:117–126.

Tsuchiya, T., Yamada, N., and Matsumota, J. J. 1978. Extraction and purification of squid myosin. *Bull. Jpn. Soc. Sci. Fish.* **44**:175–180.

Ueno, H. 1978. Cyanogen-bromide fragments of rabbit skeletal muscle α-tropomyosin. *J. Biochem.* **84**:1559–1565.

Ullrick, W. C., Toselli, P. A., Smith, J. D., and Phear, W. P. C. 1977. Fine structure of the vertebrate Z-disk. *J. Mol. Biol.* **115**:61–74.

Uyemura, D. G., Brown, S. S., and Spudich, J. A. 1978. Biochemical and structural characterization of actin from *Dictyostelium discoideum*. *J. Biol. Chem.* **253**:9088–9096.

van Eerd, J. P., and Takahashi, K. 1975. The amino acid sequence of bovine cardiac troponin-C. Comparison with rabbit skeletal troponin-C. *Biochem. Biophys. Res. Commun.* **64**:122–127.

van Eerd, J. P., and Takahashi, K. 1976. Determination of the complete amino acid sequence of bovine cardiac troponin-C. *Biochemistry* **15**:1171–1180.

van Eerd, J. P., Capony, J. P., Ferraz, C., and Pechere, J. F. 1978. The amino-acid sequence of troponin C from frog skeletal muscle. *Eur. J. Biochem.* **91**:231–242.

van Iterson, W., Hoeniger, J. F. M., and van Zanten, E. N. 1967. Microtubule in a bacterium. *J. Cell Biol.* **32**:1–10.

Vivier, E. 1965. Présence de microtubules intranucléaires chez *Metchnikovella hovassei* Vivier. *J. Microsc. (Paris)* **4**:559–562.

von Olenhusen, K. G., and Wohlfarth-Bottermann, K. E. 1979. Evidence for actin transformation during contraction–relaxation cycle of cytoplasmic actomyosin: Cycle blockade by phalloidin injection. *Cell Tissue Res.* **196**:455–470.

Wang, K., and Singer, S. J. 1977. Interaction of filamin with F-actin in solution. *Proc. Natl. Acad. Sci. USA* **74**:2021–2025.

Warner, F. D., and Mitchell, D. R. 1978. Structural conformation of ciliary dynein arms and the generation of sliding forces in *Tetrahymena* cilia. *J. Cell Biol.* **76**:261–277.

Warner, F. D., Mitchell, D. R., and Perkins, C. R. 1977. Structural conformation of the ciliary ATPase dynein. *J. Mol. Biol.* **114**:367–384.

Watanabe, T., and Flavin, M. 1976. Nucleotide-metabolizing enzymes in *Chlamydomonas* flagella. *J. Biol. Chem.* **251**:182–192.

Weatherbee, J. A., Luftig, R. B., and Weihing, R. R. 1978. *In vitro* polymerization of microtubules from HeLa cells. *J. Cell Biol.* **78**:47–57.

Weber, K. 1976. Visualization of tubulin-containing structures by immunofluorescence microscopy. *In*: Goldman, R., Pollard, T., and Rosenbaum, J., eds., *Cell Motility,* Cold Spring Harbor, N.Y., Cold Spring Harbor Laboratory, Book A, pp. 403–417.

Weber, K., Rathke, P. C., and Osborn, M. 1978. Cytoplasmic microtubular images in glutaraldehyde-fixed tissue culture cells by electron microscopy and by immunofluorescence microscopy. *Proc. Natl. Acad. Sci. USA* **75**:1820–1824.

Weeds, A., and Frank, G. 1972. Structural studies on the light chains of myosin. *Cold Spring Harbor Symp. Quant. Biol.* **37**:9–14.

Weingarten, M. D., Lockwood, A. H., Horo, S. Y., and Kirschner, M. W. 1975. A protein factor essential for microtubule assembly. *Proc. Natl. Acad. Sci. USA* **72**:1858–1862.

Weisenberg, R. C. 1972. Microtubule formation *in vitro* in solutions containing low calcium concentrations. *Science* **177**:1104–1105.

Weisenberg, R. D., Borisy, G. G., and Taylor, E. W. 1968. The colchicine binding protein of mammalian brain and its relation to microtubules. *Biochemistry* **7**:4466–4479.

Whalen, R. G., Butler-Browne, G. S., and Gros, F. 1976. Protein synthesis and actin heterogeneity in calf muscle cells in culture. *Proc. Natl. Acad. Sci. USA* **73**:2018–2122.

Wilkinson, J. M. 1976. The amino acid sequence of troponin C from chicken skeletal muscle. *FEBS Lett.* **70**:254–256.

Wilkinson, J. M. 1978. The components of troponin from chicken fast skeletal muscle: A comparison of troponin T and troponin I from breast and leg muscle. *Biochem. J.* **169**:229–238.

Wikinson, J. M., and Grand, R. J. A. 1975. The amino acid sequence of troponin I from rabbit skeletal muscle. *Biochem. J.* **149**:493–496.

Wilkinson, J. M., and Grand, R. J. A. 1978. The amino acid sequence of chicken fast skeletal muscle troponin I. *Eur. J. Biochem.* **82**:493–502.

Wilkinson, J. M., Perry, S. V., Cole, H. A., and Trayer, I. P. 1972. The regulatory proteins of the myofibril. *Biochem. J.* **127**:215–228.

Wise, B. N. 1965. Fine structure of *Euplotes*: Filaments, vesicles and kinetosomes. *J. Cell Biol.* **27**:113A–114A.

Witman, G. B. 1975. The site of *in vivo* assembly of flagellar microtubules. *Ann. N.Y. Acad. Sci.* **253**:178–191.

Witman, G. B., Carlson, K., and Rosenbaum, J. L. 1972. *Chlamydomonas* flagella. II. The distribution of tubulins 1 and 2 in the outer doublet microtubules. *J. Cell Biol.* **54**:540–555.

Witman, G. B., Cleveland, D. W., Weingarten, M. D., and Kirschner, M. W. 1976. Tubulin requires tau for growth onto microtubule initiating sites. *Proc. Natl. Acad. Sci. USA* **73**:4070–4074.

Wolff, J., and Bhattacharyya, B. 1975. Microtubules and thyroid hormone mobilization. *Ann. N.Y. Acad. Sci.* **253**:763–770.

Wuerker, R. B. 1970. Neurofilaments and glial filaments. *Tissue Cell* **2**:1–9.

Yanagisawa, T., Hasegawa, S., and Mohri, H. 1968. The bound nucleotides of the isolated microtubules of sea urchin sperm flagella and their possible role in flagellar movement. *Exp. Cell Res.* **52**:86–100.

Yen, S., Lee, V., and Shelanski, M. L. 1975. Evidence for identity of 9 nm filaments in neurons and astroglia. *J. Cell Biol.* **67**:468a.

Zigmond, S. H., Otto, J. J., and Bryan, J. 1979. Organization of myosin in a submembranous sheath in well-spread human fibroblasts. *Exp. Cell Res.* **119**:205–220

CHAPTER 3

Aikman, D. P., and Anderson, W. P. 1971. A quantitative investigation of a peristaltic model for phloem translocation. *Ann. Bot.* **35**:761–772.

Allen, N. S. 1974. Endoplasmic filaments generate the motive force for rotational streaming in *Nitella*. *J. Cell Biol.* **63**:270–287.

Allen, N. S. 1976. Undulating filaments in *Nitella* endoplasm and motive force generation. *In*: Goldman, R., Pollard, T., and Rosenbaum, J., eds., *Cell Motility*, Cold Spring Harbor, N.Y., Cold Spring Harbor Laboratory, Book B, pp. 613–621.

Allen, N. S., and Allen, R. D. 1978. Cytoplasmic streaming in green plants. *Annu. Rev. Biophys. Bioeng.* **7**:497–526.

Allen, R. D. 1961. Amoeboid movement. *In*: Brachet, J., and Mirsky, A. E., eds., *The Cell: Biochemistry, Physiology, Morphology*, New York, Academic Press, Vol. 2, pp. 135–216.

Allen, R. D. 1974. Some new insights concerning cytoplasmic transport. *Symp. Soc. Exp. Biol.* **28**:15–26.

Allen, R. D. 1977. Concluding remarks. *In*: Brinkley, B. R., and Porter, K. R., eds. *International Cell Biology, 1976–1977,* New York, Rockefeller University Press, pp. 403–406.

Allen, R. D., and Allen, N. S. 1978. Cytoplasmic streaming in amoeboid movement. *Annu. Rev. Biophys. Bioeng.* **7**:469–495.

Allen, R. D., Cooledge, J., and Hall, P. J. 1960. Streaming in cytoplasm dissociated from the giant amoeba, *Chaos chaos*. *Nature (London)* **187**:896–899.

Allen, R. D., Francis, D., and Zeh, R. 1971. Direct test of the positive gradient theory of pseudopod extension and retraction in amoebae. *Science* **174**:1237–1240.

Anderson, O. R., and Hoeffler, W. K. 1979. Fine structure of a marine proteomyxid and cytochemical changes during encystment. *J. Ultrastruct. Res.* **66**:276–287.

Anderson, W. P. 1976. Transport through roots. *In*: Lüttge, U., and Pitman, M. G., eds., *Encyclopedia of Plant Physiology,* Berlin, Springer-Verlag, n.s. Vol. 2, pp. 129–156.

Armstrong, M. T., and Armstrong, P. B. 1979. The effects of antimicrotubule agents on cell motility in fibroblast aggregates. *Exp. Cell Res.* **120**:359–364.

Augsten, H., and Finke, L. 1978. Wirkung des Lichts auf die Protoplasmaströmung in Würzelhaaren von *Hordeum vulgare*. *Biochem. Physiol. Pflanz.* **172**:181–185.

Bardele, C. F. 1972. A microtubule model for ingestion and transport in the suctorian tentacle. *Z. Zellforsch. Mikrosk. Anat.* **126**:116–134.

Bardele, C. F. 1974. Transport of material in the suctorian tentacle. *Symp. Soc. Exp. Biol.* **28**:191–208.

Bardele, C. F., and Grell, K. G. 1967. Elektronenmikroskopische Beobachtungen zur Nahrungsaufnahme bei den Suktor *Acineta tuberosa*. *Z. Zellforsch. Mikrosk. Anat.* **80**:108–123.

Bastrom, T. E., and Walker, N. A. 1976. Intercellular transport of chloride in *Chara*. *J. Exp. Bot.* **27**:347–357.

Breckheimer-Beyrich, H. 1954. Weitere Erkenntnisse über die Wirkung zentrifugaler Kräfte auf das Protoplasma von *Nitella flexilis*. *Ber. Dtsch. Bot. Ges.* **67**:86–92.

Bretscher, A., and Weber, K. 1979. Villin: The major microfilament-associated protein of the intestinal microvillus. *Proc. Natl. Acad. Sci. USA* **76**:2321–2325.

Brunser, O., and Luft, J. H. 1970. Fine structure of the apex of absorptive cells of rat small intestine. *J. Ultrastruct. Res.* **31**:291–311.

Burgess, D. R., and Schroeder, T. E. 1977. Polarized bundles of actin filaments within microvilli of fertilized sea urchin eggs. *J. Cell Biol.* **74**:1032–1038.

Canella, M. F. 1957. Studie recherche sui tentaculiferi nel quadro della biologia generale. *Ann. Univ. Ferrara* (3)**1**:1–716.

Condeelis, J. S. 1974. The identification of F-actin in the pollen tube and protoplast of *Amaryllis belladonna*. *Exp. Cell Res.* **88**:435–439.

Condeelis, J. S., and Taylor, D. L. 1977. The control of gelation, solation, and contraction in extracts from *Dictyostelium discoideum*. *J. Cell Biol.* **74**:901–927.

Corti, B. 1774. *Osservazioni microscopiche sulla tremela e sulla circulazione del fluido in una pianta acquajuola,* Luca, Italy.

Cronshaw, J., and Esau, K. 1967. Tubular and fibrillar components of mature and differentiating sieve elements. *J. Cell Biol.* **34**:801–816.

Cronshaw, J., and Esau, K. 1968. Cell division in leaves of *Nicotiana*. *Protoplasma* **65**:1–24.

Dahlström, A., Häggendal, J., Heiwall, P. O., Larsson, P. A., and Saunders, N. R. 1974. Intra-axonal transport of neurotransmitters in mammalian neurons. *Symp. Soc. Exp. Biol.* **28**:229–247.

Dellinger, O. P. 1906. Locomotion of amoebae and allied forms. *J. Exp. Zool.* **3**:337–358.

Dujardin, F. 1835. Recherches sur les organismes. *Ann. Sci. Nat. Zool. Biol. Anim.* **4**:343–377.

Dujardin, F. 1838. *Ann. Sci. Nat. Zool.* **10**:230.

Eckert, B. S., Warren, R. H., and Rubin, R. W. 1977. Structural and biochemical aspects of cell motility in amebas of *Dictyostelium discoideum*. *J. Cell Biol.* **72**:339–358.

Eddy, E. M., and Shapiro, B. M. 1976. Changes in the topography of the sea urchin egg after fertilization. *J. Cell Biol.* **71**:35–48.

Fensom, D. S., and Williams, E. J. 1974. On Allen's suggestion for long-distance translocation in phloem of plants. *Nature (London)* **250**:490–492.

Franke, W. W., Rathke, P. C., Seib, E., Trendelenburg, M. F., Osborn, M., and Weber, K. 1976. Distribution and mode of attachment of microfilamentous structures and actin in the cortex of the amphibian oocyte. *Cytobiologie Z. Exp. Zellforsch.* **14**:111–130.

Gey, G. O. 1956. Some aspects of the constitution and behavior of normal and malignant cells maintained in continuous culture. *Harvey Lect.* **50**:154–229.

Glass, A. D. M., and Perley, J. E. 1979. Cytoplasmic streaming in the root cortex and its role in the delivery of potassium to the shoot. *Planta* **145**:399–401.

Goldacre, R. J. 1952a. The folding and unfolding of protein molecules as a basis of osmotic work. *Int. Rev. Cytol.* **1**:135–164.

Goldacre, R. J. 1952b. The action of general anaesthetics on amoebae and the mechanism of the response to touch. *Symp. Soc. Exp. Biol.* **6**:128–144.

Goldacre, R. J. and Lorsch, I. J. 1950. Folding and unfolding of protein molecules in relation to cytoplasmic streaming, amoeboid movements and osmotic works. *Nature (London)* **166**:497–500.

Gray, E. G. 1975. Presynaptic microtubules and their association with synaptic vesicles, *Proc. R. Soc. London Ser. B* **190**:369–372.

Griffin, J. L., and Allen, R. D. 1960. The movement of particles attached to the surface of amoebae in relation to current theories of amoeboid movement. *Exp. Cell Res.* **20**:619–622.

Hauser, M. 1970. Elektronenmikroskopische Untersuchung an dem Suktor *Paracineta limbata* Maupas. *Z. Zellforsch. Mikrosk. Anat.* **106**:584–614.

Hepler, P. K., and Jackson, W. T. 1968. Microtubules and early stages of cell-plate formation in the endosperm of *Haemanthus katherinae* Baker. *J. Cell Biol.* **38**:437–446.

Hepler, P. K., and Palevitz, B. A. 1974. Microtubules and microfilaments. *Annu. Rev. Plant Physiol.* **25**:309–362.

Heslop, J. P. 1974. Fast transport along nerves. *Symp. Soc. Exp. Biol.* **28**:209–227.

Heslop, J. P., and Howes, E. A. 1972. Temperature and inhibitor effects on fast axonal transport in a molluscan nerve. *J. Neurochem.* **19**:1709–1716.

Hiramoto, Y. 1955. Nature of the perivitelline space in sea urchin eggs. *Jpn. J. Zool.* **11**:333–344.

Holt, P. A., and Corliss, J. O. 1973. Pattern variability in microtubular arrays associated with the tentacles of *Actinobolina* (Ciliata: Gymnostomatida). *J. Cell Biol.* **58**:213–219.

Holt, P. A., Lynn, D. H., and Corliss, J. O. 1973. An ultrastructural study of the tentacle-bearing ciliate *Actinobolina smalli*, n. sp., and its systematic and phylogenetic implications. *Protistologica* **9**:521–541.

Hull, R. W. 1961. Studies on suctorian protozoa: The mechanism of ingestion of prey cytoplasm. *J. Protozool.* **8**:351–359.

Huxley, H. E. 1969. The mechanism of muscular contraction. *Science* **164**:1356–1366.

Hyams, J. S., and Stebbings, H. 1977. The distribution and function of microtubules in nutritive tubes. *Tissue Cell* **9**:537–545.

Hyams, J. S., and Stebbings, H. 1979. The mechanism of microtubule associated cytoplasmic transport. *Cell Tissue Res.* **196**:103–116.

Ishikawa, H., Bischoff, R., and Holtzer, H. 1969. Formation of arrowhead complexes with heavy meromyosin in a variety of cell types. *J. Cell Biol.* **43**:311–316.

Iwanami, Y. 1956. Protoplasmic movement in pollen grains and tubes. *Phytomorphology* **6**:288–295.

Jahn, T. L., and Votta, J. J. 1972. Capillary suction test of the pressure gradient theory of amoeboid motion. *Science* **177**:636–637.

Jahn, T. L., Bovee, E. C., and Small, E. B. 1960. The basis for a new major dichotomy of the Sarcodina. *J. Protozool.* **7**(Suppl.):8.

Johnson, R. P. C. 1968. Microfilaments in pores between frozen etched sieve elements. *Planta* **81**:314–332.

Jurand, A., and Bomford, R. 1965. The fine structure of the parasitic suctorian *Podophrya parameciorum*. *J. Microsc.* (*Paris*) **4**:509–522.

Kamiya, N. 1959. Protoplasmic streaming. *Protoplasmatologia* **8**(3a):1–199.

Kamiya, N. 1962. Cytoplasmic streaming. *Encycl. Plant Physiol.* **17**(2):979–1035.

Kamiya, N. 1964. The motive force of endoplasmic streaming in the amoeba. *In*: Allen, R. D., and Kamiya, N., eds., *Primitive Motile Systems in Cell Biology*, New York, Academic Press, pp. 257–278.

Kamiya, N., and Kuroda, K. 1956. Velocity distribution of the protoplasmic streaming in *Nitella* cell. *Bot. Mag.* **69**:544–554.

Kamiya, N., and Kuroda, K. 1973. Dynamics of cytoplasmic streaming in a plant cell. *Biorheology* **10**:179–187.

Katsumoto, T., Takayama, H., and Takagi, A. 1978. Ultrastructural organization of cultured macrophages as shown by negative staining techniques. *J. Electron Microsc.* **27**:1–12.

Kersey, Y. M., Hepler, P. K., Palevitz, B. A. and Wessels, N. K. 1976. Polarity of actin filaments in characean algae. *Proc. Natl. Acad. Sci. USA* **73**:165–167.

Kidd, P., Schatten, G., Grainger, J., and Mazia, D. 1976. Microfilaments in the sea urchin egg at fertilization. *Biophys. J.* **16**:117a.

Koppenhöfer, E., Ode, A., Rimmel, C., Schramm, M., Schuback, P., and Schumann, H. 1977. Isolated *Nitella* protoplasm is not excitable. *J. Theor. Biol.* **68**:449–451.

Kristensson, K. 1970. Morphological studies of the neural spread of herpes simplex virus to the central nervous system. *Acta Neuropathol.* **16**:54–63.

Kumagaya, S. 1950. Cytoplasmic streaming in pollen grains. *Bot. Mag.* **63**:52.

Lewis, W. H., and Lewis, M. R. 1924. Behavior of cells in tissue cultures. *In*: Cowdry, E. V., ed., *General Cytology*, Chicago, University of Chicago Press, pp. 385–447.

Linsbauer, K. 1929. Untersuchungen über Plasma und Plasmaströmung an *Chara* Zellen. *Protoplasma* **5**:563–621.

Lushbaugh, W. V., and Pittman, F. L. 1979. Microscopic observations on the filopodia of *Entamoeba histolytica*. *J. Protozool.* **26**:186–195.

McGee-Russell, S. M. 1974. Dynamic activities and labile microtubules in cytoplasmic transport in the marine foraminiferan *Allogromia*. *Symp. Soc. Exp. Biol.* **28**:157–190.

Macgregor, H. C., and Stebbings, H. 1970. A massive system of microtubules associated with cytoplasmic movement in telotrophic ovarioles. *J. Cell Sci.* **6**:431–449.

MacRobbie, E. A. C. 1971. Phloem translocation: Facts and mechanisms, a comparative survey. *Biol. Rev.* **46**:429–482.

Mann, S., Schatten, G., Steinhardt, R., and Friend, D. S. 1976. Sea urchin sperm : oocyte interaction. *J. Cell Biol.* **70**:110a.

Marsot, P., and Couillard, P. 1978. La réaction phygocytaire chez *Amoeba proteus*. II. Phagocytose de particules non-vivantes. *Can. J. Zool.* **56**:1497–1506.

Mast, S. O. 1926. Structure, movement, locomotion, and stimulation in *Amoeba*. *J. Morphol. Physiol.* **41**:347–425.

Mast, S. O. 1931. Locomotion in *Amoeba proteus* (Leidy). *Protoplasma* **14**:321–330.

Meyer, K. H. 1929. Über Feinbau, Festigkeit und Kontraktilität tierischer Gewebe. *Biochem. Z.* **214**:253–281.

Monné, L. 1948. Functioning of the cytoplasm. *Adv. Enzymol.* **8**:1–69.

Mooseker, M. S. 1976. Brush border motility. *J. Cell Biol.* **71**:417–433.

Mooseker, M. S., and Tilney, L. G. 1975. Organization of an actin filament–membrane complex. Filament polarity and membrane attachment in the microvilli of intestinal epithelial cells. *J. Cell Biol.* **67**:725–743.

Mukherjee, T. M., and Staehelin, L. A. 1971. The fine-structural organisation of the brush border of intestinal epithelial cells. *J. Cell Sci.* **8**:573–599.

Murphy, D. B., and Tilney, L. G. 1974. The role of microtubules in the movement of pigment granules in teleost melanophores. *J. Cell Biol.* **61**:757–779.

Nachmias, V. T., and Asch, A. 1976. Regulation and polarity: Results with myxomycete plasmodium and with human platelets. *In*: Goldman, R., Pollard, T., and Rosenbaum, J., eds., *Cell Motility*, Cold Spring Harbor, N.Y., Cold Spring Harbor Laboratory, Book B, pp. 771–783.

Nachmias, V. T., Sullender, J., and Asch, A. 1977. Shape and cytoplasmic filaments in control and lidocaine-treated human platelets. *Blood* **50**:39–53.

Nagai, R., and Hayama, T. 1979. Ultrastructure of the endoplasmic factor responsible for cytoplasmic streaming in *Chara* internodal cells. *J. Cell Sci.* **36**:121–136.

Nagai, R., and Rebhun, L. I. 1966. Cytoplasmic microfilaments in streaming *Nitella* cells. *J. Ultrastruct. Res.* **14**:571–589.

Nowakowska, G., and Grebecki, A., 1978. Attachment of *Amoeba proteus* to the substrate during upside-down crawling. *Acta Protozool.* **17**:361–368.

Ochs, S., 1972. Fast transport of materials in mammalian nerve fibers. *Science* **176**:252–260.

Oplatka, A., and Tirosh, R. 1973. Active streaming in actomyosin solutions. *Biochim. Biophys. Acta* **305**:684–688.

Otto, J. J., Kane, R. E., and Bryan, J. 1979. Formation of filopodia in coelomocytes: Localization of fascin, a 58,000 dalton actin cross-linking protein. *Cell* **17**:285–293.

Palevitz, B. A. 1976. Actin cables and cytoplasmic streaming in green plants. *In*: Goldman, R., Pollard, T., and Rosenbaum, J., eds., *Cell Motility*, Cold Spring Harbor, N.Y., Cold Spring Harbor Laboratory, Book B, pp. 601–611.

Palevitz, B. A., and Hepler, P. K. 1975. Identification of actin *in situ* at the ectoplasm–endoplasm interface of *Nitella*. *J. Cell Biol.* **65**:29–38.

Palevitz, B. A., Ash, J. F., and Hepler, P. K. 1974. Actin in the green alga, *Nitella*. *Proc. Natl. Acad. Sci. USA* **71**:363–366.

Pantin, C. F. A. 1923. On the physiology of amoeboid movement. *J. Mar. Biol. Assoc. U.K.* **13**:1–69.

Pickard, W. F. 1974. Hydrodynamic aspects of protoplasmic streaming in *Chara braunii*. *Protoplasma* **82**:321–339.

Pickett-Heaps, J. D. 1967. Ultrastructure and differentiation in *Chara* sp. I. Vegetative cells. *Aust. J. Biol. Sci.* **20**:539–551.

Pitman, M. G. 1977. Ion transport into the xylem. *Annu. Rev. Plant Physiol.* **28**:71–88.

Podlubnaya, Z. A., Tskhovrebova, L. A., Zaalishvili, M. M., and Stefanenko, G. A. 1975. Electron microscopic study of α-actinin. *J. Mol. Biol.* **92**:357–359.

Pollard, T. D., and Weihing, R. R. 1974. Cytoplasmic actin and myosin and cell movement. *CRC Crit. Rev. Biochem.* **2**:1–65.

Preston, T. M., and King, C. A. 1978. An experimental study of the interaction between the soil amoeba *Naegleria gruberi* and a glass substrate during amoeboid locomotion. *J. Cell Sci.* **34**:145–158.

Rebhun, L. I. 1964. Saltatory particle movements in cells. *In*: Allen, R. D., and Kamiya, N., eds., *Primitive Motile Systems in Cell Biology*, New York, Academic Press, pp. 503–525.

Rebhun, L. I. 1972. Polarised intracellular particle transport. Saltatory movements and cytoplasmic streaming. *Int. Rev. Cytol.* **32**:93–137.

Robidaux, J., Sandborn, E. B., Fensom, D. S., and Cameron, M. L. 1973. Plasmatic filaments and particles in mature sieve elements of *Heracleum spondylium* under the electron microscope. *J. Exp. Bot.* **24**:349–359.

Rodewald, R., Newman, S. B., and Karnovsky, M. J. 1976. Contraction of isolated brush borders from the intestinal epithelium. *J. Cell Biol.* **70**:541–554.

Rudzinska, M. A. 1970. The mechanism of food intake in *Tokophrya infusionum* and ultra-structural changes in food vacuoles during digestion. *J. Protozool.* **17**:626–641.

Rudzinska, M. A. 1973. Do Suctoria really feed by suction? *BioScience* **23**:87–94.

Sabnis, D. D., and Jacobs, W. P. 1967. Cytoplasmic streaming and microtubules in the coenocytic marine alga, *Caulerpa prolifera*. *J. Cell Sci.* **2**:465–472.

Schliwa, M. 1975. Microtubule distribution and melanosome movements in fish melanophores. *In*: Borgers, M., and De Brabander, M., eds., *Microtubules and Microtubule Inhibitors*, Amsterdam, North-Holland, pp. 215–228.

Seifriz, W. 1929. The contractility of protoplasm. *Am. Nat.* **63**:410–434.

Seifriz, W. 1943 Protoplasmic streaming. *Bot. Rev.* **9**:49–123.

Seifriz, W. 1952. The rheological properties of protoplasm. *In*: Frey-Wyssling, A., ed., *Deformation and Flow in Biological Systems*, Amsterdam, North-Holland, pp. 3–156.

Spudich, J. A., and Amos, L. A. 1979. Structure of actin filament bundles from microvilli of sea urchin eggs. *J. Mol. Biol.* **129**:319–331.

Szamier, P. M., Pollard, T. D., and Fujiwara, K. J. 1975. Tropomyosin prevents the destruction of actin filaments by osmium. *J. Cell Biol.* **67**:424a.

Taylor, A. C. 1966. Microtubules in the microspikes and cortical cytoplasma of isolated cells. *J. Cell Biol.* **28**:155–168.

Taylor, A. C., and Robbins, E. 1963. Observations on microextensions from the surface of isolated vertebrate cells. *Dev. Biol.* **7**:660–673.

Taylor, D. L., and Wang, Y. L. 1978. Molecular cytochemistry: Incorporation of fluorescently labeled actin in living cells. *Proc. Natl. Acad. Sci. USA* **75**:857–861.

Taylor, D. L., Rhodes, J. A., and Hammond, S. A. 1976. The contractile basis of amoeboid movement. II. Structure and contractility of motile extracts and plasmalemma-ectoplasm ghosts. *J. Cell Biol.* **70**:123–143.

Thaine, R. 1961. Transcellular strands and particle movement in mature sieve tubes. *Nature (London)* **192**:772–773.

Thaine, R. 1969. Movement of sugars through plants by cytoplasmic pumping. *Nature (London)* **222**:873–875.

Tilney, L. G. 1977. Actin: Its association with membranes and the regulation of its polymerization. *In*: Brinkley, B. R., and Porter, K. R., eds., *International Cell Biology, 1976–1977*, New York, Rockefeller University Press, pp. 388–402.

Tilney, L. G., and Cardell, R. R. 1970. Factors controlling the reassembly of the microvillus border of the small intestine of the salamander. *J. Cell Biol.* **47**:408–422.

Tilney, L. G., and Mooseker, M. S. 1971. Actin in the brush border of epithelial cells of the chicken intestine. *Proc. Natl. Acad. Sci. USA* **68**:2611–2615.

Tilney, L. G., and Mooseker, M. S. 1976. Actin filament–membrane attachment. Are membrane particles involved? *J. Cell Biol.* **71**:402–416.

Tucker, J. B. 1978. Endocytosis and streaming of highly gelated cytoplasm alongside rows of arm-bearing microtubules in the ciliate *Nassula*. *J. Cell Sci.* **29**:213–232.

Tyree, M. T. 1970. The symplast concept. A general theory of symplastic transport according to the thermodynamics of irreversible processes. *J. Theor. Biol.* **26**:181–214.

Tyree, M. T., Fisher, R. A., and Dainty, J. 1974. A quantitative investigation of symplastic transport in *Chara corallina*. II. The symplastic transport of chloride. *Can. J. Bot.* **52**:1325–1334.

Wang, E., and Goldman, R. D. 1978. Functions of cytoplasmic fibers in intracellular movements in BHK-21 cells. *J. Cell Biol.* **79**:708–726.

Ward, P. A., and Becker, E. L. 1977. Biology of leukotaxis. *Rev. Physiol. Biochem. Pharmacol.* **77**:125–148.

Weiss, P. 1961. From cell to molecule. *In*: Allen, J. M., ed., *The Molecular Control of Cellular Activity*, New York, McGraw–Hill, pp. 1–72.

Wellings, J. V., and Tucker, J. B. 1979. Changes in microtubule packing during the stretching of an extensible microtubule bundle in the ciliate *Nassula*. *Cell Tiss. Res.* **197**:313–323.

Williamson, R. E. 1974. Actin in the alga *Chara corallina*. *Nature (London)* **248**:801–802.

Wohlman, A., and Allen, R. D. 1968. Structural organization associated with pseudopod extension and contraction during cell locomotion in *Difflugia*. *J. Cell Sci.* **3**:105–114.

Zeligs, J. D., and Wollman, S. H. 1977. Pseudopod behavior in hyperplastic thyroid follicles *in vivo*. *J. Ultrastruct. Res.* **60**:99–105.

CHAPTER 4

Adam, G. 1977. Rotation of bacterial flagella as driven by cytomembrane streaming. *J. Theor. Biol.* **65**:713–726.

Adler, J. 1976. Some aspects of the structure and function of bacterial flagella. *In*: Goldman, R., Pollard, T., and Rosenbaum, J., eds., *Cell Motility*, Cold Spring Harbor, N.Y., Cold Spring Harbor Laboratory, Book A, pp. 29–33.

Adler, J., and Dahl, M. M. 1967. A method for measuring the motility of bacteria and for comparing random and non-random motility. *J. Gen. Microbiol.* **46**:161–173.

Afzelius, B. 1959. Electron microscopy of the sperm tail: Results obtained with a new fixative. *J. Biophys. Biochem. Cytol.* **5**:269–278.

Afzelius, B. 1961. The fine structure of the cilia from ctenophore swimming-plates. *J. Biophys. Biochem. Cytol.* **9**:383–394.

Allen, C., and Borisy, G. G. 1974. Structural polarity and directional growth of microtubules of *Chlamydomonas* flagella. *J. Mol. Biol.* **90**:381–402.

Allen, R. D. 1968. A reinvestigation of cross-sections of cilia. *J. Cell Biol.* **37**:825–831.

André, J. 1961. Sur quelques details nouvellement connus de l'ultrastructure des organites vibratiles. *J. Ultrastruct. Res.* **5**:86–108.

Auclair, W., and Siegel, B. W. 1966. Cilia regeneration in the sea urchin embryo: Evidence for a pool of ciliary proteins. *Science* **154**:913–915.

Baccetti, B., and Dallai, R. 1976. The spermatozoon of Arthropoda. XXVII. Uncommon axoneme patterns in different species of cecydomyid dipterans. *J. Ultrastruct. Res.* **55**:50–69.

Baccetti, B., Burrini, A. G., Dallai, R., and Pallini, V. 1979. The dynein electrophoretic bands in axonemes naturally lacking the inner or the outer arm. *J. Cell. Biol.* **80**:334–340.

Bargmann, W. 1954. Über Feinbau und Funktion des Saccus vasculosus. *Z. Zellforsch. Mikrosk. Anat.* **40**:49–74.

Bargmann, W., and Knoop, A. 1955. Electronmikroskopische Untersuchungen der Kronichenzellen des Saccus vasculosus. *Z. Zellforsch. Mikrosk. Anat.* **43**:184–194.

Barnes, B. G. 1961. Ciliated secretory cells in the pars distalis of the mouse hypophysis. *J. Ultrastruct. Res.* **5**:453–467.

Berg, H. C. 1974. Dynamic properties of bacterial flagellar motors. *Nature (London)* **249**:77–79.

Berg, H. C. 1975. Bacterial behaviour. *Nature (London)* **254**:389–392.

Berg, H. C. 1976. Does the flagellar rotary motor step? *In*: Goldman, R., Pollard, T., and Rosenbaum, J., *Cell Motility*, Cold Spring Harbor, N.Y., Cold Spring Harbor Laboratory, Book A, pp. 47–56.

Black, F. T., Freundt, E. A., Vinther, O., and Christiansen, C. 1979. Flagellation and swimming motility of *Thermoplasma acidophilum*. *J. Bacteriol.* **137**:456–460.

Bouck, G. B. 1969. Extracellular microtubules. *J. Cell Biol.* **40**:446–460.

Bouck, G. B. 1971. The structure, origin, and composition of the tubular mastigonemes of the *Ochromonas* flagellum. *J. Cell Biol.* **50**:362–384.

Bouck, G. B., Rogalski, A., and Valaitis, A. 1978. Surface organization and composition of *Euglena*. II. Flagellar mastigonemes. *J. Cell Biol.* **77**:805–826.

Bovee, G. C., Jahn, T. L., Fonesca, J. R., and Landman, M. D. 1963. Flagellar movements in some species of mastigamoebas. *Abstr. Seventh Meet. Biphys. Soc. MD2.*

Bradfield, J. R. G. 1955. Fibre patterns in animal flagella and cilia. *Symp. Soc. Exp. Biol.* **9:**306–334.

Bradley, D. E. 1966. The ultrastructure of the flagella of the chrysomonads with particular reference to the mastigonemes. *Exp. Cell Res.* **41**:162–173.

Bradley, T. J., and Satir, P. 1979. Insect axopods. *J. Cell Sci.* **35**:165–176.

Brock, T. D. 1974. *Biology of Microorganisms*, 2nd ed., Englewood Cliffs, N.J., Prentice–Hall.

Brokaw, C. J. 1961. Movement and nucleoside polyphosphatase activity of isolated flagella from *Polytoma uvella*. *Exp. Cell Res.* **22**:151–162.

Brokaw, C. J. 1963. Movement of the flagella of *Polytoma uvella*. *J. Exp. Biol.* **40**:149–156.

Brokaw, C. J. 1965. Non-sinusoidal bending waves of sperm flagella. *J. Exp. Biol.* **43**:155–169.

Brokaw, C. J. 1972. Flagellar movement: A sliding filament model. *Science* **178**:455–462.

Brokaw, C. J., and Simonick, T. F. 1976. CO_2 regulation of the amplitude of flagellar bending. *In*: Goldman, R., Pollard, T., and Rosenbaum, J., eds., *Cell Motility*, Cold Spring Harbor, N.Y., Cold Spring Harbor Laboratory, Book C, pp. 933–940.

Brokaw, C. J., and Wright, L. 1963. Bending waves of the posterior flagellum of *Ceratium*. *Science* **142**:1169–1170.

Brown, D. L., and Rogers, K. A. 1978. Hydrostatic pressure-induced internalization of flagellar axonemes, disassembly, and reutilization during flagellar regeneration in *Polytomella*. *Exp. Cell Res.* **117**:313–324.

Brown, H. P., and Cox, A. 1954. An electron microscopic study of protozoan flagella. *Am. Midl. Nat.* **52**:106–117.

Brown, W. V., and Bertke, E. M. 1974. *Textbook of Cytology*, St. Louis, Mosby.

Burns, R. G., and Pollard, D. 1974. A dynein-like protein from brain. *FEBS Lett.* **40**:274–280.

Burton, P. R. 1973. Some structural and cytochemical observations on the axial filament complex of lung-fluke spermatozoa. *J. Morphol.* **140**:185–196.

Burton, P. R., and Silveira, M. 1971. Electron microscopic and optical diffraction studies of negatively stained axial units of certain platyhelminth sperm. *J. Ultrastruct. Res.* **36**:757–767.

Chalcroft, J. P., Bullivant, S., and Howard, B. H. 1973. Ultrastructural studies on *Selenomonas ruminantium* from the sheep rumen. *J. Gen. Microbiol.* **79**:135–146.

Chang, J. Y., DeLange, R. J., Shaper, J. H., and Glazer, A. N. 1976. Amino acid sequence of flagellin in *Bacillus subtilis* 168. I. Cyanogen bromide peptides. *J. Biol. Chem.* **251**:695–700.

Chen, L. L., and Haines, T. H. 1976. The flagellar membrane of *Ochromonas danica*. *J. Biol. Chem.* **251**:1828–1834.

Chen, L. L., Pousada, M., and Haines, T. H. 1976. The flagellar membrane of *Ochromonas danica*. Lipid composition. *J. Biol. Chem.* **251**:1835–1842.

Claybrook, J. R., and Nelson, L. 1968. Flagellar adenosine triphosphatase from sea urchin sperm: properties and relation to motility. *Science* **162**:1134–1136.

Coulton, J. W., and Murray, R. G. E. 1977. Membrane associated components of the bacterial flagellar apparatus. *Biochim. Biophys. Acta* **465**:290–310.

Coulton, J. W., and Murray, R. G. E. 1978. Cell envelope associations of *Aquaspirillum serpens* flagella. *J. Bacteriol.* **136**:1037–1049.

Davidson, B. E. 1971. The alignment of cyanogen bromide fragments from the flagellin of *Salmonella adelaide*. *Eur. J. Biochem.* **18**:524–529.

De Lange, R. J., Chang, J. Y., Shaper, J. H., Komatzus, S. K., and Glazer, A. N. 1973. On the amino-acid sequence of flagellin from *B. subtilis* 168: Comparison with other bacterial flagellins. *Proc. Natl. Acad. Sci. USA* **70**:3428–3431.

De Lange, R. J., Chang, J. Y., Shaper, J. H., and Glazer, A. N. 1976. Amino acid sequence of flagellin of *Bacillus subtilis* 68. III. Tryptic peptides, *n*-biomosuccinimide peptides, and the complete amino acid sequence. *J. Biol. Chem.* **251**:705–711.

Dentler, W. L. 1977. Structures connecting microtubules and membranes in cilia and flagella. *J. Cell Biol.* **75**:287a.

Dentler, W. L., and Rosenbaum, J. L. 1977. Flagellar elongation and shortening in *Chlamydomonas*. III. Structures attached to the tips of flagellar microtubules and their relationship to the directionality of flagellar microtubule assembly. *J. Cell Biol.* **74**:747–759.

De Pamphilis, M. L., and Adler, J. 1971a. Purification of intact flagella from *E. coli* and *B. subtilis*. *J. Bacteriol.* **105**:376–383.

De Pamphilis, M. L., and Adler, J. 1971b. Fine structure and isolation of the hook–basal body complex of flagella from *E. coli* and *B. subtilis*. *J. Bacteriol.* **105**:384–395.

De Pamphilis, M. L., and Adler, J. 1971c. Attachment of flagellar bodies to the cell envelope: Specific attachment to the outer, lipopolysaccharide membrane and the cytoplasmic membrane. *J. Bacteriol.* **105**:396–407.

de Robertis, E. 1956. Electron microscope observations on the submicroscopic organization of the retinal rods. *J. Biophys. Biochem. Cytol.* **2**:319–330.

Dillon, L. S. 1978. *The Genetic Mechanism and the Origin of Life*, New York, Plenum Press.

Dimmit, K., and Simon, M. I. 1971. Purification and thermal stability of intact *E. coli* flagella. *J. Bacteriol.* **105**:369–375.

Doetsch, R. N., and Hageage, G. J. 1968. Motility in procaryotic organisms. *Biol. Rev.* **43**:317–362.

Dute, R., and Kung, C. 1978. Ultrastructure of the proximal region of somatic cilia in *Paramecium tetraurelia*. *J. Cell Biol.* **78**:451–464.

Eckert, R. 1972. Bioelectric control of ciliary activity. *Science* **176**:473–481.

Emerson, S., and Simon, M. 1971. Variation in the primary structure of *Bacillus subtilis* flagellins. *J. Bacteriol.* **106**:949–954.

Fauré-Fremiet, E. 1958. The origin of metazoa and the stigma of the phytoflagellates. *Q. J. Microsc. Sci.* **99**:123–129.

Fawcett, D. W. 1954. The study of epithelial cilia and sperm flagella with the electron microscope. *Laryngoscope* **64**:557–567.

Fawcett, D. W. 1957. The structure of mammalian spermatozoon. *Int. Rev. Cytol.* **7**:195–234.

Fawcett, D. W. 1961. Cilia and flagella. *In*: Brachet, J., and Mirsky, A. E., eds., *The Cell*, New York, Academic Press, Vol. 2, pp. 217–297.

Follett, E. A. C., and Gordon, J. 1963. An electron microscope study of *Vibria flagella*. *J. Gen. Microbiol.* **32**:235–239.

Friend, D. S. 1966. The fine structure of *Giardia muris*. *J. Cell Biol.* **29**:317–332.

Fuerst, J. A., and Hayward, A. C. 1969. The sheathed flagellum of *Pseudomonas stizolobii*. *J. Gen. Microbiol.* **32**:235–245.

Fukuda, A., Koyasu, S., and Okada, Y. 1978. Characterization of two flagella-related proteins from *Caulobacter crescentus*. *FEBS Lett.* **95**:70–75.

Fulton, C. 1970. Amebo-flagellates as research partners: The laboratory biology of *Naegleria* and *Tetramitus*. *In:* Prescott, D. M. ed., *Methods in Cell Physiology,* New York, Academic Press, Vol. 4, pp. 341–476.

Fulton, C., and Simpson, P. A. 1976. Selective synthesis and utilization of flagellar tubulin. The multi-tubulin hypothesis. *In*: Goldman, R., Pollard, T., and Rosenbaum, J., eds., *Cell Motility*, Cold Spring Harbor, N.Y., Cold Spring Harbor Laboratory, Book C, pp. 987–1005.

Gibbons, B. H., and Gibbons, I. R. 1979. Relationship between the latent ATPase state of dynein 1 and its ability to recombine functionally with KCl-extracted sea urchin sperm flagella. *J. Biol. Chem.* **254**:197–201.

Gibbons, I. R. 1961. The relationship between fine structure and direction of beat in gill cilia of a lamellibranch mollusc. *J. Biophys. Biochem. Cytol.* **11**:179–204.

Gibbons, I. R. 1977. Structure and function of flagellar microtubules. *In*: Brinkley, B. R., and Porter, K. R., eds., *International Cell Biology, 1976–1977*, New York, Rockefeller University Press, pp. 348–357.

Gibbons, I. R., and Fronk, E. 1979. A latent ATPase form of dynein 1 from sea urchin sperm flagella. *J. Biol. Chem.* **254**:187–196.

Gibbons, I. R., and Grimstone, A. V. 1960. On flagellar structure in certain flagellates. *J. Biophys. Biochem. Cytol.* **7**:697–716.

Gibbons, I. R., and Rowe, A. J. 1965. Dynein: A protein with ATPase activity from cilia. *Science* **149**:424–425.

Gibbons, I. R., Fronk, E., Gibbons, B. H., and Ogawa, K. 1976. Multiple forms of dynein in sea urchin sperm flagella. *In*: Goldman, R., Pollard, T., and Rosenbaum, J., eds., *Cell Motility*, Cold Spring Harbor, N.Y., Cold Spring Harbor Laboratory, Book C, pp. 915–932.

Glauert, A. M., Kerridge, D., and Horne, R. W. 1963. The fine structure and mode of attachment of the sheathed flagellum of *Vibrio metchnikovii*. *J. Cell Biol.* **18**:327–336.

Glossmann, H., and Bode, W. 1972. Cyanogen bromide cleavage of *Proteus mirabilis* flagellin. *Hoppe-Seyler's Z. Physiol. Chem.* **353**:298–306.

Gordon, M., and Barnett, R. J. 1967. Fine structural cytochemical localizations of phosphatase activities of rat and guinea pigs. *Exp. Cell Res.* **48**:395–412.

Grimes, G. W., and Adler, J. A. 1976. The structure and development of the dorsal bristle complex. *J. Protozool.* **23**:135–143.

Grimes, G. W., and Adler, J. A. 1978. Regeneration of ciliary pattern in longitudinal fragments of the hypotrichous ciliate, *Stylonychia*. *J. Exp. Zool.* **204**:57–80.

Grimstone, A. V. 1963. The fine structure of some polymastigate flagellates. *Proc. Linn. Soc. London* **174**:49–52.

Guentzel, M. N., and Berry, L. J. 1975. Motility as a virulence factor for *Vibrio cholerae*. *Infect. Immun.* **11**:890–897.

Guttmann, S. G. 1978. Doctoral dissertation. University of Rochester, Rochester, N.Y.

Hanson, J., and Huxley, H. E. 1955. The structural basis of contraction in striated muscle. *Symp. Soc. Exp. Biol.* **9**:228–264.

Harris, K. 1963. Observations on *Sphaleromantis tetragona*. *J. Gen. Microbiol.* **33**:345–348.

Hendelberg, J. 1965. The different types of spermatozoa in Polycladida, Turbellaria. *Arkv. Zool.* (2)**18**:267–304.

Hershenov, B. R., Tulloch, G. S., and Johnson, A. D. 1966. The fine structure of trematode sperm tails. *Trans. Am. Microsc. Soc.* **85**:480–483.

Hibberd, D. J., and Leedale, G. F. 1972. Observations on the cytology and ultrastructure of the new algal class, Eustigmatophyceae. *Ann. Bot.* **36**:49–71.

Hilmen, M., and Simon, M. 1976. Motility and the structure of bacterial flagella. *In*: Goldman, R., Pollard, T., and Rosenbaum, J., eds., *Cell Motility*, Cold Spring Harbor, N.Y., Cold Spring Harbor Laboratory, Book A, pp 35–45.

Hilmen, M., Silverman, M., and Simon, M. 1974. The regulation of flagellar formation and function. *J. Supramol. Struct.* **2**:360–371.

Hoeniger, J. F. M., van Iterson, W., and van Zanten, E. N. 1966. Basal bodies of bacterial flagella in *Proteus mirabilis*. II. Electron microscopy of negatively stained material. *J. Cell Biol.* **31**:603–618.

Hoffmann-Berling, H. 1955. Geisselmodelle und Adeninosintriphosphat (ATP). *Biochim. Biphys. Acta* **16**:146—154.

Hopkins, J. M. 1970. Subsidiary components of the flagella of *Chlamydomonas reinhardii*. *J. Cell Sci.* **7**:823–839.

Horiguchi, T., Yamaguchi, S., Yao, K., Taira, T., and Iino, T. 1975. Genetic analysis of H1, the structural gene for phase 1 flagellin in *Salmonella*. *J. Gen. Microbiol.* **91**:139–149.

Hotani, H. 1979. Micro-video study of moving bacterial flagellar movements. *J. Mol. Biol.* **129**:305–318.

Houwink, A. L. 1953. A macromolecular mono-layer in the cell wall of *Spirillum* species. *Biochim. Biophys. Acta* **10**:360–366.

Hufnagel, L. A., and Torch, R. 1967. Intraclonal dimorphism of caudal cirri in *Euplotes vannus* cortical determination. *J. Protozool.* **14**:429–439.

Hyams, J. S., and Borisy, G. G. 1978. Isolated flagellar apparatus of *Chlamydomonas*: Characterization of forward swimming and alteration of waveform and reversal of motion by calcium ions *in vitro*. *J. Cell Sci.* **33**:235–253.

Iino, T. 1969. Genetics and chemistry of bacterial flagella. *Bacteriol. Rev.* **33**:454–475.

Iino, T. 1974. Assembly of *Salmonella* flagellin *in vitro* and *in vivo*. *J. Supramol. Struct.* **2**:372–384.

Jahn, T. L., and Bovee, E. C. 1965. Movement and locomotion of microorganisms *Annu. Rev. Microbiol.* **19**:21–58.

Jahn, T. L., and Fonseca, J. R. 1963. Mechanisms of locomotion of flagellates: V. *Trypanosoma lewisi* and *T. cruzi*. *J. Protozool.* Suppl. 10.

Jahn, T. L., Landman, M. D., and Fonseca, J. R. 1964. The mechanism of locomotion of flagellates II. Function of the mastigonemes of *Ochromonas*. *J. Protozool.* **11**:291–296.

Johnson, R. C., Walsh, M. P., Ely, B., and Shapiro, L. 1979. Flagellar hook and basal complex of *Caulobacter crescentus*. *J. Bacteriol.* **138**:984–989.

Joys, T. M., and Rankis, V. 1972. The primary structure of the phase 1 flagellar protein of *Salmonella typhimurium*. 1. The tryptic peptides. *J. Biol. Chem.* **247**:5180–5193.

Kauffmann, F. 1964. Das Kauffmann–White schema. *In*: van Oye, E. L., ed., *The World Problem of Salmonellosis,* The Hague, June, pp. 21–66.

Kaye, J. S. 1964. The fine structure of flagella in spermatids of the house cricket. *J. Cell Biol.* **22**:710–714.

Kennedy, J. R., and Brittingham, E. 1968. Fine structure changes during chloral hydrate deciliation of *Paramecium caudatum*. *J. Ultrastruct. Res.* **22**:530–545.

Kirby, H. 1941–1949. Devescovinid flagellates of termites. *Univ. Calif. Publ. Zool.* **45**:1–91.

Komeda, Y., Suzuki, H., Ishidsu, J., and Iino, T. 1975. The role of cAMP in flagellation of *Salmonella typhimurium*. *Mol. Gen. Genet.* **142**:289–298.

Komeda, Y., Silverman, M., and Simon, M. 1977. Genetic analysis of *E. coli* K-12 region I flagellar mutants. *J. Bacteriol.* **131**:801–808.

Kort, E. N., Goy, M. F., Larsen, S. H., and Adler, J. 1975. Methylation of a membrane protein involved in bacterial chemotaxis. *Proc. Natl. Acad. Sci. USA* **72**:3939–3943.

Koshland, D. E. 1977. Bacterial chemotaxis and some enzymes in energy metabolism. *Symp. Soc. Gen. Microbiol.* **27**:317–332.

Langford, G. M. and Inoué, S. 1979. Motility of the microtubular axostyle in *Pyrsonympha*. *J. Cell Biol.* **80**:521–538.

Larsen, S. H., Reader, R. W., Kort, E. N., Tso, W. W., and Adler, J. 1974. Change in direction of flagellar rotation is the basis of the chemotactic response in *E. coli*. *Nature* (*London*) **249**:74–77.

Läuger, P. 1977. Ion transport and rotation of bacterial flagella. *Nature* (*London*) **268**:360–362.

Leedale, G. F. 1967. *Euglenoid Flagellates*, Englewood Cliffs, N.J., Prentice–Hall.

Leedale, G. F., Leadbeater, B. S. C., and Massalski, A. 1970. The intracellular origin of the flagellar hairs in the Chyrsophyceae and Xanthophyceae. *J. Cell Sci.* **6**:701–719.

Lefebvre, P. A., Nordstrom, S. A., Moulder, J. E., and Rosenbaum, J. L. 1978. Flagellar elongation and shortening in *Chlamydomonas*. IV. Effects of flagellar detachment, regeneration, and resorption on the induction of flagellar protein synthesis. *J. Cell Biol.* **78**:8–27.

Lewin, R. A. 1958. The cell wall of *Platymonas*. *J. Gen. Microbiol.* **19**:87–90.

Linck, R. W. 1973. Chemical and structural differences between cilia and flagella from the lamellibranch mollusc, *Aequipecten irradians*. *J. Cell Sci.* **12**:951–981.

Linck, R. W. 1976. Fractionation of minor component proteins and tubulin from specific regions of flagellar doublet microtubules. *In*: Goldman, R., Pollard, R., and Rosenbaum, J., eds., *Cell Motility*, Cold Spring Harbor, N.Y., Cold Spring Harbor Laboratory, Book C, pp. 819–890.

Machin, K. E. 1958. Wave propagation along flagella. *J. Exp. Biol.* **35**:796–806.

Manton, I. 1952. The fine structure of plant cilia. *Symp. Soc. Exp. Biol.* **6**:306–319.

Manton, I. 1964. Further observations on the fine structure of the haptonema in *Prymnesium parvum*. *Arch. Mikrobiol.* **19**:315–330.

Manton, I., and Harris, K. 1966. Observations on the microanatomy of the brown flagellate *Sphaleromantis tetragona* Skuja with special reference to the flagellar apparatus and scales. *J. Linn. Soc. London Bot.* **59**:397–403.

Manton, I., and Leedale, G. F. 1961. Further observations on the fine structure of *Chrysochromulina ericina* Parke and Manton. *Arch. Mikrobiol.* **41**:145–155.

Manton, I., and Leedale, G. F. 1963a. Observations on the fine structure of *Prymnesium parvum* Carter. *Arch. Mikrobiol.* **45**:285–303.

Manton, I., and Leedale, G. F. 1963b. Observation on the micro-anatomy of *Chrystallolithus hyalinus* Gaarder and Markali. *Arch. Mikrobiol.* **47**:115–136.

Manton, I., and von Stosch, H. A. 1966. Observations on the fine structure of the male gamete of the marine centric diatom *Lithodesmium undulatum*. *J. R. Microsc. Soc. Lond.* **85**: 119– 134.

Manton, I., Rayns, D. G., and Ettl, H. 1965. Further observations on green flagellates with scaly flagella: The genus *Heteromastix* Korshikov. *J. Mar. Biol. Assoc. U.K.* **45**:241–255.

Marchand, B., and Mattei, X. 1976a. La spermatogenèse des acanthocéphales. I. L'appareil centriolaire et flagellaire au cours de la spermiogenèse d'*Illiosentis furcatus* var. *africana* Golvan, 1956 (Paleacanthocephala, Rhadinorhynchidae). *J. Ultrastruct. Res.* **54**:347–358.

Marchand, B., and Mattei, X. 1976b. Ultrastructure du spermatozöide de *Centrorhynchus milvus* Ward, 1956 (Paleacanthocephala, Plymorphidae). *C. R. Seances Soc. Biol. Paris* **170**:237–242.

Marchand, B., and Mattei, X. 1976c. La spermiatogenèse des acanthocéphales. II. Variation du nombre de fibres centrales dans le flagelle spermatique d'*Acanthosentis tilapiae* Baylis (Eoacanthocephala, Quadrigyridae). *J. Ultrastruct. Res.* **55**:391–399.

Marchand, B., and Mattei, X. 1977. Un type nouveau de structure flagellaire. *J. Cell Biol.* **72**:707–713.

Marchand, B., and Mattei, X. 1978. Flagellogenèse chez un Eoacanthocephala: Mise en place et désorganisation de l'axoneme spermatique. *J. Ultrastruct. Res.* **63**:41–50.

Markey, D. R., and Bouck, G. B. 1977. Mastigoneme attachment in *Ochromonas*. *J. Ultrastruct. Res.* **59**:173–177.

Maruyama, M., Lodderstaedt, G., and Schmitt, R. 1978. Purification and biochemical properties of complex flagella isolated from *Rhizobium lupini* H13-3. *Biochim. Biphys. Acta* **535**:110–124.

Massalski, A., and Leedale, G. F. 1969. Cytology and ultrastructure of the Xanthophyceae. I. Comparative morphology of the zoosphores of *Bumilleria sicula* Borzi and *Tribonema vulgare* Pascher. *Br. Phycol. J.* **4**:159–180.

Matsumura, P., Silverman, M., and Simon, M. 1977. Cloning and expression of the flagellar hook gene on hybrid plasmids in minicells. *Nature (London)* **265**:758–760.

Metzner, P. 1920. Die Bewegung und Reizbeantwortung der bipolar begeisselten Spirillen. *Jahrb. Wiss. Bot.* **59**:325–412.

Miki-Noumura, T., and Kamiya, R. 1979. Conformational change in the outer doublet microtubules from sea urchin sperm flagella. *J. Cell Biol.* **84**:355–360.

Milhaud, M., and Pappas, G. D. 1968. Cilia formation in the adult cat brain after pargyline treatment. *J. Cell Biol.* **37**:599–609.

Miller, W. H. 1958. Derivatives of cilia in the distal sense cells of the retina of *Pecten*. *J. Biophys. Biochem. Cytol.* **4**:227–228.

Norstag, K. 1967. Fine structure of the spermatozoid of *Zamia* with special reference to the flagellar apparatus. *Am. J. Bot.* **54**:831–840.

Parke, M. and Manton, I. 1965. Preliminary observations on the fine structure of *Prasinocladus marinus. J. Mar. Biol. Assoc. U.K.* **45**:525–536.

Parke, M. Lund, J. W. G., and Manton, I. 1962. Observations on the biology and fine structure of the type species of *Chrysochromulina* (*C. parva* Lackey) in the English Lake District. *Arch. Mikrobiol.* **42**:333–352.

Patterson-Delafield, J., Martinez, R. J., Stocker, B. A. D., and Yamaguchi, S. 1973. A new *fla* gene in *Salmonella typhimurium-fla*R and its mutant phenotype—superhooks. *Arch. Microbiol.* **90**:107–120.

Phillips, D. M. 1969. Exceptions to the prevailing pattern of tubules (9 + 9 + 2) in the sperm flagella of certain insect species. *J. Cell Biol.* **40**:28–43.

Pitelka, D. R. 1961. Fine structure of the silverline and fibrillar systems of three tetrahymenid ciliates. *J. Protozool.* **8**:75–89.

Pitelka, D. R. 1963. *Electron-Microscopic Structure of Protozoa,* Oxford, Pergamon Press.

Pitelka, D. R., and Schooley, C. N. 1955. Comparative morphology of some protistan flagella. *Univ. Calif. Publ. Zool.* **61**:79–128.

Pommerville, J. 1978. Analysis of gamete and zygote motility in *Allomyces. Exp. Cell Res.* **113**:161–172.

Raska, I., Mayer, F., Edlebluth, C., and Schmitt, R. 1976. Structure of plain and complex flagellar hooks of *Pseudomonas rhodos. J. Bacteriol.* **125**:679–688.

Reese, T. S. 1965. Olfactory cilia in the frog. *J. Cell Biol.* **25**:209–230.

Rosenbaum, J. L., and Child, F. M. 1967. Flagellar regeneration in protozoan flagellates. *J. Cell Biol.* **34**:345–364.

Rosenbaum, J. L., Moulder, J. E., and Ringo, D. L. 1969. Flagellar elongation and shortening in *Chlamydomonas*. The use of cycloheximide and colchicine to study the synthesis and assembly of flagellar proteins. *J. Cell Biol.* **41**:600–619.

Sager, R., and Palade, G. E. 1957. Structure and development of the chloroplast in *Chlamydomonas. J. Biophys. Biochem. Cytol.* **3**:463–487.

Satir, P. 1963. Studies on cilia: The fixation of the metachronal wave. *J. Cell Biol.* **18**:345–365.

Satir, P. 1965. Studies on cilia. II. Examination of the distal region of the ciliary shaft and the role of the filaments in motility. *J. Cell Biol.* **26**:805–834.

Satir, P. 1968. Studies on cilia. III. *J. Cell Biol.* **39**:77–94.

Satir, P. 1974. How cilia move. *Sci. Am.* **231**(4):44–52.

Schmitt, R., Raska, I., and Mayer, F. 1974a. Plain and complex flagella of *Pseudomonas rhodos*: Analysis of fine structure and composition. *J. Bacteriol.* **117**:844–857.

Schmitt, R., Bamberger, I., Acker, G., and Mayer, F. 1974b. Fine structure analysis of the complex flagella of *Rhizobium lupini* H13-3. *Arch. Microbiol.* **100**:145–162.

Schrank, G. D., and Verway, W. F. 1976. Distribution of cholera organisms in experimental *Vibrio cholerae* infections: Proposed mechanisms of pathogenesis and antibacterial immunity. *Infect. Immun.* **13**:195–203.

Shaper, J. H., DeLange, R. J., Martinez, R. J., and Glazer, A. N. 1976. Amino acid sequence of flagellin of *Bacillus subtilis* 168. II. Tryptic peptides from methylated flagellin. *J. Biol. Chem.* **251**:701–704.

Shimizu, T. 1975. Recombination of ciliary dynein of *Tetrahymena* with the outer fibers. *J. Biochem.* **78**:41–49.

Silveira, M. 1969. Ultrastructural studies on a 9 + 1 flagellum. *J. Ultrastruct. Res.* **26**:274–288.

Silveira, M. 1973. Intraaxonemal glycogen in "9 + 1" flagella of flatworms. *J. Ultrastruct. Res.* **44**:253–264.

Silveira, M. 1974. The fine structure of 9 + 1 flagella in turbellarian flatworms. *In*: Afzelius, B. A., ed., *Functional Anatomy of the Spermatozoon*, New York, Pergamon Press, pp. 289–298.

Silverman, M., and Simon, M. 1972. Flagellar assembly mutants in *E. coli*. *J. Bacteriol.* **112**:986–993.

Silverman, M., and Simon, M. 1973. Genetic analysis of flagellar mutants in *E. coli*. *J. Bacteriol.* **113**:105–113.

Silverman, M., and Simon, M. 1974a. Flagellar rotation and the mechanism of bacterial motility. *Nature (London)* **249**:73–74.

Silverman, M., and Simon, M. 1974b. Characterization of *E. coli* flagellar mutants that are insensitive to catabolite repression. *J. Bacteriol.* **120**:1196–1203.

Simon, M. I., Emerson, S. U., Shaper, J. H., Bernard, P. D., and Glazer, A. N. 1977. Classification of *Bacillus subtilis* flagellins. *J. Bacteriol.* **130**:200–204.

Sleigh, M. A. 1962. *The Biology of Cilia and Flagella*, New York, Macmillan.

Slifer, E. H., and Sekhon, S. S. 1963. Sense organs on the antennal flagellum of the milkweed bug *Lygaeus kalmii* Stal. *J. Morphol.* **112**:165–193.

Sonneborn, T. M. 1970. Gene action in development. *Proc. R. Soc. London Ser. B.* **176**:347–366.

Sorokin, S. P. 1962. Centrioles and the formation of rudimentary cilia by fibroblasts and smooth muscle cells. *J. Cell Biol.* **15**:363–377.

Steinman, R. M. 1968. An electron microscopic study of ciliogenesis in developing epidermis and trachea in the embryo of *Xenopus laevis*. *Am. J. Anat.* **122**:19–56.

Stephens, R. E. 1970. Thermal fractionation of outer fiber doublet microtubules into A- and B-subfiber components: A- and B-tubulin. *J. Mol. Biol.* **47**:353–363.

Stephens, R. S. 1975. Structural chemistry of the axoneme: Evidence for chemically and functionally unique tubulin dimers in outer fibers. *In*: Inoué, S., and Stephens, R. E., eds., *Molecules and Cell Movement*, New York, Raven Press, pp. 181–206.

Stephens, R. E. 1977. Major membrane protein differences in cilia and flagella: Evidence for a membrane associated tubulin. *Biochemistry* **16**:2047–2058.

Summers, K. E., and Gibbons, I. R. 1971. ATP-induced sliding of tubules in trypsin-treated flagella of sea-urchin sperm. *Proc. Natl. Acad. Sci. USA* **68**:3092–3096.

Summers, K. E., and Gibbons, I. R. 1973. Effects of trypsin digestion on flagellar structures and their relationship to motility. *J. Cell Biol.* **58**:618–629.

Suzuki, H., and Iino, T. 1975. Absence of mRNA specific for flagellin in non-flagellate mutants of *Salmonella*. *J. Mol. Biol.* **95**:549–556.

Suzuki, H., Iino, T., Horiguchi, T., and Yamaguchi, S. 1978. Incomplete flagellar structures in nonflagellate mutants of *Salmonella typhimurium*. *J. Bacteriol.* **133**:904–915.

Takahashi, M., and Tonomura, Y. 1978. Binding of 30 S dynein with the B-tubule of the outer doublet of axonemes from *Tetrahymena pyriformis* and ATP-induced dissociation. *J. Biochem.* **84**:1339–1355.

Tamm, S. L. 1967. Flagellar development in the protozoan *Peranema trichophorum*. *J. Exp. Zool.* **164**:163–186.

Tamm, S. L. 1976. Properties of a rotary motor in eukaryotic cells. *In*: Goldman, R., Pollard, T., and Rosenbaum, J., eds., *Cell Motility*, Cold Spring Harbor, N.Y., Cold Spring Harbor Laboratory, Book C, pp. 949–967.

Tamm, S. L. 1978. Laser microbeam study of a rotary motor in termite flagellates. *J. Cell Biol.* **78**:76–92.

Tamm, S. L. 1979. Membrane movements and fluidity during rotational motility of a termite flagellate. *J. Cell Biol.* **80**:141–149.

Tamm, S. L., and Tamm, S. 1974. Direct evidence for fluid membranes. *Proc. Natl. Acad. Sci. USA* **71**:4589–4593.

Tamm, S. L., and Tamm, S. 1976. Rotary movements and fluid membranes in termite flagellates. *J. Cell Sci.* **20**:619–639.

Tawara, J. 1965. The root of flagella of *Vibrio cholerae*. *Jpn. J. Microbiol.* **9**:49–54.

To, L. P., and Margulis, L. 1978. Ancient locomotion: Prokaryotic motility systems. *Int. Rev. Cytol.* **54**:267–293.

Tulloch, G. S., and Hershenov, B. R. 1967. Fine structure of platyhelminth sperm tails. *Nature* (London) **213**:299–300.

Turner, F. R. 1968. An ultrastructural study of plant spermatogenesis. Spermatogenesis in *Nitella*. *J. Cell Biol.* **37**:370–393.

van Iterson, W., Hoeniger, J. F. M., and van Zanten, E. N. 1966. Basal bodies of bacterial flagella in *Proteus mirabilis*. I. Electron microscopy of sectioned material. *J. Cell Biol.* **31**:585–601.

Vary, P. S., and Stocker, B. A. D. 1973. Nonsense motility mutants in *Salmonella typhimurium*. *Genetics* **73**:229–245.

von Bonsdorff, C. H., and Telkkä, A. 1965. The spermatozoon flagella in *Diphyllobothrium laterum* (fish tapeworm). *Z. Zellforsch. Mikrosk. Anat.* **66**:643–648.

Warner, F. D. 1978. Cation-induced attachment of ciliary dynein cross-bridges. *J. Cell Biol.* **77**:R19–R26.

Warner, F. D., and Mitchell, D. R. 1978. Structural conformation of ciliary dynein arms and the generation of sliding forces in *Tetrahymena* cilia. *J. Cell Biol.* **76**:261–277.

Warner, F. D., and Satir, P. 1974. The structural basis of ciliary bend formation. Radial spoke positional changes accompanying microtubule sliding. *J. Cell Biol.* **63**:35–63.

Warner, F. D., Mitchell, D. R., and Perkins, C. R. 1977. Structural conformation in the ciliary ATPase dynein. *J. Mol. Biol.* **114**:367–384.

Williams, H. R., Verwey, W. F., Schrank, G. D., and Hurry, E. K. 1973. An *in vitro* antigen–antibody reaction in relation to a hypothesis of intestinal immunity to cholera. *Proc. Joint Cholera Res. Conf.*, U.S. Department of State, pp. 161–173.

Willmer, E. N. 1955. The physiology of vision. *Annu. Rev. Physiol.* **17**:339–366.

Witman, G. B., Carlson, K., Berliner, J., and Rosenbaum, J. L. 1972. *Chlamydomonas* flagella. I. Isolation and electrophoretic analysis of microtubules, matrix, membranes, and mastigonemes. *J. Cell. Biol.* **54**:507–539.

Witman, G. B., Fay, R., and Plummer, J. 1976. *Chlamydomonas* mutants: Evidence for the roles of specific axonemal components in flagellar movement. *In*: Goldman, R., Pollard, T., and Rosenbaum, J., eds., *Cell Motility*, Cold Spring Harbor, N.Y., Cold Spring Harbor Laboratory, Book C, pp. 969–986.

Witman, G. B., Plummer, J., and Sander, G. 1978. *Chlamydomonas* flagellar mutants lacking radial spokes and central tubules. *J. Cell Biol.* **76**:729–747.

Yamaguchi, S., Iino, T., Horiguchi, T., and Ohta, K. 1972. Genetic analysis of *fla* and *mot* cistrons closely linked to *H1* in *Salmonella abortuesequi* and its derivatives. *J. Gen. Microbiol.* **70**:59—75.

Yang, G. C. H., Schrank G. D., and Freeman, B. A. 1977. Purification of flagellar cores of *Vibrio cholerae*. *J. Bacteriol.* **129**:1121–1128.

Zanetti, N. C., Mitchell, D. R., and Warner, F. D. 1979. Effects of divalent cations on dynein cross bridging and ciliary microtubule sliding. *J. Cell Biol.* **80**:573–588.

CHAPTER 5

Abel, J. H. 1969. Electron microscopic demonstration of ATP-hydrolase activity in herring gull salt glands. *J. Histochem. Cytochem.* **17**:570–584.

Anderson, E. 1955. The electron microscopy of *Trichomonas muris*. *J. Protozool.* **2**:114–124.

Anderson, E., and Dumont, J. N. 1966. A comparative study of the concrement vacuole of certain endocommensal ciliates—a so-called mechanoreceptor. *J. Ultrastruct. Res.* **15**:414–450.

Argetsinger, J. 1965. The isolation of ciliary basal bodies from *Tetrahymena*. *J. Cell. Biol.* **24**:154–157.

Beams, H. W., and Sekhon, S. S. 1969. Further studies on the fine structure of *Lophomonas blat-*

tarum with special reference to the so-called calyx axial filament and parabasal body. *J. Ultrastruct. Res.* **26**:296–315.

Bell, P. R. 1974a. The origin of the multilayered structure in the spermatozoid of *Pteridium aquilinum*. *Cytobiologie* **8**:203–212.

Bell, P. R. 1974b. Microtubules in relation to flagellogenesis in *Pteridium* spermatozoids. *J. Cell Sci.* **15**:99–111.

Bell, P. R., Duckett, J. G., and Myles, D. 1971. The occurrence of a multilayered structure in the motile spermatozoids of *Pteridium aquilinum*. *J. Ultrastruct. Res.* **34**:181–189.

Berry, S. J., and Johnson, E. 1975. Formation of temporary flagellar structures during insect organogenesis. *J. Cell Biol.* **65**:498–492.

Birbeck, J. E., Stewart, K. D., and Mattox, K. R. 1974. The cytology and classification of *Schizomeris leibleinii* (Chlorophyceae). II. The structure of quadriflagellate zoospores. *Phycologia* **13**:71–79.

Blackburn, G. R., Barrau, M. D., and Dewey, W. C. 1978. Partial purification of centrosomes from Chinese hamster ovary cells. *Exp. Cell Res.* **113**:183–187.

Bradbury, P., and Pitelka, D. R. 1965. Observations on kinetosome formation in an apostome ciliate. *J. Microsc. (Paris)* **4**:805–810.

Brinkley, B. R., and Stubblefield, E. 1970. Ultrastructure and interaction of the kinetochore and centriole in mitosis and meiosis. *Adv. Cell Biol.* **1**:119–185.

Brown, D. L., Massalski, A., and Patenaude, R. 1976. Organization of the flagellar apparatus and associated cytoplasmic microtubules in the quadriflagellate alga *Polytomella agilis*. *J. Cell Biol.* **69**:106–125.

Carothers, Z. B., and Duckett, J. G. 1977. A comparative study of the multilayered structure in developing bryophyte spermatozoids. *Bryophyt. Biblio.* **13**:95–112.

Carothers, Z. B., and Kreitner, G. L. 1967. Studies of spermatogenesis in the Hepaticae. I. Ultrastructure of the *Vierergruppe* in *Marchantia*. *J. Cell Biol.* **33**:43–57.

Carothers, Z. B., and Kreitner, G. L. 1968. Studies of spermatogenesis in the Hepaticae. II. Blepharoplast structure in the spermatid of *Marchantia*. *J. Cell Biol.* **36**:603–616.

Clark, T. B., and Wallace, F. G. 1960. A comparative study of kinetoplast ultrastructure in the Trypanosomatidae. *J. Protozool.* **7**:115–124.

de Harven, E., and Bernhard, W. 1956. Étude au microscope electronique de l'ultrastructure du centriole chez les vertébrés. *Z. Zellforsch. Mikrosk. Anat.* **45**:378–398.

Dingle, A. D., and Fulton, C. 1966. Development of the flagellar apparatus of *Naegleria*. *J. Cell Biol.* **31**:43–54.

Dippell, R. V. 1968. The development of basal bodies in *Paramecium*. *Proc. Natl. Acad. Sci. USA* **61**:461–468.

Dippell, R. V. 1976. Effects of nuclease and protease digestion on the ultrastructure of *Paramecium* basal bodies. *J. Cell Biol.* **69**:622–637.

Dirksen, E. R., and Crocker, T. T. 1966. Centriole replication in differentiating ciliated cells of mammalian respiratory epithelium. An E. M. study. *J. Microsc. (Paris)* **5**:629–644.

Dodge, J. D., and Crawford, R. M. 1968. Fine structure of the dinoflagellate *Amphidinium casteri* Hulbert. *Protistologica* **4**:231–242.

Duckett, J. G., and Bell, P. R. 1969. The occurrence of a multilayered structure in the sperm of a pteridophyte. *Planta* **89**:203–211.

Duckett, J. G., and Bell, P. R. 1977. An ultrastructural study of the mature spermatozoid of *Equisetum*. *Philos. Trans. R. Soc. London Ser. B* **277**:131–158.

DuPraw, E. J. 1968. *Cell and Molecular Biology*, New York, Academic Press.

Engelman, T. W. 1882. Über Licht- und Farbenperzeption niederster Organismen. *Pflüegers Arch. Gen. Physiol.* **29**:387–400.

Favard, P., Carasso, N., and Fauré-Fremiet, E. 1963. Ultrastructure de l'appareil adhésif des urceolaires (Ciliés péritriches). *J. Microsc. (Paris)* **2**:337–368.

Fawcett, D. W., 1961. Cilia and flagella. *In*: Brachet, J., and Mirsky, A. E., eds., *The Cell*, New York, Academic Press, Vol. 2, pp. 217–297.

Friedländer, M., and Wahrman, J. 1966. Giant centrioles in neuropteran meiosis. *J. Cell Sci.* **1**:129–144.

Friedmann, I., Colwin, A. L., and Colwin, L. H. 1968. Fine-structural aspects of fertilization in *Chlamydomonas reinhardii*. *J. Cell Sci.* **3**:115–128.

Frisch, D. 1967. Fine structure of the early differentiation of ciliary basal bodies. *Anta. Rec.* **157**:245.

Fulton, C. 1971. Centrioles. *In*: Reinert, J., and Ursprung, H., eds., *Origin and Continuity of Cell Organelles*, New York, Springer-Verlag, pp. 170–221.

Fulton, C., and Dingle, A. D. 1971. Basal bodies, but not centrioles in *Naegleria*. *J. Cell Biol.* **51**:826–836.

Gall, J. G. 1961. Centriole replication. A study of spermatogenesis in the snail *Viviparus*. *J. Biophys. Biochem. Cytol.* **10**:163–193.

Gibbons, I. R. 1961. The relationship between the fine structure and direction of beat in gill cilia of a lamellibranch mollusc. *J. Biophys. Biochem. Cytol.* **11**:179–205.

Gibbons, I. R., and Grimstone, A. V. 1960. On flagellar structure in certain flagellates. *J. Biophys. Biochem. Cytol.* **7**:697–716.

Gould, R. R. 1975. The basal bodies of *Chlamydomonas reinhardtii*. *J. Cell Biol.* **65**:65–74.

Grim, J. N. 1966. Isolated ciliary structures of *Euplotes patella*. *Exp. Cell Res.* **41**:206–226.

Grim, J. N. 1972. Fine structure of the surface and infraciliature of *Gastrostyla steinii*. *J. Protozool.* **19**:113–126.

Grimes, G. W. 1973a. Origin and development of kinetosomes in *Oxytricha fallax*. *J. Cell Sci.* **13**:43–53.

Grimes, G. W. 1973b. Morphological discontinuity of kinetosomes during the life cycle of *Oxytricha fallax*. *J. Cell Biol.* **57**:229–232.

Harris, P., and Mazia, D. 1962. The finer structure of the mitotic apparatus. *In*: Harris, R. J. C., ed. *The Interpretation of Ultrastructure*, New York, Academic Press, pp. 279–305.

Hartman, H. 1975. The centriole and the cell. *J. Theor. Biol.* **51**:501–509.

Hartmann, J. F. 1964. Cytochemical localization of ATPase in the mitotic apparatus. *J. Cell Biol.* **23**:363–370.

Heidemann, S. R., and Kirschner, M. W. 1975. Aster formation in eggs of *Xenopus laevis*: Induction by isolated basal bodies. *J. Cell Biol.* **67**:105–117.

Heidemann, S. R., Sander, G., and Kirschner, M. W. 1977. Evidence for a functional role of RNA in centrioles. *Cell* **10**:337–350.

Heitz, E. 1959. Elektronenmikroskopische Untersuchungen über zwei auffallende Strukturen an der Geisselbasis der Spermatiden von *Marchantia*. *Z. Naturforsch.* **14b:**399–401.

Henneguy, L. F. 1897. Sur les rapports des cils vibratiles avec les centrosomes. *Arch. Anat. Microsc.* **1**:481–496.

Hepler, P. K. 1976. The blepharoplast of *Marsilea*: Its *de novo* formation and spindle association. *J. Cell Sci.* **21**:361–390.

Hepler, P. K., and Myles, D. G. 1977. Spermatogenesis in *Marsilea*: an example of male gamete development in plants. *In*: Brinkley, B. R., and Porter, K. R., eds., *International Cell Biology, 1976–1977,* New York, Rockefeller University Press, pp. 569–579.

Hill, G. C., and Anderson, W. A. 1969. Effects of acriflavine on the mitochondria and kinetoplast of *Crithidia fasciculata*. *J. Cell Biol.* **41**:547–561.

Hoffman, E. J. 1965. The nucleic acids of basal bodies isolated from *Tetrahymena*. *J. Cell Biol.* **25**:217–228.

Joyon, L. 1963. Contribution a l'étude cytologique de quelques protozoaires flagellés. *Ann. Fac. Sci. Univ. Clermont* **22**:1–96.

Krijgsman, B. J. 1925. Beiträge zum Problem der Geisselbewegung. *Arch. Protistenkd.* **52**:478–488.

Kristiansen, J., and Walne, P. H. 1977. Fine structure of photokinetic systems in *Dinobryon cylindricum* var *alpinum* (Chrysophyceae). *Br. Phycol. J.* **12**:329–341.

Lang, N. J. 1963. An additional ultrastructural component of flagella. *J. Cell Biol.* **19**:631–634.

Lansing, A. I., and Lamy, F. 1961. Fine structure of the cilia of rotifers. *J. Biophys. Biochem. Cytol.* **4**:799–812.

Leedale, G. F. 1967. *Euglenoid Flagellates*, Englewood Cliffs, N.J., Prentice–Hall.

Lembi, C.A., and Herndon, W. R. 1966. Fine structure of the pseudocilia of *Tetraspora. Can. J. Bot.* **44**:710–712.

Lembi, C. A., and Lang, N. J. 1965. Electron microscopy of *Carteria* and *Chlamydomonas. Am. J. Bot.* **52**:464–477.

Lenhossek, M. 1898. Über Flimmerzellen. *Verh. Dtsch. Anat. Ges. Jena* **12**:106–128.

Lessie, P. E., and Lovett, J. S. 1968. Ultrastructural changes during sporangium formation and zoospore differentiation in *Blastocladiella emersonii. Am. J. Bot.* **55**:220–236.

Lom, J., and Corliss, J. O. 1971. Morphogenesis and cortical ultrastructure of *Brooklynella hostilis*, a dysteriid ciliate ectoparasitic on marine ciliates. *J. Protozool.* **18**:261–281.

Manton, I. 1964. Die feinere Struktur von *Pedinomonas minor*. II. Electron microscopical investigation. *Nova Hedwigia Z. Kryptogamenkd.* **3**:441–451.

Mizukami, I., and Gall, J. 1966. Centriole replication. II. Sperm formation in the fern, *Marsilea*, and the cycad, *Zamia. J. Cell Biol.* **44**:454–458.

Outka, D. E., and Kluss, B. C. 1967. The ameba-to-flagellate transformation in *Tetramitus rostratus*. II. Microtubular morphogenesis. *J. Cell Biol.* **35**:323–346.

Paolillo, D. J. 1965. On the androcyte of *Polytrichum*, with special reference to the *Dreiergruppe* and limosphere (nebenkern). *Can. J. Bot.* **43**:669–676.

Parke, M., and Manton, I. 1965. Preliminary observations on the fine structure of *Prasinocladus marinus. J. Mar. Biol. Assoc. U.K.* **45**:525–536.

Pedersen, H. 1969. Microtubules in the spermatid of the rabbit. *Z. Zellforsch. Mikrosk. Anat.* **98**:148–156.

Phillips, D. M. 1966. Observations on spermiogenesis in the fungus gnat *Sciara coprophila. J. Cell Biol.* **30**:477–497.

Phillips, D. M. 1967. Giant centriole formation in *Sciara. J. Cell Biol.* **33**:73–92.

Phillips, D. M. 1970. Insect sperm: Their structure and morphogenesis. *J. Cell Biol.* **44**:243–277.

Phillips, S. G., and Rattner, J. B. 1976. Dependence of centriole formaton on protein systhesis. *J. Cell Biol.* **70**:9–19.

Pitelka, D. R. 1961. Observations on the kinetoplast-mitochondrion and the cytostome of *Bodo. Exp. Cell Res.* **25**:87–93.

Randall, J. T., and Disbrey, C. 1965. Evidence for the presence of DNA at basal body sites in *Tetrahymena. Proc. R. Soc. London Ser. B* **162**:473–491.

Renaud, F. L., and Swift, H. 1964. The development of basal bodies and flagella in *Allomyces arbusculus. J. Cell Biol.* **23**:339–354.

Roland, J. C., Szöllösi, A., and Szöllösi, D. 1977. *Atlas of Cell Biology*, Boston, Little, Brown.

Roth, L. E., and Shigenaka, Y. 1964. The structure and formation of cilia and filaments in rumen protozoa. *J. Cell Biol.* **20**:249–270.

Satir, P., and Rosenbaum, J. L. 1965. The isolation and identification of kinetosome-rich fractions from *Tetrahymena. J. Protozool.* **12**:397–405.

Seaman, G. R. 1960. Large-scale isolation of kinetosomes from *Tetrahymena. Exp. Cell Res.* **21**:292–302.

Seaman, G. R. 1962. Protein synthesis by kinetosomes isolated from *Tetrahymena. Biochim. Biophys. Acta* **55**:889–899.

Simpson, P. A., and Dingle, A. D. 1971. Variable periodicity in the rhizoplast of *Naegleria* flagellates. *J. Cell Biol.* **51**:323–328.

Smith-Sonneborn, J., and Plaut, W. 1967. Evidence for the presence of DNA in the pellicle of *Paramecium. J. Cell Sci.* **2**:225–234.

Sorokin, S. P. 1968. Reconstructions of centriole formation and ciliogenesis in mammalian lungs. *J. Cell Sci.* **3**:207–230.

Steinman, R. 1968. An electron microscopic study of ciliogenesis in developing epidermis and trachea in embryo of *Xenopus laevis*. *Am. J. Anat.* **122**:19–56.

Stewart, J. M., and Muir, A. R. 1963. The fine structure of the cortical layers in *Paramecium aurelia*. *Q. J. Microsc. Sci.* **104**:129–134.

Stewart, K. D., Mattox, K. R., and Chandler, C. D. 1974. Mitosis and cytokinesis in *Platymonas subcordiformis*, a scaly green monad. *J. Phycol.* **10**:65–69.

Stubblefield, E., and Brinkley, B. R. 1967. Architecture and function of the mammalian centriole. *Symp. Int. Soc. Cell Biol.* **6**:175–218.

Suyama, Y., and Preer, J. R. 1965. Mitochondrial DNA from Protozoa. *Genetics* **52**:1051–1058.

Swale, E. M. F., and Belcher, J. H. 1973. A light and electron microscope study of the colorless flagellate *Aulacomonas skuja*. *Arch. Mikrobiol.* **92**:91–103.

Szöllösi, D. 1964. The structure and function of centrioles and their satellites in the jellyfish *Phialidium gregarium*. *J. Cell Biol.* **21**:465–479.

Turner, F. R. 1968. An ultrastructural study of plant spermatogenesis. Spermatogenesis in *Nitella*. *J. Cell Biol.* **37**:370–393.

Went, H. A. 1977. Can a reverse transcriptase be involved in centriole duplication? *J. Theor. Biol.* **68**:95–100.

Wheatley, D. N. 1967. Cilia and centrioles of the rat adrenal cortex. *J. Anat.* **101**:223–237.

CHAPTER 6

Abreu, I., and Santos, A. 1977. Intimate association between endoplasmic reticulum and plastids during microsporogenesis in *Lycopersicum esculentum* Mill. and *Solanum tuberosum* L. *J. Submicrosc. Cytol.* **9**:239–246.

Alley, C. D., and Scott, J. L. 1977. Unusual dictyosome morphology and vesicle formation in tetrasporangia of the marine red alga *Polysiphonia denudata*. *J. Ultrastruct. Res.* **58**:289–298.

Anttinen, H., Oikarinen, A., Ryhänen, L., and Kivirikko, K. I. 1978. Evidence for the transfer of mannose to the extension peptides of procollagen within the cisternae of the rough ER. *FEBS Lett.* **87**:222–226.

Arion, W. J., Wallin, B. K. Carlson, W., and Lange, A. J. 1972. The specificity of glucose 6-phosphatase of intact liver microsomes. *J. Biol. Chem.* **247**:2558–2565.

Avers, C. J. 1967. Distribution of cytochrome *c* peroxidase activity in wild-type and petite cells of Baker's yeast grown aerobically and anaerobically. *J. Bacteriol.* **94**:1225–1235.

Bainton, D. F., and Farquhar, M. G. 1966. Origin of granules in polymorphonuclear leukocytes. Two types derived from opposite faces of the Golgi complex in developing granulocytes. *J. Cell Biol.* **28**:277–301.

Beams, H. W., and Kessel, R. G. 1968. The Golgi apparatus: Structure and function. *Int. Rev. Cytol.* **23**:209–276.

Behnke, H. D., and Eschlbeck, G. 1979. Dilated cisternae in Capparales—An attempt towards the characterization of a specific endoplasmic reticulum. *Protoplasma* **97**:351–363.

Belcher, J. H. 1969. Some remarks upon *Mallomonas papillosa* Harris and Bradley and *M. calceolus* Bradley. *Nova Hedwigia Z. Kryptogamenkd.* **18**:257–270.

Bergeron, J. J. M., Posner, B. I., Josefsberg, Z., and Sikstrom, R. 1978. Intracellular polypeptide hormone receptors. *J. Biol. Chem.* **253**:4058–4066.

Berthold, W.-U. 1978. Ultrastrukturanalyse der endoplasmatischen Algen von *Amphistegina lessonii* d'Orbigny, Foraminifera (Protozoa) und ihre systematische Stellung. *Arch. Protistenkd.* **120**:16–62.

Birken, S., Smith, D. L. Canfield, R. E., and Boime, I. 1977. Partial amino acid sequence of human placental lactogen precursor and its mature hormone form produced by membrane-associated enzyme activity. *Biochem. Biophys. Res. Commun.* **74**:106–112.

Bisalputra, T. 1966. Electron microscope study of the protoplasmic continuity in certain brown algae. *Can. J. Bacteriol.* **44**:89–93.

Bisalputra, T., and Stein, J. R. 1966. The development of cytoplasmic bridges in *Volvox aureus*. *Can. J. Bot.* **44**:1697–1702.

Blobel, G., and Dobberstein, B. 1975. Transfer of proteins across membranes. I. Presence of proteolytically processed and unprocessed nascent immunoglobulin light chains on membrane-bound ribosomes of murine myeloma. *J. Cell Biol.* **67**:835–851.

Blobel, G., and Sabatini, D. D. 1971. Ribosome–membrane interaction in eukaryotic cells. *Biomembranes* **2**:193–195.

Bond, H. E., Cooper, J. A., Courington, D. P., and Wood, S. S. 1969. Microsome-associated DNA. *Science* **165**:705–706.

Boublik, M., and Hellmann, W. 1978. Comparison of *Artemia salina* and *E. coli* ribosome structure by electron microscopy. *Proc. Natl. Acad. Sci. USA* **75**:2829–2833.

Bouck, G. 1962. Chromatophore development, pits, and other fine structure in the red alga, *Lomentaria baileyana*. *J. Cell Biol.* **12**:553–569.

Bouck, G. 1965. Fine structure and organelle associations in brown algae. *J. Cell Biol.* **26**:523–537.

Bowles, D. J., and Northcote, D. H. 1972. The site of synthesis and transport of extracellular polysaccharides in the root tissues of maize. *Biochem. J.* **130**:1133–1145.

Bowles, D. J., and Northcote, D. H. 1974. The amount and rates of export of polysaccharides found within the membrane system of maize root cells. *Biochem. J.* **142**:139–144.

Branton, D., and Moor, H. 1964. Fine structure in freeze-etched *Allium cepa* L. root tips. *J. Ultrastruct. Res.* **11**:401–411.

Brown, D. 1978. Fenestrae in the rough ER of *Xenopus laevis* hepatocytes. *Anat. Rec.* **191**:103–110.

Brown, D. L., Massalski, A., and Patenaude, R. 1976a. Organization of the flagellar apparatus and associated cytoplasmic microtubules in the quadriflagellate alga, *Polytomella agilis*. *J. Cell Biol.* **69**:106–125.

Brown, D. L., Leppard, G. G., and Massalski, A. 1976b. Fine structure of encystment of the quadriflagellate alga, *Polytomella agilis*. *Protoplasma* **90**:139–154.

Brown, R. M. 1969. Observations on the relationship of the Golgi apparatus to wall formation in the marine chrysophycean alga, *Pleurochrysis scherffelii*. *J. Cell Biol.* **41**:109–123.

Brown, R. M., Franke, W. W., Kleinig, H., Falk, H., and Sitte, P. 1969. Cellulosic wall component produced by the Golgi apparatus of *Pleurochrysis scherffelii*. *Science* **166**:894–896.

Brown, W. V., and Bertke, E. M. 1974. *Textbook of Cytology*, 2nd ed., St. Louis, Mosby.

Bruni, C., and Porter, K. R. 1965. The fine structure of the parenchymal cell of the normal rat liver. *Am. J. Pathol.* **46**:691–755.

Buvat, R. 1958a. Nouvelles observations sur l'appareil de Golgi dans les cellules de végétaux vasculaires. *C. R. Acad. Sci. Paris* **D246**:2157–2160.

Buvat, R. 1958b. Recherches sur les infrastructures du cytoplasme, dans les cellules du méristème apical des ebauches foliares et des feuilles développées d'*Elodea canadensis*. *Ann. Sci. Nat. Bot.* **19**:121–161.

Buvat, R. 1961. Le réticulum endoplasmique des cellules végétales. *Ber. Dtsch. Bot. Ges.* **74**:261–267.

Camefort, H. 1970. Particularités structurales du gamète femelle chez la *Cryptomeria*. Formation de complexes plastes-réticulum pendant la période du maturation du gamète. *C. R. Acad. Sci. Paris* **D271**:49–52.

Campbell, P. N., and Blobel, G. 1976. The role of organelles in the chemical modification of the primary translation products of secretory proteins. *FEBS Lett.* **72**:215–226.

Carasso, N., Ovtracht, L., and Favard, F. 1971. Observation en microscopie électronique à haute tension de l'appareil de Golgi sur coupes de 0.5 à 5μ d'épasseur. *C. R. Acad. Sci. Paris* **D273**:876–879.

Cardell, R. R. 1977. Smooth endoplasmic reticulum in rat hepatocytes during glycogen deposition and depletion. *Int. Rev. Cytol.* **48**:221–279.

Cardelli, J., Long, B. H., and Pitot, H. C. 1976. Direct association of mRNA labeled in the presence of fluoroacetate with membranes of the endoplasmic reticulum in rat liver. *J. Cell Biol.* **70**:47–58.

Cardelli, J., Long, B. H., and Pitot, H. C. 1978. An examination of rat liver smooth ER membranes for the presence of associated mRNA. *Arch. Biochem. Biophys.* **191**:632–637.

Caro, L. G., and Palade, G. E. 1964. Protein synthesis, storage and discharge in the pancreatic exocrine cell. *J. Cell Biol.* **20**:473–495.

Chatterjee, S. K., Kazemie, M., and Matthaei, H. 1973. Studies on rabbit reticulocyte ribosomes. II. Separation of the ribosomal proteins by two-dimensional electrophoresis. *Hoppe-Seyler's Z. Physiol. Chem.* **354**:481–486.

Chedid, A., and Nair, V. 1972. Diurnal rhythm in endoplasmic reticulum of rat liver: Electron microscopic study. *Science* **175**:176–179.

Chrispeels, M. J. 1977. The role of the endoplasmic reticulum in the biosynthesis and transport of macromolecules in plant cells. *In*: Brinkley, B. R., and Porter, K. R., eds., *International Cell Biology, 1976–1977*, New York, Rockefeller University Press, pp. 284–292.

Christensen, A. K., and Fawcett, D. W. 1960. Fine structure of testicular interstitial cells in the opossum. *Anat. Rec.* **136**:333.

Christensen, A. K., and Fawcett, D. W. 1961. The normal fine structure of opossum testicular interstitial cell. *J. Biophys. Biochem. Cytol.* **9**:653–670.

Christie, K. N., and Stoward, P. J. 1979. Catalase in skeletal muscle fibers. *J. Histochem. Cytochem.* **27**:814—819.

Clark, C. C., and Kefalides, N. A. 1976. Carbohydrate moieties of procollagen: Incorporation of isotopically labeled mannose and glucosamine into propeptides of procollagen secreted by matrix free chick embryo tendon cells. *Proc. Natl. Acad. Sci. USA* **73**:34–38.

Clark-Walker, G. D., and Linnane, A. W. 1967. The biogenesis of mitochondria in *Saccharomyces cerevisiae*. *J. Cell Biol.* **34**:1–14.

Cleveland, L. R., and Grimstone, A. V. 1964. The fine structure of the flagellate *Mixotricha paradoxa* and its associated micro-organisms. *Proc. R. Soc. London* **B159**:668–686.

Clowes, F. A. L., and Juniper, B. E. 1968. *Plant Cells*, Oxford, Blackwell.

Conti, S. F., and Brock, T. D. 1965. Electron microscopy of cell fusion in conjugating *Hansenula wingei*. *J. Bacteriol.* **90**:524–533.

Coombs, J., Lauritis, J. A., Darley, W. M., and Volcani, B. E. 1968. Studies on the biochemistry and fine structure of silica shell formation in diatoms. *Z. Pflanzenphysiol.* **58**:124–152.

Cornick, G. G., and Kretsinger, R. H. 1977. The 30S subunit of the *E. coli* ribosomes. *Biochim. Biophys. Acta* **474**:398–410.

Coughlan, S. 1977. Isolation and characterization of a fucosyl transferase associated with dictyosomes from the brown alga *Fucus serratus* L. *FEBS Lett.* **81**:33–36.

Coughlan, S., and Evans, L. V. 1978. Isolation and characterization of Golgi bodies from vegetative tissue of the brown alga *Fucus serratus*. *J. Exp. Bot.* **29**:55–68.

Cunningham, W. P., Morré, D. J., and Mollenhauer, H. H. 1966. Structure of isolated plant Golgi apparatus revealed by negative staining. *J. Cell Biol.* **28**:169–179.

Dallman, P. R., Dallner, G., Bergstrand, A., and Ernster, L. 1969. Heterogeneous distribution of enzymes in submicrosomal membrane fragments. *J. Cell Biol.* **41**:357–377.

Dallner, G., Siekevitz, P., and Palade, G. E. 1966. Biogenesis of endoplasmic reticulum membranes. II. Synthesis of constitutive microsomal enzymes in developing rat hepatocytes. *J. Cell Biol.* **30**:97–117.

Dalton, A. J. 1961. Golgi apparatus and secretion granules. *In*: Brachet, J., and Mirsky, A. E., eds., *The Cell*, New York, Academic Press, Vol. II, pp. 603–619.

Daniels, E. W. 1964. Origin of the Golgi system in amoebae. *Z. Zellforsch. Mikrosk. Anat.* **64**:38–51.

Daniels, E. W., and Roth, L. E. 1961. X-irradiation of the giant amoeba, *Pelomyxa illinoisensis*. III. Electron microscopy of centrifuged organisms. *Radiat. Res.* **14**:66–82.

DePierre, J. W., and Dallner, G. 1975. Structural aspects of the membrane of the endoplasmic reticulum. *Biochim. Biophys. Acta* **415**:411–422.

DeRobertis, E., and Sabatini, D. D. 1960. Submicroscopic analysis of the secretory process in the adrenal medulla. *Fed. Proc. Fed. Am. Soc. Exp. Biol.* **19**, Suppl. **5**:70–78.

Diers, L. 1966. On the plastids, mitochondria, and other cell constituents during oogenesis of a plant. *J. Cell Biol.* **28**:527–543.

Dillon, L. S. 1978. *The Genetic Mechanism and the Origin of Life*, New York, Plenum Press.

Di Orio, J., and Millington, W. F. 1978. Dictyosome formation during reproduction in colchicine-treated *Pediastrum boryanum*. *Protoplasma* **97**:329–336.

Dodge, J. D., and Crawford, R. M. 1968. Fine structure of the dinoflagellate *Amphidinium casteri* Hulbert. *Protistologica* **4**:231–242.

Donaldson, R. P., and Beevers, H. 1978. Organelle membranes from germinating castor bean endosperm. I. A comparison of purified endoplasmic reticulum, glyoxysomes, and mitochondria. *Protoplasma* **97**:317–327.

Drawert, H., and Mix, M. 1963. Licht- und elektronenmikroskopische Untersuchungen an Desmidaceen. XI. Die Struktur von Nucleolus und Golgi-Apparat bei *Micrasterias denticulata*. *Port. Acta Biol.* **7**:17–28.

Duksin, D., and Bornstein, P. 1977. Impaired conversion of procollagen to collagen by fibroblasts and bone treated with tunicamycin, and inhibitor of protein glycosylation. *J. Biol. Chem.* **252**:955–962.

Dute, R. R., and Evert, R. F. 1977. Sieve-element ontogeny in the root of *Equisetum hyemale*. *Am. J. Bot.* **64**:421–438.

Dykstra, M. J. 1978. Ultrastructure of the genus *Schizoplasmodiopsis* (Protostelia). *J. Protozool.* **25**:44–49.

Elliott, A. M., and Bak, I. J. 1964. The contractile vacuole and related structures in *Tetrahymena pyriformis*. *J. Protozool.* **11**:250–261.

Elliott, A. M., and Zieg, R. G. 1968. The Golgi apparatus associated with mating in *Tetrahymena pyriformis*. *J. Cell Biol.* **36**:391–398.

Erdmann, V. A. 1980. Collection of published 5 S and 5.8 S RNA sequences and their precursors. *Nucl. Acids Res.* **8**:r31–r48.

Ericsson, J. 1966. On the fine structural demonstration of glucose-6-phosphatase. *J. Histochem. Cytochem.* **14**:361–362.

Fabre, J., Fayallas, M., Richailley, G., and Lecal, J. 1969. Ultrastructure des caelomocytes des quelques échinodermes. *Bull. Soc. Hist. Nat. Toulouse.* **105**:234–262.

Fawcett, D. W. 1955. Observations on the cytology and electron microscopy of hepatic cells. *J. Natl. Cancer Inst.* **15**:1475–1503.

Fisher, D. G., and Evert, R. F. 1979. Endoplasmic reticulum–dictyosome involvement in the origin of refractive spherules in sieve elements of *Davallia fijiensis* Hook. *Ann. Bot. (London)* **43**:255–258.

Flickinger, C. J. 1968a. The effects of enucleation on the cytoplasmic membranes of *Amoeba proteus*. *J. Cell Biol.* **37**:300–315.

Flickinger, C. J. 1968b. Cytoplasmic alterations in amebae treated with actinomycin D. *Exp. Cell Res.* **53**:241–251.

Flickinger, C. J. 1969a. The development of Golgi complexes and their dependence upon the nucleus in amoebae. *J. Cell Biol.* **43**:250–262.

Flickinger, C. J. 1969b. Fenestrated cisternae in the Golgi apparatus of the epididymis. *Anat. Rec.* **163**:39–54.

Flickinger, C. J. 1969c. The pattern of growth of the Golgi complex during the fetal and postnatal development of the rat epididymis. *J. Ultrastruct. Res.* **27**:344–364.

Flickinger, C. J. 1978. The pattern of appearance of enzymic activity during the development of the Golgi apparatus in amoebae. *J. Cell Sci.* **34**:53–63.

Freygang, W. H. 1965. Tubular ionic movement. *Fed. Proc. Fed. Am. Soc. Exp. Biol.* **24**:1135–1140.

Frey-Wyssling, A., López-Sáez, J. F., and Mühlethaler, K. 1964. Formation and development of the cell plate. *J. Ultrastruct. Res.* **10**:422–432.

Friend, D. S. 1969. Cytochemical staining of multivesicular body and Golgi vesicles. *J. Cell Biol.* **41**:269–279.

Frink, R., Krupp, P. P., and Young, R. A. 1978. The parathyroid gland of the woodchuck: A study of seasonal variation in the chief cells. *Tissue Cell* **10**:259–268.

Furthmayer, H., Timpl, R., Stark, M., Lapière, C. M., and Kühn, K. 1972. Chemical properties of the peptide extension in the *pα*1 chain of dermatosparactic skin procollagen. *FEBS Lett.* **28**:247–250.

Gantt, E., and Arnott, H. J. 1965. Spore germination and development of the young gametophyte of the ostrich fern (*Matteuccia struthiopteris*). *Am. J. Bot.* **52**:82–94.

Gantt, E., and Conti, S. E. 1965. The ultrastructure of *Porphyridium cruentum*. *J. Cell Biol.* **26**:365–381.

García, M. L., Benavides, J., Valdivieso, F., Mayor, F., Giménez-Gallego, G. 1978. Cytochromes in the rat kidney brush border membrane. *Biochem. Biophys. Res. Commun.* **82**:738–744.

Garg, B., Khandekar, J., Tuchweber, B., and Kovacs, K. 1971. Effect of D-actinomycin on phenobarbital induced ultrastructural changes in rat hepatocytes. *Pathol. Eur.* **6**:152–160.

Gatenby, J. B. 1960. Notes on the gametogenesis of a pulmonate mollusc. *La Cellule* **60**:289–303.

Gaye, P., Hue, D., Mercier, J. C., and Haze, G. 1979. Enzymatic processing of precursors of ovine lactoproteins by mammary microsomal membranes and a deoxycholate-soluble extract of rough microsomes. *FEBS Lett.* **101**:137–142.

Gemmell, R., and Stacy, B. D. 1979. Ultrastructural study of granules in the corpora lutea of several mammalian species. *Am. J. Anat.* **155**:1–14.

Giulian, D., and Diacumakos, E. G. 1976. The study of intercellular compartments by micropipette techniques. *J. Cell Biol.* **70**:332a.

Goldsmith, P. K., and Stetten, M. R. 1979. Different developmental changes in latency for two functions of a single membrane-bound enzyme. *Biochim. Biophys. Acta* **583**:133–147.

Grassé, P. P. 1957. Ultrastructure, polarité et reproduction de l'appareil de Golgi. *C. R. Acad. Sci. Paris* **D245**:1278–1281.

Grimstone, A. V. 1959. Cytoplasmic membranes and the nuclear membrane in the flagellate *Trichomympha*. *J. Biophys. Biochem. Cytol.* **6**:369–378.

Grove, S. N., Bracker, C. E., and Morré, D. J. 1968. Cytomembrane differentiation in the endoplasmic reticulum–golgi apparatus–vesicle complex. *Science* **161**:171–173.

Grunberg-Manago, M., Buckingham, R. H., Cooperman, B. S., and Hershey, J. W. B. 1978. Structure and function of the translation machinery. *Symp. Soc. Gen. Microbiol.* **28**:27–110.

Hall, W. T., and Witkus, E. R. 1964. Some effects on the ultrastructure of the root meristem of *Allium cepa* by 6-azouracil. *Exp. Cell Res.* **36**:494–501.

Hand, A. R. 1971. Morphology and cytochemistry of the Golgi apparatus of rat salivary gland acinar cells. *Am. J. Anat.* **130**:141–157.

Harwood, R., Grant, M. E., and Jackson, D. S. 1976. The route of secretion of procollagen. The influence of $\alpha\alpha'$-bipyridyl colchicine, and antimycin A on the secretion processes in embryonic chick tendon and cartilage cells. *Biochem. J.* **156**:81–90.

Hawkins, E. K. 1974a. Golgi vesicles of uncommon morphology and wall formation in the red alga, *Polysiphonia*. *Protoplasma* **80**:1–14.

Hawkins, E. K. 1974b. Growth and differentiation of the Golgi apparatus in the red alga, *Callithamnion roseum*. *J. Cell Sci.* **14**:633–655.

Hayward, A. F. 1967. Changes in fine structure of developing intestinal epithelium associated with pinocytosis. *J. Anat.* **102**:57–70.

Hedley, R. H., and Bertaud, W. S. 1962. Electron-microscopic observations of *Gromia oviformis* (Sarcodina). *J. Protozool.* **9**:79–87.

Held, W. A., Mizushima, S., and Nomura, M. 1973. Reconstitution of *E. coli* 30S ribosomal subunits from purified molecular components. *J. Biol. Chem.* **248**:5720–5730.

Hibberd, D. J., and Leedale, G. F. 1972. Observations on the cytology and ultrastructure of the new algal class, Eustigmatophyceae. *Ann. Bot.* **36**:49–71.

Higgins, J. A. 1974. The site of phospholipid synthesis in the initial phase of membrane proliferation. *J. Cell Biol.* **62**:635–646.

Higgins, J. A., and Barrnett, R. 1972. Studies on the biogenesis of smooth reticulum membranes in livers of phenobarbital-treated rats. *J. Cell Biol.* **55**:282–298.

Hildebrand, H. F. 1978. Electron microscopic investigation with developmental stages of the trophozoite of *Didymorphyes gigantea*. II. The fine structure of the deutomerite with emphasis on the nuclear division and on Golgi apparatus. *Z. Parasitenkd.* **55**:9–28.

Hinman, N. D., and Phillips, A. H. 1970. Similarity and limited multiplicity of membrane proteins from rough and smooth endoplasmic reticulum. *Science* **170**:1222–1223.

Hino, Y., Asano, A., and Sato, R. 1978a. Biochemical studies on rat liver Golgi apparatus. II. Further characterization of isolated Golgi fraction. *J. Boichem.* **83**:925–934.

Hino, Y., Asano, A., and Sato, R. 1978b. Biochemical studies on rat liver Golgi apparatus. III. Subfractionation of fragmented Golgi apparatus by counter-current distribution. *J. Biochem.* **83**:935–942.

Hohl, H. R., and Hamamoto, S. T. 1967. Ultrastructural changes during zoospore formation in *Phylophthora parasitica*. *Am. J. Bot.* **54**:1131–1139.

Hohl, H. R., Hamamoto, S. T., and Hemmes, D. E. 1968. Ultrastructural aspects of cell elongation, cellulose synthesis, and spore differentiation in *Acytostelium leptosomum*, a cellular slime mold. *Am. J. Bot.* **55**:783–796.

Huebner, E., and Anderson, E. 1976. Comparative spiralian oogenesis—Structural aspects: An overview. *Am. Zool.* **16**:315–343.

Huynh-van Tan, Delaunay, J., and Schapira, G. 1971. Eukaryotic ribosomal proteins. Two-dimensional electrophoretic studies. *FEBS Lett.* **17**:163–167.

Ito, S., and Winchester, R. J. 1963. The fine structure of the gastric mucosa in the bat. *J. Cell Biol.* **16**:541–577.

Jacob, J., and Jurand, A. 1963. Electron microscope studies on salivary gland cells of *Bradysia mycorum* Frey (Sciaridae). III. The structure of cytoplasm. *J. Ins. Physiol.* **9**:849–857.

Jamieson, J. D., and Palade, G. E. 1967a. Intracellular transport of secretory proteins in the pancreatic exocrine cell. I. Role of the peripheral elements of the Golgi complex. *J. Cell Biol.* **34**:577–596.

Jamieson, J. D., and Palade, G. E. 1967b. Intracellular transport of secretory proteins in the pancreatic exocrine cell. II. Transport to condensing vacuoles and zymogen granules. *J. Cell Biol.* **34**:597–615.

Jelsema, C. L., and Morré, D. J. 1978. Distribution of phospholipid biosynthetic enzymes among cell components of rat liver. *J. Biol. Chem.* **253**:7960–7971.

Jensen, W. A. 1965. The composition and ultrastructure of the nucellus in cotton. *J. Ultrastruct. Res.* **13**:112–128.

Jones, R. L., and Chen, R. 1976. Immunohistochemical localization of α-amylase in barley aleurone cells. *J. Cell Sci.* **20**:183–198.

Joyon, L. 1963. Contribution a l'étude cytologique de quelques protozoaires flagellées. *Ann. Fac. Sci. Univ. Clermont* **22**:1–83.

Kahan, D., Oren, R., Aaronson, S., and Behrens, V. 1978. Fine structure of the cell surface and Golgi apparatus of *Ochromonas*. *J. Protozool.* **25**:30–33.

Kayser, C., Petrovic, A., and Porte, A. 1961. Variations ultrastructurales de la parathyroide du hamster ordinaire au cours du cycle saisonnier. *C. R. Soc. Biol. Paris* **155**:2178–2181.

Kiermayer, O. 1970. Elektronmikroskopische Untersuchungen zum Problem der Cytomorphogenese von *Micrasterias denticulata*. *Protoplasma* **69**:97–132.

Kiselev, N. A., Stel'mashchuk, V. Y., Lerman, M. I., and Abakumova, O. Y. 1974. On the structure of liver ribosomes. *J. Mol. Biol.* **86**:577–586.

Kivirikko, K. I., and Risteli, L. 1976. Biosynthesis of collagen and its alterations in pathological states. *Med. Biol.* **54**:159–186.

Knudson, C. M., Stemberger, B. H., and Patton, S. 1978. Effects of colchicine on ultrastructure of the lactating mammary cell: Membrane involvement and stress on the Golgi apparatus. *Cell Tissue Res.* **195**:169–181.

Kolata, G. B. 1975. Ribosomes. I. Genetic studies with viruses. *Science* **190**:136, 183–184.

Kreibich, G., Freienstein, C. M., Pereyra, B. N., Ulrich, B. L., and Sabatini, D. D. 1978a. Proteins of rough microsomal membranes related to ribosome binding. II. Cross-linking of bound ribosomes to specific membrane proteins exposed at the binding sites. *J. Cell Biol.* **77**:488–506.

Kreibich, G., Ulrich, B. L., and Sabatini, D. D. 1978b. Proteins of rough microsomal membranes related to ribosome binding. I. Identification of ribophorins I and II, membrane proteins characteristic of rough microsomes. *J. Cell Biol.* **77**:464–487.

Kristen, U. 1978. Ultrastructure and a possible function of the intercisternal elements in dictyosomes. *Planta* **138**:29–34.

Kumagai, H., Igarashi, K., Tanaka, K., Nakao, H., and Hirose, S. 1979. A microsomal exoribonuclease from rat liver. *Biochim. Biophys. Acta* **566**:192–199.

Kupila-Ahvenniemi, S., Pihakaski, S., and Pihakaski, K. 1978. Wintertime changes in the ultrastructure and metabolism of the microsporangiate strobili of the Scotch pine. *Planta* **144**:19–29.

Kurland, C. G. 1977. Structure and function of the bacterial ribosome. *Annu. Rev. Biochem.* **46**:173–200.

Kwiatkowska, M., and Maszewski, J. 1979. Changes in the activity of the Golgi apparatus during the cell cycle in antheridial filaments of *Chara vulgaris* L. *Protoplasma* **99**:31–38.

LaClaire, J. W., and West, J. A. 1978. Light and electron-microscopic studies of growth and reproduction in *Cutleria* (Phaeophyta). I. Gametogenesis in the female plant of *C. hancocki*. *Protoplasma* **97**:93–110.

Lake, J. A. 1976. Ribosome structure determined by electron microscopy of *E. coli* small subunits, large subunits, and monomeric ribosomes. *J. Mol. Biol.* **165**:131–159.

Lake, J. A., Sabatini, D. D., and Nonomura, Y. 1974. Ribosome structure as studied by electron microscopy. *In:* Nomura, M., Tissières, A., and Lengyei, P., eds., *Ribosomes,* Cold Spring Harbor, N.Y., Cold Spring Harbor Laboratory, pp. 543–557.

Lande, M. A., Adesnik, M., Sumida, M., Tashiro, Y., and Sabatini, D. D. 1975. Direct association of mRNA with microsomal membranes in human diploid fibroblasts. *J. Cell Biol.* **65**:513–528.

Lane, N. J. 1968. Distribution of phosphatases in the Golgi region and associated structures of the thoracic ganglionic neurons in the grasshopper, *Melanoplus diffentialis*. *J. Cell Biol.* **37**:89–104.

Lang, N. J. 1963. Electron microscopy of the Volvaceae and Astrephomenaceae. *Am. J. Bot.* **50**:280–300.

Lange, L. and Olson, L. W. 1977. The zoospore of *Phlyctochytrium aestuarii*. *Protoplasma* **93**:27–43.

Lange, L., and Olson, L. W. 1978. The zoospore of *Synchytrium endobioticum*. *Can. J. Bot.* **56**:1229–1239.

Leadbeater, B. S. C. 1979. Developmental studies on the loricate choanoflagellate *Stephanoeca diplocostata* Ellis. *Protoplasma* **98**:241–262.

Leblond, C. P., and Bennett, G. 1977. Role of the Golgi apparatus in terminal glucosylation. *In:* Brinkley, B. R., and Porter, K. R., eds., *International Cell Biology, 1976–1977,* New York, Rockefeller University Press, pp. 326–336.

Leedale, G. F. 1966. Euglena: A new look with the electron microscope. *Adv. Sci.* **107**:22–37.

Leedale, G. F. 1967. *Euglenoid Flagellates,* Englewood Cliffs, N.J., Prentice–Hall, pp. 1–242.

Lehle, L., Bowles, D. J., and Tanner, W. 1978. Subcellular site of mannosyl transfer to dolichyl phosphate in *Phaseolus aureus. Plant Sci. Lett.* **11**:27–34.

Lentz, T. L. 1965. The fine structure of differentiating interstitial cells in *Hydra. Z. Zellforsch. Mikrosk. Anat.* **67**:546–560.

Lentz, T. L., and Barrnett, R. J. 1965. Fine structure of the nervous system of *Hydra. Am. Zool.* **5**:341–356.

Leskes, A., Siekevitz, P., and Palade, G. 1971. Differentiation of endoplasmic reticulum in hepatocytes. I. Glucose-6-phosphatase distribution *in situ. J. Cell Biol.* **49**:264–287.

Linnane, A. W., Vitols, E., and Nowland, P. G. 1962. Studies on the origin of yeast mitochondria. *J. Cell Biol.* **13**:345–350.

Loening, U. E. 1968. Molecular weights of ribosomal RNA in relation to evolution. *J. Mol. Biol.* **38**:355–365.

McNitt, R. E., and Shen-Miller, J. 1978. Quantitative assessment of ultrastructural changes in primary roots of corn (*Zea mays* L.) after geotropic stimulation. I. Root cap. *Plant Physiol.* **61**:644–648.

Manton, I. 1964. Observations with the electron microscope on the division cycle in the flagellate *Prymnesium parvum* Carter. *J. R. Microsc. Soc.* **83**:317–325.

Manton, I. 1966a. Observations on scale production in *Prymnesium parvum. J. Cell Sci.* **1**:375–380.

Manton, I. 1966b. Observations on scale production in *Pyramimonas amylifera. J. Cell Sci.* **1**:429–438.

Manton, I. 1967. Further observations on the fine structure of *Chrysochromulina chiton* with special reference to the haptonema 'peculiar' Golgi structure and scale production. *J. Cell Sci.* **2**:265–272.

Manton, I., and Harris, K. 1966. Observations on the microanatomy of the brown flagellate *Sphaleromantis tetragona* Skuja with special reference to flagellar apparatus and scales. *J. Linn. Soc. London Bot.* **59**:397–403.

Manton, I., and Parke, M. 1965. Observations on the fine structure of two species of *Platymonas* with special references to flagellar scales and the mode of origin of the theca. *J. Mar. Biol. Assoc. U.K.* **45**:743–754.

Marquardt, H. 1962. Der Feinbau von Hefezellen in Electronenmikroskop. II. *Saccharomyces cerevisiae. Z. Naturforsch.* **17**b:689–695.

Martini, O. H. W., and Gould, H. J. 1971. Enumeration of rabbit reticulocyte ribosomal proteins. *J. Mol. Biol.* **62**:403–405.

Massalski, A., and Leedale, G. F. 1969. Cytology and ultrastructure of the Xanthophyceae. *Br. Phycol. J.* **4**:159–180.

Mattern, C. F., Park, H. D., and Daniel, W. A. 1965. Electron microscope observations on the structure and discharge of the stenotele of *Hydra. J. Cell Biol.* **27**:621–638.

Mattox, K. R., and Stewart, K. D. 1974. A comparative study of cell division in *Trichosarcina polymorpha* and *Pseudendoclonium basiliense* (Chlorophyceae). *J. Phycol.* **10**:447–456.

Mattox, K. R., Stewart, K. D., and Floyd, G. L. 1974. The cytology and classification of *Schizomeris leibleinii* (Chlorophyceae). I. The vegetative thallus. *Phycologia* **13**:63–69.

Maurer, R. A., Gorski, J., and McKean, D. J. 1977. Partial amino acid sequence of rat pre-prolactin. *Biochem. J.* **161**:189–192.

Ménard, D., Penasse, W., Drochmans, P., and Hugon, J. S. 1974a. Glucose-6-phosphatase heterogeneity within the hepatic lobule of the phenobarbital-treated rat. *Histochemistry* **38**:229–239.

Ménard, D., Berteloot, A., and Hugon, J. S. 1974b. Action of phenobarbital on the ultrastructure and enzymatic activity of the mouse intestine and the mouse liver. *Histochemistry* **38**:241–252.

Mercier, J. C., Haze, G., Gaye, P., and Hue, D. 1978. Amino terminal sequence of the precursor of ovine β-lactoglobulin. *Biochem. Biophys. Res. Commun.* **82**:1236–1245.

Meyer, M., Bout, W. S., deVries, M., and Nanninga, N. 1974. Electron microscopic and sedimentation studies on rat-liver ribosomal subunits. *Eur. J. Biochem.* **42**:259–268.

Mignot, J. P. 1965. Étude ultrastructurale de Eugléniens. II. A. Dictyosomes et dictyocinèse chez *Distigma proteus*. B. Mastigonèmes chez *Anisonemo costatum*. *Protistologica* **1**(2):17–22.

Miller, E. J. 1976. Biochemical characteristics and biological significance of the genetically-distinct collagens. *Mol. Cell. Biochem.* **13**:165–192.

Milstein, C., Brownlee, G. C., Harrison, T. M., and Mathews, M. B. 1972. A possible precursor of immunoglobulin light chains. *Nature New Biol.* **239**:117–120.

Mims, C. W., and Rogers, M. A. 1973. An ultrastructural study of spore germination in the myxomycete, *Stemonitis virginiensis*. *Protoplasma* **78**:243–254.

Moestrup, Ø., and Walne, P. L. 1979. Studies on scale morphogenesis in the Golgi apparatus of *Pyramimonas tetrarhynchus* (Prasinophyceae). *J. Cell Sci.* **36**:437–459.

Mollenhauer, H. H. 1974. Distribution of microtubules in the Golgi apparatus of *Euglena gracilis*. *J. Cell Sci.* **15**:89–97.

Mollenhauer, H. H., and Mollenhauer, B. A. 1978. Changes in the secretory activity of the Golgi apparatus during the cell cycle in root tips of maize. *Planta* **138**:113–118.

Mollenhauer, H. H., and Morré, D. J. 1966. Golgi apparatus and plant secretion. *Annu. Rev. Plant Physiol.* **17**:27–46.

Mollenhauer, H. H., and Morré, D. J. 1978a. Structural compartmentation of the cytosol: Zones of exclusion, zones of adhesion, cytoskeletal and intercisternal elements. *Subcell. Biochem.* **5**:327–359.

Mollenhauer, H. H., and Morré, D. J. 1978b. Structural differences contrast higher plant and animal Golgi apparatus. *J. Cell Sci.* **32**:357–362.

Mollenhauer, H. H., and Whaley, W. G. 1963. An observation on the functioning of the Golgi apparatus. *J. Cell Biol.* **17**:222–225.

Mollenhauer, H. H., Evans, W., and Kogut, C. 1968. Dictyosome structure in *Euglena gracilis*. *J. Cell Biol.* **37**:579–583.

Moor, H., and Mühlethaler, K. 1963. Fine structure in frozen-etched yeast cells. *J. Cell Biol.* **17**:609–628.

Moore, R. T., and McAlear, J. H. 1963. Fine structure of Mycota. 4. The occurrence of the Golgi dictyosome in the fungus *Neobulgaria pura*. *J. Cell Biol.* **16**:131–141.

Moore, T. S., Lord, J. M., Kagawa, T., and Beevers, H. 1973. Enzymes of phospholipid metabolism in the endoplasmic reticulum of castor bean endosperm. *Plant Physiol.* **52**:50–53.

Morré, D. J., and Ovtracht, L. 1977. Dynamics of the Golgi apparatus. Membrane differentiation and membrane flow. *Int. Rev. Cytol. Suppl.* **5**:61–188.

Morré, D. J., Mollenhauer, H. H., and Bracker, C. E. 1971. Origin and continuity of Golgi apparatus. *In:* Reinert, J., and Ursprung, H., eds., *Origin and Continuity of Cell Organelles,* New York, Springer–Verlag, pp. 82–126.

Murphy, W. H., von der Mark, K., McEneany, L. S. G., and Bornstein, P. 1975. Characterization of procollagen-derived peptides unique to the precursor molecule. *Biochemistry* **14**:3243–3250.

Nair, V., and Casper, R. 1969. The influence of light on daily rhythm in hepatic drug metabolizing enzymes in rat. *Life Sci.* **8**:1291–1298.

Nair, V., Siegel, S., and Bau, D. 1970. Regulation of the diurnal rhythm in hepatic drug metabolism. *Fed. Proc. Fed. Am. Soc. Exp. Biol.* **29**:804.

Nessler, C. L., and Mahlberg, P. G. 1977. Ontogeny and cytochemistry of alkaloidal vesicles in lacticifers of *Papaver somniferum* L. (Papaveraceae). *Am. J. Bot.* **64**:541–551.

Netzel, H. 1975a. Die Entstehung der hexagonalen Schalenstruktur bei der Thekamöbe *Arcella vulgaris* var *multinucleata*. *Arch. Protistenkd.* **117**:321–357.

Netzel, H. 1975b. Morphologie und Ultrastruktur von *Centropyxis discoides*. *Arch. Protistenkd.* **117**:369–392.

Neutra, M., and Leblond, C. P. 1966a. Synthesis of the carbohydrate of mucus in the Golgi complex as shown by electron microscope radioautography of goblet cells from rats injected with glucose-H^3. *J. Cell Biol.* **30**:119–136.

Neutra, M., and Leblond, C. P. 1966b. Radioautographic comparison of the uptake of galactose-H^3 and glucose-H^3 in the Golgi region of various cells secreting glycoproteins or mucopolysaccharides. *J. Cell Biol.* **30**:137–150.

Nichols, B. 1968. A membranous apparatus in a flagellate, *Tritrichomonas augusta*. *J. Cell Biol.* **37**:559–563.

Noll, H. 1970. Organelle integration and the evolution of ribosome structure and function. *Symp. Soc. Exp. Biol.* **24**:419–447.

Northcote, D. H., and Pickett-Heaps, J. D. 1966. A function of the Golgi apparatus in polysaccharide synthesis and transport in the root-cap cells of wheat. *Biochem. J.* **98**:159–167.

Novikoff, A. B. 1976. The endoplasmic reticulum. A cytochemist's view (A review). *Proc. Natl. Acad. Sci. USA* **73**:2781–2787.

Novikoff, A. B., and Goldfischer, S. 1961. Nucleoside diphosphatase activity in the Golgi apparatus and its usefulness for cytological studies. *Proc. Natl. Acad. Sci. USA* **47**:802–810.

Novikoff, A. B., and Heus, M. 1963. A microsomal nucleoside diphosphatase. *J. Biol. Chem.* **238**:710–716.

Novikoff, A. B., Essner, E., Goldfischer, S., and Heus, M. 1962. Nucleoside-phosphatase activities of cytomembranes. *In:* Harris, R. J. C., ed., *The Interpretation of Ultrastructure,* New York, Academic Press, Vol. 1, pp. 149–192.

Novikoff, A. B., Beard, M. E., Albala, A., Sheid, B., Quintana, N., and Biemprica, L. 1971a. Localization of endogenous peroxidases in animal tissues. *J. Microsc. (Paris)* **12**:381–404.

Novikoff, P., Novikoff, A., Quintana, N., and Hauro, J. 1971b. Golgi apparatus, GERL, and lysosomes of neurons in rat dorsal root ganglia, studied by thick section and thin section cytochemistry. *J. Cell Biol.* **50**:859–886.

Novikoff, A. B., Novikoff, P. M., Ma, M., Shin, W. Y., and Quintana, N. 1974. Cytochemical studies of secretory and other granules associated with the ER in rat thyroid epithelial cells. *Adv. Cytopharmacol.* **2**:349–368.

Nunez, E. A., Whalen, J. P., and Krook, L. 1972. An ultrastructural study of the natural secretory cycle of the parathyroid gland of the bat. *Am. J. Anat.* **134**:459–480.

Oikarinen, A., Anttinen, H., and Kivirikko, K. I. 1976a. Hydroxylation of lysine and glycosylation of hydroxylysine during collagen biosynthesis in isolated chick-embryo cartilage cells. *Biochem. J.* **156**:545–551.

Oikarinen, A., Anttinen, H., and Kivirikko, K. I. 1976b. Effect of L-azetidine-2-carboxylic acid on glycosylations of collagen in chick-embryo tendon cells. *Biochem. J.* **160**:639–645.

Oikarinen, A., Anttinen, H., and Kivirikko, K. I. 1977. Further studies on the effect of the collagen triple-helix formation on the hydroxylation of lysine and the glycosylations of hydroxylysine in chick-embryo tendon and cartilage cells. *Biochem. J.* **166**:357–362.

Olsen, B. R., Guzman, N. A., Engel, J., Condit, C., and Aase, S. 1977. Purification and characterization of a peptide from the carboxy-terminal region of chick tendon procollagen type I. *Biochemistry* **16**:3030–3036.

Olson, L. W., Lange, L., and Reichle, R. 1978. The zoospore and meiospore of the aquatic phycomycete *Catenaria anguillulae*. *Protoplasma* **94**:53–71.

Orrenius, S. 1965a. On the mechanism of drug hydroxylation in rat liver microsomes. *J. Cell Biol.* **26**:713–724.

Orrenius, S. 1965b. Further studies on the induction of the drug-hydroxylating enzyme system of liver microsomes. *J. Cell Biol.* **26**:725–733.

Orrenius, S., Ericsson, J., and Ernster, L. 1965. Phenobarbital-induced synthesis of the microsomal drug-metabolizing enzyme system and its relationship to the proliferation of endoplasmic membranes. A morphological and biochemical study. *J. Cell Biol.* **25**:627–639.

Paavola, L. G. 1978a. The corpus luteum of the guinea pig. II. Cytochemical studies on the Golgi complex and lysosomes in luteal cells during maximal progesterone secretion. *J. Cell Biol.* **79**:45–58.

Paavola, L. G. 1978b. The corpus luteum of the guinea pig. III. Cytochemical studies on the Golgi complex and GERL during normal postpartum regression of luteal cells emphasizing the origin of lysosomes and autophagic vacuoles. *J. Cell Biol.* **79**:59–73.

Palade, G. E. 1975. Intracellular aspects of the process of protein synthesis. *Science* **189**:347–357.

Parry, G. 1978. Membrane assembly and turnover. *Subcell. Biochem.* **5**:261–336.

Peachey, L. D. 1965. The sarcoplasmic reticulum and transverse tubules of the frog's sartorius muscle. *J. Cell Biol.* **25**:209–233.

Peek, W. D., and McMillan, D. B. 1979. Ultrastructure of the tubular nephron of the garter snake *Thamnophis sirtalis. Am. J. Anat.* **154**:103–128.

Peek, W. D., and Youson, J. H. 1979a. Ultrastructure of chloride cells in young adults of the anadromous sea lamprey, *Petromyzon marinus L.*, in fresh water and during adaptation to sea water. *J. Morphol.* **160**:143–164.

Peek, W. D., and Youson, J. H. 1979b. Transformation of the interlamellar epithelium of the gills of the anadromous sea lamprey, *Petromyzon marinus L.*, during metamorphosis. *Can. J. Zool.* **57**:1318–1332.

Pene, J. J., Knight, E., and Darnell, J. E. 1968. Characterization of a new low molecular weight RNA in HeLa cell ribosomes. *J. Mol. Biol.* **33**:609–624.

Perry, J. W., and Evert, R. F. 1975. Structure and development of the sieve elements in *Psilotum nudum. Am. J. Bot.* **62**:1038–1052.

Pickett-Heaps, J. D. 1967. Further observations on the Golgi apparatus and its functions in cells of the wheat seedling. *J. Ultrastruct. Res.* **18**:287–303.

Pickett-Heaps, J. D. 1968a. Ultrastructure and differentiation in *Chara (fibrosa)*. IV. Spermatogenesis. *Aust. J. Biol. Sci.* **21**:655–690.

Pickett-Heaps, J. D. 1968b. Further ultrastructural observations on polysaccharide localization in plant cells. *J. Cell Sci.* **3**:55–64.

Pickett-Heaps, J. D. 1968c. Ultrastructure and differentiation in *Chara* sp. III. Formation of the antheridium. *Aust. J. Biol. Sci.* **21**:255–274.

Pitelka, D. R. 1963. *Electron-Microscopic Structure of Protozoa*, New York, Macmillan.

Policard, A., Bessis, M., Breton, J., and Thiéry, J. P. 1958. Polarité de la centrosphere et des corps de Golgi dans les leucocytes les mammifères. *Exp. Cell Res.* **14**:221–223.

Porter, K. R. 1961a. The ground substance: Observations from electron microscopy. *In:* Brachet, J., and Mirsky, A. E., eds., *The Cell,* New York, Academic Press, Vol. 2, pp. 621–675.

Porter, K. R. 1961b. *In:* Goodwin, T. W., and Lindberg, O., eds., *Biological Structure and Function*, New York, Academic Press, Vol. 1, pp. 127–155.

Porter, K. R. 1961c. The sarcoplasmic reticulum: Its recent history and present status. *J. Biophys. Biochem. Cytol.* **10**:219–225.

Porter, K. R., and Bruni, C. 1959. An electron microscope study of the early effects of 3′-Me-DAB on rat liver cells. *Cancer Res.* **19**:997–1009.

Porter, K. R., and Palade, G. E. 1957. Studies of the endoplasmic reticulum. III. Its form and distribution in striate muscle. *J. Biophys. Biochem. Cytol.* **3**:269–300.

Porter, K. R., and Yamada, E. 1960. Studies on the endoplasmic reticulum. V. Its form and differentiation in pigment epithelial cells of the frog retina. *J. Biophys. Biochem. Cytol.* **7**:167–180.

Posner, B. I., Josefsberg, Z., and Bergeron, J. J. M. 1978a. Intracellular polypeptide hormone receptors. *J. Biol. Chem.* **253**:4067–4073.

Posner, B. I., Raquidan, D., Josefsberg, Z., and Bergeron, J. J. M. 1978b. Different regulation of insulin receptors in intracellular (Golgi) and plasma membranes from livers of obese and lean mice. *Proc. Natl. Acad. Sci. USA* **75**:3302–3306.

Rambourg, A., and Chrétien, M. 1970. L'appareil de Golgi: examen en microscopie électronique de coupes épaisses (0.5–1μ) après impregnation des tissues par le tetroxyde d'osmium. *C. R. Acad. Sci. Paris* **D270**:981–983.

Rambourg, A., Hernandez, W., and Leblond, C. P. 1969. Detection of complex carbohydrates in the Golgi apparatus of rat cells. *J. Cell Biol.* **40**:395–414.

Rambourg, A., Clermont, Y., and Marraud, A. 1974. Three dimensional structure of the osmium-impregnated Golgi apparatus as seen in the high voltage electron microscope. *Am. J. Anat.* **140**:27–46.

Rambourg, A., Clermont, Y., and Hermo, L. 1979. Three dimensional architecture of the Golgi appartus in Sertoli cells of the rat. *Am. J. Anat.* **154**:455–476.

Rawson, J. R. Y., and Stutz, E. 1969. Isolation and characterization of *Euglena* cytoplasmic and chloroplast ribosomes and their ribosomal RNA components. *Biochim. Biophys. Acta* **190**:368–380.

Reid, E. 1967. Membrane systems. *In:* Roodyn, D. B., ed., *Enzyme Cytology*, New York, Academic Press, pp. 321–406.

Remmer, H., and Merker, H. 1963. Drug-induced changes in the liver endoplasmic reticulum: Association with drug-metabolizing enzymes. *Science* **142**:1657–1658.

Renaudin, S., and Capdepon, M. 1977. Association of the endoplasmic reticulum and the plastids in *Tozzia alpina* L. scale leaves. *J. Ultrastruct. Res.* **61**:303–308.

Risteli, L., Myllylä, R., and Kivirikko, K. I. 1976. Partial purification and characterization of collagen galactosyltransferase from chick embryos. *Biochem. J.* **155**:145–153.

Rubin, G. M. 1973. The nucleotide sequence of 5.8S ribosomal RNA species in *S. cerevisiae*. *Eur. J. Biochem.* **41**:197–202.

Sabatini, D. D., Bensch, K. G., and Barrnett, R. J. 1963. Cytochemistry and electron microscopy. *J. Cell Biol.* **17**:19–58.

Santos, A., Abreu, I., and Salema, R. 1979. Elaborate system of RER and degenerescence of tapetum during pollen development in some dicotyledons. *J. Submicrosc. Cytol.* **11**:99–107.

Schnepf, E. 1961. Quantitative Zusammenhänge zwischen der Sekretion des Fangschleimes und den Golgi-Strukturen bei *Drosophyllum lusitanicum*. *Z. Naturforsch.* **16b**:605–610.

Schnepf, E. 1968a. Transport by compartments. *In*: Mothes, K., Muller, E., Nelles, A., and Neumann, D., eds., *Transport and Distribution of Matter in Higher Plants,* Berlin, Akademie-Verlag, pp. 39–49.

Schnepf, E. 1968b. Zur Feinstruktur der schleimsezernierenden Drüsenhaare auf der Ochrea von *Rumex* und *Rheum*. *Planta* **79**:22–34.

Schnepf, E. 1969. Über den Feinbau von Öldrüsen. II. Die Drüsenhaare in *Calceolaria* Blüten. *Protoplasma* **67**:195–203.

Schnepf, E., and Deichgräber, G. 1969. Über die Feinstruktur von *Synura petersenii* unter besonderer Berücksichtigung der Morphogenese ihrer Kieselschuppen. *Protoplasma* **68**:85–106.

Shen-Miller, J., and McNitt, R. E. 1978. Quantitative assessment of ultrastructural changes in primary roots of corn after geotropic stimulation. II. Curving and noncurving zones of the root proper. *Plant Physiol.* **61**:649–653.

Sherton, C. C., and Wool, I. G. 1972. Determination of the number of proteins in liver ribosomes and ribosomal subunits by two-dimensional polyacrylamide gel electrophoresis. *J. Biol. Chem.* **247**:4460–4467.

Sherton, C. C., and Wool, I. G. 1974. A comparison of the proteins of rat skeletal muscle and liver ribosomes by two-dimensional polyacrylamide gel electrophoresis. *J. Biol. Chem.* **249**:2258–2267.

Sjöstrand, F. S., and Hanzon, V. 1954. Ultrastructure of Golgi apparatus of exocrine cells of mouse pancreas. *Exp. Cell Res.* **7**:415–429.

Stelten, M., and Ghosh, S. 1971. Different properties of glucose-6-phosphatase and related enzymes in rough and smooth endoplasmic reticular membranes. *Biochim. Biophys. Acta* **233**:163–175.

Stoeckel, M. E., and Porte, A. 1973. Observations ultrastructurales sur la parathyroide de mammifères et d'oiseau dans les conditions normales et experimentales. *Arch. Anat. Murosi* **62**:55–88.

Stoermer, E. F., Pankratz, H. S., and Bowen, C. C. 1965. Fine structure of the diatom *Amphipleura pellucida*. *Am. J. Bot.* **52**:1067–1078.

Stöffler, G., and Wittmann, H. G. 1977. Primary structure and three-dimensional arrangement of proteins within the *E. coli* ribosome. *In:* Weissbach, H., and Pestka, S., eds., *Molecular Mechanisms of Protein Synthesis,* New York, Academic Press, pp. 117–202.

Strauss, A. W., Bennett, C. A., Donohue, A. M., Rodkey, J. A., Boime, I., and Alberts, A. W. 1978. Conversion of rat pre-proalbumin to proalbumin *in vitro* by ascites membranes. *J. Biol. Chem.* **253**:6270–6274.

Stutz, E., and Noll, H. 1967. Characterization of cytoplasmic and chloroplast polysomes in plants: Evidence for three classes of rRNA in nature. *Proc. Natl. Acad. Sci. USA* **57**:774–781.

Szczesna, E., and Boime, I. 1976. mRNA-dependent synthesis of authentic precursor to human placental lactogen: Conversion to its mature hormone form in ascites cell-free extracts. *Proc. Natl. Acad. Sci. USA* **73**:1179–1183.

Taira, K. 1979. Studies on intramitochondrial inclusions in the pancreatic acinar cells of the Japanese newt, *Triturus pyrrogaster*. *J. Ultrastruct. Res.* **67**:89–94.

Theron, J. J., Biagio, R., Meyer, A. C., and Boekkooi, S. 1979. Microfilaments, the smooth endoplasmic reticulum and synaptic ribbon fields in the pinealocytes of the baboon (*Papio ursinus*). *Am. J. Anat.* **154**:151–162.

Threadgold, L. T. 1976. *The Ultrastructure of the Animal Cell,* 2nd ed., New York, Pergamon Press.

Thyberg, J., Nilsson, S., Moskalewski, S., and Hinek, A. 1977. Effect of colchicine on the Golgi complex and lysosomal system of chondrocytes in monolayer culture. *Cytobiologie Z. Exp. Zellforsch.* **15**:175–191.

Tissières, A. 1974. Ribosome research: Historical background. *In:* Nomura, M., Tissières, A., and Lengyel, P., eds., *Ribosomes*, Cold Spring Harbor, N.Y., Cold Spring Harbor Laboratory, pp. 3–12.

Tripodi, G., and DeMasi, F. 1977. The post-fertilization stages of red algae: The fine structure of the fusion cell of *Erythrocystis*. *J. Submicrosc. Cytol.* **9**:389–401.

Tripodi, G., and DeMasi, F. 1978. A possible vestige of a flagellum in the fusion cell of the red algae *Erythrocystis montagnei*. *J. Submicrosc. Cytol.* **10**:435–439.

Tsukahara, T., and Yamada, M. 1965. Cytological structure of *Aspergillus niger* by electron microscopy. *Jpn. J. Microbiol.* **9**:35–48.

Turner, F. R., and Whaley, W. G. 1965. Intercisternal elements of the Golgi apparatus. *Science* **147**:1303–1304.

Van Holde, K. E., and Hill, W. E. 1974. General physical properties of ribosomes. *In:* Nomura, M., Tissières, A., and Lengyel, P., eds., *Ribosomes,* Cold Spring Harbor, N.Y., Cold Spring Harbor Laboratory, pp. 53–91.

Ward, R. T., and Ward, E. 1968. The multiplication of Golgi bodies in the oocytes of *Rana pipiens*. *J. Microsc. (Paris)* **7**:1007–1020.

Warner, J. R. 1971. The assembly of ribosomes in yeast. *J. Biol. Chem.* **246**:447–454.

Warshawsky, H., Leblond, C. P., and Droz, B. 1963. Synthesis and migration of proteins in the cells of the exocrine pancreas as revealed by the specific activity determination from radioautographs. *J. Cell Biol.* **16**:1–23.

Weinberg, R. A., and Penman, S. 1970. Processing of 45S nucleolar RNA. *J. Mol. Biol.* **47**:169–178.

Weisblum, B., Herman, L., and Fitzgerald, P. J. 1962. Changes in pancreatic acinar cells during protein deprivation. *J. Cell Biol.* **12**:313–327.

Welfle, H., Stahl, J., and Bielka, H. 1971. Studies on proteins of animal ribosomes. VIII. Two-dimensional polyacrylamide gel electrophoresis of ribosomal proteins of rat liver. *Biochim. Biophys. Acta* **243**:416–419.

Welfle, H., Stahl, J., and Bielka, H. 1972. Studies on proteins of animal ribosomes. XIII. Enumeration of ribosomal proteins of rat liver. *FEBS Lett.* **26**:228–232.

Wetherbee, R., and West, J. A. 1976. Unique Golgi apparatus and vesicle formation in a red alga. *Nature (London)* **259**:566–567.

Wetherbee, R., and West, J. A. 1977. Golgi apparatus of unique morphology during early carposporogenesis in a red alga. *J. Ultrastruct. Res.* **58**:119–133.

Whaley, W. G. 1975. *The Golgi Apparatus,* Cell Biology Monograph, Vol. 2, New York, Springer–Verlag.

Whaley, W. G., and Mollenhauer, H. H. 1963. The Golgi apparatus and cell plate formation—A postulate. *J. Cell Biol.* **17**:216–221.

Whaley, W. G., Kephart, J. E., and Mollenhauer, H. H. 1959. Developmental changes in the Golgi apparatus of maize root cells. *Am. J. Bot.* **46**:743–751.

Whaley, W. G., Kephart, J. E., and Mollenhauer, H. H. 1964. The dynamics of cytoplasmic membranes during development. *In:* Locke, M., ed., *Cellular Membranes in Development,* New York, Academic Press, pp. 135–173.

Whaley, W. G., Dauwalder, M., and Kephart, J. E. 1972. Golgi apparatus: Influence on cell surfaces. *Science* **175**:596–599.

Whur, P., Herscovics, A., and Leblond, C. P. 1969. Radioautographic visualization of the incorporation of galactose-^{3}H and mannose-^{3}H by rat thyroids *in vitro* in relation to the stages of thyroglobulin synthesis. *J. Cell Biol.* **43**:289–311.

Wischnitzer, S. 1962. An electron microscopic study of the Golgi apparatus of amphibian oocytes. *Z. Zellforsch. Mikrosk. Anat.* **57**:202–212.

Wittmann, H. G. 1974. Purification and identification of *E. coli* ribosomal proteins. *In*: Nomura, M., Tissières, A., and Lengyel, P., eds., *Ribosomes*, Cold Spring Harbor, N.Y., Cold Spring Harbor Laboratory, pp. 93–114.

Wittmann, H. G., 1976. Structure, function and evolution of ribosomes. *Eur. J. Biochem.* **61**:1–13.

Wolman, M. 1955. Problems of fixation in cytology, histology, and histochemistry. *Int. Rev. Cytol.* **4**:79–102.

Wooding, R. B. P., and Northcote, D. H. 1965. Association of the endoplasmic reticulum and the plastids in *Acer* and *Pinus. Am. J. Bot.* **52**:526–531.

Wool, I. G., and Stöffler, G. 1974. Structure and function of eukaryotic ribosomes. *In*: Nomura, M., Tissières, A., and Lengyel, P., eds., *Ribosomes*, Cold Spring Harbor, N.Y., Cold Spring Harbor Laboratory, pp. 417–460.

Wujek, D. E., and Kristiansen, J. 1978. Observations on bristle and scale-production in *Mallomonas caudata* (Chrysophyceae). *Arch. Protistenkd.* **120**:213–221.

Young, O. A., and Beevers, H. 1976. Mixed function oxidases from germinating castor bean endosperm. *Phytochemistry* **15:379**–385.

Zeigel, R. F., and Dalton, A. J. 1962. Speculations based on the morphology of the Golgi system in several types of protein secreting cells. *J. Cell Biol.* **15**:45–54.

CHAPTER 7

Aaronson, S. 1973. Digestion in phytoflagellates. *In*: Dingle, J. T., ed., *Lysosomes in Biology and Pathology,* Amsterdam, North-Holland, Vol. 3, pp. 18–37.

Ackerman, G. A. 1963. Cytochemical properties of the blood basophilic granulocyte. *Ann. N.Y. Acad. Sci.* **103**:376–393.

Afzelius, B. A. 1977. Spermatozoa and spermatids of the crinoid *Antedon petasus*, with a note on primitive spermatozoa from deuterostome animals. *J. Ultrastruct. Res.* **59**:272–281.

Afzelius, B. A., and Ferraguti, M. 1978. The spermatozoon of *Priapulus caudatus* Lamarck. *J. Submicrosc. Cytol.* **10**:71–79.

Allison, A. C., and Hartree, E. F. 1970. Lysosomal enzymes in the acrosome and their possible role in fertilization. *J. Reprod. Fertil.* **21**:501–515.

Archer, G. T., and Hirsch, J. G. 1963. Isolation of granules from eosinophil leucocytes and study of their enzyme content. *J. Exp. Med.* **118**:277–285.

Archer, G. T., Air, G., Jackas, M., and Morell, D. B. 1965. Studies on rat eosinophil peroxidase. *Biochim. Biophys. Acta* **99**:96–101.

Armentrout, V. N., and Wilson, C. L., 1969. Haustorium–host interaction during mycoparasitism of *Mycotypha microspora* by *Pyptocephalis virginiana. Phytopathology* **59**:897–905.

Armentrout, V. N., Graves, L. B., and Maxwell, D. P. 1978. Localization of enzymes of oxalate biosynthesis in microbodies of *Sclerotium rolfsii. Phytopathology* **68**:1597–1599.

Arnold, G., and Holtzman, E. 1978. Microperoxisomes in the central nervous system of the postnatal rat. *Brain Res.* **155**:1–18.

Arnold, G., Liscum, L., and Holtzman, E. 1979. Ultrastructural localization of D-amino acid oxidase in microperoxisomes of the rat nervous tissue. *J. Histochem. Cytochem.* **27**:735–745.

Ashworth, J. M., and Quance, J. 1972. Enzyme synthesis in myxamoebae of the cellular slime mold *Dictyostelium discoideum* during growth in axenic culture. *Biochem. J.* **126**:601–608.

Ashworth, J. M., and Wiener, E. 1973. The lysosomes of the cellular slime mold *Dictyostelium discoideum. In:* Dingle, J. T., ed., *Lysosomes in Biology and Pathology,* Amsterdam, North-Holland, Vol. 3, pp. 38–48.

Ashworth, J. M., Duncan, D., and Rowe, A. J. 1969. Changes in fine structure during cell differentiation of the cellular slime mold *Dictyostelium discoideum. Exp. Cell Res.* **58**:73–78.

Avers, C. J., and Federman, M. 1968. The occurrence in yeast of cytoplasmic granules which resemble microbodies. *J. Cell Biol.* **37**:555–559.

Bainton, D. F. 1972. Origin, content, and fate of PMN granules. *In:* Williams, R. C., and Fudenberg, H. H., eds., *Phagocytic Mechanisms in Health and Disease,* New York, Intercontinental Medical Book Corp., p. 123.

Bainton, D. F., and Farquhar, M. G. 1970. Segregation and packaging of granule enzymes in eosinophilic leukocytes. *J. Cell Biol.* **43**:54–73.

Bainton, D. F., Ullyot, J. L., and Farquhar, M. G. 1971. The development of neutrophilic polymorphonuclear leukocytes in human bone marrow. *J. Exp. Med.* **134**:907–934.

Bainton, D. F., Nichols, B. A., and Farquhar, M. G. 1976. Primary lysosomes of blood leukocytes. *In:* Dingle, J. T., and Dean, R. T., eds., *Lysosomes in Biology and Pathology,* Amsterdam, North-Holland, Vol. 5, pp. 3–32.

Baudhuin, P. 1969. Liver peroxisomes, cytology and function. *Ann. N.Y. Acad. Sci.* **168**:214–228.

Baudhuin, P., Beaufay, H., and deDuve, C. 1965a. Combined biochemical and morphological study of particulate fractions from rat-liver. *J. Cell Biol.* **26**:219–243.

Baudhuin, P., Müller, M., Poole, B., and deDuve, C. 1965b. Non-mitochondrial oxidizing particles (microbodies) in rat liver and kidney and in *Tetrahymena pyriformis. Biochem. Biophys. Res. Commun.* **20**:53–59.

Bennuh, A., and Blum, J. J. 1966. Properties of the induced acid phosphatase and of the constitutive acid phosphatase of *Euglena. Biochim. Biophys. Acta* **128**:106–123.

Berg, C. P. 1959. Utilization of D-amino acids. *In:* Albanese, A. A., ed., *Protein and Amino Acid Nutrition,* New York, Academic Press, pp. 57–96.

Berjak, P., and Villiers, T. A. 1970. Aging in plant embryos. *New Phytol.* **69**:929–938.

Bird, J. W. C., Spanier, A. M., and Schwartz, W. N. 1978. Cathepsins B and D: Proteolytic activity and ultrastructural localization in skeletal muscle. *In:* Segal, H. L., and Doyle, D. J., eds., *Protein Turnover and Lysosome Function,* New York, Academic Press, pp. 589–604.

Blanchard, M., Green, D. E., Nocito, V., and Ratner, S. 1944. L-Amino acid oxidase of animal tissue. *J. Biol. Chem.* **155**:421–440.

Bowen, I. D., and Bryant, J. A. 1978. The fine structural localization of *p*-nitrophenyl phosphatase activity in the storage cells of pea cotyledon. *Protoplasma* **97**:241–250.

Bowen, I. D., Coakley, W. T., and James, C. J. 1979. The digestion of *Saccharomyces cerevisiae* by *Acanthamoeba castellanii*. *Protoplasma* **98**:63–71.

Breidenbach, R. W. 1969. Characterization of some glyoxysomal proteins. *Ann. N.Y. Acad. Sci.* **168**:342–347.

Breidenbach, R. W., and Beevers, H. 1967. Association of the glyoxylate cycle enzymes in a novel subcellular particle from castor bean endosperm. *Biochem. Biophys. Res. Commun.* **27**:462–469.

Brenner, D. M., and Carroll, G. C. 1968. Fine structural correlates of growth in hyphae of *Ascodesmis sphaerospora*. *J. Bacteriol.* **95**:658–671.

Claude, A. 1943. The constitution of protoplasm. *Science* **97**:451–456.

Clayton, R. K. 1959. Purified catalase from *Rhodopseudomonas sphaeroides*. *Biochim. Biophys. Acta* **36**:40–47.

Clermont, Y., and Leblond, C. P. 1955. Spermiogenesis of man, monkey, ram, and other mammals as shown by the "periodic acid–Schiff" technique. *Am. J. Anat.* **96**:229–253.

Cohn, Z. A., and Wiener, E. 1963. The particulate hydrolases of macrophages. II. Biochemical and morphological response to particle ingestion. *J. Exp. Med.* **118**:1009–1019.

Cullen, M. J., Appleyard, S. T., and Bindoff, L. 1979. Morphologic aspects of muscle breakdown and lysosomal activation. *Ann. N.Y. Acad. Sci.* **317**:440–463.

Curtis, S. K., Cowden, R. R., and Nagel, J. W. 1979. Ultrastructure of the bone marrow of the salamander, *Plethodon glutinosus*. *J. Morphol.* **159**:151–184.

de Duve, C. 1964. From cytases to lysosomes. *Fed. Proc. Fed. Am. Soc. Exp. Biol.* **23**:1045–1049.

de Duve, C. 1969. Evolution of the peroxisomes. *Ann. N.Y. Acad. Sci.* **168**:369–381.

de Duve, C. 1971. Tissue fractionation, past and present. *J. Cell Biol.* **50**:20d–55d.

de Duve, C. 1973. Biochemical studies on the occurrence, biogenesis, and life history of mammalian peroxisomes. *J. Histochem. Cytochem.* **21**:941–948.

de Duve, C., and Baudhuin, P. 1966. Peroxisomes (microbodies and related particles). *Physiol. Rev.* **46**:323–357.

de Duve, C., and Wattiaux, R. 1966. Functions of lysosomes. *Annu. Rev. Physiol.* **28**:435–492.

de Duve, C., Pressman, B. C., Gianetto, R., Wattiaux, R., and Appelmans, F. 1955. Tissue fractionation studies. 6. Intracellular distribution patterns of enzymes in rat-liver tissue. *Biochem. J.* **60**:604–617.

de Duve, C., Beaufay, H., Jacques, P., Rahman-Li, Y., Sellinger, O. Z., Wattiaux, R., and de Coninck, S. 1960. Intracellular localization of catalase and of some oxidases in rat liver. *Biochim. Biophys. Acta* **40**:186–187.

Dingle, J. T. 1972. *Lysosomes: A Laboratory Handbook,* Amsterdam, North-Holland.

Dixon, M., and Webb, E. C. 1964. *Enzymes,* 2nd ed., New York, Academic Press.

Drawert, H., and Mix, M. 1962. Licht- und elektronenmikroskopische Untersuchungen an Desmidiaceen. X. Beiträge zur Kenntnis der "Haütung" von Desmidiaceae. *Arch. Mikrobiol.* **42**:96–109.

Drum, R. W. 1963. The cytoplasmic fine structure of the diatom, *Nitzchia palea*. *J. Cell Biol.* **18**:429–440.

Eeckhout, Y. 1970. Propriétés et localisation des hydrolases acides du Trypanosomide *Crithidia luciliae*. *Arch. Int. Physiol. Biochem.* **78**:993–994.

Eeckhout, Y. 1972. Studies on acid hydrolases and on catalase of the trypanosomatid *Crithidia luciliae*. *In:* van den Bossche, H., ed., *The Comparative Biochemistry of Parasites*, New York, Academic Press.

Eeckhout, Y. 1973. Digestion and lysosomes in zooflagellates. *In:* Dingle, J. T., ed., *Lysosomes in Biology and Pathology*, Amsterdam, North-Holland, Vol. 3, pp. 3–17.

El-Hawawi, A. S. N., and King, P. E. 1978. Spermiogenesis in a pycnogonid *Nymphon gracile* (Leach). *J. Submicrosc. Cytol.* **10**:345–356.

Ericsson, J. L. E. 1969. Mechanism of cellular autophagy. *In:* Dingle, J. T., and H. B. Fell, eds., *Lysosomes in Biology and Pathology*, Amsterdam, North-Holland, Vol. 2, pp. 345–394.

Essner, E. 1966. Endoplasmic reticulum and the origin of microbodies in fetal liver. *Fed. Proc. Fed. Am. Soc. Exp. Biol.* **25**:361.

Essner, E. 1967. Endoplasmic reticulum and the origin of microbodies in fetal mouse liver. *Lab. Invest.* **17**:71–87.

Farquhar, M. G. 1969. Lysosome function in regulating secretion: Disposal of secretory granules in cells of the anterior pituitary gland. *In:* Dingle, J. T., and Fell, H. B., eds., *Lysosomes in Biology and Pathology,* Amsterdam, North-Holland, Vol. 2, pp. 462–482.

Farquhar, M. G. 1971. Processing of secretory products by cells of the anterior pituitary gland. *Mem. Soc. Endocrinol.* **19**:79–124.

Fawcett, D. W. 1966. *The Cell: Its Organelles and Inclusions*, Philadelphia, Saunders.

Franzén, Å. 1977. Sperm structure with regard to fertilization biology and phylogenetics. *Verh. Dtsch. Zool. Ges.* **1977**:123–138.

Franzén, Å. 1979. A fine structure study on spermiogenesis in the Entoprocta. *J. Submicrosc. Cytol.* **11**:73–84.

Frederick, S. E., and Newcomb, E. H. 1969. Microbody-like organelles in leaf cells. *Science* **163**:1353–1355.

Frey-Wyssling, A., and Mühlethaler, K. 1965. *Ultrastructural Plant Cytology*, New York, American Elsevier.

Frey-Wyssling, A., Greishaber, E., and Mühlethaler, K. 1963. Origin of spherosomes in plant cells. *J. Ultrastruct. Res.* **8**:506–516.

Friend, D. S., and Farquhar, M. G. 1967. Functions of coated vesicles during protein absorption in the rat vas deferens. *J. Cell Biol.* **35**:357–376.

Gahan, P. B. 1973. Plant lysosomes. *In:* Dingle, J. T., ed., *Lysosomes in Biology and Pathology,* Amsterdam, North-Holland, Vol. 3, pp. 69–85.

Gezelius, K. 1966. Acid phosphatase in *Dictyostelium discoideum*. *Physiol. Plant.* **19**:946–959.

Gibson, R. A., and Paleg, L. G. 1972. Lysosomal nature of hormonally induced enzymes in wheat aleurone cells. *Biochem. J.* **128**:367–375.

Giranello, R. D., and Axelrod, J. 1973. Genetically controlled alterations in the rate of degradation of phenylethanolamine *N*-methyl transferase. *J. Biol. Chem.* **248**:5616–5623.

Girbardt, M. 1961. Licht- und elektronmikroskopische Untersuchungen an *Polystictus versicolor*. II. Die Feinstruktur von Grundplasma und Mitochondrien. *Arch. Mikrobiol.* **39**:351–359.

Goff, L. J. 1979a. The biology of *Harveyella mirabilis* (Rhodophyceae). VI. Translocation of photoassimilated $^{14}C^{1,2}$. *J. Phycol.* **15**:82–87.

Goff, L. J. 1979b. The biology of *Harveyella mirabilis* (Cryptonemiales, Rhodophyceae). VII. Structure and proposed function of host-penetrating cells. *J. Phycol.* **15**:87–100.

Gonzalez, E., and Beevers, H. 1976. Role of the endoplasmic reticulum in glyoxysome formation in castor bean endosperm. *Plant Physiol.* **57**:406–409.

Grieshaber, E. 1964. Entwicklung und Feinbau der Sphärosomen in Pflanzenzellen. *Vierteljahrsschr. Naturforsch. Ges. Zürich* **109**:1–23.

Hamburger, V., and Hamilton, H. L. 1951. A series of normal stages in the development of the chick embryo. *J. Morphol.* **88**:49–92.

Hanstein, J. V. 1880. *Bot. Abhandl. Morph. Physiol.* **4**:1.

Hartree, E. F. 1975. The acrosome–lysosome relationship. *J. Reprod. Fertil.* **44**:125.

Hayashi, H., and Sugo, T. 1978. Some characteristics of peroxisomes in the slime mold, *Dictyostelium discoideum*. *J. Biochem.* **84**:513–520.

Heath, M. F., Gandy, G., and Jacobson, W. 1976. Lysosomes in the lung. *In:* Dingle, J. T., and Dean, R. T., eds., *Lysosomes in Biology and Pathology,* Amsterdam, North-Holland, Vol. 5, pp. 33–58.

Herbert, D., and Pinsent, J. 1948. Crystalline bacterial catalase. *Biochem. J.* **43**:193–202.

Hess, W. M. 1968. Ultrastructural comparisons of fungus hyphal cells using frozen-etched replicas and thin sections of the fungus *Pyrenochaeta terrestris. Can. J. Microbiol.* **14**:205–210.

Hieber, V., Distler, J., Myerowitz, R., Schnickel, R. D., and Jourdian, G. W. 1976. The role of glycosidically bound mannose in the assimilation of β-galactosidase by generalized gangliosidosis fibroblasts. *Biochem. Biophys. Res. Commun.* **73**:710–717.

Higashi, T., and Peters, T. 1963a. Studies on rat liver catalase. I. Combined immunochemical and enzymatic determination of catalase in liver cell fractions. *J. Biol. Chem.* **238**:3945–3951.

Higashi, T., and Peters, T. 1963b. Studies on rat liver catalase. II. Incorporation of ^{14}C-leucine into calatase of liver cell fractions *in vivo. J. Biol. Chem.* **238**:3952–3954.

Hinklemann, W. 1966. Thesis, Technische Hochschule, Braunschweig.

Hirsch, J. G. 1962. Cinemicrophotographic observations on granule lysis in polymorphonuclear leucocytes during phagocytosis. *J. Exp. Med.* **116**:827–834.

Hirsch, J. G., and Cohn, Z. A. 1960. Degranulation of polymorphonuclear leucocytes following phagocytosis of microorganisms. *J. Exp. Med.* **112**:1005–1014.

Hirsch, J. G., and Cohn, Z. A. 1964. Digestive and autolytic function of lysosomes in phagocytic cells. *Fed. Proc. Fed. Am. Soc. Exp. Biol.* **23**:1023–1025.

Hohl, H. R. 1965. Nature and development of membrane systems in food vacuoles of cellular slime molds predatory upon bacteria. *J. Bacteriol.* **90**:755–765.

Hohl, H. R., and Hamamoto, S. T. 1967. Ultrastructural changes during zoospore formation in *Phytophthora parasitica. Am. J. Bot.* **54**:1121–1139.

Holtzman, E. 1969. Lysosomes in the physiology and pathology of neurons. *In:* Dingle, J. T., and Fell, H. B., eds., *Lysosomes in Biology and Pathology,* Amsterdam, North-Holland, Vol. 1, pp. 192–216.

Holtzman, E. 1976. *Lysosomes: A Survey,* Vienna, Springer-Verlag.

Holtzman, E., Novikoff, A. B., and Villaverde, H. 1967. Lysosomes and GERL in normal and chromatolytic neurons of the rat ganglion nodosum. *J. Cell Biol.* **33**:419–436.

Hruban, Z., and Rechcigl, M. 1969. Microbodies and related particles. *Int. Rev. Cytol. Suppl.* **1**:1–265.

Hruban, Z., and Swift, H. 1964. Uricase: Localization in hepatic microbodies. *Science* **146**:1316–1318.

Hurle, J. M., Lafarga, M., and Ojeda, J. L. 1977. Cytological and cytochemical studies of the nectoric area of the bulbous of the chick embryo heart. Phagocytosis by developing myocardial cells. *J. Embryol. Exp. Morphol.* **41**:161–173.

Hurle, J. M., Lafarga, M., and Ojeda, J. L. 1978. *In vivo* phagocytosis by developing myocardial cells. An ultrastructural study. *J. Cell Sci.* **33**:363–369.

Ishikawa, T., Wicher, K., and Arbesman, C. E., 1974. *In vitro* and *in vivo* studies on uptake of antigen–antibody complexes by eosinophils. *Int. Arch. Allergy Appl. Immunol.* **46**:230–248.

Jamieson, B. G. M., and Daddow, L. 1979. An ultrastructural study of microtubules and the acrosome in spermiogenesis of Tubificidae (Oligochaeta). *J. Ultrastruct. Res.* **67**:209–224.

Jamieson, B. G. M., Daddow, L., and Bennett, J. D. 1978. Ultrastructure of the tubificid acrosome (Annelida, Oligochaeta). *Zool. Scr.* **7**:115–118.

Jarosch, R. 1961. Das Characeen-protoplasma und seine Inhaltskörper. *Protoplasma* **53**:34–56.

Kagawa, T., Lord, J. M., and Beevers, H. 1973. The origin and turnover of organelle membranes in castor bean endosperm. *Plant Physiol.* **51**:61–65.

Kaltwasser, H. 1968. Induktive Bildung partikel gebundener Uricase bei *Hydrogenomonas* H16 und anderen aeroben Bakterien. *Arch. Mikrobiol.* **60**:160–171.

Kaplan, A., Fischer, D., Achord, D. T., and Sly, W. S. 1977. Phosphohexosyl recognition is a general characteristic of pinocytosis of lysosomal glycoside bases by human fibroblasts. *J. Clin. Invest.* **60**:1088–1093.

Kaplan, A., Fischer, D., and Sly, W. S. 1978. Correlation of structural features of phosophomannans with their ability to inhibit pinocytosis of human β-glucuronidase by human fibroblasts. *J. Biol. Chem.* **253**:647–650.

Kemmler, W., Steiner, D. F., and Borg, J. 1973. Studies on the conversion of proinsulin to insulin. *J. Biol. Chem.* **248**:4544–4551.

Kerr, J. F. R. 1973. Some lysosome functions in liver cells reacting to sublethal injury. *In:* Dingle, J. T., ed., *Lysosomes in Biology and Pathology,* Amsterdam, North-Holland, Vol. 3, pp. 365–394.

Kisaki, T., and Tolbert, N. E. 1969. Glycolate and glyoxylate metabolism by isolated peroxisomes or chloroplasts. *Plant Physiol.* **44**:242–250.

Klebanoff, S. J., and Hamon, C. B. 1975. Antimicrobial systems of mononuclear leukocytes. *In:* van Furth, R., ed., *Mononuclear Phagocytes in Immunity, Infection, and Pathology*, Oxford, Blackwell, pp. 507–529.

Kleinig, H., Steinki, C., Kopp, C., and Zaar, K. 1978. Oleosomes (spherosomes) from *Daucus carota* suspension culture cells. *Planta* **140**:233–238.

Knox, R. B., and Heslop-Harrison, J. 1970. Pollen wall proteins: Localization and enzymatic activity. *J. Cell Sci.* **6**:1–27.

Köller, W., Frevert, J., and Kindl, H. 1979. Albumins, glyoxysomal enzymes and globulins in dry seeds of *Cucumis sativus:* Qualitative and quantitative analysis. *Hoppe-Seyler's Z. Physiol. Chem.* **36**:167–176.

Kozar, F., and Weijer, J. 1969. Electron-dense structures in *Neurospora crassa.* II. Lomasome-like structures, *Can. J. Genet. Cytol.* **11**:617–621.

Lamb, J. E., Riezman, H., Becker, W. M., and Leaver, C. J. 1978. Regulation of glyoxysomal enzymes during germination of cucumber. 2. Isolation and immunological detection of isocitrate lyase and catalase. *Plant Physiol.* **62**:754–760.

Lange, L., and Olson, L. W. 1977. The zoospore of *Phlyctochytrium aestuarii. Protoplasma* **93**:27–43.

Lange, L., and Olson, L. W. 1978. The zoospore of *Synchytrium endobioticum. Can. J. Bot.* **56**:1229–1239.

Lee, D. 1970. The relative permeability of lysosomes from *Tetrahymena pyriformis* to carbohydrates, lactate, and the cryoprotective nonelectrolytes, glycerol and dimethylsulfoxide. *Biochim. Biophys. Acta* **211**:550–554.

Levitt, J. 1967. The mechanism of stomatal action. *Planta* **74**:101–118.

Locke, M., and Collins, J. V. 1968. Protein uptake into multivesicular bodies and storage granules in the fat body of an insect. *J. Cell Biol.* **36**:453–483.

Locke, M., and McMahon, J. T. 1971. The origin and fate of microbodies in the fat body of an insect. *J. Cell Biol.* **48**:61–78.

Lockwood, W. R., and Allison, F. 1963. Electron micrographic studies of phagocytic cells. 1. Morphological changes of the cytoplasm and granules of rabbit granulocytes associated with ingestion of rough *Pneumococcus. Bot. J. Exp. Pathol.* **44**:593–600.

Lui, N. S. T., Roels, O. A., Trout, M. E., and Anderson, O. R. 1968. Subcellular distribution of enzymes in *Ochromonas malhamensis. J. Protozool.* **15**:536–542.

McGowan, E. B., Shafiq, S. A., and Stracher, A. 1976. Delayed degeneration of dystrophic and normal muscle cell cultures treated with pepstatin, leupeptin and antipain. *Exp. Neurol.* **50**:649–657.

Mannering, G. J., Van Harken, D. R., Makar, A. B., Tephly, T. R., Watkins, W. D., and Goodman, J. I. 1969. Role of the intracellular distribution of hepatic catalase in the peroxidative oxidation of methanol. *Ann. N.Y. Acad. Sci.* **168**:265–280.

Manocha, M. S., and Shaw, M. 1964. Occurrence of lomasomes in mesophyll cells of ''Khapli'' wheat. *Nature (London)* **203**:1402–1403.

Matile, P. 1964. Die Funktion proteolytischer Enzyme bei der Proteinaufnahme durch *Neurospora crassa*. *Naturwissenschaften* **51**:489–490.

Matile, P. 1965. Intrazelluläre Lokalisation proteolytischer Enzyme von *Neurospora crassa*. I. Funktion und subzellulare Verteilung protolytischer Enzyme. *Z. Zellforsch. Mikrosk. Anat.* **65**:884–896.

Matile, P. 1968a. Aleurone vacuoles as lysosomes. *Z. Pflanzenphysiol.* **58**:365–368.

Matile, P. 1968b. Utilization of peptides in yeasts. *Proc. 2nd Symp. Yeasts, Bratislavia*, p. 1966.

Matile, P. 1969. Plant lysosomes. *In:* Dingle, J. T., and Fell, H. B., eds., *Lysosomes in Biology and Pathology*, Amsterdam, North-Holland, Vol. 1, pp. 406–430.

Matile, P. 1975. *The Lytic Compartment of Plant Cells*. New York, Springer-Verlag.

Matile, P., and Moore, H. 1968. Vacuolation: Origin and development of the lysosomal apparatus in root-tip cells. *Planta* **80**:159–175.

Mayagkaya, G., Schellens, J. P. M., and Vreeling-Sinderlárová, H. 1979. Lysosomal breakdown of erythrocytes in the sheep placenta. *Cell Tissue Res.* **197**:79–94.

Mego, J. L. 1973. Protein digestion in isolated heterolysosomes. *In:* Dingle, J. T., ed., *Lysosomes in Biology and Pathology*, Amsterdam, North-Holland, Vol. 3, pp. 138–168.

Miller, F., de Harven, E., and Palade, G. E. 1966. Structure of eosinophil leukocytic granules in rodents and man. *J. Cell Biol.* **31**:349–362.

Moor, H., and Mühlethaler, K. 1963. Fine structure in frozen etched yeast cells. *J. Cell Biol.* **17**:609–628.

Mooradian, B. A., and Cutler, L. S. 1978. Developmental distribution of microperoxisomes in the rat submandibular gland. *J. Histochem. Cytochem.* **26**:989–999.

Moore, T., and McAlear, J. H. 1961. Fine structure of Mycota. 5. Lomasomes—Previously uncharacterized hyphal structures. *Mycologia* **53**:194–200.

Morton, D. B. 1976. Lysosomal enzymes in mammalian spermatozoa. *In:* Dingle, J. T., and Dean, R. T., eds., *Lysosomes in Biology and Pathology*, Amsterdam, North-Holland, Vol. 5, pp. 203–255.

Müller, M. 1969. Peroxisomes of Protozoa. *Ann. N.Y. Acad. Sci.* **168**:292–301.

Müller, M. 1972. Secretion of acid hydrolase and its intracellular source in *Tetrahymena pyriformis*. *J. Cell Biol.* **52**:478–487.

Müller, M. 1973. Biochemical cytology of trichomonad flagellates. *J. Cell Biol.* **57**:453–474.

Müller, M., and Møller, K. M. 1969a. Studies on some enzymes of purine metabolism in the amoebae *Chaos chaos* and *Amoeba proteus*. *C. R. Trav. Lab. Carlsberg* **36**:463–497.

Müller, M., and Møller, K. M. 1969b. Urate oxidase and its association with peroxisomes in *Acanthamoeba* sp. *Eur. J. Biochem.* **9**:424–430.

Müller, M., Baudhuin, P., and deDuve, C. 1966. Lysosmes in *Tetrahymena pyriformis*. I. Some properties and lysosomal localization of acid hydrolases. *J. Cell. Physiol.* **68**:165–175.

Müller, M., Hogg, J. F., and deDuve, C. 1968. Distribution of tricarboxylic acid cycle enzymes and of glyoxylate cycle enzymes between mitochondria and peroxisomes in *Tetrahymena pyriformis*. *J. Biol. Chem.* **243**:5385–5395.

Nakabayashi, T., and Ikezawa, H. 1978. An isoelectric focusing study of acid phosphohydrolases in rat liver lysosomes. *J. Biochem.* **84**:351–360.

Neims, A. H., and Hellerman, L. 1962. Specificity of the D-amino acid oxidase in relation to glycine oxidase activity. *J. Biol. Chem.* **237**:PC976–978.

Neufeld, E. F., Sando, G. N., Garvin, A. J., and Rome, L. H. 1977. The transport of lysosomal enzymes. *J. Supramol. Struct.* **6**:95–101.

Nicholls, P., and Schonbaum, G. R. 1963. Catalases. *In:* Boyer, P. D., Lardy, H., and Myrbäck, K., eds., *The Enzymes*, New York, Academic Press, Vol. VIII, pp. 147–225.

Nichols, B. A., and Bainton, D. F. 1973. Differentiation of human monocytes in bone marrow and blood. Sequential formation of two granule populations. *Lab. Invest.* **29**:27–40.

Nichols, B. A., and Bainton, D. F. 1975. Ultrastructure and cytochemistry of mononuclear phagocytes, *In*: van Furth, R., ed., *Mononuclear Phagocytes in Immunity, Infection, and Pathology*, Oxford, Blackwell, pp. 17–55.

Nichols, B. A., Bainton, D. F., and Farquhar, M. G. 1971. Differentiation of monocytes: Origin, nature, and fate of their azurophil granules. *J. Cell Biol.* **50**:498–515.

Noguchi, T., Takada, Y., and Fujiwara, S. 1979. Degradation of uric acid to urea and glyoxylate in peroxisomes. *J. Biol. Chem.* **254**:5272–5275.

Novikoff, A. B. 1961. Lysosomes and related particles. *In:* Brachet, J., and Mirsky, A. E., eds., *The Cell: Biochemistry, Physiology, Morphology,* Vol. II, *Cells and Their Component Parts,* New York, Academic Press, pp. 423–488.

Novikoff, A. B. 1976. The endoplasmic reticulum: A cytochemist's view (A review). *Proc. Natl. Acad. Sci. USA* **73**:2781–2787.

Novikoff, A. B., and Essner, E. 1962. Pathological changes in cytoplasmic organelles. *Fed. Proc. Fed. Am. Soc. Exp. Biol.* **21**:1130–1142.

Novikoff, A. B., and Shin, W. Y. 1964. The endoplasmic reticulum in the Golgi zone and its relation to microbodies, Golgi apparatus and autophagic vacuoles in rat liver cells. *J. Microsc. (Paris)* **3**:187–206.

Novikoff, A. B., Essner, E., and Quintana, N. 1964. Golgi apparatus and lysosomes. *Fed. Proc. Fed. Am. Soc. Exp. Biol.* **23**:1010–1023.

Novikoff, A. B., Novikoff, P. M., Davis, C., and Quintana, N. 1973. Studies on microspherosomes. V. Are microperoxisomes ubiquitous in mammalian cells? *J. Histochem. Cytochem.* **21**:737–755.

Owen, K. 1979. Biochemical studies of dystrophy in the young chicken: Lysosomal and sarcolemma enzymes. *Ann. N.Y. Acad. Sci.* **317**:247–262.

Paavola, L. G. 1979a. Cellular mechanisms involved in luteolysis. *In:* Channing, C. P., Marsh, J. M., and Sadler, W. A., eds., *Ovarian Follicular and Corpus Luteum Function*, New York, Plenum Press, pp. 527–533.

Paavola, L. G. 1979b. The corpus leuteum of the guinea pig. IV. Fine structure of macrophages during pregnancy and postpartum luteolysis, and the phagocytosis of luteal cells. *Am. J. Anat.* **154**:337–364.

Pagano, R. E., and Weinstein, J. N. 1978. Interactions of liposomes with mammalian cells. *Annu. Rev. Biophys. Bioeng.* **7**:435–468.

Palay, S. L., and Revel, J. P. 1964. The morphology of fat absorption. *In:* Meng, H. C., ed., *Lipid Transport*, Springfield, Ill., Thomas, pp. 33–43.

Papahadjopoulos, D. 1978. Liposomes and their uses in biology and medicine. *Ann. N.Y. Acad. Sci.* **308**:1–462.

Parish, R. W. 1975. Mitochondria and peroxisomes from the cellular slime mold *Dictyostelium discoideum. Eur. J. Biochem.* **58**:523–531.

Pasteels, J. J. 1973. Yolk and lysosomes. *In:* Dingle, J. T., ed., *Lysosomes in Biology and Pathology*, Amsterdam, North-Holland, Vol. 3, pp. 216–234.

Patton, G. W., and Nishimura, E. T. 1967. Developmental changes of hepatic catalase in the rat. *Cancer Res.* **27**:117–123.

Pearson, C. M., and Kar, N. C. 1979. Muscle breakdown and lysosomal activation (biochemistry). *Ann. N.Y. Acad. Sci.* **317**:465–476.

Perner, E. S. 1953. Die Vitalfärbung mit Berber insulfat und iher physiologische Wirkung und Zellen höherer Pflanzen. *Ber. Dtsch. Bot. Ges.* **65**:52–59.

Poole, A. R. 1973. Tumour lysosomal enzymes and invasive growth. *In:* Dingle, J. T., ed., *Lysosomes in Biology and Pathology*, Amsterdam, North-Holland, Vol. 3, pp. 303–337.

Poole, B. 1971. Synthesis and degradation of proteins in relation to cellular structure. *In:* Rechcigl, M., ed., *Enzyme Synthesis and Degradation in Mammalian Systems,* Baltimore, University Park Press, pp. 375–402.

Poole, B., Ohkuma, S., and Warburton, M. 1978. Some aspects of the intracellular breakdown of exogenous and endogenous proteins. *In:* Segal, H. L., and Doyle, D. L., *Protein Turnover and Lysosome Function*, New York, Academic Press, pp. 43–58.

Powell, M. J. 1976. Ultrastructure and isolation of glyoxysomes (microbodies) in zoospores of the fungus *Entophlyctis* sp. *Protoplasma* **89**:1–27.

Powell, M. J. 1978. Phylogenetic implications of the microbody–lipid globule complex in zoosporic fungi. *BioSystems* **10**:167–180.

Powell, M. J. 1979. The structure of microbodies and their associations with other organelles in zoosporangia of *Entophlyctis variabilis*. *Protoplasma* **98**:177–198.

Price, C. A. 1962. Repression of acid phophatase synthesis in *Euglena gracilis*. *Science* **135**:46.

Pringsheim, E. G. 1952. On the nutrition of *Ochromonas*. *Q. J. Microsc. Sci.* **93**:71–96.

Rechcigl, M. 1971. Intracellular protein turnover and the role of synthesis and degradation in regulation of enzyme levels. *In*: Rechcigl, M., ed., *Enzyme Synthesis and Degradation in Mammalian Systems*, Baltimore, University Park Press, pp. 237–310.

Reddy, J. K., and Kumar, N. S. 1977. The peroxisome proliferation-associated polypeptide in rat liver. *Biochem. Biophys. Res. Commun.* **77**:824–829.

Robineaux, J., and Frederic, J. 1955. Contribution a l'étude des granulations neutrophiles des polynucléaires par la microcinématographie en contraste de phase. *C. R. Seances Soc. Biol. Paris* **149**:486–489.

Robinson, J. C., Keay, L., Molinari, R., and Sizer, I. W. 1962. L-α-Hydroxyacid oxidases of hog renal cortex. *J. Biol. Chem.* **237**:2001–2010.

Rome, L. H., Garvin, A. J., Allietta, M. M., and Neufeld, E. F. 1979. Two species of lysosomal organelles in cultured human fibroblasts. *Cell* **17**:143–153.

Rosen, O. M., Rosen, S. M., and Horecker, B. L. 1965. Fate of the cell wall of *Salmonella typhimurium* upon ingestion by the cellular slime mold *Polysphondylium pallidum*. *Biochem. Biophys. Res. Commun.* **18**:270–276.

Sampson, D., and Archer, G. T. 1967. Release of histamine from human basophils. *Blood* **29**:722–736.

Sando, G. N., and Neufeld, E. F. 1977. Recognition and receptor-mediated uptake of a lysosomal enzyme, α-L-iduronidase, by cultured human fibroblasts. *Cell* **12**:619–627.

Sando, G. N., Titus-Dillon, P., Hall, C. W., and Neufeld, E. F. 1979. Inhibition of receptor-mediated uptake of a lysosomal enzyme into fibroblasts by chloroquine, procaine and ammonia. *Exp. Cell Res.* **119**:359–364.

Sapra, G. R., and Kloetzel, J. A. 1975. Programmed autophagocytosis accompanying conjugation in the ciliate *Stylonychia mytilus*. *Dev. Biol.* **42**:84–94.

Scannone, H., Wellner, D., and Novogrodsky, A. 1964. A study of amino acid oxidase specificity using a new sensitive assay. *Biochemistry* **3**:1742–1745.

Schneider, D. L., and Cornell, E. 1978. Evidence for a proton pump in rat liver lysosomes. *In*: Segal, H. L., and Doyle, D. J., eds., *Protein Turnover and Lysosome Function*, New York, Academic Press, pp. 59–66.

Seeman, P. M., and Palade, G. E. 1967. Acid phosphatase localization in rabbit eosinophils. *J. Cell Biol.* **34**:745–756.

Segal, H. L., Brown, J. A., Dunaway, G. A., Winkler, J. R., Madnick, H. M., and Rothstein, D. M. 1978. Factors involved in the regulation of protein turnover. *In*: Segal, H. L., and Doyle, D. J., eds., *Protein Turnover and Lysosome Function*, New York, Academic Press, pp. 9–28.

Seiguer, A. C., and Castro, A. E. 1972. Electron microscopic demonstration of arylsulfatase activity during acrosome formation in the rat. *Biol. Reprod.* **7**:31–42.

Sergeyeva, G. I. 1967. Activity of acid phosphatase at different stages of the life cycle of *Opalina ranarum* Ehrbg. *Tsitologiya* **9**:324–332.

Sergeyeva, G. I. 1969. An electron microscope study of lysosomes in *Opalina ranarum* Ehrbg. *Tsitologiya* **11**:714–720.

Shio, H. 1971. Master of arts thesis, Queens College, City University of New York, Flushing.

Shnitka, T. K. 1966. Comparative ultrastructure of hepatic microbodies in some mammals and birds in relation to species differences in uricase activities. *J. Ultrastruct. Res.* **16**:598–625.

Silveira, S. R., and Hadler, W. A. 1978. Catalases and peroxidases histochemical detection. Techniques suitable to discriminate these enyzmes. *Acta Histochem.* **63**:1–10.

Smith, C. G. 1974. The ultrastructural development of spherosomes and oil bodies in the developing embryo of *Crambe abyssinica*. *Planta* **119**:125–142.
Smith, R. E. 1969. Phosphohydrolases in cell organelles: Electron microscopy. *Ann. N.Y. Acad. Sci.* **166**:525–564.
Smith, R. E., and Farquhar, M. G. 1966. Lysosome function in the regulation of the secretory process in cells of the anterior pituitary gland. *J. Cell Biol.* **31**:319–336.
Smith, R. E., and van Frank, R. M. 1975. The use of amino acid derivatives of 4-methoxy-β-naphthylamine for the assay and subcellular localization of tissue proteinases. *In*: Dingle, J. T., and Dean, R. T., eds., *Lysosmes in Biology and Pathology*, Amsterdam, North-Holland, Vol. 4, p. 195.
Solomon, E. P., Johnson, E. M., and Gregg, J. H. 1964. Multiple forms of enzymes in a cellular slime mold during morphogenesis. *Dev. Biol.* **9**:314–326.
Sommer, J. R., and Blum, J. J. 1965. Cytochemical localization of acid phosphatases in *Euglena gracilis*. *J. Cell Biol.* **24**:235–248.
Sorokin, H. P. 1967. The spherosomes and the reserve fat in plant cells (*Persea americana*, *Cocos mucifera*, *Helianthus annuus*, *Brassica napus*, *Cucurbita pepo*). *Am. J. Bot.* **54**:1008–1016.
Sorokin, H. P., and Sorokin, S. 1968. Fluctuations in the acid phosphatase activity of spherosomes in guard cells of *Campanula persicifolia*. *J. Histochem. Cytochem.* **16**:791–802.
Steinman, R. M., Brodie, S. E., and Cohn, Z. A. 1976. Membrane flow during pinocytosis. A stereologic analysis. *J. Cell Biol.* **68**:665–687.
Stossel, T. P. 1974. Phagocytosis. *N. Engl. J. Med.* **290**:717–723, 774–780, 833–839.
Strickland, J. D. H., and Solorzano, L. 1966. Determination of monoesterase hydrolysable phosphate and phosphomonoesterase activity in sea water. *In*: *Some Contemporary Studies in Marine Science*, New York, Hafner, pp. 665–674.
Stroun, M., Anker, P., Charles, P., and Ledoux, L. 1966. Fate of bacterial DNA in *Lycopersicum esculentum*. *Nature (London)* **212**:397–398.
Sussman, M., and Sussman, R. 1969. Patterns of RNA synthesis and of enzyme accumulation and disappearance during cellular slime mold differentiation. *Symp. Soc. Gen. Microbiol.* **19**:403–435.
Svoboda, D., and Reddy, J. 1972. Microbodies in experimentally altered cells. IX. The fate of microbodies. *Am. J. Pathol.* **67**:541–554.
Svoboda, D., Grady, H., and Azarnoff, D. 1967. Microbodies in experimentally altered cells. *J. Cell Biol.* **35**:127–152.
Szabo, A. S., and Avers, C. J. 1969. Some aspects of regulation of peroxisomes and mitochondria in yeast. *Ann. N.Y. Acad. Sci.* **168**:302–312.
Szego, C. M. 1974. *In*: Dingle, J. T., and Dean, R. T., eds., *Lysosomes in Biology and Pathology*, Amsterdam, North-Holland, Vol. 4.
Tappel, A. L. 1968. Lysosomes. *In*: Florkin, M., and Stotz, E. H., eds., *Comprehensive Biochemistry*, Amsterdam/New York, Elsevier, Vol. 23, pp. 77–98.
Thorton, R. M. 1968. The fine structure of Phycomyces. I. Autophagic vacuoles. *J. Ultrastruct. Res.* **21**:269–280.
Thurman, R. G., and Chance, B. 1969. Inhibition of catalase in perfused rat liver by sodium azide. *Ann. N.Y. Acad. Sci.* **168**:348–353.
Tiffon, Y., Rasmont, R., deVos, L., and Bouillon, J. 1973. Digestion in lower metazoa. *In*: Dingle, J. T., ed., *Lysosomes in Biology and Pathology*, Amsterdam, North-Holland, Vol. 3, pp. 49–68.
Tolbert, N. E. 1962. Glycolate pathway. *NSF–NRC Publ.* **1145**:648–662.
Tolbert, N. E., and Yamazaki, R. K. 1969. Leaf peroxisomes and their relation to photorespiration and photosynthesis. *Ann. N.Y. Acad. Sci.* **168**:325–341.
Tolbert, N. E., Oeser, A., Kisake, T., Hageman, R. H., and Yamazaki, R. K. 1968. Peroxisomes from spinach leaves containing enzymes related to glycolate metabolism. *J. Biol. Chem.* **243**:5179–5184.

Tomlinson, G. 1967. The glyoxylate pathway in *Acanthamoeba* sp. *J. Protozool.* **14**:114–116.

Trout, J. J., Stauber, W. T., and Schottelius, B. A. 1979. Cytochemical observations of two distinct acid phosphatase-reactive structures in anterior latissimus dorsi muscle of the chicken. *Histochem. J.* **11**:223–230.

Tsukahara, T., and Yamada, M. 1965. Cytological structure of *Aspergillus niger* by electron microscopy. *Jpn. J. Microbiol.* **9**:35–48.

Tu, M., Deyoe, C. W., and Eustace, P. 1978. Exploring buffalo gourd seeds with scanning electron microscopy. *Cereal Chem.* **55**:773–778.

Ulrich, K., Mersmann, G., Weber, E., and von Figura, K. 1978. Evidence for lysosomal enzyme recognition by human fibroblasts via a phosphorylated carbohydrate moiety. *Biochem. J.* **170**:643–650.

van Furth, R., and Cohn, Z. A. 1968. The origin and kinetics of mononuclear phagocytes. *J. Exp. Med.* **128**:415–435.

Vigil, E. L. 1973. Structure and function of plant microbodies. *Subcell. Biochem.* **2**:237–285.

Walek-Czernecka, A. 1962. Mise en évidence de la phosphatase acide (Monophosphoesterase II) dans les sphérosomes des cellules épidermiques des écailles bulbaires d'*Allium cepa*. *Acta Soc. Bot. Pol.* **31**:539–543.

Wanner, G., and Theimer, R. R. 1978. Membranous appendages of spherosomes (oleosomes): Possible role in fat utilization in germinating oil seeds. *Planta* **140**:163–170.

Watts, D. J., and Ashworth, J. M. 1970. Growth of myxamoebae of the cellular slime mold *Dictyostelium discoideum* in axenic culture. *Biochem. J.* **119**:171–174.

Wellner, D., and Scannone, H. 1964. Oxidation of L-proline and L-3,4-dehydroproline by D-amino acid oxidase. *Biochemistry* **3**:1746–1749.

Wells, K. 1965. Ultrastructural features of developing and mature basidia and basidiospores of *Schizophyllum commune*. *Mycologia* **57**:236–261.

Werb, Z., and Dingle, J. T. 1976. Lysosomes as modulators of cellular functions. Influence on the synthesis and secretion of non-lysosomal materials. *In:* Dingle, J. T., and Dean, R. T., eds., *Lysosomes in Biology and Pathology*, Amsterdam, North-Holland, Vol. 5, pp. 127–156.

West, B. C., Rosenthal, A. S., Gelb, N. A., and Kimball, H. R. 1974. Separation and characterization of human neutrophil granules. *Am. J. Pathol.* **77**:41–61.

Westheide, W. 1978. Ultrastructure of the genital organs in interstitial Polychaeta. I. Structure, development, and function of the copulatory stylets in *Microphthalmus*. *Zoomorphologie* **91**:101–118.

Westheide, W. 1979. Unusual granules in the ejaculatory duct of a *Microphthalmus* species (Annelida, Polychaeta). *Cell Tissue Res.* **197**:61–68.

Wetzel, B. K. 1970. The comparative fine structure of normal and diseased mammalian granulocytes. *In*: Gordon, A. S., ed., *Regulation of Hematopoiesis*, New York, Appleton–Century–Crofts, Vol. 2, pp. 819–872.

Wetzel, B. K., Spicer, S. S., and Horn, R. G. 1967. Fine structural localization of acid and alkaline phosphatases in cells of rabbit blood and bone marrow. *J. Histochem. Cytochem.* **15**:311–334.

Wheeler, H., and Hanchey, P. 1971. Pinocytosis and membrane dilation in uranyl treated plant roots. *Science* **171**:68–71.

Wiener, E., and Ashworth, J. M. 1979. The isolation and characterization of lysosomal particles from myxamoeba of the cellular slime mold *Dictyostelium discoideum*. *Biochem. J.* **118**:505–512.

Williams, D. M., Gillett, R., and Linde, J. E. 1979. Light microscope and electron microscope alkaline phosphatase cytochemistry of rat bone marrow leukocytes. *J. Histochem. Cytochem.* **27**:665–675.

Williams, N. E., and Luft, J. H. 1968. Use of a nitrogen mustard derivative in fixation for electron microscopy and observations on the ultrastructure of *Tetrahymena*. *J. Ultrastruct. Res.* **25**:271–292.

Wilsenach, R., and Kessel, M. 1965. The role of lomasomes in wall formation in *Penicillium vermiculatum*. *J. Gen. Microbiol.* **40**:401–404.
Zachariah, K., and Fitz-James, P. C. 1967. The structure of phialides in *Penicillium claviforme*. *Can. J. Microbiol.* **13**:249–256.
Zelich, I. 1968. Investigations on photorespiration with a sensitive ^{14}C-assay. *Plant Physiol.* **43**:1829–1837.
Zeligs, J. D., and Wollman, S. H. 1977. Ultrastructure of erythrophagocytosis and red blood cell fission by thyroid epithelial cells *in vivo*. *J. Ultrastruct. Res.* **59**:57–69.
Zucker-Franklin, D., and Hirsch, J. G. 1964. Electron microscopic study on the degranulation of rabbit peritoneal leukocyte during phagocytosis. *J. Exp. Med.* **120**:569–576.

CHAPTER 8

Adman, E. T., Sieker, L. C., and Jensen, L. H. 1973. The structure of a bacterial ferredoxin. *J. Biol. Chem.* **248**:3987–3996.
Aitken, A. 1977. Purification and primary structure of cytochrome *f* from the cyanobacterium *Plectonema boryanum*. *Eur. J. Biochem.* **78**:273–279.
Ambler, R. P. 1968. The amino acid sequence of cytochrome c_3 from *Desulfovibrio vulgaris*. *Biochem. J.* **109**:47P–48P.
Ambler, R. P., and Bartsch, R. G. 1975. Amino acid sequence similarity between cytochrome *f* from a blue-green bacterium and algal chloroplasts. *Nature (London)* **253**:285–288.
Ambler, R. P., Bruschi, M., and LeGall, J. 1969. The structure of cytochrome c_3 from *Desulfovibrio gigas* (NCIB 9332). *FEBS Lett.* **5**:115–117.
Ambler, R. P., Bruschi, M., and LeGall, J. 1971. The amino acid sequence of cytochrome c_3 from *Desulfovibrio desulfuricans*. *FEBS Lett.* **18**:347–350.
Ambler, R. P., Meyer, T. E., and Kamen, M. D. 1976. Primary structure determination of two cytochromes c_2: Close similarity to functionally unrelated mitochondrial cytochrome *c*. *Proc. Natl. Acad. Sci. USA* **73**:472–475.
Andreu, J. M., Warth, R., and Muñoz, E. 1978. Glycoprotein nature of energy-transducing ATPases. *FEBS Lett.* **86**:1–5.
Apps, D. K. 1970. The NAD kinases of *Saccharomyces cerevisiae*. *Eur. J. Biochem.* **13**:223–230.
Awasthi, Y. C., Chuang, T. F., Keeman, T. W., and Crane, F. L. 1971. Tightly bound cardiolipin in cytochrome oxidase. *Biochim. Biophys. Acta* **226**:42–52.
Baccarini-Melandri, A., Jones, O. T. G., and Hauska, G. 1978. Cytochrome c_2—An electron carrier shared by the respiratory and photosynthetic electron transport chain of *Rhodopseudomonas capsulata*. *FEBS Lett.* **86**:151–154.
Baird, B. A., and Hammes, G. G. 1979. Structure of oxidative- and photophosphorylation coupling factor complexes. *Biochim. Biophys. Acta* **549**:31–53.
Barber, D., Parr, S. R., and Greenwood, C. 1976. Some spectral and steady state kinetic properties of *Pseudomonas* cytochrome oxidase. *Biochem. J.* **157**:431–438.
Barrett, J. 1956. The prosthetic group of cytochrome a_2. *Biochem. J.* **64**:626–639.
Bartsch, R. G. 1977. Cytochromes. *In*: Clayton, R. K., and Sistrom, W. R. eds., *The Photosynthetic Bacteria*, New York, Plenum Press, pp. 249–280.
Baum, H., Rieske, J. S., Sillman, H. I., and Lipton, S. H. 1967. On the mechanism of electron transfer in complex III of the electron transfer chain. *Proc. Natl. Acad. Sci. USA* **57**:798–805.
Beinert, H. 1963. Electron-transferring flavoproteins, *In*: Boyer, P. D., Lardy, H., and Myrbäck, K., eds., *The Enzymes*, 2nd ed., Vol. 7, pp. 467–476.
Bell, R. L., and Capaldi, R. A. 1976. The polypeptide composition of ubiquinone-cytochrome *c* reductase (complex III) from beef heart mitochondria. *Biochemistry* **15**:996–1001.
Berden, J. A., and Opperdoes, F. R. 1972. An antimycin-sensitive cytochrome *b* component in beef-heart mitochondria. *Biochim. Biophys. Acta* **267**:7–14.

Berg, A., Ingelman-Sundberg, M., and Gustafsson, J. A. 1979. Purification and characterization of cytochrome P-450_{meg}. *J. Biol. Chem.* **254**:5264–5271.

Bertoli, E., Parenti-Castelli, G., Sechi, M., Trigari, G., and Lenaz, G. 1978. A requirement for ubiquinone in ATPase activity and oxidative phosphorylation. *Biochem. Biophys. Res. Commun.* **85**:1–6.

Bisalputra, T., Brown, D. L., and Weier, T. E. 1969. Possible respiratory sites in a blue-green alga, *Nostoc sphaericum,* as demonstrated by potassium tellurite and TNBT reduction. *J. Ultrastruct. Res.* **27**:182–197.

Blazy, B., Thusius, D., and Baudras, A. 1976. Mechanism of yeast cytochrome b_2 action. *Biochemistry* **15**:257–261.

Boardman, N. K., and Anderson, J. M. 1967. Fractionation of the photochemical systems of photosynthesis. II. Cytochrome and carotenoid contents of particles isolated from spinach chloroplasts. *Biochim. Biophys. Acta.* **143**:187–203.

Böhme, H. 1976. Photoreactions of cytochrome b_6 and cytochrome *f* in chloroplast photosystem I fragments. *Z. Naturforsch.* **31c**:68–77.

Bonner, W. D. 1964. Plant cytochromes. *Sixth Int. Congr. Biochem. N.Y.* **4**:291–292.

Bonner, W. D., and Slater, E. C. 1970. Effect of antimycin on the potato mitochondrial cytochrome *b* system. *Biochim. Biophys. Acta* **223**:349–353.

Botelho, L. H., Ryan, D. E., and Levin, W. 1979. Amino acid composition and partial amino acid sequences of three highly purified forms of liver microsomal cytochrome P-450 from rats treated with polychlorinated biphenyls, phenobarbital, or 3-methylcholanthrene. *J. Biol. Chem.* **254**:5635–5640.

Boulter, D. 1973. The molecular evolution of higher plant cytochrome *c*. *Pure Appl. Chem.* **34**:539–552.

Boulter, D., Laycock, M. V., Ramshaw, J. A. M., and Thompson, E. W. 1970. Amino acid sequence studies of plant cytochrome *c* with particular reference to mung bean cytochrome *c*. *In*: Harbourne, J. B., ed., *Phytochemical Phylogeny,* New York, Academic Press, pp. 179–186.

Boyer, P. D. 1977. Coupling mechanisms in capture, transmission, and use of energy. *Annu. Rev. Biochem.* **46**:957–966.

Briggs, M., Kamp, P. F., Robinson, N. C., and Capaldi, R. A. 1975. The subunit structure of the cytochrome *c* oxidases complex. *Biochemistry* **14**:5123–5128.

Bright, H. J., and Porter, D. J. T. 1975. Flavoprotein oxidases. *In*: Boyer, P. D., ed., *The Enzymes,* New York, Academic Press, Vol. 12B, pp. 421–505.

Broda, E. 1971. The origins of bacterial respiration. *In*: Buvet, R., and Ponnamperuma, C., eds., *Chemical Evolution and the Origin of Life*, Amsterdam, North-Holland, pp. 446–452.

Brodie, A. F., and Gray, C. T. 1956. Activation of coupled oxidative phosphorylation in bacterial particulates by a soluble factor (S). *Biochim. Biophys. Acta* **19**:384–386.

Brown, G. G., and Beattie, D. S. 1977. Role of coenzyme Q in the mitochondrial respiratory chain. *Biochemistry* **16**:4449–4454.

Brown, R. H., and Boulter, D. 1974. The amino acid sequences of cytochrome *c* from four plant sources. *Biochem. J.* **137**:93–100.

Bruder, G., Fink, A., and Jarasch, E. D. 1978. The *b*-type cytochrome in ER of mammary gland epithelium and milk fat globule membranes consists of two components, cytochrome b_5 and cytochrome P-420. *Exp. Cell Res.* **117**:207–218.

Burton, K. 1959. Formation constants for the complexes of ADP or ATP with magnesium or calcium ions. *Biochem. J.* **71**:388–395.

Burton, S. D., Morita, R. Y., and Miller, W. 1966. Utilization of acetate by *Beggiatoa*. *J. Bacteriol.* **91**:1192–1200.

Buse, G., and Steffens, G. J. 1978. Studies on cytochrome *c* oxidase. II. The chemical constitution of a short polypeptide from the beef heart enzyme. *Hoppe-Seyler's Z. Physiol. Chem.* **359**:1005–1009.

Buvet, R., and LePort, L. 1973. Non-enzymic origin of the metabolism. *Space Life Sci.* **4**:434–447.

Callely, A. G., Rigopoulos, N., and Fuller, R. C. 1968. The assimilation of carbon by *Chloropseudomonas ethylicum*. *Biochem. J.* **106**:615–622.

Campbell, W. H., Orme-Johnson, W. H., and Burris, R. H. 1973. A comparison of the physical and chemical properties of four cytochromes *c* from *Azotobacter vinelandii*. *Biochem. J.* **135**:617–630.

Capaldi, R. A., and Briggs, M. 1976. The structure of cytochrome oxidase. *In:* Martonosi, A., ed., *The Enzymes of Biological Membranes,* New York, Plenum Press, Vol. 4, pp. 87–102.

Capaldi, R. A., Sweetland, J., and Merli, A. 1977. Polypeptides in the succinate-coenzyme Q reductase segment of the respiratory chain. *Biochemistry* **16**:5707–5710.

Carlson, S. S., Moss, G. A., Wilson, A. C., Mead, R. T., Wolin, L. D., Bowers, S. F., Foley, N. T., Muijsers, A. O., and Margoliash, E. 1977. Primary structure of mouse, rat, and guinea pig cytochrome *c*. *Biochemistry* **16**:1437–1442.

Case, G. D., Ohnishi, T., and Leigh, J. S. 1976. Intramitochondrial positions of ribiquinone and iron-sulfur centres determined by dipolar interactions with paramagnetic ions. *Biochem. J.* **160**:785–795.

Castor, L. N., and Chance, B. 1959. Photochemical determinations of the oxidases of bacteria. *J. Biol. Chem.* **234**:1587–1592.

Chance, B. 1958. The kinetics and inhibition of cytochrome components of the succinic oxidase system. III. Cytochrome *b*. *J. Biol. Chem.* **233**:1223–1229.

Chance, B. 1977. Electron transfer: Pathways, mechanisms, and controls. *Annu. Rev. Biochem.* **46**:967–980.

Chance, B., Bonner, W. D., and Storey, B. T. 1968. Electron transport in respiration. *Annu. Rev. Plant Physiol.* **19**:295–320.

Chance, B., Wilson, D. F., Dutton, P. L., and Erecińska, M. 1970. Energy-coupling mechanisms in mitochondria: Kinetic, spectroscopic, and thermodynamic properties of an energy-transducing form of cytochrome *b*. *Proc. Natl. Acad. Sci. USA* **66**:1175–1182.

Chiang, Y. L., and Coon, M. J. 1979. Comparative study of two highly purified forms of liver microsomal cytochrome P-450. *Arch. Biochem. Biophys.* **195**:178–187.

Chiang, Y. L., and King, T. E. 1979. Cytochrome c_1 complexes. *J. Biol. Chem.* **254**:1845–1853.

Clayton, R. K., and Sistrom, W. R. 1964. The importance of reaction centers for the photochemistry of photosynthesis. *Proc. Natl. Acad. Sci. USA* **52**:67–74.

Cookson, D. J., Moore, G. R., Pitt, R. C., Williams, R. J. P., Campbell, I. D., Ambler, R. P., Bruschi, M., and Le Gall, J. 1978. Structural homology of cytochromes *c*. *Eur. J. Biochem.* **83**:261–275.

Copenhaver, J. H., and Lardy, H. A. 1952. Oxidative phosphorylations: Pathways and yield in mitochondrial preparations. *J. Biol. Chem.* **195**:225–238.

Crespi, H. L., Smith, U., Gajda, L., Tisue, T., and Ammeraal, R. M. 1972. Extraction and purification of ^{1}H, ^{2}H, and isotope hybrid algal cytochrome, ferredoxin, and flavoprotein. *Biochim. Biophys. Acta* **256**:611–618.

Csonka, L. N., and Fraenkel, D. G. 1977. Pathways of NADPH formation in *E. coli*. *J. Biol. Chem.* **252**:3382–3391.

Cusanovich, M. A., and Edmundson, D. E. 1971. The isolation and characterization of *Rhodospirillum rubrum* flavodoxin. *Biochem. Biophys. Res. Commun.* **45**:327–336.

Dailey, H. A., and Strittmatter, P. 1978. Structural and functional properties of the membrane binding segment of cytochrome b_5. *J. Biol. Chem.* **253**:8203–8209.

Dalziel, K. 1975. Kinetics and mechanism of nicotinamide-nucleotide-linked dihydrogeneses. *In:* Boyer, P. D., ed., *The Enzymes,* New York, Academic Press, Vol. 11A, pp. 1–60.

Daniel, R. M. 1979. Occurrence and role of ubiquinone in electron transport to oxygen and nitrate in aerobically, anaerobically, and symbiotically grown *Rhizobium japonicum*. *J. Gen. Microbiol.* **110**:333–337.

D'Anna, J. A., and Tollin G. 1972. Studies of flavin–protein interaction in flavoproteins using protein fluorescence and circular dichroism. *Biochemistry* **11**:1073–1080.

Davenport, H. E. 1952. Cytochrome components in chloroplasts. *Nature* (*London*) **170**:1112–1114.

Davis, K. A., Hatefi, Y., Poff, K. L., and Butler, W. L. 1972. The *b*-type cytochromes of beef heart mitochondria. *Biochem. Biophys. Res. Commun.* **46**:1984–1990.

Davis, K. A., Hatefi, Y., Poff, K. L., and Butler, W. L. 1973. The *b*-type cytochromes of bovine heart mitochondria: Absorption spectra, enzymatic properties, and distribution in the electron transfer complexes. *Biochim. Biophys. Acta* **325**:341–356.

Dayhoff, M. O. 1972. *Atlas of Protein Sequence and Structure,* Vol. 5, Silver Springs, Md., National Biomedical Research Foundation.

Dayhoff, M. O. 1973. *Atlas of Protein Sequence and Structure,* Vol. 5, Suppl. 1, Washington, D.C., National Biomedical Research Foundation.

Dayhoff, M. O. 1976. *Atlas of Protein Sequence and Structure,* Vol. 5, Suppl. 2, Silver Springs, Md., National Biomedical Research Foundation.

DeLange, R. J., Glazer, A. M., and Smith, E. L. 1969. Presence and location of an unusual amino acid, ε-*N*-trimethyllysine, in cytochrome *c* of wheat germ and *Neurospora*. *J. Biol. Chem.* **244**:1385–1388.

DeLange, R. J., Glazer, A. M., and Smith, E. L. 1970. Identification and location of ε-*N*-trimethyllysine in yeast cytochrome *c*. *J. Biol. Chem.* **245**:3325–3327.

DePierre, J. W., and Ernster, L. 1977. Enzyme topology of intracellular membranes. *Annu. Rev. Biochem.* **46**:201–262.

Dickerson, R. E., and Timkovich, R. 1975. Cytochromes *c*. *In:* Boyer, P. D., ed., *The Enzymes,* New York, Academic Press, Vol. 11A, pp. 397–547.

Dickerson, R. E., Timkovich, R., and Almassy, R. J. 1976. The cytochrome fold and the evolution of bacterial energy metabolism. *J. Mol. Biol.* **100**:473–491.

Dillon, L. S. 1978. *The Genetic Mechanism and the Origin of Life,* New York, Plenum Press.

DiMaria, P., Polastro, E., DeLange, R. J., Kim, S., and Paik, W. K. 1979. Studies on cytochrome *c* methylation in yeast. *J. Biol. Chem.* **254**:4645–4652.

Dolin, M. I. 1961. Survey of microbial electron transport mechanisms. *In:* Gunsalus, I. C., and Stanier, R. Y., eds., *Bacteria: A Treatise on Structure and Function,* New York, Academic Press, Vol. 2, pp. 319–363.

Dubourdieu, M., and Fox, J. L. 1977. Amino acid sequence of *Desulfovibrio vulgaris* flavodoxin. *J. Biol. Chem.* **252**:1453–1463.

DuPraw, E. J. 1968. *Cell and Molecular Biology,* New York, Academic Press.

Dutton, P. L., Wilson, D. F., and Lee, C. P. 1970. Oxidation–reduction potentials of cytochromes in mitochondria. *Biochemistry* **9**:5077–5082.

Dutton, P. L., Wilson, D. F., and Lee, C. P. 1971. Energy dependence of oxidation–reduction potentials of the *b* and *c* cytochromes in beef heart submitochondrial particles. *Biochem. Biophys. Res. Commun.* **43**:1186–1191.

Elhammer, Å., Dallner, G., and Omura, T. 1978. Glycosylation of rat liver cytochrome b_5 on the ribosomal level. *Biochem. Biophys. Res. Commun.* **84**:572–580.

Erecińska, M., Blaise, J. K., and Wilson, D. F. 1977. Orientation of the hemes of cytochrome *c* oxidase and cytochrome *c* in mitochondria. *FEBS Lett.* **76**:235–239.

Ernster, L. 1977. Chemical and chemiosmotic aspects of electron transport-linked phosphorylation. *Annu. Rev. Biochem.* **46**:981–995.

Eytan, G. D., and Schatz, G. T. 1975. Cytochrome *c* oxidase from baker's yeast: Arrangement of the subunits in the isolated and membrane-bound enzyme. *J. Biol. Chem.* **250**:767–774.

Falmange, P., Vanderwinkle, E., and Wiane, J. M. 1965. Mise en evidence de deux malate synthases chez *Escherichia coli*. *Biochim. Biophys. Acta* **99**:246–258.

Fan, C. C., Lin, J. P. F., and Plaut, G. W. E. 1975. Effects of temperature on diphosphopyridine nucleotide-linked isocitrate dehydrogenase from bovine heart. *J. Biol. Chem.* **250**:2022–2027.

Fauque, G., Bruschi, M., and LeGall, J. 1979. Purification and some properties of cytochrome c_{553} ($_{550}$) isolated from *Desulfovibrio desulfuricans* Norway. *Biochem. Biophys. Res. Commun.* **86**:1020–1029.

Ferguson-Miller, S., Brautigon, D. L., and Margoliash, E. 1976. Correlation of the kinetics of electron transfer activity of various cytochromes *c* with binding to mitochondrial cytochrome *c* oxidase. *J. Biol. Chem.* **251**:1104–1115.

Fox, J. L., Smith, S. S., and Brown, J. R. 1972. Amino acid sequences of *Clostridium pasteurianum* flavodoxin. *Z. Naturforsch.* **27b**:1096–1100.

Fritz, I. B., and Beyer, R. E. 1969. Apparent respiratory control in uncoupled mitochondria. *J. Biol. Chem.* **244**:3075–3083.

Fry, M., and Green, D. E. 1979. Ion-channel component of cytochrome oxidase. *Proc. Natl. Acad. Sci. USA* **76**:2664–2668.

Fukumori, Y., and Yamanaka, T. 1979. Flavocytochrome *c* of *Chromatium vinosum. J. Biochem.* **85**:1405–1414.

Fuller, R. C., Smillie, F. M., Sisler, E. C., and Kornberg, H. L. 1961. Carbon metabolism in *Chromatium. J. Biol. Chem.* **236**:2140–2149.

Gellerfors, P., and Nelson, B. D. 1975. Analysis of the peptide composition of purified beef-heart complex III by dodecylsulfate electrophoresis. *Eur. J. Biochem.* **52**:433–443.

Gellerfors, P., and Nelson, B. D. 1977. Topology of the peptides in free and membrane-bound complex III (ubiquinol-cytochrome *c* reductase) as revealed by lactoperoxidase and *p*-diazonitrobenzene [^{35}S]sulfonate labeling. *Eur. J. Biochem.* **80**:275–282.

Gillespie, R. J., Maw, G. A., and Vernon, C. A. 1953. The concept of phosphate bond-energy. *Nature (London)* **171**:1147–1149.

Giorgio, N. A., Yip, A. T., Fleming, J., and Plaut, G. W. E. 1970. Diphosphopyridine nucleotide-linked isocitrate dehydrogenase from bovine heart. *J. Biol. Chem.* **245**:5469–5477.

Glass, T. L., Bryant, M. P., and Wolin, M. J. 1977. Partial purification of ferredoxin from *Ruminococcus albus* and its role in pyruvate metabolism and reduction of NAD by H_2. *J. Bacteriol.* **131**:463–472.

Gleason, F. H. 1968. Respiratory electron transport systems of aquatic fungi. I. *Leptomitus lacteus* and *Apodachyla punctata. Plant Physiol.* **43**:597–605.

Goewert, R. R., Sippel, C. J., and Olson, R. E. 1977. The isolation and identification of a novel intermediate in ubiquinone-6 biosynthesis by *S. cerevisiae. Biochem. Biophys. Res. Commun.* **77**:599–605.

Goldberg, E., Sberna, D., Wheat, T. E., Urbanski, G. J., and Margoliash, E. 1977. Cytochrome *c:* Immunofluorescent localization of the testis-specific form. *Science* **196**:1010–1012.

Gómez-Puijou, M. T., Beigel, M., and Gómez-Puijou, A. 1976. On the problem of site specific agents in oxidative phophorylation. *In:* Packer, L., and Gómez-Puijou, A., eds., *Mitochondria: Bioenergetics, Biogenesis, and Membrane Structure,* New York, Academic Press, pp. 155–165.

Green, D. E., and Blondin, G. A. 1975. Molecular mechanism of mitochondrial energy coupling. *BioScience* **28**:18–24.

Griffiths, D. E. 1976. Biochemical genetic studies of oxidative phosphorylation. *In:* Packer, L., and Gómez-Puijou, A., eds., *Mitochondria: Bioenergetics, Biogenesis, and Membrane Structure,* New York, Academic Press, pp. 265–274.

Griffiths, M. M., and Bernofsky, C. 1972. Purification and properties of reduced diphosphopyridine nucleotide kinase from yeast mitochondria. *J. Biol. Chem.* **247**:1473–1478.

Groudinsky, O. 1971. Study of heme-protein linkage in cytochrome *b. Eur. J. Biochem.* **18**:480–484.

Gudat, J., Singh, J., and Wharton, D. C. 1973. Cytochrome oxidase from *Pseudomonas aeruginosa. Biochim. Biophys. Acta* **292**:376–398.

Guerrieri, F., and Nelson, B. D. 1975. Studies on the characteristics of a proton pump in phospholipid vesicles inlayed with purified complex III from beef heart mitochondria. *FEBS Lett.* **54**:339–342.

Guiard, B., Groudinsky, O., and Lederer, F., 1974. Homology between baker's yeast cytochrome b_2 and liver microsomal cytochrome b_5. *Proc. Natl. Acad. Sci. USA* **71**:2539–2543.

Guiard, B., Lederer, F., and Jacq, C. 1975. More similarity between baker's yeast L(+)-lactate dehydrogenase and liver microsomal cytochrome b_5. *Nature (London)* **255**:422–423.

Haddock, B. A., and Jones, C. W. 1977. Bacterial respiration. *Bacteriol. Rev.* **41**:47–99.

Hägele, E., Neeff, J., and Necke, D. 1978. The malate dehydrogenase isozymes of *S. cerevisiae*. *Eur. J. Biochem.* **83**:67–76.

Hagihara, B., Sato, N., and Yamanaka, T. 1975. Type *b* cytochromes. *In:* Boyer, P. D., ed., *The Enzymes,* New York, Academic Press, Vol. 11A, pp. 549–593.

Haneishi, T., and Shirasaka, M. 1968. Comparison of yeast cytochrome *c*. *In:* Okuniki, K., Kamen, M. D., and Sekuzu, I., eds., *Structure and Function of Cytochromes,* Tokyo, University of Tokyo Press, pp. 404–412.

Harano, T., and Omura, T. 1977. Biogenesis of endoplasmic reticulum membrane in rat liver cells. *J. Biochem.* **82**:1541–1549.

Harmon, H. J., and Crane, F. L. 1976. Inhibition of mitochondrial electron transport by hydrophilic metal chelators. *Biochim. Biophys. Acta* **440**:45–58.

Harmon, H. J., Hall, J. D., and Crane, F. L. 1974. Structure of mitochondrial cristae membranes. *Biochim. Biophys. Acta* **344**:119–155.

Hartman, H. 1975. Speculations on the origin and evolution of metabolism. *J. Mol. Evol.* **4**:359–370.

Hase, T., Wada, K., and Matsubara, H. 1976a. Amino acid sequence of the major component of *Aphanothece sacrum* ferredoxin. *J. Biochem.* **79**:329–343.

Hase, T., Wada, K., Ohmiya, M., and Matsubara, H. 1976b. Amino acid sequence of the major component of *Nostoc muscorum* ferredoxin. *J. Biochem.* **80**:993–999.

Hase, T., Wakabayashi, S., Matsubara, H., Kerscher, L., Oesterhelt, D., Rao, K. K., and Hall, D. O. 1977a. *Halobacterium halobium* ferredoxin. *FEBS Lett.* **77**:308–310.

Hase, T., Wada, K., and Matsubara, H. 1977b. Horsetail (*Equisetum telmateia*) ferredoxins I and II. Amino acid sequences. *J. Biochem.* **82**:267–276.

Hase, T., Wada, K., and Matsubara, H. 1977c. Horsetail (*Equisetum arvense* ferredoxins I and II. Amino acid sequences and gene duplication. *J. Biochem.* **82**:277–286.

Hase, T., Matsubara, H., and Evans, M. C. W. 1977d. Amino acid sequence of *Chromatium vinosum* ferredoxin: Revisions. *J. Biochem.* **81**:1745–1749.

Hase, T., Wakabayashi, S., Matsubara, H., Evans, M. C. W., and Jennings, J. N. 1978a. Amino acid sequence of a ferredoxin from *Chlorobium thiosulfatophilum* strain tassajara, a photosynthetic green sulfur bacterium. *J. Biochem.* **83**:1321–1325.

Hase, T., Wakabayashi, S., Wada, K., and Matsubara, H. 1978b. Amino acid sequences of *Aphanothece sacrum* ferredoxin II. *J. Biochem.* **83**:761–770.

Hase, T., Wakabayashi, S., Matsubara, H., Kerscher, L., Oesterhelt, D., Rao, K. K., and Hall, D. O. 1978c. Complete amino acid sequence of *Halobacterium halobium* ferredoxin containing an N^{ϵ} acetyllysine residue. *J. Biochem.* **83**:1657–1670.

Hase, T., Wakabayashi, S., Matsubara, H., Ohmori, D., and Suzuki, K. 1978d. *Pseudomonas ovalis* ferredoxin: Similarity to *Azotobacter* and *Chromatium* ferredoxins. *FEBS Lett.* **91**:315–319.

Hase, T., Wakabayashi, S., Wada, K., Matsubara, H., Jüttner, F., Rao, K. K., Fry, I., and Hall, D. O. 1978e. *Cyanidium caldarum* ferredoxin: A red algae type? *FEBS Lett.* **96**:41–44.

Hatefi, Y. 1966. The functional complexes of the electron-transfer system. *Compr. Biochem.* **14**:199–231.

Hatefi, Y., and Stiggall, D. L. 1976. Metal-containing flavoprotein dehydrogenases. In: Boyer, P. D., ed., *The Enzymes,* New York, Academic Press, Vol. 13C, pp. 175–297.

Hatefi, Y., Haavik, A. G., Fowler, L. R., and Griffiths, D. E. 1962. Studies on the electron transfer system. XLII. Reconstitution of the electron transfer system. *J. Biol. Chem.* **237**:2661–2669.

Haugen, D. A., van der Hoeven, T. A., and Coon, M. J. 1975. Purified liver microsomal cytochrome P-450. Separation and characterization of multiple forms. *J. Biol. Chem.* **250**:3567–3570.

Hensagens, L. A. M., Grivell, L. A., Borst, P., and Bos, J. L. 1979. Nucleotide sequence of the mitochondrial structural gene for subunit 9 of yeast ATPase complex. *Proc. Natl. Acad. Sci. USA* **76**:1663–1667.

Hermans, L. 1979. Purification of mitochondrial NADH dehydrogenase from *Drosophila hydei* and comparisons with the 'heat shark' polypeptides. *Biochim. Biophys. Acta* **567**:125–134.

Hill, R. 1954. The cytochrome *b* component of chloroplasts. *Nature (London)* **174**:501–503.

Hill, T. J. 1979. Steady-state coupling of four membrane systems in mitochondrial oxidative phosphorylation. *Proc. Natl. Acad. Sci. USA* **76**:2236–2238.

Hind, G., and Olson, J. M. 1966. Light-induced changes in cytochrome b_6 in spinach chloroplasts. *Brookhaven Symp. Biol.* **19**:188–194.

Hinkle, P. C., and Yu, M. L. 1979. The phosphorus/oxygen ratio of mitochondrial oxidative phosphorylation. *J. Biol. Chem.* **254**:2450–2455.

Hiroshi, I., Komogata, K., and Uchino, C. 1964. Studies on the microorganisms of cereal grains. VI. *Aerobacter, Bacillus,* and *Micrococcus* isolated from rice. *Agric. Biol. Chem.* **28:**A_1.

Hoare, D. S., Hoare, S. L., and Moore, R. B. 1967. The photoassimilation of organic compounds by autotrophic blue-green algae. *J. Gen. Microbiol.* **49**:351–370.

Horio, T. 1958a. Purification of cytochromes from *Pseudomonas aeruginosa. J. Biochem.* **45**:195–205.

Horio, T. 1958b. Some physical and physiological properties of purified cytochromes of *Pseudomonas aeruginosa. J. Biochem.* **45**:267–279.

Horio, T., Higashi, T., Yamanaka, T., Matsubara, H., and Okunuki, K. 1961. Purification and properties of cytochrome oxidase from *Pseudomonas aeruginosa. J. Biol. Chem.* **236**:944–951.

Hutson, K. G., Rogers, L. J., Haslett, B. G., Boulter, D., and Cammack, R. 1978. Comparative studies on two ferredoxins from the cyanobacterium *Nostoc* strain MAC. *Biochem. J.* **172**:465–477.

Ichikawa, Y., and Yamano, T. 1965. Cytochrome 559 in the microsomes of the adrenal medulla. *Biochem. Biophys. Res. Commun.* **20**:263–268.

Ikegami, I., Katoh, S., and Takamiya, A. 1968. Nature of heme moiety and oxidation–reduction potential of cytochrome 558 in *Euglena* chloroplasts. *Biochim. Biophys. Acta* **162**:604–606.

Ito, A. 1971. Hepatic sulfite oxidase identified as cytochrome b_5-like pigment extractable from mitochondria by hypotonic treatment. *J. Biochem.* **70**:1061–1064.

Jablonski, E., and DeLuca, M. 1977. Purification and properties of NADH and NADPH specific FMN oxido-reductases from *Beneckea harveyi. Biochemistry* **16**:2932–2936.

Jacobs, E. E., and Sanadi, D. R. 1960. The reversible removal of cytochrome *c* from mitochondria. *J. Biol. Chem.* **235**:531–534.

Jacobs, N. J., and Conti, S. F. 1965. Effect of hemin on the formation of the cytochrome system of anaerobically grown *Staphylococcus epidermidis. J. Bacteriol.* **89**:675–679.

Jefcoate, C. R. E., and Gaylor, J. L. 1969. Ligand interactions with hemoprotein P-450. *Biochemistry* **8**:3464–3472.

Jeffreys, A. J., and Craig, I. W. 1977. Differences in the mitochondrially synthesized subunits of human and mouse cytochrome *c* oxidase. *FEBS Lett.* **77**:151–154.

Johnson, E. F., and Muller-Eberhard, U. 1977a. Resolution of multiple forms of rabbit liver cytochrome P-450. *Am. Chem. Soc. Symp. Ser.* **44**:72–80.

Johnson, E. F., and Muller-Eberhard, U. 1977b. Multiple forms of cytochrome P-450. Resolution and purification of rabbit liver aryl hydrocarbon hydroxylase. *Biochem. Biophys. Res. Commun.* **76**:644–651.

Johnson, E. F., and Muller-Eberhard, U. 1977c. Purification of the major cytochrome P-450 of liver microsomes from rabbits treated with 2,3,7,8-tetrachlorodibenzo-*p*-dioxin (TCDD). *Biochem. Biophys. Res. Commun.* **76**:652–659.

Johnson, E. F., Zounes, M. C., and Muller-Eberhard, U. 1979. Characterization of three forms of rabbit microsomal cytochrome P-450 by peptide mapping utilizing limited proteolysis in sodium dodecyl sulfate and analysis by gel electrophoresis. *Arch. Biochem. Biophys.* **192**:282–289.

Jones, C. W., and Redfearn, E. R. 1966. Electron transport in *Azotobacter vinelandii*. *Biochim. Biophys. Acta* **113**:467–481.

Kagawa, Y. 1976. Reconstitution of the inner mitochondrial membrane. *In*: Martonosi, A., ed., *The Enzymes of Biological Membranes*, New York, Plenum Press, Vol. 4, pp. 125–142.

Kapke, G. F., Redick, J. A., and Baron, J. 1979. Immunohistochemical demonstration of an adrenal ferredoxin-like iron-sulfur protein in rat hepatic mitochondria. *J. Biol. Chem.* **253**:8604–8608.

Kaplan, N. O. 1972. Pyridine nucleotide transhydrogenases. *Harvey Lect.* **66**:105–133.

Katoh, S. 1959. Studies on the algal cytochromes of *c*-type. *J. Biochem.* **46**:629–632.

Keilin, D., and Hartree, E. F. 1940. Succinic dehydrogenease-cytochrome system of cells. *Proc. R. Soc. London Ser. B* **129**:277–306.

Kerscher, L., and Oesterhelt, D. 1977. Ferredoxin is the coenzyme of α-ketoacid oxidoreductases in *Halobacterium halobium*. *FEBS Lett.* **83**: 197–201.

Kessler, R. J., Blondin, G. A., Van de Zande, H., Haworth, R. A., and Green, D. E. 1977. Coupling in cytochrome *c* oxidase. *Proc. Natl. Acad. Sci. USA* **74**:3662–3666.

Kikuchi, G., and Motokawa, Y. 1968. Cytochrome oxidase of *Rhodopseudomonas sphaeroides*. *In*: Okunuki, K., Kamin, M. D., and Sekuzu, I., eds., *Structure and Function of Cytochromes*, Tokyo, University of Tokyo Press, pp. 174–181.

Kita, K., Yamato, I., and Anraku, Y. 1978. Purification and properties of cytochrome b_{556} in the respiratory chain of aerobically grown *E. coli* K12. *J. Biol. Chem.* **253**:8910–8915.

Knaff, D. B. 1977. The role of cytochromes b_6 and f in cyclic electron flow in a blue-green alga. *Arch. Biochem. Biophys.* **182**:540–545.

Konstantinov, A. A., and Ruuge, E. A. 1977. Semiquinone Q in the respiratory chain of electron transport particles. *FEBS Lett.* **81**:137–141.

Korb, H., and Neupert, W. 1978. Biogenesis of cytochrome *c* in *Neurospora crassa*. *Eur. J. Biochem.* **91**:609–620.

Kornberg, H. L., and Sadler, J. R. 1961. Metabolism of C_2-compounds in microorganisms. VIII. A dicarboxylic acid cycle as a route for the oxidation of glycolate by *E. coli*. *Biochem. J.* **81**:503–513.

Krampitz, L. O. 1961. Cyclic mechanisms of terminal oxidation. *In*: Gunsalus, I. C., and Stanier, R. Y., eds., *Bacteria: A Treatise on Structure and Function*, New York, Academic Press, Vol. 2, pp. 209–256.

Kuboyama, M., Yong, F. C., and King, T. E. 1972. Studies on cytochrome oxidase. *J. Biol. Chem.* **247**:6375–6383.

Künze, U., and Junge, W. 1977. Ellipticity of cytochrome a_3 and rotational mobility of cytochrome *c* oxidase in the cristae membrane of mitochondria. *FEBS Lett.* **80**:429–434.

Kuronen, T., and Ellfolk, N. 1972. A new purification procedure and molecular properties of *Pseudomonas* cytochrome oxidase. *Biochim. Biophys. Acta* **275**:308–318.

Kuwahara, M., and Chaykin, S. 1973. Biosynthesis of pyridine nucleotides in early embryos of the mouse (*Mus musculus*). *J. Biol. Chem.* **248**:5095–5099.

Lance, C., and Bonner, W. D. 1968. The respiratory chain components of higher plant mitochondria. *Plant Physiol.* **43**:756–766.

Lawford, H. G., Cox, J. C., Garland, P. B., and Haddock, B. A. 1976. Electron transport in aerobically-grown *Paracoccus dinitrificans*. *FEBS Lett.* **64**:369–374.

Laycock, M. V. 1975. The amino acid sequence of cytochrome *f* from the brown alga *Alaria esculenta* (L.) Grev. *Biochem. J.* **149**:271–279.

Lee, C. P., Sottocasa, G. L., and Ernster, L. 1967. Use of artificial electron acceptors for abbreviated phosphorylating electron transport: Flavin–cytochrome *c*. *Methods Enzymol.* **10**:33–37.

Lemberg, R., and Barrett, J. 1973. *Cytochromes,* New York, Academic Press.

Leung, K. H., and Hinkle, P. C. 1975. Reconstitution of ion transport and respiratory control in vesicles formed from reduced CoQ-cytochrome *c* reductase and phospholipids. *J. Biol. Chem.* **250**:8467–8471.

Lindsay, J. G., Dutton, P. L., and Wilson, D. F. 1972. Energy-dependent effects on the oxidation–reduction midpoint potentials of the *b* and *c* cytochromes in phosphorylating submitochondria particles from pigeon heart. *Biochemistry* **11**:1937–1942.

Ludwig, B., Downer, N. W., and Capaldi, R. A. 1979. Labeling of cytochrome *c* oxidase with (^{35}S)diazobenzenesulfonate. Orientation of this electron transfer complex in the inner mitochondrial membrane. *Biochemistry* **18**:1401–1407.

Lundegårdh, H. 1962. The respiratory system of wheat roots. *Biochim. Biophys. Acta* **57**:352–358.

Lundegårdh, H. 1964. Actin spectra of the reducing and oxidizing systems in spinach chloroplasts. *Biochim. Biophys. Acta* **88**:37–56.

Mackey, L. N., Kuwana, T., and Hartzell, C. R. 1973. Evaluation of the energetics of cytochrome *c* oxidase in the absence of cytochrome *c*. *FEBS Lett.* **36**:326–329.

MacKnight, M. L., Gray, W. R., and Tollin, G. 1974. N-terminal amino acid sequences of *Azotobacter vinelandii* and *Rhodospirillum rubrum* flavodoxins. *Biochem. Biophys. Res. Comm.* **59**:630–635.

McLaughlin, P. J., and Dayhoff, M. O. 1973. Eukaryote evolution: A view based on cytochrome *c* sequence data. *J. Mol. Evol.* **2**:99–166.

Marks, G. S. 1969. *Heme and Chlorophyll,* London, Van Nostrand.

Martin, E., and Mukkada, A. J. 1979. Respiratory chain components of *Leishmania tropica* promastigotes. *J. Protozool.* **26**:138–142.

Masaki, R., Wada, K., and Matsubara, H. 1977. Chemical modification of spinach ferredoxin. Properties of acetylated spinach ferredoxin, *J. Biochem.* **81**:1–9.

Mason, T. L., and Schatz, G. 1973. Cytochrome *c* oxidase from baker's yeast. II. Site of translation of the protein components. *J. Biol. Chem.* **248**:1355–1360.

Massey, V., and Hemmerich, P. 1975. Flavin and pteridine monooxygenases. *In*: Boyer, P. D., ed., *The Enzymes*, New York, Academic Press, Vol. 12B, pp. 191–252.

Mathews, F. S., and Czerwinski, E. W. 1976. Cytochrome b_5 and cytochrome b_5 reductase from a chemical and x-ray diffraction viewpoint. *In*: Martonosi, A., ed., *The Enzymes of Biological Membranes*, New York, Plenum Press, Vol. 4, pp. 143–197.

Matsubara, H., Hase, T., Wakabayashi, S., and Wade, K. 1978. Gene duplications during evolution of chloroplast type ferredoxins. *In*: Matsubara, H., and Yamanaka, eds., *Evolution of Protein Molecules*, Tokyo, Japanese Scientific Society Press, pp. 209–219.

Mattoon, J. R., and Sherman, F. 1966. Reconstitution of phosphorylating electron transport in mitochondria from a cytochrome *c*-deficient yeast mutant. *J. Biol. Chem.* **241**:4330–4338.

Matuda, S. 1979a. Biochemical studies on the muscle microsomes of *Ascaris lumbricoides* var. *suum*. I. Biochemical characterization and electron transport of *Ascaris* microsomes, *J. Biochem.* **85**:343–350.

Matuda, S. 1979b. Biochemical studies on the muscle microsomes of *Ascaris lumbricoides* var. *suum*. II. Purification and characterization of *b*-type cytochrome and NADH-ferricyanide reductase from *Ascaris* muscle microsomes. *J. Biochem.* **85**:351–358.

Mayhew, S. G., and Ludwig, M. L. 1975. Flavodoxins and electron-transferring proteins. *In*: Boyer, P. D., ed., *The Enzymes*, New York, Academic Press, Vol. 12B, pp. 57–118.

Meatyard, B. T., and Boulter, D. 1974. The amino acid sequence of cytochrome *c* from *Enteromorpha intestinalis*. *Phytochemistry* **13**:2777–2782.

Mendel-Hartvig, I., and Nelson, B. D. 1978. Labeling of complex III peptides in beef heart mitochondria and submitochondrial particle by diazonium benzene (^{35}S) sulfonate. *FEBS Lett.* **92**:36–40.

Merli, A., Capaldi, R. A., Ackrell, B. A. C., and Kearney, E. B. 1979. Arrangement of complex II (succinate-ubiquinone reductase) in the mitochondrial inner membrane. *Biochemistry* **18**:1393–1400.

Mével-Ninio, M. 1972. Subunit structure of L-lactate dehydrogenase (cytochrome b_2) of *S. cerevisiae*. *Eur. J. Biochem.* **25**:254–261.

Mével-Ninio, M., Risler, Y., and Labeyrie, F. 1977. Structural studies of yeast flavocytochrome b_2: cooperative roles of the α and β globules in the formation of the flavin-binding sites. *Eur. J. Biochem.* **73**:131–140.

Mitchell, P. 1961. Coupling of phosphorylation to electron and hydrogen transfer by a chemi-osmotic type of mechanism. *Nature (London)* 191:144–148.

Mitchell, P. 1977. Vectorial chemiosmotic processes. *Annu. Rev. Biochem.* **46**:996–1005.

Mitchell, P., and Moyle, J. 1967. Proton-transport phosphorylation, some experimental tests. *In*: Slater, E. C., Kaniuga, Z., and Wojtczak, L., eds., *Biochemistry of Mitchondria,* New York, Academic Press, pp. 53–74.

Moore, A. L., and Wilson, S. B. 1978. An estimation of the proton conductance of the inner membrane of turnip (*Brassica napae* L.) mitochondria. *Planta* **141**:297–302.

Moore, A. L., Bonner, W. D., and Rich, P. R. 1978a. The determination of the proton-motive force during cyanide-insensitive respiration in plant mitochondria. *Arch. Biochem. Biophys.* **186**:298–306.

Moore, A. L., Rich, P. R., and Bonner, W. D. 1978b. Factors influencing the components of the total proton-motive force in mung bean mitochondria. *J. Exp. Bot.* **29**:1–12.

Morton, R. K., and Sturtevant, J. M. 1964. The dehydrogenation of L-lactate in the presence of and absence of ferricyanide as electron acceptor. *J. Biol. Chem.* **239**:1614–1624.

Nicholls, P. 1964. Observations on the oxidation of cytochrome *c*. *Arch. Biochem. Biophys.* **106**:25–48.

Nicholls, P. 1974. Cytochrome *c* binding to enzymes and membranes. *Biochim. Biophys. Acta* **346**:261–310.

Niece, R. L., Margoliash, E., and Fitch, W. M. 1977. Complete amino acid sequence of guanaco (*Lama guanicoe*) cytochrome *c*. *Biochemistry* **16**:68–72.

Nisimoto, Y., Takeuchi, F., and Shibata, Y. 1977. Purifications and properties of a cytochrome b_5-like hemeprotein from mitochondrial outer membranes of rat liver. *J. Biochem.* **82**:1257–1266.

Nóbrega, F. G., and Ozols, J. 1971. Amino acid sequences of tryptic peptides of cytochrome b_5 from microsomes of human, monkey, porcine, and chicken liver. *J. Biol. Chem.* **246**:1706–1717.

Nóbrega, F. G., Araujo, P. S., Pasetto, M., and Rao, I. 1969. Some properties of cytochrome *b* from liver microsomes of man, monkey, pig and chicken. *Biochem. J.* **115**:849–856.

Nochumson, S., Durban, E., Kim, S., and Paik, W. K. 1977. Cytochrome *c*-specific protein methylase III from *Neurospora crassa*. *Biochem. J.* **165**:11–18.

Ochoa, S. 1943. Efficiency of aerobic phosphorylation in cell-free heart extracts. *J. Biol. Chem.* **151**:493–505.

Ohnishi, T., Kawaguchi, K., and Hagihara, B., 1966. Preparation and some properties of yeast mitochondria. *J. Biol. Chem.* **241**:1797–1806.

Okuda, J., Nagamine, J., and Yagi, K. 1979. Exchange of free and bound coenzyme of flavin enzymes studied with [^{14}C] FAD. *Biochim. Biophys. Acta* **566**:245–252.

Omura, T., and Sato, R. 1964. The carbon monoxide-binding pigment of liver microsomes. II. Solubilization, purification, and properties. *J. Biol. Chem.* **239**:2379–2385.

Ozols, J., and Strittmatter, P. 1968. The amino acid sequence of cytochrome b_5. *J. Biol. Chem.* **243**:3376–3381.

Ozols, J., and Strittmatter, P. 1969. Correction of the amino acid sequence of calf liver microsomal cytochrome b_5. *J. Biol. Chem.* **244**:6617–6618.

Pajot, P., and Claisse, M. L. 1974. Utilization by yeast of D-lactate and L-lactate as sources of energy in the presence of antimycin A. *Eur. J. Biochem.* **49**:275–285.

Pajot, P., and Groudinsky, O. 1970. Molecular weight and quaternary structure of yeast L-lactate dehydrogenase cytochrome b_2. *Eur. J. Biochem.* **12**:158–164.

Palmer, G., Babcock, G. T., and Vickery, L. E. 1976. A model for cytochrome oxidase. *Proc. Natl. Acad. Sci. USA* **73**:2206–2210.

Papa, S. 1976. Proton translocation reactions in the respiratory chain. *Biochim. Biophys. Acta* **456**:39–84.

Parr, S. R., Wilson, M. T., and Greenwood, C. 1974. The reactions of *Pseudomonas aeruginosa* cytochrome *c* oxidase with sodium metabisulphite. *Biochem. J.* **139**:273–276.

Parr, S. R., Barber, D., Greenwood, C., Phillips, B. W., and Melling, J. 1976. A purification procedure for the soluble cytochrome oxidase and some other respiratory proteins from *Pseudomonas aeruginosa*. *Biochem. J.* **157**:423–430.

Pearce, J., Leach, C. K., and Carr, N. G. 1969. The incomplete tricarboxylic acid cycle in the blue-green alga *Anabaena variabilis*. *J. Gen. Microbiol.* **55**:371–378.

Peck, H. D. 1962. The role of adenosine 5′-phosphosulfate in the reduction of the sulfate to sulfite in *Desulfovibrio desulfuricans*. *J. Biol. Chem.* **237**:198–203.

Peck, H. D., Deacon, T. E., and Davidson, J. T. 1965. Studies on adenosine 5′-phosphosulfate reductase from *Desulfovibrio desulfuricans* and *Thiobacillus thioparus*. *Biochim. Biophys. Acta* **96**:429–446.

Perlish, J. S., and Eichel, J. H. 1971. A succinate- and DPNH-reducible *o*-type cytochrome in mitochondrial preparations from *Tetrahymena pyriformis*. *Biochem. Biophys. Res. Commun.* **44**:973–980.

Petersen, L. C., and Andréasson, L. E. 1976. The reaction between oxidized cytochrome *c* and reduced cytochrome *c* oxidase. *FEBS Lett.* **66**:52–57.

Peterson, J. A., Ebel, R. E., O'Keefe, D. H., Matsubara, T., and Estabrook, R. W. 1976. Temperature dependence of cytochrome P-450 reduction. A model for NADPH-cytochrome P-450 reductase:cytochrome P-450 interaction. *J. Biol. Chem.* **251**:4010–4016.

Peterson, J. A., White, R. E., Yasukochi, Y., Coomes, M. L., O'Keeffe, D. H., Ebel, R. E., Masters, B. S. S., Ballou, D. P., and Coon, M. J. 1977. Evidence that purified liver microsomal cytochrome P-450 is a one-electron acceptor. *J. Biol. Chem.* **252**:4431–4434.

Petrucci, D., Amicarelli, F., and Paponetti, B. 1977. NAD^+-linked malic enzyme in mitochondria of amphibian oocytes. *Int. J. Biochem.* **8**:149–157.

Pettigrew, G. W., Leaver, J. L., Meyer, T. E., and Ryle, A. P. 1975. Purification, properties, and amino acid sequence of atypical cytochrome *c* from two protozoans, *Euglena gracilis* and *Crithidia oncopelta*. *Biochem. J.* **147**:291–302.

Phan, S. H., and Mahler, H. R. 1976a. Studies on cytochrome oxidase. Partial resolution of enzymes containing 7 or 6 subunits, from yeast and beef heart, respectively. *J. Biol. Chem.* **251**:257–269.

Phan, S. H., and Mahler, H. R. 1976b. Studies on cytochrome oxidase. Preliminary characterization of an enzyme containing only four subunits. *J. Biol. Chem.* **251**:270–276.

Pinchot, G. B. 1957a. A polynucleotide coenzyme of oxidative phosphorylation. *J. Biol. Chem.* **229**:1–9.

Pinchot, G. B. 1957b. A polynucleotide coenzyme of oxidative phosphorylation. II. *J. Biol. Chem.* **229**:25–37.

Postgate, J. R., and Campbell, L. L. 1966. Classification of *Desulfovibrio* species, the nonsporulating sulfate-reducing bacteria. *Bacteriol. Rev.* **30**:732–738.

Poyton, R. O., and Schatz, G. 1975. Cytochrome *c* oxidase from baker's yeast. *J. Biol. Chem.* **250**:762–766.

Pullman, B., and Pullman, A. 1960. Electronic structure of energy-rich phosphates. *Radiat. Res. Suppl.* **2**:160–181.

Quagliariello, E., and Palmieri, F. 1968. Control of succinate oxidation by succinate-uptake by rat-liver mitochondria. *Eur. J. Biochem.* **4**:20–27.

Ragan, C. I. 1976. The structure and subunit composition of the particulate NADH-ubiquinone reductase of bovine heart mitochondria. *Biochem. J.* **154**:295–305.

Rashid, M. A., Hagihara, B., Kobayashi, M., Tani, S., and Tsugita, A. 1973. Sequential studies in human liver cytochrome b_5. *J. Biochem.* **74**:985–1002.

Reddy, V. V. R., Kupfer, D., and Caspi, E. 1977. Mechanism of C-5 double bond introduction in the biosynthesis of cholesterol by rat liver microsomes. Evidence for the participation of microsomal cytochrome b_5. *J. Biol. Chem.* **252**:2797–2801.

Rieske, J. S. 1976. Composition, structure, and function of complex III in the respiratory chain. *Biochim. Biophys. Acta* **456**:195–247.

Ritchey, T. W., and Seeley, H. W. 1976. Distribution of cytochrome-like respiration in streptococci. *J. Gen. Microbiol.* **93**:195–203.

Robbi, M., Berthet, J., Trouet, A., and Beaufay, H. 1978a. The biosynthesis of rat liver cytochrome *c*. 1. Subcellular distribution of cytochrome *c*. *Eur. J. Biochem.* **84**:333–340.

Robbi, M., Berthet, J., and Beaufay, H. 1978b. The biosynthesis of rat-liver cytochrome *c*. 2. Subcellular distribution of newly synthesized cytochrome *c*. *Eur. J. Biochem.* **84**:341–346.

Rosén, S., Brändén, R., Vänngård, T., and Malmström, B. G. 1977. EPR evidence for an active form of cytochrome *c* oxidase different from the resting enzyme. *FEBS Lett.* **74**:25–30.

Ross, E., and Schatz, G. 1976a. Cytochrome c_1 of baker's yeast. I. Isolation and properties. *J. Biol. Chem.* **251**:1991–1996.

Ross, E., and Schatz, G. 1976b. Cytochrome c_1 of baker's yeast. II. Synthesis on cytoplasmic ribosomes and influence of oxygen and heme on accumulation of the apoprotein. *J. Biol. Chem.* **251**:1997–2004.

Rubin, M. S. 1972. Subunit structure of yeast and beef cytochrome oxidase. *Fed. Proc. Fed. Am. Soc. Exp. Biol.* **31**:3896.

Rubin, M. S., and Tzagoloff, A. 1973. Assembly of the mitochondrial membrane system; Purification, characterization, and subunit structure of yeast and beef heart cytochrome oxidase. *J. Biol. Chem.* **248**:4269–4274.

Rushbrook, J. I., and Harvey, R. A. 1978. NAD-dependent isocitrate dehydrogenase from beef heart. *Biochemistry* **17**:5339–5346.

Ryrie, I. J. 1977. Yeast mitochondrial ATPase complex. Purification, subunit composition, and some effects of protease inhibitors. *Arch. Biochem. Biophys.* **184**:464–475.

Salemme, F. R. 1977. Structure and function of cytochromes *c*. *Annu. Rev. Biochem.* **46**:299–329.

Saltzgaber-Müller, J., and Schatz, G. 1978. Heme is necessary for the accumulation and assembly of cytochrome *c* oxidase subunits in *S. cerevisiae*. *J. Biol. Chem.* **253**:305–310.

Sato, N., Wilson, D. F., and Chance, B. 1971. The spectral properties of the *b* cytochromes in intact mitochondria. *Biochim. Biophys. Acta* **253**:88–97.

Saunders, G. F., Campbell, L. L., and Postgate, J. R. 1964. Base composition of DNA of sulfate-reducing bacteria deduced from buoyant density measurements in cesium chloride. *J. Bacteriol.* **87**:1073–1078.

Sebald, W., Weiss, H., and Jackl, G. 1972. Inhibition of the assembly of cytochrome oxidase in *Neurospora crassa* by chloramphenicol. *Eur. J. Biochem.* **30**:413–417.

Senior, A. E. 1973. The structure of mitochondrial ATPase. *Biochim. Biophys. Acta* **301**:249–277.

Sherman, F., Taber, H., and Campbell, W. 1965. Genetic determination of isocytochromes *c* in yeast. *J. Mol. Biol.* **13**:21–39.

Shichi, H., and Hackett, D. P. 1962. Studies on the *b*-type cytochromes from mung bean seedlings. I. Purification of cytochromes *b*-555 and *b*-561. *J. Biol. Chem.* **237**:2955–2964.

Shimazaki, K., Takamiya, K., and Nishimura, M. 1978a. Studies on electron transfer system in the marine diatom *Phaeodactylum tricornutum* I. Isolation and characterization of cytochromes. *J. Biochem.* **83**:1631–1638.

Shimazaki, K., Takamiya, K., and Nishimura, M. 1978b. Studies on electron transfer systems in the marine diatom *Phaeodactylum tricornutum*. II. Identification and determination of quinones, cytochromes and flavins. *J. Biochem.* **83**:1639–1642.

Sigal, N., Senez, J. C., LeGall, J., and Sebald, M. 1963. Base composition of the DNA of sulfate-reducing bacteria. *J. Bacteriol.* **85**:1315–1318.

Slater, E. C. 1967. The respiratory chain and oxidative phosphorylation: Some of the unsolved problems. *In*: Slater, E. C., Kaniuga, Z., and Wojtczak, L, *Biochemistry of Mitochondria*, New York, Academic Press, pp. 1–10.

Slater, E. C. 1977. Mechanism of oxidative phosphorylation. *Annu. Rev. Biochem.* **46**:1015–1026.

Slater, E. C., Lee, C. P., Berden, J. A., and Wegdan, H. J. 1970. High-energy forms of cytochrome *b*. *Nature (London)* **226**:1248–1249.

Slater, E. C., Kemp, A., van der Krann, I., Muller, J. L. M., Roveri, D. A., Verschoor, G. J., Wagenvoord, R. J., and Wielders, J. P. M. 1979. The ATP- and ADP-binding sites in mitochondrial coupling factor F and their possible roles in oxidative phosphorylation. *FEBS Lett.* **103**:7–11.

Slonimski, P. P., Acher, R., Péré, G., Sels, A., and Somlo, M. 1965. *In: Mécanismes de Régulation des Activités Cellulaires chez les Microorganismes*, Paris, Centre National Recherche Scientifique, p. 435.

Smillie, R. M. 1965. Isolation of two proteins with chloroplast ferredoxin activity from a blue-green alga. *Biochem. Biophys. Res. Commun.* **20**:621–629.

Smith, L. 1954. Bacterial cytochromes. *Bacteriol. Rev.* **18**:106–130.

Smith, L. 1968. The respiratory chain system of bacteria. *In*: Singer, F. P., ed., *Biological Oxidations*, New York, Interscience, pp. 55–122.

Sugeno, K., and Matsubara, H. 1969. The amino acid sequence of *Scenedesmus* ferredoxin. *J. Biol. Chem.* **244**:2979–2989.

Sugimura, Y., Toda, F., Murata, T., and Yakushiji, E. 1968. Studies on algal cytochromes. *In*: Okunuki, K., Kamen, M. D., and Sekuzu, I., eds., *Structure and Function of Cytochromes*, Tokyo, University of Tokyo Press, pp. 452–458.

Stadtman, E. R. 1968. The role of multiple enzymes in the regulation of branched metabolic pathways. *Ann. N.Y. Acad. Sci.* **151**:516–530.

Steffens, G. J., and Buse, G. 1979. Chemical constitution and subunit function of polypeptide II from cytochrome-*c*-oxidase. *In*: King, T. E., Orii, Y., Chance, B., and Okunuki, K., eds., *Cytochrome Oxidase*, Amsterdam, Elsevier/North-Holland Biomedical Press, pp. 79–90.

Stern, J. O., and Peisach, J. 1976. A model compound for nitrosyl cytochrome P-450: Further evidence for mercaptide sulfur ligation to heme. *FEBS Lett.* **62**:364–368.

Stiggall, D. L., Galante, Y. M., and Hatefi, Y. 1978. Preparation and properties of an ATP–P_i exchange complex (complex V) from bovine heart mitochondria. *J. Biol. Chem.* **253**:956–964.

Straub, J. P., and Colpa-Boonstra, J. P. 1962. The effect of pH on the oxidation–reduction potential of cytochrome *b* in heart-muscle preparations. *Biochim. Biophys. Acta* **60**:650–652.

Strittmatter, P., and Huntley, T. E. 1970. Protein conformation in haeme binding in cytochrome b_5. *Symp. 8th Int. Congr. Biochem.*, Switzerland **1**:21–22.

Srivastava, H. K. 1971. Carbon monoxide reactive haemoproteins in parasitic flagellate *Crithidia oncopelti*. *FEBS Lett.* **16**:189–191.

Takahashi, H., Taniguchi, S., and Egani, F. 1956. Nitrate reduction in aerobic bacteria and that in *E. coli* coupled to phosphorylations. *J. Biochem.* **43**:223–233.

Takruri, I., and Boulter, D. 1979. The amino acid sequence of ferredoxin from *Triticum aestivum*. *Biochem. J.* **179**:373–378.

Takruri, I., Haslett, B. G., Boulter, D., Andrew, P. W., and Rogers, L. J. 1978. The amino acid sequence of ferredoxin from the red alga *Porphyra umbilicalis*. *Biochem. J.* **173**:459–466.

Tanaka, M., Haniu, M., Yasunobu, K. T., Evans, M. C. W., and Rao, K. K. 1974. Amino acid sequence of ferredoxin from photosynthetic green bacterium, *Chlorobium limicola*. *Biochemistry* **13**:2953–2959.

Tanaka, M., Haniu, M., Yasunobu, K. T., Evans, M. C. W., and Rao, K. K. 1975a. The amino acid sequence of ferredoxin II from *Chlorobium limicola*, a photosynthetic green bacterium. *Biochemistry* **14**:1938–1943.

Tanaka, M., Haniu, M., Zeitlin, S., Yasunobu, K. T., Rao, K. K., and Hall, D. O. 1975b. Modification of the automated sequence determination as applied to the sequence determination of the *Spirulina maxima* ferredoxin. *Biochemistry* **14**:5535–5540.

Tanaka, M., Haniu, M., Yasunobu, K. T., Yu, C. A., Yu, L., Wei, H., and King, T. E. 1977. Amino acid sequence of the heme *a* subunit of bovine heart cytochrome oxidase and sequence homology with hemoglobin. *Biochem. Biophys. Res. Commun.* **76**:1014–1019.

Thauer, R. K., Jungermann, K., and Decker, K. 1977. Energy conservation in chemotropic anaerobic bacteria. *Bacteriol. Rev.* **41**:100–180.

Thomas, P. E., Korzeniowski, D., Bresnick, E., Bornstein, W. A., Kasper, C. B., Fahl, W. E., Jefcoate, C. R., and Levin, W. 1979. Hepatic cytochrome P-448 and epoxide hydrase: Enzymes of nuclear origin are immunochemically identical to those of microsomal origin. *Arch. Biochem. Biophys.* **192**:22–26.

Thompson, R. B., Borden, D., Tarr, G. E., and Margoliash, E. 1978. Heterogeneity of amino acid sequence in hippopotamus cytochrome *c*. *J. Biol. Chem.* **253**:8957–8961.

Thusius, D., Blazy, B., and Baudras, A. 1976. Mechanism of yeast cytochrome b_2 action. I. Thermodynamics and relaxation kinetics of the interaction between cytochrome b_2 and oxalate. *Biochemistry* **15**:250–256.

Tikhonov, A. N., Burbaev, D. S., Grigolava, I. V., Konstantinov, A. A., Ksenzenko, M. Y., and Ruuge, E. K. 1977. Interaction of the ubisemiquinone with the high-potential iron-sulfur center of succinate dehydrogenase of submitochondrial particles. ESR study at 240° and 12°K. *Biofizika* **22**:734–736.

Tilley, B. E., Watanuki, M., and Hall, P. F. 1976. Cytochrome P-450 from bovine adrenocortical mitochondria: Two species of subunit. *Biochem. Biophys. Res. Commun.* **70**:1303–1307.

Timkovich, R., and Dickerson, R. E. 1976. Amino acid sequence of *Paracoccus denitrificans* cytochrome *c* 550. *J. Biol. Chem.* **251**:2197–2206.

Ting, H. Y., Jacobson, E. L., and Jacobson, M. K. 1977. Regulation of NADP levels in yeast. *Arch. Biochem. Biophys.* **183**:98–104.

Trousil, E. B., and Campbell, L. L. 1974. Amino acid sequence of cytochrome c_3 from *Desulfovibrio vulgaris*. *J. Biol. Chem.* **249**:386–393.

Trumpower, B. L. 1976. Evidence for a proton motive Q cycle mechanism of electron transfer through the cytochrome b–c_1 complex. *Biochem. Biophys. Res. Commun.* **70**:73–80.

Tsugita, A., Kobayashi, M., Tani, S., Kyo, S., Rashid, M. A., Yoshida, Y., Kajihara, T., and Hagihara, B. 1970. Comparative study of the primary structures of cytochrome *b* from four species. *Proc. Natl. Acad. Sci, USA* **67**:442–447.

Tyree, B., and Webster, D. A. 1978a. The binding of cyanide and carbon monoxide to cytochrome *o* purified from *Vitreoscilla*. *J. Biol. Chem.* **253**:6988–6991.

Tyree, B., and Webster, D. A. 1978b. Electron-accepting properties of cytochrome *o* purified from *Vitreoscilla*. *J. Biol. Chem.* **253**:7635–7637.

Tyree, B., and Webster, D. A. 1979a. Intermediates in the reaction of reduced cytochrome *(Vitreoscilla)* (*Vitreoscilla*) with oxygen. *J. Biol. Chem.* **254**:176–179.

Tyree, B., and Webster, D. A. 1979b. Intermediates in the reaction of cytochrome *o* (*Vitreoscilla*) with oxygen. *In*: King, T. E., Orii, Y., Chance, B., and Okunuki, K., eds., *Cytochrome Oxidase*, Amsterdam, Elsevier/North-Holland Biomedical Press, pp. 21–27.

Vanderkooi, J. M., and Erecińska, M. 1976. Cytochrome *c*, a membrane-bound enzyme. *In*: Martonosi, A., ed., *The Enzymes of Biological Membranes*, New York, Plenum Press, Vol. 4, pp. 43–86.

Vanderkooi, J. M., Erecińska, M., and Chance, B. 1973a. Cytochrome *c* interaction with membranes. I. Use of a fluorescent chromophore in a study of cytochrome *c* interaction with artificial and mitochondrial membranes. *Arch. Biochem. Biophys.* **154**:219–229.

Vanderkooi, J. M., Erecińska, M., and Chance, B. 1973b. Cytochrome *c* interaction with membranes. II. Comparative study of the interaction of *c* cytochromes with the mitochondrial membrane. *Arch. Biochem. Biophys.* **157**:531–540.

van Lin, B., and Bothe, H. 1972. Flavodoxin from *Azotobacter vinelandii*. *Arch. Mikrobiol.* **82**:155–172.

Vetter, H., and Knappe, J. 1971. Flavodoxin and ferredoxin of *Escherichia coli*. *Hoppe-Seyler's Z. Physiol. Chem.* **352**:433–446.

Wachter, E., Sebald, W., and Tzagoloff, A. 1977. Altered amino acid sequence of the DCCD-binding protein in the OL1-1 resistant mutant D273-10B/A21 of *S. cerevisiae*. *In*: Bandlow, W., Schweyen, R. J., Wolf, K., and Kaudewitz, F., eds., *Mitochondria 1977*, Berlin, de Gruyter, pp. 441–449.

Wada, K., and Arnon, D. I. 1971. Three forms of cytochrome b_{559} and their relation to the photosynthetic activity of chloroplasts. *Proc. Natl. Acad. Sci. USA* **68**:3064–3068.

Wada, K., Hase, T., Tokunaga, H., and Matsubara, H. 1975. Amino acid sequence of *Spirulina platensis* ferredoxin: A far divergency of blue-green algal ferredoxins. *FEBS Lett.* **55**:102–104.

Wakabayashi, S., Hase, T., Wada, K., Matsubara, H., Suzuki, K., and Takaichi, S. 1978. Amino acid sequences of two ferredoxins from pokeweed, *Phytolacca americana*. *J. Biochem.* **83**:1305–1319.

Walasek, O. F., and Margoliash, E. 1977. Transmission of the cytochrome *c* structural gene in horse–donkey crosses. *J. Biol. Chem.* **252**:830–834.

Watanabe, H., Kamita, Y., Nakamura, T., Takimoto, A., and Yamanaka, T. 1979. The terminal oxidase of *Photobacterium phosphoreum*: A novel cytochrome. *Biochim. Biophys. Acta* **547**:70–78.

Webster, D. A., and Hackett, D. P. 1966. Respiratory chain of colorless algae. II. Cyanophyta. *Plant Physiol.* **41**:599–605.

Webster, D. A., and Orii, Y. 1977. Oxygenated cytochrome *o*. An active intermediate observed in whole cells of *Vitreoscilla*. *J. Biol. Chem.* **252**:1834–1836.

Wikström, M. K. F. 1971. Properties of three cytochrome *b*-like species in mitochondria and submitochondrial particles. *Biochim. Biophys. Acta* **253**:332–345.

Wikström, M. K. F. 1973. The different cytochrome *b* components in the respiratory chain of animal mitochondria and their role in electron transport and energy conservation. *Biochim. Biophys. Acta* **301**:155–193.

Wikström, M. K. F. 1977. Proton pump coupled to cytochrome *c* oxidase in mitochondria. *Nature (London)* **266**:271–273.

Wikström, M. K. F. 1978. Cytochrome *c* oxidase. The mechanism of a redox-coupled proton pump. *In*: Azzone, G. F., *et al.*, eds., *The Proton and Calcium Pumps*, Amsterdam, Elsevier/North-Holland Biomedical Press, pp. 215–226.

Wikström, M. K. F., and Sarri, H. T. 1977. The mechanism of energy conservation and transduction by mitochondrial cytochrome *c* oxidase. *Biochim. Biophys. Acta* **462**:347–361.

Wikström, M. K. F., Saari, H., Penttila, T., and Saraste, M. 1977. Functioning of cytochrome *c* oxidase in the mitochondrial membrane and in vesicles. *In:* Nicholls, P., *et al.*, eds., *Membrane Proteins,* New York, Pergamon Press, pp. 85–94.

Williams, R. J. P. 1978. The history and hypotheses concerning ATP-formation by energised protons. *FEBS Lett.* **85**:9–19.

Wilson, D. F., Owens, C. S., and Erecińska, M. 1979. Quantitative dependence of mitochondrial oxidative phosphorylation on oxygen concentration: A mathematical model. *Arch. Biochem. Biophys.* **195**:494–504.

Wiseman, A., Lim, T. K., and Woods, L. F. J. 1978. Regulation of the synthesis of cytochrome P-450 in brewer's yeast. Role of cyclic AMP. *Biochim. Biophys. Acta* **544**:615–623.

Wong, D. T. O., and Ajl, S. J. 1956. Conversion of acetate and glyoxylate to malate. *J. Am. Chem. Soc.* **78**:3230–3231.

Wong, D. T. O., and Ajl, S. J. 1957. Significance of the malate synthetase reaction in bacteria. *Science* **126**:1013–1014.

Yamamoto, T., and Orii, Y. 1974. The polypeptide compositions of bovine cytochrome oxidase and its proteinase treated derivative. *J. Biochem.* **75**:1081–1089.

Yamanaka, T. 1972. Evolution of cytochrome *c* molecule. *Adv. Biophys.* **3**:227–276.

Yamanaka, T. 1973. Cytochrome *c* and evolution of the energy acquiring system. *Space Life Sci.* **4**:490–504.

Yamanaka, T. 1975. A comparative study on the redox reactions of cytochromes *c* with certain enzymes. *J. Biochem.* **77**:493–499.

Yamanaka, T. 1976. The subunits of *Chlorobium* flavocytochrome *c*. *J. Biochem.* **79**:655–660.

Yamanaka, T., and Fukumori, Y. 1977. *Thiobacillus novellus* cytochrome oxidase can separate some eucaryotic cytochromes *c*. *FEBS Lett.* **77**:155–158.

Yamanaka, T., and Okunuki, K. 1963. Crystalline *Pseudomonas* cytochrome oxidase. I. Enzymatic properties with special reference to the biological specificity. *Biochim. Biophys. Acta* **67**:379–393.

Yamanaka, T., and Okunuki, K. 1968. Comparative studies of cytochrome *c* with cytochrome oxidases. *In*: Okunuki, K., Kamen, M. D., and Sekuzu, I., eds., *Structure and Function of Cytochromes*, Tokyo, University of Tokyo Press, pp. 390–403.

Yamanaka, T., Ota, A., and Okunuki, K. 1961. A nitrite reducing system reconstructed with purified cytochrome components of *Pseudomonas aeruginosa*. *Biochim. Biophys. Acta* **53**:294–308.

Yang, N. S., Sorenson, J. C., and Scandalios, J. G. 1977. Genetic control of mitochondrial malate dehydrogenases: Evidence for duplicated chromosome segments. *Proc. Natl. Acad. Sci. USA* **74**:310–314.

Yasunobu, K. T., Tanaka, M., Haniu, M., Sameshima, M., Reimer, N., Eto, T., King, T. E., Yu, C. A., Yu, L., and Wei, Y. H. 1979. Sequence studies of bovine heart cytochrome oxidase subunits. *In*: King, T. E., Orii, Y., Chance, B., and Okunuki, K., eds., *Cytochrome Oxidase*, Amsterdam, Elsevier/North-Holland Biomedical Press, pp. 91–101.

Yatscoff, R. W., Freeman, K. B., and Vail, W. J. 1977. Site of biosynthesis of mammalian cytochrome *c* oxidase subunits. *FEBS Lett.* **81**:7–9.

Yockey, H. P. 1977. On the information content of cytochrome *c*. *J. Theor. Biol.* **67**:345–376.

Yu, C. A., Yu, L., and King, T. E. 1974a. Soluble cytochrome *b–c* complex and the reconstitution of succinate-cytochrome *c* reductase. *J. Biol. Chem.* **249**:4905–4910.

Yu, C. A., Gunsalus, I. C., Katagiri, M., Suhara, K., and Takemori, S. 1974b. Cytochrome P-450 cam. I. Crystallization and properties. *J. Biol. Chem.* **249**:94–101.

Yu, L., Yu, C. A., and King, T. E. 1978. The indispensability of phospholipid and ubiquinone in mitochondrial electron transfer from succinate to cytochrome *c*. *J. Biol. Chem.* **253**:2657–2663.

Zumft, W. G., and Spiller, H. 1971. Characterization of a flavodoxin from the green alga *Chlorella*. *Biochem. Biophys. Res. Commun.* **45**:112–118.

CHAPTER 9

Afzelius, B. A. 1972. Morphological aspects of mitochondrial ontogenesis. *Proc. First John Innes Symp.* **1**, pp. 307–340.

Agsteribbe, E., Kroon, A. M., and van Bruggen, E. F. J. 1972. Circular DNA from mitochondria of *Neurospora crassa*. *Biochim. Biophys. Acta* **269**:299–303.

Allmann, D. W., Galginza, L., McCaman, R. E., and Green, D. E. 1966. The membrane systems of the mitochondrion. IV. The localization of the fatty acid oxidizing system. *Arch. Biochem. Biophys.* **117**:413–422.

Amon, H., Kühnel, W., and Petry, G. 1967. Untersuchungen an der Niere des Siebenschläfers

(*Glis glis* L.) im Winterschlafund im Sommerlichen Wachzustrand. *Z. Zellforsch. Mikrosk. Anat.* **77**:204–231.

Anderson, W. A., and Hill, G. C. 1969. Division and DNA synthesis in the kinetoplast of *Crithidia fasciculata*. *J. Cell Sci.* **4**:611–620.

André, J. 1962. Contribution à la connaissance du chondriome: Étude de ses ultrastructurales ultrastructurales pendant la spermatogénèse. *J. Ultrastruct. Res. Suppl.* **3**:1–85.

André, J. 1963. Some aspects of specialization in sperm. *In*: Mazia, D., and Tyler, A., eds., *General Physiology of Cell Specialization,* New York, McGraw-Hill, pp. 91–115.

André, J., and Marinozzi, V. 1965. Présence, dans les mitochondries, de particules ressemblant aux ribosomes. *J. Microsc. (Paris)* **4**:615–626.

Arnberg, A. C., van Bruggen, E. F. J., Ter Schegget, J., and Borst, P. 1971. The presence of DNA molecules with a displacement loop in standard mitochondrial DNA preparations. *Biochim. Biophys. Acta* **246**:353–357.

Arnberg, A. C., van Bruggen, E. F. J., Clegg, R. A., Upholt, W. B., and Borst, P. 1974. An analysis by electron microscopy of intermediates in the replication of linear *Tetrahymena* mitochondrial DNA. *Biochim. Biophys. Acta* **361**:266–276.

Arnberg, A. C., Goldbach, R. W., van Bruggen, E. F. J., and Borst, P. 1977. The structure of *Tetrahymena pyriformis* mitochondrial DNA. II. The complex structure of strain GL mitochondrial DNA. *Biochim. Biophys. Acta* **477**:51–69.

Arnold, C. G., Schimmer, O., Schötz, F., and Bathelt, H. 1972. Die Mitochondrien von *Chlamydomonas reinhardii. Arch. Mikrobiol.* **81**:50–67.

Ashwell, M., and Work, T. S. 1970. The biogenesis of mitochondria. *Annu. Rev. Biochem.* **39**:251–290.

Ashworth, C. T., Leonard, J. S., Eigenbrodt, E. H., and Wrightsman, F. J. 1966. Hepatic intracellular osmiophilic droplets. Effect of lipid solvents during tissue preparation. *J. Cell Biol.* **31**:301–318.

Attardi, G., Amalric, F., Ching, E., Constantino, P., Gelfand, R., and Lynch, D. 1976. Informational content and gene mapping of mtDNA from HeLA cells. *In*: Saccone, C., and Kroon, A. M., eds., *The Genetic Function of Mitochondrial DNA*, Amsterdam, North-Holland, pp. 37–46.

Aujame, L., and Freeman, K. B. 1979. Mammalian mt tRNAs: Chromatographic properties, size and origin. *Nucleic Acids Res.* **6**:455–469.

Aujame, L., Wallace, R. B., and Freeman, K. B. 1978. Chemical and physical properties of mammalian mitochondrial aminoacyl-tRNAs. I. Molecular weights of mitochondrial leucyl- and methionyl-tRNAs. *Biochim. Biophys. Acta* **518**:308–320.

Avadhani, N. G. 1979. Messenger RNA metabolism in mammalian mitochondria: Relationship between the decay of mitochondrial mRNA and their poly(A). *Biochemistry* **18**:2673–2676.

Avadhani, N. G., and Buetow, D. E. 1972a. Protein synthesis with isolated mitochondrial polysomes. *Biochem. Biophys. Res. Commun.* **46**:773–781.

Avadhani, N. G., and Buetow, D. E. 1972b. Isolation of active polyribosomes from the cytoplasm, mitochondria, and chloroplasts of *Euglena gracilis. Biochem. J.* **128**:353–365.

Avadhani, N. G., and Buetow, D. E. 1974. Distinguishing characteristics and a requirement for the homologous ribosomal salt-extractable fraction for protein synthesis. *Biochem. J.* **140**:73–78.

Avakyan, A. A., Kats, L. N., Mineeva, L. A., Ratner, E. N., and Gusev, M. V. 1978. Electron microscopic data on mesosomelike and myelinlike structures in blue-green algae. *Microbiologica* **47**:595–600.

Avers, C. J. 1967. Distribution of cytochrome *c* peroxidase activity in wild-type and petite cells of bakers' yeast grown aerobically and anaerobically. *J. Bacteriol.* **94**:1225–1235.

Awasthi, Y. C., Chuang, T. F., Keenan, T. W., and Crane, F. L. 1971. Tightly bound cardiolipin in cytochrome oxidase. *Biochim. Biophys. Acta* **226**:42–52.

Baker, J. E. 1962. Properties of particles derived from sweet potato mitochondria by sonic irradiation. *Plant Physiol. Suppl.* **37**:XXXI.

Baldacci, G., Falcone, C., Frontali, L., Macino, G., and Palleschi, C. 1975. tRNA and aminoacyl-tRNA synthetases from wildtype and petite yeast mitochondria. *In:* Puiseux-Dao, S., ed., *Molecular Biology of Nucleocytoplasmic Relationships,* Amsterdam, Elsevier, pp. 41–46.

Baldacci, G., Falcone, C., Frontali, L., Macino, G., and Palleschi, C. 1976. tRNA in mitochondria from *S. cerevisiae* grown in different physiological conditions. *In:* Saccone, C., and Kroon, A. M., eds., *The Genetic Function of Mitochondrial DNA,* Amsterdam, North-Holland, pp. 305–312.

Baldacci, G., Falcone, C., Frontali, L., Macino, G., and Palleschi, C. 1977. Differences in mitochondrial isoaccepting tRNAs from *S. cerevisiae* as a function of growth conditions. *In:* Bandlow, W., Schweyen, R. J., Wolf, K., and Kaudewitz, F., eds., *Mitochondria 1977,* Berlin, de Gruyter, pp. 571–574.

Barnett, W. E., and Brown, D. H. 1967. Mitochondrial tRNAs. *Proc. Natl. Acad. Sci. USA* **57**:452–458.

Barnett, W. E., Brown, D. H., and Epler, J. L. 1967. Mitochondrial-specific aminoacyl-RNA synthetases. *Proc. Natl. Acad. Sci. USA* **57**:1775–1781.

Bauer, H., and Tanaka, K. 1968. Ultrastructure of mitochondria and crystal-containing bodies in mature ballistospores of the fungus *Basidiobolus ranarum* as revealed by freeze-etching. *J. Bacteriol.* **96**:2132–2137.

Baulina, O. I., Semenova, L. R., Mineeva, L. A., and Gusev, M. V. 1978. Characteristics of cellular ultrastructural organization of the chemoheterotrophic blue-green alga *Chlorogloea fritschii. Microbiology* **47**:746–751.

Baxter, R. 1971. Origin and continuity of mitochondria. *In:* Reinert, J., and Ursprung, H., eds., *Origin and Continuity of Cell Organelles,* Berlin, Springer-Verlag, pp. 46–64.

Beattie, D. S. 1971. The synthesis of mitochondrial proteins. *Sub-Cell. Biochem.* **1**:1–23.

Beaulaton, J. A. 1968. Modifications ultrastructurales des cellules sécrétrices de la glande prothoracique de vers à soie au cours des deux derniers âges larvaires. *J. Cell Biol.* **39**:501–525.

Behnke, O. 1965. Helical filaments in rat liver mitochondria. *Exp. Cell Res.* **37**:687–698.

Bereiter-Hahn, J. 1978. Intracellular motility of mitochondria: Role of the inner compartment in migration and shape changes of mitochondria in XTH-cells. *J. Cell Sci.* **30**:99–115.

Bereiter-Hahn, J., and Morawe, G. 1972. Mitochondrial movements in epithelial cells from tadpole hearts in tissue culture. *Cytobiologie Z. Exp. Zellforsch.* **6**:447–467.

Berger, E. R. 1964. Mitochondria genesis in the retinal photoreceptor inner segment. *J. Ultrastruct. Res.* **11**:90–111.

Berk, A. J., and Clayton, D. A. 1976. Mechanism of mtDNA replication in mouse L-cells. *J. Mol. Biol.* **100**:85–102.

Bernard, U., and Küntzel, H. 1976. Physical mapping of mtDNA from *Neurospora crassa. In*: Saccone, C., and Kroon, A. M., eds., *The Genetic Function of Mitochondrial DNA,* Amsterdam, North-Holland, pp. 105–109.

Bernardi, G., Faures, M., Piperno, G., and Slonimski, P. P. 1970. Mitochondrial DNA's from respiratory-sufficient and cytoplasmic respiratory-deficient mutant yeast. *J. Mol. Biol.* **48**:23–42.

Bernardi, G., Prunell, A., Fonty, G., Kopecka, H., and Strauss, F. 1976. The mitochondrial genome of yeast: Organization, evolution, and the petite mutation. *In*: Saccone, C., and Kroon, A. M., eds., *The Genetic Function of Mitochondrial DNA,* Amsterdam, North-Holland, pp. 185–198.

Birky, C. W. 1976. The inheritance of genes in mitochondria and chloroplasts. *BioScience* **26**:26–33.

Bisalputra, T. 1966. Electron microscope study of the protoplasmic continuity in certain brown algae. *Can. J. Bot.* **44**:89–93.

Bisalputra, T., and Bisalputra, A. A. 1967. Chloroplast and mitochondrial DNA in a brown alga *Egregia menziesii. J. Cell Biol.* **33**:511–520.

Bisalputra, T., Brown, D. L., and Weier, T. E. 1969. Possible respiratory sites in a blue-green alga *Nostoc sphaericum* as demonstrated by potassium tellurite and tetranitro-blue tetrazolium reduction. *J. Ultrastruct. Res.* **27**:182–197.

Blanchy, B., Godinot, C., and Gautheron, D. C. 1978. Asymmetrical distribution of thiol groups involved in ATP–$^{32}P_i$ exchange on mitochondrial membranes. *Biochem. Biophys. Res. Commun.* **82**:776–781.

Blecher, S. R. 1967. Mitochondrial chromosomes. *Curr. Mod. Biol.* **1**:249–255.

Blomstrand, R., and Svensson, L. 1974. Studies on phospholipids with particular reference to cardiolipin of rat heart after feeding rapeseed oil. *Lipids* **9**:771–780.

Bogenhagen, D., and Clayton, D. A. 1974. The number of mitochondrial DNA genomes in mouse L and human HeLa cells. Quantitative isolation of mitochondrial DNA. *J. Biol. Chem.* **249**:7991–7995.

Boguslawski, G., Vodkin, M. H., Finkelstein, D. B., and Fink, G. R. 1974. Histidyl-tRNAs and histidyl-tRNA synthetases in wild type and cytoplasmic petite mutants of *Saccharomyces cerevisiae*. *Biochemistry* **13**:4659–4667.

Bohnert, H. J. 1973. Circular mtDNA from *Acanthamoeba castellanii* (Neff-strain). *Biochim. Biophys. Acta* **324**:199–205.

Bohnert, H. J., and Herrmann, R. G. 1974. The genomic complex of *Acanthamoeba castellanii* mitochondrial DNA, *Eur. J. Biochem.* **50**:83–90.

Borkowski, T., and Brzuszkiewicz-Zarnowski, H. 1975. Fractionation of specific mitochondrial and cytoplasmic tRNAs obtained from calf brain. *J. Neurochem.* **25**:641–644.

Borst, P. 1972. Mitochondrial nucleic acids. *Annu. Rev. Biochem.* **41**:333–376.

Borst, P. 1974. Mitochondrial nucleic acids. *Biochem. Soc. Trans.* **2**:182–185.

Borst, P., and Fairlamb, A. H. 1976. DNA of parasites, with special reference to kinetoplast DNA. *In*: Van den Bossche, H., ed., *Biochemistry of Parasites and Host–Parasite Relationships*, Amsterdam, Elsevier, pp. 169–191.

Borst, P., and Grivell, L. A. 1978. The mitochondrial genome of yeast. *Cell* **15**:705–723.

Borst, P., and Kroon, A. M. 1969. Mitochondrial DNA: Physicochemical properties, replication, and genetic function. *Int. Rev. Cytol.* **26**:107–190.

Borst, P., Fairlamb, A. H., Fase-Fowler, F., Hoeijmakers, J. H. J., and Weislogel, P. O. 1976. The structure of kinetoplast DNA. *In*: Saccone, C., and Kroon, A. M., eds., *The Genetic Function of Mitochondrial DNA,* Amsterdam, North-Holland, pp. 59–69.

Borst, P., Bos, J. L., Grivell, L. A., Groot, G. S. P., Heyting, C., Moorman, A. F. M., Sanders, J. P. M., Talen, J. L., van Kreijl, C. F., and van Ommen, G. J. B. 1977. The physical map of yeast mitochondrial DNA anno 1977 (A review). *In*: Bandlow, W., Schweyen, R. J., Wolf, K., and Kaudewitz, F., eds., *Mitochondria 1977*, Berlin, de Gruyter, pp. 213–254.

Bos, J. L., Heyting, C., Borst, P., Arnberg, A. C., and van Bruggen, E. F. J. 1978. An insert in the single gene for the large rRNA in yeast mtDNA. *Nature (London)* **275**:336–338.

Branton, D., and Moor, H. 1964. Fine structure in freeze-etched *Allium cepa* L. root tips. *J. Ultrastruct. Res.* **11**:401–411.

Braun, R., and Evans, T. E. 1969. Replication of nuclear satellite and mtDNA in the mitotic cycle of *Physarum*. *Biochim. Biophys. Acta* **182**:511–522.

Brierley, G. P., and Slautterback, D. B. 1964. Studies on ion transport. IV. An electron microscope study of the accumulation of Ca^{2+} and inorganic phosphate by heart mitochondria. *Biochim. Biophys. Acta* **82**:183–186.

Brunk, C. F., and Hanawalt, P. C. 1969. Mitochondrial DNA in *Tetrahymena pyriformis*. *Exp. Cell Res.* **54**:143–149.

Buck, C. A., and Nass, M. M. K. 1969. Studies on mitochondrial tRNA from animal cells. I. A comparison of mitochondrial and cytoplasmic tRNA and aminoacyl-tRNA synthetases. *J. Mol. Biol.* **41**:67–82.

Buetow, D. E., and Wood, W. M. 1978. The mitochondrial translation system. *Subcell. Biochem.* **5**:1–85.

Burdett, I. D. J., and Rogers, H. J. 1972. The structure and development of mesosomes studied in *Bacillus licheniformis* strain 6346. *J. Ultrastruct. Res.* **38**:113–133.

Campbell, J. W. 1973. Nitrogen metabolism. *In*: Prosser, C. L., ed., *Comparative Animal Physiology*, Philadelphia, Saunders, pp. 279–316.

Campbell, J. W., and Vorhaben, J. E. 1976. Avian mitochondrial glutamine metabolism. *J. Biol. Chem.* **251**:781–786.

Carafoli, E. 1979. The calcium cycle of mitochondria. *FEBS Lett.* **104**:1–5.

Casey, J. W., Cohen, M., Rabinowitz, M., Fukahara, H., and Getz, G. S. 1972. Hybridization of mitochondrial tRNAs with mitochondrial and nuclear DNA of *grande* (wild-type) yeast. *J. Mol. Biol.* **63**:431–440.

Casey, J. W., Hsu, H.-J., Rabinowitz, M., Getz, G. S., and Fukahara, H. 1974a. Transfer RNA genes in the mitochondrial DNA of cytoplasmic petite mutants of *S. cerevisiae*. *J. Mol. Biol.* **88**:717–733.

Casey, J. W., Hsu, H.-J., Getz, G. S., Rabinowitz, M., and Fukahara, H. 1974b. Transfer RNA genes in mitochondrial DNA of *grande* (wild-type) yeast. *J. Mol. Biol.* **88**:735–747.

Cavalier-Smith, T. 1975. The origin of nuclei and of eukaryotic cells. *Nature (London)* **256**:463–468.

Chance, B. 1964. Techniques for the assay of the respiratory enzymes. *Meth. Enzymol.* **4**:273–329.

Chance, B. 1965. Reaction of oxygen with the respiratory chain in cells and tissues. *J. Gen. Physiol.* **49**:163–188.

Chapman, M. J., Miller, L. R., and Ontko, J. A. 1973. Localization of the enzymes of ketogenesis in rat liver mitochondria. *J. Cell Biol.* **58**:284–306.

Chappell, J. B., Cohn, M., and Greville, G. D. 1963. *In*: Chance, B., ed., *Energy-Linked Functions of Mitochondria,* New York, Academic Press, pp. 219–243.

Charezinski, M., and Borkowski, T. 1973. Studies on binding of amino acids by mitochondrial and cytoplasmic tRNA from calf brain. *Acta Biochim. Pol.* **20**:153–158.

Charret, R., and André, J. 1968. La synthèse de l'ADN mitochondrial chez *Tetrahymena pyriformis*. *J. Cell Biol.* **39**:369–381.

Chèvremont, M., Chèvremont-Comhaire, S., and Baeckeland, E. 1959. Action de desoxyribonucléases neutre et acide sur les cellules somatiques vivantes cultivées *in vitro*. *Arch. Biol.* **70**:811–831.

Chi, J. C. H., and Suyama, Y. 1970. Comparative studies on mitochondrial and cytoplasmic ribosomes of *Tetrahymena pyriformis*. *J. Mol. Biol.* **53**:531–556.

Chia, L. S. Y., Morris, H. P., Randerath, K., and Randerath, E. 1976. Base composition studies on mitochondrial 4S RNA from rat liver and Morris hepatomas 5123D and 7777. *Biochim. Biophys. Acta* **425**:49–62.

Chiu, A. O. S., and Suyama, Y. 1975. The absence of structural relationships between mitochondrial and cytoplasmic leucyl-tRNA synthetases from *Tetrahymena pyriformis*. *Arch. Biochem. Biophys.* **171**:43–54.

Chiu, N., Chiu, A. O., and Suyama, Y. 1974. Three isoaccepting forms of leucyl tRNA in mitochondria. *J. Mol. Biol.* **82**:441–457.

Chiu, N., Chiu, A. O., and Suyama, Y. 1975. Native and imported tRNA in mitochondria. *J. Mol. Biol.* **99**:37–50.

Chua, N. H., and Schmidt, G. W. 1979. Transport of proteins into mitochondria and chloroplasts. *J. Cell Biol.* **81**:461–483.

Cieciura, L, Klitończyk, W., and Rydzyński, K. 1978. Stereologic analysis of mitochondria of hepatocyte from fasted rats in the course of digestion. *Folia Histochem. Cytochem.* **16**:193–204.

Cieciura, L., Rydzyński, K., and Klitończyk, W. 1979. Stereologic studies on mitochondrial configuration in different organs of the rat. *Cell Tissue Res.* **196**:347–360.

Clandinin, M. T. 1976. Fatty acid composition changes in mitochondrial membranes induced by dietary long chain fatty acids. *FEBS Lett.* **68**:41–44.

Clarke, S. 1976a. A major polypeptide component of rat liver mitochondria: Carbamyl phosphate synthetase. *J. Biol. Chem.* **251**:950–961.

Clarke, S. 1976b. Protein asymmetry in the inner membrane of rat liver mitochondria. *J. Biol. Chem.* **251**:1354–1363.

Clark-Walker, G. D., and Linnane, A. W. 1967. The biogenesis of mitochondria in *Saccharomyces cerevisiae*. *J. Cell Biol.* **34**:1–14.

Clayton, D. A., and Brambl, R. M. 1972. Detection of circular DNA from mitochondria of *Neurospora crassa*. *Biochem. Biophys. Res. Commun.* **46**:1477–1482.

Cleaves, G. R., Jones, T., and Dubin, D. T. 1976. Properties of a discrete high molecular weight poly(A)-containing mtRNA. *Arch. Biochem. Biophys.* **175**:303–311.

Clegg, R. A., Borst, P., Weijers, P. J., Arnberg, A. C., and van Bruggen, E. F. J. 1974. Replicative intermediates of *Tetrahymena pyriformis* mitochondrial deoxyribonucleic acid. *Biochem. Soc. Trans.* **2**:227–229.

Cohen, M., and Rabinowitz, M. 1972. Analysis of grande and petite yeast mitochondrial DNA by tRNA hybridization. *Biochim. Biophys. Acta* **281**:192–201.

Cohen, S. S. 1970. Are/were mitochondria and chloroplasts microorganisms? *Am. Sci.* **58**:281–289.

Cohen-Bazire, G., Pfennig, N., and Kuisawa, R. 1964. The fine structure of green bacteria. *J. Cell Biol.* **22**:207–225.

Conti, S. F., and Brock, T. D. 1965. Electron microscopy of cell fusion in conjugating *Hansenula wingei*. *J. Bacteriol.* **90**:524–533.

Coon, H. G., Horak, I., and Dawid, I. B. 1973. Propagation of both parental mitochondrial DNAs in rat–human and mouse–human hybrid cells. *J. Mol. Biol.* **81**:285–298.

Coote, J. L., Szabados, G., and Work, T. S. 1979. The heterogeneity of mtDNA in different tissues from the same animal. *FEBS Lett.* **99**:255–260.

Copeland, D. E., and Fitzjarrell, A. T. 1968. The salt absorbing cells in the gills of the blue crab with notes on modified mitochondria. *Z. Zellforsch. Mikrosk. Anat.* **92**:1–22.

Cosson, J., and Tzagoloff, A. 1979. Sequence homologies of (G + C)-rich regions of mtDNA of *S. cerevisiae*. *J. Biol. Chem.* **254**:42–43.

Cottrell, S. F., and Avers, C. J. 1970. Evidence of mitochondrial synchrony in synchronous cultures of yeast. *Biochem. Biophys. Res. Commun.* **38**:973–980.

Cottrell, S. F., Rabinowitz, M., and Getz, G. S. 1973. Mitochondrial deoxyribonucleic acid synthesis in a temperature-sensitive mutant of deoxyribonucleic acid replication of *Saccharomyces cerevisiae*. *Biochemistry* **12**:4374–4378.

Criddle, R. S., Bock, R. M., Green, D. E., and Tisdale, H. A. 1962. Physical properties of proteins of the electron transfer system and interpretation of the structure of the mitochondrion. *Biochemistry* **1**:827–842.

Cummings, D. J., Goddard, J. M., and Maki, R. A. 1976. Mitochondrial DNA from *Paramecium aurelia*. *In:* Saccone, C., and Kroon, A. M., eds., *The Genetic Function of Mitochondrial DNA*, Amsterdam, North-Holland, pp. 119–130.

Cunningham, W. P., Prezbindowski, K., and Crane, F. L. 1967. The relation between structure and function in electron transport systems. *Biochim. Biophys. Acta* **135**:614–623.

Curgy, J. J. 1968. Influence du mode de fixation sur la possibilité d'observer des structures myéliniques dans les hépatocytes d'embryons du poulet. *J. Microsc.* (*Paris*) **7**:63–80.

Curthoys, N. P., and Weiss, R. F. 1974. Subcellular localization of rat kidney glutaminase isoenzymes. *J. Biol. Chem.* **249**:3261–3266.

Davey, P. J., Yu, R., and Linnane, A. W. 1969. The intracellular site of formation of the mitochondrial protein synthetic system. *Biochem. Biophys. Res. Commun.* **36**:30–34.

Davey, P. T., Haslam, J. M., and Linnane, A. W. 1970. The effects of aminoglycoside antibiotics on the mitochondrial and cytoplasmic protein-synthesizing systems of *S. cerevisiae*. *Arch. Biochem. Biophys.* **136**:54–64.

Dawid, I. B., and Brown, D. D. 1970. The mitochondrial and ribosomal DNA components of oocytes of *Urechis caupo*. *Dev. Biol.* **22**:1–14.

Dawid, I. B., Klukas, C. K., Ohi, S., Ramirez, J. L., and Upholt, W. B. 1976. Structure and evolution of animal mitochondrial DNA. *In*: Saccone, C., and Kroon, A. M., eds., *The Genetic Function of Mitochondrial DNA*, Amsterdam, North-Holland, pp. 3–13.

de Duve, C. 1973. Origin of mitochondria. *Science* **182**:85.

DeLuca, H. F., and Engstrom, G. W. 1961. Calcium uptake by rat liver mitochondria. *Proc. Natl. Acad. Sci. USA* **47**:1744–1750.

DeMartino, C., Floridi, A., Marcante, M. L., Malorni, W., Barcellona, P. S., Bellocci, M., and Silvestrini, B. 1979. Morphological, histochemical and biochemical studies on germ cell mitochondria of normal rats. *Cell Tissue Res.* **196**:1–22.

DeRobertis, E., Nowinski, W. W., and Saez, F. A. 1970. *Cell Biology*, Philadelphia, Saunders.

Diers, L., and Schötz, F. 1965. Über den Feinbau pflanzlicher Mitochondrien. *Z. Pflanzenphysiol.* **53**:334–343.

Dillon, L. S. 1978. *The Genetic Mechanism and the Origin of Life*, New York, Plenum Press.

Dizikes, G. J., and Burke, D. D. 1978. Isolation and characterization of the mitochondrial DNA of *Allomyces macrogynus*. *Biochim. Biophys. Acta* **518**:247–256.

Dodge, J. D., and Crawford, R. M. 1968. Fine structure of the dinoflagellate *Amphidinium casteri* Hulbert. *Protistologica* **4**:231–242.

Drouet, F. 1962. *In*: Silva, P. C., ed., *Classification of Algae*, New York, Academic Press.

Dubin, D. T. 1974. Methylated nucleotide content of mitochondrial ribosomal RNA from hamster cells. *J. Mol. Biol.* **84**:257–273.

Dubin, D. T., and Friend, D. A. 1972. Comparison of cytoplasmic and mitochondrial 4S RNA from cultured hamster cells: Physical and metabolic properties. *J. Mol. Biol.* **71**:163–175.

Dubin, D. T., and Friend, D. A. 1974. Methylation properties of mitochondrion-specific tRNA from cultured hamster cells. *Biochim. Biophys. Acta* **340**:269–277.

Echlin, P. 1964. Intra-cytoplasmic membranous inclusions in the blue-green alga, *Anacystis nidulans*. *Arch. Mikrobiol.* **49**:267–274.

Edelman, M., Epstein, H. T., and Schiff, J. A. 1966. Isolation and characterization of DNA from the mitochondrial fraction of *Euglena*. *J. Mol. Biol.* **17**:463–469.

Edwards, M. R., and Stevens, R. W. 1963. Fine structure of *Listeria monocytogenes*. *J. Bacteriol.* **86**:414–428.

Eggitt, M. J. 1976. Protein synthesis in an *E. coli* system directed by yeast mitochondrial RNA. *FEBS Lett.* **61**:6–9.

Eggitt, M. J., and Scragg, A. H. 1975. Messenger activity of ribonucleic acid from yeast mitochondria. *Biochem. J.* **149**:507–512.

Englund, P. T., DiMaio, D. C., and Price, S. S. 1977. A nicked form of kinetoplast DNA in *Leishmania tarentolae*. *J. Biol. Chem.* **252**:6208–6216.

Epler, J. L., Shugart, L. R., and Barnett, W. E. 1970. *N*-Formylmethionyl tRNA in mitochondria of *Neurospora*. *Biochemistry* **9**:3575–3579.

Evans, T. E., and Suskind, D. 1971. Characterization of the mitochondrial DNA of the slime mold *Physarum polycephalum*. *Biochim. Biophys. Acta* **228**:350–364.

Fain-Maurel, M.-A., and Cassier, P. 1969. Pleomorphisme mitochondrial dans les corpora allata de *Locusta migratoria migratorioides* (R. et F.) au cours de la vie imaginale. *Z. Zellforsch. Mikrosk. Anat.* **102**:543–553.

Fauron, C. M. R., and Wolstenholme, D. R. 1976. Structural heterogeneity of mitochondrial DNA molecules within the genus *Drosophila*. *Proc. Natl. Acad. Sci. USA* **73**:3623–3627.

Federman, M., and Avers, C. J. 1967. Fine-structure analysis of intercellular and intracellular mitochondrial diversity in *S. cerevisiae*. *J. Bacteriol.* **94**:1236–1243.

Fernández-Morán, H. 1961. *In*: Edds, M. V., ed., *Macromolecular Complexes*, New York, Ronald Press, pp. 113–130.

Fernández-Morán, H., Oda, T., Blair, P. V., and Green, D. E. 1964. Macromolecular repeating unit of mitochondrial structure and function. *J. Cell Biol.* **22**:63–100.

Fitz-James, P. C. 1960. Participation of the cytoplasmic membrane in the growth and spore formation of bacilli. *J. Biophys. Biochem. Cytol.* **8**:507–528.

Flavell, R. A., and Jones, I. G. 1970. mtDNA from *Tetrahymena pyriformis* and its kinetic complexity. *Biochem. J.* **116**:811–817.

Flavell, R. A., and Jones, I. G. 1971a. Base sequence distribution in *Tetrahymena* mtDNA. *FEBS Lett.* **14**:354–356.

Flavell, R. A., and Jones, I. G. 1971b. *Paramecium* mtDNA. *Biochim. Biophys. Acta* **232**:255–260.

Fleischer, S., and Klouwen, G. 1960. The role of "soluble lipid" preparations in sub-mitochondrial enzyme systems. *Fed. Proc. Fed. Am. Soc. Exp. Biol.* **19**:32.

Foissner, W., and Simonsberger, P. 1975. Elektronenmikroskopischer Nachweis der subpelliculären Lage des Silberliniensystems bei *Colpidium colpoda* (Ciliata, Tetrahymenidae). *Protoplasma* **86**:65–82.

Fonty, G., Crouse, E. J., Stutz, E., and Bernardi, G. 1975. The mitochondrial genome of *Euglena gracilis*. *Eur. J. Biochem.* **54**:367–372.

Francisco, J. F., Vissering, F. F., and Simpson, M. V. 1977. Two aspects of mtDNA structure. *In*: Bandlow, W., Schweyen, R. J., Wolf, K., and Kandewitz, F., eds., *Mitochondria 1977*, Berlin, de Gruyter, pp. 25–37.

Fukuhara, H., Bolotin-Fukuhara, M., Hsu, H.-J., and Rabinowitz, M. 1976. Deletion mapping of mitochondrial tRNA genes in *Saccharomyces cerevisiae* by means of cytoplasmic petite mutants. *Mol. Gen. Genet.* **145**:7–17.

Gahan, P. B., and Chayen, J. 1965. Cytoplasmic DNA. *Int. Rev. Cytol.* **18**:223–247.

Gallerani, R., and Saccone, C. 1974. The DNA-dependent RNA polymerase from rat liver mitochondria. *In*: Kroon, A. M., and Saccone, C., eds., *The Biogenesis of Mitochondria*, New York, Academic Press, pp. 58–64.

Galper, J. B. 1974. Mitochondrial protein synthesis in HeLa cells. *J. Cell Biol.* **60**:755–763.

Galper, J. B., and Darnell, J. E. 1969. The presence of *N*-formyl-methionyl-tRNA in HeLa cells. *Biochem. Biophys. Res. Commun.* **34**:205–214.

Garon, C. F., Berry, K. W., and Rose, J. A. 1972. A unique form of terminal redundancy in adenovirus DNA molecules. *Proc. Natl. Acad. Sci. USA* **69**:2391–2395.

Giesbrecht, P. 1960. Über "organisierte" Mitochondrien und andere Feinstrukturen von *B. megaterium*. *Zentralbl. Bakteriol. Parasitenkd. Infektionskr. Hyg. Abt. 1 Orig.* **179**:538–581.

Goddard, J. M., and Cummings, D. J. 1975. Structure and replication of mitochondrial DNA from *Paramecium aurelia*. *J. Mol. Biol.* **97**:593–609.

Goddard, J. M., and Cummings, D. J. 1977. Mitochondrial DNA replication in *Paramecium aurelia*. Cross-linking of the initiation end. *J. Mol. Biol.* **109**:327–344.

Goldbach, R. W., Borst, P., Arnberg, A. C., and van Bruggen, E. F. J. 1976. The duplications and chewed ends of *Tetrahymena* mtDNA. *In*: Saccone, C., and Kroon, A. M., eds., *The Genetic Function of Mitochondrial DNA*, Amsterdam, North-Holland, pp. 137–142.

Goldbach, R. W., Arnberg, A. C., van Bruggen, E. F. J., Defize, J., and Borst, P. 1977. The structure of *Tetrahymena pyriformis* mitochondrial DNA. I. Strain differences and occurrence of inverted repetitions. *Biochim. Biophys. Acta* **477**:37–50.

Goldberger, R., Pumphrey, A., and Smith, A. 1962. Studies on the electron-transport system. XLVI. On the modification of the properties of cytochrome *b*. *Biochim. Biophys. Acta* **58**:307–313.

Green, D. E., and Perdue, J. F. 1966. Correlation of mitochondrial structure and function. *Ann. N.Y. Acad. Sci.* **137**:667–684.

Green, D. E., and Wharton, D. C. 1963. Stoichiometry of the fixed oxidation–reduction components of the electron transfer chain of beef heart mitochondria. *Biochem. Z.* **338**:335–348.

Green, D. E., Asai, J., Harris, R. A., and Penniston, J. T. 1968. Conformational basis of energy transformations in membrane systems. III. Configurational changes in the mitochondrial inner membrane induced by changes in functional states. *Arch. Biochem. Biophys.* **125**:684–705.

Greenawalt, J. W., and Whiteside, T. L. 1975. Mesosomes: Membranous bacterial organelles. *Bacteriol. Rev.* **39**:405–463.

Grimes, G. W., Mahler, H. R., and Perlman, P. S. 1974. Nuclear gene dosage effects on mitochondrial mass and DNA. *J. Cell Biol.* **61**:565–574.

Grimwood, B. G., and Wagner, R. P. 1976. Direct action of ethidium bromide upon mitochondrial oxidative phosphorylation and morphology. *Arch. Biochem. Biophys.* **176**:43–52.

Grivell, L. A., and Moorman, A. F. M. 1977. A structural analysis of the oxi-3 region on yeast mtDNA. *In*: Bandlow, W., Schweyen, R. J., Wolf, K., and Kaudewitz, F., eds., *Mitochondria 1977*, Berlin, de Gruyter, pp. 371–384.

Grivell, L. A., Reijnders, L., and Borst, P. 1971. Isolation of yeast mitochondrial ribosomes highly active in protein synthesis. *Biochim. Biophys. Acta* **247**:91–103.

Grivell, L. A., Arnberg, A. C., Boer, P. H., Borst, P., Bos, J. L., van Bruggen, E. F. J., Groot, G. S. P., Hecht, N. B., Heusgens, L. A. M., van Ommen, G. J. B., and Tabak, H. F. 1979. Transcripts of yeast mtDNA and their processing. *In*: Cumming, D., Borst, P., David, I., Weissman, S., and Fox, C. F., eds., *Extrachromosomal DNA*, New York, Academic Press.

Groot, G. S. P., and Poyton, R. O. 1975. Oxygen control of cytochrome *c* oxidase synthesis in isolated mitochondria from *S. cerevisiae*. *Nature (London)* **255**:238–240.

Groot, G. S. P., Flavell, R. A., and Sanders, J. P. M. 1975. Sequence homology of nuclear and mitochondrial DNAs of different yeasts. *Biochim. Biophys. Acta* **378**:186–194.

Groot-Obbink, D. J., Hall, R. M., Linnane, A. W., Lukins, H. B., Monk, B. C., Spithill, T. W., and Trembath, M. K. 1976. Mitochondrial genes involved in the determination of mitochondrial membrane proteins. *In*: Saccone, C., and Kroon, A. M., eds., *The Genetic Function of Mitochondrial DNA*, Amsterdam, North-Holland, pp. 163–173.

Grosch, G., Doxiadis, I., Lang, B., Schweyen, R. J., and Kaudewitz, F. 1977. A one-step method for *rho*$^+$ cDNA/*rho*$^-$ DNA hybridization in mapping of the mitochondrial genome of *S. cerevisiae*. *In*: Bandlow, W., Schweyen, R. J., Wolf, K., and Kaudewitz, F., eds., *Mitochondria 1977,* Berlin, de Gruyter, pp. 425–432.

Grossman, L. I., Watson, R., and Vinograd, J. 1973. The presence of ribonucleotides in mature closed-circular mitochondrial DNA. *Proc. Natl. Acad. Sci. USA* **70**:3339–3343.

Grunicke, H., Keller, H. J., Puschendorf, B., and Benaguid, A. 1975. Biosynthesis of NAD in mitochondria. *Eur. J. Biochem.* **53**:41–45.

Guérineau, M., and Paoletti, C. 1975. Rearrangement of mitochondrial DNA molecules during the differentiation of mitochondria in yeast. II. Labelling studies of the precursor product relationship. *Biochimie* **57**:931–942.

Guérineau, M., Grandchamp, C., and Slonimski, P. P. 1975. Rearrangement of mitochondrial DNA molecules during the differentiation of mitochondria in yeast. I. Electron microscopic studies of size and shape. *Biochimie* **57**:917–929.

Guttes, E. W., Hanawalt, P. C., and Guttes, S. 1967. Mitochondrial DNA synthesis and the mitotic cycle in *Physarum polycephalum*. *Biochim. Biophys. Acta* **142**:181–194.

Guttes, E. W., Guttes, S., and Devi, R. V. 1969. Division stages of the mitochondria in normal and actinomysin-treated plasmodia of *Physarum polycephalum*. *Experientia* **25**:66–68.

Guttes, S., Guttes, E. W., and Hadek, R. 1966. Occurrence and morphology of a fibrous body in the mitochondria of the slime mold *Physarum polycephalum*. *Experientia* **22**:452–454.

Hackenbrock, C. R. 1966. Ultrastructural bases for metabolically linked mechanical activity in mitochondria. *J. Cell Biol.* **30**:269–297.

Hackenbrock, C. R. 1968. Electron transport-linked ultrastructural transformations in mitochondria. *J. Cell Biol.* **37**:345–369.

Hackett, D. P. 1959. Respiratory mechanisms in higher plants. *Annu. Rev. Plant Physiol.* **10**:113–146.

Hackett, D. P. 1963. Respiratory mechanisms and control in higher plant tissues. *Annu. Symp. Soc. Gen. Physiol.* **8**:105–127.

Hall, J. D., Stiles, J. W., Awasthi, Y., and Crane, F. L. 1969. Membranifibrils on cristae and grana membranes. *Proc. Indiana Acad. Sci.* **78**:189–197.

Harris, R. A., Williams, C. H., Caldwell, M., and Green, D. E. 1969. Energized configurations of heart mitochondria *in situ*. *Science* **165**:700–702.

Hayashi, J.-I., Yomekawa, H., Gotoh, O., Motohashi, J., and Tagashira, Y. 1978a. The differences between the primary structures of mitochondrial DNAs from rat liver and ascites hepatoma (AH-130). *Cancer Lett.* **4**:125–130.

Hayashi, J.-I., Yomekawa, H., Gotoh, O., Motohashi, J., and Tagashira, Y. 1978b. Two different molecular types of rat mitochondrial DNAs. *Biochem. Biophys. Res. Commun.* **81**:871–877.

Heckman, J. E., Hecker, L. I., Schwartzbach, S. D., Barnett, W. E., Baumstark, B., and Raj-Bhandary, U. L. 1978. Structure and function of initiator $tRNA^{Met}$ from the mitochondria of *Neurospora crassa*. *Cell* **13**:83–95.

Hedley, R. H., and Bertaud, W. S. 1962. Electron-microscopic observations of *Gromia oviformis* (Sarcodina). *J. Protozool.* **9**:79–87.

Herbert, E., and Potter, V. R. 1956. Nucleotide metabolism. VI. The phosphorylation of 5′-cytosine and guanine nucleotides by cell fractions from rat liver. *J. Biol. Chem.* **222**:453–467.

Hettiarachchy, N. S., and Jones, I. G. 1974. Isolation and characterization of mitochondrial deoxyribonucleic acid of *Acanthamoeba castellanii*. *Biochem. J.* **141**:159–164.

Heyting, C., and Sanders, J. P. M. 1976. The physical mapping of some genetic markers in the 21S ribosomal region of the mitochondrial DNA of yeast. *In*: Saccone, C., and Kroon, A. M., eds., *The Genetic Function of Mitochondrial DNA*, Amsterdam, North-Holland, pp. 273–280.

Higgins, M. L., and Daneo-Moore, L. 1974. Factors influencing the frequency of mesosomes observed in fixed and unfixed cells of *Streptococcus faecalis*. *J. Cell Biol.* **61**:288–300.

Highton, P. J. 1969. An electron microscopic study of cell growth and mesosomal structure of *Bacillus licheniformis*. *J. Ultrastruct. Res.* **26**:130–147.

Hill, G. C., and Anderson, W. A. 1969. Effects of acriflavine on the mitochondria and kinetoplast of *Crithidia fasciculata*. *J. Cell Biol.* **41**:547–561.

Hird, F. J. R., and Rowsell, E. V. 1950. Additional transaminations by insoluble particle preparations of rat liver. *Nature (London)* **166**:517–518.

Hirsch, M., Spradling, A., and Penman, S. 1974. The messenger-like poly(A)-containing RNA species from the mitochondria of mammals and insects. *Cell* **1**:30–35.

Hoffman, H. P., and Avers, C. J. 1973. Mitochondrion of yeast: Ultrastructural evidence for one giant, branched organelle per cell. *Science* **181**:749–751.

Hogeboom, G. H., and Schneider, W. C. 1953. Intracellular distribution of enzymes. XI. Glutamic dehydrogenase. *J. Biol. Chem.* **204**:233–238.

Hoh, J. F. Y., McGrath, P. A., and Hale, P. T. 1977. Electrophoretic analysis of multiple forms of rat cardiac myosin: Effects of hypophysectomy and thyroxine replacement. *J. Mol. Cell. Biol.* **10**:1053–1076.

Hoh, J. F. Y., Yeoh, G. P. S., Thomas, M. A. W., and Higginbottom, L. 1979. Structural differences in the heavy chains of rat ventricular myosin enzymes. *FEBS Lett.* **97**:330–334.

Hollenberg, C. P., Borst, P., and van Bruggen, E. F. J. 1970. Mitochondrial DNA. V. A 25-μ closed circular duplex DNA molecule in wild-type mitochondria. *Biochim. Biophys. Acta* **209**:1–15.

Hootman, S. R., and Philpott, C. W. 1979. Ultracytochemical localization of Na^+, K^+-activated ATPase in chloride cells from the gills of a euryhaline teleost. *Anat. Rec.* **193**:99–130.

Horak, I., Coon, H. G., and Dawid, I. B. 1974. Interspecific recombination of mitochondrial DNA molecules in hybrid somatic cells. *Proc. Natl. Acad. Sci. USA* **71**:1828–1832.

Horovitch, S. J., Storti, R. V., Rich, A., and Pardue, M. L. 1979. Multiple actins in *Drosophila melanogaster*. *J. Cell Biol.* **82**:86–92.

Horváth, I., Zagos, P., Szabados, G., and Bauer, P. 1975. *In vitro* synthesis of guanosine polyphosphates in rat liver mitochondrial preparations. *FEBS Lett.* **56**:179–183.

Houtsmuller, U. M. T., Struijk, C. B., and Van der Beek, A. 1970. Decrease in rate of ATP synthesis of isolated rat heart mitochondria induced by dietary erucic acid. *Biochim. Biophys. Acta* **218**:564–566.

Innis, M. A., and Craig, S. P. 1978. Mitochondrial regulation in sea urchins. II. Formation of polyribosomes within the mitochondria of 4–8 cell stage embryos of the sea urchin. *Exp. Cell Res.* **111**:223–230.

Innis, M. A., Beers, T. R., and Craig, S. P. 1976. Mitochondrial regulation in sea urchins. I. Mitochondrial ultrastructure transformations and changes in the ADP : ATP ratio at fertilization. *Exp. Cell Res.* **98**:47–56.

Iwashima, A., and Rabinowitz, M. 1969. Partial purification of mitochondrial and supernatant DNA polymerases from *Saccharomyces cerevisiae*. *Biochim. Biophys. Acta* **178**:283–293.

Jackl, G., and Sebald, W. 1975. Identification of two products of mitochondrial protein synthesis associated with mitochondrial ATPase from *Neurospora*. *Eur. J. Biochem.* **54**:97–106.

Jacq, C., Kujawa, C., Grandchamp, C., and Netter, P. 1977. Physical characteristics of the difference between yeast mtDNA alleles ω^+ and ω^-. *In*: Bandlow, W., Schweyer, R. J., Wolf, K., and Kaudewitz, F., eds., *Mitochondria 1977,* Berlin, de Gruyter, pp. 255–270.

Jakovcic, S., Casey, J., and Rabinowitz, M. 1975. Sequence homology between mitochondrial DNAs of different eukaryotes. *Biochemistry* **14**:2043–2050.

Jay, G., and Kaempfer, R. 1975. Initiation of protein synthesis. Binding of mRNA. *J. Biol. Chem.* **250**:5742–5748.

Karnovsky, M. T. 1963. The fine structure of mitochondria in the frog nephron correlated with cytochrome oxidase activity. *Exp. Mol. Pathol.* **2**:347–366.

Karol, M. H., and Simpson, M. V. 1968. DNA biosynthesis by isolated mitochondria: A replicative rather than a repair process. *Science* **162**:470–473.

Kasamatsu, H., Robberson, D. L., and Vinograd, J. 1971. A novel closed-circular mtDNA with properties of a replicatory intermediate. *Proc. Natl. Acad. Sci. USA* **68**:2252–2257.

Katan, M. B., Pool, L., and Groot, G. S. P. 1976. The cytochrome bc_1 complex of yeast mitochondria. Isolation and partial characterization of the cytochrome bc_1 complex and cytochrome *b*. *Eur. J. Biochem.* **65**:95–105.

Kawakami, N. 1961. Threadlike mitochondria in yeast cells. *Exp. Cell Res.* **25**:179–181.

Keiding, J., and Westergaard, O. 1971. Induction of DNA polymerase activity in irradiated *Tetrahymena* cells. *Exp. Cell Res.* **64**:317–322.

Kellems, R. E., and Buetow, R. A. 1974. Cytoplasmic type 80S ribosomes associated with yeast mitochondria. III. Changes in the amount of bound ribosomes in response to changes in metabolic state. *J. Biol. Chem.* **249**:3304–3310.

Kellenberger, E., and Huber, L. 1953. Contribution à l'étude des équivalents des mitochondries dans les bactéries. *Experientia* **9**:289–291.

Kellenberger, E., Ryter, A., and Séchaud, J. 1958. Electron microscope study of DNA-containing plasms. II. Vegetative and mature phage DNA as compared with normal bacterial nucleoids in different physiological states. *J. Biophys. Biochem. Cytol.* **4**:671–676.

Kellerman, G. M., Biggs, D. R., and Linnane, A. W. 1969. Biogenesis of mitochondria. *J. Cell Biol.* **42**:378–391.

Kessler, D. 1969. Mitochondrial DNA in *Physarum polycephalum*. *J. Cell Biol.* **43**:68a.

Kielley, R. K., and Schneider, W. C. 1950. Synthesis of *p*-aminohippuric acid by mitochondria of liver homogenates. *J. Biol. Chem.* **185**:869–880.

King, G. A. M. 1977. Symbiosis and the origin of life. *Origins Life* **8**:39–53.

Kirschner, R. H., Wolstenholme, D. R., and Gross, N. J. 1968. Replicating molecules of circular mitochondrial DNA. *Proc. Natl. Acad. Sci. USA* **60**:1466–1472.

Kleisen, C. M., and Borst, P. 1975. Are 50% of all cellular proteins synthesized on mitochondrial ribosomes in *Crithidia luciliae? Biochim. Biophys. Acta* **390**:78–81.

Kleisen, C. M., Borst, P., and Weijers, P. J. 1976a. The structure of kinetoplast DNA. I. The mini-circles of *Crithidia luciliae* are heterogenous in base sequence. *Eur. J. Biochem.* **64**:141–151.

Kleisen, C. M., Weislagel, P. O., Fonck, K., and Borst, P. 1976b. The structure of kinetoplast DNA. II. Characterization of a novel component of high complexity present in the kinetoplast DNA network of *Crithidia luciliae*. *Eur. J. Biochem.* **64**:153–160.

Koike, K., and Wolstenholme, D. R. 1974. Evidence for discontinuous replication of circular mitochondrial DNA molecules from Novikoff rat ascites hepatoma cells. *J. Cell Biol.* **61**:14–25.

Kozar, F., and Weijer, J. 1969. Electron-dense structures in *Neurospora crassa*. I. Mesosomes. *Can. J. Gen. Cytol.* **11**:613–616.

Krawiec, S., and Eisenstadt, J. M. 1970a. RNAs from the mitochondria of bleached *Euglena gracilis*. I. Isolation of mitochondria and extraction of nucleic acids. *Biochim. Biophys. Acta* **217**:120–131.

Krawiec, S., and Eisenstadt, J. M. 1970b. RNAs from the mitochondria of bleached *Euglena gracilis*. II. Characterization of highly polymeric RNA. *Biochim. Biophys. Acta* **217**:132–141.

Krebs, H. A., Gascoyne, T., and Nolton, B. M. 1967. Generation of extramitochondrial reducing power in gluconeogenesis. *Biochem. J.* **102**:275–282.

Krebs, J. J. R., Hauser, H., and Carafoli, E. 1979. Asymmetric distribution of phospholipids in the inner membrane of beef heart mitochondria. *J. Biol. Chem.* **254**:5308–5316.

Küntzel, H. 1969. Proteins of mitochondrial and cytoplasmic ribosomes from *Neurospora crassa*. *Nature (London)* **222**:142–146.

Küntzel, H., and Schäfer, K. P. 1971. Mitochondrial RNA polymerase from *Neurospora crassa*. *Nature New Biol.* **231**:265–269.

Kuroiwa, T. 1973. Studies on mitochondrial structure and function in *Physarum polycephalum*. I. Fine structure, cytochemistry, and ^{3}H-uridine autoradiography of central body in mitochondria. *Exp. Cell Res.* **78**:351–359.

Kuroiwa, T. 1974. Studies on mitochondrial structure and function in *Physarum polycephalum*. III. Electron microscopy of large amount of DNA released from a central body in mitochondria by trypsin digestion. *J. Cell Biol.* **63**:299–306.

Kuroiwa, T., Kawano, S., and Hizume, M. 1976a. A method of isolation of mitochondrial nucleoid of *Physarum polycephalum* and evidence for the presence of a basic protein. *Exp. Cell Res.* **97**:435–440.

Kuroiwa, T., Kitane, H., Watanabe, T., and Kawano, S. 1976b. The general occurrence of mitochondrial nuclei in the true slime mold. *J. Electron Microsc.* **25**:103–105.

Kuroiwa, T., Kawano, S., and Hizume, M. 1977. Behavior of mitochondrial nucleoids throughout mitochondrial division cycle. *J. Cell Biol.* **72**:687–694.

Lambowitz, A. L., Chua, N. H., and Luck, D. J. L. 1976. Mitochondrial ribosome assembly in *Neurospora*. *J. Mol. Biol.* **107**:223–253.

Lambowitz, A. L., LaPolla, R. J., and Collins, R. A. 1979. Mitochondrial ribosome assembly in *Neurospora*. *J. Cell Biol.* **82**:17–31.

Lang, N. J. 1963. Electron microscopy of the Volvocaceae and Astrephomenaceae. *Am. J. Bot.* **50**:280–300.

Lange, L. L., and Olson, L. W. 1978. The zoospore of *Olpidium radicale*. *Trans. Br. Mycol. Soc.* **71**:43–55.

Lange, R. A., and Jacobson, M. K. 1977. Synthesis of pyridine nucleotides by mitochondrial fractions of yeast. *Biochem. Biophys. Res. Commun.* **76**:424–428.

Lasker, R., and Threadgold, L. T. 1968. "Chloride cells" in the skin of the larval sardine. *Exp. Cell Res.* **52**:582–590.

Laub-Kupersztejn, R., and Thirion, J. 1969. Étude des effects de l'acriflavine et du bromure d'ethidium sur l'incorporation de précurseurs radioactifs dans les acides nucléiques et les proteins du trypanosomide *Crithidia luciliae*. *Arch. Int. Physiol. Biochem.* **77**:566–568.

Laub-Kupersztejn, R., and Thirion, J. 1972. Effets du chloramphénicol sur le trypanosomide *Crithidia luciliae*. *Arch. Int. Physiol. Biochem.* **80**:197–198.

Laub-Kupersztejn, R., and Thirion, J. 1974. Existence of two distinct protein synthesis systems in the trypanosomatid *Crithidia luciliae*. *Biochim. Biophys. Acta* **340**:314–322.

Laurent, M., and Steinert, M. 1970. Electron microscopy of kinetoplastic DNA from *Trypanosoma mega*. *Proc. Natl. Acad. Sci. USA* **66**:419–424.

Leedale, G. F., and Buetow, D. E. 1970. Observations on the mitochondrial reticulum in living *Euglena gracilis*. *Cytobiologie Z. Exp. Zellforsch.* **1**:195–202.

Lehninger, A. L. 1966. Dynamics and mechanisms of active ion transport across the mitochondrial membrane. *Ann. N.Y. Acad. Sci.* **137**:700–707.

Leister, D. E., and Dawid, I. B. 1975. Mitochondrial ribosomal proteins in *Xenopus laevis/X. mulleri* interspecific hybrids. *J. Mol. Biol.* **96**:119–123.

Lembi, C. A., and Lang, N. J. 1965. Electron microscopy of *Carteria* and *Chlamydomonas*. *Am. J. Bot.* **52**:464–477.

Leuthardt, F., and Nielson, H. 1951. Investigation of the biological synthesis of hippuric acid. *Helv. Chim. Acta* **34**:1618–1631.

Lewin, A., Morimoto, R., Merten, S., Martin, N., Berg, P., Christianson, T., Levens, D., and Rabinowitz, M. 1977. Physical mapping of mitochondrial genes and transcripts in *S. cerevisiae*. *In*: Bandlow, W., Schweyen, R. J., Wolf, K., and Kaudewitz, F., eds., *Mitochondria 1977*, Berlin, de Gruyter, pp. 271–289.

Lewis, F. S., Rutman, R. J., and Avadhani, N. G. 1976. Messenger RNA metabolism in mammalian mitochondria. *Biochemistry* **15**:3362–3366.

Lewis, M. R., and Lewis, W. H. 1914. Mitochondria (and other cytoplasmic structures) in tissue cultures. *Am. J. Anat.* **17**:339–401.

Ley, K. D., and Murphy, M. M. 1973. Synchronization of mitochondrial DNA synthesis in Chinese hamster cells (line CHO) deprived of isoleucine. *J. Cell Biol.* **58**:340–345.

Lieberman, M., and Baker, J. E. 1965. Respiratory electron transport. *Annu. Rev. Plant Physiol.* **16**:343–382.

Lin, H. S. 1965. Microcylinders within mitochondrial cristae in the rat pinealocyte. *J. Cell Biol.* **25**:435–441.

Linnane, A. W. 1965. Aspects of the biosynthesis of the mitochondria of *S. cerevisiae*. *In*: King, T. E., Mason, H., and Morrison, M., eds., *Oxidases and Related Redox Systems*, New York, Wiley, Vol. 2, pp. 1102–1128.

Linnane, A. W., and Nagley, P. 1978. Structural mapping of mtDNA. *Arch. Biochem. Biophys.* **187**:277–289.

Linnane, A. W., Vitols, E., and Nowland, P. G. 1962. Studies on the origins of yeast mitochondria. *J. Cell Biol.* **13**:345–350.

Linnane, A. W., Haslam, J. M., Lukins, H. B., and Nagley, P. 1972. The biogenesis of the mitochondria in microorganisms. *Annu. Rev. Microbiol.* **26**:153–198.

Lizardi, P. M., and Luck, D. J. L. 1972. The intercellular site of synthesis of mitochondrial ribosomal proteins in *Neurospora crassa*. *J. Cell Biol.* **54**:56–74.

Lloyd, D. 1974. *The Mitochondria of Microorganisms*, New York, Academic Press.

Lloyd, D., and Venables, S. E. 1967. The regulation of proprionate oxidation in *Prototheca zopfi*. *Biochem. J.* **104**:639–646.

Lopez Perez, M. J., and Turner, G. 1975. Mitochondrial DNA from *Aspergillus nidulans*. *FEBS Lett.* **58**:159–163.

Lowenstein, J. M. 1971. Replenishment and depletion of citric acid cycle intermediates in muscle. *In*: Mehlman, M. A., ed., *Energy Metabolism Regulation of Metabolic Processes in Mitochondria*, New York, Academic Press, pp. 53–61.

Luck, D. J. L. 1963. Formation of mitochondria in *Neurospora crassa*. *J. Cell Biol.* **16**:483–499.

Lynch, D. C., and Attardi, G. 1976. Amino acid specificity of the tRNA species coded for by HeLa cell mitochondrial DNA. *J. Mol. Biol.* **102**:125–141.

Maccecchini, M. L., Rudin, Y., Blobel, G., and Schatz, G. 1979. Import of proteins into mitochondria. Precursor forms of the extramitochondrially made F_1-ATPase subunits in yeast. *Proc. Natl. Acad. Sci. USA* **76**:343–347.

McCully, M. 1968. Histological studies on the genus *Fucus*. III. Fine structure and possible functions of the epidermal cells of the vegetative thallus. *J. Cell Sci.* **3**:1–16.

Macino, G., and Tzagoloff, A. 1979. Assembly of the mitochondrial membrane system. The DNA sequence of a mitochondrial ATPase gene in *S. cerevisiae*. *J. Biol. Chem.* **254**:4617–4623.

Mahler, H. R., and Raff, R. A. 1975. The evolutionary origin of the mitochondrion: A nonsymbiotic model. *Int. Rev. Cytol.* **43**:1–124.

Malhotra, S. K. 1966. A study of structure of the mitochondrial membrane system. *J. Ultrastruct. Res.* **15**:14–37.

Malhotra, S. K. 1968. Mesosome-like structures in mitochondria of *Poky Neurospora*. *Nature (London)* **219**:1267–1268.

Manning, J. E., Wolstenholme, D. R., Ryan, R. S., Hunter, J. A., and Richards, O. C. 1971. Circular chlorophast DNA from *Euglena gracilis*. *Proc. Natl. Acad. Sci. USA* **68**:1169–1173.

Mantel, L. H. 1967. Asymmetry potentials, metabolism and sodium fluxes in the gills of the blue crab, *Callinectes sapidus*. *Comp. Biochem. Physiol.* **20**:743–753.

Marchant, H. J., and Fowke, L. C. 1977. Preparation, culture, and regeneration of protoplasts from filamentous green algae. *Can. J. Bot.* **55**:3080–3086.

Margulis, L. 1968. Evolutionary criteria in Thallophytes: A radical alternative. *Science* **161**:1020–1022.

Margulis, L. 1970. *Origin of Eukaryotic Cells,* New Haven, Yale University Press.

Marjanon, L. A., and Ryrie, I. J. 1976. The biogenesis of cytochrome *b* complex in yeast mitochondria. *Arch. Biochem. Biophys.* **172**:679–684.

Marquardt, H. 1962. Der Feinbau von Hefezellen im Elektronenmikroskop. II. Mitt.: *Saccharomyces cerevisiae*–Stämme. *Z. Naturforsch.* **17b**:689–695.

Martin, N. C., Rabinowitz, M., and Fukuhara, H. 1976. Isoaccepting mitochondrial glutamyl-tRNA species transcribed from different regions of the mitochondrial genome of *S. cerevisiae*. *J. Mol. Biol.* **101**:285–296.

Martin, N. C., Rabinowitz, M., and Fukuhara, H. 1977. Yeast mitochondrial DNA specifies tRNA for 19 amino acids. Deletion mapping of the tRNA genes. *Biochemistry* **16**:4672–4677.

Martin, R., Schneller, J. M., Stahl, A. J. C., and Dirheimer, G. 1976. Studies of odd bases in yeast mitochondrial tRNA. II. Characterization of rare nucleosides. *Biochem. Biophys. Res. Commun.* **70**:997–1002.

Martinucci, G. B., and Felluga, B. 1979. Mitochondria-mediated chromatin condensation and nucleus reshaping during spermiogenesis in Lumbricidae. *J. Submicrosc. Cytol.* **11**:221–228.

Matile, P., and Bahr, G. F. 1968. Biochemical and quantitative electron microscopic evidence for heterogeneity of mitochondria from *S. cerevisiae*. *Exp. Cell Res.* **52**:301–307.

Matsumoto, L., Pikó, L., and Vinograd, J. 1976. Complex mitochondrial DNA in animal thyroids. A comparative study. *Biochim. Biophys. Acta* **432**:257–266.

Matsura, H. 1973. Joint propagation of the β and β' subunits of RNA polymerase in *E. coli*. *Nature New Biol.* **244**:262–264.

Merker, H. J., Herbst, R., and Kloss, K. 1968. Elektronenmikroskopische Untersuchungen an den Mitochondrien des menschlichen Uterusepithels während der Sekretionsphase. *Z. Zellforsch. Mikrosk. Anat.* **86**:139–152.

Merli, A., Capaldi, R. A., Ackrell, B. A. C., and Kearney, E. B. 1979. Arrangement of complex II (succinate ubiquinone reductase) in the mitochondrial inner membrane. *Biochemistry* **18**:1393–1398.

Milder, R. V., and Deane, M. P. 1967. Ultrastructure of *Trypanosoma conorhini* in the crithidial phase. *J. Protozool.* **14**:65–72.

Millis, A. J. T., and Suyama, Y. 1972. Effects of chloramphenicol and cycloheximide on the synthesis of mitochondrial ribosomes in *Tetrahymena*. *J. Biol. Chem.* **247**:4063–4073.

Milner, J. 1976. The functional development of mammalian mitochondria. *Biol. Rev.* **51**:181–209.

Mockel, J. 1972. Amino acid incorporation into rat liver mitochondria. *Biochim. Biophys. Acta* **277**:628–638.

Montisano, D. F., and James, T. W. 1979. Mitochondrial morphology in yeast with and without mtDNA. *J. Ultrastruct. Res.* **67**:288–296.

Moor, H. 1964. Die Gefrier-Fixation lebender Zellen und ihre Anwendung in der Elektronenmikroscopie. *Z. Zellforsch. Mikrosk. Anat.* **62**:546–580.

Moor, H., and Mühlethaler, K. 1963. Fine structure in frozen-etched yeast cells. *J. Cell Biol.* **17**:609–628.

Moore, K. H., Johnson, P. H., Chandler, S. E. W., and Grossman, L. I. 1977. A restriction endonuclease cleavage map of mouse mitochondrial DNA. *Nucleic Acids Res.* **4**:1273–1289.

Mudd, S. 1954. Cytology of bacteria. Part I: The bacterial cell. *Annu. Rev. Microbiol.* **8**:1–22.

Mudd, S. 1956. Cellular organization in relation to function. *Bacteriol. Rev.* **20**:268–271.

Mudd, S., Winterscheid, L. C., De Lameter, E. D., and Henderson, H. J. 1951a. Evidence suggesting that the granules of mycobacteria are mitochondria. *J. Bacteriol.* **62**:459–475.

Mudd, S., Brodie, A. F., Winterscheid, L. C., Hartman, P. E., Beutner, E. H., and McLean, R. A. 1951b. Further evidence of the existence of mitochondria in bacteria. *J. Bacteriol.* **62**:729–739.

Mudd, S., Takeya, K., and Henderson, H. J. 1956. Electron-scattering granules and reducing sites in mycobacteria. *J. Bacteriol.* **72**:767–783.

Mugnaini, E. 1964. Helical filaments in astrocytic mitochondria of the corpus striatum in the rat. *J. Cell Biol.* **23**:173–182.

Mukerjee, H., and Goldfeder, A. 1973. Purification and properties of RNA polymerase from rat liver mitochondria. *Biochemistry* **12**:5096–5101.

Nagata, I., Furuya, E., Yoshida, Y., Kanaseki, T., and Tagawa, K. 1975. Development of mitochondrial membranes in anaerobically grown yeast cells. *J. Biochem.* **78**:1353–1364.

Nagley, P., Sriprakash, K. S., Rytka, J., Choo, K. B., Trembath, M. K., Lukins, H. B., and Linnane, A. W. 1976. Physical mapping of genetic markers in the yeast mitochondrial genome. *In*: Saccone, C., and Kroon, A. M., eds., *The Genetic Function of Mitochondrial DNA,* Amsterdam, North-Holland, pp. 231–242.

Nass, M. M. K. 1969. Mitochondrial DNA: Advances, problems, and goals. *Science* **165**:25–35.

Nass, M. M. K., and Buck, C. A. 1970. Studies on mitochondrial tRNA from animal cells. II. Hybridization of aminoacyl tRNA from rat liver mitochondria with H and L complementary strands of mtDNA. *J. Mol. Biol.* **54**:187–198.

Nass, M. M. K., and Nass, S. 1962. Fibrous structures within the matrix of developing chick embryo mitochondria. *Exp. Cell Res.* **26**:424–437.

Nass, M. M. K., Nass, S., and Afzelius, B. A. 1965. The general occurrence of mitochondrial DNA. *Exp. Cell Res.* **37**:516–539.

Nass, M. M. K., Schori, L., Ben-Shaul, Y., and Edelman, M. 1974. Size and configuration of mtDNA in *Euglena gracilis*. *Biochim. Biophys. Acta* **374**:283–291.

Nass, S. 1969. The significance of the structural and functional similarities of bacteria and mitochondria. *Int. Rev. Cytol.* **25**:55–129.

Nejedly, K., and Greksák, M. 1977. Sequential increase in activity of mitochondrial enzymes during respiratory adaptation of anaerobically-grown synchronous yeast. *FEBS Lett.* **77**:33–36.

Netzel, H. 1975. Struktur und Ultrastruktur von *Arcella vulgaris* var. *multinucleata* (Rhizopoda, Testacea). *Arch. Protistenkd.* **117**:219–245.

Neupert, W., Sebald, W., Schwab, A., Massinger, P., and Bücher, T. 1969. Incorporation *in vivo* of ^{14}C-labelled amino acids into the proteins of mitochondrial ribosomes from *Neurospora* sensitive to cycloheximide and insensitive to chloramphenicol. *Eur. J. Biochem.* **10**:589–591.

Newton, B. A. 1967. Isolation of DNA from kinetoplasts of *Crithidia fasciculata*. *J. Gen. Microbiol.* **48**:IV.

Newton, B. A. 1968. Biochemical peculiarities of trypanosomatid flagellates. *Annu. Rev. Microbiol.* **22**:109–130.

Nichols, J. M., and Cross, G. A. M. 1977. Isolation of mitochondria and mitochondrial RNA from *Crithidia fasciculata*. *J. Gen. Microbiol.* **99**:291–300.

Niklowitz, W. 1958. Mitochondrienäquivalente bei *Escherichia coli*. *Zentralbl. Bakteriol. Parasitenkd. Infektionskr. Hyg.* **173**:12–24.

Novikoff, A. B. 1961. Mitochondria (chondriosomes). *In*: Brachet, J., and Mirsky, A. E., eds., *The Cell,* New York, Academic Press, Vol. 2, pp. 299–421.

O'Brien, T. W. 1977. Transcription and translation in mitochondria. *In:* Brinkley, B. R., and Porter, K. R., eds., *International Cell Biology, 1976–1977,* New York, Rockefeller University Press, pp. 245–255.

Ohkawa, K. I., Vogt, M. T., and Farber, E. 1969. Unusually high mitochondrial alpha glycerophosphate dehydrogenase activity in rat brown adipose tissue. *J. Cell Biol.* **41**:441–449.

Ohnishi, T., and Ohnishi, T. 1962. Extraction of contractile protein from liver mitochondria. *J. Biochem.* **51**:380–381.

Ojala, D., and Attardi, G. 1974. Identification and partial characterization of multiple discrete poly(A)-containing RNA components coded by the HeLa cell mitochondrial DNA. *J. Mol. Biol.* **88**:205–219.

Ostroumov, S. A. 1977. Participation of chloroplasts and mitochondria in virus reproduction and the evolution of the eukaryotic cell. *J. Theor. Biol.* **67**:287–297.

Pace, N. R. 1973. Structure and synthesis of the ribosomal RNA of prokaryotes. *Bacteriol. Rev.* **37**:562–603.

Pangborn, J., Marr, A. G., and Robrish, S. A. 1962. Localization of respiratory enzymes in intracytoplasmic membranes of *Azotobacter agilis*. *J. Bacteriol.* **84**:669–678.

Parce, J. W., Cunningham, C. C., and Waite, M. 1978. Mitochondrial phospholipase A_2 activity and mitochondrial aging. *Biochemistry* **17**:1634–1639.

Parker, R. C., and Watson, R. M. 1977. Restriction endonuclease maps of rat and mouse mtDNAs. *Nucleic Acids Res.* **4**:1291–1299.

Parsons, D. F., Bonner, J. W., and Verboon, J. G. 1965. Electron microscopy of isolated plant mitochondria and plastids using both thin section and negative staining techniques. *Can. J. Bot.* **43**:647–654.

Parsons, D. F., Williams, G. R., and Chance, B. 1966. Characteristics of isolated and purified preparations of the outer and inner membranes of mitochondria. *Ann. N.Y. Acad. Sci.* **137**:643–666.

Parsons, J. A. 1965. Mitochondrial incorporation of tritiated thymidine in *Tetrahymena pyriformis*. *J. Cell Biol.* **25**:641–646.

Parsons, P., and Simpson, M. V. 1967. Biosynthesis of DNA by isolated mitochondria: Incorporation of thymidine triphosphate-2-C^{14}. *Science* **155**:91–93.

Pasquali-Ronchetti, I., Greenawalt, J. W., and Carafoli, E. 1969. On the nature of the dense matrix granules of normal mitochondria. *J. Cell Biol.* **40**:565–568.

Peek, W. D., and Youson, J. H. 1979. Ultrastructure of chloride cells in young adults of the anadromous sea lamprey, *Petromyzon marinus* L., in freshwater and during adaptation to sea water. *J. Morphol.* **160**:143–164.

Penniston, J. T., Harris, R. A., Asai, J., and Green, D. E. 1968. The conformational basis of energy transformations in membrane systems. I. Conformational changes in mitochondria. *Proc. Natl. Acad. Sci. USA* **59**:624–631.

Perlman, S., Abelson, H. T., and Penman, S. 1973. Mitochondrial protein synthesis: RNA with the properties of eukaryotic messenger RNA. *Proc. Natl. Acad. Sci. USA* **70**:350–353.

Philippe, M., and Chevaillier, P. 1978. Presence of two DNA polymerases in bull spermatozoa. *Biochem. J.* **175**:595–600.

Phillips, D. M. 1977. Mitochondrial disposition in mammalian spermatozoa. *J. Ultrastruct. Res.* **58**:144–154.

Philpott, C. W., and Copeland, D. E. 1963. Fine structure of chloride cells from three species of *Fundulus. J. Cell Biol.* **18**:389–404.

Pitelka, D. R. 1961. Observations on the kinetoplast-mitochondrion and the cytostome of *Bodo. Exp. Cell Res.* **25**:87–93.

Plattner, H., and Schatz, G. 1969. Promitochondria of anaerobically grown yeast. *Biochemistry* **8**:339–343.

Plattner, H., Salpeter, M. M., Saltzgaber, J., and Schatz, G. 1970. Promitochondria of anaerobically grown yeast. *Proc. Natl. Acad. Sci. USA* **66**:1252–1259.

Poindexter, J. S., Stove, L., and Cohen-Bazire, G. 1964. The fine structure of the stalked bacteria belonging to the family Caulobacteriaceae. *J. Cell Biol.* **23**:587–596.

Pollock, E. G., and Cassell, R. Z. 1977. An intracristal component of *Fucus* sperm mitochondria. *J. Ultrastruct. Res.* **58**:172–177.

Pontefract, R. D., Bergeron, G., and Thatcher, F. S. 1969. Mesosomes in *Escherichia coli. J. Bacteriol.* **97**:367–375.

Poyton, R. O., and Groot, G. S. P. 1975. Biosynthesis of polypeptides of cytochrome *c* oxidase by isolated mitochondria. *Proc. Natl. Acad. Sci. USA* **72**:172–176.

Price, S. S., DiMaio, D. C., and Englund, P. T. 1976. Kinetoplast DNA from *Leishmania tarentolae. In*: Saccone, C., and Kroon, A. M., eds., *The Genetic Function of Mitochondrial DNA*, Amsterdam, North-Holland, pp. 83–94.

Rabinowitz, M., Jakovcic, S., Martin, N., Hendler, F., Halbreich, A., Lewin, A., and Morimoto, R. 1976. Transcription and organization of yeast mitochondrial DNA. *In*: Saccone, C., and Kroon, A. M., eds., *The Genetic Function of Mitochondrial DNA*, Amsterdam, North-Holland, pp. 219–230.

Racker, E. 1970. The two faces of the inner mitochondrial membrane. *In*: Campbell, P. N., and Dickens, F., eds., *Essays in Biochemistry,* New York, Academic Press, Vol. 6, pp. 1–22.

Racker, E., and Conover, T. E. 1963. Multiple coupling factors in oxidative phosphorylation. *Fed. Proc. Fed. Am. Soc. Exp. Biol.* **22**:1088–1091.

Raff, R. A., and Mahler, H. R. 1972. The nonsymbiotic origin of mitochondria. *Science* **177**:575–582.

Raff, R. A., and Mahler, H. R. 1973. Origin of mitochondria. *Science* **180**:517.

Ragland, T. E., and Hackett, D. P. 1961. Intracellular localization of some oxidative activities in etiolated pea stems. *Biochim. Biophys. Acta* **54**:577–580.

Raven, P. H. 1970. A multiple origin for plastids and mitochondria. *Science* **169**:641–646.

Reaveley, D. A., and Rogers, H. J. 1969. Some enzymic activities and chemical properties of the mesosomes and cytoplasmic membranes of *Bacillus licheniformis* 6346. *Biochem. J.* **113**:67–79.

Reboul, A., and Vignais, P. 1974. Origin of mitochondrial ribosomal RNA in *Candida utilis. Biochimie* **56**:269–274.

Reid, B. D., and Parsons, P. 1971. Partial purification of mtRNA polymerase from rat liver. *Proc. Natl. Acad. Sci. USA* **68**:2830–2834.

Reijnders, L. 1975. The origin of mitochondria. *J. Mol. Evol.* **5**:167–176.

Reijnders, L., and Borst, P. 1972. The number of 4S RNA genes on yeast mitochondrial DNA. *Biochem. Biophys. Res. Commun.* **47**:126–133.

Reijnders, L., Kleisen, C. M., Grivell, L. A., and Borst, P. 1972. Hybridization studies with yeast mitochondrial RNAs. *Biochim. Biophys. Acta* **272**:396–407.

Remsen, C. C. 1966. Fine structure of frozen-etched *Bacillus cereus. Arch. Mikrobiol.* **54**:266–275.

Remsen, C. C. 1968. Fine structure of the mesosome and nucleoid in frozen-etched *Bacillus subtilis. Arch. Mikrobiol.* **61**:40–47.

Renger, H. C., and Wolstenholme, D. R. 1970. Kinetoplast deoxyribonucleic acid of the hemoflagellate *Trypanosoma lewisi*. *J. Cell Biol.* **47**:689–702.

Renger, H. C., and Wolstenholme, D. R. 1971. Kinetoplast and other satellite DNAs of kinetoplastic and dyskinetoplastic strains of *Trypanosoma*. *J. Cell Biol.* **50**:533–540.

Revel, J. P., Fawcett, D. W., and Philpott, C. W. 1963. Observations on mitochondrial structure. Angular configurations of the cristae. *J. Cell Biol.* **16**:187–195.

Richter, D., and Lipmann, F. 1970. Separation of mitochondrial and cytoplasmic peptide chain elongation factors from yeast. *Biochemistry* **9**:5065–5070.

Rigobello, M. P., Carignani, G., and Pinna, L. A. 1978. Isolation and partial characterization of a membrane-bound protein kinase from mitochondria of *S. cerevisiae*. *Biochem. Biophys. Res. Commun.* **85**:1400–1406.

Riou, G. F., and Delain, E. 1969. Electron microscopy of the circular kinetoplastic DNA from *Trypanosoma cruzi*. *Proc. Natl. Acad. Sci. USA* **62**:210–217.

Riou, G. F., and Paoletti, C. 1967. Preparation and properties of nuclear and satellite DNA of *Trypanosoma cruzi*. *J. Mol. Biol.* **28**:377–382.

Riou, G. F., and Pautrizel, R. 1969. Nuclear and kinetoplastic DNA from trypanosomes. *J. Protozool.* **16**:509–513.

Riou, G. F., and Yot, P. 1977. Heterogeneity of the kinetoplast DNA molecules of *Trypanosoma cruzi*. *Biochemistry* **16**:2390–2396.

Robertson, J. D. 1959. The ultra-structure of cell membranes and their derivatives. *Biochem. Soc. Symp.* **16**:3–43.

Rogall, G., and Wintersberger, E. 1974. Low molecular weight subunit of a rifampicin-resistant mitochondrial RNA polymerase from yeast. *FEBS Lett.* **46**:333–336.

Roodyn, D. B. 1965. The classification and partial tabulation of enzyme studies on subcellular fractions isolated by differential centrifugation. *Int. Rev. Cytol.* **18**:99–190.

Roodyn, D. B. 1967. Mitochondrion. *Enzyme Cytol.* **1967**:103–180.

Roodyn, D. B., Reis, P. J., and Work, T. S. 1961. Protein synthesis in mitochondria. Requirements for the incorporation of radioactive amino acids into mitochondrial protein. *Biochem. J.* **80**:9–21.

Rosen, D., and Edelman, M. 1976. Poly(A)-associated RNA from the mitochondrial fraction of the fungus *Trichoderma*. *Eur. J. Biochem.* **63**:525–532.

Rosen, W. G., and Siegesmund, K. A. 1961. Some observations on the fine structure of a thermophilic, acidophilic alga. *J. Biophys. Biochem. Cytol.* **9**:910–914.

Rossi, C. S., and Lehninger, A. L. 1963. Stoichiometric relations between accumulation of ions by mitochondria and the energy-coupling sites in the respiratory chain. *Biochem. Z.* **338**:698–713.

Rouslin, W. 1977. Partial independence of synthesis and membrane attachment of mitochondrial and cytoplasmic precursors of electron transfer complexes III and IV in adapting baker's yeast. *Arch. Biochem. Biophys.* **178**:408–418.

Rowe, M. J., Lansman, R. A., and Woodward, D. O. 1974. A comparison of mitochondrially synthesized proteins from whole mitochondria and cytochrome oxidase in *Neurospora*. *Eur. J. Biochem.* **41**:25–30.

Rudzinska, M. A., D'Alesandro, P. A., and Trager, W. 1964. The fine structure of *Leishmania donovani* and the role of the kinetoplast in the Leishmania–Leptomonad transformation. *J. Protozool.* **11**:166–191.

Ruska, C., and Ruska, H. 1969. Kompartimentierung und Membranbau von Herzmuskel-mitochondrien in Darstellungen durch die Gefrierätztechnik. *Z. Zellforsch. Mikrosk. Anat.* **97**:298–312.

Ruthmann, A., and Grell, K. G. 1964. Die Feinstruktur des intracapsulären cytoplasmas bei dem Radiolar *Aulacantha scolymantha*. *Z. Zellforsch. Mikrosk. Anat.* **63**:97–119.

Ryter, A. 1968. Association of the nucleus and the membrane of bacteria: A morphological study. *Bacteriol. Rev.* **32**:39–54.

Ryter, A., and Jacob, F. 1966. Étude au microscope électronique de la liaison entre noyau et mésosome chez *Bacillus subtilis*. *Ann. Inst. Pasteur* **107**:384–400.

Saccone, C., and Quagliariello, E. 1975. Biochemical studies of mitochondrial transcription and translation. *Int. Rev. Cytol.* **43**:125–165.

Saccone, C., Pepe, G., Cantatore, P., Terpstra, P., and Kroon, A. M. 1976. Mapping the transcription products of rat-liver mitochondria by hybridization. *In*: Saccone, C., and Kroon, A. M., eds., *The Genetic Function of Mitochondrial DNA*, Amsterdam, North-Holland, pp. 27–36.

Saccone, C., Pepe, G., Bakker, H., and Kroon, A. M. 1977. The genetic organization of rat liver mtDNA. *In*: Bandlow, W., Schweyen, R. J., Wolf, K., and Kaudewitz, F., eds., *Mitochondria 1977*, Berlin, de Gruyter, pp. 303–315.

Sagan, L. 1967. On the origin of mitosing cells. *J. Theor. Biol.* **14**:225–274.

Salton, M. R. J. 1978. Structure and function of bacterial plasma membranes. *Symp. Soc. Gen. Microbiol.* **28**:201–223.

Salton, M. R. J., and Owen, P. 1976. Bacterial membrane structure. *Annu. Rev. Microbiol.* **30**:451–482.

Sanders, J. P. M., Weijers, P. J., Groot, G. S. P., and Borst, P. 1974. Properties of mtDNA from *Kluyveromyces lactis*. *Biochim. Biophys. Acta* **374**:136–144.

Sanders, J. P. M., Heyting, C., DiFranco, A., Borst, P., and Slonimski, P. P. 1976a. The organization of genes in yeast mitochondrial DNA. *In*: Saccone, C., and Kroon, A. M., eds., *The Genetic Function of Mitochondrial DNA*, Amsterdam, North-Holland, pp. 259–272.

Sanders, J. P. M., Heyting, C., and Borst, P. 1976b. The variability of the mitochondrial genome of *Saccharomyces* strains. *In*: Bücher, T., Neupert, W., Sebald, W., and Weiner, S., eds., *Genetics and Biogenesis of Chloroplasts and Mitochondria*, Amsterdam, North-Holland, pp. 511–517.

Sanders, J. P. M., Heyting, C., Verbeet, M. P., Meijlink, F. C. P. W., and Borst, P. 1977. The organization of genes in yeast mtDNA. *Mol. Gen. Genet.* **157**:239–261.

Scarpa, A., and Lindsay, J. G. 1972. Maintenance of energy-linked functions in rat-liver mitochondria aged in the presence of nupercaine. *Eur. J. Biochem.* **27**:401–407.

Schäfer, K. P., Bugge, G., Grandi, M., and Küntzel, H. 1971. Transcription of mtDNA *in vitro* from *Neurospora crassa*. *Eur. J. Biochem.* **21**:478–488.

Schatz, G. 1963. The isolation of possible mitochondrial precursor structures from aerobically grown baker's yeast. *Biochem. Biophys. Res. Commun.* **12**:448–451.

Schatz, G. 1979. How mitochondria import proteins from the cytoplasm. *FEBS Lett.* **103**:203–211.

Schatz, G., and Mason, T. L. 1974. The biosynthesis of mitochondrial proteins. *Annu. Rev. Biochem.* **43**:51–87.

Schick, L., and Butler, L. G. 1969. Inorganic pyrophosphatase of rat liver mitochondria. *J. Cell Biol.* **42**:235–240.

Schmitt, H. 1970. Characterization of 72-S mitochondrial ribosome from *S. cerevisiae*. *Eur. J. Biochem.* **17**:278–283.

Schmitt, H. 1971. Core particles and proteins from mitochondrial ribosomes of yeast. *FEBS Lett.* **15**:186–190.

Schmitt, H., Grossfeld, H., and Littauer, U. Z. 1973. Mitochondrial biogenesis during differentiation of *Artemia salina* cysts. *J. Cell Biol.* **58**:643–649.

Schmitt, H., Grossfeld, H., Beckmann, J. S., and Littauer, U. Z. 1974. Biogenesis of mitochondria from *Artemia salina* cysts and the transcription *in vitro* of the DNA. *In*: Kroon, A. M., and Saccone, C., eds., *The Biogenesis of Mitochondria*, New York, Academic Press, pp. 135–146.

Schneider, W. C. 1959. Mitochondrial metabolism. *Adv. Enzymol.* **21**:1–72.

Schneller, C., Schneller, J. M., and Stahl, A. J. C. 1975. Étude comparative des methylases des acides ribonucléiques de transfert des mitochondries et du cytoplasme de la levure. *Bull. Soc. Pharm. Strasbourg* **18**:107–111.

Schneller, C., Schneller, J. M., and Stahl, A. J. C. 1976. Studies of odd bases in yeast mitochondrial tRNA. III. Characterization of the tRNA methylases associated with the mitochondria. *Biochem. Biophys. Res. Commun.* **70**:1003–1008.

Schneller, J. M., Stahl, A. J. C., and Fukuhara, H. 1975a. Coding origin of isoaccepting tRNA in yeast mitochondria. *Biochimie* **57**:1051–1057.

Schneller, J. M., Matlin, R., Stahl, A. J. C., and Dirheimer, G. 1975b. Studies on odd bases in yeast mitochondrial tRNA: Absence of the fluorescent "Y" base in mitochondrial DNA coded $tRNA^{Phe}$, absence of 4-thiouridine. *Biochem. Biophys. Res. Commun.* **64**:1046–1053.

Schneller, J. M., Accoceberry, B., and Stahl, A. J. C. 1975c. Fractionation of yeast mitochondrial $tRNA^{Tyr}$ and $tRNA^{Leu}$. *FEBS Lett.* **53**:44–48.

Schnepf, E., and Brown, R. M. 1971. On relationships between endosymbiosis and the origin of plastids and mitochondria. *In*: Reinert, J., and Ursprung, U., eds., *Origin and Continuity of Cell Organelles,* Berlin, Springer-Verlag, pp. 299–322.

Schoen, A. E., Cooks, R. G., and Wiebers, J. L. 1979. Mitochondrial DNA analyses and the origin and relative age of parthenogenetic lizards (genus *Cnemidophorus*). *Science* **203**:1247–1251.

Schötz, F. 1972. Dreidimensionale, manstabgebreue Rekonstruktion einer grünen Flagellatenzelle nach Elektronmikroskopie von Serienschnitten. *Planta* **102**:152–159.

Schötz, F., Bathelt, H., Arnold, C. G., and Schimmer, O. 1972. Die Architektur und Organisation der *Chlamydomonas*-Zelle. *Protoplasma* **75**:229–254.

Schuster, F. L. 1965. A DNA component in mitochondria of *Didymium nigripes*, a slime mold. *Exp. Cell Res.* **39**:329–345.

Schwartz, R. M., and Dayhoff, M. O. 1978. Origins of prokaryotes, eukaryotes, mitochondria, and chloroplasts. *Science* **199**:395–403.

Schweyen, R. J., Weiss-Brummer, B., Backhaus, B., and Kaudewitz, F. 1976. Localization of seven gene loci on a circular map of the mitochondrial genome of *S. cerevisiae*. *In*: Saccone, C., and Kroon, A. M., eds., *The Genetic Function of Mitochondrial DNA,* Amsterdam, North-Holland, pp. 251–258.

Schweyen, R. J., Weiss-Brummer, B., Backhaus, B., and Kaudewitz, F. 1977. The genetic map of the mitochondrial genome including the fine structure of *cob* and *oxi*3 clusters. *In*: Bandlow, W., Schweyen, R. J., Wolf, K., and Kaudewitz, F., eds., *Mitochondria 1977*, Berlin, de Gruyter, pp. 139–160.

Scragg, A. H. 1974. A mitochondria DNA-directed polymerase from yeast mitochondria. *In*: Kroon, A. M., and Saccone, C., eds., *The Biogenesis of Mitochondria,* New York, Academic Press, pp. 47–57.

Scragg, A. H. 1976. Origin of the mitochondrial RNA polymerase of yeast. *FEBS Lett.* **65**:148–151.

Scragg, A. H., Morimoto, H., Villa, V., Nekhorocheff, J., and Halvorson, H. O. 1971. Cell-free protein synthesizing system from yeast mitochondria. *Science* 171:908–910.

Sedar, A. W., and Burde, R. M. 1965. The demonstration of the succinic dehydrogenase system in *Bacillus subtilis* using TNBT combined with techniques of electron microscopy. *J. Cell Biol.* **27**:53–66.

Senior, A. E. 1973. Relationship of cysteine and tyrosine residues to ATP hydrolysis to mitochondrial ATPase. *Biochemistry* **12**:3622–3627.

Shah, D. M., and Langley, C. H. 1977. Complex mitochondrial DNA in *Drosophila*. *Nucleic Acids Res.* **4**:2949–2960.

Shearman, C. W., and Kalf, G. F. 1977. DNA replication by a membrane–DNA complex from rat liver mitochondria. *Arch. Biochem. Biophys.* **182**:573–586.

Shepherd, J. A., and Kalnitsky, G. 1951. Intracellular distribution of the phosphate-activated glutaminase of rat liver. *J. Biol. Chem.* **192**:1–7.

Siekevitz, P., and Potter, V. R. 1953. Intramitochondrial regulation of oxidative rate. *J. Biol. Chem.* **201**:1–13.

Siliprandi, D., Siliprandi, N., Scutari, G., and Zoccarato, F. 1973. Restoration of some energy linked processes lost during the ageing of rat liver mitochondria. *Biochem. Biophys. Res. Commun.* **55**:563–567.

Simpson, L. 1972. The kinetoplast of the haemoflagellates. *Int. Rev. Cytol.* **32**:139–207.

Simpson, L., and da Silva, A. 1971. Isolation and characterization of kinetoplast DNA from *Leishmania tarentolae*. *J. Mol. Biol.* **56**:443–473.

Sjöstrand, F. S. 1977. The arrangement of mitochondrial membranes and a new structural feature of the inner mitochondrial membranes. *J. Ultrastruct. Res.* **59**:292–319.

Sjöstrand, F. S. 1978. The structure of mitochondrial membranes: A new concept. *J. Ultrastruct. Res.* **64**:217–245.

Sjöstrand, F. S., and Bernhard, W. 1976. The structure of mitochondrial membranes in frozen sections. *J. Ultrastruct. Res.* **56**:233–246.

Sjöstrand, F. S., and Cassell, R. Z. 1978. Structure of inner membranes in rat heart muscle mitochondria as revealed by means of freeze-fracturing. *J. Ultrastruct. Res.* **63**:111–137.

Sjöstrand, F. S., and Kretzer, F. 1975. A new freeze-drying technique applied to the analysis of the molecular structure of mitochondrial and chloroplast membranes. *J. Ultrastruct. Res.* **53**:1–28.

Smith, D. G., Tauro, P., Schweyer, E., and Halvorson, H. O. 1968. The replication of mtDNA during the cell cycle in *Saccharomyces lactis*. *Proc. Natl. Acad. Sci. USA* **60**:936–942.

Smith, D. G., Marchant, R., Maroudas, N. G., and Wilkie, D. 1969. A comparative study of the mitochondrial structure of petite strains of *Saccharomyces cerevisiae*. *J. Gen. Microbiol.* **56**:47–54.

Sonenshein, G. E., and Holt, C. E. 1968. Molecular weight of mtDNA in *Physarum polycephalum*. *Biochem. Biophys. Res. Commun.* **33**:361–367.

Soriano, L., Smith, J., Croisille, Y., and Dastugue, B. 1974. Mitochondrial DNA polymerase, deoxyribonuclease, and ribonuclease H activities from brain of chick embryo. *Nucleic Acids Res.* **1**:1055–1097.

Sparace, S. A., and Moore, T. S. 1979. Phospholipid metabolism in plant mitochondria. *Plant Physiol.* **63**:963–972.

Staehelin, L. A. 1968. The interpretation of freeze-etched artificial and biological membranes. *J. Ultrastruct. Res.* **22**:326–347.

Stegeman, W. J., and Hoober, J. K. 1974. Mitochondrial protein synthesis in *Chlamydomonas reinhardii* y-1. *J. Biol. Chem.* **249**:6866–6873.

Steinert, M. 1969. The ultrastructure of mitochondria. *Proc. R. Soc. London Ser. B* **173**:63–70.

Steinert, M., Van Assel, S., Borst, P., Mol, J. N. M., Kleisen, C. M., and Newton, B. A. 1973. Specific detection of kinetoplast DNA in cytological preparations of trypanosomes by hybridization with complementary RNA. *Exp. Cell Res.* **76**:175–185.

Stelter, K. O., and Zillig, W. 1974. DNA-dependent RNA polymerase from *Lactobacillus curvatus*. *Eur. J. Biochem.* **48**:527–540.

Stevens, B. J. 1974. Variation in mitochondrial numbers and volume in yeast according to growth conditions. *J. Cell Biol.* **63**:336A.

Storrie, B., and Attardi, G. 1973. Heterogeneity of isolated HeLa cell mitochondria as assayed for their enzymatic and *in vivo* biosynthetic activities. *J. Biol. Chem.* **248**:5826–5834.

Sussman, R., and Rayner, E. P. 1971. Physical characterization of deoxyribonucleic acids in *Dictyostelium discoideum*. *Arch. Biochem. Biophys.* **144**:127–137.

Suter, E. R. 1969. The fine structure of brown adipose tissue. *J. Ultrastruct. Res.* **26**:216–241.

Sutfin, L. V., Holtrop, M. E., and Ogilvie, R. E. 1971. Microanalysis of individual mitochondrial granules having diameters less than 1000 Å. *Science* **174**:947–949.

Suyama, Y. 1967. The origins of mitochondrial RNAs in *Tetrahymena pyriformis*. *Biochemistry* **6**:2829–2839.

Suyama, Y., and Eyer, J. 1967. Leucyl tRNA and leucyl tRNA synthetase in mitochondria of *Tetrahymena pyriformis*. *Biochem. Biophys. Res. Commun.* **28**:746–751.

Suyama, Y., and Hamada, J. 1976. Imported tRNA: Its synthetase as a probable transport protein. *In*: Bücher, T., *et al.*, eds., *Genetics and Biogenesis of Chloroplasts and Mitochondria*, Amsterdam, Elsevier/North-Holland Biomedical Press, pp. 763–770.

Suyama, Y., and Hamada, J. 1978. The mitochondrial and cytoplasmic valyl tRNA synthetases in *Tetrahymena* are indistinguishable. *Arch. Biochem. Biophys.* **191**:437–443.

Suyama, Y., and Muira, K. 1968. Size and structural variations of mitochondrial DNA. *Proc. Natl. Acad. Sci. USA* **60**:235–242.

Suyama, Y., and Preer, J. R. 1965. Mitochondrial DNA from protozoa. *Genetics* **52**: 1051–1058.

Taira, K. 1979. Studies on intramitochondrial inclusions in the pancreatic acinar cells of the Japanese newt, *Triturus pyrrogaster*. *J. Ultrastruct. Res.* **67**:89–94.

Tait, A., Knowles, J. K. C., and Hardy, J. C. 1976. The genetic control of mitochondrial ribosomal proteins in *Paramecium*. *In*: Saccone, C., and Kroon, A. M., eds., *The Genetic Function of Mitochondrial DNA,* Amsterdam, North-Holland, pp. 131–136.

Takagi, A., and Nagata, A. 1962. Studies on the fine structure of *Candida albicans,* with special reference to the intracytoplasmic membrane system. *Jpn. J. Microbiol.* **6**:95–111.

Takagi, A., Abe, O., and Ueda, M. 1965. Reduction sites of tetrazolium salts and tellurite in *Clostridium botulinum* and *C. tetani*. *J. Gen. Appl. Microbiol.* **11**:221–231.

Tanaka, S., and Koike, K. 1977. Template specificity of rat mtDNA polymerase. *Biochim. Biophys. Acta* **479**:290–299.

Tanaka, S., and Koike, K. 1978. DNA-polymerase γ is localized in mitochondria. *Biochem. Biophys. Res. Commun.* **81**:791–797.

Tandler, B., Erlandson, R. A., Smith, A. L., and Wynder, E. L. 1969. Riboflavin and mouse hepatic cell structure and function. II. Division of mitochondria during recovery from simple deficiency. *J. Cell Biol.* **41**:477–493.

Terpstra, P., Holtrop, M., and Kroon, A. M. 1976. Restriction fragment map of *Neurospora crassa* mitochondrial DNA. *In*: Saccone, C., and Kroon, A. M., eds., *The Genetic Function of Mitochondrial DNA,* Amsterdam, North-Holland, pp. 111–118.

Terpstra, P., de Vries, H., and Kroon, A. M. 1977. Properties and genetic localization of mitochondrial tRNAs of *Neurospora crassa*. *In*: Bandlow, W., Schweyen, R. J., Wolf, K:, and Kaudewitz, F., eds., *Mitochondria 1977*, Berlin, de Gruyter, pp. 291–302.

Ter Schegget, J., and Borst, P. 1971a. DNA synthesis by isolated mitochondria. I. Effect of inhibitors and characterization of the product. *Biochim. Biophys. Acta* **246**:239–248.

Ter Schegget, J., and Borst, P. 1971b. DNA synthesis by isolated mitochondria. II. Detection of product DNA hydrogen-bonded to closed duplex circles. *Biochim. Biophys. Acta* **246**:249–257.

Thomas, R. S., and Greenawalt, J. W. 1968. Microincineration, electron microscopy, and electron diffraction of calcium phosphate-loaded mitochondria. *J. Cell Biol.* **39**:55–76.

Threadgold, L. T. 1976. *The Ultrastructure of the Animal Cell,* 2nd ed., New York, Pergamon Press.

Thyagarajan, T. R., Conti, S. F., and Naylor, H. B. 1961. Electron microscopy of yeast mitochondria. *Exp. Cell Res.* **25**:216–218.

Thyagarajan, T. R., Conti, S. F., and Naylor, H. B. 1962. Electron microscopy of *Rhodotorula glutinis*. *J. Bacteriol.* **83**:381–394.

Tobler, H., and Gut, C. 1974. Mitochondrial DNA from 4-cell stages of *Ascaris lumbricoides*. *J. Cell Sci.* **16**:593–601.

Tomasz, M., Sanno, Y., and Chambers, R. W. 1965. The chemistry of pseudouridine. V. Periodate oxidation of pseudouridylic acid and soluble RNA. *Biochemistry* **4**:1710–1719.

Trager, W. 1964. The cytoplasm of protozoa. *In*: Brachet, J., and Mirsky, A. E. eds., *The Cell*, New York, Academic Press, Vol. 6, pp. 81–137.

Travers, A. A., and Burgess, R. R. 1967. Cyclic re-use of the RNA polymerase sigma factor. *Nature (London)* **222**:536–540.

Tripodi, G., Pizzolongo, P., and Giannattacio, M. 1972. A DNase-sensitive twisted structure in the mitochondrial matrix of *Polysiphonia* (Rhodophyta). *J. Cell Biol.* **55**:530–532.

Truman, D. E. S., and Korner, A. 1962. Incorporation of amino acids into the protein of isolated mitochondria. *Biochem. J.* **83**:588–596.

Tzagoloff, A. 1977. Genetic and translational capabilities of the mitochondrion. *BioScience* **27**:18–23.

Tzagoloff, A., Rubin, M. S., and Sierra, M. F. 1973. Biosynthesis of mitochondrial enzymes. *Biochim. Biophys. Acta* **301**:71–104.

Tzagoloff, A., Akai, A., and Rubin, M. S. 1974. Mitochondrial products of yeast ATPase and cytochrome oxidase. *In*: Kroon, A. M., and Saccone, C., eds., *The Biogenesis of Mitochondria*, New York, Academic Press, pp. 405–421.

Tzagoloff, A., Foury, F., and Akai, A. 1976. Resolution of the mitochondrial genome. *In*: Saccone, C., and Kroon, A. M., eds., *The Genetic Function of Mitochondrial DNA*, Amsterdam, North-Holland, pp. 155–161.

Upholt, W. B., and Borst, P. 1974. Accumulation of replicative intermediates of mitochondrial DNA in *Tetrahymena pyriformis* grown in ethidium bromide. *J. Cell Biol.* **61**:383–397.

Uzzell, T., and Spolsky, C. 1973. Mitochondrial origins. *Science* **180**:516–517.

Vail, W. J., and Riley, R. K. 1972. Comparison of aggregated and orthodox rat heart mitochondria *in situ* using thin sections and freeze fracture techniques. *Proc. W. Va. Acad. Sci.* **44**:103–110.

van Berkel, T. J. C., and Kruijt, J. K. 1977. Different types of mitochondria in parenchymal and non-parenchymal rat-liver cells. *Eur. J. Biochem.* **73**:223–229.

Vanderwinkel, E., and Murray, R. G. E. 1962. Intracytoplasmic organelles of bacteria and the site of oxidation–reduction activity. *J. Ultrastruct. Res.* **7**:185–199.

van Iterson, W. 1965. Symposium on the fine structure and replication of bacteria and their parts. II. Bacterial cytoplasm. *Bacteriol. Rev.* **29**:299–325.

van Iterson, W., and Leene, W. 1964. A cytochemical localization of reductive sites in a Gram-positive bacterium. Tellurite reduction in *B. subtilis*. *J. Cell Biol.* **20**:361–370.

Van Ommen, G. J. G., and Groot, G. S. P. 1977. Transcription map of yeast mitochondrial DNA. *In:* Bandlow, W., Schweyen, R. J., Wolf, K., and Kaudewitz, F., eds., *Mitochondria 1977*, Berlin, de Gruyter, pp. 415–424.

Vardanis, A. 1977. Protein kinase activity at the inner membrane of mammalian mitochondria. *J. Biol. Chem.* **252**:807–813.

Vasington, F. D., and Murphy, J. V. 1962. Ca^{++} uptake by rat liver mitochondria and its dependence on respiration and phosphorylation. *J. Biol. Chem.* **237**:2670–2677.

Vickerman, K. 1977. DNA throughout the single mitochondrion of a kinetoplastid flagellate. Observations on the ultrastructure of *Cryptobia vaginalis* (Hesse, 1910). *J. Protozool.* **24**:221–233.

von Jagow, G., and Klingenberg, M. 1970. Pathways of hydrogen in mitochondria of *Saccharomyces carlsbergensis*. *Eur. J. Biochem.* **12**:583–592.

Vorhaben, J. E., and Campbell, J. W. 1972. Glutamine synthetase, a mitochondrial enzyme in uricotelics. *J. Biol. Chem.* **247**:2763–2767.

Vorhaben, J. E., and Campbell, J. W. 1977. Submitochondrial localization and function of glutamine metabolism in avian liver. *J. Cell Biol.* **73**:300–310.

Waite, M., van Deenen, L. L. M., Ruigrok, T. J. C., and Elbers, P. F. 1969. Relation of mitochondrial phospholipase A activity to mitochondrial swelling. *J. Lipid Res.* **10**:599–608.

Wallace, R. B., and Freeman, K. B. 1974a. Initiation of mammalian mitochondrial protein synthesis. The effect of methotrexate. *Biochim. Biophys. Acta* **366**:466–473.

Wallace, R. B., and Freeman, K. B. 1974b. Multiple species of methionyl-tRNA from mouse liver mitochondria. *Biochem. Biophys. Res. Commun.* **60**:1440–1445.

Wallace, R. B., Aujame, L., and Freeman, K. B. 1978. Chemical and physical properties of mammalian mitochondrial aminoacyl-tRNAs. *Biochim. Biophys. Acta* **518**:321–325.

Wallin, I. E. 1927. *Symbionticism and the Origin of Species*, Baltimore, Williams & Wilkins.

Walter, U., and Söling, H. D. 1976. Transfer of acetyl-units through the mitochondrial membrane: Evidence for a pathway different from the citrate pathway. *FEBS Lett.* **63**:260–266.

Watson, J. A., and Lowenstein, J. M. 1970. Citrate and the conversion of carbohydrate into fat. *J. Biol. Chem.* **245**:5993–6002.

Watson, K., Haslam, J. M., and Linnane, A. W. 1970. Biogenesis of mitochondria. *J. Cell Biol.* **46**:88–96.

Weissbach, A., Baltimore, D., Bollum, F., Gallo, R., and Korn, D. 1975. Nomenclature of eukaryotic DNA polymerases. *Eur. J. Biochem.* **59**:1–2.

Westergaard, O. 1970. Separation of two DNA polymerase fractions from *Tetrahymena* cells after excision-repairable damage to DNA. *Biochim. Biophys. Acta* **213**:36–44.

Westergaard, O., and Pearlman, R. E. 1969. DNA polymerase activity in methotrexate plus uridine treated *Tetrahymena. Exp. Cell Res.* **54**:309–313.

Whittaker, V. P. 1966. The ultrastructure of mitochondria. *In*: Tager, J. M., Papa, S., Quagliariello, E., and Slater, E. C., eds., *Regulation of Metabolic Processes in Mitochondria*, Amsterdam, Elsevier, pp. 1–27.

Wiebe, J. 1972. Mechanism of gonadotropin action in Amphibia: Involvement of mitochondria. *Science* **175**:67–68.

Williamson, D. H. 1976. Packaging and recombination of mitochondrial DNA in vegetatively growing yeast cells. *In*: Bandlow, W., Schweyen, R. J., Thomas, D. Y., and Wolf, K., eds., *Genetics, Biogenesis, and Bioenergetics of mitochondria,* Berlin, de Gruyter, pp. 117–123.

Williamson, D. H., and Fennell, D. J. 1975. The use of fluorescent DNA binding agent for detecting and separating yeast mitochondrial DNA. *Methods Cell Biol.* **12**:335–351.

Williamson, D. H., Johnston, L. H., Richmond, M. V., and Game, J. C. 1977. Mitochondrial DNA and the heritable unit of the yeast mitochondrial genome: A review. *In*: Bandlow, W., Schweyen, R. J., Wolf, K., and Kaudewitz, F., *Mitochondria 1977*, Berlin, de Gruyter, pp. 1–24.

Wiskich, J. T., and Bonner, W. D. 1963. Preparation and properties of sweet potato mitochondria. *Plant Physiol.* **38**:594–604.

Wolfson, J., and Dressler, D. 1972. Adenovirus-2 DNA contains an inverted terminal repetition. *Proc. Natl. Acad. Sci. USA* **69**:3054–3057.

Wolstenholme, D. R., and Gross, N. J. 1968. The form and size of mitochondrial DNA of the red bean, *Phaseolus vulgaris. Proc. Natl. Acad. Sci. USA* **61**:245–252.

Wood, D. D., and Luck, D. J. L. 1969. Hybridization of mitochondrial ribosomal RNA. *J. Mol. Biol.* **41**:211–224.

Wu, G. J., and Dawid, I. B. 1972. Purification and properties of mitochondrial DNA dependent RNA polymerase from ovaries of *Xenopus laevis. Biochemistry* **11**:3589–3595.

Wu, G. J., and Dawid, I. B. 1974. *In vitro* transcription of *Xenopus* mitochondrial DNA by homologous mitochondrial RNA polymerase. *J. Biol. Chem.* **249**:4412–4419.

Young, P. G., and Hunter, N. P. 1979. Products of mitochondrial protein synthesis in *Tetrahymena. Can. J. Biochem.* **57**:314–320.

Zachariah, K., and Fitz-James, P. C. 1967. The structure of phialides in *Penicillium claviforme. Can. J. Microbiol.* **13**:249–256.

Zuckerman, B. M., Kisiel, M., and Himmelhoch, S. 1973. Unusual mitochondrial cristae in the vinegar eelworm. *J. Cell Biol.* **58**:476–480.

CHAPTER 10

Allet, B., and Rochaix, J. D. 1979. Structure analysis of the intervening DNA sequences in the chloroplast 23S ribosomal genes of *C. reinhardii. Cell* **18**:55–60.

Allfrey, V. G. 1974. Biosynthetic reactions in the cell nucleus. *In*: Anfinsen, C. B., ed., *Aspects of Protein Biosynthesis,* New York, Academic Press, Part A, pp. 247–365.

Alscher-Herman, R., Jagendorf, A. T., and Grumet, R. 1979. Ribosome–thylakoid association in peas. Influence of anoxia. *Plant Physiol.* **64**:232–235.

Andrews, T. J., and Lorimer, G. H. 1978. Photorespiration—Still unavoidable. *FEBS Lett.* **90**:1–9.

Apel, K., and Kloppstech, K. 1978. The plastid membranes of barley (*Hordeum vulgare*). *Eur. J. Biochem.* **85**:581–588.

Argyroudi-Akoyunoglou, J. H., Kondylaki, S., and Akoyunoglou, G. 1976. Growth of grana from the "primary" thylakoids in *Phaseolus vulgaris. Plant Cell Physiol.* **17**:939–954.

Armond, P. A., Staehelin, L. A., and Arntzen, C. J. 1977. Spatial relationship of photosystems I and II, and the light-harvesting complex in chloroplast membranes. *J. Cell Biol.* **73**:400–418.

Avron, M. 1971. Biochemistry of photophosphorylation. *In*: Gibbs, M., ed., *Structure and Function of Chloroplasts*, New York, Springer-Verlag, pp. 149–168.

Avron, M., and Chance, B. 1966. The relation of light-induced oxidation–reduction changes in cytochrome *f* of isolated chloroplasts to photophosphorylation. *In*: Thomas, J. B., and Goedheer, J. C., eds., *Currents in Photosynthesis,* Rotterdam, Ad. Donker, pp. 455–463.

Bailey, J. L., Thornber, J. P., and Whyborn, A. G. 1966. The chemical nature of chloroplast lamellae. *In*: Goodwin, T. W., ed., *Biochemistry of Chloroplasts,* London, Academic Press, pp. 243–255.

Baird, B. A., and Hammes, G. G. 1979. Structure of oxidative- and photophosphorylation coupling factor complexes. *Biochim. Biophys. Acta* **549**:31–53.

Bar-Nun, S., and Ohad, I. 1974. Cytoplasmic and chloroplastic origin of chloroplast membrane proteins associated with PSII and PSI active centers in *Chlamydomonas. In*: Avron, M., ed., *Proceedings of Third International Congress on Photosynthesis,* Amsterdam, Elsevier, pp. 1627–1637.

Bartels, P. G., and Weier, T. E. 1967. Particle arrangements in proplastids of *Triticum vulgare* L. seedlings. *J. Cell Biol.* **33**:243–253.

Bassham, J. A. 1973. The role of photosynthesis in green plants. *In*: Miller, L. P., ed., *Phytochemistry*, New York, Van Nostrand Reinhold, Vol. 1, pp. 38–74.

Bedbrook, J. R., and Bogorad, L. 1976. Endonuclease recognition sites mapped on *Zea mays* chloroplast DNA. *Proc. Natl. Acad. Sci. USA* **73**:4309–4313.

Bedbrook, J. R., Kolodner, R., and Bogorad, L. 1977. *Zea mays* chloroplast rRNA genes are part of a 22,000 base pair inverted repeat. *Cell* **11**:739–749.

Belanger, F. C., and Rebeiz, C. A. 1979. Chloroplast biogenesis. XXVII. Detection of novel chlorophyll and chlorophyll precursors in higher plants. *Biochem. Biophys. Res. Commun.* **88**:365–372.

Bengis, C., and Nelson, N. 1975. Purification and properties of the photosystem I reaction center from chloroplasts. *J. Biol. Chem.* **250**:2783–2788.

Bengis, C., and Nelson, N. 1977. Subunit structure of chloroplast photosystem I reaction center. *J. Biol. Chem.* **252**:4564–4569.

Bennett, J. 1977. Phosphorylation of chloroplast membrane polypeptides. *Nature* **269**:344–346.

Bennett, J. 1979a. Chloroplast phosphoproteins. The protein kinase of thylakoid membranes is light-dependent. *FEBS Lett.* **103**:342–344.

Bennett, J. 1979b. Chloroplast phosphoproteins. Phosphorylation of polypeptides of the light-harvesting chlorophyll protein complex. *Eur. J. Biochem.* **99**:133–137.

Ben-Shaul, Y., Schiff, J. A., and Epstein, H. T. 1964. Studies of chloroplast development in Euglena. VII. Fine structure of the developing plastid. *Plant Physiol.* **39**:231–240.

Ben-Shaul, Y., Epstein, H. T., and Schiff, J. A. 1965. Studies of chloroplast development in *Euglena. Can. J. Bot.* **43**:129–136.

Berthold, W.-U. 1978. Ultrastrukuranalyse der endoplasmatischen Algen von *Amphistegina lessonii* und ihre systematische Stellung. *Arch. Protistenkd.* **120**:16–62.

Bisalputra, T., and Bisalputra, A. A. 1967. The occurrence of DNA fibrils in chloroplasts of *Laurencia spectabilis*. *J. Ultrastruct. Res.* **17**:14–22.

Bisalputra, T., and Stein, J. R. 1966. The development of cytoplasmic bridges in *Volvox aureus*. *Can. J. Bot.* **44**:1697–1702.

Bisalputra, T., Brown, D. L., and Weier, T. E. 1969. Possible respiratory sites in a blue-green alga *Nostoc sphaericum* as demonstrated by potassium tellurite and TNBT reduction. *J. Ultrastruct. Res.* **27**:182–197.

Black, C. C. 1973. Photosynthetic carbon fixation in relation to net CO_2 uptake. *Annu. Rev. Plant Physiol.* **24**:253–286.

Blair, G. E., and Ellis, R. J. 1973. Light driven synthesis of the large subunit of fraction I protein by isolated pea chloroplasts. *Biochim. Biophys. Acta* **319**:223–234.

Boatman, E. S., and Douglas, H. C. 1961. Fine structure of the photosynthetic bacterium *Rhodomicrobium vannielii*. *J. Biophys. Biochem. Cytol.* **11**:469–480.

Bogorad, L. 1975. Evolution of organelles and eukaryotic genomes. *Science* **188**:891–898.

Böhme, H. 1977. On the role of ferredoxin and ferrodoxin-$NADP^+$ reductase in cyclic electron transport of spinach chloroplasts. *Eur. J. Biochem.* **72**:283–289.

Böhme, H. 1979. Photoreactions of cytochrome b_{563} and f_{554} in intact spinach chloroplasts: Regulation of cyclic electron flow. *Eur. J. Biochem.* **93**:287–293.

Bohnert, H. J., Driesel, A. J., Crouse, E. J., Gordon, K., Herrmann, R. G., Steinmetz, H., Mubumbila, M., Keller, M., Burkard, G., and Weil, J. H. 1979. Presence of a tRNA gene in the spacer sequence between the 16S and 23S rRNA genes of spinach chloroplast DNA. *FEBS Lett.* **103**:52–56.

Bonen, L., and Doolittle, W. F. 1976. Partial sequences of 16S rRNA and the phylogeny of blue-green algae and chloroplasts. *Nature (London)* **261**:669–673.

Borowitzka, M. A., and Larkum, A. W. D. 1974. Chloroplast development in the Caulopalean alga *Halimeda*. *Protoplasma* **7**:131–144.

Bottomley, W., Higgins, T. J. V., and Whitfeld, P. R. 1976. Differential recognition of chloroplast and cytoplasmic mRNA by 70S and 80S ribosomal systems. *FEBS Lett.* **63**:120–124.

Bouck, G. B. 1962. Chromatophore development, pits, and other fine structure in the red alga, *Lomentaria baileyana*. *J. Cell Biol.* **12**:553–569.

Bouck, G. B. 1965. Fine structure and organelle associations in brown algae. *J. Cell Biol.* **26**:523–527.

Bourque, D. P., and Wildman, S. G. 1973. Evidence that nuclear genes code for several chloroplast ribosomal proteins. *Biochem. Biophys. Res. Commun.* **50**:532–537.

Bowes, J. M., Crofts, A. R., and Itoh, S. 1979. A high potential acceptor for photosystem II. *Biochim. Biophys. Acta* **547**:320–335.

Branton, D., and Park, R. B. 1967. Subunits in chloroplast lamellae. *J. Ultrastruct. Res.* **19**:283–303.

Breitenberger, C. A., Graves, M. C., and Spremulli, L. L. 1979. Evidence for the nuclear location of the gene for chloroplast elongation factor G. *Arch. Biochem. Biophys.* **194**:265–270.

Brown, R. H., and Simmons, R. E. 1979. Photosynthesis of grass species differing in CO_2 fixation pathways. I. Water-use efficiency. *Crop Sci.* **19**:375–379.

Brown, R. M., and Van Baalen, C. 1970. Comparative ultrastructure of a filamentous mutant and the wild type of *Agmenellum quadruplicatum*. *Protoplasma* **70**:87–99.

Calvert, H. E., and Dawes, C. J. 1976. Ontogenetic membrane transitions in plastids of the coenocytic alga *Caulerpa* (Chlorophyceae). *Phycologia* **15**:37–40.

Calvert, H. E., Dawes, C. J., and Borowitzka, M. A. 1976. Phylogenetic relationships of *Caulerpa* (Chlorophyta) based on comparative chloroplast ultrastructure. *J. Phycol.* **12**:149–162.

Calvin, M., and Benson, A. A. 1948. The path of carbon in photosynthesis. IV. The identity and sequence of the intermediates in sucrose synthesis. *Science* **109**:140–142.

Caple, M. B., Chow, H. C., and Strouse, C. E. 1978. Photosynthetic pigments of green sulfur bacteria. Esterifying alcohols of bacteriochlorophyll *c* from *Chlorobium limicola*. *J. Biol. Chem.* **253**:6730–6737.

Casadoro, G., and Rascio, N. 1978a. Chloroplast ontogenesis in *Helianthus annuus* L. *Protoplasma* **97**:165–172.

Casadora, G., and Rascio, N. 1978b. Thylakoid membranes in sunflower and other plants. *J. Ultrastruct. Res.* **65**:30–35.

Cerff, R. 1978a. Glyceraldehyde-3-phosphate dehydrogenase (NADP) from *Sinapis alba* L. Reversible association of the enzyme with a protein factor as controlled by pyridine nucleotides *in vitro*. *Plant Physiol.* **61**:369–372.

Cerff, R. 1978b. Glyceraldehyde-3-phosphate dehydrogenase (NADP) from *Sinapis alba* L. NAD(P)-induced conformation changes of the enzyme. *Eur. J. Biochem.* **82**:45–53.

Cerff, R. 1979. Quaternary structure of higher plant glyceraldehyde-3-phosphate dehydrogenases. *Eur. J. Biochem.* **94**:243–247.

Cerff, R., and Chambers, S. E. 1978. Glyceraldehyde-3-phosphate dehydrogenase (NADP) from *Sinapis alba* L. Isolation and electrophoretic characterization of isoenzymes. *Hoppe-Seyler's Z. Physiol. Chem.* **359**:769–772.

Chapman, K. S. R., and Hatch, M. D. 1977. Regulation of mitochondrial NAD-malic enzyme involved in C_4 pathway photosynthesis. *Arch. Biochem. Biophys.* **184**:298–306.

Chapman, K. S. R., and Hatch, M. D. 1979. Aspartate stimulation of malate decarboxylation in *Zea mays* bundle sheath cells: Possible role of C_4 photosynthesis. *Biochem. Biophys. Res. Commun.* **86**:1274–1280.

Chelm, B. K., and Hallick, R. B. 1976. Changes in the expression of the chloroplast genome of *Euglena gracilis* during chloroplast development. *Biochemistry* **15**:593–599.

Chelune, P., and Wujek, D. E. 1974. An ultrastructural study of pyrenoids in *Chaetopeltis* sp. *Phycologia* **13**:27–30.

Chua, N. H., and Luck, D. J. L. 1974. Biosynthesis of organelle ribosomes. *In:* Nomura, M., Tissières, A., and Lengyel, P., eds., *Ribosomes,* Cold Spring Harbor, N.Y., Cold Spring Harbor Laboratory, pp. 519–539.

Chua, N. H., and Schmidt, G. W. 1978. *In vitro* synthesis, transport, and assembly of ribulose 1,5-bisphosphate carboxylase. *In*: Siegelman, H. W., and Hind, G., eds., *Photosynthetic Carbon Assimilation,* New York, Plenum Press, pp. 325–347.

Chua, N. H., Matlin, K., and Bennoun, P. 1975. A chlorophyll–protein complex lacking in photosystem I mutants of *Chlamydomonas*. *J. Cell Biol.* **67**:361–377.

Clayton, R. K. 1965. The biophysical problems of photosynthesis. *Science* **149**:1346–1354.

Cohen, S. S. 1973. Mitochondria and chloroplasts revisited. *Am. Sci.* **61**:437–445.

Cohen-Bazire, G. 1971. The photosynthetic apparatus of procaryotic organisms. *In*: Harris, P. J., ed., *Biological Ultrastructure: The Origin of Cell Organelles,* Corvallis, Oreg., Oregon State University Press, pp. 65–90.

Cohen-Bazire, G., Pfennig, N., and Kunisawa, R. 1964. The fine structure of green bacteria. *J. Cell Biol.* **22**:207–225.

Conjeaud, H., Mathis, P., and Paillotin, G. 1979. Primary and secondary electron donors in photosystem II of chloroplasts. *Biochim. Biophys. Acta* **546**:280–291.

Coombs, J., and Greenwood, A. D. 1976. Compartmentation of the photosynthetic apparatus. *In*: Barber, J., ed., *The Intact Chloroplast,* Amsterdam, Elsevier, pp. 1–51.

Cramer, W. A., and Whitmarsh, J. 1977. Photosynthetic cytochromes. *Annu. Rev. Plant Physiol.* **28**:133–172.

Crawford, R. M., Dodge, J. D., and Happey, C. M. 1971. The dinoflagellate genus *Woloszynskia*. *Nova Hedwigia Z. Kryptogamenkd.* **19**:825–840.

Crawley, J. C. W. 1963. The fine structure of *Acetabularia mediterranea*. *Exp. Cell Res.* **32**:368–378.

Crick, F. H. C., Wang, J. C., and Bauer, W. R. 1979. Is DNA really a double helix? *J. Mol. Biol.* **129**:449–462.

Crill, P. A. 1977. The photosynthesis–light curve: A simple analog model. *J. Theor. Biol.* **64**:503–516.
Crowther, D., Mills, J. D., and Hind, G. 1979. Protonmotive cyclic electron flow around photosystem I in intact chloroplasts. *FEBS Lett.* **98**:386–390.
Danon, A., and Caplan, S. R. 1977. CO_2 fixation by *Halobacterium halobium. FEBS Lett.* **74**:255–258.
Davis, D. J., and San Pietro, A. 1977. Interactions between spinach ferredoxin and other electron carriers. *Arch. Biochem. Biophys.* **182**:266–277.
de Haller, G. 1959. Structure submicroscopique d'*Euglena viridis. Arch. Sci.* **12**:309–340.
Demeter, S., Herczeg, T., Droppa, H., and Horvath, G. 1979. Thermoluminescence characteristics of granal and agranal chloroplasts of maize. *FEBS Lett.* **100**:321–324.
Descomps, S. 1965. Observations sur l'infrastructure des plastes de Caulerpales (Chlorophyceés). *C. R. Acad. Sci. Paris* **261**:1061–1063.
Dillon, L. S. 1962. Comparative cytology and the evolution of life. *Evolution* **16**:102–117.
Dillon, L. S. 1972. The origins of the genetic code. *Bot. Rev.* **39**:301–345.
Dillon, L. S. 1978. *The Genetic Mechanism and the Origin of Life,* New York, Plenum Press.
Diner, B. A., and Wollman, F. A. 1979. Functional comparison of the photosystem II center–antenna complex of a phycocyaninless mutant of *Cyanidium caldarium* with that of *Chlorella pyrenoidosa. Plant Physiol.* **63**:20–25.
Dodge, J. D. 1968. The fine structure of chloroplasts and pyrenoids in some marine dinoflagellates. *J. Cell Sci.* **3**:41–48.
Dodge, J. D. 1969. A review of the fine structure of algal eyespots. *Br. Phycol. J.* **4**:199–210.
Dodge, J. D. 1973. *The Fine Structure of Algal Cells,* New York, Academic Press.
Dodge, J. D. 1975. A survey of chloroplast ultrastructure in the Dinophyceae. *Phycologia* **14**:253–263.
Dodge, J. D., and Crawford, R. M. 1968. Fine structure of the dinoflagellate *Amphidinium casteri* Hubert. *Protistologica* **4**:231–242.
Dodge, J. D., and Crawford, R. M. 1971. A fine-structural survey of dinoflagellate pyrenoids and food reserves. *J. Linn. Soc. London Bot.* **64**:105–115.
Doherty, A., and Gray, J. C. 1979. Synthesis of cytochrome *f* by isolated pea chloroplasts. *Eur. J. Biochem.* **98**:87–92.
Drum, R. W. 1963. The cytoplasmic fine structure of the diatom, *Nitzschia palea. J. Cell Biol.* **18**:429–440.
Dutton, P. L., Kaufmann, K. J., Chance, B., and Rentzepis, P. M. 1975. Picosecond kinetics of the 1250 nm band of the *Rps. sphaeroides* reaction center: The nature of the primary photochemical intermediary state. *FEBS Lett.* **60**:275–280.
Dutton, P. L., Prince, R. C., and Tiede, D. M. 1978. The reaction center of photosynthetic bacteria. *Photochem. Photobiol.* **28**:939–949.
Ellis, R. J. 1969. Chloroplast ribosomes: Stereospecificity of inhibition by chloramphenicol. *Science* **163**: 477–478.
Ellis, R. J. 1977. The synthesis of chloroplast proteins. *In*: Bogorad, L., and Weil, J. H., eds., *Nucleic Acids and Protein Synthesis in Plants,* New York, Plenum Press, pp. 195–212.
Emerson, R., and Arnold, W. J. 1932. The photochemical reaction in photosynthesis. *J. Gen. Physiol.* **16**:191–198.
Epstein, H. T., and Schiff, J. A. 1961. Studies of chloroplast development in *Euglena*. 4. Electron and fluorescence microscopy of the proplastid and its development into a mature chloroplast. *J. Protozool.* **8**:427–432.
Feher, G., and Okamura, M. Y. 1977. Reaction centers from *Rhodopsendomonas sphaeroides. Brookhaven Symp. Biol.* **28**:183–194.
Fenton, J. M., Pellin, M. J., Govindjee, and Kaufmann, K. J. 1979. Primary photochemistry of the reaction center of photosystem I. *FEBS Lett.* **100**:1–4.
Fischer-Arnold, G. 1963. Untersuchungen über die Chloroplastenbewegung bei *Vaucheria sessilis. Protoplasma* **56**:495–520.

Frenkel, A. W. 1970. Multiplicity of electron transport reactions in bacterial photosynthesis. *Biol. Rev.* **45**:569–616.

Freyssinet, G. 1977. Characterization of cytoplasmic and chloroplast ribosomal proteins of *Euglena gracilis*. *Biochimie* **59**:597–610.

Freyssinet, G. 1978. Determination of the site of synthesis of some *Euglena* cytoplasmic and chloroplast ribosomal proteins. *Exp. Cell Res.* **115**:207–219.

Freyssinet, G., Morlé, F., and Nigon, V. 1976. Chloroplast ribosomal proteins of *E. gracilis*. Immunological studies. *In:* Bücher, T., Neupert, W., Sebald, W., and Werner, S., eds., *Genetics and Biogenesis of Chloroplasts and Mitochondria,* Amsterdam, Elsevier/North-Holland Biomedical Press, pp. 653–658.

Frey-Wyssling, A., and Schwegler, F. 1965. Ultrastructure of the chromoplasts in the carrot root. *J. Ultrastruct. Res.* **13**:543–559.

Galatis, B., Katsaros, C., and Mitrakos, K. 1977. Fine structure of vegetative cells of *Sphacelaria tribuloides* Menegh. with special reference to some unusual proliferations of the plasmalemma. *Phycologia* **16**:139–151.

Gantt, E., and Arnott, H. J. 1963. Chloroplast division in the gametophyte of the fern *Matteuccia struthiopteris* (L.). *J. Cell Biol.* **19**:446–448.

Gantt, E., and Conti, S. F. 1965. The ultrastructure of *Porphyridium cruentum*. *J. Cell Biol.* **26**:365–381.

Cantt, E., and Conti, S. F. 1966a. Granules associated with the chloroplast lamellae of *Porphyridium cruentum*. *J. Cell Biol.* **29**:423–434.

Gantt, E., and Conti, S. F. 1966b. Phycobiliprotein localization in algae. *Brookhaven Symp. Biol.* **19**:393–405.

Gantt, E., and Conti, S. F. 1969. Ultrastructure of blue-green algae. *J. Bacteriol.* **97**:1486–1493.

Gantt, E., and Conti, S. F. 1975. The ultrastructure of *Porphyridium cruentum*. *J. Cell Biol.* **26**:365–381.

Giannasi, D. E. 1978. Systematic aspects of flavonoid biosynthesis and evolution. *Bot. Rev.* **44**:399–427.

Gibbs, M. 1971. Carbohydrate metabolism by chloroplasts. *In*: Gibbs, M., ed., *Structure and Function of Chloroplasts,* New York, Springer-Verlag, pp. 169–214.

Gibbs, S. P. 1962a. The ultrastructure of the pyrenoids of algae, exclusive of the green algae. *J. Ultrastruct. Res.* **7**:247–261.

Gibbs, S. P. 1962b. The ultrastructure of the pyrenoids of the green algae. *J. Ultrastruct. Res.* **7**:262–272.

Gibbs, S. P., Cheng, D., and Slankis, T. 1974a. The chloroplast nucleoid in *Ochromonas danica*. I. Three dimensional morphology in light- and dark-grown cells. *J. Cell Sci.* **16**:557–577.

Gibbs, S. P., Mak, R., Ng, R., and Slankis, T. 1974b. The chloroplast nucleoid in *Ochromonas danica*. II. Evidence for an increase in plastid DNA during greening. *J. Cell Sci.* **16**:579–591.

Gibson, K. D. 1965. Electron microscopy of chromatophores of *Rhodopseudomonas spheroides*. *J. Bacteriol.* **90**:1059–1072.

Giddings, S. J., and Young, D. A. 1974. Differences in the subcellular localization of the stimulatory effects of glucose and pyruvate on protein synthesis in rat thymus cells *in vitro*. *J. Cell Physiol.* **83**:409–417.

Giddings, T. H., and Staehelin, L. A. 1978. Plasma membrane architecture of *Anabaena cylindrica*. *Cytobiologie Z. Exp. Zellfursch.* **16**:235–249.

Giddings, T. H., and Staehelin, L. A. 1979. Changes in thylakoid structure associated with the differentiation of heterocysts in the cyanobacterium, *Anabaena cylindrica*. *Biochim. Biophys. Acta* **546**:373–382.

Gloe, A., Pfennig, N., Brockmann, H., and Trowitzsch, W. 1975. A new bacteriochlorophyll from brown-colored Chlorobiaceceae. *Arch. Microbiol.* **102**:103–110.

Golbeck, J. H., and Kok, B. 1978. Further studies of the membrane bound iron-sulfur proteins and P700 in a photosystem I subchloroplast particle. *Arch. Biochem. Biophys.* **188**:233–242.

Golbeck, J. H., and Kok, B. 1979. Redox titration of electron acceptor Q and the plastoquinone pool in photosystem II. *Biochim. Biophys. Acta* **547**:347–360.

Golbeck, J. H., Lien, S., and San Pietro, A. 1977. Isolation and characterization of a subchloroplast particle enriched in iron-sulfur protein and P700. *Arch. Biochem. Biophys.* **178**:140–150.

Goodwin, T. W. 1973. Carotenoids. *In:* Miller, L. P., ed., *Phytochemistry,* New York, Van Nostrand Reinhold, Vol. 1, pp. 112–142.

Graham, L. E., and Graham, J. M. 1978. Ultrastructure of endosymbiotic *Chlorella* in a *Vorticella. J. Protozool.* **25**:207–210.

Granick, S. 1963. The plastids: Their morphological and chemical differentiation. *In*: Locke, M., ed., *Cytodifferentiation and Macromolecular Synthesis: Proceedings,* New York, Academic Press, pp. 144–174.

Gray, B. H., Lipschultz, C. A., and Gantt, E. 1973. Phycobilisomes from a blue-green alga *Nostoc* species. *J. Bacteriol.* **116**:471–478.

Gray, P. W., and Hallick, R. B. 1978. Physical mapping of the *Euglena gracilis* chloroplast DNA and ribosomal RNA gene region. *Biochemistry* **17**:284–290.

Gray, P. W., and Hallick, R. B. 1979. Isolation of *Euglena gracilis* chloroplast 5S rRNA and mapping the 5S rRNA on chloroplast DNA. *Biochemistry* **18**:1821–1825.

Grebanier, A. E., Coen, D. M., Rich, A., and Bogorad, L. 1978. Membrane proteins synthesized but not processed by isolated maize chloroplasts. *J. Cell Biol.* **78**:734–746.

Griffiths, D. J. 1970. The pyrenoid. *Bot. Rev.* **36**:29–58.

Gruol, D., Rawson, J. R. Y., and Haselkorn, R. 1975. Ribosomal RNA genes in the nucleus and chloroplast of *Euglena. Biochim. Biophys. Acta* **414**:20–29.

Halldal, P. 1964. Phototaxis in protozoa. *In*: Hutner, S. H., ed., *Biochemistry and Physiology of Protozoa,* New York, Academic Press, Vol. 3, pp. 277–296.

Hartley, M. R. 1979. The synthesis and origin of chloroplast low-molecular-weight rRNA in spinach. *Eur. J. Biochem.* **96**:311–320.

Hartley, M. R., and Head, C. 1979. The synthesis of chloroplast high-molecular weight rRNA in spinach. *Eur. J. Biochem.* **96**:301–309.

Hatch, M. D. 1971. The C_4-pathway of photosynthesis. *Biochem. J.* **125**:425–432.

Hatch, M. D., and Kagawa, T. 1974. Activity, location, and role of NAD malic enzyme in leaves with C_4-pathway photosynthesis. *Aust. J. Plant Physiol.* **1**:357–369.

Hatch, M. D., and Kagawa, T. 1976. Photosynthetic activities of isolated bundle sheath cells in relation to differing mechanisms of C_4 pathway photosynthesis. *Arch. Biochem. Biophys.* **175**:39–53.

Hatch, M. D., and Mau, S. L. 1977. Association of NADP- and NAD-linked malic enzyme activities in *Zea mays:* Relation to C_4-pathway photosynthesis. *Arch. Biochem. Biophys.* **179**:361–369.

Hatch, M. D., and Slack, C. R. 1966. A new carboxylation reaction and the pathway of sugar formation. *Biochem. J.* **101**:103–111.

Hatch, M. D., Kagawa, T., and Craig, S. 1975. Subdivision of C_4-pathway species based on differing C_4 acid decarboxylating systems and ultrastructural features. *Aust. J. Plant Physiol.* **2**:111–128.

Haupt, W. 1963. Photoreceptor probleme der Chloroplastenbewegung. *Ber. Dtsch. Bot. Ges.* **76**:313–322.

Haupt, W. 1973. Role of light in chloroplast movement. *BioScience* **23**:289–296.

Hawkins, A. F., and Leedale, G. F. 1971. Zoospore structure and colony formation in *Pediastrum* spp. and *Hydrodictyon reticulatum. Ann. Bot.* **35**:201–211.

Heber, U., Kirk, M. R., and Boardman, N. K. 1979. Photoreactions of cytochrome *b*-559 and cyclic electron flow in photosystem II of intact chloroplasts. *Biochim. Biophys. Acta* **546**:292–306.

Heizmann, P. 1970. Propriétés der ribosomes et des RNA ribosomique d'*Euglena gracilis. Biochim. Biophys. Acta* **224**:144–154.

Heizmann, P. 1974. Maturation of chloroplast rRNA in *Euglena gracilis*. *Biochem. Biophys. Res. Commun.* **56**:112–118.

Herrmann, R. G., and Kowallik, K. V. 1970. Multiple amounts of DNA related to the size of chloroplasts. *Protoplasma* **69**:365–372.

Herrmann, R. G., Bohnert, H. J., Kowallik, K. V., and Schmitt, J. M. 1975. Size, conformation and purity of chloroplast DNA of some higher plants. *Biochim. Biophys. Acta* **378**:305–317.

Heslop-Harrison, J. 1963. Structure and morphogenesis of lamellar systems in grana-containing chloroplasts. *Planta* **60**:243–260.

Heslop-Harrison, J. 1966. Structural features of the chloroplast. *Sci. Prog. (Oxford)* **54**:519–541.

Hibberd, D. J., and Leedale, G. F. 1972. Observations on the cytology and ultrastructure of the new algal class Eustigmatophyceae. *Ann. Bot.* **36**:49–71.

Hind, G., Mills, J. D., and Slovacek, R. E. 1977. Cyclic electron transport in photosynthesis. *In*: *Proceedings of the 4th International Congress on Photosynthesis, 1977,* pp. 591–600.

Honeycutt, R. C., and Margulies, M. M. 1973. Protein synthesis in *Chlamydomonas reinhardii*. *J. Biol. Chem.* **248**:6145–6153.

Honsell, E., Avanzini, A., and Ghirardelli, L. A. 1978. Two ways of chloroplast development in vegetative cells of *Nitophyllum punctatum* (Rhodophyta). *J. Submicrosc. Cytol.* **10**:227–237.

Hori, T., and Ueda, R. 1967. Electron microscope studies on the fine structure of plastids in siphonous green algae with special reference to their phylogenetic relationships. *Sci. Rep. Tokyo Kyoiku Daigaku Sect. B* **12**:225–244.

Hutson, S. M., Van Dop, C., and Lardy, H. A. 1977. Mitochondrial metabolism of pyruvate in bovine spermatozoa. *J. Biol. Chem.* **252**:1309–1315.

Jenni, B., and Stutz, E. 1978. Physical mapping of the rDNA region of *Euglena gracilis* chloroplast DNA. *Eur. J. Biochem.* **88**:127–134.

Jenni, B., and Stutz, E. 1979. Analysis of *Euglena gracilis* chloroplast DNA. *FEBS Lett.* **102**:95–99.

Jones, B. L., Nagabhushan, N., Gulyas, A., and Zalik, S. 1972. Two-dimensional acrylamide gel electrophoresis of wheat leaf cytoplasmic and chloroplast ribosomal proteins. *FEBS Lett.* **23**:167–170.

Jones, O. T. G. 1973. Chlorophyll. *In*: Miller, L. P., ed., *Phytochemistry,* New York, Van Nostrand Reinhold, Vol. 1, pp. 75–111.

Jost, M. 1965. Die Ultrastruktur von *Oscillatoria rubescens*. *Arch. Mikrobiol.* **50**:211–245.

Joyon, L., and Fott, B. 1964. Quelques particularités infrastructurales du plaste des *Carteria* (Volvocales). *J. Microsc. (Paris)* **3**:159–166.

Kagawa, T., and Hatch, M. D. 1977. Regulation of C_4 photosynthesis. *Arch. Biochem. Biophys.* **184**:290–297.

Kaja, H. 1955. Untersuchungen über die Chromatophoren und Pyrenoide der Anthocerotales. *Protoplasma* **44**:136–153.

Kelly, G. J., Latzko, D., and Gibbs, M. 1976. Regulatory aspects of photosynthetic carbon metabolism. *Annu. Rev. Plant Physiol.* **27**:181–205.

Kirk, J. T. O. 1971. Will the real chloroplast DNA please stand up. *In*: Boardman, N. K., Linnane, A. W., and Smillie, R. M., eds., *Autonomy and Biogenesis of Mitochondria and Chloroplasts,* Amsterdam, North-Holland, pp. 267–276.

Kirk, J. T. O., and Tilney-Bassett, R. A. E. 1967. *The Plastids,* San Francisco, Freeman.

Kisaki, T., and Tolbert, N. E. 1969. Glycolate and glyoxylate metabolism by isolated peroxisomes or chloroplasts. *Plant Physiol.* **44**:242–250.

Koenig, F., Schmid, G. H., Radunz, A., Pineau, B., and Menke, W. 1976. The isolation of further polypeptides from the thylakoid membrane, their localization and function. *FEBS Lett.* **62**:342–346.

Koenig, F., Menke, W., Radunz, A., and Schmid, G. H. 1977. Localization and functional characterization of 3 thylakoid membrane polypeptides of m.w. 66,000. *Z. Naturforsch. Teil C* **32**:817–827.

Kortschak, H. P., Hartt, C. E., and Burr, G. O. 1965. CO_2 fixation in sugarcane leaves. *Plant Physiol.* **40**:209–213.
Krause, G. H., Thorne, S. W., and Lorimer, G. H. 1977. Glycolate synthesis by intact chloroplasts. *Arch. Biochem. Biophys.* **183**:471–479.
Kretzer, F., Ohad, I., and Bennoun, P. 1977. Ontogeny, insertion, and activation of two thylakoid peptides required for photosystem II activity in the nuclear, temperature sensitive T4 mutant of *Chlamydomonas reinhardii*. *In*: Bücher, T., ed., *Symposium on Genetics and Biogenesis of Chloroplasts and Mitochondria,* Amsterdam, Elsevier, pp. 25–32.
Krinsky, N. I. 1971. Function. *In*: Isler, O., ed., *Carotenoids,* Basel, Birkhäuser, pp. 669–716.
Kuehl, L. 1967. Evidence for nuclear synthesis of lactic dehydrogenase in rat liver. *J. Biol. Chem.* **242**:2299–2306.
Küster, E. 1951. *Die Pflanzenzelle,* Jena, Fischer.
Lamppa, G. K., and Bendich, A. J. 1979. Changes in chloroplast DNA levels during development of pea (*Pisum sativum*). *Plant Physiol.* **64**:126–130.
Lang, N. J. 1968a. Electron microscope studies of extraplastidic astaxanthin in *Haematococcus. J. Phycol.* **4**:12–19.
Lang, N. J. 1968b. The fine structure of blue-green algae. *Annu. Rev. Microbiol.* **22**:15–46.
Lang, N. J., and Rae, P. M. M. 1967. Structures in a blue-green alga resembling prolamellar bodies. *Protoplasma* **64**:67–75.
Lebherz, H. G., and Rutter, W. J. 1967. Glyceraldehyde-3-phosphate dehydrogenase variants in phyletically diverse organisms. *Science* **157**:1198–1200.
Leedale, G. F. 1966. Euglena: A new look with the electron microscope. *Adv. Sci.* **107**:22–37.
Leedale, G. F. 1967. *Euglenoid Flagellates,* Englewood Cliffs, N. J., Prentice–Hall.
Leggett-Bailey, J., and Whyborn, A. G. 1963. The osmiophilic globules of chloroplasts. II. Globules of the spinach-beet chloroplast. *Biochim. Biophys. Acta* **78**:163–174.
Levine, R. P. 1968. Genetic dissection of photosynthesis. *Science* **162**:768–771.
Lewin, R. A. 1962. *Physiology and Biochemistry of Algae,* New York, Academic Press.
Leyon, H. 1954. The structure of chloroplasts. III. A study of pyrenoids. *Exp. Cell Res.* **6**:497–505.
Lichtenthaler, H. K. 1969. Die Plastoglobuli von Spinat, ihre Grösse, Isolierung und Lipochinonzusammensetzung. *Protoplasma* **68**:65–77.
Link, G., Bedbrook, J. R., Bogorad, L., Coen, D. M., and Rich, A. 1978. The expression of the gene for the large subunit of ribulose 1,5-bisphosphate carboxylase in maize. *In*: Siegelman, H. W., and Hind, G., eds., *Photosynthetic Carbon Assimilation,* New York, Plenum Press, pp. 349–362.
Lonergan, T. A., and Sargent, M. L. 1979. Regulation of the photosynthesis rhythm in *Euglena gracilis*. II. Involvement of electron flow through both photosystems. *Plant Physiol.* **64**:99–103.
Löw, H., and Afzelius, B. A. 1964. Subunits of the chromatophore membranes in *Rhodospirillum rubrum*. *Exp. Cell Res.* **35**:431–434.
McBride, D. L., and Cole, K. 1972. Ultrastructural observations on germinating monospores in *Smithora naiadum* (Rhodophyceae). *Phycologia* **11**:181–192.
McCarty, R. E. 1978. AMP is converted to ADP and ATP in the medium before it is bound to coupling factor 1 in illuminated spinach chloroplast thylakoids. *FEBS Lett.* **95**:299–302.
McCrea, J. M., and Hershberger, C. L. 1976. Chloroplast DNA codes for tRNA. *Nucleic Acids Res.* **3**:2005–2018.
McEvoy, F. A., and Lynn, W. S. 1973. The peptides of chloroplast membranes. 1. The soluble coupling factor (Ca^{2+}-ATPase). *Arch. Biochem. Biophys.* **156**:335–341.
McEwen, B. S., Allfrey, V. G., and Mirsky, A. E. 1964. Dependence of RNA synthesis in isolated thymus nuclei on glycolysis, oxidative carbohydrate catabolism and a type of "oxidative phosphorylation." *Biochim. Biophys. Acta* **91**:23–28.
McKinney, D. W., Buchanan, B. B., and Wolosiuk, R. A. 1979. Association of a thioredoxin-like

protein with chloroplast coupling factor. *Biochem. Biophys. Res. Commun.* **86**:1178–1184.

Madigan, M. T., and Brock, T. D. 1977. Chlorobium-type vesicles of photosynthetically-grown *Chloroflexus aurantiacus* observed using negative staining techniques. *J. Gen. Microbiol.* **102**:279–285.

Maier, S., and Murray, R. G. E. 1965. The fine structure of *Thioploca ingrica* and a comparison with *Beggiatoa*. *Can. J. Microbiol.* **11**:645–655.

Malkin, R., and Bearden, A. J. 1978. Membrane-bound iron-sulfur centers in photosynthetic systems. *Biochim. Biophys. Acta* **505**:147–181.

Manning, J. E., and Richards, O. C. 1972. Isolation and molecular weight of circular chloroplast DNA from *Euglena gracilis*. *Biochim. Biophys. Acta* **259**:285–296.

Manton, I. 1964. Observations with the electron microscope on the division cycle of the flagellate *Prymnesium parvum* Carter. *J. R. Microsc. Soc.* **83**:317–325.

Manton, I. 1966. Some possibly significant structural relations between chloroplasts and other cell components. *In*: Goodwin, T. W., ed., *The Biochemistry of Chloroplasts*, New York, Academic Press, Vol. 1, pp. 23–47.

Manton, I. 1967. Further observations on the fine structure of *Chrysochromulina chiton* with special reference to the haptonema, "peculiar" Golgi structure and scale production. *J. Cell Sci.* **2**:265–272.

Manton, I., and Parke, M. 1965. Observations on the fine structure of two species of *Platymonas* with special reference to flagellar scales and the mode of origin of the theca. *J. Mar. Biol. Assoc. U.K.* **45**:743–754.

Margulies, M. M., and Michaels, A. 1974. Ribosomes bound to chloroplast membranes in *Chlamydomonas*. *J. Cell Biol.* **60**:65–77.

Margulies, M. M., Tiffany, H. L., and Michaels, A. 1975. Vectorial discharge of nascent polypeptides attached to chloroplast thylakoid membranes. *Biochem. Biophys. Res. Commun.* **64**:735–739.

Margulis, L. 1970. *Origin of Eucaryotic Cells,* New Haven, Conn., Yale University.

Markwell, J. P., Thornber, J. P., and Boggs, R. T. 1979. Higher plant chloroplasts: Evidence that all the chlorophyll exists as chlorophyll–protein complexes. *Proc. Natl. Acad. Sci. USA* **76**:1233–1235.

Mattheis, J. R., and Rebeiz, C. A. 1977a. Chloroplast biogenesis. *J. Biol. Chem.* **252**:8347–8349.

Mattheis, J. R., and Rebeiz, C. A. 1977b. Chloroplast biogenesis. *Arch. Biochem. Biophys.* **184**:189–196.

Mattox, K. R., Stewart, K. D., and Floyd, G. L. 1974. The cytology and classification of *Schizomeris leibleinii* (Chlorophyceae). I. The vegetative thallus. *Phycologia* **13**:63–69.

Mayer, F. 1971. Light-induced chloroplast contraction and movement. *In*: Gibbs, M., ed., *Structure and Function of Chloroplasts,* New York, Springer-Verlag, pp. 35–49.

Melis, A. 1978. Oxidation–reduction potential dependence of the two kinetic components in chloroplast system II primary photochemistry. *FEBS Lett.* **95**:202–206.

Melis, A., and Homann, P. H. 1975. Kinetic analysis of the fluorescence induction in 3-(3,4-dichlorophenyl)-1, 1-dimethylurea-poisoned chloroplasts. *Photochem. Photobiol.* **21**:431–437.

Melis, A., and Homann, P. H. 1976. Heterogeneity of the photochemical centers in system II of chloroplasts. *Photochem. Photobiol.* **23**:343–350.

Menke, W. 1962. Structure and chemistry of plastids. *Annu. Rev. Plant Physiol.* **13**:27–44.

Menke, W., and Schölzel, E. 1971. Molecular weight of polypeptides of the thylakoid membrane. *Z. Naturforsch.* **26b**:378–379.

Messer, G., and Ben-Shaul, Y. 1969. Fine structure of *Peridinium westii* Lemm, a freshwater dinoflagellate. *J. Protozool.* **16**:272–280.

Metzner, H. 1975. Water decomposition in photosynthesis? A critical reconsideration. *J. Theor. Biol.* **51**:201–231.

Michaels, A., and Margulies, M. M. 1975. Amino acid incorporation into proteins by ribosomes bound to chloroplast thylakoid membranes. Formation of discrete products. *Biochim. Biophys. Acta* **390**:352–362.

Miller, K. R. 1976. A particle spanning the photosynthetic membrane. *J. Ultrastruct. Res.* **54**:159–167.

Miller, K. R. 1978. Structural organization in the photosynthetic membrane. *In:* Akoyunoglou, G., and Argyroudi-Akoyunoglou, J. H., eds., *Chloroplast Development,* Elsevier/North-Holland Biomedical Press, pp. 17–30.

Miller, K. R., and Cushman, R. A. 1979. A chloroplast lacking photosystem II. *Biochim. Biophys. Acta* **546**:481–497.

Miller, K. R., Miller, G. J., and McIntyre, K. R. 1976. The light-harvesting chlorophyll–protein complex of photosystem II. *J. Cell Biol.* **71**:624–638.

Miller, M. M., and Lang. N. J. 1968. The fine structure of akinete formation and germination in *Cylindrospermum. Arch. Mikrobiol.* **60**:303–313.

Mills, J. D., Crowther, D., Slovacek, R. E., Hind, G., and McCarty, R. E. 1979. Electron transport pathways in spinach chloroplasts. *Biochim. Biophys. Acta* **547**:127–137.

Mitrakos, K. 1960. Feinbau und Teilung bei Plastiden einiger Florideen-Arten. *Protoplasma* **52**:611–617.

Molnar, K. E., Stewart, K. D., and Mattox, K. R. 1975. Cell division in the filamentous *Pleurastrum* and its comparison with the unicellular *Platymonas* (Chlorophyceae). *J. Phycol.* **11**:287–296.

Moore, F. D., and Shephard, D. C. 1977. Biosynthesis in isolated *Acetabularia* chloroplasts. II. Plastid pigments. *Protoplasma* **92**:167–175.

Moore, F. D., and Tschismadia, I. 1977. Biosynthesis in isolated *Acetabularia* chloroplasts. III. Complex lipids. *Progress in Acetabularia Research,* New York, Academic Press, pp. 159–173.

Morgan, J. A., and Brown, R. H. 1979. Photosynthesis in grass species differing in CO_2 fixation pathways. *Plant Physiol.* **64**:257–262.

Mühlethaler, K. 1971. The ultrastructure of plastids. *In:* Gibbs, M., ed., *Structure and Function of Chloroplasts,* New York, Springer-Verlag, pp. 7–34.

Nelson, N., Deters, D. W., Nelson, H., and Racker, E. 1973. Partial resolution of the enzymes catalyzing phosphorylation. III. *J. Biol. Chem.* **248**:2049–2055.

Nessler, C. L., and Mahlberg, P. G. 1979a. Plastids in laticifers of *Papaver.* I. Development and cytochemistry of laticifer plastids in *P. somniferum* L. *Am. J. Bot.* **66**:266–273.

Nessler, C. L., and Mahlberg, P. G. 1979b. Plastids in laticifers of *Papaver.* II. Enzyme cytochemistry of membrane-bound inclusions of laticifer plastids in *P. bracteatum* Lindl. (Papaveraceae). *Am. J. Bot.* **66**:274–279.

Newcomb, E. H. 1967. Fine structure of protein-strong plastids in bean root tips. *J. Cell Biol.* **33**:143–163.

Nichols, H. W., Ridgway, J. E., and Bold, H. C. 1966. A preliminary ultrastructural study of the freshwater red alga *Compsopogon. Ann. Mo. Bot. Gard.* **53**:17–27.

Nielsen, N. C., Smillie, R. M., Henningsen, K. W., and von Wettstein, D. 1979. Composition and function of thylakoid membranes from grana-rich and grana-deficient chloroplast mutants of barley. *Plant Physiol.* **63**:174–182.

Nigon, V., and Heizmann, P. 1978. Morphology, biochemistry, and genetics of plastid development in *Euglena gracilis. Int. Rev. Cytol.* **53**:211–290.

Oesterhelt, D., and Stoecknius, W. 1973. Functions of a new photoreceptor membrane. *Proc. Natl. Acad. Sci. USA* **70**:2853–2857.

Ohad, I. 1975. Biogenesis of chloroplast membranes. *In:* Tzagoloff, A., ed., *Membrane Biogenesis, Mitochondria, Chloroplast, and Bacteria,* New York, Plenum Press, pp. 279–350.

Ohad, I. 1977. Ontogeny and assembly of chloroplast membrane polypeptides in *Chlamydomonas reinhardii. In:* Brinkley, B. R., and Porter, K. R., eds., *International Cell Biology 1976–1977,* New York, Rockefeller University Press, pp. 193–203.

Olive, J., Wollman, F. A., Bennoun, P., and Recouvreur, M. 1979. Ultrastructure–function relationship in *Chlamydomonas reinhardii* thylakoids, by means of a comparison between wild type and the F34 mutant which lacks the photosystem II reaction center. *Mol. Biol. Rep.* **5**:139–143.

Ottaway, J. H., and Mowbray, J. 1977. The role of compartmentation in the control of glycolysis. *Curr. Top. Cell. Regul.* **12**:107–208.

Padmanabhan, U., and Green, B. R. 1978. The kinetic complexity of *Acetabularia* chloroplast DNA. *Biochim. Biophys. Acta* **521**:67–73.

Paolillo, D. J. 1970. The three-dimensional arrangement of intergranal lamellae in chloroplasts. *J. Cell Sci.* **6**:243–255.

Paolillo, D. J., and Falk, R. H. 1966. The ultrastructure of grana in mesophyll plastids of *Zea mays*. *Am. J. Bot.* **53**:173–180.

Paolillo, D. J., Mackay, N. C., and Graffius, J. R. 1969. The structure of grana in flowering plants. *Am. J. Bot.* **56**:344–347.

Park, R. B. 1965. Substructure of chloroplast lamellae. *J. Cell Biol.* **27**:151–161.

Parke, M., and Manton, I. 1965. Preliminary observations on the fine structure of *Prasinocladus marinus*. *J. Mar. Biol. Assoc. U.K.* **45**:525–536.

Parthier, B. 1973. Cytoplasmic site of synthesis of chloroplast aminoacyl-tRNA synthetases in *Euglena gracilis*. *FEBS Lett.* **38**:70–74.

Parthier, B., and Krauspe, R. 1973. Assignment to chloroplast and cytoplasm of three *Euglena gracilis* aminoacyl-tRNA synthetases with ambiguous specificity for transfer RNA. *Plant Sci. Lett.* **1**:221–227.

Radunz, A. 1977. Binding of antibodies onto the thylakoid membrane. II. Distribution of lipids and proteins at the outer surface of the thylakoid membrane. *Z. Naturforsch.* **32c**:597–599.

Rathnam, C. K. M., and Edwards, G. E. 1975. Intracellular localization of certain photosynthetic enzymes in bundle sheath cells of plants possessing C_4-pathway photosynthesis. *Arch. Biochem. Biophys.* **171**:214–225.

Rathnam, C. K. M., and Edwards, G. E. 1977. C_4 acid decarboxylation and CO_2 donation to photosynthesis in bundle sheath strands and chloroplasts representing three groups of C_4 plants. *Arch. Biochem. Biophys.* **182**:1–13.

Rawson, J. R. Y., and Boerma, C. L. 1976. A measurement of the fraction of chloroplast DNA transcribed during chloroplast development in *Euglena gracilis*. *Biochemistry* **15**:588–592.

Rawson, J. R. Y., and Stutz, E. 1969. Isolation and characterization of *Euglena gracilis* cytoplasmic and chloroplast ribosomes and their ribosomal RNA components. *Biochim. Biophys. Acta* **190**:368–380.

Raymond, J. C., and Sistrom, W. R. 1967. The isolation and preliminary characterization of a halophilic photosynthetic bacterium. *Arch. Mikrobiol.* **59**:255–268.

Rebeiz, C. A., and Castelfranco, P. A. 1973. Protochlorophyll and chlorophyll biosynthesis in cell-free systems from higher plants. *Annu. Rev. Plant Physiol.* **24**:129–172.

Reinman, S., and Thornber, J. P. 1979. The electrophoretic isolation and partial characterization of three chlorophyll–protein complexes from blue-green algae. *Biochim. Biophys. Acta* **547**:188–197.

Renger, G. 1975. Studies on the reaction mechanism of the oxidizing equivalents produced by system II. *In:* Avron, M., ed., *Proceedings, Third International Congress on Photosynthesis,* Amsterdam, Elsevier, Vol. 1, pp. 127–144.

Rijersberg, C. P., Melis, A., Amesz, J., and Swager, J. A. 1979. Quenching of chlorophyll fluorescence and photochemical activity of chloroplasts at low temperature. *Ciba Found. Symp.* **61**(n.s.):305–322.

Ris, H., and Singh, R. N. 1961. Electron microscope studies on blue-green algae. *J. Biophys. Biochem. Cytol.* **9**:63–80.

Rochaix, J. D. 1978. Restriction endonuclease map of the chloroplast DNA of *Chlamydomonas reinhardii*. *J. Mol. Biol.* **126**:597–617.

Rochaix, J. D., and Malnoe, P. 1978. Anatomy of the chloroplast rDNA of *Chlamydomonas reinhardii*. *Cell* **15**:661–670.
Rose, R. J. 1979. The association of chloroplast DNA with photosynthetic membrane vesicles from spinach chloroplasts. *J. Cell Sci.* **36**:169–183.
Rosen, W. G., and Siegesmund, K. A. 1961. Some observations on the fine structure of a thermophilic, acidophilic alga. *J. Biophys. Biochem. Cytol.* **9**:910–914.
Sabnis, D. D. 1969. Observations on the ultrastructure of the coenocytic marine alga *Caulerpa prolifera,* with particular reference to some unusual cytoplasmic components. *Phycologia* **7**:24–42.
Sagan, L. 1967. On the origin of mitosing cells. *J. Theor. Biol.* **14**:225–274.
Sakano, K., Kung, S. D., and Wildman, S. G. 1974. Identification of several chloroplast DNA genes which code for the large subunit of *Nicotiana* fraction I proteins. *Mol. Gen. Genet.* **130**:91–97.
Salvador, G., Lefort-Tran, M., Nigon, V., and Jourdan, F. 1971. Structure et evolution du corps prolamellaire dans les proplastes d'*Euglena gracilis*. *Exp. Cell Res.* **64**:457–462.
Sauer, K., and Austin, L. A. 1978. Bacteriochlorophyll–protein complexes from the light-harvesting antenna of photosynthetic bacteria. *Biochemistry* **17**:2011–2019.
Schachman, H. K., Pardee, A. B., and Stanier, R. Y. 1952. Studies on the macromolecular organization of microbial cells. *Arch. Biochem. Biophys.* **38**:245–260.
Schiff, J. A. 1970. Developmental interactions among cellular compartments in *Euglena*. *Symp. Soc. Exp. Biol.* **24**:277–301.
Schönbohm, E. 1973. Kontractile Fibrillen als aktive Elemente bei der Mechanik der Chloroplastenbewegung. *Ber. Dtsch. Bot. Ges.* **86**:407–423.
Schulz-Baldes, M., and Lewin, R. A. 1976. Fine structure of *Synechocystis* (Cyanophyta). *Phycologia* **15**:1–6.
Schwartzbach, S. D., Hecker, L. I., and Barnett, W. E. 1976. Transcriptional origin of *Euglena* chloroplast tRNAs. *Proc. Natl. Acad. Sci. USA* **73**:1984–1988.
Scott, N. S. 1973. r-RNA cistrons in *Euglena gracilis*. *J. Mol. Biol.* **81**:327–336.
Senn, G. 1908. *Die Gestalts- und Lageveränderungen der Pflanzenchromatophoren,* Leipzig, Engelmann.
Shuvalov, V. A., Bacon, K. E., and Dolan, E. 1979a. Kinetic and spectral properties of the intermediate electron acceptor A_1 in photosystem I. *FEBS Lett.* **100**:5–8.
Shuvalov, V. A., Dolan, E., and Ke, B. 1979b. Spectral and kinetic evidence for two early electron acceptors in photosystem I. *Proc. Natl. Acad. Sci. USA* **76**:770–773.
Siebert, G., and Humphrey, G. B. 1965. Enzymology of the nucleus. *Adv. Enzymd.* **27**:239–288.
Siegesmund, K. A., Rosen, W. G., and Gawlik, S. R. 1962. Effects of darkness and of streptomycin on the fine structure of *Euglena gracilis*. *Am. J. Bot.* **49**:137–145.
Simcox, P. D., and Dennis, D. T. 1978a. Isoenzymes of the glycolytic and pentose phosphate pathways in proplastids from the developing endosperm of *Ricinus communis* L. *Plant Physiol.* **61**:871–877.
Simcox, P. D., and Dennis, D. T. 1978b. 6-Phosphogluconate dehydrogenase isoenzymes from the developing endosperm of *Ricinus communis* L. *Plant Physiol.* **62**:287–289.
Simcox, P. D., Reid, E. E., Canvin, D. T., and Dennis, D. T. 1977. Enzymes of the glycolytic and pentose phosphate pathways in proplastids from the developing endosperm of *Ricinus communis* L. *Plant Physiol.* **59**:1128–1132.
Slavik, N. S., and Hershberger, C. L. 1975. The kinetic complexity of *Euglena gracilis* chloroplast DNA. *FEBS Lett.* **52**:171–174.
Slavik, N. S., and Hershberger, C. L. 1976. Internal structural organization of chloroplast DNA from *Euglena gracilis*. *J. Mol. Biol.* **103**:563–581.
Slovacek, R. E., Mills, J. D., and Hind, G. 1978. The function of cyclic electron transport in photosynthesis. *FEBS Lett.* **87**:73–76.

Slovacek, R. E., Crowther, D., and Hind, G. 1979. Cytochrome function in the cyclic electron transport pathways of chloroplasts. *Biochim. Biophys. Acta* **547**:138–148.

Smillie, R. M., Scott, N. S., and Bishop, D. G. 1973. Gene expression in chloroplasts and regulation of chloroplast differentiation. *In:* Pollak, J. K. K., and Lee, J. W., eds., *Biochemistry of Gene Expression in Higher Organisms,* Sydney, Australia and New Zealand Book Co., pp. 479–503.

Stanier, R. Y. 1974. The origins of photosynthesis in eukaryotes. *In: Evolution in the Microbial World, 24th Symposium of the Society for General Microbiology,* Cambridge, Cambridge University Press, pp. 219–240.

Stanier, R. Y., and Cohen-Bazire, G. 1957. The role of light in the microbial world: Some facts and speculations. *Symp. Soc. Gen. Microbiol.* **7**:56–89.

Steffen, K., and Walter, F. 1958. Die Chromoplasten von *Solanum capsicastrum* und ihre Genese. *Planta* **50**:640–670.

Steinbach, R. A., Sahm, H., and Schütte, H. 1978. Purification and regulation of glucose-6-phosphate dehydrogenase from obligate methanol-utilizing bacterium *Methylmonas* M15. *Eur. J. Biochem.* **87**:409–415.

Straley, S. C., Parson, W. W., Manzerall, D. C., and Clayton, R. K. 1973. Pigment content and molar extinction coefficients of photochemical reaction centers from *Rhodopseudomonas spheroides. Biochim. Biophys. Acta* **305**:597–609.

Straus, W. 1953. Chromoplasts—Development of crystalline forms, structure, state of pigments. *Bot. Rev.* **19**:147–186.

Straus, W. 1961. Studies on the chromoplasts of carrots. *Protoplasma* **53**:405–421.

Surzycki, S. J., Goodenough, U. W., Levine, R. P., and Armstrong, J. J. 1970. Nuclear and chloroplast control of chloroplast structure and function in *Chlamydomonas reinhardii. Symp. Soc. Exp. Biol.* **29**:13–35.

Taylor, D. L. 1970. Chloroplasts as symbiotic organelles. *Int. Rev. Cytol.* **27**:29–64.

Tevini, M., and Lichtenthaler, H. K. 1970. Untersuchungen über die Pigment- und Lipochinonausstattung der zwei photosynthetischen Pigment-systeme. *Z. Pflanzenphysiol.* **62**:17–32.

Tewari, K. K., and Wildman, S. G. 1966. Chloroplast DNA from tobacco leaves. *Science* **153**:1269–1271.

Tewari, K. K., and Wildman, S. G. 1970. Information content in the chloroplast DNA. *Symp. Soc. Exp. Biol.* **24**:147–179.

Tewari, K. K., Kolodner, R., Chu, N. M., and Meeker, R. R. 1977. Structure of chloroplast DNA. *In:* Bogorad, L., and Weil, L. H., eds., *Nucleic Acids and Protein Synthesis in Plants,* New York, Plenum Press, pp. 15–36.

Thompson, P., Reid, E. E., Lyttle, C. R., and Dennis, D. T. 1977. Pyruvate dehydrogenase complex from higher plant mitochondria and proplastids: Kinetics. *Plant Physiol.* **59**:849–853.

Tolbert, N. E., and Yamazaki, R. K. 1969. Leaf peroxisomes and their relation to photorespiration and photosynthesis. *Ann. N.Y. Acad. Sci.* **168**:209–381.

Toth, R. 1974. Sporangial structure and zoosporogenesis in *Chorda tomentosa* (Laminariales). *J. Phycol.* **10**:170–185.

Toth, R. 1976. The release, settlement and germination of zoospores in *Chorda tomentosa. J. Phycol.* **12**:222–233.

Trüper, H. G. 1968. *Ectothiorhodospira mobilis* Pelsh., a photosynthetic sulfur bacterium depositing sulfur outside the cells. *J. Bacteriol.* **95**:1910–1920.

Tschismadia, I., and Moore, F. E. 1979. Biosynthesis in isolated *Acetabularia* chloroplasts. IV. Plastoquinones. *In:* Bonotto, S., Kefeli, V., and Puiseux-Dao, S., eds., *Developmental Biology of Acetabularia,* Amsterdam, Elsevier/North-Holland Biomedical Press, pp. 183–194.

Turner, J. B., and Friedmann, E. I. 1974. Fine structure of capitular filaments in the coenocytic green alga *Penicillus. J. Phycol.* **10**:125–134.

Tyszkiewicz, E., Popovic, R., and Roux, E. 1977. Relationship between the redox-state of P_{700} and photosystem I-mediated proton translocation studied with chloroplasts from dark-grown *Pinus nigra* seedlings. *FEBS Lett.* **81**:65–68.

van der Rest, M., and Gindras, G. 1974. The pigment complement of the photosynthetic reaction center of *Rhodospirillum rubrum*. *J. Biol. Chem.* **249**:6446–6453.

Vandrey, J. P., and Stutz, E. 1973. Evidence for a novel DNA component in chloroplasts of *Euglena gracilis*. *FEBS Lett.* **37**:174–177.

Vatter, A. E., and Wolfe, R. S. 1958. The structure of photosynthetic bacteria. *J. Bacteriol.* **75**:480–483.

Verdier, G. 1979a. Poly-(A)-containing RNA of *Euglena gracilis* during chloroplast development. 1. Analysis of their complexity of hybridization to complementary DNA. *Eur. J. Biochem.* **93**:573–580.

Verdier, G. 1979b. Poly-(A)-containing RNA of *Euglena gracilis* during chloroplast development. 2. Transcriptional origin of the different RNA. *Eur. J. Biochem.* **93**:581–586.

Vernon, L. P. 1968. Photochemical and electron transport reactions of bacterial photosynthesis. *Bacteriol. Rev.* **32**:243–261.

Von Wettstein, D. 1959. *In:* Rudnick, D., ed., *Developmental Cytology,* New York, Ronald Press, p. 123.

Von Wettstein, D. 1961. Nuclear and cytoplasmic factors in development of chloroplast structure and function. *Can. J. Bot.* **39**:1537–1545.

Von Wettstein, D., Henningsen, K. W., Boynton, J. E., Kannangara, G. C., and Nielsen, O. F. 1971. The genic control of chloroplast development in barley. *In:* Boardman, N. K., Linnane, A. W., and Smillie, R. M., eds., *Autonomy and Biogenesis of Mitochondria and Chloroplasts,* Amsterdam, North-Holland, pp. 205–223.

Warburg, O., and Krippahl, G. 1960. Glycolsäurebildung in *Chlorella*. *Z. Naturforsch.* **15b**: 197–200.

Weier, T. E., and Thomson, W. W. 1962. The grana of starch-free chloroplasts of *Nicotiana rustica*. *J. Cell Biol.* **13**:89–108.

Weier, T. E., Engelbrecht, H. T.., Harrison, A., and Risley, E. B. 1965. Subunits in the membranes of chloroplasts of *Phaseolus vulgaris, Pisum sativum,* and *Aspidistra* sp. *J. Ultrastruct. Res.* **13**:92–111.

Weier, T. E., Stocking, C. R., and Shumway, L. K. 1966a. The photosynthetic apparatus in chloroplasts of higher plants. *Brookhaven Symp. Biol.* **19**:353–374.

Weier, T. E., Bisalputra, T., and Harrison, A. 1966b. Subunits in chloroplast membranes of *Scenedesmus quadricauda*. *J. Ultrastruct. Res.* **15**:38–56.

Weil, J. H., Burkard, G., Guillemaut, P., Jeannin, G., Martin, R., and Steinmetz, A. 1977. tRNAs and aminoacyl-tRNA synthetases in plant cytoplasm, chloroplasts and mitochondria. *In:* Bogorad, L., and Weil, J. H., eds., *Nucleic Acids and Protein Synthesis in Plants,* New York, Plenum Press, pp. 97–119.

Whitfeld, P. R., Atchinson, B. A., Bottomley, W., and Leaver, C. J. 1976. Analysis of the coding capacity of EcoRI restriction fragments of spinach chloroplast DNA. *In*: Bücher, T., Neupert, W., Sebald, W., and Werner, S., eds., *Genetics and Biogenesis of Chloroplasts and Mitochondria,* Amsterdam, Elsevier/North-Holland Biomedical Press, pp. 361–368.

Whitfeld, P. R., Herrmann, R. G., and Bottomley, W. 1978. Mapping of the rRNA genes on chloroplast DNA. *Nucleic Acids Res.* **5**:1741–1751.

Wild, M. A., and Gall, J. G. 1979. An intervening sequence in the gene coding for 25S ribosomal RNA of *Tetrahymena pigmentosa*. *Cell* **16**:565–574.

Wolken, J. J. 1973. Photodynamics: The chloroplast in photosynthesis. *In*: Miller, L. P., ed., *Phytochemistry*, New York, Van Nostrand Reinhold, pp. 15–37.

Wollman, F. A. 1979. Ultrastructural comparison of *Cyanidium caldarium* wild type and III-C mutant lacking phycobilisomes. *Plant Physiol.* **63**:375–381.

Woodcock, C. L. F., and Bogorad, L. 1971. Nucleic acids and information processing in chloroplasts. *In*: Gibbs, M., ed., *Structure and Function of Chloroplasts,* New York, Springer-Verlag, pp. 89–128.
Wujek, D. E. 1976. Ultrastructure of flagellated chrysophytes. II. *Uroglena* and *Uroglenopsis*. *Cytologia* 41:665–670.
Wujek, D. E., and Gretz, M. R. 1977. An ultrastructural study of the pyrenoid of *Schizochlamys* (Chlorophyceae: Tetrasporales). *Trans. Am. Microsc. Soc.* **96**:398–402.
Wujek, D. E., and Kristiansen, J. 1978. Observations on bristle- and scale-production in *Mallomonas caudata* (Chrysophyceae). *Arch. Protistenkd.* **120**:213–221.
Zablen, L. B., Kissil, M. S., Woese, C. R., and Buetow, D. E. 1975. Phylogenetic origin of the chloroplast and prokaryotic nature of its ribosomal RNA. *Proc. Natl. Acad. Sci. USA* **72**:2418–2422.
Zelitch, I. 1965. The relation of glycolic acid synthesis to the primary photosynthetic carboxylation reaction in leaves. *J. Biol. Chem.* **240**:1869–1876.
Zurzycki, J. 1955. Chloroplast arrangement as a factor in photosynthesis. *Acta Soc. Bot. Pol.* **24**:27–63.

CHAPTER 11

Aist, R., and Wilson, C. L. 1965. Observations on nuclear division in vegetative hyphae of *Ceratocestis fagacearum*. *Proc. Arkansas Acad. Sci.* **19**:32–36.
Aldrich, H. C. 1969. The ultrastructure of mitosis in myxamoebae and plasmodia of *Physarum flavicomum*. *Am. J. Bot.* **56**:290–299.
Ashton, M. L., and Moens, P. B. 1979. Ultrastructure of sporulation in the Hemiascomycetes *Ascoidea corymbosa, A. rubescens, Cephaloascus fragrans,* and *Saccharomycopsis capsularis*. *Can. J. Bot.* **57**:1259–1284.
Bajer, A. S. 1973. Interaction of microtubules and the mechanism of chromosome movement (zipper hypothesis). *Cytobios* **8**:139–160.
Bajer, A. S., and Allen, R. D. 1966a. Role of phragmoplast filaments in cell-plate formation. *J. Cell Sci.* **1**:455–462.
Bajer, A. S., and Allen, R. D. 1966b. Structure and organization of the living mitotic spindle of *Haemanthus* endosperm. *Science* 151:572–574.
Bajer, A. S., and Molé-Bajer, J. 1969. Formation of spindle fibers, kinetochore orientation, and behavior of the nuclear envelope during mitosis in endosperm. *Chromosoma* **27**:448–484.
Bajer, A. S., and Molé-Bajer, J. 1972. Spindle dynamics and chromosome movement. *Int. Rev. Cytol. Suppl.* **3**:1–271.
Barksdale, L., and Kim, K.-S. 1977. *Mycobacterium*. *Bacteriol. Rev.* **41**:217–372.
Bělař, K. 1926. *Formwechsel der Protistenkerne,* Jena, Fischer.
Berger, J. D. 1979. Regulation of macronuclear DNA content in *Paramecium tetraurelia*. *J. Protozool.* **26**:18–28.
Bernhard, W., and de Harven, E. 1960. L'ultrastructure du centriole et d'autres éléments de l'appareil achromatique. *Verh. Int. Kong. Elektronenmikrosk.* **4**:217–227.
Berns, M. W., and Richardson, S. M. 1977. Continuation of mitosis after selective laser microbeam destruction of the centriolar region. *J. Cell Biol.* **75**:977–982.
Berns, M. W., Rattner, J. B., Brenner, S., and Meredith, S. 1977. The role of the centriolar region in animal cell mitosis. *J. Cell Biol.* **72**:351–367.
Blum, J. J., and Padilla, G. M. 1962. Studies on synchronized cells. The time course of DNA, RNA, and protein synthesis in *Astasia longa*. *Exp. Cell Res.* **28**:512–523.
Blum, J. J., Sommer, J. R., and Kahn, V. 1965. Some biochemical, cytological, and morphogene-

tic comparisons between *Astasia longa* and a bleached *Euglena gracilis*. *J. Protozool.* **12**:202–209.

Borisy, G. G. 1978. Polarity of microtubules of the mitotic spindle. *J. Mol. Biol.* **124**:565–570.

Bouligoland, Y., Soyer, N. O., and Pusieux-Dao, S. 1968. La structure fibrillaire et l'orientation des chromosomes chez les dinoflagellées. *Chromosoma* **24**:251–287.

Bradfield, J. R. G. 1956. Bacterial anatomy. *Symp. Soc. Gen. Microbiol.* **6**:296–317.

Brenner, S., Branch, A., Meredith, S., and Berns, M. W. 1977. The absence of centrioles from spindle poles of rat kangaroo (PtK_2) cells undergoing meiotic-like reduction division *in vitro*. *J. Cell Biol.* **72**:368–379.

Brinkley, B. R., and Cartwright, J. 1975. Cold-labile and cold-stable microtubules in the mitotic spindles of mammalian cells. *Ann. N.Y. Acad. Sci.* **253**:428–439.

Brinkley, B. R., Stubblefield, E., and Hsu, T. C. 1967. The effects of colcemid inhibition and reversal on the fine structure of the mitotic apparatus of Chinese hamster cells *in vitro J. Ultrastruct. Res.* **19**:1–18.

Brown, R. M. 1969. Observations on the relationships of the Golgi apparatus to wall formation in the marine chrysophycean alga *Pleurochrysis scherffelii*. *J. Cell Biol.* **41**:109–123.

Burgdorfer, W., Brinton, L. P., and Hughes, L. E. 1973. Isolation and characterization of symbiotes from the Rocky Mountain wood tick, *Dermacentor andersoni*. *J. Invertebr. Pathol.* **22**:424–434.

Burgess, J., and Northcote, D. H. 1967. A function of the preprophase band of microtubules in *Phleum pratense*. *Planta* **75**:319–326.

Byers, B., and Goetsch, L. 1976. A highly ordered ring of membrane-associated filaments in budding yeast. *J. Cell Biol.* **69**:717–721.

Cande, W. Z., Snyder, J., Smith, D., Summers, K., and MacIntosh, J. R. 1974. A functional mitotic spindle prepared from mammalian cells in culture. *Proc. Natl. Acad. Sci. USA* **71**:1559–1563.

Carasso, N., and Favard, P. 1965. Microtubules fusoriaux dans les micro- et macronucleus de ciliés péritriches en division. *J. Microsc.* (*Paris*) **4**:395–402.

Cesana, D. 1971. Le mécanisme de la mitose gamogonique chez le forminifère *Iridia lucida*. *C. R. Acad. Sci. Paris Ser. D* **272**:3057–3060.

Chatton, E. 1920. Les Péridiniens parasitiques: Morphologie, reproduction, éthologie. *Arch. Zool. Exp. Gen.* **59**:1–475.

Cholnoky, B. von 1932. Vergleichende Studien über Kern- und Zellteilung der Fadenbildenden Conjugaten. *Arch. Protistenkd.* **78**:522–542.

Cleveland, L. R. 1953. Studies on chromosomes and nuclear division. IV. Photomicrographs of living cells during meiotic divisions. *Trans. Am. Philos. Soc.* **43**:848–869.

Cleveland, L. R. 1954. Autogamy and endomitosis in *Barbulanympha* resulting from interruption of haploid gametogenesis. *J. Morphol.* **95**:189–212.

Cleveland, L. R. 1955. Unusual behavior of gametes and centrioles of *Barbulanympha*. *J. Morphol.* **97**:511–542.

Cleveland, L. R. 1958. A factual analysis of chromosomal movements in *Barbulanympha*. *J. Protozool.* **5**:47–62.

Cleveland, L. R. 1963. Functions of flagellate and other centrioles in cell reproduction. *In:* Levine, L., ed., *The Cell in Mitosis*, New York, Academic Press, pp. 3–52.

Colwell, R. R., Smith, E. J., and Chapman, G. B. 1968. Properties of a D-quinovosamine-producing *Achromobacter*. *Can. J. Microbiol.* **14**:165–171.

Copeland, H. F. 1956. *Classification of the Lower Organisms*, Palo Alto, Calif., Pacific Books.

Cota-Robles, E. H. 1966. Internal membranes in cells of *E. coli*. *J. Ultrastruct. Res.* **16**:626–639.

Czurda, V. 1922. Zur Cytologie der Gattung *Spirogyra*. *Arch. Protistenkd.* **45**:163–199.

Danforth, H. D., and Hammond, D. M. 1972. Merogony in multinucleate merozoites of *Eimeria magna* Per. *J. Protozool.* **19**:454–457.

Daniels, E. W., and Roth, L. E. 1964. Electron microscopy of mitosis in a radiosensitive giant amoeba. *J. Cell Biol.* **20**:75–84.

Daniels, E. W., Breyer, E. P., and Kudo, R. R. 1966. *Pelomyxa palustris* Greeff. II. Its ultrastructure. *Z. Zellforsch. Mikrosk. Anat.* **73**:367–383.

de Harven, E. 1968. The centriole and the mitotic spindle. *In:* Dalton, A. J., and Haguenau, F., eds., *The Nucleus*, New York, Academic Press, pp. 197–227.

de Harven, E., and Bernhard, W. 1956. Étude au microscope électronique de l'ultrastructure du centriole chez les vertébrés. *Zellforsch. Mikrosk. Anat.* **45**:378–398.

Dillon, L. S. 1962. Comparative cytology and the evolution of life. *Evolution* **16**:102–117.

Dillon, L. S. 1963. A reclassification of the major groups of organisms based upon comparative cytology. *Syst. Zool.* **12**:71–83.

Dillon, L. S. 1978. *The Genetic Mechanism and the Origin of Life,* New York, Plenum Press.

Dodge, J. D. 1963a. Chromosome structure in the Dinophyceae. I. The spiral chromonema. *Arch. Mikrobiol.* **45**:46–57.

Dodge, J. D. 1963b. The nucleus and nuclear division in the Dinophyceae. *Arch. Protistenkd.* **106**:442–452.

Dodge, J. D. 1964. Chromosome structure in the Dinophyceae. II. Cytochemical studies. *Arch. Mikrobiol.* **48**:66–80.

Dodson, E. O. 1971. The kingdoms of organisms. *Syst. Zool.* **20**:265–281.

Doerder, F. P. 1979. Regulation of macronuclear DNA content in *Tetrahymena thermophila. J. Protozool.* **26**:28–35.

Doonan, B. B., and Jensen, T. E. 1977. Ultrastructural localization of alkaline phosphatase in the blue-green bacterium *Plectonema boryanum. J. Bacteriol.* **132**:967–973.

Drum, R. W., and Pankratz, H. S. 1963. Fine structure of a diatom centrosome. *Science* **142**:61–63.

Duda, J. J., and Slack, J. M. 1972. Ultrastructural studies on the genus *Actinomyces. J. Gen. Microbiol.* **71**:63–68.

DuPraw, E. J. 1965. The ultrastructure of human chromosomes. *Am. Zool.* **5**:648.

DuPraw, E. J. 1966. Evidence for a folded fibre organization in human chromosomes. *Nature (London)* **209**:577–581.

Echlin, P. 1964. Intra-cytoplasmic membranous inclusions in the blue-green alga, *Anacystis nidulans. Arch. Mikrobiol.* **49**:267–274.

Edmunds, L. N. 1964. Replication of DNA and cell division in synchronously dividing cultures of *Euglena gracilis. Science* **145**:266–268.

Floyd, G. L., Stewart, K. D., and Mattox, K. R. 1972. Cellular organization, mitosis and cytokinesis in the Ulotrichalean alga, *Klebsormidium. J. Phycol.* **8**:176–184.

Forer, A. 1966. Characterization of the mitotic traction system, and evidence that birefringent spindle fibers neither produce nor transmit force for chromosome movement. *Chromosoma* **19**:44–98.

Forer, A. 1976. Actin filaments and birefringent spindle fibers during chromosome movements. *In:* Goldman, R., Pollard, T., and Rosenbaum, J., eds., *Cell Motility,* Cold Spring Harbor, N.Y., Cold Spring Harbor Laboratory, Book C, pp. 1273–1293.

Forer, A., and Jackson, W. T. 1976. Actin filaments in the endosperm mitotic spindles of the higher plant *Haemanthus katherinae* Baker. *Cytobiologie Z. Exp. Zellforsch.* **12**:199–214.

Fowke, L. C., and Pickett-Heaps, J. D. 1969a. Cell division in *Spirogyra.* I. Mitosis. *J. Phycol.* **5**:240–259.

Fowke, L. C., and Pickett-Heaps, J. D. 1969b. Cell division in *Spirogyra.* II. Cytokinesis. *J. Phycol.* **5**:273–281.

Fowke, L. C., and Pickett-Heaps, J. D. 1978. Electron microscope study of vegetative cell division in two species of *Marchantia. Can. J. Bot.* **56**:467–475.

Frankel, J. 1967a. An analysis of the mechanism of resorption of developing oral structures. *J. Exp. Zool.* **164**:435–460.

Frankel, J. 1967b. The relationship of protein synthesis to cell division and oral organelle development. *J. Cell Biol.* **34**:841–858.

Frey-Wyssling, A., Lopéz-Sáez, J. F., and Mühlethaler, K. 1964. Formation and development of the cell plate. *J. Ultrastruct. Res.* **10**:422–432.

Fuge, H. 1974. The arrangement of microtubules and the attachment of chromosomes to the spindle during anaphase in tipulid spermatocytes. *Chromosoma* **45**:245–260.

Fulton, C., and Dingle, A. D. 1971. Basal bodies but not centrioles in *Naegleria*. *J. Cell Biol.* **51**:826–836.

Fuseler, J. W. 1975. Mitosis in *Tilia americana* endosperm. *J. Cell Biol.* **64**:159–171.

Gall, J. G. 1959. Macronuclear duplication in the ciliated protozoan *Euplotes*. *J. Biophys. Biochem. Cytol.* **5**:295–308.

Giese, A. C. 1973. *Blepharisma: The Biology of a Light-Sensitive Protozoan*, Stanford, Calif., Stanford University Press.

Girbardt, M. 1978. Historical review and introduction. *In*: Heath, I. B., ed., *Nuclear Division in the Fungi*, New York, Academic Press, pp. 1–20.

Glauert, A. M., Brieger, E. M., and Allen, J. M. 1961. The fine structure of vegetative cells of *Bacillus subtilis*. *Exp. Cell Res.* **22**:73–85.

Gould, R. R., and Borisy, G. G. 1978. Quantitative initiation of microtubule assembly by chromosomes from Chinese hamster ovary cells. *Exp. Cell Res.* **113**:369–374.

Green, F., and Hanson, R. P. 1973. Ultrastructure and capsule of *Mycoplasma meleagridis*. *J. Bacteriol.* **166**:1011–1018.

Grimes, G. W. 1972. Cortical structure in nondividing and cortical morphogenesis in dividing *Oxytricha fallax*. *J. Protozool.* **19**:428–445.

Grula, E. A., and Smith, G. L. 1965. Cell division in a species of *Erwinia*. IX. Electron microscopy of normally dividing cells. *J. Bacteriol.* **90**:1054–1058.

Grula, E. A., Smith, G. L., and Grula, M. M. 1968. Cell division in a species of *Erwinia*. X. Morphology of the nuclear body in filaments produced by growth in the presence of D-serine. *Can. J. Microbiol.* **14**:293–298.

Hall, R. P. 1925a. Binary fission in *Oxyrrhis marina* Dujardin. *Univ. Calif. Publ. Zool.* **26**: 281–324.

Hall, R. P. 1925b. Mitosis in *Ceratium hirudinella* O.F.M., with notes on nuclear phenomena in encysted forms and the question of sexual reproduction. *Univ. Calif. Publ. Zool.* **28**: 29–64.

Hammond, D. M. 1977. Life cycles and development of Coccidia. *In*: Hammond, D. M., and Long, P. L., eds., *The Coccidia,* Baltimore, University Park Press, pp. 45–79.

Hard, G. C. 1969. Electron microscopic examination of *Corynebacterium ovis*. *J. Bacteriol.* **97**:1480–1485.

Hayes, S. F., and Burgdorfer, W. 1979. Ultrastructure of *Rickettsia rhipicephali,* a new member of the spotted fever group rickettsiae in tissue of the host vector *Rhipicephalus sanguinea*. *J. Bacteriol.* **137**:605–613.

Heath, I. B. 1978. Experimental studies of mitosis in the fungi. *In*: Heath, I. B., ed., *Nuclear Division in the Fungi,* New York, Academic Press, pp. 89–176.

Heath, I. B., and Greenwood, A. D. 1970. Centriole replication and nuclear division in *Saprolegnia*. *J. Gen. Microbiol.* **62**:139–148.

Hepler, P. K., and Newcomb, E. H. 1967. Fine structure of cell plate formation in the apical meristem of *Phaseolus* roots. *J. Ultrastruct. Res.* **19**:498–513.

Herman, I. M., and Pollard, T. D. 1979. Comparison of purified antiactin and fluorescent heavy-meromyosin staining patterns in dividing cells. *J. Cell Biol.* **80**:509–520.

Heywood, P. 1978. Ultrastructure of mitosis in chloromonadophycean alga, *Vacuolaria virescens*. *J. Cell Sci.* **31**:37–51.

Heywood, P., and Godward, M. B. E. 1974. Mitosis in the alga *Vacuolaria virescens*. *Am. J. Bot.* **61**:331–338.

Hirumi, H., and Maramorosch, K. 1973. Ultrastructure of the aster yellows agent: Mycoplasma-like bodies in sieve tube elements of *Nicotiana rustica*. *Ann. N.Y. Acad. Sci.* **225**:201–222.

Hofstod, T., and Selvig, K. A. 1969. Ultrastructure of *Leptotrichia buccalis*. *J. Gen. Microbiol.* **56**:23–26.

Hollande, A., and Carruette-Valentin, J. 1971. Les atractophores, l'induction du fuseau et la division cellulaire chez les Hypermastigines. *Protistologica* **7**:5–100.

Hopwood, D. A., and Glauert, A. M. 1960. The fine structure of the nuclear material of a blue-green alga, *Anabaena cylindrica* Lemm. *J. Biophys. Biochem. Cytol.* **8**:813–823.

Inaba, F., and Sotokawa, Y. 1968. Electron-microscopic observation on nuclear events during binary fission in *Blepharisma wardsi*. *Jpn. J. Genet.* **43**:335–348.

Inoué, S., and Ritter, H. 1978. Mitosis in *Barbulanympha*. II. Dynamics of a two-stage anaphase nuclear morphogenesis and cytokinesis. *J. Cell Biol.* **77**:655–684.

Inoué, S., and Sato, H. 1967. Cell motility by labile association of molecules: The nature of the mitotic spindle fibers and their role in chromosome movement. *J. Gen. Physiol.* **50**:259–292.

Ivanić, M. 1936. Die Kernteilung bei *Opalina ranarum* and *O. obtrigona*. *Arch. Protistenkd.* **87**:172–184.

Jaeckel-Williams, R. 1978. Nuclear divisions with reduced numbers of microtubules in *Tetrahymena*. *J. Cell Sci.* **34**:303–319.

Jenkins, R. A. 1967. Fine structure of division in ciliate protozoans. I. Micronuclear mitosis in *Blepharisma*. *J. Cell Biol.* **34**:463–481.

Jenkins, R. A. 1977. The role of microtubules in macronuclear division of *Blepharisma*. *J. Protozool.* **24**:264–275.

Jensen, T. E. 1979. Occurrence of trilamellar bodies in isolates of *Nostoc*. *J. Phycol.* **15**:112–114.

Jensen, T. E., and Ayala, R. P. 1976. The fine structure of a microplate–microtubule array, microfilaments and polyhedral body-associated microtubules in several species of *Anabaena*. *Arch. Microbiol.* **111**:1–6.

Jensen, T. E., and Ayala, R. P. 1977. Fine structure of wall bodies in large cells of an *Anabaena* species. *Cytobios* **18**:79–88.

Jensen, T. E., and Sicko, L. M. 1974. Phosphate metabolism in blue-green algae. *Can. J. Microbiol.* **20**:1235–1239.

Johnson, U. G., and Porter, K. R. 1968. Fine structure of cell division in *Chlamydomonas reinhardii*. *J. Cell Biol.* **38**:403–425.

Joyon, L. 1963. Contribution a l'étude cytologique de quelques protozoaires flagellés. *Ann. Fac. Sci. Univ. Clermont* **22**:1–96.

Jurand, A. 1976. Some ultrastructural features of micronuclei during conjugation and autogamy in *Paramecium aurelia*. *J. Gen. Microbiol.* **94**:193–203.

Jurand, A., and Bomford, R. 1965. The fine structure of the parasitic suctorian *Podophyra parameciorum*. *J. Microsc. (Paris)* **4**:509–522.

Jurand, A., and Selman, G. G. 1970. Ultrastructure of the nuclei and intranuclear microtubules of *Paramecium aurelia*. *J. Gen. Microbiol.* **60**:357–364.

Karling, J. S. 1926. Nuclear and cell division in *Nitella* and *Chara*. *Bull. Torrey Bot. Club* **53**:319–379.

Kaufman, P. B., Yocum, C. S., and Baic, D. 1970. Ultrastructural studies on stomatal development in internodes of *Avena sativa*. *Am. J. Bot.* **57**:33–49.

Kazama, F. Y. 1974. The ultrastructure of nuclear division in *Thraustochytrium*. *Protoplasma* **82**:155–175.

Kessel, M., MacColl, R., Berns, D. S., and Edwards, M. R. 1973. Electron microscope and physical chemical characterization of C-phycocyanin from fresh extracts of two blue-green algae. *Can. J. Microbiol.* **19**:831–836.

Kimball, R. F. 1953. The structure of the macronucleus of *Paramecium aurelia*. *Proc. Natl. Acad. Sci. USA* **39**:345–347.

Kloetzel, J. A. 1970. Compartmentalization of the developing macronucleus following conjugation in *Stylonychia* and *Euplotes*. *J. Cell Biol.* **47**:395–407.

Kofoid, C. A., and Swezey, O. 1925. On the number of chromosomes and type of mitosis in *Endamoeba dysenteriae*. *Univ. Calif. Publ. Zool.* **26**:331–352.

Kubai, D. F. 1973. Unorthodox mitosis in *Trichonympha agilis:* Kinetochore differentiation and chromosome movement. *J. Cell Sci.* **13**:511–552.

Kubai, D. F. 1975. The evolution of the mitotic spindle. *Int. Rev. Cytol.* **43**:167–227.

Kubai, D. F., and Ris, H. 1969. Division in the dinoflagellate *Gyrodinium cohnii* (Schiller). *J. Cell Biol.* **40**:508–528.

Laane, M. M., Wahlstrom, R., and Mellem, T. R. 1977. Scanning electron microscopy of nuclear division stages in *Vicia faba* and *Haemanthus cinnabarinus*. *Hereditas* **86**:171–178.

Laishley, E. J., MacAlister, T. J., Clements, I., and Young, C. 1973. Isolation and morphology of native intracellular polyglucose granules from *Clostridium pasteurianum*. *Can. J. Microbiol.* **19**:991–994.

Lamont, H. C. 1969. Shear-oriented microfibrils in the mucilaginous investments of two motile oscillatoriacean blue-green algae. *J. Bacteriol.* **97**:350–361.

Lang, N. J. 1965. Electron microscopic study of heterocyst development in *Anabaena azollae* Strasburger. *J. Phycol.* **1**:127–134.

Lawry, N. H., and Jensen, T. E. 1979. Deposition of condensed phosphate as an effect of varying sulfur deficiency in the cyanobacterium *Synechococcus sp.* (*Anacystis nidulans*). *Arch. Microbiol.* **120**:1–7.

Leadbeater, B., and Dodge, J. D. 1967. An electron microscope study of nuclear and cell differentiation in a dinoflagellate. *Arch. Mikrobiol.* **57**:239–254.

Leak, L. V., and Wilson, G. B. 1965. Electron microscopic observations on a blue-green alga, *Anabaena* sp. *Can. J. Genet. Cytol.* **7**:237–249.

Leedale, G. F. 1958. Nuclear structure and mitosis in the Euglenineae. *Arch. Mikrobiol.* **32**:32–64.

Leedale, G. F. 1959a. Formation of anucleate cells of *Euglena gracilis* by miscleavage. *J. Protozool. Suppl.* **6**:26.

Leedale, G. F. 1959b. Amitosis in three species of *Euglena*. *Cytologia* **24**:213–219.

Leedale, G. F. 1967. *Euglenoid Flagellates,* Englewood Cliffs, N.J., Prentice–Hall.

Lewin, R. A., and Meinhart, J. O. 1953. Studies on the flagella of algae. III. Electron micrographs of *Chlamydomonas moewusonii*. *Can. J. Bot.* **31**:711–717.

Littlefield, L. J., and Heath, M. C. 1979. *Ultrastructure of Rust Fungi,* New York, Academic Press.

Løvlie, A., and Bråten, T. 1968. On the division of the cytoplasm and chloroplast in the multicellular green-alga *Ulva mutabilis* Føyn. *Exp. Cell Res.* **51**:211–220.

Lu, B. C. 1978. Meiosis in *Coprinis*. VIII. A time-course study in the fusion and division of the spindle pole body during meiosis. *J. Cell Biol.* **76**:761–766.

McAllister, F. 1931. The formation of the achromatic figure in *Spirogyra setiformis*. *Am. J. Bot.* **18**:838–853.

McCandless, R. G., Cohen, M., Kalmanson, G. M., and Guze, L. B. 1968. Cores, microbial organelles possibly specific to group D streptococci. *J. Bacteriol.* **96**:1400–1412.

McIntosh, J. R., and Landis, S. C. 1971. The distribution of spindle microtubules during mitosis in cultured human cells. *J. Cell Biol.* **49**:468–497.

McIntosh, J. R., Hepler, P. K., and Van Wie, D. G. 1969. Model for mitosis. *Nature* (*London*) **224**:659–663.

McIntosh, J. R., Cande, W. Z., and Snyder, J. A. 1975. Structure and physiology of the mammalian mitotic spindle. *In*: Inoué, S., and Stephens, R. E., eds., *Molecules and Cell Movement,* New York, Raven Press, pp. 31–76.

McIntosh, J. R., Cande, W. Z., Lazarides, E., McDonald, K., and Snyder, J. A. 1976. Fibrous elements of the mitotic spindle. *In*: Goldman, R., Pollard, T., and Rosenbaum, J., eds., *Cell Motility*, Cold Spring Harbor, N.Y., Cold Spring Harbor Laboratory, Book C, pp. 1261–1272.

McIntosh, J. R., Sisken, J. E., and Chu, L. K. 1979. Structural studies on mitotic spindles isolated from cultured human cells. *J. Ultrastruct. Res.* **66**:40–57.

Maier, S., and Murray, R. G. E. 1965. The fine structure of *Thioploca ingrica* and a comparison with *Beggiatoa*. *Can. J. Microbiol.* **11**:645–655.

Margulis, L. 1970. *Origin of Eukaryotic Cells,* New Haven, Conn., Yale University Press.

Markey, D. R., and Wilce, R. T. 1975. The ultrastructure of reproduction in the brown alga *Pylaiella littoralis*. *Protoplasma* **85**:219–241.

Mason, D. J., and Powelson, D. M. 1956. Nuclear division as observed in live bacteria by a new technique. *J. Bacteriol.* **71**:474–479.

Mattox, K. R., and Stewart, K. D. 1974. A comparative study of cell division in *Trichosarcina polymorpha* and *Pseudoendoclonium basiliense* (Chlorophyceae). *J. Phycol.* **10**:447–456.

Mazia, D., Harris, P. J., and Bibring, T. 1960. The multiplicity of the mitotic centers and the time course of their duplication and separation. *J. Biophys. Biochem. Cytol.* **7**:1–20.

Merriam, M. L. 1906. Nuclear division in *Zygnema*. *Bot. Gaz. (Chicago)* **41**:43–53.

Miyake, A., Tulli, M., and Nobili, R. 1979. Requirement of protein synthesis in the initiation of meiosis and other nuclear changes in conjugation of *Blepharisma*. *Exp. Cell Res.* **120**:87–93.

Moens, P. B. 1976. Spindle and kinetochore morphology of *Dictyostelium discoideum*. *J. Cell Biol.* **68**:113–122.

Molnar, K. E., Stewart, K. D., and Mattox, K. R. 1975. Cell division in the filamentous *Pleurastrum* and its comparison with the unicellular *Platymonas* (Chlorophyceae). *J. Phycol.* **11**:287–296.

Morita, R. Y., and Stave, P. W. 1963. Electron micrograph of an ultrathin section of *Beggiatoa*. *J. Bacteriol.* **85**:940–942.

Mullins, J. M., and Biesele, J. J. 1977. Terminal phase of cytokinesis in D-98S cells. *J. Cell Biol.* **73**:672–684.

Mundkur, B. 1960. Electron microscopical studies of frozen-dried yeast. I. Localization of polysaccharides. *Exp. Cell Res.* **20**:28–42.

Mundkur, B. 1961a. Electron microscopical studies of frozen-dried yeast. II. The nature of basophile particles and vesicular nuclei in *Saccharomyces*. *Exp. Cell Res.* **25**:1–23.

Mundkur, B. 1961b. Electron microscopical studies of frozen-dried yeast. III. Formation of the tetrad in *Saccharomyces*. *Exp. Cell Res.* **25**:24–40.

Nanney, D. L., and Preparata, R. M. 1979. Genetic evidence concerning the structure of the *Tetrahymena thermophila* macronucleus. *J. Protozool.* **26**:2–9.

Nanney, D. L., and Rudzinska, M. A. 1960. Protozoa. *In*: Brachet, J., and Mirsky, A. E., eds., *The Cell: Biochemistry, Physiology, Morphology,* New York, Academic Press, Vol. 4, pp. 109–150.

Nicklas, R. B. 1971. Mitosis. *In*: Prescott, D. M., Goldstein, L., and McConkey, E., eds., *Advances in Cell Biology, II,* New York, Appleton–Century–Crofts, pp. 225–297.

Novitski, C., and Bajer, A. S. 1977. Interaction of microtubules and the mechanism of chromosome movement (zipper hypothesis). *Cytobios* **18**:173–182.

Oakley, B. R., and Dodge, J. D. 1976. Mitosis and cytokinesis in the dinoflagellate *Amphidinium carteri*. *Cytobios* **17**:35–46.

Oakley, B. R., and Heath, I. B. 1978. The arrangement of microtubules in serially sectioned spindles of the alga *Cryptomonas*. *J. Cell Sci.* **31**:53–70.

Oura, G. 1953. On the mitosis of the *Spirogyra,* with special reference to the nuclear organization. *Cytologia* **18**:297–304.

Pankratz, H. S., and Bowen, C. C. 1963. Cytology of blue-green algae. I. The cells of *Symploca muscorum*. *Am. J. Bot.* **50**:387–399.

Pasternak, J. 1967. Differential genic activity in *Paramecium aurelia*. *J. Exp. Zool.* **165**:395–418.
Pearson, B. R., and Norris, R. E. 1975. Fine structure of cell division in *Pyramimonas parkeae* Norris and Pearson (Chlorophyta; Prasinophyceae). *J. Phycol.* **11**:113–124.
Peterschilka, F. 1922. Kernteilung und Pyrenoidvermehrung bei *Mougeotia*. *Arch. Protistenkd.* **45**:153–162.
Pickett-Heaps, J. D. 1967. Ultrastructure and differentiation in *Chara*. II. Mitosis. *Aust. J. Biol. Sci.* **20**:883–894.
Pickett-Heaps, J. D. 1969. Preprophase microtubules and stomatal differentiation in *Commelina cyanea*. *Aust. J. Biol. Sci.* **22**:375–391.
Pickett-Heaps, J. D., and Northcote, D. H. 1966. Organization of microtubules and endoplasmic reticulum during mitosis and cytokinesis in wheat meristems. *J. Cell Sci.* **1**:109–120.
Pitelka, D. R., and Schooley, S. N. 1955. Comparative morphology of some protistan flagella. *Univ. Calif. Publ. Zool.* **61**:79–128.
Rafalko, J. S. 1947. Cytological observations on the amoebo-flagellate, *Naegleria gruberi*. *J. Morphol.* **81**:1–44.
Rafalko, J. S. 1951. Mitotic division in the amoebo-flagellate, *Tetramitus rostratus*. *J. Morphol.* **89**:71–85.
Rao, M. V. N. 1966. Conjugation in *Kahlia* sp. with special reference to meiosis and endomitosis. *J. Protozool.* **13**:565–573.
Rattner, J. B., and Berns, M. W. 1976. Distribution of microtubules during centriole separation in rat kangaroo (*Potorous*) cells. *Cytobios* **15**:37–43.
Rieder, C. L. 1979. Ribonucleoprotein staining of centrioles and kinetochores in newt lung cell spindles. *J. Cell Biol.* **80**:1–9.
Ris, H., and Kubai, D. F. 1974. An unusual mitotic mechanism in the parasitic protozoan *Syndinium* sp. *J. Cell Biol.* **60**:702–720.
Ritter, H., Inoué, S., and Kubai, D. 1978. Mitosis in *Barbulanympha*. I. Spindle function formation, and kinetochore engagement. *J. Cell Biol.* **77**:638–654.
Rizzo, P. J., and Noodén, L. D. 1972. Chromosomal proteins in the dinoflagellate alga *Gyrodinium cohnii*. *Science* **176**:796–797.
Rizzo, P. J., and Noodén, L. D. 1973. Isolation and chemical composition of dinoflagellate nuclei. *J. Protozool.* **20**:666–673.
Robinow, C. F. 1961. Mitosis in the yeast *Lipomyces lipofer*. *J. Biophys. Biochem. Cytol.* **9**:879–892.
Robinow, C. F., and Marak, J. 1966. A fiber apparatus in the nucleus of the yeast cell. *J. Cell Biol.* **29**:129–151.
Roth, L. E., and Daniels, E. W. 1962. Electron microscopic studies of mitosis in amebae. II. The giant ameba, *Pelomyxa carolinensis*. *J. Cell Biol.* **12**:57–78.
Roth, L. E., Wilson, H. J., and Chakroborty, J. 1966. Anaphase structure in mitotic cells typified by spindle elongation. *J. Ultrastruct. Res.* **14**:460–483.
Ruffolo, J.J. 1978. Fine structure of the macronucleus during the cell division cycle of *Euplotes eurystomus,* a ciliate protozoan. *J. Morphol.* **157**:211–222.
Ryter, A., and Kellenberger, E. 1958. Étude au microscope électronique de plasmas contenant de l'acide désoxyribonucléique. *Z. Naturforsch.* **136**:597–605.
Sanger, J. W. 1975. Changing patterns of actin localization during cell division. *Proc. Natl. Acad. Sci. USA* **72**:1913–1916.
Sanger, J. W., and Sanger, J. M. 1976. Actin localization during cell division. *In*: Goldman, R., Pollard, T., and Rosenbaum, J., eds., *Cell Motility,* Cold Spring Harbor, N.Y., Cold Spring Harbor Laboratory, Book C, pp. 1295–1316.
Sawyer, H. R., and Jenkins, R. A. 1977. Stomatogenic events accompanying binary fission in *Blepharisma*. *J. Protozool.* **24**:140–149.
Schroeder, T. 1973. Actin in dividing cells: Contractile ring filaments bind heavy meromyosin. *Proc. Natl. Acad. Sci. USA* **70**:1688–1692.

Schulz-Baldes, M., and Lewin, R. A. 1976. Fine structure of *Synechocystis didemni*. *Phycologia* **15**:1–6.

Schwab, D. 1968. Nachweis von Centriolen bei *Myxotheca*. *Naturwissenschaften* **2**:88–89.

Schwab, D. 1969. Elektronenmikroskopische Untersuchung an der Foraminifere *Myxotheca arenilega* Schaudinn. *Z. Zellforsch. Mikrosk. Anat.* **96**:295–324.

Schwab, D. 1972. Electron microscopic studies on the foraminifer *Allogromia laticollaris* Arnold mitosis in agametes. *Protoplasma* **75**:79–89.

Sherline, P., and Schiavone, K. 1978. High molecular weight MAPs are part of the mitotic spindle. *J. Cell Biol.* **77**:R9–R12.

Simpson, G. G. 1945. The principles of classification and a classification of the mammals. *Bull. Am. Mus. Nat. Hist.* **85**:1–350.

Singh, A. P., Shaw, M., and Hollins, G. 1977. Preprophase bands of microtubules in developing complexes of sugarcane. *Cytologia* **42**:611–620.

Skoczylas, O. 1958. Über die Mitose von *Ceratium cornutum* und einigen anderen Peridineen. *Arch. Protistenkd.* **103**:190–228.

Sonneborn, T. M. 1938. Sex behaviour, sex determination, and the inheritance of sex in fission and conjugation in *Paramecium aurelia*. *Genetics* **23**:168–169.

Sonneborn, T. M. 1946. Inert nuclei: Inactivity of micronuclear genes in variety 4 of *Paramecium aurelia*. *Genetics* **31**:231.

Sonneborn, T. M. 1947. Recent advances in the genetics of *Paramecium* and *Euplotes*. *Adv. Genet.* **1**:263–358.

Soyer, M. 1971. Structure du noyau des *Blastodinium:* Division et condensation chromatique. *Chromosoma* **33**:70–114.

Soyer, M., and Haapala, O. K. 1973. Filaments extrachromosomiques: Variations et relations avec l'envelope nucléaire pendant la division chez les Dinoflagellés. *J. Microsc. (Paris)* **18**:267–270.

Soyer, M., and Haapala, O. K. 1974. Division and function of dinoflagellate chromosomes. *J. Microsc. (Paris)* **19**:137–146.

Spector, D. L., and Triemer, R. E., 1979. Ultrastructure of the dinoflagellate *Peridinium cinctum* f. *ovoplanum*. I. Vegetative cell ultrastructure. *Am. J. Bot.* **66**:845–850.

Srivastava, L. M., and Singh, A. P. 1972. Stomatal structure in corn leaves. *J. Ultrastruct. Res.* **39**:345–363.

Stenman, S., Wartiovaara, J., and Vaheri, A. 1977. Changes in the distribution of a major fibroblast protein, fibronectin, during mitosis and interphase. *J. Cell Biol.* **74**:453–467.

Stewart, K. D., Mattox, K. R., and Floyd, G. L. 1973. Mitosis, cytokinesis, the distribution of plasmodesmata, and other cytological characteristics in the Ulotrichales, Ulvales, and Chaetophorales: Phylogenetic and taxonomic considerations. *J. Phycol.* **9**:128–141.

Stewart, K. D., Mattox, K. R., and Chandler, C. D. 1974. Mitosis and cytokinesis in *Platymonas subcordiformis,* a scaly green monad. *J. Phycol.* **10**:65–79.

Strohl, W. R., and Larkin, J. M. 1978a. Cell division and trichome breakage in *Beggiatoa*. *Curr. Microbiol.* **1**:151–155.

Strohl, W. R., and Larkin, J. M. 1978b. Enumeration, isolation, and characterization of *Beggiatoa* from fresh water sediments. *Appl. Environ. Microbiol.* **36**:755–770.

Stump, A. B. 1943. Mitosis and cell division in *Pontigulasia vas*. *J. Elisha Mitchell Sci. Soc.* **59**:14–22.

Stump, A. B. 1959. Mitosis in the rhizopod *Lesquereuxia spiralis*. *J. Protozool.* **6**:185–189.

Szöllösi, D., Calarco, P., and Donahue, R. P. 1972. Absence of centrioles in the first and second meiotic spindles of mouse oocytes. *J. Cell Sci.* **11**:521–541.

Tartar, V. 1966. Fission after division primordium removal in the ciliate *Stentor coeruleus* and comparable experiments on reorganizers. *Exp. Cell Res.* **42**:357–370.

Taylor, J. W., and Wells, K. 1979. A light and electron microscopic study of mitosis in *Bullera alba* and the histochemistry of some cytoplasmic substances. *Protoplasma* **98**:31–62.

Thyagarajan, T. R., Conti, S. F., and Naylor, H. B. 1962. Electron microscopy of *Rhodotorula glutinis*. *J. Bacteriol.* **83**:381–394.

Tippit, D. H., and Pickett-Heaps, J. D. 1976. Apparent amitosis in the binucleate dinoflagellate *Peridinium balticum*. *J. Cell Sci.* **21**:273–289.

Tippit, D. H., and Pickett-Heaps, J. D. 1977. Mitosis in the pennate diatom *Surviella ovalis*. *J. Cell Biol.* **73**:705–727.

Tippit, D. H., Schulz, D., and Pickett-Heaps, J. D. 1978. Analysis of the distribution of spindle microtubules in the diatom *Fragilaria*. *J. Cell Biol.* **79**:737–763.

Tomasz, A., Jamieson, J. D., and Ottolenghi, E. 1964. The fine structure of *Diplococcus pneumoniae*. *J. Cell Biol.* **22**:453–467.

Tudor, J., Hashimoto, T., and Conti, S. F. 1969. Presence of nuclear bodies in some minicells of *E. coli*. *J. Bacteriol.* **98**:298–299.

Völker, H., Schweisfurth, R., and Hirsch, P. 1977. Morphology and ultrastructure of *Crenothrix polyspora* Cohn. *J. Bacteriol.* **131**:306–313.

Wada, B. 1966. Analysis of mitosis. *Cytologia* **30**(suppl.):1–158.

Wada, B. 1972. Mitotic cell studies based on *in vivo* observations. V. Evolutional aspect of the degeneration of centrioles and the development of cell walls in the plant kingdom. *Cytologia* **37**:457–468.

Wada, B. 1976. Mitotic studies based on *in vivo* observations. VIII. The evolution of mitotic spindles in eukaryotes: A negation of the breakdown of the nuclear membrane. *Cytologia* **41**:153–175.

Wenrich, D. H. 1924. Studies on *Euglenomorpha hegneri* n.g., n. sp., a euglenoid flagellate found in tadpoles. *Biol. Bull.* **47**:149–175.

Wenrich, D. H. 1940. Nuclear structure and nuclear division in the trophic stage of *Entamoeba muris*. *J. Morphol.* **66**:215–239.

Whaley, W. G., and Mollenhauer, H. H. 1963. The Golgi apparatus and cell plate formation—A postulate. *J. Cell Biol.* **17**:216–221.

Whittaker, R. H. 1969. New concepts of kingdoms of organisms. *Science* **163**:150–160.

Witt, P. L. 1977. Unequal distribution of DNA in the macronuclear division of the ciliate *Euplotes eurystomas*. *Chromosoma* **60**:59–67.

Wolfe, J. 1967. Structural aspects of amitosis: A light and electron microscope study of the isolated macronucleus of *Paramecium aurelia* and *Tetrahymena pyriformis*. *Chromosoma* **23**:59–79.

Woodard, J., Gelber, B., and Swift, H. 1961. Nucleoprotein changes during the mitotic cycle in *Paramecium aurelia*. *Exp. Cell Res.* **23**:258–264.

Woodard, J., Woodard, M., Gelber, B., and Swift, H. 1966. Cytochemical studies of conjugation in *Paramecium aurelia*. *Exp. Cell Res.* **41**:55–63.

Yamada, E. 1958. Some observations on the fine structure of centrioles in the mitotic cell. *Kurume Med. J.* **5**:36–44.

Yao, M. C., and Gall, J. G. 1979. Alteration of the *Tetrahymena* genome during nuclear differentiation. *J. Protozool.* **26**:10–13.

Yuasa, A., Soneda, M., Osumi, M., and Sumiyasu, M. 1965. Mitosis in *Wickerhamia fluorescens* Soneda. *Jpn. J. Genet.* **40**:21–24.

Zeligs, J. D., and Wollman, S. H. 1979a. Mitosis in rat thyroid epithelial cells *in vivo*. I. Ultrastructural changes in cytoplasmic organelles during the cell cycle. *J. Ultrastruct. Res.* **66**:53–77.

Zeligs, J. D., and Wollman, S. H. 1979b. Mitosis in rat thyroid epithelial cells *in vivo*. II. Centrioles and pericentriolar material. *J. Ultrastruct. Res.* **66**:97–108.

Zeligs, J. D., and Wollman, S. H. 1979c. Mitosis in thyroid follicular epithelial cells *in vivo*. III. Cytokinesis. *J. Ultrastruct. Res.* **66**:288–303.

Zimmermann, W. 1921. Zur Entwicklungsgeschichte und Zytologie von *Volvox*. *Jahrb. Wiss. Bot.* **60**:256–294.

Index

Page numbers in **bold** type refer to illustrations.